INTERNATIONAL BIBLIOGRAPHY
OF HISTORICAL SCIENCES

INTERNATIONAL BIBLIOGRAPHY OF HISTORICAL SCIENCES

INTERNATIONALE BIBLIOGRAPHIE DER GESCHICHTSWISSENSCHAFTEN
BIBLIOGRAFIA INTERNACIONAL DE CIENCIAS HISTORICAS
BIBLIOGRAPHIE INTERNATIONALE DES SCIENCES HISTORIQUES
BIBLIOGRAFIA INTERNAZIONALE DELLE SCIENZE STORICHE

VOLUME LXIX
2000

Edited by Massimo Mastrogregori

with the contribution of a number of scholars,
under the auspices of the
International Committee of Historical Sciences

K·G·SAUR MÜNCHEN 2005

The IBOHS for the years 1978 to 1992 (Vol. 47 – 61) was edited by
Michel François and Michael Keul for Vol. 47/48 (1978/1979) and
Jean Glénisson and Michael Keul for Vol. 49 – 61 (1980 – 1992)
on behalf of the International Committee of Historical Sciences
and was published by K. G. Saur Munich.

Bibliographic information published by Die Deutsche Bibliothek
Die Deutsche Bibliothek lists this publication in the Deutsche Nationalbibliografie;
detailed bibliographic data is available in the Internet at http://dnb.ddb.de.

Printed on acid-free paper / Gedruckt auf säurefreiem Papier

© 2005 by K. G. Saur Verlag GmbH, München
Printed in Germany

All Rights Strictly Reserved / Alle Rechte vorbehalten
No part of this publication may be reproduced, stored in a retrieval system,
or transmitted in any form or by any means, electronic, mechanical, photocopying,
recording, or otherwise, without permission in writing from the publisher /
Jede Art der Vervielfältigung ohne Erlaubnis des Verlags ist unzulässig

Technical partner: Dr. Rainer Ostermann, München
Managing partner and technical support: Ellediemme libri dal mondo, Roma
Printed and Bound by Strauss Offsetdruck GmbH, Mörlenbach

ISSN 0074-2015
ISBN 3-598-20431-0

General editor

Massimo MASTROGREGORI, Università di Roma 'La Sapienza'

Assistant editor

Carlo COLELLA, Roma

Advisory board

Maria Tereza AMADO, Universidad de Evora
Girolamo ARNALDI, Istituto storico italiano per il Medioevo, Roma
† Yuri BESSMERTNY, Institute of General History, Russian Academy of Sciences, Moscow
† Wiesław BIEŃKOWSKI, Polska Akademia Nauk
László BIRÓ, Hungarian Academy of sciences, Budapest
Corinne BONNET, Academia Belgica, Rome
Luciano CANFORA, Università di Bari
Alejandro CATTARUZZA, University of Buenos Aires, Argentina
Anne EIDSFELDT, University of Oslo Library
Ilse FREDERIKSEN VÄHÄKYRÖ, Turku University Library, Finland
Jean GLÉNISSON, Comité International des Sciences Historiques, Paris
Kazuhiko KONDO, University of Tokyo
Mario MAZZA, Università di Roma 'La Sapienza'
Vilém PREČAN, Institute of Contemporary History, Prague
Matjaz REBOLJ, Ljubljana
Jacques REVEL, Ecole des Hautes Etudes en Sciences Sociales, Paris
† Ruggiero ROMANO, Ecole des Hautes Etudes en Sciences Sociales, Paris
Gabrielle M. SPIEGEL, Johns Hopkins University, Baltimore
Martina STERCKEN, Universität Zürich
Natasa STERGAR, Ljubljana
Abdeljellil TEMIMI, Fondation Temimi, Tunis
Şerban TURCUŞ, Università "Babeş-Bolyai", Cluj, Napoca
Nenad VEKARIĆ, Dubrovnik
Bahaeddin YEDİYILDIZ, Hacettepe Universitesi, Ankara

Contributing editors

Kira E. AGEEVA, Institute of Archaeology of the Russian Academy of Sciences, Moscow (*Russian historiography*)
Maria Tereza AMADO, Universidad de Evora (*Portuguese historiography*)
Vassili N. BABENKO, Russian Academy of Sciences, Moscow (*Russian Historiography*)
† Wiesław BIEŃKOWSKI, Polska Akademia Nauk (*Polish historiography*)
Wolfdieter BIHL, Institut für Geschichte, Universität Wien (*Austrian historiography*)
László BIRÓ, Hungarian Academy of sciences, Budapest (*Hungarian historiography*)
Vera BRENOVA, Institute of Contemporary History, Prague (*Czech historiography*)
Alejandro CATTARUZZA, University of Buenos Aires, Argentina (*Latin American historiography*)

Carlo COLELLA (*History by countries, History of international relations*)
Laura DE GIORGI, Università di Venezia (*Chinese historiography*)
Francesca DEVESCOVI, Trieste (*History of modern culture*)
Anne EIDSFELDT, University of Oslo Library, (*Norwegian historiography*)
Ilse FREDERIKSEN VÄHÄKYRÖ, Turku University Library, Finland (*Finnish historiography*)
Anna GRUCA, Instytut Historii PAN. Pracownia Bibliografii Bieżącej (*Polish historiography*)
Alessandro GUERRA, Università di Roma 'La Sapienza' (*Modern religious history*)
Timophey GUIMON, Institute of Universal History of the Russian Academy of Sciences, Moscow (*Russian historiography*)
Libby KAHANE, The Jewish National and University Library, Jerusalem (*Historiography of Israel*)
Kazuhiko KONDO, University of Tokyo (*Japanese historiography*)
Mauro LENZI, Società Romana di Storia Patria, Roma (*Palaeography, Diplomatics, History of the book, Medieval history*)
Jean Marie MAILLEFER, Université Charles-De-Gaulle Lille 3 (*Danish and Swedish historiography*)
Massimo MASTROGREGORI, Università di Roma 'La Sapienza' (*Auxiliary sciences, General works, Modern history*)
Stjepan MATKOVIĆ, (*Croatian historiography*)
Massimo PONTESILLI, Roma (*History of modern culture*)
Matjaz REBOLJ, Ljubljana (*Slovenian historiography*)
Eszter SALGÓ, dottoranda in Storia d'Europa presso l'Universitá di Roma 'La Sapienza', Roma (*Hungarian historiography*)
Lidia SANTARELLI, Università di Roma 'La Sapienza' (*Modern Greek historiography*)
Marco SANTUCCI, Università di Roma 'La Sapienza' (*Ancient history*)
Evgeny E. SAVITSKI, Institute of Universal History of the Russian Academy of Sciences, Moscow (*Russian historiography*)
Alžbeta SEDLIAKOVÁ, (*Slovak historiography*)
A. SLIVA, Rossijskaja Akademija Nauk, Moskva (*Russian historiography*)
Natale SPINETO, Università di Torino (*History of religions*)
Natasa STERGAR, Ljubljana (*Slovenian historiography*)
Paola STIRPE, Università di Roma 'La Sapienza' (*Ancient history*)
Şerban TURCUŞ, Università "Babeş-Bolyai", Cluj, Napoca, (*Romanian historiography*)
Michaela VALENTE, Università di Roma 'La Sapienza' (*Modern religious history*)
Alec VUIJLSTEKE, Academia Belgica, Rome (*Belgian historiography*)
Galina A. YANKOVSKAYA, Perm State University, Perm (*Russian historiography*)
Bahaeddin YEDİYILDIZ, Hacettepe Universitesi, Ankara (*Historiography of Turkey*)

Consulting editors

Maurice AYMARD, Maison des sciences de l'homme, Paris
Eric BRIAN, Centre Alexandre Koyré, Paris
Louis CHATELLIER, Université de Nancy II
Sten EBBESEN, University of Copenhagen

Carlo FRANCO, Università di Venezia
Olivier GUYOTJEANNIN, Ecole nationale des Chartes, Paris
Michel MORINEAU, Paris
Brian TIERNEY, Cornell University, Ithaca
Giusto TRAINA, Università di Lecce
Pietro VANNICELLI, Università di Urbino
André VAUCHEZ, Ecole Française de Rome

Special Assistant editor

Carlo SCOGNAMIGLIO, Roma

CONTENTS

	Pages
FOREWORD	XI
SCHEME	XIII
GENERAL HISTORICAL BIBLIOGRAPHIES	XIX
BIBLIOGRAPHY	1
INDEX OF NAMES	357
GEOGRAPHICAL INDEX	429

FOREWORD

The International Bibliography of Historical Sciences (I. B. O. H. S.) is a selective and descriptive bibliography, and the works it mentions, both books and articles, are arranged according to a methodical and chronological scheme originally drawn up and established by the Bibliographical Commission of the International Committee of Historical Sciences; the scheme has been revised only in details.

An exposition of the principles which were followed in the choice of works included and of the rules which were observed for their presentation in the present volume is set out below.

A. Manner of Selection.

In agreement with the wish expressed by the Bibliographical Commission of the C. I. S. H., the selection is actuated by the twin concern of preserving for the I. B. O. H. S. its character of a general bibliography comprehending the whole field of historical sciences, and of putting at the disposal of historians as also librarians the essential facts of historical production throughout the world, in one complete volume appearing annually.

In view of the multiplication of specialized bibliographies, it has in fact appeared more than ever necessary to offer to isolated scholars and even scientific establishments unable to obtain all these bibliographies the means of keeping informed, each year, of the advancement of historical science. But it was also desirable that these bibliographies be mentioned, and this has been done in two different ways: firstly, we have listed, outside the systematic inventory and immediately preceding it, the great international or national bibliographies giving the historical production of a country and in which a conspectus of works connected with this country is given; on the other hand, in the systematic inventory, at the head of each division or subdivision are mentioned general bibliographies dealing with one of the historical disciplines or particular bibliographies devoted to one question, one author or one province or state, and which find their logical position in that division or subdivision; in the latter case, the bibliographies are preceded by an asterisk (*).

In order to justify its existence as a working instrument of a high scientific standard and of international application, the I. B. O. H. S. only mentioned books or articles with a wider scope than the narrow field of local preoccupations, and rejects also reviews which are mere presentations or of courtesies. Like-wise re-editings, translations, descriptions of research which do not include new elements of information, exhibition catalogues without commentary, typed or stencilled works and works of popularization and propaganda have been normally eliminated.

On the other hand, the contributing editors have been careful to describe those works which, through slight or of apparently only local interest, make an obvious contribution to general history or to the solution of current problems; this is the case of certain reports on excavations and of articles bearing on controversial subjects touching the history of institutions or civilization; in this case, as whenever the title of an article was too vague, it has been followed whenever possible by a brief remark or by a date in brackets for the reader's orientation. Herein can be found an effort which will not fail to be useful to those who use the I. B. O. H. S., and which can be increased in future volumes without incurring the temptation to transform this essentially *selective* and *descriptive* bibliography into an analytical and critical one, this double character being in fact reserved to specialized bibliographies.

Unlike the greater number of national bibliographies, the I. B. O. H. S. does not limit the works included by any fixed date; that is to say that works connected with the most recent history find a place in it, notably those connected with international relations (P 8); at the same time the selection had to be correspondingly strictly.

By this conception, the I. O. B. H. S. keeps a physiognomy peculiar to itself; it has no tendency towards substituting for any existing bibliography, but while avoiding as far as possible a double role, it allows to be necessary that amount or overlapping which is profitable in the scholarly world.

B. Rules of Presentation.

The volume LXIX, 2000 mentions the works published with the date: 2000. Within each division or subdivision, the works are presented in alphabetical order of their authors. Slavonic, Greek, Japanese, Hebraic and Arabic names are transcribed into Latin characters and placed according to the order of the Latin alphabet, but characters with diacritics, for instance ć, č, ś, š are considered as if ordinaries c, s. Germanic and Scandinavian names are classed according to the function of the developed value of letters of inflection: ä, ö, ø, ü become ae, oe, ue. Mc and M' are indexed as if Mac.

Anonymous or collective works are classed alphabetically according to the initial of the key word in the title, for instance: «Congress (Fourteenth) of the Learned Societies...» At the same time, in heavy type are the names of scholars who have been the object of an important biographical or historiographical study (B § 2 b) and those of Saints (G § 4, I § 13 d); in the first case, the works are indicated in the alphabetical order of the people concerned.

As it has been done for the bibliographies peculiar to a division or subdivision, the publications of texts which had their place in the alphabetical list of each division or subdivision have been extracted and transferred to the head of the alphabetical list and immediately following the list of bibliographies; these publications of texts have been distinguished by being preceded by two asterisks. Thus the reader has immediately before his eyes the bibliographies and editions of the most recent texts bearing on a particular question or period. However, concerning the texts, the procedure of two asterisks (**) has not been adopted in chapter E, F, G and H, each of which already has a division especially devoted to the texts.

When the current year has been marked by the commemoration of an important historical event, the works to which this commemoration has led are grouped separately and under a special title at the end of the subdivision where this event finds its normal place.

When a work which has been in circulation for three or four years has been the object of a review every succeeding year, only the name of the author is cited plus the essential of the title, preceded by a reference to the number of the last volume of the I. B. O. H. S. in which it was quoted; it is thus possible to follow year to year the state of the criticism which the publication of a book has provoked.

Where the «collation» of works is concerned, the unification of references to pages, plates and illustrations, etc. has been sought as far as possible by putting them into either French or English, these being the two languages which have the most words or initial letters of words of an identical meaning in common.The transferrings of works interesting for one part to a section other than their logical one, transferences indicated by «*Cf. n°*...», have been grouped at the end of each section.

In the index of names of authors and persons, the names of Classical authors, Saints and Popes are written in their Latin form.

SCHEME

GENERAL HISTORICAL BIBLIOGRAPHIES
(p. XIX-XX)

A

AUXILIARY SCIENCES
(p. 1-14)

§ 1. Palaeography. 1-33. – § 2. Diplomatics. 34-42. – § 3. History of the book (*a*. Manuscripts; *b*. Printed books). 43-96. – § 4. Chronology. 97-113. – § 5. Genealogy and family history. 114-133. – § 6. Sigillography and heraldry. 134-153. – § 7. Numismatics and metrology. 154-179.– § 8. Linguistics. 180-220. – § 9. Historical geography, travels and discoveries. 221-290. – § 10. Iconography and images. 291-316.

B

MANUALS, GENERAL WORKS AND WORKS ON LARGE PERIODS
(p. 15-58)

§ 1. Archives, libraries and museums (*a*. Archives; *b*. Libraries; *c*. Museums). 317-396. – § 2. History of historiography (*a*. General; *b*. Special studies). 397-744. – § 3. Methodology, philosophy, and teaching of history. 745-831. – § 4. Ethnology, folklore and historical anthropology. 832-895. – § 5. General history. 896-1023. – § 6. Theory of the state and of society. 1024-1084. – § 7. Constitutional and legal history. 1085-1116. – § 8. Economic and social history. 1117-1182. – § 9. History of civilization, sciences and education. 1183-1254. – § 10. History of art. 1255-1294. – § 11. History of religions (*a*. General; *b*. Special studies). 1295-1408. – § 12. History of philosophy. 1409-1428. – § 13. History of literature. 1429-1476.

C

PREHISTORY
(p. 59-64)

§ 1. General. 1477-1499. – § 2. Palaeolithic and Mesolithic. 1500-1540. – § 3. Neolithic. 1541-1564. – § 4. Bronze age. 1565-1589. – § 5. Iron age. 1590-1613.

D

THE ANCIENT EAST
(the Hellenistic States included)
(p. 65-75)

§ 1. General. 1614-1636. – § 2. The Near East. 1637-1676. – § 3. Egypt. 1677-1733. – § 4. Mesopotamia. 1734-1805. – § 5. Hittites. 1806-1830. – § 6. Jews and Semitic peoples to the end of the ancient world. 1831-1886. – § 7. Iran. 1887-1910.

E

GREEK HISTORY
(p. 77-91)

§ 1. Classical world in general. 1911-1942. – § 2. Prehellenic epoch. 1943-1956. – § 3. Sources and criticism of sources (*a*. Epigraphical sources; *b*. Literary sources). 1957-2045. – § 4. General and political history. 2046-2091. – § 5. History of law and institutions. 2092-2113. – § 6. Economic and social history. 2114-2158. – § 7. History of literature, philosophy and science (*a*. Literature; *b*. Philosophy and sciences). 2159-2292. – § 8. Religion and mythology. 2293-2313. – § 9. Archaeology and history of art. 2314-2365.

F

HISTORY OF ROME, ANCIENT ITALY AND THE ROMAN EMPIRE
(p. 93-113)

§ 1. The peoples of Italy. 2366-2388. – § 2. The Etruscans. 2389-2412. – § 3. Sources and criticism of sources. (*a*. Epigraphical sources; *b*. Literary sources). 2413-2533. – § 4. General and political history. 2534-2637. – § 5. History of law and institutions. 2638-2673. – § 6. Economic and social history. 2674-2719. – § 7. History of literature, philosophy and science (*a*. Literature; *b*. Philosophy and science). 2720-2837. – § 8. Religion and mythology. 2838-2870. – § 9. Archaeology and history of art. 2871-2943. – § 10. Late antiquity. Transformation of the Roman world. 2944-2967.

G

EARLY HISTORY OF THE CHURCH TO GREGORY THE GREAT
(p. 115-120)

§ 1. Sources. 2968-3038. – § 2. General. 3039-3047. – § 3. Special studies. 3048-3103. – § 4. Hagiography. 3104-3112.

H

BYZANTINE HISTORY
(since Justinian)
(p. 121-126)

§ 1. Sources. 3113-3161. – § 2. General. 3162-3171. – § 3. Special studies. 3172-3254.

I

HISTORY OF THE MIDDLE AGES
(p. 127-158)

§ 1. Sources and criticism of sources (*a*. Non-literary sources; *b*. Literary sources). 3255-3318. – § 2. General works. 3319-3361. – § 3. Political history (*a*. General; *b*. 476–900; *c*. 900–1300; *d*. 1300–1500). 3362-3438. – § 4. Jews. 3439-3452. – § 5. Islam. 3453-3474. – § 6. Vikings. 3475-3484. – § 7. History of law and institutions. 3485-3528. – § 8. Economic and social history. 3529-3619. – § 9. History of civilization, literature, technology and education (*a*. Civilization; *b*. Literature; *c*. Technology; *d*. Education). 3620-3771. – § 10. History of art. 3772-3823. – § 11. History of music. 3824-3845. – § 12. History of philosophy, theology and science (*a*. Sources; *b*. Studies). 3846-3893. – § 13. History of the Church and religion (*a*. General; *b*. History of the Popes; *c*. Monastic history; *d*. Hagiography; *e*. Special studies). 3894-4003. – § 14. Settlements. Place names. Town planning. 4004-4034.

K

MODERN HISTORY, GENERAL WORKS
(p. 159-214)

§ 1. General. 4035.-4121. – § 2. History by countries. 4122-5395.

L

MODERN RELIGIOUS HISTORY
(p. 215-227)

§ 1. General. 5396-5459. – § 2. Roman Catholicism (*a*. General; *b*. History of the Popes; *c*. Special studies; *d*. Religious orders; *e*. Missions). 5460-5611. – § 3. Orthodox Church. 5612-5619. – § 4. Protestantism. 5620-5678. – § 5. Non-Christian religions and sects. 5679-5728.

M

HISTORY OF MODERN CULTURE
(p. 229-265)

§ 1. General. 5729-5898. – § 2. Academies, universities and intellectual organizations. 5899-5959. – § 3. Education. 5960-6040. – § 4. The Press. 6041-6113. – § 5. Philosophy. 6114-6209. – § 6. Exact, natural, medical sciences and technique. 6210-6349. – § 7. Literature (*a*. General; *b*. Renaissance; *c*. Classicism; *d*. Romanticism and after). 6350-6500. – § 8. Art and industrial art (*a*. General; *b*. Architecture; *c*. Sculpture, painting, etching and drawing; *d*. Decorative, popular and industrial art). 6501-6634. – § 9. Music, theatre, cinema and broadcasting. 6635-6754.

N

MODERN ECONOMIC AND SOCIAL HISTORY
(p. 267-290)

§ 1. General. 6755-6819. – § 2. Political economy. 6820-6841.– § 3. Industry, mining and transportation. 6842-6913. – § 4. Trade. 6914-6956. – § 5. Agriculture and agricultural problems. 6957-6989. – § 6. Money and finance. 6990-7040. – § 7. Demography and urban history. 7041-7111. – § 8. Social history. 7112-7298. – § 9. Working-class movement and socialism. 7299-7366.

O

MODERN LEGAL AND CONSTITUTIONAL HISTORY
(p. 291-300)

§ 1. General. 7367-7395. – § 2. History of constitutional law. 7396-7437. – § 3. Public law and institutions. 7438-7494. – § 4. Civil and penal law. 7495-7565. – § 5. International law. 7566-7575.

P

HISTORY OF INTERNATIONAL RELATIONS
(p. 301-336)

§ 1. General. 7576-7641. – § 2. History of colonization and decolonization (a. General; b. Asia; c. Africa; d. America; e. Oceania). 7642-7758. – § 3. From 1500 to 1789 (a. General; b. 1500–1648; c. 1648–1789). 7759-7851. – § 4. From 1789 to 1815. 7852-7871. – § 5. From 1815 to 1910. 7872-7952. – § 6. From 1910 to 1935. The First World War. 7953-8054. – § 7. From 1935 to 1945. The Second World War (a. General; b. Diplomacy. Economy; c. Military operations; d. Resistance). 8055-8235. – § 8. From 1945. 8236-8473.

R

ASIA
(p. 337-349)

§ 1. General. 8474-8479. – § 2. Western and central Asia. 8480-8495. – § 3. South Asia and Southeast Asia. 8496-8503. – § 4. China. 8504-8738. – § 5. Japan (before 1868). 8739-8781. – § 6. Korea. 8782-8785.

S

AFRICA
(esp. to its colonization)
(p. 351-352)

N°s 8786-8809.

T

AMERICA
(esp. to its colonization)
(p. 353)

N°s 8810-8820.

U

OCEANIA
(esp. to its colonization)
(p. 355)

N°s 8821-8826.

GENERAL HISTORICAL BIBLIOGRAPHIES

I. [Austria]. Österreichische historische Bibliographie. Austrian historical bibliography 1998. [1997. Cf. Bibl. 99, n° *I.*] Hrsg. v. Günther HÖDL u. Wolfdieter BIHL.. Graz, Neugebauer u. Santa Barbara, Clio, 2000, 624 p.

II. [Belgium] Bibliographie de l'histoire de Belgique. Bibliografie van de geschiedenis van België 1998. [1997. Cf. Bibl. 99, n° *II.*]. Ed. par Romain VAN EENOO, Jean BOVESSE [et al.]. *Revue Belge de philologie et d'histoire – Belgisch Tijdschrift voor Filologie en Geschiedenis*, 2000, 78, 2*, [s. p.].

III. [Czechoslovak Republik] Bibliografie dejin Ceskych zemi za rok 1993. Praha, Historicky ustav CAV, 2000, 350 p. – Bibliografie dejin Ceskych zemi za rok 1994. Praha, Historicky ustav CAV, 2000, 372 p. – Bibliografie dejin Ceskych zemi za rok 1995. Praha, Historicky ustav CAV, 2000, 420 p. – Czech and Czechoslovak history, 1918–1999: a bibliography of select monographs, volumes of essays, and articles published from 1990 to 1999. Compiled and edited by Vera BRENOVÁ and Slavena ROHLÍKOVÁ. Prague, Institute of Contemporary History, 2000, IX-472 p. – Select bibliography on Czech history: books and articles 1990–1999. Compiled by Václava HORCÁKOVÁ, Kristina REXOVÁ and Lubos POLANSKÝ; edited and introduced by Jaroslav PÁNEK. Prague, Institute of History, 2000, XLVI-333 p. (Práce Historického ústavu AV CR. Rada D, Bibliographia, 8).

IV. [Finland] ANTIN (Kirsti). Finländsk historisk litteratur 1998 i bibliografiskt urval. [1997. Cf. Bibl. 99, n° *IV.*] (La littérature historique de la Finlande. Une sélection bibliographique. 1997). *Historisk Tidskrift (Finland)*, 2000, 85, p. 108-137.

V. [France] Bibliographie annuelle de l'histoire de France du Ve siècle à 1958. Année 1999. [1998. Cf. Bibl. 99, n° *V.*] Redigée par M. SONNET, I. HAVELANGE, B. KÉRIVEN et Cl. GHIATI. Paris, Ed. du C. N. R. S., 2000, LXXXII-1044 p.

VI. [Germany] Historische Bibliographie. Berichtsjahr 1999 [1998. Cf. Bibl. 99, n° *VI.*] Hrsg. von der Arbeitsgemeinschaft ausseruniversitärer Forschungseinrichtungen in der Bundesrepublik Deutschland. München, Oldenbourg, 2000, 1056 p. – Jahresberichte für Deutsche Geschichte. Neue Folge. 51. Jahrgang 1999. Mit Nachträgen. [50. Jahrgang 1998. Cf. Bibl. 99, n° *VI.*] Hrsg. von der Berlin-Brandenburgischen Akademie der Wissenschaften. Berlin, Akademie Verlag, 2000, 1461 p.

VII. [Great Britain] Annual bibliography of British and Irish history. Publications of 1999. [1998. Cf. Bibl. 99, n° *VII.*] Ed. by Austin GEE. Oxford, Oxford U. P., 2000, XIII-535 p.

VIII. International bibliography of historical sciences. Internationale Bibliographie der Geschichtswissenschaften. Bibliografia internacional de ciencias historicas. Bibliographie internationale des sciences historiques. Bibliografia internazionale delle scienze storiche. Vol. LXIII, 1994. Vol. LXIV, 1995 [Vol. LXII, 1993. Cf. Bibl. 99, n° *VIII.*] Ed. by Massimo MASTROGREGORI with the contribution of a number of scholars. München, K. G. Saur, 2000, 2 vol., XVI-404 p., XVIII-438 p.

IX. [Italy] Bibliografia storica nazionale. Anno LX. 1998 [Cf. Bibl. 98, n° *X.*] A cura di M. BETTONI, M. AIRES e G. PIZZORUSSO. Dir. da R. VILLARI e G. VITUCCI. Roma e Bari, Laterza, 2000, XL-470 p. (Giunta centrale per gli studi storici).

X. [Mexico] Bibliografia histórica mexicana. Publ. Anual del Centro de estudios históricos, Collegio de Mexico. Vol. 30. Mexico, Collegio de Mexico, 2000, [s. p.].

XI. [Netherlands] VAN DER PLAAT (G. N.), DE KEUNING (M.), VAN HERWIJNEN (G.), VAN VLIET (C.). Kroniek. Lijst van de voornaamste in 1999 verschenen boeken en artikelen op het terrein van de Nederlandse geschiedenis. [1998. Cf. Bibl. 99, n° *X.*] La Haye, Koninklijk Nederlands Historisch Genootschap, 2000, [s. p.].

XII. [Nigeria] AKINBODE (Rahmon O.), OLUMOROTI (Oluranti). Historical research in Nigeria: index to articles in Journal of Historical Society of Nigeria (JHSN) and bibliography of Nigerian books and monographs on history (1956–2000 A.D.). Ibadan, Options Books and Information Services, 99, 107 p.

XIII. [Norway] Norsk bokfortegnelse. (The Norwegian National Bibliography. Historical works published in Norway, 2000) [1998. Cf. Bibl. 99, n° *XII.*] Oslo, Nasjonalbiblioteket, 2000, p. 686-692.

XIV. [Poland] Bibliografia historii polskiej za rok 1999. [1998. Cf. Bibl. 99, n° *XIII.*] (Bibliographie de l'histoire polonaise pour l`année 1998). Auteurs: Wojciech FRAZIK, Stefan GĄSIOROWSKI, Anna GRUCA, Zbigniew SOLAK. Réd. Wiesław BIEŃKOWSKI. Kraków, Wydawn. Profesjonalnej Szkoły Biznesu, 2000, [s. p.]. (Pol. Akad. Nauk, Inst. Hist., Zakł. Bibliografii Bieżącej).

XV. [Romania] Bibliografia istorică a României. Vol. 9. 1994–1999 [Vol. 8. 1989–1994. Cf. Bibl. 96, n° *XIII.*] Cluj Napoca, [s. n.], 2000, 702 p.

XVI. [Slovakia] SEDLIAKOVÁ (Alžbeta). Slovenská historiografia 1998. Výberová bibliografia. [1997. Cf. Bibl. 99, n° *XIV.*] (Historiography in Slovakia. Selected bibliography). *Historický časopis*, 2000, 48, 4, p. 714-763. – Slovenská historiografia 1995–1999. Výberová bibliografia. (Slovak historiography 1995–1999. Selective bibliography). Ed. Alžbeta SEDLIAKOVÁ. Bratislava, Veda vydavateľstvo Slovenskej akadémie vied – Historický ústav Slovenskej akadémie vied, 2000, 342 p.

XVII. [Spain] Indice histórico español. Publicación semestral del centro de estudios históricos internacionales. Ed. por Pere MOLAS RIBALTA y Rosa ORTEGA CANADELL. Vol. 36, n. 110, 1998; vol. 37, n. 111, 1999. Barcelona, Publicacions de la Universitat de Barcelona, 2000, 2 vol.., 344 p., 364 p.

XVIII. [Switzerland] Bibliographie der Schweizergeschichte. Bibliographie de l'histoire suisse. 1997. [1996. Cf. Bibl. 99, n° *XVI.*] Bearb. v./Etablie par P. L. SURCHAT. Bern, Bibliothèque nationale suisse/Schweizerische landesbibliothek, 2000, XXVI-262 p.

A

AUXILIARY SCIENCES

§ 1. Palaeography. 1-33. – § 2. Diplomatics. 34-42. – § 3. History of the book (*a*. Manuscripts; *b*. Printed books). 43-96. – § 4. Chronology. 97-113. – § 5. Genealogy and family history. 114-133. – § 6. Sigillography and heraldry. 134-153. – § 7. Numismatics and metrology. 154-179.– § 8. Linguistics. 180-220. – § 9. Historical geography, travels and discoveries. 221-290. – § 10. Iconography and images. 291-316.

§ 1. Palaeography.

* 1. BMB. Bibliografia dei manoscritti in scrittura beneventana. A cura della Scuola di specializzazione per Conservatori di beni archivistici librari della civiltà medievale, Università di Cassino. Vol. 8. [Vol. 7. Cf. Bibl. 99, n° 1.] Roma, Viella, 2000, [s. p.].

* 2. Bulletin codicologique, publié par le Centre International de Codicologie. [1999. Cf. Bibl. 99, n° 3.] Ed. par Pierre Cockshaw. *Scriptorium*, 2000, 54, p. 1*-164*

3. Alfabeti: preistoria e storia del linguaggio scritto. A cura di Mario Negri. Colognola ai colli, Demetra, 2000, 255 p. (ill.). (Atlanti del pensiero).

4. ASOV (Aleksandr Igorevich). Slavianskie runy i "Boianov gimn". Moskva, Izdatel'stvo "Veche", 2000, 413 p. (ill.).

5. Autografi (Gli) di frate Francesco e di frate Leone. A cura di Attilio BARTOLI LANGELI. Turnhout, Brepols, 2000, 135 p. (plates, ill.). (Corpus Christianorum. Autographa Medii Aevi, 5).

6. AZEVEDO SANTOS (Maria José). Ler e comprender a escrita na Idade Média. Lisboa, Ediçoes Colibri, 2000, 131 p. (ill.). (Colecçao Textos pedagógicos e didácticos, 10).

7. Chartae latinae antiquiores. Facsimile-edition of the latin charters. 2nd Series: ninth century. Ed. by Guglielmo CAVALLO a. Giovanna NICOLAJ. Part 54. Italy XXVI. Ravenna, 1. Published by Giuseppe RABOTTI a. Francesca SANTONI. Part 56. Italy XXVIII. Piemonte, 1. Asti. Published by Gian Giacomo FISSORE. Dietikon, URS Graf Verlag, 2000, 2 vol., 147 p., 147 p. (facs.).

8. COLAS (Gérard). Critique et transmission des textes de l'Inde classique. *Quaderni di storia*, 2000, 26, 51, p. 203-224.

9. DOGAN (Ismail). Kafkasya'daki Göktürk (runik) isaretli yazitlar. Ankara, Atatürk Kültür, Dil ve Tarih Yüksek Kurumu, Türk Dil Kurumu, 2000, 250 p. (ill., map).

10. DORANDI (Tiziano). Le stylet et la tablette: dans le secret des auteurs antiques. Paris, Les Belles Lettres, 2000, 218 p. (L'âne d'or).

11. Drevnerusskaia knizhnost': redaktor i tekst. Experiments in investigation of written sources: archaeography, textology, codicology. S.-Peterburg, Dmitrii Bulanin, 2000, 411 p. (ill.). [English summary].

12. DULFER (Kurt), KORN (Hans-Enno). Schrifttafeln zur deutschen Paläographie des 16.–20. Jahrhunderts. Bearb. von Karsten UHDE. Marburg, Archivschule Marburg, 2000, 192 p. (Veröffentlichungen der Archivschule Marburg).

13. EGGER (Christoph), WEIGL (Herwig). Text – Schrift – Codex: Quellenkundliche Arbeiten aus dem Institut für österreichische Geschichtsforschung. Wien u. München, R. Oldenbourg, 2000, 391 p. (ill., maps). (Mitteilungen des Instituts für österreichische Geschichtsforschung, Ergänzungsband, 35).

14. Fachgebiet historische Hilfswissenschaften. Ausgewählte Aufsätze zum 65. Geburtstag von Peter Ruck. Hrsg. Erika EISENLOHR und Peter WORM. Marburg an der Lahn, [s.n.], 2000, 304 p. (ill.). (Elementa diplomatica).

15. HARRIS (Roy). Rethinking writing. London, Athlone, 2000, XVI-254 p. (ill.).

16. Hellenike (E) graphe kata tous 15. kai 16. aiones: Diethne symposia 7. Athena, Patoura, 2000, 568 p. (ill.). (Institouto Vyzantinon Ereunon. Diethne Symposia, 7).

17. HELLMANN (Martin). Tironische Noten in der Karolingerzeit: am Beispiel eines Persius-Kommentars

aus der Schule von Tours. Hannover, Hahnsche Buchhandlung, 2000, XXVIII-266 p. (pl., ill., map). (Monumenta Germaniae historica. Studien und Texte, 27).

18. HUWS (Daniel). Medieval Welsh manuscripts. Cardiff, University of Wales Press and the National Library of Wales, 2000, XVI-352 p. (plates, diagrams, tables).

19. Inventar der Handschriften des Benedektinerstiftes Melk. Teil 1. Von dern Anfängen bis ca. 1400. Katalofband-Registerband. Hrsg. v. Christine GLASSNER. Wien, Verlag des Österreichschen Akademie der Wissenschaften, 2000; 2 vol., 538 p., 183 p. (ill., CD-ROM).

20. Katalog der deutschsprachigen illustrierten Handschriften des Mittelalters, 3/3: 26. Chroniken. Hrsg. v. Hella FRÜHMORGEN-VOSS, Norbert H. OTT und Ulrike BODEMANN. München, C. H. Beck, 2000, 161-240 p. (plates, ill.).

21. KHRISTOVA (Boriana), DZHUROVA (Aksiniia), VELINOVA (Vasia). Opis na slavianskite rukopisi ot Tsentura za slaviano-vizantiiski prouchvaniia "prof. Ivan Duichev" kum SU Sv. Kliment Okhridski", XIV–XIX v. Sofiia, UI "Sv. Kliment Okhridski", 2000, 227 p. (ill., facs). (Series catalogorum, 1).

22. KOUAMÉ (Nathalie). Initiation à la paléographie japonaise: à travers les manuscrits du pèlerinage de Shikoku. Paris, Langues & Mondes l'Asiatique, 2000, 97 p., (ill., map). (Connaâitre le Japon).

23. Manoscritti (I) greci tra riflessione e dibattito: atti del V Colloquio Internazionale di Paleografia Greca (Cremona, 4–10 ottobre 1998). A cura di Giancarlo PRATO. Firenze, Gonnelli, 2000, 3 vol., XI-796 p. (pl., ill.). (Papyrologica Florentina, 31).

24. MROZOWICZ (Wojciech). Mittelalterliche Handschriften oberschlesischer Autoren in der Universitätsbibliothek Breslau (Wroclaw). Heidelberg, Palatina, 2000, 112 p.

25. MUIR (Bernard J.), KENNEDY (Nick), SMITH (Graeme). Ductus: a history of handwriting. An online course in paleography. Melbourne, Bernard J. Muir, 2000, 1 CD-ROM and user's guide.

26. New directions in later medieval manuscript studies: essays from the 1998 Harvard Conference. Ed. by Derek PEARSALL. York, York Medieval Press, in association with Boydell and Brewer and the Centre for Medieval Studies, University of York, 2000, XV-213 p. (plates, ill.).

27. PASTENA (Carlo). Breve storia dei materiali scrittori dalle origini al 15. secolo. Palermo, Regione siciliana, Assessorato dei beni culturali e ambientali e della pubblica istruzione, 2000, 146 p. (ill.).

28. Psalterium Egberti: facsimile del ms. 136. del Museo archeologico nazionale di Cividale del Friuli. A cura di Claudio BARBERI. Trieste, Ministero per i beni e le attività culturali, Soprintendenza per i beni ambientali, architettonici, archeologici, artistici e storici del Friuli-Venezia Giulia, 2000, 217 p. (1 CD-ROM.). (Relazioni della Soprintendenza per i beni ambientali e architettonici, archeologici, artistici e storici del Friuli-Venezia Giulia).

29. SCHIPKE (Renate). Scriptorium und Bibliothek des Benediktinerklosters Bosau bei Zeitz: Die Bosauer Handschriften in Schulpforte. Wiesbaden, Harrassowitz, 2000, 143 p. (plates).

30. SHUL'GINA (Emilija V.). Russkaja knizhnaja skoropis' XV v. (Russian book cursive script of the 15th century). Gos. istorichekij muzej. Sankt-Peterburg, Dmitrij Bulanin, 2000, 195 p. (tables, facs.).

31. Statut (Le) du scripteur au moyen âge. Actes du XIIe colloque scientifique du Comité international de paléographie latine (Cluny, 17–20 juillet 1998).Hrsg. v. Marie-Clotilde HUBERT, Emmanuel POULLE, Marc H. SMITH. Paris, Ecole de Chartes, 2000, 388 p. (ill.). (Matériaux pour l'Histoire, 2).

32. STOLJAROVA (Lubov' V.). Svod zapisej pistsov, khudozhnikov i perepletchikov drevnerusskikh pergamennykh kodeksov XI–XIV vekov. (Corpus of scribes', painters' and bookbinders' inscriptions on Old Russian parchment manuscripts of the 11th–14th centuries). Ed. Sergej M. KASHTANOV. RAN, In-t rossijskoj istorii. Moskva, Nauka, 2000, 543 p. (facs.; ill.; plates; bibl. p. 461-472; indices p. 473-517). [Cf. bibl. 98, n° 3285. Cf. Bibl. 99, n° 79.]

33. Studies in the Harley Manuscript: the scribes, contents, and social context of British library MS Harley 2253. Ed. by Susanna FEIN. Kalamazoo, Medieval Institute Publications, Western Michigan University, 2000, XVI-515 p. (ill., plates, tables).

Cf. n°s 34-42, 141, 3255-3318

§ 2. Diplomatics.

* 34. ANSANI (Michele). Diplomatica (e diplomatisti) nell'arena digitale. *Archivio storico italiano*, 2000, 158, 584, p. 349-380.

———

35. Atti (Gli) dell'arcivescovo e della Curia arcivescovile di Milano nel sec. XIII: Ottone Visconti (1262–1295). A cura di Maria Franca BARONI, introduzione di Grado Giovanni MERLO. Milano, Università degli studi di Milano, 2000, XCIV-462 p. (ill.).

36. Charters and the use of the written word in medieval society. Ed. by Karl HEIDECKER. Turnhout, Brepols, 2000, XI-253 p. (Utrecht studies in medieval literacy, 5).

37. Colección diplomática del Monasterio de Celanova: 842–988 p. Ed. por Emilio SÁEZ y Carlos SÁEZ. Alcalá, Universidad, 2000, 462 p. (ill.).

38. Dating undated medieval charters. Ed. by Michael GERVERS. Suffolk, Boydell Press, 2000, VII-237 p.

39. Diplomatique (La) urbaine en Europe au moyen âge: actes du congrès de la Commission internationale de diplomatique, Gand, 25–29 août 1998. Éd. Par Walter PREVENIER et Therèse DE HEMPTINNE. Leuven, Garant, 2000, XIV-6-581 p. (ill.). (Studies in urban social, economic and political history of the medieval and modern Low Countries, 9).

40. GRIGOR'EV (A. P.). Jarlyk Mukhammeda Bjuleka ot 1379 g. mitropolitu Mikhailu: Rekonstruktsija soderzhanija. (Muhammad Byulek's Yarlyk [charter] of 1379 to the Metropolitan Michail [of Russia]: reconstruction of the contents). *In*: Istoriografija i istochnikovedenie istorii stran Azii i Afriki. Vol. 19. Sankt-Peterburg, Izd-vo Sankt-Peterburgskogo un-ta, 2000, p. 3-104.

41. KÖLZER (Theo). Tra tarda antichità e medioevo: l'edizione critica dei diplomi merovingi: inaugurazione del Corso biennale, anni accademici 1998–2000, Città del Vaticano, 19 ottobre 1998. Città del Vaticano, [s. n.], 2000, 47 p. (ill.).

42. VALENTI (Filippo). Scritti e lezioni di archivistica, diplomatica e storia istituzionale. A cura di Daniela GRANA. Roma, Ufficio centrale per i beni archivistici, 2000, XIV-689 p. (ill., pl.). (Saggi, 57).

Cf. n°s 1-33, 141, 3255-3318

§ 3. History of the book.

a. Manuscripts

43. ALTURO PERUCHO (Jesus). El Llibre manuscrit a Catalunya: orígens i esplendor. Barcelona, Generalitat de Catalunya. Departament de Presidència, 2000, 291 p.

44. ALVAREZ MARQUEZ (Maria del Carmen). El libro manuscrito en Sevilla: siglo 16. Sevilla, Ayuntamiento de Sevilla, Area de cultura y fiestas mayores, 2000, 317 p. (tab., ill.).

45. AMTOWER (Laurel). Engaging words: the culture of reading in the later middle ages. New York a. Houndmills, Palgrave, 2000, XI-243 p. (ill.).

46. Bibbia (La) Amiatina / The codex Amiatinus. Complete reproduction on CD-ROM of the manuscript Firenze, Biblioteca Medicea Laurenziana, Amiatino 1. Firenze, SISMEL, Edizioni del Galluzzo, 2000, 1 CD-ROM.

47. Bibbie (Le) atlantiche: il libro delle scritture tra monumentalità e rappresentazione. A cura di Marilena MANIACI e Giulia OROFINO. Milano, CT, 2000, XXI-351 p. (ill.).

48. Biblioteca civica "Angelo Mai" (Bergamo). Bergamo, Esedra, 2000, 2 CD-ROM (Opera Omnia). (facs.).

49. BONDÉELLE-SOUCHIER (Anne). Bibliothèques de l'Ordre de Prémontré dans la France d'Ancien Régime. Paris, CNRS Éditions, 2000, 383 p. (ill., facs., maps). (Histoire des bibliothèques médiévales, 9).

50. British (The) library catalogue of additions to the manuscripts, 1956–1965: additional manuscripts 48989–53708, Egerton manuscripts 3725–3776, additional charters and rolls 75435–75577, detached seals and casts CCI.1–124. London, The British Library, 2000, 3 vol., XVI-1692 p.

51. Catalogo dei manoscritti della Biblioteca comunale di Treviso. A cura di E. LIPPI. Mss. n. 2901–3150. Treviso, Comune di Treviso, 2000, 199 p.

52. Codex: i tesori della Biblioteca Ambrosiana. Milano, Rizzoli-RCS libri, 2000, 159 p. (ill.).

53. Codice (Il) Valerio Massimo e la miniatura. Milano, Regione Lombardia, 2000, 1 CD-ROM [versione digitalizzata del ms. Les faits et les dits des romains et des autres gens di Valerio Massimo conservato presso la Biblioteca Trivulziana di Milano].

54. Codices Comboniani Aethiopici. Recensuit Osvaldus RAINERI. Citta del Vaticano, Bibliotheca Vaticana, 2000, 334 p. (tav., ill.). (Bibliothecae Apostolicae Vaticanae codices manuscripti recensiti).

55. Codici (I) della Libreria di S. Giacomo della Marca nel Museo civico di Monteprandone. A cura di Saturnino LOGGI, coordinamento scientifico di Mauro MEI. Ancona, Regione Marche, Centro beni culturali, LII-162 p. (ill.). (Fondi storici nelle biblioteche marchigiane).

56. Codici greci dell'Italia meridionale. A cura di Paul CANART e Santo LUCA. Roma, Retablo e Ministero per i beni e le attivita culturali, 2000, 191 p. (ill.).

57. CORTESE (Delia). Ismaili and other Arabic manuscripts: a descriptive catalogue of manuscripts in the library of the Institute of Ismaili studies. London a. New York, Institute of Ismaili studies, 2000, XVIII-170 p. (facs., tav.).

58. DÉROCHE (François). Manuel de codicologie des manuscrits en écriture arabe. Paris, Bibliothèque nationale de France, 2000, 413 p. (ill.). (Etudes et recherches).

59. Descriptive (A) catalogue of the medieval manuscripts of Exeter College, Oxford. Ed. by Andrew G. WATSON. Oxford, Oxford U. P., 2000, XXVIII-150 p. (tab., ill.).

60. English (The) medieval book: studies in memory of Jeremy Griffiths. Edited by A. S. G. EDWARDS, Vincent GILLESPIE and Ralph HANNA. London, The British Library, 2000, XII-264 p. (ill.).

61. GALLABARES (Giorgos). Iera Mone Iberon: eikonographemena cheirographa. Hagion Oros, Hiera Mone Iberon, 2000, 130 p. (ill.).

62. GAMESON (Richard). The manuscripts of early Norman England (c. 1066–1130). London, Oxford U. P., 99, XVIII-190 p. (ill., tables)

63. GRBIC (Dusica). Znakovi starih knjiga. Novi Sad, Biblioteka Matice srpske, 2000, 177 p. (ill., facs). (Tragovi, 9).

64. Issledovaniia knizhnykh pamiatnikov: istoriia, filologiia, istochnikovedenie. Sbornik nauchnykh statei. Redaktor-sostavitel' L. I. ILLARIONOVA. Moskva, Rossiiskaia Gosudarstvennaia Biblioteka, 2000, 195 p. (Kniga, biblioteka, kul'tura).

65. Katalog der Handschriften des Benediktinerstiftes Michaelbeuern bis 1600. Bearbeitet von Beatrix KOLL, unter mitarbeit von Josef FELDNER. Wien, Österreichische Akademie der Wissenschaften, 2000, 3 vol., 457 p. u. 16 taf., 189 p., 69 taf. (ill.). (Philosophisch-Historische Klasse. Denkschriften, 278/Veröffentlichungen der Kommission für Schrift- und Buchwesen des Mittelalters, Reihe 2: Verzeichnisse der Handschriften Österreichischer Bibliotheken, 6).

66. Katalog der mittelalterlichen deutschsprachigen Handschriften der ehemaligen Staats- und Universitätsbibliothek Königsberg. Auf der Grundlage der Vorarbeiten Ludwig Deneckes erarbeiten von Ralf G. PASLER. Hrsg. von Uwe MEVES. München, R. Oldenbourg, 2000, 284 p. (ill.). (Schriften des Bundesinstituts fur Ostdeutsche Kultur und Geschichte).

67. Libri, lettori e biblioteche dell'Italia medievale (secoli IX–XV): fonti, testi, utilizzazione del libro: atti della Tavola rotonda italo-francese (Roma, 7–8 marzo 1997). A cura di Giuseppe LOMBARDI e Donatella NEBBIAI DALLA GUARDA. Roma, ICCU, 2000, 560 p.

68. Liturgie und Buchkunst der Zisterzienser im 12. Jahrhundert: Katalogisierung von Handschriften der Zisterzienserbibliotheken. Hrsg. v. Charlotte ZIEGLER. Bern, Lang, 2000, 189 p. (taf., ill.).

69. Manoscritti (I) datati della provincia di Vicenza e della Biblioteca Antoniana di Padova. A cura di Cristiana CASSANDRO. Tavarnuzze, SISMEL, Edizioni del Galluzzo, 2000, XIX-142 p. (tav., ill.).

70. Manuscrits datés des bibliothèques de France. T. 1. Cambrai. Éd par Denis MUZERELLE. Paris, CNRS, 2000, XXVI-330 p. (ill.).

71. MARTIN (Henri-Jean), CHATELAIN (Jean-Marc), DIU (Isabelle), LE DIVIDICH (Aude), PINON (Laurent). Mise en page et mise en texte du livre français: La naissance du livre moderne (XIVe–XVIIe siècles). Paris, Editions du Cercle de la Librarie, 2000, VIII-494 p. (ill.).

72. ORNATO (Ezio). Apologia dell'apogeo. Divagazioni sulla storia del libro nel tardo medioevo. Roma, Viella, 2000, 147 p.

73. Parrhasiana. Atti della 1. Giornata di studi "Manoscritti medievali e umanistici della Biblioteca nazionale di Napoli", Napoli, 12 maggio 1999. A cura di Lucia GUALDO ROSA, Luigi MUNZI, Fabio STOK. Napoli, Arte tipografica, 2000, 73 p.

74. ROUSE (Richard H.), ROUSE (Mary A.). Illiterati et uxorati. Manuscripts and their makers: commercial book producers in medieval Paris, 1200–1500. Turnhout, Brepols, 2000, 2 vol., 424 p., 407 p. (ill., tables, maps).

75. SCHNEIDER (Wolfgang Christian). Geschlossene Bücher – offene Bücher. Das Öffnen von Sinnräumen in Schließen der Codices. *Historische Zeitschrift*, 2000, 271, 3, p. 561-592.

76. Scriptorium und Bibliothek des Benediktinerklosters Bosau bei Zeitz: die Bosauer Handschriften in Schulpforte. Beschrieben von Renate SCHIPKE. Wiesbaden, Harrassowitz, 2000, 143 p. (tav.).

77. Tesori di una biblioteca francescana: libri e manoscritti del Convento di San Nicolò in Carpi, sec. XV–XIX. A cura di Anna PRANDI, direzione scientifica di Giorgio MONTECCHI. Modena, Mucchi, 2000, 273 p. (ill.).

78. Tesori salvati: acquisizioni della Regione Lombardia, 1995–2000. Milano, Electa, 2000, 117 p. (ill.).

79. TOUATI (Houari). La dédicace des livres dans l'Islam médiéval. *In*: Texte et paratexte au Moyen Age. *Annales*, 2000, 55, 2, p. 325-354.

80. Überlieferung (Zur), Kritik und Edition alter und neuerer Texte. Beiträge des Colloquiums zum 85. Geburtstag von Werner Schröder am 12. und 13. März 1999 in Mainz. Mainz, Akademie der Wissenschaften und der Literatur. Hrsg. v. Kurt FRANZ GÄRTNER und Hans-Henril KRUMMACHER. Stuttgart, Franz Steiner, 2000, p. 308 (ill., tables).

81. VALLACQUA GUARIENTO (Maria Luisa). I codici liturgici decorati e miniati delle biblioteche della Valle d'Aosta: secoli X–XIII. Quart, Musumeci, 2000, 295 p. (ill.).

82. YILDIZ (Nuray). Eskiçağda yazı malzemeleri ve kitabın oluşumu. (Scribal instruments and formation of books in Antiquity). Ankara, Türk Tarih Kurumu, 2000, 329 p.

Cf. n°s 11, 27, 36, 3781, 3783, 3796, 3835, 3888, 3926, 3943, 3983

b. Printed books

* 83. ABHB. Annual bibliography of the history of the printed book and libraries. Vol. 28. [Vol. 27. Cf. Bibl. 99, n° 85.] Ed. by the Department of special collections of Koninklijke bibliotheek. The Hague, Martinus Nijhoff, 2000, 644 p.

* 84. Bibliographie der Buch- und Bibliotheksgeschichte (BBB). Vol. 18. 1998. [Vol. 17. 1997. Cf. Bibl. 99, n° 86.] Hrsg. v. Horst MEYER. Bad Iburg, Bibliographischer Verlag Dr. Horst Meyer, 2000, [s. p.].

85. D'AMBROSIO (Gaetano). La stampa nella città di Campagna e nella provincia di Salerno dalle origini all'Unità d'Italia. Eboli, Grafica ebolitana, 2000, 461 p. (ill.).

86. Ecrire aux XVIIe et XVIIIe siècles. Genèses de textes littéraires et philosophiques. Dir. par Jean-Louis LEBRAVE et Almuth GRÉSILLON. Paris, CNRS Editions, 2000, 240 p. (Textes et manuscrits).

87. History (A) of the book in America. Ed. by David HALL. Vol. 1. The colonial book in the Atlantic world. Ed. by Hugh AMORY and David HALL. Worcester, American Antiquarian Society a. Cambridge, Cambridge U. P., 2000, 638 p.

88. IMAŃSKA (Iwona). Druk jako wielofunkcyjny środek przekazu w czasach saskich. (Das Gedruckte als multifunktionelles Informationsmittel unter den Sachsen). Toruń, Wydaw. Uniw. M. Kopernika, 2000, 217 p.

89. LOVELL (Stephen). The Russian reading revolution: print culture in the Soviet and Post-Soviet eras. New York, St. Martin's, in association with the School of Slavonic and East European Studies, University of London, 2000, VIII-215 p. (Studies in Russia and East Europe).

90. MIX (York-Gothart). Kulturelles Kapital für 20, 50 oder 80 Pfennige. Medialisierungsstrategien Leipziger Verleger in der frühen Moderne am Beispiel der 'Universal-Bibliothek', der 'Insel-Bücherei' und der Sammlung 'Der Jüngste Tag'. *Archiv für Kulturgeschichte*, 2000, 82, p. 191-210.

91. PERÄLÄ (Anna). Suomen typografinen atlas – Finsk typografisk atlas. – Typographischer Atlas Finnlands 1642–1827. 1-2. Helsinki, Helsinki University Library, 2000, 2 vol., 702 p., 597 p. (Publ. of the Helsinki University Library, 54-55).

92. Religieux (Les) et leurs livres à l'époque moderne. Dir. par Bernard DOMPNIER et Marie-Hélène FROESCHLÉ-CHOPARD. Maringues, Presses Universitaires Blaise-Pascal, 2000, 296 p.

93. SELWYN (Pamela E.). Everyday life in the German book trade: Friedrich Nicolai as bookseller and publisher in the age of enlightenment, 1750–1810. University Park, Pennsylvania State U. P., 2000, XVI-419 p. (Penn State series on the history of the book).

94. SHARPE (Kevin). Reading revolutions: the politics of reading in early modern England. New Haven a. London, Yale U. P., 2000, XIV-358 p.

95. TOTH (Istvan György). Literacy and written culture in early Modern Central Europe. Budapest, Central European U. P., 2000, 266 p.

96. Ventes (Les) de livres et leurs catalogues XVIIe–XXe siècle. Ed. par Annie CHARON et Elisabeth PARINET. Paris, Ecole des chartes, 2000, 208 p. (Etudes et rencontres de l'Ecole des chartes, 5).

Cf. n°s 30, 32, 48, 60, 64, 348-373, 6041-6113

§ 4. Chronology.

97. BARTKY (Ian R.). Selling the true time: nineteenth-century timekeeping in America. Stanford, Stanford U. P., 2000, XVI-310 p.

98. BOURGOING (Jacqueline de). Le calendrier, maître du temps. Paris, Gallimard, 2000, 143 p. (ill.). (Découvertes. Histoire, 400).

99. BRENDECKE (Arndt). Vom Zählschritt zur Zäsur. Die Entstehung des modernen Jahrundertbegriffs. *Comparativ*, 2000, 10, 3, p. 21-37.

100. BRINCKEN (A.-D. Von den). Historische Chronologie des Abendlandes. Kalenderreformen und Jahrtausendrechnungen. Eine Einführung. Stuttgart, W. Kohlhammer, 2000, X-132 p.

101. CAPASSO (Riccardo). Elementi di cronologia e di cronogafia medievale. Roma, Media Print Edizioni, 2000, 95 p. (Accademia europea per le scienze storiche). – Il tempo nel Medioevo: rappresentazioni storiche e concezioni filosofiche, Atti del Convegno internazionale, Roma, 26–28 novembre 1998. A cura di Riccardo CAPASSO e P. PICCARI. Roma. [s. n.], 2000, [s. p.].

102. DECLERCQ (Georges). Anno Domini: The origins of the Christian era. Turnhout, Brepols, 2000, 206 p.

103. EL-SABBAN (Sherif). Temple festival calendars of ancient Egypt. Liverpool, Liverpool U. P., 2000, XII-218 p. (plates, ill.). (Liverpool monographs in archaeology & Oriental studies).

104. FREIBERG (M.) Going Gregorian, 1582–1752. A summary view. *Catholic historical review*, 2000, 86, p. 1-19.

105. GOLDBERG (Sylvie Anne). La clepsydre: Essai sur la pluralité des temps dans le judaïsme. Paris, Albin Michel, 2000, 394 p. – EADEM. Questions of times: conflicting time scales in historical perspective. *Jewish history*, 2000, 14, 3, p. 267-286.

106. HUBER (Peter J.). Babylonian short-time measurements: lunar sixes. *Centaurus: international magazine of the history of science and medicine*, 2000, 42, 3, p. 223-234.

107. Normes et usages chronologiques du moyen âge à l'époque contemporaine. Ed. par Marie-Clotilde HUBERT. Paris, Champion et Genève, Droz, 2000, 508 p. (ill., maps).

108. RICHARDS (Edward Graham). Mapping time: the calendar and its history. Oxford, Oxford U. P., 2000, XXI-438 p. (ill.).

109. RICHTER (Steffi). Syncronisierung von Welt. *Comparativ*, 2000, 10, 3, p. 7-20.

110. ROEMER (N.). Between hope and despair: conceptions of time and the German-Jewish experience in the nineteenth century. *Jewish history*, 2000, 14, 3, p. 345-363.

111. SOKOLOVSKII (Vladimir). Circle of holy feasts: the Orthodox calendar. Moskow, Raduga, 2000, 73[3]p. (ill.).

112. Synchronisation (The) of civilisations in the eastern mediterranean in the second millennium B.C.: international symposium at Schloß Haindorf, 15th–17th November 1996, and at the Austrian Academy, Vienna, 11th May 1998. Vienna, Australian Academy of Sciences Press, 2000, [s. p.]. (Contributions to the chronology of the Eastern Mediterranean,1).

113. WEINBERG (J.). Invention and convention: Jewish and Christian critique of the Jewish fixed calendar. *Jewish history*, 2000, 14, 3, p. 317-330.

Cf. n°s 141, 2578

§ 5. Genealogy and family history.

114. Beginner's guide (A) to Jewish genealogy in Great Britain. Ed. by Rosemary WENZERUL. London, The Jewish Genealogical Society of Great Britain, 2000, 80 p.

115. CRABB (Ann). The Strozzi of Florence: widowhood and family solidarity in the Renaissance. Ann Arbor, University of Michigan Press, 2000, VIII-328 p. (Studies in Medieval and Early Modern civilization).

116. DONCHEV (Doncho). Istoricheska pamet za rod i roden krai. (Mémoire historique de famille et de pays natal). Sofiia, Khristo Botev, 2000, 415 p. (Bibliotechna poreditsa, 1).

117. EDROIU (Nicolae). Studii de Genealogie. Cluj-Napoca, Academia Română, 2000, 228 p.

118. GIBSON (Jeremy Sumner Wycherley). Specialist indexes for family historians. Bury, Federation of Family History Societies, 2000, 72 p. (Guides for genealogists, family and local historians).

119. GOODY (Jack). La famiglia nella storia europea. Roma e Bari, Laterza, 2000, VIII-337 p. – IDEM. The European family: an historico-anthropological essay. Malden, Blackwell, 2000, X-209 p. (The making of Europe).

120. HERBER (Mark D). Ancestral trails: the complete guide to British genealogy and family history. Stroud, Sutton, 2000, XVI-701 p. (ill., maps, ports).

121. HEY (David). Family names and family history. London, Hambledon and London, 2000, XII-240 p. (plates, ill., maps, ports.).

122. KHAN (Iftikhar Ali). History of the ruling family of Sheikh Sadruddin Sadar-i-Jahan of Malerkotla (1449 A.D. to 1948 A.D.). Ed. by Raj Krishan GHAI. Patiala, Publication Bureau, Punjabi University, 2000, XXV-147 p.

123. KLAPISCH-ZUBER (Christiane). L'ombre des ancêtres: essai sur l'imaginaire médiéval de la parenté. Paris, Fayard, 2000, 458 p. (Esprit de la cité).

124. LANE-POOLE (Stanley). The Mohammadan dynasties. London, Routledge, 2000, XXVIII-361 p. (ill.). (Orientalism, early sources, 2).

125. Origen, nobleza y heráldica de los principales apellidos hispanos. Ed. por Endika MOGROBEJO [et al.] Bilbao, Editorial Mogrobejo-Zabala, 2000, 249 p. (plates, ill.).

126. PETTI BALBI (Giovanna). I Visconti di Genova: identità e funzioni dei Carnadino (Secoli XI–XII). *Archivio storico italiano*, 2000, 158, 586, p. 679-720.

127. SETTIPANI (Christian). Continuité gentilice et continuité familiale dans les familles sénatoriales romaines à l'époque impériale: mythe et réalité. Oxford, Unit for Prosopographical Research, Linacre College, University of Oxford, 2000, 597 p. (geneal. tables). (Occasional publications of the Oxford Unit for Prosopographical Research).

128. SOUSA (Bernardo Vasconcelos). Os Pimentéis: percursos de uma linhagem da nobreza medieval portuguesa (séculos XIII–XIV). Lisboa, Imprensa Nacional-Casa da Moeda, 2000, 358 p. (Temas portugueses).

129. SOUTHWICK (Michael). West Midlands repositories: record holdings in the West Midlands for local and family historians. Winlaton, Kingpin, 2000, 68 p. (ill., maps).

130. STARY (Giovanni). A dictionary of Manchu names: a name index to the Manchu version of the "Complete genealogies of the Manchu clans and families of the eight banners". Wiesbaden, Harrassowitz in Kommission, 2000, XVII-645 p. (Aetas Manjurica, 8).

131. Szlachta wylegitymowana w Królestwie Polskim w latach 1836–1861. (Die legitimierte Schlachta in dem Königreich Polen in den Jahren 1836–1861). Bearb. von Elżbieta SĘCZYS, Warszawa, DiG, 2000, 849 p. (Szlachta Polska).

132. WAHL (Johannes). Lebensplanung und Alltagserfahrung: württembergische Pfarrfamilien im 17. Jahrhundert. Mainz, Philipp von Zabern, 2000, VIII-286 p. (ill.). (Veröffentlichungen des Instituts für Europäische Geschichte, Mainz, 181).

133. ZAPRJANOVA (A.). Bulgarische Familien zur Zeit der Wiedergeburt. (In memoriam Panajot D. Maždrakov). *Bulgarian historical review*, 2000, 1-2, p. 98-103.

Cf. n° 141

§ 6. Sigillography and heraldry.

134. ADAMCZEWSKI (Marek). Heraldyka miast wielkopolskich do końca XVIII wieku. (Die Heraldik der Städte Großpolens [Wielkopolska] bis zum ausgehenden XVIII. Jahrhundert). Warszawa, DiG, 2000, 520 p. (Pol. Tow. Heraldyczne).

135. ANTONOV (S.). Gerbut na Juzhna Bulgarija. (Les armoiries de la Bulgarie de Sud). *Istoricheski pregled*, 2000, 1-2, p. 236-243.

136. Corpus der minoischen und mykenischen Siegel. Hrsg. V. Friedrich MATZ. Berlin, Mann, 2000, XV-368 p. (ill., maps, diagrams).

137. DEUTSCH (Robert), LEMAIRE (André). Biblical period personal seals in the Shlomo Moussaieff collection. Tel Aviv, Archaeological Center Publications, 2000, 229 p. (ill.).

138. FILIP (Václav V.). Einführung in die Heraldik. Stuttgart, Steiner, 2000, 133 p. (Historische Grundwissenschaften in Einzeldarstellungen, 3).

139. GERRING (Britta). Sphragides: die gravierten Fingerringe des Hellenismus. Oxford, Archeopress, 2000, 200 p. (plates, ill.). (BAR. International series, 848).

140. GUT (Waltraud). Schwarz auf weiss: Maske und Schrift des heraldischen Ornaments. Stuttgart, Metzler, 2000, 356 p. (ill.). (M & P Schriftenreihe für Wissenschaft und Forschung).

141. HENNING (Eckart). Auxilia Historica. Beiträge zu den Historischen Hilfswissenschaften und ihren Wechselbeziehungen. Köln, Weimar u. Wien, Böhlau, 2000, VIII-382 p.

142. Heraldický register: Slovenskej repupliky. Zv. 1. (The Heraldic Register of the Slovak Republic). Ed. by Peter KARTOUS and Ladislav VRTEL. Martin, Ministerstvo vnútra Slovenskej republiky vo vydavateľstve Matice slovenskej, 2000, 277 p. [Eng. summary].

143. Lexikón erbov šľachty na Slovensku. Zv. 1. Trenčianska stolica. Šľachta Trenčianskej stolice podľa súpisov z rokov 1596 a 1646/47. (Encyclopedia of aristocratic armorial bearings in Slovakia. Vol. 1. The County of Trenčín. The aristocracy of the County of Trenčín according to the documents from 1596 and 1646/47). Ed. by Frederik FEDERMAYER. Bratislava, Vydavateľstvo Hajko & Hajková, 2000, XXIV-303 p.

144. Manoscritti (I) araldici della Biblioteca casanatense. A cura di Isabella CECCOPIERI e Laura GIALLOMBARDO. Roma, Istituto poligrafico e zecca dell stato, 2000, XVII-246 p. (ill.). (Indici e cataloghi. Nuova serie, 13).

145. OURY (G.-M.). Les armoiries de Saint-Vanne et de Saint-Maur. *Collectanea cistercensia*, 2000, 62, p. 262-268.

146. PARRA (Iván Darío). Francisco de Miranda y los símbolos venezolanos. Maracaibo, Parra Editores, 2000, 118 p. (ill., map, ports).

147. Raccolte comunali di Assisi: monete, gettoni, medaglie, sigilli, misure e armi. A cura di Maurizio MATTEINI. Milano, Electa; Perugina, Editori umbri associati, 2000, 266 p. (ill.). (Catalogo regionale dei beni culturali dell'Umbria).

148. STIELDORF (A.). Die Siegel der Herrscherinnen. Siegelführung und Siegelbild der "deutschen" Kaiserinnen und Königinnen. *Rheinische Vierteljahrsblätter*, 2000, 64, p. 1-44 (ill.).

149. TASSINARI (G.). L'araldica veneziana a Ravenna e la rocca Brancaleone. *Ravenna*, 2000, 7, 1, p. 85-101 (ill.).

150. VASILIKOU (Dora). Mycenaean signet rings of precious metals with cult scenes. Athens, The Archaeological Society at Athens, 2000, 69 p. (Library of the Archaeological Society at Athens, 196 – Ancient sites and museums in Greece, 14).

151. WAGNER (Anthony, Sir). Heralds and heraldry in the Middle Ages: an inquiry into the growth of the armorial function of heralds. Ed. by Anthony Richard WAGNER. Oxford, Oxford U. P., 2000, IX-176 p. (ill.). (Oxford scholarly classics).

152. WOODCOCK (Thomas), ROBINSON (John Martin). Heraldry in historic houses of Great Britain. London, National Trust, 2000, 240 p. (ill.).

153. ZNAMIEROWSKI (Alfred). Flags through the ages: a guide to the world of flags, banners, standards and ensigns. London, Southwater, 2000, 128 p. (ill.).

Cf. n° 125

§ 7. Numismatics and metrology.

154. ANTONSEN (Jack). Norges grunnlovsdag i metall. Et numismatisk oppslagsverk for 17. mai Medaljene 1881–2000. (A numismatic dictionary of the 17th of May medals, 1881–2000). Oslo, Norsk numismatisk forlag, 2000, 134 p. (ill.).

155. AYKUT (Şevki Nezihi). Türkiye Selçuklu sikkeleri. (Coins of Anatolian Seljukids). İstanbul, Eren yayınları, 2000.

156. BEAN (Simon C.). The coinage of the Atrebates and Regni. Oxford, Oxford University School of Archaeology, 2000, 318 p. (ill., maps). (Studies in Celtic coinage, 4).

157. BOMPAIRE (Marc), DUMAS (Françoise). Numismatique médiévale: monnaies et documents d'origine française. Turnhout, Brepols, 2000, 687 p. (ill., map). (L'atelier du médiéviste, 7).

158. BOMPAIRE (Marc). Les monnayages d'or d'Aquitaine anglo-gasconne. Le témoignage des livres de changeurs. *Revue numismatique*, 2000, 155, p. 261-279.

159. Cartagineses (Los) y la monetización del Mediterráneo occidental. Ed. por Ma. Paz GARCÍA-BELLIDO y Laurent CALLEGARIN. Madrid, Consejo Superior de Investigaciones Científicas, Instituto de Historia, Departamento de Historia Antigua y Arqueología, Casa de Velázquez, 2000, 185 p. (ill., maps). (Anejos de Archivo Español de Arqueología, 22).

160. Concordia ditat. 50 Jahre Numismatische Kommission der Länder in der Bundesrepublik Deutschland 1950–2000. Hrsg. v. Reiner CUNZ. Regenstauf, Gietl, 2000, 287 p. (Numismatische Studien, 13).

161. GENEVIÈVE (Vincent). Monnaies et circulation monétaire à Toulouse sous l'Empire romain, Ier–Ve siècle. Toulouse, Musée Saint-Raymond, 2000, 211 p. (ill., maps).

162. GUIDO (Francesco). Monete puniche e della Sardegna romana nella collezione F. S. di Cagliari. Milano, Edizioni ennerre S.r.l., 2000, 28 p. (ill.). (Annotazioni numismatiche. Supplemento, 16).

163. HIPPEL (Wolfgang von). Mass und Gewicht im Gebiet des Königreichs Württemberg und der Fürstentümer Hohenzollern am Ende des 18. Jahrhunderts.

Stuttgart, W. Kohlhammer, 2000, XV-247 p. (map). (Veröffentlichungen der Kommission für Geschichtliche Landeskunde in Baden-Württemberg. Reihe B, Forschungen, 145).

164. JANIN (Enrico). Scritti di argomento numismatico, 1972-1999. Genova, Circolo numismatico ligure Corrado Astengo, Sezione della Società ligure di storia patria, 2000, 150-36 p. (ill.).

165. KHARITONOV (Khristo). Numizmatika na Bulgariia: entsiklopediia. Sofiia, Abagar, 2000, 486 p. (ill.).

166. MAGNELLO (Eileen). A century of measurement: an illustrated history of the National Physical Laboratory. Bath, Canopus Publishing, 2000, 223 p. (ill.).

167. MISELLI (Walter). Considerazioni per la corretta classificazione delle medaglie papali emesse sino al secolo XVIII. *Rivista Italiana di Numismatica e Scienze affini*, 2000, 101, p. 217-235.

168. Moneta (La) greca e romana. A cura di Francesco PANVINI ROSATI. Roma, "L'Erma" di Bretschneider, 2000, 161 p. (plates, ill., map). (Storia della moneta, 1). [Cf. n^{os} <scelta> 2120, 2123, 2143, 2145, 2149, 2380, 2679, 2693, 2711, 2957.]

169. Moneta i mennictwo w Małopolsce i krajach sąsiednich. Materiały z Międzynarodowego Sympozjum Naukowego z udziałem numizmatyków z Czech, Słowacji, Ukrainy i Polski, Sanok 10–12 września 1999. (Münzen und Münzwesen in Kleinpolen [Małopolska] sowie in den Anrainerstaaten. Materialien vom Internationalen Wissenschaftlichen Symposion unter Anteilnahme der Numismatiker aus der Tschechischen, Slowakischen Republik, aus der Ukraine und aus Polen, Sanok 10.–12. September 1999). Sanok, Pol. Tow. Numizmatyczne. Koło w Sanoku im. Rudolfa Mękickiego, 2000, 117 p.

170. NOESKE (Hans-Christoph.). Münzfunde aus Ägypten 1. Die Münzfunde des ägyptischen Pilgerzentrums Abu Mina und die Vergleichsfunde aus den Dioecesen Aegyptus und Oriens vom 4.–8. Jh. n. Chr.: Prolegomena zu einer Geschichte des spätrömischen Münzumlaufs in Ägypten und Syrien. Berlin, Gebr. Mann, [s. p.]. (ill.). (Studien zu Fundmünzen der Antike: SFMA).

171. Ordo et mensura VI: internationaler interdisziplinärer Kongress für Historische Metrologie vom 28. bis 31. Oktober 1999 in der Physikalisch-Technischen Bundesanstalt. Hrsg. v. Rolf C. A. ROTTLÄNDER. St. Katharinen, Scripta mercaturae, 2000, IX-343 p. (plates, ill.). (Sachüberlieferung und Geschichte, 31).

172. PASZKIEWICZ (Borys). Pieniądz górnośląski w średniowieczu. (Die Währung Oberschlesiens zur Zeit des Mittelalters). Lublin, Wydaw. Uniw. M. Curie-Skłodowskiej, 2000, 385 p.

173. Pieniądz pamiątkowy i okolicznościowy – wspólnota dziejów. Białoruś, Litwa, Łotwa, Polska, Ukraina. Materiały z IV Międzynarodowej Konferencji Numizmatycznej, Supraśl, 7–9 IX 2000. (Gedenk- und Gelegenheitsmünzen – Geschichtsgemeinschaft. Weißrussland, Litauen, Lettland, Polen, Ukraine. Materialien von der IV. Internationalen Konferenz für Numismatik, Supraśl, 7.–9. IX. 2000). Hrsg. v. Krzysztof FILIPOW. Warszawa, Pol. Tow. Numizmatyczne, 2000, 251 p.

174. ROBERTSON (Anne Strachan). An inventory of Romano-British coin hoards. London: Royal Numismatic Society, 2000, IX-520 p. (ill., maps). (Royal Numismatic Society special publication, 20).

175. STONE (Michael Edward), ERVINE (Roberta R.). The Armenian texts of Epiphanius of Salamis De mensuris et ponderibus. Leuven, Peeters, 2000, VII-144 p. (ill.). (Corpus scriptorum Christianorum Orientalium, 583, 105; Corpus scriptorum Christianorum Orientalium. Subsidia, t. 105).

176. VAGI (David L.). Coinage and history of the Roman Empire, c. 82 B.C.–A.D. 480. Vol. 2. Coinage. Chicago a. London, Fitzroy Dearborn, 2000, 656 p. (ill.).

177. VEKOV (M.). Osobenosti na predmetrichnite evropejski. Opit za istoriko-metrologichen analiz. (Particularitées des mesures européennes prémétriques. Essai d'analyse historico-métrologique). *Istorichesko budeshche*, 2000, 2, p. 95-105.

178. WANG (Victor). Die Vereinheitlichung von Mass und Gewicht in Deutschland im 19. Jahrhundert: Analyse des metrologischen Wandels im Grossherzogtum Baden und anderen deutschen Staaten 1806 bis 1871. St. Katharinen, Scripta Mercaturae, 2000, VIII-360 p. (ill.). (Sachüberlieferung und Geschichte, 32).

179. XII. Internationaler Numismatischer Kongress, Berlin 1997: Akten. Hrsg. v. Bernd KLUGE und Bernhard WEISSER. Berlin, Staatliche Museen zu Berlin, 2000, 2 vol., 1488 p. (ill., maps).

Cf. n°s 141, 291-316, 591

§ 8. Linguistics.

180. BARTHELEMY (Tiphaine). Patronimic names and "noms de terre" in the French nobility in the eighteenth and the nineteenth centuries. *History of the family. An international quarterly*, 2000, 5, 2, p. 181-197.

181. BARTOLI LANGELI (Attilio). La scrittura dell' italiano. Bologna, Il Mulino, 2000, 182 p. (tav.).

182. BERGER (Elisabeth). Name und Recht. Die Entwicklung der Familiennamen und ihre Einbeziehung in die Rechtsordnung. *Zeitschrift der Savigny-Stiftung für Rechtsgeschichte. Germanistische Abteilung*, 2000, 117, p. 564-591.

183. BIANCHI (Serge). Les "prénoms révolutionnaires" dans la Révolution française: un chantier en devenir. *Annales historiques de la Révolution française*, 2000, 322, p. 39-60.

184. BORCA (Federico). Facies locorum: morfologia, onomastica e percezione delle isole nella cultura romana. *Quaderni di storia*, 2000, 26, 51, p. 187-202.

185. CAMPBELL (Lyle). American Indian languages: the historical linguistics of Native America. New York a. Oxford, Oxford U. P., 2000, 528 p. (maps). (Oxford studies in anthropological linguistics, 4).

186. CASSAGNE (Jean-Marie), KORSAK (Mariola). Origine des noms de villes et villages de la Vendée. Saint-Jean-d'Angély, J.-M. Bordessoules, 2000, 303 p.

187. CASSELMAN (Bill). What's in a Canadian name? The origins and meanings of Canadian names. Toronto, McArthur & Co, 2000, XIII-250 p. (1 port).

188. COLE (Ann), CUMBER (Janey), GELLING (Margaret). Old English merece "wild celery, smallage" in place-names, *Nomina*, 2000, 23, p. 141-148.

189. EVERETT-HEATH (John). Place names of the world. Europe: historical context, meanings and changes. Basingstoke, Macmillan, 2000, XXII-413 p. (maps).

190. FANTI (Mario). Le vie di Bologna: saggio di toponomastica storica e di storia della toponomastica urbana. Bologna, Istituto per la Storia di Bologna, 2000, 2 vol., [s. p.]. (ill.). (Fonti per la storia di Bologna. Collana Testi, Nuova ser., 13).

191. GALMÉS DE FUENTES (Álvaro). Los topónimos: sus blasones y trofeos (la toponimia mítica). Madrid, Real Academia de la Historia, 2000, 211 p.

192. GOETZ (H.-W.). Gentes. Zur zeitgenössischen Terminologie und Wahrnehmung ostfränkischer Ethnogenese im IX. Jht.. *Mitteilungen des Instituts für österreichische Geschichtsforschung*, 2000, 108, p. 85-116.

193. GRAY (Ronald), STRUBBINGS (Derek). Cambridge street-names: their origins and associations. Cambridge, Cambridge U. P., 2000, XIX-159 p.

194. KADMON (Naftali). Toponymy: the lore, laws, and language of geographical names. New York, Vantage Press, 2000, X-333 p. (ill., maps).

195. KEMPENEERS (Paul). Toponymie van Landen. Leuven, Instituut voor Naamkunde te Leuven, 2000, XII-306 p. (ill., maps). (Nomina geographica Flandrica. Monografieë, 17).

196. KLEMPERER (Victor). The language of the Third Reich: LTI, lingua tertii imperii, a philologist's notebook. London, Athlone, 2000, 296 p.

197. LÄSSIG (Simone). Sprachwandel und Verbürgerlichung. Zur Bedeutung der Sprache im innerjüdischen Modernisierungsprozeß des frühen 19. Jahrhunderts. *Historische Zeitschrift*, 2000, 270, 3, p. 617-668.

198. LEI: lessico etimologico italiano. A cura di Elda MORLICCHIO. Wiesbaden, Reichert, 2000, [s. p.].

199. Lexicon (A) of Greek personal names. Vol. 3B: Central Greece from the Megarid to Thessaly. Ed. by P.M. FRASER and E. MATTHEWS and others. Oxford, Clarendon Press, 2000, XXI-478 p.

200. Linguistics (A) round-table on dictionaries and the history of the language. Ed. by Giulio LEPSCHY and Prue SHAW. London, University of London, Centre for Italian Studies, 2000, 127 p. (Occasional papers / University of London, Centre for Italian Studies, 4).

201. Malik al-Afdal al-`Abbas ibn `Ali. The king's dictionary: the Rasûlid Hexaglot: fourteenth century vocabularies in Arabic, Persian, Turkish, Greek, Armenian and Mongol. Ed. by Peter B. GOLDEN. Leiden a. Boston, Brill, 2000, XII-418 p. (Handbuch der Orientalistik. 8. Abt., Zentralasien, 4).

202. MINKOVA (Milena). The personal names of the Latin inscriptions in Bulgaria. New York, Peter Lang, 2000 (Studien zur klassischen Philologie, 118).

203. MONTBARBUT (Johnny). La toponymie française des États-Unis d'Amérique. Saint-Laurent, Éditions P. Tisseyre, 2000, 318 p.

204. MUSERE (Jonathan). Traditional African names. Lanham a. London, Scarecrow Press, 2000, IX-401 p.

205. Noms (Les) de famille en France: Histoires et anecdotes. Ed. par Marie-Odile MERGNAC. Préface de Jacques DUPÂQUIER. Paris, Archives & culture, 2000, 477 p. (ill., maps).

206. Noms propres. Ed. par HONORÉ, Marie-Anne PAVEAU et Gabriel PÉRIÈS. Lyon, ENS Éditions, 2000 (Mots [Special issue], 63).

207. Novum glossarium Mediae Latinitatis ab anno DCCC usque ad annum MCCC: Permachino-Pezzola. Ed. par François DOLBEAU. Genève, Droz, 2000, VII-573-948 p.

208. Onomastique et parenté dans l'Occident médiéval. K. S. B. KEATS-ROHAN et C. SETTIPANI. Oxford, Unit for Prosopographical Research, 2000, 310 p. (tables). (Prosopographica et Genealogica, 3).

209. Personennamen des Mittelalters: PMA: Namensformen für 13000 Personen gemäss den Regeln für die alphabetische Katalogisierung (RAK) = Personal names of the Middle Ages: (PMA): names of 13,000 persons according to the Regeln für die alphabetische Katalogisierung (RAK) = Nomina scriptorum medii aevi. Hrsg. v. Bayerische Staatsbibliothek/Universitätsbibliothek München. München, Saur, 2000, XXIII-696 p.

210. QUINN (Seán E.). Surnames in Ireland. Bray, Irish Genealogy Press, 2000, 183 p.

211. RÖDER (Katrin). Struktur und Verbreitung der alteuropäischen Toponymie: eine Studie am Beispiel der Wurzelformen +is- und +ur-. Berlin, Logos, 2000, 174, [9] p. (1 map).

212. ROFFIN (Raymond), ROFFIN (Françoise). L'étude des patronymes du XVIIIe au XIXe siècle: pourquoi faire?. *Archipal*, 2000, 47, p. 76-87.

213. SADHU (Shyam Lal). Place names in Kashmir. Mumbai, Bharatiya Vidya Bhavan, 2000, 196 p. (Bhavan's book university).

214. SIHLER (Andrew L.). Language history: an introduction. Amsterdam, John Benjamins, 2000, XV-298 p. (Amsterdam studies in the theory and history of linguistic science, 191, Series IV, Current issues in linguistic theory).

215. STOIANOV (Valeri). Istoriia na izuchavaneto na Codex Cumanicus: Neslavianska, kumano-pechenezhka antroponimika v bulgarskite zemi prez XV vek. (Histoire de la recherche du Codex Cumanicus. Antroponomie non slave, koumano-pétchéneque dans les terres bulgares). Sofiia, Ogledalo, 2000, 319 p. (Biblioteka "Elbegen". Bulgaro-Turcica, 3-4).

216. STOTZ (Peter). Handbuch zur lateinischen Sprache des Mittelalters. Vol. 2: Bedeutungswandel und Wortbildung. München, C. H. Beck, 2000, XXVI-482 p. (Handbuch d. Altertumswiss., 2ª, 5/2).

217. Textual parameters in older languages. Ed. by Susan C. HERRING, Pieter VAN RENEEN and Lene SCHØSLER. Amsterdam, J. Benjamins Pub., 2000, X-448 p. (Amsterdam studies in the theory and history of linguistic science, 195. Series IV, Current issues in linguistic theory).

218. TRASK (Robert Lawrence). The dictionary of historical and comparative linguistics. Edinburgh, Edinburgh U. P., 2000, VIII-403 p.

219. UTTERSTRÖM (Gudrun). Svenska adelsnamn: uppkomst och utveckling (Les noms nobles en Suède: origine et développement). *Studia anthroponymica scandinavica*, 2000 18, p. 25-63. [Résumé anglais].

220. Vocabulary (The) of English place-names: Brace-Cæster. Ed by David N. PARSONS and Tania STYLES. Nottingham, Centre for English Name-Studies, 2000, XIII-177 p.

§ 9. **Historical geography, travels and discoveries.**

* 221. BERTRAND (Giles). Bibliographie des études sur le voyage en Italie. Voyage en Italie, voyage en Europe, XVIᵉ–XVIIIᵉ siècle. *Cahiers du CRHIPA*, 2000, 2, p. 7-301.

222. ARNAUD (Vicente Guillermo). Las Islas Malvinas: descubrimiento, primeros mapas y ocupación: siglo XVI. Buenos Aires, Academia Nacional de Geografía, 2000, 255 p. (ill., maps). (Publicación especial, 13).

223. ASILTÜRK (Bâki). Osmanlı seyyahlarının gözüyle Avrupa. (Europe through the eyes of the Ottoman travelers). Ed. by Betül BILIKTÜ. İstanbul, Kaknüs yayınları, 2000, 591 p.

224. Atlas historique mondial. Ed. por Georges DUBY. Paris, Larousse, 2000, 349 p. (ill., maps).

225. BURNETT (D. Graham). Masters of all they surveyed: exploration, geography, and a British El Dorado. Chicago, University of Chicago Press, 2000, XV-298 p.

226. Cappelens historiske atlas. (Historical atlas). Ed. May Britt STAMSØ. Oslo, Cappelen, 2000, 96 p. (ill.).

227. Cassell (The) atlas of world history. Vol. 2. The Medieval and early Modern worlds. London, Cassell, 2000, 192 p. (ill., maps, ports.).

228. CHEVALLIER (Raymond). Lecture du temps dans l'espace: topographie archéologique et historique. Paris, Picard, 2000, 229 p. (ill., photos., plans, maps).

229. COURVILLE (Serge). Le Québec: genèses et mutations du territoire: synthèse de géographie historique. Sainte-Foy, Presses de l'Université Laval, 2000, XVII-508 p. (ill., maps). (Géographie historique).

230. DAVID (Robert G.). The Arctic in the British imagination, 1818–1914. Manchester, Manchester U. P., 2000, XX-278 p. (ill.). (Studies in imperialism).

231. Descobridores do Brasil: exploradores do Atlântico e construtores do Estado da Índia. Coord.: Joao Paulo OLIVEIRA E COSTA. Lisboa, Sociedade Historica da Independencia de Portugal, 2000, 478 p. (Colecçao Memória lusíada, 8).

232. Descobrimento (O) do Brasil nos textos de 1500 a 1571. Ed. por. José Manuel GARCIA. Lisboa, Fundaçao Calouste Gulbenkian, 2000, 87 p. (10 leaves of plates, ill., facsims, maps).

233. DIERKENS (Alain), SANSTERRE (Jean-Marie), KUPPER (Jean-Louis). Voyages et voyageurs à Byzance et en Occident du VIᵉ au XIᵉ siècle. Actes du colloque international organisé par la Section d'Histoire de l'Université Libre de Bruxelles en collaboration avec le Département des Sciences historiques de l'Université de Liège (5–7 mai 1994). Genève, Droz, 2000, 421 p.

234. DOLAN (Brian). Exploring European frontiers: British travellers in the age of Enlightenment. Basingstoke, Macmillan, 2000, XI-232 p. (plates, ill.).

235. DRECOLL (Carsten). Idrísí aus Sizilien: der Einfluss eines arabischen Wissenschaftlers auf die Entwicklung der europäischen Geographie. Egelsbach u. New York, Hänsel-Hohenhausen, 2000, 177 p. (ill.). (Deutsche Hochschulschriften, 1187).

236. Géographes grecs. T. 1. Introduction générale. Ed. par. Paris, Les Belles Lettres, 2000, 310 p. (ill., maps).

237. GOLVERS (N.). Jesuit cartographers in China. Francesco Brancati, S. J., and the map (1661?) of Sungchiang prefecture (Shanghai). *Imago mundi*, 2000, 52, p. 30-42 (ill.).

238. GUILLET (François). Naissance de la Normandie; genèse et épanouissement d'une image régionale en France, 1750–1850. Caen, Annales de Normandie, 2000, 586 p.

239. HEWSEN (Robert H.). Armenia: a historical atlas. Chicago, University of Chicago Press, 2000, 336 p. (ill., maps).

240. HIATT (A.). The cartographic imagination of Thomas Elmham. *Speculum*, 2000, 75, p. 859-886 (ill.).

241. Historical maps of Central Asia, 9th–19th centuries A.D. Ed. by Yuri BREGEL. Bloomington, Indiana University Research Institute for Inner Asian Studies, 2000 (leavs, maps). (Papers on Inner Asia. Special supplement).

242. HOFMANN (Catherine). La genèse de l'atlas historique en France (1630–1800): pouvoirs et limites de la carte comme "œil de l'histoire". *Bibliothèque de l'Ecole de Chartes*, 2000, 158, p. 97-128.

243. İLDEM (Arzu Etensel). Fransız gezginlerin gözüyle Türkler ve Yunanlılar: 19. yüzyılın ilk yarısında Fransız gezginlerin yapıtlarında karşılaştırmalı Türk ve Yunan imgesi. (Turks and Greeks through the eyes of French travelers: comparative Turkish and Grek images in the works of French travelogues in the first half of the XIXth century). İstanbul, Boyut Kitapları, 2000, 288 p.

244. JUZARTE (Teotônio José). Diario da navegacao. Sao Paulo, Editora da Universidade de Sao Paulo/ Imprensa Oficial do Estado, 2000, 461 p. (ill.).

245. Kartografia Królestwa Polskiego 1815–1915. Materiały XVIII Ogólnopolskiej Konferencji Historyków Kartografii, Warszawa, 21–22 listopada 1997. (Die Kartographie des Königreichs Polen 1815–1915. Materialien der XVIII. Gemeinpolnischen Konferenz der Kartographie-Historiker, Warschau, 21.–22. November 1997). Hrsg. v. Lucyna SZANIAWSKA und Jerzy OSTROWSKI. Warszawa, Bibl. Nar., 2000, 280 p. (PAN. Inst. Historii Nauki; Z Dziejów Kartografii, 10).

246. KEARNS (G.). Maps, models and registers: the historical geography of the population of England. *Journal of historical geography*, 2000, 26, 2, p. 298-304.

247. KONIAS (Andrzej). Kartografia topograficzna Śląska Cieszyńskiego i zaboru austriackiego od II połowy XVIII wieku do początku XX wieku. (Die topographische Kartographie Teschiner Schlesiens sowie des österreichischen Teilungsgebietes von der II. Hälfte des XVIII. bis zum Anfang des XX. Jahrhunderts). Katowice, Wydaw. Uniw. Śląskiego, 2000, 259 p. (Prace Naukowe Uniw. Śląskiego w Katowicach, 1866).

248. KONSTAM (Angus). Atlas of Medieval Europe. New York, Facts on File, 2000, 192 p. (ill., maps). – IDEM. Historical atlas of exploration: 1492–1600. New York, Checkmark Books, 2000, 191 p.

249. KORTE (Barbara). English travel writing from pilgrimages to postcolonial explorations. Basingstoke, Macmillan, 2000, VII-218 p.

250. KOSHAR (Rudy). German travel cultures. New York, Berg, 2000, X-241 p. (Leisure, consumption and culture).

251. KOSONEN (Katariina). Kartta ja kansakunta. Suomalainen lehdistökartografia sortovuosien protesteista Suur-Suomen kuviin 1899–1942. (The map and the nation. Finnish press cartography from the protests of oppression years to the images of Greater Finland, 1899–1942). Helsinki, SKS, 2000, 389 p. (ill. maps). (Suomal. Kirjall. Seuran toim., 793).

252. Lexicon topographicum urbis Romae. Vol. 6. A cura di Eva Margarete STEINBY. Roma, Quasar, 2000, 142 p. (ill.).

253. Lexikon mapových archivů a sbírek České republiky. (Lexicon of map's archives and collections in the Czech Republic). Ed. by Eva SEMOTANOVÁ and Robert ŠIMŮNEK. Praha, Historický ústav AV ČR, 2000, 267 p.

254. LOBATO (Mirta Zaida), SURIANO (Juan). Atlas histórico de la Argentina. Buenos Aires, Sudamericana, 2000, 587 p. (ill., maps, ports.). (Nueva historia argentina).

255. LÓPEZ-DAVALILLO (Larrea, Julio). Atlas histórico de España y Portugal: desde el paleolítico hasta el siglo XX. Madrid, Editorial Síntesis, 2000, 223 p. (maps). – IDEM. Atlas histórico mundial: Desde el Paleolítico hasta el siglo XX. Madrid, Editorial Síntesis, 2000, 255 p. (maps).

256. LÖYTÖNEN (Markku). Maapallon kartoitus. Kartläggning av jordklotet. (Mapping of the world). *In*: Terra cognita [Cf. n° 283], p. 46-72

257. LOZOVSKY (Natalia). The earth is our book: geographical knowledge in the Latin West ca. 400–1000. Ann Arbor, University of Michigan Press, 2000, VIII-182 p. (ill.).

258. MAC CONNELL (Curt). Coast to coast by automobile: the pioneering trips, 1899–1908. Stanford, Stanford U. P., 2000, XIV-349 p.

259. MARKKANEN (Tapio). Imago mundi. Kartan laadinan vaiheita. Etapper i kartframställning. (The history of mapmaking). *In*: Terra cognita [Cf. n° 283], p. 14-38. – IDEM. Karttaprojektiot. Kartprojektionerna. (Map projections). *In*: Terra cognita [Cf. n° 283], p. 39-41

260. MAYHEW (Robert J.). Enlightenment geography: the political languages of British geography, 1650–1850. Basingstoke, Macmillan, 2000, VIII-324 p. (Studies in modern history).

261. Medical geography in historical perspective. Ed. by Nicolaas A. RUPKE, London, Wellcome Trust Centre for the History of Medicine at UCL, 2000, 227 p. (plates, maps). (Medical history. Supplement, 20).

262. Mégapoles méditerranéennes: géographie urbaine rétrospective: actes du colloque organisé par l'Ecole française de Rome et la Maison méditerranéenne des sciences de l'homme (Rome, 8–11 mai 1996). Ed. par Claude NICOLET, Robert ILBERT et Jean-Charles DEPAULE. Rome, Ecole française de Rome, 2000, 1071 p. (ill.). (Collection de l'Ecole française de Rome, 261). [Cf. n[os] <sélection> 2552, 2568, 2676.]

263. MICHEL (Franck). Désirs d'ailleurs. Essai d'anthropologie des voyages. Préf. de Jean-Didier URBAIN. Paris, A. Colin, 2000, 272 p. (Chemins de traverse).

264. MORAES (Antonio Carlos Robert). Bases da formaçao territorial do Brasil: o território colonial brasileiro no "longo" século XVI. Sao Paulo, Editora Hucitec, 2000, 431 p. (Coleçao Estudos históricos, 41).

265. Osmanlı coğrafya literatürü tarihi. (The history of Ottoman geographical literature). İstanbul, İslâm Tarih, Sanat ve Kültür Araştırma Merkezi, 2000, 2 vol., [s. p.].

266. PARK (Mungo). Travels in the interior districts of Africa. London, Duke U. P., 2000, 407 p.

267. PITCHER (Edward W. R.). Discoveries in periodicals, 1720–1820: facts and fictions. Lewiston a. Lampeter, E. Mellen Press, 2000, 702 p. (Studies in British and American magazines, 7).

268. PODOSINOV (Aleksandr V.). Problemy istoricheskoj geografii Vostochnoj Evropy (antichnost' i rannee srednevekov'e). (Problems of historical geography of Eastern Europe: Antiquity and Early Middle Ages). App.: Selected bibliography of the works on source criticism and ancient history of the Northern Black Sea Littoral. Lewinston, Queenston a. Lampeter, Edwin Mellen Press, 2000, XII-378 p. (ind.). (Rossijskie issledovanija po mirovoj istorii i kul'ture, 2). [Eng. summary and table of contents].

269. RANDLES (W. G. L.). Geography, cartography and nautical science in the Renaissance: the impact of the great discoveries. Aldershot, Ashgate, 2000, 1v. (ill., maps). (Collected studies, CS689).

270. Regions and landscapes. Reality and imagination in late Medieval and Early Modern Europe. Ed. by Peter AINSWORTH and Tom SCOTT. Oxford, Bern, Berlin, Bruxelles, Frankfurt am Main, New York a. Wien, Lang, 2000, 244 p. (ill., maps, tables).

271. Représentations (Les) de la Méditerranée. Dir. par Thierry FABRE et Robert ILBERT. Paris, Maisonneuve et Larose, 2000, 737 p.

272. Romantic geographies: discourses of travel, 1775–1844. Ed. by Amanda GILROY. Manchester, Manchester U. P., 2000, XII-260 p. (ill.). (Exploring travel).

273. RUBIÉS (Joan-Pau). Travel and ethnology in the Renaissance: South India through European eyes, 1250–1625. Cambridge, Cambridge U. P., 2000, XXII-443 p. (ill., maps). (Past and present publications).

274. SALITOT (Michelle). Modes d'appropriation d'un rivage. La baie du Mont-Saint-Michel. Paris, L'Harmattan, 2000, 279 p. (Maritimes).

275. SCHEUCH (Manfred). Historischer Atlas Österreich. Wien, C. Brandstätter, 2000, 229 p (ill., maps).

276. Slovenija na vojaškem zemljevidu 1763–1787. Zv. 6. Sekcije 139, 140, 141, 142, 143, 144, 145, 146, 147, 165, 166, 167, 168, 169, 170, 171, 172, 196, 197, 198 [Kartografsko gradivo]: karte = Josephinische Landesaufnahme 1763–1787 für das Gebiet der Republik Slowenien. Band 6. Sektionen 139, 140, 141, 142, 143, 144, 145, 146, 147, 165, 166, 167, 168, 169, 170, 171, 172, 196, 197, 198: Karten. Vodja projekta, toponimija sekcij, indeks, redakcija Vincenc RAJŠP. Ljubljana, Znanstvenoraziskovalni center SAZU, Arhiv Republike Slovenije, 2000, [s. p.].

277. STEWART STOKES (Hamish I.). Del mar del Norte al mar del Sur: navegantes británicos y holandeses en el Pacífico Suroriental, 1570–1807. Valparaíso, Centro de Estudios de la Cuenca del Pacífico, Universidad de Playa Ancha, 2000, 221 p. (ill., maps).

278. Strabone e l'Asia Minore. A cura di Anna Maria BIRASCHI e Giovanni SALMERI. Napoli, Edizioni scientifiche italiane, Perugia, Università degli studi di Perugia, 2000, 650 p., (ill., map). (Incontri perugini di storia della storiografia antica e sul mondo antico, 10).

279. Studies in historical geography and biblical historiography: presented to Zecharia Kallai. Ed. by Gershon GALIL and Moshe WEINFELD. Leiden, Brill, 2000, XII-281 p. (19 p. of plates, ill., maps, 1 port.). (Supplements to Vetus Testamentum, 81).

280. Svensk militärhistorisk atlas. En guide till klassiska slagfält och krigsskådeplatser i Sverige. (Atlas historico-militaire suédois. Guide des principaux champs de bataille en Suède). Ed. par Per DAHL. Stockholm, Hjalmarson & Högberg, 2000, 249 p. (ill.).

281. 'Tableau (Le) de la géographie de la France' de Paul Vidal de La Blache: dans le labyrinthe des formes. Sous la dir. de Marie-Claire ROBIC. Paris, Ed. du CTHS, 2000, 298 p. (ill.).

282. TAKEUCHI (Keiichi). Modern Japanese geography: an intellectual history. Tokyo, Kokon Shoin, 2000, XIV-250 p.

283. Terra cognita. Maailma tulee tunnetuksi. Kännedom om världen ökar. (Discovering the world). Toim. Ed. Leena PÄRSSINEN. Helsinki, Helsinki, University Library, 2000, 197 p. (ill., maps). [Cf. n^{os} <choice> 256, 259.]

284. TISSOT (Laurent). Naissance d'une industrie touristique. Les anglais et la Suisse au XIXe siècle. Lausanne, Editions Payot, 2000, 302 p.

285. Topografia (La) antica. A cura di Pier Luigi DALL'AGLIO. Premessa di N. ALFIERI. Bologna, Clueb, 2000, 241 p. (ill., maps). (Manuali scientifici).

286. Tourisme et relations internationales. *Relations internationales*, 2000, 102, p. 141-251. [Contient: de PUYMÈGE (G.). Introduction (p. 141-146). – CHABAUD (G.). Aux origines du tourisme: les Grands Tours de l'époque moderne (p. 147-159). – RAUCH (A.). L'Orient dans l'essor du tourisme au XIXe siècle (p. 161-172). – PATIN (V.). Les diplomates découvreurs (p. 173-184). – TISSOT (L.). Agences de voyages et conquêtes touristiques, 1850–1914 (p. 185-199). – MAZUY (R.). Le tourisme idéologique en Union soviétique (p. 201-207). – CLAEYS (V.), VAN PRAET (J.). Le tourisme social dans une perspective internationale (p. 219-232). – CAZES (G.). Tourisme et relations internationales, perspective du dernier demi-siècle (p. 233-245).]

287. Trade, travel, and exploration in the Middle Ages: an encyclopedia. Ed. By John Block FRIEDMAN and Kristen MOSSLER FIGG. Associate editor, Scott D. WESTREM. Collaborating editor, Gregory G. GUZMAN. New York a. London, Garland Pub., 2000, XXXIX-715 p. (ill., maps). (Garland reference library of the humanities, 1899).

288. VAN SANT (John E.). Pacific pioneers: Japanese journeys to America and Hawaii, 1850–1880. Foreword by Roger DANIELS. Urbana a. Chicago, University of Illinois Press, 2000, XII-191 p. (Asian American experience).

289. WIGAL (Donald). Historic maritime maps used for historic exploration, 1290–1699. New York, Parkstone Press, 2000, 264 p. (ill., maps).

290. ZIFF (Larzer). Return passages: great American travel writing, 1780–1910. New Haven a. London, Yale U. P., 2000, 304 p. (ill., map).

Cf. n° 3342

§ 10. Iconography and images.

291. AMIC (Sylvain), PATRY (Sylvie). Les recueils de costumes à l'usage des peintres (XVIIIe–XIXe siècles): un genre éditorial au service de la peinture d'histoire? *Histoire de l'art*, 2000, 46, p. 39-66.

292. ANDALORO (Maria), ROMANO (Serena), FRASCHETTI (Augusto). Arte e iconografia a Roma: da Costantino a Cola di Rienzo. Milano, Jaca book, 2000, 266 p. (ill., maps, ports). (Di fronte e attraverso, 537. Storia dell'arte, 15).

293. BACCI (Michele). "Pro remedio animae": immagini sacre e pratiche devozionali in Italia centrale (secoli XIII e XIV). Pisa, ETS, 2000, 504 p. (ill.). (Piccola biblioteca Gisem,15).

294. BHATTACHARYA (Gouriswar). Essays on Buddhist, Hindu, Jain iconography and epigraphy. Dhaka, International Centre for Study of Bengal Art, 2000, 654 p. (ill., maps). (Studies in Bengal art series, 1).

295. BOECKL (Christine M.). Images of plague and pestilence iconography and iconology. Kirksville, Truman State U. P., 2000, XIV-210 p. (ill.). (Sixteenth century essays & studies, 53).

296. BOESPFLUG (François). La Trinité à l'heure de la mort. Sur les motifs trinitaires en contexte funéraire à la fin du Moyen Age (m. XIVe–déb. XVIe siècle). *Cahiers de Recherches médiévales*, 2000, 216 p.

297. BÜHNEMANN (Gudrun). The iconography of Hindu tantric deities. Groningen, Egbert Forsten, 2000, [s. p.]. (ill.). (Series, Gonda indological studies, 9).

298. DURAND (Jorge). La experiencia migrante: iconografía de la migración México-Estados Unidos. Ciudad de México, Altexto, 2000, 202 p. (ill., facims.).

299. ĒTINGOF (Ol'ga E.). Obraz Bogomatery: ocherki vizantijskoj ikonografii XI–XIII vv. (The Image of the Virgin: essays on Byzantine iconography of the 11th–13th centuries). In-t khristianskoj kul'tury srednevekov'ja. Moskva, Progress-Traditsija, 2000, 311 p. (ill., bibl.). [Eng. summary].

300. EVERSON (Paul), STOCKER (David). A newly identified figure of the Virgin from a late Anglo-Saxon rood at Great Hale, Lincolnshire. *Antiquaries journal*, 2000, 80, p. 285-294.

301. FINALDI (Gabriele). The image of Christ. London, National Gallery Company Limited; New Haven, distributed by Yale U. P., 2000, 224 p. (ill.).

302. GAPOSCHKIN (M. Cecilia). The king of France and the queen of heaven: the iconography the Porte Rouge of Notre-Dame de Paris. *Gesta*, 2000, 39, 1, p. 58-72.

303. Iconography (The) of power: ideas and images of rulership on the English Renaissance stage. Ed. by György E. SZONYI and Rowland WYMER. Szeged, JATEPress, 2000, 214 p. (ill.). (Acta Universitatis Szegediensis de Attila József Nominatae. Papers in English and American studies, 8).

304. Ikonstas: Proiskhozhdenie, razvitie, simvolika. (Iconostasis: origin, development, meaning). Ed. A. M. LIDOV. Tsentr vostochnokhristianskoj kul'tury. Moskva, Progress-Traditsija, 2000, 751 p. (ill.).

305. KUNÉ (C.). Er taufte mit Wasser. Zur Taufe Christi im deutschen religiösen Drama und in der bildenden Kunst des späten Mittelalters. *Neophilologus*, 2000, 84, p. 241-253.

306. LINI (Gabriella), GROSSENBACHER (Maya), CHRISTIE (Yves). La Bible du Roi, Daniel et Ezéchiel dans les bibles moralisées et les vitrail de la Sainte-Chapelle. *Arte medievale*, 2000, 2, 14, 1-2, p. 73-99.

307. MATCHABELI (K.). Remarques sur l'iconographie de la crucifixion sur les stèles géorgiennes du haut moyen âge. *Byzantion*, 2000, 70, p. 91-104 (ill.).

308. MICHALSKY (Tanja). Memoria und Repräsentation. Die Grabmäler des Königshauses Anjou in Italien. Göttingen, Vandenhoeck et Ruprecht, 2000, 446 p. (ill.).

309. National imaginaries, American identities: the cultural work of American iconography. Ed. by Larry J. REYNOLDS and Gordon HUTNER. Princeton, Princeton U. P., 2000

310. NOVELLI (Edoardo). C'era una volta il Pci: autobiografia di un partito attraverso le immagini della sua propaganda. Roma, Editori riuniti, 2000, 318 p. (ill.).

311. OPINEL (Annick). Du choléra morbus et de ses représentations dans la peinture française du XIXe siècle. *Histoire de l'art*, 2000, 46, p. 67-76.

312. OTT (N. H.). Ikonen deutscher Ideologie. Der Nibelungenstoff in der Bildkunst von Mittelalter bis

zur Gegenwart. *Zeitschrift für bayerische Landesgeschichte*, 2000, 43, p. 325-356 (ill.).

313. SABATIER (Gérard). La glorie du roi. Iconographie de Louis XIV de 1661 à 1672. *Histoire, économie, société*, 19, 4, p. 527-560.

314. SOUTHERN (Eileen), WRIGHT (Josephine). Images: iconography of music in African-American culture 1770s–1920s. New York a. London, Garland Pub, 2000, XXIII-299 p. (ill.). (Music in African-American culture, 1. Garland reference library of the humanities, 2089).

315. Temi di iconografia paleocristiana. A cura di Fabrizio BISCONTI. Città del Vaticano, Pontificio istituto di archeologia cristiana, 2000, 385 p. (plates, ill.). (Sussidi allo studio delle antichità cristiane, 13).

§ 10. Addenda 1999.

316. ASSAYAG (Jackie). L'Inde fabuleuse. Le charme discret de l'exotisme français. Paris, Ed. Kimé, 1999, 249 p.

Cf. n°s 154-179, 377, 1255-1294, 3086

B

MANUALS, GENERAL WORKS AND WORKS ON LARGE PERIODS

§ 1. Archives, libraries and museums (*a.* Archives; *b.* Libraries; *c.* Museums). 317-396. – § 2. History of historiography (*a.* General; *b.* Special studies). 397-744. – § 3. Methodology, philosophy, and teaching of history. 745-831. – § 4. Ethnology, folklore and historical anthropology. 832-895. – § 5. General history. 896-1023. – § 6. Theory of the state and of society. 1024-1084. – § 7. Constitutional and legal history. 1085-1116. – § 8. Economic and social history. 1117-1182. – § 9. History of civilization, sciences and education. 1183-1254. – § 10. History of art. 1255-1294. – § 11. History of religions (*a.* General; *b.* Special studies). 1295-1408. – § 12. History of philosophy. 1409-1428. – § 13. History of literature. 1429-1476.

§ 1. Archives, libraries and museums.

a. Archives

* 317. Bibliografía archivística. Ed. par España. Subdirección General de los Archivos Estatales. Madrid, Ministerio de Educación, Cultura y Deporte. Secretaría General Técnica. Centro de Publicaciones, 2000, 1 CD-ROM

* 318. Scritti di teoria archivistica italiana: rassegna bibliografica. A cura di Isabella MASSABÒ RICCI e Marco CARASSI. Roma, Ministero per i beni e le attività culturali, Ufficio centrale per i beni archivistici, 2000, 200 p.

** 319. Archivista (L') sul confine: scritti di Isabella Zanni Rosiello. A cura di Carmela BINCHI e Tiziana DI ZIO. Roma, Ministero per i beni e le attività culturali, Ufficio centrale per i beni archivistici, 2000, 453 p. (Pubblicazioni degli archivi di stato, saggi, 60).

320. ÁLVAREZ PINEDO (Francisco Javier), RODRÍGUEZ DE DIEGO (José Luis). The archives of Spain, Simancas. Barcelona, Lunwerg Editores, 2000, 296 p. (ill.).

321. BAS MARTÍN (Nicolás). Juan Bautista Muñoz (1745–1799) y la fundación del Archivo General de Indias. Valencia, Generalidad, 2000, 184 p. (ill.).

322. CARRILERO MARTÍNEZ (Ramón). Carlos V y Albacete: colección documental del emperador en el archivo histórico provincial. Albacete, Instituto de Estudios Albacetenses Don Juan Manuel, 2000, 500 p. (ill.).

323. Casperia: inventario dell'archivio 1099–1860 e studi documentari. A cura di Agostino ATTANASIO e Alfredo PELLEGRINI. Roma, Gangemi, 2000, 254 p.

324. DE RIDDER (P.). De archiven van de hospitalen, godshuizen en liefdadigheidsinstellingen en het taalgebruik te Brussel (vóór 1764). *Eigen schoon en de Brabander*, 2000, 83, p. 1-59.

325. DELSALLE (Paul). Lire et comprendre les archives des XVIe et XVIIe siècles. Besançon, Presses universitaires franc-comtoises, 2000, 232 p. (ill.). (Collection Didactiques. Histoire).

326. DUCLERT (Vincent). Le secret en politique au risque des archives? Les archives au risque du secret en politique. Une histoire archivistique française. *Matériaux pour l'histoire de notre temps*, 2000, 58, p. 9-27.

327. ELORZA MAIZTEGUI (Javier). Documentación medieval de los archivos municipales de Eibar (1409–1508) y de Soraluze/Placencia de las armas. San Sebastián, Editorial Eusko Ikaskuntza, S.A. Sociedad de Estudios Vascos, 2000, V-132-XXIII p. (Fuentes documentales medievales del País Vasco, 97).

328. FAYET-SCRIBE (Sylvie). Histoire de la documentation en France. Culture, science et technologie de l'information, 1895–1937. Paris, CNRS, 2000, 313 p.

329. FUNARO (Liana Elda). «Nelle domestiche mura». Carte dei Lorena nella Biblioteca Mediceo Laurenziana. *Archivio storico italiano*, 2000, 158, 585, p. 515-552.

330. GOTTERI (Nicole). Archives nationales. Maison de l'Empereur. Administration de l'intendance générale, an X-1815. Inventaire des articles O^2 150 à 223. Paris, Centre historique des Archives nationales, 2000, 144 p.

331. Guida agli archivi delle personalità in Toscana tra '800 e '900. L'area pisana. A cura di Emilio CAPANNELLI ed Elisabetta INSABATO. Firenze, Olschki, 2000, 378 p.

332. Guida degli archivi capitolari d'Italia. A cura di Salvatore PALESE [et al.]. Roma, Ministero per i beni e le attività culturali, Ufficio centrale per i beni archivistici, 2000 (Quaderni di "Archiva ecclesia"/Associazione archivistica ecclesiastica, 6).

333. Guide (A) to historical institutes, history departments, archives and museums in the Czech Republic. Prague, The Czech National Committee of Historians for the participants in the 19th International Congress of Historical Sciences, Oslo, 2000, 112 p. (ill.).

334. HOFFMANNOVÁ (Jaroslava), PRAŽÁKOVÁ (Jana). Biografický slovník archivářů českých zemí. (A biographic dictionary of archivists in the Czech Lands). Praha, Libri, 2000, 830 p.

335. Holocaust: sound archive oral history recordings. Ed. by Joanna LANCASTER and Richard MAC DONOUGH. London, Trustees of the Imperial Museum, 2000, III-279 p. (ill.).

336. Ipswich Borough Archives, 1255–1835: a catalogue. Ed. by David ALLEN. Woodbridge a. Rochester, Boydell and Brewer, for the Suffolk Records Society, 2000, LXIV-611 p. (ill., plates).

337. KELLERHALS-MAEDER (Andreas). Das Bundesgesetz über die Archivierung. Neue Chancen für die Zeitgeschichte [Forschungsbericht]. *Schweizerische Zeitschrift für Geschichte*, 2000, 50, p. 188-197.

338. KŘESŤAN (Jiří). Přístupnost archiválií. Právní principy a praktické zkušenosti. (Zugänglichkeit der Archivalien. Rechtsprinzipien und praktische Erfahrungen). *Archivní časopis*, 2000, 50, 2, p. 65-73.

339. MARIOTTE (Jean-Yves), METZ (Bernhard), SCHWICKER (François), GSELL (Danièle), WEIL (Brigitte). Archives municipales de Strasbourg. Les Sources manuscrites de l'histoire de Strasbourg. T. 1. des origines à 1790. Strasbourg, Archives municipales, 2000, 367 p.

340. MORONI (Andrea). L'archivio privato della famiglia Niccolini di Camugliano. *Archivio storico italiano*, 2000, 158, 584, p. 307-348.

341. NATALINI (Terzo). Archivio segreto vaticano. Roma, Gangemi e Città del Vaticano, Archivio segreto vaticano, 2000, 64 p.

342. NEJKOVA (A.). Idei i programi za izdirvane i publikuvane na pismeni izvori za bulgarskata istorija. (Idées et programmes pour la recherche et la publication de sources écrites sur l'histoire bulgare). *Godishnik na Sofijskija universitet, istoricheski fakultet*, 2000, 84-85, p. 287-325.

343. SÁNCHEZ GONZÁLEZ DE HERRERO (María Nieves), HERRERA HERNÁNDEZ (María Teresa). Diccionario español de documentos alfonsíes. Madrid, Arco Libros, 2000, 466 p.

344. SCHROLL (Heike). Spurensicherung: die Bestände des Stadtarchivs Berlin und ihr Schicksal durch den Zweiten Weltkrieg. Berlin, Gebr. Mann, 2000, 270 p. (ill.). (Schriftenreihe des Landesarchivs Berlin, 5).

345. TEMPERÁN VILLAVERDE (Elisardo), CEPEDA FANDIÑO (Antonio). Archivo Histórico Diocesano: Fondo General. Santiago de Compostela, Arzobispado de Santiago, 2000, 252 p. (ill.)

346. VIGNAL (Marie-Catherine). Des papiers d'Etat d'un ministre aux archives diplomatiques du ministère des Affaires étrangères: la destinée des dossiers politiques de Richelieu. *XVIIe Siècle*, 2000, 52, 208, p. 371-386.

347. WEISER (Johanna). Geschichte der preussischen Archivverwaltung und ihrer Leiter: von den Anfängen unter Staatskanzler von Hardenberg bis zur Auflösung im Jahre 1945. Köln, Böhlau, 2000, VII-329 p. (plates, ill.). (Veröffentlichungen aus den Archiven Preussischer Kulturbesitz. Beiheft, 7).

Cf. n° 42

b. *Libraries*

348. BLACK (Alistair). The public library in Britain, 1914–2000. London, British Library, 2000, XII-180 p. (plates, ill., map, port.).

349. CABEZA SÁNCHEZ-ALBORNOZ (María Cruz). La biblioteca universitaria de Valencia. València, Universitat de València, 2000, 241 p. (Col.lecció Cinc segles, 9).

350. DAIN (Phyllis). The New York Public Library: a universe of knowledge. New York, New York Public Library a. London, Scala Publishers, 2000, 144 p. (ill.).

351. DEL BONO (Gianna). La Biblioteca nazionale. Il modello europeo e l'Italia. *Culture del Testo e del Documento*, 2000, 1, 3, p. 19-35.

352. Essays on the history of Trinity College Library Dublin. Ed. by Vincent KINANE and Anne WALSH, Dublin, Four Courts, 2000, 206 p. (ill.).

353. FLACHOWSKY (Sören). Die Bibliothek der Berliner Universität während der Zeit des Nationalsozialismus. Berlin, Logos, 2000, X-209 p. (Berliner Arbeiten zur Bibliothekswissenschaft, 2).

354. FRIEDEL (A.). Die Bibliothek der Abtei St. Walburg zu Eichstätt. (Schriften d. Universitätsbibliothek Eichstätt, 45). Wiesbaden, O. Harrassowitz, 2000, XLII-853 p.

355. GONZALO SÁNCHEZ-MOLERO (J. L.). La biblioteca postrimera de Carlos V en España. Las lecturas del emperador. *Hispania*, 2000, 60, p. 911-943.

356. HAECKEL (Ilse). Geschichte der Universitätsbibliothek Erlangen von 1792–1844. Erlangen, Universitätsbibliothek, 2000, 183 p. (ill., ports). (Schriften der Universitätsbibliothek Erlangen-Nürnberg, 37).

357. INCH (Marcela). Bibliotecas privadas y libros en venta en Potosí y su entorno (1767–1822). Potosí, [s. n.], 2000, 241 p.

358. KALLEY (Jacqueline Audrey). Apartheid in South African libraries: the Transvaal experience. Lanham, Scarecrow Press, 2000, XVI-239 p.

359. KENDEROVA (Stoyanka). Bibliothèques et livres musulmans dans les territoires Balkaniques de l'empire Ottoman: le cas de Samokov, XVIIIe–première moitié du XIXe siècle. Villeneuve d'Ascq, Presses Universitaires du Septentiron, 2000, 585 p. (ill., maps).

360. KEUNECKE (H.-O.). Die Universitätsbibliothek Erlangen und die Bücherverbrennung von 1933. *Jahrbuch für fränkische Landesforschung*, 2000, 55, p. 634-659.

361. Librairie (La) des ducs de Bourgogne: Manuscrits conserves à la Bibliothèque royale de Belgique. 1. Textes liturgiques, ascétiques, théologiques, philosophiques et moraux. Ed. par Bernard BOUSMANNE et Céline HOOREBEECK. Turnhout, Brepols, 2000, 369 p. (ill., table).

362. Libraries and the book trade. Ed. by Robin MYERS, Michael HARRIS and Giles MANDELBROTE. New Castle, Oak Knoll Press, 2000, 192 p. (Publishing pathways).

363. Library (The) of Alexandria: centre of learning in the ancient world. Ed. by Roy MAC LEOD. London, I.B. Tauris, 2000, XII-196 p. (map).

364. MAC MULLEN (Haynes). American libraries before 1876. Westport a. London, Greenwood Press, 2000, XIV-179 p. (ill.). (Beta Phi Mu monograph series, 6).

365. MANNELLI (Goggioli, Marina). La biblioteca Magliabechiana: libri, uomini, idee per la prima biblioteca pubblica a Firenze. Firenze, L.S. Olschki, 2000, 222 p. (ill.). (Collana di monografie delle Biblioteche d'Italia, 9).

366. Registre (Le) de prêt de la bibliothèque du Collège de Sorbonne [1402–1536]: «Diarium Bibliothecae Sorbonae», Paris, Bibliothéque Mazarine, ms. 3323. Ed. par Jeanne VIELLIARD et Marie-Henriette Jullien DE POMMEROL. Paris, CNRS, 2000, 817 p. (tables). (Documents, Etudes et Répertoires, 57. Histoire des Bibliotéques Médiévales, 8).

367. RIEGER (Dietmar). Des «livres sacrés»: fiction et idée de la bibliothèque au Moyen Age. *Cahiers de Recherches médievales*, 2000, 7, p. 227-251.

368. SCHLOBACH (Jochen). Livres, lectures, envois d'auteur: catalogue de la bibliothèque de Roger Martin du Gard. Paris, H. Champion, 2000, 615 p.

369. SERRAI (Alfredo). Analecta libraria: temi di critica bibliografica e di storia bibliotecaria. A cura di Maria Grazia CECCARELLI Roma, Bulzoni, 2000, 200 p. (Il bibliotecario, 16).

370. STAIKOS (K.). The great libraries: from antiquity to the Renaissance. New Castle, Oak Knoll Press, British Library, 2000, XVI-563 p. (ill.).

371. SZOCKI (Józef). Domowy świat książek Wybrane księgozbiory polskie w XIX wieku. (Die häusliche Welt der Bücher. Die ausgewählten polnischen Büchersammlungen im XIX. Jahrhundert). Kraków, Wydaw. Nauk. AP, 2000, 167 p. (Prace Monograficzne, 279).

372. WEGMANN (Nikolaus). Bücherlabyrinthe: Suchen und Finden im alexandrinischen Zeitalter. Köln, Böhlau, 2000, XII-368 p.

373. Wissenschaftliche (Die) Stadtbibliothek und die Entwicklung kommunaler Bibliotheksstrukturen in Europa seit 1945. Hrsg. v. Jörg FLIGGE und Peter BORCHARDT. Wiesbaden, Harrassowitz, 2000, 448 p.

Cf. n°s 49, 83-96, 5037

c. Museums

374. Discovering Islamic art: scholars, collectors and collections, 1850–1950. Ed. by Stephen VERNOIT. London, I. B. Tauris, 2000, XIV-232 p. (ill.).

375. HASKELL (Francis). The ephemeral museum: old master paintings and the rise of the art exhibition. New Haven a. London, Yale U. P., 2000, XIV-200 p.

376. JOACHIMIDES (Alexis). Die Museumsreformbewegung in Deutschland und die Entstehung des modernen Museums 1880–1940. Dresden, Verlag der Kunst Dresden, 2000, 320 p.

377. Khristianskie relikvii v Moskovskom Kremle. (Christian Relics in the Kremlin of Moscow: [Catalogue of Exhibition]). Ed. Aleksej M. LIDOV. Gos. istoriko-kul'turnyj muzej-zapovednik 'Moskovskij Kreml'; Tsentr vostochno-khristianskoj kul'tury. Moskva, Radunitsa, 2000, 304 p. (ill.). [Eng. summary].

378. LERI (Jean-Marc). Musée Carnavalet: histoire de Paris. Paris, Fragments, 2000, 190 p. (ill.).

379. LEVYKIN (Konstantin G.). O zhizni muzeja v epokhu peremen... Gosudarstvennyj istoricheskij muzej. (Muzeum in the Age of Changes ... The State Historical Muzeum [Moscow]). Moskva, [s. n.], 2000, 239 p.

380. MALVERN (Sue). War, memory and museums: art and artifact in the Imperial War Museum. *History workshop*, 2000, 49, p. 177-203.

381. Medium Museum: Kommunikation und Vermittlung in Museen für Kunst und Geschichte. Hrsg. v. Thomas Dominik MEIER und Hans Rudolf REUST. Bern, P. Haupt, 2000, 190 p. (ill.).

382. MORTON (Patricia A.). Hybrid modernities: architecture and representation at the 1931 colonial exposition, Paris. Cambridge, MIT Press, 2000, IX-380 p.

383. Musée colonial (Du) au musée des cultures du monde. Actes du colloque organisé par le musée national des Arts d'Afrique et d'Océanie et le Centre Georges Pompidou, 3–6 juin 1998. Ed. par Dominique TAFFIN. Paris, Maisonneuve et Larose; Musée des Arts d'Afrique et d'Océanie, 2000, 245 p. (ill.).

384. Musei (I) della Grande guerra: guida: dall'Adamello a Caporetto. A cura di Lucio FABI. Rovereto, Osiride, 2000, 101 p. (ill., col. map).

385. Musei del XX secolo. A cura di Ersilia ALESSANDRONE PERONA. *Passato e Presente*, 2000, 18, 51, p. 15-40.

386. Museums and history in West Africa. Ed. by Claude Daniel ARDOUIN and Emmanuel ARINZE. Oxford, Smithsonian Institution Press, J. Curry, 2000, IX-182 p. (ill., maps).

387. Museums and memory. Ed. by Susan A. CRANE. Stanford, Stanford U. P., 2000, X-257 p. (ill.). (Cultural sitings).

388. New perspectives on industrial history museums. Santa Barbara, University of California Press, 2000, 175 p. (ill.). (The Public historian, 22, 3)

389. Origins (The) of museums: the cabinet of curiosities in sixteenth and seventeenth century Europe. Ed. by Oliver IMPEY and Arthur MAC GREGOR. Thirsk, House of Stratus, 2000, XX-356 p (ill., facsims, plans, ports.).

390. POULOT (Dominique). Musées menacés, musées disparus: le cas de la France post-révolutionnaire. *Musées perdus, musées retrouvés: l'expérience de l'Italie et de la France. Cahiers du CRHIPA*, 2000, 3, p. 101-125.

391. RIEKE-MÜLLER (Annelore), MÜLLER (Siegfried). Konzeptionen der Kulturgeschichte um die Mitte des 19. Jahrhunderts: das Germanische Nationalmuseum in Nürnberg und die Zeitschrift für deutsche Kulturgeschichte. *Archiv für Kulturgeschichte*, 2000, 82, p. 345-376.

392. SHEEHAN (James J.). Museums in the German art world. From the end of the Old Regime to the rise of Modernism. Oxford, Oxford U. P., 2000, 258 p. (ill.).

393. SKEATES (Robin). The collecting of origins: collectors and collections of Italian prehistory and the cultural transformation of value (1550–1999). Oxford, J. and E. Hedges, 2000, IX-143 p. (ill., map, ports.). (BAR. International series, 868).

394. SPIEGEL (Regis). Mythe et histoire, images et textes: les vies de Dominique-Vivant Denon. *Zeitschrift für Kunstgeschichte*, 2000, 63, 4, p. 562-568.

395. Storia contemporanea (La) nei musei. A cura di Massimo BAIONI. *Contemporanea*, 2000, 3, p. 495-517.

c. Museums. Addenda 1999

396. WIRTY (Emeline). Le budget des Beaux Arts sous la Restauration (1815–1830). *Cahiers d'histoire publiés par les Universités de Clermont, Lyon, Grenoble*, 1999, 44, 3, p. 399-413.

*Cf. n*os *1282, 8792*

§ 2. History of historiography.

a. General

* 397. BALARD (Michel). L'historiographie des croisades au XXe siècle (France, Allemagne et Italie). *Revue historique*, 2000, 616, p. 973-999.

* 398. Bollettino di storiografia. 2000. [1999. Cf. Bibl. 99, n° 483.] Dir. da Massimo MASTROGREGORI. Pisa e Roma, Istituti Editoriali e Poligrafici Internazionali, 2000, 82 p. (Storiografia, supplemento critico e bibliografico, 4).

* 399. DEMANTOWSKY (Marko). Das Geschichtsbewusstsein in der SBZ und DDR: historisch-didaktisches Denken und sein geistiges Bezugsfeld (unter besonderer Berücksichtigung der Sowjetpädagogik): Bibliographie und Bestandsverzeichnis 1946–1973. Berlin, Bibliothek für Bildungsgeschichtliche Forschung, 2000, VI-276 p. (Bestandsverzeichnisse zur Bildungsgeschichte, 9).

* 400. DIMITROV (S.). Ottoman studies in Bulgaria after the Second World War. *Études balkaniques*, 2000, 36, 1, p. 29-58.

* 401. KLEIN (Friz). Vierzig Jahre Weltkriegsforschung. Lüneburg, Unibuch, 2000, 40 p. (Lüneburger Universitätsreden, 1).

* 402. MORELLI (Serena). La storiografia sul Regno angioino di Napoli: una nuova stagione di studi. *Studi storici*, 2000, 41, 4, p. 1023-1046.

* 403. Siècle (Un) d'histoire du christianisme en France: bilan historiographique et perspectives. Actes du Colloque de Rennes, 15–17 septembre 1999. *Revue d'histoire de l'Eglise de France*, 2000, 86, 217, p. 321-769. [Cf. n° <sélection> 617.]

* 404. Study (The) of history: a bibliographical guide. Compiled by R. C. Roger Charles RICHARDSON. Manchester, Manchester U. P., 2000, XIV-140 p. (History and related disciplines select bibliographies).

* 405. VAN HARTESVELDT (Fred R.). The Boer war: historiography and annotated bibliography. Westport, Greenwood Press, 2000, 255 p. (Bibliographies of battles and leaders, 24).

* 406. VELICHKOVA (G.). Fashizmut v bulgarskata knizhnina (20-te–purvata polovina na 30-te godini). (Le fascisme dans la littérature bulgare, les années 20–la premiere moitié des années 30). *Munalo*, 2000, 1, p. 58-76; 3-4, p. 58-76.

** 407. Irving judgement (The): David Irving v. Penguin Books and Professor Deborah Lipstadt. London, Penguin, 2000, 348 p.

408. 8. Sjezd českých historiků. (8. Convention of the Czech historians). Ed. Jiří PEŠEK. Praha, Scriptorium, 2000, 335 p.

409. Adelige und bürgerliche Erinnerungskulturen des Spätmittelalters und der Frühen Neuzeit. Hrsg. v. Werner RÖSENER. Göttingen, Vandenhoeck & Ruprecht, 2000, 228 p. (ill., maps). (Formen der Erinnung, 8).

410. AGUIRRE ROJAS (Carlos Antonio). La réception de l'historiographie française an Amérique latine, 1870–1968. *Cahiers du monde hispanique et luso-brésilien*, 2000, 74, p. 143-158.

411. AILES (Marianne). Early French chronicle. History or literature? *Journal of medieval history*, 2000, 26, 3, p. 301-312.

412. AKSENOVA (Elena P.). Ocherki iz istorii otechestvennogo slavjanovedenija, 1930-e gody. (Essays on history of Slavonic studies in Russia, the 1930s). RAN, In-t slavjanovedenija. Moskva, [s. n.], 2000, 222 p.

413. Aktualität (Die) des Mittelalters. Hrgs. v. Hans-WERNER GOETZ. Bochum, D. Winkler, 2000, 354 p. (ill.). (Herausforderungen, 10).

414. Alfonso X el Sabio y las crónicas de España. Ed. por Inés FERNÁNDEZ-ORDÓÑEZ. Valladolid, Secretariado de Publicaciones e Intercambio Editorial, Universidad de Valladolid, Centro para la Edición de los Clásicos Espanoles, 2000, 282 p. (ill.).

415. ANDERSON (Carolyn Bernadette). Double vision: historiographers, chroniclers, romances and the invention of Royal character, 1050–1377. Ann Arbor, UMI, 2000, IX-282 p.

416. Anonim Osmanlı Tarihi (1099–1116 / 1688–1704). (Anonymous Ottoman history). Ed. by Abdülkadir ÖZCAN. Ankara, Türk Tarih Kurumu, 2000, 617 p.

417. Art, memory and family in Renaissance Florence. Ed. by Giovanni CIAPPELLI and Patricia LEE RUBIN. Cambridge, Cambridge U. P., 2000, 316 p. (ill.)

418. ARVIDSSON (Stefan). Ariska idoler. (Idoles aryennes. La mythologie indoeuropéenne entre idéologie et science). Eslöv, Brutus Östlings Bokförlag Symposion, 2000, 424 p. (ill.).

419. Assessment (An) of twentieth-century historiography: professionalism, methodologies, writings. Ed. by Rolf TORSTENDAHL, Stockholm, Kungl. Vitterhets Historie och Antikvitets Akademien, 2000, 232 p. (Konferenser – Kungl. Vitterhets, historie och antikvitets akademien –, 49).

420. BANTI (Alberto M.). Su alcuni modelli esplicativi delle origini delle nazioni. *Ricerche di storia politica*, 2000, 3, 1, p. 53-70.

421. BARCZEWSKI (Stephanie L.). Myth and national identity in nineteenth-century Britain: the legends of King Arthur and Robin Hood. Oxford, Oxford U. P., 2000, VIII-274 p.

422. BARTOŠ (Josef), SCHULZ (Jindřich), TRAPL (Miloš). Regionální dějiny. Stav, problémy a výhledy. (Local history. Situation, problems and perspectives). *Acta Universitatis Palackianae Olomucensis, Fac. Philosophica. Historica 29 Sborník prací historických*, 2000, 17, p. 21-38.

423. BEDOS-REZAK (Brigitte Miriam). Medieval identity: a sign and a concept. *American historical review*, 2000, 105, 5, p. 1489-1533.

424. BENJAMIN (Thomas). A time of reconquist: history, the Maya revival, and the Zapatista rebellion in Chiapas. *American historical review*, 2000, 105, 2, p. 417-450. – IDEM. La Revolución: Mexico's great revolution as memory, myth, and history. Austin, University of Texas Press, 2000, XI-237 p.

425. BERG (NICOLAS). Zwischen individuellen und historiographischem Gedächtnis: der Nationalsozialismus in Autobiographien deutscher Historiker nach 1945. *BIOS*, 2000, 13, 2, p. 181-207.

426. BETTS (Paul). The twilight of the idols: East German memory and material culture. *Journal of modern history*, 2000, 72, 3, p. 731-765.

427. Beyond colonialism and nationalism in the Maghrib: history, culture, and politics. Ed. by Ali Abdullatif AHMIDA. New York, Palgrave, 2000, XII-255 p.

428. BIRD (Stephen). Reinventing Voltaire: the politics of commemoration in nineteenth-century France. Oxford, Voltaire Foundation, 2000, VIII-225 p.

429. BIZZOCCHI (Roberto). Religioni, miti, nazioni. *Studi storici*, 2000, 41, 4, p. 1183-1194.

430. BLAAS (P. B. M.). Geschiedenis en nostalgie: de historiografie van een klein natie met een groot verleden: verspreide historiografische opstellen. Hilversum, Verloren, 2000, 251 p. (Publicaties van de Faculteit der Historische en Kunstwetenschappen, 31)

431. BOJE MORTENSEN (Lars), SKOVGAARD-PETERSEN (Karen), EKREM (Inger). Olavslegenden og den latinske historieskrivning i 1100-tallets Norge: en artikelsamling. (La légende de Saint Olaf et l'historiographie latine en Norvège au 12e siècle: recueil d'articles). København, Museum Tusculanum, 2000, 329 p. (ill., cartes).

432. BOONE (R.). Claude de Seyssel's translations of ancient historians. *Journal of the history of ideas*, 2000, 61, 4, p. 561-576.

433. British and German historiography, 1750–1950: traditions, perceptions, and transfers. Ed. by Benedikt STUCHTEY and Peter WENDE. Oxford, Oxford U. P. a. London, German Historical Institute of London, 2000, VIII-438 p. (Studies of the German Historical Institute London).

434. BROUCEK (Peter), PEBALL (Kurt). Geschichte der österreichischen Militärhistoriographie. Köln, Weimar u. Wien, Böhlau, 2000, XI-713 p.

435. Bulgariia i Rusiia prez XX vek: bulgaro-ruski nauchni diskusii. (La Bulgarie et la Russie au XXe siècle. Discussion scientifiques bulgaro-russe). Redaktsionna kolegiia V. TOSHKOVA, L. REVIAKINA, S. PINTEV, L. LJUBENOVA, L. STOJANOV. Sofiia, Izd. "Gutenberg", 2000, 445 p.

436. BURKERT (Martin). Die Ostwissenschaften im Dritten Reich. T. 1. Zwischen Verbot und Duldung. Die schwierige Gratwanderung der Ostwissenschaften zwischen 1933 und 1939. Wiesbaden, Harrassowitz, 2000, 771 p. (Forschungen zur osteuropäischen Geschichte, 55).

437. CARTIER (Stephan). Licht ins Dunkel des Anfangs. Studien zur Rezeption der Prähistorik in der deutschen Welt und Kulturgeschichtsschreibung des 19. Jh. Herdecke, CGA Verl., 2000, 266 p. (Forschen und Wissen – Kulturgeschichte).

438. CATTARUZZA (Alejandro), EUJANIAN (Alejandro). Héroes patricios y gauchos rebeldes. Dispositivos estatales y representaciones populares en la constitución de imágenes colectivas del pasado en la Argentina (1870–1940). *Storiografia*, 2000, 4, p. 1-22.

439. CAYLA-VARDHAN (Fabienne). Les enjeux de l'historiographie érythréennee. Bordeaux, Université Montesquieu-Bordeaux IV, 2000, 65 p. (map). (Travaux et documents / Centre d'étude d'Afrique noire, Institut d'études politiques de Bordeaux, 66-67).

440. Censoring history: citizenship and memory in Japan, Germany, and the United States. Ed. by Laura HEIN and Mark SELDEN. Armonk, M.E. Sharpe, 2000, IX-301 p. (ill.). (Asia and the Pacific).

441. CHAZAN (Mireille). La nécessité de l'Empire de Sigebert de Gembloux à Jean de Saint-Victor. *Moyen Age*, 2000, 106, 1, p. 9-36.

442. Clio's favorites: leading historians of the United States, 1945–2000. Ed. by Robert Allen RUTLAND. Columbia, University of Missouri Press, 2000, 191 p.

443. COHEN (William B.). The Algerian war and French memory. *Contemporary European history*, 2000, 9, 3, p. 489-500.

444. COLLEY (Linda). Going native, telling tales: captivity, collaborations and empire. *Past and present*, 2000, 168, p. 170-193.

445. Composing useful pasts: history as a contemporary politics. Ed. by Edmund E. JACOBITTI. Albany, State University of New York Press, 2000, XIII-176 p.

446. CONNELLY (Matthew). Taking off the Cold War lens: visions of North-South conflict during the Algerian war of independence. *American historical review*, 2000, 105, 3, p. 739-769.

447. CORRADINI (Richard). Die Wiener Handschrift Cvp 430*: Ein Beitrag zur Historiographie in Fulda im frühen 9. Jahrhundert. Frankfurt am Main, Josef Knecht, 2000, 81 p. (facsim.).

448. CROIZY-NAQUET (Catherine). Ecrire l'histoire: le choix du vers ou de la prose aux XIIe et XIIIe siècles. *Médiévales*, 2000, 38, p. 71-85.

449. DAMINA-GRINT (Peter). The new historians of the twelfth-century Renaissance: authorising history in the Vernacular revolution. Woodbridge, Boydell Press, 2000, XII-292 p.

450. DAVISON (Graeme). The use and abuse of Australian history. Crows Nest, Allen & Unwin, 2000, VII-326 p.

451. DE LA FUENTE (Ariel). Facundo and Chacho in songs and stories: oral culture and the representations of caudillos in the nineteenth-century Argentine interior. *Hispanic American historical review*, 2000, 80, 3, p. 503-536.

452. DIEHL (Gerhard). Exempla für eine sich wandelnde Welt: Studien zur norddeutschen Geschichtsschreibung im 15. und 16. Jahrhundert. Bielefeld, Verlag für Regionalgeschichte, 2000, 416 p. (Veröffentlichungen des Instituts für Historische Landesforschung der Universität Göttingen, 38).

453. Disremembering the dictatorship: the politics of memory in the Spanish transition to democracy. Ed. by Joan Ramon RESINA. Amsterdam, Rodopi, 2000, 248 p. (Portada Hispánica, 8).

454. DODD (Diane), POSTOLEC (Genevieve). Report on the survey on women in public history. *Canadian historical review*, 2000, 81, 3, p. 452-466.

455. DOOLEY (Brendan). Intelletto e mercato nella storiografia della cultura della prima età moderna. *In*: Monumento (Il): arte e storia [Cf. n° 1283], p. 603-635.

456. ËKSHTUT (Semen A.). Istorija i literatura: "polosa otchuzhdenija"? (History and literature: "No man's Land"? [19th-century Russia]). *Dialog so vremenem: Al'manakh intellektual'noj istorii*, 2000, 3, p. 47-72. [Eng. summary]

457. EMERY (Elizabeth). The «truth» about the Middle Ages: 'La revue des Deux Mondes' and the late nineteenth-century French medievalism. *Prose studies: history, theory, criticism (Londres)*, 2000, 23, 2, p. 99-114.

458. Espaces (Les) de l'historien. Ed. par Jean-Claude WAQUET, Odile GEORG et Rebecca ROGERS. Strasbourg, Presses univ. Strasbourg, 2000, 264 p.

459. FABIAN (Ann). The unvarnished truth: personal narratives in nineteenth-century America. Berkeley a. Los Angeles, University of California Press, 2000, XIII-255 p.

460. Fascismo e antifascismo. Rimozioni, revisioni, negazioni. A cura di Enzo COLLOTTI. Roma e Bari, Laterza, 2000, 546 p.

461. FISCHER (Thomas E.). Geschichte der Geschichtskultur: über den öffentlichen Gebrauch von Vergangenheit von den antiken Hochkulturen bis zur Gegenwart [1965]. Köln, Wissenschaft und Politik, 2000, 135 p. (Bibliothek Wissenschaft und Politik, 57).

462. FLEMING (K. E.). Orientalism, the Balkans, and Balkan historiography. *In*: Orientalism twenty years on. [Review essays]. *American historical review*, 2000, 105, 4, p. 1218-1233.

463. FONTANA (F.). Su alcuni modelli latini di 'Historia Regum Britannie' 137-138: Ovidio, Plauto e Virgilio. *Studi medievali*, 2000, 41, 2, p. 809-826.

464. FOUCHER (Antoine). Historia proxima poetis: l'influence de la poésie épique sur le style des historiens

latins, de Salluste à Ammiens Marcellin. Bruxelles, Latomus, 2000, 487 p. (Collection Latomus, 255).

465. FOWLER (Robert Louis). Early Greek mythography. Oxford, Oxford U. P., 2000.

466. FUREIX (Emmanuel). La cancellazione delle tracce della Rivoluzione nella Parigi della Restaurazione. *Passato e presente*, 2000, 18, 50, p. 71-90.

467. GABBA (Emilio). Roma arcaica: storia e storiografia. Roma, Edizioni di storia e letteratura, 2000, 284 p. (Storia e letteratura, 205).

468. GARCIA (Patrick). Le bicentenaire de la Révolution française: pratiques sociales d'une commémoration. Paris, CNRS Ed., 2000, 354 p.

469. GAVRISHINA (Oksana V.). Istoricheskoe soznanie v Rossii 40-kh godov XIX veka: obraz nedavnego proshlogo v biografii sovremennika. (Historical consciousness in 1840s Russia: the representation of 'Recent Past' in a contemporary's biography). Ed. in app.: I. V. MALINOVSKIJ's "O zhizni General-Majora Vol'khovskogo (On the life of Major-General Volkhovsky.)", p. 253-260. *Dialog so vremenem: Al'manakh intellektual'noj istorii*, Moskva, 2000, 3, p. 240-260. [Eng. summary]

470. GEMELLI (Giuliana). Pensiero inquieto: da Cournot a Braudel. *Intersezioni. Rivista di storia delle idee*, 2000, 20, p. 263-276.

471. Gender and history: retrospect and prospect. Ed. By Leonore DAVIDOFF, Keith MAC CLELLAND and Eleni VARIKAS. Oxford, Blackwell, 2000, X-217 p.

472. Genealogie als Denkform in Mittelalter und früher Neuzeit. Hrsg. v. Kilian HECK und Bernhard JAHN. Tübingen, Niemeyer, 2000, VIII-264 p. (ill.). (Studien und Texte zur Sozialgeschichte der Literatur, 80).

473. GENGENBACH (Heidi). Naming the past in a "scattered" land: memory and the power of women's naming practices in southern Mozambique. *International journal of African historical studies*, 2000, 33, 3, p. 523-542.

474. Geschichte (Zur) der Historiographie nach 1945. Beiträge eines Kolloquiums zum 75. Geburtstag von Gerhard Lozek. Hrsg. v. Alfred LOESDAU und Helmut MEIER. Berlin, Trafo Verl., 2000, 280 p.

475. Geschichte als Herrschaftsdiskurs: der Umgang mit der Vergangenheit in der DDR. Hrsg. v. Martin SABROW. Köln, Böhlau, 2000, 330 p. (Zeithistorische Studien, 14).

476. GIARDINA (Andrea), VAUCHEZ (André). Il mito di Roma: da Carlo Magno a Mussolini. Roma, Laterza, 2000, XI-336 p., (ill.). (Storia e società).

477. Giornata di studi bizantini. Categorie linguistiche e concettuali della storiografia bizantina: atti della quinta Giornata di studi bizantini, Napoli, 23-24 aprile 1998. A cura di Ugo CRISCUOLO e Riccardo MAISANO. Napoli, M. D'Auria, 2000, 274 p.

478. "Goldhagen effect" (The): history, memory, Nazism-facing the German past. Ed. by Geoff ELEY. Ann Arbor, University of Michigan Press, 2000, 172 p. (Social history, popular culture, and politics in Germany). [Contents: ELEY (Geoff). Ordinary Germans, Nazism, and judeocide. – BARTOV (Omer). Reception and perception: Goldhagen's holocaust and the world. – GROSSMANN (Atina). The Goldhagen effect: memory, repetition, and responsibility in the new Germany. – JUDSON (Pieter). Austrian non-reception of a reluctant Goldhagen. – CAPLAN (Jane). Reflections on the reception of Goldhagen in the United States].

479. GOLSAN (Richard Joseph). Vichy's afterlife: history and counterhistory in postwar France. Lincoln a. London, University of Nebraska Press, 2000, XI-232 p.

480. GRABSKI (Andrzej Feliks). Zarys historiografii polskiej. (Polnische Historiographie im Überblick). Poznań, Wydaw. Poznańskie, 2000, 277 p.

481. GRASMAN (Edward). All'ombra del Vasari: cinque saggi sulla storiografia dell'arte nell'Italia del Settecento. Firenze, Istituto universitario olandese di storia dell'arte, 2000, 301 p. (ill.). (Istituto universitario olandese di storia dell'arte, 14).

482. HAAR (Ingo). Historiker im Nationalsozialismus: Deutsche Geschichtswissenschaft und der "Volkstumskampf" in Osten. Göttingen, Vandenhoeck & Ruprecht, 2000, 433 p. (Kritische Studien zur Geschichtswissenschaft, 143).

483. HAROOTUNIAN (Harry). Overcome by modernity: history, culture, and community in interwar Japan. Princeton, Princeton U. P., 2000, XXXII-440 p.

484. HARTOG (François). The invention of history: the pre-history of a concept from Homer to Herodotus. *History and theory*, 2000, 39, 3, p. 384-395.

485. HAUSMANN (Frank-Rutger). "Vom Strudel der Ereignisse verschlungen". Deutsche Romanistik im "Dritten Reich". Frankfurt am Main, Klostermann, 2000, XXIII-741 p. (Analecta Romanica, 61).

486. HEAP (Ruby). The status of women in the historical profession in Canada: results of a 1998 survey. *Canadian historical review*, 2000, 81, 3, p. 436-451.

487. HEDRICK (Charles W.). History and silence. Purge and rehabilitation of memory in late antiquity. Austin, Texas U. P., 2000, XXVI-338 p. (ill.).

488. Histoires d'outre-Manche. Tendances récentes de l'historiographie britannique. Ed. par Frédérique LACHAUD, Isabelle LESCENT-GILLES et François-Joseph RUGGIU. Paris, Presses Universitaires de la Sorbonne, 2000, 360 p.

489. Historia (La) alfonsí. El modelo y sus destinos (siglos XIII–XV). Ed. por Martin GEORGES. Madrid, Casa de Velázques, 2000, 163 p.

490. Historia: the concept and genres in the middle ages. Ed. by Tuomas M. S. LEHTONEN and Paivi MEH-

TONEN. Helsinki, Societas Scientiarum Fennica, 2000, 135 p. (Commentationes Humanarum Litterarum, 116).

491. Historikerkontroversen. Hrsg. v. Hartmut LEHMANN. Göttingen, Wallstein, 2000, 189 p. (Göttinger Gespräche zur Geschichtswissenschaft, 10).

492. History after the three worlds: post-Eurocentric historiographies. Ed. by Arif DIRLIK, Vinay BAHL and Peter GRAN. Lanham, Rowman & Littlefield, 2000, VIII, 278 p.

493. HOEGEN (Saskia von). Entwicklung der spanischen Historiographie im ausgehenden Mittelalter: am Beispiel der "Crónicas de los reyes de Castilla Don Pedro I, Don Enrique II, Don Juan I y Don Enrique III" von Pero López de Ayala, der "Generaciones y semblanzas" von Fernán Pérez de Guzman und der "Crónica de los reyes católicos" von Fernando del Pulgar. Frankfurt am Main u. Oxford, Lang, 2000, XV-513 p.

494. HOWE (Stephen). Viewpoint: the politics of historical 'Revisionism': comparing Ireland and Israel/Palestine. *Past and present*, 2000, 168, p. 227-253.

495. HUMMEL (Pascale). Histoire de l'histoire de la philologie: étude d'un genre épistémologique et bibliographique. Genève, Droz, 2000, 504 p. (Histoire des idées et critique littéraire, 385).

496. HUSSON (Edouard). Comprendre Hitler et la Shoah: les historiens de la République fédérale d'Allemagne et l'identité allemande depuis 1949. Paris, Presses universitaires de France, 2000, XVII-306 p. (Perspectives germaniques).

497. Identità (L') dell'Italia repubblicana. Un dibattito sugli orientamenti storiografici. n. 210-211. A cura di Federico ROMERO. *L'Italia contemporanea*, 2000, p. 389-429.

498. IGARASHI (Yoshiyuki). Bodies of memory: narratives of war in postwar Japanese culture, 1945–1970. Princeton, Princeton U. P., 2000, X-284 p.

499. IGLESIAS (Francisco). Historiadores do Brasil: capitulos de historiografia brasileira. Rio de Janeiro, Nova Frontiera, Editora UFMG, 2000, 251 p.

500. IGOUNET (Valérie). Histoire du négationnisme en France. Paris, Seuil, 2000, 691 p. (XXe siècle).

501. Invention (L') de l'histoire. Thème coordonné par Christopher LUCKEN et Mireille SÉGUY. Saint-Denis, Presses Universitaires de Vincenne-Paris VIII, 2000, 190 p. (Médiévales, 38).

502. IRACE (Erminia). Sul «potere dei ricordi»: qualche nota di lettura. *Storiografia*, 2000, 4, p. 53-65.

503. Istorija i antiistorija: Kritika "novoj khronologii" akademika A. T. Fomenko. (History and antihistory: criticism of the 'New Chronology' by academician Anatoly T. Fomenko). Ed. Aleksej D. KOSHELEV. Moskva, Jazyki russkoj kul'tury, 2000, 523 p. (ill., maps). (Studia historica. Series minor).

504. Istoriki Rossii: Poslevoennoe pokolenie. (Historians of Russia: the generation after the Second World War: [biographical essays]). Ed. Lidija V. MAKSAKOVA. Tsentr sotsial'nogo modelirovanija CARPE DIEM; Assotsiatsija issledovatelej rossijskogo obshchestva XX v. Moskva, AIRO-XX, 2000, 238 p. (portr.).

505. JEANNENEY (Jean-Noël). La République a besoin d'histoire. Paris, Ed. du Seuil, 2000, 251 p.

506. KARSH (Efraim). Fabricating Israeli history: the "new historians". London, Frank Cass, 2000, XLIV-236 p. (Cass series in Israeli history, politics, and society).

507. KEITA (Maghan). Race and the writing of history: ridling the sphinx. Oxford, Oxford U. P., 2000, VII-214 p. (Race and American culture).

508. KIERDORF (Alexander), HASSLER (Uta). Denkmale des Industriezeitalters. Von der Geschichte des Umgangs mit Industriekultur. Tübingen u. Berlin, Wasmuth, 2000, 313 p.

509. KOCKA (Jürgen). Historische Sozialwissenschaft heute. In: Perspektiven der Gesellschaftsgeschichte [Cf. n° 806], p. 5-24.

510. KORZUN (Valentina P.). Obrazy istoricheskoj nauki na rubezhe XIX–XX v. (analiz otechestvennykh istoriograficheskikh kontseptsij). (Images of the historical science on the eve of the 21st century: an analysis of historiographical conceptions in Russia). Ekaterinburg, Izd-vo Ural'skogo un-ta. Omsk, [s. n.], 2000, 226 p.

511. KOSHAR (Rudy). From monuments to traces: artifacts of German memory, 1870–1990. Berkeley a. Los Angeles, University of California Press, 2000, 368 p. (Weimar and now: German cultural criticism, 24) (ill.).

512. KRAPAUSKAS (Virgil). Nationalism and historiography: the case of nineteenth-century Lithuanian historicism. Boulder, East European Monographs, 2000, VI-234 p.

513. KÜLZER (A.). Die Anfänge der Geschichte. Zur Darstellung des "biblischen Zeitalters" in der byzantinischen Chronistik. *Byzantinische Zeitschrift*, 2000, 93, p. 138-156.

514. KUNZ (Georg). Verortete Geschichte: Regionales Geschichtsbewusstsein in den deutschen Historischen Vereinen des 19. Jahrhunderts. Göttingen, Vandenhoeck & Ruprecht, 2000, 413 p. (Kritische Studien zur Geschichtswissenschaft, 138).

515. LAGROU (Pieter). The legacy of Nazi occupation: patriotic memory and national recovery in Western Europe, 1945–1965. Cambridge, Cambridge U. P., 2000, XIII-327 p. (Studies in the social and cultural history of modern warfare, 8).

516. LARSON (Pier M.). History and memory in the age of enslavement: becoming Merina in highland Madagascar, 1770–1822. Portsmouth, Heinemann, 2000, XXXII-414 p. (Social history of Africa).

517. LASSNER (Jacob). The Middle East remembered: forged identities, competing narratives, contested spaces. Ann Arbor, University of Michigan Press, 2000, XVII-428 p.

518. LICHTENHAM (Francine-Dominique). De l'abus de l'historiographie: approches de l'histoire russe de Herberstein à Custine. *Cahiers du monde russe*, 2000, 41, 1, p. 151-164.

519. LIM (Jie-Hyun). Shifting Marxist historiography: from hard history to soft history. *History and Culture*, 2000, 1, [s. p.].

520. LOOSER (Devoney). British women writers and the writing of history, 1670–1820. London, Johns Hopkins U. P., 2000, XI-272 p.

521. LORENZ (Stefan). Hitler und die Antike. *Archiv für Kulturgeschichte,* 2000, 82, p. 407-432.

522. LUDINGTON (Charles C.). Between myth and margin: the Huguenots in Irish history. London, Institute of Historical Research, 2000, 19 p.

523. MAC CAW (Neil). George Eliot and Victorian historiography: imagining the national past. Basingstoke, Macmillan, 2000, VIII-203 p.

524. MAC GOWAN (Margaret M.). The vision of Rome in late Renaissance France. New Haven, Yale U. P., 2000, XIV-461 p.

525. MAC KITTERICK (Rosamond). History and its audiences. Cambridge, Cambridge U. P., 2000, 73 p. (ill., maps, geneal. table).

526. MAIER (Charles S.). Consigning the twentieth century of history: alternative narratives for the modern era. *American historical review*, 2000, 105, 3, p. 807-831.

527. Mångfaldiga historien (Den). Tio historiker om forskningen inför framtiden. (L'histoire pluridisciplinaire: dix historiens et la recherche face à l'avenir). Red. par Roger QVARSELL et Bengt SANDIN. Lund, Historiska media, 2000, 232 p.

528. MANNING (Brian). Contemporary histories of the English Civil War. East Finchley, Caliban Books, 2000, 218 p.

529. MARCONE (Arnaldo). La Tarda Antichità e le sue periodizzazioni. *Rivista storica italiana*, 2000, 112, 1, p. 318-334.

530. MARKWICK (Roger D.). Rewriting history in Soviet Russia: the politics of revisionist historiography, 1956–1974. Basingstoke, Macmillan, 2000, 327 p.

531. MAZZA (Mario). "Was ist (die antike) Wirtschaftsgeschichte?" Teoria economica e storia antica prima di Bücher, Meyer e Rostovtzeff. *Mediterraneo antico*, 2000, 3, p. 499-547.

532. Meditation und Erinnerung in der Frühen Neuzeit. Hrsg. v. Gerhard KURZ. Göttingen, Vandenhoeck & Ruprecht, 2000, 405 p. (Formen der Erinnerung, 2).

533. Memorias hegemónicas, memorias disidentes: el pasado como política de la historia. Ed. por Cristóbal GNECCO y Marta ZAMBRANO. Bogotá, Universidad del Cauca, Instituto Colombiano de Antropología e Historia, Ministerio de Cultura, 2000, 349 p. (ill.).

534. Memory and justice on trial. The Papon affair. Ed. by Richard J. GOSLAN. London a. New York, Routledge, 2000, 279 p.

535. MITCHELL (Rosemary). Picturing the past: English history in text and image 1830–1870. Oxford, Oxford U. P., 2000, XI-314 p. (Oxford historical monographs).

536. MJAGKOV (German P.). Nauchnoe soobshchestvo v isotricheskoj nauke: opyt "russkoj istoricheskoj shkoly". (Scientific community in historical science: an attempt at 'Russian Historical School'). Kazan', Izd-vo Kazanskogo un-a, 2000, 297 p. (bibl.).

537. MONSON (Jamie). Claims to history and the politics of memory in southern Tanzania, 1940–1960. *International journal of African historical studies*, 2000, 33, 3, p. 543-566. – IDEM. Memory, migration and the authority of history in southern Tanzania, 1860–1960. *Journal of African history*, 2000, 41, 3, p. 347-372.

538. MUNIER (Gerald). Geschichte im Comic. Aufklärung durch Fiktion? Über Möglichkeiten und Grenzen des historisierenden Autorencomic der Gegenwart. Hannover, Unser, 2000, 258 p.

539. MUNSLOW (Alun). The Routledge companion to historical studies. London a. New York, Routledge, 2000, XV-271 p.

540. MUREŞAN (Camil). L'écriture de l'histoire: production, offre, consommation. *In*: Nouvelles Études d'Histoire. X. Publiées à l'occasion du XIXe Congrès International des Sciences Historiques. Oslo, 2000. Bucureşti, Editura Academiei Române, 2000, p. 7-16.

541. Myth (The) of the lost cause and Civil War history. Ed. by Gary W. GALLAGHER and Alan T. NOLAN. Bloomington, Indiana U. P., 2000, V-231 p.

542. Naciones, gentes y territorios: ensayos de historia e historiografía comparada de América Latina y el Caribe. Ed por Luis Javier ORTIZ MESA y Víctor Manuel URIBE URÁN. Medellín, Editorial Universidad de Antioquia, Facultad de Ciencias Humanas y Económicas de la Universidad Nacional de Colombia-Sede Medellín, 2000, XXXIV-449 p. (Colección Clío).

543. Nanjing (The) Massacre in history and historiography. Ed. by Joshua A. FOGEL. London, University of California Press, 2000, XVI-248 p. (Asia, local studies/global themes, 2).

544. NELSON (Janet L.). Gender, memory and social power. *Gender and history*, 2000, 12, p. 722-734.

545. Neue (Das) Jahrhundert. Europäische Zeitdiagnosen und Zukunftsentwürfe um 1900. Hrsg. v. Ute FREVERT. Göttingen, Vandenhoeck u. Ruprecht, 2000, 308 p.

546. NEUMANN (Klaus). Shifting memories: the Nazi past in the new Germany. Ann Arbor, University of Michigan Press, 2000, X-333 p. (ill.). (Social history, popular culture, and politics in Germany).

547. Nordic historiography in the 20th century. Ed. by Frank MEYER and Jan E. MYHRE. Oslo, Univ. Dep. of History, 2000, 352 p.

548. Ottoman past and today's Turkey. Ed. by Kemal H. KARPAT. Leiden, Brill, 2000, XXII-306 p.

549. OUDIJK (Michel R.). Historiography of the Benizaa: the postclassic and early colonial periods (1000–1600 A.D.). Leiden, Research School of Asian, African and American Studies, Universiteit Leiden, 2000, 429 p. (CNWS publications, 84).

550. PAI (Hyung Il). Constructing "Korean" origins: a critical review of archaeology, historiography, and racial myth in Korean state-formation theories. Cambridge a. London, Harvard U. P., 2000, 450 p. (ill.). (Harvard East Asian monographs).

551. Paths to power: the historiography of American foreign relations to 1941. Ed. by Michael J. HOGAN. New York, Cambridge U. P., 2000, XII-303 p.

552. *Vacat*

553. PELLING (C. B. R.). Literary texts and the Greek historian. London, Routledge, 2000, X-338 p. (Approaching the ancient world).

554. PERONNET (M.). La censure de la Faculté de théologie contre un livre: l'«Histoire philosophique et politique». *Studies on Voltaire and the Eighteenth Century*, 2000, 12, p. 273-286.

555. PERTICI (Roberto). Storici italiani del Novecento. Pisa, Istituti editoriali e poligrafici internazionali, 2000, VIII-340 p. (Storiografia, 3).

556. PETROV (E. V.). Nauchno-pedagogicheskaja dejatel'nost' russkikh istorikov-emigrantov v SShA (Pervaja polovina XX stoletija): Istochniki i istoriografija. (Scientific and pedagogical activites of Russian emigrant historians in the USA, the 1st half of the 20th century: sources and historiography). Ros. tamozhennaja akademija, Sankt-Peterburgskij filial. Pavlovsk, [s. n.], 2000, 158 p. [Cf. Bibl. 98, n° 653.]

557. PETZOLD (Joachim). Parteinahme wofür? DDR-Historiker im Spannungsfeld von Politik und Wissenschaft. Potsdam, Verlag für Berlin-Brandenburg, 2000, 397 p. (Potsdamer Studien, 15).

558. PHILLIPS (Mark). Society and sentiment: genres of historical writing in Britain, 1740–1820. Princeton, Princeton U. P., 2000.

559. PLATON (Gheorghe). La société roumaine et la gestion du passé historique. *In*: Nouvelles Études d'Histoire. X. Publiées à l'occasion du XIXe Congrès International des Sciences Historiques. Oslo, 2000. București, Editura Academiei Române, 2000, p. 43-53.

560. PLOKHII (Serhii). The city of glory: Sevastopol in Russian historical mythology. *Journal of contemporary history*, 2000, 35, 3, p. 369-383.

561. POMIAN (Krzysztof). Mariette et Winckelmann. *Revue germanique internationale*, 2000, 13, p. 11-38.

562. Portrety istorikov: Vremja i sud'by. (Portraits of historians: time and fortunes: [Biographies of Russian historians]). Vol. 1. Otechestvennaja istorija. (The Russian history). Vol. 2. Vseobshchaja istorija. (The universal history). Ed. G. N. SEVOST'JANOV, L. T. MIL'SKAJA, L. P. MARINOVICH. Moskva, Universitetskaja kniga a. Jerusalem, Gesharim, 2000, 2 vol., 431 p., 464 p. (bibl.).

563. PROCHASSON (Christophe). Une histoire culturelle de la politique. *Historical reflections. Réflexions historiques*, 2000, 26, 1, p. 93-126.

564. PRONIN (Aleksandr A.). Istoriografija rossijskoj emigratsii. (Historiography of the Russian emigration). Ural'skaja nezavisimaja obshchestvennja biblioteka. Ekaterinburg, Izd-vo Ural'skogo un-ta, 2000, 185 p. (bibl.).

565. QUINN (Sholeh Alysia). Historical writing during the reign of Shah `Abbas: ideology, imitation, and legitimacy in Safavid chronicles. Salt Lake City, University of Utah Press, 2000, XIV-197 p. (ill.)

566. RABASA (José). Writing violence on the northern frontier: the historiography of sixteenth century New Mexico and Florida and the legacy of conquest. Durham a. London, Duke U. P., 2000, XIV-359 p. (ill.). (Latin America otherwise: languages, empires, nations).

567. RAMAZANOV (Sergej P.). Krizis v rossijskoj istoriografii nachala XX veka. (The crisis in the Russian historiography in the early 20th century). Part 2. Metodologicheskie iskanija posleoktjabr'skoj istoricheskoj mysli. (Methodological search in the post-October historical thought). Volgogradskij gos. un-t; Volzhskij gumanit in-t. Volgograd, Izd-vo Volgorgadskogo un-ta, 2000, 163 p.

568. RAPHAEL (Lutz). Anstellte eines Editorials. Nationalzentrierte Sozialgeschichte in programmatischer Absicht: die Zeitschrift "Geschichte und Gesellschaft. Zeitschrift für Historische Sozialwissenschaft" in den ersten 25 Jahren ihres Bestehens. *Geschichte und Gesellschaft*, 2000, 26, 1, p. 5-37.

569. Regionalgeschichte in Europa. Methoden und Erträge der Forschung zum 16.–19. Jh. Hrsg. v. Stefan BRAKENSIEK und Axel FLÜGEL. Paderborn, Schöningh, 2000, 310 p. (Forschungen zur Regionalgeschichte, 34).

570. Reinvenzione (La) dei lumi: percorsi storiografici del Novecento. A cura di Giuseppe RICUPERATI. Firenze, L. S. Olschki, 2000, XVI-233 p. (Studi / Fondazione Luigi Einaudi, 38).

571. REITER (Reimond). Empirie und Methode in der Erforschung des "Dritten Reichs". Fallstudien zur

Inhaltsanalyse, Typusbildung, Statistik, zu Interviews und Selbstzeugnissen. Frankfurt am Main, Lang, 2000, 226 p.

572. REMBOLD (Elfie). Die festliche Nation. Geschichtsinszenierungen und regionaler Nationalismus in Großbritannien vor dem Ersten Weltkrieg. Berlin u. Wien, Philo, 267 p. (Arbeitskreis Deutsche England-Forschung, 44).

573. Répertoire des historiens français de la période moderne et contemporaine. Annuaire 2000. Ed. par Daniel ROCHE. Paris, CNRS Ed., 2000, 495 p.

574. Rethinking the foundations: historiography in the ancient world and in the Bible: essays in honour of John Van Seters. Ed. by Steven L. MAC KENZIE and Thomas RÖMER in collaboration with Hans Heinrich SCHMID. Berlin a. New York, Walter de Gruyter, 2000, XIII-304 p. (Beihefte zur Zeitschrift für die alttestamentliche Wissenschaft, 294).

575. RIBARD (Dinah). Philosophe ou écrivain? Problèmes de délimitation entre histoire littéraire et histoire de la philosophie en France, 1650–1850. In: Figures d'auteurs. Musique et littérature, France, 17e–20e siècles. Annales, 2000, 55, 2, p. 355-388.

576. RICO MORENO (Javier). Pasado y futuro en la historiografía de la Revolución mexicana. Reynosa, Universidad Autónoma Metropolitana-Unidad Azcapotzalco y Roma, Instituto Nacional de Antropología e Historia/Conaculta, 2000, 272 p. (ill.). (Colección Ensayos, 8).

577. RIO (Joseph). Mythes fondateurs de la Bretagne: aux origines de la celtomanie. Rennes, Editions Ouest-France, 2000, 351 p. (map). (De mémoire d'homme. L'histoire).

578. ROBERTS (David D.). How not to think about fascism and ideology, intellectual antecedents and historical meaning. Journal of contemporary history, 2000, 35, 2, p. 185-211.

579. ROBERTS (Richard). History and memory: the power of statist narratives. International journal of African historical studies, 2000, 33, 3, p. 513-522.

580. Romanian and British historians on the contemporary history of Romania. Ed. by George CIPAIANU and Virgiliu TÂRAU. Cluj-Napoca, Cluj U. P., 2000, 268 p.

581. ROPER (Michael). Re-remembering the soldier hero: the psychic and social construction of memory in personal narratives of the Great War. History workshop, 2000, 50, p. 181-204.

582. ROSENFELD (Gavriel D.). Munich and memory. Architecture, monuments, and the legacy of the Third Reich. Berkeley a. Los Angeles, University of California Press, 2000, XXIII-433 p.

583. ROTTER (Andrew J.). Saidism without Said: Orientalism and U. S. diplomatic history. In: Orientalism twenty years on [Review essays]. American historical review, 2000, 105, 4, p. 1205-1217.

584. ROY (Martin). Luther in der DDR: zum Wandel des Lutherbildes in der DDR-Geschichtsschreibung: mit einer dokumentarischen Reproduktion. Bochum, D. Winkler, 2000, 373 p. (Studien zur Wissenschaftsgeschichte, 1).

585. RÜSEN (Jörn). 'Cultural currence'. The nature of historical consciousness in Europe. In: Approaches to European historical consciousness: reflections and provocations. Ed. by Sharon MACDONALD. Hamburg, Edition Körber-Stiftung, 2000, p. 75-85. – IDEM. Holocaust-Erinnerungen im Wechsel der Generationen. Thesen zur Entwicklung in der Bundesrepublik. In: Das Ende der Sprachlosigkeit? Auswirkungen traumatischer Holocaust-Erfahrungen über mehrere Generationen. Hrsg. v. Liliane OPHER-COHN [et. al.]. Gießen, Psychosozial Verlag, 2000, p. 71-84. – IDEM. Was heißt und zu welchem Ende studiert man Kulturwissenschaften? Essener Universitätsreden, 2000, 4, [s. p.].

586. SCHMIDT (Hartwig). Archäologische Denkmäler in Deutschland. Rekonstruiert und wieder aufgebaut. Stuttgart, Theiss, 2000, 160 p.

587. SCHNEEBERGER (Paul). Der schwierige Umgang mit dem "Anschluss": die Rezeption in Geschichtsdarstellungen 1946–1995. Innsbruck, Studien Verlag, 2000, 568 p.

588. SCHOLZ (Birgit). Von der Chronistik zur modernen Geschichtswissenschaft. Die Warägerfrage in der russischen, deutschen und schwedischen Historiographie. Wiesbaden, Harrassowitz, 2000, 476 p.

589. SCHWARTZ (Barry). Abraham Lincoln and the forge of national memory. Chicago, University of Chicago Press, 2000, XIII-367 p.

590. SEMMEL (Stuart). British radicals and 'legitimacy': Napoleon in the mirror of history. Past and present, 2000, 167, p. 140-175.

591. Sentimento del tempo e periodizzazione della storia nel Medioevo: atti del XXXVI Convegno storico internazionale: Todi, 10–12 ottobre 1999. Spoleto, Centro italiano di studi sull'alto Medioevo, 2000, X-329 p. (Atti dei convegni del Centro italiano di studi sul basso Medioevo. Accademia tudertina e del Centro di studi sulla spiritualità medievale, 13).

592. SHAPIRO (Barbara J.). A culture of fact: England, 1550–1720. Ithaca, Cornell U. P., 2000, X-284 p.

593. SHAVIT (Jacob). History in Black: African Americans in search of an ancient past. London, Frank Cass, 2000, 422 p.

594. SIBERRY (Elizabeth). The new crusaders: images of the crusades in the nineteenth and early twentieth centuries. Aldershot, Ashgate, 2000, XII-228 p. (plates, ill.). (The nineteenth century).

595. Singular continuities: tradition, nostalgia, and identity in modern British culture. Ed. by George K.

BEHLMER and Fred M. LEVENTHAL. Stanford, Stanford U. P., 2000, 277 p.

596. Slovak Contributions to 19th International Congress of Historical Sciences. Ed. Dušan KOVÁČ. Bratislava, Veda vydavateľstvo Slovenskej akadémie vied, Historický ústav Slovenskej akadémie vied, 2000, 243 p.

597. SMITH (Anthony D.). The nation in history: historiographical debates about ethnicity and nationalism. Hanover, University Press of New England, 2000, [s. p.]. (The Menahem Stern Jerusalem lectures).

598. SOLDI RONDININI (Gigliola). Per il recupero della memoria: i primi dieci anni di vita della Società degli storici italiani. *Nuova rivista storica*, 2000, 2, p. 337-364.

599. SOLER (Leticia). Historiografía uruguaya contemporánea, 1985–2000. Montevideo, Ediciones Trilce, 2000, 127 p.

600. SOLL (Jacob). Amelot de La Houssaye (1634–1706) annotates Tacitus. *Journal of the history of ideas*, 2000, 61, 2, p. 167-188.

601. Spory o dějiny. Sborník kritických textů. Sv. 3. (Disputes over history. Tomo 3). Ed. Miloslav BEDNÁŘ. Praha, Masarykův ústav AV ČR, 2000, 190 p.

602. SRIVASTAVA (Kamal Shankar). Indian history, historians and historiography: a study. Patna, Sangeeta Prakashan, 2000, VII-258 p.

603. Städtische Geschichtsschreibung im Spätmittelalter und in der frühen Neuzeit. Hrsg. v. Peter JOHANEK. Köln, Böhlau, 2000, XXIV-356 p. (ill.). (Städteforschung. Reihe A, Darstellungen, 47).

604. ȘTEFĂNESCU (Ștefan). La science de l'histoire et la responsabilité de l'historien. *In*: Nouvelles Études d'Histoire. X. Publiées à l'occasion du XIXe Congrès International des Sciences Historiques. Oslo, 2000. București, Editura Academiei Române, 2000, p. 17-27.

605. STORA (Benjamin). Maroc, le traitement des histoires proches. *Esprit*, p. 88-102. – IDEM. Maroc-Algérie. Retour du passé et écriture de l'histoire. *Vingtième siècle. Revue d'histoire*, 2000, 68, p. 109-118.

606. SUNSERI (Thaddeus). Statist narratives and Maji ellipses. *International journal of African historical studies*, 2000, 33, 3, p. 567-584.

607. Teoría y práctica de la historiografía hispánica medieval. Ed. by Aengus WARD. Birmingham, University of Birmingham Press, 2000, 196 p.

608. TIMONEN (Asko). Cruelty and death: Roman historians' scenes of imperial violence from Commodus to Philippus Arabs. Turku, Turun Yliopisto, 2000, 273 p. (Turun yliopiston julkaisuja. Sarja B, Humaniora, 241).

609. TransAtlantic encounters: public uses and misuses of history in Europe and the United States. Ed. by David K. ADAMS and Maurizio VAUDAGNA. Amsterdam, VU U. P., 2000, 222 p. (European contributions to American studies, 46, 1).

610. TROMPF (Garry W.). Early Christian historiography: narratives of retributive justice. London, Continuum, 2000, XX-362 p.

611. Uses (The) of the past in the early Middle Ages. Ed. by Yitzhek HEN and Matthew INNES. Cambridge, Cambridge U. P., 2000, IX-283 p.

612. VAGENHEIM (Ginette). L'épigraphie: un aspect méconnu de l'histoire de la philologie classique au XVIIe siècle. *Cahiers de l'Humanisme. Revue consacrée a la littérature de langue latine dans l'Europe de la Renassaince (XIIe–XVIIIe siècles)*, 2000, 1, p. 89-115.

613. VALKEAPÄÄ (Leena). Pitäjänkirkosta kansallismonumentiksi. Suomen keskiaikaisten kivikirkkojen restaurointi ja sen tausta vuosina 1870–1920. (From a parish church to a national monument. Restoration of Finnish medieval churches and its background, 1870–1920). Helsinki, Suomen muinaismuistoyhdistys, 2000, 218 p. (SMY, A 108). (ill.).

614. VAN DÜLMEN (Richard). Historische Anthropologie. Entwicklung, Probleme, Aufgaben. Köln, Böhlau, 2000, VI-135 p.

615. VANDERPUTTEN (Steven). Clusterpatronenin de Middeleeuwse monastieke historiographie. *Revue belge de philologie et d'histoire*, 2000, 78, 3-4, p. 773-796.

616. VANSINA (J.). Historical tales (Ibiteekerezo) and the history of Rwanda. *History in Africa*, 2000, 27, p. 375-414.

617. VAUCHEZ (André). Lieux et milieux de production de l'histoire religieuse en France. *In*: Siècle (Un) d'histoire du christianisme en France [Cf. n° 403], p. 693-705.

618. Versäumte Fragen: deutsche Historiker im Schatten des Nationalsozialismus. Hrsg. v. Rüdiger HOHLS und Konrad H. JARAUSCH; unter Mitarbeit von Torsten BATHMANN [et al.]. Stuttgart, Deutsche Verlags-Anstalt, 2000, 528 p.

619. VOLDMAN (Danièle). Le témoignage dans l'histoire française du temps présent. *Bulletin de l'Institut d'histoire du temps présent*, 2000, 75, p. 41-54.

620. Was ist Militärgeschichte? Hrsg. v. Thomas KÜHNE und Benjamin ZIEMANN. Paderborn, Schöningh 2000, 359 p. (Krieg in der Geschichte, 6).

621. WAWN (Andrew). The Vikings and the Victorians: inventing the old north in nineteenth-century Britain. Cambridge, D. S. Brewer, 2000, XIII-434 p.

622. WEST (Nancy Martha). Kodak and the lens of nostalgia. Charlottesville, University Press of Virginia, 2000, XVIII-242 p. (Cultural Frames, Framing Culture).

623. WIELL (Stine). Kampen om oldtiden: nationale oldsager siden 1864. (Le combat pour le passé: les antiquités nationales depuis 1864). Aabenraa, Museum

for Sønderjylland amt, 2000, 134 p. (ill.). (Skrifter fra Museumsrådet for Sønderjyllands amt. 7).

624. WILLING (Matthias). Die DDR-Althistorie im historischen Kontext. *Quaderni di storia*, 2000, 26, 52, p. 245-276.

625. WISSELGREN (Per). Samhällets kartläggare. (Les cartographes de la société. Les débuts des sciences sociales en Suède entre 1830 et 1920). Eslöv, Östlings Bokförlag Symposion, 2000, 380 p. (ill.). (résumé ang lais).

626. WITT (Ronald G.). "In the footsteps of the ancients": the origins of humanities from Lovato to Bruni. Leiden, Boston a. Köln, Brill, 2000, XIII-562 p.

627. Wohin steuert die Osteuropaforschung? Eine Diskussion. Hrsg. Stefan CREUZBERGER, Ingo MANNTEUFEL, Alexander STEININGER und Jutta UNSER. Köln, Verlag Wissenschaft und Politik, 2000, 281 p. (Bibliothek Wissenschaft und Politik, 58).

628. WOLFRUM (Edgar). Geschichtspolitik in der Bundesrepublik Deutschland: der Weg zur bundesrepublikanischen Erinnerung, 1948–1990. Darmstadt, Wissenschaftliche Buchgesellschaft, 2000, 532 p.

629. WOOLF (D. R.). Reading history in early modern England. Cambridge, Cambridge U. P., 2000, XVI-360 p. (Cambridge studies in early modern British history).

630. WREDE (Martin). Das Reich und seine Geschichte in den Werken französischer Staatsrechtler und Historiker des 18. Jahrhuderts. *Francia*, 2000, 27, 2, p. 177-211.

631. WRIGHT (Beth S.). «That other historian, the illustrator»: voices and 'vignettes' in mid-nineteenth century France. *Oxford art journal*, 2000, 23, 1, p. 113-136.

632. WRIGHT (Donald). Gender and the professionalization of history in English Canada before 1960. *Canadian historical review*, 2000, 81, 1, p. 29-66.

633. YAMAMOTO (Masahiro). The history and historiography of the rape of Nanking. Ann Arbor, UMI Dissertation Services, 2000, XII-593 p.

634. YARDENI (Myriam). Repenser l'histoire, aspects de l'historiographie huguenote des guerres de religion à la révolution française. Paris, H. Champion, 2000, 220 p.

635. YARROW (Liv Mariah). Non-Roman portrayals of foreign affairs in the last century of the Republic: a study of the historical writings of Posidonius, Diodorus Siculus, and Pompeius Trogus. Oxford, [s. n.], 2000, 64 p. [Thesis (M. Phil.) – University of Oxford, 2000].

636. ZELLER (Joachim). Kolonialdenkmäler und Geschichtsbewußtsein. Eine Untersuchung der kolonialdeutschen Erinnerungskultur. Frankfurt am Main, Verlag für Interkulturelle Kommunikation, 2000, 325 p.

637. ZIMMER (Oliver). Competing memories of the nation: liberal historians and the reconstruction of the Swiss past 1870–1900. *Past and present*, 2000, 168, p. 194-226.

638. ZUB (Alexandru). De la istoria critică la criticism. Istoriografia română sub semnul modernității. București, Editura Enciclopedică, 2000, 383 p. – IDEM. Orizont închis. Istoriografia română sub dictatură. Iași, Institutul European din Iași, 2000, 200 p.

a. General. Addenda 1998–1999

639. CROIZY-NAQUET (Catherine). Ecrire l'histoire romaine au début du XIIIe siècle. Paris, H. Champion, 1999, 344 p.

640. DUMOULIN (Olivier). Archives au féminin, histoire au masculin. Les historiennes professionnelles en France, 1920–1965. *In*: L'Histoire sans les femmes est-elle possible? Paris, Perrin, 98, p. 334-356.

641. SCHÖTTLER (Peter). Geschichtsschreibung in einer Trümmerwelt. Reaktionen französischer Historiker auf die deutsche Historiographie während und nach dem ersten Weltkrieg. *In*: Plurales Deutschland – Allemagne plurielle. Mélanges offerts au professeur Etienne François. Göttingen, Wallstein, 99, p. 296-313.

Cf. n°s 769, 1916, 2306, 2900, 3317, 3333, 5063, 5066, 5127, 5284, 6381, 6453, 6465, 7857, 8489, 8572, 8800

b. Special studies

642. FABBRINI (Fabrizio). Silvio **Accame**: studioso del mondo antico. Roma, Istituto italiano per la storia antica, Scuola di storia antica, 2000, XV-545 p.

643. HILL (Roland). Lord **Acton**. London, Yale U. P., 2000, XXIV-548 p.

644. SEIBT (Gustav). **Anonimo romano**: Scrivere la storia alle soglie del Rinascimento. Rome, Viella, 2000, 295 p. (ill.). (I libri di Viella, 20).

645. BUCHER (G. S.). The origins, program, and composition of **Appian**'s Roman history. *Transactions of the American philological association*, 2000, 130, p. 411-458.

646. HODGES (Richard). Visions of Rome. Thomas **Ashby** archaeologist. London, British School at Rome, 2000, 134 p.

647. DECHERF (Dominique). **Bainville**, l'intelligence de l'histoire. Paris, Bartillat, 2000, 429 p.

648. PIUZ (Anne M.). Paul **Bairoch** (1930–1999). *Journal of European economic history*, 2000, 29, p. 203-208.

649. STOY (Manfred). Aus dem Briefwechsel von Wilhelm **Bauer**. T. 1. *Mitteilungen des Instituts für österreichische Geschichtsforschung*, 2000, 108, p. 376-398.

650. Ricordo di Ranuccio **Bianchi Bandinelli** (Roma, 11 maggio 2000). *Atti della Accademia Na-*

zionale dei Lincei. Rendiconti. Classe di Scienze Morali, Storiche e Filologiche, 2000, 397, 9, 11, p. 671-705.

651. BLOCH (Marc). Aus der Werkstatt des Historikers. Zur Theorie und Praxis der Geschichtswissenschaft. Frankfurt/M., Campus Verl., 2000, 361 p. – DAMAMME (Dominique). Un cas d'expertise. «L'Etrange défaite» de Marc **Bloch**. *Sociétés contemporaines*, 2000, 39, p. 95-116. – DUMOULIN (Olivier). Marc **Bloch**. Paris, Presses de Sciences Po, 2000, 336 p. – MORETTI (Mauro). Guerra e dopoguerra storiografico. Pirenne, Febvre, **Bloch**. *Società e storia*, 2000, 22, 88, p. 345-357. – PANERO (Francesco). Le nouveau servage et l'attache à la glèbe aux XIIe et XIIIe siècles: l'interprétation de Marc **Bloch** et la documentation italienne. *In*: Formes (Les) de la servitude [Cf. n° 1142], p. 555-561.

652. GASPARRI (Stefano). Gian Piero **Bognetti**, storico dei Longobardi. *La Cultura*, 2000, 38, p. 129-140.

653. DÜRRING (Norbert). Nofretete und andere Kleinigkeiten. Ludwig **Borchardt** und die frühe Ägyptologie. *Antike Welt*, 2000, 31, p. 424-426.

654. BOROZDINA (Polina A.). Zhizn' i sud'ba professora Il'i Nikolaevicha Borozdina. (Life and fortune of Prof. Ilya N. **Borozdin** [1883–1959, Russia]). Ed. M. K. BELINOV, T. M. TRUBAROV, N. F. FEDOSOV. Voronezh, Izd-vo Voronezhskogo un-ta, 2000, 406 p.

655. NOLZEN (Armin). Martin **Broszat**, der "Staat Hitlers" und die NSDAP. Einige Anmerkungen zur strukturalistischen Interpretation des "Dritten Reichs". *Revue d'Allemagne et des pays de langue allemande*, 2000, 32, 3, p. 101-118.

656. BOWERSOCK (Glen). **Burckhardt** sulla Tarda Antichità: dal "Constantin" alla "Griechische Kulturgeschichte". *Rivista storica italiana*, 2000, 112, 3, p. 1094-1108. – CHRIST (Karl). Jacob **Burckhardts** Weg zur "griechischen Kulturgeschichte". *Historia*, 2000, 49, p. 101-125. – CRÈPON (Marc). L'art de la Renaissance selon **Burckhardt** et Taine (la question des appartenances). *Revue germanique internationale*, 2000, 13, p. 187-200. – FUBINI (Riccardo). Considerazioni su **Burckhardt**. Il libro sul Rinascimento in Italia; De Sanctis e Burckhardt. *Archivio storico italiano*, 2000, 158, 583, p. 85-118. – GOSSMAN (Lionel). Basel in the age of **Burckhardt**. A study in unseasonable ideas. Chicago a. London, University of Chicago Press, 2000, XII-608 p. – GROßE (Jürgen). Letzte Stunde. Zur Analogie von Welt- und Lebensgeschichte bei Wilhelm Dilthey und Jacob **Burckhardt**. *Storia della storiografia*, 2000, 37, p. 67-99. – HINDE (John R.). Jacob **Burckhardt** and the crisis of modernity. Montreal, McGill-Queen's U. P., 2000, XII-327 p. (McGill-Queen's Studies in the History of Ideas, 29). – Jacob **Burckhardt**. Ästhetik der bildenden Kunst. Über das Studium der Geschichte. Mit dem Text der "Weltgeschichtlichen Betrachtungen" in der Fassung von 1905. München, Beck, 2000, 695 p. (Jacob Burckhardt. Werke, 10). – Jacob **Burckhardt**. Die Baukunst der Renaissance in Italien. Nach der Erstausgabe der «Geschichte der Renaissance in Italien». Hrsg. v. Maurizio GHELARDI. München, Beck, 2000, 532 p. (Jacob Burckhardt. Werke, 5). – TAUBER (Christine). Jacob **Burckhardts** "Cicerone". Eine Aufgabe zum Genießen. Tübingen, Niemeyer, 2000, VIII-314 p. (Reihe der Villa Vigoni, 13).

657. ANGELOW (Jürgen). Edmund **Burke** und die Französische Revolution. Von konservativer Kritik zur Idee des modernen Weltanschauungskrieges. *Zeitschrift für Religions- und Geistesgeschichte*, 2000, 52, 2, p. 97-114.

658. BELLAMY (Richard). A modern interpreter: Benedetto **Croce** and the politics of Italian culture. *The European legacy*, 2000, 5, 6, p. 845-862. – CINGARI (Salvatore). Dall'erudizione alla storia «sociale». Un percorso nella storiografia del giovane **Croce** (1883–1901). *Rivista storica italiana*, 2000, 112, 1, p. 235-281.

659. LAVAGNE (Henri). Lettres inédites de Franz **Cumont** à Salomon Reinach. *Académie des Inscriptions et Belles-Lettres. Comptes rendus*, 2000, 2, p. 763-774.

660. ALBERTI (Giovanni). Le tentazioni della storia: il "caso" **Cusin**. *Clio*, 2000, 36, p. 223-264.

661. Incontro di studi sull'opera di Renzo **De Felice**, Roma, Palazzo Giustiniani, 4 giugno 1997. Roma, Giunta centrale per gli studi storici, 2000, 81 p.

662. Prof. Aldo **De Maddalena**. Bibliografia. A cura di Maria ELLI. *In*: Omaggio ad Aldo De Maddalena. Per gli ottant'anni di un maestro amico. A cura di Marco CATTINI e Marzio A. ROMANI. Roma, Bulzoni, 2000, p. 227-238. (Cheiron, 34).

663. RICHE (Denyse). Un témoin de l'historiographie clunisienne à la fin du Moyen Age: le 'Chronicon' de François **de Rivo**. *Revue Mabillon*, 2000, 72, 11, p. 89-114.

664. MARCONE (Arnaldo). I libri sull'Italia antica delle "Rivoluzioni d'Italia" di Carlo **Denina**. *In*: Leo Valiani storico e politico [Cf. n° 729], p. 1072-1093.

665. PORCIANI (Leone). Gli inizi della storiografia greca nel De Thucydide di **Dionigi di Alicarnasso**: storia di una teoria. *Atti dell'Accademia delle Scienze di Torino. Classe di Scienze morali, storiche e filologiche*, 1999–2000, 133-134, p. 43-79.

666. VELCIC-CANIVEZ (M.). Histoire et intertextualité. L'écriture de George **Duby**. *Revue historique*, 2000, 124, 613, p. 187-206.

667. BOUTRY (Philippe), JULIA (Dominique). Une anthropologie religieuse du Jubilé. L'année sainte d'Alphonse **Dupront**. *Revue d'histoire de l'Eglise française*, 2000, 86, 216, p. 95-117. – BOUYSSY (Maïté). Alphonse **Dupront** et 'Le mythe de croisade'. *Revue d'histoire moderne et contemporaine*, 2000, 47, p. 616-620.

668. MAFFI (Davide). La Spagna e l'Europa: l'opera storica di sir John **Elliott**. *Rivista storica italiana*, 2000, 112, 1, p. 282-317.

669. Studi su Bartolomeo **Facio**. A cura di Gabriella ALBANESE. Pisa, ETS, 2000, XI-274 p. (16 p. of plates, ill.).

670. GREENE (Kevin). Technological innovation and economic progress in the ancient world: M. I. **Finley** re-considered. *Economic history review*, 2000, 53, 1, p. 29-59.

671. MADER (Gottfried). **Josephus** and the politics of historiography: apologetic and impression management in the Bellum Judaicum, Leiden, Brill, 2000, X-172 p. (Mnemosyne, bibliotheca classica Batava. Supplementum, 205). – THACKERAY (H. St. J.). **Flavius Josèphe**: l'homme et l'historien. Avec un appendice sur la version slavone de la 'Guerre'. Paris, Le Cerf, 2000, XIX-250 p.

672. CAPITANI (Ovidio). Ricordo di Arsenio **Frugoni**: trenta anni dopo. *La Cultura*, 2000, 38, p. 321-331.

673. DEPTUŁA (Czesław). Galla Anonima mit genezy Polski. Studium z historiozofii i hermeneutyki symboli dziejopisarstwa średniowiecznego. (Die Genese Polens **Gall Anonim**'s Überlieferung zufolge. Studium zur Historiosophie und Hermeneutik der Symbole der mittelalterlichen Geschichtsschreibung). Lublin, Inst. Europy Środkowowschodniej, 2000, 395 p.

674. FERRI (Mauro). Ricordo di Aldo **Garosci**. *Rassegna storica del Risorgimento*, 2000, 57, p. 277-280.

675. KOCZERSKA (Maria). Aleksander **Gieysztor** 17 VII 1916–9 II 1999. Szkic biograficzny. (Aleksander Gieysztor 17. VII. 1916–9. II. 1999. Biographische Skizze). *Studia Źródłoznawcze,* 2000, 37, p. 3-12.

676. BUSINO (Giovanni). Causalismo, simmetria e riflessività: una lettura dei lavori di Carlo **Ginzburg**. *L'Acropoli*, 2000, 1, p. 311-327. – SERNA ALONSO (Justo), PONS (Anaclet). Cómo se escribe la microhistoria: ensayo sobre Carlo **Ginzburg**. Madrid, Ediciones Cátedra, 2000, 288 p.

677. Bibliografia degli scritti di Edoardo **Grendi**. A cura di Osvaldo RAGGIO. *Quaderni storici*, 2000, 35, 104, p. 823-834.

678. DAWSON (S.). George **Grote** and the Ancient Greeks. *Polis*, 2000, 17, p. 187-198.

679. LAMARRIGUE (Anne-Marie). Bernard **Gui** (1261–1331): un historien et sa méthode. T. 1. Paris, H. Champion, 2000.

680. NAMER (Gérard). **Halbwachs** et la mémoire sociale. Paris, L'Hartmann, 2000, 250 p.

681. HARMSEN (Theodorus Hendrikus Bernardus Maria). Antiquarianism in the Augustan age: Thomas **Hearne**, 1678–1735. Oxford, Peter Lang, 2000, 335 p. (ill.).

682. BICHLER (Reinhold), ROLLINGER (Robert). **Herodot**. Hildesheim, Zürich u. New York, Georg Olms Verlag, 2000, 209 p. – BICHLER (Reinhold). **He**rodots Welt. Der Aufbau der Historie am Bild der fremden Länder und Völker, ihrer Zivilisation und ihrer Geschichte. Berlin, Akademie Verl., 2000, 425 p. (Antike in der Moderne). – GOULD (John). **Herodotus**. London, Bristol Classical, 2000, 164 p. (maps). (Bristol classical paperbacks). – THOMAS (Rosalind). **Herodotus** in context: ethnography, science, and the art of persuasion. Cambridge, Cambridge U. P., 2000, VIII-321 p.

683. DI COSTANZO (Giuseppe). Lo storicismo realistico di Otto **Hintze**. Prefazione di Giuseppe CACCIATORE. Bari, Palomar, 2000, 344 p. (Palomar Athenaeum, 23).

684. STRUPP (Christoph). Johan **Huinziga**. Geschichtswissenschaft als Kulturgeschichte. Göttingen, Vandenhoeck u. Ruprecht, 2000, 352 p.

685. GUYOT-BACHY (Isabelle). Le Memoriale historiarum de **Jean de Saint-Victor**: un historien et sa communauté au début du XIVe siècle. Turnhout, Brepols, 2000, 608 p. (plates, ill.). (Bibliotheca Victorina, 12).

686. ESCH (Arnold). Norbert **Kamp** als Historiker des staufischen Italien. *Quellen und Forschungen aus italienischen Archiven und Bibliotheken*, 2000, 80, p. 625-641.

687. BRUSH (Stephen G.). Thomas **Kuhn** as a historian of science. *Science & education*, 2000, 9, 1-2, p. 39-58. – CANEVA (Kenneth L.). Possible **Kuhns** in the history of science: anomalies of incommensurable paradigms. *Studies in history and philosophy of science*, 2000, 31A, 1, p. 87-124. – FULLER (Steve). Thomas **Kuhn**: a philosophical history for our times. London, University of Chicago Press, 2000, XVII-472 p.

688. LANGLOIS (Charles-Victor). Document: avertissement aux candidats à l'agrégation d'histoire (1901). *Vingtième siècle. Revue d'histoire*, 2000, 65, p. 125-136.

689. BERNARD (Jacques-Emmanuel). Le portrait chez **Tite-Live**. Essai sur une écriture de l'histoire romaine. Bruxelles, Latomus, 2000, 482 p. – CHAPLIN (J. D.). **Livy**'s exemplary history. Oxford. Oxford U. P., 2000, XII-245 p.

690. THOMAS (William). The quarrel of **Macaulay** and Croker. Politics and history in the age of reform. Oxford, Oxford U. P., 2000, 339 p.

691. *Vacat.*

692. GAMBERALE (Leopoldo), MUSTI (Domenico). Scevola **Mariotti** e la "Rivista di Filologia". *Rivista di Filologia e Istruzione classica*, 2000, 128, p. 5-20.

693. GILL (David W. J.). Collecting for Cambridge: John Hubert **Marshall** on Crete. *Annual of the British School at Athens*, 2000, 95, p. 517-526.

694. Antropologia, storia e impegno politico in Marcel **Mauss**. Interventi di Elena FASANO GUARINI e Michele BATTINI. *Passato e Presente*, 2000, 18, 50, p. 91-106.

695. Gelehrtenalltag. Der Briefwechsel zwischen Eduard **Meyer** und Georg Wissowa (1890–1927). Hrsg. v. Gert AUDRING. Hildesheim, Weidmann, 2000, VI-560 p.

696. Correspondance générale [de **Michelet**]. T. 11. 1866–1870. Textes réunis et annotés par Louis LE GUILLOU. Paris, H. Champion, 2000, 912 p.

697. DI DONATO (Riccardo). Nuovi materiali per una biografia intellettuale di Arnaldo **Momigliano**. *Atti della Accademia Nazionale dei Lincei. Rendiconti. Classe di Scienze Morali, Storiche e Filologiche*, 2000, 397, 9, 11, p. 383-398. – MOMIGLIANO (Arnaldo). Ausgewählte Schriften zur Geschichte und Geschichtsschreibung. Band 3. Die Moderne Geschichtsschreibung der Alten Welt. Hrsg. v. Glenn W. MOST, übers. v. Kai BRODERSEN und Andreas WITTENBURG. Stuttgart u. Weimar, Metzler, 2000, XIX-465 p.

698. COTTON (Hannah M.). Cassius Dio, **Mommsen** and the quinquefascales. *Chiron*, 2000, 30, p. 217-234. – KIRSTEIN (Robert), CALDER (W. M.). Agnes von Zahn-Harnack's forgotten essay on **Mommsen**. *Classical world*, 2000, 94, 1, p. 37-44. – ZANGEMEISTER (Karl). Theodor **Mommsen** als Schriftsteller. Ein Verzeichnis seiner Schriften. Bearb. u. fortgesetzt v. Emil JACOBS, neu bearb. v. Stefan REBENICH. Hildesheim, Weidmann, 2000, XXV-354 p.

699. MONTESQUIEU (Charles de Secondat, baron de La Brède et de). Considérations sur les causes de la grandeur des Romains et de leur décadence. Ed. par Françoise WEIL et Cecil P. COURTNEY. Oxford, Voltaire Foundation, 2000, XX-382 p.

700. CASINI (Simona). Bibliografia degli scritti e dei corsi universitari di Carlo **Morandi**. *Rassegna storica toscana*, 2000, 46, p. 173-220.

701. CAPONE (Alfredo). **Moscati**, Romeo e la tradizione del Risorgimento. *Clio*, 2000, 36, p. 265-279.

702. TAGLIAFERRI (Teodoro). Nazionalità territoriale e nazionalità linguistica nel pensiero storico di Lewis **Namier**. *Archivio di storia della cultura*, 2000, 13, p. 119-148.

703. MUSTÉ (Marcello), PERTICI (Roberto). Sulla quarta silloge di Adolfo **Omodeo**. *La Cultura*, 2000, 38, 2, p. 335-364.

704. MÉGIER (Elisabeth). 'Divina pagina' and the narration of history in **Orderic Vitalis**' "Historia evvlesiastica". *Revue bénédictine*, 2000, 110, 1-2, p. 106-123.

705. Paolo **Diacono**: uno scrittore fra tradizione longobarda e rinnovamento carolingio. Atti del Convegno Internazionale di Studi, Cividale del Friuli-Udine, 6–9 maggio 1999. A cura di Paolo CHIESA. Udine, Forum, 2000, 625 p. (ill., tables).

706. Vostochnaja Evropa v drevnosti i srednevekov'je: Istoricheskaja pamjat' i formy ee voploshchenija: XII chtenija pamjati V. T. Pashuto, Mat-ly konf. (Eastern Europe in Antiquity and Middle Ages: historical memory and forms of its realization. Proceedings of the 11th Conference dedicated to the memory of Vladimir T. **Pashuto**). Ed. Elena A. MEL'NIKOVA. RAN, In-t vseobshchej istorii. Moskva, [s. n.], 2000, 237 p. [Cf. Bibl. 99, n° 1043.]

707. MILLER (Peter N.). **Peiresc**'s Europe. Learning and vertue in the seventeenth century. New Haven a. London, Yale U. P., 2000, 234 p.

708. CUENCA TORIBIO (José Manuel). La obra historiográfica de Florentino **Pérez-Embid**. Sevilla, Escuela de Estudios Hispano-Americanos, CSIC, 2000, 105 p. (Colección Difusión y estudio/Escuela de Estudios Hispano-Americanos de Sevilla, 408).

709. DAGER ALVA (Joseph). Una aproximación a la historiografía del siglo XIX: vida y obra de José Toribio **Polo**, 1841–1918. Lima, Pontificia Universidad Católica del Perú, Banco Central de Reserva del Perú, 2000, 354 p. (ill.). (Publicación del Instituto Riva-Agüero, 186).

710. Proposito (A) del volume di Paolo **Prodi** "Una storia della giustizia. Dal pluralismo dei fori al moderno dualismo tra coscienza e diritto". Interventi di Reinhard ELZE, Diego QUAGLIONI, Nestore PIRILLO e Paolo POMBENI. *Annali dell'Istituto storico italo-germanico in Trento*, 2000, 26, p. 743-778.

711. DOVE (Alfred). "**Ranke**'s Verhältnis zur Biographie" von 1895. *BIOS*, 2000, 13, 2, p. 285-308.

712. POMIAN (K.). [Ruggiero] **Romano**: histoire et encyclopédie. *Revue européenne des sciences sociales*, 2000, 38, 117, p. 193-210.

713. PANEJAKH (Viktor M.). Boris Aleksandrovich **Romanov**: Tvorchestvo i sud'ba istorika. (Boris A. Romanov: work and fortune of a historian). RAN, In-t rossijskoj istorii, Sankt-Peterburgskij filial. Sankt-Peterburg, Dmitrij Bulanin, 2000, 445 p. (portr., bibl.).

714. PESCOSOLIDO (Guido). Volpe e **Romeo**: il maestro e l'allievo. *Nuova Storia contemporanea*, 2000, 4, 6, p. 97-120.

715. JOBST (Kerstin S.). Orientalism, E. W. **Said** und die Osteuropäische Geschichte. *Saeculum*, 2000, 51, 2, p. 250-266.

716. JENSSEN (Dag). **Sars**: norsk historie som eksistensiell og europeisk erfaring. (Ernst Sars [1835–1917]). European influence on Norwegian historiography). Bergen, Universitetet i Bergen, 2000, VII, 332 p.

717. SERENI (Emilio). Vita e tecniche forestali nella Liguria antica. Ambienti e storie della Liguria. Studi in ricordo di Emilio **Sereni**. Bari, Dedalo, 2000, 245 p. (Annali dell'Istituto Alcide Cervi, 19).

718. Ernesto **Sestan** 1898–1998. A cura di Emilio CRISTIANI e Giuliano PINTO. Firenze, Leo S. Olschki, 2000, 225 p. – SOLDANI (Simonetta). Ernesto **Sestan**, il mondo tedesco e le aporie del principio di nazionalità. *Passato e presente*, 2000, 18, 51, p. 95-122.

719. BURGHARTZ (Susanna). Der "große Wilde" und die "Unvergleichliche" – Figuren kolonialer Annäherung. John **Smiths** Geschichtsschreibung zu den Anfängen Virginias. *Historische Anthropologie*, 2000, 8, 2, p. 163-188.

720. BÜNZ (E.). Neues zur Biographie des Chronisten Conrad **Stolle** (1436–1501). *Deutsches Archiv für Erforschung des Mittelalters*, 2000, 56, p. 201-211.

721. RAMONDETTI (Paola). Tiberio nella biografia di **Svetonio**. Napoli, Loffredo, 2000, 93 p. (Studi latini, 40).

722. Révolution romaine (La) après Ronald **Syme**. Bilans et perspectives. Sept exposés suivis de discussions. Éd. par Adalberto GIOVANNINI, Philippe BORGEAUD et Greg ROWE. Vandoevre et Genève, Fondation Hardt, 2000, XI-342 p.

723. BIRLEY (Anthony R.). The life and death of Cornelius **Tacitus**. *Historia*, 2000, 49, p. 230-247. – O'GORMAN (Ellen). Irony and misreading in the Annals of Tacitus. Cambridge, Cambridge U. P., 2000, VII-200 p.

724. SCHARBAUM (Heike). Zwischen zwei Welten. Wissenschaft und Lebenswelt am Beispiel des deutschjüdischen Historikers Eugen **Täubler** (1879–1953). Münster, Lit, 2000, 136 p. (Münsteraner Judaistische Studien, 8).

725. TAGLIAFERRI (Teodoro). La nuova storiografia britannica e lo sviluppo del welfarismo. Ricerche su R. H. **Tawney**. Napoli, Liguori, 2000, 346 p.

726. BURK (Kathleen). Troublemaker. The life and history of A. J. P. **Taylor**. New Haven a. London, Yale U. P., 2000, XIV-491 p.

727. EASTWOOD (David). History, politics and reputation: E. P. **Thompson** reconsidered. *History*, 2000, 85, 280, p. 634-654.

728. ALONSO-NUNEZ (José Manuel). Die Archäologien des **Thukydides**. Konstanz, Universitätsverlag Konstanz, 2000, 114 p. (map). – FRIEDRICHS (Jörg). Aufschlussreiche Rhetorik: ein Versuch über die Redekultur und ihren Verfall bei **Thukydides**. Würzburg, Ergon, 2000, 154 p. (Spektrum Politikwissenschaft, 12). – JANSSENS (Emiel). **Thucydides**: een kritische bewonderaar van Pericles. *Revue belge de philologie et d'histoire*, 2000, 78, 1, p. 31-52.

729. ARA (Angelo). Leo **Valiani**, uomo e storico della Mitteleuropa. *In*: Leo Valiani storico e politico [Cf. n° 729], p. 921-998. – BUSINO (Giovanni). In memoria di Leo **Valiani** (1909–1999). *In*: Leo Valiani storico e politico [Cf. n° 729], p. 862-920. – Leo **Valiani** storico e politico. *Rivista storica italiana*, 2000, 112, 3, p. 862-1093. [Cf. n°s <scelta> 664, 729, 3919, 6167.] – Storia (Tra) e politica. Bibliografia degli Scritti di Leo **Valiani** (1926–1999). A cura di Giovanni BUSINO. Milano, Fondazione Giangiacomo feltrinelli, 2000, XLV-233 p.

730. VALDERAS (José). Francisco **Vélez de Arciniega** en la polémica de la coloquíntida. *Asclepio: archivo iberoamericano de historia de la medicina y antropología médica*, 2000, 52, 1 p. 7-35.

731. SCHMITZER (Ulrich). **Velleius Paterculus** und das Interesse an der Geschichte im Zeitalter des Tiberius. Heidelberg, C. Winter, 2000, 346 p. (Bibliothek der klassischen Altertumswissenschaften, n.F., 2. Reihe, 107).

732. DI DONATO (Riccardo), BARBAGALLO (Francesco), CANFORA (Luciano), VERNANT (Jean-Pierre). Tra passato e presente. L'impegno di Jean-Pierre **Vernant**. *Studi Storici*, 2000, 41, 1, p. 7-30.

733. CERETTI (Marinella). La diffusione della pubblicistica antigiacobina in Italia e la testimonianza delle fonti nelle «Vicende memorabili de' tempi suoi» di Alessandro **Verri**. *Rivista storica italiana*, 2000, 112, 1, p. 138-188.

734. GHISALBERTI (Carlo). Gioacchino **Volpe** e la Grande Guerra. *Clio*, 2000, 36, p. 201-222. – SASSO (Gennaro). Giovanni Gentile e Gioacchino **Volpe** dinanzi al crollo del fascismo. *La Cultura*, 2000, 38, 3, p. 381-400.

735. Art history as cultural history: **Warburg**'s projects. Ed. by Richard WOODFIELD. Abingdon, Marston, 2000, 304 p. (ill.). (Critical voices in art, theory and culture).

736. CAPOGROSSI BOLOGNESI (Luigi). Max **Weber** e le economie del mondo antico. Roma e Bari, Laterza, 2000, XX-454 p. – Max **Weber** Gesamtausgabe. Abt. 1. Schriften und Reden. Band 5. Börsenwesen. Schriften und Reden 1893–1898. Halbbd. 2. Hrsg. v. Knut BORCHARDT in Zusammenarb. mit Cornelia MEYER-STOLL. Tübingen, Mohr, 2000, 603 p. – Max **Weber** und die Stadt in Kulturvergleich. Hrsg. v. Hinnerk BRUHNS und Wilfried NIPPEL. Göttingen, Vandenhoeck u. Ruprecht, 2000, 202 p. – MOMMSEN (Wolfgang). Max **Weber**'s "Grand Sociology": the origins and composition of Wirtschaft und Gesellschaft. Sociologie. *History and theory*, 2000, 39, 3, p. 364-383.

737. BLÄNSDORF (Jürgen). Seneca und Richard von **Weizsäcker** über Geschichte und Zukunft. *Gymnasium*, 2000, 107, p. 229-246.

738. DÉCULTOT (Elisabeth). Johann Joachim **Winckelmann**: enquête sur la genèse de l'histoire de l'art. Paris, Presses Universitaires de France, 2000, X-337 p. (ill.).

b. Special studies. Addenda 1997–1999

739. CORBELLARI (Alain). Joseph **Bédier**, écrivain et philologue. Genève, Droz, 97, XXV-765.

740. HUREL (Daniel-Odon). Bibliographie et inventaire de la correspondance imprimée de dom **Bernard de Montfaucon**. *Dom Bernard de Montfaucon*, 98, 2, p. 137-212.

741. **HALÉVI** (Florence), **HALÉVI** (Elie). Six jours in URSS, septembre 1932: récit de voyage inédit. Ed.

par Sophie CŒURÉ. Paris, Presses de l'Ecole normale supériore, 98, 140 p. (ill.).

742. MÉGIER (Elisabeth). 'Cotidie operatur'. Christus und die Geschichte in der 'Historia ecclesiastica' des **Ordericus**. *Revue Mabillon*, 99, 71, 10, p. 169-204.

743. POLICHETTI (Antonio). Le "Historiae" di **Orosio** e la "Storiografia ecclesiastica" occidentale (311–471 d.C.). Napoli, Edizioni scientifiche italiane, 99, [s. p.].

744. REBÉRIOUX (Madeleine). Pierre **Vidal-Naquet** et nos guerres. *In*: Pierre Vidal-Naquet, un historien dans la cité. Paris, La Découverte, 98, p. 87-109. (Textes à l'appui. Série histoire contemporaine).

Cf. n^{os}, *1344, 1633, 2004, 2005, 2219, 2469, 2470-2472, 2518-2522, 2575, 3334, 3335*

§ 3. Methodology, philosophy, and teaching of history.

* 745. *Vacat*.

* 746. TORSTENDAHL (Rolf). Thirty-five years of theories in history. *Scandinavian journal of history*, 2000, 25, 1-2, p. 1-26.

---

747. Ad fontes. Antikkvitenskap, kildebehandling og metode. (Ad fontes. Classical studies, study of sources and method). Ed. Jon W. IDDENG. Oslo, Norsk klassisk forbund, 2000, 263 p. (ill.). (Klassisk skriftserie, 1).

748. ARNOLD (John). History: a very short introduction. Oxford, Oxford U. P., 2000, 137 p. (ill.). (Very short introductions).

749. Auf der Suche nach der verlorenen Wahrheit. Zum Grundlagenstreit in der Geschichtswissenschaft. Hrsg. v. Rainer M. KLESOW und Dieter SIMON. Frankfurt am Main, Campus Verl., 2000, 171 p.

750. BARBERI (Alessandro). Hayden White und Carlo Ginzburg. Eine Diskurs-Analyse nach Foucault. Wien, Turia u. Kant, 2000, 350 p. (Kultur als Praxis, 3).

751. BLUM (Paul Richard). Francesco Patrizi in the "Time-Sack": history and rhetorical philosophy. *Journal of the history of ideas*, 2000, 61, 1, p. 59-74.

752. BONNAUD (Robert). Tournants et périodes. Paris, Kimé, 2000, 199 p.

753. BRETONE (Mario). In difesa della storia. Roma e Bari, Laterza, 2000, VIII-89 p.

754. BUDICK (Sanford). The Western theory of tradition: terms and paradigms of the cultural sublime. London, Yale U. P., 2000, XXII-293 p. (ill.)

755. CANNADINE (David). History in our time. London, Penguin, 2000, XI-313 p.

756. CHAKRABARTY (Dipesh). Provincializing Europe: postcolonial thought and historical difference. Princeton, Princeton U. P., 2000, XII-301 p. (Princeton Studies in Culture/Power/History).

757. CONRAD (Christoph), CONRAD (Sebastian). Wie vergleicht man Historiographien? *Historical social research*, 2000, 25, 3/4, p. 266-271.

758. CONTE (Domenico). Storicismo e storia universale. Linee di un'interpretazione. Napoli, Liguori, 2000, 188 p. (La cultura storica, 14).

759. CORTIJO OCANA (Antonio). Teoría de la historia y teoría política en el siglo XVI: Sebastián Fox Morcillo, De historiae institutione dialogus = Diálogo de la enseñanza de la historia, 1557. Alcalá, Universidad de Alcalá, Servicio de Publicaciones, 2000, 311 p. (Ensayos y documentos, 35).

760. COSMANN (Peggy). Der Ausgang der Vernunft aus der Geschichte. Wieder den Versuch einer Rehabilitierung von Geschichtsphilosophie. *Tel Aviver Jahrbuch für deutsche Geschichte*, 2000, 29, p. 19-54.

761. Culture and explanation in historical inquiry. *History and theory*, 2000, 39, 3, p. 289-363. [Contents: BIERNACKI (Richard). Language and the shift from signs to practices in cultural inquiry (p. 289-310). – KANE (Anne). Reconstructing culture in historical explanation: narratives as cultural structure and practice (p. 311-330). – HALL (John R.). Cultural meanings and cultural structures in historical explanation (p. 331-347). – LORENZ (Chris). Some afterthoughts on culture and explanation in historical inquiry (p. 348-363)].

762. DETIENNE (Marcel). Comparer l'incomparable. Paris, Ed. Du Seuil, 2000, 134 p. (La librairie du XXe siècle).

763. DIPPER (Christoph). Die "Geschichtlichen Grundbegriffe". Von der Begriffsgeschichte zur Theorie der historischen Zeiten. *Historische Zeitschrift*, 2000, 270, 2, p. 281-308.

764. DIRLIK (Arif). Postmodernity's histories: the past as legacy and project. Lanham, Rowman & Littlefield, 2000, XIV-245 p. (Culture and politics)

765. DOSSE (François). L'histoire. Paris, A. Colin, 2000, 208 p. – IDEM. Paul Ricœur: les sens d'une vie. Paris, la Découverte, 2000, 789 p.

766. ENGSTROM (Eric J.). «Zeitgeschichte» as disciplinary history. On professional identity, self-reflexive narratives, and discipline-building in contemporary German History. *Tel Aviver Jahrbuch für deutsche Geschichte*, 2000, 29, p. 399-426.

767. Ernst Troeltschs "Historismus". Hrsg. v. Friedrich Wilhelm GRAF. Gütersloh, Gütersloher Verlagshaus, 2000, 304 p. (Troeltsch-Studien, 11).

768. Formatos, géneros y discursos: memoria del Segundo Encuentro de Historiografía (1998: Universidad Autónoma Metropolitana. Unidad Azcapotzalco). Ed. por José A. RONZÓN LEÓN, Saul Jerónimo ROMERO, con la colaboración de Valeria CORTÉS. Azcapotzalco: Universidad Autónoma Metropolitana Azcapotzalco, 2000, 463 p. (ill). (Colección Memorias (Universidad Autónoma Metropolitana. Unidad Azca-

potzalco. División de Ciencias Sociales y Humanidades)).

769. Formy istoricheskogo soznanija ot pozdnej antichnosti do epokhi Vozrozhdenija (Issledovanija i teksty): Sb. nauch. trudov pamjati K.D. Avdeevoj. [Forms of historical consciousness from Antiquity to the Renaissance (studies and texts): Collected papers dedicated to the memory of Klaudia D. Avdeeva]. Ed. I. V. KRIVUSHIN, V. M. TJULENEV [et al.]. Ivanovo, Ivanovskij gos. un-t, 2000, 281 p.

770. GALASSO (Giuseppe). Nient'altro che storia: saggi di teoria e metodologia della storia. Bologna, Il mulino, 2000, 379 p. (Saggi, 521)

771. Gedächtnis (Das) der Kunst: Geschichte und Erinnerung in der Kunst der Gegenwart. Hrsg. v. Kurt WETTENGL. Frankfurt, Historisches Museum, Schirn Kunsthalle, Ostfildern-Ruit, Hatje Cantz, 2000, 278 p.

772. Geschichte denken: Philosophie, Theorie, Methode. Hrsg. im Auftrag des Instituts für deutsche Geschichte von Moshe ZUCKERMANN. Gerlingen, Bleicher, 2000, 477 p. (Tel Aviver Jahrbuch für deutsche Geschichte, 29).

773. Geschichtskultur. Theorie – Empirie – Pragmatik. Hrsg. v. Bernd MÜTTER, Bernd SCHÖNEMANN und Uwe UFFELMANN. Weinheim, 2000, 369 p. (Schriften zur Geschichtsdidaktik, 11).

774. Geschichtsphilosophie: Kolloquium zum 70. Geburtstag von Wolfgang Eichhorn. Berlin, Trafo, 2000, 149 p. (Sitzungsberichte der Leibniz-Sozietät, 37, 2).

775. Geschichtswissenschaften. Eine Einführung. Hrsg. v. Christoph CORNELIßEN. Frankfurt am Main, Fischer Taschenbuch Verl., 2000, 320 p.

776. GINZBURG (Carlo). Rapporti di forza: storia, retorica, prova. Milano, Feltrinelli, 2000. 161 p. (plates, ill.). (Campi del sapere)

777. GRAS (Michel). Donner du sens à l'objet. Archéologie, technologie culturelle et anthropologie. In: Histoire (L') face à l'archéologie [Cf. n° 780], p. 601-614.

778. HARRISON (Dick). På Klios fält. Essäer om historisk forskning och historieskrivning. (Le champ de Clio. Essais sur la recherche historique et l'écriture de l'histoire). Lund, Historiska media, 2000, 190 p.

779. HARTOG (François). Le témoin et l'historien. *Gradhiva*, 2000, 27, p. 1-14.

780. Histoire (L') face à l'archéologie. Prés. de Philippe BRAUNSTEIN. *Annales*, 2000, 55, 3, p. 551-622. [Cf. n°^{os} <sélection> 777, 783, 795, 814.]

781. Histoire et nation en Europe centrale et orientale XIXe–XXe siècles. Ed. par Marie Elisabeth DUCREUX. Paris, Institut national de recherche pedagogique, 2000, 186 p. (*Histoire de l'éducation*, 2000, numéro spécial).

782. Hobbes and history. Ed. by G. A. J. ROGERS and Tom SORELL. London, Routledge, 2000, XVI-192 p. (Routledge studies in seventeenth century philosophy).

783. *Vacat*.

784. IGGERS (Georg C.). Historiography between scholarship and poetry: reflections on Hayden White's approach to historiography. *Rethinking history*, 2000, 4, 3, p. 373-390. [Cf. n° 826]

785. Interdisziplinarität (Die) der Begriffsgeschichte. Hrsg. v. Gunter SCHOLTZ. Hamburg, Meiner, 2000, 200 p. (Archiv für Begriffsgeschichte, Sonderheft).

786. JARRICK (Arne). Textual analysis in the historiography of mentalities its external and internal problem of representativity. In: Meaning and Interpretation. Ed. by Dag PRAWITZ. Stockholm, Kgl. Viterhets Historie och Antikvitets Akademien, 2000, [s. p.]. (Konferenser, 49-50).

787. Kazus: Individual'noe i unikal'noe v istorii. (Casus: the individual and unique in history: [Articles]). Vol. 3 (2000). Ed. Jurij L. BESSMERTNYJ, Mikhail A. BOJTSOV. RAN, In-t vseobshchej istorii. Moskva, Izd. Ros. gos. gumanit. un-ta, 2000, 401 p. (bibl. incl.). [Eng. summaries] [Cf. Bibl. 99, n° 844.]

788. KELLEY (Donald R.). Vico and the archeology of wisdom. *New Vico studies*, 2000, 18, [s. p.].

789. KINDT (Tom), MÜLLER (Hans H.). Dilthey gegen Scherer. Geistesgeschichte contra Positivismus. Zur Revision eines wissenschaftshistorischen Stereotyps. *Deutsche Vierteljahrschrift für Literaturwissenschaft und Geistesgeschichte*, 2000, 74, 4, p. 685-709.

790. KOCKA (Jürgen). Das Jahrhundert als Konstrukt und Befund – Erinnerung als Ressource. In: Erinnerte Zukunft. Was nehmen wir mit ins nächste Jahrtausend? Hrsg. v. Wilfried F. SCHOELLER [et al.]. Reinbek, 2000, p. 72-84.

791. KOSELLECK (Reinhart). Zeitschichten. Studien zur Historik. Frankfurt am Main, Suhrkamp, 2000, 400 p.

792. KOVALEVSKAJA (Vera B.). Komp'juternaja obrabotka massovogo arkheologicheskogo materiala iz rannesrednevekovykh pamjatnikov Evrazii. (Computer processing of the mass archaeological material from early Medieval Eurasian sites). Appendices by V. B. KOVALEVSKAJA, R. A. BAKHTADZE, V. A. GALIBIN, Z. A. L'VOVA. Moskva, [s. n.], 2000, 364 p. (ill.). (Khronologija vostchnoevropejskikh drevnostej V-IX vv., 2: "Stekljannye busy i pojasnye nabory").

793. LACAPRA (Dominick). History and reading: Tocqueville, Foucault, French studies. Toronto a. London, University of Toronto Press, 2000, 235 p. (Green College thematic lecture series).

794. LASERNA (Mario). Dos ensayos sobre la posibilidad de la historia, Carta de Heidelberg. Santafé de Bogotá, Universidad de los Andes, Ediciones Uniandes, 2000, 269 p. (Colección 50 anos)

795. LEVEAU (Philippe). Le paysage aux époques historiques: un document archéologique. *In*: Histoire (L') face à l'archéologie [Cf. n° 780], p. 555-582.

796. MAC CARNEY (Joe). Routledge philosophy guidebook to Hegel on history. London, Routledge, 2000, XI-234 p. (Routledge philosophy guidebooks).

797. MAH (Harold). Phantasies of the public sphere: rethinking the Habermas of historians. *Journal of modern history*, 2000, 72, 1, p. 153-182.

798. MAIGRET (Eric). Les trois héritages de Michel de Certeau. Un projet éclaté d'analyse de la modernité. *In*: Lectures et réceptions d'une œuvre. *Annales*, 2000, 55, 3, p. 511-550.

799. MARSDEN (William E.). Poisoned history: a comparative study of nationalism, propaganda and the treatment of war and peace in the late nineteenth and early twentieth century school curriculum. *History of Education*, 2000, 29, 1, p. 29-48.

800. MAZZARELLA (Eugenio). Nietzsche e la storia. Storicità e ontologia della vita. Napoli, Guida, 2000, 235 p.

801. MORADIELLOS (Enrique). Sine ira et studio: ejercicios de crítica historiográfica. Cáceres, Universidad de Extremadura, 2000, 221 p.

802. "Not telling": secrecy, lies, and history. Ed. by Gary MINKLEY and Martin LEGASSICK. *History and theory*, 2000, 39, 3, p. 132 p. [Contents: MINKLEY (Gary), LEGASSICK (Martin). "Not telling": secrecy, lies, and history (p. 1-10). – WHITE (Luise). Telling more: lies, secrets, and history (p. 11-22). – POHLANDT-MAC CORMICK (Helena). "I saw a nightmare ...": violence and the construction of memory (Soweto, June 16, 1976) (p. 23-44). – LALU (Premesh). The grammar of domination and the subjection of agency: colonial texts and modes of evidence (p. 45-68). – SITHOLE (Jabulani), MKHIZE (Sibongiseni). Truth or lies? Selective memories, imagings, and representations of chief Albert John Luthuli in recent political discourses (p. 69-85). – BRADFORD (Helen). Peasants, historians, and gender: a South African case study revisited, 1850–1886 (p. 86-110). – NDLOVU (Sifiso Mxolisi). Johannes Nkosi and the Communist Party of South Africa: images of "Blood river" and king Dingane in the late 1920s–1930 (p. 111-132).]

803. OSBORNE (Ken). 'Our History Syllabus Has Us Gasping': history in Canadian schools – past, present, and future. *Canadian historical review*, 2000, 81, 3, p. 404-435.

804. PAPAGNO (Giuseppe). Un modello per la storia: materiale, attività, funzione. Pref. di Maurice AYMARD. Reggio Emilia, Edizioni Diabasis, 2000, XV-331 p.

805. PARKER (Christopher). The English idea of history from Coleridge to Collingwood. Aldershot, Ashgate, 2000, 244 p.

806. Perspektiven der Gesellschaftsgeschichte. Hrsg. v. Paul NOLTE, Manfred HETTLING, Frank M. KUHLEMANN [et al.]. München, Beck, 2000. VI-170 p. [Cf. n° <Auswahl> 509.]

807. Philosophies of history: from Enlightenment to post-modernity. Ed. by Robert M. BURNS and Hugh RAYMENT-PICKARD. Oxford, Blackwell, 2000, XV-360 p.

808. RAUCH (Angelika). The hieroglyph of tradition: Freud, Benjamin, Gadamer, Novalis, Kant. Madison, Fairleigh Dickinson University Press; London, Associated University Presses, 2000, 249 p.

809. Representações; contribuição a um debate transdisciplinar. Ed. Jurandir MALERBA e Ciro FLAMARION CARDOSO. Campinas, Papirus, 2000, [s. p.].

810. RICŒUR (Paul). Aux origines de la mémoire, l'oubli de réserve. *Esprit*, 2000, p. 32-47. – IDEM. La mémoire, l'histoire, l'oubli. Paris, Ed. du Seuil, 2000, 675 p. (ill.). (L'ordre philosophique). – IDEM. L'écriture de l'histoire et la représentation du passé. *In*: Histoire et mémoire. *Annales*, 2000, 55, 4, p. 731-748.

811. ROHBECK (Johannes). Technik, Kultur, Geschichte: eine Rehabilitierung der Geschichtsphilosophie. Frankfurt am Main, Suhrkamp, 2000, 282 p. (Suhrkamp Taschenbuch Wissenschaft, 1462).

812. ROSENSTONE (Robert A.). A history of what has not yet happened. *Rethinking history*, 2000, 4, 2, p. 183-192.

813. RÜSEN (Jörn). Das Andere denken. Herausforderungen der modernen Kulturwissenschaften. Ulm, [s. n.], 2000, [s. p.]. (Bausteine zur Philosophie. Interdisziplinäre Schriftenreihe des Humboldt-Studienzentrums der Universität Ulm, 15). – IDEM. Vom Nutzen und Nachteil der Wissenschaft für das Schulbuch – am Beispiel der Geschichte. *In*: Internationale Verständigung. 25 Jahre Georg-Eckert-Institut für internationale Schulbuchforschung in Braunschweig. Hrsg. v. Ursula A. J. BECHER und Rainer RIEMENSCHNEIDER. Hannover, [s. n.], 2000, p. 49-56. (Studien zur internationalen Schulbuchforschung, 100).

814. SCHEID (John). Pour une archéologie du rite. *In*: Histoire (L') face à l'archéologie [Cf. n° 780], p. 615-622.

815. SCHLEIER (Hans). Historisches Denken in der Krise der Kultur: Fachhistorie, Kulturgeschichte und Anfänge der Kulturwissenschaften in Deutschland. Göttingen, Wallstein, 2000, 127 p. (Essener kulturwissenschaftliche Vorträge, 7).

816. SEGAL (Daniel A.). "Western civ" and the staging of history in american higher education. *American historical review*, 2000, 105, 3, p. 770-805.

817. SÉGINGER (Gisèle). Flaubert. Une poétique de l'histoire. Strasbourg, Presses universitaires de Strasbourg, 2000, 256 p.

818. SOTNIKOVA (Svetlana I.). Estestvennonauchnye metody v sovremennom istoricheskom poznanii.

(Methods of natural sciences and historical knowlege). *Dialog so vremenem: Al'manakh intellektual'noj istorii*, 2000, 2, p. 52-69. [Eng. summary]

819. Sravnitel'noe izuchenie tsivilizatsij mira: Mezhdistsiplinarnyj podkhod: Sb. st. (Comparative study of civilizations of the world: interdisciplinary approach: articles: [Theory and methodology of comparative studies; Medieval Studies]). Ed. Ksenija V. KHVOSTOVA. RAN, In-t vseobshchej istorii. Moskva, [s. n.], 2000, 352 p. [Cf. n° <choice> 3618.]

820. SÜẞMANN (Johannes). Geschichtsschreibung oder Roman? Zur Konstitutionslogik von Geschichtserzählungen zwischen Schiller und Ranke (1780–1824). Stuttgart, Steiner, 2000, 300 p. (Frankfurter Historische Abhandlungen, 41).

821. Tarihî metin çalışmalarında usul: Menâkıbu'l-kudsiyye üzerinde bir deneme. (Methodology in studies on historical texts: an essay on Menakıbu'l-kudsiyye). Ed. by Mertol TULUM. İstanbul, Deniz Kitabevi, 2000, 656 p.

822. TESSITORE (Fulvio). Contributi alla storia e alla teoria dello storicismo. Roma, Edizioni di Storia e Letteratura, 2000, 568 p. (Storia e letteratura, 204).

823. THAPAR (Romila). Narratives and the making of history: two lectures. New Delhi a. Oxford, Oxford U. P., 2000, VI-53 p.

824. THOMPSON (Willie). What happened to history? London, Pluto, 2000, XII-212 p.

825. WALIA (Shelley). Between truth and history: perspectives on culture, politics, and theory. New Delhi, Sterling, 2000, XII-220 p.

826. WERTZ (Spencer K.). Between Hume's philosophy and history: historical theory and practice. Lanham a. Oxford, University Press of America, 2000, XVI-157 p.

827. WHITE (Hayden). An old question raised again: is historiography art or science? (Response to Iggers). *Rethinking history*, 2000, 4, 3, p. 391-406. [Cf. n° 784]

828. YUZO (Mizoguchi). Sympathy, scientificity and subject in the historiography. *Journal of Japanese history*, 2000, 451, p. 41.

829. Zukunft der Geschichte. Historisches Denken an der Schwelle zum 21. Jh. Hrsg. v. Stefan JORDAN. Berlin, Trafo Verl., 2000, 187 p.

§ 3. Addenda 1999

830. CANGUILHELM (Georges). La problématique de la philosophie de l'histoire au début des années 30. *In*: Raymond Aron, la philosophie de l'histoire et les sciences sociales [Cf. n° 831], p. 9-23.

831. Raymond Aron, la philosophie de l'histoire et les sciences sociales. Textes élités par Jean-Claude CHAMBOREDON. Colloque organisé à l'Ecole normale supérieure en 1988. Paris, Ed. ENS Rue d'Ulm, 1999, 96 p. [Cf. n° <sélection> 830.]

Cf. n°s 456, 503, 510, 536, 1025, 1126, 5803

§ 4. Ethnology, folklore and historical anthropology.

** 832. Iz istorii nemtsev Kyrgyzstana, 1917–1999: Sb. dokumentov i materialov. (From the History of the Germans of Kirghizstan, 1917–1999: documents and materials). Bishkek, Sham, 2000, 780 p.

833. ADLER (Alfred). Le pouvoir et l'interdit: royauté et religion en Afrique noire: essais d'ethnologie comparative. Paris, Albin Michel, 2000, 334 p. (plates, ill., maps). (Sciences des religions).

834. AGIER (Michel). Anthropologie du carnaval. La ville, la fête et l'Afrique à Bahia. Marseille, Parenthèses, 2000, 256 p. (Eupalinos).

835. Antropologia e história: debate em regiao de fronteira. Ed. por Lilia K. Moritz SCHWARCZ e Nilma Lino GOMES. Belo Horizonte, Autêntica, 2000, 191 p. (ill.).

836. BARNARD (Alan). History and theory in anthropology. Cambridge, Cambridge U. P., 2000, XII-243 p. (ill).

837. Berättelser om ondskan: en historia genom tusen år. (Récits sur le mal: une histoire millénaire). Red. par Olav HAMMER et Catharina RAUDVERE. Stockholm, Wahlström & Widstrand, 2000, 276 p.

838. BETTINI (Maurizio). Le orecchie di Hermes: studi di antropologia e letterature classiche. Torino, Einaudi, 2000, XIII-418 p. (Biblioteca Einaudi, 96)

839. BEYOĞLU (Ağacan). Türkmen boylarının tarih ve etnografyası. (History and ethnography of Turkoman clans). İstanbul, [s. n.], 2000, 965 p.

840. BEYRU (Rauf). 19. yüzyılda İzmir'de yaşam. (Life in the XIXth century İzmir). İstanbul, Literatür yayınları, 2000, 443 p.

841. BRONZINI (Giovanni Battista). Il tabu dell'incesto da Valerio Massimo a Caravaggio. Fra scrittura, oralità e pittura. *Quaderni di storia*, 2000, 26, 51, p. 57-74.

842. BULATOVA (Angara G). Laktsy: Istoriko-etnograficheskoe issledovanie, XIX–nachalo XX v. (The Laks [Lakians]: a historical and ethnographical study, the 19th and the early 20th century). RAN, Dagestanskij nauchnyj tsentr; In-t istorii, arhkeologii i etnografii. Makhachkala, Izd-vo DNTs RAN, 2000, 387 p. (ill., portr.).

843. CAILLAVET (Chantal). Etnias del norte: etnohistoria e historia de Ecuador. Madrid, Casa de Velázquez y Lima, IFEA y Quito, Abya Yala, 2000, 499 p. (ill., maps). (Travaux de l'Institut français d'études andines, 106).

844. CHRISTINGER (Raymond). Mythologie de la Suisse ancienne: des pratiques chamaniques et du monde celtique aux métamorphoses modernes. Chêne-Bourg, Georg/Musée d'Ethnographie de la Ville de Genève, 2000, 317 p. (ill.).

845. Dizionario della fiaba italiana: simboli, personaggi, storie delle fiabe regionali. A cura di Gian Paolo CAPRETTINI. Roma, Meltemi, 2000, 518 p. (ill.). (Gli Argonauti, 70).

846. Domestiquer l'histoire: ethnologie des monuments historiques. Ed. par Daniel FABRE et Claudie VOISENAT. Paris, Éditions de la Maison des sciences de l'homme, 2000, X-222 p. (plates, ill., port.) (Collection Ethnologie de la France, cahier 15).

847. DUSSART (Françoise). The politics of ritual in an aboriginal settlement: kinship, gender, and the currency of knowledge. Washington, Smithsonian Institution Press, 2000, XVI-269 p. (ill., maps). (Smithsonian series in ethnographic inquiry).

848. Ethnologie und Nationalsozialismus. Hrsg. v. Bernhard STRECK. Gehren, Escher, 2000, 228 p. (ill., ports.). (Veröffentlichungen des Instituts für Ethnologie der Universität Leipzig. Reihe Fachgeschichte, 1).

849. Etnografía de la vida cotidiana. Ed. por Victoria NOVELO, Sergio LÓPEZ RAMOS. México, M.A. Porrúa, 2000, 165 p.

850. Extraordinaire (L') et le quotidien: variations anthropologiques: hommage au professeur Pierre Vérin. Ed. par Claude ALLIBERT, Narivelo RAJAONARIMANANA. Paris, Karthala, 2000, 607 p. (plates, ill.). (Hommes et sociétés).

851. FLORES ARROYUELO (Francisco José). Diccionario de supersticiones y creencias populares. Madrid, Alianza Editorial, 2000, 306 p. (El Libro de bolsillo. Biblioteca Temática)

852. GIOLLÁIN (Diarmuid Ó). Locating Irish folklore: tradition, modernity, identity. Sterling, Cork U. P., 2000, XI-228 p.

853. GLIOZZI (Giuliano). Adam et le Nouveau Monde: la naissance de l'anthropologie comme idéologie coloniale: des généalogies bibliques aux théories raciale (1500–1700). Lecques, Théétète, 2000, 410 p.

854. GORNENSKY (Ioann). Legendy Pamira i Gindukusha. (Legends of Pamir and Gindukush). Moskva, [s. n.], 207 p.

855. HARRISON (Dick). Mannen från Barnsdale. Historien om Robin Hood och hans legend. (L'homme de Barnsdale. Histoire de Robin des bois et de sa légende). Stockholm, Prisma, 2000, 336 p. (ill.).

856. HEISSIG (Walther). Individuelles und traditionelles Erzählen: der mongolische Erzähler Cogrub (Coyirub) aus Ordus (1912–1989). Wiesbaden, Harrassowitz, 2000, 248 p. (Asiatische Forschungen, 136).

857. HERŠAK (Emil). Iz etničke prapovijesti Evrazije: altajci, kineski izvori i Turanija. (Some details from Euroasian ethnic history – Altaic peoples, Chinese sources and Turania). *Migracijske teme: časopis za istraživanje migracija i narodnosti*, 2000, 16, 4, p. 359-392.

858. HORCAJO (Arturo), HORCAJO (Carlos). La question de l'altérité du XVIe siècle à nos jours. Paris, Ellipses, 2000, 128 p. (ill.). (Réseau. Les thématiques)

859. HUBER (Martin), LAUER (Gerhard). Nach der Sozialgeschichte: Konzepte für eine Literaturwissenschaft zwischen Historischer Anthropologie, Kulturgeschichte und Medientheorie. Tübingen, Niemeyer, 2000, XII-625 p. (ill.)

860. KLONDER (Andrzej). Wszystka spuścizna w Bogu spoczywająca. Majątek ruchomy zwykłych mieszkańców Elbląga i Gdańska w XVII wieku. (Sämtliche in Gott ruhende Hinterlassenschaft. Sachbesitz der Durchschnittseinwohner von Elbląg [Elbing] und Gdańsk [Danzig] im XVII. Jahrhundert). Warszawa, Inst. Archeologii i Etnologii PAN, 2000, 184 p. (Studia i Materiały z Historii Kultury Materialnej, 68).

861. KUPIAINEN (Jari). Tradition, trade and woodcarving in Solomon Islands. Helsinki, Höjberg, 2000, XXVII, 379 p. a. cd-rom. (ill.). (Transactions of the Finnish Anthropological Society, 45).

862. Lesbarkeit der Kultur: Literaturwissenschaften zwischen Kulturtechnik und Ethnographie. Hrsg. v. Gerhard NEUMANN und Sigrid WEIGEL. München, W. Fink, 2000, 520 p.

863. LÖNNQVIST (Minna Angelina). Between nomadism and sedentism. Amorites – a perspective of contextual archaeology. Helsinki, Juutiprint, 2000, 612 p. (ill., maps).

864. MAC KILLOP (James). A dictionary of Celtic mythology. Oxford, Oxford U. P., 2000, XXXIV-454 p. (Oxford paperback reference)

865. MACONIE (Stuart). Folklore: the official history. London, Virgin, 2000, 266 p. (plates, ports)

866. MAHAL (Bhupinder Singh). Punjab: the nomads and the mavericks. New Delhi, Sanbun Publishers, 2000, 221 p. (ill., maps).

867. Medieval folklore: an encyclopaedia of myths, legends, tales, beliefs, and customs, 1: A–K; 2: L–Z. Ed. by Carl LINDHAL, John MAC NAMARA and John LINDOW. Santa Barbara, Denver a. Oxford, ABC-CLIO, 2000, 2 vol., XXXIII-574 p., X-560 p. (ill.).

868. MITTERAUER (Michael). Von der Historischen Sozialwissenschaft zur Historischen Anthropologie? *Historical social research*, 2000, 25, 2, p. 139-148.

869. NEDKVITNE (Arnved). Beyond historical anthropology in the study of Medieval mentalities. *Scandinavian journal of history*, 2000, 25, 1-2, p. 27-52.

870. NICHOLLS (David). Conjuring the folk: forms of modernity in African America. Ann Arbor, University of Michigan Press, 2000, XI-180 p. (ill.).

4. ETHNOLOGY, FOLKLORE AND HISTORICAL ANTHROPOLOGY

871. Offenen Grenzen (Die) der Ethnologie. Schlaglichter auf ein sich Wandelndes Fach. Hrsg. v. Beatrix HEINTZE und Sylvia M. SCHOMBURG-SCHERIFF. Frankfurt/M., Lembeck, 2000, 315 p. [Cf. n° <Auswahl> 883.]

872. O'HANLON (Michael), WELSCH (Robert L.). Hunting the gatherers: ethnographic collectors, agents and agency in Melanesia, 1870s–1930s. New York a. Oxford, Berghahn Books, 2000, XVIII-286 p. (ill.). (Methodology and history in anthropology, 6).

873. PÉREZ VILATELA (Luciano). Lusitania: historia y etnología. Madrid, Real Academia de la Historia, 2000, 309 p. (ill., maps). (Publicacions del Gabinete de Antigüedades. Bibliotheca archaeologica ispana, 6).

874. PLATON (Alexandru-Florin). Societate şi mentalităţi în Europa medievală. O introducere în antropologia istorică. Iaşi, Editura Universităţii "Al. I. Cuza", 2000, 248 p.

875. POE (Marshall). A people born to slavery: Russia in early modern European ethnography, 1476–1748. Ithaca a. London, Cornell U. P., 2000, XI-293 p. (ill., maps). (Studies of the Harriman Institute).

876. Quälen (Das) des Körpers. Eine historische Anthropologie der Folter. Hrsg. v. Peter BURSCHEL. Köln, Böhlau, 2000, VIII-323 p.

877. QUIJADA (Mónica). Homogeneidad y nación: con un estudio de caso; Argentina, siglos XIX y XX. Madrid, Consejo Superior de Investigaciones Científicas, Centro de Humanidades, Instituto de Historia, 2000, 260 p. (ill.). (Colección Tierra nueva y cielo nuevo, 42).

878. RADER (Olaf B.). Prismen der Macht. Herrschaftsbrechungen und ihre Neutralisierung am Beispiel von Totensorge und Grabkulten. *Historische Zeitschrift*, 2000, 271, 2, p. 311-346.

879. Rire (Le) des Grecs: anthropologie du rire en Grèce ancienne. Ed- par Marie-Laurence DESCLOS. Grenoble, Millon, 2000, 632 p. (ill.). (Collection Horos).

880. Role (The) of migration in the history of the Eurasian steppe: sedentary civilization vs. "barbarian" and nomad. Ed. by Andrew BELL-FIALKOFF. Basingstoke, Macmillan, 2000, XI-355 p. (ill.).

881. Roots and rituals: the construction of ethnic identities. Ed by Ton DEKKER, John HELSLOOT and Carla WIJERS. Amsterdam, Het Spinhuis, 2000, 808 p. (ill.).

882. ROSLAVTSEVA (Lidija I.). Odezhda krymskikh tatar kontsa XVIII–nachala XX v.: Istoriko-etnograficheskoe issledovanie. (The clothing of the Crimean Tatars, from the late 18th to the early 20th century: a historical and ethnographical study). Moskva, Nauka, 2000, 104 p. (tables, ill.).

883. RÜSEN (Jörn). Vom Nutzen und Nachteil der Ethnologie für die Historie. Überlegungen im Anschluß an Klaus E. Müller. *In*: Offenen Grenzen (Die) der Ethnologie. Schlaglichter auf ein sich wandelndes Fach [Cf. n° 871], p. 291-309.

884. SCHIRRMACHER (Freimut). Der natürlichere Mensch: Helmuth Plessners religionsanthropologische Systematik in ihrer Bedeutung für die theologisch-anthropologische Urteilsbildung. Würzburg, Königshausen & Neumann, 2000, 203 p. (Epistemata. Reihe Philosophie).

885. SCHULZ (Monika). Magie, oder, die Wiederherstellung der Ordnung. Frankfurt am Main u. Oxford, P. Lang, 2000, 439 p. (Beiträge zur europäischen Ethnologie und Folklore, Reihe A, Texte und Untersuchungen, 5).

886. SCHUTKOWSKI (Holger). Subsistence, social status and stature. Approaches from historical anthropology to the reconstruction and significance of dietary patterns. *Jahrbuch für Wirtschaftsgeschichte*, 2000, 1, p. 13-28.

887. SJÖBLOM (Tom). Early Irish taboos. A study of cognitive history. Helsinki, Univ. Of Helsinki, Dept. of Comparative Religion, 2000, 279 p. (Helsingin yliopiston Uskontotieteen laitoksen julk., 5). (ill.).

888. Slovenija in sosednje dežele med antiko in karolinško dobo: začetki slovenske etnogeneze = Slowenien und die Nachbarländer zwischen Antike und karolingischer Epoche: Anfänge der slowenischen Ethnogenese. Uredil Rajko BRATOŽ. Ljubljana, Narodni muzej Slovenije, Slovenska akademija znanosti in umetnosti, 2000, 2 vol., 1130 p. (Situla: razprave Narodnega muzeja Slovenije = dissertationes Musei nationalis Sloveniae, 39. Razprave / Slovenska akademija znanosti in umetnosti, Razred za zgodovinske in družbene vede = Dissertationes / Academia scientiarum et artium Slovenica, Classis I: Historia et sociologia, 18).

889. STEINKOHL (Franz). Der Ethnologie und sein Objekt: Jean de Lérys edle Wilde in Brasilien. Rheinfelden, Schläuble Verlag, 2000, 95 p. (Völkerkunde, 2).

890. TILLION (Germaine). Il était une fois l'ethnographie. Paris, Editions Du Seuil, 2000, 292 p.

891. TKHAMOKOVA (Irina Kh.). Russkoe i ukrainskoe naselenie Kabardino-Balkarii. (The Russian and the Ukrainian population of Kabardino-Balkaria). In-t gumanit. issled. pravitel'stva Kabardino-Balkarskoj Respubliki; RAN, Kabardino-Balkarskij nauch. tsentr. Nal'chik, El'-Fa, 2000, 240 p.

892. TRENCHER (Susan R.). Mirrored images: American anthropology and American culture, 1960–1980. Westport, Bergin & Garvey, 2000, XIII-217 p.

893. URAZMANOVA (Raufa K.). Byt neftjanikov-tatar jugo-vostoka Tatarstana, 1950–1960: Etnosotsial'noe issledovanie. (Everyday life of the Tatar oil industry workers of South-Eastern Tatarstan, 1950–1960: an ethnographical and social study). Al'met'evsk, Fond 'Al'met'evska'ja entsiklopedija', 2000, 158 p. (ill., portr.). (Iz istorii neftjanoj promyshlennosti; 1).

894. Volkskultur und Moderne: europäische Ethnologie zur Jahrtausendwende: Festschrift für Konrad Köstlin zum 60. Geburtstag am 8. Mai 2000. Hrsg. v. Institut für Europäische Ethnologie der Universität Wien. Wien, Selbstverlag des Instituts für Europäische Ethnologie, 2000, 446 p. (ill.). (Veröffentlichungen des Instituts für Europäische Ethnologie der Universität Wien, 21).

895. ZEMON DAVIS (Natalie). The gift in sixteenth-century France. Madison, University of Wisconsin Press, 2000, X-185 p.

Cf. n°s 614, 706, 1126, 1159, 1486, 4586, 5061, 5075, 5295, 5850, 5862, 6465, 8809

§ 5. General history.

* 896. Antisemitism: an annotated bibliography. Vol. 12: 1996. Ed. by Susan Sarah COHEN. München, Saur, 2000, XXXII-527 p.

* 897. Baltische Bibliographie: Schrifttum über Estland, Lettland, Litauen, 1998, mit Nachträgen. Hrsg. v. Paul KAEGBEIN. Marburg, Verlag Herder-Institut, 2000, XVII-306 p. (Bibliographien zur Geschichte und Landeskunde Ostmitteleuropas, 24).

* 898. CRISCUOLO (Eduardo Luis). A/Z Bibliografía de la Ciudad de Buenos Aires. Buenos Aires, Instituto Histórico de la Ciudad de Buenos Aires, 2000, 475 p. (ill.).

* 899. DUCHET (Claude), KIM (In-Kyuoung), PETY (Dominique). Bibliographie du dix-neuvième siècle: lettres, arts, sciences, histoire: année 1999. Paris, SEDES, 2000, 272 p.

** 900. NERALIĆ (Jadranka). Priručnik za istraživanje hrvatske povijesti u Tajnom vatikanskom arhivu: od ranog srednjeg vijeka do sredine XVIII. stoljeća: Schedario Garampi. (Manual for the research of Croatian history in the secrete Vatican Archive: from the Early Middle Ages to the half of 18th centruy). Vol. 1–2. Zagreb, Hrvatski institut za povijest, 2000, [s. p.].

901. 140 yıllık miras Güney Afrika'da Osmanlılar. (Heritage of the 140 years: Ottomans in South Africa). Ed. by Ensar KÖSE. İstanbul, Tez yayınları, 2000, 417 p.

902. ABOUBAKER ALWAN (Daoud), MIBRATHU (Yohanis). Historical dictionary of Djibouti. Lanham a. London, Scarecrow Press, 2000, XXVIII-165 p. (map). (African historical dictionaries, 82).

903. ALAKOM (Rohat). Svensk-kurdiska kontakter under tusen år. (Les contacts suédo-kurdes au cours du dernier millénaire). Spånga, Apec-Tryck & Förlag AB, 2000, 265 p. (ill.).

904. ALARCÓN COSTTA (César). Diccionario biográfico ecuatoriano. Quito, Fundación Ecuatoriana de Desarrollo, Editorial Raíces, 2000, 1273 p. (ports.).

905. BARBERIS (Peter), MAC HUGH (John), TYLDESLEY (Mike). Encyclopedia of British and Irish political organizations: parties, groups and movements of the twentieth century. London, Pinter, 2000, VII-562 p.

906. BARDAVÍO (Joaquín), SINOVA (Justino). Todo Franco: franquismo y antifranquismo de la A a la Z. Barcelona, Plaza & Janés Editores, 2000, 701 p.

907. BARDON (Jonathan). A guide to local history sources in the Public Record Office of Northern Ireland. Belfast, Blackstaff, 2000, XII-119 p. (ill., facsims). (A Blackstaff paperback original).

908. BASS (Harold Franklin). Historical dictionary of United States political parties. Lanham a. London, Scarecrow, 2000, XXVI-389 p. (Historical dictionaries of religions, philosophies and movements, 29).

909. BENNETT (Martyn). Historical dictionary of the British and Irish civil wars, 1637–1660. Lanham a. London, Scarecrow, 2000, 253 p. (ill.). (Historical dictionaries of war, revolution & civil unrest, 14).

910. BERGGREN (Lars). En svensk historia från vikingatid till nutid. (Une histoire de la Suède des Vikings à nos jours). Lund, Studentlitteratur, 2000, 327 p. (ill.).

911. BIZZARRI (Hugo O.). Diccionario paremiológico e ideológico de la Edad Media (Castilla, siglo XIII). Buenos Aires, SECRIT, 2000, XLVIII-382 p. (Incipit publicaciones, 5).

912. BOLÒS I MASCLANS (Jordi). Diccionari de la Catalunya medieval, segles VI–XV. Barcelona, Edicions 62, 2000, 270 p. (El Cangur, 284).

913. BOOTH (Alan Rundlett). Historical dictionary of Swaziland. Lanham a. London, Scarecrow, 2000, XXXI-403 p. (maps). (African historical dictionaries, 80).

914. BORGERSRUD (Lars). Årstall og holdepunkter i norgeshistorien. (Chronology of Norwegian history). Oslo, Universitetsforlaget, 2000, 130 p.

915. BREZIANU (Andrei). Historical dictionary of the Republic of Moldova. Lanham a. London, Scarecrow Press, 2000, LXI-174 p. (maps). (European historical dictionaries, 37).

916. Bulgari i Evrei. 1–2. (Bulgares et Juifs). Sustavitel i redaktor Aleksandur FOL. Sofiia, Tangra TanNakRa Obshtobulgarska Fondatsiia, 2000, 2 vol., 290 p., 296 p. (Bulgarska vechnost, 13).

917. BURGAT (François), LARONDE (André). La Libye. Paris, Presses universitaires de France, 2000, 128 p. (Que sais-je? 1634).

918. CASSAR (Carmel). A concise history of Malta. Msida, Mireva Publications, 2000, XIV-287 p. (ill., maps).

919. CHAMBERLAIN (Muriel E.). Longman companion to formation of the European empires, 1488–1920. London, Longman, 2000, 267 p.

920. Chambers dictionary of world history. Ed by Bruce P. LENMAN and Trevor ANDERSON. Edinburgh, Chambers, 2000, X-950 p. (ill., maps).

5. GENERAL HISTORY

921. Co daly naše země Evropě a lidstvu. Díl 3, Svobodný národ na prahu třetího tisíciletí. (Was brachten unsere Länder dem Europa und der Menschheit. Tomo 3). Ed. Ivan M. HAVEL und Dušan TŘEŠTÍK. Praha, Evropský literární klub, 2000, 609 p.

922. Companion (A) to the American Revolution. Ed. by Jack P. GREENE and Jack R. POLE. Oxford, Blackwell, 2000, XVI-778 p.

923. Concise (A) history of Slovakia. Ed. by Elena MANNOVÁ. Bratislava, Historický ústav SAV, 2000, 353 p. (ill., maps, ports). (Studia historica Slovaca, 21).

924. COOK (Chris). The Longman companion to Britain since 1945. Harlow a. New York, Longman, 2000, XIV-349 p. (maps). (Longman companions to history).

925. COPPER (John Franklin). Historical dictionary of Taiwan (Republic of China). Lanham a. London, Scarecrow Press, 2000, XLIV-269 p. (Asian/Oceanian historical dictionaries, 34).

926. CRAMPTON (R. J.). A concise history of Bulgaria. Cambridge, Cambridge U. P., 2000, XV-259 p. (ill., maps). (Cambridge concise histories).

927. DE PLANHOL (Xavier). L'Islam et la mer. La mosquée et le matelot, VIIe–XXe siècle. Paris, Perrin, 2000, 658 p.

928. DELANCEY (Mark Wakeman). Historical dictionary of the Republic of Cameroon. Lanham a. London, Scarecrow Press, 2000, XXXVII-359 p. (African historical dictionaries, 81).

929. Diccionario de historia de Venezuela. Caracas, Fundación Polar, 2000 (1computer optical disc).

930. Dicionário de história de Portugal. Ed. por Joel SERRAO. Porto, Livraria Figueirinhas, 2000, s.p.

931. Dictionaire des relations internationales au XXe siècle. Sous la dir. de Maurice VAISSE. Paris, A. Colin, 2000, 298 p. (ill.).

932. Dictionary (A) of world history. Oxford, Oxford U. P., 2000, 697 p. (maps). (Oxford paperback reference)

933. Dictionary of New Zealand Biography. Vol. 5. 1941–1960. Auckland, Auckland U. P., 2000, XXV-679 p.

934. Dictionary of the ancient Near East. Ed. by Piotr BIENKOWSKI and Alan MILLARD. Philadelphia a. London, British Museum Press, 2000, X-342 p.

935. Dictionnaire historique de la France sous l'Occupation. Sous la direction de Michèle COINTET et Jean-Paul COINTET. Paris, Tallandier, 2000, 732 p.

936. Dizionario enciclopedico dell'Oriente cristiano. A cura di Edward G. FARRUGIA. Roma, Pontificio Istituto Orientale, 2000, XVI-830 p. (plates).

937. DUE-NIELSEN (Carsten), PETERSEN (Nikolaï), FELDBÆK (Ole), ALBRECTSEN (Esbern). Danmark i syv sind: tusind års dansk udenrigspolitik. (Le Danemark dans tous ses états: une politique étrangère millénaire). København, Gyldendal, 2000, 175 p. (ill.).

938. DUFFY (Seán). The concise history of Ireland. Dublin, Gill & Macmillan, 2000, 256 p. (ill., maps).

939. EDELHEIT (Hershel). History of Zionism: a handbook and dictionary. Boulder a. Oxford, Westview, 2000, XVII-653 p. (ill., map).

940. Enciclopedia dell'antichità classica: repertorio generale della civiltà greco-romana. A cura di Eugenia DOSSI. Milano, Garzanti, 2000, 1648 p. (ill., maps). (Garzantine).

941. Encyclopaedia (The) of Islam, new edition = Encyclopédie de l'Islam, nouvelle édition. Glossary and index of terms to volumes 1–9, and to the supplement, fascicules 1–6. Leiden a. New York, E.J. Brill, 2000, 427 p.

942. Encyclopedia of European social history: from 1350 to 2000. Ed. by Peter N. STEARNS. New York, Scribner, 2000, [s. p.]. (ill., maps).

943. Encyclopedia of the Korean War: a political, social, and military history. Ed. by Spencer C. TUCKER. Santa Barbara, ABC-Clio, 2000, 3 vol.

944. Encyclopedia of the Middle Ages. Ed. by André VAUCHEZ, Barrie DOBSON and Michael LAPIDGEÒ. Cambridge, James Clarke & Co., 2000, 2 vol., [s. p.].

945. FLETCHER (Stella). The Longman companion to Renaissance Europe, 1390–1530. Harlow a. New York, Longman, 2000, XIII-337 p. (ill., geneal. tables, maps). (Longman companions to history).

946. GÉRARD (Jean-Louis). Dictionnaire historique et biographique de la guerre d'Algérie. Hélette, Curutchet, 2000, 206 p.

947. GONZÁLEZ CUEVAS (Pedro Carlos). Historia de las derechas espanolas: de la Ilustración a nuestros días. Madrid, Biblioteca Nueva, 2000, 525 p. (Colección Historia Biblioteca Nueva).

948. Grands notables du Premier Empire. Sous la dir. de Louis BERGERON et Guy CHAUSSINAND-NOGARET. T. 26. Vienne. Par Guillaume LÉVÊQUE, Emmanuel DION et Sébastien JAHAN. T. 27. Somme. Par Jean-Marie WISCART. Paris, CNRS, 2000, 2 vol., 278 p., 144 p.

949. Grèce (La) pour penser l'avenir. Paris, L'Harmattan, 2000, 190 p. (L'homme et la société).

950. GREGORY (Jeremy), STEVENSON (John). The Longman companion to Britain in the eighteenth century, 1688–1820. London, Longman, 2000, IX-571 p. (maps). (Longman companions to history).

951. Große Revolutionen der Geschichte. Von der Frühzeit bis zur Gegenwart. Peter WENDE. München, Beck, 2000, 391 p.

952. HANČIČ (Damjan). Sedemsto let klaris na Slovenskem. (Seven hundred years of the poor clares in

Slovenia). Nazarje, Samostan Brezmadežne sester klaris, 2000, 101 p.

953. Handbook (A) of dates, for students of British history. Ed. by C. R. CHENEY, revised by Michael JONES. Cambridge, Cambridge U. P., 2000, XVII-246 p. (Royal historical society guides and handbooks, 4).

954. Handbook of American women's history. Ed. by Angela M. HOWARD and Frances M. KAVENIK. Thousand Oaks a. London, Sage Publications, 2000, XX-724 p. (ill.).

955. Handbuch der Geschichte Russlands. Hrsg. v. M. HELLMANN, K. ZERNACK und G. SCHRAMM. Stuttgart, A. Hiersemann, 2000.

956. HARDY (Lyda Mary). Women in U.S. history: a resource guide. Englewood, Libraries Unlimited, 2000, XVI-344 p.

957. HAUE (Henry), OLSEN (Jørgen), AARUP-KRISTENSEN (Jørn). Fra stammer til folk: Danmarks historie indtil 1840. (Des tribus au peuple: histoire du Danemark jusqu'en 1840). København, Gyldendal Uddannelse, 2000, 303 p. (ill.).

958. HEIBERG (Steffen). Danske dronninger i tusind år. (Les Reines de Danemark depuis un millénaire). København, Gyldendal, 2000, 388 p. (ill.).

959. HENDERSON (James D.), DELPAR (Helen), BRUNGARDT (Maurice P.), WELDON (Richard N.). A reference guide to Latin American history. London, M.E. Sharpe, 2000, IX-615 p. (maps).

960. HENZE (Paul B.). Layers of time: a history of Ethiopia. New York, St. Martin's Press, 2000, XXIV-372 p. (plates, ill., maps).

961. Heroic reputations and exemplary lifes. Ed. by Geoffrey CUBITT and Allen WARREN. Manchester a. New York, Manchester U. P., 2000, X-274 p.

962. HILLERBRAND (Hans Joachim). Historical dictionary of the Reformation and Counter-Reformation. London, Fitzroy Dearborn, 2000, 265 p.

963. Histoire d'un quartier de Montbéliard (Doubs): Le bourg Saint-Martin (XIIIᵉ–XXᵉ s.). Ed. par Sylvier CANTRELLE, Corinne GOY et Claudine MUNIER. Paris, Editions de la Maison des Sciences de l'Homme, 2000, 140 p. (ill.).

964. Historical encyclopedia of U.S. presidential use of force, 1789–2000. Ed by Karl R. DEROUEN, Jr. Westport a. London, Greenwood Press, 2000, IX-313 p.

965. HOOPER (R. W.). Representative chapters in ancient history. An introduction to the West's classical experience. Vol. 1. From Australopithecus to Alexander. Vol. 2. From Romulus to Justinian. Lanham, New York a. Oxford, University Press of America, 00, XV-625 p.

966. Hungarian state (The). Thousand years in Europe. Ed. by András GERGELY and Gábor MÁTHÉ. Budapest, Korona Publishing House, 2000, 531 p.

967. Illustrated guide to British history. Ed by David BOYLE [et al.]; general editor, E.J. EVANS. Bath, Parragon, 2000, 272 p. (ill., maps, ports.).

968. Ilustrirana zgodovina Slovencev. (Illustrated history of Slovenes). Urednik Marko VIDIC, sourednika Lan BRENK, Martin IVANIČ. Ljubljana, Mladinska knjiga, 2000, 526 p. (Knjižnica Enciklopedije Slovenije).

969. Istorija Peru s drevnejshikh vremen do kontsa XX veka. (A history of Peru: from the earliest times to the end of the twentieth century). Ed. S. A. SOZINA and I. I. JANCHUK. RAN, In-t vseobshchej istorii. Moskva, Nauka, 2000, 476 p. (ill., portr., bibl.).

970. JORIO (Mario). Das Historische Lexikon der Schweiz im Jahre 2000 [Forschungsbericht]. *Schweizerische Zeitschrift für Geschichte*, 2000, 50, p. 198-203.

971. KAMEN (Henry Arthur Francis). Who's who in Europe, 1450–1750. London a. New York, Routledge, 2000, X-321 p. (Who's who series).

972. KAMMEN (Carol). Encyclopedia of local history. Walnut Creek a. Oxford, AltaMira, 2000, XVI-539 p. (ill., ports.). (American Association for state and local history book series).

973. KENNY (Kevin). The American Irish: a concise history since 1700. Harlow, Longman, 2000, 352 p. (ill.). (Studies in modern history).

974. KLJASHTONYJ (S. G.), SULTANOV (T. I.). Gosudarstva i narody Evrazijskikh stepej: Drevnost' i srednevekov'e. (States and peoples of the Eurasian steppe: Antiquity and the Middle Ages). RAN, In-t vostokovedenija, Sankt-Peterburgskij filial. Sankt-Peterburg, Peterburgskoe vostokovedenie, 2000, 307 p. (bibl.). [Eng. summary]

975. Krieg und Frieden in Übergang vom Mittelalter zur Neuzeit. Theorie – Praxis – Bilder. Guerre et paix au Moyen Age aux temps modernes. Théories – pratiques – représentations. Hrsg. v. Heinz DUCHHARDT und Patrice VEIT. Mainz, von Zabern, 2000, VIII-328 p. (Veröffentlichungen des Instituts für europäische Geschichte Mainz, Abt. f. Universalgeschichte, 52).

976. Kriege entstehen (Wie). Zum historischen Hintergrund von Staatenkonflikten. Hrsg. v. Bernd WEGNER. Paderborn, München u. Wien, Schöningh, 2000, 378 p.

977. Lexikon der Renaissance. Hrsg. v. Herfried MÜNKLER und Marina MÜNKLER. München, Beck, 2000, 472 p.

978. LINIGER-GOUMAZ (Max). Historical dictionary of Equatorial Guinea. Lanham a. London, Scarecrow, 2000, XLII-567 p. (African historical dictionaries, 21).

979. LIVERMORE (Harold V.). Essays on Iberian history and literature, from the Roman Empire to the Renaissance. Aldershot, Ashgate, 2000, [s. p.]. (Collected studies series).

5. GENERAL HISTORY

980. LONGLEY (David). The Longman companion to Imperial Russia. New York, Longman, 2000, 517 p. (Longman companions to history).

981. MACLEAN (Fitzroy). Scotland: a concise history. London, Thames & Hudson, 2000, 248 p. (ill., maps, ports).

982. MAIER (Bernhard). Die Kelten. Ihre Geschichte von Anfängen bis zur Gegenwart. München, Beck, 2000, 320 p.

983. MAURER (Michael). Geschichte Englands. Leipzig, Reclam, 2000, 404 p.

984. MAXON (Robert M.), OFCANSKY (Thomas P.). Historical dictionary of Kenya. Lanham a. London, Scarecrow Press, 2000, XXVI-449 p. (African historical dictionaries, 77).

985. MIJATOVIĆ (Anđelko). The Croats and Croatia in time and space. Zagreb, Školska knjiga, 2000, 297 p.

986. Mille anni di storia: dalla città medievale all'unità dell'Europa. A cura di Rosario VILLARI. Roma, Laterza, 2000, VIII-903 p. (Storia e società).

987. MINAHAN (James). One Europe, many nations: a historical dictionary of European national groups. Westport a. London, Greenwood Press, 2000, XVII-781 p. (ill., maps).

988. NICOLSON (Colin). Longman companion to the First World War: Europe, 1914–1918. New York, Longman, 2000, 340 p. (Longman companions to history).

989. NOËL (Bernard). Dictionnaire de la Commune. Paris, Mémoire du Livre, 2000, 642 p.

990. OLSEN (Rikke Agnete). Danmark i verden. 2. Magten og æren. (Le Danemark dans le monde. 2. La puissance et l'honneur). København, Fremad, 2000, 391 p. (ill.).

991. Otechesvennaja istorija: Istorija Rossii s drevnejshikh vremen do 1917 goda. Entsiklopedija. (The history of Russia, from the most ancient time to 1917: encyclopaedia). Vol. 3. K–M. Ed. Valentin L. JANIN. Moskva, Bol'shaja rossijskaja entsiklopedija, 2000, 623 p. (ill., portr., maps, bibl.).

992. Oxford (The) companion to New Zealand military history. Ed. by Ian MAC GIBBON. Oxford, Oxford U. P., 2000, XXVI-653 p. (ill., maps, ports.).

993. OYEWOLE (A.). Historical dictionary of Nigeria. London, Scarecrow, 2000, XXX-599 p. (African historical dictionaries, 40).

994. Pakistan (Le). Ed. par Christophe JAFFRELOT. Paris, Fayard, 2000, 503 p.

995. PALKA (Joel W.). Historical dictionary of ancient Mesoamerica. London, Scarecrow Press, 2000, XVI-199 p. (plates, ill., map). (Historical dictionaries of ancient civilizations and historical eras, 2).

996. Penguin (The) dictionary of British history. Ed. by Juliet GARDINER. London, Penguin, 2000, VI-750 p. (maps).

997. POLLEY (Martin). A–Z of modern Europe since 1789. London, Routledge, 2000, XI-184 p. (maps).

998. POPESCU (J.). Bulgaria. New York, [s. n.], 2000, 96 p.

999. Problemy slavjanovedenija: Sb. nauch. st. i mat-lov. (Problems of Slavic studies: scientific articles and materials). Vol. 1, 2. Ed. Sergej I. MIKHAL'CHENKO. Brjanskij gos. ped. un-t im. I.G. Petrovskogo. Brjansk, Izd-vo BGPU, 2000, 2 vol., 272 p., 237 p. (bibl.).

1000. SALEWSKI (Michael). Geschichte Europas. Staaten und Nationen von der Antike bis zur Gegenwart. München, Beck, 2000, 1146 p.

1001. SAUL (Nigel). A companion to medieval England 1066–1485. Stroud, Tempus, 2000, 224 p. (ill., maps, ports).

1002. SAUNDERS (Christopher). Historical dictionary of South Africa. Lanham a. London, Scarecrow, 2000, XLV-373 p. (ill.). (African historical dictionaries, 78).

1003. SCHMALE (Wolfgang). Geschichte Frankreichs. Stuttgart, Ulmer, 2000, 432 p. (UTB, 2145).

1004. SCHMITZ-BERNING (Cornelia). Vokabular des Nationalsozialismus. Berlin, De Gruyter, 2000, XLI, 710 p.

1005. SCHOPPA (R. Keith). The Columbia guide to modern Chinese history. New York, Columbia U. P., 2000, XVIII-356 p. (maps). (The Columbia guides to Asian history).

1006. SCHÜLLER (Karin). Einführung in das Studium der iberischen und lateinamerikanischen Geschichte. Münster, Aschendorff, 2000, 220 p.

1007. SMITH (Joseph), DAVIS (Simon). Historical dictionary of the Cold War. Lanham, Scarecrow Press, 2000, 329 p. (Historical dictionaries of war, revolution, and civil unrest, 15).

1008. Som kongerne bød: fra trelleborge til enevælde. Festskrift til Hendes Majestæt Dronning Margrethe II i anledning af tres års dagen 16. april 2000. (Sur l'ordre des rois: de l'époque viking à l'absolutisme. Mélanges offerts à sa Majesté la reine Margrethe II pour l'anniversaire de ses 60 ans). Red. par David FAVRHOLDT, Pia GRÜNER et Flemming LUNDGREN-NIELSEN. København, Det kongelige danske Videnskabernes Selskab / C.A. Reitzel, 2000, 216 p. (ill.).

1009. Srednji vijek i renesansa: (XIII.–XVI. stoljeće). (The Middle Ages and Renaissance). Ed. by Eduard HERCIGONJA. Zagreb, Hrvatska akademija znanosti i umjetnosti i Školska knjiga, 2000, 885 p.

1010. Staat und Krieg. Vom Mittelalter bis zur Moderne. Hrsg. v. Werner RÖSENER. Göttingen, Vandenhoeck & Ruprecht, 2000, 244 p. (Sammlung Vandenhoeck).

1011. Storia d'Italia. Le regioni dall'Unità ad oggi. L'Abruzzo. A cura di Massimo COSTANTINI e Costantino FELICE. Torino, Einaudi, 2000, XX-1162 p.

1012. THANE (Pat). Old age in English history: past experiences, present issues. Oxford, Oxford U. P., 2000, X-536 p.

1013. TREADGOLD (Warren T.). A concise history of Byzantium. New York, St. Martin's Press, 2000, XII-273 p. (ill., maps).

1014. TRUHART (Peter). Regents of nations. T. 1. Antiquity worldwide. München, Saur, 2000, XXXIX-392 p.

1015. Vår skjulte kulturarv: arkeologien under overfladen. Til Hendes Majestæt Dronning Margrethe II. 16 april 2000. (Notre héritage archéologique caché. Mélanges offerts à la Reine Margrethe II de Danemark, 16 avril 2000). Red. par Steen HVASS. København, Det kongelige nordiske Oldskriftselskab et Höjberg, Jysk Arkæologisk Selskab, 2000, 239 p. (ill.). (résumé anglais).

1016. Venice reconsidered. The history and civilization of an Italian city-state, 1297–1797. Ed. by John MARTIN and Dennis ROMANO. Baltimore, Johns Hopkins U. P., 2000, 538 p.

1017. Verdämmern (Das) der Macht: vom Untergang großer Reiche. Hrsg. v. Richard LORENZ. Frankfurt am Main, Fischer Taschenbuch, 2000, 285 p.

1018. Virtuosen der Macht: Herrschaft und Charisma von Perikles bis Mao. Hrsg. Wilfried NIPPEL. München, Beck, 2000, 319 p.

1019. Vocabulaire historique du Moyen Âge: (Occident, Byzance, Islam). Ed. par François-Olivier TOUATI. Paris, Boutique de l'histoire, 2000, 331 p. (ill., plans).

1020. War and competition between states. Ed. by Philippe CONTAMINE. Oxford, Clarendon Press, 2000, XI-347 p. (The origins of the modern state in Europe, 13th to 18th centuries, theme A).

1021. WILKINSON (Endymion Porter). Chinese history: a manual. Cambridge, Harvard University Asia Center for the Harvard-Yenching Institute, 2000, XXIV-1181 p. (Harvard-Yenching Institute monograph series, 52).

1022. WINKLER (Heinrich August). Der lange Weg nach Westen. Deutsche Geschichte. Band 1. Deutsche Geschichte vom Ende des Alten Reiches bis zum Untergang der Weimarer Republik. Band 2. Deutsche Geschichte vom "Dritten Reich" bis zur Wiedervereinigung. München, Beck, 2000, 625 p.

§ 5. Addenda 1999.

1023. Seiyô-sekai no rekishi. (A history of the Western world). Ed. by Kazuhiko KONDO. Tokyo, Yamakawa Shuppansha, 99, 438 p.

Cf. n°s 476, 787, 1025, 1934, 1939, 4035-4121, 4901

§ 6. Theory of the state and of society.

** 1024. MONTESQUIEU. Réflexions sur la Monarchie universelle en Europe. Introduction et notes par Michel PORRET. Genève, Librairie Droz, 2000, 126 p.

1025. Al'ternativnye puti k tsivilizatsii. (Alternative ways to civilization: [a collective monograph]. Chapters: I. Theory of social evolution. II. Multi-linear theories of the formation of civilization. III. The emergence of state and civilization. IV. Civilizational alternatives to state. V. Nomades in the process of civilization). Ed. N. N. KRADIN, A. V. KOROTAEV, D. M. BONDARENKO, V. A. LYNSHA. Ros. gos. gumanit. un-t, etc. Moskva, Logos, 2000, 367 p. (ill.).

1026. Angst und Politik in der europäischen Geschichte. Hrsg. v. Franz BOSBACH. Dettelbach, Röll, XIX-231 p. (Bayreuther Historische Kolloquien, 13).

1027. ARMITAGE (D.). Edmund Burke and reason of state. *Journal of the history of ideas*, 2000, 61, 4, p. 617-634.

1028. ARON (Raymond). Democrazia. *Quaderni di storia*, 2000, 26, 51, p. 5-40.

1029. BARADAT (Leon P.). Political ideologies: their origins and impact. Upper Saddle River, Prentice Hall, 2000, XIV-338 p. (ill.).

1030. BOIA (Lucian). La mythologie scientifique du communisme. Paris, Les Belles Lettres, 2000, 223 p.

1031. *Vacat.*

1032. BOUCHER (David), VINCENT (Andrew). British idealism and political theory. Edinburgh, Edinburgh U. P., 2000, VIII-248 p.

1033. CAMBIANO (Giuseppe). Polis: un modello per la cultura europea. Roma e Bari, Editori Laterza, 2000, XII-492 p. (Collezione storica).

1034. Cambridge (The) history of Greek and Roman political thought. Ed. by Christopher ROWE, Malcolm SCHOFIELD, Simon HARRISON, Melissa LANE. Cambridge, Cambridge U. P., 2000, XX-745 p. [Cf. n^{os} <choice> 2053, 2060, 2077, 2080, 2084, 2091.]

1035. CLARK (Manning). The ideal of Alexis de Tocqueville. Ed. by Dymphna CLARK, David HEADON and John WILLIAMS. Carlton, Melbourne U. P., 2000, XIV-185 p. (plates, ill., ports).

1036. COLEMAN (Janet). A history of political thought: from ancient Greece to early Christianity. Oxford, Blackwell Publishers, 2000, XI-363 p.

1037. COSTA (Pietro). Storia della cittadinanza in Europa. Vol. 2. L'età delle rivoluzioni. Roma e Bari, Laterza, 770 p.

1038. Development: critical concepts in the social sciences. Ed. by Stuart CORBRIDGE. London, Routledge, 2000, 6 vol., [s. p.]. (ill., maps, charts).

1039. Dictionnaire critique du féminisme. Ed. par Helena HIRATA. Paris, Presses universitaires de France, 2000, XXX-299 p. (Politique d'aujourd'hui).

1040. EBENSTEIN (William), EBENSTEIN (Alan). Great political thinkers: Plato to the present. Fort Worth a. London, Harcourt College Publishers, 2000, XIV-1001 p.

1041. Etat (L') moderne: regards sur la pensée politique de l'Europe occidentale entre 1715 et 1848. Ed. par Simone GOYARD-FABRE. Paris, J. Vrin, 2000, 318 p. (Histoire des idées et des doctrines).

1042. FRANCISCO (Ronald A.). The politics of regime transitions. Oxford, Westview, 2000, X-177 p. (ill.).

1043. GRECO (Tommaso). Norberto Bobbio: un itinerario intellettuale tra filosofia e politica. Roma, Donzelli, 2000, XIII-272 p.

1044. GRUNERT (Frank). Normbegründung und politische Legitimität: zur Rechts- und Staatsphilosophie der deutschen Frühaufklärung. Tübingen, Niemeyer, 2000, VII-310 p. (Frühe Neuzeit, 57).

1045. HALTERN (Utz). Krone und Palast. Zur politischen Metaphorik des Bürgertums. *Archiv für Kulturgeschichte,* 2000, 82, p. 121-156.

1046. HARDT (Michael), NEGRI (Antonio). Empire. Cambridge, Harvard U. P., 2000, 478 p.

1047. HEERICH (Thomas). Transformation des Politikkonzepts von Hobbes zu Spinoza: das Problem der Souveränität. Würzburg, Königshausen & Neumann, 2000, 94 p. (Schriftenreihe der Spinoza-Gesellschaft, 8).

1048. HEINZ (Hürten). Totalitäre Herrschaft und ihre Grenzen. *Zeitschrift für Politik,* 2000, 47, 2, p. 117-130.

1049. HESS (Andreas). American social and political thought: a concise introduction. Edinburgh, Edinburgh U. P., 2000, XIII-162 p.

1050. Historia y evolución de las ideas políticas y filosóficas argentinas. Córdoba, Academia Nacional de Derecho y Ciencias Sociales de Córdoba, 2000, 378 p. (Ediciones de la Academia Nacional de Derecho y Ciencias Sociales de Córdoba,17).

1051. HORNUNG (Klaus). Politischer Messianismus: Jacob Talmon und die Genesis der totalitären Diktaturen. *Zeitschrift für Politik,* 2000, 47, 2, p. 131-172.

1052. Jewish political tradition (The). Vol. 1. Authority. Ed. by Michael WALZER, Menachem LORBERBAUM, Noam ZOHAR and Yair LORBERBAUM. New Haven a. London, Yale U. P., 2000, 578 p.

1053. KALLIS (Aristotle A.). The "Regime-Model" of fascism: a tipology. *European history quarterly,* 2000, 30, 1, p. 77-104.

1054. Klassische Politik. Politikverständnisse von der Antike bis ins 19. Jh. Hrsg. v. Hans J. LIETZMANN und Peter NITSCHKE. Opladen, Leske u. Budrich, 2000, 254 p.

1055. KOJÈVE (Alexandre). Tirannide e saggezza. *Quaderni di storia,* 2000, 26, 52, p. 5-56.

1056. LENNER (Andrew). The federal principle in American politics, 1790–1833. Madison, Madison House, 2000, XII-223 p. (A Madison House Book).

1057. LIANG (Qichao). History of Chinese political thought: during the Early Tsin period. London, Routledge, 2000, VIII-210 p. (ports.). (The international library of philosophy. Ethics and political philosophy, 3).

1058. MARRAMAO (Giacomo). Dopo il Leviatano. Individuo e comunità. Torino, Bollati Boringheri, 2000, 443 p.

1059. MEDICI (Rita). Giobbe e Prometeo: filosofia e politica nel pensiero di Gramsci. Firenze, Alinea, 2000, 240 p. (Studi e ricerche, 21).

1060. Modern political thought: a reader. Ed. by John GINGELL, Adrian LITTLE and Christopher WINCH. London a. New York, Routledge, 2000, X-293 p.

1061. Montesquieu's science of politics: essays on 'The spirit of laws'. Ed. by David W. CARRITHERS, Michael A. MOSHER and Paul A. RAHE. Lanham, Rowman & Littlefield, 2000, 458 p.

1062. NITSCHKE (Peter). Einführung in die politische Theorie der Prämoderne, 1500–1800. Darmstadt, Wissenschaftliche Buchgesellschaft, 2000, XI-178 p. (Die Politikwissenschaft).

1063. OPITZ (Peter J.). Der Weg des Himmels. Zum Geist und zur Gestalt des politischen Denkens im klassischen China. München, Fink, 2000, 313 p.

1064. Ordnung (Die) des Staates und die Freiheit des Menschen. Deutschlandpläne im Widerstand und Exil. Hrsg. v. Gerhard RINGSHAUSEN und Rüdiger von VOSS. Bonn, Bouvier, 2000, 373 p.

1065. PAVLOVICH (P.). Koncepcijata za vlastta v isljama: formirane i doktrinalni aspekti. (La conception sur le pouvoir dans l'Islam: formation et aspects doctrinaires). *Istorichesko budeshche,* 2000, 2, p. 39-63.

1066. Pensiero (Il) politico dell'età antica e medioevale: dalla polis alla formazione degli stati europei. A cura di Carlo DOLCINI. Torino, UTET libreria, 2000, XII-352 p.

1067. PETRUCCIANI (Stefano). Introduzione a Habermas. Bari e Roma, Laterza, 2000, 237 p.

1068. Politische Mythen und Rituale in Deutschland, Frankreich und Polen. Hrsg. v. Yves BIZEUL. Berlin, Duncker u. Humblot, 2000, 235 p. (Ordo Politicus, 34).

1069. Politische Theorien von der Antike bis zur Gegenwart. Wiesbaden, Fourier, 2000, 1038 p.

1070. Readings in classical political thought. Ed. by Peter J. STEINBERGER. Indianapolis, Hackett Pub. Co, 2000, XII-625 p.

1071. Republikbegriff und Republiken seit dem 18. Jahrhundert im europäischen Vergleich. Hrsg. v. Helmut REINALTER. Frankfurt am Main, Lang, 2000, 307 p. (Schriftenreihe der Internationalen Forschungsstelle "Demokratische Bewegungen in Mitteleuropa 1770–1850", 28).

1072. SARASOHN (Lisa T.). Was 'Leviathan' a patronage artifact? *History of political thought*, 2000, 21, 4, p. 606-631.

1073. SASSOON (Anne Showstack). Gramsci and contemporary politics: beyond pessimism of the intellect. London, Routledge, 2000, 173 p. (Routledge innovations in political theory, 4).

1074. SCHATZ (J.). «Imperium, pax et iustitia». Das Reich, Friedensstiftung zwischen Ordo, Regnum und Staatlichkeit. Berlin, Duncker & Humblot, 2000, 338 p. (Beitr. z. Polit. Wiss., 114).

1075. SEDDA (Lidia). Economia, politica e società sovietica nei Quaderni del carcere. [S. l.], Quaderni del Centro Studi P. Calamandrei, 2000, 142 p. (Impresa e Società, 16).

1076. SIMONETTA (Stefano). Marsilio in Inghilterra: Stato e Chiesa nel pensiero politico inglese fra XIV e XVII secolo. Milano, LED, 2000, 190 p. (Il filarete / Facoltà di lettere e filosofia dell'Università degli studi di Milano, 195).

1077. SLOMP (Gabriella). Thomas Hobbes and the political philosophy of glory. Basingstoke, Palgrave Macmillan, 2000, XI-194 p. (ill.).

1078. Stand und Probleme der Erforschung des Konservativismus. Hrsg. v. Caspar von SCHRENCK-NOTZIG. Berlin, Duncker u. Humblot, 2000, 242 p. (Studien und Texte zur Erforschung des Konservativismus, 1).

1079. TENENTI (Alberto). Sovranità e ragion di stato nell'Italia del secondo Cinquecento. *Studi veneziani*, 2000, 34, p. 97-112.

1080. THUAU (Étienne). Raison d'État et pensée politique à l'époque de Richelieu. Paris, Albin Michel, 2000, 504 p. (Bibliothèque de l'évolution de l'humanité, 35).

1081. WARBURTON (Nigel), PIKE (Jon), MATRAVERS (Derek). Reading political philosophy: Machiavelli to Mill. London, Routledge, 2000, X-382 p.

1082. WARRENDER (Howard). The political philosophy of Hobbes: his theory of obligation. Oxford, Oxford U. P., 2000, IX-346 p. (Oxford scholarly classics).

1083. WEBER-FAS (Rudolf). Über die Staatsgewalt. Von Platons Idealstaat bis zur Europäischen Union. München, Beck, 2000, VI-333 p.

1084. ZAGORIN (Perez). Hobbes without Grotius. *History of political thought*, 2000, 21, p. 16-40.

Cf. n°s 6121, 6154, 8572

§ 7. Constitutional and legal history.

1085. BAKER (John H.). The common law tradition. Lawyers, books and the law. London a. Rio Grande, Hambledon Press, 2000, XXXIV-404 p.

1086. Begründung (Die) des Rechts als historisches Problem. Hrsg. v. Dietmar WILLOWEIT und Elisabeth MÜLLER-LUCKNER. München, Oldenbourg, 2000, VI-345 p. (Schriften des Historischen Kollegs. Kolloquien, 45).

1087. BERGFELD (Christoph). Juristen-Biografien. *Zeitschrift für Neuere Rechtsgeschichte*, 2000, 22, 3/4, p. 451-502.

1088. BERTHIAU (Denis). Histoire du droit. Paris, Hachette supérieur, 2000, 192 p. (Crescendo, 19)

1089. BUCCI (Onorato). Il principio di equità nella storia del diritto. Napoli, Edizioni Scientifiche Italiane, 2000, 124 p. (Pubblicazioni del Dipartimento di Scienze giuridico-sociali e dell'amministrazione dell'Università degli studi del Molise, 3).

1090. CARBASSE (Jean-Marie). Histoire du droit pénal et de la justice criminelle. Paris, Presses universitaires de France, 2000, 445 p. (Droit fondamental. Droit pénal)

1091. Denunziation und Justiz. Historische Dimensionen eines sozialen Phänomens. Hrsg. v. Friso ROSS und Achim LANDWEHR. Tübingen, Edition discord, 2000, 283 p.

1092. Dret comú (El) i Catalunya. Actes del IX Simposi Internacional Barcelona, 4–5 de juny de 1999. la familia i el seu patrimoni. Ed. por Aquilino IGLESIA FERREIRÓS. Barcelona, Signo, 2000, 252 p. (Associació Catalana d'Història del Dret "Jaume de Montjuïc", 22).

1093. GROSSI (Paolo). Scienza giuridica italiana: un profilo storico: 1860–1950. Milano, Giuffrè, 2000, XIX-324 p.

1094. HEIRBAUT (D.). Europe and the people without legal history: on the need for a general history of non-European law. *Revue d'histoire du droit*, 2000, 68, 3, p. 269-280.

1095. HIRVONEN (Ari). Oikeuden käynti. Antigonen laki ja oikea oikeus. (Justice and law. Conflict between two laws in Antigone, legal thinking and justice). Helsinki, Loki-kirjat, 2000, 521 p.

1096. HOFMANN (Hasso). Einführung in die Rechts- und Staatsphilosophie. Darmstadt, Wissenschaftliche Buchgesellschaft, 2000, 224 p.

1097. HOLZHAUER (Heinz). Beiträge zur Rechtsgeschichte. Hrsg. v. Stefan Chr. SAAR und Andreas ROTH. Berlin, Schmidt, 2000, 275 p.

1098. KLIPPEL (Diethelm). Ideen – Normen – Lebenswelt. Exegese und Kontexterschließung in der Rechtsgeschichte. *Scientia Poetica*, 2000, 4, p. 179-191.

1099. LANGER (Stefanie). Rechtswissenschaftliche Itinerarien: Lebenswege namhafter europäischer Juristen vom 11. bis zum 18. Jahrhundert. Frankfurt am Main u. Oxford, Peter Lang, 2000, 245 p. (maps). (Rechtshistorische Reihe, 225).

1100. Lügen und Betrügen. Das Falsche in der Geschichte von der Antike bis zur Moderne. Hrsg. v. Oliver HOCHADEL und Ursula KOCHE. Köln, Weimar u. Wien, Böhlau, 2000, VI-311 p.

1101. LUPOI (Maurizio). The origins of the European legal order. Cambridge, Cambridge U. P., 2000, XIII-641 p.

1102. MARSCH (Robert M.). Weber's misunderstanding of traditional Chinese law. *American journal of sociology*, 2000, 106, 2, p. 281-302.

1103. MASS (Jeffrey P.). Family, law, and property in Japan, 1200–1350. Cambridge, Harvard University, Edwin O. Reischauer Institute of Japanese Studies, 2000, [s. p.]. (Occasional papers in Japanese studies, 2000-03).

1104. Metamorphosen der Leviathan? Staatsaufgaben um Umbruch. Hrsg. v. Irene GERLACH und Peter NITSCHKE. Opladen, Leske u. Budrich, 2000, 278 p.

1105. MOHNHAUPT (Heinz). Historische Vergleichung im Bereich von Staat und Recht: Gesammelte Aufsätze. Frankfurt am Main, V. Klostermann, 2000, VI-490 p. (Ius commune. Sonderheft, 134).

1106. MOORMAN VAN KAPPEN (O.). Lex loci. Opstellen over de Nederlandse rechtsgeschiedenis, onder redactie van E. C. COPPENS [et al.]. Nijmegen, Gerard Noodt Instituut, 2000, L-611 p. (Rechtshistorische Reeks van het Gerard Noodt Instituut, 45).

1107. Moral world (The) of the law. Ed. by Peter COSS. Cambridge, Cambridge U. P., 2000, XI-262 p. (Past and present publications).

1108. MÜLLER (Wolfgang P.). Die Abtreibung: Anfänge der Kriminalisierung, 1140–1650. Köln, Weimar u. Wien, 2000, Böhlau, 2000, VIII-355 p.

1109. Neopublikovannye stranisty tret'ego toma "Ocherkov istoricheskogo pravovedenija" P. G. Vinogradova. (Unpublished abstracts of P. G. Vinogradoff's 'Essays of the History of Law'). Ed., with intr. and comment. A. V. ANTOSHCHENKO. *Srednie veka*, Moskva, 2000, 61, p. 289-313. [Text in English; Intr. and comment. in Russian]

1110. PRODI (Paolo). Una storia della giustizia Dal pluralismo dei fori al moderno dualismo tra coscienza e diritto. Bologna, il Mulino, 2000, 499 p.

1111. Recht, Idee, Geschichte: Beiträge zur Rechts- und Ideengeschichte für Rolf Lieberwirth anlässlich seines 80. Geburtstages. Hrsg. v. Heiner LÜCK und Bernd SCHILDT. Köln, Böhlau, 2000, X-727 p. (port., map).

1112. Rechtsgeschichte(n)? Europäisches Forum Junger Rechtshistorikerinnen und Rechtshistoriker Zürich 28.–30. Mai 1999. Europäisches Forum Junger Rechtshistorikerinnen und Rechtshistoriker. Frankfurt am Main u. Oxford, P. Lang, 2000, 392 p. (ill.). (Rechtshistorische Reihe, 220).

1113. RIEDE (Judith). Die Person der Zeitgeschichte im deutschen und amerikanischen Bildnisschutz. Hamburg, Kovac, 2000, 392 p

1114. SCHMOECKEL (Mathias). Humanität und Staatsraison, die Abschaffung der Folter in Europa und die Entwicklung des gemeinen Strafprozeß- und Beweisrechts seit dem hohen Mittelalter. Köln, Weimar u. Wien, Böhlau, 2000, XI-668 p. (Norm und Struktur, Studien zum sozialen Wandel in Mittelalter und Früher Neuzeit, 14).

1115. VAN VELTEN (A. A.). Het notariaat: inderdaad een elastisch ambt, Het Nederlandse notariaat in de tweede helft van de twintigste eeuw, geschreven ter gelegenheid van het vijftigjarig bestaan van de Stichting tot bevordering van de notariële wetenschap. Amsterdam, Stichting tot bevordering van de notariële wetenschap, 2000, XXI-258 p. (Ars Notariatus, 108).

§ 7. Addenda 1999

1116. Droit (Le) de résistance, XIIe–XXe siècle. Ed. par Jean-Claude ZANCARINI. Paris, ENS Editions, 99, 339 p.

Cf. n°s 3486, 3523

§ 8. Economic and social history.

* 1117. BLAIKIE (A.). Migration and cultural identity: within and beyond the nation. A review article. *Northern Scotland*, 2000, 20, p. 179-188.

* 1118. HOWKINS (Richard A.). Bibliography in business history. *Business archives*, 2000, 80, p. 57-63.

* 1119. OTTO (Ingeborg). Frauen in den arabischen Ländern: eine Auswahlbibliographie. Hamburg, Deutsches Übersee-Institut, Übersee-Dokumentation, Referat Vorderer Orient, 2000, XXX-270 p. (Dokumentationsdienst Vorderer Orient. Reihe A, 27)

* 1120. Select bibliography 1999 of contributions to economic and social history in Scandinavian books, periodicals and year-books and Scandinavian contributions in other publications. *Scandinavian economic history review*, 2000, 48, 3, p. 96-113.

1121. Advances in agricultural economic history. Ed. by Kyle D. KAUFFMAN. Stamford, JAI Press, 2000, XIV-252 p. (ill., map) (New frontiers in agricultural history, 1).

1122. Argent (L') des villages. Comptabilités paroissiales et communales, fiscalité locale du XIIIe au XVIIIe siècle. Actes du colloque d'Angers (30–31 octobre 1998). Ed. par Antoine FOLLAIN. Rennes, Association d'histoire et sociétés rurales, 2000, 438 p. (Bibliothèque d'histoire rurale, 4).

1123. BANDHAUER-SCHÖFFMANN (Irene), BENDL (Regine). Unternehmerinnen. Geschichte und Gegenwart selbständiger Erwerbstätigkeit von Frauen. Frankfurt am Main, Lang, 2000, 399 p.

1124. BJÖRN (Ismo). Takeover: the environmental history of the coniferous forest. *Scandinavian journal of history*, 2000, 25, 4, p. 281-296.

1125. Cambridge economic history (The) of the United States. Vol. 2. The long nineteenth century. Vol. 3. The twentieth century. Ed. by Stanley L. ENGERMAN and Robert E. GALLMAN. Cambridge, Cambridge U. P., 2000, 2 vol., 1021 p., 1190 p.

1126. Chelovek v mire chuvstv: Ocherki po istorii chastnoj zhizni v Evrope i nekotorykh stranakh Azii do nachala novogo vremeni. (Man in the world of feelings: essays on history of private life in Europe and in certain countries of Asia before the modern time). Ed. Jurij L. BESSMERTNYJ [Bessmertny]. RAN, In-t vseobshchej istorii; Ros. gos. gumanit. un-t. Moskva, [s. n.], 2000, 582 p. (ill., bibl.). [Eng. summaries] [Cf. n° <choice> 1159.] – chelovek i ego blizkie na Zapade i Vostoke Evropy (do nachala novogo vremeni). (Man and his family in the west and east of Europe before the modern time: [Proceedings of a German-Russian Seminar]). Ed. Jurij L. BESSMERTNYJ, Otto G. OEXLE, Mikhail A. BOJTSOV, Pavel Sh. GABDRAKHMANOV. RAN, In-t vseobshchej istorii. Moskva, [s. n.], 2000, 268 p. (ill.). [Cf. n° <choice> 1159.]

1127. City walls: the urban enceinte in global perspective. Ed. by James D. TRACY. Cambridge, Cambridge U. P., 2000, XIX-697 p. (Studies in comparative early modern history).

1128. Comparative (A) study of thirty city-state cultures. An investigation conducted by the Copenhagen Polis Centre. Ed. by M. H. HANSEN. København, C. A. Reitzels Forlag, 2000, 636 p. (figs). [Cf. n° <choice> 2409.]

1129. Crossing slavery's boundaries. [AHR Forum]. *American historical review*, 2000, 105, 2, p. 451-484. [Contents: DAVIS (David Brion). Looking at slavery from broader perspectives. – KOLCHIN (Peter). The big picture: a comment on David Brion Davis's "Looking at slavery from broader perspectives". – SCOTT (Rebecca J.). Small-scale dynamics of large-scale processes. – ENGERMAN (Stanley L.). Slavery at different time and places].

1130. DENİZ (Bekir). Türk Dünyasında Halı Ve Düz Dokuma Yaygıları. (Carpets and looms in the Turkish world). Ed. Neval KONUK. Ankara, Atatürk Kültür Merkezi, 2000, 284 p.

1131. Deutsche Wirtschaftsgeschichte. Ein Jahrtausend im Überblick. Hrsg. v. Michael NORTH. München, Beck, 2000, 530 p.

1132. DEWEY (Scott Hamilton). Don't breathe the air: air pollution and U.S. environmental politics, 1945–1970. College Station, Texas A&M U. P., 2000, 321 p. (Environmental history series, 16).

1133. DOĞRU (Halime). XIII.–XIX yüzyıllar arasında Rumeli'de sağ kolun siyasi, sosyal, ekonomik görüntüsü ve Kozluca kazası. (Political, social and economic aspects of the right-side main trade route of Rumeli in the XIII–XIX centuries and the Kozluca District). Eskişehir, Anadolu Üniversitesi Edebiyat Fakültesi, 2000, 342 p.

1134. DOĞRUEL (Fatma). Osmanlı'dan Günümüze Tekel. (The Tekel [State Monopoly] from Ottomans to the present day). İstanbul, Türkiye Ekonomik ve Toplumsal Tarih Vakfı, 2000, 398 p.

1135. DYER (Christopher). Alternative approaches to the history of agricolture. *Past & Present*, 2000, 168, p. 254-262.

1136. Emotionalität. Zur Geschichte der Gefühle. Hrsg. v. Claudia BENTHIEN. Köln, Böhlau, 2000, 237 p.

1137. EPSTEIN (S. R.). Freedom and growth. The rise of States and markets in Europe, 1300–1750. London a. New York, Routledge, 2000, X-228 p.

1138. Europäische Sozialgeschichte: Festschrift für Wolfgang Schieder. Hrsg. v. Christoff DIPPER, Lutz KLINKHAMMER und Alexander NÜTZENADEL. Berlin, Duncker & Humblot, 2000, XIV-558 p. (port.) (Historische Forschungen, 68).

1139. FARR (James R.). Artisans in Europe 1300–1914. Cambridge, Cambridge U. P., 2000, [s. p.].

1140. FLÜGEL (Axel). Ambivalente Innovation. Anmerkungen zur Volksgeschichte. *Geschichte und Gesellschaft*, 2000, 26, 4, p. 653-671.

1141. Forest history: international studies on socioeconomic and forest ecosystem change. Ed. by M. AGNOLETTI and S. ANDERSON. Wallingford, CAB International, 2000, XIII-418 p. (IUFRO research series, 2).

1142. Formes (Les) de la servitude: esclavages et servages de la fin de l'Antiquité au monde moderne. Actes de la Table ronde de Nanterre, 12 et 13 décembre 1997, présentés par Henri BRESC. *Mélanges de l'Ecole française de Rome. Moyen Age*, 2000, 112, 2, p. 493-631. [Cf. n° <sélection> 651.]

1143. Geschichte der Ökonomie. Hrsg. v. Johannes BURKHARDT und Birger P. PRIDDAT. Frankfurt am Main, Deutscher Klassiker Verlag, 2000, 975 p. (Bibliothek der Geschichte und Politik, 21).

1144. Geschichte und Zukunft der Arbeit. Hrsg. v. Jürgen KOCKA, Claus OFFE und Beate REDSLOB. Frankfurt am Main, Campus Verl., 2000, 510 p. [Cf. n° <Auswahl> 7334.]

1145. Hauptstädte und Global Cities an der Schwelle zum 21. Jahrhundert. Hrsg. v. Andreas SOHN und Hermann WEBER. Bochum, Verlag Dr. Dieter Winckler, 2000, 496 p. (Herausforderungen, Historisch-politische Analysen, 9).

8. ECONOMIC AND SOCIAL HISTORY

1146. ILIESCU (Octavian). Aspecte ale economiei monetare în Moldova sub domnia lui Alexandru cel Bun. *Revista Istorică*, 2000, 11, 1-2, p. 59-95.

1147. JARRICK (Arne). Hamlets fråga: en svensk självmordshistoria. (Hamlet's question: a Swedish history of suicide). Stockholm, Norstedts Forlag, 2000, 239 p.

1148. JONES (Eric Lionel). Growth recurring: economic change in world history. Ann Arbor, University of Michigan Press, 2000, XLVI-247 p. (Economics, cognition, and society)

1149. KNITTLER (H.). Die europäische Stadt in der frühen Neuzeit. Institutionen, Strukturen, Entwicklungen. Wien u. Munich, Verlag für Geschichte und Politik u. R. Oldenbourg, 2000, 320 p. (Querschnitte, 4).

1150. Køn, religion og kvinder i bevægelse. Konferencerapport fra det VI. Nordiske Kvindehistorikermøde. (Sexe, religion et femmes en mouvement. Rapport des conférences des 6e Rencontres des historiennes nordiques). Ed. par Anette WARRING. Roskilde, Kvinder på Tværs, Institut for historie og samfund, 2000, 470 p. (diagr., tab., ill.).

1151. Kriminalitätsgeschichte. Beiträge zur Sozial- und Kulturgeschichte der Vormoderne. Hrsg. v. Andreas BLAUERT und Gerd SCHWERHOFF. Konstanz, UVK, Medien, 2000, 920 p.

1152. LAAKKONEN (Simo). Vesiensuojelun synty. Helsingin ja sen merialueen ympäristöhistoria 1878–1928. (The origins of water protection in Helsinki 1878–1928). Helsinki, Gaudeamus, 2000, 309 p. (ill., maps). (Hanki ja jää).

1153. LEINO-KAUKIAINEN (Pirkko). Cleaning a thousand waste-waters: water pollution in Finland, 1945–1970. *Scandinavian economic history review*, 2000, 48, 1, p. 81-99.

1154. MAC NEILL (J. R.). Something new under the sun: an environmental history of the twentieth-century world. New York, W. W. Norton, 2000, XXVI-421 p. (The global century series).

1155. Migration et migrants dans une perspective historique. Permanences et innovations. Ed. par René LEBOUTTE. Frankfurt am Main, Lang, 2000, 354 p. (Collection Europe plurielle, 12).

1156. OSTERHAMMEL (Jürgen). Sklaverei und die Zivilisation des Westens. München, Carl-Friedrich-von-Siemens-Stiftung, 2000, 72 p. (Carl-Friedrich-von-Siemens-Stiftung. Themen. 70).

1157. PETZINA (Dietmar). Die Verantwortung des Staates für die Wirtschaft. Ausgewählte Aufsätze. Essen, Klartext Verl., 2000, 236 p.

1158. POMERANZ (Kenneth). The great divergence. China, Europe and the making of the modern world economy. Princeton, Princeton U. P., 2000, 382 p.

1159. PUSHKAREVA (Natal'ja L.). "Slez ee radi ...": Opyt mikroanaliza emotsional'nykh otnoshenij sem'i "novykh russkikh" XVI stoletija. ("For Her Tears Sake" microanalysis experience of the family of emotional relations of "New Russians" in the 16th century). *In*: Sotsial'naja istorija: Ezhegodnik [Cf. n° 7277], p. 268-284. [Eng. summary] – PUSHKAREVA (Natal'ja L.). Mir chuvstv russkoj dvorjanki kontsa XVIII–nachala XIX veka: Seksual'naja sfera. (The world of feelings of a noble woman: Russia, the late 18th to the early 19th centuries: sexual sphere). *In*: Chelovek v mire chuvstv [Cf. n° 1126], p. 85-119. [Eng. summary] – PUSHKAREVA (Natal'ja L.). Obraz ideal'noj suprugi (supruga) i ego transformatsii v srednevekovoj Rusi i Rossii novogo vremeni, XII–XVII vv. (The concept of ideal wife and its transformations in Rus' and Russia from the 12th to the 17th century). *In*: Chelovek i ego blizkie na Zapade i Vostoke Evropy [Cf. n° 1126], p. 177-193.

1160. RADKAU (Joachim). Natur und Macht. Eine Weltgeschichte der Umwelt. München, Beck, 2000, 438 p.

1161. Rise (The) of the fiscal state in Europe c. 1200–1815. Ed. by Richard BONNEY. New York, Oxford U. P., 2000, XI-527 p.

1162. SANDMO (Erling). The history of violence in the Nordic Countries: a case study. *Scandinavian journal of history*, 2000, 25, 1-2, p. 53-68.

1163. SAYER (Karen). Country cottages: a cultural history. Manchester, Manchester U. P., 2000, 229 p.

1164. SCHÖBER (Peter). Wirtschaft, Stadt und Staat. Von den Anfängen bis zur Gegenwart. Köln, Böhlau, 2000, VIII-307 p.

1165. Servitude (La) dans les pays de la Méditerranée occidentale chrétienne au XIIe siècle et au-delà: déclinante ou renouvelée? Actes de la Table ronde de Rome (8 et 9 octobre 1999). Ed. par Monique BOURIN et Paul FREEDMAN. *Mélanges de l'Ecole française de Rome. Moyen Age*, 2000, 112, 2, p. 633-1085.

1166. Seta (La) in Italia dal Medioevo al Seicento: dal baco al drappo. A cura di Luca MOLA, Reinhold C. MUELLER [et al.]. Venezia, Marsilio, 2000, 568 p. (Saggi).

1167. SIEVERT (James). The origins of nature conservation in Italy. New York, Peter Lang, 2000, 298 p.

1168. Slave elites in the Middle East and Africa: a comparative study. Ed. by Miura TORA and John Edward PHILIPS. London: Kegan Paul, 2000, XII-248 p. (Islamic area studies).

1169. SMAJE (Chris). Natural hierarchies: the historical sociology of race and caste. Malden, Blackwell Publishers, 2000, IX-290 p.

1170. SMOUT (T. C.). Nature contested: environmental history in Scotland and northern England since 1600. Edinburgh, Edinburgh U. P., 2000, XII-210 p.

1171. STELLA (Alessandro). Histoire d'esclaves dans la péninsule Ibérique. Paris, Ed. de l'EHESS, 2000, 214 p.

1172. Stiftungen und Stiftungswirklichkeiten. Vom Mittelalter bis zum Gegenwart. Hrsg. v. Michael BORGOLTE. Berlin, Akademie, 2000, 343 p. (Stiftungsgeschichten, 1).

1173. STOKLUND (Bjarne). Bondefiskere og strandsiddere: studier over de store sæsonfiskerier 1350–1600. (Pêcheurs paysans et habitants de la côte: études sur les pêcheries saisonnières entre 1350 et 1600). København, Landshistorisk Selskab, 2000, 224 p. (tab., cartes, ill.).

1174. THANE (Pat). "An untiring zest for life": images and self-images of old women in England. *Journal of family history*, 2000, 25, p. 235-247.

1175. Towns in decline AD 100–1600. Hrsg. v. Terry R. SLATER. Aldershot, Ashgate, 2000, XII-325 p.

1176. TUCKER (Richard P.). Insatiable appetite: the United States and the ecological degradation of the tropical world. Berkeley, University of California Press, 2000, XIII-551 p.

1177. Unsichere Großstädte? Vom Mittelalter bis zur Postmoderne. Hrsg. v. Martin DINGERS und Fritz SACK. Konstanz, UVK, 2000, 400 p. (Konflikte und Kultur – Historische Perspektiven, 3).

1178. VERTECCHI (Giulia). Wiener Neustadt: studio di una città di formazione medievale. Roma, Bonsignori, 2000, 102 p. (ill.). (Civitates, 3; Dispense del Dottorato in "Storia delle città", 1).

1179. Vite (La) e il vino. Storia e diritto (secoli XI–XIX). A cura di Mario DA PASSANO, Antonello MATTONE, Franca MELE e Pinuccia F. SIMBULA. Introduzione di Massimo MONTANARI. Roma, Carocci editore, 2000, 1328 p.

1180. WALTERSKIRCHEN (Helene). Aristokraten. Eben zwischen Tradition und Moderne. Wien, Ueberreuter, 2000, 250 p.

1181. WILKE (Jürgen). Grundzüge der Medien- und Kommunikationsgeschichte. Von den Anfängen bis ins 20. Jahrhundert. Köln, Weimar u. Wien, Böhlau, 2000, VIII-436 p.

1182. WIRTH (John). Smelter smoke in North America: the politics of transborder pollution. Lawrence, University Press of Kansas, 2000, XX-264 p. (Development of Western resource).

Cf. n°s 114-133, 6755-6819, 8552

§ 9. **History of civilization, sciences and education.**

* 1183. FISCHER (Klaus-Dietrich). Bibliographie des textes médicaux latins. Antiquité et haut moyen âge. Premier supplément 1986–1999. Saint-Étienne Cedex, Publication de l'université de Saint-Étienne, 2000, 62 p.

* 1184. History of mathematics from antiquity to the present. A selective annotated bibliography. Ed. by Joseph W. DAUBEN and Albert C. LEWIS. Ann Arbor, American Mathematical Society, 2000, 1 CD-Rom.

* 1185. Isis, current bibliography of the history of science. Vol. 91. [Vol. 90. Cf. Bibl. 99, n° 1194]. Ed. by John Neu. *Isis*, 2000 (suppl.), 200 p.

1186. BALDINI (Massimo). Storia della scienza e della tecnica dall'antichità al '900. Roma, Armando, 2000, 447 p.

1187. BAZÁN DÍAZ (Iñaki), VÁZQUEZ GARCÍA (Francisco), MORENO MENGIBAR (Andrés). La prostitution au Pays basque entre XIVe et XVIIe siècle. *In*: Hommes et femmes. Codes amoureux et morale publique. *Annales*, 2000, 55, 6, p. 1283-1302.

1188. BODIN (Per Arne). Ryssland Idéer och identiter. (La Russie: idées et identité). Skellefteå, Norma, 2000, 214 p.

1189. BOLENS (Guillemette). La logique du corps articulaire: Les articulations du corps humain dans la littérature occidentale. Rennes, Presses Universitaires de Rennes, 2000, 249 p. (plates, ill.).

1190. Books and the sciences in history. Ed. by Marina FRASCA-SPADA and Nick JARDINE. Cambridge, Cambridge U. P., 2000, XIV-438 p. (ill., ports).

1191. BULLARD (Alice). Exile to paradise. Savagery and civilization in Paris and the South Pacific, 1700–1900. Stanford, Stanford U. P., 2000, 380 p. (ill.).

1192. CAPATTI (Alberto), MONTANARI (Massimo). La cucina italiana. Storia di una cultura. Roma e Bari, Laterza, 2000, 422 p.

1193. CARLIER (Omar). Les enjeux sociaux du corps. Le Hammam maghrébin (XIXe–XXe siècle), lieu pérenne, menacé ou recréé. *In*: Hommes et femmes. Codes amoureux et morale publique. *Annales*, 2000, 55, 6, p. 1303-1334.

1194. CARLSSON (Bo Göran). Religion, kultur och manlig homosexualitet. (Religion, culture et homosexualité masculine). Stockholm, Carlssons, 228 p.

1195. ÇIZAKÇA (Murat). A history of philanthropic foundations: the islamic world from the seventh century to the present. İstanbul, Boğaziçi U. P., 2000, 288 p.

1196. Commentaire (Le): entre tradition et innovation. Actes du colloque international de l'Institut des Traditions Textuelles (Paris et Villejuif, 22–25 septembre 1999). Ed. par Marie-Odile GOULET-CAZÉ. Paris, J. Vrin, 2000, 585 p. (ill., plates).

1197. DEMIR (Remzi). Takiyüddîn'de matematik ve astronomi: Cerîdetü'd-Dürer ve Harîdetü'l-Fiker üzerine bir inceleme. (Mathematics and astronomy in Takiyüddin: a study on Cerîdetü'd-Dürer ve Haridet'l-Fiker). Ankara, Atatürk Kültür Merkezi, 2000, 243 p.

1198. Dictionary (A) of the history of science. Ed. by Anton SEBASTIAN. Carnforth, Parthenon, 2000, 400 p. (ill.).

1199. DUBUISSON (Michel). Verba uolant. Réexamen de quelques «mots historiques» romains. *Revue belge de philologie et d'histoire*, 2000, 78, 1, p. 147-170.

1200. Encyclopedia of the scientific revolution: from Copernicus to Newton. Ed. by Wilbur APPLEBAUM. New York a. London, Garland Publishing, 2000, XXXV-758 p. (ill., ports., map). (Garland reference library of the humanities, 1800).

1201. FRENCH (Roger Kenneth). Ancients and moderns in the medical sciences: from Hippocrates to Harvey. Aldershot, Brookfield, Ashgate, 2000, XXI-280 p. (ill., port.).

1202. GRMEK (Mirko Dražen). Život, bolesti i povijest. (Life, diseases and history). [S. l.], Hrvatska akademija znanosti i umjetnosti, 2000, 127 p.

1203. Guide to the history of technology in Europe 2000. Ed. by Caroline TURNEY [et al.]. London, Science Museum, 2000, X-229 p.

1204. GUZZETTI (Luca). Science and power: the historical foundations of research policies in Europe. Luxembourg, Office for Official Publications of the European Communities, 2000, 250 p.

1205. HÄHNER-ROMBACH (Sylvelyn). Sozialgeschichte der Tuberkulose. Vom Kaiserreich bis zum Ende des Zweiten Weltkriegs unter besonderer Berücksichtigung Württembergs. Stuttgart, Steiner, 2000, 404 p. (Medizin, Gesellschaft und Geschichte, 14).

1206. Histoire des pères et de la paternité. Ed. par Jean DELUMEAU et Daniel ROCHE. Paris, Larousse, 2000, 535 p.

1207. History (The) of science and religion in the western tradition: an encyclopedia. Ed. by Gary B. FERNGREN. New York, Garland Pub., 2000, 586 p. (Garland reference library of the humanities, 1833).

1208. History of civilizations of Central Asia. Vol. 4. Age of achievement: A.D. 750 to the end of the fifteenth century. Part 2. Achievements. Ed. By C. E. BOSWORTH and M. S. ASIMOV. Paris, UNESCO Publishing, 2000, 700 p. (ill.). (Multiple history series).

1209. History of education: major themes. Ed. by Roy LOWE. London, Routledge, 2000, 4 vol., [s. p.]. (Major themes in education).

1210. History of humanity: scientific and cultural development. Vol. 4. From the seventh to the sixteenth century. Ed. by M. A. AL-BAKHIT, L. BAZIN, S. M. CISSOKO, M. S. ASIMOV [et al.]. London, Routledge, 2000, XXX-682 p. (plates, ill., maps, ports).

1211. History of Indian science, technology and culture: AD 1000–1800. Ed. by A. RAHMAN and D.P. CHATTOPADHYAYA. New Delhi, Oxford U. P., 2000, XX-445 p. (ill.). (History of science, philosophy and culture in Indian civilization, 3. Development of philosophy, science and technology in India and neighbouring civilizations, 1).

1212. Instrument – Experiment: Historische Studien. Hrsg. v. Cristoph MEINEL. Stuttgart, GNT-Verl., 2000, 423 p.

1213. International handbook on history of education. Ed. by Kadriya SALIMOVA and Nan L. DODDE. Moscow, Orbita-M, 2000, 559 p.

1214. IYANGA PENDI (Augusto). Historia de la Universidad en Europa. Valencia, Universidad de Valencia. Servicio de Publicaciones, 2000, 300 p.

1215. JONES (Ann Rosalind), STALLYBRASS (Peter). Renaissance clothing and the materials of memory. Cambridge, Cambridge U. P., 2000, XIII-386 p. (ill.).

1216. JULLIEN (François). De l'essence ou du nu. Paris, Ed. du Seuil, 2000, 157 p.

1217. KHRISTOVA (Boriana), DZHUROVA (Aksiniia), VELINOVA (Vasia). Opis na slavianskite rukopisi ot Tsentura za slaviano-vizantiiski prouchvaniia "prof. Ivan Duichev" kum SU Sv. "Kliment Okhridski": XIV–XIX v. (Inventaire des manuels slaves du Centre de recherches slave-byzantines "Prof. Ivan Doujtchev" pres l'Université d'Etat "St Climent d'Ohrid". Sofiia, UI "Sv. Kliment Okhridski", 2000, 227 p. (Series catalogorum, 1).

1218. KING (Margaret L.). Western civilization: a social and cultural history. Vol. 1. Vol. 2. Upper Saddle River, Prentice-Hall, 2000, 2 vol., XVI-512 p., VII-190 p. (ill.).

1219. Kommunikation in der ländlichen Gesellschaft vom Mittelalter bis zur Moderne. Hrsg. v. Werner RÖSENER. Göttingen, Vandenhoeck u. Ruprecht, 2000, 412 p.

1220. Körper mit Geschichte. Der menschliche Körper als Ort der Selbst- und Weltdeutung. Hrsg. v. Clemens WISCHERMANN und Stefan HAAS. Stuttgart, Steiner, 2000, 345 p. (Studien zur Geschichte des Alltags, 17).

1221. Kulturgeschichte des Salzes. 18. bis 20. Jahrhundert. Hrsg. v. Thomas HELLMUTH und Ewald HIEBL. Wien, Verlag für Geschichte und Politik u. München, Oldenbourg, 2000, 344 p.

1222. KUZMICS (Helmut), AXTMANN (Roland). Autorität, Staat und Nationalcharakter: der Zivilisationsprozess in Österreich und England, 1700–1900. Opladen, Leske und Budrich, 2000, VII-427 p. (Figurationen, 2).

1223. LANGFORD (Paul). Englishness identified: manners and character, 1650–1850. Oxford, Oxford U. P., 2000, X-389 p.

1224. LÁZARO (José), BUJOSA (Francesc). Historiografía de la psiquiatría española. Madrid, Editorial Triacastela, 2000, 194 p.

1225. LECLERC (Gérard). La mondialisation culturelle. Les civilisations à l'épreuve. Paris, Presses Universitaires de France, 2000, 486 p.

1226. LORENZ (Maren). Leibhaftige Vergangenheit. Einführung in die Körpergeschichte. Tübingen, Diskord, 2000, 239 p. (Historische Einführungen, 4).

1227. LUND (Hakon). Danmark havekunst. Bd. 1. indtil 1800. (L'art d'aménager les jardins au Danemark. Tome 1. jusqu'en 1800). København, Arkitektens forlag, 2000, 413 p. (ill.).

1228. Mahl und Repräsentation. Der Kult ums Essen. Beiträge des internationalen Symposions in Salzburg 29. April bis 1 Mai 1999. Hrsg. v. Lothar KOLMER und Christian ROHR. Paderborn, München, Wien u. Zürich, Schöningh, 2000, 288 p.

1229. Medieval (The) world and the modern mind. Ed. by Michael BROWN and Stephen H. HARRISON. Dublin a. Portland, Four Courts Press, 2000, 201 p.

1230. MERİÇ (Nevin). Osmanlı'da gündelik hayatın değişimi: âdâb-ı muâşeret 1894–1927. (Changes of everyday life in the Ottomans: the etiquette). Ed. by Mustafa HILMI BAŞ, redactor Serkan ÖZBURUN. İstanbul, Kaknüs yayınları, 2000, 486 p.

1231. MIGNOLO (Walter D.). Local histories / Global designs. Coloniality, subaltern knowledges and border thinking. Princeton, Princeton U. P., 2000, 371 p.

1232. MILLER (William Ian). The mystery of courage. Cambridge, Harvard U. P., 2000, XI-346 p.

1233. MINOIS (Georges). Histoire du rire et de la dérision. Paris, Fayard, 2000, 637 p.

1234. MOORE (Barrington). Moral purity and persecution in history. Princeton, Princeton U. P., 2000, XVI-158 p.

1235. MUCHEMBLED (Robert). Une histoire du diable, XIIe–XXe siècle. Paris, Ed. du Seuil, 2000, 403 p.

1236. NEEDHAM (Joseph). Science and civilisation in China. Vol. 6. Biology and biological technology. Cambridge, Cambridge U. P., 2000, XVIII-261 p. (ill.).

1237. NILSSON (Ingemar). Idéhistoriska perspektiv: symposium i Göteborg mars 2000. (Perspectives de l'histoire des idées: symposium tenu à Göteborg en mars 2000). Göteborg, Göteborgs universitet, Institution för idé- och lärdomshistoria, 2000, 148 p. (Arachne).

1238. Only man, studies in the history of conceptions of man. Ed. by Arne JARRICK. Stockholm, Almquist and Wiksell International, 2000, 383 p. (Acta Universitatis Stockholmiensis, 61).

1239. OUÉDRAOGO (Arouna P.). De la secte religieuse à l'utopie philanthropique. Genèse sociale du végétarisme occidental. *In*: Histoire de l'alimentation. *Annales*, 2000, 55, 4, p. 825-844.

1240. Place (The) of the dead: death and remembrance in late medieval and early modern Europe. Ed. by Bruce GORDON and Peter MARSHALL. Cambridge, Cambridge U. P., 2000, XIII-324 p. (ill., plates).

1241. Reader's guide to the history of science. Ed. by Arne HESSENBRUCH. London, Fitzroy Dearborn, 2000, XXIX-934 p.

1242. Robin Hood in popular culture: violence, transgression, and justice. Ed. by Thomas HAHN. Woodbridge a. Rochester, Boydell and Brewer, 2000, IX-278 p. (ill.).

1243. SABBAN (Françoise). Quand la forme transcende l'objet. Histoire des pâtes alimentaires en Chine (IIIe siècle av. J.-C.–IIIe siècle apr. J.-C.). *In*: Histoire de l'alimentation. *Annales*, 2000, 55, 4, p. 791-824.

1244. SCHUBRING (Gert). Kabinett – Seminar – Institut: Raum und Rahmen des forschenden Lernens. *Berichte zur Wissenschaftsgeschichte*, 2000, 23, 3, p. 269-286.

1245. Science and its times: understanding the social significance of scientific discovery. Vol. 5. 1800–1899. Ed. by Neil SCHLAGER and Josh LAUER. Detroit a. London, Gale Group, 2000, XXII-633 p. (ill., facsims, maps, ports).

1246. Self and story in Russian history. Ed. by Laura ENGELSTEIN and Stephanie SANDLER. Ithaca, Cornell U. P., 2000, IX-363 p.

1247. SWAHN (Jan-Öjvind). Fil, fläsk och falukorv. (Histoire de l'alimentation en Suède). Lund, Historiska media, 2000, 124 p.

1248. Tarihten Günümüze Anadolu'da Konut ve Yerleşme. (Housing and settlement in Anatolia: a historical perspective). Ed. by Yıldız SEY. İstanbul, Türkiye Ekonomik ve Toplumsal Tarih Vakfı, 2000, 492 p.

1249. TATON (René). Études d'histoire des sciences. Turnhout, Brepols, 2000, 544 p. (De diversis artibus, 47, 10).

1250. VIGARELLO (Georges). Passion sport. Histoire d'une culture. Paris, Ed. Textuel, 2000, 191 p.

1251. WARD-PERKINS (Bryan). Why did the Anglo-Saxons not become more British? *English historical review*, 2000, 115, 462, p. 513-533.

1252. WILSON (L. M.). A short history of science: ancient mathematics, medicine, science and technology. Cambridge, L. M. Wilson, 2000. 188 p. (ill.).

1253. Written on the body: the tattoo in European and American history. Ed. by Jane CAPLAN. Princeton, Princeton U. P., 2000, XXIII-318 p.

1254. ZILSEL (Edgar). The social origins of modern science. Dordrecht a. London, Kluwer Academic Publishers, 2000, LIX-267 p. (ill.). (Boston studies in the philosophy of science, 200).

Cf. n°s 819, 1315, 3888

§ 10. History of art.

* 1255. BHA. Bibliography of the history of Art. Bibliographie de l'histoire de l'art. Vol. 10. 1–4. 2000. [Vol. 9, 1–4, 1999. Cf. Bibl. 99, n° 1271.] Ed. by Michael RINEHART and Marise BIDEAULT. Paris, Centre National de la Recherche Scientifique a. Santa Monica, The J. Paul Getty Trust, 2000, 4 vol., [s.p.].

* 1256. MÜLLER (Frank-Bernhard), ULLMANN (Ernst). Bibliographie zur Kunstgeschichte in Sachsen.

Stuttgart, Verlag der Sächsischen Akademie der Wissenschaften zu Leipzig, 2000, XLII-497 p. (Abhandlungen der Sächsischen Akademie der Wissenschaften zu Leipzig. Philologisch-Historische Klasse, 77).

1257. Allgemeines Künstlerlexikon. Biobibliographischer Index A–Z. Vol. 4. Gaci–Hodson. Vol. 5. Hodunov–Laborier. Vol. 6. Laborim–Michallet. Vol. 7. Michallon–Pikaar. Vol. 8. Pikalov–Schintzel. Vol. 9. Schinz–Torricelli. Vol. 10. Torrico–Z. München u. Leipzig, Saur, 2000, 7 vol., [s. p.]

1258. Allgemeines Künstlerlexikon: die Bildenden Künstler aller Zeiten und Völker. Vol. 24. Damdama–Dayal. Vol. 25. Dayan–Delvoye. Vol. 26. Delwaide–Dewagut. Vol. 27. Dewailly–Dismorr. München u. Leipzig, Saur, 2000, 4 vol. [s.p.].

1259. Allgemeines Künstlerlexikon: internationale Künstlerdatenbank. 9^a ed. [1-24]. 10^a ed. [1-26]. München u. Leipzig, Saur, 2000, 2 CD-ROM.

1260. Anfang (Der) der Museumslehre in Deutschland. Das Traktat "Inscriptiones vel tituli Theatri Amplissimi" von Samuel Quiccheberg. Lateinisch-deutsch. Hrsg. v. Harriet ROTH. Berlin, Akademie, 2000, 362 p.

1261. Arti fiorentine: la grande storia dell'artigianato. Vol. 3. Il Cinquecento. A cura di Franco FRANCESCHI, Gloria FOSSI. Firenze, Giunti, 2000, 303 p. (ill.).

1262. BLAND (Kalman P.). The artless Jew: medieval and modern affirmations and denials of the visual. Princeton, Princeton U. P., 2000, IX-233 p.

1263. BREVAGLIERI (Sabina). Assistenza e patronage femminile a Venezia: la Compagnia di S. Orsola, Tintoretto e l'altare degli Incurabili. In: Committenza artistica femminile [Cf. n° 1267], p. 355-392.

1264. ÇAM (Nusret). Yunanistan'daki Türk eserleri. (Turkish monuments in Greece). Ankara, Türk Tarih Kurumu, 2000, 394 p.

1265. Cambridge Companion (The) to modern British women playwrights. Ed. by Elaine ASTON and Janette REINELT. Cambridge, Cambridge U. P., 2000, XX-276 p.

1266. Cambridge history (The) of American theatre. Vol. 3. Post-World War to the 1990s. Ed. by Don B. WILMETH and Christopher BIGSBY. Cambridge, Cambridge U. P., 2000, XVIII-582 p.

1267. Committenza artistica femminile. A cura di Sara F. MATTHEWS-GRECO e Gabriella ZARRI. *Quaderni storici*, 2000, 35, 104, p. 283-422. [Cf. n^{os} <scelta> 1263, 1270, 1279, 1284.]

1268. DIDI-HUBERMAN (Georges). Devant le temps: histoire de l'art et anachronisme des images. Paris, Ed. de Minuit, 2000, 286 p.

1269. DUARTE (Carlos F.). Diccionario biográfico documental: pintores, escultores y doradores en Venezuela: período hispánico y comienzos del período republicano. Caracas, Fundación Galería de Arte Nacional, 2000, 320 p. (ill.).

1270. EDELSTEIN (Bruce L.). Nobildonne napoletane e committenza: Eleonora d'Aragona ed Eleonora di Toledo a confronto. In: Committenza artistica femminile [Cf. n° 1267], p. 295-330.

1271. Essential guide (The) to Dutch music: 100 composers and their work. Ed. by Jolande VAN DER KLIS. Amsterdam, Amsterdam U. P., 2000, XIV-437 p. (ill.).

1272. FYFE (Gordon). Art, power and modernity: English art institutions, 1750–1950. London, Leicester U. P., 2000, IX-212 p.

1273. GABBA (Emilio). Architettura e stampa. In: Monumento (Il): arte e storia [Cf. n° 1283], p. 650-653.

1274. History (A) of art in Africa. Ed. by Monica BLACKMUN VISONÀ [et al.]. London, Thames & Hudson, 2000, 544 p. (ill., maps, ports.).

1275. Images du pouvoir. Pavements de faïence en France du XIIIe au XVIIe siècle. Ed. par Thierry CRÉPIN et Jean ROSEN. Paris, Réunion des musées nationaux et Brou, Musée de Brou, 2000, 197 p.

1276. IMORDE (J.). Dulciores sunt lacrimae orantium, quam gaudia theatrorum. Zum Wechselverhältnis von Kunst und Religion um 1600. *Zeitschrift für Kunstgeschichte*, 2000, 70, p. 1-14 (ill.).

1277. Indian art: forms, concerns and development in historical persepective. Ed. by B.N. GOSWAMY and Kavita SINGH. New Delhi, Project of History of Indian Science, Philosophy and Culture, Centre for Studies in Civilizations, 2000, XXVI-387 p. (ill.). (History of science, philosophy and culture in Indian civilization, 6, 3).

1278. KEMP (Martin). The Oxford history of Western art. New york, Oxford U. P., 2000, 564 p.

1279. KIRKHAM (Victoria). Laura Battiferra degli Ammannati benefattrice dei gesuiti fiorentini. In: Committenza artistica femminile [Cf. n° 1267], p. 331-354.

1280. KLOTZ (Heinrich). Geschichte der deutschen Kunst. Vol. 3. Neuzeit und Moderne. 1750–2000. München, Beck, 2000, 486 p.

1281. LANGMUIR (Erika). Yale dictionary of art and artists. New Haven a. London, Yale U. P., 2000, 768 p. (Nota Bene series)

1282. MARINOVICH (Ljudmila P.), KOSHELENKO (Gennadij A.). Sud'ba Parfenona. (The fortunes of the Parthenon). RAN, In-t vseobshchej istorii. Moskva, Jazyki russkoj kul'tury, 2000, 352 p. (ill.; bibl.).

1283. Monumento (Il): arte e storia. *Rivista storica italiana*, 2000, 112, 2, p. 429-702. [Cf. n^{os} <scelta> 455, 1273, 1288, 2399, 2894, 3795, 3798, 3811, 4832.]

1284. MURPHY (Caroline P.). Il teatro della vedovanza. Le vedove e il patronage pubblico delle arti visive a Bologna nel XVI secolo. In: Committenza artistica femminile [Cf. n° 1267], p. 393-422.

1285. PİLEHVARİAN (Nuran Kara). Osmanlı başkenti İstanbul'da çeşmeler. (Fountains in the Ottoman capital İstanbul). İstanbul, YEM, 2000, 216 p.

1286. PUTTFARKEN (Thomas). The discovery of pictorial composition: theories of visual order in painting 1400–1800. New Haven a. London, Yale U. P., 2000, VIII-332 p. (ill.).

1287. Renaissance (From) to Impressionism: styles and movements in Western art, 1400–1900. Ed. by Jane TURNER. London, Macmillan Reference, 2000, XII-377 p. (plates, ill.). (Grove art).

1288. SCIOLLA (Gianni Carlo). Storia dell'arte/storia della cultura. Un 'dossier' di problemi metodologici. *In*: Monumento (Il): arte e storia [Cf. n° 1283], p. 429-440.

1289. SCREECH (Timon). The Shogun's painted culture: fear and creativity in the Japanese states, 1760–1829. London, Reaktion, 2000, 311 p. (ill., ports.).

1290. Storia del cinema mondiale. Vol. 2. Gli Stati Uniti. A cura di G. Piero BRUNETTA. Torino, Einaudi, 2000, XVIII-1022 p. (ill.).

1291. Storia del teatro moderno e contemporaneo. A cura di Roberto ALONGE e Guido DAVICO BONINO. Torino, Einaudi, 2000, 2 vol., XXV-1346 p., XVIII-1248 p.

1292. TUKSAR (Stanislav). Kratka povijest hrvatske glazbe. (A short history of the Croatian music). Zagreb, Matica hrvatska, 2000, 149 p.

1293. WATSON (William). The arts of China 900–1620. New Haven a. London, Yale U. P., 2000, VI-286 p. (ill., maps). (Yale University Press Pelican history of art).

1294. ZIMMERMANN (Bernhard). Europe und die griechische Tragödie: Vom kultischen Spiel zum Theater der Gegenwart. Frankfurt am Main, Fischer Taschenbuch, 2000, 220 p. (Europäische Geschichte).

Cf. n°s 299, 1494, 1655, 3086, 3814

§ 11. History of religions.

a. General

* 1295. Chroniques bibliographiques. *Hieros*, 2000, 5, p. 50-73. [99, 4. Cf. Bibl. 99, n° 1301.]

* 1296. Critical review of books in religion. 1999. [1998. Cf. Bibl. 99, n° 1302.] Ed. by Charles PREBISH. 2000, 12, [s. p.].

* 1297. Ephemerides theologicae lovanienses. Elenchus bibliographicus. Tomus LXXVI. [Tomus LXXV. Cf. Bibl. 99, n° 1303.] Editae cura E. BRITO, L. DE FLEURQUIN, J. FAMERÉE, É. GAZIAUX, J. HAERS, A. HAQUIN, M. LAMBERIGTS, J. LUST, G. VAN BELLE, J. VERHEYDEN. Leuven, Peeters, 2000, 703 p.

* 1298. JEAN-BAPTISTE (Chantal). Index to book reviews in religion IBRR: an author, title, reviewer, series, and annual classified index o reviews of books published in and for interest to the field of religion. 1999. Evanston, American Theological Library Association, 2000, XXIII-862 p.

* 1299. MOULINET (Daniel). Guide bibliographique des sciences religieuses. Paris, Ed. Salvator, 2000, 487 p.

* 1300. *Religious Studies Review*, 2000, 26, 1-4, 407-32 p. [99, 25, 1-4. Cf. Bibl. 99, n° 1304.]

* 1301. Revue d'histoire ecclésiastique. Bibliographie. Tome 95. 2000. [Tome 94, 1999. Cf. Bibl. 99, n° 1305.] Ed. par M. HAVERALS. Louvain-la-Neuve, Bureaux de la R. H. E., Bibliothèque de l'université, 2000, 333* p.

1302. Anni (Gli) Santi nella storia: atti del Congresso Internazionale, Cagliari 16–19 ottobre 1999. A cura di Luisa D'ARIENZO. Cagliari, Edizioni AV, 2000, 692 p. (ill.).

1303. BRAUN (Willi), MAC CUTCHEON (Russell T.). Guide to the study of religion. London, Cassell Academic Press, 2000, [s. p.].

1304. DELUMEAU (Jean). Que reste-t-il du paradis? Paris, Fayard, 2000, 535 p.

1305. DOWDEN (Ken). European paganism: the realities of cult from antiquity to the Middle Ages. London a. New York, Routledge, 2000, XXI-367 p. (ill.).

1306. Enciclopedia dei papi. Roma, Istituto della enciclopedia italiana, 2000, 3 vol., 746 p., 717 p., 743 p.

1307. Encyclopédie des religions. Ed. Frédéric LENOIR et Ysé TARDAN-MASQUELIER. Nouvelle édition revue, augmentée et mise à jour. Paris, Bayard, 2000, XXVI-2513 p.

1308. FITZGERALD (Timothy). The ideology of religious studies. New York a. Oxford, Oxford U. P., 2000, XIV-276 p.

1309. GEERTZ (Armin W.). Global perspectives on methodology in the study of religions. *Method and theory in the study of religion*, 2000, 12, p. 47-73.

1310. GREISCH (Jean). La religion et les religions. *Archives de philosophie*, 2000, 63, p. 229-246.

1311. JOY (Morny). Beyond a god's eyeview. alternative perspectives in the study of religion. *Method and theory in the study of religion*, 2000, 12, p. 110-140.

1312. KLOCZOWSKI (Jerzy). A history of Polish Christianity. New York a. Cambridge U. P., 2000, XXXVIII-385 p.

1313. LEASE (Gary). The definition of religion. An analytical or hermeneutical task? *Method and theory in the study of religion*, 2000, 12, p. 287-293.

1314. MAÎTRE (Jacques). Anorexies religieuses, anorexie mentale. Essai de psychanalyse sociohistorique. De Marie de l'Incarnation à Simone Weil. Paris, Ed. du Cerf, 2000, 200 p.

1315. Nenasilie kak mirovozzrenie i obraz zhizni: istoricheskij rakurs. (Non-violence as Weltanschauung and way of life: historical aspect: [Articles]). Ed.

Tat'jana A. PAVLOVA. RAN, In-t vseobshchej istorii. Moskva, [s. n.], 2000, 300 p. [Eng. summaries]

1316. PADEN (William E.). Sacred order. *Method and theory in the study of religion*, 2000, 12, p. 207-225.

1317. Perspectives on method and theory in the study of religion. Adjunct Proceedings of the XVIIth Congress of the International Association for the History of Religions, Mexico City 1995. Ed. Armin W. GEERTZ and Russell T. MAC CUTCHEON. Leiden, E.J. Brill, 2000, VI-347 p.

1318. POUNDS (N. J. G.). A history of the English parish: the culture of religion from Augustine to Victoria. Cambridge, Cambridge U. P., 2000, XXV-593 p.

1319. Religion in Geschichte und Gegenwart. Hrsg. v. Hans Dieter BETZ, Don S. BROWNING, Berndt JANOWSKI und Eberhard JUNGEL. Band 3. F–H. Tübingen, Mohr, 2000, LXIX p.-1984 col.

1320. SALER (Benson). Conceptualizing religion. responses. *Method and theory in the study of religion*, 2000, 12, p. 353-358.

1321. SEGAL (Robert A.). Conceptualizing religion. *Method and theory in the study of religion*, 2000, 12, p. 187-194.

Cf. n°s 5396-5459

b. Special studies

* 1322. ALTHANN (Robert). Elenchus of Biblica. 1996. Roma, Ed. Istituto Pontificio Biblico, 2000, 690 p.

* 1323. Bulletin de bibliographie biblique. N. 28. Avril 2000, XXVI-118 p. N. 29. Juillet 2000, VI-172 p. N. 30. Déc. 2000, VI-153 p. [n°s 25-27. Cf. Bibl. 99, n° 1322.]

* 1324. Index Islamicus 1998. [1997. Cf. Bibl. 99, n° 1323.] Ed. by G. J. ROPER and C. H. BLEANEY. London, Melbourne, München a. New Providence, Bowker-Saur, 2000, 842 p.

* 1325. Internationale Zeitschriftenschau für Bibelwissenschaft und Grenzgebiete. International Review of Biblical Studies. Revue internationale des études bibliques. Bd. XLV 1998–1999. [Band XLIV, 1997–1998. Cf. Bibl. 99, n° 1324.] Düsseldorf, Patmos Verlag, 2000, XV-298 p.

* 1326. New Testament Abstracts. Vol. 44, 2000, 687 p. [Vol. 43. Cf. Bibl. 99, n° 1326.]

* 1327. NORTH (Robert). Elenchus of biblical bibliography. Vol. 12. 1996. [Vol. 11, 2, 1995. Cf. Bibl. 99, n° 1327.] Roma, Istituto Pontificio Biblico, 2000, [s. p.].

* 1328. Old Testament Abstracts. Vol. 23, 2000, 662 p. [Vol. 22. Cf. Bibl. 99, n° 1328.]

1329. ADAMS (Walter Randolph), SALAMONE (Frank). Anthropology and theology: gods, icons, and god-talk. Lanham, New York a. London, University Press of America, 2000, XVIII-502 p.

1330. ALDERINK (Larry J.). Walter Burkert and a natural theory of religion. *Religion*, 2000, 30, p. 211-227.

1331. Âme et corps: conceptions de la personne. *Archives de sciences sociales des religions*, 2000, 45, 112, p. 5-47.

1332. ANTES (Peter). Fundamentalism: a western term with consequences. *Method and theory in the study of religion*, 2000, 12, p. 260-266.

1333. ANTTONEN (Veikk). What is that we call "Religion"? Analyzing the epistemological status of the sacred as a scholarly category in comparative religion. *Method and theory in the study of religion*, 2000, 12, p. 195-206.

1334. Apocalyptic Time. Ed. Albert I. BAUMGARTEN. Leiden, Brill, 2000, XV-388 p. (Numen book series: studies in the history of religions, 86).

1335. APPLEBY (R. Scott). The ambivalence of the sacred: religion, violence, and reconciliation. Lanham a. London, Rowman and Littlefield, 2000, XIII-429 p.

1336. BEDUHN (Jason David). The historical assessment of speech acts: classifications of Austin and Skinne for the study of religions. *Method and theory in the study of religion*, 2000, 12, p. 477-505.

1337. BENAVIDES (Gustavo). Towards a natural history of religion. *Religion*, 2000, 30, p. 229-244.

1338. BERNER (Ulrich). Reflections upon the concept of "New Religious Movement". *Method and theory in the study of religion*, 2000, 12, p. 267-276.

1339. BOWIE (Fiona). The anthropology of religion. An introduction. Oxford a. Malden, Blackwell, 2000, XIII-284 p.

1340. BRENNER (Louis). Histories of religion in Africa. *Journal of religion in Africa*, 2000, 30, p. 143-167.

1341. BUELL (Denise Kimber). Ethnicity and religion in Mediterranean antiquity and beyond. *Religious studies review*, 2000, 26, p. 243-249.

1342. CARROLL (Brett E.). The Routledge historical atlas of religion in America. New York a. London, Routledge, 2000, 144 p.

1343. Catholicisme hier aujourd'hui demain. Encyclopédie publiée sous le patronage de l'Institut Catholique de Lille par G. MATHON et G. H. BAUDRY. Fasc. 74. Vocation–Zwingli. Paris, Letouzey & Ané, 2000, col. 1281-1572.

1344. CHAUVIN (Charles). Renan 1823–1892. Paris, DDB, 2000, 158 p.

1345. CHIDESTER (David). Material terms for the study of religion. *Journal of the American Academy of Religion*, 2000, 68, p. 367-380.

1346. CIURTIN (Eugen). La première revue d'histoire des religions en Roumanie: Zalmoxis (1938–1942) sous la direction de Mircea Eliade. *Archaeus*, 2000, 4, p. 327-364.

1347. CROCKETT (Clayton). On sublimation: the significance of psychoanalysis for the study of religion. *Journal of the American academy of religion*, 2000, 68, p. 837-855.

1348. DEMPSEY (Corinne). Religion and representation in recent ethnographies. *Religious studies review*, 2000, 26, p. 21-28.

1349. Dictionnaire d'histoire du christianisme. Paris, Albin Michel, 2000, 1174 p.

1350. Dictionnaire d'histoire et de géographie ecclésiastiques. Éd. Roger AUBERT. Tome 28. Fasc. 162. Jonopolis–Joseph-Marie. Paris, Letouzey et Ané, 2000, col. 1-256.

1351. Encyclopédie de l'Islam. Ed. P. J. BEARMAN, T. BIANQUIS, C. E. BOSWORTH, E. VAN DONZEL et W. P. HEINRICHS. Volume X. 171-172. Terengganu–Tīmūrides. 173-174. Tīmūrides–Tunbūr. Leiden, Brill, 2000, p. 449-569; p. 561-572.

1352. Esploratori del pensiero umano: Georges Dumézil e Mircea Eliade. A cura di Julien RIES e Natale SPINETO. Milano, Jaca Book, 2000, 432 p.

1353. Expériences et écriture mystiques dans les religions du livre. Actes du colloque international tenu par le Centre d'études juives, Université de Paris IC-Sorbonne, 1994. Ed. par Paul B. FENTON et Roland GOETSCHEL. Leiden, Boston et Köln, Brill, 2000, VIII-245 p. (Etudes sur le Judaïsme Médiéval, 22).

1354. Formes religieuses caractéristiques de l'ultramodernité: France, Pays Bas, États Unis, Japon, analyses globales. *Archives de sciences sociales des religions*, 2000, 45, 109, p. 5-116.

1355. Formes sensibles (Des) de la religion. *Archives de sciences sociales des religions*, 2000, 45, 111, p. 5-133.

1356. GARCÍA BAZÁN (Francisco). Aspectos inusuales de lo sagrado. Madrid, Trotta, 2000, 251 p. (Paradigmas, 26).

1357. GARDAZ (Michel). The age of discoveries and patriotism. James Darmesteter's assessment of French orientalism. *Religion*, 2000, 30, p. 353-365.

1358. GATTO TROCCHI (Cecilia). I nuovi movimenti religiosi. Brescia, Queriniana, 2000, 184 p. (Piccola biblioteca delle religioni, 19).

1359. GAUER (Werner). Muttergöttin und Vatergott. *Archiv für Religionsgeschichte*, 2000, 2, p. 255-282.

1360. GEERTZ (Armin W.), MAC CUTCHEON (Russell T.). The role of method and theory in IAHR. *Method and theory in the study of religion*, 2000, 12, p. 3-37.

1361. Gender / bodies / religions: adjunct proceedings of the XVIIth Congress of the International Association for the History of Religions. Edited by Sylvia MARCOS. Cuernavaca, ALER Publications, 2000, 363 p.

1362. GONZÁLEZ TORRE (Yolotl). The history of religion and the study of religions in Mexico. *Method and theory in the study of religion*, 2000, 12, p. 38-48.

1363. HELVE (Helena). The formation of gendered world views and gender ideology. *Method and theory in the study of religion*, 2000, 12, p. 245-259.

1364. Histoire du christianisme (Des origines à nos jours). Sous la direction de Jean-Marie MAYEUR, Charles PIETRI, Luce PIETRI, André VAUCHEZ et Marc VENARD. XIII. Crises et renouveau, de 1958 à nos jours. Sous la responsabilité de Jean-Marie MAYEUR. XIV. Anamnèsis. Sous la direction de François Laplanche. Paris, Desclée, 2000, 2 vol., 794 p.; 750 p.

1365. KALBERG (Stephen). Ideen und Interessen. Max Weber über den Ursprung ausserweltlicher Erlösungsreligionen. *Zeitschrift für Religionswissenschaft*, 2000, 8, p. 45-70.

1366. KIPPENBERG (Hans G.). Religious history, displaced by modernity. *Numen*, 2000, 47, p. 221-243.

1367. KRECH (Volkhard). From historicism to functionalism. The rise of scientific approaches in religions around 1900 and their socio-cultural context. *Numen*, 2000, 47, p. 244-265.

1368. LAWSON (E. Thomas). Towards a cognitive science of religion. *Numen*, 2000, 47, p. 338-349.

1369. LEFEBURE (Leo D.). Revelation, the religions, and violence. Maryknoll, Orbis, 2000, XVII-244 p.

1370. Lexikon für Theologie und Kirche. Neunter Bd. San–Thomas. Hrsg. v. Walter KASPER [et al.]. Freiburg, Basel, Roma u. Wien, Herder, 2000, 1538 col.

1371. LIGHT (Timothy). Orthosyncretism. An account of melding in religion. *Method and theory in the study of religion*, 2000, 12, p. 162-186.

1372. LINCOLN (Bruce). On ritual, change, and marked categories. *Journal of the American Academy of Religion*, 2000, 68, p. 487-510.

1373. MAC CAULEY (Robert N.). Overcoming barries to a cognitive psychology of religion. *Method and theory in the study of religion*, 2000, 12, p. 141-161.

1374. MAC CUTCHEON (Russell T.). Taming ethnocentrism and trans-cultural understandings. *Method and theory in the study of religion*, 2000, 12, p. 294-306.

1375. MACRAE (Clare). Myth, liminarity and a rational universe? *Studies in world Christianity*, 2000, 6, p. 245-259.

1376. Man, meaning, and mystery: 100 years of history of religions in Norway. The heritage of W. Brede Kristenses. Ed. Sigurd HJELDE. Leiden, Brill, 2000, XXII-301 p. (Numen Book Series: Studies in the History of Religions, 87).

1377. Miti e riti della preistoria. Un secolo di studi sull'origine del senso del sacro. Ed. Fiorenzo FACCHINI e Paolo MAGNANI. Milano, Jaca Book, 2000, 397 p.

1378. MÖDE (Erwin). Tabu in postmodern Zeit. *Archiv für Religionspsychologie*, 2000, 23, p. 220-230.

1379. MOLENDIJK (Arie L.). The principal religions. A landmark in the early study of religion in the Netherlands. *Nederlands Theologisch Tijdschrift*, 2000, 54, p. 18-34.

1380. MORGAN (David). Visual religion. *Religion*, 2000, 30, p. 41-53.

1381. NEZHAD (Saeid Edalat). The role of religions in minimizing the word's crises. *Studies in interreligious dialogue*, 2000, 10, p. 117-122.

1382. NYE (Malory). Religion, post-religionism, and religioning. Religious studies and contemporary cultural debates. *Method and theory in the study of religion*, 2000, 12, p. 447-476.

1383. Orient (L') dans l'histoire religieuse de l'Europe: l'invention des origines. Ed. par Mohammad Ali AMIR-MOEZZI et John SCHEID. Turnhout, Brepols, 2000, XI-234 p. (Bibliothèque de l'École des Hautes Études. Section V. Sciences religieuses, 110).

1384. PACE (Enzo), STEFANI (Piero). Il fondamentalismo religioso contemporaneo. Brescia, Queriniana, 2000, 214 p. (Piccola viblioteca delle religioni, 21).

1385. PADEN (William E.). Prototypes. Western or cross-cultural? *Method and theory in the study of religion*, 2000, 12, p. 307-313.

1386. PETERS (Frederic H.). Neurophenomenology. *Method and theory in the study of religion*, 2000, 12, p. 379-415.

1387. PINXTEN (Rik). Goddelijke fantasie. Over religie, leren en identiteit. Antwerpen, Houtekiet, 2000, 288 p.

1388. Reallexikon für Antike und Christentum. Hrsg. v. Ernst DASSMANN. 151. Iustinus II–Kain und Abel. 152. Kain und Abel–Kaiserpriester. Stuttgart, Hiersemann, 2000, col. 801-960. 961-1119.

1389. Religions et violences. Sources et interactions: Symposium. Ed. Anand NAYAK. Fribourg, Éditions universitaires, 2000, VII-364 p.

1390. RENNIE (Bryan S.). A response to Carl Olson's "Mircea Eliade, postmodernism, and the problematic nature of representational thinking". *Method and theory in the study of religion*, 2000, 12, p. 416-421.

1391. RIESEBRODT (martin). Fundamentalism and the resurgence of religion. *Numen*, 2000, 47, p. 266-287.

1392. RIFFLET (Jacques). Le mondes du sacré. Étude comparée des voies du sacré en Occident et en Orient. Bierges, Éditions Mols, 2000, 797 p.

1393. RITTER (Werner H.). Lust am Okkulten? Zwischen Verdrängung und Wiederkehr. Kulturphilosophische, theologische und pädagogische Anmerkungen. *Wege zum Menschen*, 2000, 22, p. 38-45.

1394. SCHILBRACK (Kevin). Myth and metaphisics. *International journal for phylosophy of religion*, 2000, 48, p. 11-43.

1395. SCHMIDTCHEN (Dietrich). Ökonomik der Religion. *Zeitschrift für Religionswissenschaft*, 2000, 8, p. 11-43.

1396. SIRONNEAU (Jean-Pierre). Métamorphoses du mythe et de la croyance. Paris, L'Harmattan, 2000, 288 p.

1397. SMITH (Jonathan Z.). Acknowledgements, morphology and history in Mircea Eliade's patterns in comparative religion (1949–1999). Part I. The work and its contexts. *History of religions*, 99-2000, 39, p. 315-351.

1398. STUDSTILL (Randall). Eliade, phenomenology, and the sacred. *Religious studies*, 2000, 36, p. 177-194.

1399. SULLIVAN (Lawrence E.). Values and risks of international collaborations on the study of religion. *Bulletin of the Nanzan Institute for religion and culture*, 2000, 24, p. 48-54.

1400. TANZARELLA (Sergio). René Girard e il dibattito sulla teoria mimetica. *Rassegna di teologia*, 2000, 41, p. 87-99.

1401. Theologische Realenzyclopädie. 31. Seelenwanderung Sprache / Sprachwissenschaft / Sprachphilosophie. Hrsg. v. G. MÜLLER. Berlin u. New York, de Gruyter, 2000, VI-823 p.

1402. THOMAS (Terence). Political motivations in the development of academic study of religions in Britain. *Method and theory in the study of religion*, 2000, 12, p. 74-90.

1403. TSUCHIYA (Hiroshi). "Religious Studies" in Japan and future prospects. *Bulletin of the Nanzan Institute for religion and culture*, 2000, 24, p. 8-21.

1404. URBAN (Hugh B.). Making a place to take a stand. Jonathan Z. Smith and the politics and poetics of comparison. *Method and theory in the study of religion*, 2000, 12, p. 339-378.

1405. WASSERSTROM (Steven M.). Sense and senselesness in religion. Reflections on creation of the sacred. *Religion*, 2000, 30, p. 273-281.

1406. Weltreligionen in Weltliteratur. *Religionsunterricht an höheren Schulen*, 2000, 43, p. 347-391.

1407. WIEBE (Don). Problems with the family resemblance approach to conceptualizing religion. *Method and theory in the study of religion*, 2000, 12, p. 314-322.

1408. WUNN (Ina). Beginning of religion. *Numen*, 2000, 47, p. 417-452.

Cf. n° 1282

§ 12. History of philosophy.

* 1409. Bibliografia filosofica italiana, 1998. [1997. Cf. Bibl. 99, n° 1390.] A cura di C. SCALABRIN. Firenze, L.S. Olschki, 2000, 260 p. (Biblioteca di bibliografia italiana, 160).

* 1410. Bibliography of philosophy = Bibliographie de la philosophie: a quarterly bulletin. Vol. 47, 2000. Fasc. 1–4. [Vol. 46, 1999. Cf. Bibl. 99, ° 1391.] Paris, Vrin, 2000, [s. p.].

* 1411. International philosophical bibliography = Répertoire bibliographique de la philosophie. Vol. 52, 2000. [Vol. 51, 1999. Cf. Bibl. 99, n° 1394.] Louvain, Ed. de l'Institut Supérieur de Philosophie, 2000, [s. p.].

1412. BAILLIE (James). Routledge philosophy guidebook to Hume on morality. London, Routledge, 2000, IX-226 p.

1413. Blackwell guide (The) to the modern philosophers: from Descartes to Nietzsche. Ed. by Steven M. EMMANUEL. Oxford a. Malden, Blackwell, 2000, 423 p. (Blackwell philosophy guides).

1414. Dictionary (The) of seventeenth-century British philosophers. Ed. By Andrew PYLE. Bristol, Thoemmes, 2000, 2 vol., XXI-932 p.

1415. Dictionnaire des philosophes antiques III, d'Eccélos à Juvénal. Éd. par R. GOULET. Paris, CNRS Éditions, 2000, 1071 p.

1416. GILJE (Nils), SKIRBEKK (Gunnar). History of Western thought: from Ancient Greece to the twentieth century. London, Routledge, 2000, 624 p.

1417. GOTTLIEB (Anthony). The dream of reason: a history of Western philosophy from the Greeks to the Renaissance. London, Allen Lane, 2000, IX-468 p. (ill.)

1418. Hume: general philosophy. Ed. by David W. D. OWEN. Aldershot, Ashgate, 2000, XIX-516 p. (ill.). (International library of critical essays in the history of philosophy).

1419. KAMPITS (Peter). Fra apparenza e realtà: breve storia della filosofia austriaca. Milano, Franco Angeli, 2000, 176 p. (Filosofia, 123).

1420. KORBEN (A.). Istorija na isljamskata filosofija. (Histoire de la philosophie de l'Islam). Sofiia, [s. n.], 2000, 519 p.

1421. Lithuanian philosophy: persons and ideas. Ed. by Jurate BARANOVA. Washington, Council for Research in Values and Philosophy, 2000, XII-314 p. (Cultural heritage and contemporary change. Series IVA, Eastern and Central Europe, 17. Lithuanian philosophical studies, 2).

1422. MAC CARNEY (Joseph). Routledge philosophy guidebook to Hegel on history. London, Routledge, 2000, XI-234 p. (Routledge philosophy guidebooks).

1423. Marx. Ed. by Scott MEIKLE. Aldershot, Dartmouth, 2000, XX-488 p. (International library of critical essays in the history of philosophy).

1424. MEJER (Jørgen). Überlieferung der Philosophie im Altertum: eine Einführung. København, Det Kongelige Danske videnskabernes selskab, 2000, 206 p. (Historisk-filosofiske meddelelser, 80)

1425. Nietzsche-Handbuch: Leben, Werk, Wirkung. Hrsg. v. Henning OTTMANN. Stuttgart, Metzler, 2000, 561 p.

1426. Philosophen des Mittelalters: eine Einführung. Hrsg. v. Th. KOBUSCH. Darmstadt, Wissenschaftliche Buchgesellschaft, 2000, VI-282 p.

1427. RAWLS (John). Lectures on the history of moral philosophy. Cambridge a. London, Harvard U. P., 2000, XXII-384 p.

1428. SOLARO (Giuseppe). «Uomo, perché accendi la lampada?» Sulle notti laboriose dei filosofi. *Quaderni di storia*, 2000, 26, 52, p. 227-244.

Cf. n° 1315, 6114-6209

§ 13. History of literature.

* 1429. Bibliografía en resúmenes de la literatura española.1996. Ed. por Emilio MARTÍNEZ MATA y Manuel Angel FERNÁNDEZ ALVAREZ. Oviedo, Universidad de Oviedo. Servicio de Publicaciones, 2000, 220 p.

* 1430. Bibliographie der deutschen Sprach- und Literaturwissenschaft. Band XXXIX. 1999. [Band XXXVIII. 1998. Cf. Bibl. 99, n° 1411.] Bearb. v. Doris MAREK und Susanne PRÖGER. Frankfurt am Main, Vittorio Klostermann, 2000, XXXIV-1050 p.

* 1431. Bibliographie der französischen Literaturwissenschaft. Band XXXVII. 1999. [Band XXXVI. 1998. Cf. Bibl. 99, n° 1412.] Bearb. v. Astrid KLAPP-LEHRMANN. Frankfurt am Main, Vittorio Klostermann, 2000, 1060 p.

* 1432. BIGLI. Bibliografia generale della lingua e della letteratura italiana. Vol. 8. 1998. Tomi 1–2. [Vol. 7, 1997. Tomi 1–2. Cf. Bibl. 99, n° 1413.] Diretta da Enrico MALATO. Roma, Salerno, 2000, 2 vol., [s. p.]. (Pubblicazioni del "Centro Pio Rajna").

* 1433. KOSCH (Wilhelm). Deutsches Literatur-Lexikon: das 20. Jahrhundert: biographisches und bibliographisches Handbuch. München, K.G. Saur, 2000, [s. p.].

* 1434. SEIFERT (Siegfried). Goethe-Bibliographie 1950–1990. München, Saur, 2000, 3 vol., XXIII-1565 p.

1435. ALLEN (Roger M. A.). An introduction to Arabic literature. Cambridge, Cambridge U. P., 2000, XXI-263 p.

1436. BENOIT-DUSAUSOY (Annick). History of European literature. London, Routledge, 2000, XXVIII-731 p.

1437. BOLBECHER (Siglinde). Lexikon der österreichischen Exilliteratur. Wien, Deuticke, 2000, 763 p. (ill.).

1438. Cambridge (The) companion to medieval romance. Ed. by Roberta L. KRUEGER. Cambridge, Cambridge U. P., 2000, XIX-290 p. (ill.).

1439. CANIM (Rıdvan). Latîfî tezkiretü'ş-şu'arâ ve tabsıratü'n-nuzamâ: inceleme – metin. (Latifi's tezkiretü'ş-şu'ara ve tabsıratü'n-nuzma [Latifi's biographies of poets]: study and the text). Ankara, Atatürk Kültür Merkezi, 2000, 903 p.

1440. Companion (A) to Chaucer. Ed. by Peter BROWN. London a. Malden, Blackwell, 2000, XVII-515 p. (ill.).

1441. Concise (The) Oxford companion to Irish literature. Ed. by Robert WELCH. Oxford, Oxford U. P., 2000, XXI-392 p.

1442. COX (Philip). Reading adaptation: novels and verse narratives on the stage, 1790–1840. Manchester, Manchester U. P., 2000, VII-184 p.

1443. Dante (The) Encyclopedia. Ed. by Richard LANSING. New York a. London, Garland, 2000, XXVII-1006 p. (ill., maps, tables). (Garland Reference Library of the Humanities, 1836).

1444. Dictionary of early Christian literature. Ed. by Siegmar DÖPP and Wilhelm GEERLINGS. New York, Crossroad Publishing Co., 2000, XVI-621 p.

1445. Dictionnaire de la littérature française. Paris, Albin Michel et Encyclopaedia Universalis, 2000, 894 p.

1446. Encyclopaedia of medieval literature. Ed. by Robert Thomas LAMBDIN and Laura Cooner LAMBDIN. Westport a. London, Greenwood Press, 2000, X-549 p.

1447. Encyclopedia of German literature. Ed. by Matthias KONZETT. Chicago a. London, Fitzroy Dearborn Publishers, 2000, 2 vol., [s. p.].

1448. Historia de la literatura portuguesa. Ed. por José Luis GAVILANES y António APOLINÁRIO. Madrid, Cátedra, 2000, 718 p. (ill., facs., port.). (Crítica y estudios literarios).

1449. Historia de la literatura vasca. Ed. pot Patricio URQUIZU SARASUA. Madrid, Universidad Nacional de Educación a Distancia, 2000, 685 p. (ill., facs., ports.) (Aula abierta / Universidad Nacional de Educación a Distancia, 36140).

1450. Identité littéraire de l'Europe. Ed. par Marc FUMAROLI [et al.]. Paris, Presses universitaires de France, 2000, 224 p. (ill.). (Perspectives littéraires).

1451. Interkulturelle Literatur in Deutschland: ein Handbuch. Hrsg. v. Carmine CHIELLINO. Stuttgart, Metzler, 2000, X-536 p. (ill., port.).

1452. Interpretation and allegory: antiquity to the modern period. Ed. by Jon WHITMAN. Leiden, Boston a. Köln, Brill, 2000, XV-513 p.

1453. JEWERS (Caroline A.). Chivalric fiction and the history of the novel. Gainesville, University Press of Florida, 2000, XII-206 p. (ill., map).

1454. KURNAZ (Cemâl). Osmanlı dönemi Kırım edebiyatı. (Crimean literature in Ottoman period). Ankara, Kültür Bakanlığı Yayımlar Dairesi Başkanlığı, 2000, 413 p.

1455. LINK (Eugene Perry). The uses of literature: life in the socialist Chinese literary system. Princeton, Princeton U. P., 2000, VI-387 p.

1456. Literary appropriations of the Anglo-Saxons from the thirteenth to the twentieth century. Ed. by Donald SCRAGG and Carole WEINBERG. Cambridge, Cambridge U. P., 2000, XII-242 p. (table).

1457. Littérature et religion. Ed. par Gérard FERREYROLLES. Paris, Champion, 2000, 379 p. (Littératures classiques. [Special issue], 39).

1458. MAC CONNELL (Louise). Dictionary of Shakespeare. Teddington, Peter Collin, 2000, 315 p.

1459. Nordisk kvinnolitteraturhistoria. Bd. 5. (Histoire de la littérature féminine scandinave. Tome 5). Red. Elisabeth MØLLER JENSEN. Höganäs, Bra Böcker, 2000, 396 p. (ill.).

1460. Ordbog over det norrene prosasprog/A Dictionary of old norse prose. ONP 1–2. Nøgle/Key. 2. Ban–da. Ed. by James E. KNIRK. København, Arnamagnæaske Kommission, 2000, 2 vol., 190 p., VIII-1241 p.

1461. ORIGAS (Jean-Jacques). Dictionnaire de littérature japonaise. Paris, Presses universitaires de France, 2000, XIII-366 p. (Quadrige (Presses universitaires de France). Référence).

1462. PEDRAZA JIMÉNEZ (Felipe B.), RODRÍGUEZ CÁCERES (Milagros). Manual de literatura española. Tafalla, Cenlit, 2000, 1075 p.

1463. PLOTZ (John). The crowd: British literature and public politics. Berkeley a. London, University of California Press, 2000, XII-263 p.

1464. RATH (Wolfgang). Die Novelle: Konzept und Geschichte. Göttingen, Vandenhoeck & Ruprecht, 2000, 366 p.

1465. RIEDEL (Volker). Antikerezeption in der deutschen Literatur vom Renaissance-Humanismus bis zur Gegenwart: eine Einführung. Stuttgart, Metzler, 2000, VIII-515 p.

1466. ROBERTSON (D.). Encyclopaedic dictionary of Shakespeare's works. Delhi, Ivy, 2000, 378 p.

1467. RUBIN (Rachel). Jewish gangsters of modern literature. Urbana, University of Illinois Press, 2000, XIII-189 p.

1468. Russian literature and its demons. Ed. by Pamela DAVIDSON. New York a. Oxford, Berghahn, 2000, XIV-530 p. (ill.). (Studies in Slavic literature, culture, and society, 8).

1469. SANDIFORD (Keith A.). The cultural politics of sugar: Caribbean slavery and narratives of colonialism. Cambridge, Cambridge U. P., 2000, 222 p.

1470. SARKOWICZ (Hans), MENTZER (Alf). Literatur in Nazi-Deutschland: ein biographisches Lexikon. Hamburg, Europa Verlag, 2000, 381 p. (ill., port.).

1471. SCHMIDT (Peter Lebrecht). Traditio latinitatis: Studien zur Rezeption und Überlieferung der lateinischen Literatur. Stuttgart, F. Steiner, 2000, 378 p. (ill.).

1472. SCHÖNLE (Andreas). Authenticity and fiction in the Russian literary journey, 1790–1840. Cambridge, Harvard U. P., 2000, VIII-296 p.

1473. Tradition, modernity, and postmodernity in Arabic literature: essays in honor of Professor Issa J. Boullata. Ed. by Kamal ABDEL-MALEK and Wael HALLAQ. Leiden, Brill, 2000, XV-414 p. (ill.).

1474. Victorian Gothic: literary and cultural manifestations in the nineteenth century. Ed. by Ruth ROBBINS and Julian WOLFREYS. Houndmills, Basingstoke, 2000, XX-260 p.

1475. WEBBY (Elizabeth). The Cambridge companion to Australian literature. Cambridge, Cambridge U. P., 2000, XXI-326 p.

1476. WHITE (Luise). Speaking with vampires: rumor and history in colonial Africa. Berkeley a. London, University of California Press, 2000, XVI-352 p.

Cf. n°s 1315, 6396

C

PREHISTORY

§ 1. General. 1477-1499. – § 2. Palaeolithic and Mesolithic. 1500-1540. – § 3. Neolithic. 1541-1564. – § 4. Bronze age. 1565-1589. – § 5. Iron age. 1590-1613.

§ 1. General.

1477. ANDREWS (Gill), BARRETT (John C.), LEWIS (John S. C.). Interpretation not record: the practice of archaeology. *Antiquity*, 2000, 74, p. 525-530.

1478. Arqueologia peninsular. História, teoria e prática. Actas do 3° Congresso de Arqueologia peninsular. Ed. por Jorge OLIVEIRA. Porto, Associaçao para o Desenvolvimento de Cooperaçao em Arqueologia peninsular, 2000, 547 p.

1479. BEAMISH (Matt), RIPPER (Susan). Burnt mounds in the East Midlands. *Antiquity*, 2000, 74, p. 37-38.

1480. BELINSKIJ (Andrej B.), KALMYKOV (Alexej A.), KORENEVSKIJ (Sergej N.), HÄRKE (Heinrich). The Ipatovo kurgan on the North Caucasian Steppe (Russia). *Antiquity*, 2000, 74, p. 73-74.

1481. BLACKMAN (D. J.). Is maritime archaeology on course? *American journal of archaeology*, 2000, 104, p. 591-596.

1482. BLÁZQUEZ (José María). Los pueblos de España y el Mediterráneo en la antigüedad. Madrid, Cátedra, 2000, 727 p. (ills.).

1483. BOTELLA (Miguel C.), ALEMÁN (Inmaculada), JIMÉNEZ (Sylvia A.). Los huesos humanos. Manipulación y alteraciones. Barcelona, Ediciones Bellaterra, 2000, 229 p.

1484. CHERNYI (F. N.), AVILOVA (L. I.), ORLOVSKAIA (L. B.). Metallurgical provinces and radiocarbon chronology. Moskva, Rossiiskaia Akademiia Nauk-Institut Arjeologii, 2000, 95 p.

1485. CRANE (Eva), WALKER (Penelope). Wall recesses for bee hives. *Antiquity*, 2000, 74, p. 805-811.

1486. DOLUKHANOV (Pavel M.). Istoki etnosa. (The sources of ethnos). Sankt-Peterburg, Evropejskij dom, 2000, 220 p. (ill.). [Eng. summary]

1487. DOMANSKI (Marian), WEBB (John A.). Flaking properties, petrology and use of Polish flint. *Antiquity*, 2000, 74, p. 822-823.

1488. ENDERE (María Luz). Patrimonios en disputa: acervos nacionales, investigación arcqueológica y reclamos ètnicos sobre restos humanos. *Trabajos de prehistoria*, 2000, 57, 1, p. 5-17.

1489. ETHELBERG (Per). Den sønderjyske landbrugshistorie: sten og bronzealder. (Histoire de l'agriculture du Jutland méridional: néolithique et âge du bronze). Haderslev, Haderslev Museum, 2000, 305 p. (ill.). (Skrifter udgivne af Historisk Samfund for Sønderjylland, 81).

1490. GONZÁLEZ MÉNDEZ (Matilde). Memoria, historia y patrimonio: hacia una concepción social del patrimonio. *Trabajos de prehistoria*, 2000, 57, 2, p. 9-20.

1491. HALL (David), PALMER (Rog). Ridge and furrow survival and preservation. *Antiquity*, 2000, 74, p. 29-30.

1492. HAOUR (Anne). The former Kano? Ethnoarchaeology of Kufan Kanawa, Niger. *Antiquity*, 2000, 74, p. 767-768.

1493. HIGHAM (Charles), THOSARAT (Rachanie). The origins of the civilization of Angkor. *Antiquity*, 2000, 74, p. 27-28.

1494. KIR'JAK (Dikova, Margarita A.). Drevnee iskusstvo severa Dal'nego Vostoka kak istoricheskij istochnik: Kamennyj vek. (The ancient art of the north of the far east [of Russia] as a historical source: the Stone Age). Magadan, [s. n.], 2000, 290 p. (ill.). [Eng. summary]

1495. LIPHSCHITZ (N.), BONANI (G.). Dimension of olive (Olea Europaea) stones as a reliable parameter to distinguish between wild and cultivated varieties: further evidence. *Tel Aviv*, 2000, 27, p. 23-25.

1496. Osteologisk materiale som historisk kilde. (Osteological material as historical records). Red. Audun DYBDAHL. Trondheim, Tapir, 2000, 253 p. (ill.). (Skrifter / Senter for middelalderstudier, 11).

1497. POPE (Gregory A.). Weathering of petroglyphs: direct assessment and implications for dating methods. *Antiquity*, 2000, 74, p. 833-843.

1498. SKEATES (R.). Debating the archaeological heritage. London, Duckworth, 2000, 160 p.

1499. WAINWRIGHT (Geoffrey). Time please. *Antiquity*, 2000, 74, p. 909-943.

Cf. n°s 393, 437, 792, 1025, 6453

§ 2. Palaeolithic and Mesolithic.

1500. AMBRUSTER (Barbara), PEREA (Alicia). Macizo/Hueco, soldado/fundido, morfología/tecnología. El ámbito tecnológico castreño a través de los torques con remates en doble escocia. *Trabajos de prehistoria*, 2000, 57, 1, p. 97-114.

1501. AMIRKHANOV (Khizri A.). Zarajskaja stojanka: rezul'taty kompleksnogo issledovanija 1995-1998 gg. (The Zaraisk site: results of a complex research in 1995-1998). Moskva, Nauchnyj mir, 2000, 248 p. (ill.).

1502. ANDERSON (D. G.), FAUGHT (M. K.). Palaeoindian artefact distributions: evidence and implications. *Antiquity*, 2000, 74, p. 507-513.

1503. BAHN (Paul G.). New rock-art find in Portugal. *Antiquity*, 2000, 74, p. 753-754.

1504. BIGLARI (Fereydoun), NOKANDEH (Gabriel), HEYDARI (Saman). A recent find of a possible Lower Palaeolithic assemblage from the foothills of the Zagros Mountains. *Antiquity*, 2000, 74, p. 749-740.

1505. BLASCO (Mònica), BORRELL (Mònica), BOSCH (Joseph). Las minas prehistóricas de Gavá (Barcelona): un ejemplo de estudio, conservación y presentación pública de un yacimiento arqueológico. *Trabajos de prehistoria*, 2000, 57, 2, p. 77-87.

1506. BLOCKLEY (S. P. E.), DONAHUE (R. E.), POLLARD (A. M.). Radiocarbon calibration and Late Glacial occupation in northwest Europe with reply by R. A. Housley, C. S. Gamble and P. Pettitt. *Antiquity*, 2000, 74, p. 112-121. – IIDEM. Rapid human response to Late Glacial climate change: a reply to Housley [et al.] (2000). *Antiquity*, 2000, 74, p. 427-428.

1507. BOCQUET-APPEL (Jean-Pierre), DEMARS (Pierre Yves). Neanderthal contraction and modern human colonization of Europe. *Antiquity*, 2000, 74, p. 544-562.

1508. BOSCHIAN (G.). Grotta Patrizi. Analisi petrografica di frammenti ceramici in sezione sottile. *Bullettino di paletnologia italiana*, 2000, 91, p. 111-117.

1509. CHAPMAN (Henry P.), GEAREY (Benjamin R.). Palaeoecology and the perception of prehistoric landscapes: some comments on visual approaches to phenomenology. *Antiquity*, 2000, 74, p. 316-319.

1510. CHAZAN (Michael). Flake production at the Lower Palaeolithic site of Holon (Israel): implications for the origin of the Levallois method. *Antiquity*, 2000, 74, p. 495-499.

1511. DARK (Petra). Revised 'absolute' dating of the early Mesolithic site of Star Carr, North Yorkshire, in the light of changes in the early Holocene tree-ring chronology. *Antiquity*, 2000, 74, p. 304-307.

1512. DEL LUCCHESE (A.), MARTINI (S.), NEGRINO (F.). OTTOMANO (C.). "I Ciotti" (Mortola Superiore, Ventimiglia, Imperia). Una località di approvvigionamento della materia prima per la scheggiatura durante il Paleolitico. *Bullettino di paletnologia italiana*, 2000, 91, p. 1-26.

1513. DELIBES DE CASTRO (Germán). Itinerario arqueológico de los dolmenes de Sedano (Burgos). *Trabajos de prehistoria*, 2000, 57, 2, p. 89-103.

1514. FERREIRA BICHO (Nino), HOCKETT (Bryan), HAWS (Jonathan), BELCHER (William). Hunter-gatherer subsistence at the end of the Pleistocene: preliminary results from Picareiro Cave, Central Portugal. *Antiquity*, 2000, 74, p. 500-506.

1515. GRIFONI CREMONESI (R.), RADMILLI (A. M.). La grotta Patrizi al Sasso di Furbara (Cerveteri, Roma). *Bullettino di paletnologia italiana*, 2000, 91, p. 63-110.

1516. HACHEM (Lamys). New observations on the Bandkeramik House and social organization. *Antiquity*, 2000, 74, p. 308-312.

1517. HORNOS MATA (Francisca), ZAFRA DE LA TORRE (Narciso), CASTRO LÓPEZ (Marcelo). Perspectivas, itinerarios e intersecciones: experiencias y propuestas de apropiación cultural de Marroquíes Bajos (Jaén). *Trabajos de prehistoria*, 2000, 57, 2, p. 105-118.

1518. HORROCKS (M.), JONES (M. D.), CARTER (J. A.), SUTTON (D. G.). Pollen and phytoliths in stone mounds at Pouerua, Northland, New Zealand: implications for the study of Polynesian farming. *Antiquity*, 2000, 74, p. 863-872.

1519. IRWIN (Arthur). The hooked stick in the Lascaux shaft scene. *Antiquity*, 2000, 74, p. 293-298.

1520. JACOBI (R. M.), PETTITT (P. B.). An Aurignacian point from Uphill Quarry (Somerset) and the earliest settlement of Britain by Homo sapiens sapiens. *Antiquity*, 2000, 74, p. 513-518.

1521. KARAVANIC (Ivor). Research on the Middle Palaeolithic in Dalmatia, Croatia. *Antiquity*, 2000, 74, p. 777-778.

1522. LESPEZ (Laurent), DALONGEVILLE (Rémi), NOIREL-SCHUTZ (Claudine), SUC (Jean-Pierre), KOUKOLI-CHRYSSANTHAKI (Haïdo), TREUIL (René). Les paléoenvironnements du site préhistorique de Dikili Tash (Macédoine orientale, Grèce). *Bulletin de correspondance hellénique*, 2000, 124, p. 413-434.

1523. LISITSYN (Nikolaj F.). Pozdnij paleolit Chulymo-Enisejskogo mezhdurech'ja. (The upper Paleolithic of Chulim-Yenisey region). RAN, In-t istorii material'noj kul'tury. Sankt-Peterburg, [s. n.], 2000, 232 p. (ill.). (Trudy Instituta istorii material'noj kul'tury, 3). [Eng. summary]

1524. LJUBIN (V. P.), GEDE (F. J.). Paleolit Respubliki Kot D'Ivuar: Zapadnaja Afrika. (The Paleolithic of the Republic of Côte d'Ivoire: West Africa). RAN, In-t istorii material'noj kul'tury. Sankt-Peterburg, [s. n.], 2000, 160 p. (ill.). (Trudy Instituta istorii material'noj kul'tury, 3). [Eng. summary]

1525. MAC NIVEN (Ian J.), DAVID (Bruno), BRAYER (John). Digital enhancement of Torres Strait rock-art. *Antiquity*, 2000, 74, p. 759-760.

1526. MELTZER (David J.). Renewed investigations at the Folsom Palaeoindian type site. *Antiquity*, 2000, 74, p. 35-36.

1527. MILLER-ANTONIO (S.), SCHEPARTZ (L. A.), BAKKEN (D.). Raw material utilization and new evidence for non-lithic technology at Dadong Cave, Southern China. *Antiquity*, 2000, 74, p. 372-379.

1528. MITHEN (Steven), FINLAYSON (Bill). WF16, a new PPNA site in Southern Jordan. *Antiquity*, 2000, 74, p. 11-12.

1529. MITHEN (Steven). Mesolithic sedentism on Oronsay: chronological evidence from adjacent islands in the southern Hebrides. *Antiquity*, 2000, 74, p. 298-304.

1530. MURRAY (Jane). Peau noire, masques blancs: self-image in the Mesolithic-Neolithic transition in Scotland. *Antiquity*, 2000, 74, p. 779-785.

1531. PELTENBURG (Edgar), COLLEDGE (Sue), CROFT (Paul), JACKSON (Adam), MAC CARTNEY (Carol), MURRAY (Mary Anne). Agro-pastoralist colonization of Cyprus in the $10^{l}h$ millennium BC: initial assessments. *Antiquity*, 2000, 74, p. 844-853.

1532. PETTITT (P. B.), BADER (N. O.). Direct AMS Radiocarbon dates for the Sungir mid Upper Palaeolithic burials. *Antiquity*, 2000, 74, p. 269-270.

1533. PIPERNO (M.), PELLEGRINI (E.). Risultati delle ricerche alla grotta del Pino (Sassano, Salerno): 1997/1998. *Bullettino di paletnologia italiana*, 2000, 91, p. 121-206.

1534. SCARRE (Chris), RAUX (Paul). A new decorated menhir. *Antiquity*, 2000, 74, p. 757-758.

1535. SCHOLZ (M.), TRELLISÓ CARREÑO (L.), PUSCH (C. M.). The examination of ancient DNA: guidelines on precautions, controls and sample processing. *Trabajos de prehistoria*, 2000, 57, 1, p. 115-120.

1536. SOFFER (O.), ADOVASIO (J. M.), ILLINGWORTH (J. S.), AMIRKHANOV (H. A.), PRASLOV (N. D.), STREET (M.). Paleolithic perishables made permanent. *Antiquity*, 2000, 74, p. 812-821.

1537. STEADMAN (David W.), ANTÓN (Susan C.), KIRCH (Patrick V.). Ana Manuku: a prehistoric ritualistic site on Mangaia, Cook Islands. *Antiquity*, 2000, 74, p. 873-883.

1538. STRAUS (Lawrence Guy), BICHO (Nino), WINEGARDNER (Ann C.). The Upper Palaeolithic settlement of Iberia: first-generation maps. *Antiquity*, 2000, 74, p. 533-566.

1539. WENBAN-SMITH (Francis F.). Red Barns Palaeolithic site. *Antiquity*, 2000, 74, p. 9-10.

1540. ZILHÂO (Joâo). La puesta en valor del arte rupestre del valle del Côa (Portugal). *Trabajos de prehistoria*, 2000, 57, 2, p. 57-64.

Cf. n° 1494

§ 3. Neolithic.

1541. ALONSO (Natàlia), JUNYENT (Emili), LAFUENTE (Ángel), LÓPEZ (Joan B.), TARTERA (Enric). 'La fortaleza de Arbeca. El proyecto Vilars 2000'. Investigación, recuperación y socialización del conocimiento y del patrimonio. *Trabajos de prehistoria*, 2000, 57, 2, p. 161-173.

1542. BRADLEY (Richard), BALL (Christine), CAMPBELL (Michelle), CROFT (Sharon), PHILLIPS (Tim), TREVARTHEN (David). Tomnaverie stone circle, Aberdeenshire. *Antiquity*, 2000, 74, p. 465-466.

1543. CARROBLES SANTOS (Jesus), PEREIRA SIESO (Juan), RUIZ TABOADA (Arturo). Palomar de Pintado (Villafranca de los Caballeros, Toledo): un proyecto de formación académicas, investigación y revalorización de un yacimiento arqueológico. *Trabajos de prehistoria*, 2000, 57, 2, p. 147-160.

1544. DOMÈNECH FAUS (Elisa María). Las producciones líticas del final del Epipaleolítico e inicios del Neolítico en la vertiente mediterránea española. Propuesta metodológica. *Trabajos de prehistoria*, 2000, 57, 1, p. 135-144.

1545. FUGAZZOLA DELPINO (M. A.). La piccola "dea madre" del lago di Bracciano. *Bullettino di paletnologia italiana*, 2000, 91, p. 27-45.

1546. GONZÁLEZ MORALES (Manuel R.), STRAUS (Lawrence G.). La cueva del Mirón (Ramales de la Victoria, Cantabria): excavaciones 1996–1999. *Trabajos de prehistoria*, 2000, 57, 1, p. 121-133.

1547. GRISSOM (C. A.). Neolithic statues from ʿAin Ghazal: construction and form. *American journal of archaeology*, 2000, 104, p. 25-45.

1548. KIRJUSHIN (Jurij F.), KUNGUROVA (Natal'ja Ju.), KADIKOV (Boris Kh.). Drevnejshie mogil'niki severnykh predgorij Altaja. (The earliest burial grounds of the northern foothills of Altai). Barnaul, Izd-vo Altajskogo gos. un-ta, 2000, 117 p. (ill.).

1549. KOSTYRJA (G. V.). Dol'meny Severnogo Kavkaza – gipotezy i real'nost'. (Dolmens of the northern Caucasus: hypotheses and reality). Sankt-Peterburg, [s. n.], 2000, 138 p. (ill.).

1550. KUZMIN (Yaroslav V.), ORLOVA (Lyubov A.). The Neolithization of Siberia and the Russian Far East: radiocarbon evidence. *Antiquity*, 2000, 74, p. 356-364.

1551. LARSSON (Lars). The passage of axes: fire transformation of flint objects in the Neolithic of southern Sweden. *Antiquity*, 2000, 74, p. 602-610.

1552. LILLIOS (Karina T.). A biographical approach to the ethnogeology of Late Prehistoric Portugal. *Trabajos de prehistoria*, 2000, 57, 1, p. 19-28.

1553. LONG (D.J.), TIPPING (R.), HOLDEN (T. G.), BUNTING (M. J.), MILBURN (P.). The use of henbane (Hyoscyamus niger L.) as a hallucinogen at Neolithic 'ritual' sites: a re-evaluation. *Antiquity*, 2000, 74, p. 49-53.

1554. MARTÍNEZ CORTIZAS (Antonio), FÁBREGAS VALCARCE (Ramón), FRANCO MASIDE (Susana). Evolución del paisaje y actividad humana en el área de Monte Penide (Redondela, Pontevedra): una aproximación metodológica. *Trabajos de prehistoria*, 2000, 57, 1, p. 173-184.

1555. RODRÍGUEZ ARIZA (M.ª Oliva). El paisaje vegetal de la depresión de Vera durante la Prehistoria reciente. Una aproximación desde la antracología. *Trabajos de prehistoria*, 2000, 57, 1, p. 145-156.

1556. ROLETT (Barry V.), CHEN (Wei-chun), SINTON (John M.). Taiwan, Neolithic seafaring and Austronesian origins. *Antiquity*, 2000, 74, p. 54-61.

1557. ROTTOLI (M.). Zafferanone selvatico (Carthamus lanatus) e cardo della Madonna (Silybum marianum), piante raccolte o coltivate nel Neolitico antico a "La Marmotta"? *Bullettino di paletnologia italiana*, 2000, 91, p. 47-61.

1558. RUGGLES (Clive), BARCLAY (Gordon). Cosmology, calendars and society in Neolithic Orkney: a rejoinder to Euan Mac Kie. *Antiquity*, 2000, 74, p. 62-73.

1559. SHISHLINA (N, I.), ALEXANDROVSKY (A. L.), CHICHAGOVA (O. A.), VAN DER PLICHT (J.). Radiocarbon chronology of the Kalmykia Catacomb culture of the west Eurasian steppe. *Antiquity*, 2000, 74, p. 793-799.

1560. TABOR (Richard), JOHNSON (Paul). Sigwells, Somerset, England: regional application and interpretation of geophysical survey. *Antiquity*, 2000, 74, p. 319-325.

1561. TSIRTSONI (Zoï). Les poteries du début du Néolitique Récent en Macédoine, I. Les types de récipients. *Bulletin de correspondance hellénique*, 2000, 124, p. 1-55.

1562. WAINWRIGHT (Geoffrey). The Stonehenge we deserve. *Antiquity*, 2000, 74, p. 334-342.

1563. WHITTLE (Alasdair). New research on the Hungarian Early Neolithic. *Antiquity*, 2000, 74, p. 13-14

1564. WYSOCKI (Michael), WHITTLE (Alasdair). Diversity, lifestyles and rites: new biological and archaeological evidence from British Earlier Neolithic mortuary assemblages. *Antiquity*, 2000, 74, p. 591-601.

Cf. n° 1494

§ 4. Bronze age.

1565. ANTHONY (Daid W.), BROWN (Dorcas R.). Eneolithic horse exploitation in the Eurasian steppes: diet, ritual and riding. *Antiquity*, 2000, 74, p. 75-86.

1566. ARIAS CABAL (Pablo), GONZÁLEZ SAINZ (César), MOURE ROMANILLO (Alfonso), ONTAÑÓN PEREDO (Roberto). La zona arqueológica de La Garma (Cantabria): investigación, conservación y uso social. *Trabajos de prehistoria*, 2000, 57, 2, p. 41-56.

1567. BLÁZQUEZ MARTÍNEZ (José M.). El santuario de Cancho Roano (Badajoz) y la prostitución sagrada. *Aula Orientalis*, 99-00, 17-18, p. 367-379.

1568. BRANDHERM (Dirk). El poblamiento argárico de Las Herrerías (Cuevas de Almanzora, Almería), según la documentación inédita de L. Siret. *Trabajos de prehistoria*, 2000, 57, 1, p. 157-172.

1569. BUTLER (Virginia). Resource depression on the Northwest coast of North America. *Antiquity*, 2000, 74, p. 649-661.

1570. FRANKEL (David), WEBB (Jennifer M.). Marki Alonia: a prehistoric Bronze Age settlement in Cyprus. *Antiquity*, 2000, 74, p. 763-764.

1571. GEJ (Aleksandr N.). Novotitorovskaja kul'tura. (Novotitorovo culture). Moskva, Staryj sad, 2000, 224 p. (ill.).

1572. GLUSHKOV (Igor' G.), ZAKHOZHAJA (Tat'jana M.). Keramika epokhi pozdnej bronzy Nizhnego Priirtysh'ja. (The Late Bronze Age ceramics of the Lower Irtysh Region). Surgut, RITs Surgutskogo gos. ped. in-ta, 2000, 200 p. (ill.).

1573. GOEBEL (Ted), WATERS (Michael R.), BUVIT (Ian), KONSTANTINOV (Mikhail V.), KONSTANTINOV (Aleksander V.). Studenoe-2 and the origins of microblade technologies in the Transbaikal, Siberia. *Antiquity*, 2000, 74, p. 567-575.

1574. JOHANSEN (Øystein Kock). Bronse og makt. Bronsealderen i Norge. (Bronze and power. Bronze age in Norway). Oslo, Andresen & Butenschøn, 2000, 264 p. (ill.).

1575. KNAUF (E. A.). Jerusalem in the Late Bronze and Early Iron Ages: a proposal. *Tel Aviv*, 2000, 27, p. 75-90.

1576. MACGILLIVRAY (J. A.), DRIESSEN (J. M.), SACKETT (L. H.). The Palaikastro kouros. A Minoan chryselephantine statuette and its Aegean Bronze Age context. London, British School at Athens, 2000, 195 p. (figs, pls).

1577. MEGAW (Vincent), MORGAN (Graham), STÖLLNER (Thomas). Ancient salt-mining in Austria. *Antiquity*, 2000, 74, p. 17-18.

1578. MUÑOZ IBÁÑEZ (Francisco Javier). Las puntas ligeras de proyectil del solutrense extracantábrico. Análisis tecnomorfológico funcionales. Madrid, Aula abierta, 2000, 357 p.

1579. NOWAKOWSKI (Jacek), RACZKOWSKI (Wlodzimierz). Refutation of the myth: new fortified settlement from Late Bronze Age/Early Iron Age in Wielkopolska region (Poland). *Antiquity*, 2000, 74, p. 765-766.

1580. PADDAYYA (K.), JHALDIYAL (RICHA), PETRAGLIA (M.D.). Excavation of an Acheulian workshop at Isampur, Karnataka (India). *Antiquity*, 2000, 74, p. 751-752.

1581. PÉREZ (Sebastián Celestino). Investigación, adecuación y musealización del santuario protohistórico de Cancho Roano (Zalamea de la Serena, Badajoz). *Trabajos de prehistoria*, 2000, 57, 2, p. 133-146.

1582. RAINVILLE (Lynn). Microdebris analysis in Early Bronze Age Mesopotamian households. *Antiquity*, 2000, 74, p. 291-292.

1583. RODRÍGUEZ ARIZA (M. Oliva), FRESNEDA PADILLA (Eduardo), MARTÍN MONTERO (Marcelino), MOLINA GONZÁLEZ (Fernando). Conservación y puesta en valor del yacimiento argárico de Castellón Alto (Galera, Granada). *Trabajos de prehistoria*, 2000, 57, 2, p. 119-131.

1584. SARONIO (P.). Un ripostiglio dell'età del Bronzo recente da Bettola, nell'Appennino piacentino. *Bullettino di paletnologia italiana*, 2000, 91, p. 209-221.

1585. SHAVIT (A.). Settlement patterns in the Ayalon valley in the Bronze and Iron Ages. *Tel Aviv*, 2000, 27, p. 189-230.

1586. SHELL (COLIN A.). Shell Metalworker or shaman: Early Bronze Age Upton Lovell G2a burial. *Antiquity*, 2000, 74, p. 271-272.

1587. TAKAOĞLU (Turan). Hearth structures in the religious pattern of Early Bronze Age northeast Anatolia. *Anatolian studies*, 2000, 50, p. 11-16.

1588. VAN LOKEREN (Sven). Experimental reconstruction of the casting of copper 'oxhide' ingots. *Antiquity*, 2000, 74, p. 275-276.

1589. VICENT GARCÍA (Juan M.), RODRÍGUEZ ALCALDE (Ángel L.), LÓPEZ SÁEZ (José Antonio), DE ZAVALA MORENCOS (Ignacio), LÓPEZ GARCÍA (Pilar), MARTÍNEZ NAVARRETE (M. Isabel). ¿Catástrofes ecológicas en la estepa? Arqueología del paisaje en el coplejo minerometalúrgico de Kargaly (Región de Orenburg, Rusia). *Trabajos de prehistoria*, 2000, 57, 1, p. 29-74.

§ 5. Iron age.

1590. BURTON (Nick). New tools at Avebury. *Antiquity*, 2000, 74, p. 279-280.

1591. CHIPPINDALE (Christopher), DE JONGH (Joané), FLOOD (Josephine), RUFOLO (Scott). Stratigraphy, Harris matrices & relative dating of Australian rock-art. *Antiquity*, 2000, 74, p. 285-286.

1592. CONNOR (Aileen), PALMER (Rog). An Iron Age ditched enclosure system at Limes Farm, Landbeach, Cambridgeshire. *Antiquity*, 2000, 74, p. 281-282.

1593. CURRIE (Elizabeth J.). Archaeological investigations in the Northern Highlands of Ecuador at Hacienda Zuleta. *Antiquity*, 2000, 74, p. 273-274.

1594. DRIVER (Toby), HAMILTON (Mike), LEIVERS (Matt), ROBERTS (Julia), PETERSON (Rick). New evidence from Bryn yr Hen Bobl, Llanedwen, Anglesey. *Antiquity*, 2000, 74, p. 761-762.

1595. FANTALKIN (A.). A salvage excavation at an Early Bronze Age settlement on Ha-Shophtim Street, Qiryat 'Ata. *Tel Aviv*, 2000, 27, p. 27-56.

1596. FEINMAN (Gary M.), NICHOLAS (Linda M.). The intensive survey of three hilltop terraced sites in Oaxaca, Mexico. *Antiquity*, 2000, 74, p. 21-22.

1597. JIMENO MARTÍNEZ (Alfredo). Numancia: pasado vivo, pasado sentido. *Trabajos de prehistoria*, 2000, 57, 2, p. 175-193.

1598. KENWARD (Harry), HALL (Allan). Decay of delicate organic remains in shallow urban deposits: are we at a watershed? *Antiquity*, 2000, 74, p. 519-525.

1599. LULL (Vicente). Argaric society: death at home. *Antiquity*, 2000, 74, p. 581-590. – IDEM. Death and society: a Marxist approach. *Antiquity*, 2000, 74, p. 576-580.

1600. MAC GREGOR (Arthur). An aerial relic of O.G.S. Crawford. *Antiquity*, 2000, 74, p. 87-100.

1601. MARTÍNEZ VALLE (Rafael). El parque natural de Valltorta-Gasulla (Castellón). *Trabajos de prehistoria*, 2000, 57, 2, p. 65-76.

1602. MAYORAL HERRERA (Victorino), CHAPA BRUNET (Teresa), PEREIRA SIESO (Juan), MADRIGAL BELINCHÓN (Antonio). La pesca fluvial como recurso económico en época ibérica tardía: un ejemplo procedente de Los Castellones de Céal (Hinojares, Jaén). *Trabajos de prehistoria*, 2000, 57, 1, p. 185-197.

1603. MILLER ROSEN (Arlene), CHANG (Claudia), PAVLOVICH GRIGORIEV (Fedor). Palaeoenvironments and economy of Iron Age Saka-Wusun agro-pastoralists in southeastern Kazakhstan. *Antiquity*, 2000, 74, p. 611-623.

1604. MURAIL (P.), CRUBÉZY (E.), MARTIN (H.), HAYE (L.), BRUZEK (J.), GISCARD (P. H.), TURBAT (T.), ERDENEBAATAR (D.). The man, the woman and the hyoid bone: from archaeology to the burial practices of the Xiongnu people (Egyin Gol valley, Mongolia). *Antiquity*, 2000, 74, p. 531-536.

1605. NEBELSICK (Louis Daniel). Drinking against death – drinking sets in ostentatious tombs in the Late

Bronze and Early Iron Ages in the western Carpathian basin. *Altorientalische Forschungen*, 2000, 27, p. 211-241.

1606. NICOLÁS CHECA (Elena), MÁRQUEZ MORA (Belén), RODRÍGUEZ MÉNDEZ (Jesús). Divulgación de las investigaciones en Atapuerca (Burgos). *Trabajos de prehistoria*, 2000, 57, 2, p. 21-39.

1607. PARCERO OUBIÑA (César). Tre para dos. Las formas de poblamiento en le Edad del Hierro del Noroeste Ibérico. *Trabajos de prehistoria*, 2000, 57, 1, p. 75-95.

1608. PHILLIPS (Tim), WATSON (Aaron). The living and the dead in northern Scotland 3500–2000 BC. *Antiquity*, 2000, 74, p. 786-792.

1609. SADEH (M.). Animal bones from Qiryat ᶜAta, area L. *Tel Aviv*, 2000, 27, p. 57-60.

1610. SÁNCHEZ PALENCIA (F.-Javier), FERNÁNDEZ POSSE (M.a Dolores), FERNÁNDEZ MANZANO (Julio), OREJAS (Almudena), PÉREZ (Luis Carlos), SASTRE (Inés). Las Médulas (León), un paisaje cultural patrimonio de la humanidad. *Trabajos de prehistoria*, 2000, 57, 2, p. 195-208.

1611. SOLBERG (Bergljot). Jernalderen i Norge, ca. 500 f. Kr.–1030 e. Kr. (Iron age in Norway, c. 500. BC–1030 AD). Oslo, Cappelen Akademisk, 2000, 371 p. (ill.).

1612. VOIGT (Mary M.), HENRICKSON (Robert C.). Formation of the Phrygian state: the Early Iron Age at Gordion. *Anatolian studies*, 2000, 50, p. 37-54.

1613. WEINCKE (M. H.). Lerna, a preclassical site in the Argolid: results of excavations conducted by the American School of Classical Studies at Athens. Vol. 4. Parts I e II. The architecture, stratification and pottery of Lerna III. Princeton, The American School of Classical Studies at Athens, 2000, XVII-799 p. (plans, figs).

Cf. n°s 792, 3337, 3538, 4022, 8482

D

THE ANCIENT EAST
(the Hellenistic states included)

§ 1. General. 1614-1636. – § 2. The Near East. 1637-1676. – § 3. Egypt. 1677-1733. – § 4. Mesopotamia. 1734-1805. – § 5. Hittites. 1806-1830. – § 6. Jews and Semitic peoples to the end of the ancient world. 1831-1886. – § 7. Iran. 1887-1910.

§ 1. General.

1614. BECKER (Cornelia). Tierknochenfunde – Zeugnisse ritueller Aktivitäten. *Altorientalische Forschungen*, 2000, 27, p. 167-183.

1615. BIENKOWSKI (Piotr), MILLARD (Alan). Dictionary of the ancient Near East. Philadelphia, University of Pennsylvania Press, 2000, X-342 p. (ill.).

1616. BONATZ (Dominik), NOVÁK (Mirko), OETTEL (Andreas). Totenritual und Jenseitskonzeptionen – methodische Ansätze zur Analyse von Grabbefunden. 1. Interdisziplinärer Workshop, abgehalten am Samstag, den 21. und Sonntag, den 22. März 1998. *Altorientalische Forschungen*, 2000, 27, p. 3-5.

1617. BONATZ (Dominik). Ikonographische Zeugnisse im sepulkralen Kontext. *Altorientalische Forschungen*, 2000, 27, p. 88-105.

1618. Codification (La) des lois dans l'antiquité. Actes du Colloque de Strasbourg 27–29 novembre 1997. Ed. par Edmond LÉVY. Paris, De Boccard, 2000, 344 p. [Cf. n^{os} <sélection> 1774, 1792, 1867, 1952, 1955, 2101, 2110, 2426, 2644, 2651, 2659.]

1619. FALK (Harry). Bestattungsgebräuche in der Indien betreffenden Archäologie und im vedischen Schrifttum. *Altorientalische Forschungen*, 2000, 27, p. 68-80.

1620. FALLER (S.). Taprobane im Wandel der Zeit; das Sri Lanka-Bild in griechischen und lateinischen Quellen zwischen Alexanderzug und Spätantike. Stuttgart, Franz Steiner Verlag, 2000, 243 p. (maps, pls.).

1621. FISCHER (Claudia). Die Bildsymbolik der Assyrer in der akkadischen Tradition. *Altorientalische Forschungen*, 2000, 27, p. 308-324.

1622. FREYDANK (Helmut). Mehr "Urartäisches" aus mittelassyrischen Quellen. *Altorientalische Forschungen*, 2000, 27, p. 256-259.

1623. GREEN (Anthony). "Hell's angels": zur Ikonographie des Todes in der neuassyrischen und neubabylonischen Bildkunst. *Altorientalische Forschungen*, 2000, 27, p. 242-250.

1624. KLENGEL (Evelyn), KLENGEL (Horst). Ein altbabylonischer Text betreffend die Lieferung von Schilfrohr. *Altorientalische Forschungen*, 2000, 27, p. 251-255.

1625. KRAUSE (M.). In memoriam: P. Hans Quecke SJ (1928–1998). *Orientalia*, 2000, 69, p. 189-208.

1626. LUCK (Georg). Ancient pathways and hidden pursuits: religion, morals and magic in the ancient world. Ann Arbor, University of Michigan Press, 2000, XII-314 p.

1627. MARCELLESI (Marie-Christine). Commerce, monnaies locales et monnaies communes dans les États hellénistiques. *Révue des études grecques*, 2000, 113, p. 326-358.

1628. MARGUERON (Jean-Claude). Egouts pour les eaux usées ou conduites de recuperation des eaux du pluie? *Aula Orientalis*, 1999-2000, 17-18, p. 393-405.

1629. MEYER (Jan-Waalke). Zur Möglichkeit einer kulturhistorischen Einordnung von Grabfunden. *Altorientalische Forschungen*, 2000, 27, p. 21-37.

1630. OETTEL (Andreas). Charonspfennig und Totenglöckchen. Zur Symbolik von Münzen und Glöckchen. *Altorientalische Forschungen*, 2000, 27, p. 196-220.

1631. PRUSS (Alexander), NOVÁK (Mirko). Terrakotten und Beinidole in sepulkralen Kontexten. *Altorientalische Forschungen*, 2000, 27, p. 184-195.

1632. PRUSS (Alexander). "Sie suchen den Grabhügel so groß wie möglich zu machen" – Bestattungssitten in Transkaukasien und Anatolien. *Altorientalische Forschungen*, 2000, 27, p. 38-51.

1633. Vestnik drevnej istorii. (Journal of Ancient History). Vol. [64]/2 (233): "[Issue dedicated to the memory of Igor M. Diakonoff]". Ed. Grigorij M. BONGARD-LEVIN. Moskva, [s. n.], 2000, 304 p. [Eng. summaries and table of contents]

1634. VIOLLET (Pierre-Louis). L'hydraulique dans les civilisations anciennes, 5000 ans d'histoire. Paris, Presses de l'Ecole nationale des ponts et chaussées, 2000, 374 p.

1635. WILHELM (Susanne). Formen der Bestattung und anthropologische Archäologie – ein Überblick. *Altorientalische Forschungen*, 2000, 27, p. 162-166.

1636. YENER (K. A.), EDENS (C.), HARRISON (T. P.), VERSTRAETE (J.), WILKINSON (T. J.). The Amuq Valley regional project 1995–1998. *American journal of archaeology*, 2000, 104, p. 163-220.

Cf. n°s 974, 1025, 3334

§ 2. The Near East.

** 1637. DURAND (Jean-Marie). Documents épistolaires du palais de Mari. T. 3. Littératures anciennes du Proche-Orient 18. Paris, Ed. du Cerf, 2000, 632 p. (Archives royales de Mari).

** 1638. Rendre la justice en Mésopotamie. Archives judiciaires du Proche-Orient ancien (IIIe–I^e millénaires avant J.-C.). Ed. par Francis JOANNÈS. Saint-Denis, Presses Universitaires de Vincennes, 2000, 269 p. (Temps et espaces).

1639. AKYÜREK ŞAHIN (N. E.), ŞAHIN (S.). Ein Meilenstein aus Tlos. *Klio*, 2000, 82, p. 475-482.

1640. Anatolia antiqua VIII. (Eski Anadolu VIII). Ed. par Catherine MARRO et Aksel TIBET. İstanbul, Institut Français d'Etudes Anatoliennes-Georges Dumezil et Paris, De Boccard, 2000, 383 p.

1641. ARSLAN (Murat). Antikçağ Anadolusu'nun savaşçı kavmi Galatlar. (Warrior tribe of ancient Anatolia: the Galats). Ed. by Nezih BAŞGELEN. İstanbul, Arkeoloji ve Sanat yayınları, 2000, 322 p.

1642. BAŞARAN (Sait). Aeolische Kapitelle aus Ainos (Enez). *Istanbuler Mitteilungen*, 2000, 50, p. 157-170.

1643. BELMONTE MARÍN (Juan A.). Observaciones sobre algunos topónimos recogidos en STU y DLU. *Aula Orientalis*, 1999-2000, 17-18, p. 13-22.

1644. BERGES (Dietrich), TUNA (Numan). Das Apollonheiligtum von Emecik. Bericht über die Ausgrabungen 1998 und 1999. *Istanbuler Mitteilungen*, 2000, 50, p. 171-214.

1645. BOUNEGRU (Octavian), ERDEMGIL (Selahattin). Töpfereiliste aus Pergamon-Ketiostal. *Istanbuler Mitteilungen*, 2000, 50, p. 285-295.

1646. CARRUBA (O.). Der Name der Karer. *Athenaeum*, 2000, 88, p. 49-57.

1647. COTTICA (Daniela). Unguentari tardo-antichi dal martyrion di Hierapolis, Turchia. *Mélanges de l'École Française de Rome*, 2000, 112, p. 999-1021.

1648. DARBYSHIRE (Gareth), MITCHELL (Stephen), VARDAR (Levent). The Galatian settlement in Asia Minor. *Anatolian studies*, 2000, 50, p. 75-97.

1649. DAVIES (G.). Cremna in Pisidia: a re-appraisal of the siege works. *Anatolian studies*, 2000, 50, p. 151-158.

1650. DESSE (Jean), DESSE-BERSET (Nathalie). Salaisons de poissons marins aux marges orientales du monde gréco-romain. Contributions de l'archéozoologie. *Mélanges de l'École Française de Rome*, 2000, 112, p. 119-134.

1651. DIETRICH (Manfried), LORETZ (Oswald). Baal, Leviathan and der siebenköpfige Drache Šlyt in der Rede des Todesgottes Môt (KTU 1. 5 I 1-8 ‖ 27a-31). *Aula Orientalis*, 1999-2000, 17-18, p. 55-80.

1652. DIJKSTRA (Meindert). The list of qdšm in KTU 4, 412 + ii 8ff. *Orientalia*, 1999-2000, 17-18, p. 81-89.

1653. DUPLUOY (A.). Le bûcher de Crésus. Exploitation et transformation d'une tradition iconographique et textuelle à travers le Ve siècle. *Rivista di filologia e di istruzione classica*, 2000, 128, p. 21-37.

1654. ERTUĞ (Füsun). Linseed oil and oil mills in central Turkey. Flax/Linum and Eruca, important oil plants of Anatolia. *Anatolian studies*, 2000, 50, p. 171-185.

1655. Gemmy Drevnego Vostoka: Kruglyj stol. (Gems of Ancient Near East: A 'Round Table'). Authors: S. Ja. BERZINA, E. I. KONONENKO, B. I. VAJNBERG, M. N. POGREBOVA, S. I. KHODZHASH, O. Ja. NEVEROV. *Vestnik drevnej istorii (Journal of Ancient History)*, 2000, [64], 3 (234), p. 126-141; 4 (235), p. 145-158.

1656. GILL (David W. J.). 'A rich and promising site': Winifred Lamb (1894–1963), Kusura and Anatolian archaeology. *Anatolian studies*, 2000, 50, p. 1-10.

1657. GREENEWALT (C. H. Jr.), RAUTMAN (M. L.). The Sardis campaigns of 1996, 1997 and 1998. *American journal of archaeology*, 2000, 104, p. 643-681.

1658. HÖGEMANN (Peter). Der Iliasdichter, Anatolien und der griechische Adel. *Klio*, 2000, 82, p. 7-39.

1659. HÜLDEN (O.). Pleistarchos und die Befestigungsanlagen von Herakleia am Latmos. *Klio*, 2000, 82, p. 382-408.

1660. JACOBS (Bruno). Die Reliefs der Vorfahren des Antiochos I. von Kommagene auf dem Nemrud Daği – Versuch einer Neubenennung der Frauendarstellungen in den mütterlichen Ahnenreihen. *Istanbuler Mitteilungen*, 2000, 50, p. 297-306.

1661. KAYA (M. Ali). Anadolu'daki Galatlar ve Galatya tarihi. (The Galats in Anatolia and history of Galatia). İzmir, Ege Üniversitesi Edebiyat Fakültesi, 2000, 289 p.

1662. MA (John). The epigraphy of Hellenistic Asia Minor: a survey of recent research (1992–1999). *American journal of archaeology*, 2000, 104, p. 95-121.

1663. MARTÍNEZ LACY (Ricardo). Révoltes hellénistiques. Considérations sur la lutte des classes dans le bassin de la Méditerranée du III au I siècles av. J.-Chr. *Quaderni di storia*, 2000, 52, p. 203-213.

1664. MILNER (N. P.). Notes and inscriptions on the cult of Apollo at Oinoanda. *Anatolian studies*, 2000, 50, p. 139-149.

1665. MITTAG (P. F.). Die Rolle der hauptstädtischen Bevölkerung bei den Ptolemäern und Seleukiden im 3. Jahrhundert. *Klio*, 2000, 82, p. 409-425.

1666. SANER (Turgut). Beobachtungen zur Art und Ausführung von 'Randschlag' im hellenistischen Mauerwerk Kleinasiens. *Istanbuler Mitteilungen*, 2000, 50, p. 267-283.

1667. SCHMIDT (Klaus). "Zuerst kam der Tempel, dann die Stadt". Vorläufiger Bericht zu den Grabungen am Göbekli Tepe und am Gürcütepe 1995–1999. *Istanbuler Mitteilungen*, 2000, 50, p. 5-41.

1668. SCHWARTZ (Glenn M.), CURVERS (Hans H.), STUART (Barbara). A 3rd-millennium BC élite tomb from Tell Umm el-Marra, Syria. *Antiquity*, 2000, 74, p. 771-772.

1669. SMITH (R. R. R.), RATTÉ (C.). Archaeological research at Aphrodisias in Caria, 1997 and 1998. *American journal of archaeology*, 2000, 104, p. 221-253.

1670. SOMMER (Michael). Babylonien im Seleukidenreich: Indirekte Herrschaft und indigene Bevölkerung. *Klio*, 2000, 82, p. 73-90.

1671. STROBEL (Karl), GERBER (Christoph). TAVIUM (Büyüknefes, Provinz Yozgat) – ein regionales Zentrum Anatoliens. Bericht über den Stand der Forschungen nach den ersten drei Kampagnen (1997–1999). *Istanbuler Mitteilungen*, 2000, 50, p. 215-265.

1672. SUMMERS (Geoffrey D.). The Median Empire reconsidered: a view from Kerkenes Dağ. *Anatolian studies*, 2000, 50, p. 55-73.

1673. TRAILL (David A.). 'Priam's treasure': clearly a composite. *Anatolian studies*, 2000, 50, p. 17-35.

1674. UÇANKUŞ (Hasan Tahsin). Bir insan ve uygarlık bilimi arkeoloji: tarih öncesi çağlardan Perslere kadar Anadolu. (Archeology as a science of humanity and civilization: Anatolia from prehistorical times to the Persians). Ankara, Kültür Bakanlığı Yayımlar Dairesi Başkanlığı, 2000, 1243 p.

1675. UZEL (İlter). Anadolu'da Bulunan Antik Tıp Aletleri. (Ancient medical instruments found in Anatolia). Ankara, Türk Tarih Kurumu, 2000, 318 p.

1676. YALÇIKLI (Derya). Zwei Bronzegabeln aus Zentralanatolien. *Istanbuler Mitteilungen*, 2000, 50, p. 113-130.

§ 3. Egypt.

** 1677. WILKINSON (Toby A. H.). Royal annals of ancient Egypt. The Palermo stone and its associated fragments. London a. New York, Kegan Paul International, 2000, 287 p.

1678. Amarna diplomacy: the beginnings of International relations. Ed. by R. COHEN, R. WESTBROOK. Baltimore a. London, The Johns Hopkins University Press, XVI-307 p.

1679. ANAGNOSTOU-CANAS (Barbara). La documentation judiciaire pénale dans l'Égypte romaine. *Mélanges de l'École Française de Rome*, 2000, 112, p. 753-779.

1680. ARENA (A.). La figura di Serapide nelle testimonianze degli autori latini e greci d'età romana. *Latomus*, 2000, 59, p. 57-68.

1681. BOGAERT (R.). Les opérations des banques de l'Égypte romaine. *Ancient society*, 2000, 30, p. 135-269.

1682. BREYER (Francis A. K.). Ptahhotep – von Ptahs Gnaden der Weise mit dem dreifachen Palindrom. *Die Welt des Orients*, 2000, 31, p. 19-22.

1683. CLACKSON (J.). A Greek papyrus in Armenian script. *Zeitschrift für Papyrologie und Epigraphik*, 2000, 129, p. 223-258.

1684. CLARYSSE (W.), GALLAZZI (C.), KRUIT (N.). Three joins from the Zenon archive. *Ancient society*, 2000, 30, p. 5-27.

1685. COENEN (Marc). The funerary papyri of the Bodleian library at Oxford. *Journal of Egyptian archaeology*, 2000, 86, p. 81-98.

1686. FIORE-MAROCHETTI (Elisa). Inscribed blocks from Tomb Chapels at Hawara. *Journal of Egyptian archaeology*, 2000, 86, p. 43-50.

1687. Fontes historiae Nubiorum: textual sources for the history of the Middle Nile region between the eighth century BC and the sixth century AD. Vol. 4. Corrigenda and indices. Ed. by T. EIDE, T. HÄGG, R. HOLTON PIERCE and L. TÖRÖK. Bergen, Universitetet, 2000, p. 1217-1375.

1688. GALÁN (José M.). The ancient Egyptian Sed-festival and the exemption from corvée. *Journal of Near Eastern studies*, 2000, 59, p. 255-264. – IDEM. Two old kingdom officials connected with boats. *Journal of Egyptian archaeology*, 2000, 86, p.145-150.

1689. GOPHNA (R.), BUZAGLO (E.). A note on an Egyptian pottery basin from ʿEn Besor. *Tel Aviv*, 2000, 27, p. 26-27.

1690. GOUDSMIT (Jaap), BRANDON-JONES (Douglas). Evidence from the Baboon Catacomb in North Saqqara for a West Mediterranean monkey trade route to Ptolemaic Alexandria. *Journal of Egyptian archaeology*, 2000, 86, p. 111-119.

1691. GOZZOLI (Roberto B.). The statue BM EA 37891 and the erasure of Necho II's names. *Journal of Egyptian archaeology*, 2000, 86, p. 67-80.

1692. HAMERNIK (Gottfried). An early record of the Sarcophagi of Tjaiharpata and Esshu-tefnut and the identification of some leider squeezes at the Griffith Institute Oxford. *Journal of Egyptian archaeology*, 2000, 86, p. 168-172.

1693. HAMZA (Mahmoud), KEMP (Barry). Report on a large house at Amarna, discovered near the village of el-Hagg Qandil. *Journal of Egyptian archaeology*, 2000, 86, p. 161-165.

1694. HARRELL (James H.), BROWN (V. Max), SALAH MASOUD (Masoud). An early dynastic quarry for stone vessels at Gebel Manzal el-Seyl, Eastern desert. *Journal of Egyptian archaeology*, 2000, 86, p. 33-42.

1695. HAZZARD (R. A.). Imagination of a monarchy. Studies in Ptolemaic propaganda. Toronto, Buffalo a. London, University of Toronto Press, 2000, X-244 p.

1696. HOFFMANN (F.). Ägypten. Kultur und Lebenswelt in griechisch-römischer Zeit. Eine Darstellung nach den demotischen Quellen. Oldenbourg, Akademie Verlag, 2000, 355 p. (ill.).

1697. JANSEN-WINKELN (K.). Die Fremdherrschaften in Ägypten im 1. Jahrtausend v. Chr. *Orientalia*, 2000, 69, p. 1-20.

1698. JANSSEN (Jac. J.). Idiosyncrasies in late Ramesside hieratic writing. *Journal of Egyptian archaeology*, 2000, 86, p. 51-56.

1699. KAPLONY-HECKEL (Ursula). Einweizen-Überweisungsauftrag zugunsten des (Tempel)-Wirtschafters Hor (das demotische Kalkstein-Ostrakon Leipzig ÄM 4789). *Journal of Egyptian archaeology*, 2000, 86, p. 99-109.

1700. KOLTA (K. S.), SCHWARZMANN-SCHAFHAUSER (D.). Die Heilkunde im alten Ägypten. Magie und ratio in der Krankheitsvorstellung und therapeutischen Praxis. Stuttgart, Franz Steiner Verlag, 2000, 223 p. (figs).

1701. KRUCHTEN (Jean-Marie). Assimilation and dissimilation at work in the late Egyptian verbal system: the verb forms built by means of the auxiliary iri from the second part of the nineteenth dynasty until early demotic. *Journal of Egyptian archaeology*, 2000, 86, p. 57-65.

1702. LAMBERT-ZAZULAK (Patricia). The International Ancient Egyptian Mummy Tissue Bank at the Manchester Museum. *Antiquity*, 2000, 74, p. 44-48.

1703. LAROCHE-TRAUNECKER (Françoise). Chapiteaux nabatéens, corinthiens inachevés ou simplifiés? Nouveaux exemples en Egypte. *Ktema*, 2000, 25, p. 207-213.

1704. LAYTON (Bentley). A coptic grammar with crestomathy and glossary: Sahidic dialect. Wiesbaden, Harrassowitz Verlag, 2000, XIX-520 p.

1705. LECLANT (J.), MINAULT-GOUT (A.). Fouilles et travaux en Égypte et au Soudan, 1997–1998. Seconde partie. *Orientalia*, 2000, 69, p. 141-170. – IIDEM. Fouilles et travaux en Égypte et au Soudan, 1998. *Orientalia*, 2000, 69, p. 209-329.

1706. LÓPEZ (Jesús). Los orígines de la fábula: la disputa del cuerpo y la cabeza. *Aula Orientalis*, 1999-2000, 17-18, p. 475-482.

1707. *Vacat.*

1708. MARTIN (A.). P. Mich. Inv. 5478a et le préfet d'Égypte Statilius Ammianus. *Latomus*, 2000, 59, p. 399-402.

1709. MINAS (Martina). Die hieroglyphischen Ahnenreihen der ptolemäischen Könige. Ein Vergleich mit den Titeln der eponymen Priester in der demotischen und griechischen Papyri. Mainz, von Zabern, 2000, XIV-259 p.

1710. Miscellany (A) of demotic texts and studies. Ed. by P. J. FRANDSEN and K. RYHOLT. København, Museum Tusculanum Press, 2000, VII-171 p. (plates).

1711. MÜNCH (Hans-Hubertus). Categorizing archaeological finds: the funerary material of Queen Hetepheres I at Giza. *Antiquity*, 2000, 74, p. 898-908.

1712. NIELSEN (B. E.). A catalog of duplicate papyri. *Zeitschrift für Papyrologie und Epigraphik*, 2000, 129, p. 187-214.

1713. PADRÓ (Josep). Espania en Egipto. *Aula Orientalis*, 1999-2000, 17-18, p. 483-492.

1714. QUACK (Joachim Friedrich). Die rituelle Erneuerung der Osiris-Figurinen. *Die Welt des Orients*, 2000, 31, p. 5-18.

1715. REYNARD (Jean). La géographie de l'Égypte selon Platon. Remarques sur un passage du Timée (21 e). *Révue des études grecques*, 2000, 113, p. 131-146.

1716. RIGGS (Christina). Roman mummy masks from Deir el-Bahri. *Journal of Egyptian archaeology*, 2000, 86, p. 121-144.

1717. RITNER (Robert K.). Innovations and adaptations in ancient Egyptian medicine. *Journal of Near Eastern studies*, 2000, 59, p. 107-117.

1718. RÖMER (M.). Staat und Wirtschaft im Alten Ägypten. *Orientalia*, 2000, 69, p. 407-429.

1719. RUTHERFORD (Ian). The genealogy of the Boukoloi: how Greek literature appropriated an Egyptian narrative-motif. *Journal of Hellenic studies*, 2000, 120, p. 106-121.

1720. RZEPKA (Slawomir). Reconstruction of the kilt represented on a statue of Neferefre. *Journal of Egyptian archaeology*, 2000, 86, p. 150-155.

1721. SERPICO (Margaret), WHITE (Raymond). The botanical identity and transport of incense during the Egyptian New Kingdom. *Antiquity*, 2000, 74, p. 884-897.

1722. SHAW (I.). The Oxford history of ancient Egypt. Oxford, Oxford U. P., 2000, XV-512 p. (ills, maps, pls.).

1723. SHORTLAND (Andrew). Depictions of glass in two Theban tombs and their role in the dating of early glass. *Journal of Egyptian archaeology*, 2000, 86, p. 159-161.

1724. SPIESER (Cathie). Les noms du Pharaon comme être autonomes au Nouvel Empire. Freiburg, Universitätsverlag et Göttingen, Vandenhoeck & Ruprecht, 2000, XI-398 p.

1725. STERNBERG (Myriam). Données sur le produits fabriqués dans une officine de Neapolis (Nabeul, Tunisie). *Mélanges de l'École Française de Rome*, 2000, 112, p. 135-153.

1726. STEWART (H. M.). Note on an enigmatic Shabti form. *Journal of Egyptian archaeology*, 2000, 86, p. 166-167.

1727. TAYLOR (John H.). Edward Poole's drawings of two royal coffins of the seventeenth dynasty. *Journal of Egyptian archaeology*, 2000, 86, p. 155-158.

1728. THOMAS (T. K.). Late antique Egyptian funerary sculpture. Images for this world and the next. Princeton, Princeton U. P., 2000, XXV-163 p. (ill.).

1729. TOMBER (Roberta). Indo-Roman trade: the ceramic evidence from Egypt. *Antiquity*, 2000, 74, p. 624-631.

1730. VAN MINNEN (P.). An official act of Cleopatra (with a subscription in her own hand). *Ancient society*, 2000, 30, p. 29-34.

1731. VILLAR (Francisco). El epíteto teonímico lusitano Nilaicui y la etimología del nombre del Nilo. *Aula Orientalis*, 1999-2000, 17-18, p. 469-473.

1732. WILKINSON (Toby A. H.). What a king is this: Narmer and the concept of the ruler. *Journal of Egyptian archaeology*, 2000, 86, p. 23-32.

1733. WILSON (Penelope), JEFFREYS (D.), KEMP (Barry), NICHOLSON (Paul T.), ROSE (Pamela). Fieldwork 1999–2000: Sais, Memphis, Tell El-Amarna, Tell El-Amarna glass project, Qasr Ibrim. *Journal of Egyptian archaeology*, 2000, 86, p. 1-21.

Cf. n°s 363, 8792

§ 4. Mesopotamia.

1734. AL-RAWI (Farouk), BLACK (Jeremy). A balbale of Ninurta, god of fertility. *Zeitschrift für Assyriologie und vorderasiatische Archäologie*, 2000, 90, p. 31-39.

1735. BAGG (Ariel M.). Assyrische Wasserbauten: Landwirtschaftliche Wasserbauten in Kernland Assyriens zwischen der 2. Hälfte des 1. Jahrtausends v. Chr. Mainz am Rhein, Verlag Philipp von Zabern, 2000, XXXVI-439 p. (tables).

1736. BÖCK (Barbara), MÁRQUEZ ROWE (Ignacio). MM 818: a new LB fragment of Atra-hasīs I. *Aula Orientalis*, 1999-2000, 17-18, p. 167-177.

1737. BÖCK (Barbara). Die babylonisch-assyrische Morphoskopie. Wien, Institut für Orientalistik der Universität Wien, 2000, VII-348 p.

1738. BONATZ (Dominik). Fallbeispiele aus Tall Šēh Hamad II: Spindel und Spinnwirtel. *Altorientalische Forschungen*, 2000, 27, p. 199-210.

1739. BORDREUIL (Pierre), PARDEE (Denis). Catalogue raisonné des textes ougaritiques de la Maison d'Ourtenou. *Aula Orientalis*, 1999-2000, 17-18, p. 23-38.

1740. CAMPS (Guiu). La narración de Gn 18, 1-5. ¿Un trasfondo ugarítico? *Aula Orientalis*, 1999-2000, 17-18, p. 39-43.

1741. CARROUÉ (François). Šulgi et le temple Bagara. *Zeitschrift für Assyriologie und vorderasiatische Archäologie*, 2000, 90, p. 181-193.

1742. Chronologies des pays du Caucase et de l'Euphrate aux IVe–IIIe millenaires: from the Euphrates to the Caucasus: chronologies for the IVth–IIIrd millennium B. C. vom Euphrat in den Kaukasus: vergleichende Chronologie des IV. und III. Jahrtausends v. chr. Ed. par Catherine MARRO et Harald HAUPTMANN. İstanbul, Institut Français d'Etudes Anatoliennes d'Istanbul et Paris, De Boccard, 2000, 511 p.

1743. CIVIL (Miguel). Reading Gilgameš. *Aula Orientalis*, 1999-2000, 17-18, p. 179-189.

1744. COHEN (Eran). Akkadian –ma in diachronic perspective. *Zeitschrift für Assyriologie und vorderasiatische Archäologie*, 2000, 90, p. 207-226.

1745. COLBOW (G.). Anbringungsorte neuassyrischer Felsreliefs – Bedeutsame Räume in der Königspropaganda. *Iranica antiqua*, 2000, 35, p. 1-19.

1746. COMFORT (Anthony), ABADIE-REYNAL (Catherine), ERGEÇ (Rifat). Crossing the Euphrates in antiquity: Zeugma seen from space. *Anatolian studies*, 2000, 50, p. 99-126.

1747. CUNCHILLOS (Jesús-Luis). La recensión BDFSN del texto ugarítico. *Aula Orientalis*, 1999-2000, 17-18, p. 45-53.

1748. DION (P.-E.). L'incursion d'Aššurnasirpal II Luhutu. *Orientalia*, 2000, 69, p. 133-138.

1749. DOMBRADI (Eva). Studien zu mīthārum/mithāriš und die Frage des Duplums: 1. Zum semantischen Feld von mithārum in juristischem Kontext. *Zeitschrift für Assyriologie und vorderasiatische Archäologie*, 2000, 90, p. 40-69.

1750. DREWS (Robert). Medinet Habu: oxcarts, ships and migration theories. *Journal of Near Eastern studies*, 2000, 59, p. 161-190.

1751. DURAND (Jean-Marie). Apologue sur des mauvaises herbes et un coquin. *Aula Orientalis*, 1999-2000, 17-18, p. 191-196.

1752. EGAN (V.), BIKAI (P. M.), ZAMORA (K.). Archaeology in Jordan. *American journal of archaeology*, 2000, 104, p. 561-588.

1753. ERKANAL-ÖKTÜ (Armağan). Ein Kultgefäß aus Girnavaz-Höyük bei Nusaybin/Mardin. *Istanbuler Mitteilungen*, 2000, 50, p. 143-156.

1754. FELIU (Lluís). "The Lord of the Offspring". *Aula Orientalis*, 1999-2000, 17-18, p. 197-200.

1755. FRAHM (Eckart). Wie "christlich" war die assyrische Religion? *Die Welt des Orients*, 2000, 31, p. 31-45.

1756. GARCÍA RECIO (Jesús). VAT 17256 a: fragmento del himno al templo de Keš. *Aula Orientalis*, 1999-2000, 17-18, p. 201-203.

1757. GIBBINS (David). Classical shipwreck excavation at Tektas Burnu, Turkey. *Antiquity*, 2000, 74, p. 19-20.

1758. GIBSON (McGuire), MAKTASH (Muhammad). Tell Hamoukar: early city in northeastern Syria. *Antiquity*, 2000, 74, p. 477-478.

1759. Giubileo (Il) prima del giubileo: tempo e spazio nelle civiltà mesopotamiche e nell'antico Egitto. Atti del convegno internazionale, Milano 12 febbraio 2000. A cura di L. FANTINI TERZI. Milano, Centro Studi del Vicino Oriente, 2000, 91 p.

1760. GÖRKE (Susanne), NOVÁK (Mirko). Fallbeispiele aus Tall Šēḫ Ḥamad I: die sogenannten Sekundärbeigaben. *Altorientalische Forschungen*, 2000, 27, p. 196-198.

1761. HECKER (Karl). i oder ī im Status constructus? *Altorientalische Forschungen*, 2000, 27, p. 260-268.

1762. HEIMPEL (W.). Observations on the royal letters from Mari. *Orientalia*, 2000, 69, p. 88-104.

1763. HOFTIJZER (Jacob). The opening clauses of the Lament over King Keret. *Aula Orientalis*, 1999-2000, 17-18, p. 97-104.

1764. HOROWITZ (Wayne). Astral tablets in the Hermitage, Saint Petersburg. *Zeitschrift für Assyriologie und vorderasiatische Archäologie*, 2000, 90, p. 194-206.

1765. JACOBSEN BUCKLEY (Jorunn). The evidence of women priests in Mandaeism. *Journal of Near Eastern studies*, 2000, 59, p. 93-106.

1766. KEMP (Barry). A model of Tell el-Amarna. *Antiquity*, 2000, 74, p. 15-16.

1767. KIELT COSTELLO (Sarah). Memory tools in early Mesopotamia. *Antiquity*, 2000, 74, p. 475-476.

1768. KOCH (Johannes). Neues von den babylonischen Planeten-Hypsomata. *Die Welt des Orients*, 2000, 31, p. 46-71.

1769. KOCH-WESTENHOLZ (Ulla). Babylonian liver omens. The chapters Manzāzu, Padānu and Pān tākalti of the Babylonian extispicy series mainly from Aššurbanipal's library. København, Museum Tusculanum Press, 2000, 543 p.

1770. KOHLMEYER (Kay). Der Tempel des Wettergottes von Aleppo. Münster, Thema-Verlag, 2000, 45 p. (tables).

1771. KOTTSIEPER (I.). Der Mann aus Babylonien. Steuerhinterzieher, Flüchtling, Immigrant oder Agent? Zu einem aramäischen Dekret aus neuassyrischer Zeit. *Orientalia*, 2000, 69, p. 368-392.

1772. KOUWENBERG (N. J. C.). Nouns as verbs: the verbal nature of the Akkadian stative. *Orientalia*, 2000, 69, p. 21-71.

1773. KULEMANN-OSSEN (Sabina), NOVÁK (Mirko). ᵈKūbu und das "Kind im Tof". Zur Symbolik von Topfbestattungen. *Altorientalische Forschungen*, 2000, 27, p. 121-131.

1774. LAFONT (S.). Codification et subsidiarité dans les droits du Proche-Orient ancien. *In*: Codification (La) des lois dans l'antiquité [Cf. n° 1618], p. 49-64.

1775. LEAMAN (Oliver). Eastern philosophy: key readings. London a. New York, Routledge, 2000, XVI-305 p.

1776. LIEBIG (Michael). Aššur-etel-ilāni, Sin-šum-līšir, Sin-šar-iškun und die Babylonische Chronik. *Zeitschrift für Assyriologie und vorderasiatische Archäologie*, 2000, 90, p. 281-284.

1777. LUNDSTRÖM (Steven). *Kimaḫḫu* und *Qabru*. Untersuchungen zur Begrifflichkeit akkadischer Grabbezeichnungen. *Altorientalische Forschungen*, 2000, 27, p. 6-20.

1778. MARGALIT (Baruch). New readings in Aqht. *Aula Orientalis*, 1999-2000, 17-18, p. 105-112.

1779. MARGUERON (J.). De Mari à Délos, un lien: le palmier? *Ktema*, 2000, 25, p. 55-63.

1780. MATTILA (Raija). The king's magnates: a study of the highest officials of the Neo-Assyrian empire. Helsinki, The Neo-Assyrian text corpus project, 2000, XVI-201 p.

1781. Mesopotamia and Iran in the Parthian and Sasanian periods. Rejection and revival c. 238 BC–AD 642. Proceedings of a seminar in memory of V. G. Lukonin. Ed. by J. CURTIS. London, British Museum Press, 2000, p. 80 (ills., pls.).

1782. MIGLUS (Peter A.). Altmesopotamische Stadtplanung zwischen Theorie und Wirklichkeit. *Zeitschrift für Assyriologie und vorderasiatische Archäologie*, 2000, 90, p. 123-138.

1783. MILANO (L.), DE MARTINO (S.), FALES (F. M.), LANFRANCHI (G.). Landscapes: territories, frontiers and horizons in the Ancient Near East. Papers presented to the XLIV Rencontre Assyriologique Internationale, Venezia 7–11 July 1997. Part I. Invited lectures. Part II. Geography and cultural landscapes.

Part III. Landscape in ideology, religion, literature and art. Padova, Sargon, 2000, 76 p., IV-254 p., IV-196 p.

1784. MILLET ALBÀ (Adelina). Les noms propres avec preformative en me- et mu- des archives de Mari. *Aula Orientalis*, 1999-2000, 17-18, p. 205-214.

1785. MOLINA (Manuel). Neo-Sumerian letter-orders in the British Museum. I. *Aula Orientalis*, 1999-2000, 17-18, p. 215-228.

1786. MONTERO FENOLLÓS (Jean Luis). Metales para la muerte. Costumbres funerarias en la Alta Mesopotamia durante el III milenio a.C. *Aula Orientalis*, 1999-2000, 17-18, p. 407-419.

1787. NA'AMAN (N.), ZADOK (R.). Assyrian deportations to the province of Samerina in the light of two cuneiform tablets from Tel Hadid. *Tel Aviv*, 2000, 27, p. 159-188.

1788. NOVÁK (Mirko). Das "Haus der Totenpflege". Zur Sepulkralsymbolik des Hauses im Alten Mesopotamien. *Altorientalische Forschungen*, 2000, 27, p. 132-154.

1789. OLIVA (Juan C.). Alalah VII Chronographica. Una revisión del archivo sobre la base de los textos de Yarim-Lim. *Aula Orientalis*, 1999-2000, 17-18, p. 229-239.

1790. ÖZGAN (Ramazan). "Säulensarkophage – und danach". *Istanbuler Mitteilungen*, 2000, 50, p. 365-387.

1791. RIDLEY (R. T.). The saga of an epic: Gilgamesh and the constitution of Uruk. *Orientalia*, 2000, 69, p. 341-367.

1792. ROTH (M. T.). The law collection of King Hammurabi: toward an understanding of codification and canonization. *In*: Codification (La) des lois dans l'antiquité [Cf. n° 1618], p. 9-31.

1793. SALLABERGER (Walther). Das Erscheinen Marduks als Vorzeichen: Kultstatue und Neujahrfest in der Omenserie Šumma ālu. *Zeitschrift für Assyriologie und vorderasiatische Archäologie*, 2000, 90, p. 227-262.

1794. SAX (Margaret), MEEKS (Nigel D.), COLLON (Dominique). The introduction of the lapidary engraving wheel in Mesopotamia. *Antiquity*, 2000, 74, p. 380-387.

1795. SCHWARTZ (Glenn M.), CURVERS (H. H.), GERRITSEN (F. A.), MACCORMACK (J. A.), MILLER (N. F.), WEBER (J. A.). Excavations and survey in the Jabbul Plain, Western Syria: the Umm el-Marra Project 1996–1997. *American journal of archaeology*, 2000, 104, p. 419-462.

1796. STRECK (Michael). Keilschrifttexte aus Münchener Sammlungen. *Zeitschrift für Assyriologie und vorderasiatische Archäologie*, 2000, 90, p. 263-280.

1797. TESTEN (David). Conjugating the "prefixed stative" verbs of Akkadian. *Journal of Near Eastern studies*, 2000, 59, p. 81-92.

1798. TINNEY (Steve). Notes on Sumerian sexual lyric. *Journal of Near Eastern studies*, 2000, 59, p. 23-30.

1799. VIGANÒ (Lorenzo). Rituals at Ebla II, ì-giš sag: a purification ritual or anointing of the head? *Journal of Near Eastern studies*, 2000, 59, p. 13-22.

1800. VITA (Juan-Pablo). Das Gezer-Corpus von El-Amarna: Umfang und Schreiber. *Zeitschrift für Assyriologie und vorderasiatische Archäologie*, 2000, 90, p. 70-77. – IDEM. Zur Menologie und zum Kalender von Alalah. *Altorientalische Forschungen*, 2000, 27, p. 296-307.

1801. VOLK (Konrad). Edubba'a und Edubba'a Literatur. Rätsel und Lösungen. *Zeitschrift für Assyriologie und vorderasiatische Archäologie*, 2000, 90, p. 1-30.

1802. WATSON (Wilfred G. E.). La fórmula de marcha en los textos ugaríticos. *Aula Orientalis*, 1999-2000, 17-18, p. 125-131.

1803. WITZEL (Carsten). Biologie und Taphonomie der Totenlage. *Altorientalische Forschungen*, 2000, 27, p. 155-161.

1804. WUNSCH (Cornelia). Eine Richterurkunde aus der Zeit Neriglissars. *Aula Orientalis*, 1999-2000, 17-18, p. 241-254.

1805. WYATT (N.). Just how "divine" were the kings of Ugarit? *Aula Orientalis*, 1999-2000, 17-18, p. 133-141.

§ 5. Hittites.

1806. AMADASI GUZZO (Maria Giulia). MSKT à Karatepe. *Orientalia*, 2000, 69, p. 72-80.

1807. ANDRÉ-SALVINI (Beatrice), SALVINI (Mirjo). Le liste lessicali e i vocabolari plurilingui di Ugarit. Una chiave per l'interpretazione della lingua hurrica. *La parola del passato*, 2000, 55, p. 321-348.

1808. D'ALFONSO (Lorenzo). Syro-Hittite administration at Emar: new considerations on the basis of a prosopographic study. *Altorientalische Forschungen*, 2000, 27, p. 269-295.

1809. DE MARTINO (Stefano). Il 'canto della liberazione': composizione letteraria bilingue hurrico-hittita sulla distruzione di Ebla. *La parola del passato*, 2000, 55, p. 296-320. – IDEM. Il regno hurrita di Mitanni: profilo storico-politico. *La parola del passato*, 2000, 55, p. 68-102.

1810. GIBSON (John C. L.). Death in Canaanite thinking. *Aula Orientalis*, 1999-2000, 17-18, p. 91-95.

1811. GIORGIERI (Mauro). L'onomastica hurrita. *La parola del passato*, 2000, 55, p. 278-295. – IDEM. Schizzo grammaticale della lingua hurrica. *La parola del passato*, 2000, 55, p. 171-277.

1812. GIRBAL (Christian). Das "hattische" Wort für "Kuh". *Altorientalische Forschungen*, 2000, 27, p. 373-

379. – IDEM. Ein hattischer Paragraph. *Altorientalische Forschungen*, 2000, 27, p. 367-372.

1813. GRODDEK (Detlev). Das Ende von Mythos einer verschwurdenen Gottheit: K Bo. 40, 135. *Zeitschrift für Assyriologie und vorderasiatische Archäologie*, 2000, 90, p. 285-291. – IDEM. "[Diese Angelegenheit] hörte Ištar von Ninive nicht!". Eine neue Episode einer Erzählung des Kumarbi-Kreises. *Die Welt des Orients*, 2000, 31, p. 23-30. – IDEM. Fragmenta Hethitica dispersa. *Altorientalische Forschungen*, 2000, 27, p. 359-366. – IDEM. Hethitisch ^{NA}-tahapšetai-u. ä = "Schlachtblock". *Orientalia*, 2000, 69, p. 81-85.

1814. HAAS (Volkert). Hethitische Bestattungsbräuche. *Altorientalische Forschungen*, 2000, 27, p. 52-67.

1815. MÜLLER (H.-P.). Die Tabella defixionis KAI 89 und die Magie des Fluches. *Orientalia*, 2000, 69, p. 393-406.

1816. ÖKSE (A. Tuba). Neue hethitische Siedlungen zwischen Maşat Höyük und Kuşaklı. *Istanbuler Mitteilungen*, 2000, 50, p. 87-111.

1817. PARMEGIANI (Neda). Elaborazione informatica dei testi cuneiformi in lingua hurrica. *La parola del passato*, 2000, 55, p. 366-380.

1818. PECCHIOLI DADDI (Franca). Un nuovo rituale di Muršili II. *Altorientalische Forschungen*, 2000, 27, p. 344-358.

1819. PECORELLA (Paolo Emilio). Note sulla produzione artistica hurrita e mitannica. *La parola del passato*, 2000, 55, p. 349-365.

1820. PIERALLINI (Sibilla). Observations on the lower city of Hattuša: a comparison between the epigraphic sources and the archaeological documentation. *Altorientalische Forschungen*, 2000, 27, p. 325-343.

1821. SALVINI (Mirjo). I Hurriti sulla costa orientale del Mediterraneo. *La parola del passato*, 2000, 55, p. 103-113. – IDEM. La civiltà dei Hurriti, popolo dell'Asia Anteriore Antica. Introduzione alla storia degli studi e alla documentazione testuale. *La parola del passato*, 2000, 55, p. 7-24. – IDEM. Le più antiche testimonianze dei Hurriti prima della formazione del regno di Mitanni. *La parola del passato*, 2000, 55, p. 25-67.

1822. SANMARTÍN (Joaquín). Sociedades y lenguas en el medio sirio-levantino del II milenio a.C.: Ugarit y lo hurrita. *Aula Orientalis*, 1999-2000, 17-18, p. 113-123.

1823. SICKER-AKMAN (Martina). Der Fürstensitz der späthethitischen Burganlage Karatepe-Aslantaş. *Istanbuler Mitteilungen*, 2000, 50, p. 131-142.

1824. SIPAHI (Tunç). Eine althethitische Reliefvase vom Hüseyindede Tepesi. *Istanbuler Mitteilungen*, 2000, 50, p. 63-85.

1825. SOYSAL (Oğuz). Analysis of a Hittite oracular document. *Zeitschrift für Assyriologie und vorderasiatische Archäologie*, 2000, 90, p. 85-122.

1826. SUTER (Claudia). The Hittite seal from Megiddo. *Aula Orientalis*, 1999-2000, 17-18, p. 421-430.

1827. TARACHA (Piotr). Ersetzen und Entsühnen: das mittelhethitische Ersatzritual für den Großkönig Tuthalija (CTH *448, 4) und verwandte Texte. Leiden, Boston u. Köln, Brill, 2000, XXIII-307 p.

1828. TRÉMOUILLE (Marie-Claude). La religione dei Hurriti. *La parola del passato*, 2000, 55, p. 114-170.

1829. TSAN (Tsong-Sheng). Ahnenkult aus einer ostasiatischen Sicht. *Altorientalische Forschungen*, 2000, 27, p. 81-87.

1830. YILDIRIM (Tayfun). Yörüklü/Hüseyindede: eine neue hethitische Siedlung im Südwesten von Çorum. *Istanbuler Mitteilungen*, 2000, 50, p. 43-62.

Cf. n° 1633

§ 6. Jews and Semitic peoples to the end of the ancient world.

1831. AMIT (D.), MAGNESS (J.). Not a settlement of Hermits or Essenes: a response to Y. Hirschfeld, a settlement of Hermits above ʿEn Gedi. *Tel Aviv*, 2000, 27, p. 273-285.

1832. ARNAUD (Daniel). Une bêche-de-mer antique: la langue des marchands à Tyr à la fin du XIIIe siècle. *Aula Orientalis*, 1999-2000, 17-18, p.143-166.

1833. BARJAU I RICO (M. E.), CALDERS I ARTÍS (T.). Pau de Santa Maria olim Salomó ha-Leví. *Aula Orientalis*, 1999-2000, 17-18, p. 255-259.

1834. BARRY (Catherine), FUNK (Wolf-Peter), POIRIER (Paul-Hubert), TURNER (John D.). Zostrien (NH VIII, 1). Quebec, Les Presses de l'Université Laval et Louvain, Éditions Peeters, 2000, XXIII-709 p.

1835. BEJARANO ESCANILLA (Ingrid). Qara Qūzāq: imágenes y percepciones del espacio. *Aula Orientalis*, 1999-2000, 17-18, p. 431-436.

1836. BONNET (Corinne). Brevi osservazioni comparative sull'Astarté funeraria. *Aula Orientalis*, 1999-2000, 17-18, p. 335-339.

1837. BOSSHARD-NEPUSTIL (Erich), MORENZ (Ludwig D.). Layout als Schriftspiel – zu einem edomitischen Siegel aus Horvat Qidmit. *Die Welt des Orients*, 2000, 31, p. 72-74.

1838. BOYD (Brian), CROSSLAND (Zoë). New fieldwork at Shuqba Cave and in Wadi en-Natuf, Western Judea. *Antiquity*, 2000, 74, p. 755-756.

1839. BRON (François). Divinités communes à la Syrie-Palestine et à l'Arabie du Sud préislamique. *Aula Orientalis*, 1999-2000, 17-18, p. 437-440.

1840. BROWN (J. P.). Israel and Hellas. Vol. 2. Sacred institutions with Roman counterparts. Berlin a. New York, Walter de Gruyter, 2000, XXVIII-414 p.

1841. CALLENDER (Dexter E. Jr.). Adam in myth and history. Ancient Israelite perspectives on the primal human. Winona Lake, Eisenbrauns, 2000, XVIII-244 p.

1842. CASANOVAS MIRÓ (Jorge). Dos inscripciones funerarias hebraicas del siglo XV de Móra d'Ebre (Tarragona). *Aula Orientalis*, 1999-2000, 17-18, p. 261-267.

1843. CAVALIERE (Paola). Anfore puniche utilizzate come contenitori di pesce. Un esempio olbiese. *Mélanges de l'École Française de Rome*, 2000, 112, p. 67-72.

1844. CHERIX (Pierre). Lexique analitique du parchemin Bodmer VI: version copte du Livre des Proverbes. Lausanne, Éditions du Zèbre, 2000, X-313 p.

1845. COLLINS (J. J.). Between Athens and Jerusalem: Jewish identity in the Hellenistic diaspora. Grand Rapids, William B. Eerdmans Publishing Co., 2000, XVI-327 p.

1846. CORS I MEYA (Jordi). Traces of the ancient origin of some mythic components in Philo of Biblo' *Phoenician history*. *Aula Orientalis*, 1999-2000, 17-18, p. 341-348.

1847. DÍEZ MERINO (Luis). Targum de Profeta Zacarías en la tradición sefardí. *Aula Orientalis*, 1999-2000, 17-18, p. 269-285.

1848. FANTALKIN (A.). A salvage excavations at a 6th–7th century C.E. site on Palmach Street, Beersheba. *Tel Aviv*, 2000, 27, p. 257-272.

1849. FINKELSTEIN (I.). Hazor XII–XI with an addendum on Ben-Tor's dating of Hazor X–VII. *Tel Aviv*, 2000, 27, p. 231-247.

1850. FITZMYER (J. A.). The Dead sea scrolls and Christian origins. Grand Rapids a. Cambridge, William B. Eerdmans, 2000, XVII-290 p.

1851. GEVA (Hillel). Jewish quarter excavations in the old city of Jerusalem conducted by Nahman Avigad, 1969–1982. Vol. 1. Architecture and stratigraphy: areas A, W and X-2; final report. Jerusalem, Israel Exploration Society and Institute of Archaeology, Hebrew University of Jerusalem, 2000, XI-283 p. (phot.).

1852. HAEGEMANS (Karen). Elissa, the first queen of Carthage, through Timaeus' eyes. *Ancient society*, 2000, 30, p. 277-291.

1853. HIRSCHFELD (Yizhar). The archaeology of Hermits: a reply. *Tel Aviv*, 2000, 27, p. 286-291. – IDEM. Ramat Hanadiv excavations: final report of the 1984–1998 seasons. Jerusalem, Israel Excavation Society, 2000, XV-768 p. (fig., plts.). – IDEM. A settlement of Hermits above ʿEn Gedi. *Tel Aviv*, 2000, 27, p. 103-155.

1854. HOHLFELDER (Robert L.). Beyond confidence? Marcus Agrippa and King Herod's harbor. *Journal of Near Eastern studies*, 2000, 59, p. 241-253.

1855. HUMPHREY (E. M.). Joseph and Asenath. Sheffield, Sheffield Academic Press, 2000, 121 p.

1856. JURSA (Michael), WESZELI (Michaela). Der 'Zahn' des Schreibers. Ein aramäischer Buchstabenname in akkadischer Transkription. *Zeitschrift für Assyriologie und vorderasiatische Archäologie*, 2000, 90, p. 78-84.

1857. KRAHMALKOV (Charles R.). Phoenician-Punic dictionary. Leuven, Uitgeverij Peeters en Departement Oosterse Studies, 2000, 499 p.

1858. LEMARIE (André). Les routes du Proche-Orient: des séjours d'Abraham aux caravanes de l'encens. Paris, Desclée de Brouwer, 2000, 144 p.

1859. LIPIŃSKI (Edward). The root GZR in Semitic. *Aula Orientalis*, 1999-2000, 17-18, p. 493-497.

1860. MAGDALENA NOM DE DÉU (J. Ramón). Curiosas recetas médicas en un manuscrito petropolitano de origen habraicocatalán (siglos XIV–XV). Textos y traducción. *Aula Orientalis*, 1999-2000, 17-18, p. 287-305.

1861. MARKOE (G. E.). Peoples of the past: Phoenicians. London, British Museum Press, 2000, 224 p. (maps, ills.).

1862. MARSTON SPEIGHT (R.). Narrative structures in the Hadīth. *Journal of Near Eastern studies*, 2000, 59, p. 265-271.

1863. MILEVSKI (I.). The Manahat and Gezer cylinder seals anf further Levantine glyptics. *Tel Aviv*, 2000, 27, p. 91-102.

1864. NIEMANN (H. M.). Megiddo and Solomon: a biblical investigation in relation to archaeology. *Tel Aviv*, 2000, 27, p. 61-74.

1865. NUNN (Astrid). Der figürliche Motivschatz Phöniziens, Syriens und Transjordaniens vom 6. bis zum 4. Jahrhundert v. Chr. Freiburg, Universitätsverlag u. Göttingen, Vandenhoeck & Ruprecht, 2000, IX-269 p.

1866. OTTO (Adelheid). Die Enstehung und Entwicklung der Klassisch-Syrischen Glyptik. Berlin u. New York, de Gruyter, 2000, XV-313 p.

1867. OTTO (E.). Kodifizierung und Kanonisierung von Rechtssätzen in keilschriftlichen und biblischen Rechtssammlungen. *In*: Codification (La) des lois dans l'antiquité [Cf. n° 1618], p. 77-124.

1868. RIBERA FLORIT (Josep). El género literario rabínico llamado Targum. Estudio sobre las técnicas de traducción en el Targum de Ezequiel. *Aula Orientalis*, 1999-2000, 17-18, p. 307-314.

1869. RIBICHINI (Sergio). Sacrum magnum nocturnum. Note comparative sul molchomor nelle stele N'Gaous. *Aula Orientalis*, 1999-2000, 17-18, p. 353-362.

1870. ROBERGE (Michel). La paraphrase de Sem (NH VII 1). Quebec, Les Presses de l'Université Laval et Louvain, Éditions Peeters, 2000, XXX-283 p.

1871. ROMEU FERRÉ (Pilar). Ejemplificar con el ejemplo: mešalim y ma'asiyot en el Regimiento de la vida de Moisés Almosnino. *Aula Orientalis*, 1999-2000, 17-18, p. 315-322.

1872. Sémitologie (La) aujourd'hui. Actes de la journée de l'École doctorale de l'Université de Provence du 29 Mai 1997. Ed. par Ph. CASSUTO et P. LARCHER. Aix-en-Provence, Publications de l'Université de Provence, 2000, 187 p.

1873. SHIMELMITZ (R.), BARKAI (R.), GOPHER (A.). A Canaanean blade workshop at Har Haruvim, Israel. *Tel Aviv*, 2000, 27, p. 3-22.

1874. SNYDER (H. G.). Teachers and texts in the ancient world: philosophers, Jews and Christians. Religion in the first Christian centuries. London a. New York, Routledge, 2000, XV-325 p.

1875. STEINER (Richard C.). Semitic names for utensils in the demotic word-list from Tebtunis. *Journal of Near Eastern studies*, 2000, 59, p. 191-194.

1876. STOL (M.). Birth in Babylonia and the Bible: its Mediterranean setting. Groningen, Styx Publications, 2000, X-276 p.

1877. TAL (Abraham). A dictionary of Samaritan Aramaic. Leiden, Boston a. Köln, Brill, 2000, 2 vol., XL-446 p., XLIV-521 p.

1878. TORAL-NIEHOFF (Isabel). Die Legende "Barlaam und Josaphat" in der arabisch-muslimischen Literatur: ein arabistischer Beitrag zur "Barlaam-Frage". *Die Welt des Orients*, 2000, 31, p. 110-144.

1879. USSISHKIN (D.). The credibility of the Tel Jezreel excavations: a rejoinder to Amnon Ben-Tor. *Tel Aviv*, 2000, 27, p. 248-256.

1880. VÁNDOR (Jaime). El holocausto: hacia la tipología de un nuevo género literario. *Aula Orientalis*, 1999-2000, 17-18, p. 323-333.

1881. VERNET (Joan). Cronología antiqua. *Aula Orientalis*, 1999-2000, 17-18, p. 441-443.

1882. VILADRICH (Mercè). Toponimia histórica de la primera organización de Estado omeya en el Próximo Oriente: verificando una propuesta sobre la Catalunya Vella. *Aula Orientalis*, 1999-2000, 17-18, p. 445-451.

1883. VOIGT (Rainer). Drei neue vergleichende semitistische Werke. *Die Welt des Orients*, 2000, 31, p. 165-189.

1884. WATSON-TREUMANN (Brigitte). Beyond the Cedars of Lebanon – Phoenician Timber Merchants and Trees from the "Black Mountain". *Die Welt des Orients*, 2000, 31, p. 75-83.

1885. XELLA (Paolo). I chrysoî naoí dei Cartaginesi. *Aula Orientalis*, 1999-2000, 17-18, p. 363-366.

1886. ZONTA (M.). La ricerca sulle fonti antiche della letteratura araba ed ebraica: a proposito di un libro recente. *Athenaeum*, 2000, 88, p. 597-604.

Cf. n° 691

§ 7. Iran.

1887. ABKAI-KHAVARI (Manijeh). Das Bild des Königs in der Sasanidenzeit: schriftliche Überlieferungen im Vergleich mit Antiquaria. Hildesheim, G. Olms, 2000, 307 p. (ill.). (Texte und Studien zur Orientalistik).

1888. ADIEGO (Ignasi-Xavier). Kandaules Myrsilos. *Aula Orientalis*, 1999-2000, 17-18, p. 453-454.

1889. AGUD (Ana). Lenguaje, comunidad humana y universo en la India clásica. *Aula Orientalis*, 1999-2000, 17-18, p. 455-461.

1890. EDENS (C.), WILKINSON (T. J.), BARRATT (G.). Hammat al-Qa and the roots of urbanism in southwest Arabia. *Antiquity*, 2000, 74, p. 854-862.

1891. GAUGER (J.-D.). Authentizität und Methode. Untersuchungen zum historischen Wert des persisch-griechischen Herrscherbriefes in literarischer Tradition. Hamburg, Kovac, 2000, 428 p.

1892. GEORGES (Pericles B.). Persian Ionia under Darius: the revolt reconsidered. *Historia*, 2000, 49, p. 1-39.

1893. GIUNTA (R.). Una camicia talismanica ottomana proveniente dal Museo Kircheriano. *Bullettino di paletnologia italiana*, 2000, 91, p. 225-238.

1894. GYSELEN (R.). Un dieu nimbé de flammes d'époque sassanide. *Iranica antiqua*, 2000, 35, p. 291-314.

1895. HARRISON (T.). The emptiness of Asia. Aeschylus' Persians and the history of the fifth century. London, Duckworth, 2000, 191 p.

1896. HELTZER (Michael). Some aspects of Achaemenid taxation in the V satrapy (Phoenicia). *Aula Orientalis*, 1999-2000, 17-18, p. 349-352.

1897. JAMZADEH (P.). An Achaemenid motif seen in later epic and art. *Iranica antiqua*, 2000, 35, p. 47-56.

1898. KEENAN (Jeremy). The theft of Saharan rock-art. *Antiquity*, 2000, 74, p. 287-288.

1899. MAEIR (A. M.). Sassanica varia Palaestiniensia: a Sassanian seal from T. Istaba, Israel, and other Sassanian objects from the southern Levant. *Iranica antiqua*, 2000, 35, p. 159-183.

1900. MASIA (K.). The evolution of swords and daggers in the Sasanian empire. *Iranica antiqua*, 2000, 35, p. 185-289.

1901. MEHR KIAN (J.). Un nouveau bas-relief d'Elymaide à «Shirinow», sur un passage de la migration de Baxtyaris. *Iranica antiqua*, 2000, 35, p. 57-68.

1902. MOSIG-WALBURG (K.). Die Flucht des persischen Prinzen Hormizd und sein Exil im römischen Reich – eine Untersuchung der Quellen. *Iranica antiqua*, 2000, 35, p. 69-109. – IDEM. Zu Spekulationen über den sasanidischen "Thronfolger Narsê" und seine Rolle in den sasanidisch-römischen Auseinandersetzungen im zweiten Viertel des 4. Jahrhunderts n. Chr. *Iranica antiqua*, 2000, 35, p. 111-157.

1903. PASPALAS (S. A.). On Persian-type forniture in Macedonia: the recognition and transmission of forms. *American journal of archaeology*, 2000, 104, p. 531-560.

1904. PIRART (Éric). El nombre de Ecbatana. *Aula Orientalis*, 1999-2000, 17-18, p. 463-467.

1905. SARIANIDI (V.). Neo-Assyrian parallels in late Bactrian glyptics. *Iranica antiqua*, 2000, 35, p. 21-31.

1906. SIMA (Alexander): Die Jagd im antiken Südarabien. *Die Welt des Orients*, 2000, 31, p. 84-109.

1907. THORAU (Peter). Die Ritterorden im Kampf mit Ayyubiden und Mamluken. *Die Welt des Orients*, 2000, 31, p. 145-164.

1908. VARGYAS (P.). Darius I and the Daric reconsidered. *Iranica antiqua*, 2000, 35, p. 33-46.

1909. WALSER (Gerold). Persische Teppiche als Quelle für die griechische Geschichte? *Klio*, 2000, 82, p. 54-72.

1910. ZOURNATZI (A.). Inscribed silver vessels of the Odrysian kings: gifts, tribute and the diffusion of the forms of "Achaemenid" metalware in Thrace. *American journal of archaeology*, 2000, 104, p. 683-706.

Cf. n° 3086

E

GREEK HISTORY

§ 1. Classical world in general. 1911-1942. – § 2. Prehellenic epoch. 1943-1956. – § 3. Sources and criticism of sources (*a*. Epigraphical sources; *b*. Literary sources). 1957-2045. – § 4. General and political history. 2046-2091. – § 5. History of law and institutions. 2092-2113. – § 6. Economic and social history. 2114-2158. – § 7. History of literature, philosophy and science (*a*. Literature; *b*. Philosophy and sciences). 2159-2292. – § 8. Religion and mythology. 2293-2313. – § 9. Archaeology and history of art. 2314-2365.

§ 1. Classical world in general.

* 1911. Année philologique (L'). Bibliographie critique et analytique de l'antiquité gréco-latine. Publiée par la Société Internationale de Bibliographie Classique. Tome LXIX, Bibliographie de l'année 1998 et compléments d'années antérieures. [Tome LXVIII, 1997. Cf. Bibl. 99, n° 1813.] Paris, Les Belles Lettres, 2000, LIX-1380 p.

* 1912. MALITZ (Jürgen). Gnomon. Bibliographische Datenbank. Internationales Informationssystem für die Klassische Altertumswissenschaft. Stand April 2000. München, Beck, 2000, 1 CD ROM

1913. Ancient Italy in its Mediterranean setting. Studies in honour of E. Macnamara. Ed. by D. RIDGWAY, F. SERRA RIDGWAY, M. PEARCE, E. HERRING, R. D. WHITEHOUSE and J. B. WILKINS. London, Accordia Research Institute-University of London, 2000, 336 p. (figs.).

1914. ANDERSON (G.). Fairytale in the ancient world. London a. New York, Routledge, 2000, XI-240 p.

1915. BAUMBACH (M.). Tradita et inventa. Beiträge zur Rezeption der Antike. Heidelberg, Universitätsverlag C. Winter, 2000, XI-676 p. (ill.).

1916. CHIGLINTSEV (Evgenij A.). Antichnoe rabstvo kak istoriograficheskaja problema. (The slavery of Antiquity as a historiographical problem). Kazan', Master Lajn, 2000, 135 p.

1917. CHIPPINDALE (C.), GILL (D. W. J.). Material consequences of contemporary classical collecting. *American journal of archaeology*, 2000, 104, p. 463-511.

1918. Vacat.

1919. CLARK (Gillian). Animal passions. *Greece and Rome*, 2000, 47, p. 88-93.

1920. Experts et pouvoir dans l'antiquité. [Mélanges]. *Revue historique*, 2000, 124, 615, p. 685-718. [Cont.: MOATTI (C.). Pour une histoire intellectuelle de l'Antiquité. – JOANNÈS (F.). De Babylone à Sumer: le parcours intellectuel des lettrés de la Babylone récente]

1921. FRASER (P. M.), MATTHEWS (E.). A lexicon of Greek personal names. Vol. 3 b. Central Greece from the Megarid to Thessaly. Oxford, Clarendon Press, 2000, XXI-478 p.

1922. HARDWICK (L.). Translating words, translating cultures. London, Duckworth, 2000, 160 p.

1923. HORDEN (P.), PURCELL (N.). The corrupting sea. A study of Mediterranean history. Oxford, Blackwell Publishers, 2000, XIII-761 p.

1924. HORNBLOWER (S.), MATTHEWS (E.). Greek personal names: their value as evidence. Oxford, Oxford U. P., 2000, VIII-184 p.

1925. HORNBLOWER (S.), SPAWFORTH (T.). Who's who in the classical world. Oxford, Oxford U. P., 2000, XIX-440 p.

1926. Intertextualidad en las literaturas Griega y Latina. Ed. por V. BÉCARES, F. PORDOMINGO, R. CORTÉS TOVAR y J. C. FERNÁNDEZ CORTE. Madrid, Ediciones Clásicas, 366 p.

1927. Kterismata. Philological studies dedicated to I. S. Kampitsis (1938–1990) in modern Greek. Ed. by G. M. SIFAKIS, F. I. KAKRIDIS, I. TOULOUMAKOS and O. TSAGARAKIS. Heraclion, Crete U. P., 2000, 471 p.

1928. LANZI (S.). Theos Anaitios. Storia della teodicea da Omero ad Agostino. Roma, Il calamo, 2000, 349 p.

1929. Literature in Greek and Roman worlds. A new perspective. Ed. by Oliver TAPLIN. Oxford, Oxford U. P., 2000, XV-596 p. [Cf. n[os] <choice> 2205, 2228.]

1930. MASSON (O.). Onomastica Graeca selecta. Tome 3. Genève, Librairie Droz, 2000, IX-352 p.

1931. MATZ (D.). Famous firsts in the ancient Greek and Roman world. Jefferson a. London, McFarland & Co., 2000, 154 p.

1932. Mezhdunarodnye otnoshenija i diplomatija v antichnosti. (International relations and diplomacy in Antiquity: [Articles]). Part 1. Ed. Oleg L. GABELENKO. Kazanskij gos. un-t. Kazan', [s. n.], 2000, [s. p.]. [Eng. summaries]

1933. MUSTI (Domenico). Simbologia della Vittoria dall'ellenismo a Costantino. *Rivista di filologia e di istruzione classica*, 2000, 128, p. 42-55.

1934. Neue Pauly (Der). Enzyklopädie der Antike. Hrsg v. H. CANCIK und H. SCHNEIDER. Vol. 8. Mer–Op. Vol. 9. Or-Poi. Stuttgart u. Weimar, Verlag J.B. Metzler, 2000, 2 vol., 1276 p., 1194 p.

1935. NIIMURA (Yuichiro). Kodai Suparuta shi kenkyû: Koten-ki eno michi. (Studies of the ancient Spartan history: the way to the classical period). Tokyo, Iwanami Shoten, 2000, 321p.

1936. PALMERINI (P.). Nostalgia. Dodici saggi sul mondo antico. Pisa e Roma, Istituti Editoriali e Poligrafici Internazionali, 2000, 244 p.

1937. SAUGE (André). Les degrés du verbe. Sens et formation du parfait en grec ancien. Bern, Berlin, Bruxelles, Frankfurt am Main, New York et Wien, Lang, 2000, XXII-704 p.

1938. SERGHIDOU (S.). Dorema. A tribute to the A. G. Levantis Foundation on the occasion of its 20th anniversary. Nicosia, A. G. Levantis Foundation, 2000, XIV-309 p.

1939. SPEAKE (G.). Encyclopedia of Greece and the Hellenic tradition. Vol. 1. A–K. Vol. 2. L–Z. London a. Chicago, Fitzroy Dearborn Publishers, 2000, XXXVII-XXXI-1861 p. (maps, site plans, pls.). [Cf. n° <choice> 3212.]

1940. VAN STEEN (G. A. H.). Venom in verse: Aristophanes in modern Greece. Princeton, Princeton U. P., 2000, XX-284 p.

1941. VERGER (Stéphane). Rites et espaces en pays celte et méditerranéen. Etude comparée à partir du sanctuaire d'Acy-Romance (Ardennes, France). Roma, Ecole française de Rome, 2000, 356 p. (Collection de l'Ecole française de Rome, 276).

1942. WALKER (J.). Rhetoric and poetics in antiquity. Oxford, Oxford U. P., 2000, XII-396 p.

Cf. n°s 168, 262, 692, 706, 1034, 1618, 2601, 2689

§ 2. Prehellenic epoch.

1943. AKURGAL (Ekrem). Ege batı uygarlığının doğduğu yer: Doğu Hellen kültür tarihi İ.Ö. 1050–333. (Aegean territory where the Western civilisation born: Eastern Greek cultural history 1050–333 BC.). Ed. by Mustafa ÖZTURANLİ. İzmir, İzmir Büyükşehir Belediyesi, 2000, 124 p.

1944. ALBERTI (M. Emanuela). Les poids de Malia entre les premiers et les second palais: un essai de mise en contexte. *Bulletin de correspondance hellénique*, 2000, 124, p. 57-73.

1945. DANDRAU (Alain). La peinture murale minoenne, II. Matériaux et typologie. *Bulletin de correspondance hellénique*, 2000, 124, p. 75-97.

1946. DIMOU (Élefthéria), SCHMITT (Anne), PELON (Olivier). Recherches sur les matériaux lithiques utilisés dans la construction du palais de Malia: étude géologique. *Bulletin de correspondance hellénique*, 2000, 124, p. 435-457.

1947. DUHOUX (Y.). How not to decipher the Phaistos disc: a review. *American journal of archaeology*, 2000, 104, p. 597-600.

1948. GALANIDOU (N.), TZEDAKIS (P. C.), LAWSON (I. T.), FROGLEY (M. R.). A revised chronological and palaeoenvironmental framework for the Kastritsa rockshelter, northwest Greece. *Antiquity*, 2000, 74, p. 349-355.

1949. HELLMANN (O.). Die Schlachtszesen der Ilias. Das Bild des Dichters vom Kampf in der Heroenzeit. Stuttgart, Franz Steiner Verlag, 2000, 218 p.

1950. HITCHCOCK (L. A.). Minoan architecture. A contextual analysis. Stockholm, Paul Äströms Förlag, 2000, 267 p. (ill.).

1951. KNAPPETT (Carl), SCHOEP (Ilse). Continuity and change in Minoan palatial power. *Antiquity*, 2000, 74, p. 365-371.

1952. LÉVY (Edmond). La cohérence du Code de Gortyne. *In*: Codification (La) des lois dans l'antiquité [Cf. n° 1618], p. 185-214.

1953. MOLCHANOV (Arkadij A.). Sotsial'nye struktury i obshchestvennye otnoshenija v Gretsii II tys. do n.e.: Problemy istochnikovedenija minoistiki i mikenologii. (Social structures and public relations in Greece of the second millenium B.C.: some problems of source criticism of Minoistics and Mycenology). Ed. L. P. MARINOVICH. RAN, In-t vseobshchej istorii. Moskva, [s. n.], 2000, 314 p. (ill.). [Eng. summary]

1954. PETIT (Th.). Usage de la coudée dans l'architecture palatiale de Chypre au premier millénaire. *Ktema*, 2000, 25, p. 173-188.

1955. VAN EFFENTERRE (H.), VAN EFFENTERRE (M.). La codification gortynienne, mythe ou réalité? *In*: Codification (La) des lois dans l'antiquité [Cf. n° 1618], p. 175-184.

1956. YON (M.), KARAGEORGHIS (V.), HIRSCHFELD (N.). Céramiques mycéniennes. Nicosia, Foundation A. G. Leventis, 2000, XI-251 p. (figs).

Cf. n° 268

§ 3. Sources and criticism of sources.

a. Epigraphical sources

* 1957. Bulletin épigraphique. *Revue des études grecques*, 2000, 113, p. 435-610. [99, 112. Cf. Bibl. 99, n° 1854.]

1958. BONIAS (Zissis). Une inscription de l'ancienne Bergè. *Bulletin de correspondance hellénique*, 2000, 124, p. 227-246.

1959. CASSIO (Albio Cesare). Un epigramma votivo spartano per Atena Alea. *Rivista di filologia e di istruzione classica*, 2000, 128, p. 129-134.

1960. DEL MONACO (Lavinio). Inschriften von Olympia 24. *Rivista di filologia e di istruzione classica*, 2000, 128, p. 385-405.

1961. DOBIAS-LALOU (Catherine). Le dialecte des inscriptions grecques de Cyrène. Paris, C.E.A.M., 2000, VIII-340 p.

1962. FERRARY (Jean-Louis). Les inscriptions du sanctuaire de Claros en l'honneur de Romains. *Bulletin de correspondance hellénique*, 2000, 124, p. 331-376.

1963. FEYEL (Christophe). Inscriptions inédites du Prytanée délien: dédicaces et actes d'archontes. *Bulletin de correspondance hellénique*, 2000, 124, p. 247-260.

1964. GAUTHIER (Philippe). Inscription agonistique de Messène. *Revue des études grecques*, 2000, 113, p. 631-635.

1965. Inscriptiones Graecae insularum maris Aegaei praeter Delum, consilio et auctoritate Academiae Scientiarum Berolinensis et Branderburgensis editae. Fasciculus 6. Inscriptiones Chii et Samii cum Corassiis Icariaque. Pars II. Inscriptiones Sami insulae: decreta, epistulae, sententiae, edicta imperatoria, leges, catalogi, tituli Atheniensium, tituli honorarii, tituli operum publicorum, inscriptiones ararum. Berlin, W. de Gruyter, 2000, XII-536-VIII p.

1966. Inscriptions du Sanctuaire de la Mère des Dieux Autochtone de Leukopetra (Macédoine). Ed. par P. H. M. PETSAS, M. HATZOPOULOS, L. GOUNAROPOULOU et P. PASCHIDIS. Paris, Diffusion de Boccard, 2000, 364 p. (ill., maps, pls.).

1967. PRÊTRE (Clarisse). La *Tabula* délienne de 168 av. J.-C. *Bulletin de correspondance hellénique*, 2000, 124, p. 261-271.

1968. RIAÑO RUFILANCHAS (D.). Zwei Agone in I. Priene 112, 91-95. *Zeitschrift für Papyrologie und Epigraphik*, 2000, 129, p. 89-96.

1969. ROSIVACH (V. J.). The Thracians of IG^2 1956. *Klio*, 2000, 82, p. 379-381.

1970. Supplementum Epigraphicum Graecum, 1997, 47. Ed. by H. W. PLEKET, R. S. STROUD, A. CHANIOTIS and J. H. M. STRUBBE. Amsterdam, J. C. Gieben, 2000, XXX-868 p.

b. Literary sources

1971. [Aeschines] CAREY (C.). Aeschines. Austin, University of Texas Press, 2000, XXXI-261 p.

1972. [Anonymous] COLINET (A.). Les Alchimistes Grecs. Tome 10. L'Anonyme de Zuretti ou L'art secré et divin de la chrysopée par un Anonyme. Paris, Les belles Lettres, 2000, CXVI-438 p.

1973. [Aratus] NEGRI (M.). Stelle spaventose o stelle luminose? Una nota su 'deinòs' in Arato. *Athenaeum*, 2000, 88, p. 277-280.

1974. [Aratus] NEGRI (M.). Edizioni di Arato a confronto. Un supplemento alle Concordanze dei Phenomena. *Athenaeum*, 2000, 88, p. 529-546.

1975. [Archestratus] OLSON (S. D.), SENS (A.). Archestratos of Gela: Greek culture and cuisine in the fourth century BCE. Oxford, Oxford U. P., 2000, LXXIV-261 p.

1976. [Aristophanes] GRILLI (A.). Aristofane. Le nuvole. Milano, Biblioteca Universale Rizzoli, 2000, 259 p.

1977. [Aristophanes] HENDERSON (J.). Aristophanes. Birds. Lysistrata. Women at the Thesmophoria. Cambridge a. London, Harvard U. P., 2000, 618 p.

1978. [Aristophanes] LABIANO ILUNDAIN (Juan Miguel). Estudio de las interjecciones en las comedias de Aristófanes. Amsterdam, Hakkert, 2000, 411 p.

1979. [Aristophanes] MEINECK (P.), STOREY (I. C.). Aristophanes. Clouds. Indianapolis a. Cambridge, Hackett Publishing Company, 2000, XIV-125 p.

1980. [Aristoteles] CRISP (Roger). Aristotle: Nicomachean Ethics. Cambridge, Cambridge U. P., 2000, XIII-213 p.

1981. [Aristoteles] MOREL (P. M.). Aristote. Petits traités d'histoire naturelle. Paris, G. F. Flammarion, 2000, 229 p.

1982. [Aristoteles] WEDIN (M. V.). Aristotle's theory of substance. The categories and metaphysics Zeta. Oxford, Oxford U. P., 2000, XIII-482 p.

1983. [Aristoteles] WILSON (M.). Aristotle's theory of the unity of science. Toronto, Buffalo a. London, University of Torornto Press, 2000, X-271 p.

1984. [Aristoteles, Aristophanes] HOLZHAUSEN (J.). Paideia oder Paidiá. Aristoteles und Aristophanes zur Wirkung der griechischen Tragödie. Stuttgart, Franz Steiner Verlag, 2000, 64 p.

1985. [Arrianus] AMBAGLIO (D.), OLIVA (A.). Arriano. L'India. Milano, Biblioteca Universale Rizzoli, 2000, 156 p.

1986. [Arrianus] BIFFI (Nicola). L'Indiké di Arriano. Bari, Edipuglia, 2000, 263 p.

1987. [Callimachus] ROSSI (Laura). La Chioma di Berenice: Catullo 66, 79-88, Callimaco e la propaganda di corte. *Rivista di filologia e di istruzione classica*, 2000, 128, p. 299-312.

1988. [Cassius Dio] BARZANÒ (A.), STROPPA (A.), GALIMBERTI (A.). Cassio Dione. Storia romana, libri LXIV-LXVII. Milano, Biblioteca Universale Rizzoli, 2000, 204 p.

1989. [Demosthenes] MAC DOWELL (D. M.). Demosthenes. On the false embassy (oration 19). New York, Oxford U. P., 2000, XVIII-368 p.

1990. [Demosthenes] WORTHINGTON (I.). Demosthenes. Statesman and orator. London a. New York, Routledge, 2000, XIV-289 p.

1991. [Dio Chrysostomus] SWAIN (S.). Dio Chrysostom. Politics, letters and philosophy. Oxford, Oxford U. P., 2000, X-308 p.

1992. [Dio Chrysostomus] KLAUCK (H. J.), BÄBLER (B.). Dion von Prusa. Olympische Rede oder Über die Erste Erkenntis Gottes. Darmstadt, Wissenschaftliche Buchgesellschaft, 2000, 250 p.

1993. [Euripides] CROPP (M. J.). Euripides. Iphigenia in Tauris. Westminster, Aris and Phillips, 2000, 283 p.

1994. [Euripides] MORWOOD (J.), HALL (E.). Euripides. Bacchae and other plays. Oxford, Oxford U. P., 2000, LI-227 p.

1995. [Euripides] MÜLLER (C. W.). Euripides. Philoktet. Testimonien und Fragmente. Berlin u. New York, Walter de Gruyter, 2000, 468 p.

1996. [Euripides] PADUANO (Guido). Euripide. Ippolito. Milano, Biblioteca Universale Rizzoli, 2000, 141 p.

1997. [Euripides] PREISER (C.). Euripides. Telephos. Hildesheim, Zürich u. New York, Georg Olms, 2000, 647 p.

1998. [Euripides] SULLIVAN (S. D.). Euripides' use of philological terminology. Montreal, Kingston, London a. Ithaca, McGill-Queen's U. P., 2000, XII-234 p.

1999. [Euripides] TRÉDÉ (M.). Euripide. Théâtre complet 1. Paris, G. F. Flammarion, 2000, 361 p.

2000. GERA (Deborah Levine). Two thought experiments in the Dissoi Logoi. *American journal of philology*, 2000, 121, p. 21-45.

2001. [Claudius Galenus] BOUDON (Véronique). Galien. Exhortation à l'étude de la médecine. Art médical. Paris, Les Belles Lettres, 2000, 2 tomes, 454 p. (84-117, 274-392 doppie).

2002. [Claudius Galenus] MAGNALDI (J.). Claudii Galeni Pergameni Peri psyches pathematon kai hamartematon. Roma, Istituto Poligrafico e Zecca dello Stato, 2000, LXII-131 p.

2003. [Galenus] GRANT (M.). Galen on food and diet. London a. New York, Routledge, 2000, IX-214 p.

2004. [Herodotus] DORATI (M.). Le Storie di Erodoto: etnografia e racconto. Pisa e Roma, Istituti Editoriali e Poligrafici Internazionali, 2000, 236 p.

2005. [Herodotus] MAC QUEEN (E. I.). Herodotus, book VI. Bristol, Bristol Classical Press, 2000, XVI-232 p.

2006. [Hippocrates] ANDÒ (V.). Ippocrate. Natura della donna. Milano, Biblioteca Universale Rizzoli, 2000, 317 p.

2007. [Hippocrates] JOUANNA (Jacques). L'Hippocrate de Venise (Marcianus Gr. 269; col. 533): nouvelles observationes codicologiques et histoire du texte. *Revue des études grecques*, 2000, 113, p. 193-210.

2008. [Homerus] HAMMOND (M.), GRIFFIN (J.). Homer. The Odyssey. London, Duckworth, 2000, XXVI-290 p.

2009. [Homerus] LOMBARDO (S.), MURNAGHAN (S.). Homer. Odyssey. Indianapolis a. Cambridge, Hackett Publishing Co. Inc., 2000, LXIV-414 p.

2010. [Homerus] PULLEYN (S.). Homer. Iliad I. Oxford, Oxford U. P., 2000, XI-304 p.

2011. [Homerus] TSAGARAKIS (O.). Studies in Odyssey 11. Stuttgart, Franz Steiner Verlag, 2000, 144 p.

2012. [Homerus, Hesiodus] GOTSHALK (R.). Homer and Hesiod. Myth and philosophy. Lanham, New York a. Oxford, University Press of America, 2000, XI-373 p.

2013. [Hypereides] WHITEHEAD (D.). Hypereides. The forensic speeches. Oxford, Oxford U. P., 2000, XVII-523 p.

2014. [Isocrates] MIRHADY (D. C.), TOO (Y. L.). Isocrates. Austin, University of Texas Press, 2000, XXIX-279 p.

2015. [Ister] JACKSON (S.). Istrus the Callimachean. Amsterdam, Adolf M. Hakkert, 2000, 157 p.

2016. [Lycophron] GIGANTE LANZARA (Valeria). Licofrone. Alessandra. Milano, Biblioteca Universale Rizzoli, 2000, 448 p.

2017. [Lysias] FERRANTE (D.). Lisia. Epitafio. Napoli, Ferraro, 2000, 45 p.

2018. [Lysias] IMÍZCOZ (J. M. F.). Lisias. Discursos XXVI-XXXV. Fragmentos. Madrid, Consejo Superior de Investigaciones Científicas, 2000, XXXVII-356 (30-35, 44-49, 58-62, 76-89, 102-115, 130-145, 158-162, 178-183, 200-205 doppie).

2019. [Lysias] TODD (S. C.). Lysias. Austin, University of Texas Press, 2000, XXX-402 p.

2020. [Menander] ARNOTT (W. G.). Menander. Vol. 3. Cambridge a. London, Harvard U. P., 2000, XI-639 p.

2021. [Menander] VILARDO (M.). Menandro. La donna di Samo. Milano, Biblioteca Universale Rizzoli, 2000, 179 p.

2022. [Nicolaus Damascenus, Ctesias] LENFANT (Dominique). Nicolas de Damas et le corpus des fragments de Ctésias: du fragment comme adaptation. *Ancient society*, 2000, 30, p. 293-318.

2023. [Philo Alexandrinus] NOACK (Christian). Gottesbewußtsein. Exegetische Studien zur Soteriologie und Mystik bei Philo von Alexandria. Tübingen, Mohr (Paul Siebeck), 2000, XIII-300 p.

2024. [Philostratus] BILLAULT (A.). L'univers de Philostrate. Bruxelles, Éditions Latomus, 2000, 144 p.

2025. [Plato] FERRARI (G. R. F.), GRIFFITH (Th.). Plato. The Republic. Cambridge, Cambridge U. P., 2000, XLVIII-382 p.

2026. [Plato] JOYAL (M.). The Platonic Theages. An introduction, commentary and critical edition. Stuttgart, Franz Steiner Verlag, 2000, 335 p.

2027. [Plato] SERRANO CANTARÍN (R.), DÍAZ DE CERIO DÍEZ (M.). Platón. Gorgias. Madrid, Consejo Superior de Investigaciones Científicas, 2000, CLXXII-303 p. (1-266 doppie).

2028. [Plotinus] LINGUITI (A.). La felicità e il tempo. Plotino, Enneadi I 4-I 5. Milano, Edizioni Universitarie di Lettere, Economia, Diritto, 2000, 165 p.

2029. [Plotinus] NINCI (M.). Plotino. Il pensiero come diverso dall'uno. Quinta enneade. Milano, Biblioteca Universale Rizzoli, 2000, 587 p.

2030. [Plutarchus] COMOTTI (G.), BALLERIO (R.). Plutarco. La musica. Milano, Biblioteca Universale Rizzoli, 2000, 129 p.

2031. [Plutarchus] GIEBEL (Marion). Plutarch. Die Kunst zu leben. Frankfurt am Main u. Leipzig, Insel Verlag, 2000, 195 p.

2032. [Plutarchus] ORTIZ (A. M.). Plutarco en España. Traducciones de Moralia en el siglo XVI. Murcia, Universidad de Murcia, 2000, 374 p.

2033. [Pseudo-Scymnus] MARCOTTE (D.). Géographes grecs. Introduction et Ps.-Scymnos, Circuit de la terre. Paris, Les Belles Lettres, 2000, CLXVIII-310 p.

2034. [Pseudo-Zeno] STONE (M. E.), SHIRINIAN (M. E.). Pseudo-Zeno: Anonymous philosophical treatise. Leiden, Boston a. Köln, Brill, 2000, XIV-254 p.

2035. [Claudius Ptolemaeus] BERGGREN (J. L.), JONES (A.). Ptolemy's Geography: an annotated translation of the theoretical chapters. Princeton a. Oxford, Princeton U. P., 2000, XII-192 p.

2036. [Claudius Ptolemaeus] SOLOMON (J.). Ptolemy. Harmonics. Leiden, Boston a. Köln, Brill, 2000, XXXVII-192 p.

2037. [Quintus Smyrnaeus] JAMES (A.), LEE (K.). A commentary on Quintus of Smyrna Posthomerica V. Leiden, Boston a. Köln, Brill, 2000, X-172 p.

2038. [Sextus Empiricus] SPINELLI (E.). Sesto Empirico. Contro gli astrologi. Napoli, Bibliopolis, 2000, 222 p.

2039. [Simonides] SBARDELLA (Livio). Achille e gli eroi di Platea. Simonide, frr. 10-11 W^2. *Zeitschrift für Papyrologie und Epigraphik*, 2000, 129, p. 1-11.

2040. [Sophocles] FLASHAR (H.). Sophokles, Dichter im demokratischen Athen. München, Verlag C. H. Beck, 2000, 220 p.

2041. [Sophocles] MEINECK (P.), WODRUFF (P.). Sophocles. Oedipus Tyrannus. Indianapolis a. Cambridge, Hackett Publishing Company, 2000, XXXIV-67 p.

2042. [Soranus Ephesinus] BURGUIÈRE (P.), GOUREVITCH (D.), MALINAS (Y.). Soranos d'Éphèse. Maladies des femmes. Tome 4, livre IV. Paris, Les Belles Lettres, 2000, XXVIII-197 p.

2043. [Strabo] DUECK (D.). Strabo of Amasia. A Greek man letters in Augustan Rome. London a. New York, Routledge, 2000, IX-227 p. (maps).

2044. [Themistius] PENELLA (R. J.). The private orations of Themistius. Berkeley, Los Angeles a. London, University of California Press, 2000, XIII-258 p.

2045. [Thucydides] FRIEDRICHS (J.). Aufschlussreiche Rhetorik. Ein Versuch über die Redekultur und ihren Verfall bei Thukydides. Würzburg, Ergon, 2000, 154 p.

Cf. n°s 268, 1953, 2159-2292

§ 4. **General and political history.**

2046. ABRAMENKO (A.). Der Fremde auf dem Thron. Die letze Verschwörung gegen Alexander d. Gr. *Klio*, 2000, 82, p. 361-378.

2047. Alexander the Great in fact and fiction. Ed. by A. B. BOSWORTH and E. J. BAYNHAM. Oxford, Oxford U. P., 2000, VIII-370 p.

2048. AMOURETTI (M. C.), CHRISTIEN (J.), RUZÉ (F.), SINEUX (P.). Le regard des Grecs sur la guerre. Mythes et réalités. Paris, Ellipses, 2000, 206 p.

2049. ANDERSON (Greg). Alkmeonid "homelands", political exile and the unification of Attica. *Historia*, 2000, 49, p. 387-412.

2050. BALDINI (Antonio). Considerazioni sulla cronologia di Olimpiodoro di Tebe. *Historia*, 2000, 49, p. 488-502.

2051. BETT (R.). Pyrrho, his antecedents and his legacy. Oxford, Oxford U. P., 2000, X-264 p.

2052. CAMASSA (Giorgio). Cronaca degli anni fecondi: Clistene, il demos e le eterie. *Quaderni di storia*, 2000, 26, 51, p. 41-56.

2053. CARTLEDGE (P.). Greek political thought: the historical context. *In*: Cambridge (The) history of Greek and Roman political thought [Cf. n° 1034], p. 11-22.

2054. CHÈNE (Olivier). La bataille de Symè (Thuc. VIII 42; janv. 411 a.C.). *Revue des études grecques*, 2000, 113, p. 101-130.

2055. CONSOLO LANGHER (Sebastiana Nerina). Agatocle. Da capoparte a monarca fondatore di un regno tra Cartagine e i diadochi. Messina, Dipartimento di Scienze dell'Antichità, 2000, 446 p.

2056. COZZOLI (Umberto). Aspetti emergenti nella storiografia sulla guerra deceleica. *Rivista di filologia e di istruzione classica*, 2000, 128, p. 135-143.

2057. Defining ancient Arkadia. Symposium April 1–4 1998. Ed. by T. H. NIELSEN and J. ROY. København, C. A. Reitzel, 2000, 491 p. (ill., maps).

2058. DREYER (Boris). Athen und Demetrios Poliorketes nach der Schlacht von Ipsos (301 v.Chr.). Bemerkungen zum Marmor Parium, FgrHist 239 B 27 und zur Offensive des Demetrius im Jahre 299/8 v.Chr. *Historia*, 2000, 49, p. 54-66.

2059. FUNKE (S.). Aiakidenmythos und epeirotisches Königtum. Der Weg einer hellenischen Monarchie. Stuttgart, Franz Steiner Verlag, 2000, 238 p.

2060. GOLDHILL (Simon). Greek drama and political theory. *In*: Cambridge (The) history of Greek and Roman political thought [Cf. n° 1034], p. 23-59.

2061. GRAINGER (J. D.). Aitolian prosopographical studies. Leiden, Boston a. Köln, Brill, 2000, XII-339 p.

2062. HAMMOND (Nicholas G. L.). The continuity of Macedonian institutions and the Macedonian kingdoms of the Hellenistic era. *Historia*, 2000, 49, p. 141-160.

2063. HOFER (Marc). Tyrannen, Aristokraten, Demokraten. Untersuchungen zu Staat und Herrschaft im griechischen Sizilien von Phalaris bis zum Aufstieg von Dionysios I. Bern, Berlin, Bruxelles, Frankfurt am Main, New York u. Wien, Lang, 2000, 293 p.

2064. HORSMANN (Gerhard). Athens Weg zur eigenen Währung: der Zusammenhang der metrologischen Reform Solons mit der timokratischen. *Historia*, 2000, 49, p. 259-277.

2065. KRAUTSCHICK (Stefan). Zur Entstehung eines Datums: 375 – Beginn der Völkerwandung. *Klio*, 2000, 82, p. 217-222.

2066. LINK (Stefan). Das Paros-Abenteuer des Miltiades (Hdt. 6, 132-136). *Klio*, 2000, 82, p. 40-53.

2067. MAC INERNEY (J.). The folds of Parnassos. Land and ethnicity in ancient Phokis. Austin, Univesity of Texas Press, 2000, XVI-391 p. (maps, pls).

2068. MACGREGOR MORRIS (Ian). To make a new Thermopylae: Hellenism, Greek liberation and the battle of Thermopylae. *Greece and Rome*, 2000, 47, p. 211-230.

2069. MACKAY (Christopher S.). Damon of Chaeronea: the loyalties of a Boeotian town during the Mithridatic war. *Klio*, 2000, 82, p. 91-106.

2070. MATTINGLY (Harold B.). The Athenian treaties with Troizen and Hermione. *Historia*, 2000, 49, p. 131-140.

2071. MEISSNER (Burkhard). Hofmann und Herrscher. Was es für die Griechen hieß, Freund eines Königs zu sein. *Archiv für Kulturgeschichte,* 2000, 82, p. 1-36.

2072. MITCHELL (L. G.). A new look at the election of generals at Athens. *Klio*, 2000, 82, p. 344-360.

2073. MUNN (M.). The school of history: Athens in the age of Socrates. Berkeley, Los Angeles a. London, University of California Press, 2000, XII-525 p.

2074. OSBORNE (R.). Classical Greece 500–323 BC. Oxford, Oxford U. P., 2000, IX-271 p.

2075. OWEN (Sara). New light on Thracian Thasos: a reinterpretation of the 'Cave of Pan'. *Journal of Hellenic studies*, 2000, 120, p. 139-143.

2076. PELLING (Christopher). Literary texts and the Greek historian. London a. New York, Routledge, 2000, X-338 p.

2077. PENNER (T.). Socrates. *In*: Cambridge (The) history of Greek and Roman political thought [Cf. n° 1034], p. 164-189.

2078. Polis and politics. Studies in ancient Greek history. Ed. by P. FLENSTED JENSEN, T. H. NIELSEN and L. RUBINSTEIN. København, Museum Tusculanum Press, 2000, 651 p. (maps, figs).

2079. Presenza e funzione della città di Tebe nella cultura greca. Atti del Convegno Internazionale (Urbino 7–9 luglio 1997). A cura di Paola A. BERNARDINI. Pisa e Roma, Istituti Editoriali e Poligrafici Internazionali, 2000, 378 p.

2080. RAAFLAUB (K. A.). Poets, lawgivers and the beginnings of political reflection in archaic Greece. *In*: Cambridge (The) history of Greek and Roman political thought [Cf. n° 1034], p. 23-59.

2081. RICHER (N.). Nouvelles recherches sur Sparte archaïque. *L'antiquité classique*, 2000, 69, p. 277-284.

2082. RUSCHENBUSCH (E.). Weitere Untersuchungen zu Pherekydes von Athen (FgrHist 3). *Klio*, 2000, 82, p. 335-343.

2083. SANCISI-WEERDENBURG (H.). Peisistratos and the tyranny. A reappraisal of the evidence. Amsterdam, J. C. Gieben, 2000, XII-183 p.

2084. SCHOFIELD (M.). Approaching the Republic. *In*: Cambridge (The) history of Greek and Roman political thought [Cf. n° 1034], p. 190-232.

2085. SCHOLTEN (J. B.). The politics of plunder. Aitolians and their koinon in the early Hellenistic era, 279–217 BC. Berkeley, Los Angeles a. London, University of California Press, 2000, XXVI-339 p.

2086. SHIPLEY (G.). The Greek world after Alexander 323–30 a.C. London a. New York, Routledge, 2000, XXXI-568 p. (ills, maps).

2087. SORDI (Marta). Trasibulo tra politica e religione. *Rivista di filologia e di istruzione classica*, 2000, 128, p. 182-191.

2088. SURIKOV (Igor' E.). Iz istorii grecheskoj aristokratii pozdnearkhaicheskoj i ranneklassicheskoj epokh: Rod Alkmeonidov v politicheskoj zhizni Afin VII–V vv. do n.e. (From the history of Greek late archaic and early classical aristocracy: the family of Alkmeonides in the political life of Athens, 7th–5th centuries B.C). RAN, In-t vseobshchej istorii. Moskva, [s. n.], 2000, 283 p. (tables).

2089. THOMMEN (Lukas). Spartas Umgang mit der Vergangenheit. *Historia*, 2000, 49, p. 40-53.

2090. VIDAL NAQUET (P.). Les Grecs, les historiens, la démocratie: le grand écart. Paris, Éditions La Découverte, 2000, 284 p.

2091. WINTON (R.). Herodotus, Thucydides and the sophists. *In*: Cambridge (The) history of Greek and Roman political thought [Cf. n° 1034], p. 89-121.

§ 5. History of law and institutions.

2092. ALLEN (Danielle S.). The world of Prometheus. The politics of punishing in democratic Athens. Princeton, Princeton U. P., 2000, XIII-449 p.

2093. *Vacat*.

2094. CANTARELLA (Eva). Les peines de mort en Grèce et à Rome. Origines et fonctions des supplices capitaux dans l'Antiquité classique. Paris, Albin Michel, 2000, 320 p.

2095. CARNEY (Elizabeth Donnelly). Women and monarchy in Macedonia. Norman, University of Oklahoma Press, 2000, XIII-367 p. (ill.).

2096. COHEN (E. E.). The Athenian nation. Priceton, Princeton U. P., 2000, XX-250 p.

2097. CROMEY (R. D.). Kleisthenes' 700 Epistia. *L'antiquité classique*, 2000, 69, p. 43-63.

2098. DEISSLER (J.). Antike Sklaverei und Deutsche Aufklärung. Im Spiegel von Johann Friedrich Reitemeiers 'Geschichte und Zustand der Sklaverey und Leibeigenschaft in Griechenland' (1789). Stuttgart, Franz Steiner Verlag, 2000, VI-507 p.

2099. FOLLET (Simone). Les deux archontes Pamménès du Ier siècle a.C. à Athènes. *Revue des études grecques*, 2000, 113, p. 188-192.

2100. GEHRKE (H.-J.). Verschriftung und Verschriftlichung sozialer Normen im archaischen und klassischen Griechenland. *In*: Codification (La) des lois dans l'antiquité [Cf. n° 1618], p. 141-159.

2101. HAMILTON (R.). Treasure map: a guide to the Delian inventories. Ann Arbor, The University of Michigan Press, 2000, XIII-479 p.

2102. HESK (J.). Deception and democracy in classical Athens. Cambridge, Cambridge U. P., 2000, VIII-336 p.

2103. L'HOMME (M.-L.). La notion de patrie dans la pensée politique de Solon. *L'antiquité classique*, 2000, 69, p. 21-41.

2104. LANZILLOTTA (Eugenio). Elementi di diritto costituzionale nelle iscrizioni greche del IV secolo a.C. *Rivista di filologia e di istruzione classica*, 2000, 128, p. 144-154.

2105. Law and social status in classical Athens. Ed. by V. HUNTER and J. EDMONDSON. Oxford, Oxford U. P., 2000, XIII-206 p.

2106. LAZZARINI (M. Letizia). Atene, gli alleati e il tesoro di Atena. Considerazioni su alcuni inventari del IV secolo a.C. *Rivista di filologia e di istruzione classica*, 2000, 128, p. 155-169.

2107. LUPI (M.). L'ordine delle generazioni. Classi di età e costumi matrimoniali nell'antica Sparta. Bari, Edipuglia, 2000, 228 p.

2108. MIRON (Dolores). Transmitters and representatives of power: royal women in ancient Macedonia. *Ancient society*, 2000, 30, p. 35-52.

2109. OSTWALD (Martin). Oligarchia. The development of a constitutional form in ancient Greece. Stuttgart, Franz Steiner Verlag, 2000, 96 p.

2110. PIÉRART (M.). Qui étaient les nomothètes à Athènes à l'epoque du Dèmosthène? *In*: Codification (La) des lois dans l'antiquité [Cf. n° 1618], p. 229-256.

2111. Production and public powers in classical antiquity. Ed. by E. LO CASCIO and D. W. RATHBONE. Cambridge, Cambridge Philological Society, 2000, 99 p.

2112. RUSSELL (F. S.). Information gathering in Classical Greece. Ann Arbor, University of Michigan Press, 2000, VIII-267 p.

2113. WILSON (Peter). The Athenian institution of the khoregia: the chorus, the city and the stage. Cambridge, Cambridge U. P., 2000, XVI-435 p. (figs.).

Cf. n° 1128

§ 6. Economic and social history.

2114. ARNOULD (D.). Les grenouilles de Sériphos. *L'antiquité classique*, 2000, 69, p. 257-260.

2115. Art (The) of ancient spectacle. Ed. by B. BERGMANN and C. KONDOLEON. New Haven a. London, Yale U. P., 2000, 374 p. (ills, pls.).

2116. BARAKO (T. J.). The Philistine settlement as mercantile phenomenon? *American journal of archaeology*, 2000, 104, p. 513-530.

2117. BRAND (Helmut). Griechische Musikanten im Kult von der Frühzeit bis zum Beginn der Spätklassik. Dettelbach, Döll, 2000, 258 p.

2118. BRESSON (A.). La cité merchande. Bordeaux, Éditions Ausonius, 2000, 343 p.

2119. BRUN (J.-P.). The production of perfumes in antiquity: the cases of Delos and Paestum. *American journal of archaeology*, 2000, 104, p. 277-308.

2120. CAHN (H. A.). La moneta greca dalle guerre persiane ad Alessandro Magno. *In*: Moneta (La) greca e romana [Cf. n° 168], p. 33-43.

2121. CARTER (J. C.), CRAWFORD (M.), LEHMAN (P.), NIKOLAENKO (G.), TRELOGAN (J.). The chora of Chersonesos in Crimea, Ukraine. *American journal of archaeology*, 2000, 104, p. 707-741.

2122. CASSIO (Albio Cesare), MUSTI (Domenico), ROSSI (Luigi Enrico). Synaulia. Cultura musicale in Grecia e contatti mediterranei. Napoli, A.I.O.N., 2000, 320 p.

2123. CUBELLI (V.), FORABOSCHI (D.). Caratteri generali della monetazione ellenistica. *In*: Moneta (La) greca e romana [Cf. n° 168], p. 61-76.

2124. DE CAZANOVE (O.). Bacanal ou citerne? *L'antiquité classique*, 2000, 69, p. 237-253.

2125. Death and disease in the ancient city. Ed. by V. M. HOPE and E. MARSHALL. London a. New York, Routledge, 2000, XII-194 p.

2126. Economies beyond agriculture in the classical world. Ed. by D. J. MATTINGLY and J. SALMON. London a. New York, Routledge, 2000, XII-324 p. (ills).

2127. FOURNET (Jean-Luc). Un nom rare de boulanger: artokolletes. *Revue des études grecques*, 2000, 113, p. 392-412.

2128. GARRISON (D. H.). Sexual culture in ancient Greece. Norman, University of Oklahoma Press, 2000, X-331 p. (figs).

2129. GOETTE (Hans Rupprecht). Ho axiologos demos Sounion. Landeskundliche Studien in Südost-Attika. Rhaden/West., Leidorf, 2000, VI-160 p.

2130. HAENTJENS (A. M. E.). Reflections on female infanticide in the Greco-Roman world. *L'antiquité classique*, 2000, 69, p. 261-164.

2131. Hellenistic economies. Ed. by Z. H. ARCHIBALD, J. DAVIES, V. GABRIELSEN and G. J. OLIVER. London a. New York, Routledge, 2000, XVI-400 p.

2132. HODKINSON (S.). Property and wealth in classical Sparta. London, Duckworth, 2000, XIII-498 p.

2133. JOHNSON (William A.). Musical evenings in the early empire: new evidence from a Greek papyrus with musical notation. *Journal of Hellenic studies*, 2000, 120, p. 57-85. – IDEM. Toward a sociology of reading in classical antiquity. *American journal of philology*, 2000, 121, p. 593-627.

2134. JORDAN (B.). The crews of Athenian triremes. *L'antiquité classique*, 2000, 69, p. 81-101.

2135. KORANJAK (M.). Publikum und Redner. Ihre Interaktion in der sophistischen Rhetorik der Kaiserzeit. München, C. H. Beck, 2000, 254 p.

2136. MAIULLARI (Franco). Sul concetto di omertà a partire dalla Grecia antica. *Quaderni di storia*, 2000, 51, p. 75-109.

2137. MAYOR (Adrienne). The first fossil hunters. Paleontology in Greek and Roman times. Princeton, Princeton U. P., 2000, XX-361 p.

2138. MENU (M.). Jeunes et vieux chez Lysias. L'*akolasia* de la jeunesse au IVe siècle av. J.-C. Rennes, Presses Universitaires de Rennes, 2000, 249 p.

2139. MÖLLER (A.). Naukratis. Trade in archaic Greece. Oxford, Oxford U. P., 2000, XVII-290 p. (ill., pls.).

2140. MORAWETZ (T.). Der Demos als Tyrann und Banause. Aspekte antidemokratischer Polemik im Athen des 5. und 4. Jahrhunderts v. Chr. Frankfurt am Main, Peter Lang, 2000, 186 p.

2141. MUSTI (Domenico). Un bilancio sulla questione dei Nikephoria di Pergamo. *Rivista di filologia e di istruzione classica*, 2000, 128, p. 257-298.

2142. NORMAN (A. F.). Antioch as a centre of Hellenic culture as observed by Libanius. Liverpool, Liverpool U. P., 2000, XXII-198 p.

2143. PARISE (Nicola Franco). Lineamenti di preistoria monetaria greca. *In*: Moneta (La) greca e romana [Cf. n° 168], p. 11-18.

2144. PARISINOU (Eva). 'Lighting' the world of women: lamps and torches in the hands of women in the late Archaic and Classical periods. *Greece and Rome*, 2000, 47, p. 19-43.

2145. PÉBARTHE (Ch.). Fiscalité, empire athénien et écriture: retour sur les causes de la guerre du Péloponnese. *Zeitschrift für Papyrologie und Epigraphik*, 2000, 129, p. 47-76.

2146. RADNOTI ALFÖLDI (M.). Gli inizi della monetazione nel Mediterraneo fino alle guerre persiane. *In*: Moneta (La) greca e romana [Cf. n° 168], p. 19-31.

2147. RAVEN (J. E.). Plants and plant lore in ancient Greece. Oxford, Oxbow Books, 2000, XXVII-106 p.

2148. ROSIVACH (Vincent J.). The audiences of New Comedy. *Greece and Rome*, 2000, 47, p. 169-171.

2149. RUTTER (N. K.). Magna Grecia e Sicilia. *In*: Moneta (La) greca e romana [Cf. n° 168], p. 45-60.

2150. SAMONS II (L. J.). Empire of the owl. Athenian imperial finance. Stuttgart, Franz Steiner Verlag, 2000, 358 p.

2151. SPINA (L.). La forma breve del dolore. Ricerche sugli epigrammi funerari greci. Amsterdam, Hakkert, 2000, 101 p.

2152. TOO (Y. L.). The pedagogical contract. The economies of teaching and learning in the ancient world. Ann Arbor, University of Michigan Press, 2000, 176 p.

2153. TROISI (F. F.). La donna nella società ellenistica. Testimonianze epigrafiche. Bari, Edipuglia, 2000, 120 p.

2154. TURLEY (D.). Slavery. Oxford, Blackwell, 2000, VIII-174 p.

2155. UGOLINI (Gherardo). Sofocle e Atene: vita politica e attività teatrale nella Grecia classica. Roma, Carocci, 2000, 275 p.

2156. VAN LIEFFERINGE (C.). Auditions et conférences à Delphes. *L'antiquité classique*, 2000, 69, p. 149-164.

2157. VANDENSTEENDAM (Gh.). Les concours musicaux et les musiciens d'Argos du VIIe siècle avant J.-C. au IIIe siècle après J.-C. *Maia*, 2000, 4, p. 105-114.

2158. ZELNICK-ABRAMOVITZ (R.). Did patronage exist in classical Athens? *L'antiquité classique*, 2000, 69, p. 65-80.

Cf. n° 1916

§ 7. History of literature, philosophy and science.

a. Literature

2159. ALLAN (W.). The Andromache and Euripidean tragedy. Oxford, Oxford U. P., 2000, XII-310 p.

2160. AUERBERGER (Janick). «Du prince au berger, tout homme a son content de fromage...», Odyssée 4, 87-88. *Revue des études grecques*, 2000, 113, p. 1-41.

2161. AX (W.). Lexis und Logos. Studien zur antike Grammatik und Rhetorik. Stuttgart, Franz Steiner Verlag, 2000, 229 p.

2162. BELFIORE (E. S.). Murder among friends. Violation of philia in Greek tragedy. New York a. Oxford, Oxford U. P., 2000, XIX-282 p.

2163. BENARDETE (S.). The argument of the action. Essays on Greek poetry and philosophy. Chicago a. London, The University of Chicago Press, 2000, XXI-434 p.

2164. BERNAYS (L.). Ars Poetica. Studien zu formalen Aspekten der antiken Dichtung. Frankfurt am Main, Peter Lang, 2000, 237 p.

2165. BICHLER (R.). Herodots Welt. Der Aufbau der Historie am Bild der Fremden Länder und Völker, ihrer Zivilisation und ihrer Geschichte. Oldenbourg, Akademie Verlag, 2000, 424 p.

2166. BILLAULT (A.). La littérature grecque. Paris, Hachette Supérieur, 2000, 283 p.

2167. BILLERBECK (M.), ZUBLER (C.). Das Lob der Fliege von Lukian bis L. B. Alberti. Gattungsgeschichte Texte, Übersetzungen und Kommentar. Bern, Peter Lang, 2000, 264 p.

2168. BIZZARRO (F. C.). Poetica e critica letteraria nei frammenti dei poeti comici greci. Napoli, M. D'Auria, 2000, 208 p.

2169. BOST POUDERON (Cécile). Le ronflement des Tarsiens: l'interprétation du Discours XXXIII de Dion de Pruse. *Revue des études grecques*, 2000, 113, p. 636-651.

2170. BOSWORTH (A. B.). The historical context of Thucydides' funeral oration. *Journal of Hellenic studies*, 2000, 120, p. 1-16.

2171. BRISSON (Luc). Sur une lecture récente des Lois de Platon. *Revue des études grecques*, 2000, 113, p. 230-250.

2172. BURIAN (Peter). Translation, the profession and the poets. *American journal of philology*, 2000, 121, p. 299-307.

2173. CALAME (Claude). Le récit en Grèce ancienne. Paris, Belin, 2000, 301 p. (L'antiquité au présent).

2174. CANDAU MORÓN (José María). Plutarch's Lysander and Sulla: integrated characters in Roman historical perspective. *American journal of philology*, 2000, 121, p. 453-478.

2175. CASTELLI (C.). Meter sophiston. La tragedia nei trattati greci di retorica. Milano, Ediz. Universitarie di Lettere, Economia, Diritto, 2000, 187 p.

2176. CLARKE (M.). Flesh and spirit in the songs of Homer. A study of words and myths. Oxford, Clarendon Press, 2000, XV-378 p.

2177. COOK (Brad L.). Theopompos not Theophrastos: correcting an attribution in Plutarch Demosthenes 14,4. *American journal of philology*, 2000, 121, p. 537-547.

2178. DEAN ANDERSON (R.). Glossary of Greek rhetorical terms connected to methods of argumentation. Figures and tropes from Anaximenes to Quintilian. Leuven, Peeters, 2000, 130 p.

2179. DEE (J. H.). Epitheta hominum apud Homerum. Hildesheim, [s. n.], 2000, XXXIV-610 p.

2180. DEFORGE (Bernard). À propos du vers 843 des Bacchantes d'Euripide. *Revue des études grecques*, 2000, 113, p. 611-615.

2181. DEPEW (M.), OBBINK (D.). Matrices of genre. Authors, canons and society. Cambridge a. London, Harvard U. P., 2000, VI-346 p.

2182. Desde los poemas homéricos hasta la prosa griega del siglo IV d.C. Vientiséis estudios filológicos. Ed. por J. A. LOPEZ FÉREZ. Madrid, Ediciones Clásicas, 2000, VIII-523 p.

2183. DI MARCO (Massimo). La tragedia greca. Forma, gioco scenico, tecniche drammatiche. Roma, Carocci, 2000, 356 p.

2184. DIETZ (G.). Menschenwürde bei Homer. Vorträge und Aufsätze. Heidelberg, C. Winter, 2000, 246 p.

2185. DONIGER (W.). The bedtrick. Tales of sex and masquerade. Chicago a. London, University of Chicago Press, 2000, XXXI-598 p.

2186. DURAN (M.). El jefe de la fiesta en el fragmento 114 West de Arquiloco. *L'antiquité classique*, 2000, 69, p. 201-204.

2187. DYSON (M.), KEE (K. H.). The funeral of Astyanax in Euripides' Troades. *Journal of Hellenic studies*, 2000, 120, p. 17-33.

2188. EDWARDS (M. J.). In defense of Euthyphro. *American journal of philology*, 2000, 121, p. 213-224.

2189. ERCOLANI (A.). Il passaggio di parola sulla scena tragica. Didascalie interne e struttura delle rheseis. Stuttgart e Weimar, Verlag J. B. Metzler, 2000, 252 p.

2190. FARIOLI (Marcella). Mito e satira politica nei Chironi di Cratino. *Rivista di filologia e di istruzione classica*, 2000, 128, p. 406-431.

2191. FEDERSPIEL (Michel). Notes linguistiques et critiques sur le livre II des Croniques d'Apollonius de Perge. *Revue des études grecques*, 2000, 113, p. 359-391.

2192. FRANCESIO (M.). A proposito di un passo dell'Antiochikos di Libanio. *Athenaeum*, 2000, 88, p. 580-584.

2193. FUCHS (A.). Dramatische Spannung: moderner Begriff – antikes Konzept. Stuttgart u. Weimar, Verlag J. B. Metzler, 2000, X-349 p.

2194. GÖDDE (S.), HEINZE (T.). Skenika. Beiträge zum antiken Theater und seiner Rezeption. Festschrift zum 65. Geburtstag von H.-D. Blume. Darmstadt, Wissenschaftliche Buchgesellschaft, 2000, XIII-462 p. (ill., pls.).

2195. GÖDDE (S.). Das Drama der Hikesie. Ritual und Rhetorik in Aischylos' Hiketiden. Münster, Aschendorff, 2000, VIII-300 p.

2196. GOLDHILL (Simon). Civic ideology and the problem of difference: the politics of Aeschylean tragedy, once again. *Journal of Hellenic studies*, 2000, 120, p. 34-56.

2197. HALL STERNBERG (Rachel). The nurturing male: bravery and bedside manners in Isocrates' Aegineticus (19, 24-29). *Greece and Rome*, 2000, 47, p. 172-185.

2198. HARDIN (R. E.). Love in a green shade. Idyllic romances ancient to modern. Lincoln a. London, University of Nebraska Press, 2000, X-279 p.

2199. HAUROLD (Johannes). Homer's people: epic poetry and social formation. Cambridge, Cambridge U. P., 2000, XVI-240 p.

2200. IGLESIAS ZOIDO (J. C.). ¿Se pronunciaron realmente las arengas de Tucídides? El testimonio de Thuc. VII 61-70. *Athenaeum*, 2000, 88, p. 515-528.

2201. Intratextuality. Greek and Roman textual relations. Ed. by A. SHARROCK and H. MORALES. Oxford, Oxford U. P., 2000, XII-363 p.

2202. JACQUEMIN (Anne). Pausanias à Délos ou un chapitre recomposé du livre imaginé des Kikladika. *Ktema*, 2000, 25, p. 19-36.

2203. KATZ (Joshua T.), VOLK (Katharina). 'Mere bellies'? A new look at Theogony 26-28. *Journal of Hellenic studies*, 2000, 120, p. 122-131.

2204. KIM (J.). The pity of Achilles. Oral style and the unity of the Iliad. Lanham, Boulder, New York a. Oxford, Rowman & Littlefield, 2000, X-203 p.

2205. KURKE (L.). The strangeness of 'song culture': archaic Greek poetry. *In*: Literature in Greek and Roman worlds [Cf. n° 1929], p. 58-87.

2206. LOURENÇO (Frederico). An interpolated song in Euripides? Helen 229-252. *Journal of Hellenic studies*, 2000, 120, p. 132-139.

2207. MARSILIO (Maria S.). Farming and poetry in Hesiod's works and days. Lanham, New York a. Oxford, University Press of America, 2000, XXII-112 p.

2208. Medea nella letteratura e nell'arte. A cura di Bruno GENTILI e Franca PERUSINO. Venezia, Marsilio, 2000, 215 p. (figs.).

2209. MILLER (Dean D. A.). The epic hero. Baltimore a. Oxford, Johns Hopkins U. P., 2000, XVI-501 p.

2210. MILLS (Sophie). Achilles, Patroclus and parental care in some Homeric similes. *Greece and Rome*, 2000, 47, p. 3-18.

2211. MITFORD (Tim). Thalatta, thalatta: Xenophon's view of the Black Sea. *Anatolian studies*, 2000, 50, p. 127-131.

2212. MUREDDU (P.), NIEDDU (G. F.). Furfanterie sofistiche: omonimie e falsi ragionamenti tra Aristofane e Platone. Bologna, Pàtron Editore, 2000, 77 p.

2213. MURPHY (E. M.), MALLORY (J. P.). Herodotus and the cannibals. *Antiquity*, 2000, 74, p. 388-394.

2214. MUSTI (Domenico). Il tema dell'autonomia nelle Elleniche di Senofonte. *Rivista di filologia e di istruzione classica*, 2000, 128, p. 170-181.

2215. NIETO HERNÁNDEZ (Pura). Back in the cave of the Cyclops. *American journal of philology*, 2000, 121, p. 345-366.

2216. OKA (M.). An interpretation of Oidipus Tyrannos (in japanese). *Journal of classical studies*, 2000, 48, p. 1-18.

2217. PADILLA (Mark). Gifts of humiliation: charis and tragic experience in Alcestis. *American journal of philology*, 2000, 121, p. 179-211.

2218. PERNOT (L.). La rhétorique dans l'antiquité. Paris, Le livre de Poche, 2000, 337 p.

2219. Polybios-Lexicon. Band I. Lieferung I (α-γ). Bearbeitet von Arno MAUERSBERGER, verbesserte Aufl. von Christian-Friedrich COLLATZ, Hadwig HELMS und Melsene SCHÄFER. Berlin, Akademie Verlag, 2000, 392 p.

2220. RAEBURN (David). The significance of stage properties in Euripides' Electra. *Greece and Rome*, 2000, 47, p. 149-168.

2221. Rhetorical theory and praxis in Plutarch. Acta of the IVth International Congress of the International Plutarch Society, Leuven July 3–6, 1996. Ed. by L. VAN DER STOCKT. Louvain a. Namur, Éditions Peeters and Société des Études Classiques, 2000, XII-562 p.

2222. ROLLEY (Claude). Le casque (d'Archiloque?). *Bulletin de correspondance hellénique*, 2000, 124, p. 217-219.

2223. ROSSLYN (F.). Tragic plots. A new reading from Aeschylus to Lorca. Aldershot, Burlington, Singapore a. Sydney, Ashgate, 2000, 248 p.

2224. ROWE (Galen O.). Anti-Isocratean sentiment in Demosthenes' Against Androtion. *Historia*, 2000, 49, p. 278-302.

2225. SCHMITZ (Thomas A.). Plausibility in Greek orators. *American journal of philology*, 2000, 121, p. 47-77.

2226. SILK (M. S.). Aristophanes and the definition of comedy. Oxford, U. P., 2000, 456 p.

2227. SKOCZYLAS POWNALL (F.). Shifting viewpoints in Xenophon's Hellenica: the Arginusae episode. *Athenaeum*, 2000, 88, p. 499-513.

2228. TAPLIN (Oliver). The spring of the Muses: Homer and related poetry. *In*: Literature in Greek and Roman worlds [Cf. n⁰ 1929], p. 22-57.

2229. VOX (Onofrio). Lisia 'solonico'. *Quaderni di storia*, 2000, 52, p. 191-201.

2230. WARTELLE (André). Brèves remarques de vocabulaire grec. *Revue des études grecques*, 2000, 113, p. 211-229.

2231. WILKINS (J.). The boastful chef. The discourse of food in ancient Greek comedy. Oxford, Oxford U. P., 2000, XXVIII-465 p.

2232. WOLFRAM (A. X.). Lexis und Logos. Studien zur antiken Grammatik und Rhetorik. Stuttgart, Franz Steiner Verlag, 2000, 229 p.

Cf. n⁰ˢ 682, 1971-2045

b. Philosophy and sciences

* 2233. Bibliography on Plato's Laws. Ed. by T. J. SAUNDERS and L. BRISSON. Sankt Augustin, Akademia Verlag, 2000, 141 p.

2234. ALESSE (F.). La Stoa e la tradizione socratica. Napoli, Bibliopolis, 2000, 387 p.

2235. ANNAS (J.). Platonic ethics, old and new. Ithaca a. London, Cornell U. P., 2000, VIII-196 p.

2236. BALTUSSEN (H.). Theophrastus against the Presocratics and Plato. Peripatetic dialect in the De sensibus. Leiden, Boston a. Köln, Brill, 2000, XIV-285 p.

2237. BENSON (H. H.). Socratic wisdom. The model of knowledge in Plato's early dialogues. New York a. Oxford, Oxford U. P., 2000, IX-292 p.

2238. BRISSON (L.). Lectures de Platon. Paris, Librairie Philosophique J. Vrin, 2000, 270 p.

2239. CASERTANO (G.). La struttura del dialogo platonico. Napoli, Loffredo editore, 2000, 331 p.

2240. CHARLES (D.). Aristotle on meaning and essence. Oxford, Clarendon Press, 2000, XIV-410 p.

2241. COMOTH (Katharina). Vom Grunde der Idee: Konstellationen mit Platon. Heidelberg, Universitätsverlag C. Winter, 2000, 48 p. (ill.).

2242. Constructions du temps dans le monde grec ancien. Ed. par Catherine DARBO-PESCHANSKI. Paris, CNRS Editions, 2000, 493 p.

2243. DIXSAUT (M.). Platon et la question de la pensée. Paris, Librairie Philosophique J. Vrin, 2000, 330 p.

2244. GASTALDI (Silvia). La giustizia e la forza. Le tesi di Callicle nel Gorgia di Platone. *Quaderni di storia*, 2000, 52, p. 85-105.

2245. Geschichte der Mathematik und der Naturwissenschaften in der Antike. Vol. 2. Geographie und verwandte Wissenschaften. Hrsg. v. G. WÖHRLE. Stuttgart, Franz Steiner Verlag, 2000, 258 p.

2246. GOURINAT (J. B.). La dialectique des Stoïciens. Paris, Librairie Philosophique J. Vrin, 2000, 386 p.

2247. GUARDASOLE (A.). Tragedia e medicina nell'Atene del V secolo a. C. Napoli, M. D'Auria, 2000, 296 p.

2248. GUNDERT (B.). Soma and psyche in Hippocratic medicine. *In*: Psyche and soma [Cf. n° 2272], p. 13-35.

2249. HOBBS (A.). Plato and the hero. Courage, manliness and the impersonal good. Cambridge, Cambridge U. P., 2000, XVII-280 p.

2250. JEDAN (Christoph). Willensfreiheit bei Aristoteles? Göttingen, Vandenhoeck & Ruprecht, 2000, 203 p.

2251. KALIMTZIS (K.). Aristotle on political enmity and disease: an inquiry into stasis. New York, State University of New York Press, 2000, XVII-233 p.

2252. KENNY (A.). Essays on the Aristotelian tradition. Oxford, Clarendon Press, 2000, VI-174 p.

2253. KONSTAN (David). La pitié comme émotion chez Aristote. *Revue des études grecques*, 2000, 113, p. 616-630.

2254. KÜHN (W.). La fin du Phèdre de Platon. Critique de la rhétorique et de l'écriture. Firenze, Leo S. Olschki, 2000, 137 p.

2255. Lengua (La) científica griega: origines, desarollo y influencia en las lenguas modernas europeas. Ed. por J. A. LOPEZ FÉREZ. Madrid, Ediciones Clásicas, 2000, 2 vol., 200 p., 305 p.

2256. Letteratura scientifica e tecnica greca e latina. Atti del Seminario Internazionale di Studi (Messina 29–31 ottobre 1997). A cura di P. RADICI COLACE, A. ZUMBO. Messina, Edizioni dr. A. Sfameni, 2000, VII-481 p. [Cf. n° <scelta> 2834.]

2257. MAC FARLANE (John). Aristotle's definition of anagnōrisis. *American journal of philology*, 2000, 121, p. 367-383.

2258. MAC PHERRAN (M. L.). Recognition, remembrance and reality. New essays on Plato's epistemology and metaphysics. Kelowna, Academic Printing and Publishing, 2000, IX-157 p.

2259. MANN (W. R.). The discovery of things: Aristotle's Categories and their context. Princeton, Princeton U. P., 2000, XII-231 p.

2260. MICHELINI (Ann N.). Socrates plays the buffoon: cautionary protreptic in Euthydemus. *American journal of philology*, 2000, 121, p. 509-535.

2261. MOES (M.). Plato's dialogue form and the care of the soul. New York, Peter Lang, 2000, XIV-222 p.

2262. MÖLLENDORFF (P. von). Lukian Hermotimos oder Lohnt es sich, Philosophie zu studieren? Darmstadt, Wissenschaftliche Buchgesellschaft, 2000, 226 p.

2263. MONOSON (S. Sara). Plato's democratic entanglements. Athenian politics and the practice of philosophy. Princeton, Princeton U. P., 2000, 252 p.

2264. MONTIGLIO (Silvia). Silence in the land of Logos. Princeton, Princeton Unversity Press, 2000, X-313 p. – EADEM. Wandering philosophers in classical Greece. *Journal of Hellenic studies*, 2000, 120, p. 86-105.

2265. MOREL (Pierre-Marie). Atome et nécessité: Démocrite, Épicure, Lucrèce. Paris, Presses Universitaires de France, 2000, 136 p.

2266. MORGAN (K. A.). Myth and philosophy from the Presocratics to Plato. Cambridge, Cambridge U. P., 2000, VIII-313 p.

2267. NAGLE (B.). Alexander and Aristotle's Pambasileus. *L'antiquité classique*, 2000, 69, p. 117-132.

2268. NEWELL (W. R.). Ruling passion. The erotics of statecraft in Platonic political philosophy. Lanham, Boulder, New York a. Oxford, Rowman & Littlefield, 2000, VII-205 p.

2269. Oxford studies in ancient philosophy. Vol. 18. Summer 2000. Ed. by D. SEDLEY. Oxford, Oxford U. P., 2000, 374 p.

2270. PALERNE (J. C.). Études sur l'aspect chez Platon. Saint Étienne, Publications de l'Université de Saint Étienne, 2000, 381 p.

2271. PENDER (E. E.). Images of persons unseen. Plato's metaphors for the gods and the soul. Sankt Augustin, Akademia Verlag, 2000, XI-278 p.

2272. Psyche and soma. Physicians and metaphysicians on the mind-body problem from antiquity to Enlightenment. Ed. by J. P. WRIGHT and P. POTTER. Oxford, Clarendon Press, 2000, XII-298. [Cf. n°s <choice> 2248, 2280.]

2273. RIEDENAUER (Markus). Orexis und Eupraxia. Ethikbegründung im Streben bei Aristoteles. Würzburg, Königshausen & Neumann, 2000, 374 p.

2274. SAFFREY (H. D.). Le Néoplatonisme après Plotin. Paris, Librairie Philosophique J. Vrin, 2000, IX-318 p.

2275. SALAZAR (C. F.). The treatment of war wounds in Graeco-Roman antiquity. Leiden, Boston a. Köln, Brill, 2000, XXVII-299 p. (ills).

2276. SCHAMP (J.). L'homme sans visage. Pour una lecture politique du Charmide. *L'antiquité classique*, 2000, 69, p. 103-116.

2277. SIRINELLI (Jean). Plutarque de Chéronée. Un philosophe dans le siècle. Paris, Fayard, 2000, [s. p.].

2278. SLAATTÉ (H. A.). Plato's dialogues and ethics. Lanham, New York a. Oxford, University Press of America, 2000, 146 p.

2279. SMITH (Martin Ferguson). Elementary, my dear Lycians: a pronouncement on physics from Diogenes of Oinoanda. *Anatolian studies*, 2000, 50, p. 133-137.

2280. STADEN (H. von). Body, soul and nerves: Epicurus, Herophilus, Erasistratus, the Stoics and Galen. *In*: Psyche and soma [Cf. n° 2272], p. 79-116.

2281. STAUFFER (D.). Plato's introduction to the question of justice. Albany, State University of New York Press, 2000, 14 p.

2282. STEVENS (Annick). L'ontologie d'Aristote: au carrefour du logique et du réel. Paris, Librairie Philosophique J. Vrin, 2000, 301 p.

2283. SZABÓ (A.). L'aube des mathématiques grecques. Paris, Librairie Philosophique J. Vrin, 2000, 367 p.

2284. TANAKA (R.). The problem of relativism in Sextus Empiricus (in japanese). *Journal of classical studies*, 2000, 48, p. 64-75.

2285. TARRANT (H.). Plato's first interpreters. London, Duckworth, 2000, VIII-263 p.

2286. VEGETTI (Mario). Normale, naturale, normativo in Aristotele. *In*: Realismo politico (Sul) [Cf. n° 6186], p. 73-84.

2287. WARDY (Robert). Aristotle in China: language, categories and translation. Cambridge, Cambridge U. P., 2000, X-170 p.

2288. WARREN (James). Diogenes Epikourios: keep taking the tablets. *Journal of Hellenic studies*, 2000, 120, p. 144-148.

2289. Who speaks for Plato? Studies in Platonic anonymity. Ed. by G. A. PRESS. Lanham, Boulder, New York a. Oxford, Rowman & Littlefield, 2000, VI-245 p.

2290. WLODARCZYK (Marta Anna). Pyrrhonian inquiry. Cambridge, Cambridge Philological Society, 2000, X-72 p.

2291. ZAGDOUN (M. A.). La philosophie stoïcienne de l'art. Paris, CNRS Éditions, 2000, 308 p.

2292. ZANKER (Graham). Aristotle's poetics and the painters. *American journal of philology*, 2000, 121, p. 225-235.

Cf. n°s 1971-2045, 2835

§ 8. Religion and mythology.

* 2293. Bibliographie analytique de la prière grecque et romaine (1898–1998). Ed. par G. FREYBURGER et L. PERNOT. Turnhout, Brepols, 2000, 361 p.

2294. ALBINUS (L.). The house of Hades. Studies in ancient Greek eschatology. Aarhus. Aarhus U. P., 2000, 247 p.

2295. CALAME (Claude). Poétique des mythes dans la Grèce antique. Paris, Hachette, 2000, 288 p.

2296. CHANDEZON (Christophe). Le culte d'Héra à Pérachora (VIIIe–VIe s.), essai de bilan. *Revue des études grecques*, 2000, 113, p. 70-100.

2297. Cultes (Les) des cités phocéennes. Actes du colloque international: Aix-en-Provence/Marseille 4–5 juin 1999. Ed. par A. HERMARY et H. TRÉZINY. Aix-en-Provence, Études Massaliètes, 2000, 204 p. (ill.).

2298. DEMONT (Paul). Lois héroïques: remarques sur le tirage au sort de l'Iliade aux Sept contre Thèbes d'Eschyle. *Revue des études grecques*, 2000, 113, p. 299-325.

2299. FOUNTOLAKIS (A.). The artists of Aphrodite. *L'antiquité classique*, 2000, 69, p. 133-147.

2300. GAUER (W.). Die Spuren eines ungesühnten, der Zorn des Zeus und zwei Schatzhäuser der Athener in Delphi. *Ktema*, 2000, 25, p. 75-85.

2301. HANSEN (William). Foam-born Aphrodite and the mythology of transformation. *American journal of philology*, 2000, 121, p. 1-19.

2302. HARRISON (T.). Divinity and history: the religion of Herodotus. Oxford, Clarendon Press, 2000, XII-320 p.

2303. Héros et héroïnes dans le mythe et les cultes grecs. Ed. par V. PIRENNE DELFORGE et E. SUÁREZ DE LA TORRE. Liège, Centre International d'Étude de la Religion Grecque Antique, 2000, XXIII-447 p.

2304. LERNOULD (Alain). Plutarque, E de Delphes 387 d2-9. Une interprétation philosophique de l'épisode de l'enlèvement du trépied par Héraclès: une erreur de jeunesse. *Revue des études grecques*, 2000, 113, p. 147-171.

2305. LÉVY (Edmond). Peut-on parler d'une religion grecque? *Ktema*, 2000, 25, p. 11-18.

2306. LINCOLN (B.). Theorizing myth: narrative, ideology and scholarship. Chicago a. London, University of Chicago Press, 2000, XV-298 p.

2307. MARINATOS (N.). The goddess and the warrior. The naked goddess and mistress of animals in early Greek religion. London a. New York, Routledge, 2000, XIII-162 p. (ills).

2308. MORETTI (Jean-Charles). Le théâtre du sanctuaire de Dionysos Éleuthéreus à Athènes au Ve s. av. J.-C. *Revue des études grecques*, 2000, 113, p. 275-298.

2309. MUELLER-GOLDINGEN (Chr.). Tradition und Innovation. Zu Stesichoros' Umgang mit dem Mythos. *L'antiquité classique*, 2000, 69, p. 1-19.

2310. NOVARO-LEFÈVRE (Daniela). Le culte d'Héra à Pérachora (VIIIe–VIe s.), essai de bilan. *Revue des études grecques*, 2000, 113, p. 42-69.

2311. SCHEER (Tanja S.). Die Gottheit und ihr Bild: Untersuchungen zur Funktion griechischer Kultbilder in Religion und Politik. München, Verlag C. H. Beck, 2000, X-338 p.

2312. TALALAY (L. E.). Cultural biographies of the Great Goddess. *American journal of archaeology*, 2000, 104, p. 789-792.

2313. VEYNE (Paul). Inviter les dieux, sacrifier, banqueter. Quelques nuances de la religiosité gréco-romaine. *Annales*, 2000, 55, 1, p. 3-42.

Cf. n° 1282

§ 9. Archaeology and history of art.

2314. ALEMDAR (S.). Le monument de Lysicrate et son trépied. *Ktema*, 2000, 25, p. 199-206.

2315. Antichnoe Prichernomor'e· Sb. st. po klassi-chekoj arkheologii. (Ancient Black Sea littoral: papers in classic archaeology). Ed. S. L. SOLOV'EV. Sankt-Peterburg, Izd-vo Gos. Ermitazha, 2000, 236 p. [Eng. summary]

2316. AUPERT (Pierre), HERMARY (Antoine), FOURRIER (Sabine), SCHMID (Martin), PETIT-AUPERT (Catherine), MARANGOU (Antigone). Rapport sur les travaux de l'École française à Amathonte en 1999. *Bulletin de correspondance hellénique*, 2000, 124, p. 527-546.

2317. BADAL'JANETS (Jurij S.). Ellinisticheskij Rodos: keramicheskie klejma IOSPE III – kak istoricheskij istochnik: analiz problemy, reshenija. (Hellenistic Rodos: ceramic stamps IOSPE III as a historical source: analysis of the problem and decisions). Moskva, [s. n.], 2000, 343 p.

2318. BENEDIKTSON (D. Th.). Literature and the visual arts in ancient Greece and Rome. Norman, University of Oklahoma Press, 2000, XII-259 p. (ill).

2319. BLONDÉ (Francine), PICON (Maurice). Autour de la céramique du VIe siècle dans le Nord-Est de l'Égée: quelques approches différentes. *Bulletin de correspondance hellénique*, 2000, 124, p. 161-188.

2320. BOMMELAER (Jean-François). Rapport sur les travaux de l'École française d'Athènes en 1999. Delphes. *Bulletin de correspondance hellénique*, 2000, 124, p. 499-500.

2321. BONACASA (Nicola), ENSOLI (Serena). Cirene. Milano, Electa, 2000, 224 p.

2322. BRÉCOULAKI (Hariclia). Sur la technè de la peinture grecque ancienne d'après les monuments funéraires de Macédonie. *Bulletin de correspondance hellénique*, 2000, 124, p. 189-216.

2323. BRILLIANT (Richard). My Laocoön: alternative claims in the interpretation of artworks. Berkeley, Los Angeles a. London, University of California Press, 2000, XVI-146 p. (figs., ill.).

2324. CABANES (Pierre), LAMBOLEY (Jean-Luc), VREKAJ (Bashkim). Travaux menés en collaboration avec l'École Française en 1999. Apollonia d'Illyrie (Albanie). *Bulletin de correspondance hellénique*, 2000, 124, p. 620-630.

2325. CAPUTO (Riccardo), HELLY (Bruno). Travaux menés en collaboration avec l'École Française en 1999. Archéosismicité de l'Égée: étude des failles actives de la Thessalie. *Bulletin de correspondance hellénique*, 2000, 124, p. 560-588.

2326. CHANKOWSKI (Véronique), FOUACHE (Éric). Travaux menés en collaboration avec l'École Française en 1999. Pistiros (Bulgarie). *Bulletin de correspondance hellénique*, 2000, 124, p. 643-654.

2327. Corpus vasorum antiquorum: Danemark, Copenhague, Thorwaldsen Museum, Collection des Antiquités classiques. Fasc. 1. Ed. par Torben MELANDER. Copenhague, Thorwaldsen Museum, 2000, 94 p. (ills., tab.).

2328. COULIÉ (Anne). Les vases à reliefs thasiens de l'epoque archaïque. *Bulletin de correspondance hellénique*, 2000, 124, p. 99-160.

2329. DARLAS (Andréas), DE LUMLEY (Henry). Fouilles franco-helléniques de la grotte de Kalamakia (Aréopolis, Péloponnèse). *Bulletin de correspondance hellénique*, 2000, 124, p. 659-663.

2330. DENTI (Mario). Nuovi documenti di ceramica orientalizzante della Grecia d'Occidente. Stato della questione e prospettive della ricerca. *Mélanges de l'École Française de Rome*, 2000, 112, p. 781-842.

2331. DOĞER (Lale). İzmir Arkeoloji Müzesi Örnekleriyle Kazıma Dekorlu Ege-Bizans Seramikleri. (Decorative Aegean-Byzantine ceramics with its examples from İzmir Archeology Museum). İzmir, Ege Üniversitesi Edebiyat Fakültesi, 2000, 144 p.

2332. EMPEREUR (Jean-Yves). Travaux menés en collaboration avec l'École Française en 1999. Alexandrie (Égypte). *Bulletin de correspondance hellénique*, 2000, 124, p. 595-619.

2333. ÉTIENNE (Roland). Les activités de l'EFA. *Bulletin de correspondance hellénique*, 2000, 124, p. 461-488.

2334. FREITAG (K.). Der Golf von Korinth. Historisch-topographische Untersuchungen von der Archaik bis in das I. Jh. v. Chr. München, Tuduv, 2000, 504 p. (maps).

2335. GIESS BEVILACQUA (Valérie). Le nu féminin dans la peinture de vases à figures rouges de la fin du VIe siècle à la fin du Ve siècle avant J.C. *Ktema*, 2000, 25, p. 101-114.

2336. GRANDJEAN (Yves), SALVIAT (François), BLONDÉ (Francine), MULLER (Arthur), MULLIEZ (Dominique). Rapport sur les travaux de l'École française d'Athènes en 1999. Thasos. *Bulletin de correspondance hellénique*, 2000, 124, p. 506-521.

2337. GRECO (Emanuele), KALPAXIS (Thanassis), SCHNAPP (Alain), VIVIERS (Didier). Travaux menés en collaboration avec l'École Française en 1999. Itanos (Crète orientale). *Bulletin de correspondance hellénique*, 2000, 124, p. 547-559.

2338. Greek-Swedish excavations (The) at the Agia Aikaterini Square Kastelli, Khania, 1970–1987. Vol. 2. The Late-Minoan III C settlement. Ed. by E. HALLAGER, B. P. HALLAGER. Stockholm, Paul Åströms Förlag, 2000, 229 p. (figs., pls).

2339. GUILAINE (Jean), BRIOIS (François), VIGNE (Jean-Denis), CARRÈRE (Isabelle), WILLCOX (George), DUCHESNE (Sylvie). Travaux menés en collaboration avec l'École Française en 1999. L'habitat néolitique de Shillourokambos (Parekklisha, Chypre). *Bulletin de correspondance hellénique*, 2000, 124, p. 589-594.

2340. HADJISAVVAS (Sophoklis). Chronique des fouilles et découvertes archéologiques à Chypre en 1999. *Bulletin de correspondance hellénique*, 2000, 124, p. 665-699.

2341. HOLLOWAY (R. R.). The archaeology of ancient Sicily. London a. New York, Routledge, 2000, XVIII-211 p.

2342. HOLTZMANN (Bernard). Note sur la frise de l'Érechtheion: le sujet de MAcr 1073. *Bulletin de correspondance hellénique*, 2000, 124, p. 221-226.

2343. KARAGEORGHIS (V.). Ancient art from Cyprus. The Cesnola collection in the Metropolitan Museum of

Art. New York, The Metropolitan Museum, 2000, 305 p. (ills, maps).

2344. KNELL (H.). Athen im 4. Jahrhundert v. Chr. Eine Stadt verändert ihr Gesicht. Archäologisch-kulturgeschichtliche Betrachtungen. Darmstadt, Wissenschaftliche Buchgesellschaft, 2000, 219 p.

2345. KOURINOU (E.). Sparta. Contribution to its monumental topography. Athina, Horos, 2000, XVI-296 p. (plans, pls.).

2346. KUZNETSOV (Vladimir D.). Organizatsija obshchestvennogo stroitel'stva v Drevnej Gretsii. (Organization of public building in Ancient Greece). RAN, In-t arkheologii. Moskva, Jazyki russkoj kul'tury, 2000, 536 p.

2347. LERA (Pétrika), PRENDI (Frano), TOUCHAIS (Gilles). Travaux menés en collaboration avec l'École Française en 1999. Sovjan (Albanie). *Bulletin de correspondance hellénique*, 2000, 124, p. 631-642.

2348. MANGOLD (M.). Kassandra in Athen. Die Eroberung Trojas auf attischen Vasenbildern. Berlin, Reimer, 2000, 257 p.

2349. MARC (Jean-Yves). Combien y avait-il d'agoras à Délos? *Ktema*, 2000, 25, p. 41-45.

2350. MARCHETTI (Patrick). Recherches sur les mythes et la topographie d'Argos, V. Quelques mises au point sur les rues d'Argos. À propos de deux ouvrages récents. *Bulletin de correspondance hellénique*, 2000, 124, p. 273-289.

2351. MĂRGINEANU-CÂRSTOIU (Monica), SEBE (Andrei). Remarques sur le tracé des volutes ioniques hellénistiques. Observations sur leurs corrélations géométriques dans la composition. *Bulletin de correspondance hellénique*, 2000, 124, p. 291-330.

2352. MEDWID (L. M.). The makers of classical archaeology. A reference work. Amherst, Humanity Books, 2000, 352 p.

2353. MERKER (G. S.). Corinth: results of the excavations conducted by the American School of Classical Studies at Athens. Vol. 18. The sanctuary of Demeter and Kore. Terracotta figurines of the Classical, Hellenistic and Roman periods. Princeton, American School of Classical Studies at Athens, 2000, XXVII-394 p.

2354. MORETTI (Jean-Charles), FRAISSE (Philippe), ALABE (Françoise). Rapport sur les travaux de l'École française d'Athènes en 1999. Délos. *Bulletin de correspondance hellénique*, 2000, 124, p. 522-526.

2355. MÜLLER (Christel), ABRAMOV (Andrei), ABRAMZON (Mikaïl), ALEXEIEVA (Ekaterina), BEZRUCHENKO (Igor), EFIMOV (Alexei), GORLOV (Youri), KUZNETSOV (Vladimir), MASLENNIKOV (Alexandre), MALYSHEV (Alexei), SAVOSTINA (Elena), TOLSTIKOV (Vladimir), ZAVOIKINE (Alexei). Chronique des fouilles et découvertes archéologiques dans le Bosphore cimmérien (mer Noire septentrionale). *Bulletin de correspondance hellénique*, 2000, 124, p. 701-751.

2356. MÜLLER (Christel), FOUACHE (Éric), GORLOV (Youri), POROTOV (Alexei). Travaux menés en collaboration avec l'École Française en 1999. Péninsule de Taman (Russie méridionale). *Bulletin de correspondance hellénique*, 2000, 124, p. 655-657.

2357. MÜLLER-CELKA (Sylvie). Rapport sur les travaux de l'École française d'Athènes en 1999. Malia. *Bulletin de correspondance hellénique*, 2000, 124, p. 501-505.

2358. Not the classical ideal. Athens and the construction of the other in Greek art. Ed. by B. COHEN. Leiden, Boston a. Cologne, Brill, 2000, XVI-559 p.

2359. PHILIPPA-TOUCHAIS (Anna), TOUCHAIS (Gilles), PIÉRART (Marcel), MARCHETTI (Patrick), MARCHETTI-LAKAKI (Maria), RIZAKIS (Yvonne). Rapport sur les travaux de l'École française d'Athènes en 1999. Argos. *Bulletin de correspondance hellénique*, 2000, 124, p. 489-498.

2360. Proceedings. The 101st annual meeting of the Archaeological Institute of America. *American journal of archaeology*, 2000, 104, p. 311-371.

2361. RIDGWAY (B. S.). Hellenistic sculpture II. The styles of ca. 200–100 a.C. Madison, University of Wisconsin Press, 2000, XIX-374 p.

2362. TOUCHAIS (Gilles), HUBER (Sandrine), PHILIPPA-TOUCHAIS (Anna). Chronique des fouilles et découvertes archéologiques en Grèce en 1999. *Bulletin de correspondance hellénique*, 2000, 124, p. 753-1023.

2363. TSETSKHLADZE (G. R.), PRAG (A. J. N. W.), SNODGRASS (A. M.). Periplous: Papers on Classical art and archaeology presented to Sir John Boardman. London, Themes & Hudson, 2000, 416 p. (ills).

2364. VAN GELDER (K.). Attic Geometric: Athenian and provincial. *L'antiquité classique*, 2000, 69, p. 269-275.

2365. WILSON JONES (M.). Doric measure and architectural design 1: the evidence of the relief from Salamis. *American journal of archaeology*, 2000, 104, p. 73-93.

Cf. n° 1282

F

HISTORY OF ROME, ANCIENT ITALY AND THE ROMAN EMPIRE

§ 1. The peoples of Italy. 2366-2388. – § 2. The Etruscans. 2389-2412. – § 3. Sources and criticism of sources. (*a*. Epigraphical sources; *b*. Literary sources). 2413-2533. – § 4. General and political history. 2534-2637. – § 5. History of law and institutions. 2638-2673. – § 6. Economic and social history. 2674-2719. – § 7. History of literature, philosophy and science (*a*. Literature; *b*. Philosophy and science). 2720-2837. – § 8. Religion and mythology. 2838-2870. – § 9. Archaeology and history of art. 2871-2943. – § 10. Late antiquity. Transformation of the Roman world. 2944-2967.

§ 1. The peoples of Italy.

2366. Adriatico tra IV e III sec. a. C. Vasi alto-adriatici tra Piceno, Spina e Adria. Atti del convegno di studi, Ancona, 20-21 giugno 1997. A cura di Maurizio LANDOLFI. Roma, 'L'Erma' di Bretschneider, 2000, XVI-169 p. (ill.) [Cf. n^{os} <scelta> 2371, 2376, 2377, 2378.]

2367. BAGOLAN (M.), LEONARDI (G.). Il bronzo finale nel Veneto. *In*: Protovillanoviano (II) al di qua e al di là dell'Appennino [Cf. n° 2381], p. 15-46.

2368. BARTOLONI (Piero). La necropoli di Monte Sirai. I. Roma, CNR, Istituto per la civiltà fenicia e punica 'Sabatino Moscati', 2000, 199 p. (ill., tav.).

2369. BELLINTANI (P.). Il medio Polesine tra la tarda età del bronzo e l'inizio dell'età del ferro. *In*: Protovillanoviano (II) al di qua e al di là dell'Appennino [Cf. n° 2381], p. 47-84.

2370. BISPHAM (E. H.), BRADLEY (G. J.), HAWTHORNE (J. W. J.), KANE (S.). Towards a phenomenology of Samnite fortified centres. *Antiquity*, 2000, 74, p. 23-24.

2371. BRACCESI (L.). Dorica Ancon e problemi connessi. *In*: Adriatico tra IV e III sec. a. C. [Cf. n° 2366], p. 3-9.

2372. BRADLEY (G.). Ancient Umbria. State, culture, and identity in central Italy from the iron age to the Augustan era. Oxford, Oxford U. P., 2000, XIV-333 p. (ill., maps).

2373. COARELLI (Filippo). Il Lucus Pisaurensis e la romanizzazione dell'Ager Gallicus. *In*: Roman middle republic (The) [Cf. n° 2612], p. 195-205.

2374. DE GUIO (A.). Ex occidente lux: linee di un percorso critico di rivisitazione del bronzo finale nel Veneto. *In*: Protovillanoviano (II) al di qua e al di là dell'Appennino [Cf. n° 2381], p. 259-357.

2375. DI GENNARO (F.), GUIDI (A.). Il bronzo finale dell'Italia centrale. Considerazioni e prospettive di indagine. *In*: Protovillanoviano (II) al di qua e al di là dell'Appennino [Cf. n° 2381], p. 99-131.

2376. GUZZO (P. G.). Perché i Piceni non erano alla battaglia di Cuma? *In*: Adriatico tra IV e III sec. a. C. [Cf. n° 2366], p. 11-18.

2377. HARARI (M.). Modelli etnico-culturali e ceramografia. I vasi alto-adriatici. *In*: Adriatico tra IV e III sec. a. C. [Cf. n° 2366], p. 161-169.

2378. LANDOLFI (M.). I Galli e l'Adriatico. *In*: Adriatico tra IV e III sec. a. C. [Cf. n° 2366], p. 19-46.

2379. MATTHÄUS (H.). Die Rolle Zyperns und Sardiniens im mittelmeerischen Interaktionsprozeß während des späten zweiten und frühen ersten Jahrtausends v. Chr. *In*: Akten des Kolloquiums zum Thema 'Der Orient und Etrurien' [Cf. n° 2389], p. 41-75.

2380. PANVINI ROSATI (F.). Monetazione preromana in Italia. Gli inizi della monetazione romana in Italia e la monetazione romano-campana. *In*: Moneta greca e romana (La) [Cf. n° 168], p. 79-93.

2381. Protovillanoviano (II) al di qua e al di là dell'Appennino. Atti della giornata di studio. Pavia, Collegio Ghislieri, 17 giugno 1995. A cura di Maurizio HARARI e Mark PEARCE. Como, New press, 2000, 395 p. (ill.). (Biblioteca di Athenaeum, 38). [Cf. n^{os} <scelta> 2367, 2369, 2374, 2375, 2387.]

2382. RIDGWAY (D.). The orientalizing phaenomenon in Campania: sources and manifestations. *In*: Akten des Kolloquiums zum Thema 'Der Orient und Etrurien' [Cf. n° 2389], p. 233-244.

2383. RIX (Helmut). «Tribù», «stato», «città» e «insediamento» nelle lingue italiche. *Archivio glottologico italiano*, 2000, 85, p. 196-231.

2384. SPADONI (Maria Carla). I Sabini nell'antichità: dalle origini alla romanizzazione. Rieti, D.E.U.I., 2000, 199 p.

2385. Storia del Trentino. 2. L'età romana. A cura di Ezio BUCHI. Bologna, Il Mulino, 645 p.

2386. UNTERMANN (J.). Wörterbuch des Oskisch-Umbrischen. Heidelberg, Winter, 2000, 902 p.

2387. ZANINI (A.). Il bronzo finale in Toscana. *In*: Protovillanoviano (Il) al di qua e al di là dell'Appennino [Cf. n° 2381], p. 201-212.

§ 1. Addenda 1999.

2388. Vigilia di romanizzazione. Altino e il Veneto Orientale tra II e I sec. a.C. Atti del Convegno, Venezia, San Sebastiano, 2–3 dicembre 1997. A cura di G. BANDELLI e M. VERZÁR-BASS. Roma, edizioni Quasar, 99, 326 p. (ill.).

§ 2. The Etruscans.

2389. Akten des Kolloquiums zum Thema 'Der Orient und Etrurien'. Zum Phänomen des 'Orientalisierens' im westlichen Mittelmeerraum (10.–6. Jh. v. Chr.), Tübingen, 12.–13. Juni 1997. Hrsg. v. Friedhelm PRAYON und Wolfgang RÖLLIG. Pisa u. Roma, Istituti editoriali e poligrafici internazionali, 2000, 329 p. (Taf.). [Cf. n°ˢ <Auswahl> 2379, 2382, 2393, 2394, 2395, 2398, 2400.]

2390. ALBINI (Pierluigi). L'Etruria delle donne. Vita pubblica e privata delle donne etrusche. Valentano, Scipioni, 2000, 128 p. (ill.).

2391. AMANN (Petra). Die Etruskerin. Geschlechterverhältnis und Stellung der Frau im frühen Etrurien (9.–5. Jh. v. Chr.). Wien, Verlag der Österreichischen Akademie der Wissenschaften, 2000, 332 p. (ill.).

2392. BARKER (Graeme), RASMUSSEN (Tom). The Etruscans. Oxford, Blackwell, 2000, XII-379 p.

2393. BONGHI JOVINO (Maria). Funzioni, simboli e potere. I bronzi del "complesso" di Tarquinia. *In*: Akten des Kolloquiums zum Thema 'Der Orient und Etrurien' [Cf. n° 2389], p. 287-298.

2394. BRIQUEL (D.). Pélasges et Tyrrhènes en zone égéenne. *In*: Akten des Kolloquiums zum Thema 'Der Orient und Etrurien' [Cf. n° 2389], p. 19-36.

2395. CAMPOREALE (Giovannangelo). Gli Etruschi. Storia e civiltà. Torino, UTET, 2000, XXXIX-603 p. (ill.). – IDEM. Sopravvivenze villanoviane nell'orientalizzante vetuloniese. *In*: Akten des Kolloquiums zum Thema 'Der Orient und Etrurien' [Cf. n° 2389], p. 153-170.

2396. DI FAZIO (M.). Porsenna e la società di Chiusi. *Athenaeum*, 2000, 88, p. 393-412.

2397. FACCHETTI (Giulio M.). Frammenti di diritto privato etrusco. Firenze, L. S. Olschki, 2000, 114 p. (Biblioteca dell'Archivum Romanicum. Serie II: Linguistica, 50).

2398. GAUER (W.). Olympia, der Orient und Etrurien. *In*: Akten des Kolloquiums zum Thema 'Der Orient und Etrurien' [Cf. n° 2389], p. 113-127.

2399. HARARI (Maurizio). Due vecchi sposi di Volterra e la questione del realismo: un dibattito italiano. *In*: Monumento (Il): arte e storia [Cf. n° 1283], p. 636-649.

2400. HASE (F.-W. von). Die goldene Prunkfibel aus der Tomba Regolini-Galassi in Cerveteri – Überlegungen zu ihrer Genese und Funktion. *In*: Akten des Kolloquiums zum Thema 'Der Orient und Etrurien' [Cf. n° 2389], p. 129-152.

2401. HAYNES (S.). Etruscan civilization: a cultural history. London, British Museum Press, 2000, XIX-432 p. (ill., maps)

2402. HEMPHILL (P.). Archaeological investigations in southern Etruria. Vol. 1. The Civitella Cesi survey. Stockholm, Paul Åströms Förlag, 2000, 150 p. (ill., maps).

2403. IZZET (Vedia). The Etruscan sanctuary at Cerveteri, Sant'Antonio: preliminary report of excavations 1995–1998. *Papers of the British School at Rome*, 2000, 68, p. 321-336. – IDEM. Tuscan order: the development of Etruscan sanctuary architecture. *In*: Religion in archaic and republican Rome and Italy [Cf. n° 2864], p. 34-53.

2404. MIARI (Monica). Stipi votive dell'Etruria padana. Roma, Giorgio Bretschneider, 2000, 398 p. (ill.).

2405. PALMUCCI (A.). Odisseo in Etruria: mito e archeologia. *Aufidus*, 2000, 42, p. 7-38.

2406. PAPI (Emanuele). L'Etruria dei romani. Opere pubbliche e donazioni private in età imperiale. Roma, Quasar, 2000, 278 p. (ill.).

2407. PITTAU (Massimo). Tabula Cortonensis. Lamine di Pirgi e altri testi etruschi tradotti e commentati. Sassari, Editrice democratica sarda, 2000, 131 p.

2408. Tabula Cortonensis. A cura di L. AGOSTINIANI e F. NICOSIA. Roma, 'L'Erma' di Bretschneider, 2000, 175 p. (tav.).

2409. TORELLI (Mario).The Etruscan city-state. *In*: Comparative study (A) of thirty city-state cultures [Cf. n° 1128], p. 189-208.

2410. VITALI (Daniele), KAENEL (Gilbert). Un Helvète chez les Étrusques vers 300 av. J.-C. *Archäologie der Schweiz*, 2000, 23, p. 115-122.

2411. WARDEN (P. Gregory), THOMAS (Michael L.), GALLOWAY (Jess). The Etruscan settlement of Poggio

Colla (1995-1998 excavations). *Journal of Roman archaeology*, 2000, 12, p. 231-246.

2412. WYLIN (K.). Il verbo etrusco: ricerca morfosintattica delle forme usate in funzione verbale. Roma, 'L'Erma' di Bretschneider, 2000, 348 p. (ill., tav.). (Studia philologica, 20).

§ 3. **Sources and criticism of sources**.

a. Epigraphical sources

* 2413. Année épigraphique (L'). 1997. [1996. Cf. Bibl. 99, n° 2307.] Revue des publications épigraphiques relatives à l'antiquité romaine. Éd. par M. CORBIER, P. LE ROUX et S. DARDAINE. Paris, Presses universitaires de France, 2000, 799 p.

2414. ALFÖLDY (G.). Zur Lage und zu den Inschriften des Diana-Heiligtums von Saguntum, *Zeitschrift für Papyrologie und Epigraphik*, 2000, 129, p. 275-280.

2415. ANASTASIADIS (Vasilis I.), SOURIS (George A.). An index to Roman imperial constitutions from Greek inscriptions and papyri 27 BC to 284 AD. Berlin a. New York, de Gruyter, 2000, VII-225 p.

2416. BARTELS (Klaus). Roms sprechende Steine. Inschriften aus zwei Jahrtausenden gesammelt, übersetzt und erläuert. Zürich, Verlag Neue Zürcher Zeitung u. Mainz, von Zabern, 2000, 307 p.

2417. BLECKMANN (B.). Silvanus und seine Anhänger in Italien: Zur Deutung zweier Kampanischen Inschriften für den Usurpator Silvanus (CIL X 6945 und 6946). *Athenaeum*, 2000, 88, p. 477-483.

2418. Catalogo delle iscrizioni latine del Museo Nazionale di Napoli (ILMN). Vol. 1. Roma e Latium. A cura di G. CAMODECA. Napoli, Loffredo, 2000, 399 p. (ill.).

2419. Corpus inscriptionum Latinarum consilio et auctoritate Academiae scientiarum Berolinensis et Brandenburgensis editum. Vol. VI. Inscriptiones urbis Romae Latinae. Pars VIII. Titulos et imagines collegit schedasque comparavit Silvio PANCIERA. Fasc. III. Titulos magistratuum populi Romani ordinum senatorii equestrisque thesauro schedarum imaginumque ampliato edidit Géza ALFÖLDY adiuvantibus Maria Letizia CALDELLI, Laura CHIOFFI, Fritz MITTHOF, Heike NIQUET, Silvia ORLANDI, Cecilia RICCI, Andrea SCHEITHAUER, Manfred G. SCHMIDT, Gabriele WESCH-KLEIN, Christian WITSCHEL itemque Claudia KRAMER, Jens-Uwe KRAUSE, Peter KRUSCHWITZ. Berlin u. New York, de Gruyter, 2000, XXXII, p. 4667-5294.

2420. D'ARMS (J. H.). Memory, money, and status at Misenum: three new inscriptions from the Collegium of the Augustales. *Journal of Roman studies*, 2000, 90, p. 126-144.

2421. DE CAZANOVE (O.). I destinatari dell'iscrizione di Tiriolo e la questione del Senatoconsulto *De Bacchanalibus*. *Athenaeum*, 2000, 88, p. 59-68.

2422. DEL MASTRO (G.). Secondo supplemento al Catalogo dei papiri ercolanesi. *Cronache ercolanesi*, 2000, 30, p. 157-242.

2423. ECK (W.). Monumente der Virtus. Kaiser und Heer im Spiegel epigraphischer Denkmäle. *In*: Kaiser, Heer und Gesellschaft in der Römischen Kaiserzeit [Cf. n° 2695], p. 483-496.

2424. Epigraphic landscape (The) of Roman Italy. Ed. by A. E. COOLEY. London, Institute of classical studies, School of advanced study, University of London, 2000, XIV-212 p.

2425. FABRE (G.), SILLIÈRES (P.). Inscriptions latines d'Aquitaine (ILA) Lectoure. Paris, de Boccard, 2000, 254 p. (ill.).

2426. FEISSEL (D.). Une constitution de l'empereur Julien. Entre texte épigraphique et codification (CIL III, 459 et Cth I, 16, 8). *In:* Codification (La) des lois dans l'antiquité [Cf. n° 1618], p. 315-337.

2427. FLOWER (Harriet I.). Damnatio Memoriae and Epigraphy. *In*: Caligula (From) to Constantine, [Cf. n° 2876], p. 58-69.

2428. GASCOU (J.). L'inscription de Saint-Jean-de-Garguier en l'honneur du sévir augustal Q. Cornelius Zosimus. *Mélanges de l'Ecole française de Rome. Antiquité*, 2000, 112, p. 279-295.

2429. LIOU (Bernard). Les inscriptions peintes des amphores du Pecio Gandolfo (Almería). *Mélanges de l'Ecole française de Rome. Antiquité*, 2000, 112, p. 7-25.

2430. MROZEK (S.). Die epigraphische Streuung des Denars und Sesterzes in Italien und den westlichen Provinzen der frühen römischen Kaiserzeit. *Ancient society*, 2000, 30, p. 115-134.

2431. NIQUET (H.). Monumenta virtutum titulique. Senatorische Selbstdarstellung im spätantiken Rom im Spiegel der epigraphischen Denkmäler. Stuttgart, Franz Steiner Verlag, 2000, 350 p.

2432. PANCIERA (Silvio). Corpus Inscriptionum Latinarum Supplementa Italica. Supplementa Italica – nuova serie Unione Accademica Nazionale. Roma, Quasar, 2000, 394 p. (ill.).

2433. STEINBY (E. M.). Herbert Bloch and the new CIL XV.1. *Journal of Roman archaeology*, 2000, 13, p. 201-206.

2434. VISMARA (C.), CALDELLI (M. L.). Epigrafia anfiteatrale dell'Occidente romano. 5. Alpes Maritimae, Gallia Narbonensis, Tres Galliae, Germaniae, Britannia. Roma, Quasar, 2000, 263 p. (tav.).

2435. WIEGELS (R.). Lopodunum II. Inschriften und Kultdenkmäler aus dem römischen Ladenburg am Neckar. Stuttgart, Theiss, 2000, 300 p. (Abb.).

Cf. n° 202

b. Literary sources

2436. [Agrimensores Romani] Writings (The) of the Roman land surveyors. Introduction, text, translation an commentary. Ed. by Brian CAMPBELL. London, Society for the promotion of Roman studies, 2000, LXI-566 p. (Journal of Roman studies monographs, 9).

2437. [Ammianus Marcellinus] LEPORE (P.). In margine ad Ammiano Marcellino 26. 5. 8-14. *Athenaeum*, 2000, 88, p. 585-597.

2438. [Ammianus Marcellinus] TOUGHER (Shaun). Ammianus Marcellinus on the empress Eusebia: a split personality? *Greece and Rome*, 2000, 47, p. 94-101.

2439. [Appianus] RICHARDSON (J. S.). Appian. Wars of the Romans in Iberia. Iberike. With an introduction, translation and commentary. Warminster, Aris & Phillips, 2000, VIII-184 p.

2440. [Apuleius] Apuleio. Storia del testo e interpretazioni. A cura di G. MAGNALDI e G. F. GIANOTTI, Torino, Edizioni dell'Orso, 2000, 332 p.

2441. [Apuleius] Apuleio neosofista. Discorso per la sua statua a Cartagine (Floridum 16). Introduzione, testo, traduzione e commento. A cura di Alberto TOSCHI. Firenze, Opus libri, 2000, 133 p. (Università degli Studi di Parma, Dipartimento di Filologia classica e medievale).

2442. [Apuleius] MARANGONI (C.). Il mosaico della memoria. Studi sul Florida e sulle Metamorfosi di Apuleio. Padova, Imprimitur, Dipartimento di scienze dell'antichità e Verona, Dipartimento di linguistica, letteratura e scienze della comunicazione, 2000, 112 p.

2443. [Apuleius] REGEN (F.). Il De deo Socratis di Apuleio (II parte). *Maia*, 2000, 52, p. 41-66.

2444. [Apuleius] ZIMMERMAN (M.). Apuleius Madaurensis. Methamorphoses. Book X. Text, introduction and commentary. Groningen, Egbert Forsten, 2000, 487 p.

2445. [Aulus Gellius] VARDI (Amiel D.). Brevity, conciseness, and compression in Roman poetic criticism and the text of Gellius' Noctes Atticae 19.9.10. *American journal of philology*, 2000, 121, p. 291-298.

2446. [Caesar] MARTIN (Paul M.). La guerre des Gaules, la guerre civile. César, l'actuel. Paris, Ellipses, 2000, 192 p.

2447. [Cassius Dio] MOSCOVICH (M. J.). Ausonia: the context of Cassius Dio fr. 2.1. *Historia*, 2000, 49, p. 358-377.

2448. [Cato] CHURCHILL (J. Bradford). Cato Orationes 66 and the case against M'. Acilius Glabrio in 189 B.C.E. *American journal of philology*, 2000, 121, p. 549-557.

2449. [Catullus] BRACKE (W.). À propos d'un commentaire sur Catulle datant du XVe siècle. *Latomus*, 2000, 59, p. 414-426.

2450. [Catullus] ROSSI (L.). La Chioma di Berenice: Catullo 66, 79-88, Callimaco e la propaganda di corte. *Rivista di filologia e di istruzione classica*, 2000, 128, p. 299-312.

2451. [Cicero] GRAZZINI (S.). La synkrisis fra Pompeo ed Alessandro nel Somnium Scipionis: a proposito di Cicerone, De re publica VI 22. *Museum Helveticum*, 2000, 57, p. 220-236.

2452. [Cicero] KROSTENKO (Brian A.). Beyond (dis)belief: rhetorical form and religious symbol in Cicero's de divinatione. *Transactions of the American philological association*, 2000, 130, p. 353-391.

2453. [Cicero] OPPERMANN (Irene). Zur Funktion historischer Beispiele in Ciceros Briefen. München u. Leipzig, Saur, 2000, 338 p.

2454. [Cicero] WALSH (P. G.). Cicero, On Obligations (De Officiis). Translated with an introduction and explanatory notes. Oxford, Oxford U. P., 2000, LX-218 p.

2455. [Claudianus] Claudien. Œuvres. Tome II. Vol. 1 et 2. Poèmes politiques (395-398). Éd. par J. L. CHARLET. Paris, Les Belles Lettres, 2000, LXXXVIII-222, 226 p. (Collection des Universités de France, Association Guillaume Budé)

2456. [Claudianus] HAJDU (P.). Die Menschen in De raptu Proserpinae des Claudius Claudianus. *Acta antiqua Academiae scientiarum Hungaricae*, 2000, 40, p. 133-150.

2457. [Ennius] FLORES (E.). Quinto Ennio, Annali (Libri I-VIII) Vol. I. Napoli, Liguori Editore, 2000, 91 p.

2458. [Ennius] MASIÁ (Andrés). Ennio, Tragedias. Alcmeo. El ciclo troyano. Amsterdam, Hakkert, 2000, 667 p.

2459. [Florus] BERTÉ (M.). Floro (3, 10, 26) e Petrarca: un esempio di riscrittura. *Rivista di filologia e di istruzione classica*, 2000, 128, p. 56-91.

2460. [Horatius] AKBAR KHAN (H.). Ut mater iuuenem ... sic … quaerit patria Caesarem: a note on the simile at Horace, Odes 4, 5, 9-16. *Latomus*, 2000, 59, p. 549-551.

2461. [Horatius] CITTI (F.). Studi oraziani. Tematica e intertestualità. Bologna, Pàtron editore, 2000, 275 p.

2462. [Horatius] MAGNO (P.). Le beatae arces di Orazio (Carm. II, 6, 21-22): metafora di una precisa entità geografica. *Latomus*, 2000, 59, p. 547-548.

2463. [Horatius] PARKER (Holt N.). Horace Epodes 11.15-18: what's shame got to do with it? *American journal of philology*, 2000, 121, p. 559-570.

2464. [Horatius] SCHLEGEL (C.). Horace and his fathers: Satires 1.4 and 1.6. *American journal of philology*, 2000, 121, p. 93-119.

2465. [Iulianus] SARDIELLO (R.). Giuliano imperatore. Simposio I Cesari. Edizione critica, traduzione e commento. Galatina, Mario Congedo, 2000, LIX-202 p.

2466. [Iuvenalis] SCHMITZ (Christine). Das Satirische in Juvenals Satiren. Berlin u. New York, de Gruyter, 2000, XII-305 p.

2467. KULIKOWSKI (Michael). The Notitia Dignitatum as historical source. *Historia*, 2000, 49, p. 358-377.

2468. [Libanius] MALOSSE (P. L.). Libanios, ses 'témoins oculaires', Eusèbe et Praxagoras: le travail préparatoire du sophiste et la question des sources dans L'éloge de Constance et Constant. *Revue des études grecques*, 2000, 113, p. 172-187.

2469. [Livius] BRIQUEL (D.). Tite-Live II, 44-48 – Denys d'Halicarnasse IX, 6-13. Essai d'analyse d'un récit de bataille. *Latomus*, 2000, 59, p. 858-872.

2470. [Livius] BURTON (Paul J.). The last republican historian: a new date for the composition of Livy's first pentad. *Historia*, 2000, 49, p. 429-446.

2471. [Livius] MINEO (B.). L'année 207 dans le récit livien. *Latomus*, 2000, 59, p. 512-540.

2472. [Livius] YARDLEY (J. C.). Livy: The dawn of the Roman Empire. Books 31-40. Oxford, Oxford U. P., 2000, XXXVI-612 p. (maps).

2473. [Lucanus] LEIGH (Matthew). Lucan and the libyan tale. *Journal of Roman studies*, 2000, 90, p. 95-109.

2474. [Lucanus] ROSSI (Andreola). The Aeneid revisited: the journey of Pompey in Lucan's Pharsalia. *American journal of philology*, 2000, 121, p. 571-591.

2475. [Lucanus] SALEMME (G.). Sul senso della storia nella Pharsalia di Lucano. *Bollettino di studi latini*, 2000, 30, p. 514-529.

2476. [Martialis] DI GIOVINE (C.). Per il testo e l'esegesi di Marziale 10,48,18-24. *Rivista di filologia e di istruzione classica*, 2000, 128, p. 454-466.

2477. [Martialis] FUSI (A.). Marziale e la fama di Ovidio (nota a Mart. 5,10). *Rivista di filologia e di istruzione classica*, 2000, 128, p. 313-322.

2478. [Ovidius] GILDENHARD (I.), ZISSOS (A.). Ovid's Narcissus (Met. 3.339-510): echoes of Oedipus. *American journal of philology*, 2000, 121, p. 129-147.

2479. [Ovidius] HEYWOOD (T.). Art of love: the first complete english translation of Ovid's Ars amatoria. Ann Arbor, University of Michigan Press, 2000, XII-173 p.

2480. [Ovidius] MUSGROVE (Margaret Worsham). The student's Ovid: selections from the Metamorphoses. Norman, University of Oklahoma Press, 2000, XII-211 p.

2481. [Ovidius]. Ovid. Metamorphoses book XIII. Ed. by N. HOPKINSON. Cambridge, Cambridge U. P., 2000, VII-252 p.

2482. [Ovidius] Ovid. Metamorphoses XIII–XV and indexes to Metamorphoses I-XV. Ed. by D. E. HILL. Warminster, Aris & Phillips Ltd, 2000, VI-250 p.

2483. [Ovidius] Publio Ovidio Nasón. Cartas desde el Ponto. Introducción, texto, traducción y notas. Ed. por Ana PÉREZ VEGA y Francisco SOCAS GAVILÁN. Madrid, Consejo Superior de Investigaciones Científicas, 2000, LXVIII-252 p.

2484. [Ovidius] VIARRE (S.). Ovide, Tristes II: l'héritage gréco-latin. *Latomus*, 2000, 59, p. 552-563.

2485. [Petronius] MURGATROYD (P.). Petronius, Satyricon 132. *Latomus*, 2000, 59, p. 346-352.

2486. [Petronius] RUDEN (S.). Petronius Satyricon, translation with notes and topical comments. Indianapolis a. Cambridge, Hackett Publishing Company Inc., 2000, IX-193 p.

2487. [Phaedrus] OBERG (Eberhard). Phaedrus-Kommentar. Stuttgart, Steiner, 2000, 287 p.

2488. [Plautus] OWENS (William M.). Plautus' Stichus and the political crisis of 200 B.C. *American journal of philology*, 2000, 121, p. 385-407.

2489. [Plautus] Plautus, Amphitruo. Ed. by David M. CHRISTENSON. Cambridge, Cambridge U. P., 2000, X-339 p.

2490. [Plinius minor] BIRLEY (A. R.). Onomasticon to the Younger Pliny. Letters and Panegyric. München a. Leipzig, Saur, 2000, XI-111 p.

2491. [Plinius senior] MÉTHY (N.). Neron: Mage ou monstre? Sur un passage de Pline l'Ancien (NH 30, 14-17). *Rheinisches Museum*, 2000, 143, p. 381-399.

2492. [Pomponius Mela] BATTY (Roger). Mela's phoenician geography. *Journal of Roman studies*, 2000, 90, p. 70-94.

2493. [Propertius] BECK (M.). Properzens Elegie 2,7 und die augusteische Ehegesetzgebung. Philologus, 2000, 144, p. 303-324.

2494. [Propertius] JANAN (M.). The politics of desire: Propertius IV. Berkeley, Los Angeles a. London, University of California Press, 2000, X-244 p.

2495. [Propertius] O'NEILL (Kerill). Propertius 4.2: slumming with Vertumnus? *American journal of philology*, 2000, 121, p. 259-277.

2496. [Propertius] TRAINA (A.). Cinzia come Corinna. Una crux properziana: 2, 3a, 22. *Rivista di filologia e di istruzione classica*, 2000, 128, p. 38-41.

2497. [Pseudo-Claudianus] GUEX (Sophie). Ps.-Claudien, Laus Herculis. Introduction, texte, traduction et commentaire. Bern, Berlin, Bruxelles, Frankfurt am Main, New York et Wien, Lang, 2000, 244 p.

2498. [Pseudo-Quintilianus] SCHNEIDER (C.). Quelques réflexions sur la date de publication des Grandes déclamations pseudo-quintiliennes. *Latomus*, 2000, 59, p. 614-632.

2499. [Quintilianus] CHURCHILL (J. Bradford). Dice and Facie: Quintilian, Institutio Oratoria 1.7.23 and

9.4.39. *American journal of philology*, 2000, 121, p. 279-289.

2500. [Sallustius] Sallustio. Sugli dèi e il mondo. A cura di Riccardo DI GIUSEPPE. Milano, Adelphi, 2000, 265 p.

2501. [Seneca] DOGNINI (C.). Il De situ Indiae di Seneca e i commerci fra Roma e l'India all'epoca di Claudio e Nerone. *Invigilata lucernis*, 2000, 22, p 47-59.

2502. [Seneca] GRANT (M.). Seneca's tragic geography. *Latomus*, 2000, 59, p. 88-95.

2503. [Seneca] HACHMANN (E.). Der fortuna-Begriff in Senecas Epistulae morales. *Gymnasium*, 2000, 107, p. 295-319.

2504. [Seneca] HINE (H. M.). Seneca. Medea. Warminster, Aris & Phillips Ltd, 2000, VI-218 p.

2505. [Seneca] Lettere a Lucilio. Libro III: lettere 22 e 23. Testo, introduzione, traduzione e commento. A cura di Giovanni LAUDIZI. Galatina, Congedo, 2000, 129 p.

2506. [Seneca] MAURACH (G.). Seneca: Leben und Werk. Darmstadt, Wissenschaftliche Buchgesellschaft, 2000, X-234 p.

2507. [Seneca] ROSSI (E.). Seneca. Ercole sul monte Eta. Milano, Biblioteca Universale Rizzoli, 2000, 188 p.

2508. [Seneca] SCHRÖDER (St.). Beiträge zur Kritik und Interpretation von Senecas 'Oidipus'. *Hermes*, 2000, 128, p. 65-90.

2509. [Seneca] SOVERINI (P.). La clemenza dei potenti. Considerazioni sul primo libro del De clementia di Seneca. *Bollettino di studi latini*, 2000, 30, p. 48-61.

2510. [Sidonius Apollinaris] ROUSSEAU (Ph.). Sidonius and Majorian: the censure in Carmen V. *Historia*, 2000, 49, p. 251-257.

2511. [Silius Italicus] DEVILLERS (O.). Sur deux noms de Puniques chez Silius Italicus (*Pun.*, V, 627; XV, 700). *L'antiquité classique*, 2000, 69, p. 213-215.

2512. [Silius Italicus] FRÖHLICH (Uwe). Regulus, Archetyp römischer Fides. Das sechste Buch als Schlüssel zu den Punica des Silius Italicus. Interpretation, Kommentar und Übersetzung. Tübingen, Stauffenburg Verlag, 2000, XI-447 p.

2513. [Statius] P. Papinius Statius, Thebais. Die Sieben gegen Theben. Lateinischer Text mit Einleitung, Übersetzung im Versmaß des Originals, kurzen Erläuterungen, Eigennamenverzeichnis und Nachwort. Hrsg. v. Hermann RUPPRECHT. Mitterfels, Verlag Stolz, 2000, 474 p.

2514. [Statius] PAGÁN (Victoria E.). The mourning after: Statius Thebaid 12. *American journal of philology*, 2000, 121, p. 423-452.

2515. [Suetonius] GRATWICK (A. S.). ISTO VILIVS (Suetonius fr. 112, Terence, Ad. 981). *American journal of philology*, 2000, 121, p. 79-92.

2516. [Suetonius] Suetonius, Lives of the Caesars. Ed. by C. EDWARDS. Oxford, Oxford U. P., 2000, XXXVII-392 p. (maps).

2517. [Suetonius] Suetonius: Vespasian. Ed. by B. W. JONES. Bristol, Bristol Classical Press, 2000, XVIII-142 p.

2518. [Tacitus] AVELINE (J.). Tacitus, Annales 12, 65: clarifying a difficult passage. *Historia*, 2000, 49, p. 126-127.

2519. [Tacitus] BOVEY (M.). Le Dialogus de oratoribus de Tacite et les manuels de rhétorique. *Latomus*, 2000, 59, p. 353-363.

2520. [Tacitus] GIUA (M. A.). Tra storiografia e comunicazione ufficiale. *Athenaeum*, 2000, 88, p. 253-275.

2521. [Tacitus] KRAGELUND (P.). Nero's luxuria in Tacitus and in the Octavia. *Classical quarterly*, 2000, 50, p. 494-515.

2522. [Tacitus] PELLEGRINO (C.). Una crux tacitiana: Ann. XV, 44, 7. *Latomus*, 2000, 59, p. 105-108.

2523. [Terentius] MAUGER-PLICHON (B.). Térence et le problème de l'éducation: réflexions sur les Adelphes et l'Héautontimoroumenos. *Latomus*, 2000, 59, p. 802-818.

2524. [Terentius] Terence. The Eunuch. Ed. by A. J. BROTHERS. Warminster, Aris & Phillips, 2000, VI-213 p.

2525. [Valerius Flaccus] WIJSMAN (H. J. W.). Valerius Flaccus, Argonautica, Book VI. Leiden, Boston und Köln, Brill, 2000, XII-310 p. (Mnemosyne supplement, 204).

2526. [Valerius Maximus] SHACKLETON-BAILEY (D. R.). Valerius Maximus. Memorable doings and sayings, I. Books I-V; II. Books VI-IX. Cambridge a. London, Harvard U. P., 2000, 547, 462 p. (Loeb Classical Library 492-493).

2527. [Varro] KRENKEL (Werner A.). Varro. Menippeische Satiren. Wissenschaft und Technik. Hamburg, Joachim Jungius-Gesellschaft der Wissenschaften u. Göttingen, Vandenhoeck & Ruprecht, 2000, 68 p.

2528. [Varro] Marcus Terentius Varro, Saturae Menippeae. Lateinisch-deutsch, mit Anmerkungen. Hrsg. v. W. A. KRENKEL. I. 1-332; II. 333-591. Rostock, Universität Rostock, Institut für Altertumswissenschaften, 2000, 2 vol., IV-569, LI-573 p.

2529. [Varro Atacinus] PORTE (D.). En marge de la guerre des Gaules: le Bellum Sequanicum de Varron d'Atax. *Latomus*, 2000, 59, p. 276-288.

2530. [Velleius Paterculus] KOBER (Michael). Die politischen Anfänge Octavians in der Darstellung des Velleius und dessen Verhältnis zur historiographischen Tradition. Ein philologischer Quellenvergleich: Nikolaos von Damaskus, Appianos von Alexandria, Velleius Paterculus. Würzburg, Königshausen & Neumann, 2000, 468 p.

2531. *Vacat.*

2532. [Vergilius] HORSFALL (N.). Virgil, Aeneid 7. A commentary. Leiden, Boston a. Köln, Brill, 2000, XLIV-567.

2533. [Vergilius] Virgil. Aeneid VII–XII; Appendix Vergiliana. Ed. by H. R. FAIRCLOUGH. Cambridge a. London, Harvard U. P., 2000, 590 p.

Cf. n°s 268, 645, 689, 723, 769, 2720-2837

§ 4. General and political history.

* 2534. Bulletin analytique d'histoire romaine. Nouvelle série. Vol. 9. Strasbourg, 2000, Université Marc Bloch, 2000, 501 p.

2535. ALFÖLDY (Géza). Provincia Hispania Superior. Heidelberg, Winter, 2000, VII-72 p.

2536. ANDO (C.). Imperial ideology and provincial loyalty in the Roman empire. Berkeley, Los Angeles a. London, University of California Press, 2000, XXI-494 p.

2537. BALL (W.). Rome in the east. The transformation of an empire. London a. New York, Routledge, 2000, XIX-523 p. (ill.).

2538. BARDEN DOWLING (Melissa). The clemency of Sulla. *Historia*, 2000, 49, p. 303-340.

2539. BARRETT (Anthony A.). Caligula. The corruption of power. London a. New York, Routledge, 2000, XXVI-334 p.

2540. BESSONE (L.). Le problème de la première conjuration de Catilina. *Patavium*, 2000, 15, p. 23-36.

2541. BIRLEY (Anthony R.). Hadrian. The restless emperor. London a. New York, Routledge, 2000, XVII-399 p. – IDEM. Senators as generals. *In*: Kaiser, Heer und Gesellschaft in der Römischen Kaiserzeit [Cf. n° 2695], p. 97-119.

2542. BOATWRIGHT (Mary T.). Hadrian and the cities of the Roman empire. Princeton, Princeton U. P., 2000, XVIII-243 p.

2543. BOÏADJIEV (D.). Les relations ethno-linguistiques en Thrace et en Mésie pendant l'époque romaine. Sofia, Presses Universitaires 'St Kliment Ohridski', 2000, p. 174.

2544. BORCA (Federico). Terra mari cincta: insularità e cultura romana. Roma, Carocci editore, 2000, 217 p.

2545. BRILLIANT (R.). The Pax Romana: bridge or barrier between Romans and Barbarians. *In*: Gegenwelten zu den Kulturen Griechenlands und Roms in der Antike [Cf. n° 2570], p. 391-408.

2546. BRUUN (Ch.). 'What every man in the street used to know': M. Furius Camillus, Italic legends and Roman historiography. *In*: Roman middle republic (The) [Cf. n° 2612], p. 41-68.

2547. BURTON (G. P.). The resolution of territorial disputes in the provinces of the Roman empire. *Chiron*, 2000, 30, p. 195-215.

2548. Cambridge ancient history (The). Vol. 11. The high empire, A.D. 70-192. Ed. by Alan K. BOWMAN, Peter GARNSEY and Dominic RATHBONE. Cambridge, Cambridge U. P., 2000, XXI-222 p. (ill., cartes).

2549. CAMPANILE (M. D.). Del bere sangue di toro e della morte di Annibale. *Chiron*, 2000, 30, p. 117-129.

2550. CARSANA (C.). Considerazioni sulla fondazione di Lione alla luce di una rilettura dell'epistolario ciceroniano. *Athenaeum*, 2000, 88, p. 203-217.

2551. CHAUSSON (François). De Didius Julianus aux Nummii Albini. *Mélanges de l'Ecole française de Rome. Antiquité*, 2000, 112, p. 843-879.

2552. COARELLI (Filippo). Roma, la città come cosmo. *In*: Mégapoles méditerranéennes [Cf. n° 262], p. 288-310.

2553. Cultural identity in the Roman empire. Ed. by R. LAURENCE and J. BERRY. London a. New York, Routledge, 2000, XI-205 p. (ill.).

2554. DAHLHEIM (Werner). An der Wiege Europas. Städtische Freiheit im antiken Rom. Frankfurt am Main, Fischer Taschenbuch Verlag, 2000, 238 p.

2555. DAVID (Jean-Michel). La république romaine de la deuxième guerre punique à la bataille d'Actium 218–31. Crise d'une aristocratie. Paris, Éditions du Seuil, 2000, 309 p.

2556. DETTENHOFER (Maria H.). Herrschaft und Widerstand im augusteischen Principat. Die Konkurrenz zwischen Res publica und Domus Augusta. Stuttgart, Steiner, 2000, 234 p. (Historia Einzelschriften 140).

2557. DIETZ (K.). Kaiser Julian in Phönizien. *Chiron*, 2000, 30, p. 807-855.

2558. DRÄGER (Michael). Überlegungen zu den Reisen Hadrians durch Kleinasien. *Klio*, 2000, 82, p. 208-216.

2559. DRINKWATER (J. F.). The revolt and ethnic origin of the usurper Magnentius (350–353), and the rebellion of Vetranio (350). *Chiron*, 2000, 30, p. 131-159.

2560. DRUMMOND (Andrew). Rullus and the Sullan possessores. *Klio*, 2000, 82, p. 126-153.

2561. Elites municipales (Les) de l'Italie péninsulaire, de la mort de César à la mort de Domitien, entre continuité et rupture: classes sociales dirigeantes et pouvoir central. Actes du Colloque de Naples (6–8 février 1997). Sous la dir. de Mireille CÉBEILLAC-GERVASONI. Roma, Ecole française de Rome, 2000, 528 p. (Collection de l'Ecole française de Rome, 271).

2562. ELSNER (Jaś). The Itinerarium burdigalense: politics and salvation in the geography of Constantine's empire. *Journal of Roman studies*, 2000, 90, p. 181-195.

2563. ESMONDE CLEARY (A. S.). The ending of Roman Britain. London a. New York, 2000, XI-242 p. (ill., tables).

2564. Experiencing Rome. Culture, identity and power in the Roman empire. Ed. by J. HUSKINSON. London a. New York, Routledge, 2000, XV-378 p. (ill.).

2565. FAULKNER (N.). The decline and fall of Roman Britain. Stroud, Tempus, 2000, 192 p. (ill.).

2566. FERRARY (J.-L.). Les gouverneurs des provinces romaines d'Asie Mineure (Asie et Cilicie), depuis l'organisation de la province d'Asie jusqu'à la première guerre de Mithridate (126–88 av. J.-C.). *Chiron*, 2000, 30, p. 161-193.

2567. FERRIÈS (M.-C.). La légende noire de P. Canidius Crassus. *Athenaeum*, 2000, 88, p. 413-430.

2568. FRASCHETTI (A.). La città di Roma in epoca augustea. Amministrare sorvegliando. *In*: Mégapoles méditerranéennes [Cf. n° 262], p. 724-731.

2569. GABBA (E.), LAFFI (U.). Sociedad y política en la Roma republicana (siglos III–I a. C.). Pisa, Pacini editore, 2000, 334 p.

2570. Gegenwelten zu den Kulturen Griechenlands und Roms in der Antike. Hrsg. v. Tonio HÖLSCHER. München u. Leipzig, Saur, 2000, 498 p. (Abb) [Cf. n[os] <Auswahl> 2545, 2859.]

2571. Grenzen der Macht. Zur Rolle der römischen Kaiserfrauen. Hrsg. v. Ch. KUNST und U. RIEMER. Stuttgart, Steiner, 2000, IX-174 p.

2572. GÜNTHER (R.). Sexuelle Diffamierung und politische Intrigen in der Republik: P. Clodius Pulcher und Clodia. *In*: Frauenwelten in der Antike [Cf. n° 2689], p. 227-241.

2573. GURY (F.). L'idéologie impériale et la lune: Caligula. *Latomus*, 2000, 59, p. 564-595.

2574. HALLOF (K.), HEIL (M.). Ein neues senatorisches Epitheton. *Klio*, 2000, 82, p. 483-491.

2575. HATSCHER (Christoph R.). Charisma und res publica. Max Webers Herrschaftssoziologie und die römische Republik. Stuttgart, Steiner, 2000, 263 p. (Historia. Einzelschriften, 136).

2576. Histoire romaine. T. 1. Des origines à Auguste. Ed. par François HINARD. Paris, Fayard, 2000, 1075 p. (ill.).

2577. HÖLKESKAMP (K. J.), STEIN-HÖLKESKAMP (E.). Von Romulus zu Augustus. Grosse Gestalten der römischen Republik. Munich, Verlag C. H. Beck, 2000, 393 p.

2578. HUMM (M.). Spazio e tempo civici: riforma delle tribù e riforma del calendario alla fine del IV secolo a. C. *In*: Roman middle republic (The) [Cf. n° 2612], p. 91-119.

2579. HURLET (Frédéric). Pouvoir des images, images du pouvoir impérial. La province d'Afrique aux deux premiers siècles de notre ère. *Mélanges de l'Ecole française de Rome. Antiquité*, 2000, 112, p. 297-364.

2580. HUTTNER (Ulrich). Zur Zivilisationskritik in der frühen Kaiserzeit: die Diskreditierung der pax Romana. *Historia*, 2000, 49, p. 447-466.

2581. JANVIER (Y.). Sur une erreur à éviter dans le calcul des durées en histoire ancienne. *Latomus*, 2000, 59, p. 766-772.

2582. JUNKELMANN (M.). Römische Helme. Mainz, von Zabern, 2000, 208 p. (Abb.).

2583. KÖNIG (I.). Q. Sertorius. Ein Kapitel des frühen römischen Bürgerkriegs. *Klio*, 2000, 82, p. 441-458.

2584. *Vacat.*

2585. Légions (Les) de Rome sous le haut-empire. Actes du congrès de Lyon (17–19 septembre 1998). Vol. I–II. Éd. par Yann LE BOHEC. Lyon et Paris, De Boccard, 2000, 2 vol., 410 p., 337 p. (tables).

2586. LENSKY (N.). The election of Jovian and the role of the late imperial guards. *Klio*, 2000, 82, p. 492-515.

2587. Lexicon of the Greek and Roman cities and place names in antiquity ca. 1500 B.C. – ca. A.D. 500. Ed. by O. BOUNEGRU, K. BRANIGAN, M. J. DEARNE, N. DEMAND, F. FELTEN, R. NOUWEN, E. OLSHAUSEN, M. K. PHELPS, G. WALDHERR, K. A. WORP and M. ZAHARIADE. Fasc. 5: Apollonia-Arados. Amsterdam, Hakkert, 2000, p. 41-800.

2588. MACKAY (Christopher S.). Sulla and the monuments: studies in his public persona. *Historia*, 2000, 49, p. 161-210.

2589. MACMULLEN (R.). Romanization in the time of Augustus. New Haven a. London, Yale U. P., 2000, XI-222 p. (ill.).

2590. MARCONE (Arnaldo). Costantino il Grande. Roma e Bari, Laterza, 2000, 142 p.

2591. MASIER (A.). I Sodales nell'età di Antonino Pio. *Patavium*, 2000, 15, p. 53-80.

2592. MILLER (John F.). Triumphus in palatio. *American journal of philology*, 2000, 121, p. 409-422.

2593. MONTI (Enrico). Marc Aurel. Kaiser aus Pflicht. Regensburg, Pustet, 2000, 207 p.

2594. MORGAN (Gwyn). Clodius Macer and Calvia Crispinilla. *Historia*, 2000, 49, p. 467-487.

2595. MORLEY (Neville). Trajan's engines. *Greece and Rome*, 2000, 47, p. 197-210.

2596. Mos Maiorum. Untersuchungen zu den Formen der Identitätsstiftung und Stabilisierung in der römischen Republik. Hrsg. v. B. LINKE und M. STEMMLER. Stuttgart, Steiner, 2000, 319 p. (Historia Einzelschriften, 141).

2597. MUNDUBELTZ (G.). Octavien et son armée au lendemain de la guerre de Sicile (36–35 av. J.-C.). *Athenaeum*, 2000, 88, p. 169-201.

2598. Onomasticon provinciarum Europae Latinarum (OPEL). Vol. III. Labareus–Pythea. Hrsg. v. A. MÓCSY, R. FELDMANN, E. MARTON und M. SZILÁGYI. Wien, Forschungsgesellschaft Wiener Stadtarchäologie, 2000, 190 p.

2599. Oxford classical dictionary (The) on CD-ROM. Oxford, Oxford U. P., 2000, CD-ROM.

2600. PATTERSON (John R.). Political life in the city of Rome. London, Bristol classical press a. Duckworth, 2000, VII-90 p.

2601. Paulys Realencyclopädie der classischen Altertumswissenschaft. Gesamtregister II: Systematisches Sach- und Suchregister. Erarbeitet v. Christa FRATEANTONIO und Markus E. FUCHS. Stuttgart, Metzler, 2000, CD-ROM.

2602. POHL (Walter). Die Germanen. München, Oldenbourg, 2000, X-159 p.

2603. POUCET (J.). Les rois de Rome. Tradition et histoire. Bruxelles, Classe des Lettres Académie Royale de Belgique, 2000, 517 p.

2604. RAMAGE (Edwin S.). Augustus' propaganda in Africa. *Klio*, 2000, 82, p. 171-207.

2605. RÉMY (Bernard). Loyalisme politique et culte impérial dans les provinces des Alpes occidentales (Alpes cottiènnes, graies, maritimes et poenines) au Haut-Empire. *Mélanges de l'Ecole française de Rome. Antiquité*, 2000, 112, p. 881-924.

2606. RENUCCI (P.). Les idées politiques et le gouvernement de l'empereur Julien. Bruxelles, Latomus, 2000, 537 p. (Collection Latomus, 259).

2607. REYNOLDS (J. M.). New letters of Hadrian to Aphrodisias: trials, taxes, gladiators and an aqueduct. *Journal of Roman archaeology*, 2000, 13, p. 5-20.

2608. RHOR VIO (Francesca). Le voci del dissenso. Ottaviano Augusto e i suoi oppositori. Padova, Il Poligrafo, 2000, 399 p.

2609. RICCI (C.). Domo Roma. Il contributo della capitale all'esercito di confine e alle milizie urbane (età imperiale). *In*: Kaiser, Heer und Gesellschaft in der Römischen Kaiserzeit [Cf. n° 2695], p. 193-205.

2610. RIDLEY (Ronald T.). The dictator's mistake: Caesar's escape from Sulla. *Historia*, 2000, 49, p. 211-229.

2611. Roma antica. A cura di A. GIARDINA. Roma e Bari, Laterza, 2000, XXXI-430 p.

2612. Roman middle republic (The). Politics, religion, and historiography c. 400–133 B.C. Papers from a conference at the Institutum Romanum Finlandiae, september 11–12, 1998. Ed. by Christer BRUUN. Roma, Institutum Romanum Finlandiae, 2000, IX-310 p. (Acta Instituti Romani Finlandiae, 23). [Cf. n^{os} <choice> 2373, 2546, 2578, 2632, 2643, 2645, 2655, 2667, 2670, 2869, 2870.]

2613. Romanità orientale e Italia meridionale dall'antichità al medioevo. Paralleli storici e culturali. Atti del II convegno di studi italo-romeno (Bari 19–22 ottobre 1998). A cura di Stefania SANTELIA. Bari, Edipuglia, 2000, 262 p. (ill.).

2614. Römer (Die) zwischen Alpen und Nordmeer. Zivilisatorisches Erbe einer europäischen Militärmacht. Katalog-Handbuch zur Landesausstellung des Freistaates Bayern Rosenheim 2000. Hrsg. v. Ludwig WAMSER, Christof FLÜGEL und Bernward ZIEGAUS. Mainz, von Zabern, 2000, XXII-466 p.

2615. RUMSCHEID (J.). Kranz und Krone. Zu Insignien, Siegespreisen und Ehrenzeichen der römischen Kaiserzeit. Tübingen, Ernst Wasmuth Verlag, 2000, XI-270 p.

2616. SABIN (Philip). The face of Roman battle. *Journal of Roman studies*, 2000, 90, p. 1-17.

2617. SAUTON (Gilles). L'histoire végétalisée. Ornements et politique à Rome. Paris, Picard, 2000, 249 p. (Antiqua).

2618. SCHÄFER (N.). Die Einbeziehung der Provinzialen in den Reichsdienst in augusteischer Zeit. Stuttgart, Steiner, 2000, 181 p.

2619. SCHIAVONE (A.). The end of the past. Ancient Rome and the modern West. Cambridge and London, Harvard U. P., 2000, VIII-278 p.

2620. SCHULZ (Raimund). Caesar und das Meer. *Historische Zeitschrift*, 2000, 271, 2, p. 281-310.

2621. Serta antiqua et medievalia. II. Tradizione enciclopedica e divulgazione in età imperiale. Roma, Giorgio Bretschneider, 2000, VII-222 p.

2622. SHALIMOV (Oleg A.). Obraz ideal'nogo praviteljа v Drevnem Rime v seredine I–nachale II veka n.e. (The concept of the ideal ruler in Ancient Rome in the middle of 1st–the early 2nd century A.D). RAN, In-t vseobshchej istorii. Moskva, [s. n.], 2000, 183 p. [Eng. summary]

2623. SHOTTER (David C. A.). Agrippina the Elder – a woman in a man's world. *Historia*, 2000, 49, p. 341-357.

2624. SION-JENKINS (Karin). Von der Republik zum Prinzipat. Ursachen für den Verfassungswechsel in Rom im historischen Denken der Antike. Stuttgart, Steiner, 2000, 250 p.

2625. STEMMLER (Michael). Die Römische Manipularordnung und der Funktionswandel der Zenturionen. *Klio*, 2000, 82, p. 107-125.

2626. STROTHMANN (M.). Augustus Vater der res publica. Zur Funktion der drei Begriffe restitutio, saeculum, pater patriae im augusteischen Principat, Stuttgart, Franz Steiner Verlag, 2000, 320 p.

2627. Studia dacica. Collected papers. Ed. by Lajos SZABÓ. Debrecen, University of Debrecen, Institut of history, 2000, p. IX-226 p.

2628. TANNER (Jeremy). Portraits, power, and patronage in the late Roman republic. *Journal of Roman studies*, 2000, 90, p. 18-50.

2629. TEICHNER (Felix), NEVILLE (Ann). Romanization, christianization and islamicization in southern Lusitania. *Antiquity*, 2000, 74, p. 33-34.

2630. THOME (Gabriele). Zentrale Wertvorstellungen der Römer I. II. Texte, Bilder, Interpretationen. Bamberg, Buchners Verlag, 2000, 2 vol., 151 p., 152 p. (ill.).

2631. TODD (Malcolm). Die Germanen. Von den frühen Stammesverbänden zu den Erben des Weströmischen Reiches. Stuttgart, Theiss, 2000, 270 p. (Abb.).

2632. UNGERN-STERNBERG (J. von). Eine Katastrophe wird verarbeitet: Die Gallier in Rom. *In*: Roman middle republic (The) [Cf. n° 2612], p. 207-222.

2633. WALDHERR (G. H.). 'Punica fides' – Das Bild der Karthager in Rom. *Gymnasium*, 2000, 107, p. 193-222.

2634. WITSCHEL (C.). Zwei neue Bücher zum römischen Städtewesen in Nordafrika. *Klio*, 2000, 82, p. 516-522.

2635. WOLTERS (R.). Die Römer in Germanien. München, Verlag C. H. Beck, 2000, 128 p. (Abb.).

2636. WOODS (D.). Caligula's seashells. *Greece and Rome*, 2000, 47, p. 80-87.

2637. ZHMODIKOV (A.). Roman republican heavy infantrymen in battle (IV–II centuries B.C.). *Historia*, 2000, 49, p. 67-78.

Cf. n°s 467, 608, 722

§ 5. History of law and institutions.

2638. BAUMAN (R. A.). Did Cicero want to abolish the jury-courts? *Latomus*, 2000, 59, p. 842-849.

2639. BRENNAN (T. C.). The praetorship in the Roman republic. Vol. I-II. Oxford, Oxford U. P., 2000, 2 vol., XVII-354, VII-615 p.

2640. CALORE (Antonello). 'Per Iovem lapidem'. Alle origini del 'sacro' nell'esperienza giuridica romana. Milano, Giuffrè, 2000, VIII-197 p.

2641. CASCIONE (C.). Una norma dimenticata delle XII tavole? Dion. Hal. 10.60.6. *Index*, 2000, 28, p. 187-201.

2642. CAVALIERI MANASSE (G.). Un documento catastale dell'agro centuriato veronese. *Athenaeum*, 2000, 88, p. 5-48.

2643. CELS SAINT-HILAIRE (J.). Citoyenneté de droit de vote: à propos du procès des Scipions. *In*: Roman middle republic (The) [Cf. n° 2612], p. 177-194.

2644. CORIAT (P.). Consolidation et précodification du droit impérial à la fin du principat. *In*: Codification (La) des lois dans l'antiquité [Cf. n° 1618], p. 273-284.

2645. CORNELL (T. J.). The lex Ovinia and the emancipation of the Senate. *In*: Roman middle republic (The) [Cf. n° 2612], p. 69-89.

2646. Corpus der römischen Rechtsquellen zur antiken Sklaverei (CRRS). Teil IX. Irrtümlich als Sklaven gehaltene freie Menschen und Sklaven in unsicheren Eigentumsverhältnissen – homines liberi et servi alieni bona fide servientes. Hrsg. v. J. Michael RAINER. Stuttgart, Steiner, 2000, XIV-157 p.

2647. D'IPPOLITO (Federico M.). Del fare diritto nel mondo romano. Torino, Giappichelli, 2000, VII-146 p.

2648. DIOUF (Eu.). Magie et droit chez Pline l'Ancien. *In*: Magie (La) [Cf. n° 2854], p. 71-84.

2649. ERMANN (Joachim). Strafprozess öffentliches Interesse und private Strafverfolgung. Untersuchungen zum Strafrecht der römischen Republik. Köln u. Weimar, Böhlau, 2000, XI-134 p.

2650. FERRETTI (Paolo). Le donazioni tra fidanzati nel diritto romano. Milano, Giuffrè, 2000, XIV-324 p.

2651. GASCOU (J.). Fragment d'un codex juridique du bas-empire (P. Strasb. L 9). *In*: Codification (La) des lois dans l'antiquité [Cf. n° 1618], p. 285-291.

2652. GIOVANNINI (A.). Le droit fécial et la déclaration de guerre de Rome à Carthage en 218 avant J.-C. *Athenaeum*, 2000, 88, p. 69-116.

2653. GNOLI (Tommaso). Roma, Edessa e Palmira nel III sec. d. C. Problemi istituzionali. Uno studio sui papiri dell'Eufrate. Pisa e Roma, Istituti editoriali e poligrafici internazionali, 2000, 190 p.

2654. GUTIÉRREZ MASSON (L.). La ritualización de la violencia en el derecho romano arcaico. *Index*, 2000, 28, p. 253-272.

2655. HÖLKESKAMP (K.-J.). Fides – deditio in fidem – dextra data et accepta: Recht, Religion und Ritual in Rom. *In*: Roman middle republic (The) [Cf. n° 2612], p. 223-250.

2656. JONES (H.). Le détournement d'institution, instrument technique du droit romain. *Latomus*, 2000, 59, p. 15-35.

2657. LORETO (Luigi). Il comando militare nelle province procuratorie 30 a. C.–280 d. C. Dimensione militare e dimensione costituzionale. Napoli, Jovene, 2000, 92 p.

2658. LOVISI (Claire). Contribution à l'étude de la peine de mort sous la République romaine (509–133 avant J.-C.). Paris, De Boccard, 2000, 393 p.

2659. MANTOVANI (Dario). Aspetti documentari del processo criminale nella Repubblica. Le tabulae publicae. *Mélanges de l'Ecole française de Rome. Antiquité*,

2000, 112, p. 651-691. – IDEM. L'édit comme code. *In*: Codification (La) des lois dans l'antiquité [Cf. n° 1618], p. 257-271.

2660. MAROTTA (Valerio). Ulpiano e l'impero. I. Napoli, Loffredo, 2000, 214 p.

2661. MOATTI (Claudia). Recherches sur l'administration romaine: le cas des archives judiciaires pénales. *Mélanges de l'Ecole française de Rome. Antiquité*, 2000, 112, p. 647-649.

2662. MOREAU (Philippe). Quelques aspects documentaires de l'organisation du procès pénal républicain. *Mélanges de l'Ecole française de Rome. Antiquité*, 2000, 112, p. 693-721.

2663. NETTIS (A. V.). Padroni, sesso e schiavi. *Index*, 2000, 28, p. 155-172.

2664. PERGAMI (Federico). L'appello nella legislazione del tardo impero. Milano, Giuffrè, 2000, IX-522 p.

2665. QUERZOLI (Serena). I testamenta e gli officia pietatis. Tribunale centumvirale, potere imperiale e giuristi tra Augusto e i Severi. Napoli, Loffredo, 2000, 268 p.

2666. RODRÍGUEZ-ALMEIDA (Emilio). Breve nota sulla lex di una furcatio aquaria. *Mélanges de l'Ecole française de Rome. Antiquité*, 2000, 112, p. 231-236.

2667. SANDBERG (K.). Tribunician and non-tribunician legislation in mid-republican Rome. *In*: Roman middle republic (The) [Cf. n° 2612], p. 121-140.

2668. STERN (Yaakov). The testamentary phenomenon in ancient Rome. *Historia*, 2000, 49, p. 413-428.

2669. SUÁREZ PIÑEIRO (A. M.). Les leges iudiciariae ante la crisis de la República Romana (133–44 a. C.). *Latomus*, 2000, 59, p. 253-275.

2670. TORELLI (M.). C. Genucio(s) Clousino(s) prai(fectos). La fondazione della praefectura Caeritum. *In*: Roman middle republic (The) [Cf. n° 2612], p. 141-176.

2671. TRAVERSO (M.). Il centurionato nelle legioni romane: la legio II Augusta. *Athenaeum*, 2000, 88, p. 219-252.

2672. WELWEI (K.-W.). Sub corona vendere. Quellenkritische Studien zu Kriegsgefangenschaft und Sklaverei in Rom bis zum Ende des Hannibalkrieges. Unter Berücksichtigung des Nachlasses von Gottfried Prachner. Stuttgart, Steiner, 2000, VIII-181 p

§ 5. Addenda 1999.

2673. ARENDS OLSEN (L.). La femme et l'enfant dans les unions illégitimes à Rome. L'évolution du droit jusqu'au début de l'Empire. Bern, Peter Lang, 99, XIV-247 p. (tables).

Cf. n°s 2094, 2624

§ 6. Economic and social history.

2674. ALFÖLDY (G.). Das Heer in der Sozialstruktur des Römischen Kaiserreiches. *In*: Kaiser, Heer und Gesellschaft in der Römischen Kaiserzeit [Cf. n° 2695], p. 33-57.

2675. BAHARAL (D.). Public image and women at court in the era of the adoptive emperors (AD 98-180). The case of Faustina the Younger. *In*: Studies in Latin literature and Roman history [Cf. n° 2806], p. 328-344.

2676. BOITEAUX (M.). Lieux de fête et lieux de pouvoir dans l'espace public romain. Le palais, la ville, le peuple. *In*: Mégapoles méditerranéennes [Cf. n° 262], p. 311-350.

2677. BRADLEY (Keith). Animalizing the slave: the truth of fiction. *Journal of Roman studies*, 2000, 90, p. 110-125.

2678. Caesaren und Gladiatoren. Die Macht der Unterhaltung im antiken Rom. Hrsg. v. Eckart KÖHNE und Cornelia EWIGLEBEN. Mainz, von Zabern u. Hamburg, Museum für Kunst und Gewerbe, 2000, 160 p.

2679. CALABRIA (P.). La moneta romana da Augusto a Settimio Severo. *In*: Moneta greca e romana (La) [Cf. n° 168], p. 105-125.

2680. CARRERAS MONFORT (C.). Economía de la Britannia romana; la importacíon de alimentos. Barcelona, Publicacions Universitat de Barcelona, 2000, p. 344 (ill.).

2681. COCHET (A.). Le plomb en Gaule romaine: techniques de fabrication et produits. Montagnac, Editions Monique Mergoil, 2000, VIII-223 p. (ill.).

2682. DALBY (A.). Empire of pleasures. Luxury and indulgence in the Roman world. London and New York, Routledge, 2000, X-335 p. (ill.).

2683. DE BLOIS (L.). Army and society in the late Roman republic: professionalism and the role of the military middle cadre. *In*: Kaiser, Heer und Gesellschaft in der Römischen Kaiserzeit [Cf. n° 2695], p. 11-31.

2684. DE GROSSI MAZZORIN (Jacopo). État de nos connaissances concernant le traitement et la consommation du poisson dans l'antiquité, à la lumière de l'archéologie. L'exemple de Rome. *Mélanges de l'Ecole française de Rome. Antiquité*, 2000, 112, p. 155-167.

2685. DE LIGT (L.). Studies in legal and agrarian history II: tenancy under the republic. *Athenaeum*, 2000, 88, p. 377-391.

2686. DUŠANIĆ (S.). Army and mining in Moesia superior. *In*: Kaiser, Heer und Gesellschaft in der Römischen Kaiserzeit [Cf. n° 2695], p. 343-363.

2687. ÉTIENNE (R.), MAYET (F.), TAVERES DA SILVA (C.). Chronologie des usines de salaison de Lusitanie. *Mélanges de l'Ecole française de Rome. Antiquité*, 2000, 112, p. 99-117.

2688. FEAR (A. T.). Status symbol or leisure pursuit? Amphitheatres in the Roman world. *Latomus*, 2000, 59, p. 82-87.

2689. Frauenwelten in der Antike. Geschlechterordnung und weibliche Lebenspraxis. Hrsg. v. Th. SPÄTH und B. WAGNER-HASEL. Stuttgart u. Weimar, Metzler, 2000, XXVI-494 p. (Abb.) [Cf. n^{os} <Auswahl> 2572, 2793, 2843, 2888, 3200.]

2690. FUTRELL (Alison). Blood in the arena. The spectacle of Roman power. Austin, Texas U. P., 2000, XII-338 p.

2691. GÖBL (Robert). Die Münzprägung der Kaiser Valerianus I./Gallienus/Saloninus (253/268)/Regalianus (260) und Macrianus/Quietus (260/262). Text- und Tafelband. Wien, Verlag der Österreichischen Akademie der Wissenschaften, 2000, 306, 160 p.

2692. HÄNNINEN (Marja-Leena). Women as worshippers of Juno from the mid-republican to the Augustan era. Helsinki, M.-L. Hänninen, 2000, 160 p.

2693. HUVELIN (H.). La moneta romana dalla morte di Alessandro Severo all'avvento di Diocleziano. *In*: Moneta greca e romana (La) [Cf. n° 168], p. 127-142.

2694. JUNKELMANN (M.). Das Spiel mit dem Tod. So kämpften Roms Gladiatoren. Mainz, von Zabern, 2000, IV-196 p. (Abb.).

2695. Kaiser, Heer und Gesellschaft in der römischen Kaiserzeit. Hrsg. v. G. ALFÖLDY, B. DOBSON u. W. ECK, Stuttgart, Steiner, 2000, 509 p. [Cf. n^{os} <Auswahl> 2423, 2541, 2609, 2674, 2683, 2686, 2698, 2699, 2715, 2719.]

2696. KOLB (Anne). Transport und Nachrichtentransfer im Römischen Reich. Berlin, Akademie Verlag, 2000, 380 p.

2697. LA PENNA (A.). Eros dai cento volti. Modelli etici ed estetici nell'eta dei Flavi. Venezia, Marsilio, 2000, 221 p.

2698. LE BOHEC (Y.). Le rôle social et politique de l'armée romaine dans les provinces d'Afrique. *In*: Kaiser, Heer und Gesellschaft in der Römischen Kaiserzeit [Cf. n° 2695], p. 207-226.

2699. LE ROUX (P.). Armée et société en Hispanie sous l'empire. *In*: Kaiser, Heer und Gesellschaft in der Römischen Kaiserzeit [Cf. n° 2695], p. 261-278.

2700. LO CASCIO (Elio). Il princeps e il suo impero. Studi di storia amministrativa e finanziaria romana. Bari, Edipuglia, 2000, 372 p.

2701. ŁOŚ (Andrzej). Les affaires «industrielles» des élites des villes campaniennes sous les Julio-Claudiens et les Flaviens. *Mélanges de l'Ecole française de Rome. Antiquité*, 2000, 112, p. 243-277.

2702. MEISSNER (Burkhard). Über Zweck und Anlaß von Diokletians Preisedikt. *Historia*, 2000, 49, p. 79-100.

2703. Mercati permanenti e mercati periodici nel mondo romano. Atti degli incontri capresi di storia dell'economia antica (Capri, 13–15 ottobre 1997). A cura di Elio LO CASCIO. Bari, Edipuglia, 2000, [s. p.].

2704. Metallanalytische Untersuchungen an Münzen den Römischen Republik. Hrsg. v. W. HOLLSTEIN. Berlin, Mann, 2000, 190 p. (Abb., Taf.).

2705. MOATTI (Claudia). Le côntrole de la mobilité des personnes dans l'empire romain. *Mélanges de l'Ecole française de Rome. Antiquité*, 2000, 112, p. 925-958.

2706. MOLLO (S.). La mobilità sociale a Brescia romana. Milano, Edizioni Universitarie di Lettere Economia Diritto, 2000, X-401 p.

2707. MORELLI (A. M.). La vita beata del pastore ed un passo del Culex. *Rivista di filologia e di istruzione classica*, 2000, 128, p. 432-453.

2708. MROZEK (S.). Le fonctionnement des fondations dans les provinces occidentales et l'économie de crédit à l'époque du haut-empire romain. *Latomus*, 2000, 59, p. 327-345.

2709. NELIS-CLÉMENT (J.). Les beneficiarii: militaires et administrateurs au service de l'empire (1er s. a. C. – VIe s. p. C.). Bordeaux, Éditions Ausonius, 2000, 557 p.

2710. NICOLET (C.). Censeurs et publicains. Économie et fiscalité dans la Rome antique. Paris, Fayard, 2000, 500 p.

2711. PANVINI ROSATI (F.). La moneta romana dall'introduzione del denario ad Augusto. *In*: Moneta greca e romana (La) [Cf. n° 168], p. 95-103.

2712. SCHULZ (R.). Zwischen Kooperation und Konfrontation. Die römische Weltreichsbildung und die Piraterie. *Klio*, 2000, 82, p. 426-440.

2713. SIMPSON (C. J.). Musicians and the arena: dancers and the hippodrome. *Latomus*, 2000, 59, p. 633-639.

2714. Sordes urbis. La eliminación de residuos en la ciudad romana. Actas de la reunión de Roma (15–16 de noviembre de 1996). Ed. por X. DUPRÉ RAVENTÓS y J.-A. REMOLÀ. Roma, 'L'Erma' di Bretschneider, 2000, XV-150 p.

2715. SPEIDEL (M. A.). Sold und Wirtschaftslage der römischen Soldaten. *In*: Kaiser, Heer und Gesellschaft in der Römischen Kaiserzeit [Cf. n° 2695], p. 65-94.

2716. VAN OSSEL (P.), OUZOULIAS (P.). Rural settlement economy in northern Gaul in the late empire: an overview. *Journal of Roman archaeology*, 2000, 13, p. 133-160.

2717. VEYNE (Paul). La «plèbe moyenne» sous le Haut-Empire romain. *In*: Catégories sociales de l'espace politique. *Annales*, 2000, 55, 6, p. 1169-1200.

2718. WEISS (P.). Euergesie oder römische Prägegenehmigung? Aitesamenou-Formular auf Städtemünzen der Provinz Asia, Roman provincial coinage

(RPC) II und persönliche Aufwendungen im Münzwesen. *Chiron*, 2000, 30, p. 235-254.

2719. WILKES (J. J.). Army and society in Roman Dalmatia. *In*: Kaiser, Heer und Gesellschaft in der Römischen Kaiserzeit [Cf. n° 2695], p. 327-341.

Cf. n°s 127, 1916

§ 7. History of literature, philosophy and science.

a. Literature

2720. ACHARD (G.), LEDENTU (M.). Orateur, auditeurs, lecteurs: à propos de l'éloquence romaine à la fin de la République et au début du Principat. Paris, Diffusion De Boccard, 2000, 117 p.

2721. ADKIN (Neil). Jerome, Seneca, Juvenal. *Revue belge de philologie et d'histoire*, 2000, 78, 1, p. 119-128. – IDEM. Vergilian Etymologizing: the case of Acestes. *L'antiquité classique*, 2000, 69, p. 205-207.

2722. BAJONI (M. G.). Sul gesto di Telefrone in Apul., Met., II, 21 p. 42 Helm. *L'antiquité classique*, 2000, 69, p. 209-212.

2723. BANNON (C. J.). Naming brother and sister in Catullus 100. *Latomus*, 2000, 59, p. 541-546.

2724. BARRETT (A. A.). The Laus Caesaris: its history and its place in Latin literature. *Latomus*, 2000, 59, p. 596-606.

2725. BINDER (Gerhard). Dido und Aeneas: Vergils Dido-Drama und Aspekte seiner Rezeption. Trier, Wissenschaftlicher Verlag Trier, 2000, 321 p. (Abb.).

2726. BIRD (T. A.). One wedding and two funerals. An undetected aspect of Aeneid IV. *In*: Studies in Latin literature and Roman history [Cf. n° 2806], p. 197-208.

2727. BÖMER (F.). Ovid als Erzähler. Interpretationen zur poetischen Technik der Metamorphosen. *Gymnasium*, 2000, 107, p. 1-23.

2728. BONVICINI (M.). Le forme del pianto. Catullo nei Tristia di Ovidio. Bologna, Pàtron, 2000, 194 p.

2729. BRACHET (J.-P.). Recherches sur les préverbes de- et ex- du latin. Bruxelles, Latomus, 2000, 400 p. (Collection Latomus 258).

2730. CAVARZERE (Alberto). Oratoria a Roma. Storia di un genere pragmatico. Roma, Carocci, 2000, 288 p.

2731. CITRONI MARCHETTI (S.). Soffrire come e più di Ulisse: Teognide, Plauto e le origini di un paragone ovidiano (trist. 1.5.58). *Prometheus*, 2000, 26, p. 119-136.

2732. CLÉMENT (S.). Fama et le poète: pour une poétique de la monstruosité dans l'Énéide. *Bulletin de l'association Guillaume Budé*, 2000, 4, 309-328.

2733. COLTON (Robert E.). Some literary influences on Sidonius Apollinaris. Amsterdam, Hakkert, 2000, 241 p.

2734. CRACA (C.). Le possibilità della poesia. Lucrezio e la madre frigia in De rerum natura II 598-660. Bari, Edipuglia, 2000, 181 p.

2735. Création lexicale (La) en latin. Actes de la Table ronde du IXème colloque internationale de linguistique latine à Madrid le 16 avril 1997. Textes réunis. Éd. par Michèle FRUYT et Christian NICOLAS. Paris, Presses de l'université de Paris-Sorbonne, 2000, 181 p.

2736. DAINTREE (D.). Non omnis moriar: the re-emergence of the Horatian lyrical tradition in the early middles ages. *Latomus*, 2000, 59, p. 889-902.

2737. DE NADAÏ (J. C.). Rhétorique et poétique dans la Pharsale de Lucain. La crise de la représentation dans la poésie antique. Louvain et Paris, Éditions Peeters, 2000, 363 p. (Bibliothèque d'Études Classiques).

2738. DELARUE (F.). Stace, poète épique. Originalité et cohérence. Louvain et Paris, Éditions Peeters, 2000, VIII-453 p. (Bibliothèque d'Études Classiques).

2739. DEROUX (C.). En remontant du borain au latin (à propos de l'adverbe munde). *Latomus*, 2000, 59, p. 505-511. – IDEM. Helvius Cinna et ses huit porteurs bithyniens (Catulle 10, 29–30: grauis et non Gaius). *Latomus*, 2000, 59, p. 850-857.

2740. DIMUNDO (R.). L'elegia allo specchio. Studi sul I libro degli Amores di Ovidio. Bari, Edipuglia, 2000, 363 p.

2741. DOMINIK (William J.), WEHRLE (William T.). Roman verse satire. Lucilius to Juvenal. A selection with an introduction, text, translations, and notes. Wauconda, Bolchazy-Carducci, 2000, XVII-221 p.

2742. Dramatische Wäldchen. Festschrift für Eckard Lefèvre zum 65. Geburtstag. Hrsg. v. Ekkehard STÄRK und Gregor VOGT-SPIRA. Hildesheim, Zürich u. New York, Olms, 2000, X-717 p. (Abb.) [Cf. n° <Auswahl> 2818.]

2743. DUBUISSON (M.). Art de la voltige et 'code-switching' (Apulée, Métamorphoses I,1, 5-6). *Latomus*, 2000, 59, p. 607-613.

2744. EICH (Armin). Politische Literatur in der römischen Gesellschaft. Studien zum Verhältnis von politischer und literarischer Öffentlichkeit in der späten Republik und frühen Kaiserzeit. Köln, Weimar u. Wien, Böhlau, 2000, VIII-413 p.

2745. FITZGERALD (William). Slavery and the Roman literary imagination. Cambridge, Cambridge U. P., 2000, XII-159 p.

2746. FÖGEN (Th.). Patrii sermonis egestas. Einstellungen lateinischer Autoren zu ihrer Muttersprache. Ein Beitrag zum Sprachbewußtsein in der römischen Antike. München u. Leipzig, Saur, 2000, 279 p.

2747. FOUCHER (A.). Nature et formes de l'«histoire tragique» à Rome. *Latomus*, 2000, 59, p. 773-801.

2748. FRAKES (R. M.). Ammianus Marcellinus and his intended audience. *In*: Studies in Latin literature and Roman history [Cf. n° 2806], p. 392-442.

2749. GALE (M. R.). Virgil on the nature of things: the Georgics, Lucretius and the didactic tradition. Cambridge, Cambridge U. P., 2000, XII-321 p.

2750. GÄRTNER (T.). Drei Konjekturen zu hochmittelalterlichen Drames. *Latomus*, 2000, 59, p. 647-651.

2751. GILDENHARD (Ingo), ZISSOS (Andrew). Inspirational fictions: autobiography and generic reflexivity in Ovid's proems. *Greece and Rome*, 2000, 47, p. 67-79.

2752. GILDENHARD (Ingo), ZISSOS (Andrew). Ovid's Narcissus (Met. 3.339-510): echoes of Oedipus. *American journal of philology*, 2000, 121, p. 129-147.

2753. GREBE (S.). Kriterien für die Latinitas bei Varro und Quintilian. *In*: Hortus litterarum antiquarum [Cf. n° 2756], p. 191-210.

2754. HAVERLING (Gerd). On sco- verbs, prefixes and semantic functions. A study in the development of prefixed and unprefixed verbs from early to late latin. Göteborg, Acta Universitatis Gothoburgensia, 2000, XII-533 p.

2755. HOLZBERG (N.). Lesbia, the poet, and the two faces of Sappho: 'womanufacture' in Catullus. *Proceedings of the Cambridge philological society*, 2000, 46, p. 28-44.

2756. Hortus litterarum antiquarum. Festschrift für Hans Armin Gärtner zum 70. Gerburtstag. Hrsg. v. Andreas HALTENHOFF und Fritz-Heiner MUTSCHLER. Heidelberg, Winter, 2000, XII-560 p. [Cf. n[os] <Auswahl> 2753, 2783, 2828.]

2757. Identität und Alterität in der frühromischen Tragödie. Hrsg. v. G. MANUWALD. Würzburg, Ergon Verlag, 2000, 372 p.

2758. JOHNSON (W. R.). Lucretius and the modern world. London, Duckworth, 2000, X-163 p.

2759. KEITH (A. M.). Engendering Rome. Women in Latin epic. Cambridge, Cambridge U. P., 2000, XII-149 p.

2760. KEITH DIX (T.). The library of Lucullus. *Athenaeum*, 2000, 88, p. 441-464.

2761. LAIGNEAU (S.). Ovide, Amores 1.6: un paraklausithyron très ovidien. *Latomus*, 2000, 59, p. 317-326.

2762. LANDOLFI (L.). Scribentis imago. Eroine ovidiane e lamento epistolare. Bologna, Pàtron, 2000, 259 p.

2763. LAVENCY (M.). Propositions relatives compléments adjoints de noms en latin. *L'antiquité classique*, 2000, 69, p. 179-199. – IDEM. Syntagmes à l'ablatif en latin classique: conditionnement et valeurs. *Latomus*, 2000, 59, p. 819-841.

2764. Letteratura e propaganda nell'Occidente latino da Augusto ai regni romanobarbarici. A cura di F. E. CONSOLINO. Roma, L'Erma di Bretschneider, 2000, 227 p.

2765. LOBE (M.). Die Gebärden in Vergils Aeneis. Zur Bedeutung und Funktion von Körpersprache im römischen Epos. Frankfurt am Main, Peter Lang, 2000, 230 p. (Classica et Neolatina Studien zur lateinischen Literatur, 1)

2766. *Vacat*.

2767. MAC CARTHY (K.). Slaves, masters and the art of authority in Plautine comedy. Princeton a. Oxford, Princeton U. P., 2000, XI-231 p.

2768. MADER (G.). Quis queat digne eloqui? Speech, gesture and the grammar of the mundus inversus in Seneca's Thyestes. *Antike und Abendland*, 2000, 46, p. 153-172.

2769. MAGNALDI (G.). La forza dei segni. Parolespia nella tradizione manoscritta dei prosatori latini. Amsterdam, Adolf M. Hakkert, 2000, 176 p.

2770. MALEUVRE (J.-Y.). Violence et ironie dans les Bucoliques de Virgile. Paris, Jean Touzot Libraire-Editeur, 2000, 445 p. (Textes et Images de l'Antiquité, 7)

2771. MARCHETTI (S. C.). Amicizia e potere nelle lettere di Cicerone e nelle elegie ovidiane dall'esilio. Firenze, Università degli Studi di Firenze, 2000, X-405 p.

2772. MARSHALL (P.-K.). The Spangenberg bifolium of Servius: the manuscript and the text. *Rivista di filologia e di istruzione classica*, 2000, 128, p. 192-209.

2773. MAXIA (C.). Seneca e l'età dell'oro. *Bollettino di studi latini*, 2000, 30, p. 87-105.

2774. MAZZOCCHINI (P.). Forme e significati della narrazione bellica nell'epos virgiliano. I cataloghi degli uccisi e le morti minori nell'Eneide. Fasano, Schena Editore, 2000, 436 p.

2775. Meisterwerke der antiken Literatur. Von Homer bis Boethius. Hrsg. v. Martin HOSE. München, Beck, 2000, 188 p.

2776. MERLI (E.). Arma canant alii. Materia epica e narrazione elegiaca nei Fasti di Ovidio. Firenze, Università degli Studi di Firenze, 2000, 356 p.

2777. MÉTHY (Nicole). Éloge rhétorique et propagande politique sous le Haut-Empire. L'exemple du Panégyrique de Trajan. *Mélanges de l'Ecole française de Rome. Antiquité*, 2000, 112, p. 365-411. – EADEM. Rome, «ville éternelle»? À propos de deux verses de Tibulle (II, 5, 23-24). *Latomus*, 2000, 59, p. 69-81.

2778. MOLYVIATI (O.). Narrative sequence and closure in Aeneid, XII, 931-952. *L'antiquité classique*, 2000, 69, p. 165-177.

2779. MORELLI (A. M.). L'epigramma latino prima di Catullo. Cassino, Edizioni dell'Università, 2000, 399 p.

2780. MORGAN (Llewelyn). The autopsy of C. Asinius Pollio. *Journal of Roman studies*, 2000, 90, p. 51-69.

2781. Moribus antiquis res stat Romana. Römische Werte und römische Literatur im 3. und 2. Jh. v. Chr.

Hrsg. v. Maximilian BRAUN, Andreas HALTENHOFF und Fritz-Heiner MUTSCHLER. München u. Leipzig, Saur, 2000, 374 p.

2782. NADEAU (Y.). The death of Aeneas – Vergil's version (and Ovid's). An insight into the politics of Vergil's poetry. *Latomus*, 2000, 59, p. 289-316.

2783. PEGLAU (M.). Varro und die angeblichen Schriften des Numa Pompilius. *In*: Hortus litterarum antiquarum [Cf. n° 2756], p. 441-450.

2784. PEROTTI (P. A.). La locuzione quod si, e altre simili. *Latomus*, 2000, 59, p. 8-14.

2785. PERUTELLI (A.). Il tema della casa nella Mostellaria. *Maia*, 2000, 52, p. 19-34.

2786. PETROVA (Marina S.). Martsian Kapella (prosograficheskij ocherk). (Martianus Capella: A prosopographical essay). *Dialog so vremenem: Al'manakh intellektual'noj istorii*, 2000, 2, p. 110-141. [Eng. summary]

2787. PLAZA (M.). Laughter and derision in Petronius' Satyrica. A literary study. Stockholm, Almqvist & Wiksell International, 2000, XII-227 p. (Acta Universitatis Stockholmiensis. Studia Latina Stockholmiensia, 46).

2788. PUCCINI DELBEY (G.). L'amour conjugal à l'épreuve de l'exil dans l'œuvre d'Ovide. *Bulletin de l'association Guillaume Budé*, 2000, 4, p. 329-352.

2789. RABINOWITZ (J.). Joyous dread. Erotic masochism and initiatory symbolism in the poetry of Propertius. *In*: Studies in Latin literature and Roman history [Cf. n° 2806], p. 209-231.

2790. RACCANELLI (R.). Parenti e amici a confronto. Per un sistema degli affetti nelle declamazioni latine (Ps.-Quint. decl. mai. 9 e 16; decl. min. 321). *Bollettino di studi latini*, 2000, 30, p. 106-133.

2791. Rede und Redner. Bewertung und Darstellung in den antiken Kulturen. Kolloquium Frankfurt am Main, 14.–16. Oktober 1998. Hrsg. v. Christoff NEUMEISTER und Wulf RAECK. Möhnesee, Bibliopolis, 2000, XI-312 p.

2792. RICOTTILLI (L.). Gesto e parola nell'Eneide. Bologna, Pàtron, 2000, 246 p.

2793. ROHWEDER (Ch.). Eine römische Dichterin: Sulpicia. *In*: Frauenwelten in der Antike, [Cf. n° 2689], p. 147-161.

2794. ROSKAM (G.). Mariage ou virginité? Le carmen 62 de Catulle et la lutte entre deux idéaux de vie. *Latomus*, 2000, 59, p. 41-56.

2795. ROSSI (Andreola). The tears of Marcellus: history of a literary motif in Livy. *Greece and Rome*, 2000, 47, p. 56-66.

2796. SCHAMP (J.). Le poète et les Claudii de la République (Jean le Lydien, Mag. I, 39-44). *Latomus*, 2000, 59, p. 109-128.

2797. SCHEITHAUER (A.). Kaiserliche Bautätigkeit in Rom. Das Echo in der antiken Literatur. Stuttgart, Franz Steiner Verlag, 2000, 338 p.

2798. SCHIEVENIN (R.). Amicizia perfetta e amicizia comune nel Laelius ciceroniano. *Bollettino di studi latini*, 2000, 30, p. 447-465.

2799. SCHINDLER (C.). Untersuchungen zu den Gleichnissen im römischen Lehrgedicht. Lucrez, Vergil, Manilius. Göttingen, Vandenhoeck and Ruprecht, 2000, 315 p.

2800. SCHMIDT (E. A.). Aparte. Das dramatische Verfahren und Senecas Technik. *Rheinisches Museum*, 2000, 143, p. 400-429.

2801. SCHWINDT (Jürgen Paul). Prolegomena zu einer 'Phänomenologie' der römischen Literaturgeschichtsschreibung. Von den Anfängen bis Quintilian. Göttingen, Vandenhoeck & Ruprecht, 2000, 249 p.

2802. Seneca e il suo tempo. Atti del convegno internazionale di Roma-Cassino, 11–14 novembre 1998. A cura di P. PARRONI. Roma, Salerno, 2000, 514 p. (ill., tav.).

2803. Seneca in performance. Ed. by George W. M. HARRISON. London, Duckworth, 2000, XI-260 p.

2804. SETAIOLI (Aldo). Facundus Seneca. Aspetti della lingua e dell'ideologia senecana. Bologna, Pàtron, 2000, 480 p.

2805. SOLANA PUJALTE (J.). El hexámetro del Aeneidos Liber XIII de Maffeo Vegio y sus modelos clásicos (II). *Latomus*, 2000, 59, p. 652-665.

2806. Studies in Latin literature and Roman history. Ed. by Carl DEROUX. Bruxelles, Latomus, 2000, 501 p. [Cf. n^{os} <choice> 2675, 2726, 2748, 2789, 2807.]

2807. TUPLIN (C.). Nepos and the origins of political biography. *In*: Studies in Latin literature and Roman history [Cf. n° 2806], p. 124-161.

2808. VAINIO (R.). Latinitas and barbarisms according to the Roman grammarians. Attitudes towards language in the light of grammatical examples. Turku, University of Turku, 2000, 180 p.

2809. VONS (J.). L'image de la femme dans l'œuvre de Pline l'Ancien. Bruxelles, Latomus, 2000, 480 p. (Collection Latomus, 256).

2810. WHEELER (S. M.). Narrative dinamics in Ovid's Metamorphoses. Tübingen, Gunter Narr Verlag, 2000, VII-174 p. (Classica Monacensia 20).

2811. Why Vergil? A collection of interpretations. Ed. by Stephanie QUINN. Wauconda, Illinois Bolchazy-Carducci, 2000, XXV-451 p.

2812. WOYTEK (E.). 'In medio et mihi Caesar erit ...'. Vergilimitationen im Zentrum von Ovids Remedia amoris. *Wiener Studien*, 2000, 113, p. 181-213.

2813. ZWIERLEIN (O.). Antike Revisionen des Vergil und Ovid. Wiesbaden, Wesdeutscher Verlag, 2000, 86 p.

Cf. n°s 464, 2436-2533

b. Philosophy and science

2814. BÄCK (A.). Logic in imperial Rome. *Apeiron*, 2000, 33, p. 75-85.

2815. BRANCACCI (A.). La filosofia in età imperiale. Le scuole e le tradizioni filosofiche. Roma, Bibliopolis, 2000, 326 p.

2816. CENTRONE (B.). Platonism and Pythagoreanism in the early empire. *In*: Cambridge (The) history of Greek and Roman political thought. [Cf. n° 1034], p. 559-584.

2817. DALFEN (J.). Marc Aurel. 'Werde so, wie die Philosophie dich haben will'. *In*: Philosophen des Altertums [Cf. n° 2832], p. 128-144.

2818. EIGLER (U.). Cicero und die römische Tragödie. Eine Strategie zur Legitimation philosophischer Literatur im philosophischen Spätwerk Ciceros. *In*: Dramatische Wäldchen [Cf. n° 2742], p. 619-636.

2819. Epikureismus in der späten Republik und der Kaiserzeit. Akten der 2. Tagung der Karl und Gertrud Abel Stiftung vom 30. September–3. Oktober 1998 in Würzburg. Hrsg. v. M. ERLER. Stuttgart, Franz Steiner Verlag, 2000, 316 p.

2820. Filosofia (La) in età imperiale. Le scuole e le tradizioni filosofiche. Atti del colloquio, Roma 17–19 giugno 1999. A cura di Aldo BRANCACCI. Napoli, Bibliopolis, 2000, 326 p.

2821. FLEMMING (R.). Medicine and the making of Roman women. Gender, nature and authority from Celsus to Galen. Oxford, Oxford U. P., 2000, XII-453 p.

2822. GEE (Emma). Ovid, Aratus and Augustus. Astronomy in Ovid's Fasti. Cambridge, Cambridge U. P., 2000, XI-226 p.

2823. Genie und Wahnsinn. Konzepte psychischer 'Normalität' und 'Abnormität' im Altertum. Hrsg. v. Bernd EFFE und Reinhold F. GLEI. Trier, Wissenschaftlicher Verlag Trier, 2000, 260 p.

2824. GIESECKE (Annette Lucia). Atoms, ataraxy, and allusion: cross-generic imitation of the De rerum natura in early augustan poetry. Hildesheim, Zürich a. New York, Georg Olms Verlag, 2000, VI-202 p.

2825. HARRISON (S. G.). Apuleius. A latin sophist. Oxford, Oxford U. P., 2000, VI-281 p.

2826. HEALY (John F.). Pliny the Elder on science and technology. Oxford, Oxford U. P., 2000, XV-467 p.

2827. HOLZHAUSEN (J.). Hadrians nous und seine animula. *Rheinisches Museum*, 2000, 143, p. 96-109.

2828. KLUG (W.). Geschichtsphilosophische Perspektiven in der Germania des Tacitus. Ein Beitrag zur Discussion der literarischen Form. *In*: Hortus litterarum antiquarum [Cf. n° 2756], p. 269-288.

2829. KÖHLER (H.). Die Sonnenfinsternis aus der Fußgängerperspektive: Mehrfache Ursachen in Lukrez' De rerum natura. *In*: 'Stürmend auf finsterem Pfad ...' [Cf. n° 2836], p. 113-124.

2830. LANGSLOW (D. R.). Medical latin in the Roman empire. Oxford, Oxford U. P., 2000, XV-517 p.

2831. LUCIANI (S.). L'éclair immobile dans la plaine. Philosophie et poétique du temps chez Lucrèce. Louvain et Paris, Éditions Peeters, 2000, 343 p.

2832. Philosophen des Altertums. Eine Einführung. 1. Von der Frühzeit bis zur Klassik. 2. Vom Hellenismus bis zur Spätantike. Hrsg. v. M. ERLER und A. GRAESER. Darmstadt, Wissenschaftliche Buchgesellschaft, Primus Verlag, 2000, VIII-224, VII-235 p. [Cf. n° <Auswahl> 2817.].

2833. RUNIA (David T.). The idea and the reality of the city in the thought of Philo of Alexandria. *Journal of the history of ideas*, 2000, 61, 3, p. 361-380.

2834. SERIO (A.). Un principio della fisica epicurea nell'ode oraziana II 5: l'isonomia. *In*: Letteratura scientifica e tecnica greca e latina [Cf. n° 2256], p. 275-299.

2835. SHIROKOVA (Nadezhda S.). Kul'tura kel'tov i nordicheskaja traditsija antichnosti. (The Celtic culture and the 'Nordic' tradition of the Antiquity). Sankt-Peterburg, Evrazija, 2000, 351 p. (maps, ill.). (Barbaricum).

2836. 'Stürmend auf finsterem Pfad...'. Ein Symposion zur Sonnenfinsternis in der Antike. Hrsg. v. H. KÖHLER, H. GÖRGEMANNS und M. BAUMBACH. Heidelberg, Winter, 2000, VII-169 p. [Cf. n° <Auswahl> 2829.]

2837. Textes médicaux latins (Les) comme littérature. Actes du VIe colloque international sur les texte médicaux latins du 1er au 3 septembre 1998 à Nantes. Ed. par A. J. PIGEAUD. Nantes, Université de Nantes, 2000, 389 p.

Cf. n°s 1183, 1415, 2436-2533

§ 8. Religion and mythology.

2838. Antycznego świata (Z): Religio, cultus, homines. Ed. by W. APPEL and P. WOJCIECHOWSKIEGO. Toruń, Uniwersytet Mikołaja Kopernika, 2000, 271 p.

2839. BAKHOUCHE (B.). Magie, mystère et philosophie: le livre II de Noces de Philologie et Mercure. *In*: Magie (La) [Cf. n° 2854], p. 109-127.

2840. BECK (Roger). Ritual, myth, doctrine, and initiation in the mysteries of Mythras: new evidence from a cult vessel. *Journal of Roman studies*, 2000, 90, p. 145-180.

2841. BELAYCHE (N.). Deae Suriae Sacrum. La romanité des cultes «orientaux». *Revue historique*, 2000, 124, 615, p. 565-592.

2842. CABOURET (B.). L'empereur Julien an prière. *In*: Prières méditerranéennes hier et aujourd'hui [Cf. n° 2863], p. 115-123.

2843. CANCIK-LINDEMAIER (H.). Die vestalischen Jungfrauen. *In*: Frauenwelten in der Antike [Cf. n° 2689], p. 111-123.

2844. CIAPPI (M.). La narrazione ovidiana del mito di Fetonte e le sue fonti: l'importanza della tradizione tragica. *Athenaeum*, 2000, 88, p. 117-178.

2845. CIFARELLI (Francesco). Il culto di Ercole a Segni e l'assetto topografico del suburbio meridionale. *Mélanges de l'Ecole française de Rome. Antiquité*, 2000, 112, p. 173-216.

2846. CUMONT (F.). Astrologie et religion chez les grecs et les romains. Bruxelles et Roma, Institut historique belge de Rome, Brepols, 2000, 175 p.

2847. DE CAROLIS (E.). Dei ed eroi nella pittura pompeiana. Roma, 'L'Erma' di Bretschneider, 2000, 78 p. (ill.).

2848. DEREMETZ (A.). Mythe et science dans l'anthropologie lucrétienne. *In*: Uranie [Cf. n° 2868], p. 97-107.

2849. Divination and portents in the Roman world. Ed. by R. L. WILDFAG and J. ISAGER. Odense, Odense U. P., 2000, 79 p.

2850. DORIVAL (G.). Païens en prière. *In*: Prières méditerranéennes hier et aujourd'hui [Cf. n° 2863], p. 87-110.

2851. EGELHAAF-GAISER (Ulrike). Kulträume im römischen Alltag. Das Isisbuch des Apuleius und der Ort von Religion im kaiserzeitlichen Rom. Stuttgart, Steiner, 2000, 631 p. (Abb., Taf.)

2852. INGEMARK (Dominic), GERDING (Henrik). Liv och död i antikens Rom. (La vie et la mort dans la Rome antique). Lund, Historiska media, 2000, 223 p. (ill.).

2853. JOLIVET (J.-Ch.). Une anthropologie virgilienne: Hercules Empedocleus? *In*: Uranie [Cf. n° 2868], p. 109-123.

2854. Magie (La). Actes du colloque international de Montpellier, 25–27 mars 1999. Études rassemblées. Éd. par A. MOREAU et J.-C. TURPIN. Tome I. Du monde babylonien au monde hellénistique. Tome II. La magie dans l'antiquité grecque tardive. Les mythes. Tome III. Du monde latin au monde contemporain. Tome IV. Bibliographie. Montpellier, Université Paul Valéry, Montpellier III, 2000, 4 vol., 328 p., 336 p., 358 p., 169 p. (tables) [Cf. n^{os} <sélection> 2648, 2839, 2856, 2860, 2866.]

2855. MARCO SIMÓN (Francisco), PINA POLO (Francisco). Mario Gratidiano, los compita y la religiosidad popular a fines de la república. *Klio*, 2000, 82, p. 154-170.

2856. MÉTHY (N.). Magie, religion et philosophie au second siècle de notre ère. À propos du dieuroi d'Apulée. *In*: Magie (La) [Cf. n° 2854], p. 85-107.

2857. MEULDER (M.). Le feu et la source à Rome. *Latomus*, 2000, 59, p. 749-765.

2858. MEURANT (Alain). L'idée de gémellité dans la légende des origines de Rome. Bruxelles, Académie royale de Belgique, 2000, 335 p.

2859. MUTH (S.). Gegenwelt als Glückswelt – Glückswelt als Gegenwelt? Die Welt der Nereiden, Tritonen und Seemonster in der römischen Kunst. *In*: Gegenwelten zu den Kulturen Griechenlands und Roms in der Antike, [Cf. n° 2570], p. 467-498.

2860. NOVARA (A.). Magie, amour et humour chez Properce. *In*: Magie (La) [Cf. n° 2854], p. 15-44.

2861. OTTO (Cl.). Lat. Lūcus, nemus «bois sacré» et les deux formes de sacralité chez les Latins. *Latomus*, 2000, 59, p. 3-7.

2862. PONESSA-SALATHÉ (J.). Venus and the representation of virtue on third-century sarcophagi. *Latomus*, 2000, 59, p. 873-888.

2863. Prières méditerranéennes hier et aujourd'hui. Études réunis. Actes du colloque organisé par le Centre Paul-Albert Février à Aix-en-Provence les 2 et 3 avril 1998. Éd. par G. DORIVAL et D. PRALON. Aix-en-Provence Cedex, Publications de l'université de Provence, 2000, 340 p. (tables). [Cf. n^{os} <sélection> 2842, 2850, 3105.]

2864. Religion in archaic and republican Rome and Italy. Evidence and experience. Ed. By E. BISPHAM and C. SMITH. Edinburgh, Edinburgh U. P., 2000, XIV-199 p. [Cf. n° <choice> 2403.]

2865. SECCI (E.). Non movent divos preces (Phaedr. 1242): aspetti delle invocazioni agli dèi nelle tragedie di Seneca. *Prometheus*, 2000, 26, p. 52-70.

2866. SEGAL (Ch.). Tantum medicamina possunt: la magie dans les Métamorphoses d'Ovide. *In*: Magie (La) [Cf. n° 2854], p. 45-70.

2867. TANSEY (Patrick). The inauguration of Lentulus Niger. *American journal of philology*, 2000, 121, p. 237-258.

2868. Uranie. Mythe et/ou philosophie dans les textes grecs et latins sur les origines de l'humanité. Actes des journées d'études des 13 et 14 novembre 1998. Éd. par Jacqueline FABRE-SERRIS. Villeneuve-d'Ascq, Université Charles-de-Gaulle-Lille 3, Centre de recherche interdisciplinaire 'Myths et littératures', 2000, 135 p. [Cf. n^{os} <sélection> 2848, 2853.].

2869. VAAHTERA (J. E.). Roman religion and the Polybian politeia. *In*: Roman middle republic (The) [Cf. n° 2612], p. 251-264.

2870. WISEMAN (T. P.). Liber: myth, drama and ideology in republican Rome. *In*: Roman middle republic (The), [Cf. n° 2612], p. 265-299.

Cf. n°s 2313, 2835

§ 9. Archaeology and history of art.

2871. ACCARDO (S.). Villae Romanae nell'ager Bruttius. Il paesaggio rurale calabrese durante il dominio romano. Roma, 'L'Erma' di Bretschneider, 2000, 237 p. (ill.).

2872. AURIEMMA (Rita). Le anfore del relitto di Grado e il loro contenuto. *Mélanges de l'Ecole française de Rome. Antiquité*, 2000, 112, p. 27-51.

2873. BESTE (H.-J.). The construction and phases of development of the wooden arena flooring of the Colosseum. *Journal of Roman archaeology*, 2000, 13, p. 79-92.

2874. BRADLEY (Richard). Vera Collum and the excavation of a 'Roman' megalithic tomb. *Antiquity*, 2000, 74, p. 39-43.

2875. BROWN (A. G.), MEADOWS (I.). Roman vineyards in Britain: finds from the Nene valley and new research. *Antiquity*, 2000, 74, p. 491-492.

2876. Caligula (From) to Constantine. Tyranny and transformation in Roman portraiture. Ed. by E. R. VARNER. Atlanta, Michael C. Carlos Museum, 2000, 251 p. (ill.). [Cf. n° <choice> 2427.]

2877. CARROLL (M.), GODDEN (D.). The sanctuary of Apollo at Pompeii: reconsidering chronologies and excavation history. *American journal of archaeology*, 2000, 104, p. 743-754.

2878. CASTORIO (J.-N.). Sculptures funéraires gallo-romaines de Toul (Meurthe-et-Moselle). *Latomus*, 2000, 59, p. 364-398.

2879. CHANSON (H.). Hydraulics of Roman acqueducts: steep chutes, cascades, and dropshafts. *American journal of archaeology*, 2000, 104, p. 47-72.

2880. COARELLI (Filippo). The column of Trajan. Roma, Editore Colombo a. the German archaeological institute, 2000, VIII-273. (ill.).

2881. Colonne Aurélienne (La). Autour de la colonne Aurélienne. Geste et image sur la colonne de Marc Aurèle à Rome. Ed. par J. SCHEID et V. HUET. Turnhout, Brepols, 2000, 446 p. (tables).

2882. D'ARMS (J. H.). P. Lucilius Gamala's feasts for the Ostians and their Roman models. *Journal of Roman archaeology*, 2000, 13, p. 192-200.

2883. DAVIES (P. J. E.). Death and the emperor: Roman imperial funerary monuments from Augustus to Marcus Aurelius. Cambridge, Cambridge U. P., 2000, XIV-265 p. (ill., maps).

2884. DE LA BARRERA (José Luis). La decoración arquitectónica de los foros de Augusta Emerita. Roma, 'L'Erma' di Bretschneider, 2000, 480 p. (Bibliotheca archaeologica, 25).

2885. DE POLIGNAC (François). Archives de l'archéologie romaine du XVIIIe siècle. I. Documents inédits sur les fouilles d'Anzio, de la via Latina et du Palatin (1711–1730). *Mélanges de l'Ecole française de Rome. Antiquité*, 2000, 112, p. 611-646.

2886. DELUSSU (Fabrizio). Le conserve di pesce. Alcuni dati da contesti italiani. *Mélanges de l'Ecole française de Rome. Antiquité*, 2000, 112, p. 53-65.

2887. DESSE-BERSET (N.), DESSE (J.). Salsamenta, garum et autres préparations de poisson. Ce qu'en disent les os. *Mélanges de l'Ecole française de Rome. Antiquité*, 2000, 112, p. 73-97.

2888. DIERICHS (A.). Erotik in der Bildenden Kunst der Römischen Welt. *In*: Frauenwelten in der Antike [Cf. n° 2689], p. 394-411.

2889. DILLON (S.). Subject selection and viewer reception of Greek portraits from Herculaneum and Tivoli. *Journal of Roman archaeology*, 2000, 13, p. 21-40.

2890. DORSCH (Klaus-Dieter), SEELIGER (Hans Reinhard). Römische Katakombenmalereien im Spiegel des Photoarchivs Parker. Dokumentation von Zustand und Erhaltung 1864–1994. Münster, Aschendorff, 2000, 272 p.

2891. DUCATÉ (S.). Un couple votif en terre cuite provenant de la région de Santa Severa (province de Rome). *Latomus*, 2000, 59, p. 36-40.

2892. FISHWICK (D.). A new forum at Corduba. *Latomus*, 2000, 59, p. 96-104.

2893. GANZERT (Joachim). Im Allerheiligsten des Augustusforums. Fokus 'Oikoumenischer Akkulturation'. Mainz, von Zabern, 2000, V-120 p. (Abb.).

2894. GIULIANO (Antonio). L'arco di Costantino come documento storico. *In*: Monumento (Il): arte e storia [Cf. n° 1283], p. 441-474.

2895. GOUDINEAU (CH.), GUICHARD (V.), REDDÉ (M.), SIEVERS (S.), SOULHOL (H.). Caesar und Vercingetorix. Mainz, von Zabern, 2000, 71 p. (Abb.). (Zaberns Bildbände zur Archäologie).

2896. *Vacat.*

2897. HARTER (Gabriele). Römische Gläser des Landesmuseums Mainz. Wiesbaden, Reichert, 2000, 482 p. (Abb., Taf.).

2898. HARVEY (P. B., jr.). The inscribed bowl from the Garigliano (Minturnae): local diversity and romanization in the 4th c. B.C. *Journal of Roman archaeology*, 2000, 13, p. 164-174.

2899. HEINZELMANN (Michael). Die Nekropolen von Ostia. Untersuchungen zu den Gräberstraßen vor der Porta Romana und an der Via Laurentina. München, Pfeil, 2000, 368 p. (Abb.).

2900. HINGLEY (R.). Roman officers and English gentlemen. The imperial origins of Roman archaeology. London a. New York, Routledge, 2000, XV-224 p. (ill.).

2901. İNAN (Jale). Perge'nin Roma Devri Heykeltraşlığı, I. (Sculpture of Perge in the Roman times). Ed.

by Nezih BAŞGELEN. İstanbul, Arkeoloji ve Sanat, 2000, 54 p.

2902. JAPP (Sarah). Die Baupolitik Herodes' des Großen. Die Bedeutung der Architektur für die Herrschaftslegitimation eines römischen Klientelkönigs. Rahden, Leidorf, 2000, XII-169 p. (Taf.). (Internationale Archäologie, 64).

2903. KESSENER (P.). The aqueduct at Aspendos and its inverted siphon. *Journal of Roman archaeology*, 2000, 13, p. 104-132.

2904. Klassische Archäologie. Eine Einführung. Hrsg. v. Adolf BORBEIN, Tonio HÖLSCHER und Paul ZANKER. Berlin, Reimer, 2000, 382 p.

2905. KLYNNE (Allan), LILJENSTOLPE (Peter). Where to put Augustus? A note on the placement of the Prima Porta statue. *American journal of philology*, 2000, 121, p. 121-128.

2906. KNOPF (Thomas). Das römische Sindelfingen. Stuttgart, Theiss, 2000, 134 p. (Abb.).

2907. LANCASTER (L.). Building Trajan's Markets 2: the construction process. *American journal of archaeology*, 2000, 104, p. 755-785.

2908. LANDWEHR (Christa). Die römischen Skulpturen von Caesarea Mauretaniae. Band II. Idealplastik. Männliche Figuren. Mainz, von Zabern, 2000, XVIII-131 p. (Abb., Taf.).

2909. LAVAGNE (Henri). Recueil général des mosaïques de la Gaule. III. Province de Narbonnaise. 3. Partie sud-est: Cités des Allobroges, Vocontii, Bodiontici, Reii, Salluuii, Oxubii, Deciates, Veinatii. Paris, CNRS éditions, 2000, 420 p.

2910. LECIEJEWICZ (Lech). Torcello. Nuove ricerche archeologiche. Roma, Giorgio Bretschneider, 2000, 153 p.

2911. LÖNNQVIST (K.). Pontius Pilate – An aqueduct builder? – Recent findings and new suggestions. *Klio*, 2000, 82, p. 459-474.

2912. MARCHESI (Henri). L'habitat rural de Musoleo: contribution à la connaissance des céramiques antiques en Corse (Ier-Ve s.). *Mélanges de l'Ecole française de Rome. Antiquité*, 2000, 112, p. 959-997.

2913. MARCONE (A.). Tra archeologia e storia economica: il mausoleo dei Secundini a Igel. *Athenaeum*, 2000, 88, p. 485-497.

2914. MESSINEO (Gaetano). La tomba dei Nasonii. Roma, 'L'Erma' di Bretschneider, 2000, 88 p.

2915. MEYER (Hugo). Prunkkameen und Staatsdenkmäler römischer Kaiser. Neue Perspektiven zur Kunst der frühen Prinzipatszeit. München, Biering & Brinkmann, 2000, 144 p. (Abb.).

2916. NOY (David). 'Half-burnt on an emergency pyre': Roman cremations which went wrong. *Greece and Rome*, 2000, 47, p. 186-196.

2917. OBMANN (Jürgen). Studien zu römischen Dolchscheiden des 1. Jahrhunderts n. Chr. Archäologische Zeugnisse und Bildliche Überlieferung. Rahden, Leidorf, 2000, VII-45 p.

2918. PATTERSON (H.), DI GENNARO (F.), DI GIUSEPPE (H.), FONTANA (S.), GAFFNEY (V.), HARRISON (A.), KEAY (S. J.), MILLETT (M.), RENDELI (M.), ROBERTS (P.), STODDART (S.), WITCHER (R.). The Tiber Valley project: the Tiber and Rome through two millennia. *Antiquity*, 2000, 74, p. 395-403.

2919. POTTER (T. W.), ROBINSON (B.). New Roman and prehistoric aerial discoveries at Grandford, Cambridgeshire. *Antiquity*, 2000, 74, p. 31-32.

2920. RAECK (Wulf), GORYS (Andrea), GOSSEL-RAECK (Berthild). Untersuchungen zur Vorgängerbebauung des Trajaneums von Pergamon II: Der Bereich westlich des Tempels, untere Hangstufe. *Istanbuler Mitteilungen*, 2000, 50, p. 307-364.

2921. REA (R.). Studying the valley of the Colosseum (1970–2000): achievements and prospects. *Journal of Roman archaeology*, 2000, 13, p. 93-103.

2922. RICHARDSON (L.). A catalog of identifiable figure painters of ancient Pompeii, Herculaneum and Stabiae. Baltimore a. London, Johns Hopkins U. P., 2000, XVII-190 p.

2923. RIEMER (Ellen). Romanische Grabfunde des 5.–8. Jahrhunderts in Italien. Rahden, Leidorf, 2000, 485 p. (Abb., Taf.).

2924. RODRÍGUEZ-ALMEIDA (Emilio). A proposito della Forma marmorea e di altre formae. *Mélanges de l'Ecole française de Rome. Antiquité*, 2000, 112, p. 217-230.

2925. RUTGERS (L. V.). Subterranean Rome. In search of the roots of Christianity in the catacombs of the eternal city. Leuven, Peeters, 2000, 164 p.

2926. SALETTI (C.). Ritratti di Augusto in Cisalpina: il togato velato capite di Aquileia. *Athenaeum*, 2000, 88, p. 431-439.

2927. Scavi archeologici di Tusculum rapporti preliminari delle campagne 1994–1999. Ed. by X. DUPRÉ. Roma, Escuela española de historia y arqueología en Roma, 2000, XVI-540 p. (ill.).

2928. SCHINGO (G.). A history of earlier excavations in the arena. *Journal of Roman archaeology*, 2000, 13, p. 69-78.

2929. SCHLEGEL (Oliver). Germanen im Quadrat. Die Neckarsweben im Gebiet von Mannheim, Ladenburg und Heidelberg während der frühen römischen Kaiserzeit. Rahden, Leidorf, 2000, 266 p. (Abb., Taf.). (Internationale archäologie, 34).

2930. SCOTT (R. T.). The triple arch of Augustus and the Roman triumph. *Journal of Roman archaeology*, 2000, 13, p. 183-191.

2931. SERRANO RAMOS (Encarnacion). Ceramica comun romana: siglos 2. a. C. al 7. d. C.: materiales importados y de producción local en el territorio malacitano. Portico, Servicio de Publicaciones e Intercambio Cientifico de la Universidad de Malaga, 2000, 175 p. (ill.).

2932. STANCO (E.). Ricerche sulla topografia dell'Umbria. *Mélanges de l'Ecole française de Rome. Antiquité*, 2000, 112, p. 237-242.

2933. THEODORESCU (Răzvan). Visualité et langage dans l'ancienne civilisation des Roumains. *In*: Nouvelles Études d'Histoire. X. Publiées à l'occasion du XIXe Congrès International des Sciences Historiques. Oslo, 2000. Bucureşti, Editura Academiei Române, 2000, p. 55-72.

2934. TOMEI (Maria Antonietta). I resti dell'arco di Ottavio sul Palatino e i portico delle Danaidi. *Mélanges de l'Ecole française de Rome. Antiquité*, 2000, 112, p. 557-610.

2935. TRIMBLE (J.). Replicating the body politic: the Herculaneum women statue types in early imperial Italy. *Journal of Roman archaeology*, 2000, 13, p. 41-68.

2936. TUCK (S.). A new identification for the 'Porticus Aemilia'. *Journal of Roman archaeology*, 2000, 13, p. 175-182.

2937. VARONE (Antonio). L'erotismo a Pompei. Roma, 'L'Erma' di Bretschneider, 2000, 115 p.

2938. VOLPE (Rita). Paesaggi urbani tra Oppio e Fagutal. *Mélanges de l'Ecole française de Rome. Antiquité*, 2000, 112, p. 511-556.

2939. WESTGATE (R.). Pavimenta atque emblemata vermiculata: regional styles in hellenistic mosaic and the first mosaics at Pompeii. *American journal of archaeology*, 2000, 104, p. 255-275.

2940. WILSON (R. J. A.). Campanaio, an agricultural settlement in Roman Sicily. *Antiquity*, 2000, 74, p. 289-290.

2941. WILSON JONES (M.). Principles of Roman architecture. New Haven a. London, Yale U. P., 2000, XII-270 p. (ill., tables).

2942. ZEVOLINO (G.). La tomba del Calzolaio dalla necropoli monumentale romana di Nocera Superiore. Roma, L'Erma di Bretschneider, 2000, 80 p. (ill.).

§ 9. Addenda 1999.

2943. Proceedings of the XVth international congress of classical archaeology, Amsterdam, july 12–17, 1998. Classical archaeology towards the third millennium: reflections ande perspectives. I. Text. II. Plates. Ed. by Roald F. DOCTER and Eric M. MOORMANN. Amsterdam, Allard Pierson Stichting, Publication department, 99, XVII-469 p., XV-177 p. (ill., tables).

Cf. n° 3177

§ 10. Late antiquity. Transformation of the Roman world.

2944. BOWERSOCK (Glen W.). Selected papers on late antiquity. Bari, Edipuglia, 2000, 211 p.

2945. Cambridge ancient history. Vol. 14. Late antiquity: empire and successors AD 425–600. Ed. by A. CAMERON, B. WARD-PERKINS and M. WHITBY. Cambridge, Cambridge U. P., 2000, XX-1166 p. (ill., maps).

2946. COATES (Simon). Venantius Fortunatus and the image of episcopal authority in late Antique and early Merovingian Gaul. *English historical review*, 2000, 115, 464, p. 1109-1137.

2947. Early medieval Rome and the Christian West: essays in honour of Donald A. Bullough. Ed. by Julia M. H. SMITH. Leiden, Brill, 2000, XXXII-446 p. (ill.). (The Medieval Mediterranean)

2948. Ethnicity and culture in late antiquity. Ed. by S. MITCHELL and G. GREATEX. London, Duckworth, 2000, 343 p. (ill.).

2949. Greek biography and panegyric in late antiquity. Ed. by T. HÄGG and Ph. ROUSSEAU. Berkeley, Los Angeles a. London, California U. P., 2000, XII-288 p. [Cf. n[os] <choice> …].

2950. LASSANDRO (Domenico). Sacratissimus imperator. L'immagine del princeps nell'oratoria tardoantica. Bari, Edipuglia, 2000, 167 p.

2951. LEE (A. D.). Pagans and Christians in late antiquity. London a. New York, Routledge, 2000, XXI-328 p. (ill.).

2952. Lexicon mediae et infimae Latinitatis Polonorum. Lexicon compilatur editur moderante Christina WEYSSENHOFF BROZKOWA. Vol. VII, fasc. 10 (61): protectio – quaero. Kraków, Polska Akademia Nauk, 2000, p. 1441-1600.

2953. Long (The) eight century. Production, distribution and demand. Ed. by Inge Lyse HANSEN and Chris WICHKAM. Leiden, Boston a. Köln, Brill, 2000, X-388 p. (ill., maps). (The transformation of the Roman world, 11).

2954. MAAS (M.). Readings in late antiquity. A sourcebook. London a. New York, Routledge, 2000, LXVIII-375 p. (ill.).

2955. MARCONE (Arnaldo). Tardoantico. Antologia delle fonti. Roma, Carocci, 2000, 250 p.

2956. MURRAY (Alexander Callander). From Roman to Merovingian Gaul: a reader. Peterborough a. Orchard park, Broadview Press, 2000, XVI-679 p. (ill.).

2957. RADNOTI ALFÖLDI (M.). La monetazione romana in età tardo-antica (284/476 d.C.). *In*: Moneta greca e romana (La) [Cf. n° 168], p. 143-157.

2958. Religions of late antiquity in practice. Ed. by R. VALANTASIS. Princeton a. Oxford, Princeton U. P., 2000, XVI-511 p.

2959. Rituals of power: from late antiquity to the early middle ages. Ed. by Frans THEUWS and Janet L. NELSON. Leiden, Boston a. Cologne, Brill, 2000, IX-503 p.

2960. SCOTT (S.). Art and society in fourth-century Britain. Villa mosaics in context. Oxford, Oxbow Books, 2000, 192 p. (ill.).

2961. SKEGRO (Ante). Je li rimski Delminij bio biskupsko sjedište. (Was the Roman Delminium a Bishop's residence). *Povijesni prilozi*, 2000, 19, p. 9-85.

2962. SOUTHERN (Pat); DIXON (Karen R.). The late Roman army. London, Routledge, 2000, XVII-206 p.

2963. Tardoantico (Il) alle soglie del duemila. Diritto, religione, società. Atti del quinto convegno nazionale dell'associazione di studi tardoantichi. A cura di Giuliana LANATA. Pisa, Edizioni ETS, 2000, 352 p. (ill.). (Associazione di studi tardoantichi. Atti dei convegni. 5.). [Cf. n° <scelta> 3178.]

2964. Towns and their territories between late antiquity and the early Middle Ages. Ed. by G. P. BROGIOLO, N. GAULTHIER and N. CHRISTIE. Leiden, Boston and Köln, Brill, 2000, XVII-403 p.

2965. Transformation (The) of frontiers: from late Antiquity to the Carolingians. Ed. by Walter POHL, Ian WOOD and Helmut REIMITZ. Leiden, E. J. Brill, 2000, 299 p. (The tranformation of the Roman world, 10).

2966. WEBER (G.). Träume und Visionen in Prinzipat und Spätantike. *Historische Zeitschrift*, 2000, 270, p. 99-117.

2967. *Vacat.*

Cf. n°s 487, 529, 2968-3112

G

EARLY HISTORY OF THE CHURCH TO GREGORY THE GREAT

§ 1. Sources. 2968-3038. – § 2. General. 3039-3047. – § 3. Special studies. 3048-3103. – § 4. Hagiography. 3104-3112.

§ 1. Sources.

* 2968. ALTHANN (Robert). Elenchus of Biblica. 1996. Roma, Pontificio Istituto Biblico, 2000, 690 p. (Elenchus of biblical bibliography, 12).

* 2969. LUTZKA (Carolina). Bibliographie. *Ostkirchliche Studien*, 2000, 49, 4, p. 345-366.

2970. [Ambrosius] MORETTI (P. F.). Non harundo sed calamus. Aspetti letterari della 'Explanatio psalmorum XII' di Ambrogio. Milano, Edizioni Universitarie di Lettere Economia Diritto, 2000, 263 p.

2971. Année (L') en fêtes: les Pères commentent la liturgie de la Parole. Ed. par A.-G. HAMMAN et Y. FROT. Paris, Migne et La Ferrière, Littéral, 2000, 560 p.

2972. [Arnobius] Arnobio. Difesa della vera religione. Introduzione, traduzione e note. A cura di Biagio AMATA. Roma, Città nuova, 2000, 438 p. (Collana di testi patristici, 153).

2973. [Augustinus] Augustine: political writings. Ed. By E. M. ATKINS and R. J. DODARO. Cambridge, Cambridge U. P., 2000, 229 p.

2974. [Augustinus] Augustinus von Hippo. Predigten zum Buch Genesis (Sermones 1–5): Einleitung, Text, Übersetzung und Anmerkungen. Hrsg. v. Hubertus R. DROBNER. Bern u. Frankfurt am Main, Lang, 2000, 228 p. (ill.). (Patrologia, 7).

2975. [Augustinus] Augustinus von Hippo. Sermones ad populum. Überlieferung und Bestand. Bibliographie – Indices. Hrsg. v. Hubertus R. DROBNER. Leiden, Boston u. Köln, Brill, 2000, XX-226 p.

2976. [Augustinus] MADER (G.). Blocked eyes and ears: the eloquent gestures at Augustine, conf., VI, 8, 13. *L'antiquité classique*, 2000, 69, p. 217-220.

2977. [Augustinus] Sancti Augustini opera. Contra sermonem Arrianorum (praecedit Sermo Arrianorum). Hrsg. v. Max Josef SUDA. De correptione et gratia. Hrsg. v. Georges FOLLIET. Wien, Verlag der Österreichischen Akademie der Wissenschaften, 2000, 284 p. (Corpus scriptorum ecclesiasticorum Latinorum, 92).

2978. [Augustinus] SCHULTE-KLÖCKER (Ursula). Das Verhältnis von Ewigkeit und Zeit als Widerspiegelung der Beziehung zwischen Schöpfer und Schöpfung. Eine textbegleitende Interpretation der Bücher XI-XIII der 'Confessiones' des Augustinus. Bonn, Borengässer, 2000, XVI-396 p.

2979. [Augustinus] Specimina eines Lexicon Augustinianum. SLA. 14. Hrsg. v. Werner HENSELLECK und Peter SCHILLING. Wien, Verlag der Österreichischen Akademie der Wissenschaften, 2000, 52 p.

2980. [Augustinus] WIESER (Marie-Therese). Die handschriftliche Überlieferung der Werke des heiligen Augustinus. Band VIII/1. Belgien, Luxemburg und Niederlande. Werkverzeichnis. Band VIII/2. Belgien, Luxemburg und Niederlande. Verzeichnis nach Bibliotheken. Hrsg. v. Marie Therese WIESER. Wien, Verlag der Österreichischen Akademie der Wissenschaften, 2000, 414, 338 p.

2981. [Ausonius] D. Magnus Ausonius, Mosella. Mit Texten von Symmachus und Venantius Fortunatus. Lateinisch/Deutsch. Hrsg. v. Otto SCHÖNBERGER. Stuttgart, Reclam, 2000, 111 p.

2982. [Ausonius] Decimo Magno Ausonio, Ordo urbium nobilium. Introduzione, testo critico, traduzione e note di commento. A cura di Lucia DI SALVO. Napoli, Loffredo, 2000, 341 p.

2983. [Basilius Caesariensis] HENKE (Rainer). Basilius und Ambrosius über das Sechstagewerk: eine vergleichende Studie. Basel, Schwabe, 2000, 500 p.

2984. [Cyrillus Alexandrinus] FERRARI TONIOLO (Chiara). Cyrilliana in Psalmos. I frammenti del Commento ai Salmi di Cirillo di Alessandria nel codice Laudiano greco 42. Catania, Centro di studi sul'antico cristianesimo, Università di Catania, 2000, VIII-123 p.

2985. [Cyrillus Alexandrinus] RUSSELL (Norman). Cyril of Alexandria. London a. New York, Routledge, 2000, X-271 p.

2986. [Cyrillus Alexandrinus] WESSEL (Susan). Cyril of Alexandria and the Nestorian controversy. [S. l.], [s. n.], 2000, 445 p.

2987. [Didache] DRAPER (J. A.). Ritual process and ritual symbol in Didache 7–10. *Vigiliae Christianae*, 2000, 54, p. 121-158.

2988. [Dracontius] GRILLONE (A.). Note critiche all'Orestis Tragoedia di Draconzio. *L'antiquité classique*, 2000, 69, p. 221-236.

2989. [Endelechius] BARTON (Monika). Spätantike Bukolik zwischen paganer Tradition und christlicher Verkündigung. Das Carmen De mortibus boum des Endelechius. Trier, Wissenschaftlicher Verlag Trier, 2000, 243 p.

2990. Enfant (L') a naître. Tertullien, Grégoire, Augustin, Maxime, Cassiodore, Ps. Augustin. Ed. par M.-H. CONGOURDEAU et D. SICARD. Paris, Migne et La Ferrière, Littéral, 2000, 211 p.

2991. [Ennodius] FINI (Carla). Il censimento dei codici di Ennodio. Pisa e Roma, Istituti editoriali e poligrafici internazionali, 2000, 69 p.

2992. [Eusebius Caesariensis] Eusebio di Cesarea. Dimostrazione evangelica. Introduzione, traduzione e note. A cura di Paolo CARRARA. Milano, Paoline, 2000, 874 p.

2993. [Evagrius] WHITBY (M.). The ecclesiastical history of Evagrius Scholasticus. Liverpool, Liverpool U. P., 2000, LXIII-390 p. (maps).

2994. [Gregorius Magnus] Grégoire le Grand (Pierre de Cava). Commentaire sur le Premier Livre des Rois. 4, (IV, 79-217). Introduction, texte, traduction et notes. Ed. par Adalbert DE VOGÜÉ. Paris, Ed. du Cerf, 2000, 340 p. (Sources chrétiennes, 432).

2995. [Gregorius Nazianzenus] Corpus Nazianzenum. 8. Studia Nazianzenica 1. Ed. par Bernard COULIE. Turnhout, Brepols, 2000, XI-290 p.

2996. [Gregorius Nazianzenus] Corpus Nazianzenum. 9. Sancti Gregorii Nazianzeni opera. Versio iberica. 2. Orationes XV, XXIV, XIX. Ed. par Helene METREVELI. Turnhout, Brepols, 2000, X-223 p.

2997. [Gregorius Nazianzenus] GÓMEZ VILLEGAS (N.). Gregorio de Nazianzo en Constantinopla. Orthodoxia, heterodoxia y régimen teodosiano en una capital cristiana. Madrid, Consejo superior de investigaciones científicas, 2000, XXII-234 p.

2998. [Gregorius Nazianzenus] Gregorio di Nazianzo. Tutte le orazioni. A cura di Claudio MORESCHINI, Chiara SANI, Maria VINICELLI e Carmelo CRIMI. Milano, Bompiani, 2000, 1526 p.

2999. [Gregorius Nyssenus] BANDINI (M.). Note sulla tradizione e sul testo di Gregorio Nisseno. *Rivista di filologia e di istruzione classica*, 2000, 128, p. 323-337.

3000. [Gregorius Nyssenus] Gregory of Nyssa. Homilies on the beatitudes. An English version with commentary and supporting studies. Proceedings of the eight international colloquium on Gregory of Nyssa (Paderborn, 14–18 september 1998). Ed. by Hubertus R. DROBNER and Albert VICIANO. Leiden a. Köln, Brill, 2000, XXVIII-680 p.

3001. [Gregorius Nyssenus] Lexicon Gregorianum: Wörterbuch zu den Schriften Gregors von Nyssa. Hrsg. v. der Forschungsstelle Gregor von Nyssa an der Westfälischen Wilhelms-Universität. 2. *Babaí-dorophoría*. Leiden u. New York, Brill, 2000, XII-555 p.

3002. [Hieronymus] CANELLIS (A.). Saint Jérôme et les passions sur les 'Quattuor perturbationes' des Tusculanes. *Vigiliae Christianae*, 2000, 54, p.178-203.

3003. [Hieronymus] D'ANNA (G.). A proposito di una recente edizione degli Additamenta di Girolamo ai Chronica di Eusebio. *Athenaeum*, 2000, 88, p. 547-558.

3004. [Hilarius Pictaviensis] Hilaire de Poitiers. La Trinité. 2 (livres IV-VIII). Texte critique et traduction. Ed. par P. SMULDERS, G. M. DE DURAND, Ch. MOREL et G. PELLAND. Paris, Ed. du Cerf, 2000, 483 p. (Sources chrétiennes, 448).

3005. [Iohannes Chrysostomus] Jean Chrysostome. De l'incompréhensibilité de Dieu: homélies 1–5 contre les Anoméens. Traduction, préfation et notes. Ed. par Pierre MARÉCHAUX. Paris, Rivages, 2000, 151 p.

3006. [Isidorus Hispalensis] Isidori Hispalensis versus. Cura et studio José María SÁNCHEZ MARTÍN. Turnhout, Brepols, 2000, 274 p. (Corpus Christianorum. Series Latina, 113).

3007. [Isidorus Pelusiota] Isidore de Péluse, Lettres. Tome II: Lettres 1414–1700. Texte critique, traduction et notes. Ed. par Pierre ÉVIEUX. Paris, Ed. du Cerf, 2000, 521 p.

3008. [Iustinus Martyr] BINGHAM (D. J.). Justin and Isaiah 53. *Vigiliae Christianae*, 2000, 54, p. 248-261.

3009. [Licentius] CUTINO (M.). Licentii carmen ad Augustinum. Catania, Università di Catania, Centro di Studi sull'antico Cristianesimo, 2000, 227 p.

3010. [Liturgica] La preghiera dei cristiani. Introduzione, traduzione e commento. A cura di Salvatore PRICOCO e Manlio SIMONETTI. Roma, Fondazione Lorenzo Valla e Milano, Mondadori, 2000, XXX-671 p.

3011. [Maximus Tauriniensis] DE FILIPPIS CAPPAI (Ch.). Riflessioni sul concubinato nella società romana tardoimperiale. In margine al sermone LXXXVIII di Massimo di Torino: leggi divine e leggi civili. *Bollettino di studi latini*, 2000, 30, p. 561-576.

3012. [Origenes] Origene. Dizionario: la cultura, il pensiero, le opere. A cura di Adele MONACI CASTAGNO. Roma, Città nuova, 2000, XXI-489 p.

3013. [Origenes] Origène. Exégèse spirituelle. 2. L'Exode, le Lévitique. Ed. par A. ÉGRON. Paris, Ed. du Cerf, 2000, V-339 p. (ill.).

3014. [Paulinus Nolanus] CONYBEARE (C.). Paulinus Noster: self and simbols in the letters of Paulinus of Nola. Oxford, Oxford U. P., 2000, XI-187 p.

3015. [Paulinus Nolanus] Paulini Nolani operum concordantiae. 3 voll. Hrsg. v. Matthias SKEB. Hildesheim, Olms-Weidmann, 2000, XI-2 159 p.

3016. [Prudentius] ENCUENTRA ORTEGA (A.). El hexámetro de Prudencio. Estudio comparado de métrica verbal. Logroño, Instituto de Estudios Riojanos, 2000, 498 p.

3017. [Prudentius] GNILKA (Christian). Prudentiana I. Critica. München u. Leipzig, Saur, 2000, 762 p.

3018. [Prudentius] PARTOENS (G.). The influence of the historian Florus on Prudentius' Libri contra Symmachum. *Ancient society*, 2000, 30, p. 331-347.

3019. [Pseudo-Macarius] Pseudo-Makarios, Reden und Briefe. Hrsg. v. Klaus FITSCHEN. Stuttgart, Hiersemann, 2000, VII-549 p.

3020. Sénèque et Saint Paul. Lettres. Traduction et préfation. Ed. par Paul AIZPURUA. Paris Le promeneur, 2000, 76 p.

3021. SKOGOREV (Aleksandr P.). Apokrificheskie dejanija apostolov. Arabskoe Evangelie detstva Spasitelja: Issled., perevody, komment. (The apocryphal acts of Apostles. the Arabic gospel of the childhood of the saviour: study, transl. and comment). Ed. N. S. SVENTSITSKAJA. Sankt-Peterburg, Aletejja, 2000, 480 p. (ill.). (Antichnoe khristianstvo. Istochniki). [Text partly in Latin]

3022. [Synesius Cyrenaicus] Synésios de Cyrène. Tome II. Correspondance, Lettres I–LXIII. Tome III. Correspondance, Lettres LXIV–CLVI. Ed. par Antonio GARZYA et Denis ROQUES. Paris, Belles Lettres, 2000, 2 vol., CXLVII-172, 311 p. (Collection des universités de France)

3023. [Testamentum novum] BURTON (P.). The Old latin Gospels: a study of their texts and language. Oxford, Oxford U. P., 2000, X-232 p.

3024. [Testamentum novum] CLASSEN (Carl Joachim). Rhetorical criticism of the New Testament. Tübingen, Mohr, 2000, VII-195 p.

3025. [Testamentum novum] DAHL (Nils Alstrup). Studies in Ephesians. Ed. by David HELLHOLM, Vermund BLOMKVIST and Tord FORNBERG. Tübingen, Mohr, 2000, XII-548 p.

3026. [Testamentum novum] Diglossia and other topics in New Testament linguistics. Ed. by Stanley E. PORTER. Sheffield, Sheffield Academic Press, 2000, 305 p. (Journal for the Study of the New Testament Supplement Series 193).

3027. [Testamentum novum] FREY (Jörg). Die johanneische Eschatologie. Band III. Die eschatologische Verkündingung in den johanneischen Texten. Tübingen, Mohr, 2000, XVII-600 p.

3028. [Testamentum novum] Handbuch zum Neuen Testament. 15, 1. Der Jakobusbrief. Hrsg. v. Christoph BURCHARD. Tübingen, Mohr, 2000, 217 p.

3029. [Testamentum novum] Handbuch zum Neuen Testament. 9, 1. Der erste Korintherbrief. Hrsg. v. Andreas LINDEMANN. Tübingen, Mohr, 2000, IX-389 p.

3030. [Testamentum novum] LABAHN (Michael). Offenbarung in Zeichen und Wort. Untersuchungen zur Vorgeschichte von Joh 6,1–25a und seiner Rezeption in der Brotrede. Tübingen, Mohr, 2000, XII-358 p. (Wissenschaftliches Untersuchungen zum Neuen Testament. 2. Reihe).

3031. [Testamentum Novum] MALHERBE (A. J.). The Letters to the Thessalonians. New York, London, Toronto, Sydney a. Auckland, The Anchor Bible a. Doubleday, 2000, XX-508 p.

3032. [Testamentum novum] Vetus Latina. Die Reste der altlateinischen Bibel. 26. Neues Testament. 2. Apocalypsis Johannis. 1. Einleitung. 2. Einleitung (Fortsetzung und Schluss); Apc. 1, 1–2, 7. Hrsg. v. Roger GRYSON. Freiburg im Breisgau, Herder, 2000, 80, 81-160 p.

3033. [Testamentum vetus] Älteren griechischen Katenen (Die) zum Buch Hiob. Band III: Fragmente zu Hiob 23,1–42,17. Hrsg. v. Ursula und Dieter HAGEDORN. Berlin u. New York, de Gruyter, 2000, IX-415 p. (Patristische Texte und Studien, 53).

3034. [Testamentum vetus] ECKART (Otto). Das Deuteronomium im Pentateuch und Hexateuch. Studien zur Literaturgeschichte von Pentateuch und Hexateuch im Lichte des Deuteronomiumrahmens. Tübingen, Mohr Siebeck, 2000, 250 p.

3035. [Venantius Fortunatus] Venanzio Fortunato. Epitaphium Vilithutae (IV 26). Introduzione, traduzione e commento. A cura di Paola SANTORELLI. Napoli, Liguori, 2000, 130 p.

3036. [Vergilius] FREUND (S.). Vergil im frühen Christentum. Untersuchungen zu den Vergilzitaten bei Tertullian, Minucius Felix, Novatian, Cyprian und Arnobius. Paderborn, Münich, Vienna u. Zürich, Schöningh, 2000, 430 p.

§ 1. Addenda 1999.

3037. [Gregorius Nazianzenus] Corpus Nazianzenum. 6. Sancti Gregorii Nazianzeni opera. Versio armeniaca. 2. Orationes IV et V. 7. Sancti Gregorii Nazianzeni opera. Versio armeniaca. 3. Orationes XXI, VIII, VII. Ed. par Anna SIRINIAN et Bernard COULIE. Turnhout, Brepols, 99, 2 vol., XXXIX-384 p., XVI-270 p.

3038. [Gregorius Nyssenus] Lexicon Gregorianum: Wörterbuch zu den Schriften Gregors von Nyssa. Hrsg.

v. der Forschungsstelle Gregor von Nyssa an der Westfälischen Wilhelms-Universität. 1. *Abarés-áoros.* Leiden u. New York, Brill, 99, XVIII-664 p.

Cf. n° 769

§ 2. General.

3039. BEDOUELLE (Guy). Die Geschichte der Kirche. Orig.-Ausg. Paderborn, Bonifatius, 2000, 316 p. (ill.).

3040. BROCK (E. A. G.). Authority, politics, and gender in early Christianity. Mary, Peter and the portrayal of leadership, [S. l.], [s. n.], 2000, 242 p.

3041. Dictionnaire de l'histoire du Christianisme. Paris, Albin Michel, 2000, 173 p. (Carte, plans, index)

3042. Early Christian world (The). Voll. I. II. Ed. by Philip F. ESLER. London a. New York, Routledge, 2000, 2 vol., XXVI-689, IX-649 p.

3043. Handbuch der Kirchengeschichte. Hrsg. v. Hubert JEDIN. Berlin, Directmedia publ., 2000, 1 CD-ROM.

3044. Histoire du christianisme (des origines à 250). 1. Le nouveau peuple: (des origines à 250). Ed. par L. PIETRI. Paris, Desclée, 2000, 938 p. (cartes).

3045. Kairos. Studi di letteratura cristiana antica per l'anno 2000. A cura di Pietro GIANNINI e Maria Pia CICCARESE. Galatina, Congedo, 2000, 223 p.

3046. Metzler-Lexikon christlicher Denker: 700 Autorinnen und Autoren von den Anfängen des Christentums bis zur Gegenwart. Hrsg. v. Markus VINZENT. Stuttgart, Metzler, 2000, X-821 p.

3047. Reallexicon für Antike und Christentum. Sachwörterbuch zur Auseinandersetzung des Christentums mit der antiken Welt. Hrsg. v. E. DASSMANN, H. BRAKMANN, C. COLPE, A. DIHLE, J. ENGEMANN, K. HOHEISEL, W. SPEYER und K. THRAEDE. Register der Bände I bis XV. Stuttgart, Hiersemann, 2000, 454 p.

Cf. n° 1401

§ 3. Special studies.

3048. BRENNECKE (H. Ch.). Imitatio – reparatio – continuatio. Die Judengesetzgebung im Ostgotenreich Theoderichs des Großen als reparatio imperii? *Zeitschrift für antikes Christentum*, 2000, 4, p. 133-148.

3049. BYRSKOG (Samuel). Story as history – history as story. The Gospel tradition in the context of ancient oral history. Tübingen, Mohr, 2000, XIX-386 p.

3050. CAMPIONE (Ada). La Basilicata paleocristiana. Diocesi e culti. Bari, Edipuglia, 2000, 221 p. (ill.).

3051. CAVALIERI (M.). La basilica in Italia: decorazione scultorea e sue valenze politico-culturali. *Athenaeum*, 2000, 88, p. 465-476.

3052. Comunità (La) cristiana di Roma: la sua vita e la sua cultura dalle origini all'Alto Medioevo. A cura di Letizia PANI ERMINI e Paolo SINISCALCO. Città del Vaticano, Libreria editrice Vaticana, 2000, 476 p. (ill., cartes).

3053. Concilio (Dal) di Calcedonia (451) a Giovanni Damasceno (750): i Padri orientali. A cura Angelo DI BERARDINO. Genova, Marietti, 2000, XIII-719 p.

3054. CURRAN (J.). Pagan city and christian capital. Rome in the fourth century. Oxford, Clarendon Press, 2000, XX-389 p. (ill.).

3055. DRAKE (H. A.). Costantine and the bishops. The politics of intolerance. Baltimore a. London, Johns Hopkins U. P., 2000, XX-609 p. (ill.).

3056. DÜNZL (Franz). Die Absetzung des Bischofs Meletius von Antiochien 361 n.C. *Jahrbuch für Antike und Christentum*, 2000, 43, p. 71-93. – IDEM. Pneuma. Funktionen des theologischen Begriffs in frühchristlicher Literatur. Münster, Aschendorff, 2000, 451 p.

3057. DUVAL (Y.). Chrétiens d'Afrique l'aube de la paix constantinienne. Les premiers échos de la grande persécution. Paris, Institut d'études augustiniennes, 2000, 524 p. (ill.).

3058. Église (L') et la mission au VIe siècle: la mission d'Augustin de Cantorbéry et les Églises de Gaule sous l'impulsion de Grégoire le Grand. Actes du colloque d'Arles de 1998. Ed. par Christophe DE DREUILLE et Louis-Marie BILLÉ. Paris, Ed. du Cerf, 2000, p. XII-427 p.

3059. Encyclopedia of monasticism. Ed. by M. JOHNSTON. London a. Chicago, Dearborn, 2000, 2 vol., XXXIV-1556 p., ill cartes.

3060. FIEDROWICZ (Michael). Apologie im frühen Christentum. Die Kontroverse um den christlichen Wahrheitsanspruch in den ersten Jahrhunderten. Paderborn, München, Wien u. Zürich, Schöningh, 2000, 361 p.

3061. FÖRSTER (Hans). Die Feier der Geburt Christi in der Alten Kirche: Beiträge zur Erforschung der Anfänge des Epiphanie- und des Weihnachtsfest. Tübingen, Mohr, 2000, X-218 p.

3062. FOSCHIA (L.). La réutilisation des sanctuaires païens par les chrétiens en Grèce continentale (IVe–VIIe s.). *Revue des études Grecques*, 2000, 113, p. 413-434.

3063. FRANCK (Georgia). The memory of the eyes: pilgrims to living saints in Christian late antiquity. Berkeley, Los Angeles a. London, University of California Press, 2000, XIII-219 p. (The Transformation of the Classical Heritage, 30).

3064. GEHRING (Roger W.). Hausgemeinde und Mission: die Bedeutung antiker Häuser und Hausgemeinschaften: von Jesus bis Paulus. Giessen, Brunnen, 2000, 582 p.

3065. GUYON (Jean). Les premiers baptistères des Gaules, 4.–8. siècles. Roma, Unione internazionale degli istituti di archeologia e storia dell'arte in Roma, 2000, 82 p. (ill.). (Conferenze).

3066. HARRISON (C.). Augustine christian truth and fractured humanity. Oxford, Oxford U. P., 2000, XIII-242 p.

3067. HARTENSTEIN (Judith). Die zweite Lehre. Erscheinungen des Auferstandenen als Rahmenerzählungen frühchristlicher Dialoge. Berlin, Akademie Verlag, 2000, 394 p. (Texte und Untersuchungen zur Geschichte der altchristlichen Literatur, 146).

3068. HOFFMANN (Andreas). Kirchliche Strukturen und Römisches Recht bei Cyprian von Karthago. Paderborn, München, Wien u. Zürich, Schöningh, 2000, 345 p.

3069. KENNELL (S. A. H.). Magnus Felix Ennodius. A gentleman of the church. Ann Arbor, Michigan U. P., 2000, XIII-256 p.

3070. KLEIN (Richard). Die Haltung der kappadokischen Bischöfe Basilius von Caesarea, Gregor von Nazianz und Gregor von Nyssa zur Sklaverei. Stuttgart, Steiner, 2000, VIII-306 p.

3071. LEPPIN (Hartmut). Die Kirchenväter und ihre Zeit. München, Beck, 2000, 125 p. (ill.).

3072. LEYSER (Conrad). Authority and asceticism from Augustine to Gregory the Great. Oxford, Clarendon Press, 2000, XII-221 p.

3073. LIZZI TESTA (Rita). Privilegi economici e definizione di 'status': il caso del vescovo tardoantico. *Rendiconti dell'Accademia nazionale dei Lincei*, 2000, 11, p. 55-103.

3074. MAC DONALD (Lee Martin), PORTER (Stanley E.). Early Christianity and its sacred literature. Peabody, Hendrickson, 2000, XXVII-708 p. (ill., cartes)

3075. MANZI (F.). La fede degli uomini e la singolare relazione filiale di Gesù con Dio nell'Epistola agli Ebrei. *Biblica*, 2000, 81, p. 32-62.

3076. MARTIN (Wallraff). Die Ursprünge der christlichen Gebetsostung. *Zeitschrift für Kirchengeschichte*, 2000, 111, p. 170-184.

3077. MEIER (J. P.). The historical Jesus and the historical Samaritans: what can be said? *Biblica*, 2000, 81, p. 202-232.

3078. MESA SANZ (J. F.). Grafías y sonidos del sarcófago paleocristiano de Santa Engracia (Zaragoza). *Latomus*, 2000, 59, p. 403-413.

3079. MITCHELL (Margaret M.). The heavenly trumpet. John Chrysostom and the art of Pauline interpretation. Tübingen, Mohr, 2000, XXXIV-564 p.

3080. MYSZOR (Wincenty). Europa: pierwotne chrześcijaństwo, idee i życie społeczne chrześcijan (II i III wiek). (L'Europe: la chrétienté primitive, les idées et la vie sociale des chrétiens, IIe et IIIe siècles). Warszawa, Studium Generale Europa, 2000, 252 p.

3081. Orígenes (Los) del cristianismo en Valencia y su entorno. Valencia, Ajuntament de Valencia, 2000, 296 p. (ill.). (Grandes temas arqueologicos)

3082. Orthodoxie, Christianisme, histoire. Ed. par S. ELM, É. REBILLARD et A. ROMANO. Paris, de Boccard, 2000, XXV-404 p. (École française de Rome).

3083. Papato (Il) di san Simmaco (498–514). Atti del convegno internazionale di studi, Oristano 19–21 novembre 1998. A cura di G. MELE, N. SPACCAPELO e M. LORENZANI. Cagliari, Pontificia facoltà teologica della Sardegna, 2000, XX-505 p.

3084. PERROT (Charles). Après Jésus: le ministère chez les premiers chrétiens. Paris, Éditions de l'Atelier, 2000, 271 p.

3085. PESARINO (ASTRID). Contributo allo studio del tipo della 'Virgo Lactans': il papiro PSI XV 1574 dell'istituto papirologico G. Vitelli di Firenze. *Latomus*, 2000, 59, 640-646 p.

3086. PODOSINOV (Aleksandr V.). Simvoly chetyrekh evangelistov: Ikh proiskhozhdenie i znachenie. (Symbols of four gospel-writers: their origin and meaning). Moskva, Jazyki russkoj kul'tury, 2000, 171 p. (ill.). [Zusammenfassung]

3087. PORTER (Stanley E.). The criteria for authenticity in historical-Jesus research. Previous discussion and new proposals. Sheffield, Sheffield academic press, 2000, 299 p.

3088. REISER (M.). Numismatik und Neues Testament. *Biblica*, 2000, 81, p. 457-488.

3089. Rom und das himmlische Jerusalem. Die frühen Christen zwischen Anpassung und Ablehnung. Hrsg. v. Raban von HAEHLING. Darmstadt, Wissenschaftliche Buchgesellschaft, 2000, XII-308 p.

3090. ROMANIELLO (Giuseppe). La scrittura apocrifa dei primi due capitoli del Vangelo secondo Matteo e controversione della datazione della nascita di Cristo. Latina, Selbstverlag, 2000, 207 p.

3091. Romanité et cité chrétienne: permanences et mutations, intégration et exclusion du Ier au VIe siècle. Mélanges en honneur d'Yvette Duval. Ed. par F. PRÉVOT. Paris, de Boccard, 2000, 445 p.

3092. RONCHEY (Silvia). Les procès-verbaux des martyres chrétiens dans les Acta Martyrum et leur fortune. *Mélanges de l'école française de Rome. Antiquité*, 2000, 112, p. 723-752.

3093. SABW KANYANG (Jean-Anatole). Episcopus et plebs: l'évêque et la communauté ecclésiale dans les conciles africains (345-525). Bern et Frankfurt am Main, Lang, 2000, XIV-393 p.

3094. SCHOLTEN (C.). Verändert sich Gott, wenn er die Welt erschafft? Die Auseinandersetzung der Kirchenväter mit einem philosophischen Dogma. *Jahrbuch für Antike und Christentum*, 2000, 43, p. 25-43.

3095. SORABJI (Richard). Emotion and peace of mind. From stoic agitation to Christian temptation. Oxford, Oxford U. P., 2000, XI-499 p.

3096. SPICKERMANN (Wolfgang). Der Subdiakonat, ein Amt der spätantiken Kirchenverwaltung. *Zeitschrift für Kirchengeschichte*, 2000, 111, p. 313-341.

3097. TANTILLO (I.). Filostorgio e la tradizione sul testamento di Costantino. *Athenaeum*, 2000, 88, p. 559-563.

3098. THURÉN (Lauri). Derhetorizing Paul. A dynamic perspective on Pauline theology and the law. Tübingen, Mohr, 2000, IX-213 p.

3099. Topographie chrétienne des cités de la Gaule des origines au milieu du VIIIe siécle. Ed. par N. GAUTHIER. Paris, de Boccard, 85 p. (ill.).

3100. UBIÑA (José Fernández). Cristianos y militares. La iglesia antigua ante el ejército y la guerra. Granada, Universidad de Granada, Caja general de ahorros de Granada, 2000, 730 p.

3101. VAGGIONE (Richard Paul). Eunomius of Cyzicus and the Nicene Revolution. Oxford a. New York, Oxford U. P., 2000, XXV- 425 p.

3102. VIELBERG (Meinolf). Klemens in den pseudoklementinischen Rekognitionen. Studien zur literarischen Form des spätantiken Romans. Berlin, Akademie Verlag, 2000, 236 p. (Texte und Untersuchungen zur Geschichte der altchristlichen Literatur, 145).

3103. ZLINSZKY (J.). Der Prozeß Jesu aus der Sicht des antiken Prozeßrecht. *Acta antiqua Academiae scientiarum Hungaricae*, 2000, 40, p. 505-517.

Cf. n°s 377, 8792

§ 4. Hagiography.

3104. BEAUJARD (Brigitte). Le culte des saints en Gaule. Les premiers temps. D'Hilaire de Poitiers à la fin du VIe siècle. Paris, Ed. du Cerf, 2000, IV-613 p.

3105. BOULHOL (P.). La prière dangereuse. Imprécations et malédictions dans l'hagiographie antique (IIe–VIe siècles). *In*: Prières méditerranéennes hier et aujourd'hui [Cf. n° 2863], p. 177-199.

3106. DE BOER (Esther). Maria Maddalena. Oltre il mito: alla ricerca della sua vera identità. Torino, Claudiana, 2000, p. 176 p. (ill.).

3107. MOULINIER (J.-C.). Autour de la tombe de Saint Victor de Marseille: textes et monuments commémoratifs d'un martyre. Ed. par Jean GUYON. Marseille, Tacussel, 2000, 198 p. (ill.).

3108. PEDREGAL (Amparo). Las mártires cristianas: género, violencia y dominación en el cuerpo femenino. *Studia Historica. Historia antigua*, 2000, 18, p. 277-294.

3109. TODA (Satoshi). La Vie de S. Macaire l'Égyptien: état de la question. *Analecta Bollandiana*, 2000, 118, p. 267-290.

3110. VAN DER STRAETEN (Joseph). Catalogues de manuscrits latins: inventaire hagiographique. 23. *Analecta Bollandiana*, 2000, 118, p. 399-440.

3111. VIGGIANI (Maria Carmen). Santa Marcellina: una nobile romana a Milano: vita, opere e devozione della sorella di S. Ambrogio. Genova, De Ferrari, 2000, 173 p. (ill.).

§ 4. Addenda 1999.

3112. VAN DER STRAETEN (Joseph). Catalogues de manuscrits latins: inventaire hagiographique. 22. *Analecta Bollandiana*, 99, 117, p. 394-419.

H

BYZANTINE HISTORY
(since Justinian)

---

§ 1. Sources. 3113-3161. – § 2. General. 3162-3171. – § 3. Special studies. 3172-3254.

§ 1. Sources.

3113. Anonymi Historia Imperatorum, parte prima. Introduzione, testo critico, versione italiana, note e indici. A cura di F. IADEVAIA. Messina, EDAS, 2000, 293 p.

3114. Anonymi professoris Epistulae. Hrsg. v. Athanasios MARKOPOULOS. Berlin u. New York, de Gruyter, 2000, IX-165 p. (Corpus fontium historiae Byzantinae. Series Berolinensis, 37).

3115. ANTONOPOULOU (Theodora). Lexicographical addenda from the homelies of the Emperor Leo VI. *Byzantion*, 2000, 70, p. 9-24.

3116. [Basilius Minimus] SCHMIDT (Th. S.). Les Commentaires de Basile le Minime: liste révisée des manuscrits et des éditions. *Byzantion*, 2000, 70, p. 155-181.

3117. BEIHAMMER (Alexander Daniel). Nachrichten zum byzantinischen Urkundenwesen in arabischen Quellen (565–811). Bonn, Dr. Rudolf Habelt, 2000, LXXXVII-514 p. (Poikila Byzantina, 17).

3118. [Boethius] Boethius. De consolatione philosophiae. Opuscula theologica. Hrsg. v. C. MORESCHINI. München u. Leipzig, Saur, 2000, XXI-262 p.

3119. [Christodoros] TISSONI (Francesco). Cristodoro. Un'introduzione e un commento. Alessandria, Edizioni dell'Orso, 2000, 257 p.

3120. Cinque poeti bizantini [Elia Sincello, Ignazio Diacono, Leone Magistro, Giovanni di Gaza, 'Giorgio Grammatico']. Anacreontee del Barberiniano greco 310. Testo critico, introduzione, traduzione e note. A cura di F. CICCOLELLA. Alessandria, Edizioni dell'Orso, 2000, LXIII-295 p.

3121. [Constantinus Porphyrogenetus] DAGRON (G.), BINGGELI (A.), FEATHERSTONE (M.), FLUSIN (B.). L'organisation et le déroulement des courses d'après le Livre des Cérémonies. *Travaux et mémoires*, 2000, 13, p. 3-200.

3122. [Constantinus Porphyrogenetus] HALDON (J. F.). Theory and practice in the tenth-century military administration. Chapters II, 44 et 45 of the Book of Ceremonies. *Travaux et mémoires*, 2000, 13, p. 201-352.

3123. [Constantinus Porphyrogenetus] KRESTEN (O.). Beobachtungen zu Kapitel I 96 des 'Zeremonienbuches'. *Byzantinische Zeitschrift*, 2000, 93, p. 474-489.

3124. [Constantinus Porphyrogenetus] KRESTEN (Otto). 'Staatsempfänge' im Kaiserpalast von Konstantinopel um die Mitte des 10. Jahrhunderts. Beobachtungen zu Kapitel II 15 des sogenannten 'Zeremonienbuches'. Wien, Verlag der Österreichischen Akademie der Wissenschaften, 2000, 61 p.

3125. *Vacat.*

3126. [Eustathius Thessalonicensis] Eustathii Thessalonicensis opera minora magnam parte inedita. Hrsg. v. Peter WIRTH. Berlin u. New York, de Gruyter, 2000, 409 p. (Corpus fontium historiae Byzantinae. Series Berolinensis, 32).

3127. [Eustathius Thessalonicensis] Eustazio di Tessalonica. Introduzione al commentario a Pindaro. A cura di M. NEGRI. Brescia, Paideia, 2000, 310 p.

3128. [Georgios Amiroutzes] Jorge Ameruzes de Trebisonda. El diálogo de la fe con el sultán de los Turcos. Ed. por Óscar DE LA CRUZ PALMA. Madrid, Consejo superior de investigaciones científicas, Universidad autónoma de Barcelona, Departamento de Ciencias de la antigüedad y de la edad media, 2000, XXIX-194 p.

3129. [Iohannes Kameniates] FRENDO (D.), FOTIOU (A.). John Kaminiates. The capture of Thessaloniki. Translation, introduction and notes. Perth, 00,189 p.

3130. [Iohannes Kinnamos] GENTILE MESSINA (R.). Un caso di mimesis al servizio della propaganda nell'Epitome di Giovanni Cinnamo. *Byzantion*, 2000, 70, p. 408-421.

3131. [Iohannes Malalas] THURN (I.). Ioannis Malalae Chronographia. Berlin u. New York, de Gruyter, 2000, VII, 30*-551 p. (Corpus fontium historiae Byzantinae. Series Berolinensis, 35).

3132. [Kekaumenos] Cecaumeno. Consejos de un aristócrata bizantino. Introducción, traducción y notas. Ed. por J. SIGNES CODOÑER. Madrid, Alianza editorial, 2000, 158 p.

3133. KIOURTZIAN (G.). Recueil des inscriptions grecques chrétiennes des Cyclades. De la fin du IIIe au VIIe siècle après J.-C. Paris, de Boccard, 2000, 316 p.

3134. LITAVRIN (Gennadij G.). Vizantija, Bolgarija, Drevnjaja Rus' (IX–nachalo XII v.). (Byzantium, Bulgaria and Old Rus' in the 9th–the early 12th century). Sankt-Peterburg, Aletejja, 2000, 415 p. (Vizantijskaja biblioteka). – IDEM. K izucheniju problemy dokhodnosti krest'janskogo khozjajstva v Vizantii v X-XI vv. (On the problem of profitableness of peasant's household in Byzantium in the 10th and the 11th centuries). *In*: Vizantijskij vremennik = BYZANTINA XPONIKA [Cf. n° 3161], p. 5-23.

3135. [Michael Psellus] PAPAIOANNOU (E.). Michael Psellos' rhetorical gender. *Byzantine and modern Greek studies*, 2000, 24, p. 133-146.

3136. [Nicetas Choniata] BOSSINA (L.). La bestia e l'enigma. Tradizione classica e cristiana in Niceta Coniata. *Medioevo greco*, 2000, p. 35-68.

3137. NICOLOTTI (A.). Sul metodo per lo studio dei testi liturgici. In margine alla liturgia eucaristica bizantina. *Medioevo greco*, 2000, p. 143-179.

3138. [Nonnus Panopolitanus] FAYANT (M. C.). Nonnos de Panopolis. Les Dionysiaques. Tome XVII, chant XLVII. Paris, Les belles lettres, 2000, 204 p. (Collection des Universités de France).

3139. [Nonnus Panopolitanus] Nonno di Panopoli. Parafrasi del vangelo di San Giovanni. Canto B. Introduzione, testo critico, traduzione e commento. A cura di E. LIVREA. Bologna, Edizioni dehoniane, 2000, 317 p.

3140. Perceptions of Byzantium and its neighbours (843–1261). Ed. by Olenka Z. PEVNY. New York, Metropolitan Museum of Art, 2000, XII-197 p. (ill.). (Metropolitan Museum of Art Symposia).

3141. [Photius] CANFORA (Luciano). Il rogo dei libri di Fozio. *In*: Fozio [Cf. n° 3144], p. 17-28.

3142. [Photius] DANEK (G.). Iamblichs Babyloniaka und Heliodor bei Photios: Referattechnik und Handlungsstruktur. *Wiener Studien*, 2000, 113, p.113-134.

3143. [Photius] ELEUTERI (Paolo). I manoscritti greci della biblioteca di Fozio. *Quaderni di storia*, 2000, 26, 51, p. 109-156.

3144. [Photius] Fozio. Tra crisi ecclesiale e magistero letterario. A cura di Giovanni MENESTRINA. Brescia, Morcelliana, 2000, 82 p. [Cf. n° <scelta> 3141.].

3145. [Photius] SCHAMP (J.). Les Vies des dix orateurs attiques. Fribourg, Éditions universitaires, 2000, 252 p.

3146. [Procopius] Procopio de Cesarea. Historia de las Guerras, Libros I-II, Guerra Persa. Introducción, traducción y notas. Ed. por F. A. GARCÍA ROMERO. Madrid, Editorial Gredos, 2000, 323 p.

3147. [Procopius] Procopio de Cesarea. Historia de las Guerras, Libros III-IV, Guerra Vándala. Introducción, traducción y notas. Ed. por J. A. FLORES RUBIO. Madrid, Editorial Gredos, 2000, 359 p.

3148. [Procopius] Procopio de Cesarea. Historia secreta. Introducción, traducción y notas. Ed. por J. SIGNES CODOÑER. Madrid, Editorial Gredos, 2000, 350 p.

3149. [Procopius] Prokopios, Hemlig historia. Översättning från grekiskan, med inledning och kommentar av S. LINNÉR. Stockholm, Wahlström & Widstrand, 2000, 154 p.

3150. [Procopius] Prokopios, Vandalkriegen. Översättning från grekiskan, med inledning och kommentar av S. LINNÉR. Stockholm, Wahlström & Widstrand, 2000, 204 p.

3151. [Rufus de Shotep] LUCCHESI (E.). La langue originale des commentaires sur les Évangiles de Rufus de Shotep, *Orientalia*, 2000, 69, p. 86-87.

3152. SCHAMP (J.). Photios et Jean Philopon: sur la date du De opificio mundi. *Byzantion*, 2000, 70, p. 135-154.

3153. SCHEMBRA (R.). La tradizione manoscritta della I redazione degli Homerocentones. *Byzantinische Zeitschrift*, 2000, 93, p. 162-175.

3154. SEROV (Vadim V.). Finansovaja politika rannevizantijskogo imperatora: analiz meroprijatij Anastasija I. (Financial policy of an early Byzantine Emperor: an analysis of policy of Anastasius I). Barnaul, OAO 'Altajskij poligraficheskij kombinat', 2000, 208 p.

3155. STAHL (A. M.). The documents of Angelo de Carturia and Donato Fontanella, Venetian notaries in fourteenth-century Crete. Washington, Dumbarton Oaks, 2000, XXI-295 p.

3156. TARAGNA (A. M.). Logoi historias. Discorsi e lettere nella prima storiografia retorica bizantina. Alessandria, Edizioni dell'Orso, 2000, 278 p.

3157. [Theodorus II Ducas Lascaris] Theodorus II Ducas Lascaris. Opuscula rhetorica. Hrsg. v. A. TARTAGLIA. München u. Leipzig, Saur, 2000, XX-221 p.

3158. [Theodorus Methochites] FEATHERSTONE (Jeffrey Michael). Theodore Metochites's poems 'To himself'. Introduction, text, and translation. Wien, Verlag der Österreichischen Akademie der Wissenschaften, 2000, 156 p.

3159. [Theophanes] YANNOPOULOS (P.). Les vicissitudes historiques de la Chronique de Théophane. *Byzantion*, 2000, 70, p. 527-553.

3160. Through the looking glass: Byzantium through British eyes. Papers from the twenty-ninth spring symposium of Byzantine studies, London, March 1995. Ed. by Robin CORMACK and Elizabeth JEFFREY. Aldershot, Ashgate, 2000, XII-258 p. (ill.). (Society for the promotion of Byzantine studies, publications, 7).

3161. Vizantijskij vremennik = BYZANTINA XPONIKA. (Byzantine Review). Vol. 59 (84). Ed. Gennadij G. LITAVRIN. RAN, In-t vseobshchej istorii. Moskva, Nauka, 2000, 304 p. [Cf. n° <choice> 3134, 3234.]

Cf. n°s 56, 69, 268, 769, 819, 8489

§ 2. General.

* 3162. Byzantinische Zeitschrift. Band 93. [Band 92. Cf. Bibl. 99, n° 2976.] Hrsg. v. Peter SCHREINER. München u. Leipzig, Saur, 2000, 810 p. (Taf.).

3163. 26th Annual Byzantine studies conference abstracts. 26–29 October 2000. Cambridge, Harvard U. P., 2000, 135 p.

3164. BĂNESCU (N.). Istoria imperiului bizantin. Vol. I. Imperial creștin și asaltul invaziilor (310–610): Histoire de l'empire byzantin. L'empire chrétien et l'assaut des invasions. Vol. I. București, Ediție îngrijită de Tudor Teoteoi, 2000, 771 p.

3165. Bizantino-Sicula III. Miscellanea di scritti in memoria di Bruno Lavagnini. Palermo, 2000, L-358 p. (Istituto Siciliano di studi bizantini e neoellenici. Quaderni, 14).

3166. BOUREAU (Alain). Des politiques tirées de l'Ecriture. Byzance et l'Occident (note critique). In: Gilbert Dagron et le césaro-papisme. La sacralité du pouvoir impérial. Annales, 2000, 55, 4, p. 879-888.

3167. Byzanz als Raum. Zu Methoden und Inhalten der historischen Geographie des östlichen Mittelmeerraumes. Hrsg. v. K. BELKE, F. HILD, J. KODER und P. SOUSTAL. Wien, Verlag der Österreichischen Akademie der Wissenschaften, 2000, 316 p. [Cf. n° <Auswahl> 3211.]

3168. HALDON (J.). Byzantium. A history. Brimscombe Port, Tempus, 2000, 192 p.

3169. ΛΙΘΟΣΤΡΩΤΟΝ. Studien zur byzantinischen Kunst und Geschichte. Festschrift für Marcell Restle. Hrsg. v. B. BORKOPP und Th. STEPPAN. Stuttgart, Hiersemann, 2000, 353 p. (Abb.).

3170. PATLAGEAN (Evelyne). Byzance et la question du roi-prête (note critique). In: Gilbert Dagron et le césaro-papisme. La sacralité du pouvoir impérial. Annales, 2000, 55, 4, p. 871-878.

3171. Polypleuros nous. Miscellanea für Peter Schreiner zu seinem 60. Geburtstag. Mit einem Geleitwort von Herbert Hunger. Hrsg. C. SCHOLZ und G. MAKRIS. München u. Leipzig, Saur, 2000, XII-434 p. (Byzantinisches Archiv, 19). [Cf. n^{os} <Auswahl> 3205, 3215, 3219, 3227, 3230, 3248.]

§ 3. Special studies.

3172. ALTRIPP (Michael). Beobachtungen zu Synthronoi und Kathedren in byzantinischen Kirchen Griechenlands. Bulletin de correspondance héllenique, 2000, 124, p. 377-412.

3173. ANDRÉS SANTOS (F. J.). La jurisdicción de los magistrados municipales en el Digesto de Justiniano y su relación con las leyes municipales hispanas. Hispania antiqua, 2000, 24, p. 277-297.

3174. Anna Komnene and her times. Ed. by T. GOUMA-PETERSON. New York, Garland Publishing, 2000, XIV-193 p.

3175. AVENARIUS (A.). Die byzantinische Kultur und die Slawen. Zum Problem der Rezeption und Transformation (6. bis 12. Jahrhundert). Wien, Oldenbourg, 2000, 264 p. (ill.).

3176. BARTLETT (W. B.). An ungodly war. The Sack of Constantinople and the fourth crusade. Stroud, Sutton, 2000, XVIII-229 p.

3177. BAYLISS (Richard). The fortifications and water supply systems of Constantinople. Antiquity, 2000, 74, p. 25-26.

3178. BEAUCAMP (J.). Donne, patrimonio, chiesa (Bisanzio, IV–VII secolo). In: Tardoantico (II) alle soglie del duemila [Cf. n° 2693], p. 249-265.

3179. BEIHAMMER (A. D.). Nachrichten zum byzantinischen Urkundenwesen in arabischen Quellen (565–811). Bonn, Habelt, 2000, LXXVII, 514 p. – IDEM. Quellenkritische Untersuchungen zu den ägyptischen Kapitulationsverträgen der Jahre 640–646. Wien, Verlag der Österreichischen Akademie der Wissenschaften, 2000, 71 p.

3180. BENNETT (D.). Medical practice and manuscripts in Byzantium. Social history of medicine, 2000, 13, p. 279-291.

3181. BERGER (Albrecht). Streets and public spaces in Constantinople. Dumbarton Oaks Papers, 2000, 54, p. 161-172.

3182. BRAND (Charles M.). Did Byzantium have a free market? Byzantinische Forschungen, 2000, 26, p. 63-72.

3183. BRANDES (W.). Liudprand von Cremona (Legatio cap. 39-41) und eine bisher unbeachtete westöstliche Korrespondenz über die Bedeutung des Jahres 1000 A.D. Byzantinische Zeitschrift, 2000, 93, p. 435-463.

3184. BÜHL (G.). Die Regelmäßigkeit des Unregelmäßigen. Byzantinische Zeitschrift, 2000, 93, p. 23-36.

3185. Byzantine Macedonia: identity, image and history. Papers from the Melbourne conference July 1995. Ed. By J. BURKE and R. SCOTT. Melbourne, 2000, XV-231 p.

3186. Byzantinische Malerei. Bildprogramme, Ikonographie, Stil. Symposion in Marburg vom 25.–29.6. 1997. Hrsg. v. G. KOCH. Wiesbaden, Reichert, 2000, 456 p.

3187. Centres (Les) proto-urbains russes entre Scandinavie, Byzance et Orient. Ed. par M. KAZANSKI, A. NERCESSIAN et C. ZUCKERMAN. Paris, Lethielleux, 2000, 440 p. [Cf. n° <sélection> 3242.]

3188. COSENTINO (S.). Prosopografia dell'Italia bizantina (493–804). II. G-O. Bologna, Lo scarabeo, 2000, 547 p.

3189. CRISCUOLO (U.). Oriente e Occidente fra tardoantico e bizantino: la questione religiosa. Acme, 2000, 53, p. 3-22.

3190. DEMICHELI (A. M.). L'editto XIII di Giustiniano. In tema di amministrazione e fiscalità dell'Egitto bizantino. Torino, Giappichelli, 2000, 168 p.

3191. EKONOMOU (A. J.). Byzantium on the Palatine: eastern influences on Rome and the Papacy, 590–752 A.D. [S. l.], [s. n.], 2000, 559 p. [Thesis (Phil. Diss.) – Emory University, 2000].

3192. EVANS (J. A. S.). The age of Justinian. The circumstances of imperial power. London a. New York, Routledge, 2000, X-345 p.

3193. FALCONE (Luigi). Tradizione giuridica bizantina e prassi canonica latina nella diocesi di Bisignano. La formazione del patrimonio normativo fra X e XVI secolo. Soveria Mannelli, Rubbettino, 2000, III-232 p.

3194. FATOUROS (G.). Das Grab des Kaisers Manuel I. Komnenos. Byzantinische Zeitschrift, 2000, 93, p. 108-112.

3195. FINE (John V. A.). Croats and Slavs: theories about the historical circumstances of the Croats' appearance in the Balkans. Byzantinische Forschungen, 2000, 26, p. 205-218.

3196. FRYDE (E.). The early Paleologan renaissance (1261c.–1360). Leiden, Brill, 2000, XXI-423 p.

3197. GANCHOU (Th.). La date de la mort du Basileus Jean IV Komnènos de Trebizonde. Byzantinische Zeitschrift, 2000, 93, p. 113-124.

3198. GARZYA (A.). La cultura letteraria bizantina in Italia. Ellenika menymata, 2000, 3, p. 10-26.

3199. GONZÁLEZ BLANCO (Antonino). La función de los columbarios en la antigüedad tardía: ex Oriente lux. Aula Orientalis, 1999-2000, 17-18, p. 381-391.

3200. HARDER (R. E.). Die Aristokratin als Mäzenin und Autorin im Byzanz der Komnenenzeit. In: Frauenwelten in der Antike [Cf. n° 2689], p. 183-197.

3201. HERRIN (Judith). The imperial feminine in Byzantium. Past and present, 2000, 169, p. 3-35.

3202. HINTERBERGER (M.). Autobiography and hagiography in Byzantium. Symbolae Osloenses, 2000, 75, p. 139-164.

3203. HOLO (Joshua). A Genizah letter from Rhodes evidently concerning the Byzantine reconquest of Crete. Journal of near eastern studies, 2000, 59, p. 1-13.

3204. IVISON (Eric). Urban renewal and imperial revival 730–1025. Byzantinische Forschungen, 2000, 26, p. 1-46.

3205. JACOBY (D.). Diplomacy, trade, shipping and espionage between Byzantium and Egypt in the twelfth century. In: Polypleuros nous [Cf. n° 3171], p. 83-102.

3206. KARAMBOULA (D.). Der byzantinische Kaiser als Politiker, Philosoph und Gesetzgeber (Politikos – Philosophos – Nomothetes). Jahrbuch der Österreichischen Byzantinistik, 2000, 50, p. 5-50.

3207. KOLBABA (Tia M.). The Byzantine lists: errors of the Latins. Urbana a. Chicago, University of Illinois Press, 2000, XIII-231 p. (Illinois Medieval Studies).

3208. KOUNTOURA (E.). The origins of the Genesios family. Byzantinische Zeitschrift, 2000, 93, p. 464-473.

3209. KRESTEN (Otto). Die Beziehungen zwischen den Patriarchaten von Konstantinopel und Antiocheia unter Kallistos I. und Philotheos Kokkinos im Spiegel des Patriarchatsregisters von Konstantinopel. Stuttgart, Steiner, 2000, 85 p.

3210. LAIOU (A.). Le débat sur les droits du fisc et les droits régaliens au début du 14e siècle. Revue des études Byzantines, 2000, 58, p. 97-122.

3211. LAMBROPOULOU (A.). Le Péloponnèse occidental à l'époque protobyzantine (IVe–VIIe siècles). Problèmes de géographie historique d'un espace à reconsidérer. In: Byzanz als Raum [Cf. n° 3167], p. 95-113.

3212. LITTLEWOOD (A. R.). Romance, Byzantine. The reign of Manuel I Komnenos (1143–1180) saw the revival of both satire and romance, the latter dormant since the 4th century AD. In: Encyclopedia of Greece and the Hellenic tradition [Cf n° 1939], p. 1459-1462.

3213. MAGDALINO (Paul). The maritime neighborhoods of Constantinople: commercial and residential functions, sixth to twelfth centuries. Dumbarton Oaks Papers, 2000, 54, p. 209-226.

3214. MAHÉ (J. P.). Norme écrite et droit coutumier en Arménie du Ve au XIIIe siècle. Travaux et mémoires, 2000, 13, p. 683-705.

3215. MAKSIMOVIĆ (L.). Byzantinische Herrscherideologie und Regierungsmethoden im Falle Serbien. Ein Beitrag zum Verständnis des byzantinischen Commonwealth. In: Polypleuros nous [Cf. n° 3171], p. 174-192.

3216. MALAMUT (E.). Les reines de Milutin. Byzantinische Zeitschrift, 2000, 93, p. 490-507.

3217. MANGO (Cyril). The triumphal way of Constantinople and the golden gate. Dumbarton Oaks Papers, 2000, 54, p. 173-188.

3. SPECIAL STUDIES

3218. MANIATIS (George C.). The organizational setup and functioning of the fish market in tenth-century Constantinople. *Dumbarton Oaks Papers*, 2000, 54, p. 13-42.

3219. MATSCHKE (K.-P.). Der Fall von Konstantinopel 1453 in den Rechnungsbüchern der genuesischen Staatsschuldenverwaltung. *In*: Polypleuros nous [Cf. n° 3171], p. 204-222.

3220. MAXFIELD (St.). The causes of the Iconoclasm. *Hermeneia*, 2000, p. 7-18.

3221. MAXWELL (Kathleen). Paris, Bibliothèque Nationale de France, Codex Grec 54: modus operandi of scribes and artists in a Palaiologan gospel book. *Dumbarton Oaks Papers*, 2000, 54, p. 117-138.

3222. MELLUS (M.). La schiavitù nell'età giustinianea. Besançon, Presses universitaires Franc-Comtoises, 2000, 339 p.

3223. MEUNIER (F.). La rhétorique dans les romans byzantins du XIIe siècle: besogne ou plaisir? *Erytheia*, 2000, 21, p. 51-71.

3224. MEYER (C.). Bir Umm Fawakhir survey project 1993: a Byzantine gold-mining town in Egypt. Chicago, Oriental institute of the University of Chicago, 2000, XVII, 92 p.

3225. MUNDELL MANGO (Marlia). The commercial map of Constantinople. *Dumbarton Oaks Papers*, 2000, 54, p. 189-208.

3226. NARDI (E.). Donne a Bisanzio. Nuove prospettive storiografiche. *Quaderni medievali*, 2000, 49, p. 44-61.

3227. NECIPOĞLU (N.). Constantinopolitan merchants and the question of their attitudes towards Italians and Ottomans in the late Palaiologan period. *In*: Polypleuros nous [Cf. n° 3171], p. 251-263.

3228. NEVILLE (Leonora). The Marcian treatise on taxation and the nature of bureaucracy in Byzantium. *Byzantinische Forschungen*, 2000, 26, p. 47-62.

3229. NOETHLICHS (K. L.). Quid possit antiquitas nostris legibus abrogare? Politische Propaganda und praktische Politik bei Justinian I. im Lichte der kaiserlichen Gesetzgebung und der antiken Historiographie. *Zeitschrift für antikes Christentum*, 2000, 4, p. 116-132.

3230. ORIGONE (S.). Liguria bizantina. *In*: Polypleuros nous [Cf. n° 3171], p. 538-643, 272-289.

3231. PAPASOTERIOU (Ch.). Βυζαντινή υψηλή στρατηγική: 6^{ος}- 11^{ος} αιώνα. Athen, Ποιότητα, 2000, 323 p.

3232. PATLAGEAN (E.). Les États d'Europe centrale et Byzance ou l'oscillation des confins. *Revue historique*, 2000, 302-4, p. 827-868.

3233. PITSAKIS (C. G.). Législation et stratégies matrimoniales. Parenté et empêchements de mariage dans le droit byzantin. *L'Homme*, 2000, 154-155, p. 677-696.

3234. POPOVA (O. S.). Vizantijskoe iskusstvo v Italii: Mozaiki Torchello. (Byzantine art in Italy: the mosaics of Torcello). *In*: Vizantijskij vremennik = BYZANTINA XPONIKA [Cf. n° 3161], p. 152-165.

3235. Protobyzantine Eleutherna, vol. I. Ed. by P. G. THEMELES. Rhethymnon, University of Crete, 2000, 319 p. (ill.).

3236. Roman (Der) im Byzanz der Komnenenzeit. Referate des Internationalen Symposiums an der Freien Universität Berlin, 3. bis 6. April 1998. Hrsg. v. P. A. AGAPITOS und D. R. REINSCH. Frankfurt am Main, Beerenverlag, 2000, XI-146 p.

3237. RONCHEY (Silvia). Bisanzio veramente «volle cadere»? Realismo politico e avventura storica da Alessio I Comneno al Mediterraneo di Braudel. *In*: Realismo politico (Sul) [Cf. n° 6186], p. 137-158. – EADEM. Un'alleanza dinastica per rifondare Bisanzio. *Byzantinische Zeitschrift*, 2000, 93, p. 521-567.

3238. SAVVIDES (A. G. C.). Can we refer to a concerted action among Rapsomates, Caryces and the Emir Tzachas between A. D. 1091 and 1093? *Byzantion*, 2000, 70, p. 122-134.

3239. SIGNES CODONER (J.). Bizancio y sus circunstancias: la evolución de la ideologia imperial en contacto con las culturas de su entorno. *Minerva*, 2000, 14, p. 129-175.

3240. SKINNER (Alexander). The birth of a 'Byzantine' senatorial perspective. *Arethusa*, 2000, 33, p. 363-377.

3241. SKRŽINSKAJA (E. Ch.). Rus', Italija i Bizantija v Srednevekov'e. (Russia, Italy and Byzantium in the Middle Ages). Sankt-Peterburg, Aleteja, 2000, 283 p.

3242. SORLIN (I.). Voies commerciales, villes et peuplement de la Kôsia au Xe siècle d'après le De Administrando Imperio de Constantin Porphyrogénète. *In*: Centres (Les) proto-urbains russes entre Scandinavie, Byzance et Orient [Cf. n° 3187], p. 337-364.

3243. SPECK (P.). Die Frauen und Söhne Konstantins V. *Byzantinische Zeitschrift*, 2000, 93, p. 568-585.

3244. STEPHENSON (P.). Byzantium's Balkan frontier. A political study of the northern Balkans, 900–1204. Cambridge, Cambridge U. P., 2000, XII-353 p.

3245. STICHEL (R. H. W.). Ein byzantinischer Kaiser als Sensenmann. *Byzantinische Zeitschrift*, 2000, 93, p. 586-608.

3246. Strangers to themselves: the Byzantine outsider. Papers from the thirty-second Spring Symposium of Byzantine Studies. University of Sussex, Brighton, March 1998. Ed. by D. C. SMYTHE. Aldershot, Ashgate, 2000, X-269 p.

3247. SVENSHON (Helge), STICHEL (Rudolf H. W.). Neue Beobachtungen an der ehemaligen Kirche der

Heiligen Sergios und Bakchos (Küçük Ayasofya Camisi) in Istanbul. *Instanbuler Mitteilungen*, 2000, 50, p. 389-409.

3248. TINNEFELD (F.). Zur Entstehung von Briefsammlungen in der Palaiologenzeit. *In*: Polypleuros nous [Cf. n° 3171], p. 365-381.

3249. TOKO (H.). John Oxeites' criticism of Charistike. An aspect of monastic reform in Byzantium, 11th century. *Kirisutokyo-Shigaku*, 2000, 54, p. 80-92.

3250. Traditions (Les) apocalyptiques au tournant de la chute de Constantinople. Actes de la table ronde d'Instanbul, 13–14 avril 1996. Ed. par B. LELLOUCH et S. YERASIMOS. Paris, L'harmattan, 2000, 192 p.

3251. VAKALOUDI (Anastasia). Religion and magic in Syria and Wider Orient in the early Byzantine period. *Byzantinische Forschungen*, 2000, 26, p. 255-280.

3252. VAN TRICHT (F.). «La gloire de l'Empire». L'idée impériale de Henri de Flandre-Hainaut, deuxième empereur latin de Constantinople (1206–1216). *Byzantion*, 2000, 70, p. 211-241.

3253. WRIGHT (G. R. H.). Some Byzantine bronze objects from Beycesultan. *Anatolian studies*, 2000, 50, p. 159-170.

3254. ZUCKERMAN (C.). Le voyage d'Olga et la première ambassade espagnole à Constantinople en 946. *Travaux et mémoires*, 2000, 13, 647-672.

Cf. n°s 299, 377, 1282, 8792

I

HISTORY OF THE MIDDLE AGES

§ 1. Sources and criticism of sources (*a*. Non-literary sources; *b*. Literary sources). 3255-3318. – § 2. General works. 3319-3361. – § 3. Political history (*a*. General; *b*. 476–900; *c*. 900–1300; *d*. 1300–1500). 3362-3438. – § 4. Jews. 3439-3452. – § 5. Islam. 3453-3474. – § 6. Vikings. 3475-3484. – § 7. History of law and institutions. 3485-3528. – § 8. Economic and social history. 3529-3619. – § 9. History of civilization, literature, technology and education (*a*. Civilization; *b*. Literature; *c*. Technology; *d*. Education). 3620-3771. – § 10. History of art. 3772-3823. – § 11. History of music. 3824-3845. – § 12. History of philosophy, theology and science (*a*. Sources; *b*. Studies). 3846-3893. – § 13. History of the Church and religion (*a*. General; *b*. History of the Popes; *c*. Monastic history; *d*. Hagiography; *e*. Special studies). 3894-4003. – § 14. Settlements. Place names. Town planning. 4004-4034.

§ 1. Sources and criticism of sources.

a. Non-literary sources

* 3255. KOCH (Walter), GLASER (Maria), BORNSCHLEGEL (Franz-Albrecht). Literaturbericht zur mittelalterlichen und neuzeitlichen Epigraphik (1992–1997). Hannover, Hahnsche Buchhandlung, 2000, 765 p. (M.G.H., Hilfsmittel, 19).

3256. 294 numaralı Hınıs Livası mufassal tahrir defteri (963/1556): dizin, transkribe ve tıpkıbasım. (Detailed tax register of the Hınıs district (no. 294): index, transcription, and facsimile). Project director, Yusuf SARINAY; Project executive, Necati AKTAŞ, Necati GÜLTEPE, Mustafa KAPLAN; ed. by Ahmet ÖZKİLİNÇ [et al.]. Ankara, Devlet Arşivleri Genel Müdürlüğü Osmanlı Arşivi Daire Başkanlığı [General Directorate of State Archives], 2000, 90 p.

3257. 373 numaralı 'Ayntab Livası mufassal tahrir defteri (950/1543): dizin, transkribe ve tıpkıbasım. (Tax register of the Ayıntab district, no. 373 dated 950/1543: index, transcription, and facsimile). Project director Yusuf SARINAY; Project executive Necati AKTAŞ, Necati GÜLTEPE, Mustafa KAPLAN; ed. by Ahmet ÖZKİLİNÇ [et al.]. Ankara, Devlet Arşivleri Genel Müdürlüğü Osmanlı Arşivi Daire Başkanlığı, [General Directorate of State Archives] 2000, 90 p.

3258. Acta (The) of Hugh of Wells, bishop of Lincoln, 1209–1235. Ed. by David M. SMITH. Woodbridge a. Rochester, Boydell and Brewer, for the Lincoln Record Society, 2000, LIII-256 p. (tables). (Lincoln Record Society, 88).

3259. ALBERTUS (Bohemus). Das Brief- und Memorialbuch des Albert Behaim. Hrsg. v. Thomas FRENZ und Peter HERDE. München, Monumenta Germaniae Historica, 2000, XVI-655 p. (maps).

3260. Carteggio degli oratori mantovani alla corte sforzesca (1450–1500). 3. 1461. A cura di Isabella LAZZARINI. Roma, Ministero per i Beni e le Attività Culturali, Ufficio Centrale per i Beni Archivistici, 2000, 470 p.

3261. Catàleg dels Protocols notarials dels antics districtes de Falset i Gandesa. Ed. por Roser PUIG I TARRECH. Barcelona, Fundació Noguera, 2000, 202 p. (Inventaris d'arxius notarials de Catalunya, 22).

3262. Chartrier (Le) de l'abbaye prémontrée de Saint-Yved de Braine (1134–1250). Ed. par Olivier GUYOTJEANNIN [et al.]. Paris, Ecole des Chartes, 2000, 455 p. (plates, ill., table, maps).

3263. Charts of Abingdon Abbey, 1. Ed. by K E. KELLY. Oxford a. New York, Oxford U. P., for the British Academy, 2000, XXXXXI-205 p. (ill., maps).

3264. Colección diplomática de Santa María la Mayor de Calatayud. Ed. por Herminio LAFOZ RABAZA. Zaragoza, Institución Fernando el Católico, 2000, 330 p. (Fuentes históricas aragonesas, 32).

3265. Colección diplomática medieval de la Orden de Alcántara (1157?–1494). Bonifacio PALACIOS MARTÍN y Carlos de AYALA MARTÍNEZ [et al.]. Madrid, Editorial Complutense, 2000, [s. p.].

3266. Colección documental del monasterio de Santa Maria del Carbajal (1093–1461). Ed. por Santiago DOMÍNGUEZ SÁNCHEZ. León, Centro de Estudios e Investigación "San Isidoro", Caja España de Inversiones y Archivo Histórico Diocesano, 2000, 510 p. (Fuentes y estudios de historia leonesa, 87).

3267. Congresso nazionale di archeologia medievale: Musei civici, Chiesa di Santa Giulia, Brescia, 28 settembre–1 ottobre 2000. A cura di Gian Pietro BROGIOLO. Firenze, All'insegna del giglio, 2000, 489 p. (ill.).

3268. COSTA (Maria). Parchemins valdôtains du Moyen Âge (XIIe–XVe siècles). Pergamene medievali della Valle d'Aosta (secc. XII–XV). Aosta, Imprimerie Valdôtaine, 2000, 99 p. (ill.).

3269. Deutschen (Die) Königspfalzen. T. 2, Thüringen. Hrsg. v. Michael GOCKEL. Göttingen, Vandenhoeck & Ruprecht, 2000, XXI-295 p.

3270. Documents (The) of Angelo de Cartura and Donato Fontanella: Venetian notaries in fourteenth-century Crete. Ed. by Alan M. STAHL. Washington, Dumbarton Oaks Research Library and Collection, 2000, XXI-295 p.

3271. Documents per l'història de la cultura catalana mig-eval. Ed. por Antoni RUBIO Y LLUCH. Barcelona, Institut d'Estudis Catalans, 2000, 2 vol., s. p.

3272. DUBY (Georges). Recueil des pancartes de l'abbaye de La Ferté-sur-Grosne, 1113–1178. Bruxelles, De Boeck Université, 2000, 260 p. (table, maps).

3273. English Episcopal Acta, 20: York, 1154–1181. Ed. by Marie LOVATT. Oxford, Oxford U. P., for the British Academy, 2000, LXIX-224 p. (plates).

3274. Enquête à Chypre au XVe siècle. Le «sindicamentum» de Napoleone Lomellini, capitaine gênois de Famagouste (1459). Ed. par Catherine OTTEN-FROUX. Nicosie, Centre de recherche scientifique, 2000, 310 p. (tables, ill.).

3275. Estatutos y Actos municipales de Jaca y sus montañas. Ed. por Manuel GÓMEZ DE VALENZUELA. Zaragoza, Institución Fernando el Católico, 2000, 538 p. (Fuentes históricas aragonesas, 33).

3276. GRILLON (Louis), REVIRIEGO (Bernard). Le cartulaire de l'abbaye Notre-Dame de Chancelade. Périgueux, Archives départementales de la Dordogne, 2000, 279 p. (ill.). (Archives en Dordogne. Études et documents, 2).

3277. Het Goudse hofstedengeldregister van ca. 1397 en andere bronnen voor de vroege stadsontwikkeling van Gauda. Hrsg. v. K. GOUDRIAN, B. J. IBELINGS, J. C. VISSER und A. W. HASSELINK. Hilaversum, Historische Vereniging Holland u. Verloren, 2000, XCII-99 p. (ill., tables, maps).

3278. Izvori za srednovekovnata istoriia na Bulgariia (VII–XV v.) v avstriiskite rukopisni sbirki i arkhivi. T. 2. Italianski, latinski i nemski izvori. [Sources sur l'histoire médiévale de la Bulgarie (VIIe–XVe siècles) dans les archives manuscrits autrichiennes. T. 2. Sources italiennes, latines et allemandes]. Sustavitel Vasil GIUZELEV. Sofiia, Univ. izdatelstvo "Sv. Kliment Okhridski", 2000, 271 p. (Arkhivite govoriat, 9).

3279. JANIN (Valentin L.), ZALIZNJAK (Anatolij A.). Novgorodskie gramoty na bereste: Iz raskopok 1990–1996 gg. (The Birch-Bark documents of Novgorod: from the excavations of 1984–1989 [and new comments to some of the documents published earlier]). RAN, Otd. istorii. Moskva, Russkie slovari, 2000, 430 p. (ill.). [Cf. n^{o} <choice> 3279.] – ZALIZNJAK (Anatolij A.). Paleografija berestjanykh gramot i ikh vnestratigraficheskoe datirovanie. (Palaeography of Birch-Bark Documents and Methods of Their Non-Stratigraphic Dating). *In*: Novgorodskie gramoty na bereste [Cf. n° 3279], p. 133-429.

3280. Liber Officii provvisionis Romanie: Genova, 1424–1428. A cura di Laura BALLETTO. Genova, Università degli studi di Genova, 2000, LXXVI-515 p. (Collana di fonti e studi, 6).

3281. Libri (I) iurium della Repubblica di Genova, 1/6. A cura di Eleonora PALLAVICINO. Roma, Ministero per i Beni e le attività Culturali, Ufficio Centrale per i Beni Archivistici, 2000, XLVI-559 p. (Pubblicazioni degli Archivi di Stato, Fonti, 32; Fonti per la storia della Liguria, 13).

3282. Libros de apeos del Monasterio de Piedra (1344). Libro de cuentas de la Bolsería del Monasterio de Piedra (1307–1348). Ed. por Concepción DE LA FUENTE COBOS. Zaragoza, Institución Fernando el Católico, 2000, 281 p. (Fuentes históricas aragonesas, 31).

3283. Lletres reials a la ciutat de Girona: 1293–1515. Ed. por M. Josepa ARNALL I JUAN. Barcelona, Fundació Noguera; Girona, Ajuntament de Girona, 2000, 1021 p. (Diplomataris, 22-23).

3284. Llivre (El) Verd de la ciutat de Girona 1144–1533. Ed. por Christian GUILLERÉ, Barcelona, Fundació Noguera; Girona, Ajuntament, 2000, 737 p.

3285. MEDYNTSEVA (Al'bina A.). Gramotnost' v Drevnej Rusi: Po pamjatnikam epigrafiki X–pervoj poloviny XIII veka. (Literacy in Old Rus': inscriptions of the 10th–the 1st half of 13th century). RAN, In-t arkheologii. Moskva, Nauka, 2000, 290 p.

3286. NORTIER (Michel). Documents normands du règne de Charles V (8 avril 1364–16 septembre 1380) et complément pour le règne de Jean II le Bon conservés au Département des manuscrits. Paris, Bibliothèque nationale de France. Rouen, Société de l'Histoire de Normandie, 2000, 430 p.

3287. PLETNEVA (Svetlana A.). Ocherki khazarskoj arkheologii. (Essays on the Archeology of the Khazars). Moskva a. Jerusalem, [s. n.], 2000, 365 p. (ill.).

3288. Plus (Les) anciens documents originaux de l'abbaye de Cluny. 2: documents n. 31 à 60: Paris, Bibliothèque nationale de France, Collection de Bourgogne, vol. 77, nos 33 à 61. Ed. par Hartmut ATSMA, Sébastien BARRET et Jean VEZIN. Turnhout, Brepols, 2000, 158 p. (Monumenta Paleographica Medii Aevi, Series Gallica).

3289. Port (The) of medieval Dublin: archaeological excavations at the civic offices, Winetaven Street, Dublin, 1993. Ed. by Andrew HALPIN [et al.]. Dublin a.

Portland, Four Courts Press, 2000, 189 p. (plates, ill., tables).

3290. "Regesta Imperii" (Die) im Fortschreiten und Fortschritt. Hrsg. v. Harald ZIMMERMANN. Köln, Weimar u. Vienne, Böhlau, 2000, VI-158 p.

3291. Regesten Kaiser Friedrichs III. T. 14, Die Urkunden und Briefe aus Archiven und Bibliotheken der Stadt Nürnberg, 1^e part., 1440–1449. Hrsg. v. Dieter RÜBSAMEN. Wien, Weimar u. Köln, Böhlau, 2000, 371 p.

3292. Register (The) of William Bateman, Bishop of Norwich, 1344–1355, 2. Ed. by Phyllis E. POBST. Woodbridge a. Rochester, Boydell and Brewer, 2000, VIII-248 p.

3293. Registros de la Casa de Francia. Ed. por Juan CARRASCO y Pascual TAMBURRI. Pamplona, Departemento de Educación y Cultura, 2000, [s. p.].

3294. ROFFE (David). Domesday: the inquest and the book. Oxford a. New York, Oxford U. P., 2000, XIX-282 p. (tables).

3295. Sermon (The). Ed. by Beverly Mayne KIENZLE. Turnhout, Brepols, 2000, 998 p. (Typologie des sources du moyen âge occidental, 81-83).

3296. SYMEON OF DURHAM. Libellus de exordio atque procursu istius, hoc est Dunhelmensis, ecclesie: tract on the origins and progress of this the church of Durham. Ed. by David ROLLASON. Oxford, Clarendon Press, 2000, XCVI-353 p.

Cf. n°s 1-42

b. Literary sources

** 3297. Kronika anonymného notára kráľa Bela. Gesta Hungarorum. (The chronicle of Anonymous Notary of the King Bel). Ed. Vincent MÚCSKA. Budmerice, Vydavateľstvo Rak v spolupráci s nadáciou Osudy predkov, 2000, 159 p.

3298. ANDERSSON (Theodore M.), GADE (Kari Ellen). Morkinskinna: the earliest Icelandic chronicle of the Norwegian kings (1030–1157). Ithaca a. London, Cornell U. P., 2000, XV-556 p. (maps, ill.). (Islandica, 51).

3299. ANDREA (Alfred J.), WHALEN (Brett E.). Contemporary sources for the fourth crusade. Leiden, Boston a. Köln, Brill, 2000, XII-330 p.

3300. Anglo-Saxon (The) Chronicle: a collaborative edition, 8: MS F. Ed. by Peter S. BAKER. Woodbridge a. Rochester, Boydell and Brewer, 2000, CXIII-158 p. (ill.).

3301. BERENGUER DE PUIGPARDINES. Sumari d'Espanya. A cura de Joan IBORRA. València, Universitat de València, 2000, 266 p. (Fonts històriques valencianes, 3).

3302. BOJTSOV (Mikhail A.). Pogrebenie imperatora Fridrikha III v 1493 godu. (The burial of the Emperor Friedrich III in 1493: Edited from MS. Bayerisches Haupstaatsarchiv München. Fürstensachen 296, with Transl. and Comment.). Srednie veka, Moskva, 2000, 61, p. 254-288. [Text in German; intr., transl. and comment. in Russian]

3303. Chronicon Sanctae Sophiae (cod. Vat. Lat. 4939). Ed. par Jean-Marie MARTIN. Roma, Istituto storico italiano per il Medio Evo, 2000, 2 vol., 897 p. (ill.). (Fonti per la storia dell'Italia medievale-Rerum italicarum scriptores, 3).

3304. Cronaca del Templare di Tiro (1243–1314): La caduta degli Stati Crociati nel racconto di un testimone oculare. A cura di Laura MINERVINI. Napoli, Liguori, 2000, X-490 p. (maps).

3305. GIPPIUS (Aleksej A.). "Povest' vremennykh let": O vozmozhnom proiskhozhdenii i znachenii nazvanija. ("Povest' vremennykh let": on the possible origin and meaning of the title [translated as 'Tale of times and years']). In: Iz istorii russkoj kul'tury (From the history of Russian culture). Vol. 1. Drevnjaja Rus' (Old Rus'). Moskva, Jazyki russkoj kul'tury, 2000, p. 448-460.

3306. GUTOT-BACHY (Isabelle). Le "Memoriale historicarum" de Jean de Saint-Victor. Un historien et sa communauté au début du XIVe siècle. Turnhout, Brepols, 2000, 608 p. (plates, ill., tables, maps). (Bibliotheca Victorina, 12).

3307. JACKSON (Tat'jana N.). Islandskie korolevskie sagi o Vostochnoj Evrope (Seredina XI–seredina XIII v.): Texty, perevod, commentarij. [Icelandic kings' sagas on Eastern Europe (the 11th–13th centuries): texts, translations, comments]. RAN, In-t vseobshchej istorii. Moskva, Ladomir, 2000, 364 p. (Drevneyshie istochniki po istorii narodov Vostochnoy Evropy). [Eng. summary] – EADEM. Drevnjaja Rus' glazami srednevekovykh islandtsev. (Old Rus' through the eyes of Medieval Icelanders). Lewinston, Queenston a. Lampeter, Edwin Mellen Press, 2000, XX-389 p. (ill., maps). (Rossijskie issledovanija po mirovoj istorii i kul'ture, 3). [Eng. summary and table of contents].

3308. KEY (Newton E.), WARD (Joseph P.). 'Divided into parties': exclusion crisis origins in Monmouth. English historical review, 2000, 115, 464, p. 1159-1183.

3309. MAC KITTERICK (Rosamond). The illusion of royal power in the Carolingian annals. English historical review, 2000, 115, 460, p. 1-20.

3310. MANDEVILLE (Jean de). Le livre des merveilles du monde. Ed. par Christian DELUZ. Paris, CNRS, 2000, 528 p. (ill, tables, graph., maps). (Sources d'histoire médiévale).

3311. Monumenta Arroasiensia. Ed. par Benoît-Michel TOCK et Ludo MILLIS. Turnhout, Brepols, 2000, LXV-811 p.

3312. Polnoe sobranie russkikh letopisej. (The complete edition of the Russian Chronicles). Vol. 6. Part 1. Sofijskaya pervaja letopis' starshego izvoda (The 1st Sophian chronicle: the elder version). Text prepared by S. N. KISTEREV, L. A. TIMOSHINA; intr. by B. M. KLOSS; Indices by L. A. TIMOSHINA. RAN, In-t rossiyskoj istorii; Federal'naja arkhivnaja sluzhba, Ros. gos. arkhiv drevnikh aktov. Moskva, Jazyki russkoj kul'tury, 2000, 581 p.

3313. RICHER VON SAINT-REMI. Historiae. Ed. par Hartmut HOFFMANN. Hanover, Hahnsche Buchhandlung, 2000, V-433 p. (M.G.H., Scriptores, 38).

3314. SENDEROVICH (S. Ja.). Metod Shakhmatova, rannee russkoe letopisanie i problema nachala russkoj istoriografii. (Aleksey Shakhmatov's method, early Russian chronicle-writing and the problem of the beginning of the Russian historiography). *In*: Iz istorii russkoj kul'tury (From the history of Russian culture). Vol. 1. Drevnjaja Rus' (Old Rus'). Moskva, Jazyki russkoj kul'tury, 2000, p. 461-499.

3315. SNORRI STURLUSON. Histoire des rois de Norvège Heimskringla, 1. Des origines mythiques de la dynastie à la bataille de Svold. Paris, Gallimard, 2000, 706 p. (ill., maps).

3316. TANZINI (Lorenzo). Le due redazioni del «Liber de origine civitatis Florentie et eiusdem famosis civibus». Osservazioni sulla recente edizione. *Archivio storico italiano*, 2000, 158, 583, p. 141-160.

3317. TOLSTOJ (N. I.). Etnicheskoe samopoznanie i samosoznanie Nestora Letopistsa, avtora "Povesti vremennykh let". (Ethnic self-cognition and self-perception of Nestor the Chronicler, the author of the [Old Russian] "Povest' vremennykh let"). *In*: Iz istorii russkoj kul'tury (From the history of Russian culture). Vol. 1. Drevnjaja Rus' (Old Rus'). Moskva, Jazyki russkoj kul'tury, 2000, p. 441-447.

3318. World (The) of El Cid: Chronicles of the Spanish reconquest. Ed. by Simon BARTON and Richard FLETCHER. Manchester a. New York, Manchester U. P., 2000, XIV-281 p. (map).

Cf. n°s 1-42, 268, 706, 769, 3285, 3333, 3334, 5066, 8489

§ 2. General works.

* 3319. Bibliographie annuelle du Moyen-Âge tardif. Auteurs et textes latin. T. 10. [T. 9. Cf. Bibl. 99, n° 3184.] Rassemblée et compilée à la section latine de l'Institut de recherche et d'histoire des textes. Coordination et rédaction de Jean-Pierre ROTHSCHILD, co-rédaction Frédéric DUVAL, avec la collaboration de Pascale BARMON, Christine GEORGELIN et Patrice SIARD. Paris et Turnhout, Brepols, 2000, IX-580 p.

* 3320. International Medieval Bibliography (450–1500). T. 33. Part 1. January–June 1999. Part 2. July–December 1999. [T. 32, part 1 and 2. Cf. Bibl. 99, n° 3185.] Ed. by Alan MURRAY. Leeds, International Medieval Institute, University of Leeds a. Turnhout, Brepols, 2000, 2 vol., XLVIII-476 p., XLIX-479 p.

* 3321. Medioevo latino. Bollettino bibliografico della cultura europea da Boezio a Erasmo (secc. VI–XV).

Vol. 21. [Vol. 20. Cf. Bibl. 99, n° 3186.] A cura di Claudio LEONARDI e Lucia PINELLI. Firenze, SISMEL – Edizioni del Galluzzo, 2000, XXXVII-1284 p.

* 3322. OSHCHEPKOVA (M. M.). Srednie veka i rannee novoe vremja: Ukazatel' literatury, opublikovannoj v Rossii v 1996 g. (Middle Ages and early modern time: bibliography of works published in Russia in 1996). *Bjulleten' Vserossijskoj assotsiatsii medievistov i istorikov rannego novogo vremeni*, 2000, 1, p. 77-88, p. 92-165.

3323. Aetas media aetas moderna. Studia ofiarowane profesorowi Henrykowi Samsonowiczowi w osiemdziesiątą rocznicę urodzin. (Aetas media aetas moderna. Professor Henryk Samsonowicz gewidmete Studien anlässlich des 80. Geburtstages). Hrsg. v. Halina MANIKOWSKA, Agnieszka BARTOSZEWICZ und Wojciech FAŁOWSKI. Warszawa, Inst. Historii Uniw. Warszawskiego, 2000, 732 p.

3324. Area (Un') di strada: l'Emilia occidentale nel Medioevo. Ricerche storiche e riflessioni metodologiche. Atti dei convegni di Parma e Castell'Arcuato, novembre 1997. A cura di Roberto GRECI. Bologna, CLUEB, 2000, VIII-373 p.

3325. BARBERO (Alessandro). Valle d'Aosta medievale. Napoli, Liguori, 2000, 271 p. (graph.). (Bibliothèque d'Archivum Augustanum, 27).

3326. BARRELL (A. D. M.). Medieval Scotland. Cambridge, Cambridge U. P., 2000, XIII-296 p. (ill., maps).

3327. BENITO RUANO (Eloy). Tópicos y realidades de la Edad media (I). Madrid, Real Academia de la Historia, 2000, 301 p. (ill.). (Estudios, 2).

3328. BOOCKMANN (Hartmut). Wege ins Mittelalter: historische Aufsätze. München, C. H. Beck, 2000, XI-481 p. (ill.).

3329. CASSAGNES-BROUQUET (Sophie). Histoire de l'Angleterre médiévale. Gap et Paris, Ophrys, 2000, 301 p. (ill., maps, tables). (Synthèse et Histoire).

3330. CHIBNALL (Marjorie). Piety, power and history in medieval England and Normandy. Aldershot, Ashgate, XIV-336 p. (ill., maps).

3331. Courts and regions in medieval Europe. Ed. by Sarah REES JONES, Richard MARKS and A. J. MINNIS. York, York Medieval Press, in association with Boydell and Brewer and the Centre for Medieval Studies, University of York, 2000, XI-226 p. (ill., plates, tables).

3332. DAILEADER (Philip). True citizens. Violence, memory and identity in medieval Perpignan, 1162–1397. Leiden, Boston a. Köln, Brill, 2000, XIV-280 p. (ill.).

3333. DANILEVSKIJ (Igor' N.). Russkie zemli glazami sovremennikov i potomkov (XII–XIV vv.): Kurs lektsij. (Russian Lands through the eyes of contemporaries and descendants, 12th–14th centuries: a lecture course). Moskva, Aspekt-press, 2000, 389 p. [Cf. Bibl. 98, n° 3330.]

3334. Drevnejshie gosudarstva Vostochnoj Evropy (The earliest states of Eastern Europe), 1998 g.: Pamjati A. P. Novosel'tseva. (Articles devoted to the memory of Anatoly P. Novoseltsev); Rus', Kavkaz i Zakavkaz'e v vostochnykh istochnikakh (Rus', Caucasus and Transcaucasia in Eastern sources: A. P. Novosel'tsev's selected papers on early history of Eastern Europe, p. 264-477). Ed. Tat'jana M. KALININA. RAN, In-t vseobshchej istorii, In-t rossijskoj istorii. Moskva, Vostochnaja literatura, 2000, 494 p.

3335. Drugie srednie veka: K 75-letiju A. Ja. Gurevicha. (Other Middle Ages: articles devoted to the 75th birthday of Aron Ya. Gurevich). Ed. I. V. DUBROVSKIJ, S. V. OBOLENSKAJA, M. Ju. PARAMONOVA. Moskva a. Sankt-Peterburg, Universitetskaja kniga, 2000, 464 p.

3336. Europas mittle um 1000. Hrsg. v. Alfried WIECZOREK und Hans-Martin HINZ. Stuttgart, Theiss, 2000, [s.p.].

3337. Evrazijskaja step' i lesostep' v epokhu rannego srednevekov'ja: Sb., posvjashch. 60-letiju A.V. Vinnikova. (The steppe and the forest-steppe of Eurasia in the Early Middle Ages: articles dedicated to the 60th anniversary of A. V. Vinnikov). Ed. A.D. PRJAKHIN. Voronezh, Izd-vo Voronezhskogo gos. un-ta, 2000, 152 p. (ill.). (Arkheologija Vostochnoevropejskoj Lesostepi, 14).

3338. Extraordinary women of the medieval and renaissance world: a biographical dictionary. Ed. by Carole LEVIN. Westport a. London, Greenwood Press, 2000, XXI-329 p. (ill.).

3339. GARBARINO (Osvaldo). Monaci, milites e coloni: materiali scritti e costruiti per una storia del Tigullio altomedioevale. Genova, De Ferrari, 2000, 316 p. (ill.). (Sestante).

3340. GUIDETTI (Massimo). Il Mediterraneo e la formazione dei popoli europei: V–X secolo. Milano, Jaca book, 2000, 160 p. (EDM: un'enciclopedia del Mediterraneo).

3341. Italie (L') au Moyen Age: V^e–XVe siècle. Ed. par Jean-Pierre DELUMEAU et Isabelle HEULLANT-DONAT. Paris, Hachette, 2000, 319 p. (ill.). (Carré histoire).

3342. KONOVALOVA (Irina G.), PERKHAVKO (Valerij B.). Drevnjaja Rus' i Nizhnee Podunav'je. (Old Rus' and the Lower Danube region). RAN, In-t vseobshchej istorii, In-t rossijskoj istorii. Moskva, Pamjatniki istoricheskoj mysli, 2000, 272 p. (ill., maps). [Table of contents in English].

3343. KURTÉN-LINDBERG (Birgitta). Medeltida människor. (Les hommes du Moyen Age). Lund, Historiska media, 2000, 231 p.

3344. MAC KEE (Sally). Uncommon dominion: Venetian Crete and the myth of ethnic purity. Philadelphia, University of Pennsylvania Press, 2000, XIII-273 p. (tables).

3345. Manipulus florum: aus Mittelalter, Landesgeschichte, Literatur und Historiographie. Festschrift für Peter Johanek zum 60. Geburtstag. Hrsg. v. Ellen WIDDER, Mark MERSIOWSKY und Maria-Theresia LEUKER. Münster, Waxmann, 2000, 395 p. (ill.).

3346. Medieval (The) state: essays presented to James Campbell. Ed. by J. R. MADDICOTT and D. M. PALLISER. London a. Rio Grande, Hambledon Press, 2000, XLII-262 p. (ill.).

3347. Medioevo Mezzogiorno Mediterraneo: studi in onore di Mario Del Treppo. A cura di Gabriella ROSSETTI e Giovanni VITOLO. Pisa, GISEM, 2000, 2 vol., XXXII-363 p., 375 p.

3348. Monferrato (Il): crocevia politico, economico e culturale tra Mediterraneo e Europa: atti del Convegno internazionale, Ponzone, 9–12 giugno 1998. A cura di Gigliola SOLDI RONDININI. Ponzone, 2000, 352 p. (ill.). (Collana di fonti e studi).

3349. Nihil superfluum esse. Prace z dziejów średniowiecza ofiarowane Profesor Jadwidze Krzyżaniakowej. (Nihil superfluum esse. Professorin Jadwiga Krzyżaniakowa gewidmete Arbeiten zur Geschichte des Mittelalters). Hrsg. v. Jerzy STRZELCZYK und Józef DOBOSZ. Poznań, Wydaw. Inst. Historii Uniw. A. Mickiewicza, 2000, 596 p. (Publikacje Inst. Historii Uniw. A. Mickiewicza, 33).

3350. Normandie (La) vers l'an Mil. Études et documents. Ed. par Fr. DE BEAUREPAIRE et J. P. CHALINE. Rouen, Societé de l'Histoire de Normandie, 2000, 222 p. (maps, plans, ill.).

3351. OCCHIPINTI (Elisa). L' Italia dei comuni: secoli XI–XIII. Roma, Carocci, 2000, 159 p. (Studi superiori).

3352. Occident et Orient, IXe–XVe siècles. Histoire et Archéologie. Actes du colloque d'Amiens, 8, 9 et 10 octobre 1998. Amiens, CAHMER, 2000, 286 p. (tables).

3353. Peace, negotiation and reciprocity: strategies for coexistence in the Middle Ages and Renaissance. Ed. by Diane WOLFTHAL. Turnhout, Brepols, 2000, XXX-265 p.

3354. Penser le pouvoir au Moyen Âge (VIIIe–XVe siècle). Études d'histoire et de littérature offertes à François Autrand. Ed. par Dominique BOUTET et Jacques VERGER. Paris, Rue d'Ulm, 2000, 443 p. (ill., tables).

3355. PETRUKHIN (Vladimir Ja.). Drevnjaja Rus': Narod. Knjaz'ja. Religija. (Old Rus': the people, princes and religion). *In*: Iz istorii russkoj kul'tury (From the History of Russian Culture). Vol. 1. Drevnjaja Rus' (Old Rus'). Moskva, Jazyki russkoj kul'tury, 2000, p. 11-410.

3356. Polska i jej sąsiedzi w późnym średniowieczu. (Polen und Anrainerstaaten im späten Mittelalter). Hrsg. v. Krzysztof OŻÓG und Stanisław SZCZUR. Kraków, Societas Vistulana, 2000, 431 p.

3357. Postcolonial (The) Middle Ages. Ed. by Jeffrey Jerome COHEN. New York, St. Martin's Press, 2000, VII-286 p. (ill.).

3358. Prestige, authority and power in late medieval manuscripts and texts. Ed. by Felicity RIDDY. York, York Medieval Press, 2000, VIII-199 p. (ill.).

3359. Reich, Regionen und Europa in Mittelalter und Neuzeit. Festschrift für Peter Moraw. Hrsg. v. Paul-Joachim HEINIG, Sigrid JAHNS, Hans-Joachim SCHMIDT, Reiner Christoph SCHWINGES und Sabine WEFERS. Berlin, Duncker & Humblot, 2000, XVII-759 p. (Historische Forschungen, 67).

3360. TABACCO (Giovanni). Dai re ai signori: forme di trasmissione del potere nel Medioevo. Torino, Bollati Boringhieri, 2000, 163 p. (Nuova didattica).

3361. TRAMONTANA (Salvatore). Il Mezzogiorno medievale: normanni, svevi, angioini, aragonesi nei secoli XI–XV. Roma, Carocci, 2000, 284 p.

Cf. n°s 268, 792, 819, 974, 1025, 1126, 3538, 5061, 5066

§ 3. Political history.

a. General

* 3362. CROSBY (Everett U.). Medieval warfare: a bibliographical guide. New York a. London, Garland, 2000, XVII-215 p.

3363. BROOKS (Nicholas). Anglo-Saxon myths: state and church, 400–1066. London a. Rio Grande, Hambledon Press, 2000, XVII-308 p. (ill.). – IDEM. Community and warfare, 700–1400. London a. Rio Grande, Hambledon Press, 2000, XVII-298 p. (ill.).

3364. DEBORD (André). Aristocratie et pouvoir: Le rôle du château dans la France médiévale. Ed par André BAZZANA and Jean-Michel POISSON. Paris, Picard, 2000, 238 p. (plates, ill., tables). (Espaces Médiévaux).

3365. DICHTBURN (David). Scotland and Europe: the medieval kingdom and its contacts with Christendom, c. 1215–1545. 1. Religion, culture and commerce. East Linton, Tuckwell Press, 2000, XIII-335 p. (ill.).

3366. FÖßEL (Amalie). Die Könige im mittelalterlichen Reich: Herrschaftsausübung, Herrschaftsrechte, Handlungsspielräume. Stuttgart, Jan Thorbecke, 2000, 443 p. (ill., tables).

3367. Guerre, pouvoir et noblesse au Moyen Âge. Mélanges en l'honneur de Philippe Contamine. Ed. par Jacques PAVIOT et Jacques VERGIER. Paris, Presses de l'Université de Paris-Sorbonne, 2000, 691 p. (ill.).

3368. KINTZINGER (Martin). Westbindungen im spätmittelalterlichen Europa. Auswärtige Politik zwischen dem Reich, Frankreich, Burgund und England in der Regierungszeit Kaiser Sigmunds. Stuttgart, Thorbecke, 2000, X-485 p. (ill., tables). (Mittelalter-Forschungen, 2).

3369. Ordines coronationis Franciae. Texts and ordines for the coronation of Frankish and French kings and Queens in the Middle Ages. Vol. 2. Ed. by Richard A. JACKSON. Philadelphia, University of Pennsylvania Press, 2000, XII-436 p.

3370. PORTER (Pamela). Medieval warfare in manuscripts. Toronto, University of Toronto Press, 2000, 64 p. (ill.). (The medieval world in manuscripts).

3371. War, government and power in late medieval France. Ed. par Christopher ALLMAND. Liverpool, Liverpool U. P., 2000, XVI-238 p.

Cf. n°s 1020, 3916

b. 476–900

3372. BARBERO (Alessandro). Carlo Magno. Un padre dell'Europa. Roma e Bari, Laterza, 2000, 451 p.

3373. DİVİTÇİOĞLU (Sencer). Kök Türkler: kut, küç, ülüg. (The Gokturks: the sovereignty, the power, the fate). Ed. Korkut TANKUTER. İstanbul, Yapı Kredi yayınları, 2000, 360 p.

3374. INNES (Matthew). State and society in the early middle ages: the middle Rhine Valley, 400–1000. Cambridge, Cambridge U. P., 2000, XVI-316 p. (ill.). (Cambridge Studies in Medieval Life and Thought, 4th ser., 47).

3375. KRAH (Adelheid). Die Entstehung der potestas regia im Westfrankreich während der ersten Regierungsjahre Kaiser Karls II (840–877). Berlin, Akademie Verlag, 2000, 346 p. (tables).

3376. STOCKING (Rachel L.). Bishops, councils, and consensus in the Visigothic Kingdom, 589–633. Ann Arbor, University of Michigan Press, 2000, XII-217 p. (map).

Cf. n° 3287

c. 900–1300

3377. ALTHOFF (Gerd). Die Ottonen. Königsherrschaft ohne Staat. Stuttgart, Berlin, u. Köln, Kohlhammer, 2000, 283 p.

3378. ASBRIDGE (Thomas S.). The creation of the principality of Antioch, 1098–1130. Woodbridge a. Rochester, Boydell and Brewer, 2000, XII-233 p. (maps).

3379. BALAKIN (Vasilij D.). Ital'janskaja politika imperii Ottonov. (The Italian policy of the Ottonian empire). Moskovskij ped. un-t. Moskva, Narodnyj uchitel', 2000, 287 p. – IDEM. Srednevekovaja Rimskaja imperija: ideja i real'nost'. (Medieval Roman empire: idea and reality). *Dialog so vremenem: Al'manakh intellektual'noj istorii*, 2000, 2, p. 14-35. [Eng. summary]

3380. BARLOW (Frank). William Rufus. New Haven a. London, Yale U. P., 2000, XXIII-486 p. (ill., plates). (Yale English Monarchs).

3381. BARTLETT (Robert). England under the Norman and Angevin Kings, 1075–1225. Oxford, Clarendon Press, 2000, XXX-772 p. (ill., maps, plates). (The New Oxford History of England).

3382. BENSCH (Stephen Paul). Barcelona i els seus dirigents: 1096–1291. Barcelona, Seminari d'Història de Barcelona, Proa, 2000, XIX-424 p. (ill.). (Bcn Biblioteca històrica, 6).

3383. Cour (La) Plantagenêt (1154–1204). Actes du Colloque tenu à Thouars du 30 avril au 2 mai 1999. Ed. par Martin AURELL. Poitiers, Centres d'Études supérieures de civilisation médiévale, 2000, 364 p. (ill., plans). (Civilisation médiévale, 8).

3384. CROUCH (David). The reign of King Stephen, 1135–1154. Harlow, Pearson Education, 2000, XIV-384 p. (maps, table).

3385. EHLERS (Joachim). Die Kapetinger. Stuttgart, Berlin u. Köln, Kohlhammer, 2000, 310 p.

3386. GARNIER (Claudia). Amicus amicis, inimicus inimicis: Politische Freundschaft und fürstliche Netzwerke im 13. Jahrhundert. Stuttgart, Anton Hiersemann, 2000, IX-375 p. (Monographien zur Geschichte des Mittelalters, 46).

3387. GILLINGHAM (John). The English in the twelfth century: imperialism, national identity and political values. Woodbridge a. Rochester, Boydell and Brewer, 2000, XXVI-289 p.

3388. HAMILTON (Bernard). The leper king and his heirs. Baldwin IV and the crusader kingdom of Jerusalem. Cambridge, Cambridge U. P., 2000, XXV-288 p. (ill., maps).

3389. HERMANSON (Lars). Släkt, vänner och makt: en studie av elitens politiska kultur i 1100-talets Danmark. (Kindred, friends, and power: a study of the elite's political culture in twelfth-century Denmark). Göteborg, Avhandlingar från Historiska institutionen i Göteborg, 2000, 280 p. (Avhandlingar från Historiska institutionen i Göteborg, 1).

3390. HUFFMANN (Joseph P.). The social politics of medieval diplomacy: Anglo-German relations (1066–1307). Ann Arbor, University of Michigan Press, 2000, XI-361 p. (Studies in Medieval and Early Modern Civilization).

3391. IMSEN (Steinar). Earldom and Kingdom. Orkney in the Realm of Norway 1195–1379. *Historisk tidsskrift*, 2000, 79, 2, p. 163-180.

3392. JACKSON (Tat'jana N.). Chetyre norvezhskikh konunga na Rusi: Iz istorii russko-norvezhskikh otnoshenij poslednej treti X – pervoj poloviny XI v. (Four Norwegian kings in Rus': from the history of Russian-Norwegian political relations of the late 10th and the first half of the 11th century.). Moskva, Jazyki russkoj kul'tury, 2000, 188 p. (ill.). (Studia historica. Series minor). [Eng. summary] [Cf. Bibl. 93, n° 3050. Cf. Bibl. 94, n° 3002.]

3393. JESSEE (W. Scott). Robert the Burgundian and the counts of Anjou, ca. 1025–1098. Washington, Catholic University of America Press, 2000, XII-206 p. (ill., map).

3394. KAUFHOLD (Martin). Deutsches Iterregnum und europäische Politik. Konfliktlösungen und Entscheidungsstrukturen, 1230–1280. Hanover, Hahnsche Buchhandlung, 2000, XXXIX-485 p. (M.G.H., Schriften, 49).

3395. KING (Edmund). Stephen of Blois, Count of Mortain and Boulogne. *English historical review*, 2000, 115, 461, p. 271-296.

3396. KRAG (Claus). Norges historie fram til 1319. (History of Norway to 1319). Oslo, Universitetsforlaget, 2000, 316 p. (ill.).

3397. KURZE (Wilhelm). Federico II e l'Italia: le grandi signorie monastiche tra Chiesa e Impero (Italia centrale). *Archivio storico italiano*, 2000, 158, 584, p. 215-254.

3398. LOUD (Graham). The age of Robert Guiscard: southern Italy and the Norman conquest. Harlow, Pearson Education, 2000, XII-329 p.

3399. MACÉ (Laurent). Les comtes de Toulouse et leur entourage, XIIe–XIIIe siècles. Rivalités, alliances et jeux de pouvoir. Toulouse, Privat, 2000, 445 p. (tabl., ill, maps). (Bibliothéque historique Privat).

3400. MAZZONI (Vieri). Note sulla confisca dei beni dei ghibellini a Firenze nel 1267 e sul ruolo della Parte Guelfa. *Archivio storico italiano*, 2000, 158, 583, p. 3-28.

3401. MOORE (R. I.). The first European revolution, c. 970–1215. Oxford and Malden, Blackwell, 2000, XVII-237 p. (ill., maps).

3402. MURRAY (Alan V.). The crusader kingdom of Jerusalem. A dynastic history. 1099–1125. Oxford, Unit for Prosopographical Research, 2000, VII-280 p. (ill., maps, tables). (Prosopographica et Genealogica, 4).

3403. NAZARENKO (Aleksandr V.). Porjadok prestolonaslednija na Rusi X–XII vv.: nasledstvennye razdely, sen'orat i popytki designatsii (tipologicheskie nabljudenija). (Succession of the throne in Rus' from the 10th to the 12th century: hereditary divisions, seigneurat and attempts of designation: typological observations). *In*: Iz istorii russkoj kul'tury (From the History of Russian Culture). Vol. 1. Drevnjaja Rus' (Old Rus'). Moskva, Jazyki russkoj kul'tury, 2000, p. 500-519.

3404. Neighbours of Poland in the 10th century. Ed by Przemysław URBAŃCZYK. Warsaw, Scientia, 2000, 203 p. (Inst. of Archaeology and Ethnology Pol. Acad. of Sciences).

3405. Normans (The) in Europe. Ed. by Elisabeth VAN HOURS. Manchester a. New York, Manchester U. P., 2000, XII-308 p. (tables, maps).

3406. PAGE (Mark). Cornwall, Earl Richard, and the Barons' war. *English historical review*, 2000, 115, 460, p. 21-38.

3407. Pierre II de Savoie. «Le petit Charlemagne» † 1298. Ed. par Bernard ANDENMATTEN, Agostino PARAVICINI BAGLIANI et Eva PIBIRI. Lausanne, [s. n.], 2000, 444 p. (Cahiers lausannois d'histoire médiévale, 27).

3408. RABAN (Sandra). England under Edward I and Edward II, 1259–1327. Oxford a. Malden, Blackwell, 2000, XII-204 p. (plates, tables, maps).

3409. ROGERS (Clifford J.). War cruel and sharp: English strategy under Edward III, 1327–1360. Woodbridge a. Rochester, Boydell and Brewer, 2000, XXII-458 p. (maps, tables).

3410. SCHNEIDMÜLLER (Bernd). Die Welfen. Herrschaft und Erinnerung (819–1252). Stuttgart, Berlin u. Köln, Kohlhammer, 2000, 378 p.

3411. TURNER (Ralph V.), HEISER (Richard R.). The reign of Richard Lionheart, ruler of the Angevin Empire, 1189–1199. Harlow, Longman, 2000, XII-292 p.

3412. VOGT (Lino). Laug, slægt og stat: den politiske udvikling i Firenze 1281–1295. (Corporation, famille et Etat: l'évolution politique à Florence de 1281 à 1295). København, Museum Tusculanum, 2000, 150 p. (Studier for sprog- og oldtidsforskning. 336).

3413. WARREN (Michelle R.). History on the edge: Excalibur and the borders of Britain, 1100–1300. Minneapolis a. London, University of Minnesota Press, 2000, XIII-303 p. (ill.).

3414. WHITE (Graeme J.). Restoration and Reform, 1153–1165: recovery from Civil War in England. Cambridge, Cambridge U. P., 2000, XVII-248 p. (Cambridge Studies in Medieval Life And Thought, 4th ser., 46).

3415. WOLFRAM (Herwig). Konrad II. 990–1039. Kaiser dreier Reiche. München, Beck, 2000, 464 p.

3416. Ziemie polskie w X wieku i ich znaczenie w kształtowaniu się nowej mapy Europy. (Die polnischen Gebiete im X. Jahrhundert und deren Bedeutung bei der Gestaltung einer neuen Landkarte Europas). Hrsg. v. Henryk SAMSONOWICZ. Kraków, Universitas, 2000, 475 p.

Cf. n°s 3134, 3287, 3307

d. 1300–1500

3417. Agincourt, 1415. Henry V, Sir Thomas Erpingham and the triumph of English archers. Ed. by Anne CURRY. Brimscombe Port Stroud, Tempus, 2000, 160 p.

3418. BOGATYREV (Sergei). The sovereign and his counsellors. Ritualised consultations in Muscovite political culture, 1350s–1570s. Saarijärvi, Gummerus, 2000, 297 p. (Suomalaisen Tiedeakatemian toimituksia. Annales Academiae Scentiarum Fennicae, Humaniora Series, 307).

3419. CARL (Horst). Der Schwäbische Bund 1488–1534. Landfrieden und Genossenschaft im Übergang von Spätmittelalter zur Reformation. Leinfelden-Echterdingen, DRW-Verlag, 2000, XII-592 p. (Schriften zur südwestdeutschen Landeskunde, 24).

3420. CASTOR (Helen). The king, the crown, and the Duchy of Lancaster. Public authority and private power, 1399–1461. Oxford, Oxford U. P., 2000, XIII-343 p.

3421. COHN (Samuel K.). Creating the Florentine state: peasants and rebellion, 1348–1434. Cambridge, Cambridge U. P., 2000, XIII-308 p.

3422. COLLINS (Hugh E.L.). The order of the Garter, 1348–1461: chivalry and politics in late medieval England. Oxford, Clarendon Press; New York, Oxford U. P., 2000, XI-327 p.

3423. CONNELL (William J.). La città dei crucci: fazioni e clientele in uno stato repubblicano del '400. Firenze, Nuova Toscana, 2000, X-318 p.

3424. CURRY (Anne). The battle of Agincourt. Sources and interpretations. Woodbridge, Boydell Press, 2000, 474 p.

3425. FLEET (Kate). Dentro il «buco nero»: la Turchia occidentale nel Trecento. *Quaderni di storia*, 2000, 26, 52, p. 57-70.

3426. Florentine Tuscany: structures and practices of power. Ed. by William J. CONNELL and Andrea ZORZI. Cambridge, Cambridge U. P., 2000, XII-357 p. (ill., maps, tables).

3427. Fourteenth century England, 1. Ed. by Nigel SAUL. Woodbridge a. Rochester, Boydell and Brewer, 2000, XII-210 p. (ill., graphs, tables).

3428. GRABARCZYK (Tadeusz). Piechota zaciężna Królestwa Polskiego w XV wieku. (Das Söldnerfussvolk des Königreichs Polen im XV. Jahrhundert). Łódź, Ibidem, 2000, 313 p.

3429. HAUG (Eldbjørg). Margrete – den siste dronning i Sverreætten. Nordens fullmektige frue og rette husbonde. (Margrete (1353–1412), Queen of Denmark, Norway and Sweden). Oslo, Cappelen, 2000, 399 p. (ill.).

3430. IBLER (Mladen), STRČIĆ (Petar). Hrvatsko skandinavske veze u prvoj polovici XV. stoljeća (Ivan VI. Frankapan i Erik VII. Pomeranski). (Croatian-Scandinavian relations in the first half of the 15th century), *Hrvatska akademija znanosti i umjetnosti*, 2000, 61, p. 123-145.

3431. LADERO QUESADA (Miguel Ángel). Andalucía a fines de la Edad Media. Estructuras, valores, sucesos. Cadix, Universidad de Cádiz-Servicio de publicacions, 2000, 368 p. (maps.).

3432. LEROY (Béatrice). Histoire et politique en Castille au XVe siècle. Biographies et portraits de Fernán Pérez de Guzmán (1380–1460). Limoges, PULIM, 2000, 127 p.

3433. MUTGÉ VIVES (Josefina). Contribució a l'estudi de les relacions politíques i comercials a la Mediterrània occidental: anàlisi d'un procés de l'any 1442, conservat a l'Arxiu de la Corona d'Aragó de Barcelona. Atti del "XVI Congresso Internazionale di Storia della Corona d'Aragona" (1997). Napoli, [s. n.], 2000, 2 vol., 1042 p. – EADEM. La corona d'Aragona ai tempi di Alfonso il Magnanimo: i modelli politico-istituzionali; la circolazione degli uomini, delle idee, delle merci; gli influssi della società sul costume. Napoli, Caserta e Ischia, 18–24 settembre 1997. Napoli, Paparo, 2000, 2 vol., XXIX-1910 p.

3434. New Cambridge (The) medieval history. 6. C. 1300–c.1415. Ed. by Michael JONES. Cambridge, Cambridge U. P., 2000, XXX-1110 p. (ill.).

3435. POLLARD (A.J.). Late medieval England, 1399–1509. Harlow, Pearson Education, 2000. XVIII-454 p. (plates, maps, table).

3436. PRYDS (Darleen N.). The king embodies the word: Robert d'Anjou and the politics of preaching. Leiden, Boston a. Köln, Brill, 2000, XIII-143 p.

3437. ROWELL (S. C.). Lithuania ascending. A pagan empire within east-central Europe, 1295–1345. Cambridge, Cambridge U. P., 2000, XXXIX-375 p. (maps, tables, ill.).

3438. RUSSELL (Peter). Prince Henry "the Navigator": a life. New Haven a. London, Yale U. P., 2000, XVI-448 p. (ill., tables, map).

Cf. n°s 40, 3302

§ 4. Jews.

3439. BRAND (Paul). Jews and the law in England, 1275–1290. *English historical review*, 2000, 115, 464, p. 1138-1159.

3440. CHAZAN (Robert). God, humanity and history: the Hebrew first crusade narratives. Berkley, Los Angeles a. London, University of California Press, 2000, XI-270 p.

3441. CLAMAN (Henry N.). Jewish images in the Christian church: art as the mirror of the Jewish-Christian conflict, 2000–1250 C.E. Macon, Mercer U. P., 2000, XI-212 p. (plates, ill.).

3442. CLUSE (Christoph). Studien zur Geschichte der Juden in den Mittelalterlichen Niederlanden. Hannover, Hahnsche Buchhandlung, 2000, VII-495 p. (ill., tables, map).

3443. DEBBY (N. B-A.). Jews and Judaism in the rhetoric of popular preachers: the Florentine sermons of Giovanni Dominici (1356–1419) and Bernardino da Siena (1380–1444). *Jewish history*, 2000, 14, 2, p. 175-200.

3444. JAEGER (Achim). Ein jüdischer Artusritter: Studien zum jüdisch-deutschen „Widuwilt" ("Artushof") und zum "Wigalois" des Wirnt von Gravenberc. Tübingen, Max Niemeyer, 2000, IX-465 p. (ill.).

3445. Jewish poet in Muslim Egypt: Moses Dar'ī's Hebrew Collection. Ed. by Leon J. WEINBERGER. Leiden, Boston a. Köln, Brill, 2000, XI-532 p.

3446. Jewish texts on the visual arts. Ed. by Vivian B. MANN. Cambridge, Cambridge U. P., 2000, XIX-236 p. (ill.).

3447. KANARFOGEL (E.). Progress and tradition in medieval Ashkenaz. *Jewish history*, 2000, 14, 3, p. 287-316.

3448. Medieval English Jews and royal officials: entries of Jewish interest in the English memoranda rolls, 1266–1293. Ed. by Zefira Entin ROKÉAH. Jerusalem, Hebrew University Magnes Press, 2000, XXXVII-514 p.

3449. MILLER (Elaine R.). Jewish multiglossia: Hebrew, Arabic, and Castillan in medieval Spain. Newark, Juan de la Cuesta, 2000, 160 p.

3450. Movimientos migratorios y expulsiones en la diaspora occidental: terceros encuentros judaicos de Tudela, 14–17 de julio de 1998. Universidad Publica de Navarra: Gobierno de Navarra, Departamento de Educacion y Cultura, dep.leg. 2000, 211 p.

3451. RUDAVSKY (T.M.). Time Matters: time, creation, and cosmology in medieval Jewish philosophy. Albany, State University of New York Press, 2000, XVIII-287 p. (ill.).

3452. Studies in medieval Jewish history and literature, 3. Ed. by Isadore TWERSKY and Jay M. HARRIS. Cambridge, Harvard University Centre for Jewish Studies, 2000, VII-298 p.

Cf. n° 3287

§ 5. Islam.

3453. ALGAZI (Gadi), DRORY (Rina). L'amour à la cour des Abbassides. Un code de compétence sociale. *In*: Hommes et femmes. Codes amoureux et morale publique. *Annales*, 2000, 55, 6, p. 1255-1282.

3454. BASHA (Ali Musa). Islami në Shqipëri gjatë shekujve. (L'Islam in Albania durante i secoli). Tiranë, [s. n.], 2000, 275 p. (Biblioteka Islame).

3455. BRAMON (Dolors). De quan érem o no musulmans. Textos del 713 al 1000. Barcelona, Institut Universitari d'Història Jaume Vicens, 2000, 427 p.

3456. CRONE (Patricia). Ninth century Muslim anarchists. *Past and present*, 2000, 167, p. 3-28.

3457. DEL LUNGO (Stefano). Bahr as Sham: la presenza musulmana nel Tirreno centrale e settentrionale nell'Alto Medioevo. Oxford, John and Erica Hedges, British archaeological reports, 2000, VII-102 p.

3458. DURAND (Robert). Musulmans et chrétiens en Méditerranée occidentale, 10.–13. siècles: contacts et échanges. Rennes, Presses universitaires de Rennes, 2000, 265 p.

3459. GARCIN (Jean Claude). États, sociétés et cultures du monde musulman médiéval, X^e–XVe siècle. T. 2. Sociétés et cultures. T. 3. Problèmes et prospectives de recherche. Paris, PUF, 2000, 2 vol., CXXIV-554 p., 288 p. (maps).

3460. Grandes ville méditerranéennes du monde musulman médiéval. Ed. par Jean-Claude GARCIN, Jean-Luc ARNAUD et Sylvie DENOIX. Roma, École française de Rome, 2000, 325 p. (ill., plans).

3461. GUTAS (Dimitri). Greek philosophers in the Arabic tradition. Aldershot a. Burlington, Ashgate, 2000, XII-322 p. (ill., tables).

3462. ISTANBULI (Yasin). Diplomacy and diplomatic practices in the early Islamic era. Karachi a. Oxford, Oxford U. P., 2000, VIII-171 p.

3463. KUKKONEN (T.). Plenitude, possibility, and the limits of reason: a Medieval Arabic debate on the metaphysics of nature. *Journal of the history of ideas*, 2000, 61, 4, p. 539-560.

3464. MARCONI (Silvio). Il giardino-paradiso: l'influsso iranico negli orti-giardino e in altri ambiti materiali e immateriali della Sicilia islamica. Roma, I versanti, 2000, 151 p.

3465. MARÍN (Manuela). Mujeres en al-Andalus. Madrid, Consejo Superior de Investigaciones Científicas, 2000, 781 p. (Estudios onomástico-biográficos de al-Andalus, 11).

3466. ORTEGA (Pascual). Musulmans en Cataluña. Las comunidades musulmanas de las encomiendas templarias y hospitalarias de Ascó y Miravet (siglos XII–XIV). Barcelona, CSIC-Institución Milà y Fontanals-Departamento d'estudios medievales, 2000, 192 p. (tables).

3467. REDFORD (Scott). Landscape and the state in medieval Anatolia: Seljuk gardens and pavilions of Alanya, Turkey. Oxford, Arxhaeopress, 2000, XIV-309 p. (ill., graphs).

3468. ROBINSON (Chase F.). Empire and elites after the Muslim conquest: the transformation on Northern Mesopotamia. Cambridge, Cambridge U. P., 2000, XV-206 p. (Cambridge studies in Islamic civilization).

3469. RUGGLES (D. Fairchild). Gardens, landscape, and vision in the palaces of Islamic Spain. University Park, Pennsylvania State U. P., 2000, XVI-275 p. (ill., tables, plates).

3470. SAFRAN (Janina M.). The second Umayyad caliphate: the articulation of caliphal legitimacy in al-Andalus. Cambridge a. London, Center for Middle Eastern Studies of Harvard University, 2000, XI-272 p.

3471. Salâhaddin Eyyûbi ve devri. (Saladin the Ayyubid and his time). Ed. by Ramazan ŞEŞEN; preface by Ekmeleddin İHSANOĞLU. İstanbul, İslam Tarih, Sanat ve Kültürünü Araştırma Vakfı, 2000, 658 p.

3472. TOUATI (Houari). Islam et voyage au Moyen Age. Histoire et anthropologie d'une pratique lettrée. Paris, Ed. Du Seuil, 2000, 344 p. (L'univers historique).

3473. ZOMEÑO (Amalia). Dote y matrimonio en al-Andalus y el Norte de Africa: estudios sobre la jurisprudencia islámica medieval. Madrid, Consejo Superior de Investigaciones Científicas, 2000, 302 p.

§ 5. Addenda 1998.

3474. Genèse de la ville islamique en al-Andalus et au Maghreb occidental. Ed par Patrice CRESSIER et Mercedes GARCIA-ARENAL. Madrid, Casa de Velazquez, 1998, 402 p. (ill.).

Cf. n°s 40, 58, 974, 3021, 3334, 4022, 8482, 8489

§ 6. Vikings.

** 3475. Morkinskinna. The earliest Icelandic chronicle of the Norwegian kings (1030–1157). Ed. by Theodore M. ANDERSSON and Kari Ellen GADE. Ithaca a. Londres, Cornell U. P., 2000, XVI-556 p. (maps). (Islandica, 51).

3476. ANDERSEN (Knud-Erik). På sporet af vikingerne. Nye fynd, nye viden. (Sur les traces des Vikings. Nouvelles découvertes, nouvelles connaissances). Valby, Borgen, 2000, 109 p. (ill., cartes).

3477. Cultures in contact: Scandinavian settlement in England in the ninth and tenth centuries. Ed. by Dawm M. HADLEY. Turnhout, Brepols, 2000, VIII-331 p. (ill.). (Studies in the early middle ages, 2).

3478. DUMÉZIL (Georges). Mythes et dieux de la Scandinavie ancienne. Paris, Gallimard, 2000, XX-376 p. (table).

3479. HADLEY (D. M.). The Northern danelaw: its social structure, c. 800–1100. London a. New York, Leicester U. P., 2000, X-374 p. (ill.).

3480. INGELMAN-SUNDBERG (Catharina). Boken om vikingar. (Le Livre des Vikings). Stockholm, Prisma, 245 p. (ill.).

3481. LUND (Niels). Viking og Hvidekrist. Et internationelt symposium på Nationalmuseet om Norden og Europa i den sene vikingetid og tidligste middelalder. (Viking et Christ blanc. Colloque international au Musée National sur la Scandinavie et l'Europe à la fin de l'époque viking et au début du Moyen Age). København, C. A. Reitzel, 2000, 205 p. (ill.).

3482. SAWYER (Birgit). The Viking-Age Rune-Stones: custom and commemoration in early medieval Scandinavia. Oxford a. New York, Oxford U. P., 2000, XXI-269 p. (plates, ill., tables, maps).

3483. SVANBERG (Fredrik). Vikingatiden i Skåne. (L'epoque Viking en Scanie). Lund, Historiska media, 2000, 108 p. (ill.).

3484. Volsunga saga: the saga of the Volsungs. Ed. by Kaaren GRIMSTAD. Saarbrücken, AQ.Verlag, 239 p. (ill., tables).

Cf. n°s 621, 3392, 3916

§ 7. History of law and institutions.

** 3485. D'URSEL (Caroline), GENICOT (Luc-Francis), SPÈDE (Raphaël), WEBER (Philippe). Donjours médiévaux de Wallonie. Vol. 1. Province de Brabant, Arrondissement de Nivelles. Bruxelles, Ministère de la Région wallonne-Division du Patrimoine DGATLP, 2000, 106 p. (ill., maps). (Inventaires thématiques).

** 3486. EMCHENKO (Elena B.). Stoglav: Issledovanie i tekst. (The Stoglav [Muscovite canon law code, 1551]: study and text). RAN, In-t rossijskoj istorii. Moskva, Indrik, 2000, 499 p. (facs.).

** 3487. Law code (The) ["Datastanagirk' "] of Mxit'ar Goš. Ed. by Robert W. THOMSON. Amsterdam a. Atlanta, Rodopi, 2000, 359 p.

** 3488. Leyes de los Adelantados Mayores: regulations, attributed to Alfonso X of Castile, concerning the King's vicar in the Judiciary and in territorial administration. Ed. by Robert A. MAC DONALD. New York, Hispanic Seminary of Medieval Studies, 2000, VIII-180 p.

** 3489. MATTHÄUS VON KRAKAU. De contractibus. Matthias NUDING. Heidelberg, C. Winter, 2000, 171 p. (ill.). (Editiones Heidelbergenses, 28).

3490. ASCHERI (Mario). I diritti del Medioevo italiano. Secoli XI–XIV. Roma, Carocci editore, 2000, 452 p. (Università. Argomenti di storia medievale, 193).

3491. BARNWELL (P. S.). Emperors, jurists and kings: law and custom in the late Roman and early medieval West. Past and present, 2000, 168, p. 6-29.

3492. BLAUERT (Andreas). Das Urfehdewesen im deutschen Südwesten im Spätmittelalter und in der Frühen Neuzeit. Tübingen, Bibliotheca academica, 2000, 199 p. (Frühneuzeit-Forschungen, 7).

3493. BOUREAU (Alain). How law came to the monks: the use of law in English society at the beginning of the thirteenth-century. Past and present, 2000, 167, p. 29-74.

3494. Code (The) of Cuenca. Municipal law on the twelfth-century Castilian frontier. Ed. by James F. POWER. Philadelphia, University of Pennsylvania Press, 2000, VIII-245 p.

3495. Coniugi nemici. La separazione in Italia dal XII al XVIII secolo. I processi matrimoniali degli archivi ecclesiastici italiani. A cura di Silvana SEIDEL MENCHI e Diego QUAGLIONI. Bologna, Il Mulino, 2000, 570 p. (Istituto trentino di cultura. Annali dell'Istituto storico italo-germanico in Trento. Quaderni, 53).

3496. DUMOLYN (J.). The legal repression of revolts in late medieval Flanders. Revue d'histoire du droit, 2000, 68, 4, p. 479-521.

3497. Esclavitud (De l') a la llibertat: esclaus I lliberts a l'edat mitjana. Actes del Col.loqui Internacional celebrat a Barcelona, del 27 al 29 de maig de 1999. Ed. por Maria Teresa FERRER I MALLOL y Josefina MUTGÉ VIVES. Barcelona, Consell Superior d'Investigacions Científiques, Institució Milà i Fontanals, Departement d'Estudios Medievals, 2000, XII-751 p. (ill. tables, graphs, map). (Annuario de Estudios Medievales, Annex 38).

3498. Excerptiones iuris: Studies in Honour of André Gouron. Ed. par Bernard DURAND et Laurent MAYALI. Berkeley, Robbins Collections, 2000, VII-827 p. (ill.).

3499. FENGER (Ole). Notarius publicus: notaren i latinsk middelalder. (Notarius publicus: le notaire dans le Moyen Age latin). Århus, Århus Universitets forlag, 2000,203 p. (ill.). [Résumé français].

3500. FIREY (A.). Ghostly recensions in early medieval canon law: the problem of the Collectio Dacheriana and its shades. Revue d'histoire du droit, 2000, 68, 1-2, p. 63-82.

3501. GERMIVNIK (Franciscus, c. m.). Indices ad Corpus iuris canonici. Ed. by Michel THÉRIAULT. Ottawa, Faculty of Canon Law, University of St. Paul, 2000, IX-497 p.

3502. GIESE (Wolfgang). Designative Nachfolgeregelungen in germanischen Reichen der Völkerwanderungszeit. Zeitschrift der Savigny-Stiftung für Rechtsgeschichte. Germanistische Abteilung, 2000, 117, p. 39-121.

3503. GOURON (André). Juristes et droits savants: Bologne et la France médiévale. Aldershot, Ashgate, 2000, X-304 p. (Variorum Collected Studies Series, 679).

3504. HART (A. R.). History of the King's serjeants at law in Ireland: honour rather than advantage? Dublin a. Portland, Four Courts Press, 2000, XVI-213 p. (ill., tables).

3505. HEIRBAUT (Dirk). Over lenen en families: Een studie over de vroegste geschiedenis van het zakelijk leenrecht in hit graafschap Vlaanderen (ca. 1000–1305). Bruxelles, Koninklijke Vlaamse Academie van België voot Wetenschappen en Kunsten, 2000, 258 p .

3506. HYAMS (P.R.). 'Ius commune' et 'common law' au moyen âge. Les scélérats et les honnêtes gens. Bibliothèque de l' École des Chartes, 2000, 158, p. 407-430.

3507. Integration (Die) des südlichen Ostseeraumes in das Alte Reich. Hrsg. v. Nils JÖRN und Michael NORTH. Köln, Weimar u. Wien, Böhlau, 2000, VIII-554 p. (Quellen und Forschungen zur höchsten Gerichtsbarkeit im Alten Reich, 35).

3508. KIM (Keechang). Aliens in medieval law. The origins of modern citizenship. Cambridge, Cambridge

U. P., 2000, XII-250 p. (Cambridge studies in English legal history).

3509. LUCCHESI (Marzia). Stefano Costa e il Tractatus de Ludo (1471), prime note. *Rivista di storia del diritto italiano*, 2000, 73, p. 19-64.

3510. Medieval woman and the law. Ed. by Noël MENUGE. Woodbridge, Boydell, 2000, XIII-169 p. (graphs).

3511. MEUTEN (Ludger). Die Erbfolgeordnung des Sachsenspiegels und des Magdeburger Rechts. Frankfurt am Main, Berlin, Bern, Bruxelles, New York, Oxford u. Wien, Lang, 2000, 351 p. (Rechtshistorische Reihe, 218).

3512. MEYER (Andreas). Felix et inclitus notarius: Studien zum italienischen Notariat vom 7. bis zum 13. Jahrhundert. Tübingen, M. Niemeyer, 2000, XI-857 p. (Bibliothek des Deutschen Historischen Instituts in Rom).

3513. MEYER (Christoph H.F.). Die Distinktionstechnik in der Kanonistik des 12. Jahrhunderts: Ein Beitrag zur Wissenschaftsgeschichte des Hochmittelalters. Leuven, Leunven U. P., 2000, VIII-363 p. (Mediaevalia Lovaniensia, 1/29).

3514. MOSCATI (Laura). Un'inedita vita d'Irnerio. *Rivista di storia del diritto italiano*, 2000, 73, p. 5-18.

3515. MURRAY (A. V.). The judicial inquest into the death of Count Charles of Flanders (1127), location and chronology. *Revue d'histoire du droit*, 2000, 68, 1-2, p. 47-61.

3516. NIETO SORIA (José Manuel). Legislar y gobernar en la Corona de Castilla: el Ordenamiento de Medina del Campo de 1433. Madrid, [s. n.], 2000, 276 p.

3517. PIÑOL ALABART (Daniel). El notariat públic al Camp de Tarragona: història, activitat, escriptura i societat: segles XIII–XIV. Barcelona, Fundació Noguera, 2000, 411 p. (Estudis, 27).

3518. Podestà (I) dell'Italia comunale. 1. Reclutamento e circolazione degli ufficiali forestieri (fine XII sec.– metà XIV sec.). A cura di Jean-Claude MAIRE VIGUEUR. Roma, Ecole Française de Rome, 2000, 2 vol., VI-690 p., IV-539 p. (graphs, maps, tables). (Collection de l' Ecole Française de Rome, 268. Istituto storico italiano per il medio evo, Nuovi Studi Storici, 51).

3519. Pouvoir, justice et société. Actes des XIXe journées d'histoire du droit (9–11 juin 1999). Ed. par Jacqueline HOAERAU-DODINAU et Pascal TEXIER. Limoges, PULIM, 2000, 599 p. (ill., graphs).

3520. Rites (Les) de la justice. Gestes et rituels judiciaires au Moyen Age occidental. Ed. par Claude GAUVARD et Robert JACOB. Paris, Cahiers du Léopard d'or, 2000, 239 p.

3521. ROSSO (Paolo). Catone Sacco. Problemi biografici. La tradizione delle opere. *Rivista di storia del diritto italiano*, 2000, 73, p. 237-338.

3522. STOCLET (Alain J.). La Clausula de unctione Pippini regis, vingt ans après. *Revue belge de philologie et d'histoire*, 2000, 78, 3-4, p. 719-772.

3523. Sudebnik 1497 g. v kontekste istorii rossijskogo i zarubezhnogo prava XI–XIX vv. (The Law Code of 1497 in the history of Russian and foreign law from the 11th to the 19th century: [Articles]). Ed. Andrej N. SAKHAROV. Moskva, Parad, 2000, 243 p.

3524. TUBBS (J.W.). The common law mind: medieval and early modern conceptions. Baltimore a. London. Johns Hopkins U. P., 2000, XV-253 p.

3525. VAN CAENEGEM (R. C.). The modernity of medieval law. *Revue d'histoire du droit*, 2000, 68, 3, p. 313-329.

3526. VINOGRADOV (Pavel Gavrilovich). Derecho romano en la Europa medieval. Proceso formativo, Francia, Inglaterra y Alemanna. Derecho romano en Escocia. Traducciones a cargo de Manuel J. PELÁEZ, Universidad de Málaga, Cátedra de Historia del Derecho y de las Instituciones, 2000, 177 p.

3527. WICKHAM (Chris). Legge, pratiche e conflitti. Tribunali e risoluzione delle dispute nella Toscana del XII secolo. Roma, Viella, 2000, 521 p. (maps). (I libri di Viella, 23).

3528. WINROTH (Anders). The making of Gratian's "Decretum". Cambridge, Cambridge U. P., 2000, XVI-245 p. (ill., tables).

Cf. n°s 40, 1109, 1114, 3193, 3403, 3660

§ 8. Economic and social history.

** 3529. FOSSIER (Robert). Sources de l'histoire économique et sociale du Moyen Âge occidental. Questions, sources, documents commentés. Turnhout, Brepols, 2000, 408 p.

3530. ABULAFIA (David). Mediterranean encounters, economic, religious, political, 1100–1550. Aldershot, Ashgate, 2000, XVIII-352 p. (ill.).

3531. AINSWORTH (Peter). The image of the city in peace and war in a Burgundian manuscript of Jean Froissart's Chronicles. *In*: Carrefour (Au) des époques [Cf. n° 3548], p. 295-314.

3532. AIRÒ (Anna). Per una storia dell'universitas di Taranto nel Trecento. *Archivio storico italiano*, 2000, 158, 583, p. 29-84.

3533. ARLIGHAUS (Franz-Josef). Zwischen Notiz und Bilanz: zur Eigendynamik des Schriftgebrauchs in der kaufmännischen Buchführung am Beispiel der Datini/di Berto-Handelsgesellschaft in Avignon (1367–1373). Frankfurt am Main, Peter Lang, 2000, 531 p. (ill., graphs).

3534. Aziende agrarie nel Medioevo: forme della conduzione fondiaria nell'Italia nord-occidentale (secoli 9.–15.). A cura di Rinaldo COMBA e Francesco

PANERO. Cuneo, Società per gli studi storici, archeologici ed artistici della provincia di Cuneo, 2000, 309 p.

3535. BACKHOUSE (Janet). Medieval rural life in the Lutrell Psalter. Toronto, University of Toronto Press, 2000, 64 p. (ill.).

3536. BARRAQUÉ (J.-P.). Du bon usage du pacte: les passeries dans les Pyrénées occidentales à la fin du Moyen-Age. *Revue historique*, 2000, 124, 614, p. 307-338.

3537. BAYKARA (Tuncer). Türkiye'nin sosyal ve iktisadî tarihi (XI–XIV. yüzyıllar). (Social and economic history of Turkey [XI–XIVth centuries]). Ankara, Türkiye Diyanet Vakfı, 2000, 221 p.

3538. BELAVIN (Andrej M.). Kamskij torgovyj put'. Srednevekovoe Predural'e v ego ekonomicheskikh i etnokul'turnykh svjazjakh. (The Kama trade route: Medieval pre-Urals region in its economic and ethnocultural relations, [the 11th–14th centuries]). Perm', Izd-vo Perskogo gos. ped. un-ta, 2000, 200 p. (ill.). [Eng. summary].

3539. BOIS (Guy). La grande dépression médiévale: XIVe–XVe siècles. Le précédent d'une crise systémique. Paris, PUF, 2000, 214 p. (Actuel Marx Confrontation).

3540. BONACINI (Pierpaolo). Il Monastero di San Benedetto Polirone: formazione del patrimonio fondiario e rapporti con l'aristocrazia italica nei secoli XI e XII. *Archivio storico italiano*, 2000, 158, 586, p. 623-678.

3541. BONNASSIÉ (Pierre). Les sociétés de l'an mil. Un monde entre deux âges. Bruxelles, Se Boeck Université, 2000, 517 p. (tables, maps).

3542. BOTTICINI (Maristella). A tale of "Benevolent" governments: private credit markets, public finance, and the role of Jewish lenders in Medieval and Renaissance Italy. *Journal of economic history*, 2000, 60, 1, p. 164-189.

3543. BOURIN (Monique), DURAND (Robert). Vivre au village au Moyen âge. Les solidarités paysannes du XIe au XIIIe siècle. Rennes, P.U. Rennes, 2000, 207 p.

3544. BROWN (Andrew). Urban jousts in the Later Middle Ages: the white bear of Bruges. *In*: Carrefour (Au) des époques [Cf. n° 3548], p. 315-330.

3545. BULLÓN-FERNÁNDEZ (María). Fathers and daughters in Gower's "Confessio Amantis": authority, family, state, and writing. Woodbridge a. Rochester, Boydell and Brewer, 2000, VIII-241 p.

3546. CAMPBELL (Bruce M.S.). English seigniorial agricolture, 1250-1450. Cambridge, Cambridge U. P., 2000, XXVI-517 p. (ill., tables, maps).

3547. CARPENTER (D. A.). The second century of English feudalism. *Past and present*, 2000, 168, p. 30-71.

3548. Carrefour (Au) des époques: la ville bourguignonne sous le ducs Valois. Actes du colloque inaugural du Centre de Recherche Francophones Belges (University of Edinburgh) 24–26 mai 1996. Ed. et intr. par Philip E. BENNETT. *Revue belge de philologie et d'histoire*, 2000, 78, 2, p. 289-444. [Cf. n[os] <sélection> 3531, 3544, 3552, 3577.]

3549. CASTÁN LANASPA (Guillermo). Política económica y poder político: moneda y fisco en el reinado de Alfonso X el Sabio. Valladolid, Consejería de Educación y Cultura, 2000, 234 p. (studios de historia).

3550. Castrum 6. Maisons et espaces domestiques dans le monde méditerranéen au Moyen Âge. Ed. par André BAZZANA et Étienne HUBERT. Roma, École française de Rome et Madrid, Casa de Velásquez, 2000, 271 p. (ill.). (Coll. Histoire).

3551. Chasse (La) au moyen âge: société, traités, symboles. Agostino PARAVICINI BAGLIANI, Baudoin VAN DEN ABEELE. Firenze, SISMEL-Edizioni del Galluzzo, 2000, 266 p. (ill., tables). (Micrologus' Library).

3552. COCKSHAW (Pierre). L'image de la ville dans les miniatures des manuscrits présentés aux ducs de Bourgogne. *In*: Carrefour (Au) des époques [Cf. n° 3548], p. 331-338.

3553. COSMAN (Madeleine Pelner). Women at work in medieval Europe. New York, Facts on File, 2000, XV-192 p. (ill.).

3554. CRISTIAN (Luca). Depozitul lui Radu Mihnea la Zecca Veneției. *Studii și materiale de istorie medie*, 2000, 18, p. 189-195.

3555. CROUCH (David J.F.). Piety, fraternity and power: religious gilds in late medieval Yorkshire, 1389–1547. York, York Medieval Press, 2000, XI-331 p. (ill., tables).

3556. DALENA (Pietro). Ambiti territoriali, sistemi viari e strutture del potere nel Mezzogiorno medievale. Bari, M. Adda, 2000, IV-249 p. (ill.).

3557. DERVILLE (Alain). La société française au Moyen Âge. Villeneuve d'Ascq, P.U. du Septentrion, 2000, 273 p. (Histoire de civilisations).

3558. DEVROEY (Jean-Pierre). Men and women in early medieval serfdom: the ninth-century morth Frankish evidence. *Past and present*, 2000, 166, p. 3-30.

3559. DOCKRAY-MILLER (Mary). Motherhood and Mothering in Anglo-Saxon England. New York, St. Martin's Press, 2000, XIV-161 p. (The New Middle Ages).

3560. DUBROVIN (Gennadij E.). Vodnyj i sukhoputnyj transport srednevekovogo Novgoroda X–XV vv.: po arkheologicheskim dannym. (Waterway and land transport of Medieval Novgorod, the 10th–15th centuries: archaeological data). Vol. 1. [Text and catalogue]. Vol. 2. [Ill.]. Moskva, [s. n.], 2000, 445 p. [Eng. summary].

3561. DYER (Christopher). Everyday life in medieval England. London a. New York, Hambledon a. London, 2000, XVI-336 p. (ill., tables).

3562. EWERT (Ulf Christian), HIRSCHBIEGEL (Jan). Gabe und Gegengabe. Das Erscheinungsbild einer Sonderform höfischer Repräsentation am Beispiel des französisch/burgundischen Gabentausches zum neuen Jahr um 1400. *Vierteljahrschrift für Sozial- und Wirtschaftsgeschichte*, 2000, 87, p. 5-37.

3563. Family, marriage and property devolution in the Middle Ages. Ed. Lars Ivar HANSEN. Tromsø, Department of History, University of Tromsø, 2000, 165 p.

3564. FEBRER Y ROMAGUERA (Manuel Vicente). Dominio y explotación territorial en la Valencia foral. Valencia, Universitat, 2000, 365 p.

3565. FENIELLO (A.). Marchandises et charges publiques: la fortune des d'Afflitto, hommes d'affaires napolitains du XVe siècle. *Revue historique*, 2000, 124, 613, p. 55-120.

3566. Feudalesimo (Il) nell'alto Medioevo: settimane di studio del Centro italiano di studi sull'alto Medioevo, 47. 8–12 aprile 1999. Spoleto, presso la sede del Centro, 2000, 2 vol., XIV-1060 p.

3567. FILIPPOV (I. S.). Sredizemnomorskaia Frantsiia v rannee srednekov'e: Problema stanovleniia feodalizma. Moscow, Izdatel'stvo Skriptorii, 2000, 800 p. (plates, maps).

3568. Fondazione (La) di Bobbio nello sviluppo delle comunicazioni tra Langobardia e Toscana nel Medioevo. Atti del Convegno internazionale, Bobbio, Auditorium di S. Chiara, 1-2 ottobre 1999. A cura di Flavio G. NUVOLONE. Bobbio, Associazione culturale Amici di Archivum Bobiense, 2000, 301 p. (ill., maps). (Archivum Bobiense, Studia, 3).

3569. FOSSIER (Robert). Le travail au Moyen Âge. Paris, Hachette, 2000, 316 p.

3570. Frauenzimmer (Das). Die Frau bei Hofe in Spätmittelalter un Früher Neuzeit. Hrsg. v. Jan HIRSCHBIEGEL und Werner PARAVICINI. Stuttgart, Thorbecke, 2000, 485 p. (Residenzforschung, 11).

3571. GRINDER-HANSEN (Keld). Kongemagtens krise: det danske møntvæsen 1241–ca 1340: den pengebaserede økonomi og møntcirkulation i Danmark i perioden 1241–ca 1340. (Crise du pouvoir royal: la monnaie danoise de 1241 à 1340. Economie et circulation monétaires au Danemark). København, Museum Tusculanum, 311 p.

3572. HAMMEL-KIESOW (Rolf). Die Hanse. München, Beck, 2000, 128 p. (Beck Wissen in der Beck'schen Reihe, 2131).

3573. HAREN (Michael). Sin and society in fourteenth century England: a study of the "Memoriale presbiteriorum". Oxford, Clarendon Press a. New York, Oxford U. P., 2000, XVIII-254 p. (ill.).

3574. HINOJOSA MONTALVO (José). Esclavos, nobles y corsarios en el Alicante medieval. Alicante, Fundación de Estudios Medievales Jaime II, Universitat d'Alacant, 2000, [s.p.]. (Cuadernos de la Fundación de Estudios Medievales Jaime II, 1).

3575. JACOB (Robert). Bannissement et rite de la langue tirée au Moyen Age. Du lien des lois et de sa rupture. *In*: Bannir au Moyen Age, naturaliser à l'époque moderne. *Annales*, 2000, 55, 5, p. 1039-1080.

3576. JAHNKE (Carsten). Das Silber des Meeres. Fang und Vertrieb von Ostseehering zwischen Norwegen und Italien (12.–16. Jahrhundert). Köln, Weimar u. Wien, Böhlau, 2000, IX-452 p. (Quellen und Darstellungen zur hansischen geschichte, 49).

3577. JONES (Michael). Small-town life in a late medieval Burgundy: the case of Cluny. *In*: Carrefour (Au) des époques [Cf. n° 3548], p. 359-378.

3578. JUSSEN (Bernhard). Spiritual kinship as social practice: god parenthood and adoption in the early middle ages. Newark, University of Delaware Press a. London, Associated U. P., 2000, 362 p. (The Family in Interdisciplinary Perspective).

3579. KAUFHOLD (Martin). Weibliche Hochzeitschancen und soziale Zwänge auf dem Florentiner Heiratsmarkt im Quattrocento am Beispiel von Caterina Tanagli und Filippo Strozzi. *Vierteljahrschrift für Sozial- und Wirtschaftsgeschichte*, 2000, 87, p. 423-441.

3580. KAYE (Joel). Economy and nature in the fourteenth century. Money, market exchange and the emergence of scientific thought. Cambridge, Cambridge U. P., 2000, X-273 p.

3581. KESSLER (Herbert L.), ZACHARIAS (Johanna). Rome, 1300: on the path of the pilgrim. New Haven a. London, Yale U. P., 2000, IX-237 p. (ill.).

3582. KIM (Hyonim). The Knight without the sword: a social landscape of Malorian chivalry. Woodbridge a. Rochester, Boydell and Brewer, 2000, VIII-155 p. (Arthurian Studies, 45).

3583. LE BIÉVE (Daniel). La part du pauvre: l'assistance dans les pays du Bas-Rhône du XIIe siècle au milieu du XVe siècle. Rome, Ecole Française de Rome, 2000, IV-436 p. (ill., maps, plates, tables). (Collection de l'Ecole Française de Rome, 265).

3584. LENZI (Mauro). La terra e il potere: gestione delle proprietà e rapporti economico-sociali a Roma tra alto e basso Medioevo (secoli X–XII). Roma, presso la Società alla Biblioteca Vallicelliana, 2000, 170 p. (Miscellanea della Società romana di storia patria).

3585. LEROY (Béatrice). Hommes et milieux en Espagne médiévale. Navarrais Castillans du XIIIe au XVe siècle. Biarritz, Atlantica, 2000, 306 p. (maps).

3586. MARTÍNEZ MARTÍNEZ (María). Las mujeres en la organización de una sociedad de frontera: la etapa colonizadora-repobladora de Murcia, 1266–1272. Murcia, Universidad, 2000, [s.p.].

3587. Moneda y monedas en la Europa medieval (siglos XII–XV). XXVI Semana de estudios medievales, Estella, 19 a 23 de julio de 1999. Pampelune, Gobierno de Navarra-Departemento de Educación y Cultura, 2000, 517 p. (tables, ill.).

3588. Montagne (La) dans le texte médiéval. Entre mythe et réalité. Ed. par Claude THOMASSET et Danièle JAMES-RAOUL. Paris, Presses de l'Université de Paris-Sorbonne, 2000, 348 p. (maps, ill.). (Cultures et civilisations médiévales, 19).

3589. MORSEL (Joseph). La noblesse contre le prince. L'espace social des Thüngen à la fin du Moyen Age (Franconie, v. 1250–1525). Stuttgart, Thorbecke, 2000, 757 p.

3590. MOXÓ Y ORTIZ DE VILLAJOS (Salvador de). Feudalismo, señorío y nobleza en la Castilla medieval. Madrid, Real Academia de la Historia, 2000, 370 p.

3591. NIGHTINGALE (Pamela). Knights and merchants: trade and the gentry in late Medieval England. *Past and present*, 2000, 169, p. 36-62.

3592. Nobles and nobility in medieval Europe: concepts, origins, transformations. Ed. by Anne J. DUGGAN. Woodbridge a. Rochester, Boydell and Brewer, 2000, XII-285 p. (plates, tables).

3593. Noblesse (La) dans les territoires angevins à la fin du moyen âge. Actes du colloque international organisée par l'Université d'Angers, Angers-Saumur 3–6 juin 1998. Ed. par Noël COULET et Jean-Michel MATZ. Roma, Ecole Française de Rome, 2000, IV-841 p. (plates, ill., maps, tables).

3594. NOLTE (Paul). Die Ordnung der deutschen Gesellschaft. Selbstentwurf und Selbstbeschreibung im 10. Jahrhundert. München, Beck, 2000, 520 p.

3595. Ordering medieval society. Perspectives on intellectual and practical modes of shaping social relations. Ed. by Bernhard JUSSEN. Philadelphia, University of Pennsylvania Press, 2000, VII-328 p. (ill). (The Middle Ages series).

3596. Origines (Les) de la féodalité. Hommage à Claudio Sánchez Albornoz. Ed. par Joseph PÉREZ et Santiago AGUADÉ NIETO. Madrid, Casa de Velázquez-Universidad de Alcalá, 2000, 253 p. (maps).

3597. ORTI GOST (Pere). Renda i fiscalitat en una ciutat medieval: Barcelona, siegles XII–XIV. Barcelona, CSIC-Institución Milà y Fontanals-Departamento d'estudios medievales, 2000, 736 p. (tables, graph.).

3598. PAUK (Marcin, Rafał). Działalność fundacyjna możnowładztwa czeskiego i jej uwarunkowania społeczne (XI–XIII wiek). (Die Stiftungstätigkeit der tschechischen Machthaber und deren gesellschaftliches Bedingen, XI.–XIII. Jahrhundert). Kraków u. Warszawa, Societas Vistulana, 2000, 278 p.

3599. Per Vito Fumagalli: terra, uomini, istituzioni medievali. A cura di Massimo MONTANARI e Augusto VASINA. Bologna, CLUEB, 2000, 567 p.

3600. Problems (The) of Labour in Fourteenth-Century England. Ed. by James BOTHWELL, P. J. P. GOLDBERG and W. M. ORMROD. York, York Medieval Press, 2000, VIII-153 p. (ill.).

3601. RADY (Martyn). Nobility, land and service in medieval Hungary. Houndmills, Palgrave, in association with the School of Slavonic and East European Studies, University College London, 2000, XV-231 p.

3602. RAUKAR (Tomislav). Jadranski gospodarski sustavi: Split 1475.–1500. godine. (Adriatic economical systems: Split, 1475–1500). *Rat*, 2000, 38, 480, p. 49-125.

3603. Région (Une) frontalière au Moyen Age: les vallées du Turano et du Salto entre Sabine et Abruzzes. Ed. par Etienne HUBERT. Roma, École française de Rome, 2000, 472 p. (ill.). (Collection de l'Ecole française de Rome Recherches d'archéologie médiévale en Sabine).

3604. REINLE (Christine). Exempla weiblicher Stärke? Zu den Ausprägungen des mittelalterlichen Amazonenbildes. *Historische Zeitschrift*, 2000, 270, 1, p. 1-38.

3605. RHEUBOTTOM (David). Age, marriage, and politics in fifteenth-century Ragusa. Oxford, Oxford U. P., 2000, XIII-220 p. (ill., tables, maps).

3606. SCHRÖTER (Harm G.). Der Verlust der "europäischen Form des Zusammenspiels von Ordnung und Freiheit": von Untergang der deutschen Konsumgenossenschaften. *Vierteljahrschrift für Sozial- und Wirtschaftsgeschichte*, 2000, 87, p. 442-467.

3607. Señores (Los) de la guerra y de la tierra: nuevos textos para el estudios de los Parientes Mayores guipuzcoanos, 1265–1548. Ed. por José Angel LEMA PUEYO [et al.]. Donostia/San Sebastian, Diputación Foral de Guipuzkoa, 2000, 364 p.

3608. SETTIA (Aldo). Viabilità e corti regie nell'Italia occidentale: Marengo e le vie marenche. *Archivio storico italiano*, 2000, 158, 585, p. 439-460.

3609. SHATZMILLER (Joseph). Shylock revu et corrigé: les juifs, les chrétiens et le prêt d'argent dans la société médiévale. Paris, Les belles lettres, 2000, 327 p.

3610. Social identity in early medieval Britain. Ed. by William O. FRAZER and Andrew TYRREL. London a. New York, Leicester U. P., 2000, XIV-283 p. (ill.). (Studies in the Early History of Britain).

3611. STAHL (Alan M.). Zecca: the mint of Venice in the Middle Ages. Baltimore a. London, Johns Hopkins U. P., in association with the American Numismatic Society, 2000, XV-497 p. (ill., graphs, tables, maps).

3612. Survey (The) of archbishop preacher's Kentish Manors, 1283–1285. Ed. by Kenneth WITNEY. Maidstone, Kent Archaeological Society, 2000, LXXXIV-390 p. (ill., maps).

3613. TOGNETTI (Sergio). Uno scambio diseguale. Aspetti dei rapporti commerciali tra Firenze e Napoli nella seconda metà del Quattrocento. *Archivio storico italiano*, 2000, 158, 585, p. 461-490.

3614. Transformació (La) de la frontiera al segle XI. Reflexions des Guissona arran del IX Centenari de la consagració de l'església de Santa Maria. Ed. por Flocel SABATÉ CURULL. Universitat de Lleida, Institut d'Estudis Ilerdencs, 2000, 199 p. (IV Fòrum d'Arqueologia i Història de Guissona).

3615. Treasure in the medieval West. Ed. par Elizabeth M. TYLER. Woodbridge, York Medieval Press, 2000, XI-175 p. (ill.).

3616. URSO (Carmela). Donne e potere nella Gallia merovingia e carolingia. Catania, CULC, 2000, 226 p. (Studi e ricerche dei Quaderni catanesi).

3617. Vie (Le) del Medioevo: atti del Convegno internazionale di studi, Parma, 28 settembre–1 ottobre 1998. A cura di Arturo Carlo QUINTAVALLE. Milano, Electa, 2000, 493 p. (ill.). (I convegni di Parma).

3618. VIN (Jurij Ja.). Mezhdistsiplinarnoe issledovanie srednevekovoj sel'skoj obshchiny: Sotsio-kul'turnyj i istoriko-psikhologicheskij aspekty. (Interdisciplinary study of Medieval rural community: socio-cultural and historical-psychological aspects). *In:* Sravnitel'noe izuchenie tsivilizatsij mira: Mezhdistsiplinarnyj podkhod [Cf. n° 819], p. 230-258.

3619. Violence in medieval society. Ed. by Richard W. KAEUPER. Woodbridge a. Rochester, Boydell a. Brewer, 2000, XIII-226 p.

Cf. n°s 869, 1126, 1159, 3134, 3560, 7237

§ 9. History of civilization, literature, technology and education.

a. Civilization

3620. AURELL I CARDONA (J.). Culture marchande et culture nobilaire à Barcelone au XVe siècle. *Revue historique*, 2000, 124, 613, p. 33-54.

3621. BARCELÓ CRESPÍ (Maria), ENSENYAT I PUJOL (Gabriel). Els nous horitzons culturals a Mallorca aò final de l'Edat Mitjana. Palma, Documenta Balear, 2000, 228 p. (Menjavents, 36).

3622. BASCHET (Jerôme). Le sein du père: Abraham et la paternité dans l'Occident médiéval. Paris, Gallimard, 2000, 413 p. (ill.). (Le temps des images).

3623. BELOVA (Ol'ga V.). Slavjanskij bestiarij: Slovar' nazvanij i simvoliki. ([Medieval] Slavonic bestiary: a glossary of names and symbols). RAN, In-t slavjanovedenija. Moskva, Indrik, 2000, 318 p. (ill.).

3624. BLUME (Dieter). Regenten des Himmels: Astrologische Bilder in Mittelalter und Renaissance. Berlin, Akademie, 2000, X-486 p., (plates, ill.). (Studien aus dem Warburg-Haus, 3).

3625. Body (The) in late medieval and early modern culture. Ed. by Darryl GRANTLEY and Nina TAUNTON. Aldershot a. Burlington, Ashgate, 2000, XX-388 p. (ill., tables).

3626. CAMILLE (Michael). L'art de l'amour au Moyen Age. Objets et sujets du désir. Köln, Könemann, 2000, 176 p.

3627. CLOUZOT (M.). Le son et le pouvoir en Bourgogne au XVe siècle. *Revue historique*, 2000, 124, 615, p. 615-628.

3628. Concepts and patterns of service in the later Middle Ages. Ed. by Anne CURRY and Elizabeth MATTHEW. Woodbridge a. Rochester, Boydell and Brewer, 2000, XXIII-195 p. (ill.).

3629. COOTE (Lesley A.). Prophecy and public affairs in later medieval England. York, York Medieval Press, in association with Boydell and Brewer, and the Centre for Medieval Studies, University of York, 2000, IX-301 p.

3630. Crusade propaganda and ideology: model sermons for the pacing of the Cross. Ed. by Christoph T. MAIER. Cambridge, Cambridge U. P., 2000, VIII-280 p.

3631. Cultures italiennes (XIIe–XVe siècle). Ed. par Isabelle HEULLANT-DONAT, Gian Mario ANSELMI, Enrico ARTIFONI et Alessandro BARBERO. Paris, Editions du Cerf, 2000, III-394 p.

3632. Etranger (L') au Moyen Age. Actes du congrès de la SHMESP. Paris, Publications de la Sorbonne, 2000, 308 p.

3633. FONTAINE (Jacques). Isidore de Séville. Genèse et originalité de la culture hispanique au temps des Wisigoths. Turnhout, Brepols, 2000, 486 p. (Témoins de notre histoire).

3634. FRÉDÉRIC II DE HOHENSTAUFEN. Frédéric II de Hohenstaufen, "L'art de chasser avec les oiseaux": Le traité de fauconnerie, «De arte venandi cum avibus». Ed. par Anne PAULUS et Baudoin VAN DEN ABEELE. Nogent-le-Roi, J. Laget, 2000, 563 p. (ill.). (Bibliotheca Cynegetica, 1).

3635. GANTET (C.). La dimension «sainte» du Saint-Empire romain germanique. Les représentations du pouvoir en Allemagne entre paix et guerre. *Revue historique*, 2000, 124, 615, p. 629-654.

3636. GREEN (Monika H.). Women's healthcare in the medieval West: text and context. Aldershot a. Burlington, Ashgate, 2000, XX-388 p. (ill., tables).

3637. GUREVIC (Aron Jakovlevic). Contadini e santi: problemi della cultura popolare nel Medioevo. Torino, Einaudi, 2000, XVI-383 p. (Biblioteca Einaudi).

3638. HIGGIT (John). The Murthly Hours: devotion, literacy and luxury in Paris, England and the Gaelic West. London, British Library a. Toronto a. Buffalo, University of Toronto, 2000, XXII-362 p. (plates, ill., tables, CD-ROM).

3639. Inscribing the Hundred Years' War in French and English Cultures. Ed. by Denise N. BAKER. Albany, State University of New York Press, 2000, X-227 p. (SUNY Series in Medieval Studies).

3640. KELLER (Hildegard Elisabeth). My secret is mine: studies on religion and eros in the German middle ages. Leuven, Peeters, 2000, IX-297 p. (Studies in spirituality, supplement, 4).

3641. KLEINSCHMIDT (Harald). Understanding the middle ages: the transformation of ideas and attitudes in the medieval world. Woodbridge a. Rochester, Boydell a. Brewer, 2000, XIX-401 p. (ill.).

3642. Liber para ser: mujeres creadoras e cultura en la Europa medieval. Ed. por Marirí MARTINENGO. Madrid, Narcea, 2000, [s. p.].

3643. Listening to Heloise: the voice of a twelfth-century woman. Ed. by Bonnie WHEELER. New York, St. Martin's Press, 2000, XXII-394 p.

3644. Mariage et sexualité au Moyen Age. Accord ou crise? Colloque international de Conques. Dir. par Michel ROUCHE. Paris, Presses Universitaires de Paris-Sorbonne, 2000, 365 p.

3645. Medieval Futures: Attitudes to the future in the Middle Ages. Ed. by J. A. BURROW and Ian P. WEI. Woodbridge a. Rochester, Boydell and Brewer, 2000, XIV-188 p.

3646. Memory and medieval tomb. Ed. by Elizabeth VALDEZ DEL ALAMO. Aldershot a. Brookfield, Ashgate, 2000, XVII-317 p. (ill., tables, map).

3647. MILLER (Maureen C.). Religion makes a difference: clerical and lay cultures in the courts of northern Italy, 1000–1300. *American historical review*, 2000, 105, 4, p. 1095-1130.

3648. MÖHRING (Hannes). Der Weltkreise der Endzeit: Entstehung, Wandel und Wirkung einer tausendjährigen Weissagung. Stuttgart, Jan Thorbecke, 2000, 526 p. (ill.). (Mittelalter-Forschungen, 3).

3649. Mondo (Il) animale. The world of animals. Firenze, SISMEL-Edizioni del Galluzzo, 2000, 2 vol., 350 p., 697 p. (Micrologus, VIII, 1 e 2).

3650. MOREIRA (Isabel). Dreams, visions, and spiritual authority in Merovingian Gaul. Ithaca a. London, Cornell U. P., 2000, XIII-262 p.

3651. MORPURGO (Piero). L'armonia della natura e l'ordine dei governi (secoli XII–XIV). Firenze, SISMEL-Edizioni del Galluzzo, 2000, 346 p. (Micrologus Library, 4).

3652. MURRAY (Alexander). Suicide in the middle ages. 2. The curse on self-murder. Oxford, Oxford U. P., 2000, XXIV-620 p. (ill.).

3653. NAGY (Piroska). Le don des larmes au moyen âge: un instrument spirituel en quête d'institution (V^e–XIIIe siècle). Paris, Albin Michel, 2000, 447 p. (tables).

3654. NEWHAUSER (Richard). The early history of greed: the sin of avarice in early medieval thought and literature. Cambridge, Cambridge U. P., 2000, XIV-246 p.

3655. NIEDERMAN (Carry J.). Worlds of difference: European discourses of toleration, c. 1100–c. 1550. University Park, Pennsylvania State U. P., 2000, X-157 p.

3656. Occident et Proche-Orient: contacts scientifiques au temps des Croisades: actes du colloque de Louvain-la-Neuve, 24 et 25 mars 1997. Ed. par Isabelle DRAELANTS, Anne TIHON, Baudouin VAN DEN ABEELE. Turnhout, Brepols, 2000, V-404 p. (ill.). (Reminiscences).

3657. PETERS (Christine). Gender, sacrament and ritual: the making and meaning of marriage in late Medieval and early Modern England. *Past and present*, 2000, 169, p. 63-96.

3658. PIPONNIER (Françoise), MANE (Perrine). Dress in the middle ages. New Haven a. London, Yale U. P., 2000, VII.168 p. (ill.).

3659. ROLLO (David). Glamorous sorcery: magic and literacy in the high middle ages. Minneapolis a. London, University of Minnesota Press, 2000, XXV-235 p.

3660. UVAROV (Pavel Ju.). Istorija intellektualov i intellektual'nogo truda v srednevekovoj Evrope: Spetskurs. (A history of intellectuals and intellectual labour in Medieval Europe: a lecture course). Appendix <choice> by Pavel Ju. UVAROV: Universitetskij intellektual u parizhskogo notariusa: k voprosu o 'normal'nom iskljuchenii'. (A University Intellectual at a Parisian Notary: a 'Normal Exeption'). RAN, In-t vseobshchej istorii: Kazanskij gos. un-t. Moskva, [s. n.], 2000, 98 p.

3661. VOISENET (Jacques). Bêtes et hommes dans le monde médiéval. Le bestiaire des clercs du Ve au XIIe siècle. Préf. de Jacques LE GOFF. Turnhout, Brepols, 2000, 535 p.

3662. WILSON (Sephen). The magical universe: everyday ritual magic in pre-modern Europe. London, Hambledon Press, 2000, XXX-546 p.

3663. ZHIVOV (Vladimir M.). Osobennosti retseptsii vizantijskoj kul'tury v Drevnej Rusi. (Features of reception of the Byzantine culture in Old Rus'). *In*. Iz istorii russkoj kul'tury (From the History of Russian Culture). Vol. 1. Drevnjaja Rus' (Old Rus'). Moskva, Jazyki russkoj kul'tury, 2000, p. 586-617.

b. Literature

* 3664. Chaucer's "Pardoner's Prologue" and "Tale": an annotated bibliography, 1900 to 1995. Ed. by Marilyn SUTTON. Toronto a. London, University of Toronto Press, in association with University of Rochester, 2000, LII-445 p. (ill.).

* 3665. Compensum Auctorum Latinorum Medii Aevi (500–1500). Fasc. 1. (Abelardus Petrus-Agobardus Lugdunensis archep.) et Elenchus adbreviationum. Firenze, SISMEL-Edizioni del Galluzzo, 2000, 2 fasc., XLI-86 p., 47 p.

* 3666. Études de littérature médiévale. Recherches actuelles en Hongrie. Ed. par. Katalin HALASZ, Debreceni Egyetem, 2000, 178 p.

** 3667. BRUNI (Leonardo). Laudatio Florentine urbis. A cura di Stefano U. BALDASSARRI. Firenze, SISMEL-Edizioni Galluzzo, 2000, C-48 p. (ill.). (Millennio Medievale, 16).

** 3668. Canzonieri (I) della lirica italiana delle origini. 1. Il Canzoniere Vaticano, Biblioteca Apostolica Vaticana, Vat. Lat. 3793; 2: Il Canzoniere Laurenziano, Firenze, Biblioteca Medicea Laurenziana, Redi 9; 3: Il Canzoniere Palatino, Biblioteca Nazionale Centrale di Firenze, Banco Rari 217, ex Palatino 418. A cura di Lino LENONARDI. Firenze, SISMEL-Edizioni del Galluzzo, 2000, CXV (tables, plates).

** 3669. GÓMEZ SIERRA (Esther). Diálogo entre el prudente rey el sabio aldeano (olim Libro de los pensamientos variables). London, Department of hispanic studies, Queen Mary and Westfield College, 2000, 117 p. (ill.).

** 3670. GOWER (John). Confessio Amantis, 1. Kalamazoo, Medieval Institute Publications, Western Michigan University, for TEAMS in association with University of Rochester, 2000, XII-363 p. (ill.).

** 3671. HEINRICH VON DEM TÜRLIN. Die Krone (Verse 1.12281), nach der Handschrift 2779 der Österreichischen Nationalbibliothek nach Vorarbeiten von Alfred EBENBAUER, Klaus ZATLOUKAL et Horst P. PÜTZ. Hrsg. v. Fritz Peter KNAPP und Manuela NIESNER. Tübingen, Niemeyer, 2000, XXXII-381 p. (Altdeutsche Textbibliothek, 112).

** 3672. «Histoire d'Erec» (L') en prose: Roman du XVe siècle. Ed. par Maria Colombo TIMELLI, Genève, Droz, 2000, 347 p. (ill., tables). (Textes Littéraires Français, 524).

** 3673 ISIDORE DE SÉVILLE. Versus. Ed. por José María SÁNCHEZ MARTÍN. Turnhout, Brepols, 2000, 274 p.

** 3674. JØRGENSEN (John Gunnar). Det tapte håndskriftet Kringla. (The lost manuscript Kringla). Oslo, Det historisk-filosofiske fakultet, Universitetet i Oslo, Unipub, 2000, 253 p. (ill.). (Acta humaniora, 80).

** 3675. Kundrun, nach der ausgabe von Karl BARTSCH. Ed. par Karl STACKMANN. Tübingen, Niemeyer, 2000, XXIX-362 p. (Altdeutsche Textbibliothek, 115).

** 3676. Manawydan uab Llyr. Text from the diplomatic edition of the White Book of Rhydderch, by Gwenogvryn EVANS. Ed. by Patrick K. FORD. Belmont, Ford & Bailie, 2000, XXXI-60 p. (ill.). [text in Middle Welsh with introduction and commentary in English].

** 3677. Mystère (Le) de Jour de Jugement: texte original du XIVe siècle. Ed. par Jean-Pierre PERROT et Jean-Jacques NONOT. Chambéry, Comp'Act, 2000, 281 p.

** 3678. Nibelungenlied (Das) nach der Handschulbibliothek Darmstadt. Ed. par Jürgen VORDERSTEMANN. Tübingen, Niemeyer, 2000, XXXV-161 p. (Altdeutsche Textbibliothek, 114).

** 3679. Óláfs saga Tryggvasonar en mesta, 3. Ed. by ÓLAFUR HALLDÓRSSON. København, C. A. Reitzel, 2000, CCCLII-156 p. (ill., table).

** 3680. Old (The) English poem "Judgement day II": A critical edition with editions of "De die iudicii" and the Hatton 113 Homily "Be domes draeg". Ed. by Graham D. CAIE. Woodbridge a. Rochester, Boydell and Brewer, 2000, XVI-161 p.

** 3681. PETRARCA (Francesco). Il codice degli abbozzi. Edizione e storia del manoscritto vaticano latino 3196. A cura di Laura PAOLINO. Milano, Ricciardi, 2000, X-314 p.

** 3682. PHILETICUS (Martinus). In corruptores latinitatis. A cura di Maria Agata PINCELLI. Roma, Edizioni di Storia e Letteratura, 2000, XLVI-133 p. (Edizione nazionale dei testi umanistici, 12-13).

** 3683. Roman (Le) de Guillaume d'Orange. Ed. par Madeleine TYSSENS, Nadine HENRAD et Louis GEMENNE. Paris, Honoré Champion, 2000, XII-589 p.

** 3684. Serta mediaevalia: textus varii saeculorum X–XIII. 1. Tractatus et epistulae. 2. Poetica. Indices. Ed. by R. B. C. HUYGENS. Turnhout, Brepols, 2000, 2 vol., 652 p., 304 p.

** 3685. Siège (Le) de Barbastre. Ed. par Bernard GUIDOT. Paris, Honoré Champion, 2000, 485 p.

** 3686. WAUQUELIN (Jehan). Les faits et les conquêtes d'Alexandre le Grand de Jehan Wauquelin (XVe siècle). Ed. par Sandrine HÉRICHÉ. Geneva, Droz, 200, 705 p. (ill.).

3687. AERS (David). Faith, ethics and church: writing in England, 1360–1409. Woodbridge a. Rochester, Boydell and Brewer, 2000, XII-153 p.

3688. Alexanderdichtungen im Mittelalter. Kulturelle Selbstbestimmung im Kontext literarischer Beziehungen. Ed. par Jan CÖLLIN, Susanne FRIEDE et Hartmut WULFRAM. Göttingen, Wallstein, 2000, 484 p.

3689. Arthur (The) of Germans. The Arthurian legend in medieval German and Dutch literature. Ed. by W. H. JACKSON and S. A. RANAWAKE. Cardiff, University of wales Press, 2000, XII-337 p. (ill.). (Arthurian literature in middle ages, 3).

3690. BACHORSKI (Hans-Jürgen). Dreams that have never been dreamt at all: interpreting dreams in Medieval literature. *History workshop*, 2000, 49, p. 95-127.

3691. BALDWIN (John W.). Aristocratic life in medieval France: the romances of Jean Renart and Gerbert de Montreuil, 1190-1230. Baltimore a. London, Johns Hopkins U. P., 2000, XIX-359 p.

9. HISTORY OF CIVILIZATION, LITERATURE, TECHNOLOGY AND EDUCATION

3692. BARANSKI (Zygmunt G.). Dante e i segni: saggi per una storia intellettuale di Dante Alighieri. Napoli, Liguori, 2000, XI-231 p.

3693. BENROSE (Stephen). A new life of Dante. Exeter, University of Exeter Press, 2000, XXI-249 p.

3694. BORRA (Antonello). Guittone d'Arezzo e la maschera del poeta: la lirica cortese tra ironia e palinodia. Ravenna, Longo, 2000, 113 p. (Memoria del tempo, 19).

3695. BOYDE (Patrick). Human vices and human worth in Dante's "Comedy". Cambridge, Cambridge U. P., 2000, X-323 p. (ill.).

3696. CHANCE (Jane). Medieval mythography, 2: From the School of Chartres to the Court of Avignon, 1177–1350. Gainesville, University Press of Florida, 2000, XXVI-517 p. (ill., tables).

3697. Charles d'Orléans in England (1415–1440). Ed. by Mary-Jo ARN. Woodbridge a. Rochester, Boydell and Brewer, 2000, X-231 p.

3698. CHILDRESS (Diana). Chaucer's England. North Haven, Shoe String Press, 2000, XVII-137 p. (ill., tables, map).

3699. CINGOLANI (Stefano Maria). Politica, societat i letteratura, claus per a una reinterpretació de "Lo somni" de Bernat Metge. Barcelona, Institució de les Lletres Catalanes, 2000, 106 p.

3700. COLDIRON (A.E.B.). Canon, period, and the poetry of Charles of Orleans: found in translation. Ann Arbour, University of Michigan Press, 2000, XVI-336 p. (ill., tables).

3701. CORNISH (Alison). Reading Dante's stars. New Haven a. London, Yale U. P., 2000, XI-226 p. (ill.).

3702. COULSON (Frank T.), ROY (Bruno). Incipitarium Ovidianum: a finding guide for texts in Latin related to the study of Ovid in the Middle Ages and Renaissance. Turnhout, Brepols, 2000, 208 p. (Publications of the Journal of medieval Latin).

3703. Crossing the bridge: comparative essays on medieval European and Heian Japanese women writers. Ed. by Barbara STEVENSON and Cynthia HO. New York a. Basingstoke, Palgrave, 2000, XIV-234 p. (ill.).

3704. CÜNNEN (Janina). Fiktionale Nonnenwelten: Angelsächsische Frauenbriefe des 8. und 9. Jahrhunderts. Heidelberg, C. Winter, 2000, XI-364 p. (ill., tables).

3705. DE CESARE (Giovanni Battista). Letteratura spagnola medioevale. Salerno, Edizioni del Paguro. 2000, 382 p.

3706. "Decameron" (The) and the "Canterbury Tales": new essays on an old question. Ed. by Leonard Michael KOFF and Brenda Deen SCHILDGEN. Madison a. Teaneck, Fairleigh Dickinson U. P. a. London, Associated University Presses, 2000, 352 p.

3707. DELCOURT (Thierry). La littérature arthurienne. Paris, Presses Universitaires de France, 2000, 127 p.

3708. DODWELL (Charles Reginald). Anglo-Saxon gestures and the Roman stage. Cambridge, Cambridge U. P., 2000, 171 p. (tav., ill.). (Cambridge studies in Anglo-Saxon England).

3709. Dutsch Romances. 1. Roman van Walewein. 2. Ferguut. Ed. by David F. JOHNSON and Geert H. M. CLAASSENS. Woodbridge a. Rochester, Boydell and Brewer, 2000, IV-541 p.

3710. FOEHR-JANSSENS (Yasmina). La veuve en majesté: Deuil et savoir au féminin dans la littérature médiévale. Genève, Droz, 2000, 301 p. (Publications Romanes et Françaises, 226).

3711. GALDERISI (Claudio). Une poétique des enfances. Fonctions de l'incongru dans la littérature française médiévale. Orléans, Paradigme, 2000, 259 p. (Medievalia, 34).

3712. GASTALDELLI (Ferruccio). Scritti di letteratura, filologia e teologia medievali. Spoleto, Centro Italiano di Studi sull'Alto Medioevo, 2000, XIV-405 p. (plate, diagrams).

3713. Geistliches in weltlicher und Weltliches in geistlicher Literatur des Mittelalters. Hrsg. v. Christoph HUBER, Burghart WACHINGER und Hans-Joachim ZIEGLER. Tübingen, Max Niemeyer, 2000, VI-348 p.

3714. GOUTTEBROZE (Jean-Guy). Le précieux sang de Fédecamp: origine et développement d'un mythe chrétien. Paris, Honoré Champion, 2000, 111 p. (Essais sur le Moyen Age, 23).

3715. GRIGSBY (John L.). The "Gab" as a latent genre in Medieval French literature: drinking and boasting in the Middle Ages. Cambridge, Medieval Academy of America, 2000, IX-255 p. (tables). (Medieval Academy Books, 103).

3716. Guerres, voyages et quêtes au Moyen Âge. Mélanges de littérature médiévale offerts à Jean-Claude Faucon. Ed. A. LABBÉ, D. W. LACROIX et D. QUÉREL. Paris, Champion, 2000, 472 p.

3717. Heinrich Hallers Übersetzung von „De spiritualibus ascensionibus" des Gerald Zerbold von Zutphen; James Hogg, Kartäuserhandschriftenbestände in öffentliche Bibliotheken Frankreichs. Hrsg. v. Erika BAUER. Salzburg, Institut für Anglistik und Amerikanistik, Universität Salzburg, 2000, 120 p. (ill.). (Analecta Cartusiana, 165).

3718. HILL (John M.), SINNREICH-LEVI (Deborah M.). The rhetorical poetics of the Middle Ages: reconstructive polyphony. Essays in honor of Robert O. Payne. London, Madison and Teaneck, Farleigh Dickinson U. P. a. Associated University Presses, 2000, 304 p. (ill.).

3719. HUBNER (Gert). Lobblumen: Studien zur Genese und Funktion der Geblumten Rede. Tübingen, A. Francke, 2000, X-504 p. (Bibliotheca germanica).

3720. KELLERMANN (Karina). Abschied vom "historischen Volkslied": Studien zu Funktion, Ästhetik und Publizität der Gattung historisch-politische Ereignisdichtung. Tübingen, Max Niemeyer, 2000, IX-416 p.

3721. King Arthur in the medieval Low Countries. Ed. by Geert H. M. CLAASSENS and David F. JOHNSON. Leuven, Leuven U. P., 2000, XIII-274 p. (ill., tables). (Mediaevalia Lovaniensia, 1/28).

3722. Literature (The) of Andalus. Ed. by María Rosa MENOCAL, Raymond P. SCHEINDLIN and Michael SELLS. Cambridge, Cambridge U. P., 2000, IX-507 p. (ill.).

3723. LIUZZA (R.M.). Beowulf. Peterbourgh, Broadview Press, 2000, 242 p.

3724. LYNDE-RECCHIA (Molly). Prose, verse, and truth-telling in the thirteenth century: an essay on form and function in selected texts, accompanied by an edition of the prose "Thèbes" as found in the "Histoire ancienne jusqu'à César". Lexington, French Forum, 2000, 206 p. (Esward C. Armstrong Monographs on Medieval Literature, 10).

3725. MADDOX (Donald). Fictions of identity in medieval France. Cambridge, Cambridge U. P., 2000, XX-295 p. (ill., tables).

3726. Manuscript, narrative, lexicon: essays on literary and cultural transmission in honour of Whitney F. Bolton. Ed. by Robert BOENIG, Lewisburg, Bucknell University Press a. London, Associated U. P., 2000, 261 p. (ill., tables).

3727. Medieval literature and historical inquiry. Essays in honor of Derek Pearsall. Ed. by David AERS. Woodbridge a. Roachester, Boydell and Brewer, 2000, XV-212 (ill.).

3728. NECHUTOVA (Jana). Latinska literatura ceskeho stredoveku do roku 1400. Vysehrad, Typografie Clara Istlerova, 2000, 365 p.

3729. NELSON (Janet L.). Early medieval biography. *History workshop journal*, 2000, 50, p. 129-136.

3730. OGAWA (Hiroshi). Studies in the history of old English prose. Tokyo, Nan'un-do, 2000, X-295 p. (tables).

3731. Old Icelandic literature and society. Ed. by Margaret CLUNIES ROSS. Cambridge, Cambridge U. P., 2000, XII-336 p. (ill., tables).

3732. Petrarca e i suoi lettori. A cura di Vittorio CARATOZZOLO e Georges GÜNTERT. Ravenna, Longo, 2000, 211 p. (ill, tables).

3733. POE (Elizabeth W.). Compilatio: lyric texts and prose commentaries in troubadour manuscript H (Vat. Lat. 3207). Lexington, French Forum, 2000, 307 p. (Edward C. Armstrong Monographs on Medieval Literature, 11).

3734. Poésie lyrique latine du Moyen Age. Ed. par Pascale BOURGAIN. Paris, Librairie générale française, 2000, 352 p. (Le livre de poche. Lettres gothiques).

3735. Proceedings of the Ninth Colloquium. Ed. by Andrew M. BERESFORD and Alan DEYERMOND. London, Department of Hispanic Studies, Queen Mary and Westfield College, 2000, 238 p. (Papers of the Medieval Hispanic Research Seminar, 26).

3736. Proceedings of the tenth colloquium. Ed. by Alan DEYERMOND. London, Department of Hispanic studies, Queen Mary and Westfield College, 2000, 160 p. (ill., tables).

3737. RAFFA (Guy P.). Divine dialectic: Dante's incarnational poetry. Toronto, Buffalo a. London, University of Toronto Press, 2000, XII-254 p.

3738. Rewriting Old English in the twelfth century. Ed. by Mary SWAN and Elaine M. TREHARNE. Cambridge, Cambridge U. P., 2000, X-213 p. (Cambridge studies in Anglo-Saxon England).

3739. "Richard the Redeless" and "Mum and the Sothsegger". Ed. by James M. DEAN. Kalamazoo, Medieval Institute Publications, Western Michigan University, 2000, VIII-175 p.

3740. ROBEY (David). Sound and structure in the "Divine Comedy". Oxford, Oxford U. P., 2000, X-204 p. (tables).

3741. ROUBAUD-BÉNICHOU (Sylvia). Le roman de chevalerie en Espagne. Entre Arthur et Don Quichotte. Paris, Champion, 2000, 404 p.

3742. SHERWOOD-SMITH (Maria C.). Studies in the reception of the "Historia scholastica" of Peter Comestor: the "Schwarzwälder Predigten", the "Weltchronik" of Rudolf von Ems, the "Scholastica" of Jacob van Maerlant and the "Historiebijbel van 1360". Oxford, Society for the Study of Medieval Languages and Literature, 2000, IX-181 p. (tables).

3743. Spirit (The) of medieval English popular Romance. Ed. by Ad PUTTER and Jane GILBERT. Harlow, Pearson Education, 2000, VBIII-304 p.

3744. SUMMIT (Jennifer). Lost property: the woman writer and English literary history, 1380–1589. Chicago u. London, University of Chicago Press, 2000, X-274 p. (ill.).

3745. TERRY (Arthur). Three fifteenth-century Valencian poets. London, Department of Hispanic studies, Queen Mary and Westfield College, 2000, 64 p.

3746. Testimonianze dantesche nella Biblioteca Estense universitaria, sec. XIV–XX. A cura di Ernesto MILANO. Modena, Il bulino, 2000, 351 p. (ill.) (Il giardino delle Esperidi).

3747. Textos épicos castellanos: Problemas de edición y crítica. Ed. por David G. PATTISON. London, Department of hispanic Studies, Queen Mary and Westfield College, 2000, 138 p. (ill., table).

3748. Tradizioni patristiche nell'umanesimo. Atti del Convegno Istituto Nazionale Studi sul Rinascimento, Biblioteca Medicea Laurenziana. Firenze, 6–8 febbraio 1997. A cura di Mariarosa CORTESI e Claudio LEO-

NARDI. Firenze, SISMEL-Edizioni del Galluzzo, 2000, 425 p. (Millennio Medievale, 17 – Atti di convegni, 4).

3749. VEGLIA (Marco). "La vita lieta": una lettura del "Decameron". Ravenna, Longo, 2000, XII-388 p. (Memoria del Tempo, 17).

3750. WAITE (Greg). Old English prose translations of King Alfred's reign. Woodbridge a. Rochester, Boydell and Brewer, 2000, XIV-394 p. (diagrams).

3751. WHITE (Hugh). Nature, sex and goodness in a medieval literary tradition. Oxford a. New York, Oxford U. P., 2000, IX-278 p.

3752. YAMAMOTO (Dorothy). The boundaries of the human in medieval English literature. Oxford a. New York, Oxford U. P., 2000, XI-257 p. (diagrams).

Cf. n° 36

c. Technology

3753. Artisan (L') au village dans l'Europe médiévale et moderne. Actes des XIXe Journées internationales d'Histoire de l'Abbaye de Flaran, 5–7 septembre 1997. Toulouse, [s. n.], 2000, 336 p.

3754. Brique (La) antique et médiévale. Production et commercialisation d'un matériau. Actes du colloque international organisé par le Centre d'histoire urbaine de l'école normale supérieure de Fontenay/Saint Cloud et l'école française de Rome (Saint-Cloud, 16–18 novembre 1995). Ed. par Patrick BOUCHERON, Henri BROISE et Yvon THÉBERT. Roma, École française de Rome, 2000, X-486 p. (tables, ill., graph.).

3755. Fortifications (Le) dans les domaines Plantagenêt. XIIe–XIVe siècles. Actes du Colloque international tenu à Poitiers du 11 au novembre 1994. Ed. par Marie-Pierre BAUDRY. Poitiers, Centre d'études supérieures de Civilisation médiévale, 2000, 138 p. (maps, ill). (Civilisation médiévale, 10).

3756. GIOSTRA (Caterina). L' arte del metallo in età longobarda: dati e riflessione sulle cinture ageminate. Spoleto, Centro italiano di studi sull'Alto Medioevo, 2000, 136 p. (ill.). (Studi e ricerche di archeologia e storia dell'arte).

3757. HOWLETT (David). Caledonian craftsmanship. The Scottish Latin tradition. Dublin, Four Courts Press, 2000, VIII-208 p.

3758. Laterizi (I) nell'alto medioevo italiano. A cura di Sauro GELICHI e Paola NOVARA. Ravenna, Società di Studi Ravennati, 2000, 221 p. (plates, ill., table). (Biblioteca di "Ravenna Studi e Ricerche," 3).

3759. LOHRMANN (Dietrich). Technischer Austausch zwischen Ost und West zur Zeit der Kreuzzüge. *Archiv für Kulturgeschichte,* 2000, 82, p. 319-344.

3760. Working with water in Medieval Europe: technology and resource-use. Ed. by Paolo SQUATRITI. Leiden, Boston a. Köln, Brill, 2000, XX-446 p. (ill.).

d. Education

3761. BOUREAU (Alain). La censure dans les universités médiévales (note critique). *In*: Texte et paratexte au Moyen Age. *Annales,* 2000, 55, 2, p. 313-324.

3762. CLARAMUNT (Salvador), CONDE (Rafael). Privilegi de fundació de la Universitat de Barcelona, 1450. Barcelona, Publicacions de la Universitat de Barcelona, Copp., 2000, [s. p.].

3763. Enseñanza (La) en la Edad media: X Semana de Estudios Medievales: Nájera, 1999. Logroño, Instituto de Estudios Riojanos, 2000, 520 p.

3764. Gakumon eno tabi: Yôroppa chûsei. (A Journey for learning: Medieval Europe). Ed. by Shozaburo KIMURA. Tokyo, Yamakawa Shuppansha, 2000, 281p.

3765. Księga promocji Wydziału Sztuki Uniwersytetu Krakowskiego z XV wieku. (Das Promotionsbuch der Fakultät für Kunst der Krakauer Universität vom XV. Jahrhundert). Hrsg. v. Antoni GĄSIOROWSKI. Kraków, PAU, 2000, 249 p.

3766. ROEST (Bert). A history of Franciscan education (c.1210–1517). Leiden, Boston a. Köln. Brill, 2000, X-405.

3767. SCHEMMANN (Ulrike). Confessional literature and lay education: the "Manuel dé pechez" as a book of good conduct and guide to personal religion. Düsseldorf, Droste, 2000, VI-361 p.

3768. SCHMUTZ (Jürg). Juristen für das Reich. Die deutschen Rechtsstudenten an der Universität Bologna 1265–1425. Basel, Schwabe & Co., 2000, 2 vol., 312 p., 800 p.

3769. Scienze matematiche e insegnamento in epoca medioevale: atti del Convegno internazionale di studio, Chieti, 2–4 maggio 1996. A cura di Paolo FREGUGLIA, Luigi PELLEGRINI e di Roberto PACIOCCO. Napoli, Edizioni scientifiche italiane, 2000, 325 p. (Biblioteca di Studi medievali e moderni Sezione medievale).

3770. Universities and schooling in medieval society. Ed. by William J. COURTENAY and Jürgen MIETHKE. Leiden, Boston a. Köln, Brill, 2000, VI-244 p.

3771. WHEATLEY (Edward). Mastering Aesop: medieval education, Chaucer, and his followers. Gainesville, University Press of Florida, 2000, IX-278 p.

Cf. n° 17

§ 10. History of art.

3772. Affreschi in Val Comino e nel Cassinate. A cura di Giulia OROFINO. Cassino, Edizioni dell'Università degli studi, 2000, 277 p. (ill.).

3773. ANTIPOV (Il'ja V.). Drevnerusskaja arkhitektura vtoroj poloviny XIII–pervoj treti XIV v.: Katalog pamjatnikov. (Old Russian architecture, c. 1239–1330: catalogue of monuments). Sankt-Peterburgskij gos. un-

t. Ed. Val. A. BULKIN. Sankt-Peterburg, Izd-vo Sankt-Peterburgskogo un-ta, 2000, 203 p. (ill.). [Eng. summary]

3774. Art, politics, and civil religion in central Italy, 1261–1352. Ed. by Joanna CANNON and Betj WILLIAMSON. Aldershot a. Brookfield, Ashgate, 2000, IX-317 p. (ill., tables).

3775. Attila (From) to Charlemagne: Arts of the early medieval period in the Metropolitan Museum of Art. Ed. by Katharine Reynolds BROWN, Dafydd KIDD and Charles T. LITTLE. New York, Metropolitan Museum of Art a. New Haven, Yale U. P., 2000, XVI-395 p. (ill.).

3776. BELJAEV (Leonid A.). Obshcheevropejskie elementy v drevnerusskom iskusstve, X–XII v. (General European elements in the Old Russian art, 10th–12th centuries). In: Iz istorii russkoj kul'tury (From the History of Russian Culture). Vol. 1. Drevnjaja Rus' (Old Rus'). Moskva, Jazyki russkoj kul'tury, 2000, p. 732-755.

3777. CABAÑERO SUBIZA (Bernabé). La techumbre mudéjar de la sala capitular del Monasterio de Sijena (Huesca): nuevos datos para el estudio de la evolución durante el siglo XII de los modelos de tableros geométricos de la Aljafería de Zaragoza. Tarazona, centro de Estudios Turiasoenenses, 2000, 109 p. (ill.).

3778. CADOGAN (Jean K.). Domenico Ghirlandaio: artist and artisan. New Haven a. London, Yale U. P., 2000, XII-425 p. (plates).

3779. CHRISTE (Yves). Il giudizio universale nell'arte del Medioevo. Milano, Jaca Book, 2000, 371 p. (ill.). (Complementi alla storia dell'arte europea).

3780. CIRANNA (Simonetta). Spolia e caratteristiche del reimpiego nella basilica di San Lorenzo fuori le mura a Roma. Roma, Edizioni Librerie Dedalo, 2000, 172 p. (ill.).

3781. CLARK (Gregory T.). Made in Flanders: the master of the Ghent Privileges and manuscript painting in the Southern Netherlands in the time of Philip the Good. Turnhout, Brepols, 2000, 499 p. (ill., tables).

3782. CLAUDE (Sandrine). Le château de Gréoux-les-Bains (Alpes-de-Haute-Provence): une résidence seigneuriale du moyen âge à l'époque moderne. Paris, Maison des Sciences de l'Homme, 2000, 187 p. (tables). (Documents d'Archéologie Française, 80).

3783. COLLINS (Minta). Medieval herbals: the illustrative traditions. London, British Library a. Toronto, University of Toronto Press, 2000, 334 p. (ill.). (British Library studies in medieval culture).

3784. Congreso de Archiveros de la iglesia en España (14°. 1998. Barcelona). Arte y archivos de la iglesia. Santoral ispano-mozárabe en las diócesis de España: actas del XIV Congreso, Barcelona, 1998. Ed por Augustín BALLINA. Oviedo, Asociación de Archiveros de la Iglesia en España, 2000, [s. p.]. (Memoria Ecclesiae, 16).

3785. COOMANS (Thomas). L'abbaye de villers-en-Brabant: construction, configuration et signification d'une abbaye cistercienne gothique. Brussel e. Brecht, Cîteaux commentarii cistercienses, 2000, 622 p. (ill., plans, maps).

3786. CORLEY (Brigitte). Painting and patronage in Cologne, 1300–1500. Turnhout, Harvey Miller, 2000, 342 p.(ill.).

3787. CORONEO (Roberto). Scultura mediobizantina in Sardegna. Nuoro, Poliedro, 2000, 287 p. (ill.).

3788. DIEBOLD (William J.). Word and image: an introduction to early medieval art. Boulder a. Oxford, Westview Press, 2000, XII-160 p. (ill.).

3789. DUBY (Georges). L'art et l'image: une anthologie. Marseille, Parentheses, 2000, 203 p. (ill.). (Eupalinos).

3790. EMERY (Anthony). Greater medieval houses of England and Wales, 1300–1500. 2. East Anglia, Central England, and Wales. Cambridge, Cambridge U. P., 2000, XV-724 p. (ill., plates).

3791. ENNABLI (Liliane). La Basilique de Carthagenna et le locus des sept moines de Gafsa: nouveaux édifices chrétiens de Carthage. Paris, CNRS ed., 2000, 150 p. (ill.). (Etudes d'antiquités africaines).

3792. Fabric (The) of images: European paintings on textile supports in the fourteenth and fifteenth centuries. Ed. by Caroline VILLERS. London, Archetype Publications, 2000, X-117 p. (ill., table).

3793. FERNIE (Eric). The architecture of Norman England. Oxford a. New York, Oxford U. P., 2000, XVIII-352 p. (plates, ill., tables).

3794. FIGGE (Valerie). Das Bild des Bischofs: Bischofsviten in Bilderzählungen des 9. bis 13. Jahrhunderts. Weimar, VDG, 2000, 234 p. (ill.). (Marburger Studien zur Kunst- und Kulturgeschichte).

3795. GARCÍA MARSILLA (Juan Vicente). Le immagini del potere e il potere delle immagini. I mezzi iconici al servizio della monarchia aragonese nel basso medioevo. In: Monumento (Il): arte e storia [Cf. n° 1283], p. 569-602.

3796. GATHERCOLE (Patricia May). The depiction of women in medieval French manuscript illumination. Lewiston, Mellen Press, 2000, IV-139 p. (ill.).

3797. Grabmäler. Tendenzen der Forschung an Beispielen aus Mittelalter und früher Neuzeit. Hrsg. Wilhelm MAIER, Wolfgang SCHMID und Michael Viktor SCHWARZ. Berlin, Gebr. Mann, 2000, 262 p.

3798. GUERRINI (Roberto). Dulci pro libertate. Taddeo di Bartolo: il ciclo di eroi antichi nel Palazzo pubblico di Siena (1413–1414). Tradizione classica ed iconografia politica. In: Monumento (Il): arte e storia [Cf. n° 1283], p. 510-568.

3799. HAMANN (Matthias). Die Burgundische Prioratskirche von Anzy-le-Duc und die romanische

Plastik im Brionnas. Würzburg, Deutscher Wissenschafts-Verlag, 2000, 2 vol., 380 p., 286 p.

3800. HAMBURGER (Jeffrey F.). Peindre au couvent. La culture visuelle d'un couvent médiéval. Paris, Gérard Monfort, 2000, 254 p. (ill.).

3801. HISCOCK (Nigel). Wise (The) Master Builder: platonic geometry in plans of medieval abbeys and Cathedrals. Aldershot a. Brookfield, Ashgate, 2000, XVIII-341 p. (plates, ill.).

3802. HOURIHANE (Colum). Virtue & vice: the personifications in the Index of Christian art. Princeton, Princeton U. P., 2000, XVIII-456 p. (ill.). (Index of Christian art resources).

3803. HOWARD (Deborah). Venice and the East: the impact of the Islamic world on Venetian architecture, 1100–1500. New Haven a. London, Yale U. P., 2000, XVI-283 p. (ill., maps).

3804. IKEGAMI (Shunnichi). Bannô-jin to Medhichike no seiki: Runesansu saikô. (A century of Uomo universale and the Medicis: a reconsideration of the Renaissance). Tokyo, Kodansha, 2000, 284 p.

3805. Index (An) of images in English manuscripts from the time of Chaucer to Henry VIII, c. 1380–c. 1509: the Bodleian library, Oxford, 1: MSS additional-digby. Ed. by Ann Eljenholm NICHOLS, Michael T. ORR, Kathleen L. SCOTT and Lynda DENNISON. Turnhout, Brepols, 2000, 144 p. (ill.).

3806. KESSLER (Herbert L.). Spiritual seeing: picturing God's invisibility in medieval art. Philadelphia, University of Pennsylvania Press, 2000, XV-267 p. (ill.). (The Middle Ages Series).

3807. LAZAREV (Viktor Nikitic). Studies in early Russian art. London, Pindar Press, 2000, VI-638 p. (ill.).

3808. LEPORE (Giuseppe). Edifici di culto cristiano nella valle del Cesano. Pesaro-Ancona: la documentazione storica e archeologica tra tardo antico e Medioevo. Imola, University press Bologna, 2000, 254 p. (ill.). (Studi e scavi).

3809. MAEKAWA (Kumiko). Narrative and experience. Innovations in thirteenth-century picture books. Frankfurt am Main, Berlin, Bruxelles, Wien a. New York, 2000, 349 p. (ill., tables).

3810. MANNELLI (Maria Francesca), PUNTONI (Gabriella). Catalogo critico delle chiese medievali della Versilia. Pisa, ETS, 2000, 266 p. (ill.).

3811. MARTINCIC (Lorena). Iconografia del Davide musico nella miniatura medievale dall'VIII al XIII secolo. In: Monumento (Il): arte e storia [Cf. n° 1283], p. 475-509.

3812. Medieval mosaics: light, color, materials. Ed. by Eve BORSOOK, Fiorella GIOFFREDI SUPERBI and Giovanni PAGLIARULO. Milano, Silvana; Firenze, Harvard University Center for Italian Renassaince Studies at Villa I Tatti. 2000. 328 p. (plates, ill., tables).

3813. MILLER (Maureen Catherine). The bishops palace: architecture and authority. Ithaca, London, 2000, XV-307 p. (ill.). (Conjunctions of religion & power in the medieval past).

3814. Novgorodskij istoricheskij sbornik. (The Novgorod historical review). Vol. 8 (18). 800-letiju Nereditsy posvjashchaetsja (To the 800th Anniversary of the Saviour Church in Nereditsa [articles on history, architecture and wall-paintings of the church]). Eds. Valentin L. JANIN, Elisa A. GORDIENKO [et al.]. Sankt-Peterburg, Dmitry Bulanin, 2000, 390 p. (ill., maps). [Cf. n° <choice> 4024.]

3815. Opere (Le) e i nomi: prospettive sulla firma medievale: in margine ai lavori per il corpus delle opere firmate del Medioevo italiano. A cura di Maria Monica DONATO e Monia MANISCALCHI. Pisa, Scuola normale superiore, Centro di ricerche informatiche per i beni culturali, 2000, 69 p. (ill.).

3816. PACE (Valentino). Arte a Roma nel Medioevo: committenza, ideologia e cultura figurativa in monumenti e libri. Napoli, Liguori, 2000, IX-509 p. (ill.). (Biblioteca. Nuovo Medioevo).

3817. PARDI (Renzo). Architettura religiosa medievale in Umbria. Spoleto, Centro italiano di studi sull'Alto Medioevo, 2000, X-554 (ill.).

3818. PICCININI (Chiara). Capitelli a foglie nella Firenze del Due e Trecento: "Fogliame rustico e barbaro". Florence, Leo S. Olschki, 2000, X-96 p. (plates, ill.). (Fondazione Carlo Marchi, Studi, 13).

3819. PRACHE (Anne). Cathedrals of Europe. Ithaca a. London, Cornell U. P., 2000, 279 p.

3820. Rome: art et archéologie. Ed. par Andrea AUGENTI. Paris, Hazan, 2000, 223 p. (ill.).

3821. ROSARIO (Iva). Art and propaganda: Charles IV of Bohemia, 1346–1378. Woodbridge a. Rochester, Boydell and Brewer, 2000, XVII-155 p. (plates, ill.).

3822. TURCUŞ (Veronica). La storiografia sull'architettura cistercense europea. Tradizione e nuovi lineamenti. *Studia Universitatis Babeş-Bolyai. Historia*, 2000, 45, 1-2, p. 195-204.

3823. ULUÇAM (Abdüsselam). Ortaçağ ve sonrasında Van Gölü çevresi mimarlığı. (Architecture of the Van lake and its environs in the middle ages and aftermath). Ankara, Kültür Bakanlığı Yayımlar Dairesi Başkanlığı, 2000, [s. p.].

Cf. n°s 47, 53, 61, 81, 299, 377, 792, 1282, 3838, 3999

§ 11. History of music.

** 3824. Catalogo del fondo musicale del duomo di Castelfranco Veneto. A cura di Franco ROSSI. Venezia, Fondazione Levi, 2000, LXXX-441 p. (Studi musicologici. Cataloghi e bibliografie).

** 3825. Catalogo del fondo musicale di Santa Maria in Trastevere nell'Archivio storico del Vicariato di Roma: tre secoli di musica nella Basilica romana di

Santa Maria in Trastevere. A cura di Eleonora SIMI BONINI. Roma, IBIMUS, 2000, 447 p.

** 3826. Fondo (Il) musicale della Basilica di San Prospero a Reggio nell'Emilia. A cura di Stefania RONCROFFI. Firenze, L. S. Olschki, 2000, 129 p. (tav.).

** 3827. Fondo (Il) musicale della Cappella regia sabauda. A cura Enrico DEMARIA; introduzione di Marie-Therese BOUQUET BOYER. Lucca, Libreria musicale italiana, 2000, XLIX-429 p.

3828. AUBREY (Elizabeth). The music of the troubadours. Bloomington a. Indianapolis, Indiana U. P., 2000, XXIV-327 p. (tables, map). (Music: Scholarship and Performance).

3829. BAROFFIO (Giacomo). Ipsi canamus gloriam: i frammenti liturgici latini dell'Archivio storico comunale di Nonantola. Comune di Nonantola, 2000, 67 p., 1 CD-ROM (ill.).

3830. Cobras e son: papers on the text, music and manuscripts of the "cantigas de Santa Maria". Ed. by Stephen PARCKINSON. Oxford, Modern Humanities Research Association, 2000, XIII-246 p. (ill., tables).

3831. Coventry (The) corpus Christi plays. Ed. by Pamela M. KING and Clifford DAVIDSON. Kalamazoo, Medieval Institute publications, 2000, XII-326 p. (ill., map). (Early Drama, Art, and Music, Monograph Series, 27).

3832. CROCKER (Richard L.). An introduction to Gregorian Chant. New Haven a. London, Yale U. P., 2000, VIII-248 p. (CD-ROM, plates, ill.).

3833. CUNNINGHAM (Martin G.). Alfonso X el Sabio: Cantigas de loor. Dublin, University College Dublin Press, 2000, VIII-280 p.

3834. DAVIS (Lisa Fagin). The Gottschalk antiphonary: music and liturgy in twelfth-century Lambach. Cambridge, Cambridge U. P., 2000, XV-316 p. (ill., tables). (Cambridge studies in Paleography and Codicology, 8).

3835. Vacat.

3836. FRITZ (Jean-Marie). Paysages sonores du Moyen Âge. Paris, Champion, 2000, 477 p.

3837. HENTSCHEL (Frank). Sinnlichkeit und Vernunft in der mittelalterlichen Musiktheorie: Strategien der Konsonanzwertung und der Gegenstand der "Musica sonora" um 1300. Stuttgart, Franz Steiner, 2000, 368 p. (ill., diagrams).

3838. Jubilate Deo: miniature e melodie gregoriane. Testimonianze della Biblioteca L. Feininger. A cura di Giacomo BAROFFIO, Danilo CURTI e Marco GOZZI. Trento, Provincia autonoma, Servizio beni librari e archivistici, 2000, 383 p. (ill.).

3839. Oral (The) epic: performance and music. Ed. by Karl REICHL. Berlin, Verlag für Wissenschaft und Bildung, 2000, VIII-248 p. (ill., tables).

3840. PASQUINI (Elisabetta). Libri di musica a Firenze nel Tre-Quattrocento. Firenze, L. S. Olschki, 2000, 195 p. (tav., ill.). (Studi e testi per la storia della musica).

3841. Performer's (A) guide to medieval music. Ed. by Martin J. DUFFIN. Bloomington a. Indianapolis, Indiana U. P., 2000, XIII-599 p. (ill.). (Early Music America: Performer's Guides to Early Music).

3842. Segno e musica: codici miniati e musicali nel millenario della nascita di Guido d'Arezzo. A cura di Giacomo BAROFFIO. Milano, Mazzotta, 2000, 130 p. (ill.).

3843. STEYN (Frances Caroline). Three Unknown Carthusian liturgical manuscripts, with music of the 14th to the 16th centuries in the Grey Collection, South African Library, Cape Town. Salzburg, Institut für Anglistik und Amerikanistik, Universität Salzburg, 2000, XI-195 p. (ill., tables).

3844. Sussex. Ed. by Cameron LOUIS. Turnhout a. Toronto, University of Toronto Press, 2000, CIX-403 p. (maps). (Records of Early English Drama).

3845. TAMMEN (Björn R.). Musik und Bild im Chorraum mittelalterlichen Kirchen, 1100–1500. Berlin, Reimer, 2000, 553 p. (ill., tables).

§ 12. History of philosophy, theology and science.

a. Sources

3846. ABŪ MA'ŠAR. Abū Ma'šar on historical astrology: the book of religions and dynasties (On the great conjunctions). 1. The Arabic original. 2. The Latin versions. Ed. by Keiji YAMAMOTO and Charles BURNETT. Leiden, Boston a. Köln, Brill, 2000, 2 vol., XXVII-620 p., XXXIII-578 p. (Islamic Philosophy, Theology and Science: texts and studies, 33-34).

3847. Aristoteles Latinus, 17/2/1/1: De historia animalium, translatio Guillelmi de Morbeke. 1. Lib. I–V. Ed. by Pieter BEULLENS and Fernand BOSSIER. Leiden, Boston a. Köln, E.J.Brill, 2000, XCIII-172 (ill.).

3848. ARNAU DE VILANOVA. Tractatus de intenzione medicorum. Ed. por Michael MAC VAUGH. Barcelona, Publicacions de la Universitat de Barcelona, 2000, 222 p. (Opera medica omnia).

3849. AUGUSTINUS DE FERRARIA. Quaestiones super librum Praedicamentorum Aristotelis. Ed. by Robert ANDREWS. Stockholm, Almqvist & Wiksell, 2000, XXXIX-308 p. (ill.).

3850. AVERROES (Ibn Rushd). Averroes' Middle Commentary on Aristotle's «Poetics». South Bend, St. Augustine Press, 2000, XXI-161 p. (tables).

3851. BENZO D'ALESSANDRIA. Il "Chronicon" di Benzo d'Alessandria e i classici latini all'inizio del XIV secolo. Edizione critica del libro XXIV: "De moribus et vita philosophorum". A cura di Marco PETOLETTI. Milano, Vita e Pensiero, 2000, XXIII-389 p. (plates).

3852. BURLEY (Walter). On the purity of the art of logic: the shorter and the longer treaties. New Haven a. London, Yale U. P., 2000, XXXV-323 p. (tables). – IDEM. Quaestiones super librum Posteriorum. Ed. by Mary Catherine SOMMERS. Toronto, Pontifical Institute of Medieval Studies, 2000, X-214 p. (table). (Studies and Texts, 136).

3853. Giardino (Il) magico degli alchimisti: un erbario illustrato trecentesco della Biblioteca universitaria di Pavia e la sua tradizione. Introduzione, edizione critica e commento di Vera SEGRE RUTZ. Milano, Il polifilo, 2000, XCI-338 p. (ill.).

3854. GUILLELMUS DURANTIUS. Guillelmi Duranti Rationale divinorum officiorum VII–VIII: Praefatio, indices. Ed. by A. DAVRIL, O.S.B. and T. M. THIBODEAU. Turnhout, Brepols, 2000, 470 p.

3855. RUSBROCHIUS (IOANNES). Ioannis Rusbrochii Ornatus spiritualis desponsationis, Gerardo Magno interprete. Ed. by Rijcklof HOFMAN. Turnhout, Brepols, 2000, XCVII-230 p. (ill.).

3856. LULLUS (Raimundus). Raimundi Lulli opera Latina, 92–96: In civitate Maiorcensi anno MCCC composita. Ed. por Fernando Domínguez REBOIRAS. Turnhout, Brepols, 2000, XXXVII-424 p. (ill.). (Corpus Christianorum, Continuatio Mediaeualis, 112).

3857. MARSILIUS VON INGHEN. Quaestiones super quattouor libros sententiarum Vol. 1. super primum, quaestiones 1–7. Ed. by Manuel Santos NOYA. Leiden, Boston a. Köln, Brill, 2000, LV-318 p. (Studies in the history of Christian thought, 87).

3858. SYDRAC LE PHILOSOPHE. Le livre de la fontaine de toutes sciences. Edition des enzyklopädischen Lehrdialogs aus dem XIII. Jahrhundert. Ed. Par Ernstpeter RUHE. Wiesbaden, Dr. Ludwig Reichert Verlag, 2000, XVI-490 p. (ill.).

b. Studies

3859. BILLER (Peter). The measure of multitude. Population in medieval thought. Oxford, Oxford U. P., 2000, XXI-476 p. (ill.).

3860. BILLOTTE (Denis). Le vocabulaire de traduction par Jean de Meun de la « Consolatio Philosophiae » de Boèce. Paris, Champion, 2000, 2 vol., 6/10 p., 592 p. (Nouvelle Bibliothèque du Moyen age, 54).

3861. BLACK (Robert). La consolazione della filosofia nel Medioevo e nel Rinascimento italiano: libri di scuola e glosse nei manoscritti fiorentini. Tavarnuzze, Impruneta, XXII-362 p. (tav., ill.) (Biblioteche e archivi).

3862. BOJE (Lars Mortensen). The Anchin manuscript of Passio Olaui (Douai 295), William of Jumièges, and Theodoricus Monachus. New evidence for intellectual relations between Norway and France in the 12th century. *Symbolae Osloenses*, 2000, 75, p. 165-189.

3863. BOYLE (Leonard E., O.P.). Facing history: a different Thomas Aquinas. Louvain-la-Neuve, Féderation Internationale des Instituts d'Etudes Médiévales, 2000, XXXIV-170 p. (ill., tables).

3864. CARABINE (Deidre). John Scottus Eriugena. New York a. Oxford, Oxford U. P., 2000, XI-131 p. (Great Medieval thinkers).

3865. CAROL (Harrison). Augustine: Christian thruth and fractured humanity. Oxford, Oxford, 2000, XVI-242 p. (Christian Theology in context).

3866. CASTILLEJO GORRÁIZ (Miguel). Averroes: el aquinatense islámico. Córdoba, Publicaciones Obra Social y Cultural CajaSur, 2000, 215 p. (ill.). (Colección mayor).

3867. CHABÁS (José), GOLDSTEIN (Bernhard R.). Astronomy in the Iberian peninsula: Abraham Zakut and the transition from manuscript of print. Philadelphia, American Philosophical Society, 2000, XII-196 p. (ill., tables).

3868. CIFUENTES COMAMALA (Lluís). La medicina en las galeras de la Corona de Aragón a finales de la Edad Media: la caja del barbero y sus libros. Barcelona, J. Uriach, 2000, 16 p. (ill.). (Medicina e Historia, 4).

3869. CLANCHY (Michael). Abélard. Paris, Flammarion, 2000, 487 p. (maps).

3870. COHEN (Adam S.). The Uta codex: art, philosophy, and reform in eleventh-century Germany. University Park, Pennsylvania U. P., 2000, XV-276 p. (ill., plates).

3871. COHEN (Esther). The animated pain of the body. *American historical review*, 2000, 105, 1, p. 36-68.

3872. CRUZ HERNANDEZ (Miguel). Il pensiero in al-Andalus, secoli 9.–14. Brescia, Paideia, 2000, p. 400-750.

3873. EVANS (G.R.). Bernard of Clairvaux. New York a. Oxford, Oxford U. P., 2000, IX-220 p. (Great Medieval Things).

3874. FAUPEL-DREVS (Kirstin). Vom rechten Gebrauch der Bilder im liturgischen Raum: Mittelalterliche Funktionsbestimmungen bildender Kunst im "Rationale divinorum officiorum" des Durandus von Mende (1230/1–1296). Leiden, Boston u. Köln, Brill, 2000, XVII-432 p. (ill., tables). (Studies in the history of Christian Thought, 89).

3875. Gioacchino da Fiore tra Bernardo di Clairvaux e Innocenzo III. Atti del 5° Congresso internazionale di studi gioachimiti, San Giovanni in Fiore – 16–21 settembre 1999. A cura di Roberto RUSCONI. Roma, Viella, 2000, 378 p. (ill.). (Opere di Gioacchino da Fiore: testi e strumenti, 13).

3876. GRONDEUX (Anne). Le «Graecismus» d'Evrard de Béthune à travers ses gloses: entre grammaire positive et grammaire spéculative du XIIIe au XVe siècle. Turnhout, Brepols, 2000, VII-553 p. (ill., tables).

3877. HASSE (Dag Nikolaus). Avicenna's "De anima" in the Latin West: the formation of a peripatetic philosophy of the Soul, 1160–1300. Ed. by Dag Nikolaus HASSE. London, Warburg Institute; Torino, Nino Aragno, 2000, X-350 p.

3878. LAGERLUNF (Henrik). Modal syllogistics in the Middle Ages. Leiden, Boston a. Köln, Brill, 2000, XVII-261 p. (ill., tables). (Studien und Texte zur Geistesgeschichte des Mittelalters, 70).

3879. MAC EVOY (James). Robert Grosseteste. Oxford, Oxford U. P., 2000, XX-219 p. (ill.).

3880. MARENBON (John). Aristotelian logic, Platonism, and the context of early Medieval philosophy in the West. Aldershot a. Burlington, Ashgate, 2000, VIII-384 p. (table).

3881. Medicina (La) monastica: atti del Convegno di studi, Istituto di studi politici S. Pio. 5., Roma: Roma, 24 marzo 2000. Roma, APES, 2000, 110 p. (Documenti del nostro tempo).

3882. Philosophie und Theologie des ausgehenden Mittelalters: Marsilius von Inghen und das Denken seiner Zeit. Hrsg. v. Maarten HOENEN und Paul J. J. M BAKKER. Leiden, Boston u. Köln, Brill, 2000, X-322 p.

3883. PODSKALSKY (Gerhard). Theologische Literatur des Mittelalters in Bulgarien und Serbien, 865–1459. München, C.H. Beck, 2000, X-578 p.

3884. REEVES (Marjorie), GOULD (Warwick). Gioacchino da Fiore e il mito dell'Evangelo eterno nella cultura europea. Roma, Viella, 2000, XXVII-373 p. (ill.). (Opere di Gioacchino da Fiore: testi e strumenti, 12).

3885. RYAN (Thomas F.). Thomas Aquinas as reader of the Psalms. Notre Dame, University of Notre Dame Press, 2000, IX-233 p. (tables, diagram).

3886. SCHABEL (Chris). Theology at Paris, 1316–1345: Peter Auriol and the problem of divine foreknowledge and future contingents. Aldershot a. Burlington, Ashgate, 2000, XI-368 p.

3887. SCHLOSSER (Marianne). Lucerna in caliginoso loco: Aspekte des Prophetie-Begriffes in der scholastischen Theologie. Paderborn, Ferdinand Schöningh, 2000, XXXII-317 p. (Münchener Universitätsschriften, Katholisch-Theologische Fakultät; Veröffentlichungen des Gehabmahn-Institutes zur Erforschung der mittelalterlichen Theologie und Philosophie, 43).

3888. Science antique, science medievale: Autour d'Avranches, 235. Actes du Colloque international, Mont-Saint-Michel 4–7 septembre 1998. Éd. par Louis CALLEBAT et Olivier DESBORDES. Hildesheim, Olms, 2000, 469 p. (tav., ill.).

3889. TODESCHINI (Giacomo). «Ecclesia» e mercato nei linguaggi dottrinali di Tommaso d'Aquino. *In*: Etiche economiche [Cf. n° 6826], p. 585-622.

3890. TRIFOGLI (Cecilia). Oxford-Physics in the thirteenth century (ca. 1250–1270): motion, infinity, place and time. Leiden, Boston a. Köln, Brill, 2000, VII-289 p.

3891. TÜRK (Monika). Lucidaire de grant sapientie. Untersuchung und Edition der altfranzösischen Übersetzung 1 des "Elucidarium" von Honorius Augustidunensis. Tübingen, Niemeyer, 2000, X-448 p.

3892. UBL (Karl). Engelbert von Admont. Ein Gelehrter im Spannungsfeld von Aristotelismus und christlicher Überlieferung. Wien u. München, Oldenbourg, 2000, 260 p.

3893. WIPPEL (John F.). The metaphysical thought of Thomas Aquinas: from finite being to uncreated being. Washington, Catholic University of America Press, 2000, XXVII-630 p.

Cf. n°s 1426, 3072

§ 13. History of the Church and religion.

a. General

** 3894. Province ecclésiastique de Mayence (Germania Prima). Ed. par Nancy GAUTHIER, Brigitte BEAUJARD, Rollins GUILD et Marie-Pierre TERRIEN. Paris, De Boccard, 2000, 89 p. (ill., tables, maps).

3895. ANDRIĆ (Stanko). Mogućnosti istraživanja crkvene povijesti Slavonije u srednjem vijeku. (The possibilities of research on the Medieval church history of Slavonia). *Historijski zbornik*, 2000, 53, p. 15-22.

3896. ARCHETTI GIAMPAOLINI (Elisabetta). San Pier Damiani: il coraggio di un riformatore (e altro). Roma, Viella, 2000, 240 p.

3897. Bishofs Burchard von Worms, 1000–1025. Hrsg. v. Wilfried HARTMANN. Mayence, Selbstverlag der Gesellschaft für mittelrheinische Kirchengeschichte, 2000, XII-389 p. (tables).

3898. CAMERON (Ewan). Waldensess: rejections of holy church in medieval Europe. Oxford a. Malden, Blackwell, 2000, XI-336 p. (ill., maps).

3899. CANTERA MONTENEGRO (Santiago). Los cartujos en la religiosidad y la sociedad españolas, 1390–1563. Salzburg, Institut für Anglistik und Amerikanistik, Universität Salzburg, 2000, 2 vol., XXII-490 p., III-219 p. (ill., tables, maps). (Analecta Cartusiana, 166).

3900. CHARLES-EDWARDS (T. M.). Early Christian Ireland. Cambridge, Cambridge U. P., 2000, XIX-707 p.

3901. COHN (Norman). Europe's inner demons: the demonization of Christians in medieval Christendom. Chicago, University of Chicago Press, 2000, XIII-274 p.

3902. DICKINSON (Gary). Religious enthusiasm in the medieval West: revivals, crusades, saints. Aldershot, Ashgate, 2000, XVI-304 p. (ill.). (Variorum Collected Studies, 695).

3903. Divine (The) office in the Latin Middle Ages: methodology and source studies, regional developments, hagiography. Ed. by Margor E. FASSLER and Rebecca A. BALTZER. Oxford, Oxford U. P., 2000, XXIV-632 p. (ill.).

3904. Eglise et culture en France méridionale (XIIe–XIVe siècle). Toulouse, Privat, 2000, 554 p. (Cahiers de Fanjeaux, 35).

3905. ESDERS (Stefan), MIERAU (Heike Johanna). Der althochdeutsche Klerikereid. Bischöfliche Diözesangewalt, kirchliches Benefizalwesen und volkssprachliche Rechtspraxis im frühmittelalterlichen Baiern. Hannover, Hahnsche Buchhandlung, 2000, L-317 p. (ill.).

3906. FOOT (Sarah). Veiled Women. 1. The disappearance of Nuns from Anglo-Saxon England. 2. Female Religious Communities in England, 871–1066. Aldershot a. Burlington, Ashgate, 2000, 2 vol., XVII-228 p., XII-274 p. (map).

3907. HAMES (Harvey J.). The Art of conversion: Christianity and Kabbalah in the thirteenth century. Leiden, Boston a. Köln, Brill, 2000, XIII-332 p. (The Medieval Mediterranean: Peoples, Economies and Cultures, 400 – 1453, 26).

3908. LOYN (H. R.). The English church, 940–1154. Harlow, Pearson Education, 2000, X-174 p. (The medieval world).

3909. MARGETIĆ (Lujo). Hrvatska i Crkva u srednjem vijeku: pravnovijesne i povijesne studije. (Croatia and church in the Middle Ages: legal and historical studies). Rijeka, Pravni fakultet Sveučilišta u Rijeci, 2000, 534 p.

3910. MOREY (James H.). Book and verse: a guide to Middle English biblical literature. Urbana a. Chicago, University of Illinois Press, 2000, XXI-429 p.

3911. Mystical (The) gesture: essays on medieval and early modern spiritual culture in honour of Mary E. Giles. Ed. by Robert BOENIG. Aldershot, Ashgate, 2000, IX-226 p.

3912. OFFLER (H.S.). Church and crown in the fourteenth century: studies in European history and political thought. Aldershot, Ashgate, 2000, XXXVIII-336 p. (ill.).

3913. PALAZZO (Éric). Liturgie e société au Moyen Âge. Paris, Aubier, 2000, 276 p. (ill.). (Coll. Historique).

3914. PASZTOR (Edith). Donne e sante: studi sulla religiosità femminile nel Medio Evo. Roma, Studium, 2000, XII-311 p. (Religione e società).

3915. Religion et société urbaine au Moyen Age. Etudes offertes à Jean-Louis Biget par ses anciens élèves. Ed. par Patrick BOUCHERON et Jacques CHIFFOLEAU. Paris, Publications de la Sorbonne, 2000, 567 p.

3916. STEINSLAND (Gro). Den hellige kongen. Om religion og herskermakt fra vikingtid til Middelalder. (The holy king. Religion and power from the Viking age to the Middle Ages). Oslo, Pax, 2000, 233 p. (ill.).

3917. WEIß (Bardo). Ekstase und Liebe: die Unio mystica bei den deutschen Mystikerinnen des 12. und 13. Jahrhunderts. Paderborn, Ferdinand Schöningh, 2000, IX-987 p.

Cf. n°s 40, 81, 293, 1353, 3486, 3355

b. History of the Popes

3918. Bonifacio VIII e il suo tempo: anno 1300 il primo giubileo. A cura di Marina RIGHETTI TOSTI-CROCE. Milano, Electa, 2000, 261 p. (ill.).

3919. CAROCCI (Sandro). Feudo, vassallaggi e potere papale nello Stato della Chiesa (metà XI sec.–inizio XIII sec.). In: Leo Valiani storico e politico [Cf. n° 729], p. 999-1035.

3920. COWDREY (H.E.J.). Popes and church reform in the 11th century. Aldershot, Ashgate, 2000, VII-310 p. (Tables). (Variorum Collected Studies Series, 674).

3921. GOEZ (Werner). Gregor VII., Mathilde von Canossa und die Kosten des Investiturstreits. *Archiv für Kulturgeschichte*, 2000, 82, p. 303-318.

3922. HERKLOTZ (Ingo). Gli eredi di Costantino: il papato, il Laterano e la propaganda visiva nel XII secolo. Roma, Viella, 2000, 243 p. (ill.). (La Corte dei Papi, 6).

3923. SCHUCHARD (Christiane). Die päpstlichen Kollektoren im späten Mittelalter. Tübingen, Niemeyer, 2000, X-430 p. (maps).

Cf. n° 6396

c. Monastic history

** 3924. Cistercian Lay Brothers: twelfth-century usages, with related texts. Latin text with concordance of Latin terms, English translations and notes. Ed. by Chrysogonus WADDEL, OCSO. Brecht, Cîteaux, Commentarii Cistercienses, 2000, 232 p. (ill.). (Studia et Documenta, 10).

** 3925. MASSER (Achim). Regula Benedicti des Cod. 915 der Stiftsbibliothek von St. Gallen. Die Korrektur vorlage der lateinisch-althochdeutschen Benediktinerregel. Göttingen, Vandenhoeck & Ruprecht, 2000, 120 p. (ill.).

** 3926. Mittelalterlichen (Die) nichtliturgischen Handschriften des Zisterzienserklosters Salem. Beschrieben von Wilfried WERNER. Wiesbaden, L. Reichert, 2000, LXV-428 p. (taf.). (Kataloge der Universitätsbibliothek Heidelberg).

3927. Beiträge zur Geschichte des Paulinerordens. Hrsg. v. Kaspar ELM. Berlin, Duncker & Humblot, 2000, 333 p. (maps, tables). (Berliner historische Studien, 32 – Ordensstudien).

3928. BERMAN (Constance Hoffmann). The Cistercian evolution: the invention of a religious order in twelfth-century Europe. Philadelphia, University of Pennsylvania Press, 2000, XXIV-382 p. (ill., tables). (The Middle Ages Series).

3929. BUSUIOC VON HASSELBACH (Dan Nicolae). Țara Făgărașului în sec. al XIII-lea. I-II. Cluj-Napoca, Mănăstirea cisterciană Cârța, 2000, 2 vol., 338 p., 383 p.

3930. CABY (Cécile). De l'érémitisme rural au monarchisme urbain. Les Camaldules en Italie à la fin du Moyen Age. Roma, Ecole française de Rome, 2000, 873 p. (BEFAR, 305).

3931. Clercs (Des) au service de la reforme: études et documents sur les chanoines réguliers de la province de Rouen. Ed. par Mathieu ARNOUX. Turnhout, Brepols, 2000, 404 p. (Bibliotheca Victorina).

3932. CONSTABLE (Giles). Cluny from the tenth to the twelfth centuries: further studies. Aldershot, Ashgate, 2000, [s. p.]. (ill.). (Collected studies series).

3933. DAVRIL (Anselme), PALAZZO (Eric). La vie des moines au temps des grandes abbayes: 10.–13. siècles. Paris, Hachette littératures, 2000, 344 p. (ill.). (La vie quotidienne).

3934. DUNN (Marylin). The emergence of monasticism: from the desert fathers to the early Middle Ages. Oxford a. Malden, Blackwell, 2000, VIII-280 p.

3935. JOSÉ DE SIGÜENZA. Historia de la Orden de San Jerónimo. Ed. por Francisco J. CAMPOS Y FERNÁNDEZ DE SEVILLA. Vallaloid, Junta de Castilla y Leòn, 2000, 2 vol., [s. p.]. (Libros recuperados, 2).

3936. KOLSRUP (Inge-Lise). Aspekter af dansk klostervæsen i middelalderen. (Aspects du monachisme danois). Aarhus, 2000, 124 p. (ill.).

3937. LAMBERTINI (Roberto). La povertà pensata: evoluzione storica della definizione dell'identità minoritica da Bonaventura ad Ockham. Modena, Mucchi, 2000, 327 p. (Collana di Storia Medievale, 1).

3938. LOUD (G.A.). Montecassino and Benevento in the middle ages: essays in South Italian church history. Aldershot, Ashgate, 2000, XII-322 p. (ill., tavles, maps).

3939. Mission und Christianisierung am Hoch- und Oberrhein (6.–8. Jahrhundert). Hrsg. v. Walter BERSCHIN, Dieter GEUENICH und Heiko STEUER. Stuttgart, J. Thorbecke, 2000, 222 p. (ill.). (Archäologie und Geschichte).

3940. Monks and nuns, saints and outcast religion in medieval society: essays in honor of Lester K. Little. Ed. by Sharon FARMER and Barbara H. ROSENWEIN. Ithaca a. London, Cornell U. P., 2000, 249 p.

3941. NYBERG (Tore). Monasticism in North-Western Europe, 800–1200. Aldershot, Ashgate, 2000, XI-295 p. (ill., maps, table).

3942. PATZOLD (Steffen). Konflikte im Kloster. Studien zur Auseinandersetzungen in monastischen Gemeinschaften des ottonischen-salischen Reichs. Husum, Matthiesen Verlag, Historische, 2000, 426 p. (Studien, 463).

3943. Rayonnement (Le) spirituel et culturel de l'abbaye de Saint-Gall. Colloque tenu au Centre culturel Suisse, Paris, 12 octobre 1993. Actes publiés sous la direction de Carol HEITZ, Werner VOGLER et François HEBER-SUFFRIN. Paris, Picard, 2000, 142 p., (tav., ill.). (Les cahiers du Centre de recherche sur l'Antiquité tardive et le haut Moyen Âge).

3944. RICHE (Denyse). L'Ordre de Cluny à la fin du Moyen Age. «Le vieux pays clunisien», XIIe–XVe siècles. Saint-Etienne, Publications de l'Université de Saint-Etienne, 2000, 765 p. (CERCOR, Travaux et Recherches, 13).

3945. TURCUȘ (Șerban). Misiunea cisterciană în Transilvania. *Acta Mvsei Napocensis*, 2000-2001, 37-38, 2, p. 49-74.

3946. TURCUȘ (Veronica). La nascita dell'ordine cistercense e la sua espansione. *Acta Mvsei Napocensis*, 2000-2001, 37-38, 2, p. 19-48.

3947. Unanimité et diversité cisterciennes. Filiations-Réseaux-Relectures du XIIe au XVIIe siècle. Actes du quatrième colloque international du C.E.R.C.O.R., Dijon, 23–25 septembre 1998. Saint-Étienne, Publications de l'Université, 2000, 715 p. (ill., tables).

3948. VATIN (Nicolas). Rhodes et l'ordre de Saint-Jean-de-Jérusalem. Paris, CNRS, 2000, 119 p. (ill., maps). (Patrimonie de la Méditerranée).

3949. VOGTHERR (Thomas). Die Reichsabteien der Benediktiner und das Königtum im hohen Mittelalter (900–1125). Stuttgart, Jan Thorbecke, 2000, VIII-361 p. (tables). (Mittelalter-Forschungen, 5).

3950. WEBSTER (Jill Rosemary). Els Franciscans Catalans a l'Edat mitjana: els primers menorets I menoretes a la Corona d'Aragó. Lleida, Pagès, 2000, 372 p. (Seminari (Pagès. Catalònia), 14).

3951. WENTA (Jaroslaw). Studien über die Ordensgeschichtsschreibung am Beispiel Preußens. Torún, Wydawnictwo Uniwersytetu Mikolaia Kopernika, 2000, 287 p. (Subsidia historiographica, II).

Cf. n°s 68, 3814, 4024

d. Hagiography

** 3952. Chaste passions: medieval English virgin martyr legends. Ed. by Karen A. WINSTEAD. Ithaca a. London, Cornell U. P., 2000, XI-201 p. (ill.).

** 3953. DE JOINVILLE (Jean). Storia di San **Luigi**. A cura di Armando LIPPIELLO. Roma, Il Cigno Galileo Galilei, 2000, 240 p.

13. HISTORY OF THE CHURCH AND RELIGION

** 3954. Old (The) Norse-Icelandic legend of Saint **Barbara**. Ed. by Kirsten WOLF. Toronto, Pontifical Institute of Mediaeval Studies, 2000, IX-187 p.

3955. AIGRAIN (René). L'hagiographie: Ses sources ses méthodes son histoire. Avec un complément bibliographique par Rober GODDING. Bruxelles, Société des Bollandistes, 2000, VIII-539 p. (Subsidia Hagiographica, 80).

3956. ANDRIĆ (Stanko). The miracles of St. **John Capistran**. Budapest, Central European U. P., 2000, 454 p.

3957. ASCHTON (Gail). The generation of identity in late medieval hagiography: speaking the saint. London a. New York, Routledge, 2000, VIII-176 p. (ill.). (Routledge Research in Medieval Studies, 1).

3958. CROOK (John). The architectural setting of the cult of saints in the early Christian West, c.300–1200. Oxford, Clarendon Press, 2000, XXV-308 p. (ill.).

3959. Febronia e Trofimena: agiografia latina nel Mediterraneo altomedievale: atti della Giornata di studio, Patti, 18 luglio 1998. A cura di Reginald GREGOIRE. Cava de' Tirreni, Avagliano, 2000, 138 p. (ill.). (Schola Salernitana. Studi e testi).

3960. GALERA PEDROSA (Andreu). Sant **Celdoni** i Sant **Ermenter**. Cardona, Confraria dels Sants Màrtirs, Manuel Sala i Queralt, 2000, 179 p.

3961. GRANDE QUEJIGO (Francisco Javier). Hagiografia y diffusión en la Vida de San **Millán de la Cogolla de Gonzalo de Berceo**. Logroño, Instituto de Estudios Riojanos, 2000, 372 p. (Centro de Estudios Gonzalo de Berceo, 18).

3962. Hagiographie im Kontext: Wirkungsweisen und Möglichkeiten historischer Auswertung. Hrsg. v. Dieter R. BAUER und Klaus HERBERS. Stuttgart, Franz Steiner, 2000, XXVIII-288 p. (ill., map).

3963. HENRIET (Patrick). La parole et la prière au Moyen Age. Le Verbe efficace dans l'hagiographie monastique des XIe et XIIe siècles. Bruxelles, De Boeck Université, 2000, 482 p. (Bibliothèque du Moyen Age, 16).

3964. HILKEN (Charles, FSC). The necrology of San **Nicola della Cicogna** (Montecassino, Archivio della Badia 179, Pp. 1–64). Toronto, Pontifical Institute of Medieval Studies, 2000, IX-178 p.

3965. JANKULAK (Karen). The medieval cult of St. **Petroc**. Woodbridge a. Rochester, Boydell and Brewer, 2000, 272 p. (maps).

3966. LABUDA (Gerard). Święty Wojciech biskup-męczennik, patron Polski, Czech i Węgier. (Sankt **Adalbert** – Bischof-Märtyrer, Beschützer von Polen, Böhmen und Ungarn). Wrocław, Funna, 2000, 333 p. (Monografie Fundacji na Rzecz Nauki Polskiej. Seria Humanistyczna).

3967. LEWIS (Katherin J.). The cult of St. **Katherine of Alexandria** in late medieval England. Cambridge, Cambridge U. P., 2000, XXVIII-471 p. (ill., tables).

3968. LODI (Enzo). San **Petronio**: patrono della città e Diocesi di Bologna. Bologna, Renografica, 2000, XIII-298 p. (tav., ill.).

3969. MARNER (Dominic). St. **Cuthbert**: his life and cult in medieval Durham. Toronto, University of Toronto Press, 2000, 112 p. (ill.).

3970. MASCANZONI (Leardo). San **Giacomo**: il guerriero e il pellegrino: il culto iacobeo tra la Spagna e l'Esarcato, secc. XI–XV. Spoleto, Centro italiano di studi sull'alto Medioevo, 2000, X-572 p.

3971. Medieval Hagiography: an anthology. Ed. by Thomas HEAD. New York a. London, Garland, 2000, XLIX-834 p. (Garland Reference Library of the Humanities).

3972. Miracle et karāma: hagiographies médiévales comparées, 2. Ed. by Denise AIGLE. Turnhout, Brepols, 2000, 690 p. (ill.).

3973. MUELLER (Joan). **Francis**: the Saint of Assisi. Allen, Thomas More, 2000, 296 p.

3974. ORME (Nicholas). The saints of **Cornwall**. Oxford, Oxford U. P., 2000, XVII-302 p. (maps, tables).

3975. PÉRICARD-MÉA (Denise). Compostelle et cultes de saint **Jacques** au Moyen Age. Paris, Presses Universitaires de France, 2000, 385 p. (Le nœud gordien).

3976. Pubblico (Il) dei santi. Forme e livelli di ricezione dei messaggi agiografici. Atti del III Convegno di studio dell'Associazione italiana per lo studio della santità, dei culti e dell'agiografia, Verona 22–24 ottobre 1998. A cura di Paolo GOLINELLI. Roma, Viella, 2000, 489 p. (ill.).

3977. Santità ed eremitismo nella Toscana medievale: atti delle Giornate di studio, 11–12 giugno 1999. A cura di Alessandra GIANNI. Siena, Cantagalli, 2000, 150 p. (ill.).

3978. Supplementary lives in some manuscripts of the "Gilte Legende". Ed. by Richard HAMER and Vida RUSSELL. Oxford, Oxford U. P., for the Early English Text Society, 2000, XXXIV-566 p. (facsim., tables). (Early English Text Society, O.S., 315).

3979. VOGEL (Lothar). Vom Werden eines Heiligen. Eine Untersuchung der «Vita Corbiniani» des Bishofs Arbeo von Freising. Berlin u. New York, Walter de Gruyter, 2000, XI-542 p. (Arbeiten zur Kirchengeschichte, 77).

3980. WÜNSCH (Thomas). Der heilige Bischof – Zur politischen Dimension von Heiligkeit im Mittelalter und ihrem Wandel. *Archiv für Kulturgeschichte*, 2000, 82, p. 261-302.

e. Special studies

** 3981. AMERUZES DE TREBISONDA (Jeorge). El diálogo de la fe con el sultán de los turcos. Ed. por Oscar DE LA CRUZ PALMA. Madrid, Consejo Superior de Investigaciones Cientificas, Universidad Autónoma de Barcelona, Departamento de Ciencias de la Antigüedad y de la Edad Media, 2000, XXIX-195 p.

** 3982. Quaderno (Il) della fraternità di Santa Maria di Tricesimo: ms. 147, Fondo Joppi. A cura di Federico VICARIO. Udine, Comune di Udine, Biblioteca Civica V. Joppi, 2000, 199 p. (Quaderni della Biblioteca civica V. Joppi. Fonti e documenti).

3983. BOSCH (Lynette M. F.). Art, liturgy, and legend in Renaissance Toledo: the Mendoza and the Iglesia Primada. University Park, Pennsylvania State U. P., XII-292 p. (ill.).

3984. BYNUM (Caroline Walker), FREEDMAN (Paul). Last things: and the Apocalypse in the Middle Ages. Philadelphia, University of Pennsylvania Press, 2000, VIII-365 p. (The Middle Ages Series).

3985. FLORY (David A.). Marian representations in the miracle tales of thirteenth-century Spain and France. Washington, Catholic University of America Press, 2000, XIX-156 p. (ill.).

3986. FRIEDLANDER (Alan). The hammer of the inquisitors: Bernard Délicieux and the struggle against the inquisition in fourteenth-century France. Leiden, Boston a. Köln, Brill, 2000, XX-328 p. (maps). (Cultures, Beliefs and Traditions: Medieval and Early Modern Peoples, 9).

3987. HAMILTON (Bernard). Crusaders, cathars and the holy places. Aldershot, Ashgate, 2000, [s.p.] (Collected studies series, CS 656).

3988. Hildegard von Bingen in ihrem historischen Umfeld. Actes du congrès scientifique international tenu à Bingen du 13 au 19 septembre 1998. Hrsg. v. A. HAVERKAMP. Mayence, [s. n.], 2000, 637 p.

3989. JANSEN (Katherine Ludwig). The making of the Magdalen: preaching and popular devotion in the later middle ages. Princeton, Princeton U. P., 2000, XVII-389 p. (ill.).

3990. LINTHOE NÆSHAGEN (Ferdinand). Medieval Norwegian religiosity: historical sources and modern social science. *Scandinavian journal of history*, 2000, 25, 4, p. 297-316.

3991. MAC KINNON (James). The advent project: the later-seventh-century creation of the Roman mass proper. Berkeley, Los Angeles a. London, University of California Press, 2000, XIV-466 p. (tables).

3992. NOODT (Birgit). Religion und Familie in der Hansestadt Lübeck anhand der Bürgertestamente des 14. Jahrhunderts. Lübeck, Schmidt-Römhild, 2000, XI-618 p. (ill., tables).

3993. Pays (Le) cathare: les religions médiévales et leurs expressions méridionales. Ed. par Jacques BERLIOZ. Paris, Editions du Seuil, 2000, 319 p. (ill., maps). (Histoire, 279).

3994. PEIKOLA (Matti). Congregation of the Elect: patterns of self-fashioning in English lollard writings. Turku, University of Turku, 2000, XI-362 p. (tables). (Anglicana Turkuensia, 21).

3995. PELLEGRINI (Luigi). "Che sono queste novità?" Le "religiones novae" in Italia meridionale (secoli XIII e XIV). Napoli, Liguori, 2000, X-403 p. (tables, maps). (Mezzogiorno Medievale e Moderno, 1).

3996. Peregrinatio ad loca sancta: testimonianze del passaggio dei pellegrini lungo i percorsi viari a sud-est di Ravenna. A cura di Paola NOVARA. Ravenna, Danilo Montanari, 2000, 110 p., (ill.). (Millennium).

3997. Prima della Francigena: itinerari romei nel Regnum Langobardorum. A cura di Renato STOPANI. Firenze, Le lettere, 2000, 106 p. (ill.). (Le vie della storia).

3998. Quellen zur Geschichte der Waldeinser von Freiburg im Üchtland (1399–1439). Hrsg. v. Kathrin UTZ TREMP. Hannover, Hahnsche Buchhandlung, 2000, X-837 p. (tables).

3999. Vangeli (I) dei popoli: la parola e l'immagine del Cristo nelle culture e nella storia. A cura di Francesco D'AIUTO, Giovanni MORELLO e Ambrogio M. PIAZZONI. Città del Vaticano, Rinnovamento dello Spirito Santo e Biblioteca Apostolica Vaticana, 2000, XXVII-493 p. (ill.).

4000. VÉSTEINSSON (Orri). The Christianization of Iceland: priests, power, and social change 1000–1300. Oxford, Oxford U. P., 2000, XV-318 p.

4001. VINCENT (C.). Discipline du corps et de l'esprit chez les Flagellants au Moyen Age. *Revue historique*, 2000, 124, 615, p. 593-614.

4002. Worlds of difference: European discourses of toleration, c. 1100–c. 1550. Ed. by Cary J. NEDERMAN. University Park, Pennsylvania State U. P., 2000, X-157 p.

4003. Writing religious women: female spiritual and textual practices in late medieval England. Ed. by Denis RENEVEY and Christiania WHITEHEAD. Toronto a. Buffalo, University of Toronto Press, 2000, XI-270 p.

Cf. n[os] *299, 377, 879*

§ 14. Settlements. Place names. Town planning.

4004. Aurea Roma: dalla città pagana alla città cristiana. A cura di Serena ENSOLI ed Eugenio LA ROCCA. Roma, L'Erma di Bretschneider, 2000, 711 p. (ill.).

4005. BENITO MARTIN (Félix). La formación de la ciudad medieval: la red urbana en Castilla y León. Valladolid, Universidad, Secretariado de Publicaciones e Intercambio Editorial, 2000, 293 p. (ill.). (Arquitectura y urbanismo, 33).

4006. BONARIA URBAN (Maria). Cagliari aragonese: topografia e insediamento. Cagliari, Istituto sui rapporti italo-iberici, 2000, 387 p. (ill.). (Collana di studi italo-iberici, 25).

4007. Cambridge urban history (The) of Britain. Vol. 1. 600–1540. Ed. by David M. PALLISER. Cambridge, Cambridge U. P., 2000, XXVI-841 p.

4008. Castelli: storia e archeologia del potere nella Toscana medievale. 1. A cura di Riccardo FRANCOVICH e Maria GINATEMPO. Firenze, All'insegna del giglio, 2000, 297 p., (ill.). (Biblioteca del Dipartimento di archeologia e storia delle arti, Sezione archeologica, Universita di Siena).

4009. Christiana Loca: lo spazio cristiano nella Roma del primo millennio. A cura di Letizia PANI ERMINI. Roma, Palombi, 2000, s. p. (ill.).

4010. CIGLENEČKI (Slavko). Tinje nad Loko pri Žusmu: poznoantična in zgodnjesrednjeveška naselbina = Tinje oberhalb von Loka pri Žusmu: Spätantike und frühmittelalterliche Siedlung. Ljubljana, ZRC SAZU, Založba ZRC, 2000, 196 p. (Opera Instituti archaeologici Sloveniae, 4).

4011. Destruction et reconstruction des villes. T. 2. Destruction par le pouvoir seigneurial, les troubles internes et les guerres. T. 3. Rapport final. Ed. par Martin KÖRNER. Bern, Stuttgart, Wien, 2000, 2 vol., 459 p., 196 p.

4012. Gorod v srednevekovoj tsivilizatsii Evropy. (Town in Medieval civilization of Europe). Ed. Ada A. SVANIDZE. RAN, In-t vseobshchej istorii. In 4 vols. Vol. 2. Extra muros. Gorod, obshchestvo, gosudarstvo (Town, Society and State). Ed. Ol'ga I. VAR'JASH. Vol. 3. Celovek vnutri gorodskokh sten. Formy obshchestvennykh svjazej (Man inside the urban walls. Forms of Social Relations). Ed. Pavel Ju. UVAROV. Moskva, Nauka, 2000, 2 vol., 354 p., 378 p. (ill.; bibl.; geogr. ind.). [Eng. summary]. [Cf. Bibl. 99, n° 3914.]

4013. HODGES (Richard). Towns and trade in the age of Charlemagne. London, Duckworth, 2000, 144 p. (Duckworth Debates in Archaeology).

4014. JAWOR (Grzegorz). Osady prawa wołoskiego i ich mieszkańcy na Rusi Czerwonej w późnym średniowieczu. (Die kraft des Siedlungsrechts der Walachai gegründeten Siedlungen und deren Einwohner in Rotreußen [Ruś Czerwona] im späten Mittelalter). Lublin. Wydaw. Uniw. M. Curie-Skłodowskiej, 2000, 223 p.

4015. JEAN-MARIE (Laurence). Caen aux XIe et XIIe siècles. Espace urbain, pouvoirs et societé. Caen, La Mandragore, 2000, 349 p. (ill., maps, tables).

4016. Jornades d'història i d'arqueologia medieval del Maresme (1. 1999. Mataró), L'arquitectura militar medieval: Jornades d'Història i d'Arqueologia Medei-

vale del Maresme: actes del 13 al 30 d'octobre de 1999. Mataró, Grup d'Història del Casal i Argentona, La Comarcal, 2000, 162 p. (ill.).

4017. MARTÍN VISO (Iñaki). Poblamiento y structuras sociales en la norte de la Península Ibérica: siglos VI–XIII. Salamanca, Universidad, 2000, 397 p. (Acta salamaticensa. Estudios históricos y geográficos, 111).

4018. Medieval Dublin I. Proceedings of the Friends of Medieval Dublin Symposium, 1999. Ed. by Seán DUFFY. Dublin a. Portland, Four Courts Press, 2000, 237 p. (ill., tables).

4019. Medieval practices of space. Ed. by Barbara A. HANAWALT and Michal KOBIALKA. Minneapolis a. London, University of Minnesota Press, 2000, XVIII-269 p. (ill., tables).

4020. MERÇİL (Erdoğan). Türkiye Selçukluları'nda meslekler. (Occupations in the Anatolian Seljuqids). Ankara, Türk Tarih Kurumu, 2000, 233 p.

4021. Monastero e castello nella costruzione del paesaggio: 1. Seminario di geografia storica: Cassino, 27–29 ottobre 1994. A cura di Gabriella ARENA, Andrea RIGGIO e Paola VISOCCHI. Perugia, Rux, 2000, 470 p. (ill.).

4022. NEDASHKOVSKIJ (Leonard F.). Zolotoordynskij gorod Ukek i ego okruga. (Ukek: a city of the Golden Horde and its environs). Moskva, Vostochnaja literatura, 2000, 224 p. (ill.). [Eng. summary]

4023. NEDKVITNE (Arnved), NORSENG (Per). Middelalderbyen ved Bjørvika, Oslo 1000–1536. (The medieval town at Bjørvika, Oslo 1000–1536). Oslo, Cappelen, 2000, 475 p. (ill.).

4024. PETROVA (L. I.), ANKUDINOV (I. Ju.), POPOV (V. A.), SILAEVA (T. V.). Topografija prigorodnykh monastyrej Novgoroda Velikogo. (The topography of the suburban monasteries of Novgorod the Great). *In*: Novgorodskij istoricheskij sbornik. Vol. 8 (18) [Cf. n° 3814], p. 95-157. (ill., maps).

4025. PRINGLE (Denys). Fortification and settlement in crusader Palestine. Aldershot, Ashgate, 2000, X-334 p. (ill.).

4026. ROIG (Alanyà). Urbanisme i vida a la Morella medieval: s. XIII-XV. Morella, Ajuntament, Associació d'Amicis de Morella i Comarca, 2000, 528 p. (ill.).

4027. Sector (El) nord de la ciutat de Girona: de l'inici al segle XIV. Ed. por Josep CANAL [et al.]. Girona, Ajuntament de Girona, 2000, 169 p. (ill.).

4028. STOPAR (Ivan). Grajske stavbe v osrednji Sloveniji. 1, Gorenjska. Knj. 5. Med Goričanami in Gamberkom. (Castels in central Slovenia). Ljubljana a. Viharnik, Znanstveni inštitut Filozofske fakultete, 2000, 159 p. (Zbirka Grajske stavbe, 10).

4029. STOUFF (Louis). Arles au Moyen Âge. Marseille, La Thune, 2000, 256 p.

4030. SUZIKI (Seiichi). The quoit brooch style and Anglo-Saxon settlement: a casting and recasting of cultural identity symbols. Woodbridge a. Rochester, Boydell and Brewer, 2000, XIV-218 p. (plates, ill, tables).

4031. Territori i societat a l'Edat Mitjana. 3. Història, arqueologia, documentació. Ed. por Jordi BOLÒS y Joan J. BUSQUETA. Lleida, Universitat, 2000, 454 p.

4032. Towns (The) of Italy in the later Middle Ages. Ed. by Trevor DEAN. Manchester a. New York, Manchester U. P., 2000, X-252 p. (map). (Manchester Medieval Sources Series).

4033. TURCUŞ (Şerban). Teritoriul românesc ca spaţiu-frontieră în secolul al XIII-lea. *Studia Universitatis Babeş-Bolyai. Historia*, 2000, 45, 1-2, p. 3-14.

4034. Ville (La) médiévale en deçà et au-delà de ses murs. Mélanges Jean-Pierre Leguay. Ed. par Philippe LARDIN et Jean-Louis ROCH. Rouen, Publications de l'Université de Rouen, 2000, 432 p. (maps, tables).

Cf. n°s 180-220, 792, 1175, 3342, 8482

K

MODERN HISTORY, GENERAL WORKS

§ 1. General. 4035-4121. – § 2. History by countries. 4122-5395.

§ 1. General.

* 4035. Bibliographie zur Zeitgeschichte. *Vierteljahrshefte für Zeitgeschichte*, 2000, 48, 160 p. [47, 1999. Cf. Bibl. 99, n° 3922.]

4036. 1848 – A European revolution? International ideas and national memories of 1848. Ed. by Axel KÖRNER. New York, St. Martin's, 2000, XI-232 p.

4037. ABERNETHY (David R.). The dynamics of global dominance: European overseas empires, 1415–1980. New Haven a. London, Yale U. P., 2000, VIII-524 p.

4038. BABEROWSKI (Jörg). Nationalismus aus dem Geist der Inferiorität. Autokratische Modernisierung und die Anfänge muslimischer Selbstvergewisserung im östlichen Transkaukasien 1828–1914. *Geschichte und Gesellschaft*, 2000, 26, 3, p. 371-406.

4039. BADE (Klaus J.). Europa in Bewegung. Migration vom späten 18. Jahrhundert bis zur Gegenwart. München, Beck, 2000, 510 p.

4040. BAMBERGER-STEMMANN (Sabine). Der Europäische Nationalitätenkongress 1925 bis 1938. Nationale Minderheiten zwischen Lobbyistentum und Großmachtinteressen. Marburg, Herder-Institut, 2000, VIII-619 p. (Materialen und Studien zur Ostmitteleuropa-Forschung, 7).

4041. BARTOLINI (Stefano). The political mobilisation of the European Left, 1860–1980: the class cleavage. Cambridge, Cambridge U. P., 2000, XXIV-637 p. (Cambridge studies in comparative politics).

4042. BARTOV (Omer). Mirrors of destruction: war, Genocide, and modern identity. Oxford, Oxford U. P., 2000, VIII-302 p.

4043. BERNIER (Olivier). The world in 1800. New York, John Wiley & Sons, 2000, XI-452 p.

4044. Between past and future. The Revolutions of 1989 and their aftermath. Ed. by Sorin ANTOHI and Vladimir TISMANEANU. Budapest, Central European University Press, 2000, [s. p.].

4045. BEYRAU (Dietrich). Schachtfeld der Diktatoren. Osteuropa im Schatten von Hitler und Stalin. Göttingen, Vandenhoeck & Ruprecht, 2000, 158 p.

4046. BIDUSSA (David). Antifascismo e «vie nazionali». A proposito del VII Congresso del Comintern. *In*: Novecento italiano [Cf. n° 4843], p. 137-159.

4047. BOSCO (Anna). Comunisti: trasformazioni di un partito in Italia, Spagna e Portogallo. Bologna, Il Mulino, 2000, 334 p. (Ricerca).

4048. BRENDON (Piers). The dark valley: a panorama of the 1930s. London, Jonathan Cape, 2000, XVIII-701 p.

4049. Challenge (The) of change: military institutions and new realities, 1918–1941. Ed. by Harold R. WINTON and David R. METS. Lincoln, University of Nebraska Press, 2000, XIX-246 p. (Studies in war, society, and the military).

4050. Chelovek XVI stoletija. (Man of the 16th century: [Articles]). Ed. Ada A. SVANIDZE, Vladimir A. VEDJUSHKIN. RAN, In-t vseobshchej istorii. Moskva, [s. n.], 2000, 227 p.

4051. Crisis (La) del estado liberal en la Europa del Sur: II Encuentro de Historia de la Restauración. Ed. por Manuel SUÁREZ CORTINA. Santander, Sociedad Menéndez Pelayo, 2000, 431 p. (Serie Estudios de literatura y pensamiento hispánicos, 17).

4052. CURTIN (Philip D.). The world and the west. The European challenge and the overseas response in the age of empire. Cambridge, New York a. Melbourne, Cambridge U. P., 2000, XIV-294 p.

4053. DEMEL (Walter). Europäische Geschichte des 18. Jahrhunderts. Ständische Gesellschaft und europäisches Mächtesystem im beschleunigten Wandel (1689/1700–1789/1800). Stuttgart, Berlin u. Köln, 2000, 300 p.

4054. DETTI (Tommaso), GOZZINI (Giovanni). Storia contemporanea. Vol. 1. L'Ottocento. Milano, Bruno Mondadori, 2000, 416 p.

4055. Enciclopedia della sinistra europea nel XX secolo. Diretta da Aldo AGOSTI, con la collaborazione di Luciano MARROCU e Leonardo RAPONE. Roma, Editori Riuniti, 2000, XLV-1344 p.

4056. Enclaves territoriales (Les) aux temps modernes (XVIe–XVIIe siècles). Ed. par Paul DELSALLE et André FERRER. Besançon, Presses universitaires franc-comtoises, 2000, 448 p.

4057. Erste Weltkrieg (Der) und die europäische Nachkriegsordnung: sozialer Wandel und Formveränderung der Politik. Hrsg. v. Hans MOMMSEN. Köln, Böhlau, 2000, 246 p. (Industrielle Welt, 60).

4058. European democratization since 1800. Ed. by John G. GARRARD, Vera TOLZ and Ralph WHITE. New York, St. Martin's Press, 2000, X-292 p.

4059. FERTIG (Georg). Lokales Leben, atlantische Welt. Die Entscheidung zur Auswanderung vom Rhein nach Nordamerika im 18. Jahrhundert. Osnabrück, Rasch, 2000, 466 p. (Studien zur historischen Migrationsforschung, 7).

4060. FOWKES (Ben). Eastern Europe, 1945–1969: from Stalinism to stagnation. Harlow, Longman, 2000, XV-164 p. (Seminar studies in history).

4061. FRIEDRICH (Karin). The other Prussia: royal Prussia, Poland and liberty, 1569–1772. Cambridge, Cambridge U. P., 2000, XIX-280 p. (Cambridge studies in early modern history).

4062. Frontiéres (Les) et l'espace national en Europe du Centre-Est: exemples de quatre pays – Biélorussie, Lituanie, Pologne et Ukraine = The borders and national space in East-Central Europe: the example of the following four countries – Belarus, Lithuania, Poland, and Ukraine. Sous la diréction de Jerzy KLOCZOWSKI, Piotr PLISIECKI et Hubert LASZKIEWICZ. Lublin, Instytut Europy Srodkowo-Wschodniej, 2000, 214 p. (ill., maps).

4063. GAWRECKI (Dan). K rozdílným pohledům české, polské a německé historiografie na národnostní vývoj v rakouském Slezsku 1848–1918. (Die nationale Entwicklung in Österreichisch-Schlesien 1848–1918. Ansichten der tschechischen, polnischen und deutschen Historiographie). *In*: Evropa mezi Německem a Ruskem. Ed. Miroslav ŠESTÁK. Praha, 2000, p. 239-254.

4064. Gendered nations/nationalisms in the long 19th century Europe and beyond. Ed. by Ida BLOM, Karen HAGEMANN and Catherine HALL. Oxford a. New York, Berg, 2000, [s. p.]. [Cf. n° <choice> 7221.]

4065. GREENE (Molly). A shared world. Christians and Muslims in the early modern Mediterranean. Princeton, Princeton U. P., 2000, XII-228 p.

4066. Grenzen in Ostmitteleuropa im 19. und 20. Jahrhundert. Aktuelle Forschungsprobleme. Hrsg. v. Hans LEMBERG. Marburg, Verlag Herder-Institut, 2000, 291 p. (Tagungen zur Ostmitteleuropa-Forschung, 10).

4067. Große Revolutionen der Geschichte. Von der Frühzeit bis zur Gegenwart. Hrsg. v. Peter WENDE. München, Beck, 2000, 391 p.

4068. GUSTAFSSON (Harald). Gamla riken, nya stater. Statsbildning, politisk kultur och identiter under Kalmarunionens upplössningsskede 1512–1541. (Vieux royaumes, nouveaux Etats. Formation de l'Etat, culture politique et identité à l'époque de la dissolution de l'Union de Kalmar, 1512–1541). Stockholm, Atlantis, 2000, 381 p.

4069. HINRICHS (Ernst). Fürsten und Mächte. Zum Problem das europäischen Absolutismus. Göttingen, Vandenhoeck & Ruprecht, 2000, 279 p.

4070. HURD (Madeleine). Public spheres, public mores, and democracy: Hamburg and Stockholm, 1870–1914. Ann Arbor, University of Michigan Press, 2000, VIII-316 p. (Social History, Popular Culture, and Politics in Germany).

4071. Iran and beyond: essays in Middle Eastern history in honor of Nikki R. Keddie. Ed. by Rudi MATTHEE and Beth BARON. Costa Mesa, Mazda Publishers, 2000, XII-291 p. (ill.).

4072. Istorija Evropy: S drevnejshikh vremen do nashikh dnej. (History of Europe: from the earliest times to nowadays). Vol. 5. Ot Frantsuzskoj revoljutsii kontsa XVIII veka do Pervoj mirovoj vojny. (From the French Revolution of 1789 to the First World War). Ed. Alla S. NAMAZOVA, Svetlana P. POZHARSKAJA. Moskva, Nauka, 2000, 667 p. (maps, ill., portr.).

4073. JANOS (Andrew C.). East Central Europe in the modern world: the politics of the borderlands from pre- to postcommunism. Stanford, Stanford U. P., 2000, XVI-488 p.

4074. KORUNIĆ (Petar). Fenomen nacije: porijeklo, integracija i razvoj. (Phenomena of nation: origin, integration and development). *Historijski zbornik*, 2000, 53, p. 49-100.

4075. KOVÁČ (Dušan). 20. storočie – storočie svetla, storočie temna. (20th century – century of light, century of darkness). Bratislava, Vydavateľstvo Q111, 2000, 208 p.

4076. LIM (Jie-Hyun). Fascismui Jinjeejungwa Hapuidokjae. (Fascism as 'a War of Position' and the 'Consensus dictatorship'). *Contemporary criticism*, 2000, 12, [s. p.].

4077. LINDEMANN (Thomas). Die Macht der Perzeptionen und Perzeptionen von Mächten. Berlin, Duncker & Humblot, 2000, 317 p. (Beiträge zur Politischen Wissenschaft, 118).

4078. MANČEV (Krastjo). Národné stereotypy a politika balkánskych štátov. (National stereotypes and the politics of Balkan States). *Historický časopis*, 2000, 48, 2, p. 217-330.

4079. MARINO (Giuseppe Carlo). Eclissi del principe e crisi della storia: apogeo e tramonto della democrazia rivoluzionaria nel XX secolo. Milano, F. Angeli, 2000, 256 p. (Storia, 257).

4080. MAYER (Arno). The furies: violence and terror in the French and Russian revolutions. Princeton, Princeton U. P., 2000, 716 p.

4081. Middle East monarchies: the challenge of modernity. Ed. by Joseph KOSTINER. Boulder, Lynne Rienner Publishers, 2000, VII-341 p.

4082. Militär und Gesellschaft in der frühen Neuzeit. Hrsg. v. Stefan KROLL und Kersten KRÜGER. Münster, LIT, 2000, 392 p. (Herrschaft und soziale Systeme in der frühen Neuzeit, 1).

4083. Minderheiten, Regionalbewußtsein und Zentralismus in Ostmitteleuropa. Hrsg. v. Heinz-Dietrich LÖWE, Günther H. TONTSCH und Stefan TROEBST. Köln, Weimar u. Wien, Böhlau, 2000, VIII-237 p. (Siebenbürgisches Archiv, 35).

4084. Napoleon's legacy: problems of government in Restoration Europe. Ed. by David LAVEN and Lucy RIALL. Oxford, Berg, 2000, XIII-291 p.

4085. Negros, mulatos, zambaigos. Derroteros africanos en los mundos ibéricos. Ed. par Berta ARES QUEIJA y Alessandro STELLA. Sevilla, Escuela de Estudios Hispanico-Americanos, Consejo Superior de Investigaciones Científicas, 2000, 389 p.

4086. New frontiers: imperialism's new communities in East Asia, 1842–1933. Ed. by Robert BICKERS and Christian HENRIOT. Manchester, Manchester U. P., 2000, 290 p. (Studies on imperialism).

4087. OFFEN (Karen M.). European feminisms, 1700–1950: a political history. Stanford, Stanford U. P., 2000, XXVIII-554 p.

4088. Opium regimes: China, Britain and Japan, 1839–1952. Ed. by Timothy BROOK and Bob Tadashi WAKABAYASHI. Berkeley a. Los Angeles, University of California Press, 2000, XIV-444 p.

4089. ORLOW (Dietrich). Common destiny: a comparative history of the Dutch, French, and German Social Democratic Parties, 1945–1969. New York, Berghahn, 2000, XI-370 p.

4090. PASZYN (Danuta). The Soviet attitude to political and social change in Central America, 1979–1990: case-studies on Nicaragua, El Salvador and Guatemala. Basingstoke, Macmillan in association with School of Slavonic and East European Studies, University of London, 2000, IX-161 p. (Studies in Russia and East Europe).

4091. PAULMANN (Johannes). Pomp und Politik. Monarchenbegegnungen in Europa zwischen Ancien Régime und Ersten Weltkrieg. Paderborn, München, Wien u. Zürich, Schöning, 2000, 482 p.

4092. Polen und Österreich im 18. Jahrhundert. Hrsg. v. Walter LEITSCH und Stanisław TRAWKOWSKI; unter Mitwirkung von Wojciech KRIEGSEISEN. Warszawa, Wydawnictwo Naukowe Semper, 2000, 152 p.

4093. Pomorze Zachodnie w Tysiącleciu. (Westpommern [Pomorze Zachodnie] im Laufe des Jahrtausends). Praca zbiorowa. Hrsg. v. Paweł BARTNIK und Kazimierz KOZŁOWSKI. Szczecin, Dokument, 2000, 634 p. (Regiony w Dziejach Polski).

4094. PROCACCI (Giuliano). Storia del XX secolo. Milano, Bruno Mondadori, 2000, 598 p.

4095. PUGH (Martin). The march of the women: a revisionist analysis of the campaign for women's suffrage, 1866–1914. Oxford, Oxford U. P., 2000, 303 p.

4096. PUȘCAȘ (Vasile). Central Europe since 1989: concepts and developments. Cluj-Napoca, Editura Dacia, 2000, 271 p.

4097. Revoluția de la 1848–1849 în Europa centrală. Perspectivă istorică și istoriografică. Cluj-Napoca, Editura Presa Universitară Clujeană, 2000, 472 p.

4098. Revolution (A) from above? The power state of he 16th and 17th century Scandinavia. Ed. by Leon JESPERSEN. Odense, Odense U. P., 2000, 383 p. [Cf. n[os] <choice> 4950, 5223.]

4099. Revolutions in the Americas. [AHR Forum]. *American historical review*, 2000, 105, 1, p. 92-152. [Cf. n[os] <Choice> 4101, 4729, 4914, 5316.]

4100. Rivoluzione liberale (La) e le nazioni divise. A cura di Pier Luigi BALLINI. Venezia, Istituto veneto di scienze, lettere ed arti, 2000, VI- 370 p.

4101. RODRÍGUEZ O. (Jaime E.). The emancipation of America. *In*: Revolutions in the Americas [Cf. n° 4099], p. 131-152.

4102. Rossija i Frantsija, XVIII–XX veka. (La Russie et la France, XVIIIe–XXe siècles: [articles]). Vol. 3. Ed. Petr P. CHERKASOV. RAN, In-t vseobshchej istorii. Moskva, Nauka, 2000, 283 p. (bibl. incl. and p. 270-279). [Cf. Bibl. 98, n° 4206.]

4103. Rückblicke auf das 20. Jahrhundert. Hrsg. v. Hartmut LEHMANN. Göttingen, Wallstein, 2000, 118 p. (Göttinger Gespräche zur Geschichtswissenschaft, 11).

4104. SCHWARZ (Hans-Peter). Fragen an das 20. Jahrhundert. *Vierteljahrshefte für Zeitgeschichte*, 2000, 48, 1, p. 1-36.

4105. Search (In) of peace and prosperity: new German settlements in eighteenth-century Europe and America. Ed. by Hartmut LEHMANN, Hermann WELLENREUTHER and Renate WILSON. Assisted by John B. FRANTZ and Carola WESSEL. University Park, Pennsylvania State U. P., 2000, XII-332 p.

4106. Siècle (Le) des communismes. Ed. par Michel DREYFUS, Bruno GROPPO, Claudio INGERFLOM, Rolando LEW, Claude PENNETIER, Bernard PUDAL et Serge WOLIKOW. Paris, Ed. de l'Atelier, 2000, 542 p.

4107. SPETH (Rudolf). Nation und Revolution. Politische Mythen in 19. Jahrhundert. Opladen, Leske-Budrich, 2000, 502 p.

4108. STOLLBERG-RILINGER (Barbara). Europa im Jahrhundert der Aufklärung. Stuttgart, Reclam, 2000, 407 p.

4109. TANDETER (Enrique). Tradición y modernidad en América Latina. Aportes para una discusión. *In*: Proceedings of the 19th International Congress of Historical Sciences. Oslo, University of Oslo, 2000, p. 357-369.

4110. TAYLOR (Miles). The 1848 revolutions and the British empire. *Past and present*, 2000, 166, p. 146-180.

4111. THODY (Philip Malcolm Waller). Europe since 1945. London a. New York, Routledge, 2000, VIII-328 p.

4112. Tsentral'no-Vostochnaja Evropa vo vtoroj polovine XX veka. (Central and Eastern Europe in the second half of the 20th century: [A collective monograph]). Ed. Aleksandr D. NEKIPELOV. Vol. 1. Stanovlenie 'real'nogo sotsializma', 1945-1965. (The emergence of 'Real Socialism', 1945-1965). Ed. Igor' I. ORLIK. Moskva, Nauka, 2000, 484 p. (tables).

4113. VIOLA (Paolo). L'Ottocento. Torino, Einaudi, 2000, 386 p. – IDEM. Il Novecento. Torino, Einaudi, 2000, 545 p.

4114. VOLKOV (Vladimir K.). Uzlovye problemy novejshej istorii stran Tsentral'noj i Jugo-Vostochnoj Evropy. (The focal problems of the recent history of Central and Eastern European countries). RAN, In-t slavjanovedenija. Moskva, [s. n.], 2000, 367 p.

4115. VYKOUKAL (Jiří), LITERA (Bohuslav), TEJCHMAN (Miroslav). Východ. Vznik, vývoj a rozpad sovětského bloku 1944-1989. (East. The creation, development and disintegration of the Soviet block). Praha, Libri, 2000, 860 p. (photogr.).

4116. War, institutions, and social change in the Middle East. Ed. by Steven HEYDEMANN. Berkeley a. Los Angeles, University of California Press, 2000, VII-327 p.

4117. WEINDLING (Paul Julian). Epidemics and genocide in Eastern Europe 1890-1945. Oxford, Oxford U. P., 2000, XXI-463 p.

4118. WHITE (George W.). Nationalism and territory: constructing group identity in Southeastern Europe. Lanham a. Oxford, Rowman & Littlefield, 2000, XV-311 p. (Geographical perspectives on the human past).

4119. WILSON (Peter H.). Absolutism in Central Europe. London a. New York, Routledge, 2000, 173 p.

4120. WUNDERLICH (Dieter). Vernetzte Karrieren. Friedrich der Große, Maria Theresia, Katharina die Große. Regensburg, Pustet, 2000, 286 p.

4121. Zentralismus und Föderalismus im 19. und 20. Jahrhundert: Deutschland und Italien im Vergleich. Hrsg. v. Oliver JANZ, Pierangelo SCHIERA und Hannes SIEGRIST. Berlin, Duncker & Humblot, 2000, 282 p. (ill.). (Schriften des Italienisch-Deutschen Historischen Instituts in Trient, 15).

Cf. n°s 420, 896-1023, 1053, 3322, 5575, 6770, 7384, 7576-7641

§ 2. History by countries.

Afghanistan

** 4122. DIXIT (Jyotindra Nath). An Afghan diary: Zahir Shah to Taliban. New Delhi, Konark Publishers, 2000, XVI-525 p.

4123. Afghanen im Exil: Identität und politische Verantwortung. Iserlohn, Institut für Kirche und Gesellschaft, 2000, 112 p. (Tagungsprotokolle / Institut für Kirche und Gesellschaft der Evangelischen Kirche von Westfalen).

4124. BRENTJES (Burchard), BRENTJES (Helga). Taliban: a shadow over Afghanistan. Varanasi, Rishi Publications, 2000, XI-195 p.

4125. EWANS (Martin). Afghanistan: a new history. Richmond, Curzon, 2000, 320 p. (ill.).

4126. GIUSTOZZI (Antonio). War, politics and society in Afghanistan, 1978-1992. London, C. Hurst, 2000, XIV-320 p.

4127. JALALZAI (Musa Khan). The pipeline war in Afghanistan. Lahore, Mobile Institute of International Affairs, 2000, 166 p.

4128. RASHID (Ahmed). Taliban: Islam, oil and the new great game in Central Asia. London a. New York, I.B. Tauris, 2000, XI-274 p.

Albania

4129. Albaniia i albanskite identichnosti: izsledvaniia. (Albania and the Albanian identities). Sustavitel Antonina ZHELIAZKOVA. Sofiia, Mezhdunaroden tsentur po problemite na maltsinstvata i maltsinstvata i kulturnite vzaimodeistviia, 2000, 303 p. (Sudbata na miusiulmanskite obshtnosti na Balkanite, 5).

4130. KOLA (Paulin). Albania, its isolation and the Albanian national question, with particular emphasis on Kosova, 1941-1992. London, [s. n.], 2000, 365 p. [Thesis (Ph. D.), London, LSE, 2000].

4131. NASKA (Kaliopi). Keshilli Kombetar 1920: parlamenti i parë shqiptar. (Il Consiglio nazionale del 1920: il primo parlamento albanese). Tiranë, Drejtoria e Pergjithshme e Arkivave, 2000, 178 p. (ill.).

Cf. n° 4248

Algeria

** 4132. BOVE (Roger). L'Algérie: chroniques d'une guerre oubliée: mémoire d'une géneration sacrifiée: Correspondances algériennes de juillet à décembre 1962. Paris, Editions des Ecrivains, 2000, 243 p.

4133. BELHOCINE (Mabrouk). Le courrier Alger-le Caire, 1954–1956: le congrès de la soummam dans la révolution. Alger, Casbah editions, 2000, 349 p.

4134. BOUGUESSA (Kamel). Aux sources du nationalisme algérien: les pionniers du populisme révolutionnaire en marche. Alger, Casbah éditions, 2000, 383 p. (ill.).

4135. Immigration algérienne (L') en France: des origines à l'indépendance. Sous la direction de Jacques SIMON. Paris, Paris-Méditerranée, 2000, 412 p. (ill.). (Collection Documents, témoignages et divers).

4136. KAYANAKIS (Nicolas). Algérie 1960: la victoire trahie: guerre psychologique en Algérie. Friedberg, Edition Atlantis, 2000, 282 p. (Collection France-Algérie, 16).

4137. SAADALLAH (Aboul-Kassem). al-Harakah al-wataniyah al-Jaza'iriyah. Al-Juz' 1. 1860–1900. (Mouvement national algérien). Bayrut, Dar al-Gharb al-Islami, 2000, 655 p.

Angola

** 4138. LARA (Lúcio). Documentos e comentários para a história do MPLA [Movimento Popular de Libertaçao de Angola]. Lisboa, Dom Quixote, 2000, [s. p.]. (Caminhos da memória, 22).

4139. DO AMARAL (Ilídio). O Consulado de Paulo Dias de Novais: Angola no último quartel do século XVI e primeiro do século XVII. Lisboa, Ministério da Ciência e da Tecnologia, Instituto de Investigaçao Científica Tropical, 2000, 278 p.

4140. HEYWOOD (Linda Marinda). Contested power in Angola, 1840s to the present. Rochester, University of Rochester Press, 2000, XVIII-305 p. (ill., maps). (Rochester studies in African history and the diaspora, 6).

Argentina

** 4141. Nomeolvides: memoria de la Resistencia Peronista, 1955-1972. Prefacio de Antonio Francisco CAFIERO. Buenos Aires, Editorial Biblos, 2000, 378 p.

** 4142. Periodismo (El) de la resistencia peronista 1955–1972: anos de luchas y de victorias. Ed. por Miguel Angel MOYANO LAISSUÉ. [S. l.], Asociación de la Resistencia Peronista, 2000, 100 p.

4143. ALONSO (Paula). Between revolution and the ballot box: the origins of the Argentine Radical Party in the 1890s. New York, Cambridge U. P., 2000, XIII-242 p.

4144. AUYERO (Javier). Poor people's politics: Peronist survival networks and the legacy of Evita. Durham, Duke U. P., 2000, XIV-257 p.

4145. BALCEDO (Antonio). Miedos, broncas y esperanzas: 1979–1993, mirando el país desde el sindicalismo. Buenos Aires, Corregidor, 2000, 494 p.

4146. Cordobazo (El): una rebelión popular. Compilación e introducción, Juan Carlos CENA. Buenos Aires, La Rosa Blindada, 2000, 399 p. (ill.). (Colección de ensayos Emilio Jáuregui).

4147. DE LA FUENTE (Ariel). Children of Facundo: Caudillo and Gaucho insurgency during the Argentine state-formation process (La Rioja, 1853–1870). Durham, Duke U. P., 2000, XIII-249 p.

4148. DE RIZ (Liliana). La política en suspenso, 1966-1976. Buenos Aires, Paidós, 2000, 203 p. (Colección Historia argentina, 8).

4149. DELEIS (Mónica), DE TITTO (Ricardo), ARGUINDEGUY (Diego L.). El libro de los presidentes argentinos del siglo XX. Prólogo de Félix LUNA. Buenos Aires, Aguilar, 2000, 485 p.

4150. DUARTE CASANUEVA (Felipe R.). Argentina: pasado y presente en la construccion de la sociedad y el Estado. Buenos Aires, Editorial Universitaria de Buenos Aires Eudeba, 2000, 313 p.

4151. GALMARINI (Hugo R.). Los negocios del poder: reforma y crisis del Estado: 1776–1826. Buenos Aires, Corregidor, 2000, 332 p.

4152. Historia (Una) de la censura: violencia y proscripción en la Argentina del siglo XX. Compilada por Fernando FERREIRA; prólogo de Néstor RUIZ. Buenos Aires, Grupo Editorial Norma, 2000, 429 p. (Colección Biografías y documentos).

4153. JAMES (Daniel). Doña María's story: life history, memory, and political identity. Durham, Duke U. P., 2000, XV-316 p. (Latin America otherwise: languages, empires, nations).

4154. LETTIERI (Alberto Rodolfo). La república de las instituciones: proyecto, desarrollo y crisis del régimen político liberal en la Argentina en tiempos de la organización nacional (1852–1880). Buenos Aires, Quijote Editorial, 2000, 398 p. (Colección Faro de la historia).

4155. LUNA (Félix). Conflictos en la Argentina próspera: de la Revolución del Parque a la restauración conservadora. Buenos Aires, Planeta, 2000, 158 p. (ill.). (Momentos clave de la historia integral de la Argentina, 6).

4156. MANSON (Enrique). Argentina en el mundo del siglo XX. Buenos Aires, Ediciones Caligraf, 2000, 451 p.

4157. Nueva Historia Argentina. Volumen 2. La sociedad colonial. Ed. por Enrique TANDETER. Buenos Aires, Sudamericana, 2000, [s. p.].

4158. POZZI (Pablo A.), SCHNEIDER (Alejandro). Los setentistas: izquierda y clase obrera 1969–1976. Buenos Aires, Editorial Universitaria de Buenos Aires-EUDEBA, 2000, 458 p. (Mundo contemporáneo).

4159. Region and nation: politics, economics, and society in twentieth-century Argentina. Ed. by James P. BRENNAN and Ofelia PIANETTO. New York, St. Martin's Press, 2000, XVII-233 p.

4160. Revolución Libertadora (De la) al menemismo: historia social y política argentina. Compiladores: Hernán CAMARERO, Pablo POZZI y Alejandro SCHNEIDER. Buenos Aires, Ediciones Imago Mundi, 2000, 328 p. (Colección Bitácora argentina).

4161. SPINELLI (Maria Estela). La conformacion de las identidades politicas en la Argentina del siglo XX. Cordoba, Universidad Nacional de Cordoba, 2000, 373 p.

Cf. n°s 4284, 4917, 5375, 5684, 5717

Australia

4162. BOOTH (Robert R.). Warring tribes: the story of power development in Australia. West Perth, Bardak Group, 2000, XV-258 p. (ill.).

4163. HANCOCK (Ian). National and permanent? The federal organisation of the Liberal Party of Australia, 1944–1965. Carlton, Melbourne U. P., 2000, 317 p. (ill.).

4164. HIRST (John). The sentimental nation: the making of the Australian Commonwealth. South Melbourne a. Oxford, Oxford U. P., 2000, XI-388 p.

4165. KINGSTON (Beverley). The Oxford history of Australia. Vol. 3. 1860–1900: glad, confident morning. Melbourne a. Oxford, Oxford U. P., 2000, [s. p.]

4166. KNIGHTLEY (Phillip). Australia: a biography of a nation. London, Jonathan Cape, 2000, VII-372 p.

4167. Rise (The) and fall of one nation. Ed. by Michael Leach, Geoffrey Stokes and Ian Ward. St. Lucia, University of Queensland Press, 2000, XI-272 p.

Austria

* 4168. BUIKE (Bruno). Türkische Osmanen und österreichische Habsburger: bibliographischer Behelf in forschungsstrategischer Absicht. T. 1. Einzelwissenschaften. Marburg, Tectum, 2000, [s. p.]. (Edition Wissenschaft. Reihe Geschichte, 58).

** 4169. Alfred Klahr Gesellschaft (Die) und ihr Archiv: Beiträge zur österreichischen Geschichte des 20. Jahrhunderts. Hrsg. im Auftrage der Alfred Klahr Gesellschaft von Hans HAUTMANN. Wien, Alfred Klahr Gesellschaft, 2000, 389 p. (Quellen & Studien 2000).

** 4170. Protokolle des Ministerrates der Ersten Republik: 1918–1938. Hrsg. v. Rudolf NECK. Abt. 9. 29. Juli 1934 bis 11. März 1938. 4. Kabinett Dr. Kurt Schuschnigg: 2. Dezember 1935 bis 6. März 1936. Bearb. v. Gertrude ENDERLE-BURCEL. Wien, Verl. d. österr. Staatsdruckerei, 2000, LXXX-482 p.

4171. BRUNNTHALER (Adolf). Strom für den Führer: der Bau der Ennskraftwerke und die KZ-Lager Ternberg, Grossraming und Dipoldsau. Weitra, Verlag Publikation PNo1, 2000, 126 p. (ill.).

4172. BUKEY (Evan Burr). Hitler's Austria: popular sentiment in the Nazi era, 1938–1945. Chapel Hill a. London, University of North Carolina Press, 2000, XVI-320 p.

4173. CHORHERR (Thomas). Die roten Bürger: 30 Jahre sozialistisches Österreich: Gedanken eines Konservativen. Wien, Molden, 2000, 223 p.

4174. COLE (Laurence). "Für Gott, Kaiser und Vaterland": nationale Identität der deutschsprachigen Bevölkerung Tirols, 1860–1914. Frankfurt am Main, Campus Verlag, 2000, 552 p. (Studien zur Historischen Sozialwissenschaft, 28).

4175. DIPPELREITER (Michael). Niederösterreich: Land im Herzen, Land an der Grenze. Wien, Böhlau, 2000, 780 p. (ill., maps). (Geschichte der österreichischen Bundesländer seit 1945).

4176. ERBE (Michael). Die Habsburger 1493–1918: eine Dynastie im Reich und in Europa. Stuttgart, Kohlhammer, 2000, 292 p. (Kohlhammer Urban-Taschenbücher, 454).

4177. FEICHTLBAUER (Hubert). Der Fall Österreich: Nationalsozialismus, Rassismus: eine notwendige Bilanz. Wien, Holzhausen, 2000, 386 p.

4178. GOLOUBEVA (Maria). The glorification of emperor Leopold I in image, spectacle and text. Mayence, Philipp von Zabern, 2000, XI-254 p. (Veröffentlichungen des Instituts für europäische Geschichte Mainz, Abteilung für Universalgeschichte, 184).

4179. GRUNER (Wolf). Zwangsarbeit und Verfolgung: Österreichische Juden im NS-Staat 1938–1945. Innsbruck, Studien-Verl, 2000, 356 p.

4180. Habsburger (Die) im deutschen Südwesten: neue Forschungen zur Geschichte Vorderösterreichs. Hrsg. v. Franz QUARTHAL und Gerhard FAIX. Stuttgart, Thorbecke, 2000, 470 p.

4181. Habsburgermonarchie (Die) 1848–1918. Band 7. Verfassung und Parlamentarismus. 1. Verfassungsrecht, Verfassungswirklichkeit, zentrale Repräsentativkörperschaften. 2. Die regionalen Repräsentativkörperschaften. Im Auftrag der Kommission für die Geschichte der Österreichisch-Ungarischen Monarchie (1848–1918). Hrsg. v. Helmut RUMPLER und Peter URBANITSCH. Wien, Verlag der Österreichischen Akademie der Wissenschaften, 2000, 2 vol., 1310 p., 2695 p.

4182. HINTERSTOISSER (Hermann), JUNG (Peter). Geschichte der Gendarmerie in Österreich-Ungarn: Adjustierung 1816–1918: Einsätze im Felde 1914–

1918. Wien, Stöhr, 2000, 171 p. (Österreichische Militärgeschichte. Sonderband, 2000-2).

4183. LEWIS (Jill). Austria 1950: strikes, 'putsch' and the political context. *European history quarterly*, 2000, 30, 4, p. 533-552.

4184. LOHRMANN (Klaus). Zwischen Finanz und Toleranz: das Haus Habsburg und die Juden: ein historischer Essay. Graz, Styria, 2000, 232 p. (ill.).

4185. Marshall Plan (The) in Austria. Ed. by Günter BISCHOF, Anton PELINKA und Dieter STIEFEL. New Brunswick a. London, Transaction Publishers, 2000, 588 p. (Contemporary Austrian studies, 8).

4186. MENASSE (Robert). Erklär mir Österreich: Essays zur österreichischen Geschichte. Frankfurt am Main, Suhrkamp, 2000, 175 p. (Suhrkamp Taschenbuch, 3161).

4187. OKEY (Robin). The Habsburg monarchy, c. 1765–1918: from enlightenment to eclipse. Basingstoke, Macmillan, 2000, 456 p. (European studies).

4188. Österreich: Berichte aus Quarantanien. Hrsg. v. Isolde CHARIM und Doron RABINOVICI. Frankfurt am Main, Suhrkamp, 2000, 174 p. (Edition Suhrkamp, 2184).

4189. Österreichs Kirche und Widerstand 1938–1945. Hrsg. v. Jan MIKRUT. Wien, Dom Verlag, 2000, 296 p.

4190. PELINKA (Anton), ROSENBERGER (Sieglinde). Österreichische Politik: Grundlagen, Strukturen, Trends. Wien, WUV, 2000, 266 p.

4191. PICK (Hella). Guilty victim: Austria from the Holocaust to Haider. London, I.B. Tauris, 2000, XV-246 p.

4192. ROZENBLIT (Marsha L.). Reconstructing a national identity: the Jews of Habsburg Austria during World War I. New York, Oxford U. P., 2000, XIV-252 p. (Studies in Jewish history).

4193. SCHOBER (Richard). Von der Revolution zur Konstitution. Tirol in der Ära des Neoabsolutismus (1849/51–1860). Innsbruck, Wagner, 2000, 381 p. (Veröffentlichungen des Tiroler Landesarchivs, 9).

4194. SCHORN-SCHÜTTE (Luise). Karl V. Kaiser zwischen Mittelalter und Neuzeit. München, Beck, 2000, 110 p.

4195. TRIESSNIG (Simon). Der Kärntner slowenische Klerus und die nationale Frage, 1920–1932. Klagenfurt, Mohorjeva, 2000, 172 p. (Studia Carinthiaca, 17).

4196. VALENTIN (Hellwig). Nationalismus oder Internationalismus? Arbeiterschaft und nationale Frage; mit besonderer Berücksichtigung Kärntens 1918–1934. Klagenfurt, Verlag des Geschichtsvereines für Kärnten, 2000. 480 p. (Archiv für vaterländische Geschichte und Topographie, 83).

4197. VOCELKA (Karl). Geschichte Österreichs: Kultur, Gesellschaft, Politik. Graz, Styria, 2000, 408 p.

4198. VOITHOFER (Richard). Drum schliesst Euch frisch an Deutschland an: die Grossdeutsche Volkspartei in Salzburg 1920–1936. Wien, Böhlau, 2000, 485 p. (Schriftenreihe des Forschungsinstituts für politisch-historische Studien der Dr.-Wilfried-Haslauer-Bibliothek, 9).

4199. WALTERSKIRCHEN (Gudula). Blaues Blut für Österreich: Adelige im Widerstand gegen den Nationalsozialismus. Wien, Amalthea, 2000, 336 p. (ill).

4200. WASSERMANN (Heinz P.). "Zuviel Vergangenheit tut nicht gut!" Nationalsozialismus im Spiegel der Tagespresse der Zweiten Republik. Innsbruck, Studien Verlag, 2000, 582 p.

4201. WINKLER (Eduard). Wahlrechtsreformen und Wahlen in Triest 1905–1909: eine Analyse der politischen Partizipation in einer multinationalen Stadtregion der Habsburgermonarchie. München, R. Oldenbourg, 2000, 405 p. (ill., maps). (Südosteuropäische Arbeiten, 105).

Cf. n^{os} *4534, 4734, 4938, 7967*

Belarus

4202. DEAN (Martin). Collaboration in the Holocaust: crimes of the local police in Belorussia and Ukraine, 1941–1944. Basingstoke, Macmillan, 2000, XX-241 p.

4203. KUZNETSOV (Igor' Nikolaevich), MAZETS (Valentin Genrikhovich). Istoriia Belarusi: v dokumentakh i materialakh. Minsk, "Amalfeia", 2000, 672 p.

4204. RADZIK (Ryszard). Między zbiorowością etniczną a wspólnotą narodową. Białorusini na tle przemian w Europie Środkowo-Wschodniej XIX stulecia. (Zwischen ethnischer Zusammengehörigkeit und nationaler Gemeinschaft. Weißrussen in Anbetracht des Wandels in Mittel-Ost-Europa des XIX. Jahrhunderts). Lublin, Wydaw. Uniw. M. Curie-Skłodowskiej, 2000, 301 p.

Belgium

4205. FIERS (Stefaan). Vijftig jaar volksvertegenwoordiging: de circulatie onder de Belgische parlementsleden 1946–1995. Brussel, KVAB, 2000, 257 p. (ill). (Verhandelingen van de Koninklijke Vlaamse Academie van België voor Wetenschappen en Kunsten. Nieuwe reeks, 1).

4206. LATTEUR (Nicolas). La gauche en mal de la gauche. Bruxelles, De Boeck Université, 2000, 215 p. (ill.). (Pol-his).

4207. REBÉRIOUX (Madeleine). Parti Ouvrier belge et socialisme français. *Jean Jaurés – cahiers trimestriels*, 2000, 158, p. 21-33.

4208. STENGERS (Jean). Histoire du sentiment national en Belgique des origines à 1918. Tome 1. Les racines de la Belgique, jusqu'à la révolution de 1830. Bruxelles, Racine, 2000, 348 p. – IDEM. Sur l'histoire du gouvernement belge de Londres. *Revue belge de philologie et d'histoire*, 2000, 78, 3-4, p. 1009-1022.

4209. VANDENHOUTE (Thierry). La réforme des polices en Belgique. Brussles, Bruylant, 2000, 290 p. (Collection des travaux et monographies / Université libre de Bruxelles; Ecole des sciences criminologiques Léon Cornil, 22).

4210. WITTE (Els), CRAEYBECKX (Jan), MEYNEN (Alain). Political history of Belgium from 1830 onwards. Antwerpen, Standaard Uitgeverij a. Brussels, VUB U. P., 2000, 297 p.

Cf. n° 4938

Benin

4211. FRÈRE (Marie-Soleil). Presse et démocratie en Afrique francophone: les mots et les maux de la transition au Bénin et au Niger. Préface de Patrick QUANTIN. Paris, Karthala, 2000, 540 p. (ill.). (Hommes et sociétés).

Bolivia

** 4212. Documentos del Partido Comunista de Bolivia: seis congresos y el manifiesto de 1950. La Paz, Editorial Roalva, 2000, 620 p.

4213. ARZE CUADROS (Eduardo). La independencia de Bolivia: orígenes económicos y estructura territorial, 1492–1825. La Paz, Editorial "Los Amigos del Libro", 2000, 374 p.

4214. ASTVALDSSON (Astvaldur). The dynamics of Aymara duality: change and continuity in sociopolitical structures in the Bolivean Andes. *Journal of Latin American studies*, 2000, 32, 1, p. 145-174.

4215. ECHAZÚ (Carlos). Estado y clases dominantes en Bolivia: analisis comparativo de las post-guerras en la historia de Bolivia. La Paz, C&C Editores, 2000, 81 p.

4216. GORDILLO (José M.). Campesinos revolucionarios en Bolivia: identidad, territorio y sexualidad en el valle alto de Cochabamba 1952–1964. La Paz, Plural Editores, PROMEC, Universidad de la Cordillera, CEP, 2000, 281 p.

4217. IBAÑEZ ROJO (Enrique). The Unión Democrática y popular government and the crisis of the Boliviam left (1982–1985). *Journal of Latin American studies*, 2000, 32, 1, p. 175-206.

4218. IRUROZQUI (Marta). The sound of the Pututos: politicisation and indigenous rebellions in Bolivia, 1826–1921. *Journal of Latin American studies*, 2000, 32, 1, p. 85-114.

4219. PARRA DÁVILA (Alvaro). El pensamiento político del libertador Bolívar y la Constitución de Bolivia. Caracas, Ediciones El Centauro, 2000, 122 p.

4220. PERALTA RUIZ (Víctor), IRUROZQUI VICTORIANO (Marta). Por la concordia, la fusión y el unitarismo: estado y caudillismo en Bolivia, 1825–1880. Madrid, Consejo Superior de Investigaciones Científicas, 2000, 277 p. (Colección Tierra nueva e cielo nuevo, 41).

4221. VÁZQUEZ-VIANA (Humberto). Historia de la guerrilla del Che en Bolivia. Vol. 1. Antecedentes: una guerrilla para el Che. Santa Cruz de la Sierra, Editorial R.B., 2000, [s. p.].

Cf. n° 4290

Bosnia

Cf. n° 5391

Botswana

4222. FAWCUS (Peter), TILBURY (Alan). Botswana: the road to independence. Gaborone, Pula Press a. Botswana Society, 2000, 240 p.

Brazil

** 4223. Defesa (Em) dos trabalhadores e do povo brasileiro: documentos do PC do Brasil de 1960 a 2000. Sao Paulo, Anita Garibaldi, 2000, 532 p.

** 4224. DEZEM (Rogério). Shindô-Renmei: terrorismo e repressao. Organizadora: Maria Luiza TUCCI CARNEIRO. Sao Paulo, Editora, Arquivo do Estado, 2000, 203 p. (Inventario Deops, Módulo 3 – Japoneses).

4225. Acuarela de Brasil, 500 años después: seis ensayos sobre la realidad histórica y económica brasilena: II Coloquio Internacional de Historia de América: Area de Historia de América, Dpto. de Historia Medieval, Moderna y Contemporánea: Salamanca, 10–12 de noviembre 1999. Ed. por Jonathan Irvine ISRAEL y J. Manuel SANTOS PÉREZ. Salamanca, Ediciones Universidad de Salamanca, 2000, 95 p. (ill., maps). (Aquilafuente, 15). [Cf. n° <selección> 7797.]

4226. BARONOV (David). The abolition of slavery in Brazil: the "liberation" of Africans through the emancipation of capital. Westport a. London, Greenwood Press, 2000, 236 p. (ill.). (Contributions in Latin American studies, 17).

4227. BATALHA (Claudio H. M.). O movimento operário na Primeira República. Rio de Janeiro, Jorge Zahar Editor, 2000, 78 p. (ill.). (Descobrindo o Brasil).

4228. BENNASSAR (Bartolomé), MARIN (Richard). Histoire du Brésil: 1500–2000. Paris, Fayard, 2000, III-629 p.

4229. Brasil 1701–1824: formaçao histórica da nacionalidade brasileira: seminário internacional em comemoraçao aos 500 anos do descobrimento: passagem para o século XXI realizado de 20–22 de outubro de 1999 na Fundaçao Joaquim Nabuco. Organizators: Manuel CORREIA DE OLIVEIRA ANDRADE, Eliane MOURY FERNANDES, Sandra MELO CAVALCANTI. Brasilia e Recife, Conselho Nacional de Desenvolvimento

2. HISTORY BY COUNTRIES

Cientifico e Tecnologico, Fundaçao Joaquim Nabuco, Editora Massangana, 2000, 224 p. (Série Descobrimentos, 14).

4230. CHASIN (José). A miséria brasileira: 1964–1994. Do golpe militar à crisis social. Santo Andre, Estudos e Edicoes Ad Hominem, 2000, XXXIV-367 p. (Livros (Santo André, Brazil).

4231. Colonial Brazil: foundations, crises, and legacies. *Hispanic American historical review*, 2000, 80, 4, p. 681-944. [Cf. n^{os} <Choice> 4237, 4239.]

4232. DA SILVA ROQUETTE LOPREATO (Christina). O espírito da revolta: a greve geral anarquista de 1917. Sao Paulo, Annablume, 2000, 224 p. (ill.).

4233. DE ALENCASTRO (Luiz Felipe). O trato do viventes. Formação do Brasil no Atlântico Sul. São Paulo. Companhia das Letras, 2000, 525 p.

4234. Dicionário do Brasil colonial, 1500–1808. Direçao, Ronaldo VAINFAS. Rio de Janeiro, Objetiva, 2000, 594 p. (ill.).

4235. FERNANDES (Bernardo Mançano). A formaçao do MST [Movimento dos Trabalhadores Rurais sem Terra] no Brasil. Petropolis, Editora Vozes, 2000, 319 p. (ill., maps).

4236. MALERBA (Jurandir). A corte no exílio: civilizaçao e poder no Brasil às vésperas da Independência (1808 a 1821). Sao Paulo, Companhia das Letras, 2000, 412 p. (ill.).

4237. MOSHER (J. C.). Political mobilization, party ideology, and Lusophobia in nineteenth-century brazil: Pernambuco, 1822–1850. *In*: Colonial Brazil: foundations, crises, and legacies [Cf. n° 4231], p. 881-912.

4238. PAULO (Heloisa). "Aqui também é Portugal": a colónia portuguesa do Brasil e o Salazarismo. Coimbra, Quarteto Editora, 2000, 624 p. (Coleccao Teses, 10).

4239. PEDREIRA (J. M.). From growth to collapse: Portugal, Brazil, and the breakdown of the old colonial system (1750–1830). *In*: Colonial Brazil: foundations, crises, and legacies [Cf. n° 4231], p. 839-864.

4240. PIVA (Luiz Guilherme). Ladrilhadores e semeadores: a modernizaçao brasileira no pensamento político de Oliveira Vianna, Sérgio Buarque de Holanda, Azevedo Amaral e Nestor Duarte (1920–1940). Sao Paulo, Departamento de Ciencia Politica da USP e Editora 34, 2000, 262 p.

4241. REIS FILIIO (Danicl Aarao). Ditadura militar, esquerdas e sociedade. Rio de Janeiro, Jorge Zahar Editor, 2000, 84 p. (ill.). (Descobrindo o Brasil). – IDEM. Intelectuais, história e política: séculos XIX e XX. Rio de Janeiro, 7 Letras, 2000, 289 p.

4242. ROSE (R. S.). One of the forgotten things: Getúlio Vargas and Brazilian social control, 1930–1954. Westport a. London, Greenwood Press, 2000, XVI-227 p. (Contributions in Latin American studies, 15).

4243. SERBIN (Kenneth P.). Secret dialogues: church-state relations, torture, and social justice in authoritarian Brazil. Pittsburgh, University of Pittsburgh Press, 2000, XX-312 p. (ill.). (Pitt Latin American series).

4244. TREECE (Dave). Exiles, allies, rebels: Brazil's indianist movement, indigenist politics, and the imperial nation-state. Westport a. London, Greenwood Press, 2000, VIII-271 p. (Contributions in Latin American studies, 16).

4245. VIOTTI DA COSTA (Emilia). The Brazilian empire: myths and histories. Chapel Hill, University of North Carolina Press, 2000, XXVII-320 p. (ill.).

Cf. n°s 264, 7916, 8144

Bulgaria

** 4246. Bulgarskoto vustanie ot 1835 g. (Velchovata zavera): materiali i dokumenti. [L'insurrection bulgare du 1835. (La conjuration de Veltcho et ses adeptes). Matériaux et documents]. Sustavitel, avtor na predgovora i na ochertsite Ivan RADEV. Veliko Turnovo, Slovo, 2000, 275 p.

** 4247. Drugata Bulgariia: dokumenti za organizatsiite na bulgarskata politicheska emigratsiia 1944–1989. (L'autre Bulgarie 1944–1989. Des documents sur les organisations de l'émigration politique bulgare 1944–1989). Sustaviteli Elena STATELOVA [et al.]. Sofiia, Anubis, 2000, 318 p.

** 4248. ELDUROV (Svetlozar). Bulgarite v Albaniia 1913–1939: izsledvane i dokumenti. Sofiia, "Ivrai", 2000, 361 p. (Bulgarska istoricheska biblioteka).

4249. BELOKONSKI (Petko). Istina po vreme na voina: propagandata v Bulgariia prez 1941–1944. Sofiia, Sv. Kliment Okhridski, 2000, 237 p.

4250. BÜCHSENSCHÜTZ (Ulrich). Maltsinstvenata politika v Bulgariia: politikata na BKP kum evrei, romi, pomatsi i turtsi 1944–1989. (La politique des minorités en Bulgarie. La politique du Parti Communiste Bulgare à l'égard de Juifs, de Romes, de Pomaks et de turcs 1944–1989]. Prevod ot nemski ezik Ivo GEORGIEV. Sofiia, IMIR, 2000, 247 p.

4251. Bulgarski durzhavnitsi i polititsi, 1918–1947. (Hommes d'Etat et hommes politiques bulgares 1918–1947). Sustavitel Mariia RADEVA. Sofiia, Universitetsko izd-vo "Sv. Kliment Okhridski", 2000, 201 p.

4252. Bulgarskoto revoliutsionno dvizhenie v navecherieto na suzdavaneto na BRTSK 1869 g. Dokladi i suobshteniia ot nauchnata sesiia, posvetena na 130-godishninata ot nachaloto na revoliutsionnata propaganda na Vasil Levski v Bulgariia, Pleven, 13 mai 1999 g. (Le mouvement révolutionnaire bulgare à la veille de la fondation du Comité bulgare révolutionnaire central du 1869. rapports et communications de la session scientifique consacrée au 130ème anniversaire du début de la propagande révolutionnaire de Vassil Levski en Bulgarie. Pleven, 13 mai 1999). Sustavitelstvo i redaktsiia Mikhail GRUNCHAROV. Sofiia, Goreks-Pres, 2000, 179 p.

4253. Dubrovnishki izvori za bulgarskata istoriia. (Des sources de Doubrovnic sur l'histoire bulgare). Sustavitelstvo, prevod i komentar Ioanna D. SPISAREVSKA. Sofiia, Glavno upravlenie na arkhivite, 2000, 318 p. (Arkhivite govoriat, 10).

4254. Evropa i Bulgariia: sbornik v pamet na prof. Khristo Gandev. (L'Europe et la Bulgarie. Recueil en mémoire du Prof. Khristo Gandev). Sustavitel Milen SEMKOV. Sofiia, UI "Sv. Kl. Ohridski", 2000, 361 p.

4255. GALCHEV (Ilia). Bulgarskoto samosuznanie na naselenieto v Makedoniia prez Vuzrazhdaneto. (La conscience bulgare nationale de la population en Macedonie pendant la Renaissance). Sofiia, "Gutenberg", 2000, 359 p.

4256. ISUSOV (Mito). Politicheskiiat zhivot v Bulgariia 1944–1948. (Vita politica in Bulgaria, 1944–1948). Sofiia, Akademichno izdatelstvo "Prof. Marin Drinov", 2000, 422 p.

4257. KACUNOV (V.). Etnichesko samosuznanie na bulgarite prez XV–XVII v. (La conscience nationale des Bulgares aux XVe–XVIIe siècles). *Godishnik na Sofijskija universitet, istoricheski fakultet*, 2000, 88, p. 37-68.

4258. KOLEV (V.). Borbata na demokraticheskija sgovor s opozicijata za upravlenie na obshchinite (1926–1931). (La lutte de l'Entente démocratique avec l'opposition pour le gouvernement des communes, 1926–1931). *Godishnik na Sofijskija universitet, istoricheski fakultet*, 2000, 89-90, p. 189-242.

4259. MANOLOVA (Mariia G.). Parlamentarnoto upravlenie v Bulgariia, 1894–1912. (Le gouvernement parlamentaire en Bulgarie 1894–1912). Sofiia, Akademichno izdatelstvo "Prof. Marin Drinov", Izdatelstvo SIELA, 2000, 208 p.

4260. MARCHEVA (Iliiana). Todor Zhivkov – putiat kum vlastta: politika i ikonomika v Bulgariia 1953–1964 g. (Todor Zhivkov – la voie vers le pouvoir. Politique et économie en Bulgarie, 1953–1964). Sofiia, In-t po istoriia – BAN, 2000, 281 p.

4261. NEDELCHEV (Mikhail). Radikaldemokratizmut: opit za liberalno politichesko povedenie v Bulgariia. (Radicaldemocratisme. Essai de conduit libérale politique en Bulgaria). Sofiia, D-vo Grazhdanin, 2000, 227 p. (Liberalizmut pred tretoto khiliadoletie).

Cf. n°s 916, 6805

Burkina Faso

4262. GERVAIS (Raymond R.), MANDÉ (Issiaka). From crisis to national identity: migration in mutation, Burkina faso, 1930–1960. *International journal of African historical studies*, 2000, 33, 1, p. 59-80.

Cambodia

Cf. n° 5390

Cameroon

4263. ETABA OTOA (Didier). Le Cameroun libre avec les Français libres, 1940. Yaounde, Presses de l'UCAC, 2000, 161 p. (ill.).

Canada

4264. CASTONGUAY (René). Rodolphe Lemieux et le Parti libéral, 1866–1937: le chevalier du roi. Quebec, Presses de l'Universite Laval, 2000, XVII-238 p.

4265. CHRISTIE (Nancy). Engendering the state: family, work, and welfare in Canada. Toronto a. London, University of Toronto Press, 2000, XIV-459 p.

4266. KETTUNEN-HUJANEN (Eija). Elämän pakkoraossa vai matkalla vaurauteen? Kuopion ja Mikkelin lääneistä sekä Kainuusta vuosina 1918–1930 muuttaneiden siirtolaisten sijoittuminen työelämään Kanadassa 1920- ja 1930-luvuilla sekä heidän paluumuuttonsa Suomeen. (A wretched life or a journey to wealth? Adaptation of immigrants from Savo, North Karelia and Kainuu to Canada in 1918–1930). Helsinki, SKS, 2000, 235 p. (ill., maps, tables). (Bibliotheca historica, 58).

4267. KOLENEKO (Vadim A.). Katolicheskij sindikalizm v Kanade: Teorija i praktika, 1920–1960. (Catholic Syndicalism in Canada: Theory and Practice, 1920–1960). RAN, In-t vseobshchej istorii. Moskva, [s. n.], 2000, 221 p.

4268. MAC KAY (I.). The liberal order framework: a prospectus for a reconnaissance of Canadian history *Canadian historical review*, 2000, 81, 4, p. 617-645.

4269. MAC NAIRN (Jeffrey). The capacity to judge: public opinion and deliberative democracy in upper Canada, 1791–1854. Buffalo, University of Toronto Press, 2000, XI-460 p.

4270. MARGOLIAN (Howard). Unauthorized entry: the truth about Nazi war criminals in Canada, 1946–1956. Toronto a. London, University of Toronto Press, 2000, VIII-327 p.

4271. Parlementarisme canadien (Le). Sous la dir. de Manon TREMBLAY, Réjean PELLETIER et Marcel R. PELLETIER. Sainte-Foy, Presses de l'Université Laval, 2000, XVIII-461 p.

4272. STURGIS (James), BIRD (Margaret). Canada's imperial past: the life of F.J. Ney, 1884–1973. Edinburgh, University of Edinburgh, Centre of Canadian Studies, 2000, IV-325 p. (ill.).

Chile

** 4273. Izquierda chilena (La) (1969–1973): documentos para el estudio de su línea estratégica. Compilación de Víctor FARÍAS. Berlin, Wissenschaftlicher Verlag Berlin, 2000, 7 vol., XIII-5062 p. (ill.).

** 4274. Verdugo (Patricia). La caravana de la muerte: pruebas a la vista. Santiago de Chile, Editorial Sudamericana, 2000, 262 p. (Colección Crónicas y testimonios).

4275. BRAVO VALDIVIESO (Germán). La sublevación de la escuadra y el período revolucionario 1924–1932. Santiago, Ediciones Altazor, 2000, 213 p. (ill.).

4276. CARDOSO RUÍZ (Patricio). Formación y desarrollo del Estado nacional en Chile: de la independencia hasta 1930. Toluca, Estado de Mexico, Universidad Autonoma dal Estado de Mexico, 2000, 200 p.

4277. CORVALÁN MARQUÉZ (Luis). Los partidos políticos y el golpe del 11 de Septiembre: contribución al estudio del contexto histórico. Santiago, Ediciones ChileAmerica-CESOC, 2000, 405 p.

4278. DE LA CERDA (María Soledad). Chile y los hombres del Tercer Reich. Santiago de Chile, Editorial Sudamericana, 2000, 490 p. (Colección Crónicas y testimonios).

4279. DÍAZ NIEVA (José). Chile, de la Falange Nacional a la Democracia Cristiana. Madrid, Universidad Nacional de Educacion a Distancia, 2000, 249 p. (Varia).

4280. ENSALACO (Mark). Chile under Pinochet: recovering the truth. Philadelphia, University of Pennsylvania Press, 2000, XV-280 p. (Pennsylvania studies in human rights).

4281. FARÍAS (Víctor). Los nazis en Chile. Santiago, Editorial Seix Barral, 2000, 586 p. (Tres mundos. Historia).

4282. HUNEEUS (Carlos). El régimen de Pinochet. Edición al cuidado de Jorgelina MARTÍN. Providencia, Santiago de Chile, Editorial Sudamericana, 2000, 670 p. (Colección Crónicas y testimonios). – IDEM. Technocrats and politicians in an authoritarian regime. The 'ODELPLAN boys' and the 'Gremialists' in Pinochet's Chile. *Journal of Latin American studies*, 2000, 32, 2, p. 461-502.

4283. ROSEMBLATT (Karin Alejandra). Gendered compromises: political cultures and the state in Chile, 1920–1950. Chapel Hill a. London, University of North Carolina Press, 2000, XIV-346 p.

4284. WAYLEN (Georgina). Gender and democratic politics: a comparative analysis of consolidation in Argentina and Chile. *Journal of Latin American studies*, 2000, 32, 3, p. 765-794.

China

Cf. n°s 8130, 8504-8738

Colombia

** 4285. Antología del pensamiento y programas del Partido Liberal, 1820–2000. Compilación y prólogo por Fernando JORDÁN FLÓREZ. Santafé de Bogotá, Partido Liberal Colombiano, 2000, 3 vol., [s. p.].

** 4286. SANTOS MOLANO (Enrique). Documentos para entender la historia de Colombia. Santafe de Bogota, Planeta, 2000, 266 p. (Línea del horizonte).

4287. ALVAREZ LLANOS (Jaime), COLPAS GUTIÉRREZ (Jaime), GONZÁLEZ CHAMORRO (Ever). Prensa, desarrollo urbano y política en Barranquilla, 1880–1930. Barranquilla, Fondo de Publicaciones de la Universidad del Atlantico, 2000, 189 p. (Colección de ciencias económicas y sociales Rodrigo Noguera Barreneche).

4288. ANGELERI (Sandra). Guerrillas y búsqueda de paz en Colombia. Caracas, Jose Agustin Catala, El Centauro Ediciones, 2000, 241 p.

4289. LAROSA (Michael). De la derecha a la izquierda: la iglesia católica en la Colombia contemporánea. Santafe de Bogota, Planeta, 2000, 272 p. (Línea del horizonte).

4290. LEE VAN SCOTT (Donna). A political analysis of legal pluralism in Bolivia and Colombia. *Journal of Latin American studies*, 2000, 32, 1, p. 207-234.

4291. PONCE MURIEL (Alvaro). De clérigos y generales: crónicas sobre la Guerra de los Mil Días. Santafe de Bogota, Panamericana Editorial, 2000, 232 p. (Cuadernillos de historia).

4292. SALGADO (Carlos), PRADA (M. Esmeralda). Campesinado y protesta social en Colombia 1980–1995. Santafe de Bogota, Cinep, 2000, 309 p.

4293. STEINER (Claudia). Imaginación y poder: el encuentro del interior con la costa en Urabá, 1900–1960. Medellin, Editorial Universidad de Antioquia, 2000, XXIX-159 p. (Colección Clío).

4294. TOCANCIPÁ (Jairo). La formación del estado-nación y las disciplinas sociales en Colombia. Popayan, Colombia, Taller Editorial Universidad del Cauca, 2000, 321 p.

4295. TORRES DEL RÍO (César). Fuerzas armadas y seguridad nacional. Santafe de Bogota, Planeta Colombiana Editorial, 2000, 294 p. (Temas de hoy).

4296. URIBE-URAN (Victor M.). Honorable lives: lawyers, family, and politics in Colombia, 1780–1850. Pittsburgh, University of Pittsburgh Press, 2000, XII-276 p. (Pitt Latin American series).

Comores

4297. IBRAHIME (Mahmoud). La naissance de l'élite politique comorienne (1945–1975). Paris, l'Harmattan, 2000, 204 p.

Congo

4298. DE WITTE (Ludo). L'assassinat de Lumumba. Paris, Karthala, 2000, 415 p. (Collection Les Afriques).

4299. KOUVIBIDILA (Gaston-Jonas). Histoire du multipartisme au Congo-Brazzaville. Vol. 1. La marche à rebours, 1940–1991. Vol. 2. Les débuts d'une crise attendue, 1992–1993. Paris, L'Harmattan, 2000, 2 vol., 318 p., 336 p. (Etudes africaines).

4300. O'BALLANCE (Edgar). The Congo-Zaire experience, 1960–1998. Basingstoke, Macmillan a. New York, St. Martin's Press, 2000, XXV-205 p.

4301. TURNER (Thomas). Ethnogenèse et nationalisme en Afrique Centrale: aux racines de Patrice Lumumba: Préfacé par Crawford YOUNG. Paris, L'Harmattan, 2000, 459 p. (Collection Congo-Zaïre – histoire et société).

Costa Rica

4302. OBREGÓN (Clotilde María). El proceso electoral y el poder ejecutivo en Costa Rica: 1808–1998. San José, Editorial de la Universidad de Costa Rica, 2000, 467 p. (ill.).

Croatia

4303. KARAMAN (Igor). Hrvatska na pragu modernizacije: (1750.–1918.). (Croatia on the threshold of modernization, 1750–1918). Zagreb, Naklada Ljevak, 2000, 319 p.

4304. MARKUS (Tomislav). Hrvatski politički pokret 1848.–1849. godine: ustanove, ideje, ciljevi, politička kultura. (Croatian political movement, 1848–1849: institutions, ideas, goals, and political culture). Zagreb, Dom i svijet, 2000, 449 p.

4305. MATIJEVIĆ (Zlatko). Hrvatski katolički seniorat i politika (1912.–1919.). (The Croatian Catholic seniorate and politics, 1912–1919). *Croatica Christiana periodica*, 2000, 24, 46, p. 121-162.

4306. OREŠKOVIĆ (Luc). Luj XIV. i Hrvati. (Louis XIV and the Croats). Zagreb, Dom i svijet, 2000, 321 p.

4307. RADELIĆ (Zdenko). Križari: ustaška gerila 1945.–1950. – problemi istraživanja. (The Crusaders: the Ustasha guerilla in the period 1945–1950 – the research problems). *Časopis za suvremenu povijest*, 2000, 32, 1, p. 5-28.

4308. SPEHNJAK (Katarina). Vlast i javnost u Hrvatskoj 1945.–1952. (The authority and the public in Croatia 1945–1952). *Časopis za suvremenu povijest*, 2000, 32, 3, p. 507-514.

Croatia. Addenda 1999

4309. LONZA (Nella). Izborni postupak Dubrovačke Republike. (The electoral procedure in the Republic of Dubrovnik). *Anali zavoda za povijesne znanosti*, 99, 38, p. 9-50.

Cuba

4310. AMERINGER (Charles D.). The Cuban democratic experience: the auténtico years, 1944–1952. Gainesville, University Press of Florida, 2000, 229 p.

4311. AMORES (Juan Bosco). Cuba en la época de Ezpeleta, 1785-1790. Pamplona, Ediciones Universidad de Navarra, 2000, XXVI-571 p. (Colección histórica).

4312. IBARRA GUITART (Jorge Renato). El fracaso de los moderados en Cuba: las alternativas reformistas de 1957 a 1958. La Habana, Editora Politica, 2000, 356 p.

4313. MORENO FRAGINALS (Manuel), [et al.]. Cien años de historia de Cuba (1898–1998). Madrid, Editorial Verbum, 2000, 248 p. (Ensayo).

4314. WHITNEY (Robert). The architect of the Cuban state: Fulgencio Batista and populism in Cuba, 1937–1940. *Journal of Latin American studies*, 2000, 32, 2, p. 435-460.

4315. ZEUSKE (Michael). Insel der Extreme: Kuba im 20. Jahrhundert. Zurich, Rotpunktverlag, 2000, 278 p.

Cyprus

** 4316. Sources for the history of Cyprus. Ed. by Paul W. WALLACE and Andreas G. ORPHANIDES. Vol. 9. The final days of British rule in Cyprus: dispatches and diaries of Consul General Taylor Belcher and Edith Belcher. Selected and edited by David W. MARTIN and Paul W. WALLACE. Altamont, Greece and Cyprus Research Center, 2000, XVIII-364 p.

4317. BOROWIEC (Andrew). Cyprus: a troubled island. Westport, Praeger, 2000, XVI-193 p.

4318. ÇEVİKEL (Nuri). Kıbrıs Eyâleti: yönetim, kilise, ayân ve halk (1750–1800): bir değişim döneminin anatomisi. (Province of Cyprus: the government, the church, the notables and the subjects [1750–1800]). Gazimağusa, Doğu Akdeniz Üniversitesi, 2000, 407 p.

4319. KATSIAOUNES (Rolandos). He diaskeptike, 1946–1948: me anaskopese tes periodou 1878–1945. (Considerazioni, 1946–1948: con rivisitazione del periodo 1878–1915). Leukosia, Kentro Epistemonikon Ereunon, 2000, 578 p. (Peges kai meletes tes Kypriakes historias, 28).

Czech Republik

** 4320. Formování československého zahraničního odboje v letech 1938–1939 ve světle svědectví Jana Opočenského. (Jan Opočenský's 'The President's Sojourn in the United States'). Ed. Milan HAUNER, collab. Marek ĎURČANSKÝ, Václav PODANÝ and Michal ŠULC. Praha, Archiv AV ČR, 2000, 368 p. (photogr.).

** 4321. JIRÁSEK (Miroslav). Svědectví jednoho života. Vzpomínky z dětství, prezidentské kanceláře a komunistických kriminálů. (Testimony by one life. Memories of the childhood, office of the President and communist camps). Praha, Maroli, 2000, 170 p. (photogr.).

** 4322. KOUDELKA (František). Husákův pád 1987. Dokumenty k oddělení funkcí prezidenta ČSSR a generálního tajemníka KSČ a k nástupu Miloše Jakeše do čela KSČ. (Husák's Fall, 1987. Documents on the Separation of the Office of the President of the CSSR from the Office of CPCz General Secretary and the Rise of Miloš Jakeš to the Head of the CPCz). *Soudobé dějiny. Suppl. Archiv soudobých dějin*, 2000, 7, 3, p. 473-525.

** 4323. MÍŠKOVÁ (Alena), ŠUSTEK (Vojtěch). Josef Pfitzner a protektorátní Praha v letech 1939–1945. Svazek 1: Deník Josefa Pfitznera. Úřední korespondence Josefa Pfitznera s karlem Hermannem Frankem. (Josef Pfitzner und Prag protektorat in den Jahren 1939–1945. Band 1. Das tagebuch von Josef Pfitzner. Die Amtskorrespondenz Josef Pfitzners mit Karl Hermann Frank). Praha, Scriptorium, 2000, 654 p. (Documenta Pragensia, Monographia, 11/1).

** 4324. PALEČEK (Pavel). Ministr Hubert Ripka a jeho osobní archiv. Inventář osobního fondu. Dokumenty. (Minister Hubert Ripka and his privat archive). Brno, Prius, 2000, 141 p. (Prameny a studie k dějinám československého exilu 1948–1989, 4).

** 4325. PEKAŘ (Josef). Deníky Josefa Pekaře 1916–1933. (Josef Pekař's Diaries, 1916–1933). Ed. Josef HANZAL. Praha, Vyšehrad, 2000, 195 p. (Historica).

** 4326. Polické programy Ceských národních stran 1860–1890. Editor, Pavel CIBULKA. Praha, Historicy Ustav av CR, 2000, 343 p.

** 4327. Securitas imperii. Denní situační zprávy StB z listopadu a prosince 1989. Sv. 6/I-III. (Securitas imperii. Daily situation reports of the state security forces, November–December 1989. Tomo 6/1-3). Ed. Patrik BENDA. Praha, Themis, 2000, 1256 p.

4328. BLODIG (Vojtěch). Genocida českých a moravských Židů v letech nacistické okupace. (Genocide of Czech and Moravian Jews during the Nazi occupation of Czechoslovakia). *In*: Židovská Morava – Židovské Brno. Ed. Jan KRATOCHVIL. Brno, 2000, p. 90-99.

4329. BORÁK (Mečislav). Fenomén tzv. vojenských táborů nucené prace v Československu a jeho mezinárodní souvislosti. (The phenomenon of the so called military camps of hard labour in Czechoslovakia and their international continuities). *Slezský sborník*, 2000, 98, 1/2, p. 78-92.

4330. BROKLOVÁ (Eva). T. G. Masaryk v politickém systému první republiky. (T. G. Masaryk in political system of prewar Czechoslovakia). *In*: Evropa mezi Německem a Ruskem. Ed. Miroslav ŠESTÁK. Praha, 2000, p. 193-199.

4331. CABADA (Ladislav). Český stranický systém 1890–1939. (Czech party system, 1890–1939). Plzeň, Západočeská univerzita, 2000, 100 p.

4332. ČECHUROVÁ (Jana), ČECHURA (Jaroslav). Edvard Beneš – diplomat na cestách. Depeše z padesáti zahraničních cest ministra Beneše 1919–1928. (Edvard Beneš – der Diplomat auf der Reisen). Praha, Karolinum, 2000, 212 p.

4333. ČECHUROVÁ (Jana). České politické elity po vzniku Československa. (Die tschechische politische Elite nach der Entstehung der Tschechoslowakei). *In*: Společnost v přerodu. Češi ve 20. století. Praha, 2000, p. 221-230.

4334. ČELOVSKÝ (Bořivoj). Politici bez moci. (Politicians without power). Šenov u Ostravy, Tilia, 2000, 319 p. (photogr.).

4335. CHIRIBIRI (Alessandro). Storia della Cecoslovacchia, 1918–1948. Torino, Celid, 161 p.

4336. FELCMAN (Ondřej). Vláda a prezident. Období pražského jara (prosinec 1967–srpen 1968). (The government and the president. Prague Spring, 1967–1968). Brno, Doplněk, 2000, 501 p. (Prameny k dějinám československé krize 1967-1970, 8/1).

4337. GRUNTOVÁ (Jitka), VAŠEK (František). Vlakové transporty vězňů z koncentračního tábora Osvětim v lednu 1945 přes východní Čechy. (Eisenbahntransporte von Häftlingen aus dem Konzentrationslager Auschwitz in Januar 1945 durch Ostböhmen). *Pomezí Čech a Moravy*, 2000, 4, p. 179-198.

4338. HARNA (Josef). Specifika výzkumu dějin Moravy první poloviny 20. století. (Das Spezikum der Erforschung der Geschichte Mährens in der ersten Hälfte des 20. Jahrhunderts). *In*: Dějiny Moravy a Matice moravská. Problémy a perspektivy. Ed. Libor JAN. Brno, 2000, p. 223-231.

4339. HAVLÍK (Bohumil). Perzekuce a odsun Němců na jihozápadní Moravě v letech 1945–1946. (Persekution und Vertreibung der Deutschen im westöstlichen Mähren in den Jahren 1945–1946). *Vlastivědný sborník Vysočiny. Odd. věd společenských*, 2000, 12, p. 145-170.

4340. Historie okupovaného pohraničí 1938–1945. Sv. 5-6. (Die Geschichte des Okkupationsgrenzgebiet 1938–1945. Tomo 5-6). Ed. Zdeněk RADVANOVSKÝ. Ústí nad Labem, Univerzita J. E. Purkyně, 2000, 2 vol., 183 p., 242 p.

4341. Jan Patočka, Češi a Evropa. Sborník prací z konference uspořádané k 50. výročí působení Jana Patočky na Pedagogické fakultě Masarykovy univerzity v Brně. (Jan Patočka, Czechs and Europe). Ed. Erika VONKOVÁ. Brno, Masarykova univerzita, 2000, 119 p. (photogr.).

4342. JANÁK (Dušan). Politické a legislativní aspekty táborů nucené práce. 1-3. (The political and legal aspects of camps of hard labour. 1-3). *Slezský sborník*, 2000, 98, 1/2-4, p. 93-109; p. 171-190; p. 300-315.

4343. KAPLAN (Karel). Kořeny československé reformy 1968. (The roots of the Czechoslovak reform of 1968). Brno, Doplněk, 2000, 323 p. (Knihy a dokumenty).

4344. KÁRNÍK (Zdeněk). České země v éře První republiky 1918–1938. Díl 1, Vznik, budování a zlatá léta republiky 1918–1929. (Czech Lands in the age epoch of the First Republic, 1918–1938. Tomo 1). Praha, Libri, 2000, 571 p. (photogr.). (Dějiny českých zemí).

4345. KLIMEK (Antonín). Strany a stranictví v meziválečném Československu. (Die tschechoslowakischen politischen Parteien in der Zwischenkriegszeit). *In*: Společnost v přerodu. Češi ve 20. století. Praha, 2000, p. 28-41. – IDEM. Velké dějiny zemí Koruny české. Vol. 13, 1918–1929. (The great histories of the Czech Lands. Tomo 13, 1918–1929). Praha, Paseka, 2000, 821 p. (photogr.).

4346. KREJČOVÁ (Helena). Židovská komunita v Sudetech a její osudy po Mnichovu 1938. (Jewish community in the Sudetenland and their fate after the Munich Agreement of 1938). *In*: Židé v Sudetech. Praha, 2000, p. 129-139.

4347. LEIKERT (Jozef). Čierny piatok sedemnásteho novembra. (The Black Friday – November 17th). Bratislava, Veda vydavateľstvo Slovenskej akadémie vied – Historický ústav Slovenskej akadémie vied, 2000, 454 p.

4348. MAREK (Pavel). Přehled politického stranictví na území českých zemí a Československa v letech 1861–1998. (Survey of the political party system in the Czech Lands and Czechoslovakia, 1861–1998). Rosice u Brna, Gloria, 2000, 403 p.

4349. NOSKOVÁ (Helena), VÁCHOVÁ (Jana). Reemigrace Čechů a Slováků z Jugoslávie, Rumunska a Bulharska 1945–1954. (Re-emigration of Czechs and Slovaks from Yugoslavia, Romania and Bulgaria, 1945–1954). Praha, Ústav pro soudobé dějiny AV ČR, 2000, 297 p. (Studijní materiály ÚSD AV ČR, 3).

4350. OLIVOVÁ (Věra). Dějiny první republiky. (History of the First Republic). Praha, Karolinum, 2000, 355 p.

4351. PERNES (Jiří). Snahy o překonání politickohospodářské krize v Československu v roce 1953. (Efforts to getting over political-economic crisis in the Year 1953). Brno, Prius, 2000, 70 p. (Krize komunistického systému v Československu 1953–1957, 1).

4352. PETRŮV (Helena). Právní postavení židů v Protektorátu Čechy a Morava 1939–1941. (The legal status of the Jews in Protectorate of Bohemia and Moravia, 1939–1941). Praha, Federace židovských obcí Sefer, 2000, 175 p.

4353. PODANÝ (Václav), BARVÍKOVÁ (Hana). Emigrace z Ruska v meziválečném Československu. Prameny v českých, moravských a slezských archivech. (The Emigration from Russia to Czechoslovakia between the two world wars. Sources in the Czech, Moravian and Silesian Archives). Praha, Archiv AV ČR, 2000, 490 p. (Práce z dějin Akademie věd. Serie B, 15).

4354. RYCHLÍK (Jan). Situace v Protektorátu Čechy a Morava v letech 1941–1942 ve zprávách bulharských diplomatů. (Circumstances in the Protectorate Bohemia and Moravia in the Years 1941–1942 in the Reports of Bulgarian diplomats). *Lidé města*, 2000, 4, p. 142-159. – IDEM. Slovensko a Slováci v české politice 19. a 20. století. (Slovakia and Slovaks in Czech politics during the nineteenth and twentieth centuries). *Česko-slovenská historická ročenka*, 2000, p. 59-73.

4355. SANDER (Rudolf). Abecední přehled dislokace československé mírové armády v letech 1918–1939. (Alphabetical survey of the stationing of the Czechoslovak peace corps during the period 1918–1939). *Sborník archivních prací*, 2000, 50, 1, p. 205-319.

4356. TEJCHMANOVÁ (Ladislava). Státněbezpečnostní zájmy orgánů politické správy a sledování politických stran za předmnichovské republiky 1918–1938. (The State-and-security interests of the authorities of political administration and the monitoring of political parties in the first Czechoslovak Republic, 1918–1938). *In*: Evropa mezi Německem a Ruskem. Ed. Miroslav ŠESTÁK. Praha, 2000, p. 211-227.

4357. Tomáš Garrigue Masaryk a Slovensko. Tomáš Garrigue Masaryk – 150 rokov od narodenia. Pamätné vydanie k výročiu. (Tomáš Garrigue Masaryk. 150th anniversary). Ed. Dušan KOVÁČ. Bratislava, Nakladateľstvo Nestor, 2000, 76 p.

4358. UHLÍŘ (Jan B.). 20. červenec 1944 v českém protektorátním tisku. (20th July 1944 reflected in the protectorate and exile press). *Moderní dějiny*, 2000, 8, p. 83-104.

4359. VONDROVÁ (Jitka), NAVRÁTIL (Jaromír). Komunistická strana Československa. Konsolidace (květen–srpen 1968). (The Communist Party of Czechoslovakia. Consolidation, May–August 1968). Brno, Doplněk, 2000, 485 p. (Prameny k dějinám československé krize 1967–1970, 9/2).

4360. ŽÁČEK (Pavel). Staatssicherheit versus Státní bezpečnost. K bezpečnostní terminologii dvou satelitních tajných služeb sovětského bloku. (Staatssicherheit vs Státní bezpečnost. Concerning the terminology of the security forces of two Soviet-bloc states). *Soudobé dějiny*, 2000, 7, 3, p. 357-361.

Cf. n° 5091

Dahomey

4361. EDGERTON (Robert Breckenridge). Warrior women: the Amazons of Dahomey and the nature of war. Boulder, Westview Press, 2000, VIII-196 p. (ill., map).

4362. LE CORNEC (Jacques). La calebasse dahoméenne, ou, les errances du Bénin. T. 1. Du Bénin au Dahomey. T. 2. Du Dahomey au Bénin. Paris, L'Harmattan, 2000, 2 vol., [s. p.]

Denmark

** 4363. Kancelliets brevbøger vedrørende Danmarks indre forhold. Bd 32. 1653. (Minutes de la

Chancellerie danoise concernant la situation intérieure. Tome 32. Année 1653). Ed. par Ole DEGN. København, Rigsarkivet, 2000, 554 p.

4364. BIRKELUND (Peter). De loyale oprørere: den nationalt-borgerlige modstandsbevægelses opståen og udvikling 1940–1945. En undersøgelse af de illegale organisation De frie Danske, Studenternes Efterretningstjeneste og Hjemmefronten. (Les insurgés loyaux: genèse et développement de la résistance nationaliste de 1940 à 1945. Etude des organisations illégales 'les Danois Libres', le 'Service de renseignement des étudiants' et le 'Front National'). Odense, Odense universitets forlag, 472 p. (ill.). (Odense University studies in History and Social Sciences. 233).

4365. BJØRN (Claus). Modern Denmark: a synthesis of converging developments. *Scandinavian journal of history*, 2000, 25, 1-2, p. 119-130.

4366. DAMSHOLT (Tine). Fædrelandskærlighed og borgerdyd: patriotisk diskurs og militære reformer i Danmark i sidste del af 1700-tallet. (Amour de la patrie et vertu bourgeoise: discours patriotique et réformes militaires dans la dernière partie du 18e siècle). København, Museum Tusculanum, 2000, 369 p. (ill.) (Etnologiske studier 6).

4367. Danmark og ØMU'en. Politiske aspekter. (Le Danmark et l'Union Monétaire européenne. Aspects politiques). Ed. par Søren DOSENRODE. Århus, Systime, 2000, 287 p.

4368. Dansk forvaltningshistorie: stat, forvaltning og samfund. (Histoire de l'administration danoise: Etat, administration et société). Bd. 1. Fra middelalderen til 1901. (Tome 1. Du Moyen Age à 1901). Ed. por Leon JESPERSEN et E. LADEWIG PETERSEN. Bd. 2. Folkestyrets forvaltning fra 1901 til 1950. (Tome 2. L'administration de la démocratie de 1901 à 1950). Ed. par Tim KNUDSEN. København, Jurist- og Økonomforbundets forlag, 2000, 2 vol., 986 p., 960 p. (ill.).

4369. DEHN-NIELSEN (Henning). Christian VII: den gale konge. (Christian VII de Danemark: le roi fou). København, Sesam, 2000, 203 p. (ill.).

4370. FALDBORG (Henne). Kvinder og magt: samfundskrønike 1915-2000. (Les Femmes et le pouvoir: chroniques sociales, 1915–2000). Ferritslev, Septem, 2000, 207 p. (ill.).

4371. FINK (Jørgen). Storindustri eller middelstand: det ideologiske opgør i Det Konservative Folkeparti 1918–1920. (Grande industrie ou classes moyennes: le débat idéologique dans le Parti conservateur danois 1918–1920). Århus, Aarhus Universitets forlag, 2000, 357 p. (ill.).

4372. FRIISBERG (Claus). Orla Lehmann. Danmarks første moderne politiker: Orla Lehmann, de nationalliberale og Danmark 1810–1849. En politisk biografi. (Orla Lehmann. Le premier homme politique moderne du Danmark: Orla Lehmann, les Nationaux-Libéraux et le Danemark, 1810–1849. Une biographie politique). Varde, Vestjysk Kulturforlag, 2000, 331 p.

4373. GØBEL (Erik). De styrede rigene: embedsmændene i den dansk-norske civile administration 1660–1814. (Le Gouvernement des deux royaumes: les fonctionnaires de l'administration civile dano-norvégienne de 1660 à 1814). Odense, Odense universitets forlag, 2000, 270 p. (ill., tab., diagr.).

4374. HANSEN (Peer Erik), SØRENSEN (Jakob). Påskekrisen 1948: dansk dobbeltspil på randen af den kolde krig. (La Crise de Pâques 1948: le double jeu danois à l'orée de la guerre froide). København, Høst, 2000, 263 p. (ill.).

4375. HELSTRUP (Søren). Truslen mod Danmark: regeringen og de militære chefers trusselopfattelse 1938–1940. (La menace contre le Danemark: la perception de la menace par le gouvernement et les chefs militaires danois entre 1938 et 1940). København, Lindhardt og Ringhof, 2000, 115 p.

4376. JANSEN (Trine S.). 'Komintern og dannelsen af de skandinaviske kommunistpartier'. *Arbeiderhistorie*, 2000, 3, p. 20-24. [English summary].

4377. KELM-HANSEN (Christian). Det koster at være solidarisk: socialdemokratisk u-landspolitik 1945–2000. (Le prix de la solidarité: la politique social-démocrate d'aide au Tiers monde de 1945 à 2000). København, Fremad, 2000, 300 p. (ill.).

4378. Lederskab i dansk industri og samfund 1888–1960. (Les Dirigeants dans l'industrie et la société danoises, 1888–1960). Ed. par Marianne ROSTGAARD et Michael F. WAGNER. Aalborg, Historiestudiet, Aalborg universitet, 2000, 271 p. [English summ.].

4379. MONRAD PETERSEN (Andreas). Schalburgkorpset: historien om korpset og dets medlemmar 1943–1945. (La Milice Schalburg: histoire de la Milice danoise et de ses membres, 1943–1945). Odense, Odense Universitets forlag, 2000, 225 p. (ill.). (Odense University studies in history and social sciences, 231).

4380. NOERGAARD (Asbjoern Sonne). Party politics and the organization of the Danish welfare state, 1890–1920: the bourgeois roots of the modern welfare state. *Scandinavian political studies*, 2000, 23, 3, p. 183-217.

4381. Nordiske protestantisme (Den) og vælfærdsstaten. (Le Protestantisme nordique et l'Etat-providence). Ed. par Tim KNUDSEN. Århus, Århus universitetsforlag, Center for europæisk kirkeret og kirkekundskab, 2000, 153 p. (résumé anglais).

4382. PEDERSEN (Andreas Monrad). Schalburgkorpset: historien om korpset og dets medlemmer 1943–1945. Odense, Odense Universitetsforlag, 2000, 225 p. (ill., ports). (Odense University studies in history and social sciences, 231).

4383. PETERSEN (Klaus). Velfærdsstat. Principiel politik og politisk pragmatism: studier i socialdemokratiet og velfærdsstaten i efterkrigstidens Danmark med særlig henblik på 1960'erne og 1970'erne. (L'Etat-providence. Principes et pragmatisme politques: études sur la social-démocratie et l'Etat-providence dans le Danemark de l'après-guerre, particulièrement dans les années 1960 et 1970). København, Institut for Historie, Københavns Universitet, 2000, 401 p.

4384. RÜNITZ (Lone). Danmark og den jødiske flygtinge 1933–1940: en bog om flygtinge og menneskerettigheder. (Le Danemark et les réfugiés juifs de 1933 à 1940: un livre sur les réfugiés et les droits de l'homme). København, Museum Tusculanum, 2000, 239 p.

Dominican Republic

4385. CRUZ INFANTE (José Abigail). La campana de Hipólito. Santo Domingo, Editora Manatí, 2000, 148 p. (ill.).

4386. GARCÍA (Juan Manuel). Gobierno del PLD [Partido de la Liberación Dominicana]: crónica de una moral sin memoria, Santo Domingo, Editora Manatí, 2000, XII-484 p.

4387. GRAFENSTEIN GAREIS (Johanna von). República Dominicana, una historia breve. San Juan Mixcoac, Instituto Mora, 2000, 178 p. (Perfiles. América latina).

4388. SAGÁS (Ernesto). Race and politics in the Dominican Republic. Gainesville, University Press of Florida, 2000, XII-164 p.

Cf. n° 8339

Ecuador

4389. DE LA TORRE (Carlos). Populist seduction in Latin America: the Ecuadorian experience. Athens, Ohio University Center for International Studies, 2000, XIX-185 p. (Research in international studies. Latin America series, 32).

4390. POLONI-SIMARD (Jacques). La mosaique indienne: mobilité, stratification sociale et métissage dans le corregimiento de Cueca (Equateur) du XVIe au XVIIIe siècle. Paris, Editions de l'Ecole des hautes etudes en sciences sociales, 2000, 514 p. (Civilisations et sociétés, 99).

4391. ROBERTS (Lois J.). The Lebanese immigrants in Ecuador: a history of emerging leadership. Boulder, Westview Press, 2000, XII-243 p.

4392. RODAS CHAVES (Germán). La izquierda ecuatoriana en el siglo 20: (aproximación histórica). Quito, Abya-Yala, 2000, 200 p.

Egypt

4393. ABDO (Geneive). No God but God: Egypt and the triumph of Islam. Oxford, Oxford U. P., 2000, XI-223 p.

4394. BADRAWI (Malak). Political violence in Egypt, 1910–1924: secret societies, plots and assassinations. Richmond, Curzon, 2000, XII-250 p.

4395. BEATTIE (Kirk J.). Egypt during the Sadat years. New York a. Basingstoke, Palgrave, 2000, XI-340 p.

4396. HATINA (Meir). On the margins of consensus: the call to separate religion and state in modern Egypt. *Middle eastern studies*, 2000, 36, 1, p. 35-67.

4397. HUNTER (F. Robert). State society relations in nineteenth-century Egypt: the years of transition, 1848–1879. *Middle eastern studies*, 2000, 36, 3, p. 145-159.

El Salvador

4398. ANDRADE-EEKHOFF (Katharine). Gobernabilidad urbana y exclusión social en San Salvador. San Salvador, FLACSO, Programa El Salvador, 2000, 201 p.

4399. WOOD (Elisabeth Jean). Forging democracy from below: insurgent transitions in South Africa and El Salvador. Cambridge, Cambridge U. P., 2000, XIV-247 p.

Estonia

4400. KASEKAMP (Andres). The radical right in interwar Estonia. New York, St. Martin's Press a. London, University of London, School of Slavonic and East European studies, 2000, XI-218 p. (Studies in Russia and East Europe).

Cf. n° 5075

Ethiopia

** 4401. Ethiopian records of the Menelik era: selected Amharic documents from the Nachlass of Alfred Ilg 1884–1900. Ed. by Bairu TAFLA. Wiesbaden, Harrassowitz, 2000, 563 p. (ill.). (Aethiopistische Forschungen, 54).

** 4402. Nakfa documents (The): the despatches, memoranda, reports and correspondence describing and explaining the stories of the feudal societies of the Red Sea littoral from the Christian-Muslim wars of the sixteenth century to the establishment 1885–1901 of the Italian colony of Eritrea. Ed. and ann. by Anthony D'AVRAY, in collaboration with Richard PANKHURST. Wiesbaden, Harrassowitz, 2000, VIII-307 p. (Äthiopistische Forschungen, 53).

4403. KISSI (Edward). The politics of famine in U.S. relations with Ethiopia, 1950–1970. *International journal of African historical studies*, 2000, 33, 1, p. 113-132.

4404. LE HOUÉROU (Fabienne). Ethiopie-Erythrée: frères ennemis de la Corne de l'Afrique. Paris, L'Harmattan et Saint-Rémy-de-la-Vanne, Les Nouvelles d'Addis, 2000, 159 p. (ill., maps).

Cf. n° 7911

Finland

4405. AHTOKARI (Reijo). Salat ja valat. Vapaamuurarit suomalaisessa yhteiskunnassa ja julkisuudessa

2. HISTORY BY COUNTRIES

1756–1996. ("Hemligheterna och ederna". Frimurarna i det finska samhället och i offentligheten 1756–1996. – Secrets and oaths. Freemasonry in Finnish society and public record 1756–1996 Geheimnis und Gelöbnis. die Freimaurer in der finnischen Gesellschaft und in der Öffentlichkeit 11756–1996). Helsinki, SKS, 2000, 389 p. (ill., maps). (Bibliotheca historica, 54).

4406. JUNILA (Marianne). Kotirintaman aseveljeyttä. Suomalaisen siviiliväestön ja saksalaisen sotaväen rinnakkaiselo Pohjois-Suomessa 1941–1944. (Das Zusammenleben der finnischen Zivilbevölkerung und der deutschen Truppen in Nordfinnland in den Jahren 1941–1944. Coexistence between the Finnish civilian population and the German troops in the Northern Finland in 1941–1944). Helsinki, SKS, 2000, 385 p. (Bibliotheca historica, 61).

4407. KATAINEN (Elina). Communist Women in Finland, 1944–1948: Olga Virtanen's story. *In*: When the war was over: women, war and peace in Europe, 1940–1956. London, Leicester U. P., 2000, [s. p.].

4408. KUJALA (Antti). The breakdown of a society: Finland in the Great Northern war 1700–1714. *Scandinavian journal of history*, 2000, 25, 1-2, p. 69-86.

4409. PELTONEN (Matti). Between landscape and language: the Finnish national self-image in transition. *Scandinavian journal of history*, 2000, 25, 4, p. 265-280.

4410. PELTOVUORI (Risto). Sankarikansa jakavaltajat: Suomi Kolmannen valtakunnan lehdistössä 1940–1944. (Heroic nation and traitors: Finland in the press of the Third Reich 1940–1944). Helsinki, Suomalaisen Kirjallisuuden Seura, 2000, 262 p. (Historiallisia Tutkimuksia, 208).

4411. POHJONEN (Juha). Maanpetturin tie. Maanpetoksesta Suomessa vuosina 1945–1972 tuomitut. (High treason in Finland, 1945–1972. Doomed spyes). Helsinki, Otava, 2000, 543 p. (ill.).

4412. SAHLSTRÖM (André). Under blåsvarta färger. Den estniska konstitutionella krisens verkningar i de finsk-estniska relationerna åren 1934–1938. (Die Ende der legalen Tätigkeit der Freiheitskrieger. Partei und die finnisch-estnischen Relationen, 1934–1938). Helsinki, SKS, 2000, 322 p. (Bibliotheca historica, 51). (ill.).

4413. SCREEN (John Ernest Oliver). Mannerheim: the Finnish years. London, C. Hurst, 2000, X-288 p. (ill., maps, ports.).

4414. SMOLANDER (Jyrki). Suomalainen oikeisto ja "kansankoti". Kansallisen Kokoomuksen suhtautuminen pohjoismaiseen hyvinvointivaltiomalliin jälleenrakennuskaudelta konsensusajan alkuun. (The Finnish Right Wing and "Folkhemmet". Attitudes of the National Coalition Party toward the Nordic welfare model from the period of reconstruction to the beginning of consensus). Helsinki, SKS, 2000, 342 p. (Bibliotheca historica, 63). (ill.).

France

** 4415. DE GAULLE (Charles). Mémoires. Introduction par Jean-Louis CRÉMIEUX-BRILHAC; édition présentée, établie et annotée par Marius-François GUYARD; chronologie et relevé de variantes par Jean-Luc BARRÉ. Paris, Gallimard, 2000, CXL-1505 p. (ill., facs., maps). (Bibliothèque de la Pléiade, 465).

** 4416. DE GAULLE (Philippe). Mémoires accessoires, 1946–1982. Paris, Plon, 2000, 329 p.

** 4417. NOGUÈRES (Louis). Vichy, juillet 40. Paris, Fayard et Cherche-Midi, 2000, 152 p. (Pour une histoire du XXe siècle).

** 4418. RAJSFUS (Maurice). De la victoire à la débâcle: (juin 1919–juin 1940). Paris, Le Cherche midi, 2000, 271 p. (Collection "Documents").

4419. Action française (L') et ses amis étrangers, sources. Actes de la table ronde organisée à l'Université de Nantes, 27 novembre 1999. Ed. par Claude HAUSER et Catherine POMEYROLS. Paris, Histoire du présent, 2000, 112 p.

4420. AGRIKOLIANSKY (Eric). Les partis politiques en France au 20e siècle. Paris, A. Colin, 2000, 96 p. (Collection Synthèse. Série Histoire, 111).

4421. AGULHON (Maurice). De Gaulle, histoire, symbole, mythe. Paris, Plon, 2000, 165 p.

4422. ANCEAU (Eric). Les députés du Second Empire: prosographie d'une élite du XIXe siècle. Paris, Champion, 2000, 1018 p. (ill., maps). (Bibliothèque d'histoire moderne et contemporaine, 4).

4423. ANDRESS (David). Massacre at the Champ de Mars: popular dissent and political culture in the French Revolution. Rochester, Boydell & Brewer, 2000, X-239 p. (Studies in History, new series, 17).

4424. ANTIER (Chantal). La Grande Guerre: un arrêt à l'émancipation politique des Françaises? *In*: Femmes (Les) et la guerre [Cf. n° 7164], p. 67-78.

4425. ANTOMARCHI (Véronique). Politique et famille sous la IIIe République, 1870–1914. Paris, L'Harmattan, 2000, 219 p.

4426. APRILE (Sylvie). La IIe République et le Second Empire, 1848–1870: du Prince Président à Napoléon III. Paris, Pygmalion/Gerard Watelet, 2000, 399 p. (Histoire politique de la France).

4427. AZÉMA (Jean-Pierre), WIEVORKA (Olivier). Vichy, 1940–1944. Paris, Ed. Perrin, 2000, 374 p.

4428. AZIMI (Vida). Les premiers sénateurs français: Consulat et premier empire 1800–1814. Préface de Christian PONCELET. Paris, Picard, 2000, 271 p.

4429. BANNISTER (Mark). Condé in context. Ideological change in seventeenth-century France. Oxford, European Humanities research Centre of the University of Oxford, 2000, XIII-240 p.

4430. BEN-AMOS (Avner). Funerals, politics, and memory in modern France, 1789–1996. Oxford a. New York, Oxford U. P., 2000, IX-425 p. (ill.).

4431. BIDOUZE (Frédéric). Les remontrances du Parlement de Navarre au XVIIIe siècle: essai sur une culture politique en province au Siècle des Lumières. Biarritz, Atlantica, 2000, 733 p.

4432. BODINIER (Bernard), TEYSSIER (Eric), ANTOINE (François). L'événement le plus important de la Révolution: la vente des biens nationaux. Paris, Société des Etudes robespierristes, Comité des Travaux historiques et scientifiques, 2000, 501 p.

4433. BORGETTO (Michel), LAFORE (Robert). La république sociale: contribution à l'étude de la question démocratique en France. Paris, Presses universitaires de France, 2000, 367 p. (Politique éclatée).

4434. BOUDON (Jacques-Olivier). Histoire du Consulat et de l'Empire, 1799–1815. Paris, Perrin, 2000, 511 p.

4435. CHILDS (Nick). A political academy in Paris 1724–1731: the Entresol and its members. Oxford, Voltaire Foundation, 2000, XI-289 p. (Studies on Voltaire and the eighteenth century, 2000, 10).

4436. COMBEAU (Yvan), NIVET (Philippe). Histoire politique de Paris au XXe siècle: une histoire locale et nationale. Paris, Presses universitaires de France, 2000, 351 p. (Politique éclatée).

4437. Commune (La) de 1871: utopie ou modernité? Sous la dir. de Gilbert LARGUIER et Jérôme QUARETTI. Perpignan, Presses universitaires de Perpignan, 2000, 447 p.

4438. CORNETTE (Joël). Absolutisme et Lumières, 1652–1783. Paris, Hachette, 2000, 287 p. – IDEM. Les années cardinales: chronique de la France, 1599–1652. Paris, SEDES, 2000, 445 p.

4439. DE WAELE (Michel). Les relations entre le Parlement de Paris et Henri IV. Paris, Publisud, 2000, 456 p. (La France au fil des siècles).

4440. DE WAILLY (Henri). 1940: l'effondrement. Paris, Perrin, 2000, 409 p.

4441. DECAUX (Alain). Morts pour Vichy: Darlan, Pucheu, Pétain, Laval. Paris, Perrin, 2000, 436 p.

4442. DEHAUDT (C.). Le Comité de la Guerre (1781–1784): une institution méconnue de la fin de l'Ancien Régime. *Revue historique*, 2000, 124, 616, p. 869-894.

4443. Démocratie (La) en France. Vol. 1. Idéologies. Vol. 2. Limites. Sous la direction de Marc SADOUN. Paris, Gallimard, 2000, 2 vol., [s. p.].

4444. DESFORGES (Michel). La chouannerie, 1794–1832. Saint-Sulpice-les-Feuilles, Saint-Sulpice, 2000, 189 p. (Tranches d'histoire).

4445. Development (The) of the radical right in France: from Boulanger to Le Pen. Conference on The Extreme Right Wing in France from 1880 to the present (Proceedings: 1998: Trinity College, Dublin). Ed. by Edward J. ARNOLD. Basingstoke, Macmillan, 2000, XXI-288 p.

4446. Directoire (Du) au Consulat. Vol. 2. L'intégration des citoyens dans la Grande Nation. Actes des Tables rondes de Valenciennes et de Lille. Ed. par J. BERNET, Jean-Pierre JESSENNE et Hervé LEUWERS. Lille, CHRENO-Université de Lille III, 2000, 316 p.

4447. DUCCINI (Hélène). Histoire de la France: XVIIe siècle. Paris, SEDES, 2000, 191 p.

4448. DUHAMEL (Éric), FORCADE (Olivier). Histoire et vie politique en France depuis 1945. Sous la dir. de André ZYSBERG. Paris, Nathan, 2000, VII-294 p. (ill., maps).

4449. DUHAMEL (Eric). Histoire politique de la IVe République. Paris, La Découverte, 2000, 121 p.

4450. ECKERT (Hans-Wilhelm). Konservative Revolution in Frankreich? Die Nonkonformisten der Jeune Droite und des Ordre Nouveau in der Krise der 30er Jahre. München, Oldenbourg, 2000, VI-267 p. (Studien zur Zeitgeschichte, 58).

4451. Elites locales (Les) dans la tourmente: du Front populaire aux années cinquante. Sous la direction de Gilles LE BÉGUEC et Denis PESCHANSKI. Paris, CNRS éditions, 2000, 460 p. (CNRS histoire).

4452. ENGELS (Jens Ivo). Königsbilder. Sprechen, Singen und Schreiben über den französischen König in der ersten Hälfte des achtzehnten Jahrhunderts. Bonn, Bouvier, 2000, VIII-333 p. (Pariser historische Studien, 52).

4453. Etat et société en France aux XVIIe et XVIIIe siècles: Mélanges offerts à Yves Durand. Sous la direction de Jean-Pierre BARDET [et al.]. Paris, Presses de l'Université de Paris-Sorbonne, 2000, 548 p. (Collection du Centre Roland Mousnier).

4454. FORTESCUE (William). The Third Republic in France, 1870–1940: conflicts and continuities. London a. New York, Routledge, 2000, VIII-258 p. (Routledge sources in history).

4455. France (La) au XIXe siècle: approches du politique. *Revue historique*, 2000, 124, 614, p. 339-434. [Cf. n^{os} <sélection> 4484, 4501.]

4456. France at war in the twentieth century: propaganda, myth and metaphor. Ed. by Valerie HOLMAN and Deborah KELLY. Oxford a. New York, Berghahn Books, 2000, XII-164 p. (ill.). (Contemporary France, 3).

4457. French history since Napoleon. Ed. by Martin S. ALEXANDER. New York, Oxford U. P. a. London, Arnold, 2000, XIII-434 p.

4458. GOLDSTEIN (Claudine). République et républicains en France de 1848 à nos jours: aspects culturels, idéologiques, institutionnels, politiques et sociaux. Paris, Ellipses, 2000, 239 p. (ill.).

4459. GRAHAM (Lisa Jane). If the king only knew: seditious speech in the reign of Louis XV. Charlottesville a. London, University Press of Virginia, 2000, XI-324 p. (ill.).

4460. GUENIFFEY (Patrice). La politique de la Terreur: essai sur la violence révolutionnaire (1789–1794). Paris, Fayard, 2000, 376 p.

4461. GUICHARD (Jean Pierre). De Gaulle face aux crises (1940–1968). Paris, Le Cherche Midi Editeur, 2000, 452 p. (Collection "Documents").

4462. HARAN (Alexandre Yali). Le lys et le globe. Messianisme dynastique et rêve impérial à l'aube des temps modernes. Seyssel, Champ Vallon, 2000, 377 p.

4463. HELFT-MALZ (Véronique), LÉVY (Paule H.). Les femmes et la vie politique française. Paris, Presses universitaires de France, 2000, 127 p. (Que sais-je? 3550).

4464. HILDESHEIMER (Françoise). Relectures de Richelieu. Paris, Publisud, 2000, 276 p.

4465. Images civiles de la France en guerre. Intr. de Jean-Claude ALLAIN. *Guerres mondiales et conflits contemporaine*, 2000, 50, 197, p. 3-172. [Cf. n^{os} <sélection> 4487, 4505, 4508, 5770, 5913.]

4466. JANKOWSKI (Paul). Cette vilaine affaire Stavisky. Histoire d'un scandale politique. Paris, Fayard, 2000, 467 p.

4467. JOURDAN (Annie). L'empire de Napoléon. Paris, Flammarion, 2000, 351 p. (Champs-Université).

4468. KENNEDY (Michael J.). The Jacobin clubs in the French Revolution, 1793–1795. New York, Berghahn, 2000, 312 p.

4469. KLEßMANN (Eckart). Napoleon. Ein Charakterbild. Köln, Weimar u. Wien, Böhlau, 2000, 162 p.

4470. KROEN (Sheryl). Politics and theater: the crisis of legitimacy in restoration France, 1815–1830. Berkeley a. London, University of California Press, 2000, XIV-394 p. (ill., map). (Studies on the history of society and culture, 40).

4471. LACAM (Jean-Patrice). La France, une République de mandarins? Les hauts fonctionnaires et la politique. Bruxelles, Editions Complexe, 2000, 189 p. (Collection Théorie politique).

4472. LEFÈVRE (Marianne). Géopolitique de la Corse: le modèle républicain en question. Paris, L'Harmattan, 2000, 335 p. (ill., maps). (Collection Histoire et perspectives méditerranéennes).

4473. MARNOT (Bruno). Les ingénieurs au Parlement sous la IIIe République. Paris, CNRS Editions, 2000, 322 p. (ill., maps, charts). (CNRS histoire).

4474. MAYAFFRE (Damon). Le poids des mots. Le discours de gauche et de droite dans l'entre-deux-guerres. Maurice Thorez, Léon Blum, Pierre-Etienne Flandin et André Tardieu (1928–1939). Paris, H. Champion, 2000, 798 p.

4475. Modérés (Les) dans la vie politique française (1870–1965): colloque organisé par l'Université de Nancy 2, 18–20 novembre 1998. Sous la direction de François ROTH; actes réunis par Jean EL GAMMAL, Gilles LE BÉGUEC et François ROTH. Nancy, Presses universitaires de Nancy, 2000, 532 p.

4476. Monarchie (La) entre Renaissance et Révolution, 1515–1792. Volume dirigé par Joël CORNETTE. Paris, Ed. du Seuil, 2000, 503 p. (Histoire de la France politique).

4477. NEWTON (William Ritchey). L'espace du roi: la cour de France au Château de Versailles, 1682–1789. Préface de Jean-Pierre BABELON. Paris, Fayard, 2000, 588 p.

4478. NICKLAS (Thomas). Charles de Gaulle. Held im demokratischen Zeitalter. Göttingen u. Zürich, Muster-Schmidt, 2000, 135 p. (Persönlichkeit und Geschichte, 158/159).

4479. NICOLET (Claude). Histoire, nation, République. Paris, O. Jacob, 2000, 342 p.

4480. Parti socialiste (Le) entre résistance et République. Sous la dir. de Serge BERSTEIN [et al.]. Paris, Publications de la Sorbonne, 2000, 357 p. (ill.). (Publications de la Sorbonne. Série science politique, 2).

4481. PAS (N.). Six heures pour le Vietnam. Histoire des Comités Vietnam français 1965–1968. *Revue historique*, 2000, 124, 613, p. 157-186.

4482. PESCHOT (Bernard). La chouannerie en Anjou, de la Révolution à l'Empire. Montpellier, Université Paul-Valéry, 2000, 312 p.

4483. PILBEAM (Pamela M.). The constitutional monarchy in France, 1814–1848. Harlow a. New York, Longman, 2000, VII-140 p. (ill.). (Seminar studies in history).

4484. PLOUX (F.). L'imaginaire social et politique de la rumeur dans la France du XIXe siècle (1815–1870). *In*: France (La) au XIXe siècle: approches du politique [Cf. n° 4455], p. 395-434.

4485. Prince (Le), le peuple et le droit: autour des plébiscites de 1851 et 1852. Sous la direction de Frédéric BLUCHE. Paris, Presses universitaires de France, 2000, 318 p. (ill.). (Léviathan).

4486. RAPPORT (Michael). Nationality and citizenship in Revolutionary France: the treatment of foreigners 1789–1799. Oxford, Clarendon Press, 2000, VIII-382 p.

4487. RICHARD (Thibault). 1940–1944. L'administration française face à l'occupant en Seine-et-Oise: des relations conflictuelles et ambiguës. *In*: Images civiles de la France en guerre [Cf. n° 4465], p. 149-172.

4488. ROSANVALLON (Pierre). La Démocratie inachevée: histoire de la souveraineté du peuple en France. Paris, Gallimard, 2000, 440 p. (ill.). (Bibliothèque des histoires).

4489. ROTH (François). Raymond Poincaré: un homme d'État républicain. Paris, Fayard, 2000, 715 p.

4490. RUBINSTEIN (Nina). Die französische Emigration nach 1789. Ein Beitrag zur Soziologie der politischen Emigration. Ausgewählte Schriften. Hrsg. u. eingel. v. Dirk RAITH. Mit Beitr. v. Hanna PAPANEK und David KETTLER. Graz u. Wien, Nausner & Nausner, 2000, 252 p. (Bibliothek sozialwissenschaftlicher Emigranten, 6).

4491. SCHAUB (Jean-Frédéric). Révolutions sans révolutionnaires? Acteurs ordinaires et crises politiques sous l'Ancien Régime (note critique). *In*: Conflits politiques d'Ancien Régime (France et monarchie hispanique). *Annales*, 2000, 55, 3, p. 645-654.

4492. Serviteurs de l'État. Une histoire politique de l'administration française, 1875–1945. Sous la dir. de Olivier BARUCH et Vincent DUCLERT. Paris, La Découverte, 2000, 587 p. (Collection "L'espace de l'histoire").

4493. Sharl' de Goll': K 110-letiju so dnja rozhdenija. (Charles de Gaulle: articles dedicated to his 110th anniversary). Ed. Marina Ts. ARZAKANJAN, Aleksandr O. CHUBAR'JAN. RAN, In-t vseobshchej istorii. Moskva, [s. n.], 2000, 203 p.

4494. SHIGAKI (Yoshio). Furansu zettai-ôsei to ryôshusaiban-ken. (Absolutism and seigneurial rights in early modern France). Fukuoka, Kyusyu U. P., 2000, 329 p.

4495. SIROT (Stéphane). Maurice Thorez. Paris, Presses de Sciences Po, 2000, 301 p. (Références/ Facettes).

4496. SOWERWINE (Charles). France since 1870: culture, politics and society. New York, St. Martin's Press, 2000, XXV-505 p.

4497. SPITZ (Jean-Fabien). L'amour de l'égalité: essai sur la critique de l'égalitarisme républicain en France, 1770–1830. Paris, J. Vrin, Editions de l'Ecole des hautes etudes en sciences sociales, 2000, 286 p. (Contextes).

4498. STERNHELL (Zeev). La droite révolutionnaire, 1885–1914: les origines françaises du fascisme. Paris, Fayard, 2000, 436 p. – IDEM. Ni droite, ni gauche: l'idéologie fasciste en France. Bruxelles, Edition Complexe, 2000, 539 p.

4499. TACKETT (Timothy). Conspiracy obsession in a time of revolution: French elites and the origins of the Terror, 1789–1792. *American historical review*, 2000, 105, 3, p. 691-713.

4500. TEYSSIER (Arnaud). Le Ier Empire, 1804–1815: de Napoléon à Louis XVIII. Paris, Pygmalion, 2000, 347 p.

4501. TORT (O.). La dissolution de la Chambre des députés sous la Restauration: la difficile apprivoisement d'une pratique institutionnelle ambiguë. *In*: France (La) au XIXe siècle: approches du politique [Cf. n° 4455], p. 339-366.

4502. TRANVOUEZ (Yvon). Catholiques et communistes: la crise du progressisme chrétien (1950–1955). Paris, Ed. du Cerf, 2000, 363 p.

4503. TUMBLETY (Joan). 'Civilwars of the mind': the commemoration of 1789 Revolution in the Parisian Press of the Radical Right, 1939. *European history quarterly*, 2000, 30, 3, p. 389-430.

4504. VAJDA (Sarah). Maurice Barrès. Paris, Flammarion, 2000, 437 p. (ill.). (Grandes biographies).

4505. VALADE (Jean-Michel). La lutte contre le francs-maçons sous le régime de Vichy (1940–1944). L'exemple de la Corrèze. *In*: Images civiles de la France en guerre [Cf. n° 4465], p. 117-128.

4506. VERRIÈRE (Jacques). Genèse de la nation française. Paris, Flammarion, 2000, 363 p. (Champs, 435).

4507. WHALEY (Leigh Ann). Radicals: politics and republicanism in the French Revolution. Stroud, Sutton, 2000, 212 p. (ill.).

4508. YAGIL (Limoré). Le mouvement «Redressement français» et la Révolution nationale de Vichy. *In*: Images civiles de la France en guerre [Cf. n° 4465], p. 129-148.

4509. ZINK (Anne). Pays ou circonscriptions: les collectivités territoriales de la France du Sud-Ouest sous l'Ancien Régime. Préface d'Emmanuel LE ROY LADURIE. Paris, Publications de la Sorbonne, 2000, 374 p.

Cf. n[os] *4098, 5145, 5192, 7662, 7833, 7852, 8147, 8177*

Gabon

4510. ROSSATANGA-RIGNAULT (Guy). L'État au Gabon: histoire et institutions. Libreville, Editions Raponda-Walker, 2000, 485 p.

Germany

** 4511. Adenauer: Briefe, 1957–1959. Bearb. von Hans Peter MENSING. Paderborn, Ferdinand Schöning 2000, XVII-580 p. (ill.).

** 4512. Arbeiterführer, Parlamentarier, Parteiveteran. Die Tagebücher des Sozialdemokraten Hermann Molkenbuhr 1905 bis 1927. Hrsg. v. Bernd BRAUN und Joachim EICHLER. München, Oldenbourg, 2000, 405 p. (Schriftenreihe der Stiftung Reichspräsident-Friedrich-Ebert-Gedenkstätte, 8).

** 4513. BRANDT (Willy). Berliner Ausgabe. Band 2. Zwei Vaterländer. Deutsch-Norweger im schwedischen Exil – Rückkehr nach Deutschland 1940–1947. Bearb. v. Einhart LORENZ. Band 4. Auf dem Weg nach vorn. Willy Brandt und die SPD 1947–1972. Bearb. v. Daniela MÜNKEL. Bonn, Dietz, 2000, 2 vol., 424 p., 659 p.

** 4514. Deutschland 1949 bis 1999. Band. 1. Apr. 1948–Sep. 1953. Band. 2. Sep. 1953–Okt. 1957. Band. 3. Okt. 1957–Mai 1962. Band. 4. Mai 1962

2. HISTORY BY COUNTRIES

Okt. 1966. Band. 5. Okt. 1966–Apr. 1970. Band. 6. Apr. 1970–Nov. 1973. Band. 7. Nov. 1973–Jun. 1979. Band. 8. Jun. 1979–Dez. 1985. Band. 9. Jan. 1986–Jun. 1994. Band. 10. Jul. 1994–Dez. 1999 [und] Register. Sankt Augustin, Siegler, 2000, 10 vol., [s. p.]. (Archiv der Gegenwart).

** 4515. Friedrich I. von Sachsen-Gotha und Altenburg. Die Tagebücher 1667–1686. Band 1. Tagebücher 1667–1677. Band 2. Tagebücher 1678–1686. Bearb. v. Roswitha JACOBSEN unt. Mitarb. v. Juliane BRANDSCH. Weimar, Böhlaus Nachf., 2000, 2 vol., 479 p., 487 p. (Veröffentlichungen aus Thüringischen Staatsarchiven, 4).

** 4516. Kabinettsprotokolle (Die) der Bundesregierung. Kabinettsausschuß für Wirtschaft. Band 2. 1954–1955. Bearb. v. Michael HOLLMANN. München, Oldenbourg, 2000, 791 p.

** 4517. Kabinettsprotokolle (Die) der Bundesregierung. Band 10. 1957. Bearb. v. Ulrich ENDERS und Josef HENKE. München, Oldenbourg, 2000, 611 p.

** 4518. Kabinettsprotokolle (Die) der Hessischen Landesregierung. Kabinett Geiler 1945–1946. Hrsg. v. Andreas HEDWIG in Zusammenarb. mit Jutta SCHOLL-SEIBERT. Wiesbaden, Historische Kommission für Nassau, 2000, LXXXIV-900 p. (Vorgeschichte und Geschichte des Parlamentarismus in Hessen, 20).

** 4519. Nazism 1919–1945: a documentary reader. Vol. 2. State, economy and society 1933–1939. Ed. by J. NOAKES and G. PRIDHAM. Exeter, University of Exeter, 2000, 433 p. (Exeter studies in history).

** 4520. Protokolle (Die) des Preußischen Staatsministeriums 1817–1934/38. Hrsg. v. der Berlin Brandenburgischen Akademie der Wissenschaften (vormals Preußische Akademie der Wissenschaften) unter der Leitung v. Jürgen KOCKA und Wolfgang NEUGEBAUER. Band 3. 9. Juni 1840 bis 14. März 1848. Bearb. v. Bärbel HOLTZ. Hildesheim, Zürich u. New York, Olms-Weidmann, 2000, IX-555 p. (Acta Borussica, NF., Rh. 1, 3).

** 4521. Quellen zur Geschichte des Deutschen Bundes. Für die Historische Kommission bei der Bayerischen Akademie der Wissenschaften hrsg. v. Lothar GALL. Abt. 1. Quellen zur Entstehung und Frühgeschichte des Deutschen Bundes 1813–1830. Band 1. Die Entstehung des Deutschen Bundes 1813–1815. Bearb. v. Eckhardt TREICHEL. München, Oldenbourg, 2000, CLXXVI-1671 p.

** 4522. SPD (Die) unter Kurt Schumacher und Erich Ollenhauer 1946 bis 1963. Sitzungsprotokolle der Spitzengremien. Band 1. 1946 bis 1948. Hrsg. u. bearb. v. Willy ALBRECHT. Bonn, Dietz, 2000, 555 p.

4523. Adenauer und die Wiederbewaffnung. Hrsg. v. Wolfgang KRIEGER. Bonn, Bouvier, 2000, VIII-160 p. (Rhöndorfer Gespräche, 18).

4524. ALBRECHT (Dieter). König Ludwig II. von Bayern und Bismarck. *Historische Zeitschrift*, 2000, 270, 1, p. 39-64.

4525. ALEXANDER (Matthias). Die Freikonservative Partei, 1890–1918: gemässigter Konservatismus in der konstitutionellen Monarchie. Düsseldorf, Droste, 2000, 421 p. (Beiträge zur Geschichte des Parlamentarismus und der politischen Parteien, 123).

4526. ALLINSON (Mark). Politics and popular opinion in East Germany 1945–1968. Manchester, Manchester U. P., 2000, XI-178 p.

4527. ANDERSON (Margaret Lavinia). Practicing democracy: elections and political culture in Imperial Germany. Princeton a. Chichester, Princeton U. P., 2000, XIX-483 p.

4528. Ausbeutung, Vernichtung, Öffentlichkeit. Neue Studien zur nationalsozialistischer Lagerpolitik. Hrsg. im Auftrag des Instituts für Zeitgeschichte von Norbert FREI, Sybille STEINBACHER und Bernd C. WAGNER. München, K. G. Saur, 2000, 335 p. (Darstellungen und Quellen zur Geschichte von Auschwitz, 4).

4529. BACKES (Uwe). Liberalismus und Demokratie: Antinomie und Synthese: zum Wechselverhältnis zweier politischer Strömungen im Vormärz. Düsseldorf, Droste, 2000, 570 p. (ill.). (Beiträge zur Geschichte des Parlamentarismus und der politischen Parteien, 120).

4530. BADIA (Gilbert). Ces Allemands qui ont affronté Hitler. Paris, Atelier, 2000, 254 p.

4531. BAJOHR (Frank). Verfolgung aus gesellschaftsgeschichtlicher Perspektive. Die wirtschaftliche Existenzvernichtung der Juden und die deutsche Gesellschaft. [Diskussionsforum]. *Geschichte und Gesellschaft*, 2000, 26, 4, p. 629-652.

4532. BENSER (Günter). DDR – gedenkt ihrer mit Nachsicht. Berlin, K. Dietz, 2000, 479 p.

4533. BENZ (Wolfgang). Geschichte des Dritten Reiches. München, C. H. Beck, 2000, 288 p. (ill.).

4534. BRAUNEDER (Wilhelm). Deutsch-Österreich 1918. Die Republik entsteht. Wien, München, Amalthea, 2000, 368 p.

4535. BREDOHL (Thomas M.). Class and religious identity: the Rhenish Center Party in Wilhelmine Germany. Milwaukee, Marquette U. P., 2000, 288 p. (Marquette studies in theology, 18).

4536. BROWNING (Christopher R.). Nazi policy, Jewish workers, German killers. Cambridge a. New York, Cambridge U. P., 2000, XI-185 p.

4537. BRYSAC (Shareen Blair). Resisting Hitler: Mildred Harnack and the Red Orchestra. Oxford, Oxford U. P., 2000, XI-498 p. (ill.).

4538. BURLEIGH (Michael). The Third Reich: a new history. London, Macmillan, 2000, XXV-965 p.

4539. BUSCHFORT (Wolfgang). Parteien im Kalten Krieg: die Ostbüros von SPD, CDU und FDP. Berlin, Ch. Links, 2000, 260 p. (Analysen und Dokumente: wissenschaftliche Reihe des Bundesbeauftragten für die Unterlagen des Staatssicherheitsdienstes der ehemaligen Deutschen Demokratischen Republik, 19).

4540. BUSYGINA (Irina M.). Regiony Germanii. (Regions of Germany). RAN, In-t Evropy. Moskva, ROSSPEN, 2000, 351 p. (maps).

4541. CLARK (Christopher). Kaiser Wilhelm II. London, Longman, 2000, XVI-271 p.

4542. CLINGAN (C. Edmund). The budget debate of 1926: a case study in Weimar democracy. *European history quarterly*, 2000, 30, 1, p. 33-48.

4543. CONNOR (Ian). German refugees and the Bonn Government's resettlement programme: the role of the Trek Association in Schleswig-Holstein, 1951–1953. *German history*, 2000, 18, 3, p. 337-361.

4544. DAVIS (Belinda J.). Home fires burning: food, politics, and everyday life in World War I Berlin. Chapell Hill, University of North Carolina Press, 2000, XIV-349 p.

4545. Demokratie und Arbeiterbewegung in der deutschen Revolution von 1848/1849: Beiträge des Kolloquiums zum 150. Jahrestag der Revolution von 1848/49 am 6. und 7. Juni 1998 in Berlin. Hrsg. v. Helmut BLEIBER, Rolf DLUBEK und Walter SCHMIDT. Berlin, Trafo, 2000, 284 p. (Gesellschaft, Geschichte, Gegenwart, 22).

4546. Demokratie, Agrarfrage und Nation in der bürgerlichen Umwälzung in Deutschland: Beiträge des Ehrenkolloquiums zum 70. Geburtstag von Helmut Bleiber am 28. November 1998. Hrsg. v. Walter SCHMIDT. Berlin, Trafo, 2000, 298 p. (Gesellschaft, Geschichte, Gegenwart, 29).

4547. Demokratisches Denken in der Weimarer Republik. Hrsg. v. Christoph GUSY. Baden-Baden, Nomos, 2000, 680 p. (Interdisziplinäre Studien zu Recht und Staat, 16).

4548. DENNIS (Mike). The rise and fall of the German Democratic Republic, 1945–1990. Harlow, Longman/Pearson, 2000, XVIII-334 p.

4549. Deutsche Umbrüche im 20. Jahrhundert. Hrsg. v. Dietrich PAPENFUSS und Wolfgang SCHIEDER; Redaktion Petra TERHOEVEN. Köln, Böhlau, 2000, X-691 p. [Tagungsbeiträge eines Symposiums der Alexander von Humboldt-Stiftung, Bonn-Bad Godesberg, veranstaltet vom 14.–18. März 1999].

4550. EHRENFREUND (Jacques). Mémoire juive et nationalité allemande. Les juifs berlinois à la Belle Epoque. Paris, Presses Universitaires de France, 2000, 304 p.

4551. ENGEL (David). The Holocaust: the Third Reich and the Jews. Harlow a. New York, Longman, 2000, VI-148 p. (maps). (Seminar studies in history).

4552. FAHNENSCHMIDT (Willi). DDR-Funktionäre vor Gericht: die Strafverfahren wegen Amtsmißbrauch und Korruption im letzten Jahr der DDR und nach der Vereinigung. Berlin, Berlin-Verl. Spitz, 2000, XXII-347 p. (Berliner juristische Universitätsschriften / Strafrecht, 5).

4553. Föderative Nation: Deutschlandkonzepte von der Reformation bis zum Ersten Weltkrieg. Hrsg. v. Dieter LANGEWIESCHE und Georg SCHMIDT. München, R. Oldenbourg, 2000, 428 p.

4554. FRICKE (Karl Wilhelm). Der Wahrheit verpflichtet: Texte aus fünf Jahrzehnten zur Geschichte. Herausgegeben von der Stiftung zur Aufarbeitung der SED-Diktatur und vom Deutschlandfunk; wissenschaftlicher der DDR. Bearbeiter: Ilko-Sascha KOWALCZUK. Berlin, Ch. Links, 2000, 636 p.

4555. GDR (The) and its history: Rückblick und Revision: die DDR im Spiegel der Enquête-Kommissionen. Ed. by Peter BARKER. Amsterdam a. Atlanta, Rodopi, 2000, 217 p. (ill.). (German monitor, 49).

4556. Germanija i Rossija: Sobytija, obrazy, ljudi. (Germany and Russia: events, images, people). Vol. 3. Materialy mezhdunar. nauch. konf. 'Rossija i Germanija: Opyt i uroki otnoshenij v XIX-XX vv.' (Proceedings of the International Conference 'Russia and Germany: experience and lessons of the relations, 19th–20th centuries', the State University of Voronezh, 13–14 October, 1999). Ed. V. A. ARTEMOV. Voronezh, [s. n.], 2000, 235 p.

4557. Gestapo (Die) im Zweiten Weltkrieg: "Heimatfront" und besetztes Europa. Hrsg. v. Gerhard PAUL, Klaus-Michael MALLMANN. Darmstadt, Wissenschaftliche Buchgesellschaft, 2000, VIII-674 p.

4558. GIESEKE (Jens). Die hauptamtlichen Mitarbeiter der Staatssicherheit: Personalstruktur und Lebenswelt 1950–1989/90. Berlin, Ch. Links, 2000, 615 p. (Analysen und Dokumente, 20).

4559. GODA (Norman J. W.). Black marks: Hitler's bribery of his senior officers during World War II. *Journal of modern history*, 2000, 72, 2, p. 413-452.

4560. GONEN (Jay Y.). The roots of Nazi psychology: Hitler's utopian barbarism. Lexington, University Press of Kentucky, 2000, 224 p.

4561. "Gott mit uns": Nation, Religion und Gewalt im 19. und frühen 20. Jahrhundert. Hrsg. v. Gerd KRUMREICH und Hartmut LEHMANN. Göttingen, Vandenhoeck & Ruprecht, 2000, VI-322 p. (ill.). (Veröffentlichungen des Max-Planck-Instituts für Geschichte, 162).

4562. GRESCHAT (Martin). "Mehr Wahrheit in der Politik!" Das Tübinger Memorandum von 1961. *Vierteljahrshefte für Zeitgeschichte*, 2000, 48, 3, p. 491-514.

4563. GREVELHÖRSTER (Ludger). Kleine Geschichte der Weimarer Republik 1918–1933: ein problemgeschichtlicher Überblick. Munster, Aschendorff, 2000, 220 p. (ill.).

4564. GRIESSMER (Axel). Massenverbände und Massenparteien im wilhelminischen Reich: zum Wandel der Wahlkultur, 1903–1912. Düsseldorf, Droste, 2000, 338 p. (ill., ports). (Beiträge zur Geschichte des Parlamentarismus und der politischen Parteien, 124).

4565. GRIX (Jonathan). The role of the masses in the collapse of the GDR. Basingstoke, Macmillan, 2000, XIII-213 p. (ill.). (New perspectives in German studies).

4566. GRÜNBAUM (Robert). Deutsche Einheit Opladen, Leske-Budrich, 2000, 165 p. (Beiträge zur Politik und Zeitgeschichte).

4567. GRUNER (Wolf). Die NS-Judenverfolgung und die Kommunen. Zur wechselseitigen Dynamisierung von zentraler und lokaler Politik 1933–1941. *Vierteljahrshefte für Zeitgeschichte*, 2000, 48, 1, p. 75-126.

4568. Handbuch der baden-württembergischen Geschichte. Band 1. Allgemeine Geschichte. T. 2. Vom Spätmittelalter bis zum Ende des Alten Reiches. Im Auftrag der Kommission für geschichtliche Landeskunde in Baden-Württemberg. Hrsg. v. Meinrad SCHAAB und Hansmartin SCHWARZMAIER in Verb. mit Gerhard TADDEY. Redaktion: Michael KLEIN. Stuttgart, Klett-Cotta, 2000, XXI-872 p.

4569. HOFFMANN (Dierk). Im Laboratorium der Planwirtschaft. Von der Arbeitseinweisung zur Arbeitskräftewerbung in der SBZ/DDR (1945–1961). *Vierteljahrshefte für Zeitgeschichte*, 2000, 48, 4, p. 631-666.

4570. HOFFMANN (Stefan-Ludwig). Brothers or strangers? Jews and Freemasons in nineteenth-century Germany. *German history*, 2000, 18, 2, p. 143-161.

4571. HOFMANN (Daniel). "Verdächtige Eile". Der Weg zur Koalition aus SPD und F.D.P. nach der Bundestagswahl vom 28. September 1969. *Vierteljahrshefte für Zeitgeschichte*, 2000, 48, 3, p. 515-564.

4572. JAMES (Harold). A German identity: 1770 to the present day. London, Phoenix Press, 2000, IX-270 p.

4573. JANSEN (Christian). Einheit, Macht und Freiheit: die Paulskirchenlinke und die deutsche Politik in der nachrevolutionären Epoche 1849–1867. Düsseldorf, Droste, 2000, 687 p. (ill.). (Beiträge zur Geschichte des Parlamentarismus und der politischen Parteien, 119).

4574. JONES (Larry Eugene). Catholic conservatives in the Weimarer Republic: the politics of the Rhenish-Westphalian aristocracy, 1918–1933. *German history*, 2000, 18, 1, p. 60-85.

4575. Juden, Bürger, Deutsche: zur Geschichte von Vielfalt und Differenz 1800–1933. Hrsg. v. Andreas GOTZMANN, Rainer LIEDTKE und Till VAN RAHDEN. Tübingen, Mohr Siebeck, 2000, IX-444 p. (ill.). (Schriftenreihe wissenschaftlicher Abhandlungen des Leo Baeck Instituts, 63).

4576. KAUDERS (Anthony). Catholics, the Jews and democratization in post-war Germany, Munich 1945–1965. *German history*, 2000, 18, 4, p. 461-484.

4577. KERSHAW (Ian). Hitler, 1936–1945: nemesis. London, Allen Lane The Penguin Press, 2000, XLVI-1115 p. (plates, ill., maps, ports).

4578. KIM (Young Tae). Politik und Parteienkonkurrenz im vereinigten Deutschland. Wiesbaden, Deutscher Universitats-Verlag, 2000, XIV-211 p. (ill.). (DUV Sozialwissenschaft).

4579. KNATZ (Christian). "Ein Heer im grünen Rock"? Der Mitteldeutsche Aufstand 1921, die preussische Schutzpolizei und die Frage der inneren Sicherheit in der Weimarer Republik. Berlin, Duncker & Humblot, 2000, 434 p. (Quellen und Forschungen zur brandenburgischen und preussischen Geschichte, 19).

4580. KREUTZER (Heike). Das Reichskirchenministerium im Gefüge der nationalsozialistischen Herrschaft. Düsseldorf, Droste, 2000, IX-390 p. (Schriften des Bundesarchivs, 56).

4581. LAUTERBACH (Ansgar). Im Vorhof der Macht: die nationalliberale Reichstagsfraktion in der Reichsgründungszeit (1866–1880). Frankfurt am Main u. Oxford, Lang, 2000, 389 p. (Europäische Hochschulschriften. Reihe III, Geschichte und ihre Hilfswissenschaften, 873 = Publications universitaires européennes, 873 = European university studies, 873).

4582. LEHMANN (Axel). Der Marshall-Plan und das neue Deutschland. Die Folgen amerikanischen Besatzungspolitik in den Westzonen. Münster, New York, München u. Berlin, Waxmann, 2000, 528 p. (Internationale Hochschulschriften, 335).

4583. LENGEMANN (Jochen). Das Deutsche Parlament (Erfurter Unionsparlament) von 1850: ein Handbuch: Mitglieder, Amtsträger, Lebensdaten, Fraktionen. München, Urban & Fischer, 2000, 447 p. (ill., ports). (Veröffentlichungen der Historischen Kommission für Thüringen. Grosse Reihe, 6).

4584. Letzte Jahr (Das) der SBZ: politische Weichenstellungen und Kontinuitäten im Prozeß der Gründung der DDR. Hrsg. v. Dierk HOFFMANN. München, Oldenbourg, 2000, 296 p. (Schriftenreihe der Vierteljahrshefte für Zeitgeschichte).

4585. LEVINGER (Matthew). Enlightened nationalism: the transformation of Prussian political culture 1806–1848. Oxford, Oxford U. P., 2000, XIV-317 p.

4586. LEVINSON (Kirill A.). Chinovniki v gorodakh Juzhnoj Germanii XVI–XVII vv.: Opyt istoricheskoj antropologii bjurokratii. (The bureaucracy in the cities of Southern Germany, 16th–17th centuries: an attempt at historical anthropology of the bureaucracy). RAN, In-t vseobshchej istorii. Moskva, [s. n.], 2000, 287 p.

4587. LEWY (Günter). The nazi persecution of the gypsies. Oxford, Oxford U. P., 2000, IX-306 p.

4588. LÖSCHE (Peter), WALTER (Franz). Katholiken, Konservative und Liberale: Milieus und Lebenswelten bürgerlicher Parteien in Deutschland während des 20. Jahrhunderts. [Diskussionsforum]. *Geschichte und Gesellschaft*, 2000, 26, 3, p. 471-492.

4589. LUCHTERHANDT (Martin). Der Weg nach Birkenau. Entstehung und Verlauf der nationalsozialistischen Verfolgung der "Zigeuner". Lübeck, Schmidt-Römhild, 2000, 344 p. (Schriftenreihe der Deutschen Gesellschaft für Polizeigeschichte, 4).

4590. MALYCHA (Andrea). Die SED: Geschichte ihrer Stalinisierung, 1946–1953. Paderborn, F. Schoningh, 2000, 541 p.

4591. MARCUS (Kenneth H.). The politics of power: elites of an early modern state in Germany. Mainz, P. von Zabern, 2000, X-288 p. (ill.). (Veröffentlichungen des Instituts für Europäische Geschichte Mainz, 177).

4592. MATTHIESEN (Helge). Greifswald in Vorpommern. Konservatives Milieu im Kaiserreich, in Demokratie und Diktatur 1900–1990. Düsseldorf, Droste, 2000, 764 p. (Beiträge zur Geschichte des Parlamentarismus und der politischen Parteien, 122).

4593. MEINL (Susanne). Nationalsozialisten gegen Hitler: die nationalrevolutionäre Opposition um Friedrich Wilhelm Heinz. Berlin, Siedler, 2000, 446 p.

4594. MOMMSEN (Hans). Alternative zu Hitler. Studien zur Geschichte des deutschen Widerstandes. München, Beck, 2000, 424 p.

4595. NANZKA (Martin). Spionage der ehemaligen DDR gegen die Bundesrepublik Deutschland: verfassungsrechtliche Grenzen der Strafverfolgung wegen Landesverrates, geheimdienstlicher Agententätigkeit und damit in Zusammenhang stehender Straftaten nach der Herstellung der Einheit Deutschlands. Frankfurt am Main, Lang, 2000, XII-248 p. (Europäische Hochschulschriften, 2).

4596. NEEMANN (Andreas). Landtag und Politik in der Reaktionszeit. Sachsen 1848/50–1866. Düsseldorf, Droste, 2000, 543 p. (Beiträge zur Geschichte des Parlamentarismus und der politischen Parteien, 126).

4597. NICHOLLS (David). Adolf Hitler: a biographical companion. Santa Barbara a. Oxford, ABC-CLIO, 2000, XXX-357 p. (ill.).

4598. NIEMANN (Mario). Mecklenburgischer Großgrundbesitz im Dritten Reich. Soziale Struktur, wirtschaftliche Stellung und politische Bedeutung. Köln, Weimar u. Wien, Böhlau, 2000, 421 p. (Mitteldeutsche Forschungen, 116).

4599. NIPPERT (Klaus). Nachbarschaft der Obrigkeiten. Zur Bedeutung frühneuzeitlicher Herrschaftsvielfalt am Beispiel des Hannoverschen Wendlands im 16. und 17. Jahrhundert. Hannover, Hahnsche Buchhandlung, 2000, 367 p. (Veröffentlichungen der Historischen Kommission für Niedersachsen und Bremen, 196).

4600. NS-Verbrechen und der militärische Widerstand gegen Hitler. Hrsg. v. Gerd. R. UEBERSCHÄR. Darmstadt, Primus-Verl., 2000, X-213 p. (Schriftenreihe des Fritz-Bauer-Instituts, 18).

4601. PANAYI (Panikos). Ethnic minorities in nineteenth and twentieth century Germany. Jews, Gypsies, Poles, Turks and others. Harlow, Pearson Education, 2000, XVI-288 p.

4602. POHL (Dieter). Holocaust. Die Ursachen, das Geschehen, die Folgen. Freiburg, Basel u. Wien, Herder, 2000, 188 p.

4603. POLLACK (Detlef). Politischer Protest: politisch alternative Gruppen in der DDR. Opladen, Leske-Budrich, 2000, 282 p.

4604. PÖTZSCH (Hansjörg). Antisemitismus in der Region. Antisemitische Erscheinungsformen in Sachsen, Hessen, Hessen-Nassau und Braunschweig 1870–1914. Wiesbaden, Kommission für die Geschichte der Juden in Hessen, 2000, IX-413 p. (Schriften der Kommission für die Geschichte der Juden in Hessen, 17).

4605. PREISENDÖRFER (Bruno). Staatsbildung als Königskunst: Ästhetik und Herrschaft im preussischen Absolutismus. Berlin, Akademie Verlag, 2000, 432 p.

4606. PRITCHARD (Gareth). The making of the GDR, 1945–1953: from antifascism to Stalinism. Manchester, Manchester U. P., 2000, IX-244 p.

4607. Problem (The) of revolution in Germany, 1789–1989. Ed. by Reinhard RÜRUP. Oxford, Berg, 2000, 204 p.

4608. RABEHL (Bernd). Feindblick: der SDS im Fadenkreuz des "Kalten Krieges". Berlin, Philosophischer Salon, 2000, 155 p.

4609. RAITHEL (Thomas), STRENGE (Irene). Die Reichstagsbrandverordnung. Grundlegung der Diktatur mit den Instrumenten des Weimarer Ausnahmezustands. *Vierteljahrshefte für Zeitgeschichte*, 2000, 48, 3, p. 413-460.

4610. RÖLL (Wolfgang). Sozialdemokraten im Konzentrazionslager Buchenwald 1937–1945. Unter Einbeziehungen biographischer Skizzen. Hrsg. von der Stiftung Gedenkstätten Buchenwald und Mittelbau-Dora. Göttingen, Wallstein, 2000, 357 p.

4611. ROSS (Corey). Constructing socialism at the grass-roots: the transformation of East Germany, 1945–1965. Basingstoke, Macmillan, 2000, XII-262 p. (ill.).

4612. RÖSSEL (Jörg). Soziale Mobilisierung und Demokratie: die preussischen Wahlrechtskonflikte 1900 bis 1918. Wiesbaden, Deutscher Universitats-Verlag, 2000, XI-402 p. (ill.). (DUV. Sozialwissenschaft).

4613. SCHÄFER (Annette). Zwangsarbeiter und NS-Rassenpolitik. Russische und polnische Arbeitskräfte in Württemberg 1939–1945. Stuttgart, Kohlhammer, 2000, XXVII-262 p. (Veröffentlichungen der Kommission für Geschichtliche Landeskunde in Baden-Württemberg, Rh. B: Forschungen, 143).

4614. SCHEUERMANN (Martin). Minderheitenschutz contra Konfliktverhütung? Die Minderheitenpolitik des Völkerbundes in den zwanziger Jahren. Marburg, Herder-Institut, 2000, XIII-516 p. (Materialien und Studien zur Ostmitteleuropa-Forschung, 6).

4615. SCHLEMMER (Thomas). Grenzen der Integration. Die CSU und der Umgang mit der nationalsozialistischen Vergangenheit – der Fall Dr. Max Frauendorfer. *Vierteljahrshefte für Zeitgeschichte*, 2000, 48, 4, p. 675-742.

4616. SCHMIECHEN-ACKERMANN (Detlef). Der "Blockwart". Die unteren Parteifunktionäre im nationalsozialistischen Terror- und Überwachungsapparat. *Vierteljahrshefte für Zeitgeschichte*, 2000, 48, 4, p. 575-602.

4617. SCHMÖLDERS (Claudia). Hitlers Gesicht. Eine physiognomische Biographie. München, Beck, 2000, 264 p.

4618. SCHÜLER (Barbara). "Im Geiste der Gemordeten ..." Die "Weiße Rose" und ihre Wirkung in der Nachkriegszeit. Paderborn, München u. Wien, Schöningh, 2000, 548 p. (Politik- und kommunikationswissenschaftliche Veröffentlichungen der Görres-Gesellschaft, 19).

4619. SCOLZ (Michael F.). Skandinavische Erfahrungen erwünscht? Nachexil und Remigration. Die ehemaligen KPD-Emigranten in Skandinavien und ihr weiteres Schicksal in der SBZ/DDR. Stuttgart, Steiner, 2000, 415 p. (Historische Mitteilungen, 37).

4620. SEIBT (Ferdinand). Das alte böse Lied: Rückblick auf die deutsche Geschichte 1900 bis 1945. München, Piper, 2000, 403 p.

4621. SELIGMANN (Matthew S.), MAC LEAN (Roderick R.). Germany from Reich to Republic, 1871–1918: politics, hierarchy and elites. Basingstoke, Macmillan a. New York, St. Martin's, 2000, XII-195 p. (European history in perspective).

4622. SIEG (Uwe). Sozialpolitische Zielsetzungen Theobald von Bethmann Hollwegs in den Jahren 1907 bis 1909. *Vierteljahrschrift für Sozial- und Wirtschaftsgeschichte*, 2000, 87, p. 144-165

4623. SMYSER (W. R.). From Yalta to Berlin: the Cold War struggle over Germany. Basingstoke, Macmillan, 2000, 496 p.

4624. SS (Die). Elite unter dem Totenkopf. Hrsg. v. Ronald SMELSER und Enrico SYRING. Paderborn, Schöningh, 2000, 462 p.

4625. Staat, Demokratie und Innere Sicherheit in Deutschland. Hrsg. v. Hans-Jürgen LANGE. Opladen, Leske-Budrich, 2000, 436 p. (Studien zur Inneren Sicherheit, 1).

4626. STALMANN (Volker). Die Partei Bismarcks: die Deutsche Reichs- und Freikonservative Partei, 1866–1890. Düsseldorf, Droste, 2000, 544 p. (ill.). (Beiträge zur Geschichte des Parlamentarismus und der politischen Parteien, 121).

4627. Standort- und Kommandaturbefehle des Konzentrationslager Auschwitz 1940-1945. Hrsg. im Auftrag des Instituts für Zeitgeschichte von Norbert FREI, Thomas GROTUM, Jan PARCER, Sybille STEINBACHER und Bernd C. WAGNER. München, K. G. Saur, 2000, XII-604 p. (Darstellungen und Quellen zur Geschichte von Auschwitz, 1).

4628. STEINBACHER (Sybille). "Musterstadt" Auschwitz. Germanisierungspolitik und Judenmord in Ostoberschlesien. München, K. G. Saur, 2000, 419 p. (Darstellungen und Quellen zur Geschichte von Auschwitz, 2).

4629. STEINSDORFER (Helmut). Die Liberale Reichspartei (LRP) von 1871. Stuttgart, Steiner, 2000, 502 p.

4630. STÖBER (Gunda). Pressepolitik als Notwendigkeit: zum Verhältnis von Staat und Öffentlichkeit im Wilhelminischen Deutschland 1890–1914. Stuttgart, F. Steiner, 2000, 303 p. (Historische Mitteilungen. Beiheft, 38).

4631. STOCKERT (Harald). Adel im Übergang: die Fürsten und Grafen von Löwenstein-Wertheim zwischen Landesherrschaft und Standesherrschaft, 1780–1850. Stuttgart, W. Kohlhammer, 2000, XXXV-330 p. (ill., maps). (Veröffentlichungen der Kommission für Geschichtliche Landeskunde in Baden-Württemberg. Reihe B, Forschungen, 144).

4632. STÜRMER (Michael). The German Empire, 1871–1919. London, Weidenfeld & Nicolson, 2000, XXI-121 p.

4633. TEICHERT (Carsten). Chasak! Zionismus im nationalsozialistischen Deutschland, 1933–1938. Köln, ELEN-Verlag, 2000, 560 p.

4634. THOMPSON (Alastair P.). Left liberals, the state, and popular politics in Wilhelmine Germany. Oxford, Oxford U. P., 2000, XII-425 p.

4635. TIMMER (Karsten). Vom Aufbruch zum Umbruch: die Bürgerbewegung in der DDR 1989. Göttingen, Vandenhoeck & Ruprecht, 2000, 416 p. (Kritische Studien zur Geschichtswissenschaft, 142).

4636. VERHEY (Jeffrey). The spirit of 1914: militarism, myth and mobilization in Germany. Cambridge, Cambridge U. P., 2000, XIV-268 p. (ill.). (Studies in the social and cultural history of modern warfare, 10).

4637. WAGNER (Bernd C.). IG Auschwitz. Zwangsarbeit und Vernichtung von Häftlingen des Lagers Monowitz 1941–1945. München, K. G. Saur, 2000, 378 p. (Darstellungen und Quellen zur Geschichte von Auschwitz, 3).

4638. WELSKOPP (Thomas). Das Banner der Brüderlichkeit: die deutsche Sozialdemokratie vom Vormärz bis zum Sozialistengesetz. Bonn, J.H W. Dietz Nachf., 2000, 839 p. (Reihe Politik- und Gesellschaftsgeschichte / Historisches Forschungszentrum der Friedrich-Ebert-Stiftung, 54).

4639. WIRSCHING (Andreas). Die Weimarer Republik: Politik und Gesellschaft. München, R. Oldenbourg, 2000, X-160 p. (Enzyklopädie deutscher Geschichte, 58).

Cf. n°s 475, 478, 832, 4089, 5047, 5060, 5091, 5221, 5503, 5568, 5636, 5850, 5862, 6759, 8262, 8466

Ghana

4640. AFFRIFAH (Kofi). The Akyem factor in Ghana's history: 1700–1875. Accra, Ghana Universities Press, 2000, 259 p.

4641. ASIAMAH (Alfred Effah Aduenum). The mass factor in rural politics: the case of the Asafo revolution in Kwahu political history. Accra, Ghana Universities Press, 2000, 189 p.

4642. GOCKING (Roger). A chiftancy dispute and ritual murder in Elmina, Ghana, 1945–1946. *Journal of African history*, 2000, 41, 2, p. 197-220.

4643. LENZIN (Rene). "Afrika macht oder bricht einen Mann": soziales Verhalten und politische Einschätzung einer Kolonialgesellschaft am Beispiel der Schweizer in Ghana (1945–1966). Basel, Basler Afrika Bibliographien, 2000, 272 p. (ill.).

4644. OSEI-TUTU (John Kwadwo). The asafoi (sociomilitary groups) in the history and politics of Accra (Ghana) from the 17th to the mid-20th century. Trondheim, Department of History, Norwegian University of Science and Technology NTNU, 2000, XXXI-413 p. (ill., maps, ports.). (Trondheim studies in history, 31, 3).

Great Britain

* 4645. Igirisu shi no shin-chôryû: Shûsei-shugi no kinsei-shi. (New currents of the study of British history: revisionism and the early modern history). Ed. by Jun IWAI and Akihiro SASHI. Tokyo, Sairyusha, 2000, 247 p.

** 4646. ASHDOWN (Paddy). The Ashdown diaries. Vol. 1. 1988–1997. London, Allen Lane, Penguin Press, 2000, [s. p.].

** 4647. CLARK (Alan). Diaries into politics. Edited with introduction and notes by Ion TREWIN. London: Weidenfeld & Nicolson, 2000, XXIV-389 p.

** 4648. Daring to hope: the diaries and letters of Violet Bonham Carter, 1946–1969. Ed. by Mark POTTLE. London, Weidenfeld & Nicolson, 2000, XXXI-431 p.

** 4649. Diaries (The) of Samuel Bamford. Ed. by Martin Hewitt and Robert Poole. Stroud, Sutton, 2000, XXXI-383 p.

** 4650. LEVITT (Jan). Scottish papers submitted to the Cabinet, 1945–1966: a guide to records held at the Public Record Office and National Archive of Scotland. *Scottish economic and social history*, 2000, 20, 1, p. 58-125.

** 4651. Parliament and politics in the age of Churchill and Attlee: the Headlam diaries, 1935–1951. Ed. by Stuart BALL. New York, Cambridge U. P., 2000, [s. p.]. (Camden fifth series, 14).

** 4652. Proceedings in the opening session of the Long Parliament, House of Commons. Vol. 1. 3 November–19 December 1640. Vol. 2. 21 December 1640–20 March 1641. Ed. by Maija JANSSON; assisted by Jennifer Klein MORRISON, Alisa PLANT and Shawn SMITH. Woodbridge, University of Rochester Press, 2000, 2 vol., LX-678 p., 836 p.

4653. BRADDICK (Michael J.). State formation in early modern England, c.1550–1700. Cambridge, Cambridge U. P., 2000, 448 p.

4654. BUSSFELD (Christina). 'Democracy versus dictatorship': die Herausforderung des Faschismus und Kommunismus in Großbritannien, 1932–1937. Paderborn, München, Wien u. Zürich, Schöning, 2000, 336 p.

4655. CALLOW (John). The making of king James II: the formative years of a fallen king. New York, Sutton Publishing, 2000, IX-373 p.

4656. COPSEY (Nigel). Anti-fascism in Britain. Basingstoke, Macmillan, 2000, IX-229 p.

4657. DAVIES (Joan). The secretariat of Henri I, Duc de Montmorency, 1563–1614. *English historical review*, 2000, 115, 463, p. 812-842.

4658. DOLL (Peter M.). Revolution, religion, and national identity: imperial Anglicanism in British North America, 1745–1795. Cranbury, Fairleigh Dickinson U. P., 2000, 336 p.

4659. DUFFY (Michael). The Younger Pitt. Harlow, London a. New York, Longman, 2000, XIV-247 p.

4660. English Civil War (The). Ed. by Peter GAUNT. Oxford, Blackwell, 2000, 360 p.

4661. EVANS (Eric J.). Parliamentary reform in Britain, c. 1770–1918. harlow a. New York, Longman, 2000, IV-154 p. (Seminar studies in history).

4662. FRASER (W. Hamish). Scottish popular politics: from radicalism to Labour. Edinburgh, Polygon at Edinburgh, 2000, XI-176 p.

4663. GERMAIN (Lucienne). Réflexes identitaires et intégration: les Juifs en Grande-Bretagne de 1830 à 1914. Préface de Roland MARX. Paris, Champion, 2000, 245 p. (Bibliothèque d'études juives, 11. Série histoire, 8).

4664. GOTTLIEB (Julie V.). Feminine fascism: women in Britain's fascist movement 1923–1945. London, I. B. Tauris, 2000, 378 p.

4665. GOULD (Eliga H.). The persistence of empire: British political culture in the age of the American Revolution. Chapel Hill, Published for the Omohundro Institute of Early American History and Culture, Williamsburg, Virginia, by the University of North Carolina Press, 2000, XXIV-262 p. (ill., maps).

4666. GRAHAM (Jenny). The nation, the law, and the king: reform politics in England, 1789–1799. Lanham, University Press of America, 2000, 2 vol., XIX-1093 p. (ill.).

4667. HALL (Catherine), MAC CLELLAND (Keith) RENDALL (Jane). Defining the Victorian nation: class, race, gender and the British Reform Act of 1867. Cambridge, Cambridge U. P., 2000, XIII-303 p.

4668. HART (Peter). 'Operations abroad': the IRA in Britain, 1919–1923. *English historical review*, 2000, 115, 460, p. 71-102.

4669. HOPPIT (Julian). A land of liberty? England, 1689–1727. Oxford, Clarendon Press, 2000, XXII-580 p.

4670. JAKOVLEV (Nikolaj N.). Britanija i Evropa. (Britain and Europe, [the 18th and the early 19th century: collected papers]). RAN, In-t vseobshchej istorii. Moskva, [s. n.], 2000, 298 p.

4671. JOHNSON (Odai). Rehearsing the revolution: radical performance, radical politics in the English Restoration. Newark, University of Delaware Press a. London, Associated University Presses, 2000, 184 p. (ill.).

4672. JONES (Whitney R. D.). The tree of Commonwealth, 1450–1793. Madison, Fairleigh Dickinson University Press a. London, Associated University Presses, 2000, 394 p.

4673. KAPITALNIAK (Tomasz). Historia Szkocji. (Geschichte Schottlands). Łódź, Ibidem, 2000, 125 p.

4674. KEOHANE (Dan). Security in British politics, 1945–1999. Basingstoke, Macmillan Press, 2000, IX-240 p.

4675. Labour Party (The): a centenary history. Ed. by Brian BRIVATI and Richard HEFFERNAN. Basingstoke, Macmillan, 2000, XXII-512 p.

4676. Labour's first century. Ed. by Duncan TANNER, Pat THANE and Nick TIRATSOO. Cambridge a. New York, Cambridge U. P., 2000, X-418 p.

4677. LAWES (Kim). Paternalism and politics: the revival of paternalism in early nineteenth-century Britain. Basingstoke, Macmillan, 2000, X-229 p. (Studies in modern history).

4678. LAYBOURN (Keith). A century of Labour: a history of the Labour Party. Stroud, Sutton, XVIII-204 p.

4679. LINEHAN (Thomas). British fascism, 1918–1939: parties, ideology and culture. Manchester, Manchester U. P., 2000, XIII-306 p. (Manchester studies in modern history).

4680. LUNGER KNOPPERS (Laura). Constructing Cromwell: ceremony, portrait and print, 1645–1661. New York, Cambridge U. P., 2000, XIII-249 p.

4681. MACHIN (Ian). The rise of democracy in Britain, 1830–1918. Basingstoke, Macmillan, 2000, IX-182 p. (British studies series).

4682. MARWICK (Arthur). A history of the modern British Isles, 1914–1999. Circumstances, events and outcomes. Oxford, Blackwell, 2000, XVIII-414 p.

4683. METZDORF (Jens). Politik – Propaganda – Patronage. Francis Hare und die englische Publizistik im spanischen Erbfolgekrieg. Mainz, von Zabern, 2000, XV-566 p. (Veröffentlichungen des Instituts für Europäische Geschichte Mainz, Abt. Universalgeschichte, 179).

4684. MILLER (John). After the Civil Wars: English politics and government in the reign of Charles II. Harlow, Longman, 2000. IX-318 p.

4685. MILLMAN (Brock). Managing domestic dissent in First World War Britain. London, Frank Cass, 2000, X-335 p. (ill., maps). (Cass series – British politics and society).

4686. MORI (Jennifer). Britain in the age of the French Revolution, 1785–1820. Harlow, Longman, X-259 p.

4687. Nineteenth century (The): the British Isles, 1815–1901. Ed. by Colin MATTHEW. Oxford, Oxford U. P., 2000, XIV-342 p. (ill., maps, ports).

4688. PEDEN (G. C.). The Treasury and British public policy, 1906–1959. Oxford, Oxford U. P., 2000, XIV-581 p.

4689. PIERRI (Bruno). Gran Bretagna 1945: le elezioni generali e la politicizzazione della stampa. Manduria, Lacaita, 2000, 150 p.

4690. Politics (The) of the excluded, c.1500–1850. Ed. by Tim HARRIS. Basingstoke, Palgrave, 2000, IX-295 p. (Themes in focus).

4691. POOLE (Steve). The politics of regicide in England, 1760–1850. Manchester, Manchester U. P., 2000, VIII-232 p.

4692. POZDEEVA (L. V.). London – Moskva: Britanskoe obshchestvennoe mnenie i SSSR, 1939–1945. (London – Moscow: british public opinion and the USSR, 1939–1945). RAN, In-t vseobshchej istorii. Moskva, [s. n.], 2000, 305 p. [Table of contents in English]

4693. PROCHASKA (Frank). The republic of Britain, 1760–2000. London, Allen Lane, 2000, XIX-292 p. (ill.).

4694. Radicalism and revolution in Britain, 1775–1848: essays in honour of Malcolm I. Thomis. Ed. by Michael T. Davis. Basingstoke, Macmillan a. New York, St. Martin's Press, 2000, XV-242 p.

4695. RENTON (Dave). Fascism, anti-fascism, and Britain in the 1940s. Basingstoke, Macmillan, 2000, IX-203 p.

4696. ROYLE (Edward). Revolutionary Britannia? Reflections on the threat of revolution in Britain 1789–1848. Manchester, Manchester U. P., 2000, IX-214 p.

4697. SCOTT (David). The Barwis affair: political allegiance and the Scots during the British civil wars. *English historical review*, 2000, 115, 463, p. 843-863.

4698. SELF (Robert). The evolution of the British party system, 1885–1940. Harlow, Longman, 2000, IX-217 p.

4699. SHARPE (Kevin). Remapping early modern England. The culture of seventeenth-century politics. Cambridge, Cambridge U. P., 2000, 475 p.

4700. SMITH (Jeremy). The Tories and Ireland, 1920-1914: Conservative Party politics and the home rule crisis. Dublin a. Portland, Irish Academic Press, 2000, 251 p.

4701. SOLOV'EVA (Tat'jana S.). Religioznaja politika liberal'nykh tori v Anglii, 20-e gody XIX veka. (Religious politics of liberal Tories in England in the 1820s). Moskva, Izd-vo Moskovskogo un-ta, 222 p.

4702. SPURR (John). England in the 1670s. "This masquerading age". Oxford, Blackwell, 2000, XVII-350 p.

4703. THORPE (Andrew). The British Communist Party and Moscow, 1920-1943. Manchester, Manchester U. P., XI-308 p.

4704. THURLOW (Richard C.). Fascism in modern Britain. Stroud, Sutton, 2000, XIV-192 p. (Sutton modern British history, 9).

4705. TREGIDGA (Garry). The Liberal Party in southwest Britain since 1918: political decline, dormancy and rebirth. Exeter, University of Exeter Press, 2000, XII-271 p. (ill.).

4706. VICKERS (Rhiannon). Manipulating hegemony: state power, Labour and the Marshall Plan in Britain. Foreword by Andrew GAMBLE. Basingstoke, Macmillan, 2000, XVI-185 p. (International political economy series).

4707. WALKER (Simon). Rumour, sedition and popular protest in the reign of Henry IV. *Past and present*, 2000, 166, p. 31-65.

4708. WARNICKE (Retha M.). The marrying of Anne of Cleves. Royal protocol in early modern England. Cambridge a. Melbourne, Cambridge U. P., 2000, XIV-343 p.

4709. WASSON (Ellis). Born to rule: British political elites. Stroud, Sutton, 2000, XVI-224 p.

4710. WHITING (Richard). The Labour Party and taxation: party identity and political purpose in twentieth-century Britain. Cambridge, Cambridge U. P., 2000, XII-294 p.

4711. WILLIAMSON (Philip). Christian conservatives and the totalitarian challenge, 1933-1940. *English historical review*, 2000, 115, 462, p. 607-642.

4712. Women in British politics, 1760-1860: the power of the petticoat. Ed. by Kathryn GLEADLE and Sarah RICHARDSON. Basingstoke, Macmillan, 2000, XII-179 p.

4713. WORLEY (Matthew). The Communist International, the Communist party of Great Britain, and the 'third period', 1928-1932. *European history quarterly*, 2000, 30, 2, p. 185-208.

4714. ZARET (David). Origins of democratic culture: printing, petitions, and the public sphere in early-modern England. Princeton, Princeton U. P., 2000, XV-291 p. (Princeton studies in cultural sociology).

4715. ZWEINIGER-BARGIELOWSKA (Ina). Austerity in Britain: rationing, controls, and consumption, 1939-1955. Oxford, Oxford U. P., 2000, XIII-286 p. (ill.).

Cf. n°s 5363, 7402, 7775, 8147

Greece

4716. After the war was over: reconstructing the family, nation, and state in Greece, 1943-1960. Ed. by Mark MAZOWER. Princeton, Princeton U. P., 2000, X-312 p. (Princeton modern Greek studies).

4717. KALOUDIS (George Stergiou). Modern Greek democracy: the end of a long journey? Lanham, University Press of America, 2000, IX-119 p. (ill.).

4718. KARIPSIADES (Giorgos). He Hellada hos diadocho kratos: themata diadoches kraton kata ten hidryse tou Hellenikou Kratous kai ten henose ton Heptaneson. (Le cas grec de succession d'état: la fondation de l'Etat hellénique et l'union des îles ioniennes). Prologos, A. A. PHATOUROU. Athena, Ekdoseis Ant. N. Sakkoula, 2000, 594 p. (Seira Syntagmatiko dikaio sten Europe, 10).

4719. PETTIFER (James). The Greeks: the land and people since the war. London, Penguin, 2000, XXVII-255 p.

4720. VERVENIOTI (Tasoula). The adventure of women's suffrage in Greece. *In*: When the war was over: women, war and peace in Europe, 1940-1956. London, Leicester U. P., 2000, [s. p.].

Cf. n° 1282

Guatemala

4721. ALDA MEJÍAS (Sonia). La participación indígena en la construcción de la República de Guatemala, S. XIX. Madrid, Universidad Autónoma de Madrid Ediciones, 2000, 285 p. (Colección de estudios, 69).

4722. GRANDIN (Greg). The blood of Guatemala: a history of race and nation. Durham, Duke U. P., 2000, XVIII-343 p. (ill., maps). (Latin America otherwise: languages, empires, nation).

Guinea

4723. OSBORN (Emily Lynn). Power, authority and gender in Kankan-Bate, 1650-1920. Ann Arbor, UMI dissertation Services, 2000, 367 p. [Thesis (Ph. D) – Stanford university, Department of History].

Guyana

4724. BISNAUTH (D. A.). The settlement of Indians in Guyana, 1890-1930. Leeds, Peepal Tree, 2000, 296 p. (maps).

Haiti

4725. CORTEN (André). Diabolisation et mal politique. Haïti: misère, religion et politique. Montréal, Editions Cidihca et Paris, Karthala, 2000, 245 p.

4726. DIXON (Chris). African America and Haiti: emigration and Black nationalism in the nineteenth century. Westport a. London, Greenwood Press, 2000, XII-249 p. (ill.). (Contributions in American history, 186).

4727. HECTOR (Michel). Crises et mouvements populaires en Haïti. Montréal, Éditions du Cidihca, 2000, 206 p. (ill., portr.).

4728. JEAN-PIERRE (Jean Reynold). ... et Toussaint Louverture émerge: 1793–1802. [S. l.], Editions Presses Nationales d'Haïti, 2000, 208 p. (ill.).

4729. KNIGHT (Franklin W.). The Haitian revolution. *In*: Revolutions in the Americas [Cf. n° 4099], p. 103-115.

4730. SHELLER (Mimi). Democracy after slavery: black publics and peasant radicalism in Haiti and Jamaica. Gainesville, University Press of Florida, 2000, XV-270 p. (ill., maps).

Honduras

** 4731. DE OYUELA (Leticia). De la corona a la libertad: documentos comentados para la historia de Honduras, 1778–1870. Obispado de Choluteca, Ediciones Subirana, 2000, 341 p. (ill.). (Colección Padre Manuel Subirana, 11).

4732. CONTRERAS (Carlos A.). Hacia la dictadura cariísta: la campaña presidencial de 1932. Tegucigalpa, Editorial Iberoamericana, 2000, 253 p.

4733. Significado de los movimientos populares en la gestación del estado y la identidad nacional en Honduras: memoria del seminario de historia, 1996. Instituto Hondureno de Antropología e Historia. Tegucigalpa, Alin Editora, 2000, 69 p. (Estudios antropológicos e históricos, 12).

Hungary

4734. CORNWALL (Mark). The undermining of Austria-Hungary: the battle for hearts and minds. Basingstoke, Macmillan a. New York, St. Martin's Press, 2000, XVI-485 p.

4735. DEMIRKAN (Tarık). Macar turancıları. (Hungarian Turanists). Ed. by. Ayşen ANADOL. İstanbul, Türkiye Ekonomik ve Toplumsal Tarih Vakfı, 2000, 158 p.

4736. DUCZMAL (Małgorzata). Izabela Jagiellonka królowa Węgier. (Jagiellonin Isabella [Izabela Jagiellonka] Königin Ungarns). Warszawa, Rytm, 2000, 473 p.

4737. FREIFELD (Alice). Nationalism and the crowd in Liberal Hungary, 1848–1914. Washington, Woodrow Wilson Center Press a. Baltimore a. London, Johns Hopkins U. P., 2000, XII-398 p.

4738. HADJÚ (Tibor). La Hongrie dans les années de crise après la Première Guerre mondiale (1918–1920). *In*: Hongrie (La) dans les conflits du XXe siècle [Cf. n° 7611], p. 37-52.

4739. ORMOS (Mária). Introduction: la Hongrie au XXe siècle. *In*: Hongrie (La) dans les conflits du XXe siècle [Cf. n° 7611], p. 3-24.

4740. TILKOVSZKY (Loránt). Ľudovít Žigmund Szeberényi a Slováci v Maďarsku. (Ludwig Sigismund Szeberényi und die Slowaken in Ungarn). *Historický časopis*, 2000, 48, 3, p. 417-434. [Deutsche Zfassung].

4741. TILKOVSZKY (Loránt). Német nemzetiségmagyar hazafiság: Tanulmányok a magyarországi németség történetéböl. (German nationality-Hungarian patriotism: studies concerning the history of the Hungarian-Germans). Budapest, JPTE U. P., 2000, 261 p.

Cf. n°s 4182, 7611, 8132

Iceland

4742. GUNNAR (Karlsson). The history of Iceland. Minneapolis, University of Minnesota Press, 2000, XIII-418 p. (ill., maps).

4743. HÁLFDANARSON (Guðmundur). Iceland: a peaceful secession. *Scandinavian journal of history*, 2000, 25, 1-2, p. 87-100.

India

** 4744. Selected works of Jawaharlal Nehru. General editor S. GOPAL. Vol. 26. 1 June 1954–30 September 1954. Ed. by Ravinder KUMAR and H.Y. Sharada PRASAD. New Delhi, Jawaharlal Nehru Memorial Fund, 650 p.

4745. BENICHOU (Lucien D.). From autocracy to integration: political developments in Hyderabad State, 1938-1948. Chennai, Orient Longman, 2000, XI-313 p. (ill., maps).

4746. CAUDHARI (Amita). Bharatapura Rajya evam Ista Indiya Kampani: eka adhyayana, 1803–1857. [Political history of Bharatpur, Princely State with special reference to the role of East India Company, a study; covers the period, 1803–1857]. Jayapura, Klasika Pablikesansa, 2000, 212 p.

4747. CORBRIDGE (Stuart), HARRISS (John). Reinventing India: liberalization, Hindu nationalism and popular democracy. Cambridge, Polity Press, 2000, XX-313 p. (maps).

4748. D'ORAZI FLAVONI (Francesco). Storia dell'India: società e sistema dall'indipendenza ad oggi. Venezia, Marsilio, 2000, XI-386 p. (Saggi. Critica).

4749. Inventing boundaries: gender, politics, and the partition of India. Ed. by Mushirul HASAN. New Delhi a. Oxford, Oxford U. P., 2000, VIII-393 p.

4750. KHAN (Shaharyar M.). The begums of Bhopal: a dynasty of women rulers in Raj India. London a. New York, I. B. Tauris, 2000, X-276 p.

4751. MALHOTRA (Bimal). Reform, reaction and nationalism in Western India: 1885–1907. Mumbai, Himalaya Pub. House, 2000, 285 p.

4752. MISHRA (Shreegovind). Democracy in India. New Delhi, Sanbun Publishers, 2000, 777 p.

4753. Seed-time (The) of communist movement in India, 1919–1926. Ed. By Ladli Mohon RAYCHAUDHURY. Calcutta, National Book Agency, 2000, XV-194 p.

4754. WARREN (Alan). Waziristan, the Faqir of Ipi, and the Indian Army: the north west frontier revolt of 1936–1937. Oxford, Oxford U. P., 2000, XXXII-324 p.

Indonesia

4755. East Timor question (The): the struggle for independence from Indonesia. Ed. by Paul HAINSWORTH and Stephen MAC CLOSKEY; foreword by John PILGER; preface by José RAMOS-HORTA. London, Tauris, 2000, XV-222 p.

4756. Indonésie (L'): un demi-siècle de construction nationale. Textes réunis par Françoise CAYRAC-BLANCHARD, Stéphane DOVERT et Frédéric DURAND. Paris, L'Harmattan, 2000, 352 p. (maps). (Recherches asiatiques).

4757. LEV (Daniel S.). Legal evolution and political authority in Indonesia: selected essays. The Hague a. London, Kluwer Law International, 2000, 349 p. (London-Leiden series on law administration and development, 4).

4758. MICHEL (Franck). L'Indonésie éclatée mais libre. De la dictature à la démocratie (1998–2000). Paris, L'Harmattan, 2000, 160 p. (Points sur l'Asie).

Iran

* 4759. CRONIN (Stephanie). The Left in Iran: illusion and disillusion. [Review article] *Middle eastern studies*, 2000, 36, 3, p. 231-243.

4760. BEHROOZ (Maziar). Rebels with a cause: the failure of the left in Iran. London, I. B. Tauris, 2000, XV-239 p. (ill.).

4761. CHELKOWSKI (Peter J.), DABASHI (Hamid). Staging a revolution: the art of persuasion in the Islamic Republic of Iran. London, Booth-Clibborn Editions, 2000, 312 p. (ill.).

4762. DANIEL (Elton L.). The history of Iran. Westport, Greenwood Press, 2000, [s. p.]. (Greenwood histories of the modern nations).

4763. GENIS (Vladimir L.). Krasnaja Persija: Bolsheviki v Giljane, 1920–1921 gg.: Dokumental'naja khronika. (Red Persia: the bolsheviks in Gilan [the Persian Soviet Republic], 1920–1921: a documentary chronicle). Moskva, MNPI, 2000, 559 p. (ill.).

4764. JAFARI (Reza). Centre-periphery relations in Iran: the case of the Southern Rebellion in 1946. Oxford, [s. n.], 2000, 275 p. [Thesis (D. Phil.) – University of Oxford, 2000].

4765. KATOUZIAN (Homa). State and society in Iran: the eclipse of the Qajars and the emergence of the Pahlavis. London a. New York, I. B. Tauris, 2000, XII-351 p. (Library of modern Middle East studies, 28).

4766. MAJD (Mohammad Gholi). Resistance to the Shah: landowners and ulama in Iran. Gainesville, University Press of Florida, 2000, XIV-415 p. (ill., maps).

4767. MARTIN (Vanessa). Creating an Islamic state: Khomeini and the making of a new Iran. London, I. B. Tauris, 2000, XV-248 p. (Library of modern Middle East studies, 24).

Iraq

4768. SHIELDS (Sarah). Mosul before Iraq: like bees making five-sided cells. Albany, State University of New York Press, 2000, XIV-278 p.

4769. TRIPP (Charles). A history of Iraq. Cambridge, Cambridge U. P., 2000, XVII-311 p. (ill., maps).

Ireland

4770. FAREWELL (A) to arms? From 'long war' to long peace in Northern Ireland. Ed. by Michael COX, Adrian GUELKE and Fiona STEPHEN. Manchester, Manchester U. P., 2000, XX-360 p.

4771. FARRELL (Sean). Rituals and riots. Sectarian violence and political culture in Ulster, 1784–1886. Lexington, Universuty Press of Kentucky, 2000, IX-252 p.

4772. Ireland: the politics of independence, 1922–1949. Ed. by Mike CRONIN and John M. REGAN. New York, St. Martin's Press, 2000, X-237 p.

4773. Irish diaspora (The). Ed. Andy BIELENBERG. New York, Longman, 2000, VI-368 p.

4774. KEE (Robert). The green flag: a history of Irish nationalism. London, Penguin, 2000, 877 p.

4775. KENNEDY (Michael J.). Division and consensus: the politics of cross-border relations in Ireland, 1925–1969. Dublin, Institute of Public Administration, 2000, X-422 p.

4776. MAC LAUGHLIN (Jim). Reimagining the nation state: the contested terrains of nation-building. London, Pluto Press, 2000, 289 p.

4777. MAGENNIS (Eoin). The Irish political system, 1740–1765: the golden age of the undertakers. Dublin, Four Courts Press, 2000, 231 p.

4778. MULHOLLAND (Marc). Northern Ireland at the cross-roads: Ulster unionism in the O'Neill years

1960–1969. New York, St. Martin's Press, 2000, XI-287 p.

4779. Ní AOLÁIN (Fionnuala). The politics of force: conflict management and state violence in Northern Ireland. Foreword by John WADHAM. Belfast, Blackstaff Press, 2000, XII-336 p.

4780. Ó CIARDHA (Éamonn). Ireland and the Jacobite cause, 1685–1766: a fatal attachment. Dublin, Four Courts, 2000, 468 p. (ill., maps, ports).

4781. Political ideas in eighteenth-century Ireland. Ed. by Sean J. CONNOLLY. Dublin, Four Courts Press, 2000, 236 p.

4782. Rebellion and remembrance in modern Ireland. Ed. by Laurence M. GEARY. Dublin, Four Courts, 2000, 240 p.

4783. Revolution, counter-revolution and union: Ireland in the 1790s. Cambridge, Cambridge U. P., 2000, XII-245 p.

4784. ROSE (Peter). How the Troubles came to Northern Ireland. Basingstoke, Macmillan in association with Institute of Contemporary British History, 2000, XVIII-216 p.

4785. WALKER (Brian Mercer). Past and present: history, identity and politics in Ireland. Belfast, Institute of Irish Studies, Queen's University Belfast, 2000, X-148 p.

4786. WHELAN (Bernadette). Ireland and the Marshall Plan, 1947–1957. Portland, Four Courts Press, 2000, 426 p.

Cf. n° 8075

Israel

4787. DERORI (Ze'ev). Utopyah be-madim: terumat Tsahal le-hityashvut, li-kelitat ha-`aliyah ule-hinukh be-reshit yeme ha-medinah. [Utopia in uniform]. Kiryat Sedeh-Boker, ha-Merkaz le-moreshet Ben-Guryon a. Beersheba, Hotsa'at ha-sefarim shel Universitat Ben-Guryon ba-Negev, 2000, 213 p.

4788. HELLER (Joseph). The birth of Israel, 1945–1949: Ben-Gurion and his critics. Gainesville, University Press of Florida, 2000, XVII-379 p

4789. Israel: the first hundred years. Vol. 1. Israel's transition from community to state. Vol. 2. From war to peace? Ed. by Efraim KARSH. London, Frank Cass, 2000, 2 vol., VI-253 p., 280 p. (Israeli history, politics, and society).

4790. Israel's first fifty years. Ed. by Robert Owen FREEDMAN. Gainesville, University Press of Florida, 2000, XIX-290 p.

4791. Lo `al magash shel kesef: toldot medinat Yiþsra'el me-reshit ha-hityashvut `ad `idan ha-shalom. [Not on a silver platter: a history of Israel, 1900–2000]. `Orkhim Yehudah VALAKH, Pirhiyah KOHEN. Tel Aviv, Miþsrad ha-bitahon, ha-hotsa'ah le-or a. Yerushalayim, Karta, 2000, 240 p. (ill.).

4792. PARSONS (Laila Helen). The Druze between Palestine and Israel, 1947–1949. Basingstoke, Macmillan, 2000, XVII-180 p. (St Antony's series).

Cf. n° 8265

Italy

* 4793. COLLOTTI (Enzo), KLINKHAMMER (Lutz). Zur Neubewertung des italienischen Faschismus. [Diskussionsforum]. *Geschichte und Gesellschaft*, 2000, 26, 2, p. 285-306.

** 4794. ALBERTINI (Luigi). I giorni di un liberale: diari 1907–1923. A cura di Luciano MONZALI. Bologna, Il Mulino, 2000, 434 p. (Storia/memoria).

** 4795. Deliberazioni (Le) del Comune di Pescia (1526–1532): Regesti. A cura di Massimo BRACCINI. Roma, Ministero per i beni e le attivita culturali, Ufficio centrale per i beni archivistici, 2000, XI-547 p. (Pubblicazioni degli archivi di Stato, Strumenti, 144).

** 4796. Hoi anaphores ton Veneton Provlepton tes Zakynthou: 16os–18os ai. (Le relazioni dei Provveditori veneziani di Zante). Ekdose Demetres ARVANITAKES. Venezia, Helleniko Institouto Vyzantinon kai Metavyzantinon Spoudon Venetias, 2000, 546 p. (Graecolatinitas nostra. Peges, 2).

** 4797. Primer (A) of Italian fascism. Edited and with an introduction by Jeffrey T. SCHNAPP; translated by Jeffrey T. SCHNAPP, Olivia E. SEARS, and Maria G. STAMPINO. Lincoln a. London, University of Nebraska Press, 2000, XXI-325 p. (European horizons).

** 4798. ROPA (Rossella). L'antisemitismo nella Repubblica sociale italiana: repertorio delle fonti conservate all'Archivio centrale dello Stato. Bologna, Pàtron, 2000, 89 p. (Proposte di storia, 12).

** 4799. TANUCCI (Bernardo). Epistolario. Vol. 16. 1765–1766. A cura e con introduzione di Maria Grazia MAIORINI. Napoli, Società napoletana di storia patria, 2000, XXXIII-563 p.

** 4800. TORI (Giorgio). Lucca giacobina: primo governo democratico della Repubblica lucchese (1799). 1. Saggio introduttivo. 2. Regesti degli atti. Roma, Ministero per i beni e le attivita culturali, Ufficio centrale per i beni archivistici, 2000, 2 vol., [s. p.]. (Pubblicazioni degli archivi di Stato. Strumenti, 142, 143).

** 4801. VIVARELLI (Roberto). La fine di una stagione. Memoria 1943–1945. Bologna, Il Mulino, 2000, 125 p.

4802. AEBISCHER (Tullio). Le ipotesi territoriali nella Questione Romana dal 1870 al 1929. *Rassegna storica del Risorgimento*, 2000, 3, p. 411-430.

4803. ANTONACCI (Nicola). Dalla repubblica napoletana alla monarchia italiana. Politica e società in Terra di Bari (1799–1860). Bari, Edipuglia, 2000, 143 p.

4804. APPUHN (Karl). Inventing nature: forests, forestry, and state power in Renaissance Venice. *Journal of modern history*, 2000, 72, 4, p. 861-889.

4805. ARMANDO (David), CATTANEO (Massimo), DONATO (Maria Pia). Una rivoluzione difficile: la Repubblica Romana del 1798–1799. Roma, Istituti editoriali e poligrafici internazionali, 2000, 285 p. (Storia, 3).

4806. BANTI (Alberto). La nazione del Risorgimento. Parentela, santità e onore alle origini dell'Italia unita. Torino, Einaudi, 2000, XIII-214 p.

4807. BELLASSAI (Sandro). La morale comunista: pubblico e privato nella rappresentazione del PCI (1947–1956). Roma, Carocci editore, 2000, 382 p. (Istituto Gramsci Emilia-Romagna, 3).

4808. BELOUSOV (Lev S.). Rezhim Mussolini i massy. (The Mussolini's regime and the society). Moskovskij gos. un-t im. M.V. Lomonosova, Istor. f-t. Moskva, MGU, 2000, 367 p. (Trudy Istoricheskogo fakul'teta MGU, 14; Ser. 2: "Istoricheskie issledovanija", 3).

4809. BIOCCA (Dario), CANALI (Mauro). L'informatore: Silone, i comunisti e la polizia. Milano, Luni, 2000, 275 p. (Biblioteca di storia contemporanea, 12).

4810. BOSWORTH (R. J. B.). Per necessità familiare: hypocrisy and corruption in Fascist Italy. *European history quarterly*, 2000, 30, 3, p. 357-388.

4811. CALONACI (Stefano). «Accordar lo spirito col mondo». Il cardinal Ferdinando de Medici a Roma durante i pontificati di Pio V e Gregorio XIII. *Rivista storica italiana*, 2000, 112, 1, p. 5-74.

4812. CARRINO (Annastella). La città aristocratica, linguaggi e pratiche della politica a Monopoli fra Cinque e Seicento. Bari, Edipuglia, 2000, 324 p.

4813. COLARIZI (Simona). Storia del Novecento italiano. Cent'anni di entusiasmo, di paure, di speranza. Milano, Rizzoli, 453 p.

4814. Comune democratico (Il): Riccardo Dalle Mole e l'esperienza delle giunte nel Veneto giolittiano, 1900–1914. A cura di Renato CAMURRI. Venezia, Marsilio, 2000, XIII-277 p. (Ricerche).

4815. CONTI (Fulvio). L'Italia dei democratici: sinistra risorgimentale, massonismo e associazionismo fra Otto e Novecento. Milano, F. Angeli, 2000, 361 p. (Collana della Fondazione di studi storici Filippo Turati, 18).

4816. CRAINZ (Guido). L'Italia, repubblicana. Firenze, Giunti, 2000, 123 p. (ill., some col.). (Collana XX secolo).

4817. DALLA CASA (Brunella). Attentato al duce: le molte storie del caso Zamboni. Bologna, Il mulino, 2000, VI-291 p. (ill.). (Biblioteca storica).

4818. DE GRAND (Alexander). The hunchback's tailor: Giovanni Giolitti and liberal Italy from the challenge of mass politics to the rise of fascism, 1882–1922. Westport a. London, Praeger, 2000, X-294 p. (Italian and Italian American studies).

4819. FENTRESS (James). Rebels and mafiosi: death in a Sicilian landscape. Ithaca, Cornell U. P., 2000, 297 p.

4820. FINO (Antonio). Dalla Costituzione allo Stato repubblicano: momenti e problemi dell'Italia degasperiana. Galatina, Congedo, 2000, 197 p. (Itinerari di ricerca storica. Supplementi, 17).

4821. FOLLINI (Marco). La DC. Bologna, Il Mulino, 2000, 161 p.

4822. GABACCIA (Donna R.). Italy's many diasporas. Forewords by Robin COHEN. Seattle, University of Washington Press, 2000, XV-264 p. (Global diasporas series).

4823. GENTA (Enrico). Una rivoluzione liberale mancata: il progetto Cavour-Santarosa sull'amministrazione comunale e provinciale (1858). Torino, Deputazione subalpina di storia patria, 2000, 370 p. (Biblioteca di storia italiana recente, 26).

4824. GENTILE (Emilio). Fascismo e antifascismo: i pariti italiani fra le due querre. Firenze, Felice Le Monnier, 2000, 545 p. (Quaderni di Storia, Fondati da Giovanni Spadolini).

4825. GRINER (Massimiliano). La «Banda Koch». Il reparto speciale di polizia: 1943–1944. Torino, Bollati Boringhieri, 2000, 432 p.

4826. IVETIC (E.). Oltremare. L'Istria nell'ultimo dominio veneto. Venezia, Istituto Veneto di Scienze, Lettere ed Arti, 2000, VIII-470 p. (Memorie – Classe di scienze morali, lettere ed arti, 89).

4827. KREILE (Michael). Die Republik Italien 1946–1996. *Geschichte und Gesellschaft*, 2000, 26, 2, p. 255-284.

4828. LARIZZA (Vincenzo). Mafia, politica, antimussoliniani e corruzione in Italia: conseguenze della sconfitta militare nella Seconda Guerra Mondiale 1939–1945. [S. l.], Grafica Enotria, 2000, 121 p. (ill.).

4829. LIVI (Angelo). Massoneria e fascismo. Presentazione di Luigi PRUNETI; prefazione di Luigi ALFIERI. Foggia, Bastogi, 2000, 206 p. (ill.).

4830. LÜHE (Marion). Der venezianische Adel nach dem Untergang der Republik (1797–1830). Köln, SH-Verlag, 2000, 193 p. (Italien in der Moderne, 7).

4831. LUPO (Salvatore). Il fascismo: la politica in un regime totalitario. Roma, Donzelli, 2000, VI-456 p. (Saggi. Storia e scienze sociali).

4832. LUZZATTO (Sergio). La repubblica pietrificata. Corpo e tomba di Mazzini al cimitero genovese di Staglieno. *In*: Monumento (Il): arte e storia [Cf. n° 1283], p. 654-702.

4833. MARTELLONE (Anna Maria). Nel Seicento della decadenza italiana: Lelio Marretti, o della lezione delle storie tra tacitismo e pratica politica alla corte pontificia. *Archivio storico italiano*, 2000, 158, 584, p. 255-306.

4834. MASTELLONE (Salvo). La democrazia etica di Mazzini (1837–1847). Roma, Archivio Guido Izzi: Istituto per la storia del Risorgimento italiano, 2000, X-185 p. (Biblioteca scientifica. Serie 2. Memorie / Istituto per la storia del Risorgimento italiano, 45).

4835. MENIGHETTI (Romolo), NICASTRO (Franco). L'eresia di Milazzo. Crisi del cattolicesimo politico in Sicilia e ruolo del PCI 1958–1960. Caltanisetta, Sciascia, 2000, 215 p.

4836. MERIGGI (Marco). Soziale Klassen, Institutionen und Nationalisierung im liberalen Italien. *Geschichte und Gesellschaft*, 2000, 26, 2, p. 201-218.

4837. MERLOTTI (Andrea). L'enigma delle nobiltà: stato e ceti dirigenti nel Piemonte del Settecento. Firenze, Leo S. Olschki editore, 2000, XIII-348 p. (Studi e testi, 14).

4838. MISSAGGIA (Maria Giovanna). La manipolazione dei risultati elettorali: la convalida delle elezioni nella IX e nella X Legislatura del Regno d'Italia. *Rivista storica italiana*, 2000, 112, 1, p. 189-234.

4839. MULLEN (Anne). Inquisition and inquiry: Sciascia's inchiesta. Market Harborough, Troubador, 2000, 78 p. (Hull Italian Texts).

4840. MUSI (A.). L'Italia dei viceré. Integrazione e resistenza nel sistema imperiale spagnolo. Cava dei Tirreni, Avagliano ed., 2000, 254 p.

4841. Napoleonic Italy: historical, literary and cultural perspectives. Ed. with an introduction by Eileen A. MILLER. Glasgow, University of Glasgow Press, 2000, XXII-137 p. (Italian research studies).

4842. NATOLI (Claudio). Fascismo, democrazia, socialismo: comunisti e socialisti tra le due guerre. Milano, F. Angeli, 2000, 336 p. (Studi e ricerche storiche, 268).

4843. Novecento italiano: studi in ricordo di Franco De Felice. A cura di Silvio PONS. Roma, Carocci, 2000, 333 p. (Studi storici Carocci, 1). [Cf. n° <scelta> 4046.]

4844. NOVELLI (Claudio). Il Partito d'azione e gli italiani: moralità, politica e cittadinanza nella storia repubblicana. Milano, La nuova Italia, 2000, XXIII-296 p. (Biblioteca di storia. Dall'azionismo agli azionisti, 7).

4845. Origini (Alle) del Risorgimento: la repubblica bresciana dal 18 marzo al 20 novembre 1797: atti della Giornata di studio, Brescia, 18 marzo 1997. A cura di Luigi Amedeo BIGLIONE DI VIARIGI. Brescia, Ateneo di Brescia, 2000, 174 p.

4846. OSTI GUERRAZZI (Amedeo). Grande industria e legislazione sociale in età giolittiana. Torino, Paravia scriptorium, 2000, 334 p. (Collana del Dipartimento di storia dell'Università di Torino).

4847. PAGANO (Emanuele). Pro e contro la repubblica: cittadini schedati dal governo cisalpino in un'inchiesta politica del 1798. Milano, UNICOPLI, 2000, 216 p. (Storia lombarda, 8).

4848. PARISELLA (Antonio). Cattolici e Democrazia Cristiana nell'Italia repubblicana: analisi di un contesto politico. Roma, Gangemi, 2000, 174 p. (Studi storici e sociali).

4849. PARLATO (Giuseppe). La sinistra fascista: storia di un progetto mancato. Bologna, Il Mulino, 2000, 404 p. (Ricerca).

4850. PEDIO (Alessia). La cultura del totalitarismo imperfetto: il Dizionario di politica del Partito nazionale fascista: 1940. Milano, UNICOPLI, 2000, 290 p. (Biblioteca di storia contemporanea; 2).

4851. POLECRITTI (Cynthia L.). Preaching peace in Renaissance Italy: Bernardino of Siena and his audience. Washington, Catholic University of America Press, 2000, XI-271 p.

4852. REINHARDT (Nicole). Macht und Ohnmacht der Verflechtung. Rom und Bologna unter Paul V. Studien zur frühneuzeitlichen Mikropolitik im Kirchenstaat. Tübingen, Bibliotheca Academica Verlag, 2000, [s. p.]. (Frühneuzeit-Forschungen, 8).

4853. Repubblica romana (La) nel movimento europeo tra il 1848 e il 1849. Atti del Convegno internazionale di studi, Roma 30 giugno–1 luglio 1999. *Rassegna storica del Risorgimento*, 2000, p. 151-396.

4854. Riforme (Le) del 1847 negli Stati italiani. Atti del Convegno di studi. Firenze, 20–21 marzo 1998. *Rassegna storica toscana*, 99, 2, [s. p.].

4855. ROCHAT (Giorgio). Ufficiali e soldati. L'esercito italiano dalla prima alla seconda guerra mondiale. Udine, Gasparri, 2000, 222 p.

4856. RODÉN (Marie-Louise). Church politics in seventeenth-century Rome: Cardinal Decio Azzolino, Queen Christina of Sweden, and the Squadrone Volante. Stockholm, Almqvist & Wiksell International, 2000, 327 p. (Stockholm studies in history, 60).

4857. ROSSI (Andrea). Fascisti toscani nella Repubblica di Salò, 1943–1945. Pisa, BFS Edizioni, 2000, 159 p. (ill., ports). (Biblioteca di cultura storica, 17).

4858. ROSSI-DORIA (Anna). Italian women enter politics. *In*: When the war was over: women, war and peace in Europe, 1940–1956. London, Leicester U. P., 2000, p. 89-102.

4859. SABETTI (Filippo). The search for good government: understanding the paradox of Italian democracy. Montreal, McGill-Queen's U. P., 2000, XII-313 p.

4860. SANTOMASSIMO (Gianpasquale). La marcia su Roma. Firenze, Giunti, 2000, 127 p.

4861. SARFATTI (Michele). Gli ebrei nell'Italia fascista. Vicende, identità, persecuzione. Torino, Einaudi, 2000, 377 p.

4862. SCROCCARO (Mauro). Dall'aquila bicipite alla croce uncinata: l'Italia e le opzioni nelle nuove provincie Trentino, Sudtirolo, Val Canale, 1919–1939. Trento, Museo storico in Trento, 2000, 380 p. (ill.). (Collana di pubblicazioni del Museo storico in Trento).

4863. SHAW (Christine). The politics of exile in Renaissance Italy. Cambridge, Cambridge U. P., 2000, X-257 p. (Cambridge studies in Italian history and culture).

4864. Sidney Sonnino e il suo tempo. A cura di Pier Luigi BALLINI. Firenze, L. S. Olschki, 2000, [s. p.]. (Biblioteca storica toscana. Sezione di storia del Risorgimento, 28).

4865. TULLIO-ALTAN (Carlo). La nostra Italia: clientelismo, trasformismo e ribellismo dall'unità al 2000. Prefazione di Roberto CARTOCCI. Milano, EGEA, Universita Bocconi, 2000, VII-278 p.

Cf. n°s 181, 5536, 5551, 6396

Jamaica

Cf. n° 4730

Japan

4866. AKIYAMA (T.). The Hojo Clan and Tokuso (Main Hojo Line) government. *Journal of Japanese history*, 2000, 458, p. 23-50.

4867. BANNO (Junji). Democracy in pre-war Japan: concepts of government, 1871–1937: collected essays. London, Routledge, 2000, 200 p. (Routledge/Ominato modern Japanese studies).

4868. GOW (Ian). Military intervention in pre-war Japanese politics: Admiral Kato Kanji and the 'Washington system'. Richmond, Curzon, 2000, X-358 p.

4869. HARADA (Keiichi). Meibouka to Seiji: Ôsaka-hu Toyoshima-gun Toyonaka-mura Okuno Kumaichirô no Baai. (Okuno Kumaichirô, man of high reputation and politic). *Oryô Shigaku*, 2000, 26, p. 213-244.

4870. HIGUCHI (Hidemi). Shûsenshi jô no «Sengo»: Takagi Sôkichi no shûsen kôsaku to sengo kôsô. (Takagi Sôkichi's role in ending the Pacific War and his post-war views). *Gunji Shigaku*, 2000, 32, 2, p. 4-19.

4871. INEDA (Masahiro). Ziyûminken undô no bunkashi: Atarashii seiji-bunka no tanjô. (Cultural history of the movement for the freedom of the people's rights: birth of a new political culture). Tokyo, Chikuma Shobo, 2000, 362 p.

4872. LEWIS (Michael). Becoming apart: national power and local politics in Toyama, 1868–1945. Cambridge, Harvard U. P., 2000, XVIII-340 p. (Harvard East Asian monographs, 192).

4873. LONE (Stewart). Army, empire and politics in Meiji Japan: the three careers of General Katsura Tarō. New York, St. Martin's Press, 2000, VII-247 p.

4874. MAC CARGO (Duncan). Contemporary Japan. Basingstoke, Macmillan, 2000, XII-223 p. (ill.). (Contemporary states and societies).

4875. MAC NELLY (Theodore). The origins of Japan's democratic constitution. Lanham, University Press of America, 2000, XIII-221 p.

4876. MATSUZAWA (Yûsaku). Meiji 17nen no Chihô seido kaikaku: Saitama-ken no jirei wo chûshin ni. (Local government reform in 1884: the case of Saitama prefecture). *Shigaku Zasshi*, 2000, vol.109, no.7, p. 1-35 [English summ.].

4877. MURASE (Shinichi). Nomura Yasushi naishô no Yûutsu. (Gloom of the Home Secretary, Nomura Yasushi, 1894–1896). *Nihon Rekishi*, 2000, 630, p. 73-76.

4878. NISHIKAWA (Makoto). Kagami no seiritsu: Kindai kessai kaigi bunsho seiritsu-kô. (The formation of "Kagami": a study on the formation of the modern documents of sanction and circulars in Japan). *Nihon Rekishi*, 2000, 628, p. 4-21.

4879. OBINATA (Sumio). Kindai Nihon no keisatsu to chiiki syakai. (The police and community in modern Japan). Tokyo, Chikuma Shobo, 2000, 387 p.

4880. PRESTON (Peter Wallace). Understanding modern Japan: a political economy of development, culture and global power. London, SAGE, 2000, VI-239 p.

4881. Preussens Weg nach Japan: Japan in den Berichten von Mitgliedern der preussischen Ostasienexpedition 1860–1861. Herausgegeben, kommentiert und mit einer Einführung versehen von Holmer Stahncke. München, Indicium, 2000, 262 p. (ill.).

4882. SAALER (Sven). Zwischen Demokratie und Militarismus: die Kaiserlich- Japanische Armee in der Politik der Taishô-Zeit (1912–1926). Bonn, Bier'sche Verlagsanstalt, 2000, XIII-587 p. (ill.). (Bonner Japanforschungen, 21).

4883. SWALE (Alistair). The political thought of Mori Arinori: a study of Meiji conservatism. Richmond, Japan Library, 2000, X-249 p. (Meiji Japan series, 7).

4884. TSUZUKI (Chushichi). The pursuit of power in modern Japan 1825–1995. Oxford, Oxford U. P., 2000, X-550 p.

Cf. n°s 8739-8781

Jordan

4885. ALON (Yoav Zvi). State, tribe, and mandate in Transjordan, 1918–1946. Oxford, [s. n.], 2000, XVII-382 p. [Thesis (D. Phil.) – University of Oxford, 2000].

4886. FISCHBACH (Michael R.). State, society and land in Jordan. Leiden, Brill, 2000, IX-236 p. (ill., maps). (Social, economic, and political studies of the Middle East and Asia, 75).

Kazakhstan

Cf. n° 7070

Kenya

4887. GROEN (Gerrit). The Afrikaners in Kenya 1903-1969. Ann Arbor, UMI Dissertation Services, 2000, [s. p.].

4888. Kenya: the making of a nation: a hundred years of Kenya's history, 1895-1995. Ed. by Bethwell A. OGOT and W. R. OCHIENG'. Maseno, Institute of Research and Postgraduate Studies, Maseno University, 2000, 228 p. (maps).

4889. LEWIS (Joanna). Empire state-building: war and welfare in Kenya 1925-1952. Oxford, J. Currey, 2000, XVII-393 p. (Eastern African studies).

4890. ODED (Arye). Islam and politics in Nenya. Boulder a. London, Lynne Rienner Publishers, 2000, 236 p.

Cf. n° 4905

Kirghizstan

Cf. n° 832

Korea

4891. HONG (Yong-pyo). State security and regime security: President Syngman Rhee and the insecurity dilemma in South Korea, 1953-1960. Basingstoke, Macmillan, 2000, 208 p. (St. Antony's series).

Cf. n°s 8458, 8782-8785

Kuwait

4892. KOCH (Christian). Politische Entwicklung in einem arabischen Golfstaat: die Rolle von Interessengruppen im Emirat Kuwait. Berlin, Klaus Schwarz, 2000, II-287 p. (Islamkundliche Untersuchungen, 232).

Cf. n° 8167

Laos

4893. DEUVE (J.). Le Laos 1945-1949: contribution a l'histoire du mouvement Lao Issala. Montpellier, Universite Paul Valery, 2000, 394 p. (ill.). (Etudes militaires, 27).

Latvia

4894. LABSVIRS (Janis). Latvijas lauksaimniecibas kolektivizacija, 1944-1956 (Collectivization of Latvian agriculture, 1944-1956). Riga, Zinatne, 2000, 215 p.

Cf. n° 5075

Lebanon

** 4895. ASSAF (Antoine-Joseph). Terre blanche: journal d'un otage au Liban. Préface de Régine PERNOUD. Paris, Sarment, 2000, 497 p. (ill.).

4896. EL-KHAZEN (Farid). The breakdown of the state in Lebanon, 1967-1976. London, I. B. Tauris, 2000, 432 p.

4897. ZAMIR (Meir). Lebanon's quest: the road to statehood, 1926-1939. London, I. B. Tauris, 2000, XII-313 p. (ill.).

4898. ZISSER (Eyal). Lebanon: the challenge of independence. London, I. B. Tauris, 2000, XIII-297 p.

Liberia

4899. NASS (I. A.). A study in internal conflicts: the Liberian crisis and the West African peace initiative. Enugu, Fourth Dimension, 2000, VII-354 p. (ill., maps).

Lithuania

4900. LALKOÙ (Ihar). Aperçu de l'histoire politique du Grand-Duché de Lithuanie. Ouvrage préfacé par Bruno DRWESKI. Paris, L'Harmattan, 2000, 125 p. (Collection Biélorussie).

Cf. n° 5075

Macedonia

* 4901. Macedonian question (The): culture, historiography, politics Ed. by Victor ROUDOMETOF. Boulder, East European Monographs, 2000, IV-304 p. (East European monographs, 553).

Macedonia. Addenda 1999

4902. New Macedonian question (The). Ed. by James PETTIFER. Basingstoke, Macmillan, 99, XXXVI-311 p.

Madagascar

** 4903. RABEMANANJARA (Raymond William). Correspondance de Madagascar 1950-1999. Paris, l'Harmattan, 2000, 2 vol., 204 p., 246 p. – IDEM. Madagascar: l'affaire de mars 1947. Paris, l'Harmattan, 2000, 128 p.

4904. RAZAFINDRANALY (Jacques). Les soldats de la Grande île: d'une guerre à l'autre, 1895-1918. Paris, L'Harmattan, 2000, 373 p. (ill.). (Repères pour Madagascar et l'océan Indien).

Malawi

4905. DE JONG (Albert). Mission and politics in Eastern Africa: Dutch missionaries and African nationalism in Kenya, Tanzania and Malawi, 1945-1965. Nairobi, Paulines Publications Africa, 2000, 352 p.

4906. HARRIGAN (Jane). From dictatorship to democracy: economic policy in Malawi 1964-2000. Aldershot, Ashgate, 2000, 376 p. (The making of modern Africa).

Mali

4907. Democracy and development in Mali. Ed. by R. James BINGEN, David ROBINSON and John M.

STAATZ. East Lansing, Michigan State U. P., 2000, XX-380 p.

Mexico

4908. CARRILLO CÁZARES (Alberto). El debate sobre la guerra chichimeca, 1531–1585: derecho y política en la Nueva España. Zamora, Colegio de Michoacán y San Luis Potosí, Colegio de San Luis, 2000, 2 vol., 761 p. (ill.). (Colección Fuentes).

4909. Consenso y coacción: estado e instrumentos de control político y social en México y América Latina (siglos XIX y XX). Coordinadores Ricardo FORTE y Guillermo GUAJARDO; con la colaboración de María LUNA. México, Centro de Estudios Históricos, El Colegio de México, Colegio Mexiquense, 2000, XXVIII-469 p. (ill.).

4910. ESPARZA VALDIVIA (Ricardo Cuauhtémoc). El fenómeno magonista en México y en Estados Unidos 1905–1908. Zacatecas, Centro de Investigaciones Historicas, Universidad Autonoma de Zacatecas, 2000, 196 p. (ill.).

4911. Estudios sobre el zapatismo. Coord. Laura ESPEJEL LÓPEZ. Mexico, Instituto Nacional de Antropología e Historia, 2000, 477 p. (ill.). (Colección Biblioteca del INAH. Serie Historia).

4912. FOWLER (Will). Tornel and Santa Anna: the writer and the Caudillo, Mexico 1795–1853. Westport, Greenwood, 2000, XV-308 p. (Contributions in Latin American studies, 14).

4913. Gobierno y administración pública en el Estado de México: una mirada a 175 años de historia. Coord. Alfonso X. IRACHETA CENECORTA. México, El Colegio Mexiquense, 2000, 456 p.

4914. GUEDEA (Virginia). The process of Mexican independence. In: Revolutions in the Americas [Cf. n° 4099], p. 116-130.

4915. HENDERSON (Peter V. N.). In the absence of Don Porfirio: Francisco León de la Barra and the Mexican revolution. Wilmington, Scholarly Resources, 2000, XIII-338 p. (Latin American silhouettes).

4916. MARTÍNEZ VALLE (Adolfo). El Partido Acción Nacional: una historia política. México, Editorial Porrúa, 2000, XIV-110 p.

4917. NEGRETTO (Gabriel L.), AGUILAR-RIVERA (José Antonio). Rethinking the legacy of the liberal state in Latin America: the cases of Argentina (1835–1916) and Mexico (1857–1910). *Journal of Latin American studies*, 2000, 32, 2, p. 361-398.

4918. OCEGUERA RAMOS (Rafael). De cara a la democracia: los desafíos del PRI. México, Miguel Angel Porrúa, 2000, 204 p.

4919. OCHOA (Enrique C.). Feeding Mexico: the political uses of food since 1910. Wilmington, Scholarly Resources, 2000, XIII-267 p. (Latin American silhouettes).

4920. OLCOTT (Jocelyn Harrison). Las hijas de la Malinche: women's organizing and state formation in postrevolutionary Mexico, 1934–1940. Ann Arbor, Bell & Howell Information and Learning Company, 2000, VI-466 p.

4921. PIETSCHMANN (Horst). Mexico zwischen Reform und Revolution. Vom bourbonischen Zeitalter zur Unabhängigkeit. Hrsg. v. Jochen MEIßNER, Renate PIEPER und Peer SCHMIDT. Stuttgart, Steiner, 2000, XII-299 p. (Beiträge zur Kolonial- und Überseegeschichte, 80).

4922. TOPIK (Steven C.). When Mexico had the blues: a transatlantic tale of bonds, bankers, and nationalists, 1862–1910. *American historical review*, 2000, 105, 3, p. 714-738.

4923. WASSERMAN (Mark). Everyday life and politics in nineteenth century Mexico: men, women, and war. Albuquerque, University of New Mexico Press, 2000, XIII-248 p. (ill., maps). (Diálogos).

Cf. n° 7961

Moldova

4924. KING (Charles). The Moldovans: Romania, Russia, and the politics of culture. Stanford, Hoover Institution Press, 2000, XXIX-303 p. (Studies of nationalities).

Montenegro

4925. RASTODER (Serbo). Politicke stranke i Crnoj Gori 1918–1929. Podgorica, Conteco-Bar, 2000, 727 p.

Morocco

4926. LÓPEZ GARCÍA (Bernabé). Marruecos político: cuarenta años de procesos electorales, 1960–2000. Madrid, Centro de Investigaciones Sociológicas, Siglo XXI de España Editores, 2000, XXII-335 p. (ill., maps). (Colección Monografías, 176).

4927. PENNELL (C. R.). Morocco since 1830: a history. New York, New York U. P., 2000, XXXIV-442 p. (ill., maps).

4928. TAHTAH (M.). Entre pragmatisme, réformisme et modernisme: le rôle politico-religieux des Khattabi dans le Rif (Maroc) jusqu'à 1926. Leuven, Uitgeverij Peeters en Departement Oosterse Studies, 2000, 288 p. (Orientalia Lovaniensia analecta, 91).

Mozambique

4929. BOWEN (Merle). The state against the peasantry: rural struggles in colonial and postcolonial Mozambique. Charlottesville, University Press of Virginia, 2000, XIV-256 p.

4930. CABRITA (Joao M.). Mozambique: the tortuous road to democracy. Basingstoke, Palgrave, 2000, XII-311 p.

Namibia

4931. GEWALD (Jan-Bart). "We thought we would be free ...": socio-cultural aspects of Herero history in Namibia 1915–1940. Köln, Köppe, 2000, 273 p. (ill.). (History, cultural traditions, and innovations in Southern Africa, 8).

4932. IDOWU (Stephen Babatunde). Namibia from colonisation to statehood: the paradoxical relationship between power and law in international society. London, [s. n.], 2000, 359 p. [Thesis: (Ph. D) – University of London 2000].

Netherlands

4933. KOSSMANN (Ernst Heinrich). Political thought in the Dutch Republic: three studies. Amsterdam, Koninklijke Nederlandse Akademie van Wetenschappen, 2000,1 97 p. (Verhandelingen der Koninklijke Nederlandse Akademie van Wetenschappen, Afd. Letterkunde; nieuwe reeks, 179).

4934. KUNZE (Rolf-Ulrich). 'Vader des Vaderlands'. Protorevolutionär oder toleranter Fürst? Zur Rolle Wilhelms von Oranien im Aufstand der Niederlande, 1566–1584. *Archiv für Kulturgeschichte,* 2000, 82, p. 93-120.

4935. MUNNICHS (Geert). Publiek ongenoegen en politieke geloofwaardigheid: democratische legitimiteit in een ontzuilde samenleving. Assen, Van Gorcum, 2000, 127 p.

4936. Politieke veranderingen in Nederland 1971–1998: kiezers en de smalle marges van de politiek. Onder redactie van Jacques THOMASSEN, Kees AARTS en Henk VAN DER KOLK. Den Haag, Sdu, 2000, 255 p. (ill.).

4937. VAN ZANDEN (J. L.), VAN RIEL (Arthur). Nederland 1780–1914: staat, institutië en economische ontwikkeling. Amsterdam, Balans, 2000, 480 p.

4938. ZEDINGER (Renate). Die Verwaltung der Österreichischen Niederlande in Wien (1714–1795): Studien zu den Zentralisierungstendenzen des Wiener Hofes im Staatswerdungsprozess der Habsburgermonarchie. Wien, Böhlau, 2000, 237 p. (Schriftenreihe der Österreichischen Gesellschaft zur Erforschung des 18. Jahrhunderts, 7).

Cf. n° 4089

Nicaragua

4939. CARDENAL TELLERÍA (Marco A.). Nicaragua y su historia, 1502–1936: cronología del acontecer histórico y construcción de la nación nicaragüense. Managua, Banco Mercantil, 2000, V-624 p.

4940. ZIMMERMANN (Matilde). Sandinista: Carlos Fonseca and the Nicaraguan revolution. Durham, Duke U. P., 2000, X-277 p.

Niger

4941. MAIGNAN (Jean-Claude). La difficile démocratisation du Niger. Ed. par Ginette FABRE et Jean-François LIONNET. Paris, Centre des hautes études sur l'Afrique et l'Asie modernes, 2000, 191 p.

Cf. n° 4211

Nigeria

4942. AGBESE (Dan). Fellow Nigerians: turning points in the political history of Nigeria. Ibadan, Umbrella Books, 2000, XXIV-167 p.

4943. VAUGHAN (Olufemi). Nigerian chiefs: traditional power in modern politics, 1890s–1990s. Rochester, University of Rochester Press, 2000, XIV-293 p. (Rochester studies in African history and the diaspora, 7).

4944. VICKERS (Michael). Ethnicity and subnationalism in Nigeria: movement for a mid-west state. Oxford, WorldView Publishing, 2000, XVIII-410 p. (WorldView African studies series, 3).

Norway

* 4945. NEUMANN (Iver B.). State and nation in the nineteenth century. Recent research on the Norwegian case. *Scandinavian journal of history,* 2000, 25, 3, p. 239-260.

4946. BORGERSRUD (Lars). Konspirasjon og kapitulasjon. Nytt lys på forsvarshistorien fra 1814 til 1940. (Conspiracy and capitulation. New perspectives on the Norwegian history of defence 1814–1940). Oslo, Oktober, 2000, 384 p. (ill.).

4947. ELIÆSON (Sven), BJÖRK (Ragnar). Union och secession. (Union et sécession. Les identités nationales et l'union Suède-Norvège). Stockholm, Carlssons, 2000, 336 p. (ill.).

4948. LINTHOE NÆSHAGEN (Ferdinand). Norway's democratic and conservative tradition in policing. *Scandinavian journal of history,* 2000, 25, 3, p. 177-196.

4949. NAKKEN (Alfhild). Sentraladministrasjonen i København og sentralorganer i Norge 1660–1814. (Central administration in Copenhagen and Norway 1660–1814). Oslo, Tano Aschehoug, Riksarkivaren, 2000, 143 p. (ill.). (Administrasjon og arkiver, 1, Skriftserie / Riksarkivaren, 9).

4950. ØYSTEIN (Rian). State, elite and peasant power in a Norwegian region, Bratsberg County, in the 17th century. *In:* Revolution (A) from above? [Cf. n° 4098], p. 183-243.

4951. TAMELANDER (Michael), ZETTERLING (Niklas). Nionde April. (Le 9 Avril 1940. L'invasion de la Norvège). Lund, Historiska media, 2000, 320 p.

4952. WEIDLING (Tor). Eneveldets menn i Norge, sivile sentralorganer og embetsmenn 1660–1814. (Absolute monarchy, central administration and officers of the crown in Norway 1660–1814). Oslo, Riksarkivaren / Messel forlag, 2000, 333 p. (Skriftserie / Riksarkivaren, 7).

Cf. n° 4373

Oman

** 4953. Sultanate of Oman (The), 1914–1918. Ed. by Raghid EL-SOLH. Reading, Ithaca, 2000, X-482 p. – The Sultanate of Oman, 1918–1939. Administrative affairs. Ed. by Raghid EL-SOLH. Reading, Ithaca, 2000, XX-402 p. – The Sultanate of Oman, 1918–1939. Domestic affairs. Ed. by Raghid EL-SOLH. Reading, Ithaca, 2000, XXII-297 p. – The Sultanate of Oman, 1939–1945. Ed. by Raghid EL-SOLH; with a foreword by Issam AL-RAWWAS. Reading, Ithaca, 2000, XII-378 p.

4954. AL-HINAI (Abdulmalik Abdullah). State formation in Oman: 1861–1970. London, [s. n.], 2000, 225 p. [Thesis (Ph. D.). – LSE, 2000].

4955. ALLEN (Calvin H.), RIGSBEE (W. Lynn). Oman under Qaboos: from coup to constitution, 1970–1996. London, Frank Cass, 2000, XIX-251 p, (ill.).

4956. COOPER (Tristan). The Popular Front for the Liberation of Oman and the Arab Gulf: the role of external forces in its political development (1963–1975). Oxford, [s. n.], 2000, 67 p. [Thesis (M. Phil.) – University of Oxford, 2000].

Pakistan

4957. BANERJEE (Mukulika). The Pathan unarmed: opposition and memory in the North West frontier. Oxford, James Currey, 2000, X-238 p, (ill.). (World anthropology).

4958. HASAN (Burhanuddin). Uncensored: an eyewitness account of abuse of power and media in Pakistan. Karachi, Royal Book Co., 2000, XXV-362 p. (ill.).

4959. KHALIL (Jehanzeb). Mujahideen movement in Malakand and Mohmand Agencies, 1900–1940. ed. by M. Y. EFFENDI. Peshawar, Area Study Centre University of Peshawar a. Islamabad, Hanns Seidel Foundation, 2000, XXIX-458 p. (ill., maps).

4960. Pakistan: government and politics. Ed. by Verinder GROVER. New Delhi, Deep & Deep, 2000, XII-1024 p. (ill.). (Government and politics of Asian countries, 12).

4961. SAFDAR (Mahmood). Pakistan: political roots and development, 1947–1999. Karachi a. Oxford, Oxford U. P., 2000, 500 p. (The millennium series).

4962. TALHA (Naureen). Economic factors in the making of Pakistan (1921–1947). Oxford, Oxford U. P., 2000, X-220 p.

Cf. n° 4749

Palestine

4963. BERNSTEIN (Deborah). Constructing boundaries: Jewish and Arab workers in mandatory Palestine. Albany, State University of New York Press, 2000, XVI-277 p. (ill.). (SUNY series in Israeli studies).

4964. FABRIZIO (Daniela). La questione dei Luoghi santi e l'assetto della Palestina: 1914–1922. Milano, Franco Angeli, 2000, 255 p.

4965. FRIEDMAN (Isaiah). Palestine: a twice-promised land? Vol. 1. The British, the Arabs and Zionism, 1915–1920. New Brunswick, Transaction, 2000, LXXVII-411 p.

4966. GLASS (Joseph B). From new Zion to old Zion: American Jewish immigration and settlement in Palestine, 1917–1939. Detroit, Wayne State U. P., 2000, [s. p.]. (America-Holy Land monographs).

4967. KIMMERLING (Baruch). The formation of Palestinian collective identities: the Ottoman and mandatory periods. *Middle eastern studies*, 2000, 36, 2, p. 48-81.

4968. PERRIN (Dominique). Palestine: une terre, deux peuples. Villeneuve-d'Ascq, Presses universitaires du Septentrion, 2000, 346 p., (ill.). (Histoire et civilisations).

4969. YAZBAK (Mahmoud). From poverty to revolt: economic factors in the outbreak of the 1936 rebellion in Palestine. *Middle eastern studies*, 2000, 36, 3, p. 93-113.

4970. YOUNIS (Mona). Liberation and democratization: the South African and Palestinian national movements. Minneapolis, University of Minnesota Press, 2000, XV-264 p. (Social movements, protest, and contention, 11).

Panama

4971. DAL BONI (Diego). Panamá, Italia y los italianos: en la época de la construcción del Canal (1880–1915). Panamá, República de Panamá, Crucero de Oro, 2000, 157 p. (ill.)

Paraguay

4972. PANGRAZIO (Miguel Ángel). Historia política del Paraguay. Vol. 2. Asunción, Intercontinental Editora, 2000, [s. p.].

Peru

4973. COTLER (Julio), GROMPONE (Romeo). El fujimorismo: ascenso y caída de un régimen autoritario. Lima, Instituto de Estudios Peruanos, 2000, 178 p. (Serie Ideología y política, 15).

4974. DRINOT DE ECHAVE (Paulo). Workers, the state, and radical politics in Peru in the early 1930s. Oxford, [s. n.], 2000, 332 p. [Thesis (D. phil.) – University of Oxford, 2000].

4975. GUTIÉRREZ SÁNCHEZ (Tomás). El 'Hermano' Fujimori: evangélicos y poder político en el Perú del '90. Lima, Perú, Archivo Histórico del Protestantismo Latinoamericano, 2000, 201 p, (ill.).

4976. KLARÉN (Peter F.). Peru: society and nationhood in the Andes. New York a. Oxford, Oxford U. P., 2000, XVI-494 p, (ill.). (Latin American histories).

4977. LORA CAM (Jorge). Los orígenes coloniales de la violencia política en el Perú. Puebla, Benémerita Universidad Autónoma de Puebla y Tlaxcal, Universidad Autónoma de Tlaxcala, 2000, 377 p.

4978. PELLEGRINO SOARES (Gabriela). Projetos políticos de modernizaçao e reforma no Peru: 1950–1975. Sao Paulo, FAPESP, 2000, 198 p. (Selo universidade, 122, História).

4979. PERALTA RUIZ (Víctor). Sendero Luminoso y la prensa, 1980–1994: la violencia política peruana y su representación en los medios. Cuzco, Centro de Estudios Regionales Andinos "Bartolomé de las Casas" y Lima, Casa de Estudios del Socialismo, 2000, 334 p. (Temas de actualidad, 5).

4980. RIZO-PATRÓN BOYLAN (Paul). Linaje, dote y poder: la nobleza de Lima de 1700 a 1850. Lima, Pontifica Universidad Católica del Perú, Fondo Editorial, 2000, 400 p, (ill.).

Cf. n° 969

Philippines

4981. VITUG (Marites Danguilan), GLORIA (Glenda M.). Under the crescent moon: rebellion in Mindanao. Quezon City, Ateneo Center for Social Policy & Public Affairs, Institute for Popular Democracy, 2000, IX-327 p.

Poland

** 4982. SZPILMAN (Władysław). Pianista. Warszawskie wspomnienia 1939-1945. (Der Pianist. Warschauer Erinnerungen 1939–1945). Bearb. von Andrzej SZPILMAN. Kraków, Znak, 2000, 216 p.

4983. Cenzura w PRL. Relacje historyków. (Die Zensur in der VR Polen. Berichterstattungen der Historiker). Bearb. v. Zbigniew ROMEK. Warszawa, Inst. Historii PAN, 2000, 286 p.

4984. CHWALBA (Andrzej). Historia Polski 1795–1918. (Geschichte Polens 1795–1918). Kraków, Wydaw. Literackie, 2000, 671 p.

4985. DANOWSKA (Bogumiła). Grudzień 1970 roku na Wybrzeżu Gdańskim. Przyczyny, przebieg, reperkusje. (Dezemberereignisse 1970 – Wybrzeże Gdańskie [Danziger Küste]. Hintergründe, Verlauf, Widerhall). Pelplin, Bernardinum, 2000, 311 p.

4986. DOMINICZAK (Henryk). Organy bezpieczeństwa PRL w walce z Kościołem katolickim 1944–1990 w świetle dokumentów MSW. (Der Sicherheitsdienst der VR Polen im Kampf gegen katholische Kirsche 1944–1990 angesichts der Dokumente des Innenministeriums [MSW]). Warszawa, Bellona, 2000, 543 p.

4987. DOROSZEWSKI (Jerzy). Oświata i życie kulturalne społeczności ukraińskiej na Lubelszczyźnie w latach 1918–1939. (Bildung und Kulturleben der ukrainischen Gemeinschaft auf dem Lubliner Boden [Lubelszczyzna] in den Jahren 1918–1939). Lublin, LTN, 2000, 338 p. (Studia z Dziejów Oświaty i Szkolnictwa w Latach II Rzeczypospolitej na Lubelszczyźnie, 4).

4988. FAŁOWSKI (Janusz). Parlamentarzyści mniejszości niemieckiej w Drugiej Rzeczypospolitej. (Parlamentarier der deutschen Minderheit in der Zweiten Polnischen Republik). Częstochowa, Wydaw. WSP, 2000, 425 p.

4989. HOFMANN (Andreas R.). Die Nachkriegszeit in Schlesien: Gesellschafts- und Bevölkerungspolitik in den polnischen Siedlungsgebieten, 1945–1948. Köln, Böhlau, 2000, XV-476 p. (Beiträge zur Geschichte Osteuropas, 30).

4990. KAMIŃSKI (Łukasz). Polacy wobec nowej rzeczywistości 1944–1948. Formy pozainstytucjonalnego, żywiołowego oporu społecznego. (Polen gegen neue Wirklichkeit 1944–1948. Formen des außerinstitutionellen, regen Widerstandes der Gesellschaft). Toruń, Wydaw. Adam Marszałek, 2000, 283 p.

4991. KARPIŃSKI (Andrzej). W walce z niewidzialnym wrogiem. Epidemie chorób zakaźnych w Rzeczypospolitej w XVI–XVIII wieku i ich następstwa demograficzne, społeczno-ekonomiczne i polityczne. (Im Kampf gegen einen heimlichen Feind. Ausbrüche der ansteckenden Krankheiten in der Republik Polen im XVI.–XVIII. Jahrhundert und deren demographische, gesellschaftlich-ökonomische und politische Folgen). Warszawa, Neriton, 2000, 447 p. (Inst. Historii).

4992. LANGE (Tadeusz Wojciech). Zakon Maltański w Drugiej Rzeczypospolitej 1919–1939. (Der Malteser-Orden in der Zweiten Polnischen Republik 1919–1939). Poznań, Wydaw. Poznańskie, 2000, 309 p.

4993. LIM (Jie-Hyun). Gdaedului Jayoo, Wooridului Jayoo. (For your freedom and our freedom: a history of Polish irredentist movement). Seoul, Acanet, 2000, [s. p.].

4994. MICH (Włodzimierz). Ideologia polskiego ziemiaństwa 1918–1939. (Die Ideologie der polnischen Gutsbesitzer 1918–1939). Lublin, Wydaw. Uniw. M. Curie-Skłodowskiej, 2000, 367 p.

4995. NITSCHKE (Bernadetta). Wysiedlenie czy wypędzenie? Ludność niemiecka w Polsce w latach 1945–1949. (Aussiedlung oder Vertreibung? Deutsche Bevölkerung in Polen in den Jahren 1945–1949). Toruń, Wydaw. Adam Marszałek, 2000, 306 p.

4996. PALKIJ (Henryk). Sejmy 1736 i 1738 roku. U progu nowej sytuacji politycznej w Rzeczypospolitej. (Sejm der Jahre 1736 und 1738. An der Schwelle einer neuen politischen Situation in der Republik Polen).

Kraków, PAU, 2000, 256 p. (Rozprawy Wydz. Historyczno-Filozoficznego).

4997. PORTER (Brian A.). When nationalism began to hate: imagining modern politics in nineteenth century Poland. New York a. Oxford, Oxford U. P., 2000, X-307 p.

4998. ŚLESZYŃSKI (Wojciech). Białystok w sowieckiej fotografii propagandowej 1939–1941. Proces aneksji i polityczno-prawnej sowietyzacji Białostocczyzny. (Białystok in Bildern der sowjetischen Propaganda. Der Prozess der Annexion sowie der politisch-rechtlichen Sowjetisierung der Białostocczyzna). Białystok, BTN, 2000, 221 p. (Prace Białostockiego Tow. Naukowego, 46).

4999. SOWIŃSKI (Paweł). Komunistyczne święto. Obchody 1 maja w latach 1948–1954. (Kommunistisches Fest. 1. Mai-Feierlichkeiten in den Jahren 1948–1954). Warszawa, Trio, 2000, 130 p. (W Krainie PRL).

5000. SZLACHTA (Bogdan). Polscy konserwatyści wobec ustroju politycznego do 1939 roku. (Die polnischen konserwativen Kreise angesichts der politischen Ordnung bis 1939). Kraków, Wydaw. Uniw. Jagiellońskiego, 2000, 472 p. (Biblioteka Myśli Politycznej, 29).

5001. SZWAGRZYK (Krzysztof). Winni? Niewinni? Dolnośląskie podziemie niepodległościowe (1945–1956) w świetle dokumentów. (Schuldig? – Frei von Schuld? Niederschlesiens geheime Freiheitsbewegungen [1945–1956] in Anbetracht der Dokumente). Wrocław, Zarząd Gł. Stow. Społ.-Kombatanckiego "Wolność i Niezawisłość", 2000, 430 p. (Biblioteka Zeszytów Historycznych WiN-u).

5002. TUREK-KWIATKOWSKA (Lucyna). Życie codzienne w Szczecinie w latach 1800–1939. (Das Alltagsleben in Szczecin [Stettin] in den Jahren 1800–1939). Szczecin, Wydaw. Nauk. Uniw. Szczecińskiego, 2000, 210 p. (Rozprawy i Studia, 363).

5003. Vertreibung (Die) der Juden aus Polen 1968. Antisemitismus und politisches Kalkül. Hrsg. v. Beate KOSMALA. Berlin, Metropol, 2000, 191 p. (Zentrum für Antisemitismusforschung Berlin, Reihe Dokumente, Texte, Materialen, 34).

5004. WALASZEK (Adam). Poland as the 'Promised Land': Polish American corporations and Poland after World War I. In: Migrants and the homeland. images, symbols, and realities. Ed. by Harald RUNBLOM. Uppsala, Center for Multiethnic Research, 2000, p. 265-286. (Uppsala multiethnic papers, 44).

Cf. n°s 6010, 7967

Portugal

5005. COSTA PINTO (Antonio). The blue shirts. Portuguese fascist and the New State. New York, Columbia U. P., 2000, XVI-272 p.

5006. DE LA TORRE GÓMEZ (Hipólito), CERVELLÓ SÁNCHEZ (Josep). Portugal en la edad contemporánea: 1807–2000: historia y documentos. Madrid, Universidad Nacional de Educación a Distancia, 2000, 574 p. (Varia).

5007. DE SOUSA (Marcelo Rebelo). A revolução e o nascimento do PPD. Vol. 1. Abril de 1974–maio de 1975. Vol. 2. Maio de 1975–dezembro de 1975. Venda Nova, Bertrand Editora, 2000, 2 vol., 1133-LX p., (ill.). (Ensaios e documentos).

5008. FARIA (Cristina Isabel). As lutas estudantís contra a ditadura militar (1926–1932). Lisboa, Ediçoes Colibri, 2000, 357 p. (Coleccao Colibri história, 27).

5009. FARIA (Telmo). Debaixo de fogo! Salazar e as Forças Armadas (1935–1941). Lisboa, Ediçoes Cosmos, Instituto de Defensa Nacional, 2000, 286 p, (ill.). (Coleccao Atena, 7).

5010. GOLDNER (Loren). Ubu saved from drowning: worker insurgency and statist containment in Portugal and Spain, 1974–1977. Cambridge, Queequeg Publications, 2000, 113 p. (Marx/Third Millennium series).

5011. HERMANN (Christian), MARCADÉ (Jacques). Les royaumes ibériques au XVIIe siècle. Paris, SEDES, 2000, 191 p. (Regards sur l'histoire. Histoire moderne, 143).

5012. MARINHO (José da Silva). Construction d'un gouvernement municipal: Élites, élections et pouvoir à Guimaraes entre Absolutisme et Liberalisme (1753–1834). Braga, Universidade do Minho, 2000, 438 p. (ill.).

5013. MARTINS (Manuel Gonçalves). O Estado novo e a oposiçao (1933–1974). Sintra, P. Ferreira, editor, 2000, 432 p.

5014. MEDINA (Joao). Salazar, Hitler e Franco: estudos sobre Salazar e a dictadura. Lisboa, Livros Horizonte, 2000, 308 p, (ill.). (Coleccao Horizonte histórico).

5015. PATRIARCA (Fátima). Sindicatos contra Salazar: a revolta de 18 de Janeiro de 1934. Lisboa, Imprensa de Ciênas Sociais, 2000, 556 p.

5016. PIMENTEL (Irene Flunser). História das organizaçoes femininas no Estado Novo. Lisbon, Círculo de Leitores, 2000, 473 p. (ill.).

5017. Primeira República portuguesa (A): entre o liberalismo e o autoritarismo. Coordenação de Nuno SEVERIANO TEIXEIRA e António COSTA PINTO. Lisboa, Ediçoes Colibri, Instituto de História Contemporânea, Universidade Nova de Lisboa, 2000, 222 p. (ill.). (Cursos de Verao, 3).

5018. VENTURA (António). Anarquistas republicanos e socialistas em Portugal: as convergências possíveis (1892–1910). Lisboa, Ediçoes Cosmos, 2000, 334 p, (ill.). (Coleccao de História moderna e contemporânea, 5).

Cf. n°s 4238, 5939

Puerto Rico

5019. CAMUNAS MADERA (Ricardo R.). Desplazamiento y revolución en el Puerto Rico de siglo XIX.

San Juan, Instituto de Cultura Puertorriquena, 2000, XIII-152 p.

5020. GARCÍA COLÓN (Pablo) [et al.]. Tras las huellas del pasado: mosaico de historia de Puerto Rico: siglos XIX y XX. San Juan, Isla Negra y Humacao, Decanato de Asuntos Académicos, Recinto de Humacao, Universidad de Puerto Rico, 2000, 177 p. (ill.). (Colección Visiones y cegueras).

Qatar

5021. RATHMELL (Andrew), SCHULZE (Kirsten). Political reform in the Gulf: the case of Qatar. *Middle eastern studies*, 2000, 36, 4, p. 47-62.

Reunion

5022. GAUVIN (Gilles). Le parti communiste de la Réunion (1946–2000). XXe *siècle*, 2000, 68, p. 73-94.

Romania

** 5023. IANCU (Carol). La Shoah en Roumanie: les Juifs sous le regime d'Antonescu, 1940–1944: documents diplomatiques français inédits. Montpellier, Université Paul Valéry, 2000, 205 p. (Collection Sem).

5024. Anii 1954–1960: fluxurile şi refluxurile stalinismului: comunicari prezentate la Simpozionul de la Sighetu Marmatiei (2–4 iulie 2000). Editor Romulus RUSAN. Bucureşti, Fundaţia Academia Civică, 2000, 1006 p. (Analele Sighet, 8).

5025. BERINDEI (Dan). Les représentants consulaires français et les Roumains au cours du second quart du XIXe siècle. *In*: Nouvelles Études d'Histoire. X. Publiées à l'occasion du XIXe Congrès International des Sciences Historiques. Oslo, 2000. Bucureşti, Editura Academiei Române, 2000, p. 133-153.

5026. BULEI (Ion). Conservatori şi conservatorism în România. Bucureşti, Editura Enciclopedică, 2000, 699 p.

5027. CALAFETEANU (Ion). Politică şi exil. Din istoria exilului românesc, 1946–1950. Bucureşti, Editura Enciclopedică, 2000, 386 p.

5028. CÂMPEANU (Remus). Elitele româneşti din Transilvania veacului al XVIII-lea. Cluj-Napoca, Editura Presa Universitară Clujeană, 2000, 457 p.

5029. CÂRJA (Ion). Les roumains de Transylvanie et l'Empire des Habsbourgs dans la période 1848–1851 – entre réalité et imaginaire. *Revista Bistriţei*, 2000, 14, p. 231-244.

5030. CEAUŞU (Mihai-Ştefan). Bucovina în sistemul parlamentar al monarhiei de Habsburg. 1848–1918. *Anuarul Institutului de Istorie «A. D. Xenopol» Iaşi*, 2000, 37, p. 145-160.

5031. GABANYI (Anneli Ute). The Ceausescu cult: propaganda and power policy in communist Romania. Bucureşti, Romanian Cultural Foundation Publishing House, 2000, 432 p. (Istorie).

5032. IOANID (Radu). The Holocaust in Romania: the destruction of Jews and Gypsies under the Antonescu regime, 1940–1944. Forewords by Elie WIESEL and Paul A. SHAPIRO. Chicago, Ivan R. Dee a. Washington, United States Holocaust Memorial Museum, 2000, XXIV-352 p.

5033. KUNZE (Thomas). Nicolae Ceausescu: eine Biographie. Berlin, Links, 2000, 463 p.

5034. MAMINA (Ion). Monarhia constituţională în România: enciclopedie politică, 1866–1938. Bucureşti, Editura Enciclopedică, 2000, 423 p.

5035. SUCIU (Dumitru). Antecedentele dualismului austro-ungar şi mişcarea naţională a românilor din Transilvania 1848–1867. (I precedenti del dualismo austro-ungarico e il movimento nazionale dei romeni della Transilvania, 1848–1867). Bucureşti, Editura Albatros, 2000, 319 p. (Historia).

5036. TANASESCU (Florian). Parlamentul şi viaţa parlamentară din România (1930–1940). Bucureşti, Lumina Lex, 2000, 464 p.

Cf. n°s 7956, 8132

Russia (USSR)

* 5037. Beloe dvizhenie: Katalog kollektsii listovok, 1917–1920. (The White Movement: catalogue of collection of leaflets, 1917–1920). Ed. Galina V. MIKHEEVA [et al.]. Sankt-Peterburg, Izd-vo Rossijskoj natsional'noj biblioteki, 2000, 504 p.

* 5038. GULAG: Glavnoe upravlenie lagerej, 1918–1960. (The GULAG: the Central Direction of the Camps, 1918–1960). Ed. Aleksandr N. JAKOVLEV; Compiled by Aleksandr I. KOKURIN, Nikolaj V. PETROV; Scientific ed. Vjacheslav N. SHOSTAKOVSKIJ. Moskva, Mezhdunar. fond 'Demokratija', 2000, 886 p. (Rossija. XX vek. Dokumenty).

* 5039. IL'IN (Pavel V.). 14 dekabrja 1825 goda: Istochniki, issledovanija, istoriografija, bibliografija. (14 December 1825: sources, studies, historiography, bibliography). RAN, In-t rossijskoj istorii. Parts 2, 3. Sankt-Peterburg a. Kishinev, Nestor, 2000, 2 vol., 306 p., 338 p. (Izvestija vuzov, 1-2).

* 5040. LYKOVA (L. A.) [et al.]. Politbjuro TsK RKP(b) – VKP(b): Povestki dnja zasedanij, 1919–1952: Katalog. (Catalogue of agenda of sessions of the Politburo of the Central Committee of the RKP(b) – VKP(b), 1919–1952) Vol. 1. 1919–1929. Ed. Grant M. ADIBEKOV, Kirill M. ANDERSON. Moskva, ROSSPEN, 2000, 831 p.

** 5041. DIMITROV (Georgi). Tagebücher 1933–1943. Hrsg. v. Bernhard H. BAYERLEIN. Kommentare und Materialien zu den Tagebüchern 1933–1943. Hrsg. v. Bernhard H. BAYERLEIN und Wladislaw HEDELER unt. Mitarb. v. Birgit SCHLIEWENZ und Maria MATUSCHEK. Berlin, Aufbau-Verlag, 2000, 2 vol., 712 p., 773 p.

** 5042. Dimitrov and Stalin, 1934–1943. Letters from the Soviet archives. Ed. by Alexander DALLIN and Fridrich I. FIRSOV. New Haven a. London, Yale U. P., 2000, XXX-278 p. (Annals of communism).

** 5043. DOLGOVA (S. R.), LAPTEVA (T. A.). Povsednevnye zapiski delam knjazja A.D. Menshikova, 1717–1720, 1726–1727. (The diary of Prince Alexander D. Menshikov, 1717–1720, 1726–1727). Moskva, Pedaktsija Al'manakha "Rossijskij arkhiv", 2000, 647 p. (ill.). (Rossijskij arkhiv. Istorija Otechestva v svidetel'stvakh i dokumentakh, 10).

** 5044. Moskva poslevoennaja, 1945–1947: Arkhivnye dokumenty i materialy. (Moscow after the War, 1945–1947: documents and materials from archives). Ed. A. S. KISELEV. Compiled by M. M. GORINOV, A. N. PONOMAREV, E. V. TARANOV. Moskva, Izd-vo ob'edinenija 'Mosgorarkhiv', 2000, 767 p. (ill., tables).

** 5045 Na korme vremeni: Interv'ju s leningradtsami 1930-kh godov. (Interviews with Inhabitants of Leningrad of the 1930s). Ed. M. VITUKHNOVSKAJA. Intr. by M. VITUKHNOVSKAJA, K. GERASIMOVA. Comment. by A. N. CHISTIKOV. Sankt-Peterburg, Zhurnal 'Neva', 2000, 383 p. (ill.).

** 5046. Togliatti negli anni del Comintern: 1926–1943: documenti inediti dagli archivi russi. A cura di Aldo AGOSTI; in collaborazione con Marina LITRI. Roma, Carocci, 2000, 228 p. (Annali / Fondazione Istituto Gramsci, 10).

5047. ARTEMOV (Viktor A.). Karl Radek: Ideja i Sud'ba. (Karl Radek: idea and fortune). Voronezh, Tsentral'no-Chernozemnoe knizhnoe izdatel'stvo, 2000, 175 p.

5048. BOOBBYER (Philip). Truth-telling, conscience and dissent in late Soviet Russia: evidence from oral histories. *European history quarterly*, 2000, 30, 4, p. 554-586.

5049. DI BIAGIO (Anna). Moscow, the Comintern and the War Scare, 1926–1928. *In*: Russia in the age of wars, 1914–1945 [Cf. n° 5077], p. 83-102.

5050. DULLIN (Sabine). Litvinov and the People's Commissariat of Foreign Affairs: the fate of an administration under Stalin, 1930–1939. *In*: Russia in the age of wars, 1914–1945 [Cf. n° 5077], p. 121-146.

5051. Grande terrore (Il). *Storica*, 2000, 6, 18, p. 7-62. [Con scritti di: GRAZIOSI (Andrea). Cosa stiamo imparando. – CHLEVNJUK (Oleg). I nuovi dati. – MARTIN (Terry). Un'interpretazione contestuale alla luce delle nuove ricerche].

5052. HILDERMEIER (Manfred). The Russian Socialist Revolutionary Party before the First World War. New York, St. Martin's Press a. Munster, Lit Verlag, 2000, 385 p. (ill.).

5053. HILLYAR (Anna), MAC DERMID (Jane). Revolutionary women in Russia, 1870–1917: a study in collective biography. Manchester a. New York, Manchester U. P., 2000, 232 p.

5054. "Ihr verreckt hier bei ehrlicher Arbeit!" Deutsche im GULAG 1936–1956. Anthologie des Erinnerns. Hrsg. v. Eva DONGA-SYLVESTER, Günter CZERNETZKY und Hildegard TOMA. Graz u. Stuttgart, Stocker, 2000, 366 p.

5055. IL'INA (Irina N.). Obshchestvennye organizatsii Rossii v 1920-e gody. (Public organizations in Russia in the 1920s). RAN, In-t rossijskoj istorii. Moskva, [s. n.], 2000, 215 p.

5056. JĘDRYCHOWSKA (Barbara). Polscy zesłańcy na Syberii (1830–1882). Działalność pedagogiczna, oświatowa i kulturalna. [Polen in der Verbannung in Sibirien (1830–1882). Das Wirken im Bereich der Pädagogik, Bildung und Kultur]. Wrocław, Wydaw. Uniw. Wrocławskiego, 2000, 240 p. (Acta Universitatis Wratislaviensis, 2187).

5057. JOSEPHSON (Paul R.). Red atom: Russia's nuclear power program from Stalin to today. New York a. Basingstoke, W.H. Freeman, 2000, X-352 p. (ill., map, ports.).

5058. KANG (Yoonhee). The Leningrad party organisation during the first Five Year Plan. Glasgow, [s. n.], 2000, 392 p. [Thesis (Ph. D.) – University of Glasgow, 2000].

5059. KANGASPURO (Markku). Neuvosto-Karjalan taistelu itsehallinnosta. Nationalismi ja suomalaiset punaiset Neuvostoliiton vallankäytössä vuosina 1920–1939. (The role of nationalism and the question of nationalities regarding the state-building progress of the Soviet Union and the Karelian Republic in the 1920s and 30s). Helsinki, SKS, 2000, 402 p. (ill., maps). (Bibliotheca historica, 60).

5060. LEVIN (Leonid I.). Rossijskij generalissimus gertsog Anton Ul'rikh: Istorija "Braunshvejgskogo semejstva" v Rossii. (Russian Duke and Generalissimus Anton Ulrich: history of the 'Braunschweig family' in 18th-century Russia). Sankt-Peterburg, Peterburgskij pisatel' a. BLITs, 2000, 355 p. (ill.). [Zusammenfassung]

5061. LUKIN (Pavel V.). Narodnye predstavlenija o gosudarstvennoj vlasti v Rossii XVII veka. (Popular concepts of the state power in 17th-century Russia). RAN, In-t rossijskoj istorii. Moskva, Nauka, 2000, 294 p. [Eng. summary]

5062. MAIN (Steven J.). The Red Army and the future war in Europe. *In*: Russia in the age of wars 1914–1945 [Cf. n° 5077], p. 181-185.

5063. MALYSHEVA (Svetlana Ju.). Vremennoe pravitel'stvo Rossii. Sovremennaja otechestvennaja istoriografija. (The Provisional Governement of Russia, [1917]: contemporary Russian historiography). Kazan', Kheter, 2000, 208 p. [Cf. Bibl. 99, n° 5013.]

5064. MAWDSLEY (Evan), WHITE (Stephen). The Soviet elite from Lenin to Gorbachev: the Central

Committee and Its Members, 1917–1991. New York, Oxford U. P., 2000, XXIII-323 p.

5065. MIRONOV (Boris Nikolaevich), EKLOF (Ben). A social history of Imperial Russia, 1700–1917. Boulder a. Oxford, Westview Press, 2000, XXXIV-562 p. (maps).

5066. MOROZOVA (Ljudmila E.). Smuta nachala XVII veka glazami sovremennikov i potomkov. (The time of troubles in early 17th-century Russia through the eyes of contemporaries and descendants). RAN, In-t rossijskoj istorii. Moskva, [s. n.], 2000, 464 p. (tables).

5067. Moskau (Von) nach St. Petersburg: das russische Reich im 17. Jahrhundert. Hrsg. v. Hans-Joachim TORKE. Wiesbaden, Harrassowitz, 2000, 301 p. (ill.). (Forschungen zur osteuropäischen Geschichte, 56).

5068. NORDHOF (Anton Wilhelm). Die Geschichte der Zerstörung Moskaus im Jahre 1812. Hrsg. v. Claus SCHARF unt. Mitw. v. Jürgen KESSEL. München, Boldt bei Oldenbourg, 2000, 343 p. (Deutsche Geschichtsquellen des 19. und 20. Jahrhunderts, 61).

5069. O'ROURKE (Shane). Warriors and peasants: the Don Cossacks in late imperial Russia. New York, St. Martin's, 2000, XIV-200 p. (St. Antony's Series).

5070. PENTER (Tanja). Odessa 1917. Revolution an der Peripherie. Köln, Weimar u. Wien, Böhlau, 2000, X-469 p. (Beiträge zur Geschichte Osteuropas, 32).

5071. PEREGUDOVA (Zinaida I.). Politicheskij sysk Rossii, 1880–1917. (Political detection in Russia, 1880–1917). Gos. arkhiv Rossijskoj Federatsii. Moskva, ROSSPEN, 2000, 431 p. (ill.).

5072. Petrograd na perelome epokh: Gorod i ego zhiteli v gody revoljutsii i Grazhdanskoj vojny. (Petrograd between the epochs: the city and its inhabitants during the Revolution and the Civil War, [1917–1921]). Ed. V. A. SHISHKIN. Authors: E. M. BALASHOV, V. I. MUSAEV, A. M. RUNASOV, A. N. CHISTJAKOV, V. A. SHISHKIN. Sankt-Peterburg, Dmitrij Bulanin, 2000, 349 p. (ill.).

5073. PETRONE (Karen). Life has become more joyous, comrades: celebrations in the time of Stalin. Bloomington, Indiana U. P., 2000, X-266 p. (Indiana-Michigan series in Russian and East European studies).

5074. PHILLIPS (Steve). Lenin and the Russian revolution. Oxford, Heinemann, 2000, IV-177 p. (Heinemann advanced history).

5075. Rossija i Baltija: Narody i strany, vtoraja polovina XIX–30-e gody XX v. (Russia and Baltia: peoples and countries, from the 2nd half of the 19th century to the 1930s). Ed. Aleksandr O. CHUBAR'JAN. RAN, In-t vseobshchej istorii. Moskva, [s. n.], 2000, 172 p.

5076. RUDNEVA (Svetlana E.). Demokraticheskoe soveshchanie (sentjabr' 1917 g.): Istorija foruma. (The democratic consultation, September 1917: a history of the forum). RAN, In-t rossijskoj istorii. Moskva, Nauka, 2000, 255 p.

5077. Russia in the age of wars: 1914–1945. Ed. by Silvio PONS and Andrea ROMANO. Milano, Feltrinelli, 2000, XXVIII-322 p. (Annali / Fondazione Giangiacomo Feltrinelli, 34). [Cf. n^{os} <choice> 5049, 5050, 5062, 8078, 8092, 8098.]

5078. Russian modernity: politics, knowledge and practices, 1800–1950. Ed. by David L. HOFFMANN and Yanni KOTSONIS. Basingstoke, Macmillan Press, 2000, VIII-279 p..

5079. SAMUELSON (Lennart). Plans for Stalin's war machine: Tukhachevsky and military-economic planning, 1925–1941. Foreword by Vitalii SHLYKOV. New York, St. Martin's Press a. University of Birmingham, Centre for Russian and East European Studies, 2000, XV-267 p.

5080. SANDAG (Shagdariin), KENDALL (Harry H.). Poisoned arrows: The Stalin-Choibalsan Mongolian massacres, 1921–1941. Boulder, Westview Press, 2000, XX-228 p. (ill.).

5081. SERVICE (Robert). Lenin: a biography. Cambridge, Harvard U. P., 2000, XXV-561 p.

5082. Sovetskoe obshchestvo: budni kholodnoj vojny: materialy "kruglogo stola", 29 marta 2000 g., Moskva. (Soviet society: realities of the Cold War: Proceedings of a 'round table', 29 March 2000, Moscow). Ed. Vitalij S. LEL'CHUK, Garij Sh. SAGATELJAN. RAN, In-t rossijskoj istorii; Arzamasskij gos. ped. in-t im. A.P. Gajdara. Moskva a. Arzamas, [s. n.], 2000, 338 p. (tables).

5083. Soviet defence-industry complex (The) from Stalin to Khrushchev. Ed. by John BARBER and Mark HARRISON. New York, St. Martin's Press a. University of Birmingham, Centre for Russian and East European studies, 2000, XVIII-283 p.

5084. Stalinism. New directions. Ed. by Sheila FITZPATRICK. London a. New York, Routledge, 2000, XVIII-377 p.

5085. STONE (David R.). Hammer and rifle: the militarization of the Soviet Union, 1926–1933. Lawrence, University Press of Kansas, 2000, VII-287 p. (Modern war studies).

5086. WADE (Rex A.). The Russian Revolution, 1917. Cambridge, New York a. Melbourne, Cambridge U. P., 2000, XVII-337 p.

5087. WHITE (James D.). Lenin: the practice and theory of revolution. Basingstoke, Palgrave, 2000, X-262 p. (European history in perspective).

5088. WORTMAN (Richard S.). Scenarios of power. Myth and ceremony in Russian monarchy. Vol. 2. From Alexander II to the abdication of Nicholas II. Princeton, Princeton U. P., 2000, 586 p.

Cf. n°s 377, 379, 412, 456, 469, 504, 536, 556, 562, 564, 567, 654, 713, 832, 842, 882, 891, 893, 1159, 3486, 4692, 4763, 5616, 5737, 5831, 5850, 5851, 5862, 5865, 5961, 6010, 6045, 6048, 6057,

6070, 6453, 6465, 6758, 6787, 6957, 6958, 7070, 7083, 7180, 7242, 7249, 7277, 7298, 7302, 7363, 7543, 7562, 7641, 7747, 7852, 7916, 7956, 7967, 8011, 8067, 8078, 8092, 8098, 8121, 8125, 8130, 8132, 8183, 8262, 8265, 8266

Rwanda

Cf. n° 616

Senegal

5089. FAYE (Cheikh Faty). La vie sociale à Dakar (1945–1960). Paris, l'Harmattan, 2000, 332 p. – IDEM. Les enjeux politiques à Dakar (1945–1960). Paris, l'Harmattan, 2000, 400 p.

Slovakia

** 5090. Informatívny sprievodca štátnych archívov Slovenskej republiky. (Information guide of the State Archives of the Slovak Republic). Zv. 1. Ed. Mária STIEBEROVÁ. Bratislava, Ministerstvo vnútra Slovenskej republiky – Odbor archívnictva a spisovej služby, 2000, 212 p. [Deutsche Zfassung].

** 5091. Odsun. Die Vertreibung der Sudetendeutschen, Vyhnáuí sudetských Němců. Dokumentation zu Ursachen, Planung und Realisierung einer "ethnischen Säuberung" in der Mitte Europas 1848/49–1945/46. Band 1. Vom Völkerfrühling und Völkerzwist 1848/49 bis zum Münchner Abkommen 1938 und zur Errichtung des "Protektorats Böhmen und Mähren" 1939. Berab. v. Roland J. HOFFMANN und Alois HARASKO. München, Sudetendeutsches Archiv, 2000, 944 p.

5092. BOBRÍK (Miroslav). Činnosť nemeckého telocvičného a športového zväzu v období Slovenského štátu (1939–1945). (Die Tätigkeit des Deutschen Turn- und Sportverbandes in der Zeit des Slowakischen Staates, 1939–1945). *Historický časopis*, 2000, 48, 3, p. 505-516. [Deutsche Zfassung].

5093. Dejiny Slovenska. (History of Slovakia). Aut. Dušan ČAPLOVIČ, Viliam ČIČAJ, Dušan KOVÁČ, Ľubomír LIPTÁK and Ján LUKAČKA. Bratislava, Academic Electronic Press, 2000, 309 p.

5094. DULOVIČ (Erik). Osudy a postavenie legionárov na východnom Slovensku v rokoch 1919–1929. (Schicksale und die Stellen der Legionäre in der Ostslowakei in den Jahren 1919–1929). *Historický časopis*, 2000, 48, 1, p. 54-74. [Deutsche Zfassung]

5095. DVOŘÁKOVÁ (Daniela). Šľachtici z Ludaníc a kráľ Žigmund Luxemburský. (Die Adeligen von Ludanice und der König Sigismund von Luxemburg). *Historický časopis*, 2000, 48, 1, p. 35-43. [Deutsche Zfassung].

5096. FUNDÁRKOVÁ (Anna). Politické pomery v predmoháčskom Uhorsku a kráľovná Mária Habsburská (1521–1526). (Die politischen Verhältnisse in Ungarn vor Mohács und Königin Maria Habsburg, 1521–1526). *Historický časopis*, 2000, 48, 2, p. 231-255. [Deutsche Zfassung].

5097. HARBUĽOVÁ (Ľubica). Pôsobenie ruskej porevolučnej emigrácie na Slovensku v rokoch 1920–1939. (Das Wirken der russischen Emigration nach der Revolution in der Slowakei in den Jahren 1920–1939). *Historický časopis*, 2000, 48, 3, p. 435-461. [Deutsche Zfassung].

5098. Historik v čase a priestore. Laudatio Ľubomírovi Liptákovi. (The historian in the time and space. Laudatio for Ľubomír Lipták). Ed. by Ivan KAMENEC, Elena MANNOVÁ and Eva KOWALSKÁ. Bratislava, Veda vydavateľstvo Slovenskej akadémie vied a. Historický ústav Slovenskej akadémie vied, 2000, 323 p.

5099. HOENSCH (Jörg Konrad). Studia Slovaca: Studien zur Geschichte der Slowaken und der Slowakei. Hrsg. v. Hans LEMBERG [et al.]. München, R. Oldenbourg, 2000, XIII-423 p. (Veröffentlichungen des Collegium Carolinum, 93).

5100. KAMENEC (Ivan). Hľadanie a blúdenie v dejinách. Úvahy, štúdie a polemiky. (Searching and wandering through the history. Reflections, studies and polemics). Bratislava, Kalligram, 2000. 404 p.

5101. KIPKE (Rüdiger), VODICKA (Karel). Slowakische Republik: Studien zur politischen Entwicklung. Munster, LIT, 2000, 209 p. (Tschechien und Mitteleuropa, 3).

5102. LETZ (Róbert). Dejiny Slovenskej ligy na Slovensku (1920–1948). (The history of the Slovak League in Slovakia). Martin, Matica slovenská 2000, 294 p. [Eng. summary].

5103. LICHNER (Vlastimil Patrick). Z Brezovej do Ameriky. (From Brezová to America). Ed. Slavomír MICHÁLEK. Bratislava, Vydavateľstvo Pozsony a. Bratislava, Pressburg, 2000, 176 p. [Eng. summary].

5104. LIPTÁK (Ľubomír). Slovensko v dvadsiatom storočí. (Slovakia in the 20th century). Bratislava, Kalligram, 2000, 376 p.

5105. MAREČKOVÁ (Sylvie). K sjezdu přátel Slováků 13.8.1905 v Hodoníně. (On the meeting of friends of Slovaks in Hodonín on August 13th, 1905). *Historický časopis*, 2000, 48, 2, p. 331-339.

5106. November 1989 na Slovensku. Súvislosti, predpoklady a dôsledky. Štúdie a úvahy. (November 1989 in Slovakia. Context, assumptions and consequences. studies and reflections. Ed. by Jan PEŠEK and Soňa SZOMOLÁNYI. Bratislava, Nadácia Milana Šimečku a. Historický ústav Slovenskej akadémie vied, Katedra politológie Filozofickej fakulty Univerzity Komenského, 2000, 180 p.

5107. PEŠEK (Jan). Nástroj represie a politickej kontroly. Štátna bezpečnosť na Slovensku 1953–1970. (The tool of repression and political control. The state security in Slovakia in the years 1953-1970). Bratislava, Veda vydavateľstvo Slovenskej akadémie vied,

2000, 258 p. [Eng. summary]. – IDEM. Štátna bezpečnosť na Slovensku v období pokusu o spoločenskú reformu a jeho likvidácie (1968–1970). (Die Staatssicherheit in der Slowakei in der Zeit des Versuches um eine Gesellschaftsreform und ihre Liquidation 1968–1970). *Historický časopis*, 2000, 48, 3, p. 478-494. [Deutsche Zfassung].

5108. ŠIMEKOVÁ (Slávka). Židia v právnych pamiatkach Uhorska za Arpádovcov. (Juden in den Rechtsdenkmälern von Ungarn in der Zeit der Herrschaft des Hauses Arpad). *Historický časopis*, 2000, 48, 3, p. 401-416. [Deutsche Zfassung].

5109. ŠTEFÁNIK (Martin). Pokusy Benátskej republiky o atentát na uhorského kráľa Žigmunda Luxemburského. (Die Versuche Venezianische Republik um ein Attentat auf den ungarischen König Sigismund von Luxemburg). *Historický časopis*, 2000, 48, 2, p. 209-230. [Deutsche Zfassung].

5110. STEINHÜBEL (Ján). Uhorské kráľovstvo a Nitrianske kniežatstvo za vlády Štefana I. (Das ungarische Königtum und Nitrauer Fürstentum unter der Herrschaft von Stephan I). *Historický časopis*, 2000, 48, 1, p. 3-34. [Deutsche Zfassung].

5111. TAJTÁK (Ladislav). K niektorým otázkam interpretácie revolučných rokov 1848–1849 na Slovensku. (On some issues of interpretation of revolutionary years 1848–1849 in Slovakia). *Historický časopis*, 2000, 48, 1, p. 113-130.

5112. Židovská komunita na Slovensku. Obdobie autonómie. Porovnanie s vtedajšími udalosťami v Rakúsku. (Jewish community in Slovakia. The period of autonomy. The comparison with the events in Austria). Ed. by Eduard NIŽŇANSKÝ. Bratislava, Inštitút judaistiky Univerzity Komenského, 2000, 142 p.

Cf. nos 923, 8177

Slovenia

** 5113 Dokumenti ljudske revolucije v Sloveniji. Knj.1. Marec 1941–marec 1942. (Documents on Slovene people's revolution. Vol. 1. March 1941–March 1942). Uredniški odbor Tone FERENC [et al.]. Ljubljana, Inštitut za zgodovino delavskega gibanja, 2000, 353 p. FERENC (Tone)

** 5114. DRNOVŠEK (Darinka). Zapisniki politbiroja CK KPS/ZKS 1945–1954. (Minutes of the Politburo of the Slovene Communist Party Central Committee/ Slovene Communist Union 1945–1954). Ljubljana, Arhivsko društvo Slovenije, 2000, 386 p. (Viri / Arhivsko društvo Slovenije, 15).

** 5115. RAB – Arbe – Arbissima: confinement, police raids, internment in the Ljubljana Province: 1941–1943: documents. Compiled by Tone FERENC. Ljubljana, Društvo piscev zgodovine NOB = Society of the Writers of the History of the Liberation War in Slovenia, Inštitut za novejšo zgodovino = Institute for Contemporary History, 2000, 499 p.

5116. Brez milosti: ranjeni, invalidni in bolni povojni ujetniki na Slovenskem. (Without mercy: wounded, invalid, and sick prisoners in Slovenia after the Second World War). Uredil Lovro ŠTURM. Ljubljana, Nova revija, 2000, 382 p.

5117. GOW (James), CARMICHAEL (Cathie). Slovenia and the slovenes: a small state and the new Europe. London, Hurst & Company, 2000, XI-234 p.

5118. KACIN-WOHNIZ (Milica), PIRJEVEC (Jože). Zgodovina Slovencev v Italiji 1866–2000. (The history of Slovenes in Italy 1866–2000). Ljubljana, Nova revija, 2000, 336 p. (Zbirka Korenine / Nova revija).

5119. MREVLJE (Božidar). Evropa in slovenske dežele v obdobju 1815–1848: priročnik za učitelje zgodovine v osnovni šoli. (Europe and Slovene lands between 1815 and 1848). Ljubljana, Zavod Republike Slovenije za šolstvo, 2000, 99 p. (Modeli poučevanja in učenja. Zgodovina).

5120. Novejša vojaška zgodovina na Slovenskem: (posvet z dne 1. marca 2000). (Conference on the modern history of Slovene military, March 1, 2000). Zbrala in uredila Viktor KRAJNC in Zvezdan MARKOVIĆ. Ljubljana, Generalštab Slovenske vojske, Center za vojaškozgodovinsko dejavnost, 2000, 74 p.

5121. PLETERSKI (Janko). Avstrija in njeni Slovenci: 1945–1976. Ljubljana, Inštitut za narodnostna vprašanja = Institut for Ethnic Studies, 2000, 263 p. (Ethnicity, 4).

Cf. nos 7425, 7426, 7431

Somalia

5122. DRYSDALE (John Gordon Stewart). Stoics without pillows: a way forward for the Somalilands. London, HAAN Associates, 2000, X-203 p.

5123. FOX (Mary-Jane). Political culture in Somalia: tracing paths to peace and conflict. Uppsala, Uppsala University / Department of Peace and Conflict Research, 2000, 178 p. (Rapport / Uppsala Universitet, Avdelningen för freds- och konfliktforskning, 56 = Report / Uppsala University, Dept.of Peace and Conflict Research, 56).

South Africa

5124. ALEXANDER (Peter). Workers, war, and the origins of Apartheid: labour and politics in South Africa 1939–1948. Athens, Ohio U. P. a. Oxford, James Currey, 2000 X-214 p.

5125. Bad behaviour in urban South Africa. *Journal of African history*, 2000, 41, 2, p. 221-290. [Cf. nos <choice> 5130, 5920, 7181.]

5126. CARTON (Benedict). Blood from your children: the colonial origins of generational conflict in South Africa. Charlottesville, University Press of Virginia, 2000, XIII-215 p. (Reconsiderations in southern African history).

5127. CHRISTIE (Kenneth). The South African truth commission. New York, St. Martin's Press, 2000, XI-215 p.

5128. Comrades (From) to citizens: the South African civics movement and the transition to democracy. Ed. by Glenn ADLER and Jonny STEINBERG. Basingstoke, Macmillan, 2000, XVI-253p. (International political economy series).

5129. DREW (Allison). Discordant comrades: identities and loyalties on the South African left. Aldershot, Ashgate, 2000, VII-309p.

5130. KYNOCH (Gary). Politics violence in the 'Russian Zone': conflict in Newclare south, 1950–1957. *In*: Bad behaviour in urban South Africa [Cf. n° 5125], p. 267-290.

5131. MAWBY (Arthur Andrew). Gold mining and politics: Johannesburg, 1900–1907: the origins of the old South Africa? Lewiston a. Lampeter, Edwin Mellen Press, 2000, 2 vol., XVI-995 p. (ill.). (African studies, 58b).

5132. POHLANDT-MAC CORMICK (Helena). Controlling woman: Winnie Mandela and the 1976 Soweto uprising. *International journal of African historical studies*, 2000, 33, 3, p. 585-614.

5133. SEEKINGS (Jeremy). The UDF: a history of the United Democratic Front in South Africa, 1983–1991. Athens, Ohio U. P., 2000, XIII-371 p.

5134. South Africa's resistance press: alternative voices in the last generation under apartheid. Ed. by Les SWITZER and Mohamed ADHIKARI. Athens, Ohio University Center for international studies, 2000, XXIV-505 p. (Research in international studies, Africa series, 74).

5135. SPITZ (Richard), CHASKALSON (Matthew). The politics of transition: a hidden history of South Africa's negotiated settlement. Oxford, Hart Publishing, 2000, XVIII-461 p.

5136. Trade union and democratization in South Africa. Ed. by Glenn ADLER and Eddie WEBSTER. New York, St. Martin's Press, 2000, XVII-238 p.

5137. VAN KESSEL (Ineke). "Beyond our wildest dreams": the United Democratic Front and the transformation of South Africa. Charlottesville, University Press of Virginia, 2000, XVIII-367 p. (Reconsiderations in Southern African history).

Cf. n°s 4399, 4970

Spain

5138. AIZPURU (Mikel). El Partido Nacionalista Vasco en Guipúzcoa, 1893–1923: orígenes, organización y actuación política. Bilbao, Universidad del Pais Vasco, Servicio Editorial = Euskal Herriko Unibertsitatea, Argitalpen Zerbitzua, 2000, 510 p. (ill., maps). (Historia contemporánea).

5139. ALAMO MARTELL (María Dolores). El Capitán General de Canarias en el siglo XVIII. Prólogo del Dr. José Antonio ESCUDERO. Las Palmas de Gran Canaria, Universidad de Las Palmas de Gran Canaria, 2000, 301 p.

5140. ALTMAN (Ida). Transatlantic ties in the Spanish empire: Brihuega, Spain, and Puebla, Mexico, 1560–1620. Stanford, Stanford U. P., 2000, VIII-254 p.

5141. ANGUERA (Pere). Els precedents del catalanisme: catalanitat i anticentralisme, 1808–1868. Barcelona, Editorial Empuries, 2000, 360 p. (Biblioteca universal Empúries, 146).

5142. ARREGUI ZAMORANO (Pilar). Monarquía y señoríos en la Castilla moderna: los adelantamientos de Castilla, León y Campos, 1474–1643. [S. l.], Junta de Castilla y Leon, Consejeria de Educacion y Cultura, 2000, 390 p. (Estudios de historia).

5143. BALFOUR (Sebastiano), LA PORTE (Pablo). Spanish military cultures and the Moroccan wars, 1909–1936. *European history quarterly*, 2000, 30, 3, p. 307-333.

5144. BARRAGÁN MORIANA (Antonio). Córdoba, 1898/1905: crisis social y regeneracionismo político. Cordoba, Servicio de Publicaciones, Universidad de Cordoba, 2000, 239 p.

5145. BERCÉ (Yves-Marie), DURAND (Yves), LE FLEM (Jean-Paul). Les monarchies espagnole et française du milieu du XVIe siècle à 1714. Paris, SEDES, 2000, 320 p.

5146. BÖCKER (Manfred). Antisemitismus ohne Juden: die Zweite Republik, die antirepublikanische Rechte und die Juden; Spanien 1931 bis 1936. Frankfurt am Main, Lang, 2000, 392 p. (Hispano-Americana, 23).

5147. BUENO MADURGA (Jesús Ignacio). Zaragoza, 1917–1936: de la movilización popular y obrera a la reacción conservadora. Zaragoza, Institucion "Fernando el Catolico", C.S.I.C., Excma, Diputacion de Zaragoza, 2000, 370 p. (Publicación número 2.069 de la Institución "Fernando el Católico").

5148. CASTELLS (Irene), MOLINER (Antonio). Crisis del Antiguo Régimen y Revolución Liberal en España, (1789–1845). Barcelona, Editorial Ariel, 2000, 226 p. (Ariel practicum).

5149. CASTILLO (Santiago) [et al.]. Temps de postguerra: estudis sobre les comarques gironines (1939–1955). Girona, Cercle d'Estudis Historics i Socials, 2000, 264 p. (ill.). (Quaderns del Cercle, 16).

5150. CAZORLA SÁNCHEZ (Antonio). Las políticas de la victoria: la consolidación del Nuevo Estado franquista (1938–1953). Madrid, Marcial Pons, 2000, 266 p. (Historia).

5151. CLEMENTE (Josep Carles). Cuestiones carlistas y otras reflexiones históricas. Madrid, Editorial Fundamentos, 2000, 270 p. (Colección Ciencia. Serie Sociología, 244).

5152. Conflictos y represiones en el Antiguo Régimen. Valencia, Universitat de Valencia, 2000, 222 p. (Monografías y fuentes, 20).

5153. Corte (La) de Carlos V. Dir. por José MARTÍNEZ MILLÁN. Vol. 1–2. Corte y gobierno. Coordinadores, José MARTÍNEZ MILLÁN y Carlos Javier DE CARLOS MORALES. Vol. 3. Los consejos y los consejeros de Carlos V. Coordinador, Carlos Javier DE CARLOS MORALES. Vol. 4-5. Los servidores de las Casas Reales. Coordinador, Santiago FERNÁNDEZ CONTI. Madrid, Sociedad Estatal para la Conmemoración de los Centenarios de Felipe II y Carlos V, 2000, 5 vol., [s. p.]. (ill.).

5154. CRUZ (Jesus). Los notables de Madrid: las bases sociales de la revolución liberal española. Madrid, Alianza Editorial, 2000, 327 p. (Libro universitario. Historia y geografía. Ensayo, 150).

5155. Cuestión vasca (La): una mirada desde la historia. Ed. por Mercedes ARBAIZA VILALLONGA. Bilbao, Universidad del País Vasco, Serivio Editorial = Euskal Herriko Unibertsitatea, Argitalpen Zerbitzua, 2000, 250 p. [Trabajos presentados en el V Simposium de Historia, celebrado en Vitoria-Gasteiz en julio de 1999].

5156. DE LA FUENTE MONGE (Gregorio). Los revolucionarios de 1868: élites y poder en la España liberal. Madrid, Marcial Pons, Ediciones de Historia, 2000, 291 p. (Historia).

5157. DE LA GRANJA SAINZ (José Luis). El nacionalismo vasco, 1876–1975. Madrid, Arco/Libros, 2000, 95 p. (Cuadernos de historia, 81).

5158. DE RIQUER (Borja), CULLA I CLARÀ (Joan B.). El franquisme i la transició democràtica, 1939–1988. Nota preliminar de Pierre VILAR. Barcelona, Edicions 62, 2000, 471 p. (Història de Catalunya, 7).

5159. DEL RÍO ALDAZ (Ramón). Revolución liberal, expolios y desastres de la primera guerra carlista en Navarra y en el Frente del Norte. Pamplona, Gobierno de Navarra, Departamento de Educacion y Cultura, 2000, 442 p. (Serie Historia, 101).

5160. DUARTE (Angel), GABRIEL (Pere). El Republicanismo español. Madrid, Asociacion de Historia Contemporanea, Marcial Pons, Ediciones de Historia, 2000, 273 p. (Ayer, 39).

5161. Epoca (La) de la Restauración (1875–1902). Vol. 1. Estado, política e islas de ultramar. Vol. 2. Civilización y cultura. Coordinación e introducción por Manuel ESPADAS BURGOS. Madrid, Espasa Calpe, 2000, [s. p.]. (Historia de España, 36).

5162. ESCUDERO ANDÚJAR (Fuensanta). Lo cuentan como lo han vivido (república, guerra y represión en Murcia). Murcia, Universidad de Murcia, Servicio de Publicaciones, 2000, 300 p. (ill.).

5163. FERNÁNDEZ TERRICABRAS (Ignasi). Felipe II y el clero secular: la aplicación del Concilio de Trento. Madrid, Sociedad Estatal para la Conmemoracion de los Centenarios de Felipe II y Carlos V, 2000, 441 p. (Colección Historia).

5164. FEROS (Antonio). Kingship and favoritism in the Spain of Philip III, 1598–1621. Cambridge a. New York, Cambridge U. P., 2000, XVI-299 p. (ill.). (Cambridge studies in early modern history).

5165. FRANCH BENAVENT (Ricardo). La sedería valenciana y el reformismo borbónico. Valencia, Institucio Alfons el Magnanim, Diputacio de Valencia, 2000, 182 p. (Estudios universitarios, 85).

5166. FUSI AIZPURÚA (Juan Pablo). España: la evolución de la identidad nacional. Madrid, Ediciones Temas de Hoy, 2000, 309 p. (Historia, Temas de Hoy).

5167. GARCÍA SCHMIDT (Armando). Die Politik der Gabe: Handlungsmuster und Legitimationsstrategien der politischen Elite der frühen spanischen Restaurationszeit 1876–1902. Saarbrücken, Verlag für Entwicklungspolitik Saarbrücken, 2000, 202 p. (ill.). (Forschungen zu Spanien, 22).

5168. GIL PECHARROMÁN (Julio). "Sobre España inmortal, sólo Dios": José María Albinana y el Partido Nacionalista Español (1930–1937). Madrid, Universidad Nacional de Educacion a Distancia, 2000, 226 p. (Aula abierta, 36157).

5169. Haciendo historia: homenaje a Ma. Ángeles Larrea. Ed. por Rafael Ma. MIEZA Y MIEG y Juan GRACIA CÁRCAMO. Bilbao, Universidad del País Vasco, Servicio Editorial = Euskal Herriko Unibertsitatea, Argitalpen Zerbitzua, 2000, XIV-480 p.

5170. HARGREAVES (John). Freedom for Catalonia? Catalan nationalism, Spanish identity, and the Barcelona Olympic Games. New York a. Cambridge, Cambridge U. P., 2000, IX-178 p.

5171. HILLGARTH (Jocelyn N.). The mirror of Spain, 1500–1700. The formation of a myth. Ann Arbor, University of Michigan Press, 2000, 584 p.

5172. Historia (La) de ETA. Coor. de Antonio ELORZA; epílogo de Patxo UNZUETA. Madrid, Temas de Hoy, 2000, 447 p. (Historia).

5173. INIGO FERNÁNDEZ (Luis). La derecha liberal en la segunda república española. Madrid, Universidad Nacional de Educacion a Distancia, 2000, 634 p. (Aula abierta, 36149).

5174. Institucionismo y reforma social en España: el Grupo de Oviedo. Coor. de Jorge URÍA. Madrid, Talasa Ediciones, 2000, 330 p. (Agora, 11).

5175. Liberales, agitadores y conspiradores: biografías heterodoxas del siglo XIX. Coor. de Isabel BURDIEL y Manuel PÉREZ LEDESMA. Madrid, Espasa Calpe, 2000, 365 p. (ill.). (Espasa Biografías).

5176. LÓPEZ-CORDÓN (María Victoria), MARTÍNEZ DE SAS (María Teresa), PÉREZ SAMPER (María Angeles). La casa de Borbón: familia, corte y política. Madrid,

Alianza Editorial, 2000, 2 vol., 751 p. (ill.). (El libro de bolsillo, Historia, H 4191-H 4192).

5177. MAC CLANCY (Jeremy). The decline of Carlism. Reno, University of Nevada Press, 2000, XIX-349 p. (Basque series).

5178. MACARRO VERA (José Manuel). Socialismo, república y revolución en Andalucía. Sevilla, Universidad de Sevilla, Secretariado de Publicaciones, 2000, 497 p. (Serie Historia y geografía, 51).

5179. MADRAZO (Santos). Estado débil y ladrones poderosos en la España del siglo XVIII: historia de un peculado en el reinado de Felipe V. Madrid, Los Libros de la Catarata, 2000, 200 p.

5180. MARÍN I CORBERA (Martí). Catalanisme, clientelisme i franquisme: Josep Maria de Porcioles. Barcelona, Societat Catalana d'Estudis Historics, Institut d'Estudis Catalans, 2000, 129 p. (Institut d'Estudis Catalans. D'ahir per avui, 1).

5181. Memòria de la transició a Espanya i a Catalunya. Ed. por Rafael ARACIL i Antoni SEGURA. Barcelona, Edicions Universitat de Barcelona, 2000, [s. p.].

5182. Monarquía (La) de Felipe II a debate. Coordinador, Luis A. RIBOT GARCÍA. Madrid, Sociedad Estatal para la Conmemoración de los Centenarios de Felipe II y Carlos V, 2000, 527 p.

5183. MORADIELLOS (Enrique). La España de Franco (1939–1975): política y sociedad. Madrid, Editorial Sintesis, 2000, 319 p. (ill.). (Historia de España, 33).

5184. NÚÑEZ FLORENCIO (Rafael). Las Españas de 1898: de la guerra en ultramar a la crisis nacional. San Juan de Puerto Rico, Editorial LEA, 2000, 256 p. (Cuadernos del 98, 6).

5185. ORTIZ (David). Paper liberals: press and politics in Restoration Spain. Westport a. London, Greenwood Press, 2000, VIII-139 p. (Contributions to the study of world history, 73).

5186. PAYNE (Stanley G.). Fascism in Spain 1923–1977. Madison, University of Wisconsin Press, 2000, XII-601 p.

5187. PELAZ LÓPEZ (José-Vidal). Caciques, apóstoles y periodistas: medios de comunicación, poder y sociedad en Palencia, 1898–1939. Valladolid, Universidad de Valladolid, Secretariado de Publicaciones e Intercambio Editorial, Diputación Provincial de Palencia, Departamento de Cultura, 2000, 526 p. (Serie Historia y sociedad, 87).

5188. Pluma (La), la mitra y la espada: estudios de historia institucional en la edad moderna. Ed. por Juan Luis CASTELLANO, Jean Pierre DEDIEU y María Victoria LÓPEZ-CORDÓN. Madrid, Marcial Pons, Ediciones de Historia y Bordeaux, Universidad de Burdeos, 2000, 365 p. (Historia).

5189. RAMOS ROVI (María José). Andalucía en el Parlamento Español (1876–1902). Prólogo del Dr. D. José Manuel CUENCA TORIBIO. Córdoba, Publicaciones de la Universidad de Córdoba, Obra Social y Cultural Cajasur, 2000, 427 p. (Colección mayor).

5190. RODRÍGUEZ JIMÉNEZ (José Luis). Historia de Falange Española de las JONS [Juntas Ofensivas Nacional-Sindicalistas]. Madrid, Alianza Editorial, 2000, 552 p.

5191. ROMERO MAURA (Joaquín). La romana del diablo: ensayos sobre la violencia política en España, 1900–1950. Madrid, Marcial Pons, 2000, 252 p. (Historia. Estudios).

5192. SABATIER (Gérard), EDOUARD (Sylvène). Les monarchies de France et d'Espagne (milieu XVIe siècle–fin XVIIe siècle). Paris, A. Colin, 2000, 320 p.

5193. SALMERÓN GIMÉNEZ (Francisco Javier). Caciques murcianos: la construcción de una legalidad arbitraria, 1891–1910. Murcia, Universidad de Murcia, 2000, 310 p.

5194. SECO SERRANO (Carlos). Historia del conservadurismo español: una línea política integradora en el siglo XIX. Madrid, Temas de Hoy, 2000, 343 p. (Colección Historia).

5195. SEIDMAN (Michael). Agrarian collectives during the Spanish revolution and Civil War. *European history quarterly*, 2000, 30, 2, p. 209-236.

5196. SELLÉS I QUINTANA (Magda). El Foment del Treball Nacional, 1914–1923. Presentació de Jordi CASASSAS; pròleg de Francesc CABANA I VANCELLS. Barcelona, Abadía de Montserrat, 2000, 418 p. (Biblioteca Abat Oliba, 224).

5197. SESMERO CUTANDA (Enriqueta). Clases populares y carlismo en Bizkaia, 1850–1872. Bilbao, Universidad de Deusto, 2000, 253 p. (Serie Historia, 15).

5198. SMERDOU ALTOLAGUIRRE (Luis). Carlos IV en el exilio. Pamplona, Ediciones Universidad de Navarra, 2000, 357 p. (Astrolabio, 285, Serie Historia).

5199. Spanish history since 1808. Ed. by José Alvarez JUNCO and Adrian SHUBERT. London, Arnold, 2000, IX-389 p.

5200. SUÁREZ CORTINA (Manuel). El gorro frigio: liberalismo, democracia y republicanismo en la Restauración. Madrid, Biblioteca Nueva, Sociedad Menendez Pelayo, 2000, 375 p. (Colección Historia Biblioteca Nueva).

5201. TERMES (Josep). Història del catalanisme fins el 1923. Barcelona, Portic, 2000, 802 p. (Monografies, 10).

5202. TOWNSON (Nigel). The crisis of democracy in Spain: centrist politics under the Second Republic, 1931–1936. Brighton a. Portland, Sussex Academic Press, 2000, XV-444 p. (ill., maps).

Spain. Addenda 1999

5203. CRUZ ARTACHO (Salvador), COBO ROMERO (Francisco). Potere politico e Stato nella storia dell'Andalusia contemporanea. Verso una necessaria reinterpretazione storiografica del ruolo dei poteri locali nella costruzione politica della nazione (1890–1939). *Società e storia*, 99, 84, p. 359-396.

Cf. n°s 906, 947, 5010, 5011

Sri Lanka

5204. GOONERATNE (John). A decade of confrontation: Sri Lanka and India in the 1980s. Pannipitiya, Stamford Lake, 2000, 239 p.

5205. MELEGODA (Nayani Samarasinghe). The policies of three prime ministers of Ceylon: 1948–1956, with special reference to relations with Great Britain. Colombo, Wijesooriya Grantha Kendraya, 2000, 290 p.

Sweden

** 5206. LINDH (Björn), ULATE SEGURA (Bodil). Sverige i franska arkiv: guide till franska källor om svensk historia. (La Suède dans les archives françaises: guide des sources françaises sur l'histoire suédoise). Stockholm, Riksarkivet, 2000, 245 p. (Skrifter utgivna av Riksarkivet. 15).

5207. AGRELL (Wilhelm). Fred och Fruktan. Sveriges säkerhetspolitiska historia 1918–2000. (La paix et la peur. Histoire de la politique de sécurité suédoise depuis 1918). Lund, Historiska media, 2000, 368 p.

5208. ANSHELM (Jonas). Mellan frälsning och domedag. (Entre salut et jugement dernier: la politique atomique suédoise depuis 1945). Eslöv, Brutus Östlings bokförlag Symposion, 2000, 519 p.

5209. APPELQVIST (Örjan). Bruten brygga. Gunnar Myrdal och Sveriges ekonomiska efterkrigspolitik 1943–1947. (Gunnar Myrdal et la politique économique de l'après-guerre, 1943–1947). Stockholm, Santérus förlag, 2000, 502 p. (Sverige under det kalla kriget. 8).

5210. DAHLBERG (Hans). Hundra år i Sverige. (Cent ans en Suède). Albert Bonniers förlag, 2000, 384 p. (ill.).

5211. ENGLUND (Peter). Den oövervinnerlige. Om den svenska stormaktstiden och en man i dess mitt. (Charles X de Suède: le roi invincible et l'Époque de la grandeur suédoise). Stockholm, Atlantis, 2000, 797 p. (ill.).

5212. ENGMAN (Max). Lejonet och dubbelörnen. (Le lion et l'aigle bicéphale: le Grand duché de Finlande dans l'empire russe, 1830–1890). Stockholm, Atlantis, 2000, 402 p.

5213. HALL (Patrik). Den svenkaste historian. Nationalismen i Sverige under 6 sekler. (Une histoire très suédoise. Le nationalisme en Suède depuis six siècles). Stockholm, Carlssons, 2000, 328 p. (ill.).

5214. HUHTAMIES (Mikko). Sijaissotilasjärjestelmä ja väenotot. Taloudellis-sosiaalinen tutkimus sijaissotilaiden käytöstä Ala-Satakunan väenotoissa vuosina 1631–1648. (The substitute system and conscriptions in imperial Sweden. Ersatzmänner, Zwangaushebungen und Bauergemeinschaft im 17. Jahrhundert in Schweden). Helsinki, 2000, 199 p. (ill., maps).

5215. KURUNMÄKI (Jussi). Representation, nation and time: the political rhetoric of the 1866 parliamentary reform in Sweden. Jyväskyla, Jyväskylän Yliopisto, 2000, 253 p. (Jyväskyla studies in education, psychology and social research, 170).

5216. LEKEBY (Kjell). Kung Kristina, drottningen som ville byta kön. (Le 'Roi' Christine, la reine qui voulut changer de sexe). Stockholm, Vertigo förlag, 2000, 120 p.

5217. LILJEGREN (Bengt). Karl XII: en biografi. (Biographie du roi Charles XII de Suède). Lund, Historiska media, 2000, 442 p. (ill.).

5218. LINDER (Jan). Dans på slak lina: svensk neutralitetspolitik 1939–1999. (Exercice d'équilibriste: la politique de neutralité suédoise de 1939 à 1999). Stockholm, Infomanager, 2000, 224 p. (ill.).

5219. LUNDIN (Ingvar). Baltiska Judar. Fördrivna, förföljda, förintade. (Les Juifs baltes: expulsés, pourchassés, exterminés). Sävedalen, Warne förlag, 2000, 237 p. (ill.).

5220. NORDIN (Jonas). Ett fattigt men fritt folk. (Un peuple pauvre mais libre: l'identité nationale en Suède de 1660 à 1772). Eslöv, Brutus Östlings bokförlag Symposion, 2000, 528 p. [English summary].

5221. SCHÖNHAGEN (Anne Renate). "Die Brüder schauen sehnsüchtig nach einer ausgestreckten Bruderhand von der anderen Seite des Meeres": das Verhältnis der Schwedischen Staatskirche zum nationalsozialistischen Deutschland. Frankfurt am Main, Lang, 2000, 254 p. (Europäische Hochschulschriften / 31, 394).

5222. SÖDERBERG (Johan). Den moderna människans uppkomst och andra historiska uppsatser. (La naissance de l'homme moderne et autres essais d'histoire contemporaine suédoise). Stockholm, Carlssons, 2000, 250 p.

5223. VILLSTRAND (Nils Erik). Adaptation or protestation: local community facing the conscription of infantry for the Swedish armed forces, 1620–1679. *In*: Revolution (A) from above? [Cf. n° 4098], [s. p.].

Cf. n° 4947

Switzerland

5224. ALBERS-SCHÖNBERG (Heinz). Die Schweiz und die jüdischen Flüchtlinge 1933–1945: eine unabhängige Studie. Stäfa, Th. Gut, 2000, 255 p.

5225. GAL (Stéphane). Grenoble au temps de la Ligue: étude politique, sociale et religieuse d'une cité en crise (vers 1562–vers 1598). Préface de Denis CROUZET. Grenoble, Presses universitaires de Grenoble, 2000, 629 p. (Collection "La pierre et l'écrit").

5226. KOBI (Silvia). Des citoyens suisses contre l'élite politique: le cas des votations fédérales, 1979–1995. Paris, Harmattan, 2000, 306p (Collection Logiques politiques).

5227. Making (The) of modern Switzerland, 1848–1998. Ed. by Michael BUTLER, Malcolm PENDER and Joy CHARNLEY. Basingstoke, Macmillan, 2000, XIII-163 p. (New perspectives in German studies).

5228. MEYERHOFER (Ursula). Von Vaterland, Bürgerrepublik und Nation: nationale Integration in der Schweiz 1815–1848. Zürich, Chronos, 2000, 237 p.

5229. RAUBER (André). Histoire du mouvement communiste suisse. Vol. 2. Genève, Slatkine, 2000, [s. p.]. (Suisse ev événements, 2).

5230. STÄMPFLI (Regula). Die Nationalisierung der Schweizer Frauen. Frauenbewegung und Geistige Landesverteidigung 1933–1939. *Schweizerische Zeitschrift für Geschichte*, 2000, 50, p. 155-180.

5231. Suisse (La) et les réfugiés à l'époque du national-socialisme. Paris, Fayard, pour la Commission indépendante d'experts Suisse – Seconde Guerre mondiale, 2000, 471 p.

Syria

5232. DOUWES (Dick). The Ottomans in Syria: a history of justice and oppression. London, I. B. Tauris, 2000, VIII-244 p.

5233. MOUBAYED (Sami M.). Damascus between democracy and dictatorship. Lanham, University Press of America, 2000, XXV-212 p.

Tanzania

5234. ERIKSSON SKOOG (Gun). The soft budget constraint: the emergence, persistence, and logic of an institution. Boston a. London, Kluwer Academic, 2000, XIX-400 p.

5235. KEBEDE (John Admassu). The changing face of rural policy in Tanzania, from collectivism to capitalism. London, Minerva, 2000, 86 p, (ill.).

Cf. n° 4905

Thailand

5236. HAZRA (Kanai Lal). Political history. New Delhi, Decent Books, 2000, XIV-337 p. (Thailand: political history and Buddhist cultural influences, 1).

5237. Thailand: government and politics. Ed. by Verinder GROVER. New Delhi, Deep & Deep, 2000, X-346 p. (Government and politics of Asian countries, 15).

Trinidad and Tobago

5238. FIGUEIRA (Daurius). A spy in the houses of hate: the ironies and paradox of black on black racism in post colonial Trinidad and Tobago. Trinidad, D. Figueira, 2000, 148 p.

5239. MEIGHOO (Kirk Peter). Politics in Trinidad and Tobago, 1956–2000: toward an understanding of politics in a 'half-made society'. Hull, [s. n.], 2000, [s. p.]. [Thesis (Ph. D.) –University of Hull, 2000]

Turkey

* 5240. Atatürk ve Türkiye Cumhuriyeti konusunda yurtdışında yayınlanmış kitaplar bibliyografyası. (Atatürk and the Turkish Republic: Bibliography of books published abroad. / Atatürk et la Republique de Turquie: bibliographie des livres publies a l'etranger. / Atatürk und die Türkische Republik: Bibliographie der im Ausland erschienenen Bücher). Project director Azmi SÜSLÜ; yed. by Azmi SÜSLÜ [et al.]. Ankara, Atatürk Araştırma Merkezi, 2000, 885 p.

** 5241. 82 numaralı mühimme defteri (1026–1027/1617–1618): özet – transkripsiyon – indeks ve tıpkıbasım. (Register of outgoing imperial decrees under the number of 82 dated 1026–1027/1617–1618). Project director Yusuf SARINAY; Project executive Necati AKTAŞ [et al.]; ed. by Hacı Osman YILDIRIM [et al.]. Ankara, Devlet Arşivleri Genel Müdürlüğü Osmanlı Arşivi Daire Başkanlığı [General Directorate of State Archives], 2000, 176 p.

** 5242. İBN KEMAL (Kemalpaşazâde). Tevârih-i Âl-i Osman IV. Defter: Metin Ve Transkripsiyon. (History of the House of Osman, Fourth Register: text and transcription) Ed. by Koji IMAZAWA. Ankara, Türk Tarih Kurumu, 2000, 548 p.

** 5243. ILHAN (M. Mehdi). Amid (Diyarbakır): 1518 detailed register. Ankara, Turkish Historical Society, 2000, 201 p.

** 5244. KESKİN (Mustafa). Kayseri tarihinin kaynakları: H. 1247–1277 / M. 1831–1860 tarihli Kayseri nüfus müfredat defteri. (Sources of the History of Kayseri: Kayseri Census Register dated 1247–1277/ 1831–1860). Kayseri, Kayseri Büyükşehir Belediyesi, 2000, 232 p.

** 5245. KURT (Yılmaz). Çukurova tarihinin kaynakları IV: Adana evkaf defteri. (The sources of Çukurova History IV: Waqf Register of Adana) Ankara, Türk Tarih Kurumu, 2000, 139 p.

5246. AKŞİN (Sina). Essays in Ottoman-Turkish political history. İstanbul, Isis, 2000, 243 p.

5247. ÇAGLAR (Gazi). Staat und Zivilgesellschaft in der Türkei und im Osmanischen Reich. Frankfurt am Main u. Oxford, Lang, 2000, 610 p. (Europäische Hochschulschriften. Reihe XXXI, Politikwissenschaft = Publications universitaires européennes. Série XXXI, Sciences politiques = European university studies. Series XXXI, Political science, 415).

5248. ÇALIK (Ramazan). Alman kaynaklarına göre II. Abdülhamit devrinde Ermeni olayları. (Armenian incidents at the time of Abdülhamid II according to German sources). Ankara, Kültür Bakanlığı Yayımlar Dairesi Başkanlığı, 2000, 226 p.

5249. DAVAZ (Kemal Özcan). Atatürk, Bangladeş, Kazi Nazrul İslam. (Atatürk, Bangladesh, Islam). Ankara, Atatürk Araştırma Merkezi, 2000, 112 p.

5250. GÖKBEL (Ahmet). Kıpçak Türkleri: siyasî ve dinî tarihi. (The Kıpcak Turks: political and religious history). İstanbul, Ötüken yayınları, 2000, 384 p

5251. GÜLSOY (Ufuk). Osmanlı gayrimüslimlerinin askerlik serüveni. (Military experience of Ottoman non-Muslims). Ed. by Erhan AFYONCU and Gültekin YILDIZ. İstanbul, Simurg yayınları, 2000, 227 p.

5252. HANIOGLU (M. Sükrü). Preparation for a revolution: the Young Turks, 1902–1908. New York, Oxford U. P., 2000, XVI-538 p. (Studies in Middle Eastern history).

5253. KANSU (Aykut). Politics in post-revolutionary Turkey, 1908–1913. Leiden, Brill, 2000, VIII-521 p. (Social, economic, and political studies of the Middle East and Asia, 70).

5254. KIESER (Hans-Lukas). Der verpasste Friede: Mission, Ethnie und Staat in den Ostprovinzen der Türkei 1839–1938. Zürich, Chronos, 2000, 642 p.

5255. KIRIŞCI (Kemal). Disaggregating Turkish citizenship and immigration practices. *Middle eastern studies*, 2000, 36, 3, p. 1-22.

5256. LANGENSIEPEN (Bernd). 1828–1923 Osmanlı donanması. (The Ottoman navy 1828–1923). İstanbul, Kaptan Yayıncılık, 2000, 248 p.

5257. PARUSHEV (Parashkev). Atatiurk: diktatorut demokrat. (Atatjurk. Le dictateur démocrate). Sofiia, Akademichno izd-vo "Prof. Marin Drinov", 2000, 275 p. (ill.).

5258. ŞAŞMAZ (Musa). British policy and the application of reforms for the Armenians in Eastern Anatolia 1877–1897. Ankara, Turkish Historical Society, 2000, 306 p.

5259. Seventy-five years of the Turkish Republic. Ed. by Sylvia KEDOURIE. London a. Portland, OR, Frank Cass, 2000, 237 p.

5260. SHELDON (Garrett Ward). Jefferson and Atatürk: political philosophies. New York, P. Lang, 2000, XI-139 p. (Major concepts in politics and political theory, 21).

5261. SONYEL (Salâhi Ramadan). The Great War and the tragedy of Anatolia: Turks and Armenians in the maelstrom of major powers. Ankara, Turkish Historical Society, 2000, X-221 p. (Publications of Turkish Historical Society. Serial 16, 88).

5262. TURFAN (Mehmed Naim). Rise of the young Turks: politics, the military and Ottoman collapse. London, I. B. Tauris, 2000, XIX-490 p.

5263. YAVUZ (Celalettin). Osmanlı Bahriyesi'nde Yabancı Misyonlar: Çeşme Faciası'ndan Birinci Dünya Harbi'ne Kadar Osmanlı Bahriyesi'nde Çağdaşlaşma Gayretleri. (Foreign missions in Ottoman Navy: Modernization efforts from the Çeşme disaster to the First World War). İstanbul, [s. n.], 2000, 286 p.

Cf. n° 1282

Uganda

** 5264. BARRETT-GAINES (K.), KHADIAGALA (L.). Finding what you need in Uganda's archives. *History in Africa*, 2000, 27, p. 455-470.

5265. OCITTI (Jim). Political evolution and democratic practice in Uganda, 1952–1996. Lewiston a. Lampeter, E. Mellen Press, 2000, XIV-452 p. (African studies, 51).

5266. ODONGO (Onyango). A political history of Uganda: Yoweri Museveni's referendum 2000. London, WiCU, 2000, VIII-101 p.

5267. ROBERTSHAW (Peter), TAYLOR (David). Climate change and the rise of political complexity in Western Uganda. *Journal of African history*, 2000, 41, 1, p. 1-28.

5268. TRIPP (Aili Mari). Women and politics in Uganda. Madison, University of Wisconsin Press, 2000, XXVII-277 p.

Ukraine

5269. CHENTSOV (Viktor Vasil'evich). Political repressions in the Soviet Ukraine in the 20th. Ternopil', "Zbruch", 2000, 479 p, (ill.).

5270. Ukraine, renaissance d'un mythe national. Actes publiés sous la direction de Georges NIVAT, Vilen HORSKY et Miroslav POPOVITCH. Genève, Institut européen de l'Université de Genève, 2000, VIII-274 p. (Euryopa. Etudes, 10-2000).

5271. WILSON (Andrew). The Ukrainians: unexpected nation. New Haven a. London, Yale U. P., 2000, XVIII-366 p.

Cf. n°s 882, 891, 7967

United States

** 5272. Lincoln (With) in the White House: letters, memoranda, and other writings of John G. Nicolay, 1860–1865. Ed. by Michael BURLINGAME. Carbondale, Southern Illinois U. P., 2000, XXI-274 p.

** 5273. Papers (The) of Benjamin Franklin. May 1 through October 31, 1781. Ed. by Barbara B. OBERG. New Haven a. London, Yale U. P., 2000, 832 p. (ill.).

** 5274. Papers (The) of James Madison. 16 May–31 October, 1803 (Secretary of state series). Ed. by David B. MATTERN. Charlottesville, University Press of Virginia, 2000, XXXV-643 p.

** 5275. Papers (The) of Ulysses S. Grant. Vol. 23. February 1–December 31, 1872. Vol. 24. 1872. Ed. by John Y. SIMON. Carbondale, Southern Illinois U. P., 2000, 2 vol., XXIII-536 p., XXIII-557 p.

** 5276. Presidential documents: the speeches, proclamations and policies that have shaped the nation from Washington to Clinton. Ed. by J. F. WATTS and Fred L. ISRAEL. New York a. London, Routledge, 2000, XI-396 p.

** 5277. STODDARD (William Osborn). Inside the White House in war times: memoirs and reports of Lincoln's secretary. Ed. by Michael BURLINGAME. Lincoln, University of Nebraska Press, 2000, XXI-226 p.

5278. ADAMS (Willi Paul). Die USA im 20. Jahrhundert. München, Oldenbourg, 2000, XIV-296 p. (Oldenbourg Grundriß der Geschichte, 29).

5279. ALLGOR (Catherine). Parlor politics: in which the ladies of Washington help build a city and a government. Charlottesville a. London, University Press of Virginia, 2000, 299 p. (ill.). (Jeffersonian America).

5280. ALTSCHULER (Glenn C.), BLUMIN (Stuart M.). Rude republic: Americans and their politics in the nineteenth century. Princeton a. Oxford, Princeton U. P., 2000, XII-316 p. (ill.).

5281. American presidents (The): critical essays. Ed. by Melvin I. UROFSKY. New York a. London, Garland Pub., 2000, XIII-530 p. (ill.).

5282. American radicalism. Ed. by Daniel POPE. Oxford a. Malden, Blackwell, 2000, [s. p.]. (Blackwell readers in American social and cultural history).

5283. American social and political movements, 1900–1945: pursuit of progress. Ed. by by Robert J. ALLISON. Detroit a. London, St. James Press, 2000, XXII-293 p. (ill.). (History in dispute, 3). – American social and political movements, 1945–2000: pursuit of liberty. Ed. by by Robert J. ALLISON. Detroit a. London, St. James Press, 2000, XXIV-323 p. (ill.). (History in dispute, 2).

5284. Amerikanskie issledovanija v Sibiri. (American Studies in Siberia). Vol. 4. Materialy regional'noj nauchnoj konferentsii (Proceedings of a Regional Scientific Conference), 10–11 December 1999. Ed. Mikhail Ja. PELIPAS'. Tomsk, Izd-vo Tomskogo un-ta, 2000, 145 p.

5285. APPLEBY (Joyce). Inheriting the revolution: the first generation of Americans. Cambridge, Belknap Press of Harvard U. P., 2000, VIII-322 p.

5286. Articulating America: fashioning a national political culture in early America: essays in honor of J. R. Pole. Ed. by Rebecca STARR. Lanham a. Oxford, Rowman & Littlefield, 2000, XI-276 p.

5287. AYERS (Edward L.), [et al.]. American passages: a history of the United States. Fort Worth a. London, Harcourt College Publishers, 2000, XXXI-1147 p. (ill., maps).

5288. BENNETT (G. H.). The American presidency 1945–2000: illusions of grandeur. Stroud, Sutton, 2000, XIV-274 p.

5289. BENNETT (Lerone). Forced into glory: Abraham Lincoln's white dream. Chicago, Johnson Pub. Co., 2000, 652 p.

5290. BRADLEY (Richard). American political mythology from Kennedy to Nixon. New York a. Oxford, Peter Lang, 2000, 267 p. (ill.). (Modern American history, 3).

5291. BROWN (William O.), BURDEKIN (Richard C. K.). Turning points in the U.S. civil war: a British perspective. *Journal of economic history*, 2000, 60, 1, p. 216-231.

5292. BUTLER (Jon). Becoming American: the revolution before 1776. Cambridge, Harvard U. P., 2000, X-324 p.

5293. BYRNES (Mark S.). The Truman years, 1945–1953. Harlow a. New York, Longman, 2000, XVII-162 p. (ill., maps). (Seminar studies in history).

5294. CHAN (Sewell). No race to look to: the politics of U.S. immigration reform, 1964–1965. Oxford, [s. n.], 2000, 110 p. [Thesis (M. Phil.) – University of Oxford, 2000].

5295. CHERTINA (Zoja S.). Plavil'nyj kotel? Paradigmy etnicheskogo razvitija SShA. (A smelting furnace? Paradigms of ethnic development of the USA). RAN, In-t vseobshchej istorii. Moskva, [s. n.], 2000, 163 p.

5296. CLIFT (Eleanor), BRAZAITIS (Tom). Madam President: shattering the last glass ceiling. New York a. London, Scribner, 2000, 349 p. (ill.).

5297. Cold war constructions: the political culture of United States imperialism, 1945–1966 Ed. by Christian G. APPY. Amherst, University of Massachusetts Press, 2000, 340 p. (ill.). (Culture, politics, and the cold war).

5298. Continuity and change in House elections. Ed. by David W. BRADY, John F. COGAN and Morris P. FIORINA. Stanford, Stanford U. P., Hoover Institution Press, 2000, XV-297 p.

5299. DANIEL (Pete). Lost revolutions: the South in the 1950s. Chapel Hill, University of North Carolina Press, for the Smithsonian National Museum of American History, 2000, XII-378 p.

5300. DIGGINS (John P.). On hallowed ground: Abraham Lincoln and the foundations of American history. New Haven a. London, Yale U. P., 2000, XXI-330 p.

5301. DOLLINGER (Marc). Quest for inclusion: Jews and liberalism in modern American. Princeton a. Oxford, Princeton U. P., 2000, 296 p. (ill.).

5302. DUDZIAK (Mary L.). Cold War civil rights: race and the image of American democracy. Princeton a. Chichester, Princeton U. P., 2000, XII-330 p. (ill.). (Politics and society in twentieth-century America).

5303. DUNN (Charles W.). The scarlet thread of scandal: morality and the American presidency. Lanham a. Oxford, Rowman & Littlefield, 2000, 209 p.

5304. EDLING (Max M.). A revolution in favour of government: the American constitution and ideas about state formation, 1787–1788. Stockholm, Stockholm University, Dept. of Political Science, 2000, X-321 p. (Stockholm studies in politics, 71).

5305. EDSFORTH (Ronald). The New Deal. America's response to the Great Depression. Malden a. Oxford, Blackwell, 2000, 324 p.

5306. EISNER (Marc Allen). From warfare state to welfare state: World War I, compensatory state-building, and the limits of the modern order. University Park, Pennsylvania State U. P., 2000, 371 p.

5307. EVANS (Harold), BUCKLAND (Gail), BAKER (Kevin). The American century. New York, Alfred A. Knopf, 2000, XXIII-710 p. (ill.).

5308. FERLING (John). Setting the world ablaze: Washington, Adams, Jefferson, and the American Revolution. New York a. Oxford, Oxford U. P., 2000, XXIV-392 p.

5309. FREEMAN (Jo). A room at a time: how women entered party politics. Lanham a. Oxford, Rowman & Littlefield, 2000, XII-353 p.

5310. FRIEDBERG (Aaron L.). In the shadow of the garrison state: America's anti-statism and its Cold War grand strategy. Priceton a. Chichester, Princeton U. P., 2000, XIV-362 p. (ill.). (Princeton studies in international history and politics).

5311. GARGARELLA (Roberto). The scepter of reason: public discussion and political radicalism in the origins of constitutionalism. Dordrecht a. Boston, Kluwer Academic, 2000, XXVIII-140 p. (Law and philosophy library, 48).

5312. George Washington and the origins of the American presidency. Ed. by Mark J. ROZELL, William D. PEDERSON and Frank J. WILLIAMS. Westport, Praeger, 2000, XV-210 p.

5313. GLAUDE (Eddie S.). Exodus! Religion, race, and nation in early nineteenth-century Black America. Chicago a. London, University of Chicago Press, 2000, X-216 p.

5314. GRANT (Susan-Mary). North over South: northern nationalism and American identity in the antebellum era. Lawrence, University Press of Kansas, 2000, XIII-250 p.

5315. GREENBAUM (Fred). Men against myths: the progressive response. Westport a. London, Praeger, 2000, 223 p.

5316. GREENE (Jack P.). The American Revolution. *In*: Revolutions in the Americas [Cf. n° 4099], p. 93-102.

5317. GREENSTEIN (Fred I.). The presidential difference: leadership style from FDR [Franklin Delano Roosevelt] to Clinton. New York, Martin Kessler Books, 2000, V-282 p. (ill.).

5318. HANNAH (Matthew G.). Governmentality and the mastery of territory in nineteenth- century America. Cambridge, Cambridge U. P., 2000, XIII-245 p. (ill.). (Cambridge studies in historical geography, 32).

5319. HERMAN (Arthur). Joseph McCarthy: re-examining the life and legacy of America's most hated senator. New York a. London, Free Press, 2000, 404 p. (ill.).

5320. HORSMAN (Reginald). The new republic: the United States of America, 1789–1815. Harlow a. New York, Longman, 2000, IX-275 p. (ill.). (The Longman history of America).

5321. ISSERMAN (Maurice), KAZIN (Michael). America divided: the civil war of the 1960s. Oxford, Oxford U. P., 2000, X-358 p.

5322. ITON (Richard). Solidarity blues: race, culture, and the American left. Chapel Hill a. London, University of North Carolina Press, 2000, XV-335 p.

5323. JAFFA (Harry V.). A new birth of freedom: Abraham Lincoln and the coming of the Civil War. Lanham a. Oxford, Rowman & Littlefield Publishers, 2000, XIV-550 p.

5324. JAMES (Scott Curtis). Presidents, parties, and the state: a party system perspective on Democratic regulatory choice, 1884–1936. Cambridge, Cambridge U. P., 2000, X-307 p. (ill.).

5325. JENNINGS (Francis). The creation of America: through revolution to empire. Cambridge a. New York, Cambridge U. P., 2000, XII-340 p.

5326. KARAAGAC (John). Between promise and policy: Ronald Reagan and conservative reformism. Lanham, Lexington Books, 2000, [s. p.].

5327. KERSH (Rogan). Dreams of a more perfect union. Ithaca, Cornell U. P., 2000, XI-358 p.

5328. KEYSSAR (Alexander). The right to vote: the contested history of democracy in the United States. New York, BasicBooks, 2000, XXIV-467 p.

5329. KING (Desmond). Making Americans: immigration, race, and the origins of the diverse democracy. Cambridge, Harvard U. P., 2000, X-388 p.

5330. KLEINMAN (Mark L.). A world of hope, a world of fear: Henry A. Wallace, Reinhold Niebuhr, and American liberalism. Columbus, Ohio State U. P., 2000, XVII-370 p.

5331. KOHUT (Andrew), [et al.]. The diminishing divide: religion's changing role in American politics. Washington, Brookings Institution Press, 2000, XI-178 p. (ill.).

5332. KORNBLUH (Mark Lawrence). Why America stopped voting: the decline of participatory democracy and the emergence of modern American politics. New

York a. London, New York U. P., 2000, XV-243 p. (ill.). (American social experience series).

5333. KRYDER (Daniel). Divided arsenal: race and the American state during World War II. New York a. Cambridge, Cambridge U. P., 2000, XV-301 p. (ill., maps).

5334. LAMMERS (William W.), GENOVESE (Michael A.). The presidency and domestic policy: comparing leadership styles, FDR [Freanklin Delano Roosevelt] to Clinton. Washington, CQ Press, 2000, XII-383 p.

5335. LEAB (Daniel J.). I was a communist for the F.B.I: the unhappy life and times of Matt Cvetic. University Park, Pennsylvania State U. P., 2000, XII-170 p. (ill., ports).

5336. LIEBERMAN (Robbie). The strangest dream: communism, anticommunism, and the U.S. Peace Movement, 1945–1963. Syracuse, Syracuse U. P., 2000, XVII-244 p. (Syracuse Studies on Peace and Conflict Resolution).

5337. Lincoln's side (At): John Hay's Civil War correspondence and selected writings. Ed. by Michael BURLINGAME. Carbondale, Southern Illinois U. P., 2000, XXVII-294 p. (ill.).

5338. LIPSET (Seymour Martin), MARKS (Gary). It didn't happen here: why socialism failed in the United States. New York, W. W. Norton, 2000, 379 p.

5339. MAC WILLIAMS (Wilson C.). Beyond the politics of disappointment? American elections, 1980–1998. New York a. London, Chatham House Publishers/Seven Bridges Press, 2000, VIII-165 p.

5340. MALSBERGER (John W.). From obstruction to moderation: the transformation of Senate conservatism, 1938–1952. Cranbury, Susquehanna U. P. a. London, Associated University Presses, 2000, 320 p.

5341. MIDDLETON (Catherine M.). Opposition to Indian removal and the emergence of the second party system in the United States, 1828–1834. Oxford, [s. n.], 2000, 372 p. [Thesis (D. Phil.) – University of Oxford, 2000].

5342. MITCHELL (Gordon R.). Strategic deception: rhetoric, science, and politics in missile defense advocacy. East Lansing, Michigan State U. P., 2000, XIX-390 p. (ill., map, ports.). (Rhetoric and public affairs series).

5343. MOORE (Nina M.). Governing race: policy, process, and the politics of race. Westport a. London, Praeger, 2000, XXVI-216 p. (ill.).

5344. MOREL (Lucas E.). Lincoln's sacred effort: defining religion's role in American self-government. Lanham, Lexington Books, 2000, [s. p.].

5345. NEIBERG (Michael S.). Making citizen soldiers: ROTC and the ideology of American military service. Cambridge, Harvard U. P., 2000, VIII-264 p.

5346. Neither separate nor equal: Congress in the 1790s. Ed. by Kenneth R. BOWLING and Donald R. KENNON. Athens, Ohio U. P. for the United States Capitol Historical Society, 2000, XI-344 p. (ill.). (Perspectives on the history of Congress, 1789–1801).

5347. NICOLSON (Colin). The "infamas Govener" Francis Bernard and the origins of the American Revolution. Boston, Northeastern U. P., 2000, XIII-326 p. (ill.).

5348. ONUF (Peter S.). Jefferson's empire: the language of American nationhood. Charlottesville a. London, University Press of Virginia, 2000, XI-250 p. (Jeffersonian America).

5349. ØVERLAND (Orm). Immigrant minds, American identities: making the United States home, 1870–1930. Urbana a. Chicago, University of Illinois Press, 2000, X-243 p. (Statue of Liberty-Ellis Island centennial series).

5350. PISTON (William Gerret), HATCHER (Richard W. III). Wilson's Creek: the second battle of the Civil War and the men who fought it. Chapell Hill, University of North Carolina Press, 2000, XIX-408 p. (Civil war Americas).

5351. POLENBERG (Richard). The era of Franklin D. Roosevelt, 1933–1945: a brief history with documents. Boston, Bedford/St.Martin's, 2000, XIV-251 p. (Bedford series in history and culture).

5352. POWE (Lucas A., jr.). The Warren Court and American politics. Cambridge, Belknap Press of Harvard U. P., 2000, XVI-566 p.

5353. READ (James H.). Power versus liberty: Madison, Hamilton, Wilson, and Jefferson. Charlottesville, University Press of Virginia, 2000, XI-201 p.

5354. RICHARDS (Leonard L.). The slave power: the free North and southern domination, 1780–1860. Baton Rouge, Louisiana State U. P., 2000, X-228 p.

5355. Richmond campaign (The) 1862: the Peninsula and the Seven Days. Ed. by Gary W. GALLAGHER. Chapell Hill, University of North Carolina Press, 2000, XV-272 p. (Military campaigns of the Civil War).

5356. RUDGERS (David F.). Creating the secret state: the origins of the Central Intelligence Agency, 1943–1947. Lawrence, University Press of Kansas, 2000, 244 p.

5357. SCHMIDT (Regin). Red scare: FBI and the origins of anticommunism in the United States, 1919–1943. København, Museum Tusculanum Press, University of Copenhagen, 2000, 391 p.

5358. SCHUH (Anna Marie). Timing successful policy change: lessons from the civil service. Lanham, University Press of America, 2000, XIII-202 p. (ill.).

5359. STEPHAN (Alexander). "Communazis": FBI surveillance of German emigré writers. New Haven a. London, Yale U. P., 2000, 384 p. (ill.).

5360. STIVERS (Camilla). Bureau men, settlement women: constructing public administration in the progressive era. Lawrence, University Press of Kansas,

2000, XI-187 p. (Studies in government and public policy).

5361. STROUTHOUS (Andrew). US labor and political action, 1918–1924: a comparison of independent political action in New York, Chicago and Seattle. Basingstoke, Macmillan Press a. St. Martin's Press, 2000, XI-208 p.

5362. SUMMERS (Mark Wahlgren). Rum, Romanism and rebellion: the making of a president, 1884. Chapel Hill, University of North Carolina Press, 2000, XV-377 p.

5363. SYKES (Patricia Lee). Presidents and prime ministers: conviction politics in the Anglo-American tradition. Lawrence, University Press of Kansas, 2000, XIII-399 p.

5364. TEIXEIRA (Ruy A.), ROGERS (Joel). America's forgotten majority: why the white working class still matters. New York, BasicBooks, 2000, XIV-215 p.

5365. TESTI (Arnaldo). Trionfo e declino dei partiti politici negli Stati Uniti, 1860–1930. Torino, Otto, 2000, 208 p.

5366. WEIGLEY (Russell Frank). A great Civil War: a military and political history, 1861–1865. Bloomington, Indiana U. P., 2000, [s. p.].

5367. WEISBERGER (Bernard A.). America afire: Jefferson, Adams, and the revolutionary election of 1800. New York, William Morrow, 2000, VI-345 p.

5368. WHITBY (Kenny J.). The color of representation: congressional behavior and Black interests. Ann Arbor, University of Michigan Press, 2000, 189 p. (ill.).

5369. WOLFENSBERGER (Donald R.). Congress and the people: deliberative democracy on trial. Washington, Woodrow Wilson Center Press a. Baltimore a. London, Johns Hopkins U. P., 2000, XI-308 p. (ill.).

5370. Women and the unstable state in nineteenth-century America. Ed. by Alison M. PARKER and Stephanie COLE. College Station, Texas A&M U. P., 2000, X-164 p. (Walter Prescott Webb memorial lectures, 33).

Cf. n°s 556, 4910, 5103, 5260, 5623, 7747, 7833, 8374, 8375

Uruguay

5371. ALDRIGHI (Clara) [et al.]. Antisemitismo en Uruguay: Raíces, discursos, imágenes, 1870–1940. Prólogo Teresa PORZECANSKI. Montevideo, Ediciones Trilce, 2000, 224 p. (Colección Desafíos).

5372. GIMÉNEZ RODRÍGUEZ (Alejandro). Por mi honor: vida de los presidentes uruguayos. Montevideo, Arca, 2000, 192 p.

5373. JOHNSON (Charlotte Nicola). 'The right to have rights': gender politics, citizenship and the state in Uruguay. London, [s. n.], 2000, 381 p. [Thesis (Ph. D) – University of London, 2000]

5374. PELÚAS (Daniel) [et al.]. Coparticipación y coalición: 164 años de acuerdo entre Blancos y Colorados. Montevideo, arca-HUMUS, 2000, 119 p. (Humus, 1).

5375. ROCK (David), LÓPEZ-ALVES (Fernando). State-building and political systems in nineteenth-century Argentina and Uruguay. *Past and present*, 2000, 167, p. 176-202.

5376. Uruguayos (Los) del centenario: nación, ciudadanía, religión y educación (1910–1930). Coordinador, Gerardo CAETANO. Montevideo, Ediciones Santillana, Taurus, 2000, 275 p.

Venezuela

5377. CABALLERO (Manuel). La gestación de Hugo Chávez: 40 años de luces y sombras en la democracia venezolana. Madrid, Libros de la Catarata, 2000, 167 p. (Serie Desarrollo y cooperación, 106).

5378. CHARIER (Alain). Le mouvement noir au Venezuela: revendication identitaire et modernité. Préface d'Yvon LE BOT. Paris, L'Harmattan, 2000, 350 p. (Collection recherches et documents. Amériques latines. Recherches & documents. Amérique latine).

5379. GÓMEZ (Carlos Alarico). Los sesenta: historia de la hegemonía andina, 1899–1945. [S. l.], Editorial Plain Art, 2000, 317 p.

5380. GOTT (Richard). In the shadow of the liberator: Hugo Chávez and the transformation of Venezuela. London, Verso, 2000, VIII-246 p.

5381. IRWIN G. (Domingo). Relaciones civiles-militares en el siglo XX. Caracas, El Centauro Ediciones, 2000, 224 p.

5382. LANGUE (Frédérique). Aristocrátas, honor y subversión en la Venezuela del siglo XVIII. Caracas, Academia Nacional de la Historia, 2000, 339 p. (Biblioteca de la Academia Nacional de la Historia, 252. Fuentes para la historia colonial de Venezuela).

5383. MEZA DORTA (Norelky). Estudio sobre relaciones compartidas en poblados rurales, 1919–1939. Barinas, Ediciones de la Universidad Ezequiel Zamora, 2000, 143 p. (Colección Ciencias sociales. Colección Docencia universitaria).

5384. PÉREZ LECUNA (Roberto). Apuntes para la historia militar de Venezuela: 1º de Enero de 1936, 18 de Octubre de 1945. [S. l.], Editorial El Viaje de Pez, 2000, 1243 p.

5385. PINO ITURRIETA (Elías). Fueros, civilización y ciudadania: estudios sobre el siglo XIX en Venezuela. Caracas, Universidad Católica Andrés Bello, 2000, 198 p.

5386. PLAZA (Elena). Versiones de la tiranía en Venezuela: el último régimen del General José Antonio Páez, 1861–1863. [S. l.], Facultad de Ciencias Jurídicas y Políticas, Universidad Central de Venezuela, 2000, 253 p.

5387. STRAKA (Tomás). La voz de los vencidos: ideas del Partido Realista de Caracas, 1810–1821. Caracas, Comisión de Estudios de Postgrado, Facultad de Humanidades y Educación-Universidad Central de Venezuela, 2000, XIII-262 p. (Colección Monografías).

5388. URQUIJO GARCÍA (José Ignacio). El movimiento obrero de Venezuela. Lima, OIT y Caracas, UCAB, INAESIN, 2000, 264 p.

Vietnam

5389. CHAPUIS (Oscar). The last emperors of Vietnam: from Tu Duc to Bao Dai. Westport a. London, Greenwood Press, 2000, XIII-185 p. (Contributions in Asian studies, 7).

5390. ELLIOTT (David W. P.). The Vietnamese war: revolution and social change in the Mekong Delta, 1930–1975. Armonk, M.E. Sharpe, 2000, 2 vol., [s. p.]. (Pacific Basin Institute book).

Yugoslavia

5391. Nacionalni aspekt kadrova u BiH 1945–1991 = Ethnical structure of management personnel in Bosnia and Hercegovina between 1945 and 1991. Uredio Arif ZULIC. Sarajevo, Vijece Kongresa bosnjackih intelektualaca, 2000, 107 p. (Biblioteka Posebna izdanja, 66).

5392. NIKOLIC (Kosta). Komunisti u Kraljevini Jugoslaviji: od socijal-demokratije do stalinizma: 1919–1941. Gornji Milanovac, LIO a. Beograd, Centar za savremenu istoriju Jugoistocne Evrope, 2000, 195 p. (Biblioteka Monografija, 2).

5393. PALMER (Peter). The Communists and the Roman Catholic Church in Yugoslavia, 1941–1946. Oxford, [s. n.], 2000, 320 p. [Thesis (D.Phil.). University of Oxford, Faculty of Modern History].

Zanzibar

5394. Political plight (The) of Zanzibar. Ed. by T. L. MALIYAMKONO. Dar es Salaam, TEMA Publishers, 2000, XIV-255 p.

Zimbabwe

5395. MAKUMBE (John), COMPAGNON (Daniel). Behind the smokescreen: the politics of Zimbabwe's 1995 general elections. Harare, University of Zimbabwe Publications, 2000, VIII-340 p.

L

MODERN RELIGIOUS HISTORY

§ 1. General. 5396-5459. – § 2. Roman Catholicism (*a*. General; *b*. History of the Popes; *c*. Special studies; *d*. Religious orders; *e*. Missions). 5460-5611. – § 3. Orthodox Church. 5612-5619. – § 4. Protestantism. 5620-5678. – § 5. Non-Christian religions and sects. 5679-5728.

§ 1. General.

5396. Acceptation (L') de l'autre: de l'Edit de Nantes à nos jours [Actes du colloque national organisé les 16 et 17 décembre 1998 à Paris au Carrousel du Louvre]. Sous la dir. de Jean DELUMEAU. Paris, Fayard, 2000, 246 p.

5397. Adventure (The) of religious pluralism in early modern France: papers from the Exeter conference, April 1999. Ed. by Keith CAMERON, Mark GREENGRASS and Penny ROBERTS. Oxford a. New York, Peter Lang, 2000, 321 p.

5398. ALLIEVI (Stefano), BIDUSSA (David), NASO (Paolo). Il libro e la spada. Le sfide dei fondamentalismi religiosi; ebraismo, cristianesimo, islam. Torino, Claudiana, 2000, 205 p.

5399. Anticlericalism in Britain, c. 1500-1914. Ed. by Nigel ASTON and Matthew CRAGOE. Phoenix Mill, Sutton Pub., 2000, XXII-225 p.

5400. ASTON (Nigel). Religion and revolution in France, 1780–1804. Washington D.C., Catholic University of America Press, 2000, XII-435 p.

5401. Aufklärung, Revolution, Restauration (1750–1830). Hrsg. v. Bernard PLONGERON und Thomas BREMER. Freiburg, Herder, 2000, XXIII-880 p. (Die Geschichte des Christentums, 10)

5402. BAUBÉROT (Jean) - ZUBER Valentine) Une haine oubliée: l'antiprotestantisme avant le «pacte laïque», 1870–1905. Paris, A. Michel, 2000, 332 p.

5403. BECHT (Michael). Pium consensum tueri. Studien zum Begriff consensus im Werk von Erasmus von Rotterdam, Philipp Melanchthon und Johannes Calvin. Münster, Aschendorff, 2000, XIV-590p.

5404. BELLINI (Piero). La coscienza del principe. Prospettazione ideologica e realtà politica delle interposizioni prelatizie nel governo della cosa pubblica. Torino, Giappichelli, 2000, 2 vol., [s. p.].

5405. BÉRENGER (Jean). Tolérance ou paix de religion en Europe centrale. 1415–1792. Paris, Champion, 2000, 282 p.

5406. BESANÇON (Alain). L'image interdite: une histoire intellectuelle de l'iconoclasme. Paris, Gallimard, 2000, 722 p.

5407. BEYER (Renate). Interreligiöser Dialog-Schlagwort oder Chance? Christentum-Islam-Buddhismus. Gütersloh, Gütersloher Verl.-Haus, 2000, 127 p.

5408. BLASCHKE (Olaf). Das 19. Jahrhundert: ein Zweites Konfessionelles Zeitalter? *Geschichte und Gesellschaft*, 2000, 26, 1, p. 38-75.

5409. BLÜCKERT (Kjell). The church as nation: a study in ecclesiology and nationhood. Frankfurt am Main a. New York, P. Lang, 2000, 363 p.

5410. BORRUSO (Paolo). L'ultimo impero cristiano: religione e nazione nell'Etiopia di Hailé Selassie (1916–1974). Roma, [s. n.], 2000, X-277 p.

5411. BURKARD (Dominik). Staatskirche Papstkirche, Bischofskirche: die Frankfurter Konferenzen und die Neuordnung der Kirche in Deutschland nach der Säkularisation. Roma, Herder, 2000, 832 p.

5412. CAFFIERO (Marina). Religione e modernità in Italia (secoli XVII–XIX). Pisa e Roma, IEPI, 2000, 301 p.

5413. České církevní dějiny ve druhé polovině 20. století. Sborník příspěvků ze sekce církevních dějin na VIII. sjezdu českých historiků. (Czech church history in the second half of the 20th century). Ed. Libor JAN. Brno, Centrum pro studium demokracie a kultury, 2000, 119 p. (Historia ecclesiastica, 5).

5414. COLONOMOS (Ariel). Eglises en réseaux. Trajectoires politiques entre Europe et Amerique. Paris, Presses de sciences, 2000, 315 p.

5415. COSTA I BOU (Joan). Nacio i nacionalismes: una reflexio en el marc del magisteri pontifici contemporanei. Barcelona, Euroeditors, 2000, 254 p.

5416. Cristianità ed Europa. Miscellanea di studi in onore di Luigi Prosdocimi. A cura di Cesare ALZATI. Roma, Herder, 2000, 399 p.

5417. DAVIE (Grace). Religion in modern Europe: a memory mutates. Oxford, Oxford U. P., 2000, VIII-218 p.

5418. DEOL (Harnik). Religion and nationalism in India: the case of the Punjab. London a. New York, Routledge, 2000, 200 p.

5419. DUIJZINGS (Ger). Religion and the politics of identity in Kosovo. New York, Columbia U. P., 2000, XV-238 p.

5420. Early modern witches. Witchcraft cases in contemporary writing. Ed. By Marion GIBSON. London, Routledge, 2000, XIV-338 p.

5421. FIORE (Fabio). Fede e tolleranza: problemi religiosi e riforme della cristianità in Europa fra XVI e XVII secolo. Torino, Paravia, 2000, 125 p.

5422. FRIEDER (Ludwig). Zwischen Kolonialismuskritik und Kirchenkampf. Interaktionen afrikanischer, indischer und europäischer Christen während der Weltmissionskonferenz in Tambaram 1938. Göttingen, Vandenhoeck & Ruprecht, 2000, 352 p.

5423. GALIMBERTI (Umberto). Orme del sacro: il cristianesimo e la desacralizzazione del sacro. Milano, Feltrinelli, 2000, 356 p.

5424. Gottes ist der Orient! Gottes ist der Occident! Goethe und die Religionen der Welt. Hrsg. v. Wolfgang BEUTIN. Frankfurt am Main, Lang, 2000, 225 p.

5425. GREYERZ (Kaspar). Religion und Kultur: Europa, 1520-1820. Göttingen, Vandenhoeck & Ruprecht, 2000, 395 p.

5426. HAZELGROVE (Jenny). Spiritualism and British society between the wars. Manchester, Manchester U. P., 2000, 294 p.

5427. HOCHLEITNER (Janusz). Religijność potrydencka na Warmii (1551–1655). (Die Religiosität im Ermland (Warmia) nach dem Trienter Konzil, 1551–1655). Olsztyn, Wydaw. Uniw. Warmińsko-Mazurskiego, 401 p.

5428. HOMZA (Lu Ann). Religious authority in the Spanish Renaissance. Baltimore a. London, Johns Hopkins U. P., 2000, XXIII-312 p.

5429. Johannes a Lasco (1499–1560): polnischer Baron, Humanist und europäischer Reformator. Beiträge zum internationalen Symposium vom 14.–17. Oktober 1999 in der Johannes a Lasco Bibliothek Emden. Tübingen, Mohr Siebeck 2000 VIII, 390 p.

5430. Journalisme et religion (1685–1785). Ed. par Jacques WAGNER. New York, Oxford, Lang, 2000, XVIII-413 p.

5431. Katholiken und Protestanten in den Aufbaujahren der Bundesrepublik. Hrsg. v. Thomas SAUER. Mit Beitr. von Martin GRESCHAT. Stuttgart, Kohlhammer, 2000, 222 p.

5432. Katholischer Antisemitismus im 19. Jahrhundert: Ursachen und Traditionen im internationalen Vergleich. Hrsg. v. Olaf BLASCHKE. Zürich, Orell Füssli, 2000, VII-383 p.

5433. Lebenswege und Religion: Biographie in Bibel, Dogmatik, und Religionspädagogik. Hrsg. v. Detlev DORMEYER, Herbert MÖLLE und Thomas RUSTER. Münster, Lit, 2000, 309 p.

5434. MACIOTTI (Maria Immacolata). Pellegrinaggi e giubilei. I luoghi del culto. Roma e Bari, Laterza, 2000, 219 p.

5435. MACK (Phyllis). Religious dissenters in Enlightenment England. History workshop, 2000, 49, p. 1-23.

5436. MARGRY (P. J.). Teedere quaesties. Religieuze rituelen in conflict. confrontaties tussen katholieken en protestanten rond de processiecultuur in 19de-eeuws Nederland. Hilversum, Verloren, 2000, 688 p.

5437. O'CONNELL (Michael). The idolatrous eye. iconoclasm and theater in Early-Modern England. New York a. Oxford. Oxford U. P., 2000 VIII-198 p.

5438. Parliament and the church 1529–1960. Ed. by J. P. PERRY and Stephen TAYLOR. Edinburgh, Edinburgh U. P., 2000, 193 p.

5439. Pèlerins et pèlerinages dans l'Europe moderne: acte de la table ronde organisée par le Département d'Histoire et Civilisation de l'Institut Universitaire Européen de Florence et l'École Francaise de Rome (Rome, 4–5 juin 1993). Sous la direction de Philippe BOUTRY. Rome, École Francaise de Rome, 2000, 518 p.

5440. POOLE (Kristen). Radical religion from Shakespeare to Milton. Figures of nonconformity in early modern England. Cambridge, Cambridge U. P., 2000, 312 p.

5441. Reform and renewal in the Middle Ages and the Renaissance: studies in honor of Louis Pascoe, S.J. Ed. by Thomas M. IZBICKI and Christopher M. BELLITTO. Leiden a. Boston, Brill, 2000, VIII-263 p.

5442. Religions, droit et sociétés dans l'Europe communautaire. Actes du XIIIe colloque de l'Institut de Droit et d'Histoire Religieuse (IDHR) Aix-en-Provence, 19–20 mai 1999. Aix-en-Provence, Presses Univ. d'Aix-Marseille, 2000, 294 p.

5443. ROCK (David). Voices in times of change: the role of writers, opposition movements and the churches in the transformation of East Germany. New York a. Oxford, Berghahn Books, 2000, VIII-262 p.

5444. SÉGUENNY (André). Les spirituels: philosophie et religion chez les jeunes humanistes allemands au seizième siècle. Baden-Baden, Bouxwiller-V. Koerner, 2000, 287 p.

5445. SEIFERT (Katharina). Glaube und Politik: die Ökumenische Versammlung in der DDR. Leipzig, Benno, 2000 XXIX-378 p.

5446. SHARKANSKY (Ira). The politics of religion and the religion of politics. Lanham, Lexington Books, 2000, XI-161 p.

5447. SHARMA (Arvind). Religion and secularism in India: after Ayodhya. New York, St. Martin's Press, [s.p.]

5448. Société et religion en France et aux Pays-Bas, XVe–XIXe siècle. Mélanges en l'honneur d'Alain Lottin. Par Gilles DEREGNAUCOURT. Arras, Artois Presses Université, 2000, 648 p.

5449. SPINI (Giorgio). La libertà religiosa in Italia e Europa. Torino, Claudiana, 2000, 48 p.

5450. Stát a církev v roce 1950. Sborník příspěvků z konference. (State and Church in the year 1950). Ed. by Jiří HANUŠ and Jan STŘÍBRNÝ. Brno, Centrum pro studium demokracie a kultury, 2000, 143 p. (Historia Ecclesiastica, 4).

5451. SUNDKLER BENGT (Gustaf Malcolm), STEED (Christopher). A history of the Church in Africa. Cambridge, Cambridge U. P., 2000, XIX-1232 p.

5452. Textes judéophobes (Les) et judéophiles dans l'Europe chrétienne à l'époque moderne: actes du colloque organisé par le Centre d'études juives à l'Université de Paris IV-Sorbonne [le 23 mai 1995]. Textes réunis par Daniel TOLLET; préf. de Pierre CHAUNU. Paris, Presses universitaires de France, 2000, XIII-246 p.

5453. Toleration in Enlightenment Europe. Ed. by Ole Peter GRELL and Roy PORTER. Cambridge, Cambridge U. P., 2000, IX-270 p.

5454. TRAPL (Miloš). Postavení katolické církve a politický katolicismus v zemích střední Evropy v letech 1918–1938. (Position of the Catholic Church and political Catholicism in Central Europe in the years 1918–1938). *Moderní dějiny*, 2000, 8, p. 169-179.

5455. Union, Konversion, Toleranz. Dimensionen der Annäherung zwischen den christlichen Konfessionen im 17. und 18. Jahrhundert. Hrsg. v. Heinz DUCHHARDT und Gerhard MAY, Mainz, von Zabern, 2000, 365 p.

5456. WARD (Keith). Religion and community. Oxford, Clarendon Press a. Oxford U. P., 2000, 366 p.

5457. Witchcraft in early modern Scotland. James VI's Demonology and the North Berwick witches. Edited by Lawrence NORMAND and Gareth ROBERTS. Exeter, University of Exeter Press, 2000, XIII-454 p.

5458. WOODING (Lucy E. C.). Rethinking Catholicism in Reformation England. Oxford a. New York, Clarendon Press, 2000, VIII-305 p.

5459. ZARRI (Gabriella). Recinti: donne, clausura e matrimonio nella prima età moderna. Bologna, Il mulino, 2000, 498 p.

Cf. n^{os} *626, 1282, 1295-1321, 4701, 7462*

§ 2. Roman Catholicism.

a. General

* 5460. DEGLER-SPENGLER (Brigitte). Helvetia Sacra. Arbeitsbericht 1999 [Forschungsbericht]. *Schweizerische Zeitschrift für Geschichte*, 2000, 50, p. 181-187.

** 5461. Bibbia (La). Edizioni del XVI secolo. A cura di Antonella LUMINI. Firenze, L.S. Olschki, 2000 XXXIX-327 p.

** 5462. FIRPO (Massimo), MERCATTO (Dario). I processi inquisitoriali di Pietro Carnesecchi (1557–1567). Volume 2. Il processo sotto Pio V (1566–1567). Tomo 1. Giugno 1566–ottobre 1566. Tomo 2. Novembre 1566–gennaio 1567. Tomo 3. Gennaio 1567–agosto 1567. Città del Vaticano, Archivio segreto Vaticano, 2000, [s. p.].

5463. BERGHAHN (Klaus L.). Grenzen der Toleranz. Juden und Christen im Zeitalter der Aufklärung. Köln, Weimar u. Wien, Böhlau, 2000, 304 p.

5464. BESIER (Gerhard). Kirche, Politik und Gesellschaft im 20. Jahrhundert. München, Oldenbourg, 2000, X-180 p. (Enzyklopädie deutscher Geschichte, 56).

5465. Beyond the mainstream. The emergence of religious pluralism in Finland, Estonia, and Russia. Ed. by Jeffrey KAPLAN. Helsinki, SKS, 2000, 386 p. (Studia historica, 63).

5466. BLASCHKE (Olaf), ALTERMATT (Urs). "Katholizismus und Antisemitismus". Eine Kontroverse. *Schweizerische Zeitschrift für Geschichte*, 2000, 50, p. 204-236.

5467. CALLAHAN (William J.). The Catholic church in Spain, 1875–1998. Washington, Catholic University of America Press, 2000, XVI-695 p.

5468. CASSESE (Michele). La Chiesa cattolica del Nord-Est italiano ed il suo rapporto con gli zingari. *Ricerche di storia sociale e religiosa*, 2000, 39, 57, p. 43-86.

5469. Catholicism contending with modernity. Roman Catholic modernism and anti-modernism in historical context. Ed. by Darrell JODOCK. Cambridge, Cambridge U. P., 2000, XIV-345 p.

5470. CONGAR (Yves). La chiesa cattolica di fronte alla questione razziale. Roma, Accademia degli Incolti, 2000, 91 p.

5471. DAVIS (Derek H.). Religion and the continental congress 1774–1789: contributions to original intent. Oxford, Oxford U. P., 2000, XIV-309 p. (Religions in American series).

5472. Exemte Bistum Bamberg (Das). 3. Die Bischofsreihe von 1522 bis 1693. Bearb. v. Dieter J. WEIß. Berlin u. New York, de Gruyter, 2000, XVI-682 p. (Germania Sacra, 38/1).

5473. Frontiers of faith. Religious exchange and the constitution of religious identities, 1400–1700. Ed. by Eszter ANDOR and Istvàn György TOTH, with introduction by Robert MUCHEMBLED. Budapest, Central U. P., 2000, [s. p.].

5474. FULLER (Robert C.). Stairways to heaven: drugs in American religious history. Boulder, Westview, 2000, IX-237 p.

5475. Gender and vocation: women, religion and social change in the Nordic Countries, 1830–1940. Ed. by Pirjo MARKKOLA. Helsinki, Suomalaisen Kirjallisuuden Seura, 2000, 245 p. (Studia Historica, 64).

5476. Geschichte des kirchlichen Lebens in den deutschsprachigen Ländern seit dem Ende des 18. Jahrhunderts – Die Katholische Kirche. Band 6. Die Kirchenfinanzen. Hrsg. v. Erwin GATZ. Freiburg, Basel u. Wien, Herder, 2000, 511 p.

5477. HARLINE (Craig), PUT (Eddy). A Bishop's tale: Mathias Hovius among his flock in seventeenth-century Flanders. New Haven, Yale U. P., 2000, X-387 p.

5478. HERZIG (Arno). Der Zwang zum wahren Glauben. Rekatholisierung vom 16. bis zum 18. Jahrhundert. Göttingen, Vandenhoeck & Ruprecht, 2000, 266 p. – IDEM. Die Rekatholisierung in deutschen Territorien im 16. und 17. Jahrhundert. *Geschichte und Gesellschaft*, 2000, 26, 1, p. 76-104.

5479. HOLZEM (Andreas). Religion und Lebensformen. Katholische Konfessionalisierung im Sendgericht der Fürstbistums Münster 1570–1800. Paderborn, Schöningh, 2000, XI-570 p. (Forschungen zur Regionalgeschichte, 33).

5480. JENDORFF (Alexander). Reformatio catholica. Gesellschaftliche Handlungsspielräume kirchlichen Wandels im Erzstift Mainz 1514–1630. Münster, Aschendorf, 2000, XII-594 p. (Reformationsgeschichtliche Studien und Texte, 142).

5481. KITTELSON (James). Toward an established church. Strasbourg from 1500 to the dawn of the seventeenth century. Mainz, von Zabern, 2000, 280 p. (Veröffentlichungen des Instituts für Europäische Geschichte Mainz, Abt. Abendländische Religionsgeschichte, 182).

5482. LENNON (Colm). Archbishop Richard Creagh of Armagh, 1523–1586: an Irish prisoner of Conscience of the Tudor era. Portland, Four Courts Press, 2000, 166 p.

5483. LOTTIN (Alain). Être et croire à Lille et en Flandre XVIe–XVIIIe siècle. Arras, Artois Université, 2000, 546 p.

5484. O'MALLEY (John W.). Trent and all that: renaming catholicism in the early modern era. Cambridge, Harvard U. P., 2000, 219 p.

5485. PATROUCH (Joseph F.). A negotiated settlement: the Counter-Reformation in upper Austria under the Habsburgs. Boston, Humanities Press, 2000, XI-283 p. (Studies in Central European histories).

5486. PÉROUAS (Louis). Le catholicisme en Limousin aux XIXe et XXe siècles à travers sa presse. Treignac, Editions Les Monédières, 2000, 187 p. (Les semaines religieuses).

5487. PETERS (Shawn Francis). Judging Jehovah's witnesses: religious persecution and the dawn of the rights revolution. Lawrence, University Press of Kansas, 2000, X-342 p.

5488. PHAYER (Michael). The Catholic Church and the Holocaust, 1930–1965. Bloomington, Indiana U. P., 2000, XVIII-301 p.

5489. PIERONI (Geraldo). Os excluídos do reino: a Inquisiçao portuguesa e o degredo para o Brasil colônial. Brasilia e Sao Paulo, Editora Universidade de Brasilia, 2000, 308 p. (ill.).

5490. Religions and transition dans la seconde moitié du XVIII siècle. Ed. par Louis CHÂTELLIER. Oxford, Voltaire Foundation, 2000, 300 p. (Studies on Voltaire and the eighteenth century, 2).

5491. Rendre ses vœux. Les identités pèlerines dans l'Europe moderne (XVIe–XVIIIe siècle). Ed. par Philippe BOUTRY, Pierre-Antoine FABRE et Dominique JULIA. Paris, Ed. de l'EHESS, 2000, 587 p.

5492. RICCARDI (Andrea). Vescovi d'Italia. Storie e profili del Novecento. Cinisello Balsamo, San Paolo, 2000, 231 p.

5493. ROPER (Lyndal). 'Evil imaginings and fantasies': child-witches and the end of the witch craze. *Past and present*, 2000, 167, p. 107-139.

5494. ROSSI (Ernesto). Il Sillabo e dopo. A cura di Giuseppe ARMANI. Milano, Kaos, 2000, 245 p.

5495. SCHIMDT (Jürgen Michael). Glaube und Skepsis. Die Kurpfalz und die abendländische Hexenverfolgung 1446–1685. Bielefeld, Verlag für Regionalgeschichte, 2000, 510 p. (Hexenforschung, 5).

5496. SHATTUCK (Gardiner H.). Episcopalians and race: Civil War to civil rights. Lexington, University Press of Kentucky, 2000, XIII-298 p. (Religion in the South).

5497. SPAETH (Donald A.). The church in an age of danger: parsons and parishioners, 1660–1740. Cambridge, Cambridge U. P., 2000, XIII-279 p. (Cambridge studies in early modern British history).

5498. TACCOLINI (Mario). Per il pubblico bene. La soppressione di monasteri e conventi nella Lombardia austriaca del secondo Settecento. Roma, Bulzoni, 2000, 337 p.

5499. TALLON (Alain). Le concile de Trente. Paris, Ed. du Cerf, 2000, 135 p.

5500. TARIC-ZUMSTEG (Fabienne). Les sorciers à l'assaut du village. Gollion (1615–1631). Lausanne, Ed. du Zèbre, 2000, 363 p.

5501. TIPPETT-SPIRTOU (Sandy). French Catholicism: church, state and society in a changing era. Basingstoke, McMillan press, 2000, XVIII-238 p.

5502. VAN RAHDEN (Till). Juden und andere Breslauer. Die Beziehungen zwischen Juden, Protestanten und Katholiken in einer deutschen Großstadt von 1860–1925. Göttingen, Vandenhoeck & Ruprecht, 2000, 382 p. (Kritische Studien zur Geschichtswissenschaft, 139).

5503. Vatican (The) and the Holocaust: the Catholic church and the Jews during the Nazi era. Ed. by Randolph BRAHAM. New York, Rosenthal institute for Holocaust studies, 2000, VI-137 p.

5504. ZUCCOTTI (Susan). Under his very windows: the Vatican and the Holocaust. New Haven, Yale U. P., 2000, 408 p.

Cf. n°s 4580, 5638, 7201

b. History of the Popes

5505. ANDERSON (Robin). Papa Pio VII (Barnaba Chiaromonti). La vita, il regno e il conflitto con Napoleone nel periodo seguente alla Rivoluzione francese (1742–1823). Roma, Benedectina, 2000, 221 p.

5506. BALDAN (Sergio). Il conclave di Venezia. L'elezione di papa Pio VII, 1 dicembre 1799–14 marzo 1800. Venezia, Marsilio, 2000, 280 p.

5507. BENIGNI (Mario), ZANCHI (Goffredo). Giovanni XXIII. Biografia ufficiale. Cinisello Balsamo, San Paolo, 2000, 453 p.

5508. BURKLE-YOUNG (Francis). Papal elections in the age of transition, 1878–1922. Lanham, Lexington Books, 2000, XIII-185 p.

5509. DE LUCA (Vittorio). Papa Giovanni nunzio apostolico, patriarca di Venezia, papa del Concilio, santo del nuovo secolo. Venezia, Marsilio, 2000, 221 p.

5510. Episcopato e società tra Leone XIII e Pio X: direttive romane ed esperienze locali in Emilia-Romagna e Veneto. A cura di Daniele MENOZZI. Bologna, il Mulino, 2000, 266 p.

5511. FELDKAMP (Michael F.). Pius XII und Deutschland. Göttingen, Vandenhoeck & Ruprecht, 2000, 236 p.

5512. GATTONI (Maurizio). Leone X e la geopolitica dello Stato pontificio 1513–1521. Città del Vaticano, ASV, 2000, 367 p.

5513. GUIDI BRUSCOLI (Francesco). Benvenuto Olivieri. I mercatores fiorentini e la Camera apostolica nella Roma di Paolo III Farnese, 1543–1549. Firenze, Olschki, 2000, XXVI-362 p.

5514. HERMANN DE FRANCESCHI (Sylvio). L'autorité pontificale face au legs de l'antiromanisme catholique et régaliste des Lumières. *Archivum Historiae Pontificiae*, 38, 2000, p. 119-164.

5515. MENOZZI (Daniele). I papi del '900. Firenze, Giunti, 2000, 127 p.

5516. Papes et papauté au XVIIIe siècle: VIe colloque Franco-Italien, Société française d'étude du XVIIIe siècle, Université de Savoie, Chambéry, 21–22 septembre 1995. Etudes réunies par Philippe KOEPPEL. Paris, H. Champion, 2000, 317 p.

5517. Pontificati (I) di Pio VI e Pio VII. Atti del Convegno (Cesena, 9 ottobre 1999). A cura di Marino MENGOZZI. Cesena, Diocesi di Cesena-Sarsina e Stilgraf, 2000, 338 p.

5518. Storia d'Italia. Annali, 16. Roma città del papa. Vita civile e religiosità dal giubileo di Bonifacio VIII al giubileo di papa Wojtyla. A cura di Adriano PROSPERI e Luigi FIORANI. Torino, Einaudi, 2000, p.

Cf. n°s 4852, 6396

c. Special studies

** 5519. Investigations into magic: Martin del Rio. Ed. by P. G. MAXWELL-STUART. Manchester, Manchester U. P., 2000, XI-290 p.

** 5520. KUDELIĆ (Zlatko). Izvješće zagrebačkoga biskupa Benedikta Vinkovića apostolskom nunciju Casparu Mattheiju o Marčanskoj biskupiji i Vlasima iz 1640. godine. (Report by Zagreb Bishop Benedikt Vinković to the Apostolic Delegate Caspar Matthei about the Marča Diocese and the Vlachs from 1640). *Povijesni prilozi*, 2000, 19, p. 153-179.

** 5521. Modernismo italiano (Il). Le «lettere» di Buonaiuti e le obiezioni di Prezzolini. A cura di Daniele ROLANDO. Genova, Name, 2000, 247 p.

** 5522. *Vacat.*

** 5523. PUCCI (Francesco). De praedestinatione. Introd., testo, note e nota crit. a cura di Mario BIAGIONI. Firenze, Olschki, 2000, X-354 p.

** 5524. TELLECHEA IDIGORAS (J. I.). Los conflictos de Milán (1567–1570). Cartas de S. Carlo Borromeo al Nuncio de España Mons. Juan Bautista Castagna, Arzobispo de Rossano. *Scriptorium Victoriense*, 2000, 67, p. 47-127.

** 5525. VALENTINI (Filippo). Il principe fanciullo. Trattato inedito dedicato a Renata e Ercole II d'Este. A cura di Lucia FELICI. Firenze, Olschki, 2000, IX-307 p.

5526. 1968: fra utopia e vangelo. Contestazione studentesca e mondo cattolico. A cura di Agostino GIOVAGNOLI. Roma, Ave, 2000, 188 p.

5527. BOCHNAK (Władysław). Religijne stowarzyszenia i bractwa katolików świeckich w diecezji wrocławskiej od XVI wieku do 1918 roku. (Religiöse Gesellschaften und Bruderschaften der katholischen Laien in der Diözese von Wrocław [Breslau] vom XVI. Jahrhundert bis 1918). Wrocław u. Legnica, Alta2, 2000, 311 p.

5528. BONECHI (Simone). L'impossibile restaurazione: i vescovi filonapoleonici nell'Italia francese tra "servilismo" e primato di Pietro (1801–1814). *Cristianesimo nella Storia*, 21, 2000, p. 343-381.

5529. BOURDIN (Philippe). Le noir et le rouge: itinéraire social, culturel et politique d'un prêtre patriote, 1736–1799. Clermont-Ferrand, Presses universitaires Blaise Pascal, 2000, 520 p.

5530. BRAMBILLA (Elena). Alle origini del Sant'Uffizio. Penitenza, confessione e giustizia spirituale dal Medioevo al XVI secolo. Bologna, il Mulino, 2000, 591 p.

5531. Bücherzensur – Kurie – Katholizismus und Moderne: Festschrift für Herman H. Schwedt. Hrsg. v. Peter WALTER. Frankfurt am Main, Lang, 2000, 376 p.

5532. CANESSA (ANDREW). Contesting hybridity: evangelistas y kataristas in Highland Bolivia. *Journal of Latin America studies*, 32, 2000, p. 115-144.

5533. CARNEMOLLA (Piero Antonio). Un cristiano siciliano. Rassegna degli studi su Giorgio La Pira (1978-1998). Caltanissetta, Sciascia, 2000, 385 p.

5534. Catholic (The) question in Ireland, 1762–1829. Edited and introduced by Nicholas LEE. Bristol, Thoemmes, 2000, 8 v. 3420p.

5535. Chiesa cattolica (La) e gli zingari: storia di un difficile rapporto. A cura di Gabriele DE ROSA. Roma, Centro di studi zigani, 2000, 232 p.

5536. CIAMPANI (Andrea). Cattolici e liberali durante la trasformazione dei partiti. La «Questione romana» tra politica nazionale e progetti vaticani (1876–1883). Roma, Archivio Guido Izzi, 2000, 426 p.

5537. COVA (Anne). Au service de l'Eglise, de la patrie et de la famille, femmes catholiques et maternité sous la IIIe République. Paris, Montréal, Budapest, l'Harmattan, 2000, 221 p.

5538. DAVIES (Adrian). The Quakers in English society, 1655–1725. Oxford, Clarendon Press, 2000, 262 p.

5539. DE SPIRITO (Angelomichele). Il 1799 a San Giorgio del Sannio tra rivoluzionari e insorgenti. *Ricerche di storia sociale e religiosa*, 29, 2000, 57, p. 117-142.

5540. DE VRIES ZANUCCOLI (Luciana). Cattolici d'Olanda ai confini della Chiesa di Roma. Firenze, Polistampa, 2000, 213 p.

5541. DOYLE (William). Jansenism: Catholic resistance to authority from the Reformation to the French Revolution. New York, Macmillan Press, St. Martin's Press, 2000, X-109 p.

5542. ERRERA (Andrea). Processus in causa fidei. L'evoluzione dei manuali inquisitoriali nei secoli XVI-XVIII e il manuale inedito di un inquisitore perugino. Bologna, Monduzzi, 2000, XVIII-427 p.

5543. Erudizione e devozione. Le raccolte di vite di santi in età moderna e contemporanea. A cura di Gennaro LUONGO. Roma, Viella, 2000, 367 p.

5544. FALZONE (Maria Teresa). Da questo vi riconosceranno: Chiesa e poveri in Sicilia in età contemporanea. Caltanissetta, Sciascia, 2000, 283 p.

5545. FIRPO (Massimo). Giorgio Siculo. Discussione del volume di Adriano Prosperi. *Storica*, 2000, 6, 18, p. 143-152. [Cf. n° 5566].

5546. FORNACA (Remo). La politica scolastica della Chiesa: dal Risorgimento al dibattito contemporaneo. Roma, Carocci, 2000, 257 p.

5547. FRAGNITO (Gigliola). "Dichino corone e rosari": censura ecclesiastica e libri di devozione. *Cheiron*, 17, 2000, p. 135-158.

5548. FRIAS GARCIA (Maria del Carmen). Iglesia y constitucion: la jerarquía catolica entre la II Repubblica. Madrid, Centro de estudios politicos, 2000, XXXIX-778 p.

5549. GIBERTI (Gian Matteo). Le Costituzioni per il clero (1542). A cura di Roberto PASQUALI. Vicenza, Istituto per le ricerche di storia sociale, 2000, CX-765 p.

5550. GIOVANNUCCI (Pierluigi). La canonizzazione del card. Gregorio Barbarigo. *Ricerche di storia sociale e religiosa*, 2000, 29, 57, p. 21-41.

5551. GIUNTELLA (Maria Cristina). La Fuci tra modernismo, Partito Popolare e fascismo. Roma, Studium, 2000, 296 p.

5552. GLEASON (Elisabeth). Religione e società nella Venezia di Lorenzo Lotto. *Cheiron*, 2000, 17, p. 79-88.

5553. GODMAN (Peter). The Saint as censor. Robert Bellarmine between Inquisition and Index. Leiden, Boston a. Köln, Brill, 2000, 503 p. (Studies in Medieval and Reformation thought, 80).

5554. History of the Catholic Diocese of Dublin. Ed. by James KELLY and Dáire KEOGH. Dublin, Four courts press, X-390 p.

5555. HOURS (Bernard). L'Eglise et la vie religieuse dans la France moderne, XVIe–XVIIIe siècle. Paris, Presses universitaires de France, 2000, XVII-384 p.

5556. Inquisizione (L') e gli storici: un cantiere aperto. Roma, Accademia Nazionale dei Lincei, 2000, 447 p.

5557. Lexikón katolíckych kňazských osobností Slovenska. (Thesaurus of Catholic priest personalities of Slovakia). Autors Július PAŠTEKA a kol. Bratislava, Lúč, 2000, 1550-LIV p.

5558. LUXMOORE (Jonathan), BABIUCH (Jolanta). The Vatican and the red flag: the struggle for the soul of Eastern Europe. London, Chapman, 2000, XIV-351 p.

5559. MAYER (Thomas F.). Cardinal Pole in European context: a via media in the Reformation. Aldershot Hants, Variorum, 2000, 350 p.

5560. Modernismo (Il) tra cristianità e secolarizzazione. Atti del convegno di Urbino 1–4 ottobre 1997. A cura di Alfonso BOTTI e Rocco CERRATO. Urbino, Quattro Venti, 2000, 925 p.

5561. NANNI (Stefania). Il mondo nuovo: l'edificazione della Chiesa universale. Pisa, Iepi, 2000, 274 p.

5562. PAIANO (Maria). Liturgia e società nel Novecento: percorsi del movimento liturgico di fronte ai processi di secolarizzazione. Roma, Edizioni di Storia e Letteratura, 2000, 314 p.

5563. PAZZAGLIA (Luciano). Scuola e religione nell'Italia giolittiana. Milano, ISU, 2000, 241 p.

5564. Pietismus (Der) im neunzehnten und zwanzigsten Jahrhundert. Hrsg. v. Gustav Adolf BENRATH, Martin SALLMANN und Ulrich GÄBLER. Göttingen, Vandenhoeck & Ruprecht, 2000, xii-605 p.

5565. POGGI (Vincenzo). Per la storia del Pontificio Istituto Orientale. Saggi sull'istituzione, i suoi uomini e l'Oriente Cristiano. Roma, Pontificio Istituto Orientale, 2000, 448 p.

5566. PROSPERI (Adriano). L'eresia del Libro grande. Storia di Giorgio Siculo e della sua setta. Milano, Feltrinelli, 2000, 490 p. (Campi del sapere. Culture).

5567. RICCI (Saverio). Giordano Bruno nell'Europa del Cinquecento. Roma, Salerno, 2000, 649 p.

5568. RICHTER (Reinhard). Nationales Denken im Katholizismus der Weimarer Republik. Münster, Lit, 2000, 428 p.

5569. RODRIGUEZ BESNE (Jose Ramon). El consejo de la suprema inquisición: perfil juridico de una institución. Madrid, Complutense, 2000, 282 p.

5570. ROZZO (Ugo). A proposito del «Thesaurus de la littérature interdite au XVIe siècle». La bibliofilia, 2000, 102, p. 325-337.

5571. TURBANTI (Giovanni). Un Concilio per il mondo moderno. La redazione della Costituzione pastorale «Gaudium et spes» del Vaticano II. Bologna, il Mulino, 2000, 829 p.

5572. Ut bene regantur. Politica e amministrazione periferica nello Stato ecclesiastico. A cura di Paola MONACCHIA. Modena, Mucchi, 2000, 278 p.

5573. VENARD (Marc). Le catholicisme à l'épreuve dans la France du XVIe siècle. Paris, Éd. du Cerf, 2000, 290 p.

5574. Volti di fine Concilio. Studi di storia e teologia sulla conclusione del Vaticano II. A cura di Joseph DORÉ e Alberto MELLONI. Bologna, il Mulino, 2000, 445 p.

5575. WARNER (Carolyn M.). Confessions of an interest group: the Catholic Church and political parties in Europe. Princeton, Princeton U. P., 2000, XVI-249 p.

Cf. n° 4267

d. Religious orders

* 5576. Bibliographie sur l'Histoire de la Compagnie de Jésus. [Cf. Bibl. 99, n° 5430.] Ed. par Nicoletta BASILOTTA et Lászlo POLGÁR. Archivum historicum societatis Iesu, 2000, 89, [s. p.].

5577. Cystersi w społeczeństwie Europy Środkowej. Materiały z konferencji naukowej odbytej w klasztorze OO. Cystersów w Krakowie Mogile z okazji 900 rocznicy powstania Zakonu Ojców Cystersów. Poznań-Kraków-Mogiła – 10 października 1998. (Zisterzienser in der Gesellschaft des Mittel-Europas. Materialien von der wissenschaftlichen Konferenz in dem Zisterzienser-Kloster in Kraków-Mogiła anlässlich des 900. Jahrestages der Gründung des Zisterzienser-Ordens. Poznań-Kraków-Mogiła – 10. Oktober 1998). Hrsg. v. Andrzej Marek WYRWA und Józef DOBOSZ. Poznań, Wydaw. Poznańskie, 2000, 825 p.

5578. DE NOBILI (Roberto). Preaching wisdom to the wise: three treatises; translated and introduced by Anand AMALADASS and Francis X. CLOONEY. Saint Louis, Institute of Jesuit Sources, 2000, XXI-345 p.

5579. GUERRA (Alessandro). Per un'archeologia della strategia missionaria dei gesuiti: le Indipetae e il sacrificio nella "vigna del Signore". Archivio italiano per la storia della pietà, 2000, 13, p. 109-192.

5580. Jesuit Ratio studiorum (The). 400th anniversary perspectives. Ed. by Vincent J. DUMINUCO. New York, Fordham U. P., 2000, XII-307 p.

5581. KANIOR (Marian). Polska Kongregacja Bernardyńska Świętego Krzyża 1709–1864. (Die polnische Bernhardiner-Kongregation zum Heiligen Kreuz 1709–1864). Kraków, Tyniec, 2000, 321 p.

5582. MARTIN (Catherine). Les compagnies de la Propagation de la foi (1632–1685). Paris, Grenoble, Aix, Lyon, Montpellier. Etude d'un réseau d'associations fondé en France au temps de Louis XIII pour lutter contre l'hérésie des origines à la révocation de l'Edit de Nantes. Genève, Droz, 2000, 547 p.

5583. Moines (Les) du Der, 673–1790. Actes du Colloque international d'histoire, Joinville-Montier-en-Der, 1er–3 octobre 1998. Publiés par Patrick CORBET, avec le concours de Jackie LUSSE et Georges VIARD. Langres, Ed. Dominique Guéniot, 2000, 728 p. (ill.).

5584. PÉREZ FERREIRO (Elvira). El tratado de Uceda contra los estatutos de limpieza de sangre. Una reacción ante el establecimiento del estatuto de limpieza en la Orden Franciscana. Madrid, Aben Ezra, 2000, 182 p.

5585. SCHIAMONE (Pietro). La SS. Trinità negli Esercizi spirituali di S. Ignazio di Loyola. Roma, ADP, 2000, 340 p.

5586. TAYLOR (Bruce). Structures of reform. The Mercedarian Order in the Spanish Golden Age. Leiden a. Boston, Brill, 2000, XIX-506 p.

5587. TORRES SÁNCHEZ (Concha). La clausura imposible. Conventualismo femenino y expansión contrarreformista. Madrid, Asociación Cultural Al-Mudayna, 2000, 200 p.

5588. TRAMPUS (Antonio). I gesuiti e l'Illuminismo. Politica e religione in Austria e nell'Europa centrale (1773–1798). Firenze, Olschki, 2000, 386 p.

5589. Vecchio (Tra) e nuovo mondo. Retorica, immagini, prassi missionaria. Pistoia, Provincia romana dei frati predicatori, 2000, 529 p.

e. Missions

5590. BILLINGTON HARPER (Susan). In the shadow of the Mahatma: Bishop V. S. Azariah and the travails of Christianity in British India. Grand Rapids, William B. Eerdmans, 2000, XIX-462 p. (Studies in the history of Christian missions).

5591. *Vacat.*

5592. Church mission society (The) and world christianity, 1799–1999. Ed. by Kevin WARD and Brian STANLEY. Grand Rapids, William B. Eerdmans, 2000, XVIII-382 p. (Studies in the history of Christian mission).

5593. DE CASTELNAU-L'ESTOILE (Charlotte). «Les ouvriers d'une vigne stérile». Les jésuites et la conversion des Indiens au Brésil (1580–1620). Lisboa, Fondation Calouste Gulbekian, 2000, 548 p.

5594. Espaço & arqueologia nas missões jesuíticas. O caso de São João Batista Barcelos. Por Artur Henrique FRANCO. Porto Alegre, Edipucrs, 2000, 408 p.

5595. Formaciones religiosas en la América colonial. Coordinadoras: María Alba PASTOR y Alicia MAYER. México, Facultad de Filosofía y Letras, Dirección General de Asuntos del Personal Académico, Universidad Nacional Autónoma de México, 2000, 262 p. (Colección Seminarios).

5596. GIRARD (Pascale). Les religieux occidentaux en Chine à l'époque moderne. Essai d'analyse textuelle comparée. Lisboa, Fundação Calouste Gulbenkian et Paris, Centre culturel Calouste Gulbenkian, 2000, 619 p.

5597. Gramáticas misioneras (Las) de tradición hispánica (siglos XVI–XVII). Bajo la dirección de Otto ZWARTJES. Amsterdam, Rodopi, 2000, 309 p.

5598. GREGORY (Brad). Salvation at stake: Christian martyrdom in early Modern Europe. Cambridge, Harvard U. P., 2000, XVI-528 p. (Harvard historical studies, 134).

5599. HAUSBERGER (Bernd). Für Gott und König. Die Mission der Jesuiten im Kolonialen Mexiko. Wien, Verlag für Geschichte und Politik u. München, Oldenbourg, 2000, 648 p. (Studien zur Geschichte und Kultur der Iberischen und Iberoamerikanischen Länder, 6).

5600. HUBERT (Olivier). Sur la terre comme au ciel. La gestion des rites par l'Eglise catholique du Québec (fin XVIIe–mi-XIXe siècle). Sainte-Foy, Les Presses de l'Université Laval, 2000, 341 p.

5601. INGLEBY (Jonathan C.). Missionaries, education, and India. Issues in Protestant missionary education in the long nineteenth century. Delhi, Ispck, 2000, XVIII-400 p.

5602. Invented identities: the interplay of gender, religion and politics in India. Ed by Julia LESLIE and Mary MAC GEE. New Delhi a. Oxford, Oxford U. P., XIII-309 p.

5603. JACKSON (Robert H.). From savages to subjects. Missions in the history of the American Southwest. Armonk a. London, Sharpe, 2000, XVII-151 p.

5604. LAUX (Claire). Les théocraties missionnaires en Polynésie (Tahiti, Hawaii, Cook, Tonga, Gambier, Wallis et Futuna) au XIXe siècle. Des cités de Dieu dans les mers du Sud. Paris, Harmattan, 2000, 382 p.

5605. Lettres édifiantes et curieuses des jésuites de l'Inde au XVIIIe siècle. présentées par Isabelle VISSIERE et Jean-Louis VISSIERE. Saint-Etienne, Publications de l'Université de Saint- Etienne, 2000, 201 p.

5606. LUDWIG (Frieder). Zwischen Kolonialismuskritik und Kirchenkampf. Interaktionen afrikanischer, indischer und europäischer Christen während der Weltmissionskonferenz in Tambaram 1938. Göttingen, Vandenhoeck & Ruprecht, 2000, 352 p.

5607. Martino Martini S.J. (1614–1661) und die Chinamission im 17. Jahrhundert. Hrsg. v. Roman MALEK und Arnold ZINGERLE. Nettetal, Steyler, 2000, 260 p.

5608. NORRIS (Jim). After "The Year Eighty": the demise of Franciscan power in Spanish New Mexico. Albuquerque, University of New Mexico Press, 2000, X-212 p.

5609. Portugueses (Os) no Tibete. Os primeiros relatos dos Jesuítas (1624–1635). Estudo histórico de Hugues DIDIER. Coordenacão e fixação dos textos da edição portuguesa por Paulo LOPES MATOS. Tradução de Lourdes JÚDICE LISBOA. Lisboa, Comissão Nacional para as Comemorações dos Descobrimentos Portugueses, 2000, 287 p.

5610. "… usque ad ultimum terrae": die Jesuiten und die transkontinentale Ausbreitung des Christentums, 1540–1773. Hrsg. v. Johannes MEIER. Göttingen, Vandenhoeck & Ruprecht, 2000, 211 p.

5611. WEN-CHAO (Li). Die christliche China-Mission im 17. Jahrhundert. Verständnis, Unverständnis, Missverständnis: eine geistesgeschichtliche Studie zum Christentum, Buddhismus und Konfuzianismus. Stuttgart, Franz Steiner, 2000, 648 p.

§ 3. Orthodox Church.

5612. DEGIEL (Rafał). Protestanci i prawosławni. Patronat wyznaniowy Radziwiłłów birżańskich nad Cerkwią prawosławną w księstwie słuckim w XVII w. Warszawa. (Protestanten und Orthodoxe. Russisch-

Orthodoxe Kirche unter Fittichen der Familie Radziwiłł von Birża im Fürstentum Sluzk [Słuck] im XVII. Jhdt). Warszawa, Neriton, 2000, 153 p.

5613. DELLA VALLE (Antimo). L'internazionale ortodossa. Religioni e conflitti nei Balcani. Torino, Ananke, 2000, 125 p.

5614. FELMY (Karl Christian). Vom urchristlichen Herrenmahl zur Göttlichen Liturgie der Orthodoxen Kirche. Ein historischer Kommentar. Erlangen, Oikonomia, 2000, IV-132 p.

5615. HINTIKKA (Kaisamari). The Romanian Orthodox Church and the World Council of Churches, 1961–1977. Helsinki, Luther-Agricola Society, 2000, 207 p. (Schriften der Luther-Agricola-Gesellschaft, 48).

5616. LEONT'EVA (T. G.). Zhizn' i perezhivanija sel'skogo svjashchennika, 1861–1904. (The life and the emotional peasantry of the [Russian] countrypriest, 1861–1904). In: Sotsial'naja istorija: Ezhegodnik [Cf. n° 7277], p. 34-56. [Eng. summary].

5617. OELDEMANN (Johannes). Die Apostolizität der Kirche im Ökumenischen Dialog mit der Orthodoxie: der Beitrag russischer orthodoxer Theologen zum ökumenischen Gespräch über die apostolische Tradition und die Sukzession in der Kirche. Paderborn, Bonifatius, 2000, 434 p.

5618. VALLIERE (Paul). Modern Russian theology. Bukharev, Soloviev, Bulgakov. Orthodox theology in a new key. Edinburgh, T & T Clark, 2000, 448 p.

5619. VAPORIS (Nomikos Michael). Witnesses for Christ: Orthodox Christian neomartyrs of the Ottoman period, 1437–1860. Crestwood, St. Vladimir's Seminary Press, 2000, XIV-377 p.

Cf. n°s 377, 3486

§ 4. Protestantism.

** 5620. Italian Reformation (The) of the sixteenth century and the diffusion of Renaissance culture: a bibliography of the secondary literature, ca. 1750–1997 Ed. by John TEDESCHI. With an historiographical introduction by Massimo FIRPO. Ferrara, Panini, 2000, LXIII-1047 p.

** 5621. Registers of the Consistory of Geneva in the time of Calvin. Ed. by Thomas A. LAMBERT and Isabella M. WATT; with the assistance of Jeffrey R. WATT. Grand Rapids, Eerdmans, 2000,

5622. ALBRECHT (Christian) Historische Kulturwissenschaft neuzeitlicher Christentumspraxis: Klassische Protestantismustheorien in ihrer Bedeutung für das Selbstverständnis der Praktischen Theologie. Tübingen, Mohr Siebeck, XIII-364 p.

5623. ANDREWS (Dee E.). The Methodists and revolutionary America, 1760–1800: the shaping of an evangelical culture. Princeton, Princeton U. P., 2000, XV-367 p.

5624. BAYROU (François). Ils portaient l'écharpe blanche: l'aventure des premiers réformés, des Guerres de religion à l'Edit de Nantes, de la Révocation à la Révolution. Paris, Librairie générale française, 2000, 381 p.

5625. BENINI CLEMENTI (Enrica). Riforma religiosa e poesia popolare a Venezia nel Cinquecento: Alessandro Caravia. Firenze, Olschki, 2000, IX-315 p.

5626. BIAGETTI (Stefania). Dal Bayle al Gerdes: la "riforma italiana" nella storiografia tra la fine del XVII secolo e l'età illuministica. Clio, 2000, 36, p. 91-128.

5627. BOISSON (Didier). Les protestants de l'ancien colloque du Berry de la révocation de l'Edit de Nantes à la fin de l'Ancien Régime (1679–1789) ou l'inégale résistance de minorités religieuses. Paris, Honoré Champion, 2000, 799 p.

5628. BROWN (Keith M.). Noble society in Scotland: wealth, family and culture from Reformation to Revolution. Edinburgh, Edinburgh U. P., 2000, X-369 p.

5629. CABANEL (Patrick). Les protestants et la République. Bruxelles, Editions Complexe, 2000, 272 p. (Les dieux dans la cité).

5630. CAPONETTO (Salvatore). Melantone e l'Italia. Torino, Claudiana, 2000, 102 p.

5631. CASHDOLLAR (Charles D.). A spiritual home: life in British and American reformed congregations, 1830–1915. University Park, Pennsylvania State U. P., 2000, XIV-336 p.

5632. CLARK (James G.). Reformation and reaction at St Albans Abbey, 1530–1558. English historical review, 2000, 115, 461, p. 297-328.

5633. COFFEY (John). Persecution and toleration in Protestant England: 1558–1689. Harlow, Longman, 2000, X-256 p.

5634. Continuity and change: the harvest of late medieval and Reformation history. Essays presented to Heiko A. Oberman on his 70th birthday. Ed. by Robert James GOW and Andrew COLIN. Leiden a. Boston, Brill, 2000, XV-459 p.

5635. CROUZET (Denis). Jean Calvin. Vies parallèles. Paris, Fayard, 2000, 481 p.

5636. DENTAN (Paul-Emile). Impossible de se taire: des protestants suisses face au nazisme. Genève, Labor et Fides, 2000, 133 p.

5637. DEVINE (T. M.). Scotland's shame? Bigotry and sectarianism in modern Scotland. Edinburgh, Mainstream, 2000, 281 p.

5638. Flugschriften gegen die Reformation (1525–1530). Hrsg. u. bearb. v. Adolf LAUBE unt. Mitarb. v. Ulman WEIß. Berlin, Akademie, 2000, 2 vol., VIII-1332 p.

5639. Four (The) horsemen of the apocalypse: religion, war, famine and death in Reformation Europe.

Ed. By Andrew CUNNINGHAM and Ole Peter DRGRELL. Cambridge a. New York, Cambridge U. P., 2000, XIII-360 p.

5640. GREEN (Ian M.). Print and protestantism in early modern England. Oxford a. New York, Oxford U. P., 2000, XXIII-691 p.

5641. GREGORY (Jeremy). Restoration, reformation and reform, 1660–1828: Archbishops of Canterbury and their Diocese. New York, Clarendon Press, 2000, [s. p.]. (Oxford historical monographs).

5642. HAMMANN (Konrad). Universitätsgottesdienst und Aufklärungspredigt: die Göttinger Universitätskirche im 18. Jahrhundert und ihr Ort in der Geschichte des Universitätsgottesdienstes im deutschen Protestantismus. Tübingen, Mohr Siebeck, X-433 p.

5643. HAUDE (Sigrun). In the shadow of "Savage Wolves": anabaptist Münster and the German reformation during the 1530s. Boston, Humanities Press, 2000, XIII-192 p. (Studies in Central European histories).

5644. HOTSON (Howard). Johann Heinrich Alsted 1588–1638. Between Renaissance, reformation and universal reform. Oxford a. New York, Oxford U. P., 2000, XIV-271 p.

5645. HUDSON (D. Dennis). Protestant origins in India: Tamil evangelical Christians, 1706–1835. Grand Rapids, William B. Eerdmans, 2000, XI-220 p.

5646. Vacat.

5647. KETOLA (Mikko). The nationality question in the Estonian evangelical Lutheran church, 1918–1939. Helsinki, Suomen kirkkohistoriallinen seura, 2000, 361 p.

5648. KOOI (Christine). Liberty and religion: church and state in Leiden's Reformation. Leiden, Brill, 2000, XII-243 p.

5649. KOSLOFSKY (Craig). The Reformation of the dead. Death and ritual in early modern Germany, 1450–1720. Basingstoke, Macmillan, 2000, XIII-223 p.

5650. KUNKLER (Stephan). Zwischen Humanismus und Reformation: der Humanist Joachim Camerarius (1520–1574) im Wechselspiel von pädagogischem Pathos und theologischem Ethos. Hildesheim, Olms, 2000, 259 p.

5651. KUNZE (Johannes). Erasmus und Luther. Der Einfluss des Erasmus auf die Kommentierung des Galaterbriefes und der Psalmen durch Luther 1519–1521. Hamburg, Lit, 2000, 337 p.

5652. Landgraf Moritz der Gelehrte. Ein Kalvinist zwischen Politik und Wissenschaft. Hrsg. v. Gerhard MENK. Marburg, Trautvetter & Fischer, 2000, 264 p. (Beiträge zur Hessischen Geschichte, 15).

5653. MAC CULLOCH (Diarmand). Tudor church militant: Edward VI and the Protestant Reformation. London, Penguin, 2000, XVII-284 p.

5654. MANETSCH (Scott M.). Theodore Beza and the quest for peace in France, 1562–1598. Leiden a. Boston, Brill, 2000, XII-380 p.

5655. MAU (Rudolf). Evangelische Bewegung und frühe Reformation 1521 bis 1532. Leipzig, Evangelische Verlagsanstalt, 2000, 254 p. (Kirchengeschichte in Einzeldarstellungen, 2, 5).

5656. MOORE (Rosemary). The light in their consciences: early Quakers in Britain 1646–1666. University Park, Pennsylvania State U. P., 2000, XIII-314 p.

5657. MULLER (Richard Alfred). The unaccommodated Calvin: studies in the foundation of a theological tradition. New York, Oxford U. P., XII-308 p.

5658. MURDOCK (Graeme). Calvinism on the Frontier, 1620–1660. International Calvinism and the Reformed Church in Hungary and Transylvania. Oxford, Clarendon Press, 2000, XII-359 p.

5659. NONOSE (Koji). Doitsu-nômin-sensô to Syûkyô-kaikaku: Kinsei Suisu shi no ichi-danmen. (The German peasant war and the Reformation: an aspect of the early modern Swiss history), Tokyo, Keio U. P., 2000, 385 p.

5660. OSTRANDER (Rick). The life of prayer in a world of science: protestants, prayer, and American culture 1870–1930. New York, Oxford U. P., 2000, 232 p. (Religion in America Series).

5661. PARISH (Helen L.). Clerical marriage and the English Reformation: precedent, policy, and practice. Aldershot, Ashgate, 2000, X-276 p.

5662. Penitence in the age of Reformations. Ed. by Jackson LUALDI and Katharine THAYER. Burlington, Ashgate, 2000, XVI-276 p.

5663. Pier Paolo Vergerio il Giovane, un polemista attraverso l'Europa del Cinquecento. Convegno internazionale di studi, Cividale del Friuli, 15–16 ottobre 1998, a cura di Ugo ROZZO. Udine, Forum, 2000 VIII-375 p.

5664. Protestantizem – zatočišče izgnanih na Petanjcih (Nádasdyjev dvorec): [zbornik znanstvenega srečanja v Radencih in na Tišini, (28. in 29. oktober 1999)]. (Protestantism – Exiles' refuge in Petanjci [the Nádasdy Mansion]. Collection of Papers from the Conference in Radenci and in Tišina, October 28–29, 1999). Glavni urednik Jože VUGRINEC. Petanjci, Ustanova dr. Šiftarjeva fundacija a. Ljubljana, Znanstvenoraziskovalni center SAZU, 2000, 313 p.

5665. Reformation, revolt and civil war in France and the Netherlands 1555–1585. Ed. by Philip BENEDICT, Guido MARNEF, Henk VAN NIEROP and Marc VENARD. Amsterdam, Koninklijke Nederlandse Akademie van Wetenschappen, 2000, 298 p.

5666. REJ (Krzysztof Jan). Ewangelicka służba duszpasterska w wojsku polskim 1919–1950. (Die evangelische Seelsorge bei dem polnischen Militär

1919–1950). Hrsg. v. Elżbieta ALABRUDZIŃSKA. Warszawa, Ewangelickie Duszpasterstwo Wojskowe, 2000, 235 p.

5667. RONCHI DE MICHELIS (Laura). Eresia e Riforma nel Cinquecento. La dissidenza religiosa in Russia. Torino, Claudiana, 2000, 256 p.

5668. RUMMEL (Erika). The confessionalization of humanism in Reformation Germany. Oxford a. New York, Oxford U. P., 2000, 211 p.

5669. SAARD (Riho). Eesti rahvusest luterliku pastorkonna väljakujunemine ja vaba rahvakiriku projekti loomine, 1870–1917. (Virolaisen luterilaisen papiston muotoutuminen ja vapaan kansankirkon ohjelman synty 1870–1917. – Die Herausbildung der lutherischen Pfarrerschaft estnischer Nationalität und die Gründung des Projekts der freien Volkskirche in den Jahren 1870 bis 1917. – The formation of a Lutheran clergy of Estonian descent and the establishment of a programme for a free people's church, 1870–1917). Helsinki, SKS, 2000, 379 p. (ill.). (Suomen Kirkkohist. Seur. toim., 184).

5670. SCHMIDT (Christoph). Auf Felsen gesät: die Reformation in Polen und Livland. Göttingen, Vandenhoeck & Ruprecht, 2000, 341 p.

5671. STAYER (James M.). Martin Luther, German saviour: German evangelical theological factions and the interpretation of Luther, 1917–1933. Montreal a. Ithaca, McGill-Queen's U. P., 2000, XIV-177 p. (McGill-Queen's studies in the history of religion).

5672. STUNT (Timothy C. F.). From awakening to secession. Radical evangelicals in Switzerland and Britain, 1815–1835. Edinburg, T&T Clark, 2000, 402 p.

5673. Sveriges kyrkohistoria. Band 3. Reformationstid. (Histoire ecclésiastique de la Suède. Tome 3. L'Epoque de la Réforme). Ed. par Åke ANDRÉN. Stockholm: Verbum, 2000, 360 p.

5674. TICHY (Christiane). Deutsche evangelische Auslandsgemeinden in Frankreich 1918–1944. Stuttgart, Berlin u. Köln, Kohlhammer, 2000, IX-275 p. (Konfession und Gesellschaft, 17).

5675. TORZINI (Roberto). I labirinti del libero arbitrio: la discussione tra Erasmo e Lutero. Firenze, Olschki, 2000, XVII-267 p.

5676. VOOGT (Gerrit). Constraint on trial. Dirk Volckertsz Coornhert and religious freedom. Kirksville, Truman State U. P., 2000, 268 p.

5677. WAITE (Gary K.). Reformers on stage: popular drama and religious propaganda in the Low Countries of Charles V, 1515–1556. Buffalo, University of Toronto Press, 2000, XXII-364 p.

5678. ZUBER (Valentine). Le Congrès pour la tolérance (Genève, août 1953). Histoire et mémoire chez les protestants libéraux. *Bulletin de la Société de l'histoire du protestantisme français*, 2000, 146, p. 487-521.

Cf. n°s 4381, 5702, 5731

§ 5. Non-Christian religions and sects.

5679. ALI (Azra Asghar). The emergence of feminism among Indian Muslim women, 1920–1947. Oxford, New York, Oxford U. P., 2000, XXI-291 p.

5680. ALLÈS (Elisabeth). Musulmans de Chine. Une anthropologie des Hui de Henan. Paris, Ed. de l'EHESS, 2000, 334 p.

5681. AMAYA BANEGAS (Jorge Alberto). Los judíos en Honduras. Tegucigalpa, Editorial Guaymuras, 2000, 158 p. (ill.).

5682. ANDERSSON (Lars M.). En jude är en jude är en jude ...: Representationen av "juden" i svensk skämtpress omkring 1900–1930. (A Jew is a Jew is a Jew ...: representations of "Jews" in the Swedish humor press ca. 1900–1930). Lund, Nordic Academic Press, 2000, 622 p.

5683. BELINKI (Karmela). Shylock i Finland. Judarna och Finlands litteratur 1920–1970. (Shylock in Finland. Jews and Finnish literature, 1920–1970). Abo, Abo Akademi U. P., 2000, 91 p. (Religionvetenskapliga studier).

5684. BEN-DROR (Graciela). ha-Kenesiyah ha-Katolit veha-Yehudim, Argentinah 1933–1945. (Catholic Church and the Jews, Argentina 1933–1945). Yerushalayim, Merkaz Zalman Shazar le-toldot Yiḃsra'el, ha-Hevrah ha-historit ha-Yiḃsre'elit: ha-Merkaz ha-benle'umi la-heker ha-antishemiyut `a. sh. Vidal þSaḃson, ha-Universitah ha-'Ivrit, 2000, 320 p. (Sidrah le-toldot ha-antishemiyut).

5685. BERKOWITZ (Michael). The Jewish self-image in the West. New York, New York U. P., 2000, 176 p.

5686. Black Zion: African American religious encounters with Judaism. Ed. by Yvonne CHIREAU and Nathaniel DEUTSCH. New York, Oxford U. P., 2000, XII-241p.

5687. CARLEBACH (E.). Jews, Christians, and the endtime in early modern Germany. *Jewish history*, 2000, 14, 3, p. 331-344.

5688. CHIH (Rachida). Le soufisme au quotidien. Confréries d'Egypte au XXe siècle. Arles et Paris, Sinbad/Actes Sud, 2000, 362 p.

5689. Contemporary debates in Islam. An anthology of modernist and fundamentalist thougth. Ed. by Mansoor MOADDEL and Kamran TALATTOF. Basingstoke, Macmillan, 2000, X-382p.

5690. COOK (Michael). Commanding right and forbidding wrong in Islamic thought. Cambridge, Cambridge U. P., 2000, XVII-702 p.

5691. COPLEY (A. R. H.). Gurus and their followers: new religious reform movements in colonial India. New Delhi a. Oxford, Oxford U. P., 2000, XXII-235 p.

5692. DAVIES (Horton), DAVIES (Marie-Hélène). French Huguenots in English-speaking lands. New

York, Peter Lang, 2000, 147 p. (Studies in Church history, 11).

5693. DIANTEILL (Erwan). Des dieux et des signes. Initiation, écriture et divination dans les religions afrocubaines. Paris, Ed. de l'EHESS, 2000, 381 p.

5694. DINER (Hasia R.). Lower East Side memories. A Jewish place in America. Princeton, Princeton U. P., 2000, XIII-219 p.

5695. DITHMAR (Christiane). Zinzendorfs nonkonformistische Haltung zum Judentum. Heidelberg, Winter, 2000, 335 p. (Schriften der Hochschule für Jüdische Studien Heidelberg, 1).

5696. EATON (Richard Maxwell). Essays on Islam and Indian history. New Delhi a. Oxford, Oxford U. P., X-275 p.

5697. EIDE (Oyvind). Revolution and religion in Ethiopia: the growth and persecution of the Menane Yesus Church 1974–1985. Oxford, Currey, 2000, XX-320 p.

5698. FREIBERGER (Oliver). Der Orden in der Lehre: zur religiösen Deutung des Sangha im frühen Buddhismus. Wiesbaden, Harrassowitz, 2000, 278 p.

5699. Geschichte (Zur) und Kultur der Juden in Ost- und Westpreußen. Hrsg. v. Michael BROCKE. Hildesheim, Olms, 2000, 663 p.

5700. GRIFFEL (Frank). Apostasie und Toleranz im Islam: die Entwicklung zu al-Gazalis Urteil gegen die Philosophie und die Reaktionen der Philosophen. Leiden, Brill, 2000, X-521 p.

5701. Haß, Verfolgung und Toleranz: Beiträge zum Schicksal der Juden von der Reformation bis in die Gegenwart. Hrsg. v. Thomas SIRGES. Frankfurt am Main, Lang, 2000, 211 p.

5702. HEINRICHS (Wolfgang E.). Das Judenbild im Protestantismus des Deutschen Kaiserreichs. Ein Beitrag zur Mentalitätsgeschichte des deutschen Bürgertums in der Krise der Moderne. Köln, Rheinland-Verlag, 2000, XIII-851 p. (Schriftenreihe des Vereins für Rheinische Kirchengeschichte, 145).

5703. Identità dissimulata (L'). Giudaizzanti iberici nell'Europa cristiana dell'età moderna. A cura di Pier Cesare IOLY ZORATTINI, Firenze, Olschki, 2000, 385 p.

5704. Jewish and Christian doctrines: the classics compared. Ed. by Jacob NEUSNER and Bruce CHILTON. London. New York, Routledge, 2000, XI-240 p.

5705. Jews and gender. The challenge to hierarchy. Ed. by Jonathan FRANKEL. Oxford, Institute of Contemporary Jewry a. Jerusalem, Hebrew Univerity of Jerusalem, 2000, 395 p. (Studies in contemporary jewry, 16).

5706. Judaism and Islam: boundaries, communications and interaction; essays in honor of William M. Brinner. Ed. by Benjamin H. HARY. Leiden, Brill, 2000, XLIII-438 p.

5707. Juifs (Les) et la ville. Ed. par Chantal Bordes-BENAYOUN et Patrick CABANEL. Toulouse, Presses universitaires du Mirail, 2000, 308 p. (Tempus).

5708. KAPLAN (Yosef). An alternative path to modernity: the Sephardi diaspora in Western Europe. Leiden, Brill, 2000, IX-309 p.

5709. KAPSTEIN (Matthew). The Tibetan assimilation of Buddhism. Conversion, contestation, and memory. Oxford, Oxford U. P., 2000, XX-316 p.

5710. KEUM (Jang-tae). Confucianism and Korean thoughts. Seoul, Jimoondang, 2000, VII-248 p.

5711. KIEVAL (Hillel J.). Languages of community: the Jewish experience in the Czech lands. Berkeley a. Los Angeles, University of California Press, 2000, XI-311 p.

5712. KLIER (Dzh. D.). Rossiia sobiraet svoikh evreev: Proiskhozhdenie evreiskogo voprosa v Rossii; 1772–1825. (Russia gathers her Jews, the origins of the "Jewish Question" in Russia; 1772–1825). Moscow, Mosty kul'tury. Gesharim, Jerusalem, 2000, 351 p. (Sovmestnyi izdatel'skii proekt; Bibliotheca Judaica; Seriia "Sovremennye Issledovaniia").

5713. LONDON (Louise). Whitehall and the Jews, 1933–1948: British immigration policy, Jewish refugees, and the Holocaust. Cambridge, Cambridge U. P., 2000, XIII-313 p.

5714. "Machen Sie doch unseren Islam nicht gar zu schlecht". Der Briefwechsel der Islamwissenschaftler Ignaz Goldziher und Martin Hartmann, 1894–1914. Hrsg. und kommentiert von Ludmila HANISCH. Wiesbaden, Harrassowitz, 220000, XXVII-465 p.

5715. MAKDISI (Ussama). The culture of sectarianism: community, history, and violence in nineteenth century Ottoman Lebanon. Berkeley, Univ. of California Press, XV-259 p.

5716. MASSEY (Irving). Philo-semitism in nineteenth-century German literature. Tübingen, Niemeyer, 2000, VI-199 p.

5717. MELAMED (Diego). Los judíos y el menemismo: un reflejo de la sociedad argentina. Buenos Aires, Editorial Sudamericana, 2000, 284 p.

5718. MICHAŁOWSKA (Anna). Między demokracją a oligarchią. Władze gmin żydowskich w Poznaniu i Swarzędzu (od połowy XVII do końca XVIII wieku). (Zwischen Demokratie und Oligarchie. Die Behörden der jüdischen Gemeinden in Poznań und Swarzędz, von der Mitte des XVII. bis zum Ende des XVIII. Jahrhunderts). Warszawa, Dialog, 2000, 335 p.

5719. NIEZEN (Ronald). Spirit wars, native North Americain religions in the age of nation building. Berkeley a. Los Angeles, University of California Press, 2000, 256 p.

5720. NOACK (Christian). Muslimischer Nationalismus im russischen Reich: Nationsbildung und Nationalbewegung bei Tataren und Baschkiren: 1861–1917. Stuttgart, Franz Steiner Verlag, 2000, 614 p.

5721. RIVOAL (Isabelle). Les maîtres du secret. Ordre mondain et ordre religieux dans la communauté druze en Israël. Paris, Editions de l'EHESS, 2000, 427 p.

5722. RO'I (Yaacov). Islam in the Soviet Union: from the second World War to Gorbachev. New York, Columbia U. P., 2000, XXVII-764 p.

5723. ROBINSON (David). Paths of accommodation: Muslim societies and French colonial authorities in Senegal and Mauritania, 1880–1920. Athens, Ohio U. P., 2000, XVI-361 p.

5724. RUDERMAN (David B.). Jewish enlightenment in an English key: Anglo-Jewry's construction of modern Jewish thought. Princeton a. Oxford, Princeton U. P., 2000, XV-291 p.

5725. SONNENBERG-STERN (Karina). Emancipation and poverty: the Ashkenazi Jews of Amsterdam, 1796–1850. Basingstoke, Macmillan, 2000, XVI-236 p.

5726. TARADEL (Ruggero), RAGGI (Barbara). La segregazione amichevole. La «Civiltà Cattolica» e la questione ebraica 1850–1945. Roma, Editori Riuniti, 2000, XIV-265 p.

5727. WYRWA (Ulrich). «Perché i moderni rabbini pretendono di dare ad intendere una favola chimerica ...». L'illuminismo toscano e gli ebrei. *Quaderni storici*, 2000, 35, 103, p. 139-162.

5728. ZAVOS (John). The emergence of Hindu nationalism in India. Delhi, Oxford U. P., 2000, VIII-245 p.

Cf. n^{os} *916, 4192, 4550, 5032, 5432, 5452, 5466, 7232*

M

HISTORY OF MODERN CULTURE

§ 1. General. 5729-5898. – § 2. Academies, universities and intellectual organizations. 5899-5959. – § 3. Education. 5960-6040. – § 4. The Press. 6041-6113. – § 5. Philosophy. 6114-6209. – § 6. Exact, natural, medical sciences and technique. 6210-6349. – § 7. Literature (*a*. General; *b*. Renaissance; *c*. Classicism; *d*. Romanticism and after). 6350-6500. – § 8. Art and industrial art (*a*. General; *b*. Architecture; *c*. Sculpture, painting, etching and drawing; *d*. Decorative, popular and industrial art). 6501-6634. – § 9. Music, theatre, cinema and broadcasting. 6635-6754.

§ 1. General.

* 5729. Bibliografia italiana di studi sull'Umanesimo e il Rinascimento. 1998. [1997. Cf. Bibl. 99, n° 5559.] Firenze, Olschki, 2000, [s. p.]. (Rinascimento, XXXVIII. Supplemento).

* 5730. CONLON (Pierre M.). Le siècle des lumières: bibliographie chronologique. Part: T. 20. 1782–1783. Genève, Droz, 2000, XXVIII-426 p. (Histoire des idées et critique littéraire, 384).

* 5731. TEDESCHI (John A.). The Italian Reformation of the sixteenth century and the diffusion of Renaissance culture: a bibliography of the secondary literature, ca. 1750–1997. Modena e Ferrara, F.C. Panini e ISR, 2000, LXIII-1047 p. (Strumenti / Istituto di studi rinascimentali).

* 5732. WATANABE-O'KELLY (Helen), SIMON (Anne). Festivals and ceremonies: a bibliography of works relating to court, civic, and religious festivals in Europe 1500–1800. London, Mansell, 2000, XIX-533 p.

** 5733. Brüder Grimm, Briefwechsel mit Ludwig Hassenpflug (einschließlich der Briefwechsel zwischen Ludwig Hassenpflug und Dorothea Grimm, geb. Wild, Charlotte Hassenpflug, geb. Grimm, ihren Kindern und Amalie Hassenpflug). Hrsg. und bearb. v. Ewald GROTHE. Kassel, Brüder Grimm-Gesellschaft, 2000, 448 p. (Bruder Grimm, Werke und Briefwechsel, Kasseler Ausgabe, 2).

** 5734. PANIZZA (Giorgio), COSTA (Barbara). L'Archivio Verri. Parte seconda, la «raccolta verriana». Milano, Fondazione Raffaele Mattioli, 2000, 346 p.

5735. 18. Jahrhundert (Das). Vernunft und Träume. Hrsg. v. Michael JEISMANN. München, Beck, 2000, 92 p. (Beck'sche Reihe, 4118).

5736. 1949/1989 cultural perspectives on division and unity in East and West. Ed. by Clare FLANAGAN and Stuart TABERNER. Amsterdam a. Atlanta, Rodopi, 2000, VI-297 p.

5737. ABASHEV (Vladimir V.). Perm' kak tekst: Perm' v russkoj kul'ture i literature XX v. (Perm as a text: the city of Perm in Russian culture and literature of the 20th century). Perm', Izd-vo Permskogo un-ta, 2000, 403 p.

5738. ANDERMANN (Kurt). Die geistlichen Staaten am Ende des Alten Reiches. *Historische Zeitschrift*, 2000, 271, 3, p. 593-620.

5739. Années 68 (Les): le temps de la contestation. Ed. par Geneviève DREYFUS-ARMAND. Bruxelles, Editions Complexe, 2000, 525 p. (Histoire du temps présent).

5740. ARCHER-STRAW (Petrina). Negrophilia: Avant-Garde Paris and black culture in the 1920s. London, Thames and Hudson, 2000, 200 p. (ill.).

5741. ARMITAGE (David). The ideological origins of the British empire. Cambridge, Cambridge U. P., 2000, XII- 240 p. (Ideas in context, 59).

5742. BATTISTINI (Andrea). Il Barocco. Cultura, miti, immagini. Roma, Salerno Editrice, 2000, 330 p.

5743. BEN-GHIAT (Ruth). La cultura fascista. Bologna, Mulino, 2000, 354 p. (Biblioteca storica).

5744. BEßLICH (Barbara). Wege in den "Kulturkrieg". Zivilisationskritik in Deutschland 1890–1914. Darmstadt, Wissenschaftliche Buchgesellschaft, 2000, IX-416 p.

5745. BLITZ (Hans M.). Aus Liebe zum Vaterland. Die Idee der Nation im 18. Jh. Hamburg, Hamburger Edition, 2000, 436 p.

5746. BLOK (Frans F.). Isaac Vossius and his circle: his life until his farewell to Queen Christina of Sweden, 1618–1655. Göningen, Egbert Forsten P., 2000, 520 p.

5747. BOEHM (Laetitia). Studium, Büchersammlung, Bildungsreise: Elemente gelehrter Allgemeinbildung und individueller Ausprägung historisch-politischer Weltanschauung im konfessionellen Zeitalter. *Acta historica Leopoldina*, 2000, 31, p.117-151.

5748. BOUDOU (Bénédicte). Mars et les muses dans l'Apologie pour Hérodote d'Henri Estienne. Genève, Droz, 2000, 686 p. (Travaux d'humanisme et Renaissance, 334).

5749. BOUWSMA (William J.). The waning of the Renaissance 1550–1640. New Haven, Yale U. P., 2000, XI-288 p. (Yale Intellectual History of the West).

5750. BROOKS (Jeffrey). Thank you, comrade Stalin! Soviet public culture from Revolution to Cold War. Princeton, Princeton U. P., 2000, XX-319 p.

5751. BUCK-MORSS (Susan). Dreamworld and catastrophe: the passing of mass utopia in East and West. Cambridge a. London, MIT, 2000, XVI-368 p. (ill.).

5752. Bürgerliche Wertehimmel (Der). Innenansichten des 19. Jahrhunderts. Hrsg. v. Manfred HETTLING und Stefan-Ludwig HOFFMANN. Göttingen, Vandenhoeck & Ruprecht, 2000, 307 p.

5753. CABADA (Ladislav). Intelektuálové a idea komunismu v českých zemích 1900–1939. (Intellectuals and the idea of communism in the Bohemian Lands, 1900–1939). Praha, Institut pro středoevropskou kulturu a politiku, 2000, 197 p.

5754. CAPLAN (Jay). In the king's wake: Post-Absolutist culture in France. Chicago, University of Chicago Press, 2000, 222 p.

5755. CERASI (Laura). Gli ateniesi d'Italia. Associazioni di cultura a Firenze nel primo Novecento. Milano, Franco Angeli, 2000, 234 p.

5756. CHOCANO MENA (Magdalena). La fortaleza docta. Elite letrada y dominación social en México colonial, siglos XVI–XVII. Barcelona, Ediciones Bellaterra, 2000, 415 p.

5757. Civil histories: essays presented to Sir Keith Thomas. Ed. by Peter BURKE, Brian HARRISON and Paul SLACK. Oxford, Oxford U. P., 2000, 399 p.

5758. CLARK (Peter). British clubs and societies 1580–1800. The origins of an associational world. Oxford, Clarendon Press, 2000, XII-516 p.

5759. COLLARD (Patrick). Cambio de siglo: ideas, mentalidades, sensibilidades en España hacia 1900. Amsterdam, Rodopi, 2000, 148 p. (Foro hispánico, 18).

5760. Commercium: scambi culturali italo-tedeschi nel 18. secolo = deutsch-italienischer Kulturaustausch im 18. Jahrhundert. A cura di Federica LA MANNA. Firenze, L. S. Olschki, 2000, IX-230 p. (Villa Vigoni, 5).

5761. CONDETTE (J.-F.). La translation des cendres d'Emile Zola au Panthéon: la difficile et posthume revanche de l'intellectuel dreyfusard (juillet 1906–juin 1908). *Revue historique*, 2000, 124, 615, p. 655-684.

5762. COSANDEY (Fanny). La reine de France. Symbole et pouvoir XVe–XVIIIe siècle. Paris, Gallimard, 2000, 414 p.

5763. CRUNDEN (Robert M.). Body and soul: the making of American modernism. New York, Basic Books, 2000, XVII-475 p.

5764. Currents in contemporary French intellectual life. Ed. by Christopher FLOOD and Nick HEWLETT. Basingstoke, Macmillan a. New York, St. Martin's Press, 2000, IX-248 p.

5765. DARNTON (Robert). An early information society: news and the media in eighteenth-century Paris. *American historical review*, 2000, 105, 1, p. 1-35.

5766. DARROW (Margaret M.). French women and the First World War: war stories on the home front. New York, Berg, 2000, IX-341 p. (Legacy of the Great War).

5767. DAUPHIN (Cécile). Prête moi ta plume ... Les manuels épistolaires au XIXe siècle. Paris, Kimé, 2000, 199 p.

5768. DAVIES (Peter). Divided loyalties: East German writers and the politics of German division 1945–1953. Leeds, Maney Publishing for the Modern Humanities Research Association and the Institute of Germanic Studies, University of London School of Advanced Study, 2000, VI-277 p. (MHRA texts and dissertations, 49).

5769. DELON (Michel). Le savoir-vivre libertin. Paris, Hachette, 2000, 348 p.

5770. DEMM (Eberhard). Barbusse et son Feu: la dernière cartouche de la propagande de guerre française. *In*: Images civiles de la France en guerre [Cf. n° 4465], p. 43-64.

5771. Deutsche Frauen der frühen Neuzeit. Dichterinnen, Malerinnen, Mäzeninnen. Hrsg. v. Kerstin MERKEL und Heide WUNDER. Darmstadt, Primus-Verlag, 2000, 294 p.

5772. DOGLIO (Maria Luisa). L'arte delle lettere. Idea e pratica della scrittura epistolare tra Quattro e Seicento. Bologna, Bologna, Il Mulino, 2000, 236 p.

5773. «Dolce dono graditissimo». La lettera privata dal Settecento al Novecento. A cura di Maria Luisa BETRI e Daniela MALDINI CHIARITO. Milano, Franco Angeli, 2000, 474 p.

5774. D'ORSI (Angelo). La cultura a Torino tra le due guerre. Torino, Einaudi, 2000, XV-377 p. (Biblioteca Einaudi, 87).

5775. EAGLES (R. D. E.). Francophilia in English society, 1748–1815. Basingstoke, Macmillan, 2000, X-229 p. (ill.).

5776. ESPAGNE (Michel). Le creuset allemand: histoire interculturelle de la Saxe, XVIIIe–XIXe siècles. Paris, Presses universitaires de France, 2000, VII-328 p. (Perspectives germaniques).

5777. Europa und die Türken in der Renaissance. Hrsg. v. Bodo GUTHMÜLLER und Wilhelm KÜHLMANN. Tübingen, Max Niemeyer Verlag, 2000, VII-451 p.

5778. Europäische Identität? Versuch, kulturelle Aspekte eines Phantoms zu beschreiben. Hrsg. v. Martin KUTZ und Petra WEYLAND. Bremen, Edition Temmen, 2000, 304 p. (Schriftenreihe des Wissenschaftlichen Forums für Internationale Sicherheit, 15).

5779. FAZLHASHEMI (Mohammad). Exemplets makt. (Le pouvoir de l'exemple. Les représentations de l'Occident en Iran de 1850 à 1980). Eslöv, Brutus Östlings Bokförlag Symposion, 2000, 245 p.

5780. Female communities, 1600–1800: literary vision and cultural realities. Ed. by Rebecca D'MONTE and Nicole POHL. Houndmills, Basingstoke a. London, Macmillan, 2000, XVI-264 p.

5781. Femmes en toutes lettres: les épistolières du XVIIIe siècle. Ed. par Marie-France SILVER et Marie-Laure GIROU SWIDERSKI. Oxford, Voltaire Fondation, 2000, 277 p. (Studies on Voltaire and the Eighteenth century, SVEC-04).

5782. Festive culture in Germany and Europe from the sixteenth to the twentieth century. Ed. by Karin FRIEDRICH. Lewiston, Mellen Press, 2000, 373 p.

5783. Fin (The) de siècle. A reader in cultural history, 1880–1900. Ed. by Sally LEDGER and Roger LUCKHURST. Oxford, Oxford U. P., 2000, 392 p.

5784. FISCHER (Joachim). Das Deutschlandbild der Iren 1890–1939: Geschichte, Form, Funktion. Heidelberg, Universitätsverlag C. Winter, 2000, XV-680 p. (ill.). (Anglistische Forschungen, 284).

5785. FLASCH (Kurt). Die geistige Mobilmachung. Die deutschen Intellektuellen und der Erste Weltkrieg: ein Versuch. Berlin, Alexander Fest, 2000, 447 p. (ill.).

5786. FOUSEK (John). To lead the free world: American nationalism and the cultural roots of the Cold War. Chapel Hill a. London, University of North Carolina Press, 2000, XIV-253 p. (ill.).

5787. FOX (Adam). Oral and literate culture in England, 1500–1700. Oxford, Clarendon Press, 2000, XII-498 p.

5788. FRÄNGSMYR (Tore). Svensk idéhistoria. Bd. 2. År 1809–2000. (Histoire des idées en Suède. Tome 2. 1809–2000). Stockholm, Natur och Kultur, 2000, 422 p. (ill.).

5789. FREITÄGER (Andreas). Johannes Cincinnius von Lippstadt (ca. 1485–1555). Bibliothek und Geisteswelt eines westfälischen Humanisten. Münster, Aschendorff, 2000, 438 p. (Veröffentlichungen der Historischen Kommission für Westfalen, 18. Westfälische Biographien, 10).

5790. FREVERT (Ute). Das neue Jahrhundert: Europäische Zeitdiagnosen und Zukunftsentwürfe um 1990. Göttingen, Vandenhoeck & Ruprecht, 2000, 308 p. (Geschichte und Gesellschaft, 18).

5791. GABRIELLI (Patrizia). Mondi di carta. Lettere, autobiografie, memorie. Siena, Protagon Editori Toscani, 2000, 222 p.

5792. GEORGIEV (Velichko). Bulgarskata inteligentsiia i natsionalnata kauza v Purvata svetovna voina: Suiuzut na bulgarskite ucheni, pisateli i khudozhnitsi (1917–1918). (Bulgarian intellectuals and the national cause in World War One: the Union of the Bulgarian scholars, writers and artists, 1917–1918). Sofiia, Makedonski nauchen institut, 2000, 112 p.

5793. GERBOD (Paul). La Restauration hors foyer en Europe. Paris, H. Champion, 2000, 150 p. (Histoire culturelle de l'Europe, 1).

5794. GERVASONI (Marco). L'intellettuale come eroe: Piero Gobetti e le culture del Novecento. Firenze, La nuova Italia, 2000, IX-482 p. (Biblioteca di storia, 81).

5795. Geschichte der österreichischen Humanwissenschaften. Band 3. Teil II. Menschliches Verhalten und gesellschaftliche Institutionen – Wirtschaft, Politik und Recht. Hrsg. v. Karl ACHAM. Wien, Passagen-Verlag, 2000, 520 p.

5796. Gesichter der Weimarer Republik. Eine physiognomische Kulturgeschichte. Hrsg. v. Claudia SCHMÖLDERS und Sander GILMAN. Köln, Dumont, 2000, 335 p.

5797. GRELL (Chantal). Histoire intellectuelle et culturelle de la France du Grand Siècle, 1654–1715. Paris, Nathan, 2000, 304 p. (Fac. Histoire).

5798. GRÖSSING (Helmuth). Frühling der Neuzeit. Wissenschaft, Gesellschaft und Weltbild in der frühen Neuzeit. Wien, Erasmus, 2000, 182 p. (Perspektiven der Wissenschaftsgeschichte, 12).

5799. GUGGISBERG (Daniel). Das Bild der "Alten Eidgenossen" in Flugschriften des 16. bis Anfang 18. Jh. (1531–1712). Tendenzen und Funktionen eines Geschichtsbildes. Frankfurt am Main, Peter Lang, 2000, XVI-845 p.

5800. GUNDLE (Stephen). Between Hollywood and Moscow: the Italian communists and the challenge of mass culture, 1943–1991. Durham a. London, Duke U. P., 2000, 269 p. (ill.). (American encounters/global interactions).

5801. GUNN (Simon). The public culture of the Victorian middle class: ritual and authority and the English industrial city, 1840–1914. Manchester, Manchester U. P., 2000, X-207 p.

5802. HÄRMÄNMAA (Marja). Un patriota che sfidò la decadenza. F. T. Marinetti e l'immagine dell'uomo nuovo fascista, 1929–1944. Helsinki Suomalainen tiedeakatemia, 2000, 379 p. (ill.). (A. Acad. Sci. Fennicae. Ser Humana, 310).

5803. HAYWARD (Rhodri). Policing dreams: history and the moral uses of the unconscious. *History workshop*, 2000, 49, p. 142-160.

5804. HEIN (Rudolf Branko). Gewissen bei Adrian von Utrecht (Hadrian VI), Erasmus von Rotterdam und Thomas More: ein Beitrag zur systematischen Analyse des Gewissensbegriffs in der katholischen nordeuropäischen Renaissance. Münster, Lit, 2000, IX-539 p.

5805. HENNING (Joseph M.). Outposts of civilization: race, religion, and the formative years of American-Japanese relations. New York, New York U. P., 2000, XII-243 p.

5806. HEWITSON (Mark). National identity and political thought in Germany: Wilhelmine depictions of the French Third Republic, 1890–1914. Oxford, Oxford U. P., 2000, X-288 p. (Oxford historical monographs).

5807. HEXELSCHNEIDER (Erhard). Kulturelle Begegnungen zwischen Sachsen und Russland 1790–1849. Köln, Böhlau, 2000, 617 p. (Geschichte und Politik in Sachsen, 13).

5808. HILLEBRECHT (Frauke). Göteborg in den nordischen Kulturideologie. Heidelberg, Universitätsverlag C. Winter, 2000, 239 p. (Skandinavistische Arbeiten, 16).

5809. Historicism, psychoanalysis, and early modern culture. Ed. by Carla MAZZIO and Douglas TREVOR. New York a. London, Routledge, 2000, VIII-417 p. (ill.). (Culture work).

5810. History, religion and culture: British intellectual history, 1750–1950. Ed. by Stefan COLLINI, Richerd WHATMORE and Brian YOUNG. Cambridge, Cambridge U. P., 2000, VIII-289 p.

5811. HOURMANT (François). Au pays de l'avenir radieux. Voyages des intellectuels français en U.R.S.S., à Cuba et en Chine populaire. Paris, Aubier, 2000, [s. p.]. – IDEM. La croisière rouge, entre simulacre et théâtrocratie. Le système de privilèges des voyageurs aux pays de l'Avenir Radieux. *Revue historique*, 2000, 124, 613, p. 121-156.

5812. IANZITI (Gary). A life in politics: Leonardo Bruni's Cicero. *Journal of the history of ideas*, 2000, 61, 1, p. 39-58.

5813. Intellectuals, identities and popular movements: ten case studies from France, Britain, Germany and the Balkans. Ed. by Clive E. HILL. London, Middlesex U. P., 2000, 334 p.

5814. Intellektuelle im 20. Jahrhundert in Deutschland: ein Forschungsreferat. Hrsg. von Jutta SCHLICH. Tübingen, M. Niemeyer, 2000, IX-395 p. (Internationales Archiv für Sozialgeschichte der deutschen Literatur. Sonderheft, 11).

5815. Intellektuelle im Nationalsozialismus. Hrsg. von Wolfgang BIALAS und Manfred GANGL. Frankfurt am Main a. New York, P. Lang, 2000, 363 p. (Schriften zur politischen Kultur der Weimarer Republik, 4).

5816. Intellettuali (Gli) e la Grande Guerra. A cura di Vincenzo CALÌ, Gustavo CORNI e Giuseppe FERRANDI. Boligna, Il Mulino, 2000, 425 p. (Annali dell'Istituto storico italo-germanico di Trento. Quaderni, 54).

5817. Italija i russkaja kul'tura XV–XX vekov. (Italy and Russian culture, 15th–20th centuries: [Articles and Publications]). Part 2. Ed. Nelli P. KOMOLOVA. RAN, In-t vseobshchej istorii. Moskva, [s. n.], 2000, 135 p. – Rossija i Italija. (Russia and Italy). Vol. 4. Vstrecha kul'tur. (Meeting of the cultures). Ed. Nelli P. KOMOLOVA. RAN, In-t vseobshchej istorii; Assots. kul'turnogo i delovogo sotrudnichestva s Italijej. Moskva, Nauka, 2000, 361 p. (ill.).

5818. JENNINGS (Lawrence C.). French anti-slavery: the movement for the abolition of slavery in France, 1802–1848. Cambridge, Cambridge U. P., 2000, X-320 p.

5819. JONAS (Raymond). France and the cult of the sacred heart: an epic tale for modern times. Berkeley and Los Angeles, University of California Press, 2000, XV-308 p. (Studies on the History of Society and Culture, 29).

5820. KAISER (Thomas). The evil empire? The debate on Turkish despotism in eighteenth-century French political culture. *Journal of modern history*, 2000, 72, 1, p. 6-34.

5821. KAPLAN (Alice). The collaborator: the trial and execution of Robert Brasillach. Chicago, University of Chicago Press, 2000, XVI-308 p.

5822. KENT (Dale). Cosimo de'Medici and the Florentine Renaissance: the patron's oeuvre. New Haven a. London, Yale U. P., 2000, XIV-538 p. (ill.).

5823. KLEIN (Natalie). "L'humanité, le christianisme, et la liberté": die internationale philhellenische Vereinsbewegung der 1820er Jahre. Mainz, von Zabern, 2000, XII-382 p. (Veröffentlichungen des Instituts für europäische Geschichte Mainz, Abt. für Universalgeschichte, 178).

5824. KLEPSCH (Michael). Romain Rolland im Ersten Weltkrieg. Ein Intellektueller auf verlorenem Posten. Stuttgart, Berlin u. Köln, Kohlhammer, 2000, 312 p.

5825. KLÖTZER (Wolfgang). Kleine Schriften zur Frankfurter Kulturgeschichte. Vol. 2. Keine liebere Stadt als Frankfurt. Frankfurt am Main, Kramer, 2000, 458 p. (Studien zur Frankfurter Geschichte, 45).

5826. KNAPÍK (Jiří). Kdo spoutal naši kulturu. Portrét stalinisty Gustava Bareše. (Who did bind our culture. Portrait of Stalin's man Gustav Bareš). Přerov, Šárka, 2000, 205 p.

5827. KUKLÍK (Jan), HASIL (Jan). Výbor z textů k dějinám českého myšlení. (Texts concerning the history of Czech idea). Praha, Karolinum, 2000, 249 p.

5828. Kultur und Wissenschaft beim Übergang ins "Dritte Reich". Hrsg. v. Carsten KÖNNEKER, Arnd FLORACK und Peter GEMEINHARDT. Marburg, Tectum, 2000, 186 p. (ill.).

5829. Kulturgeschichte Ostpreußens in der Frühen Neuzeit. Akten des interdisziplinären Kolloquiums in Rauschen/Swetlogorsk vom 18–25 September 1994. Hrsg. v. Klaus GRABER, Manfred KOMOROWSKI und Axel WALTER. Tübingen, Max Niemeyer Verlag, 2000, 980 p. (Frühe Neuzeit, 56).

5830. LAKE (Peter), QUESTIER (Michael). Puritans, papists, and the "Public sphere" in early modern England: the Edmund Campion affair in context. *Journal of modern history*, 2000, 72, 3, p. 587-627.

5831. LEONOVA (Lira S.). "Ja ne mogu ujti v odnu nauku ...": Obshchestvenno-politicheskie vzgljady V.I. Vernadskogo. (Social and Political Views of Vladimir I. Vernadsky). Sankt-Peterburg, Aletejja, 2000, 394 p. (ill.).

5832. LLANQUE (Marcus). Demokratisches Denken im Krieg: die deutsche Debatte im Ersten Weltkrieg. Berlin, Akademie Verlag, 2000, 365 p. (Politische Ideen, 11).

5833. LÓPEZ CRESPÍ (Miquel). Cultura i antifranquisme. Barcelona, Edicions de 1984, 2000, 280 p. (Assaig, 1).

5834. LOWE (K. J. P.). Cultural links between Portugal and Italy in the Renaissance. Oxford a. New York, Oxford U. P., 2000, XVIII-330 p. (ill.).

5835. MAC CARTHY (Conor). Modernisation: crisis and culture in Ireland, 1969–1992. Dublin, Four Courts Press, 2000, 240 p.

5836. Mao's unfilial sons. Cultural revolution and the democratic critics of the early modernization period. *In*: China @com. Näkökulmia 2000-luvun Itä-Aasiaan. Toim. Eevamaria MIELONEN. Turku, Turun yliopisto, 2000, p. 122-151.

5837. MATHY (Jean-Philippe). French Resistance: the French-American culture wars. Minneapolis, University of Minnesota Press, 2000, XI-211 p.

5838. Medien und Weltbilder im Wandel der Frühen Neuzeit. Hrsg. v. Franz MAUELSHAGEN und Benedikt MAUER. Augsburg, Wißner, 2000, 218 p. (Documenta Augustana, 5).

5839. MEHLMAN (Jeffrey). Émigré New York: French intellectuals in wartime Manhattan, 1940–1944. Baltimore, Johns Hopkins U. P., 2000, 209 p.

5840. MOMMSEN (Wolfgang J.). Bürgerliche Kultur und politische Ordnung: Künstler, Schriftsteller und Intellektuelle in der deutschen Geschichte 1830–1933. Frankfurt am Main, Fischer Taschenbuch Verlag, 2000, 272 p.

5841. MOOGK (Peter N.). La Nouvelle France. The making of French Canada. A cultural history. East Lansing, Michigan State U. P., 2000, XIX-340 p.

5842. MORSE (David). The age of virtue: British culture from the Restoration to Romanticism. Houndmills, Basingstoke a. London, Macmillan, 2000, VIII-330 p.

5843. MORTIER (Roland). Les combats des Lumières: recueil d'études sur le dix-huitième siècle. Ferney-Voltaire, Centre international d'étude du XVIIIe siècle, 2000, XLIX-423 p. (Publications du Centre international d'étude du XVIIIe siècle, 5).

5844. MÜLLER (Jan-Werner). Another country: German intellectuals, unification and national identity. New Haven, Yale U. P., 2000, 310 p.

5845. MUNCK (Thomas). The Enlightenment. A comparative social history 1721–1794. London, Arnold a. New York, Oxford U. P., 2000, XII-249 p.

5846. Nation transformed (A) by information: how information has shaped the United States from colonial times to the present. Ed. by Alfred D. CHANDLER Jr. and James W. CORTADA. Oxford, Oxford U. P., 2000, XII-380 p.

5847. NELLES (Paul). Sainte-Beuve between the Renaissance and Enlightenment. *Journal of the history of ideas*, 2000, 61, 3, p. 453-492.

5848. NETHERCOTT (Frances). Russia's Plato: Plato and the platonic tradition in Russian education, science and ideology (1840–1930). Aldershot, Ashgate, 2000, IX-233 p.

5849. Nicola Chiaromonte, Ignazio Silone: l'eredità di "Tempo presente". A cura di Goffredo FOFI, Vittorio GIACOPINI e Monica NONNO. Roma, Fahrenheit 451, 2000, 133 p.

5850. OBOLENSKAJA (Svetlana V.). Germanija i nemtsy glazami russkikh, XIX vek. (Germany and Germans through the eyes of 19th-century Russians). RAN, In-t vseobshchej istorii. Moskva, [s. n.], 2000, 209 p.

5851. Ocherki russkoj kul'tury XIX v. (Essays on Russian 19th-century culture). Vol. 2. Vlast' i kul'tura. (Power and culture). Ed. Lidija V. KOSHMAN. Moskva, Izd-vo Moskovskogo un-ta, 2000, 479 p. (ill.).

5852. PARIS (Michael). Warrior nation: images of war in British popular culture, 1850–2000. London, Reaktion, 2000, 303 p. (ill.).

5853. PELIZZARI (Maria Rosaria). La penna e la zappa. Alfabetizzazione, culture e generi di vita nel Mezzogiorno moderno. Salerno, P. Laveglia editore, 2000, 274 p.

5854. PENDERGAST (Tom). Creating the modern man: American magazines and consumer culture, 1900–1950. Columbia a. London, University of Missouri Press, 2000, X-289 p. (ill.).

5855. Petrus Canisius SJ (1521–1597). Humanist und Europäer. Hrsg. v. Rainer BERNDT. Berlin, Akademie, 2000, 500 p. (Erudiri Sapientia. Studien zum Mittelalter und zu seiner Rezeptionsgeschichte, 1).

5856. Political thought in seventeenth-century Ireland: kingdom or colony. Ed. by Jane H. OHLMEYER. Cambridge, Cambridge U. P., 2000, XVII-290 p.

5857. PORTER (Roy). Enlightenment: Britain and the making of the modern world. London, Allen Lane, 2000, [s. p.].

5858. Renaissance civic humanism: reappraisals and reflections. Ed. by James HANKINS. New York, Cambridge U. P., 2000, X-314 p. (Ideas in context, 57).

5859. Rethinking Victorian culture. Ed. by Juliet JOHN and Alice JENKINS. Houndmills, Basingstoke, 2000, XVI-244 p.

5860. REUTHER (Thomas). Die ambivalente Normalisierung. Deutschlanddiskurs und Deutschlandbilder in den USA, 1941–1955. Stuttgart, Steiner, 2000, 476 p.

5861. ROSATI (Massimo). Il patriottismo italiano: culture politiche e identità nazionale. Roma e Bari, Laterza, 2000, XIX-197 p.

5862. Russkie i nemtsy v XVIII v.: vstrecha kul'tur. (Russians and Germans in the 18th century: meeting of the cultures: [Proceedings of the 2nd Russian-German Conference on the 18th century, 16–18 May 1996]). Ed. S. Ja. KARP. Moskva, Nauka, 2000, 310 p. (ill.). [German summary and table of contents]

5863. RYAN (Mary P.). "A laudable pride in the whole of us": City Halls and civic materialism. *American historical review*, 2000, 105, 4, p. 1131-1170.

5864. SALMON (John Hearsey McMillan). Ideas and contexts in France and England from the Renaissance to the Romantics. Aldershot, Ashgate, 2000, XII-304 p. (ill). (Variorum collected studies series).

5865. Samodejatel'noe khudozhestvennoe tvorchestvo v SSSR: ocherki istorii, 1917–1932 gg. (Essays on history of amateur talent activities in the USSR, 1917–1932). Ed. L. P. SOLNTSEVA, S. Ju. RUMJANTSEV. RAN, In-t iskusstvoznanija Ministerstva kul'tury RF. Sankt-Peterburg, Dmitrij Bulanin a. Moskva, GIIS, 2000, 535 p. – Samodejatel'noe khudozhestvennoe tvorchestvo v SSSR: ocherki istorii, 1930–1950 gg. (Essays on history of amateur talent activities in the USSR, 1930–1950). Ed. L. P. SOLNTSEVA, S. Ju. RUMJANTSEV. RAN, In-t iskusstvoznanija Ministerstva kul'tury RF. Sankt-Peterburg, Dmitrij Bulanin a. Moskva, GIIS, 2000, 550 p.

5866. SAUERMANN (Eberhard). Literarische Kriegsfürsorge. Österreichische Dichter und Publizisten im Ersten Weltkrieg. Köln, Weimar u. Wien, Böhlau, 2000, 403 p. (Literaturgeschichte in Studien und Quellen, 4).

5867. SCHIKEDANZ (Hans-Joachim). Ästhetische Rebellion und rebellische Ästheten: eine kulturgeschichtliche Studie über den europäischen Dandyismus. Frankfurt am Main u. Bern, 2000, Peter Lang, 245 p.

5868. SCHLEIFER (Ronald). Modernism and time: the logic of abundance in literature, science and culture, 1880–1930. Cambridge, Cambridge U. P., 2000, XVIII-278 p.

5869. SCHMIDT (Jens). Sich hart machen, wenn es gilt: Männlichkeitskonzeptionen in Illustrierten der Weimarer Republik. Münster, LIT, 2000, 196 p. (ill.). (Geschlecht, Kultur, Gesellschaft, 3).

5870. SCHOONOVER (Thomas D.). The French in Central America: culture and commerce, 1820–1930. Wilmington, Scholarly Resources, 2000, XXV-244 p. (Latin American Silhouettes, Studies in History and Culture).

5871. SCHWARTZ (Richard Alan). Cold War culture: media and the arts, 1945–1990. New York, Checkmark Books, 2000, VII-376 p. (ill.).

5872. SCOTT-SMITH (Giles). A radical democratic political offensive: Melvin J. Lasky, "Der Monat", and the Congress for cultural freedom. *Journal of contemporary history*, 2000, 2, p. 263-280.

5873. SIGMIREAN (Cornel). Istoria formării intelectualității românești din Transilvania și Banat în epoca modernă. Cluj-Napoca, Editura Presa Universitară Clujeană, 2000, 807 p.

5874. SIMMONS (Clare A.). Eyes across the Channel: French revolutions, party history and British writing, 1830–1882. Australia, Harwood Academic, 2000, 240 p. (ill.). (Interdisciplinary nineteenth-century studies, 1).

5875. SIMONINI (Gian Luca). Il giardino italiano allo specchio. Il revival tra Ottocento e Novecento. Rimini, IdeaLibri, 2000, 167 p.

5876. SMITH (Angela K.). The second battlefield: women, modernism, and the First World War. Manchester, Manchester U. P., 2000, 214 p.

5877. SMITH (Paul Julian). The moderns: time, space, and subjectivity in contemporary Spanish culture. Oxford, Oxford U. P., 2000, XII-206 p. (ill.).

5878. SPANG (Rebecca L.). The invention of the restaurant: Paris and modern gastronomic culture. Cambridge, Harvard U. P., 2000, VII-325 p. (Harvard historical studies, 135).

5879. Späthumanismus. Studien über das Ende einer kulturhistorischen Epoche. Hrsg. v. Notker HAMMERSTEIN und Gerrit WALTHER. Göttingen, Wallstein, 2000, 312 p.

5880. ŠULC (Zdislav). Psáno inkognito. Doba v zrcadle samizdatu 1968–1989. (Written incognito. Twenty-one years in Czechoslovakia reflected in Samizdat, 1968–1989). Praha, Ústav pro soudobé dějiny AV ČR, 2000, 288 p. (Svědectví o době a lidech, 7).

5881. Technologie und Kultur. Europas Blick auf Amerika vom 18. bis zum 20. Jh. Hrsg. v. Michael WALA und Ursula LEHMKUHL. Köln, Böhlau, 2000, XX-245 p. (Beiträge zur Geschichte der Kulturpolitik, 7).

5882. TERÁN (Oscar). Vida intelectual en el Buenos Aires fin-de-siglo (1880–1910): derivas de la "cultura científica". Buenos Aires, Fondo de Cultura Económica, 2000, 309 p. (Sección de obras de historia).

5883. Toleration in Enlightenment Europe. Ed. by Ole Peter GRELL and Roy PORTER. Cambridge, Cambridge U. P., 2000, IX-270 p.

5884. Transactions, transgressions, transformations. American culture in Western Europe and Japan. Ed. by Heide FEHRENBACH and Uta G. POIGER. New York a. Oxford, Berghahn, 2000, XL-258 p.

5885. TUCKER (Avezier). The philosophy and politics: the theory and practice of Czech dissidence from Patočka to Havel. Pittsburgh, University of Pittsburgh Press, 2000, XIV-295 p. (Pitt series in Russian and East European studies).

5886. VÖLCKER (Lars). Tempel für die Großen der Nation. Das kollektive Nationaldenkmal in Deutschland, Frankreich und Großbritannien im 18. und 19. Jh. Frankfurt am Main, Peter Lang, 2000, 377 p. (Europäische Hochschulschriften. R.3.877).

5887. Voltaire en Europe. Hommage à Christiane Mervaud. Ed. par Michel DELON et Catriona SETH. Oxford, Voltaire Foundation, 2000, XXII-383 p.

5888. WALTON (Whitney). Eve's proud descendants: four women writers and republican politics in nineteenth-century France. Stanford, Stanford U. P., 2000, 308 p.

5889. WALZ (Robin). Pulp surrealism: insolent popular culture in early twentieth-century Paris. Berkeley, University of California Press, 2000, XII-206 p. (ill.).

5890. WAQUET (Françoise). Le prince et son lecteur. Avec l'édition de Charles Dantal, Les délassements littéraires ou l'heures de lecture de Frédéric II. Paris, Honoré Champion, 2000, 82 p.

5891. WAQUET (Jean-Claude). La conjuration des dictionnaires. Vérité des mots et vérités de la politique dans la France moderne. Strasbourg, Presses universitaires de Strasbourg, 2000, 270 p.

5892. WEISBROD (Bernd). Military violence and male fundamentalism: Ernst Jünger's contribution to the conservative revolution. *History workshop*, 2000, 49, p. 68-94.

5893. WILLIAMSON (George S.). What killed August von Kotzebue? The temptations of virtue and the political theology of German nationalism, 1789–1819. *Journal of modern history*, 2000, 72, 4, p. 890-943.

5894. Windows on the sixties: exploring key texts of media and culture. Ed. by Anthony ALDGATE, James CHAPMAN and Arthur MARWICK. London a. New York, I. B. Tauris, 2000, XXI-194 p. (ill.).

5895. Women in Italian Renaissance culture and society. Ed. by Letizia PANIZZA. Oxford, Legenda, 2000, XXI-512 p. (ill.).

5896. WOOTTON (David). Unhappy Voltaire, or 'I Shall Never Get Over it as Long as I Live'. *History workshop*, 2000, 50, p. 137-155.

5897. Writing in the Irish Republic: literature, culture, politics, 1949–1999. Ed. by Ray RYAN. Basingstoke, Macmillan and New York, St Martin's Press, 2000, X-289 p.

5898. YU-JOSE (Lydia N). Building Cultural Bridges: Japan and the Philippines in the 1930s. International Conference on Japanese Studies, 1999, Japan-Southeast Asia Relations. Singapore, Department of Japanese Studies, National University of Singapore, 2000, [s. p.].

Cf. n°s 181, 896, 899, 1062, 1072, 1084, 3660, 5061, 5961, 6045

§ 2. Academies, universities and intellectual organizations.

* 5899. GRAZZINI (Giovanni). Di Crusca in Crusca: per una bibliografia dell'Accademia. Ospedaletto, Pacini Editore, 2000, 295 p. (ill.).

** 5900. Istruzione (L') universitaria, 1859–1915. A cura di Gigliola FIORAVANTI, Mauro MORETTI e Ilaria PORCIANI. Roma, Ministero per i beni culturali e ambientali, 2000, 376 p. (Fonti per la storia della scuola, 5).

** 5901. Philipps-Universität Marburg (Die) im Nationalsozialismus. Dokumente zu ihrer Geschichte. Hrsg. v. Anne Christine NAGEL. Stuttgart, Steiner, 2000, X-563 p. (Pallas Athene, 1).

** 5902. Protocollum Contubernii. Visitation und Rechnungsprüfung von 1568–1615. Hrsg. v. Gerhard MERKEL. Heidelberg, Winter, 2000, XXIV-375 p. (Die Amtsbücher der Universität Heidelberg, Rh. C: Die Amtsbücher der Collegien und Bursen).

5903. Academia in upheaval: origins, transfers and transformations of the communist academic regime in Russia and East Central Europe. Ed. by Michael DAVID-FOX and György PÉTERI. Westport a. London, Bergin & Garvey, 2000, XI-334 p. (ill.).

5904. AHSMANN (Margreet). J.A.M. Collegium und Kolleg. Der juristische Unterricht an der Universität Leiden 1575–1630 unter besonderer Berücksichtigung der Disputationen. Frankfurt am Main, Klostermann, 2000, XI-762 p. (Studien zur Europäischen Rechtsgeschichte, 138).

5905. ANGELETTI (Luciana Rita). Il ruolo del Lancisi e del Baglivi all'interno delle Accademie mediche romane. *Medicina nei secoli*, 2000, 12, 1, 29-47 p.

5906. ASCHE (Matthias). Von der reichen hansischen Bürgeruniversität zur armen mecklenburgischen Landeshochschule. Das regionale und soziale Besucherprofil der Universitäten Rostock und Bützow in der frühen Neuzeit (1500–1800). Stuttgart, Steiner, 2000, XIV-635 p. (Contubernium, 52).

5907. ASHMANN (Margreet J. A. M.). Collegium und Kolleg, der juristische Unterricht an der Universität Leiden, 1575–1630, unter besonderer Berücksichtigung der Disputationen. Aus dem Niederländischen übersetzt von Irene SAGEL-GRANDE. Frankfurt am Main, Klostermann, 2000, XI-762 p. (Ius Commune, Sonderhefte: Studien zur Europäischen Rechtsgeschichte, 138).

5908. BARZMAN (Karen). The Florentine Academy and the early modern state. The discipline of Disegno. Cambridge, Cambridge U. P., 2000, 380 p. (ill.).

5909. BENDALL (A. Sarah), BROOKE (Christopher Nugent Lawrence), COLLINSON (Patrick). A history of Emmanuel College, Cambridge. Woodbridge, Boydell, 2000, XVII-741 p.

5910. BOJADŽIEVA (E.). Germanskata akademija v Mjunkhen i nejnite lektori v Bălgarija (1925–1944). (L'Academie allemande à Munich et ses lecteurs en Bulgarie, 1925–1944). *Istoricheski pregled*, 2000, 3-4, p. 9-111.

5911. CHAUBET (F.), LOYER (E.). L'Ecole libre des hautes études de New-York: exil et résistance intellectuelle (1942–1946). *Revue historique*, 2000, 124, 616, p. 939-972.

5912. COCO (Antonio), LONGHITANO (Adolfo), RAFFAELE (Silvana). La Facoltà di medicina e l'Università di Catania, 1434–1860. Catania, [s. n.], 2000, 285 p.

5913. CONDETTE (Jean-François). L'Université de Lille dans la Première Guerre mondiale, 1914-1918. *In*: Images civiles de la France en guerre [Cf. n° 4465], p. 83-102.

5914. CONNELLY (John). Captive University. The Sovietization of East German, Czech, and Polish higher education, 1945–1946. Chapell Hill, University of North Carolina Press, 2000, XVIII-432 p.

5915. CUNHA (Norberto Ferreira da). Elites e académicos na cultura portuguesa setecentista. [Lisboa], Imprensa Nacional-Casa da Moeda, 2000, 249 p. (Temas portugueses).

5916. CZERNIAKOWSKA (Małgorzata). Związki króla Stanisława Augusta i uczonych z jego kręgu z Royal Society w Londynie. (Connections of king Stanislaus Augustus and scientists from his environment with Royal Society in London). Gdańsk, Oddz. Gdański Tow. Przyj. Książki u. Komisja Hist. Gdańskiego Tow. Nauk., 32 p.

5917. DEFRANCE (Corinne). Les Alliés occidentaux et les universités allemandes 1945–1949. Préf. de Hans-Peter SCHWARZ. Paris, CNRS Ed. 2000, 406 p.

5918. DEL TACCA (MARIO). Storia della medicina nello studio generale di Pisa dal XIV al XX secolo. Pisa, Primula, 2000, 203 p. (ill.).

5919. DONATO (Maria Pia). Accademie romane. Una storia sociale, 1671–1824. Napoli, Edizioni Scientifiche Italiane, 2000, 309 p.

5920. DURRILL (Wayne K.). Shaping a settler elite: students, competition and leadership at South African College, 1829–1895. *In*: Bad behaviour in urban South Africa [Cf. n° 5125], p. 221-240.

5921. DYBIEC (Julian). Uniwersytet Jagielloński 1918–1939. (Jagiellonen-Universität 1918–1939). Kraków, PAU, 2000, 759 p.

5922. FROST (Dan R.). Thinking Confederates: Academia and the idea of progress in the new South. Knoxville, University of Tennessee Press, 2000, XIV-207 p.

5923. GATTO (Romano). Storia di una "anomalia": le facoltà di scienze dell'Università di Napoli tra l'Unità d'Italia e la Riforma Gentile, 1860–1923. Napoli, Fridericiana editrice universitaria, 2000, 612 p. (Fridericiana historia, 4).

5924. Gelehrte Gesellschaften im mitteldeutschen Raum (1650–1820). Band 1. Hrsg. v. Detlef DÖRING und Kurt NOWAK. Leipzig u. Stuttgart, Verlag der Sächsischen Akademie der Wissenschaften in Leipzig, 2000, 237 p. (Abhandlungen der Sächsischen Akademie der Wissenschaften zu Leipzig, Philologisch-Historische Klasse, 76, 2).

5925. GOETZ (Helmut). Il giuramento rifiutato. I docenti universitari e il regime fascista. Milano, La Nuova Italia, 2000, XXIII-314 p.

5926. GRADMANN (Christoph). Money and microbes: Robert Koch, tuberculin and the foundation of the Institute for Infectious Diseases in Berlin in 1891. *History and philosophy of life sciences*, 2000, 22,1, p. 59-79.

5927. HAMMERSTEIN (Notker). Res publica litteraria. Ausgewählte Aufsätze zur frühneuzeitlichen Bildungs-, Wissenschafts- und Universitätsgeschichte. Berlin, Duncker u. Humblot, 2000, 409 p. (Historische Forschungen, 69).

5928. HEWITT (Steve). 'Information Believed True': RCMP security intelligence activities on Canadian university campuses and the controversy surrounding them, 1961–1971. *Canadian historical review*, 2000, 81, 2, p. 191-228.

5929. History (The) of the University of Oxford. Vol. 7. Part 2. Nineteenth-century Oxford. Ed. by Michael G. BROCK and Mark C. CURTHOYS. Oxford, Oxford U. P., 2000, IV-993 p.

5930. History of universities. Vol. 15. 1997–1999. Ed. by Peter DENLEY and Mordechai FEINGOLD. Oxford, Oxford, U.P., 2000, 380 p. (ill.).

5931. Hitotsubashi University, 1875–2000: a hundred and twenty-five years of higher education in Japan. Ed. by Makoto IKEMA. Basingstoke, Macmillan, 2000, XIX-296 p. (ill.).

5932. HOFSTETTER (Michael J.) The romantic idea of a university: England and Germany, 1770–1850. Basingstoke, Palgrave, 2000, XIV-162 p. (Romanticism in perspective).

5933. Influencias (Las) de las culturas académicas alemana y española desde 1898 hasta 1936. Ed. por Jaime DE SALAS y Dietrich BRIESEMEISTER. Madrid, Iberoamericana y Frankfurt am Main, Vervuert, 2000, 286 p. (ill.).

5934. IVES (Eric William), SCHWARZ (Leonard D.), DRUMMOND (Diane K.). The first civic university: Birmingham 1880–1980. An introductory history. Birmingham, University of Birmingham Press, 2000, XVIII-462 p.

5935. KAASCH (Michael), KAASCH (Joachim). Vom Werden, Wirken, Widerstehen. Die Deutsche Akademie der Naturforscher Leopoldina im Wandel der Zeiten. *Naturwissenschaftliche Rundschau*. 53, 2000, 5/6, p. 231-239, p. 289-297.

5936. KAYA (Mehmet Ali). Edebiyat Fakültesi Tarihi. (The history of faculty of literature). İzmir, Ege Üniversitesi, 2000, 178 p.

5937. LONGO (Gisella). L'Istituto nazionale fascista di cultura: da Giovanni Gentile a Camillo Pellizzi, 1925–1943. Roma, Pellicani, 2000, 320 p. (Fascismo/fascismi, 5).

5938. MÄHRLE (Wolfgang). Academia Norica. Wissenschaft und Bildung an der Nürnberger Hohen Schule in Altdorf (1575–1623). Stuttgart, Steiner, 2000, IX-592 p. (Contubernium, 54).

5939. MARQUILHAS (Rita). A faculdade das letras. Leitura e escrita em Portugal no séc. XVII. Lisboa, Imprensa Nacional-Casa da Moeda, 2000, 367 p. (Filologia portuguesa).

5940. NICHOLLS (Christine Stephanie). The history of St Antony's College, Oxford, 1950–2000. Basingstoke, Macmillan a. Oxford, St Antony's College, 2000, XII-316 p.

5941. Origini (Alle) dell'Università dell'Aquila. Cultura, Università, collegi gesuitici all'inizio dell'età moderna in Italia meridionale. A cura di F. IAPPELLI e U. PARENTE. Roma, IHSI, 2000, 824p.

5942. PERKOWSKA (Urszula). Jubileusze Uniwersytetu Jagiellońskiego. (Die Jubiläen der Jagiellonen-Universität). Kraków, Secesja, 2000, 294 p.

5943. Preußische Akademie der Wissenschaften (Die) zu Berlin 1914–1945. Hrsg. v. Wolfram FISCHER unt. Mitarb. v. Rainer HOHLFELD und Peter NÖTZOLDT. Berlin, Akademie, 2000, XI- 594 p. (Interdisziplinäre Arbeitsgruppen, Forschungsberichte, 8).

5944. PULLAN (Brian), ABENDSTERN (Michele). A history of the University of Manchester, 1951–1973. Manchester, Manchester U. P., 2000, XI-281 p.

5945. Rektoren (Die) der Universität Rostock 1419–2002. Hrsg. v. Angela HARTWIG, Tilmann SCHMIDT. Rostock, Universität Verlag, 2000, 229 p. (Beiträge zur Geschichte der Universität Rostock, 23).

5946. ROBERTS (Jon H.), TURNER (James). The sacred and secular university. Intr. by John F. WILSON. Princeton, Princeton U. P., 2000, 184 p.

5947. SALADIN (Jean Christophe). Léon X philologue: le collège des jeunes Grecs du Quirinal (1514–1521). *Quaderni di storia*, 2000, 26, 51, p. 157-186.

5948. SCHUBERT (Ernst). Wissenschaftliche Unabhängigkeit und gesellschaftliche Verantwortung: Der Wandel von Leitbildern in der Geschichte der Akademie. *Jahrbuch der Akademie der Wissenschaften in Göttingen*, 2000, p. 71-124.

5949. SCHWEIGARD (Jörg). Aufklärung und Revolutionsbegeisterung. Die katholischen Universitäten in Mainz, Heidelberg und Würzburg im Zeitalter der Französischen Revolution (1789–1792/93–1803). Frankfurt am Main, Berlin u. Bern, Lang, 2000, 559 p. (Schriftenreihe der internationalen Forschungsstelle "Demokratische Bewegungen in Mitteleuropa 1770–1850", 29).

5950. SEE (Klaus von). Die Göttinger Sieben. Kritik einer Legende. Heidelberg, Winter, 2000, 124 p. (Beiträge zur Neueren Literaturgeschichte, Folge 3, 155).

5951. Storia della Facoltà di lettere e filosofia de La Sapienza. A cura di Lidia CAPO e Maria Rosa DI SIMONE; prefazione di Emanuele PARATORE. Roma, Viella, 2000, XIX-707 p.

5952. Storia della Facoltà di lettere e filosofia dell'Università di Torino. A cura di Italo LANA; prefazione di Nicola TRANFAGLIA. Firenze, Olschki, 2000, XI-570 p. (Storia / Facoltà di lettere e filosofia, Fondo di studi Parini Chirio, Università degli studi di Torino, 4).

5953. Studenti e dottori nelle università italiane (origini–20. secolo): atti del Convegno di studi: Bologna, 25–27 novembre 1999. A cura di Gian Paolo BRIZZI e Andrea ROMANO. Bologna, CLUEB, 2000, 402 p. (Centro interuniversitario per la storia delle università italiane. Studi, 1).

5954. University, city and state: the University of Glasgow since 1870. Ed. by Michael MOSS, J. FORBES MUNRO and Richard H. TRAINOR. Edinburgh, Edinburgh U. P., 2000, 382 p.

5955. VALERO GARCÍA (Pilar). La Universidad de Salamanca en la época de Carlos I/V. *In*: Aspectos históricos y culturales bajo Carlo V = Aspekte der Geschichte und Kultur unter Karl V. Madrid, Iberoamericana, 2000, p. 47-67.

5956. WATSON (Nigel). And their works do follow them: the story of North London Collegiate School 1850–2000. London, James & James, 2000, 140 p.

5957. WETTMANN (Andrea). Heimatfront Universität. Preußische Hochschulpolitik und die Universität Marburg im Ersten Weltkrieg. Köln, SH-Verlag, 2000, 515 p.

5958. ZIMMERMANN (Susanne). Die medizinische Fakultät der Universität Jena während der Zeit des Nationalsozialismus. Berlin, Verlag für Wissenschaft und Bildung, 2000, 223 p. (Ernst Haeckel Haus Studien. Monographien zur Geschichte der Biowissenschaften und Medizin, 2).

§ 2. Addenda 1999.

5959. TURI (Gabriele). L'Università di Firenze e la persecuzione razziale. *Italia contemporanea*, 99, 219, p. 227-247.

Cf. n°s 412, 5862, 5960-6040

§ 3. Education.

* 5960. Historical sources in U.S. reading education, 1900–1970: an annotated bibliography. Ed. by Richard D. ROBINSON. Newark, International Reading Association, 2000, 93 p.

5961. ANDREEV (Andrej Ju.). Moskovskij universitet v obshchestvennoj i kul'turnoj zhizni Rossii nachala XIX veka. (The University of Moscow in public and cultural life of early 19th-century Russia). Moskva, Jazyki russkoj kul'tury, 2000, 310 p. (ill.). (Studia Historica. Series minor).

5962. ANTOHI (Sorin). Commuting to Castalia. Noica's school, culture and power in Ceausescu's Romania. Introduction to LIICEANU (Gabriel). The Paltinis diary [Cf. n° 6003], p. VII-XXIV.

5963. BABINI (Valeria). Science, feminism and education: the early work of Maria Montessori. *History workshop*, 2000, 49, p. 44-67.

5964. BAŁCZEWSKI (Marian). Gry i zabawy Turków osmańskich na podstawie polskich i obcych źródeł od schyłku średniowiecza do końca Oświecenia. (Ludisches Leben der osmanischen Türken unter Anlehnung an die polnischen sowie fremden Quellen vom ausgehenden Mittelalter bis zum Ende der Aufklärung). Warszawa, Dialog, 2000, 237 p. (Świat Orientu).

5965. BALUKIEWICZ (Małgorzata). Protektoraty lwowskie. Początki i rozwój praktyki opiekuńczowychowawczej we Lwowie i na ziemi lwowskiej od końca XVII stulecia do wybuchu II wojny światowej [Lwower Förderungstätigkeit. Anfänge und Entwicklung des Betreuungs- und Erziehungs-Handelns in Lwow sowie auf dem Lwower Boden vom Ende des XVII. Jahrhunderts bis zum Ausbruch des II. Weltkrieges]. Katowice, Wydaw. Uniw. Śląskiego, 2000, 156 p. (Prace Naukowe Uniw. Śląskiego w Katowicach, 1862).

5966. BARREAU (Jean-Michel). Vichy, contre l'école de la République: Théoriciens et théories scolaires de la "Révolution nationale". Paris, Flammarion, 2000, 334 p.

5967. BAUER (Franz J.). Geschichte des Deutschen Hochschulverbandes. München, Saur, 2000, 234 p.

5968. BRISTON (Adrian). Dame schools. Chester, Imogen, 2000, XIII-102 p.

5969. BRODY (Judit). A Victorian physician's struggle for the introduction of physiology, hygiene and Swedish gymnastics into the elementary schools. *History of Education Society Bulletin*, 2000, 66, p. 80-88.

5970. CAFFERA (Hugo R.). Educación y luchas populares: historia crítica de la educación en la Argentina. Buenos Aires, Ediciones Cinco, 2000, 333 p.

5971. CASELLA (Francesco). Il Mezzogiorno d'Italia e le istituzioni educative salesiane (1879-1922). Roma, Istituto storico salesiano, 2000, 830 p.

5972. CLARK (Charles E.). Uprooting otherness: the literacy campaign in NEP-era Russia. Cranbury, Susquehanna U. P., 2000, 235 p.

5973. CURTIS (Sarah A.). Educating the faithful: religion, schooling in nineteenth-century France. DeKalb, Northern Illinois U. P., 2000, XII-255 p.

5974. CUTLER (William W.). Parents and school: the 150-year struggle for control in American education. Chicago, Chicago U. P., 2000, XIII-290 p. (ill.).

5975. DEPAEPE (Marc). Order in progress: everyday educational practice in primary schools, Belgium, 1880–1970. Leuven, Leuven U. P., 2000, 265 p. (Studia paedagogica, 29).

5976. DOUGLASS (John Aubrey). The California idea and American higher education. Stanford, Stanford U. P., 2000, XIII-460 p. (ill.).

5977. Education in Germany since unification. Ed. by David PHILLIPS. Wallingford, Symposium Books, 2000, 172 p. (Oxford studies in comparative education, 10).

5978. Education: institutions of Scotland. Ed. by Heather HOLMES. East Linton, Tuckwell, 2000, XVIII-558 p.

5979. EDWARDS (Elizabeth). Women principals, 1900–1960: gender and power. *History of education*, 2000, 29, 5, p. 405-14.

5980. EGANA BARAONA (María Loreto). La educación primaria popular en el siglo XIX en Chile: una práctica de política estatal. Santiago, Lom Ediciones, 256 p. (Colección sociedad y cultura, 22).

5981. Elementarschulverhältnisse im Niederstift Münster im 18. Jahrhundert. Die Schulvisitationsprotokolle Bernard Overbergs für die Ämter Meppen, Cloppenburg und Vechta 1783/84. Hrsg. v. Alwin HANSCHMIDT. Mit Beitr. v. Franz BÖLSKER-SCHLICHT, Alwin HANSCHMIDT und Hubert STEINHAUS. Münster, Aschendorff, 2000, XIII-353 p. (Veröffentlichungen der Historischen Kommission für Westfalen, 22 B. Geschichtliche Arbeiten zur Westfälischen Landesforschung, Geistesgeschichtliche Gruppe, 3).

5982. ELLEDGE (Paul). Lord Byron at Harrow School: speaking out, talking back, acting up, bowing

3. EDUCATION

out. Baltimore a. London, Johns Hopkins U. P., 2000, XIII-221 p.

5983. FISCHER (Didier). L'histoire des étudiants en France de 1945 à nos jours. Paris, Flammarion, 2000, 612 p.

5984. FREELY (John). A history of Robert College: the American College for girls, and Boğaziçi University (Bosphorus University). Ed. by Ayşe ERDEM. İstanbul, Yapı Kredi yayınları, 2000, 2 vol., [s. p.].

5985. GAILLARD (Jean Michel). Un siècle d'école républicaine. Paris, Editions du Seuil, 2000, 197 p.

5986. GEISSLER (Gert). Geschichte des Schulwesens in der Sowjetischen Besatzungszone und in der Deutschen Demokratischen Republik, 1945 bis 1962. Frankfurt am Main u. Oxford, Peter Lang, 2000, 590 p.

5987. GINIO (Ruth). Marshal Petain spoke to school children: Vichy propaganda in French West Africa, 1940–1943. *International journal of African historical studies*, 2000, 33, 2, p. 291-312.

5988. Grammaire et enseignement du français, 1500–1700. Ed. par Jan DE CLERCQ, Nico LIOCE, Pierre SWIGGERS. Leuven, Peeters, 2000, XXXIV-671 p. (ill.). (Orbis: monographies publiées par le Centre International de Dialectologie Générale [Louvain], 16, Supplementa).

5989. GRANTLEY (Darryll). Wit's Pilgrimage: drama and the social impact of education in early modern England. Aldershot, Ashgate, 2000, VIII-270 p.

5990. GREENWOOD (Keith M.). Robert College: the American founders. İstanbul, Boğaziçi U. P., 2000, 266 p.

5991. GROMADA (Barbara). Szkoły Sióstr Nazaretanek w okresie Polski Ludowej. (Die Ordensschulen des Schwesternordens der Heiligen Familie von Nazareth). Lublin, Tow. Nauk. KUL, 2000, 48 p. (Prace z Historii Szkolnictwa w Polsce, 3).

5992. GUEST (Harriet). Small change: women, learning, patriotism, 1750–1810. Chicago, Chicago U. P., 2000, X-350 p.

5993. HEATHORN (Stephen J.). For home, country, and race: constructing gender, class, and Englishness in the elementary school, 1880–1914. Toronto a. London, University of Toronto Press, 2000, XII-300 p.

5994. Hochschuloffiziere und Wiederaufbau des Hochschulwesens in Deutschland 1945–1949. Die Sowjetische Besatzungszone. Hrsg. v. Manfred HEINEMANN unt. Mitarb. v. Alexandr HARITONOW, Berit HARITONOW, Matthias JUDT [et al.]. Berlin, Akademie, 2000, XV-478 p. (Edition Bildung und Wissenschaft, 4).

5995. KOSIŃSKI (Krzysztof). O nową mentalność. Życie codzienne w szkołach 1945–1956. (Einer neuen Geistes- und Gemütsart halber. Schulalltag 1945–1956). Warszawa, Trio, 2000, 32 p. (W Krainie PRL).

5996. KUMAR (Nita). Lessons from schools: the history of education in Benaras. New Delhi a. London, Sage, 2000, 232 p.

5997. KÜTÜKOĞLU (Mübahat S.). XX. Asra Erişen İstanbul Medreseleri. (İstanbul Medreses reaching to the XXth century). Ankara, Türk Tarih Kurumu, 2000, 390 p.

5998. LAAMANEN (Hilkka). Kenen koulu, sen kansa. Lehdistökeskustelu kansakoulun muodosta ja sisällöstä kansakouluasetuksesta piirijakoasetukseen (1866–1898). (The political wrangling over the form, duration and curriculum of the common school, 1866–1898). Helsinki, [s. n.], 2000, 240 p. (Suomen kouluhistoriallisen seuran julk., 2000).

5999. LASCARIDES (Celia). History of early childhood education. New York a. London, Falmer, XXVI-662 p. (Garland reference library of social science, 982).

6000. LAUFER (Ulrike). Technik und Bildung. Bürgerliche Initiativen und Staatliche Reglementierung im Beruflich-technischen Schulwesen Bayerns und der bayerischen Pfalz 1789–1848. Mannheim, Palatium, 2000, 438 p. (Mannheimer historische Forschungen, 19).

6001. LEE DOWNS (Laura). Municipal communism and the politics of childhood: Ivry-sur-Seine 1925–1960. *Past and present*, 2000, 166, p. 205-241.

6002. LEŃ (Kazimierz). Jezuickie kolegium św. Jana w Jarosławiu 1573–1773. (St. Johann-Jesuiten-Kolleg in Jarosław 1573–1773). Kraków, Wyższa Szkoła Filoz.-Pedagog. "Ignatianum", 2000, 154 p. (Studia i Materiały do Dziejów Jezuitów, Polskich, 4).

6003. LIICEANU (Gabriel). The Paltinis diary: a paideic model in humanist culture. Intr. by Sorin ANTOHI [Cf. n° 5962]. Budapest a. New York, Central European U. P., 2000, XXXII-227 p. (Central European library of ideas).

6004. LOST (Christine). Sowietpädagogik: Wandlungen, Wirkungen, Wertungen in der Bildungsgeschichte der DDR. Baltmannsweiler, Schneider Verlag Hohengehren, 2000, 278 p.

6005. MAC GRATH (Michael). The Catholic Church and Catholic schools in Northern Ireland. Dublin, Irish Academic Press, 2000, XX-330 p. (ill.).

6006. MACLURE (Stuart). The inspectors' calling: HMI and the shaping of educational policy, 1945–1992. London, Hodder and Stoughton, 2000, XXXIV-350 p.

6007. MAIORCA (Bruno). La cattedra del duce: vita della scuola elementare fascista tra cronaca, liturgia e ideologia. Cagliari, Tema, 2000, 319 p. (ill.).

6008. MANGAN (James Anthony). Athleticism in the Victorian and Edwardian public school: the emergence and consolidation of an educational ideology. London, Cass, 2000, VI-346 p.

6009. MARTIN (Jane). Working for the people? Mrs Bridges Adams and the London School Board, 1897–1904. *History of education*, 2000, 29, 1, p. 49-62.

6010. MIKHAL'CHENKO (Sergej I.). Juridicheskij fakul'tet Varshavskogo universiteta, 1869–1917 gg.: Kratkij istoricheskij ocherk. (The Faculty of Law of the University of Warsaw, 1869–1917: a brief history). Brjansk, Izd. Brjanskogo ped. un-ta, 2000, 155 p. (portr). (Trudy Tsentra Slavjanovedenija).

6011. MIYAKOSHI (Eiichi). 19 seiki Eikoku no kikinritsu bunpô-gakkô: Charitî no dentô to henyô. (The reconstruction of the Endowed Grammar Schools: change and continuity in charitable provision in nineteenth century England). Tokyo, Sobunsha, 2000, 350 p.

6012. MOLINA (Iván), PALMER (Steven). Educando a Costa Rica: alfabetización popular, formación docente y género (1880–1950). San José, Editorial Porvenir, 2000, 180 p.

6013. MORAN (Jeffrey P.). Teaching sex: the shaping of adolescence in the 20th century. Cambridge, Harvard U. P., 2000, X-281 p.

6014. MORDECHAI (Eliav). Jüdische Erziehung in Deutschland im Zeitalter der Aufklärung und der Emanzipation. Münster, Waxmann, 2000, 488 p. (Jüdische Bildungsgeschichte in Deutschland, 2).

6015. MÜLLER (Michael). Die Entwicklung des höheren Bildungswesens der französischen Jesuiten im 18. Jahrhundert bis zur Aufhebung, 1762–1764: mit besonderer Berücksichtigung der Kollegien von Paris und Moulins. Frankfurt am Main u. Oxford, P. Lang, 2000, 526 p. (Mainzer Studien zur Neueren Geschichte, 4).

6016. Naissance, enfance et éducation dans la France méridionale du XVIe au XXe siècle. Hommage à Mireille Laget. Actes du colloque des 15 et 16 mars 1996, Montpellier III. Textes recueillis par Roland ANDRÉANI, Henri MICHEL et Elie PÉLAQUIER. Montpellier, Publications de l'Université Paul-Valéry, 2000, 453 p.

6017. NOSKOVÁ (Helena). Minority v minoritě a národnostní školství. (Minorities as a minority and national school system). *Lidé města*, 2000, 3, p. 79-126.

6018. OSGOOD (Robert L.). For "children who vary from the normal tupe". Special education in Boston, 1830–1930. Washington, Gallaudet U. P., VIII-214 p.

6019. PAIGE (John Rhodes). Preserving order amid chaos: the survival of schools in Uganda, 1971–1986. New York a. Oxford, Berghahn Books, 2000, XVI-208 p.

6020. PAVLIČ (Slavica). Sto znamenitih osebnosti v šolstvu na Slovenskem. (One hundred prominent personalities in Slovene education). Ljubljana, Prešernova družba, 2000, 239 p. (Koledarska zbirka).

6021. Practical visionaries: women, education and social progress, 1790–1930. Ed. by Mary HILTON and Pam HIRSCH. Harlow, Longman, 2000, XIII-252 p. (ill.).

6022. PRIETO (Víctor Manuel). El ginnasio moderno y la formación de la élite liberal bogotana, 1914–1948. Bogotá, Universidad Pedagógica Nacional, 138 p.

6023. Prinz Albert und die Entwicklung der Bildung in England und Deutschland im 19. Jahrhundert. Hrsg. v. Franz BOSBACH, William FILMER-SANKEY und Hermann HIERY. München, Saur, 2000, 256 p. (ill.) (Prinz-Albert-Studien, 18).

6024. Prometheus's fire: a history of scientific and technological education in Ireland. Ed. by Norman MAC MILLAN. Carlow, Tyndall, 2000, 610 p.

6025. RĘDZIŃSKI (Kazimierz). Żydowskie szkolnictwo świeckie w Galicji w latach 1813–1918. (Das jüdische weltliche Schulwesen in Galizien in den Jahren 1813–1918). Częstochowa, Wydaw. WSP, 2000 283 p.

6026. REMY (Johannes). Higher education and national identity. Polish student activism in Russia 1832–1863. Helsinki, SKS, 2000, 380 p. (ill., maps). (Bibliotheca historica 57).

6027. ROSSETTO (Nicola). Chiesa e istruzione popolare nel Risorgimento. L'opera di mons. Andrea Charvaz precettore di Vittorio Emanuele II, nella diocesi di Pinerolo (1834–1847). Pinerolo, Alzani, 2000, 223 p.

6028. SAINE (Harri). Uskonnonopetus Suomen oppivelvollisuuskouluissa 1900-luvulla. (Religious teaching in Finland in the 20th century). Turku, 2000, 285 p. (ill.). (A. Acad. Turkuensis, Ser C 165).

6029. SCHNEIDER (Barbara). Die Höhere Schule im Nationalsozialismus. Zur Ideologisierung von Bildung und Erziehung. Köln, Weimar u. Wien, Böhlau, 2000, XI-474 p. (Beiträge zur historischen Bildungsforschung, 21).

6030. SEEBERG (Vilma). The rhetoric and reality of mass education in Mao's China. Lewiston, Edwin Mellen Press, 2000, XVI-562 p.

6031. SIMONTON (Deborah). Schooling the poor: gender and class in eighteenth-century England. *British journal for eighteenth-century studies*, 2000, 23, 2, p. 183-202.

6032. STACK (Annetta). Relinquishing educational dominance: the Catholic church and Irish secondary schools, 1829–1989. *History studies: University of Limerick History Society journal*, 2000, 2, p. 69-77.

6033. STEWART (Mary Lynn). For health and beauty: physical culture for Frenchwomen, 1880s–1930s. Baltimore, Johns Hopkins U. P., 2000, XII-274 p. (ill.).

6034. SUCHMIEL (Jadwiga). Działalność naukowa kobiet w Uniwersytecie we Lwowie do roku 1939. (Die wissenschaftliche Tätigkeit der Frauen an der Universität zu Lwow bis 1939). Częstochowa, Wydaw. WSP, 2000, 367 p.

6035. SZABÓ (Anikó). Vertreibung, Rückkehr, Wiedergutmachung. Göttinger Hochschullehrer im Schatten des Nationalsozialismus. Mit einer biographischen Dokumentation der entlassenen und verfolgten Hochschullehrer: Universität Göttingen – TH Braunschweig – TH Hannover – Tierärztliche Hochschule Hannover. Göttingen, Wallstein, 2000, 765 p. (Veröffentlichungen des Arbeitskreises Geschichte des Landes Niedersachsen, 15).

6036. TOMUSK (Voldemar). The blinding darkness of the Enlightenment. Towards the understanding of post state-socialist higher education in Eastern Europe. Turku, Univ. of Turku, 2000, 311 p. (tables). (RUSE Research report 53).

6037. TROUVÉ (Susan). L'école élémentaire et le sport (1870-1940). *Revue française de civilisation britannique*, 2000, 10, 4, p. 77-88.

6038. TYERMAN (Christopher). A history of Harrow School, 1324-1991. Oxford, Oxford U. P., 2000, XI-599 p.

6039. WELS (Volkhard). Triviale Künste: die humanistische Reform der grammatischen, dialektischen und rhetorischen Ausbildung an der Wende zum 16. Jahrhundert. Berlin, Weidler, 2000, 332 p. (Studium litterarum, 1).

6040. Women, educational policy-making and administration in England: authoritative women since 1880. Ed. by Joyce GOODMAN and Sylvia HARROP. London, Routledge, 2000, XII-210 p.

Cf. n°s 556, 5546, 5563, 5862, 5899-5959

§ 4. The Press.

* 6041. DUMAN (Hasan). Başlangıcından Harf Devrimine Kadar Osmanlı-Türk Süreli Yayınlar ve Gazeteler Bibliyografyası ve Toplu Kataloğu, 1828-1928. (A bibliography and catalogue of Ottoman-Turkish serials and newspapers from the beginning to the introduction of the modern Turkish alphabet, 1828-1928). Ankara, Enformasyon ve Dokümantasyon Hizmetleri Vakfı, 2000, 3 vol., [s. p.].

* 6042. Editoria libraria in Italia dal Settecento a oggi: bibliografia 1980-1998. A cura di Luca CLERICI, Bruno FALCETTO e Gianfranco TORTORELLI. Milano, Il Saggiatore e Fondazione Arnoldo e Alberto Mondadori, 2000, 272 p. e 1 computer laser optical disc

* 6043. HAGELWEIDE (Gert). Literatur zur deutschsprachigen Presse. Eine Bibliographie. Von den Anfängen bis 1970. München, Saur, 2000, XXI-385 p. (Dortmunder Beiträge zur Zeitungsforschung, 35).

* 6044. KANELLOS (Nicolás). Hispanic periodicals in the United States, origins to 1960: a brief history and comprehensive bibliography. Houston, Arte Público Press, 2000, 359 p. (ill.).

* 6045. MEZ'ER (Avgusta V.). Slovar' russkikh tsenzorov: Materialy k bibliografii po istorii russkoj tsenzury. (A dictionnary of Russian censors: materials to bibliography of the history of Russian censorship). Intr. by Oleg R. KHROMOV. Gos. publ. ist. biblioteka Rossii. Moskva a. Pjatigorsk, [s. n.], 2000, 144 p.

* 6046. Mighty engine (The): the printing press and its impact. Ed. by Peter Charles Gordon ISAAC and Barry MAC KAY. Winchester a. New Castle, St. Paul's Bibliographies and Oak Knoll Press, 2000, XI-205 p.

* 6047. WEGEHAUPT (Heinz). Alte deutsche Kinderbücher III: Bibliographie, 1524-1900. Stuttgart, Hauswedell, 2000, 488 p. (ill.).

** 6048. VOLKOV (A. I.), PUGACHEVA (M. G.). JARMOLJUK (S. F.). Pressa v obshchestve (1959-2000): Otsenki zhurnalistov i sotsiologov. Dokumenty. (The press in the [Russian] society, 1959-2000: interviews with journalists and sociologists, documents). RAN, Int sotsiologii. Moskva, Moskovskaja shkola politicheskikh issledovanij, 2000, 613 p. (Biblioteka Moskovskoj shkoly politicheskikh issledovanij).

6049. ALBERT (Pierre), KOCH (Ursula E.). Les médias en Allemagne. Paris, Presses universitaires de France, 2000, 126 p.

6050. ANDREWS (Stuart). The British periodical press and the French Revolution, 1789-1799. Basingstoke, Palgrave, 2000, XI-280 p.

6051. ARCHER (Caroline). The Kynoch Press: the anatomy of a printing house, 1876-1981. London a. New Castle, British Library and Oak Knoll Press, 2000, XI-222 p.

6052. BALMUTH (Daniel). The Russian bulletin, 1863-1917: a liberal voice in tsarist Russia. New York, Peter Lang, 2000, 462 p. (American universities studies, 194).

6053. BARKER (Hannah). Newspapers, politics and English society, 1695-1855. Harlow, Longman, 2000, 246 p.

6054. BAUDINO (Isabelle). La presse et le développement du marché de l'art à Londres dans la première moitié du XVIIIe siècle. *Bulletin de la société d'études anglo-américaines des XVIIe et XVIIIe siècles*, 2000, 50, p. 233-45.

6055. BENITEZ (José Antonio). Los orígines del periodismo en nuestra América. Buenos Aires, Grupo Editoriale Lumen, 2000, 190 p.

6056. BERTAUD (Jean Paul). La presse et le pouvoir de Louis XIII à Napoléon Ier. Paris, Perrin, 2000, 277 p.

6057. BLJUM (Arlen V.). Sovetskaja tsenzura v epokhu total'nogo terrora, 1929-1953. (Soviet censorship in the age of total terror, 1929-1953). Sankt-Peterburg, Akademicheskij proekt, 2000, 312 p.

6058. BOULARD (Claire). Presse et socialisation féminine en Angleterre de 1690 à 1750: conversations à l'heure du thé. Paris, L'Harmattan, 2000, 536 p. (ill.).

6059. BRAIDA (Ludovica). Stampa e cultura in Europa tra XV e XVI secolo. Roma e Bari, Laterza, 2000, 162 p.

6060. BRUMETT (Palmira). Image and imperialism in the Ottoman revolutionary press. Albany, State University of New York Press, 2000, 470 p. (ill.).

6061. BRUNI (Domenico Maria). Per uno studio della censura in Toscana: appunti sulla legge del 6 maggio 1847. *Rassegna storica toscana*, 2000, 1, p. 43-60.

6062. Deutsche Publizistik im Exil, 1933 bis 1945: Personen – Positionen – Perspektiven. Festschrift für Ursula Koch. Hrsg. v. Markus BEHMER. Münster, Lit, 2000, 433 p. (ill.) (Kommunikationgeschichte, 11).

6063. DRÁPALA (Milan). Na ztracené vartě Západu. Antologie české nesocialistické publicistiky z let 1945–1948. (Last bastion of the West. Anthology of Czech non-socialist journalism, 1945–1948). Praha, Prostor, 2000, 683 p. (photogr.). (Obzor, 31).

6064. Einblattdrucke des 15. und frühen 16. Jahrhunderts. Hrsg. v. Volker HONEMANN und Sabine GRIESE. Tübingen, Max Niemeyer Verlag, 2000, 460 p.

6065. FEUCHTWANGER-SARIG (Naomi). How Italian are the Venice "Minhagim" of 1593? A chapter in the history of Yiddish printing in Italy. *In*: Schöpferische Momente des europäischen Judentums in der frühen Neuzeit. Heidelberg, Winter, 2000, p. 177-205.

6066. FEYEL (Gilles). L'annonce et la nouvelle: la presse d'information en France sous l'Ancien Régime (1630-1788). Oxford, Voltaire Foundation, 2000, VII-1387 p. (ill.). (Histoire du livre).

6067. FORTE (Isabel). A censura de Salazar no Jornal de notícias: da actuação da Comissão de Censura do Porto no Jornal de notícias durante o governo de António de Oliveira Salazar. Coimbra, Minerva, 2000, 193 p. (ill.). (Colecção Comunicação, 16).

6068. FRANKEL (J.). Jewish politics and the press: The "reception" of the Alliance Israélite Universelle (1860). *Jewish history*, 2000, 14, 1, p. 29-50.

6069. GERMANN (Martin). Zwischen Konfiskation, Zerstreuung und Zerstörung. Schicksale der Bücher und Bibliotheken in der Reformationszeit in Basel, Bern und Zürich. *Zwingliana*, 2000, 27, p. 63-77

6070. GORJAEVA (Tat'jana M.). Radio Rossii: Politicheskij kontrol' sovetskogo veshchanija v 1920–1930 gg.: Dokumentirovannaja istorija. (The radio of Russia: political control over Soviet broadcasting in 1920–1930: a documented history). Moskva, ROSSPEN, 2000, 175 p. (ill.). (Kul'tura i vlast' ot Stalina do Gorbacheva: Issledovanie).

6071. GRANT (Alfred). The American Civil War and the British press. Jefferson, McFarland, 2000, VIII-197 p.

6072. GUSTAFSSON (Karl-Erik), RYDÉN (Per). Den svenska pressens historia. Bd. 1 (-1830). (Histoire de la presse suédoise. Tome 1. Jusqu'en 1830). Stockholm, Ekerlids förlag, 2000, 341 p.

6073. HARRISON (Henrietta). Newspapers and nationalism in rural China 1890–1929. *Past and present*, 2000, 166, p. 181-204.

6074. HARTSOCK (John C.). A history of American literary journalism: the emergence of a modern narrative forme. Amherst, University of Massachusetts, 2000, XIV-294 p.

6075. HEISE (Joachim S.). Für Firma, Gott und Vaterland. Betriebliche Kriegszeitschriften im Ersten Weltkrieg. Das Beispiel Hannover. Hannover, Hahnsche Buchhandlung, 2000, 452 p. (Hannoversche Studien, 9).

6076. JEFFREY (Robin). India's newspaper revolution: capitalism, politics, and the Indian-language press, 1977–1999. London, Hurst, 2000, XXI-234 p. (ill.).

6077. Journalisme et religion (1685–1785). Ed. par Jacques WAGNER. Bern u. Berlin, Peter Lang, 2000, 432 p. (Eighteenth century French intellectual history, 6).

6078. KABACALI (Alpay). Başlangıcından günümüze Türkiye'de matbaa, basın ve yayın. (Printing-press, and publication history in Turkey from the beginning). General editor Kenan KOCATÜRK; ed. by Öner CIRAVOĞLU. İstanbul, Literatür yayınları, 2000, 279 p.

6079. KEITSCH (Christine). Der Fall Struensee – ein Blick in die Skandalpresse des ausgehenden 18. Jahrhunderts. Hamburg, Krämer, 2000, 336 p. (Beiträge zur deutschen und europäischen Geschichte, 26).

6080. LEDIN (Per). Veckopressens historia. Del 2. (Histoire de la presse hebdomadaire. Tome 2). Lund, Lunds universitet, Institutionen för nordiska språk, 2000, 199 p. (ill.).

6081. LEHUU (Isabelle). Carnival on the page: popular print media in antebellum America. Chapel Hill, University of North Carolina Press, 2000, XI-244 p.

6082. LESZCZYŃSKI (Adam). Sprawy do załatwienia. Listy do "Po prostu" 1955–1957. (Zu Erledigendes. Briefe an "Po prostu" 1955–1957). Warszawa, Trio, 2000, 221 p. (W Krainie PRL).

6083. LUSTOSA (Isabel). Insultos impressos: a guerra dos jornalistas na Independência, 1821–1823. Sao Paulo, Companhia das Letras, 2000, 497 p. (ill.).

6084. MANN (Alastair J.). The Scottish book trade, 1500–1720: print commerce and print control in early modern Scotland. East Linton, Tuckwell Press, 2000, 308 p.

6085. MELIS (François). Zeitungsdruck. Die Entwicklung der Technik vom 17. zum 20. Jh. München, Saur, 2000, 102 p. (Dortmunder Beiträge zur Zeitungsforschung, 58).

6086. MICHELS (T.). "Speaking to Moyshe": The early socialist Yiddish press and its readers. *Jewish history*, 2000, 14, 1, p. 51-82.

6087. MITTLMEIER (Christine). Publizistik im Dienste antijüdischer Polemik. Spätmittelalterliche und frühneuzeitliche Flugschriften und Flugblätter zu Hostienschändungen. Frankfurt am Main, Peter Lang, 2000, 184 p. (Mikrokosmos, 56).

6088. Nineteenth century media and the construction of identities. Ed. by Laurel BRAKE, Bill BELL and David FINKELSTEIN. New York, St. Martin's Press, 2000, XV-387 p. (ill.).

6089. NORDBLOM (Pia). Für Glaube und Volkstum. Die katholische Wochenzeitung "Der Deutsche in Polen" (1934–1939) in der Auseinandersetzung mit dem Nationalsozialismus. Paderborn, München u. Wien, Saur u. Schöning, 2000, 758 p. (Veröffentlichungen der Kommission für Zeitgeschichte, Rh. B: Forschungen, 87).

6090. Northcliffe's legacy: aspects of the British popular press, 1896–1996. Ed. by Peter CATTERALL, Colin SEYMOUR-URE and Adrian SMITH. Basingstoke, Macmillan in association with the Institute of Contemporary British History, 2000, XII-237 p.

6091. O'MALLEY (Tom), SOLEY (Clive). Regulating the press (1945–1999). London, Pluto, 2000, 244 p.

6092. ONSLOW (Barbara). Women of the press in nineteenth-century Britain. Basingstoke, Macmillan, 2000, XII-297 p.

6093. OPPEN VON (Karoline). The role of the writer and the press in the unification of Germany, 1989–1990. New York, Peter Lang, 2000, XIV-277 p. (German life and civilization, 31).

6094. POTTER (Simon James). Nationalism, imperialism and the press in Britain and the Dominions, c. 1898–1914. Oxford, [s. n.], 2000, 331 p. [Thesis (D. Phil.), University of Oxford, 2000].

6095. Presse et événement: journaux, gazettes, almanachs (XVIIIe–XIXe siècles). Actes du Colloque international "La perception de l'événement dans la presse de langue allemande et française" (Université de la Sarre, 12–14 mars 1998). Ed. par Hans-Jürgen LÜSEBRINK et Jean-Yves MOLLIER. Bern, Lang, 2000, XII-323 p. (ill.). (Convergences, 16).

6096. PROSS (Harry). Zeitungsreport: Deutsche Presse im 20. Jahrhundert. Weimar, Hermann Böhlaus Nachfolger, 2000, VIII-333 p. (ill.).

6097. Rivista all'avanguardia (Una). La «Riforma sociale» 1849–1935. Politica, società, istituzioni, statistica. A cura e con introduzione di Corrado MALANDRINO, presentazione di Gian Mario BRAVO. Firenze, Olschki, 2000, XXXVI-430 p. (Fondazione Luigi Einaudi, Studi, 39).

6098. ROBINE (Nicole). Lire des livres en France: des années 1930 à 2000. Paris, Editions du Cercle de la librairie, 2000, 260 p.

6099. SÁNCHEZ LLAMA (Inigo). Galería de escritoras isabelinas: la prensa periódica entre 1833 y 1895. Madrid, Ediciones Cátedra, 2000, 417 p. (Feminismos, 61).

6100. SCHULZ (Andreas). Der Aufstieg der "vierten Gewalt". Medien, Politik und Öffentlichkeit im Zeitalter der Massenkomunikation. *Historische Zeitschrift*, 2000, 270, 1, p. 65-97.

6101. SPREE (Ulrike). Das Streben nach Wissen. Eine vergleichende Gattungsgeschichte der populären Enzyklopädie in Deutschland und Großbritannien im 19. Jahrhundert. Tübingen, Max Niemeyer Verlag, 2000, VI-372 p. (Communicatio, 24).

6102. STEIN (S. A.). Creating a taste for news: Historicizing Judeo-Spanish periodicals of the Ottoman Empire. *Jewish history*, 2000, 14, 1, p. 9-28.

6103. STÖBER (Rudolf). Deutsche Pressegeschichte. Einführung, Systematik, Glossar. Konstanz, UVK-Medien, 2000, 370 p. (Uni-Papers, 8).

6104. THOMPSON (J. Lee). Northcliffe: press baron in politics, 1865–1922. London, John Murray, 2000, XVI-642 p.

6105. THÝFNER (Margit). Princely Pieties: the 1598–1617 accessions of the Royal Library in Brussels. *Quaerendo*, 2000, 30, 2, p.130-153

6106. TRANFAGLIA (Nicola), VITTORIA (Albertina). Storia degli editori italiani. Roma e Bari, Laterza, 2000, VIII-574 p.

6107. VAN BRUAENE (Anne L.). Printing plays: the publication of the Ghent Plays of 1539 and the reaction of the authorities. *Dutch Crossing*, 2000, 24, 2, p. 265-284.

6108. VERVLIET (Hendrik D. L.). Printing types of Pierre Haultin (ca. 1510–1587). Vol. 1. Roman types. Vol. 2. Italic, greek and music types. *Quaerendo*, 2000, 30, 2, p. 87-129; 3, p. 173-227.

6109. VOLMER (Annette). Presse und Frankophonie im 18. Jahrhundert: Studien zur französischsprachigen Presse in Thüringen, Kursachsen und Russland. Leipzig, Leipziger Universitätsverlag, 2000, 318 p.

6110. WALSHAM (Alexandra). "Domme Preachers"? Post-Reformation English Catholicism and the culture of print. *Past and Present*, 2000, 168, p.72-123.

6111. War (The) for public mind: political censorship in nineteenth-century Europe. Ed. by Robert Justin GOLDSTEIN. Westport, Praeger, 2000, XIII-277 p.

6112. WEHDE (Susanne). Typographische Kultur. Eine zeichentheoretische und kulturgeschichtliche Studie zur Typographie und ihrer Entwicklung. Tübingen, Max Niemeyer Verlag, 2000, X- 496 p. (Studien und Texte zur Sozialgeschichte der Literatur, 69).

6113. ZELKA (Luan). Tirana: objekt dhe subjekt i shtypit shqiptar: shtypi shqiptar për Tiranën: fillimet e shtypit Tiranas. (Tirana: oggetto e soggetto della stampa albanese. Stampa albanese per Tirana: inizi della stampa di Tirana). Tiranë, Botimet Enciklopedike, 2000, 168 p. (ill.).

Cf. n°s 83-96, 4211, 5134, 5765, 5862

§ 5. Philosophy.

* 6114. DELFOSSE (Heinrich P.). Kant-Index. Hrsg. v. Norbert HINSKE. Bd. 15. Stellenindex und Konkordanz zur "Grundlegung zur Metaphysik der Sitten". Section 2. Indices zum Ethik Korpus. Stuttgart, Frommann u. Bad Cannstatt, Holzboog, 2000, CX-487 p. (ill.). (Forschungen und Materialien zur deutschen Aufklärung, 3,22).

* 6115. Paul Ricoeur: bibliographie primaire et secondaire, 1935-2000. Ed. par F. D. VANSINA. Leuven, Leuven U. P., 2000, XXV-544 p.

** 6116. Vita (La) degli studi. Carteggio Gioele Solari-Norberto Bobbio 1931-1952. A cura di Angelo D'ORSI. Milano, Franco Angeli, 2000, 234 p.

6117. ABBEY (Ruth). Charles Taylor. Teddington, Acumen, 2000, VI-250 p. (Philosophy now).

6118. ADKINS (Gregory Matthew). When ideas matter: the moral philosophy of Fontenelle. *Journal of the history of ideas*, 2000, 61, 3, p. 433-452.

6119. BALAKRISHNAN (Gopal). The enemy: an intellectual portrait of Carl Schmitt. New York, Verso, 2000, VI-312 p.

6120. BAST (Rainer A.). Problem, Geschichte, Form: das Verhältnis von Philosophie und Geschichte bei Ernst Cassirer im historischen Kontext. Berlin, Duncker & Humblot, 2000, 603 p. (Philosophische Schriften, 40).

6121. BERGER WALDENEGG (Georg Christoph). Krieg und Expansion bei Machiavelli. Überlegungen zu einem vernachlässigten Kapitel seiner "politischen Theorie". *Historische Zeitschrift*, 2000, 271, 1, p. 1-56.

6122. BONGIOVANNI (Bruno). Da Marx alla catastrofe dei comunismi. Traiettorie e antinomie del socialismo. Milano, Unicopli, 368 p.

6123. BRIENT (Elizabeth). Hans Blumenberg and Hannah Arendt on the "Unworldly Worldliness" of the Modern Age. *Journal of the history of ideas*, 2000, 61, 3, p. 513-532.

6124. BURGIO (Alberto). «Dal mondo del pensiero al mondo del reale». Sul realismo politico di Marx. *In*: Realismo politico (Sul) [Cf. n° 6186], p. 107-136.

6125. BURKHARD (Bud). French Marxism between the wars: Henri Lefebvre and the "Philosophies". Amherst, Prometheus, 2000, 278 p.

6126. BURROW (John Wyon). The crisis of reason: European thought, 1848-1914. New Haven a. London, Yale U. P., 2000, XV-271 p. (ill.). (Yale intellectual history of the West).

6127. CHANDE (M. B.). Indian philosophy in modern times. New Delhi, Atlantic, 2000, XVI-405 p.

6128. DELACAMPAGNE (Christian). La philosophie politique aujourd'hui: idées, débats, enjeux. Paris, Éd. du Seuil, 2000, 245 p. (ill.).

6129. Descartes im Diskurs der Neuzeit. Hrsg. v. Wilhelm F. NIEBEL, Angelica HORN und Herbert SCHNÄDELBACH. Frankfurt am Main, Suhrkamp, 2000, 369 p. (Suhrkamp Taschenbuch Wissenschaft, 1436).

6130. Descartes' natural philosophy. Ed. by Stephen GAUKROGER, John SCHUSTER and John SUTTON. London, Routledge, 2000, X-797 p. (Routledge studies in seventeenth-century philosophy, 3).

6131. DILTHEY (Wilhelm). Gesammelte Schriften. Band 23. Allgemeine Geschichte der Philosophie: Vorlesungen 1900-1905. Hrsg. v. Gabriele GEBHARDT und Hans-Ulrich LESSING. Göttingen, Vandenhoeck & Ruprecht, 2000, XXVII-418 p.

6132. DUSTDAR (Farah). Vom Mikropluralismus zu einem makropluralistischen Politikmodell. Kants wertgebundener Liberalismus. Berlin, Duncker & Humblot, 2000, 239 p. (Beiträge zur Politischen Wissenschaft, 115).

6133. Elemente moderner Wissenschaftstheorie: zur Interaktion von Philosophie, Geschichte und Theorie der Wissenschaften. Hrsg. von Friedrich STADLER. Wien u. New York, Springer, 2000, XXVI-220 p. (ill.). (Veröffentlichungen des Instituts Wiener Kreis, 8).

6134. English philosophy in the age of Locke. Ed. by Alexander STEWART. Oxford, Oxford U. P., 2000, VIII-326 p. (Oxford studies in the history of philosophy).

6135. ESDERS (Michael). Begriffs-Gesten: Philosophie als Kurze Prosa von Friedrich Schlegel bis Adorno. Frankfurt am Main, Peter Lang, 2000, 381 p. (Literatur als Sprache: Literaturtheorie, Interpretation, Sprachkritik, 14).

6136. EZE (E. C.). Hume, race, and human nature. *Journal of the history of ideas*, 2000, 61, 4, p. 691-698.

6137. FATTORI (Marta). Linguaggio e filosofia nel Seicento europeo. Firenze, L. S. Olschki, 2000, XXXIV-430 p. (ill.). (Lessico intellettuale europeo, 83).

6138. FELICE (Domenico). Oppressione e libertà: filosofia e anatomia del dispotismo nel pensiero di Montesquieu. Pisa, ETS, 2000, 261 p. (Filosofia, 31).

6139. FILONI (Marco). Filosofia e politica. Attualità di Eric Weil. Urbino, Università degli studi di Urbino, 2000, 143 p. (Serie di filosofia pedagogia psicologia, 8).

6140. Filosofia italiana (La) di fronte al fascismo. Gli anni Trenta: contrasti e trasformazioni. A cura di Amedeo VIGORELLI e Marzio ZANANTONI. Milano, UNICOPLI, 2000, 241 p. (Biblioteca di cultura filosofica, 15).

6141. Filosofia e cultura nel Settecento britannico. A cura di Antonio SANTUCCI. Bologna, Il Mulino, 2000, 2 vol., 474 p., 526 p. (Percorsi).

6142. FLIKSCHUH (Katrin). Kant and modern political philosophy. Cambridge, Cambridge U. P., 2000, X-216 p.

6143. FONTANA (Benedetto). Logos and Kratos: Gramsci and the ancients on hegemony. *Journal of the history of ideas*, 2000, 61, 2, p. 305-326.

6144. GAIGER (Jason). Schiller's theory of landscape depiction. *Journal of the history of ideas*, 2000, 61, 1, p. 115-132.

6145. GARFF (Joakim). SAK: Søren Aabye Kierkegaard. En Biografi. (Biographie de Søren Kierkegaard). København, Gad, 738 p. (pl., ill.).

6146. German 20th-century philosophy: the Frankfurt school. Ed. by Wolfgang SCHIRMACHER. New York, Continuum, 2000, XX-244 p. (The German library, 78).

6147. GIUSTINO VITOLO (Angela). Storia e filosofia in N.A. Berdjaev. Milano, FrancoAngeli, 2000, 297 p. (Filosofia, 113).

6148. GOOCH (Todd A.). The numinous and modernity: an interpretation of Rudolf Otto's philosophy of religion. Berlin, W. de Gruyter, 2000, VIII-233 p. (Beihefte zur Zeitschrift für die alttestamentliche Wissenschaft, 293).

6149. GORNER (Paul). Twentieth century German philosophy. Oxford a. New York, Oxford U. P., 2000, 225 p.

6150. GUYER (Paul). Kant on freedom, law, and happiness. Cambridge, Cambridge U. P., 2000, 452 p.

6151. HACOHEN (Malachi Haim). Karl Popper – The formative years, 1902–1945. Politics and philosophy in interwar Vienna. Cambridge a. New York, Cambridge U. P., 2000, XIII-610 p.

6152. HAMLIN (William M.). On continuities between skepticism and early ethnography, or Montaigne's providential diversity. *Sixteenth century journal*, 2000, 31, 2, p. 361-379.

6153. HÄUSSLING (Roger). Nietzsche und die Soziologie: zum Konstrukt des Übermenschen, zu dessen anti-soziologischen Implikationen und zur soziologischen Reaktion auf Nietzsches Denken. Würzburg, Königshausen & Neumann, 2000, 290 p. (Epistemata, Reihe Philosophie, 289).

6154. HEERICH (Thomas). Transformation des Politikkonzepts von Hobbes zu Spinoza: das Problem der Souveränität. Würzburg, Königshausen & Neumann, 2000, 94 p. (Schriftenreihe der Spinoza-Gesellschaft, 8).

6155. Hermeneutik (Die) im Zeitalter der Aufklärung. Hrsg. v. Manfred BEETZ und Giuseppe CACCIATORE. Köln, Böhlau, 2000, VI-337 p.

6156. HERZBERG (Guntolf). Aufbruch und Abwicklung: Neue Studien zur Philosophie in der DDR. Berlin, Ch. Links, 2000, 267 p. (Forschungen zur DDR-Gesellschaft).

6157. História do pensamento filosófico português. Vol. 5. O síecolo XX. Ed. de Pedro CALAFATE. Lisboa, Caminho, 2000, 2 vol., 964 p.

6158. HOEGES (Dirk). Niccolò Machiavelli. Die Macht und der Schein. München, Beck, 2000, 290 p.

6159. HOWARD (Thomas A.). Religion and the rise of the Historicism. Historicist thought in the shadow of theology: W. M. L. De Wette, Jacob Burckhardt and the shaping of the nineteenth century. New York, Cambridge U. P., 2000, XI-250 p.

6160. Humanism and early modern philosophy. Ed. by Jill KRAYE and M. W. F. STONE. London, Routledge, 2000, XIII-270 p. (ill.). (London studies in the history of philosophy).

6161. HUNTER (I.). Christian Thomasius and the desacralization of philosophy. *Journal of the history of ideas*, 2000, 61, 4, p. 595-616.

6162. ILIFFE (Rob). The masculine birth of time: temporal frameworks of early modern natural philosophy. *British journal for the history of science*, 2000, 33, 119, p. 427-453.

6163. Jüdische Traditionen in der Philosophie des 20. Jahrhunderts. Hrsg. v. Joachim VALENTIN und Saskia WENDEL. Darmstadt, Primus, 2000, VI-298 p.

6164. JURIST (Elliot L.). Beyond Hegel and Nietzsche: philosophy, culture, and agency. Cambridge a. London, MIT Press, 2000, XII-355 p. (Studies in contemporary German social thought).

6165. KEMP (C.). Two meanings of the term "Idea": acts and contents in Hume's treatise. *Journal of the history of ideas*, 2000, 61, 4, p. 675-690.

6166. KEUTH (Herbert). Die Philosophie Karl Poppers. Tübingen, Mohr Siebeck, 2000, XXI-444 p. (UTB für Wissenschaft, 2156).

6167. LANDUCCI (Sergio). Il punto sul "De tribus impostoribus". *In*: Leo Valiani storico e politico [Cf. n° 729], p. 1036-1071.

6168. LE GALL (J.-M.). Lectures méditerranéennes d'Erasme au XVIe siècle. *Revue historique*, 2000, 124, 614, p. 435-444.

6169. LÉVY (Bernard Henri). Le siècle de Sartre: enquête philosophique. Paris, B. Grasset, 2000, 663 p.

6170. LEVY (Carl). Gramsci and the anarchists. New York, Berg, 2000, XII-272 p.

6171. LEVY (Lia). L'automate spirituel: la subjectivité moderne d'après l'Ethique de Spinoza. Assen, Van Gorcum, 2000, XII-365 p. (Philosophia Spinozae perennis – Spinoza's philosophy and its relevance, 10).

6172. Libertà (La) nella filosofia classica tedesca: politica e filosofia tra Kant, Fichte, Schelling e Hegel. A cura di Giuseppe DUSO e Gaetano RAMETTA. Milano, F. Angeli, 2000, 185 p. (Per la storia della filosofia politica, 10).

6173. LONG (Eugene Thomas). Twentieth-century Western philosophy of religion, 1900–2000. Dordrecht a. London, Kluwer Academic, 2000, XII-538 p. (Handbook of contemporary philosophy of religion, 1).

6174. MAC CUMBER (John). Philosophy and freedom: Derrida, Rorty, Habermas, Foucault. Bloomington, Indiana U. P., 2000, XI-191 p. (Studies in continental thought).

6175. MARTIN (Raymond). Naturalization of the soul: self and personal identity in the eighteenth century. London a. New York, Routledge, 2000, XI-203 p. (Routledge studies in eighteenth century philosophy, 1).

6176. MASROORI (C.). European thought in nineteenth-century Iran: David Hume and others. *Journal of the history of ideas*, 2000, 61, 4, p. 657-674.

6177. MASSEAU (Didier). Les ennemis des philosophes: l'antiphilosophie au temps des Lumières. Paris, Albin Michel, 2000, 451 p. (Bibliothèque Albin Michel des idées).

6178. MEIN (Georg). Die Konzeption des Schönen: der ästhetische Diskurs zwischen Aufklärung und Romantik: Kant, Moritz, Hölderlin, Schiller. Bielefeld, Aisthesis, 2000, 224 p.

6179. MICHEL (Andreas). Die französische Heidegger-Rezeption und ihre sprachlichen Konsequenzen: ein Beitrag zur Untersuchung fachsprachlicher Varietäten in der Philosophie. Heidelberg, C. Winter, 2000, XX-754 p. (Studia Romanica, 91).

6180. MIQUEL (Bastien). Joseph de Maistre: un philosophe à la cour du Tsar. Paris, Albin Michel, 2000, 252 p.

6181. Monismus um 1900. Wissenschaftskultur und Weltanschauung. Hrsg. v. Paul ZICHE. Berlin, Verlag für Wissenschaft und Bildung, 2000, 200 p.

6182. Nietzsche's postmoralism: essays on Nietzsche's prelude to philosophy's future. Ed. by Richard SCHACHT. Cambridge, Cambridge U. P., 2000, 296 p.

6183. ÖÇAL (Şamil). Kemal Paşazâde'nin felsefî ve kelâmî görüşleri. (The Philisophical and theological views of Kemal Paşazade). Ankara, Kültür Bakanlığı Yayımlar Dairesi Başkanlığı, 2000, 481 p.

6184. Potentia dei: l'onnipotenza divina nel pensiero dei secoli XVI e XVII. A cura di Guido CANZIANI, Miguel A. GRANADA e Yves Charles ZARKA. Milano, F. Angeli, 2000, 688 p. (Filosofia e scienza nel Cinquecento e nel Seicento, I,52).

6185. RAMBALDI (Enrico I.). Millenarismi nella cultura contemporanea: con un'appendice su yovel ebraico e giubileo cristiano. Milano, Franco Angeli, 2000, 247 p. (Filosofia e scienza nel Cinquecento e nel Seicento).

6186. Realismo politico (Sul). Relazioni del seminario tenuto presso la Biblioteca Cantonale di Locarno (I parte). *Quaderni di storia*, 2000, 26, 52, p. 71-190. [Cf. n^{os} <scelta> 2286, 3237, 6124.]

6187. Reception (The) of Kant's critical philosophy: Fichte, Schelling, and Hegel. Ed. by Sally SEDGWICK. Cambridge, Cambridge U. P., 2000, X-338 p.

6188. Representations of the self from the Renaissance to Romanticism. Ed. by Patrick COLEMAN, Jayne LEWIS and Jill KOWALIK. Cambridge, Cambridge U. P., 2000, XII-284 p.

6189. REY (Roselyne). Naissance et développement du vitalisme en France de la deuxième moitié du XVIIIe siècle à la fin du Premier Empire. Oxford, Voltaire Foundation, 2000, XII-472 p. (ill.). (Studies on Voltaire and the eighteenth century, 381).

6190. RICUPERATI (Giuseppe). Non Swedenborg, ma Giannone. Sulla scoperta di un autografo parziale del «Triregno» nell'archivio dell'Inquisizione. *Rivista storica italiana*, 2000, 112, 1, p. 75-137.

6191. ROCCHI (Jean). Giordano Bruno après le bûcher. Bruxelles, Complexe, 2000, 304 p. (Questions à l'histoire).

6192. SCARAVELLI (Luigi). Lezioni su Leibniz (1953–1954). A cura di Gianfranco BRAZZINI. Soveria Mannelli, Rubbettino, 2000, 258 p. (ill.). (Biblioteca di studi filosofici, 11).

6193. Schelling: zwischen Fichte und Hegel. Hrsg. von Christoph ASMUTH, Alfred DENKER und Michael VATER. Amsterdam, B.R. Grüner, 2000, VII-423 p. (Bochumer Studien zur Philosophie, 32).

6194. Scottish Enlightenment (The): essays in reinterpretation. Ed. by Paul WOOD. Rochester, University of Rochester Press, 2000, X-399 p. [Cf. n^o <choice> 6325.]

6195. Sens moral (Le): une histoire de la philosophie morale de Locke à Kant. Ed. par Laurent JAFFRO. Paris, P.U.F., 2000, 140 p. (Débats philosophiques).

6196. Skizzen zur österreichischen Philosophie. Hrsg. von Rudolf Haller. Amsterdam, Rodopi, 2000, XII-572 p. (Special issue of Grazer philosophische Studien, 58/59).

6197. SORELL (Tom). Descartes, a very short introduction. Oxford, Oxford U. P., 2000, 116 p. (ill.). (Very short introductions, 30).

6198. Spätphilosophie (Die) J.G. Fichtes: Tagung der Internationalen J.G.-Fichte-Gesellschaft (15.–27. September 1997) in Schulpforte in Verbindung mit der Landesschule Pforta und dem Istituto Italiano per gli Studi Filosofici (Napoli). Hrsg. v. Wolfgang H. SCHRADER. Amsterdam, Rodopi, 2000, VII-231 p. (Fichte-Studien, 17–18).

6199. SWENSON (James). On Jean-Jacques Rousseau: considered as one of the first authors of the Revolution. Stanford, Stanford U. P., 2000, XIII-320 p. (Atopia: philosophy, political thought, aestetics).

6200. TARANTINO (G.). Martin Clifford 1624–1677. Deismo e tolleranza nell'Inghilterra della Restaurazione. Firenze, L. S. Olschki, 2000, [s. p.]. (Studi e testi per la storia della tolleranza in Europa nei secoli XVI–XVIII, 3).

6201. TREIBER (Gerhard). Philosophie der Existenz. Das Entscheidungsproblem bei Kierkegaard, Jaspers, Heidegger, Sartre, Camus: literarische Erkundungen bei Kundera, Céline, Broch, Musil. Frankfurt am Main u. Oxford, P. Lang, 2000, 236 p. (Europäische Hochschulschriften, 20, Philosophie, 610).

6202. VAN RULER (Han). Mind, forms, and spirits: the nature of Cartesian disenchantment. *Journal of the history of ideas*, 2000, 61, 3, p. 381-396.

6203. VECCHIOTTI (Icilio). Introduzione alla filosofia di Giordano Bruno. Urbino, Quattro Venti, 2000, 307 p. (Filosofia e storia delle idee).

6204. VERDICCHIO (Massimo). Naming things: aesthetics, philosophy and history in Benedetto Croce. Napoli, La città del Sole, 2000, 245 p.

6205. "Voll Verdienst, doch dichterisch wohnet der Mensch auf dieser Erde": Heidegger und Hölderlin. Hrsg. von Peter TRAWNY. Frankfurt, Klostermann, 2000, 235 p. (Schriftenreihe / Martin-Heidegger-Gesellschaft, 6).

6206. WEINER (Thomas). Die Philosophie Arthur Schopenhauers und ihre Rezeption. Hildesheim u. New York, G. Olms, 2000, 175 p. (Europaea memoria: Studien und Texte zur Geschichte der europäischen Ideen, 1, Studien, 15).

6207. WITTE (Egbert). Logik ohne Dornen: die Rezeption von A. G. Baumgartens Ästhetik im Spannungsfeld von logischem Begriff und ästhetischer Anschauung. Hildesheim, Georg Olms, 2000, 255 p. (Studien und Materialien zur Geschichte der Philosophie, 53).

6208. WOLTERSTORFF (Nicholas). Thomas Reid and the story of epistemology. Cambridge a. New York, Cambridge U. P., 2000, 280 p. (Modern European philosophy).

6209. ZOTTA (Franco). Immanuel Kant: Legitimität und Recht: eine Kritik seiner Eigentumslehre, Staatslehre und seiner Geschichtsphilosophie. Freiburg, Verlag K. Alber, 2000, 320 p. (Alber-Reihe praktische Philosophie, 58).

Cf. *n°s 575, 751, 1024-1084, 1409-1428, 1043, 1418, 5831*

§ 6. Exact, natural, medical sciences and technique.

6210. ABBRI (Ferdinando). Alchemy and chemistry: chemical discourses in the seventeenth century. *Early science and medicine*, 2000, 5, 2, p. 214-226.

6211. ADAMS (Steve). Frontiers: twentieth-century physics. London, Taylor & Francis, 2000, X-507 p. (ill.).

6212. ARNOLD (David). The new Cambridge history of India. III. 5. Science, technology and medicine in colonial India. Cambridge, Cambridge U. P., 2000, XII-234 p.

6213. AUNESLUOMA (Juhana). The engineering profession and civil society in Finland. Towards a political history of technology. *Tekniikan vaiheita – Teknik i tiden*, 2000, 18, 4. p. 20-31.

6214. BARTLEY (Paula). Prostitution: prevention and reform in England, 1860–1914. London a. New York, Routledge, 2000, XI-229 p. (Women's and gender history).

6215. BATES (D.). Machina Ex Deo: William Harvey and the meaning of instrument. *Journal of the history of ideas*, 2000, 61, 4, p. 577-594.

6216. Between the natural and the artificial: dyestuffs and medicines. Proceedings of the XXth International Congress of History of Science (Liège, 20–26 July 1997). Ed. by Gérard EMPTOZ and Patricia Elena ACEVES PASTRANA. Turnhout, Brepols, 2000, 192 p. (De diversis artibus, 42).

6217. BEWELL (Alan). Romanticism and colonial desease. Baltimore, Johns Hopkins U. P., 2000, XV-373 p.

6218. BIERBRODT (Johannes). Naturwissenschaft und Ästhetik 1750–1810. Würzburg, Königshausen & Neumann, 2000, 371 p. (Epistemata, Rh.: Literaturwissenschaft, 279).

6219. BLEKER (Johanna). Ärztinnen aus dem Kaiserreich: Lebensläufe einer Generation. Weinheim, Deutscher Studien Verlag, 2000, 348 p. (ill.).

6220. BLUM (Alain). La purge de 1924 à la Direction centrale de la statistique. *In*: Fabriquer la statistique (Urss et France). *Annales*, 2000, 55, 2, p. 249-282.

6221. Borders (At the) of the human: beasts, bodies and natural philosophy in the early modern period. Ed. by Erica FUDGE, Ruth GILBERT and Susan WISEMAN. New York, St. Martin's Press, 2000, XII-269 p.

6222. BORRELLI (Antonio). Istituzioni scientifiche, medicina e società. Biografia di Domenico Cotugno (1736–1822). Firenze, Olschki, 2000, XII-270 p. (Studi e testi, 38).

6223. BRADSHAW (Ann). The nurse apprentice, 1860–1977. Aldershot, Ashgate, 2000, IX-267 p.

6224. BURGUIÈRE (André). La centralisation monarchique et la naissance des sciences sociales. Voyageurs et statisticiens à la recherche de la France à la fin du 18e siècle. *In*: Enquêtes et ethnographie, France et Russie (fin 18e–fin 19e siècles). *Annales*, 2000, 55, 1, p 199-218.

6225. BURNEY (Jan). Body of evidence: medicine and the politics of the English inquest, 1830–1926. Baltimore, Johns Hopkins U. P., 2000, X-245 p.

6226. CHABOT (Hugues). Le tribunal de la science. Les rapports négatifs à l'Académie des sciences comme illustrations d'un scientifiquement (in)correct (1795–1835). *Annales historiques de la Révolution Française*, 2000, 320, p. 173-182.

6227. CHAUVAUD (Frédéric). Les experts du crime: la médicine légale en France au XIXe siècle. Paris, Aubier, 2000, 301 p.

6228. Chemistry, society and environment: a new history of the British chemical industry. Ed. by Colin A. RUSSELL. Cambridge, Royal Society of Chemistry, 2000, XVI-372 p. (ill.).

6229. CHITTY (Dennis). Do lemmings commit suicide? Beautiful hypotheses and ugly facts. New York, Oxford U. P., 2000, 290 p.

6230. CHRISTIANSON (John Robert). On Tycho's island: Tycho Brahe and his assistants, 1570–1601. Cambridge, Cambridge U. P., 2000, XII-451 p.

6231. CLIFF (Andrew D.), HAGGETT (Peter), SMALLMAN-RAYNOR (Matthew R.). Island epidemics. Oxford, Oxford U. P., 2000, XXI-563 p.

6232. Competing for the Sugar Bowl. Ed. by Roger MUNTING and Tamás SZMRECSÁNYI. St. Katharinen, Scripta Mercaturae Verlag, 2000, XIV-193 p. [Cf. n° <choice> 6337.]

6233. CURTIS (Bruce). Social investment in medical forms: the 1866 cholera scare and beyond. *Canadian historical review*, 2000, 81, 3, p. 347-379.

6234. D'AGOSTINO (Salvo). A history of the ideas of theoretical physics. Essays on the nineteenth and twentieth century physics. Dordrecht a. London, Kluwer, 2000, XVIII-381 p. (Boston studies in the philosophy of science, 213).

6235. DANIEL (Thomas). Pioneers of medicine and their impact on tubercolosis. Rochester, University of Rochester Press, 2000, XIII-255 p. (ill.).

6236. DARRIGOL (Olivier). Electrodynamics from Ampère to Einstein. Oxford, Clarendon Press, 2000, 552 p.

6237. Development of mathematics 1950–2000. Ed. by Jean-Paul PIER. Basel, Birkhäuser Verlag, 2000, X-1372 p. (ill.).

6238. DI SIMPLICIO (Oscar). Inquisizione, stregoneria, medicina: Siena e il suo Stato (1580–1721). Siena, Il leccio, 2000, 226 p. (ill.). (Documenti di storia, 35).

6239. DILG (Peter). Die Apotheke als Forschungsstätte. *Berichte zur Wissenschaftsgeschichte*, 2000, 23, 3, p.303-316.

6240. DINGWALL (Helen M.). To be insert in the Mercury: medical practitioners and the press in eighteenth-century Edinburgh. *Social history of medicine*, 2000, 13, p. 23-44.

6241. DORN (Matthias). Das Problem der Autonomie der Naturwissenschaften bei Galilei. Stuttgart, Steiner, 2000, 193 p. (Sudhoffs Archiv. Beih., 43).

6242. DRAYTON (Richard). Nature's government: science, imperial Britain, and the 'Improvement' of the world. New Haven a. London, Yale U. P., 2000, XXI-346 p.

6243. DUCK (Ian). 100 years of Planck's Quantum. Singapore a. London, World Scientific, 2000, XII-545 p. (ill.).

6244. Einstein: the formative years 1879–1909. Ed. by Don HOWARD and John STACHEL. Boston, Birkhauser, 2000, VIII-258 p. (ill.). (Einstein studies, 8).

6245. ELIASSON (Per), LISBERG JENSEN (Ebba). Naturens nytta. (L'utilité de la nature. La science en Suède de Linné à nos jours). Lund, Historiska media, 2000, 224 p.

6246. EVENDEN (Doreen). The midwives of seventeenth century London. Cambridge, Cambridge U. P., 2000, XI-260 p. (ill.).

6247. FALTIN (Thomas). Heil und Heilung: Geschichte der Laienheilkunde und Struktur antimodernistischer Weltanschauungen in Kaiserreich und Weimarer Republik am Beispiel Eugen Wenza (1856–1945). Stuttgart, Franz Steiner, 2000, 458 p. (Medizin, Gesellschaft und Geschichte, 15).

6248. FASSIN (Didier). Les enjeux politiques de la santé. Etudes sénégalaises, équatoriennes, françaises. Paris, Karthala, 2000, 344 p.

6249. FELL (Ulrike). Disziplin, Profession und Nation: die Ideologie der Chemie in Frankreich vom Zweiten Kaiserreich bis in die Zwischenkriegszeit. Leipzig, Leipziger Universitätsverlag, 2000, 384 p. (Deutsch-französische Kulturbibliothek, 14).

6250. FISHER (Kate). 'She was quite satisfied with the arrangements I made': gender and birth control in Britain 1920–1950. *Past and present*, 2000, 169, p. 161-193.

6251. FREWER (Andreas). Medizin und Moral in Weimarer Republik und Nationalsozialismus: die Zeitschrift "Ethik" unter Emil Abderhalden. Frankfurt am Main u. New York, Campus Verlag, 2000, 318 p. (ill.).

6252. GALERA (Andrés). Los guisantes mágicos de Darwin y Mendel. *Asclepio: archivo iberoamericano de historia de la medicina y antropología médica*, 2000, 52, 2, p. 213-222.

6253. GALPERIN (Charles). Aspects du développement dans la biologie du XXe siècle: introduction. *Revue d'histoire des sciences*, 2000, 53, 3-4, p. 331-338.

6254. GAZI (Effi). Scientific national history: the Greek case in comparative perspective (1850–1920). Frankfurt am Main a. New York, Peter Lang, 2000, 179 p. (European university studies. Series III, History and allied studies, 871).

6255. GEYER-KORDESCH (Johanna). Pietismus, Medizin und Aufklärung in Preußen im 18. Jahrhundert. Tübingen, Max Niemeyer Verlag, 2000, 300 p. (Hallesche Beiträge zur Europäischen Aufklärung, 13).

6256. GIGLIONI (Guido). Immaginazione e malattia. Saggio su Jan Baptiste Van Helmont. Milano, Franco Angeli, 2000, 187 p.

6257. GRAFTON (Anthony). Cardano's cosmos: the worlds and works of a Renaissance astrologer. Cambridge, Harvard U. P., 2000, XII-284 p.

6258. GRAY (Jeremy). The Hilbert challenge. Oxford a. New York, Oxford U. P., 2000, XII-315 p. (ill.).

6259. GRIMOULT (Cédric). Histoire de l'évolutionnisme contemporain en France, 1945–1995. Genève, Droz, 2000, 616 p. (ill.). (Travaux de sciences sociales, 186).

6260. GRONEMAN (Carol). Nymphomania: a history. New York, W. W. Norton, 2000, XXIII-238 p.

6261. HART (Mitchell B.). Social science and the politics of modern Jewish identity. Stanford, Stanford U. P., 2000, VIII-340 p. (Stanford studies in Jewish history and culture).

6262. HAWKINS (Thomas). The emergence of the theory of Lie groups: an essay in the history of mathematics, 1869–1926. New York a. London, Springer, 2000, XIII-564 p. (ill.). (Sources and studies in the history of mathematics and physical sciences).

6263. HAZEN (Crai James). The village enlightenment in America: popular religion and science in the nineteenth century. Urbana a. Chicago, University of Illinois Press, 2000, 194 p.

6264. HECHT (Jennifer Michael). Vacher de Lapouge and the rise of Nazi science. *Journal of the history of ideas*, 2000, 61, 2, p. 285-304.

6265. HENIG (Robin Marantz). A monk and two peas: the story of Gregor Mendel and the discovery of genetics. London, Weidenfeld & Nicolson, 2000, VI-278 p.

6266. HEPLER (Alison L.). Women in labor: mothers, medicine and occupational health in the United States, 1890–1980. Columbus, Ohio State U. P., 2000, XII-177 p. (Women and health).

6267. HIBNER KOBLITZ (Ann). Science, women and Revolution in Russia. Amsterdam, Harwood Academic, 2000, XV-211 p. (Women in science).

6268. HILAIRE-PÉREZ (Liliane). L'invention technique au siècle des Lumières. Paris, A. Michel, 2000, 443 p. (L'évolution de l'humanité).

6269. Histoire de la pensée médicale en occident. Vol. 3. Du romantisme à la science moderne. Dir. par Mirko D. GRMEK. Paris, Ed. du Seuil, 2000, 428 p.

6270. HODGE (Jonathan). Canguilhem and the history of biology. *Revue d'histoire des sciences*, 2000, 53, 1, p. 65-81.

6271. HODGKISS (Andrew). From lesion to metaphor: chronic pain in British, French and German medical writings, 1800–1914. Amsterdam a. Atlanta, Rodopi, 2000, III-218 p. (Clio medica, 58).

6272. HUANG (H. T.). Science and civilisation in China. Vol. 6. Biology and biological technology. Part 6. Fermentations and food science. Cambridge, Cambridge U. P., 2000, XXVIII-741 p.

6273. HÜBNER (Klaus). Eugen Goldstein und die frühe Verwertung der Röntgenschen Entdeckung in Berlin. Berlin, ERS-Verlag, 2000, 118 p. (ill.). (Berliner Beiträge zur Geschichte der Naturwissenschaften und der Technik, 26).

6274. HUDEMANN-SIMON (Calixte). Die Eroberung der Gesundheit, 1750–1900. Frankfurt am Main, Fischer Taschenbuch Verl., 2000, 255 p. – IDEM. La conquête de la santé en Europe, 1750–1900. Quetigny, De Boeck, 2000, 176 p.

6275. Instruments and experimentation in the history of chemistry. Ed. by Frederic L. HOLMES and Trevor H. LEVERE. Cambridge a. London, MIT Press, 2000, XXI-415 p. (ill.). (Dibner Institute studies in the history of science and technology).

6276. JAGGI (Om Prakash). Medicine in India: modern period. Oxford, Oxford U. P., 2000, XXVI-355 p. (History of science, philosophy and culture in Indian civilization, 9/1).

6277. JANKOVIC (V.). The place of nature and the nature of place: the chorographic challenge to the history of British provincial science. *History of Science*, 2000, 38, 1, p. 79-113.

6278. JEANMONOD (Gilles), HELLER (Geneviève). Eugénisme et contexte socio-politique. L'exemple de l'adoption d'une loi sur la stérilisation des handicapés et malades mentaux dans le canton de Vaud en 1928. *Schweizerische Zeitschrift für Geschichte*, 2000, 50, p. 20-44.

6279. JONES (Colin). Pulling teeth in eighteenth-century Paris. *Past and present*, 2000, 166, p. 100-145.

6280. KÂHYA (Esin). Bilimin Işığında Osmanlıdan Cumhuriyete Tıp Ve Sağlık Kurumları. (Medicine and medical institutions in Turkey from the Ottoman to Republic in the light of science). Ankara, Türkiye Diyanet Vakfı, 2000, 466 p.

6281. KAY (Lily E.). Who wrote the book of life? A history of the genetic code. Stanford, Stanford U. P., 2000, XXIV-441 p. (ill.).

6282. KIRCHBERGER (Ulrike). Deutsche Naturwissenschaftler im britischen Empire. Die Erforschung der außereuropäischen Welt in Spannungsfeld zwischen deutschem und britischem Imperialismus. *Historische Zeitschrift*, 2000, 271, 3, p. 621-660.

6283. LASZLO (Pierre). Miroir de la chimie. Paris, Ed. Seuil, 2000, 328 p.

6284. LEINKAUF (Thomas). Der Natur-Begriff des 17. Jh. und zwei seiner Interpretamente: "res extensa" und "intima rerum". *Berichte zur Wissenschaftsgeschichte*, 2000, 23, 4, p.399-432

6285. LENGWILER (Martin). Zwischen Klinik und Kaserne. Die Geschichte der Militärpsychiatrie in

Deutschland und der Schweiz 1870–1914. Zürich, Chronos, 2000, 432 p.

6286. LEYS (Ruth). Trauma: a genealogy. Chicago, University of Chicago Press, 2000, X-318 p.

6287. LÓPEZ PÉREZ (Miguel). La alquimia: un problema social en la España del siglo XVII. *Llull: boletín de la Sociedad Española de Historia de las Ciencias*, 2000, 23, 48, p. 643-659.

6288. LUDMERER (Kenneth M.). Time to heal – American medical education from the turn of the century to the era of managed care. New York, Oxford U. P., 2000, 540 p.

6289. MAC ALLISTER (William B.). Drug diplomacy in the twentieth century: an international history. London a. New York, Routledge, 2000, XVII-344 p.

6290. MAC GREGOR (JoAnn), RANGER (Terence). Displacement and disease: epidemics and ideas about malaria in Matabeleland, Zimbabwe, 1945–1996. *Past and present*, 2000, 167, p. 203-237.

6291. MARQUIS (J. P.). The other warriors: American social science and nation building in Vietnam. *Diplomatic history*, 2000, 24, 1, p. 79-105.

6292. Medizin und Pharmazie im 18. und 19. Jahrhundert: Beiträge zur Geschichte der Wissenschaftsbeziehungen zwischen Deutschland und dem Russischen Reich. Hrsg. v. Ingrid KÄSTNER und Regine PFREPPER. Aachen, Shaker, 2000, VIII-201 p. (ill.). (Deutsch-Russische Beziehungen im Medizin und Naturwissenschaften, 2).

6293. METZLER (Gabriele). Internationale Wissenschaft und nationale Kultur: deutsche Physiker in der internationalen Community 1900–1960. Göttingen, Vandenhoeck & Ruprecht, 2000, 304 p.

6294. MORAN (Bruce T.). Alchemy, chemistry and the history of science. *Studies in history and philosophy of science*, 2000, 31, 4, p. 711-720.

6295. New chemistry (The). Ed. by Nina HALL. Cambridge, Cambridge U. P., 2000, XI-493 p. (ill.).

6296. NIEMI (Marjaana). Public helath discourses in Birmingham and Gothenbur, 1890–1920. *In*: Body and city: a cultural history of urban public health. Ed. by Sally SHEARD and Helen POWER. Aldershot, Ashgate, 2000, p. 123-142.

6297. NILSSON (Ingemar), PETERSSON (Hans-Inge). Medicinens idéhistoria. (Histoire des idées médicales). Stockholm, SNS Förlag, 2000, 250 p. (ill.).

6298. O'CONNOR (Erin). Raw material: producing pathology in Victorian culture. Durham, Duke U. P., 2000, XIII-272 p. (ill.).

6299. OOSTERHUIS (Harry). Stepchildren of nature. Krafft-Ebing, psychiatry, and the making of sexual identity. Chicago, University of Chicago Press, 2000, X-321 p.

6300. PARISI (Giovanni). Il problema della eredità biologica: dai Fattori mendeliani al riconoscimento della natura molecolare del gene. Napoli, Accademia Pontaniana, 2000, 124 p. (ill.). (Quaderni dell'Accademia Pontaniana, 29).

6301. PEARD (Julyan G.). Race, place, and medicine: the idea of the tropics in nineteenth-century Brazilian medicine. Durham, N.C., Duke U. P., 2000, X-315 p.

6302. PFEIFER (Klaus). Medizin der Goethezeit: Christoph Wilhelm Hufeland und die Heilkunst des 18. Jahrhunderts. Köln, Böhlau, 2000, X-293 p. (ill.).

6303. PICKSTONE (John). Ways of knowing: a new history of science, technology and medicine. Manchester, Manchester U. P., 2000, XII-271 p.

6304. PINET (Patrice). Robert Boyle (1627–1691) et la pharmacologie. *Revue d'histoire de la pharmacie*, 2000, 48, 328, p. 471-484.

6305. PLANERT (Ute). Der dreifache Körper des Volkes. Sexualität, Biopolitik und die Wissenschaften vom Leben. *Geschichte und Gesellschaft*, 2000, 26, 4, p. 539-576.

6306. REDFIELD (Peter). Space in the tropics: from convicts to rockets in French Guiana. Berkeley a. Los Angeles, University of California Press, 2000, [s. p.].

6307. REINHARDT (Carsten). Heinrich Caro and the creation of modern chemical industry. Dordrecht a. London, Kluwer Academic, 2000, XXII-453 p. (ill.). (Chemists and chemistry, 19).

6308. RENNEVILLE (Marc). Le langage des crânes. Une histoire de la phrénologie. Paris, Institut d'Edition Sanofi-Synthélabo, 2000, 356 p. (Les empêcheurs de penser en rond).

6309. REPP (Kevin). "More corporeal, more concrete": liberal humanism, eugenics, and German progressives at the last Fin de Siècle. *Journal of modern history*, 2000, 72, 3, p. 683-730.

6310. Rethinking the scientific revolution. Ed. by Margaret OSLER. Cambridge, Cambridge U. P., 2000, 352 p.

6311. ROBERTS (Harry E.). From Kodachrome to instant color: the history of silver halide emulsion chemistry. Old Forge, H. Roberts, 2000, VII-119 p.

6312. RODRÍGUEZ NOZAL (Raúl). Orígenes, desarrollo y consolidación de la industria farmacéutica española (ca. 1850–1936). *Asclepio: archivo iberoamericano de historia de la medicina y antropología médica*, 2000, 52, 1, p. 127-159.

6313. ROSENBAND (Leonard N.). Papermaking in eighteenth-century France. Management, labor and revolution at the Montgolfier mill, 1761–1805. Baltimore, Johns Hopkins U. P., 2000, 210 p.

6314. ROUSSEAU (George Sebastian), HAYCOCK (D.). The jew of Crane Court: Emanuel Mendes da

Costa (1717–1791), natural history and natural excess. *History of Science*, 2000, 38, 2, p. 127-70.

6315. SALOMON (Christine). Savoirs et pouvoirs thérapeutiques kanaks. Paris, Presses Universitaires de France, 2000, 160 p. (Ethnologies).

6316. SANDELOWSKI (Margarete). Devices and desires: gender, technology, and American nursing. Chapell Hill, University of North Carolina Press, 2000, XVII-295 p. (Studies in social medicine).

6317. SANDL (Marcus). Raumvorstellungen und Erkenntnismodelle im 18. Jh. *Berichte zur Wissenschaftsgeschichte*, 2000, 23, 4, p. 389-398.

6318. SCHMIDT (Leigh Eric). Hearing things: religion, illusions, and the American enlightenment. Cambridge, Harvard U. P., 2000, XIII-318 p.

6319. SCHNEIDER (Ivo). Der Einfluß der griechischen Mathematik auf Inhalt und Entwicklung der mathematischen Produktion deutscher Rechenmeister im 16. und 17. Jh. *Berichte zur Wissenschaftsgeschichte*, 2000, 23, 2, p.203-218.

6320. SCHÜTT (Hans W.). Alchemie im Zeitalter der Aufklärung. *Berichte zur Wissenschaftsgeschichte*, 2000, 23, 2, p. 157-166. – IDEM. Auf der Suche nach dem Stein der Weisen. Die Geschichte der Alchemie. München, Beck, 2000, 602 p.

6321. Science and society in Southern Africa. Ed. by Saul DUBOW. Manchester, Manchester U. P., 2000, X-241 p. (Studies in imperialism).

6322. SECORD (James A.). Victorian sensation: the extraordinary publication, reception, and secret authorship of "Vestiges of the natural history of creation". Chicago, University of Chicago Press, 2000, XIX-624 p.

6323. Sedemnásty zborník z dejín fyziky. (17th miscellany about the history of physics). Ed. Ingrid HYMPÁNOVÁ, Miroslav Tibor MOROVICS. Bratislava, Odborná skupina dejín a metodológie fyziky SFS a Slovenská spoločnosť pre dejiny vied a techniky pri Slovenskej akadémii vied, 2000,107 p.

6324. SEIDLER (Eduard). Kinderärzte 1933–1945. Entrechtet – geflohen – ermordet. Bonn, Bouvier, 2000, 494 p.

6325. SHER (Richard B.). Science and medicine in the Scottish Enlightenment: the lessons of book history. *In:* Scottish Enlightenment (The): essays in re-interpretation [Cf. n° 6194]. p. 99-156.

6326. SOMERVILLE (Christopher). The twentieth century trajectory of plant biology. *Cell*, 2000, 100, 1, p. 13-25.

6327. SPARY (E. C.). Utopia's garden. French natural history from Old Regime to Revolution. Chicago, University of Chicago Press, 2000, XV-321 p.

6328. SPILLANE (Joseph F.). Cocaine: from medical marvel to modern menace in the United States, 1884–1920. Baltimore, Johns Hopkins University in association with the Hagley Museum and Library, Wilmington, Del, 2000, X-214 p. (Studies in Industry and Society).

6329. STANZIANI (Alessandro). Les enquêtes orales en Russie, 1861–1914. *In*: Enquêtes et ethnographie, France et Russie (fin 18e–fin 19e siècles). *Annales*, 2000, 55, 1, p. 219-242.

6330. Statistica ufficiale e storia d'Italia: gli "Annali di Statistica" dal 1871 al 1997. A cura di Paola GERETTO. Roma, Istituto Nazionale di Statistica, 2000, 446 p. (Annali di statistica, Anno 129, Serie X, 21).

6331. STEPHENSON (Bruce), BOLT (Marvin), FRIEDMAN (Anna Felicity). The universe unveiled: instruments and images through history. Cambridge, Cambridge U. P., 2000, 152 p.

6332. STOLBERG (Michael). Self-pollution, anxiety and the body in the Eighteenth century. *Social history of medicine*, 2000, 13, p.1-21

6333. STRATHERN (Paul). Mendeleyev's dream: the quest for the elements. London, Hamish Hamilton, 2000, 320 p. (ill.).

6334. STRICK (James Edgar). Sparks of life: Darwinism and the Victorian debates over spontaneous generation. Cambridge a. London, Harvard U. P., 2000, XI-283 p. (ill.).

6335. STUBHAUG (Arild). Det var mine tankers djervhet. Matematikeren Sophus Lie. [The mathematician Sophus Lie (1842–1899)]. Oslo, Aschehoug, 2000, 624 p. (ill.).

6336. STUURMAN (Siep). François Bernier and the invention of racial classification. *History workshop*, 2000, 50, p. 1-21.

6337. SZMRECSÁNYI (Tamás), PELAEZ ALVAREZ (Victor Manuel). The search for a perfect substitute: technological and economic trajectories of synthetic sweeteners, from Saccharin to Aspartame (c. 1880–1980). *In*: Competing for the Sugar Bowl [Cf. n° 6232], p. 172-193.

6338. SZMRECSÁNYI (Tamás). Por una história econômica de ciência e da tecnologia. *Economia aplicada*, 2000, 4, 2, p. 399-407

6339. TOGNOTTI (Eugenia). Il mostro asiatico. Storia del colera in Italia. Roma e Bari, Laterza, 2000, 282 p.

6340. Trobades d'història de la ciència i de la tècnica. Actes de les V Trobades d'història de la ciència i de la tècnica, Roquetes, 11–13 desembre 1998. Ed. por Josep BATLLÓ ORTIZ, Pere DE LA FUENTE COLLELL i Roser PUIG AGUILAR. Barcelona, Societat Catalana d'Història de la Ciència i de la Tècnica, 2000, 562 p. (ill.).

6341. Věda v Československu v letech 1953–1963. (Science and scholarship in Czechoslovakia, 1953–1963). Praha, Archiv Akademie věd ČR, 2000, 591 p. (Práce z dějin vědy, 1).

6342. WEAR (Andrew). Knowledge and practice in English medicine, 1550–1680. Cambridge, Cambridge U. P., 2000, VIII-496 p.

6343. WEATHERALL (Mark). Gentlemen, scientists, and doctors: medicine at Cambridge, 1800–1940. Rochester, Boydell Press, 2000, X-341 p. (History of the University of Cambridge. Texts and studies, 3).

6344. WEBER (Wolfhard), ENGELSKIRCHEN (Lutz). Streit um die Technikgeschichte in Deutschland, 1945–1975. Münster u. New York, Waxmann, 2000, 446 p. (ill.). (Cottbuser Studien zur Geschichte von Technik, Arbeit und Umwelt, 15).

6345. WETZELL (Richard F.). Inventing the criminal. A history of German criminology, 1880–1945. Chapell Hill, University of North Carolina Press, 2000, XIV-348 p.

6346. WHORTON (James C.). Inner hygiene – Constipation and the pursuit of health in modern society. New York, Oxford U. P., 2000, 338 p.

6347. WILSON (Renate). Pious traders in medicine: a German pharmaceutical network in eighteenth-century north America. University Park, Pennsylvania State U. P., 2000, XIV-258 p.

6348. WOLLOCH (Nathaniel). Christiaan Huygen's attitude towards animals. *Journal of the history of ideas*, 2000, 61, 3, p. 415-432.

6349. WORBOYS (Michael). Spreading germs. Disease theories and medical practice in Britain, 1865–1900. Cambridge a. New York, Cambridge U. P., 2000, XVI-327 p.

Cf. n°s 592, 5831

§ 7. Literature.

a. General

6350. BELLIN (Joshua David). The demon of the continent: Indians and the shaping of American literature. Philadelphia, University of Pennsylvania Press, 2000, 247 p.

6351. Courts, patrons and poets. Ed. by David MATEER. New Haven a. London, Yale U. P., 2000, XI-383 p. (ill.).

6352. DENIS (Benoît). Littérature et engagement: de Pascal à Sartre. Paris, Ed. de Seuil, 2000, 319 p. (Points Essais, 407).

6353. DUVAL (Sophie), MARTINEZ (Marc). La satire (littérature française et anglaise). Paris, A. Colin, 2000, 272 p.

6354. HAFEZ-ERGAUT (Agnès). Le vertige du vide: Huysmans, Céline, Sartre. Lewiston, Edwin Mellen Press, 2000, 323 p. (Studies in French literature, 42).

6355. HELFRICH (Cornelia). Die Rezeption von Gestalt und Werk der hl. Therese von Avila in der französischen Literatur des 19./20. Jahrhunderts: der Einfluss der spanischen Mystik auf die französische Literatur. Frankfurt am Main u. Bern, Peter Lang, 2000, 216 p.

6356. Histoire (L') dans la littérature. Ed. par Laurent ADERT et Eric EIGENMANN. Genève, Groz, 2000, 349 p. (Recherches et rencontres, 15).

6357. History (A) of women's writing in France. Ed. by Sonya STEPHENS. Cambridge, Cambridge UP, 2000, 340 p.

6358. History of European literature. Ed. by Annick BENOIT-DUSAUSOY and Guy FONTAINE. London a. New York, Routledge, 2000, XXVIII-732 p.

6359. HOGAN (Patrick). Colonialism and cultural identity: crises of tradition in the anglophone literatures of India, Africa and the Caribbean. New York, State University of New York Press, 2000, XX-354 p.

6360. JOUHAUD (Christian). Les pouvoirs de la littérature: histoire d'un paradoxe. Paris, Gallimard, 2000, 450 p.

6361. KIRSCH (Fritz Peter). Epochen des französischen Romans. Wien, Wiener Universitätsverlag, 2000, 311 p. (UTB für Wissenschaft, 2125).

6362. LINDEMANN (Uwe). Die Wüste. Terra Incognita – Erlebnis – Symbol. Heidelberg, C. Winter, 2000, 451 p. (Beiträge zur neueren Literaturgeschichte, 175).

6363. MATHIEU-CASTELLANI (Gisèle). La rhétorique des passions. Paris, Presses universitaires de France, 2000, 202 p.

6364. NEUMAYR (Anton). Dichter und ihre Leiden: Jean-Jacques Rousseau, Friedrich Schiller, August Strindberg, Georg Trakl. München, Deuticke, 2000, 400 p.

6365. PATRIZI (Giorgio). Narrare l'immagine. La tradizione degli scrittori d'arte. Roma, Donzelli, 2000, 138 p.

6366. PAUVERT (Jean-Jacques). La littérature érotique. Paris, Flammarion, 2000, 128 p. (Dominos, 219).

6367. RABAU (Sophie). Fictions de présence: la narration orale dans le texte romanesque du roman antique au XXe siècle. Paris, H. Champion, 2000, 488 p. (Bibliothèque de littérature générale et comparée, 25).

6368. RANZATO SANTIN (Federica). Réalisme: alle origini di una parola moderna. Padova, CLEUP Editrice, 2000, 194 p.

6369. Representation of the self from the Renaissance to Romanticism. Ed. by Patrick COLEMAN, Jayne LEWIS and Jill KOWALIK. Cambridge, Cambridge U. P., 2000, XII-284 p.

6370. RIVA (Silvia). Rulli di tam-tam dalla torre di Babele. Storia della letteratura del Congo-Kinshasa. Milano, LED, 2000, 452 p.

6371. Roman (Le) historique: récit et histoire. Ed. par Daniel COUÉGNAS et Dominique PEYRACHE-LEBORGNE. Nantes, Pleins feux, 2000, 359 p.

6372. Russian literature, Modernism and the visual arts. Ed. by Catriona KELLY and Stephen LOVELL. New York, Cambridge U. P., 2000, 314 p. (ill.).

6373. Schriftsteller als Intellektuelle. Politik und Literatur im Kalten Krieg. Hrsg. v. Svan HANUSCHEK, Therese HÖRNIGK und Christine MALENDE. Tübingen, Max Niemeyer Verlag, 2000, 340 p. (Studien und Texte zur Sozialgeschichte der Literatur, 73).

Cf. n°s 769, 1465

b. Renaissance

6374. ARMSTRONG (Adrian). Technique and Technology: script, print and poetics in France, 1470–1550. Oxford, Clarendon Press, 2000, CII-246 p.

6375. BERGER (Harry). The absence of grace: sprezzatura and suspicion in two Renaissance courtesy books. Stanford, Stanford U. P., 2000, XIV-267 p.

6376. BORRIS (Kenneth). Allegory and epic in English Renaissance literature. Heroic form in Sidney, Spenser and Milton. Cambridge, Cambridge U. P., 2000, 332 p.

6377. BROOKS (Douglas A.). From playhouse to printing house: drama, authorship in early modern England. Cambridge, Cambridge U. P., 2000, XVIII-294 p. (Cambridge studies in Renaissance literature and culture, 36).

6378. CAPATA (Alessandro). Semper truffare paratus. Guerre e ideologia nel "Baldus" di Folengo. Roma, Bulzoni, 2000, 222 p. (Studi italiani, 9).

6379. CLOSE (Anthony). Cervantes and the comic mind of his age. Oxford, Oxfrod U. P., 2000, VIII-375 p.

6380. CLOSSON (Marianne). L'imaginaire démonique en France (1550–1650): genèse de le littérature fantastique. Genève, Droz, 2000, 544 p. (Travaux d'Humanisme et Renaissance, 341).

6381. CONCONI (Bruna). Le prove del testimone: scrivere di storia, fare letteratura nella seconda metà del Cinquecento: l'Histoire memorable de Jean de Léry. Bologna, Pàtron, 2000, 236 p. (Biblioteca del Dipartimento di lingue e letterature straniere moderne dell'Università degli studi di Bologna, 25).

6382. CORTINI (Maria Antonietta), MULAS (Luisa). Selva di vario narrare. Schede per lo studio della narrazione breve nel Seicento. Roma, Bulzoni, 2000, 587 p.

6383. DAUVOIS (Nathalie). Le sujet lyrique à la Renaissance. Aris, Presses universitaires de France, 2000, 128 p.

6384. DE MIRANDA (Girolamo). Una quiete operosa. Forma e pratiche dell'Accademia napoletana degli Oziosi (1611–1645). Napoli, Fredericiana Editrice Universitaria, 2000, 393 p.

6385. DESA WIGGINS (Peter). Donne, Castiglione and the poetry of courtliness. Bloomington a. Indianapolis, Indiana U. P., 2000, VIII-174 p.

6386. DOTOLI (Giovanni). Littérature et société en France au XVIIe siècle. Fasana, Schena et Paris, Didier Erudition, 2000, 371 p.

6387. EGGERT (Katherine). Showing like a queen: female authority and literary experiment in Spenser, Shakespeare and Milton. Philadelphia, University of Pennsylvania Press, 2000, 290 p.

6388. GRAY (Floyd). Gender, rhetoric and print culture in French Renaissance writing. Cambridge, Cambridge UP, 2000, 240 p. (Cambridge studies in French, 63).

6389. HACKETT (Helen). Women and romance fiction in the English Renaissance. Cambridge, Cambridge U. P., 2000, VII-236 p.

6390. LESTRIGANT (Frank), RIEU (Josiane), TARRÊTE (Alexandre). Littérature française du XVIe siècle. Paris, Presses universitaires de France, 2000, XIII-503 p.

6391. Satire (Fra) e Rime ariostesche. Atti del IV Seminario di letteratura italiana. Gargnano del Garda, 14–16 ottobre 1999. A cura di Claudia BERRA. Milano, Cisalpino, 2000, 572 p. (Quaderni di "Acme", 43).

6392. SCHWARZ (Kathryn). Tough love: Amazon encounters in the English Renaissance. Durham, Duke U. P., 2000, XVI-284 p.

6393. TOSCANO (Tobia R.). Letterati corti accademie. La letteratura a Napoli nella prima metà del Cinquecento. Napoli, Loffredo, 2000, 369 p. (Le ricerche di "critica letteraria", 4).

6394. Vernacular literature and current affairs in the early Sixteenth century: France, England and Scotland. Ed. by Jennifer BRITNELL and Richard BRITNELL. Aldershot, Ashgate, 2000, XXVI-212 p.

6395. VUILLEUMIER LAURENS (Florence). La raison des figures symboliques à la Renaissance et à l'âge classique. Genève, Droz, 2000, 544 p.

6396. ZARETSKIJ (Jurij P.). Renessansnaja avtobiografija i samosoznanie lichnosti: Enea Silvio Pikkolomini (Pij II): Monografija. (Renaissance Autobiography and Individual Self-Consciousness: Enea Silvio Piccolomini (Pius II): Monograph). Ed. Viktorija I. UKOLOVA. Ministerstvo obrazovanija RF, Nizhegorodskij gos. un-t. Nizhnij Novgorod, Izd. Nizhegor. un-ta, 2000, 154 p. (ill.). [Eng. summary]

Cf. n° 769

c. Classicism

* 6397. GIAMBONINI (Francesco). Bibliografia delle opere a stampa di Giambattista Marino. Firenze, Olschki, 2000, 2 vol., 920 p.

** 6398. PINDEMONTE (Ippolito). Lettere a Isabella (1784–1828). A cura di Gilberto PIZZAMAGLIO. Firenze, Olschki, 2000, LXXVIII-413 p. (Biblioteca di letteratura italiana. Studi e testi, 45).

6399. Attualità di Giuseppe Parini: poesia e impegno civile. Atti del convegno internazionale di Milano, 27–30 settembre 1999. A cura di Giorgio BARONE. Pisa e Roma, Istituti Editoriali e Poligrafici Internazionali, 2000, 710 p.

6400. BANNET TAVOR (Eve). The domestic revolution: Enlightenment feminisms and the novel. Baltimore, Johns Hopkins U. P., 2000, 288 p.

6401. BARRELL (John). Imaging the King's death: figurative treason, fantasies of regicide, 1793–1796. Oxford, Oxford U. P., 2000, XVIII-738 p.

6402. Beaumarchais: homme de lettres, homme de société. Ed. par Philip ROBINSON. Bern u. Berlin, Peter Lang, 2000, 293 p. (French studies of the 18th and 19th centuries, 8).

6403. BLOCH (Olivier). Molière / Philosophie. Paris, Albin Michel, 2000, 189 p. (Idées).

6404. BOYLE (Frank). Swift as nemesis: modernity and its satirist. Stanford, Stanford U. P., 2000, XIV-242 p.

6405. BURGESS (Miranda J.). British fiction and the production of the social order, 1740–1830. Cambridge, Cambridge U. P., 2000, XII-307 p.

6406. CAMBOU (Pierre). Le traitement voltairien du conte. Paris, H. Champion, 2000, 702 p. (Les Dix-huitèmes siècles, 45).

6407. CHEMELLO (Adriana) RICALDONE (Luisa). Geografie e genealogie letterarie. Erudite, biografe, croniste, narratrici, epistolières, utopiste tra Settecento e Ottocento. Padova, Il Poligrafo, 2000, 252 p.

6408. DE ROJAS (Fernando). La Celestina. Tragicomedia de Calisto e Melibea. Ed. por Francisco J. LOBERA, Guillermo SERÉS, Paloma DIAZ-MAS, Carlos MOTA, Inigo RUIZ ARZALLUZ y Francisco RICO. Barcelona, Critica, Bibliotrca clásica, 2000, 874 p.

6409. GREINER (Frank). Les métamorphoses d'Hermès: tradition alchimique et esthétique littéraire dans la France de l'âge baroque (1583–1646). Paris, H. Champion, 2000, 663 p. (Bibliothèque littéraire de la Renaissance, 42).

6410. LALLEMAND (Marie-Gabrielle). La lettre dans le récit: étude de l'œuvre de M.lle de Scudéry. Tübingen, G. Narr, 2000, 446 p. (Biblio 17, 120).

6411. MADRIGNANI (Carlo Alberto). All'origine del romanzo in Italia, 1753–1785. Napoli, Liguori, 2000, 354 p.

6412. MARTIN (Dieter). Barock um 1800. Bearbeitung und Aneignung deutscher Literatur des 17. Jahrhunderts von 1700 bis 1830. Frankfurt am Main, Klostermann, 2000, XI-702 p. (Das Abendland, 26).

6413. MOENNIGHOFF (Burkhard). Goethes Gedichttitel. Berlin u. New York, de Gruyter, 2000, VI-207 p. (Quellen und Forschungen zur Literatur- und Kulturgeschichte, 16).

6414. NIETHAMMER (Ortrun). Autobiographien von Frauen im 18. Jahrhundert. Tübingen u. Basel, Francke, 2000, 307 p.

6415. PARMENTIER (Bérengère). Le siècle des moralistes: de Montaigne à La Bruyère. Paris, Editions de Seuil, 2000, 349 p. (Points, 406).

6416. PAULSON (Ronald). The life of Henry Fielding. Oxford, Blackwell, 2000, XIV-400 p.

6417. PECH (Thierry). Conter le crime: droit et littérature sous la Contre-Réforme: les histoires tragiques (1559–1644). Paris, H. Champion, 2000, 480 p. (Lumière classique, 24).

6418. RICCÒ (Laura). "Parrebbe un romanzo". Polemiche editoriali e linguaggi teatrali ai tempi di Goldoni, Chiari, Gozzi. Roma, Bulzoni, 2000, 302 p.

6419. SCHLICK (Werner). Goethe's Die Wahlverwandtschaften. Heidelberg, C. Winter, 2000, X-538 p. (Beiträge zur neueren Literaturgeschichte, 172).

6420. SCHMITZ-BURGARD (Sylvia). Das Schreiben des anderen Geschlechts: Richardson, Rousseau, Goethe. Würzburg, Königshausen & Neumann, 2000, 295 p.

6421. SELLIER (Philippe). Port-Royale et la littérature. Vol. 2. Le siècle de Saint Augustin, La Rochefoucauld, Mme de Lafayette, Sacy, Racine. Paris, H. Champion, 2000, 294 p.

6422. SGARD (Jean). Le roman français à l'âge classique, 1600–1800. Paris, Librairie générale française, 2000, 254 p. (Le Livre de poche. Références. Littérature, 571).

6423. SPERA (Lucinda). Il romanzo italiano del tardo Seicento (1670–1700). Milano, La Nuova Italia, 223 p.

6424. STEINER (Uwe). Poetische Theodizee. Philosophie und Poesie in der lehrhaften Dichtung im achtzehnten Jahrhundert. München, Funk, 2000, 384 p.

6425. TERNAUX (Jean-Claude). Lucain et la littérature de l'âge baroque en France: citation, imitation et création. Paris, H. Champion, 2000, 461 p. (Bibliothèque littéraire de la Renaissance, 43).

6426. WINGROVE (Elizabeth Rose). Rousseau's republican romance. Princeton, Princeton U. P., 268 p.

6427. Women and literature in Britain, 1700–1800. Ed. by Vivien JONES. Cambridge, Cambridge U. P., 2000, 320 p.

6428. WUTHENOW (Ralph-Rainer). Die gebändigte Flamme. Zur Wiederentdeckung der Leidenschaften im Zeitalter der Vernunft. Heidelberg, Winter, 2000, 151 p. (Beiträge zur neueren Literaturgeschichte, 178).

6429. ZEUCH (Ulrike). Umkehr der Sinnshierarchie. Herder und die Aufwertung des Tastsinns seit der frühen Neuzeit. Tübingen, Max Niemeyer Verlag, 2000, X-332 p. (Communicatio. Studien zur europäischen Literatur- und Kulturgeschichte, 22).

d. *Romanticism and after*

** 6430. DE LAMARTINE (Alphonse). Correspondance d'Alphonse de Lamartine (1830–1867). Vol. 1. 1830–1832. Ed. par Christian CROISILLE. Paris, H. Champion, 2000, 693 p. (Textes de littérature moderne et contemporaine, 42).

6431. ALT (Peter André). Schiller. Leben – Werk – Zeit. München, Beck, 2000, 2 vol., 783 p., 686 p.

6432. ATTRIDGE (Derek). Joyce effects. On language, thory and history. Cambridge, Cambridge U. P., 2000, XVIII-208 p.

6433. BECKER-CANTARINO (Barbara). Schriftstellerinnen der Romantik. Epoche – Werke – Wirkung. München, Beck, 320 p.

6434. Beiträge zur Rezeption der britischen und irischen Literatur des 19. Jahrhunderts im deutschsprachigen Raum. Hrsg. v. Norbert BACHLEITNER. Amsterdam, Rodopi, 2000, X-534 p. (Internationale Forschungen zur allgemeinen und vergleichenden Literaturwissenschaft, 45).

6435. BENUCCI (Elisabetta). Paolina Leopardi. Viaggio notturno intorno alla mia camera. Traduzione dal francese dell'opera di X. de Maistre e altri scritti. Venosa, Edizioni Osanna, 2000, 198 p.

6436. BREARTON (Fran). The Great War in Irish poetry: W. B. Yeats to Michael Longley. Oxford, Oxford U. P., 2000, XII-316 p.

6437. BRILLI (Attilio). In viaggio con Leopardi. Bologna, Il Mulino, 2000, 130 p.

6438. CARTER (William). Marcel Proust: a life. New Haven a. London, Yale U. P., 2000, 960 p.

6439. CASADEI (Alberto). Romanzi di Finisterre. Narrazione della guerra e problemi del realismo. Roma, Carocci, 2000, 290 p.

6440. CERISOLA (Pier Luigi). Giovanni Pascoli tra estetica ed ermeneutica. Firenze, La nuova Italia, 2000, 205 p.

6441. CHAMPEAU (Sthéphanie). La notion d'artiste chez les Goncourt (1852–1870). Paris, H. Champion, 2000, 557 p. (Romantisme et modernités, 33).

6442. CIPRIANI (Fernando). Il romanzo d'infanzia in Francia (1913–1929). Problematiche e protagonisti. Pescara, Edizioni Campus, 2000, 303 p. (Finnegans, 7).

6443. COUSSY (Denise). La littérature africaine moderne au Sud du Shara. Paris, Editions Karthala, 208 p.

6444. DANNA (Bianca). Dal taccuino alla lanterna magica. De Amicis reporter e scrittore di viaggi. Firenze, Olschki, 2000, 192 p.

6445. DI BENEDETTO (Arnaldo). Dal tramonto dei Lumi al Romanticismo. Valutazioni. Modena, Mucchi, 2000, 292 p. (Il vaglio, 46).

6446. DIAFANI (Laura). La "stanza silenziosa". Studio sull'epistolario di Leopardi. Firenze, Le Lettere, 2000, 264 p.

6447. DRYDEN (Linda). Joseph Conrad and the imperial romance. Houndmills, Basingstoke and London, Macmillan, 2000, XII-228 p.

6448. ELSAGHE (Yahya). Die imaginäre Nation. Thomas Mann und das "Deutsche". München, Fink, 2000, 429 p.

6449. EMERSON (Everett). Mark Twain. A literary life. Philadelphia, University of Pennsylvania Press, 2000, XIV-386 p.

6450. FERRIS (David S.). Silent urns. Romanticism, Hellenism, Modernity. Stanford, Stanford U. P., 2000, XX-248 p.

6451. FINCH (Alison). Women's writing in Nineteenth century France. Cambridge, Cambridge U. P., 2000, 332 p.

6452. FOLLI (Anna). Penne leggère. Neera, Ada Negri, Sibilla Aleramo. Scritture femminili italiane fra Otto e Novecento. Milano, Guerini e Associati, 2000, 256 p.

6453. FORMOZOV (Alexandr A.). Puskhkin i drevnosti: Nabljudenija arkheologa. (Alexander Pushkin and antiquities: observations by an archaeologist). Moskva, Jazyki russkoj kul'tury, 2000, 144 p. (ill.). (Studia historica. Series minor).

6454. FRIEDRICH (Hans-Edwin). Deformierte Lebensbilder. Erzählmodelle der Nachkriegsautobiographie (1945–1960). Tübingen, Niemeyer, 2000, VI-456 p. (Studien und Texte zur Sozialgeschichte der Literatur, 74).

6455. GENEVRAY (Françoise). George Sand et ses contemporaines russes: audience, échoes, reécriture. Paris, L'Harmattan, 2000, 412 p.

6456. GIOVIALE (Fernando). Scenari del racconto. Mutazioni di scrittura nell'Otto/Novecento. Caltanisetta e Roma, Sciascia, 326 p.

6457. GUY (Josephine M.), SMALL (Jan). Oscar Wilde's profession: writing and the culture industry in the late Nineteenth century. Oxford, Oxford U. P., 2000, X-314 p.

6458. HALLENSLEBEN (Markus). Else Lasker-Schüler. Avantgardismus und Kunstinszenierung. Tübingen u. Basel, Francke, 2000, 367 p.

6459. HARZER (Friedmann). Erzählte Verwandlung. (Eine Poetik epischer Metamorphosen). Tübingen, Max Niemeyer Verlag, 2000, VIII-231 p. (Studien zur deutschen Literatur, 157).

6460. HATHAWAY (Heather). Caribbean waves: relocating Claude McKay and Paule Marshall. Bloomington and Indianapolis, Indiana U. P., 2000, XII-200 p.

6461. HAYES (Kevin J.). Poe and the printed word. Cambridge, Cambridge U. P., 2000, XVIII-146 p. (Cambridge studies in American literature and culture, 124).

6462. JONES (Steven E.). Satire and Romanticism. New York, St. Martin's Press, 2000, 262 p.

6463. KESTNER (Joseph A.). The Edwardian detective, 1901-1915. Aldeshot, Ashgate, 2000, VIII-416 p.

6464. KONZETT (Matthias). The rhetoric of national dissent in Thomas Bernhard, Peter Handke and Elfriede Jelinek. Rochester, Camden House, 2000, XII-164 p.

6465. KOSHELEV (Vjacheslav A.). Pushkin: Istorija i predanie: Ocherki. (Alexander Pushkin, history and legendary tradition: essays). Sankt-Peterburg, Akademicheskij proekt, 2000, 359 p.

6466. KROBB (Florian). Selbstdarstellungen. Untersuchungen zur deutsch-jüdischen Erzählliteratur im neunzehnten Jahrhundert. Würzburg, Königshausen & Neumann, 2000, 206 p.

6467. Leonardo Sciascia e la tradizione dei siciliani. A cura di Rosario CASTELLI. Roma e Caltanisetta, Salvatore Sciascia, 2000, 222 p.

6468. Leopardi e il libro nell'età romantica. Atti del Convegno Internazionale di Birmingham, 29-31 ottobre 1998. A cura di Michael CAESAR e Franco D'INTINO. Roma, Bulzoni, 2000, 390 p.

6469. LUCAS LAWSON (Ann). La ricerca dell'ignoto. I romanzi di avventura di Emilio Salgari. Firenze, Olschki, 2000, 206 p.

6470. MAC CANN (Sean). Gumshoe America: hardboiled crime fiction and the rise and fall of New Deal liberalism. Durham, Duke U. P., 2000, VIII-370 p. (New Americanists).

6471. MAC KEE (Patricia). Producing American races. Henry James, William Faulkner, Toni Morrison. Durham, Duke U. P., 2000, X-242 p.

6472. Mary Shelley in her times. Ed. by Betty BARNETT and Stuart CURRAN. Baltimore, Johns Hopkins U. P., 2000, XIV-312 p.

6473. MELLOR (Anne K.). Mothers of the nation: women's political writings in England, 1780-1830. Bloomington a. Indianapolis, Indiana U. P., 2000, 172 p.

6474. MIDGLEY (David). Writing Weimar. Critical realism in German literature, 1918-1933. Oxford, Oxford U. P., 2000, 390 p.

6475. Naturalismus, Fin de siècle, Expressionismus 1890-1918. Hrsg. v. York-Gothart MIX. München, Hanser, 2000, 760 p. (Hansers Sozialgeschichte der deutschen Literatur vom 16. Jahrhundert bis zur Gegenwart, 7).

6476. NEWLYN (Lucy). Reading, writing and Romanticism: the anxiety of reception. Oxford, Oxford U. P., 2000, XX-398 p.

6477. OGEDE (Ode). Achebe and the politics of representation: form against itself, from colonial conquest and occupation to post-independence disillusionment. Trenton, Africa World Press, 2000, 164 p.

6478. PALERMO (Antonio). Ottocento italiano. L'idea civile della letteratura. Cattaneo, Tenca, De Sanctis, Carducci, Imbriani, Capuana. Napoli, Liguori, 2000, 173 p.

6479. Paysages romantiques. Ed. par Gérard PEYLET. Pessac, Université Michel de Montaigne, 2000, 497 p. (Eidôlon, 54).

6480. PENZEN STADLER (Franz). Romantische Lyrik und klassizistische Tradition: Ode und Elegie in der französischen Romantik. Stuttgart, F. Steiner, 2000, 353 p.

6481. PEPPIS (Paul). Literature, politics and the English Avant-Garde: nation and empire, 1901-1918. Cambridge, Cambridge U. P., 2000, X-236 p.

6482. PFATSCHBACHER (Klaus). Jules Verne und der Populärroman. Frankfurt am Main u. Bern, Peter Lang, 2000, 203 p.

6483. PIPPIN (Robert B.). Henry James and modern moral life. Cambridge, Cambridge U. P., 2000, XII-194 p.

6484. RAITT (Suzanne). May Sinclair: a modern Victorian. Oxford, Clarendon Press, 2000, XVI-308 p.

6485. RAWLINSON (Mark). British writing of the Second World War. Oxford, Clarendon Press, 2000, VIII-248 p.

6486. SACCENTI (Mario). Bacchelli. Memoria e invenzione. Firenze, Le Lettere, 2000, 309 p.

6487. SAUVÉ (Rachel). De l'éloge à l'exclusion: les femmes auteurs et leurs préfaciers au XIXe siècle. Saint Denis, Presse universitaires de Vincennes, 2000, 250 p.

6488. SCHRAMM (Jan Melissa). Testimony and advocacy in Victorian law, literature and theology. Cambridge, Cambridge U. P., 2000, 262 p. (Cambridge studies in Nineteenth century literature and culture, 27).

6489. SÉGINGER (Gisèle). Flaubert, une étique de l'art pur. Paris, Editions Sedes, 2000, 220 p.

6490. SHULMAN (Robert). The power of political art: the 1930s literary left reconsidered. Chapel Hill a. London, University of North Carolina Press, 2000, 340 p.

6491. SIMONIN (Anne). La mise à l'épreuve du nouveau roman. Six cent cinquante fiches de lecture d'Alain Robbe-Grillet (1955-1959). In: Figures d'auteurs. Musique et littérature, France, 17e-20e siècles. Annales, 2000, 55, 2, p. 415-438.

6492. SPECTOR (Scott). Prague territories. National conflict and cultural innovation in Franz Kafka's Fin de Siècle. Berkeley a. Los Angeles, University of California Press, 2000, XIV-331 p. (Weimar and now: German cultural criticism, 21).

6493. STEEN (Inken). Parodie und parodistische Schreibweise in Thomas Manns "Doktor Faust". Tübingen, Max Niemeyer Verlag, 2000, VIII-208 p. (Untersuchungen zur deutschen Literaturgeschichte, 105).

6494. STROBEL (Jochen). Entzauberung der Nation. Die Repräsentation Deutschlands im Werk Thomas Manns. Dresden, Thelem, 2000, 394 p. (Arbeiten zur neueren deutschen Literaturwissenschaft, 1).

6495. THOMAS (Helen). Romanticism and slave narratives: transatlantic testimonies. Cambridge, Cambridge U. P., 2000, XII-332 p. (Cambridge studies in Romanticism, 38).

6496. THOMAS (Ronald R.). Detective fiction and the rise of forensic science. Cambridge, Cambridge U. P., 2000, XVIII-342 p. (Cambridge studies in Nineteenth century literature and culture, 26).

6497. TOMASELLO (Dario). Oltre il Futurismo. Percorsi delle Avanguardie in Sicilia. Con lettere inedite di F. T. Marinetti, L. Russolo, P. Buzzi, C. Alvaro. Roma, Bulzoni, 2000, 238 p.

6498. TROTTER (David). Cooking with mud: the idea of mess in Nineteenth century art and fiction. Oxford, Oxford U. P., 2000, 350 p.

6499. YEE (Jennifer). Clichés de la femme exotique: un regard sur la littérature coloniale française entre 1871 et 1914. Paris, L'Harmattan, 368 p.

6500. Zlatá léta šedesátá. Česká literatura a společnost v letech tání, kolotání a zklamání. (Czech literature in the golden sixties). Ed. Radka DENEMARKOVÁ. Praha, Ústav pro českou literaturu AV ČR, 2000, 420 p. (photogr.).

Cf. nos 456, 1472, 5737, 6045, 6057

§ 8. Art and industrial art.

a. General

6501. AMES-LEWIS (Francis). The intellectual life of the early Renaissance artist. New Haven a. London, Yale U. P., 2000, X-322 p.

6502. ANDREWS (Malcolm). Landscape and western art. Oxford a. New York, Oxford U. P., 2000, 248 p.

6503. Art (L') en Espagne et au Portugal. Ed. par Jean Louis AUGÉ. Paris, Citadelles et Mazenod, 2000, 620 p. (ill.).

6504. BEYER (Andreas). Parthenope. Neapel und der Süden der Renaissance. München u. Berlin, Deutscher Kunstverlag, 2000, 240 p. (ill.). (Kunstwissenschaftliche Studien, 84).

6505. BOCOLA (Sandro). The art of Modernism. Art, culture and society from Goya to the present day. New York, Prestel, 2000, 624 p. (ill.).

6506. BRAUN (Emily). Mario Sironi and Italian Modernism. Art and politics under Fascism. Cambridge, Cambridge U. P., 2000, 316 p. (ill.).

6507. CAVANAUGH (Jan). Out looking in. Early modern Polish art, 1890–1918. Berkeley, University of California Press, 2000, 328 p.

6508. České umění 1939–1999. Programy a impulzy. (Czech art, 1939–1999. Programmes and impulses). Praha, Akademie výtvarných umění, 2000, 159 p.

6509. COHEN (Sarah). Art, dance and the body in French culture of the Ancien Régime. New York a. Cambridge, Cambridge U. P., 2000, 352 p.

6510. CORN (Wanda). The great American thing: modern art and national identity, 1915–1935. Berkeley, University of California Press, 2000, 448 p. (ill.).

6511. DA PAZ (Alfredo). Dal realismo al simbolismo. Vicende e figure dell'arte postromantica europea. Bologna, CLUEB, 2000, 534 p. (ill.).

6512. Early modern visual culture: representation, race and empire in Renaissance England. Ed. by Peter ERICKSON and Clark HULSE. Philadelphia, University of Pennsylvania Press, 2000, VI-404 p.

6513. FLINT (Kate). The Victorians and the visual imagination. Cambridge, Cambridge U. P., 2000, 427 p. (ill.).

6514. FRASCINA (Francis). Art, politics and dissent: aspects of the art left in Sixties America. New York a. Manchester, Manchester U. P., 2000, 248 p.

6515. GREEN (Christopher). Art in France: 1900–1940. New Haven a. London, Yale U. P., 2000, 322 p. (ill.).

6516. HARBISON (Robert). Reflections on Baroque. London, Reaktions Books, 2000, 260 p. (ill.).

6517. HELGERSON (Richard). Adulterous alliances: home, state, and history in early modern European drama and painting. Chicago, University of Chicago Press, 2000, 238 p.

6518. HEUSS (Anja). Kunst- und Kulturgutraub. Eine vergleichende Studie zur Besatzungspolitik der Nationalsozialisten in Frankreich und der Sowjetunion. Heidelberg, Winter, 2000, 385 p.

6519. HOPKINS (David). After Modern art: 1945–2000. Oxford, Oxford U. P., 2000, 288 p. (ill.).

6520. Ideal (Das) der Schönheit. Rheinische Kunst in Barock und Rokoko. Hrsg. v. Frank G. ZEHNDER und Werner SCHÄFKE. Köln, DuMont, 2000, 447 p. (Der Riss im Himmel, 6).

6521. Imagination und Wirklichkeit. Zum Verhältnis von mentalen und realen Bildern in der Kunst der frühen Neuzeit. Hrsg. v. Klaus KRÜGER und Alessandro NOVA. Mayence, Philipp von Zabern, 196 p. (ill.).

6522. IZENBERG (Gerald N.). Modernism and masculinity: Mann, Wedekind, Kandinsky through World

War I. Chicago, University of Chicago Press, 2000, XII-257 p.

6523. JACHEE (Nancy). The philosophy and politics of abstract Expressionism, 1940–1960. Cambridge, Cambridge U. P., 2000, 268 p.

6524. KASFIR LETTELFIELD (Sidney). Contemporary African art. New York, Thames and Hudson, 2000, 220 p. (ill.).

6525. KEß (Bettina). Kunstleben und Kulturpolitik in der Provinz. Würzburg 1919 bis 1945. Würzburg, Bayerische Blätter für Volkskunde, 2000, 506 p. (Veröffentlichungen zur Volkskunde und Kulturgeschichte, 76).

6526. KINNA (Ruth). William Morris: art, work, leisure. *Journal of the history of ideas*, 2000, 61, 3, p. 493-512.

6527. LEMARIE (Gérard-Georges). L'univers des Orientalistes. Paris, Ed. Place des Victoires, 2000, 360 p. (ill.).

6528. LONGONI (Ana), MESTMAN (Mariano). Del Di Tella a "Tucumán Arde": vanguardia artística y política en el '68 argentino. Buenos Aires, El Cielo por Asalto, 2000, 383 p. (ill.). (Colección La cultura argentina).

6529. MILNER (John). Art, war, and Revolution in France, 1870–1871: myth, reportage, and reality. New Haven a. London, Yale U. P., 2000, XI-243 p.

6530. PETROPOULOS (Jonathan). The Faustian bargain: the art world in Nazi Germany. Oxford, Oxford U. P., 2000, XVII-395 p.

6531. PINELLI (Antonio). Nel segno di Giano. Passato e futuro nell'arte europea tra Sette e Ottocento. Roma, Carocci, 404 p.

6532. RUBIN LEE (Patricia), WRIGHT (Alison). Renaissance Florence: the art of the 1470s. New Haven a. London, Yale U. P., 2000, 352 p. (ill.).

6533. SCHIMMEL (Annemarie). Im Reich der Großmoguln. Geschichte, Kunst, Kultur. München, Beck, 2000, 432 p.

6534. SIEGEL (Jonah). Desire and excess: the Nineteenth century culture of art. Princeton, Princeton U. P., 2000, 328 p. (ill.).

6535. SPIEGEL (Régis). Dominique-Vivant Denon et Benjamin Zix: témoins et acteurs de l'épopée napoléonienne: 1805–1812. Paris, l'Harmattan, 2000, 220 p. (ill.).

6536. STOICHITA (Victor), BRÈVE histoire de l'ombre. Genève, Droz, 2000, 300 p. (ill.).

6537. Storia delle arti in Toscana. Il Cinquecento. A cura di Roberto Paolo CIARDI e Antonio NATALI. Firenze, Edifir, 2000, 303 p. (ill.).

6538. TAUSCH (Harald). Entfernung der Antike. Tübingen, Max Niemeyer Verlag, 2000, 340 p. (Studien zur deutschen Literatur, 156).

6539. TICKNER (Lisa). Modern life & modern subjects: British art in the early twentieth century. New Haven, Yale U. P., 2000, 256 p. (ill.).

6540. Utopi och verklighet. Svensk modernism 1900–1960. Konst, arkitektur, fotografi, film, formgivning. (Utopie et réalité. Le modernisme suédois de 1900 à 1960. Art, architecture, photographie, cinéma, design). Ed. par Cecilia WIDENHEIM. Stockholm, Nordstedts, 2000, 320 p. (ill.). (Catalogue de l'exposition du Moderna Museet, 07. 10. 2000-14. 01. 2001).

Cf. n°s 738, 1272, 1282

b. Architecture

6541. ANDERSON (Stanford). Peter Behrens and a new architecture for the Twentieth century. Cambridge, MIT Press, 2000, 393 p. (ill.).

6542. ARNOLD (Dana). Re-presenting the metropolis: architecture, urban experience and social life in London, 1800–1840. Aldershot, Ashgate, 2000, 151 p. (ill.).

6543. BASTEA (Eleni). The creation of modern Athens. New York, Cambridge U. P., 2000, 280 p.

6544. BENSA (Alban). Ethnologie et architecture. Le Centre culturel Tjibaou. Paris, Adam Biro, 2000, 207 p.

6545. BERGDOLL (Barry). European architecture 1750–1890. Oxford, Oxford U. P., 2000, 326 p. (ill.).

6546. BERGERON (Louis), MAIULLARI PONTOIS (Maria Teresa). Industry architecture and engineering: American ingenuity 1750–1950. New York, Abrams, 2000, 288 p. (ill.).

6547. CALVESI (Maurizio). Gli incantesimi di Bomarzo. Il Sacro Bosco tra arte e letteratura. Milano, Bompiani, 305 p. (ill.).

6548. CHÂTELET-LANGE (Liliane). Die Catherinenburg. Residenz des Pfalzgrafen Johann Casimir von Zweibrücken. Ein Bau der Zeitenwende, 1619–1622. Stuttgart, Jan Thorbecke, 2000, 240 p. (Residenzenforschung, 12) (ill.).

6549. Edilizia pubblica (L') nell'età dell'Illuminismo. A cura di Giorgio SIMONCINI. Firenze, Olschki, 2000, 3 vol., 1081 p. (ill.).

6550. ENGEL (Ute). Die Kathedrale von Worcester. München, Deutscher Kunstverlag, 2000, 368 p. (Kunstwissenschaftliche Studien, 88).

6551. FARA (Giovanni Maria). Albrecht Dürer teorico dell'architettura. Una storia italiana. Firenze Olschki, 2000, 223 p. (ill.). (Studi, 181).

6552. GALLO (Luigi). Variazioni sul Classico. L'architettura francese dal Rinascimento alla Rivoluzione. Roma, Lithos, 2000, 285 p. (ill.).

6553. Hans Poelzig in Breslau, Architektur und Kunst 1900–1916. Hrsg. v. Beate STÖRTKUHL und ILKOSZ Jerzy. Delmenhorst, Aschenbeck u. Holstein, 2000, 600 p.

6554. HOPKINS (Andrew). Santa Maria della Salute: architecture and ceremony in Baroque Venice. Cambridge, Cambridge U. P., 2000, 271 p. (ill.).

6555. JASKOT (Paul B.). The architecture of oppression: the SS, forced labor and the Nazi monumental building economy. London a. New York, Routledge, 2000, XV-207 p. (The Architext series).

6556. KAGAN (Richard). Urban images of the Hispanic World, 1493–1793. New Haven, Yale U. P., 2000, 240 p. (ill.).

6557. KING (Ross). Brunelleschi's dome. New York, Walker & Co., 2000, 192 p. (ill.).

6558. LEINWEBER (Luise). Bologna nach dem Tridentinum. Private Stiftungen und Kunstaufträge im Kontext der katholische Konfessionalisierung: das Beispiel San Giacomo Maggiore. Hildesheim u. Zürich, Georg Olms, 327 p. (ill.).

6559. Leo von Klenze. Architekt zwischen Kunst und Hof, 1784–1864. Hrsg. v. Winfried NERDINGER. München, Prestel Verlag, 2000, 540 p. (ill.).

6560. MÄKINEN (Anne). Suomen valkoinen sotilasarkkitehtuuri 1926–1939. ("White" military architecture in Finland, 1926–1939). Helsinki, SKS, 2000, 257 p. (ill.). (Bibliotheca historica, 53).

6561. MUMFORD (Eric). The CIAM [Congrès Internationaux d'Architecture moderne] discourse on urbanism. 1928–1960. Cambridge, MIT Press, 2000, XV-375 p.

6562. PÉROUSE DE MONTCLOS (Jean-Marie). Philibert De L'Orme. Architecte du roi (1514–1570). Paris, Mengès, 2000, 388 p. (ill.).

6563. PETZET (Michael). Claude Perrault und die Architektur des Sonnenkönigs. Der Louvre König Ludwigs XIV. Und das Werk Claude Perraults. München u. Berlin, Deutscher Kunstverlag, 2000, 596 p. (ill.).

6564. RABREAU (Daniel). Claude-Nicolas Ledoux (1736–1806). L'architecture et les fastes du temps. Bordeaux, Willaim Blake & Co, 2000, 435 p. (ill.).

6565. SALOMON (Frank). Building on ruins. The rediscovery of Rome and English architecture. Aldershot, Ashgate, 2000, 264 p. (ill.).

6566. SCHÜTZ (Bernhard). Die kirchliche Barockarchitektur in Bayern und Oberschwaben, 1580–1780. München, Hirmer Verlag, 2000, 196 p. (ill.).

6567. STIGLMAYR (Cristina María). Der Palast Karls V. in Granada. Frankfurt am Main, Berlin, Bern, Bruxelles, New York u. Wien, Lang, 2000, XI-274 p.

6568. Storia dell'architettura italiana. Il Settecento. A cura di Giovanna CURCIO e Elisabeth KIEVEN. Milano, Electa, 2000, 2 vol., 734 p. (ill.).

6569. TALENTI (Simona). L'histoire de l'architecture en France. Emergence d'une discipline (1863–1914). Paris, Picard, 292 p. (ill.).

6570. WILSON (Richard), MACKLEY (Alan). Creating paradise: the building of the English country house, 1660–1880. London a. New York, Hambledon Press, 2000, XVIII-428 p.

6571. WOLTERS (Wolfgang). Architektur und Ornament: venezianischer Bauschmuck der Renaissance. München, Beck, 2000, 320 p. (ill.).

Cf. n°s 382, 511

c. Sculpture, painting, etching and drawing

6572. Antonio Canova e il suo ambiente artistico fra Venezia, Roma e Parigi. A cura di Giuseppe PAVANELLO. Venezia, Istituo Veneto di Scienze, Lettere e Arti, 2000, XII-574 p. (ill.). (Studi di arte veneta, 1)

6573. APPUHN-RADTKE (Sibylle). Visuelle Medien im Dienst des Gesellschaft Jesu. Johann Christoph Storer (1620–1671) als Maler der katholischen Reform. Regensburg, Schnell und Steiner, 2000, 411 p. (ill.).

6574. AULICH (James), SYLVESTROVA (Marta). Political posters in Central and Eastern Europe, 1945–1995. New York, St. Martin's Press, 2000, 228 p. (ill.).

6575. BAKER (Malcolm). Figured in marble. The making and viewing of Eighteenth century sculpture. London, V&A Publications, 2000, 192 p. (ill.).

6576. BAUER (Hermann), DER MÜLBE (Wolf-Christian von). Barocke Deckenmalerei in Süddeutschland. München u. Berlin, Deutscher Kunstverlag, 2000, 240 p. (ill.).

6577. BERGSTEIN (Mary). The scuplture of Nanni di Banco. Princeton, Princeton U. P., 2000, 230 p. (ill.).

6578. BRINK (Claudia). Arte e Marte. Kriegskunst und Kunstliebe im Herrscherbild des 15. und 16. Jahrhunderts in Italien. München u. Berlin, Deutscher Kunstverlag, 2000, 224 p. (ill.).

6579. CAGLIOTI (Francesco). Donatello e i Medici: storia del Davide e della Giuditta. Firenze, Olshki, 2000, 530 p. (ill.).

6580. CALI (Maria). La pittura del Cinquecento. Torino, UTET, 2000, 2 vol., 682 p. (ill.).

6581. CALLEN (Anthea). The art of Impressionism. Painting technique and the making of Modernity. New Haven a. London, Yale U. P., 2000, 245 p. (ill.).

6582. COHEN-SOLAL (Annie). Un jour ils auront des peintres. L'avènement des peintres américains, Paris 1867 – New York 1948. Paris, Gallimard, 2000, 465 p.

6583. CRANSTON (Jody). The poetics of portraiture in the Italian Renaissance. Cambridge, Cambridge U. P., 2000, XIII-258 p. (ill.).

6584. CROPPER (Elizabeth), DEMPSEY (Charles). Nicolas Poussin: friendship and the love of painting. Princeton, Princeton U. P., 2000, 374 p. (ill.).

6585. CURTIS (Penelope). Sculpture 1900–1945. New York, Oxford U. P., 2000, 286 p. (ill.).

6586. DOMÉNECH (F. Benito). Joan de Joanes. Un maestro del Renacimiento. Madrid, Fundación Santander Central Hispano, 2000, 203 p.

6587. GIBSON (Walter S.). Pleasant places. The rustic landscape from Brügel to Ruisdael. Berkeley a. Los Angeles, University of California Press, 2000, 292 p. (ill.).

6588. GOLDING (John). Paths to the absolute. Mondrian, Malevich, Kandinsky, Pollock, Newman, Rothko, Still. London, Thames and Hudson, 2000, 240 p. (ill.).

6589. HAARDT DE LA BAUME (Caroline). Alexandre Iacovleff. L'artiste voyageur. Paris, Flammarion, 2000, 160 p. (ill.).

6590. Italia en el horizonte de las artes plásticas. Argentina, siglos XIX y XX. Ed. por Diana Beatriz WECHSLER. Buenos Aires, Instituto Italiano de Cultura, 2000, 223 p.

6591. JACOBY (Joachim). Hans von Aachen (1552–1615). München u. Berlin, Deutscher Kunstverlag, 2000, 342 p. (ill.).

6592. KECKS (Ronald G.). Domenico Ghirlandaio und die Malerei der florentiner Renaissance. München u. Berlin, Deutscher Verlag, 2000, 438 p. (ill.).

6593. KORZILIUS (Jean Loup). La peinture abstraite en Allemagne 1933–1955. Paris, L'Harmattan, 2000, 310 p. (ill.).

6594. KUHN (Rudolf). Erfindung und Komposition in der monumentalen zyklischen Historienmalerei des 14. und 15. Jahrhunderts in Italien. Frankfurt am Main u. Bern, Peter Lang, 2000, 663 p.

6595. LOYRETTE (Henri), PANTAZZI (Michael). Daumier. New Haven, Yale U. P., 2000, 600 p. (ill.).

6596. MAC WILLIAM (Neil). Jean Baffier. A nationalist sculptor in Fin-de-Siècle France. University Park, Pennsylvania U. P., 2000, 326 p. (ill.).

6597. MANCA (Joseph). Cosmè Tura. The life and art of a painter in Estense Ferrara. Oxford, Oxford U. P., 2000, 286 p. (ill.).

6598. MANRIQUE (María Elena). Jusepe Martínez, un pintor zaragosano en la Roma del Seicento. Zaragoza, Institución Fernando el Católico, 2000, 217 p. (ill.).

6599. MEILMAN (Patricia). Titian and the altarpiece in Renaissance Venice. Cambridge a. New York, Cambridge U. P., 2000, XIV-260 p. (ill.).

6600. NAGEL (Alexander). Michelangelo and the reform of art. Cambridge, Cambridge U. P., 2000, XVI-303 p. (ill.).

6601. OLSON (Roberta). The Florentine tondo. Oxford, Oxford U. P., 2000, 408 p. (ill.).

6602. PARKER (Deborah). Bronzino: Renaissance painter as poet. Cambridge, Cambridge U. P., 2000, X-233 p. (ill.).

6603. PHILLIPS (Quitman E.). The practices of painting in Japan, 1475–1500. Stanford, Stanford U. P., 2000, 267 p. (ill.).

6604. PLAX (Julie Anne). Watteau and the cultural politics of Eighteenth century France. Cambridge, Cambridge U. P., 2000, 260 p. (ill.).

6605. ROBERTS (Warren). Jacques-Louis David and Jean-Louis Prieur, revolutionary artists: the public, the populace, and images of the French Revolution. Albany, State University of New York Press, 2000, XX-370 p.

6606. ROSENBERG (Pierre). From drawing to painting: Poussin, Watteau, Fragonard, David and Ingres. Princeton, Princeton U. P., 2000, 265 p. (ill.).

6607. ROSENBERG (Rafael). Beschreibungen und Nachzeichnungen der Skulpturen Michelangelos. Eine Geschichte der Kunstbetrachtung. München u. Berlin, Deutscher Kunstverlag, 2000, 298 p. (ill.).

6608. SCHALHORN (Andreas). Historienmalerei und Heiligsprechung. Pierre Subleyras (1699–1749) und das Bild für den Papst im 17. und 18. Jh. München, Scaneg, 2000, 322 p. (Akádemos, 3).

6609. SIDLAUSKAS (Susan). Body, place and self in Nineteenth century painting. Cambridge, Cambridge U. P., 2000, 230 p. (ill.).

6610. THOMAS (Greg). Art and ecology in Nineteenth century France: the landscapes of Théodore Rousseau. Princeton, Princeton U. P., 2000, 276 p. (ill.).

6611. VOGT (Marion). Zwischen Ornament und Natur. Edgar Degas als Maler und Photograph. Hildesheim u. Zürich, Georg Olms, 2000, 237 p. (ill.).

6612. WIND (Edgar). The religious symbolism of Michelangelo: the Sistine Ceiling. Oxford, 2000, LIII-240 p.

6613. ZEIT (Lisa). Tizian, teurer Freund. Tizian und Federico Gonzaga. Kunstpatronage in Mantua im 16. Jahrhundert. Petersburg, Michael Imhof Verlag, 2000, 256 p. (ill.).

d. Decorative, popular and industrial art

** 6614. HERBERT (Robert L.). Nature's workshop. Renoir's wrtings on decorative arts. New Haven a. London, Yale U. P., 2000, 278 p.

6615. ARWAS (Victor). Art Nouveau. From Mackintosh to Liberty. London, Papadakis, 2000, 200 p. (ill.).

6616. AYNSLEY (Jeremy). Graphic design in Germany, 1890–1945. Berkeley a. Los Angeles, University of California Press, 2000, 240 p. (Weimar and now: German cultural criticism, 28) (ill.).

6617. BAKER (Fiona). 20th century furniture. London, Carlton Book, 256 p. (ill.).

6618. BLASZCZYK LEE (Regina). Imagining consumers. Design and innovation from Wedgwood to Corning. Baltimore, Johns Hopkins U. P., 2000, XIII-380 p. (ill.).

6619. BRUNET (François). La naissance de l'idée de la photographie. Paris, Presses universitaires de France, 2000, 326 p.

6620. EDWARDS (Clive). Encyclopedia of furniture: materials, trades and techniques. Aldershot, Ashgate, 2000, 288 p. (ill.).

6621. GRÜNEWALD (Dietrich). Comics. Tübingen, Max Niemeyer Verlag, 2000, 112 p. (Grundlagen der Medienkommunikation, 8).

6622. GUIDOT (Raymond). Histoire du design, 1940–2000. Paris, Hazan 386 p. (ill.).

6623. HANKS (David A.), HOY (Anne). Design for living: 1950–2000. Paris, Flammarion, 2000, 200 p. (ill.).

6624. HORBAS (Claudia), MÖLLER (Renate). Silber. Von der Renaissance zur Moderne. München u. Berlin, Deutscher Kunstverlag, 2000, 168 p. (ill.).

6625. JARDINE (Lisa), BROTTON (Jerry). Global interests: Renaissance arte between East and West. Ithaca, Cornell U. P., 2000, 220 p.

6626. LEDDEROSE (Lothar). Ten thousand things: module and mass production in Chinese art. Princeton, Princeton U. P., 265 p. (ill.).

6627. Medaillensammlung Goethes (Die). Band 1. Bestandkatalog. Band 2. Quellen. Bearb. v. Jochen KLAUß. Berlin, Deutsche Gesellschaft für Medaillenkunst/Stiftung Weimarer Klassik, 2000, 2 vol., 482 p., 258 p. (Die Kunstmedaille in Deutschland, 13/1-2).

6628. Mobile barocco (Il) in Italia. Arredi e decorazioni d'interni dal 1600 al 1738. A cura di Enrico COLLE. Milano, Electa, 2000, 448 p. (ill.).

6629. Pietra dipinta. Tesori nascosti del '500 e del '600 da una collezione privata milanese. A cura di Marco BONA CASTELLOTTI. Milano, Federico Motta, 250 p. (ill.).

6630. PILE (John F.). A history of interior design. London, Laurence King, 2000, 400 p. (ill.).

6631. REDHEAD (David). Products of our time. Basel a. Berlin, August/Birkhauser, 2000, 144 p. (ill.).

6632. RENZI (Giovanni), RENZI (Chiara). Curve e biondi riccioli viennesi. Mobili in faggio curvati, da Michael Thonet ad Antonio Volpe. Cinisello Balsamo, Silvana, 124 p. (ill.).

6633. SMITH (Erin A.). Hard-boiled: working-class readers and pulp magazines. Philadelphia, Temple U. P., 2000, XI-215 p.

6634. Style and socialism: modernity and material culture in post-war eastern Europe. Ed. by Susan E. REID and David CROWLEY. Oxford, Berg, 2000, XIII-213 p.

Cf. n° 5865

§ 9. Music, theatre, cinema and broadcasting.

6635. ALLMAN (Eileen). Jacobean revenge tragedy and the politics of virtue. London, Associated Universities Presses, 2000, 212 p.

6636. ARCANGELI (Alessandro). Davide o Salomè? Il dibattito europeo sulla danza nella prima età moderna. Treviso, Fondazione Benetton studi e ricerche e Roma, Viella, 2000, 390 p. (Ludica, 5).

6637. ARECCO (Sergio). Ingmar Bergman. Segreti e magie. Recco, Le Mani, 2000, 241 p.

6638. ARONSON (Arnold). American Avant-Garde theater: a history. London a. New York, Routledge, 2000, 242 p.

6639. ASKÉNAZI-SEKNADJE (Enrique). Roberto Rossellini et la Seconde Guerre Mondiale. Un cinéaste entre propagande et réalisme. Paris, L'Harmattan, 2000, 264 p.

6640. Aspects of Twentieth century theatre in French. Ed. by Michael CARDY and Derek CONNON. Bern, Peter Lang, 2000, 243 p.

6641. BAGH (Peter von). Drifting shadows. A guide to Finnish cinema. Helsinki, Otava, 2000, 127 p. (ill.).

6642. BARROT (Olivier). Noir et blanc: 250 acteurs su cinéma français, 1930–1960. Paris, Flammarion, 2000, 595 p. (ill.).

6643. Bellezza (La) e lo sguardo. Il cinematrografo di Robert Bresson. A cura di Luciano DE GIUSTI. Milano, Il Castoro, 2000, 254 p.

6644. BENESCELLI (Alberto). Felicità sognate. Il teatro di Metastasio. Genova, Il Melangolo, 2000, 182 p.

6645. BERTRAND (Michael T.). Race, rock, and Elvis. Urbana and Chicago, University of Illinois Press, 2000, XII-327 p. (Music in American Life).

6646. BLASIUS (Dirk). Die Ausstellung "Entartete Musik" von 1938. *Archiv für Kulturgeschichte*, 2000, 82, p. 391-406.

6647. BLÜDNIKOW (Bent). Tele Danmark – et erhverveventyr. (Histoire économique de la Télévision danoise). København, Tele Danmark, 2000, 138 p. (ill.).

6648. BOHRN (Patricia). André Antoine und seine Théâtre libre: eine spezifische Ausformung des naturalistischen Theaters. Frankfurt am Main u. Bern, Peter Lang, 2000, 251 p. (Wiener Beiträge zu Komparatistik und Romanistik, 8).

6649. BOISITS (Barbara). Historismus und Musikwissenschaft um 1900. *Archiv für Kulturgeschichte*, 2000, 82, p. 377-390.

6650. British cinema of the 90s. Ed. by Robert MURPHY. Berkeley a. Los Angeles, University of California Press, 2000, 264 p. (ill.).

6651. BROOKS (Jeanice). Courtly song in late sixteenth century France. Chicago a. London, University of Chicago Press, XVI-560 p.

6652. BRYANT-BERTAIL (Sarah). Space and time in epic theatre. The Brechtian legacy. Rochester, Camden House, 2000, 245 p.

6653. BURLING (William J.). Summer theatre in London, 1661–1820, and the rise of Haymarket Theatre. London, Associated Universities Presses, 2000, 326 p.

6654. BURROUGHS (Catherine). Women in British Romantic theatre: drama, performance and society, 1790–1840. Cambridge, Cambridge U. P., 2000, XVI-344 p.

6655. CAPUA (Michelangelo). Montgomery Clift. Vincitore e vinto. Torino, Lindau, 2000, 185 p.

6656. CARNEY (Ray), QUART (Leonard). The films of Mike Leigh. Cammbridge, Cambridge U. P., 2000, 304 p.

6657. CASCETTA (Annamaria). Il tragico e l'umorismo: studio sulla drammaturgia di Samuel Beckett. Firenze, Le Lettere, 2000, 403 p. (Storia dello spettacolo. Saggi, 4).

6658. CERNUSCHI (Alain). Penser la musique dans l'Encyclopédie: étude sur les enjeux de la musicographie des Lumières et sur ses liens avec l'encyclopedisme. Paris, H. Champion, 2000, 789 p. (Les Dix-huitième siècle, 47).

6659. CHATELAIN (Violaine). La télévision publique française: une fabrique politique? Les intervensions de Georges Pompidou Premier Ministre, 1962–1968. *Revue d'histoire moderne et contemporaine*, 2000, 4, p. 768-782.

6660. CHEGAI (Andrea). L'esilio di Metastasio: forme e riforme dello spettacolo d'opera fra Sette e Ottocento. Firenze, Le Lettere, 2000, 325 p. (Storia dello spettacolo. Saggi, 2).

6661. Cinema (Il) nella scrittura. A cura di Benvenuto CUMINETTI e Stefano GHISLOTTI. Bergamo, Bergamo U. P., 2000, 270 p.

6662. Cinema italiano 1930–1995. Le imprese di produzione. A cura di Aldo BERNARDINI. Roma, Anica, 2000, 527 p.

6663. CITRON (Marcia J.). Opera on screen. New Haven a. London, Yale U. P., 00 295 p.

6664. COOPER (Darius). The cinema of Satyajit Ray. Between tradition and modernity. Cambridge, Cambridge U. P., 2000, 288 p.

6665. CRAIG (Douglas B.). Fireside politics: radio and political culture in the United States, 1920–1940. Baltimore, Johns Hopkins U. P., 2000, XX-362 p.

6666. CREMONINI (Giorgio). Playtime. Viaggio non organizzato nel cinema comico. Torino, Lindau, 2000, 212 p.

6667. CRITTENDEN (Camille). Johann Strauss and Vienna. Cambridge, Cambridge U. P., 2000, 332 p.

6668. CRUSELLS (Magi). La Guerra civil española: cine y propaganda. Barcelona, Editorial Ariel, 2000, 299 p.

6669. DA COSTA (Cláudio). Cinema brasileiro, anos 60–70: dissimetria, oscilaçao e simulacro. Rio de Janeiro, 7Letras, 2000, 162 p.

6670. DAGRADA (Elena). La rappresentazione dello sguardo nel cinema dalle origini in Europa. Nascita della soggettività. Bologna, Cleub, 2000, 339 p.

6671. DAVIS (Tracy C.). The economics of the British stage, 1800–1914. Cambridge, Cambridge U. P., 2000, XVIII-506 p.

6672. DE VINCENTI (Giorgio). Il concetto di modernità nel cinema. Milano, Lampi di stampa, 2000, 270 p.

6673. DEHEE (Yannick). Mythologie politique du cinéma français, 1960–2000. Paris, Presses universitaires de France, 2000, 312 p.

6674. Diderot, l'invention du drame. Actes de la journéè d'études du 14 octobre 2000 à l'Université VII – Denis Diderot. Ed. par Marc BUFFET. Paris, Klincksieck, 2000, 192 p. (Actes et colloques, 61).

6675. DILLON (Janette). Theatre, court and city, 1595–1610: drama and social space in London. Cambridge, Cambridge U. P., 2000, X-188 p.

6676. ELSAESSER (Thomas). Weimar and cinema and after. Germany's historical imaginary. London a. New York, Routledge, 2000, 480 p.

6677. FALCO (Raphael). Charismatic authority in early modern English tragedy. Baltimore, Johns Hopkins U. P., 2000, 244 p.

6678. FARASSINO (Alberto). Tutto il cinema di Luis Buñuel. Milano, Baldini & Castoldi, 2000, 370 p.

6679. FERGUSON (Ronnie). The theatre of Angelo Beolco (Ruzante), text, context and performance. Ravenna, Longo, 2000, 249 p. (Memoria del tempo, 20).

6680. FERRAZZI (Marialuisa). Commedie e comici dell'arte italiana alla corte russa (1731–1738). Roma, Bulzoni, 2000, 339 p.

6681. FLETCHER (Alan). Drama, performance and polity in Pre-Cromwellian Ireland. Cork, Cork U. P., 2000, XVI-520 p.

6682. FOX (Jo). Filming women in the Third Reich. Oxford a. New York, Berg, 2000, 224 p.

6683. FREEDMAN (Richard). The chansons of Orlando di Lasso: music, piety and print in sixteenth century France. Rochester, Rochster U. P., 2000, XXIV-259 p. (ill.).

6684. FULCHER (Jane F.). Concert et propagande politique en France au début du 20e siècle. *In*: Figures d'auteurs. Musique et littérature, France, 17e–20e siècles. *Annales*, 2000, 55, 2, p. 389-414.

6685. GABERSCEK (Carlo). Sentieri del Western. Vol. 2. Gemona del Friuli, La Cineteca del Friuli, 2000, 240 p.

6686. GARDIES (André). Cinéma d'Afrique noire francophone. Paris, L'Harmattan, 2000, 192 p.

6687. GLENN (Susan A.). Female spectacle: the theatrical roots of modern feminism. Cambridge, Harvard U. P., 2000, X-294 p.

6688. GOLDZINK (Jean). Comique et comédie au siècle des Lumières. Paris, L'Harmattan, 2000, 381 p.

6689. GOTTSCHILD DIXON (Brenda). Waltzing in the dark: African American vaudeville and race politics in the swing era. New York, St. Martin's Press, 2000, 270 p. (ill.).

6690. GREENE (John C.). Theatre in Belfast, 1736–1800. London, Associated Universities Presses, 2000, 400 p.

6691. GREINER (Bernhard). Kleists Dramen und Erzählungen. Experimente zum "Fall" der Kunst. Tübingen u. Basel, Francke, 2000, 460 p.

6692. HALLINGBERG (Gunnar). Tidens tusende tungor. (Les mille langues de l'époque: histoire de la radio suédoise depuis les années 50). Stockholm, Atlantis, 2000, 519 p.

6693. HAMILTON (Marybeth). Sexuality, authenticity and the making of the Blues tradition. *Past and present*, 2000, 169, p. 132-160.

6694. Histoire (Une) du cinéma français. Ed. par Claude BEYLIE. Paris, Larousse, 2000, 278 p.

6695. HOLLEDGE (Julie), TOMPKINS (Joanne). Women's intercultural performance. New York, Routledge, 227 p. (ill.).

6696. HORÁK (Antonín). Světla a stíny zlínského filmu. (Light and shadows of Zlín cinema). *Iluminace*, 2000, 12, 2 (38), p. 121-140.

6697. JUDD COLLINS (Cristle). Reading Renaissance music theory: hearing with the eyes. Cambridge, Cambridge U. P., 2000, XXXIII-339 p. (ill.).

6698. KATER (Michael H.). Composers of the Nazi era: eight portraits. New York, Oxford U. P., 2000, XIII 399 p.

6699. KRUGLER (David F.). The voice of America and the domestic propaganda battles, 1945–1953. Columbia, University of Missouri Press, 2000, XII-246 p.

6700. KURKINEN (Marjaana). The spectre of the Orient. French mime and traditional Japanese theatre in the 1930s. Helsinki, M. Kurkinen, 2000, 227-12 p. (ill.).

6701. KUZNIAR (Alice). The queer German cinema. Stanford, Stanford U. P., 2000, 314 p.

6702. LANDY (Marcia). Italian film. Cambridge, Cambridge U. P., 2000, 458 p.

6703. LAURA (Ernesto G.). Le stagioni dell'aquila. Storia dell'Istituto Luce. Roma, Ente dello spettacolo, 2000, 432 p.

6704. LAURENT (Natacha). L'oeil du Kremlin: cinéma et censure en URSS s sous Stalin (1928–1953). Toulouse, Privat, 2000, 284 p. (ill.).

6705. Luci (Le) dei maestri dimenticati: cinema georgiano, 1908–1999. A cura di Fiorano RANCATI e Giulio ROSSINI. Roma, Unione circoli del cinema Arci, 2000, 143 p. (ill.).

6706. MAC VEAGH (John). Thomas Durfey and Restoration drama: the work of a forgotten writer. Aldershot, Ashgate, 2000, VIII-220 p.

6707. MAISONNEUVE (Sophie). Dischi e socialità: la nuova cultura dell'ascolto musicale negli anni venti e trenta del Novecento. *Quaderni storici*, 2000, 35, 104, p. 437-468.

6708. MALLY (Lynn). Revolutionary acts: amateur theater and the Soviet State, 1917–1938. Ithaca, Cornell U. P., 2000, X-250 p.

6709. MARTINELLI (Vittorio). Cuor d'oro e muscoli d'acciaio: il cinema francese degli anni Venti e la critica italiana. Gemona del Friuli, La cineteca del Friuli, 2000, 137 p. (ill.).

6710. MAY (Lary). The big tomorrow: Hollywood and the politics of the American way. Chicago, University of Chicago Press, 2000, XV-348 p.

6711. MAZDON (Lucy). Encore Hollywood. Remaking French cinema. Berkeley a. Los Angeles, University of California Press, 2000, 240 p. (ill.).

6712. MICELI (Sergio). Musica e cinema nella cultura del Novecento. Milano, Sansoni, 2000, 539 p.

6713. Milano città della Radiotelevisione 1945–1958. A cura di Ada FERRARI e Gaia GIUSTO. Milano, Franco Angeli, 2000, 139 p.

6714. MOSCATI (Italo). 2001, un'altra odissea. Quando il futuro sedusse il cinema. Venezia, Marsilio, 2000, 176 p.

6715. Music and British culture, 1785–1914. Ed. by Christina BASHFORD and Leanne LANGLEY. Oxford, Oxford U. P., 2000, 422 p.

6716. Music and the culture of print. Ed. by Kate VAN ORDEN. New York, Garland, 2000, XXI-354 p. (Critical Garland reference library of the humanities, 2027).

6717. PAECH (Anne). Menschen im Kino. Film und Literatur erzählen. Stuttgart, Metzler, 2000, 342 p.

6718. PARANAGUÁ (Paulo Antonio). Le cinéma en Amérique Latine. Le miroir éclaté, historiographie et comparatisme. Paris, L'Harmattan, 2000, 288 p.

6719. PAVESIO (Monica). Calderón in Francia: ispanismo e italianismo nel teatro francese del XVII secolo. Alessandria, Edizioni dell'Orso, 2000, 250 p. (Medusa, 1).

6720. PÉROT (Nicolas). Discours sur la musique à l'époque de Chateaubriand. Paris, Presses universitaires de France, 2000, 343 p.

6721. PETERS STONE (Julie). Theatre of the book, 1480–1880: print, text and performance in Europe. Oxford, Oxford U. P., 2000, XII-494 p.

6722. PETRIE (Duncan). Screening Scotland. Berkeley a. Los Angeles, University of California Press, 2000, 240 p. (ill.).

6723. PIEKARSKI (Stanisław). Teatr obozowy jako forma życia kulturalnego jeńców narodowości polskiej w latach I wojny światowej (1914–1918). (Das Lagertheater als Form des Kulturlebens der Gefangenen polnischer Nationalität während des I. Weltkrieges, 1914–1918). *Łambinowicki Rocznik Muzealny,* 2000, 23, p. 27-43.

6724. PINEAU (Joseph). Le théâtre de Molière. Une dynamique de la libert. Paris, Minard, 2000, 223 p. (Situation, 54).

6725. PIOTROWSKA (Magdalena). Lubownicy sceny, czyli polskie teatry amatorskie w Wielkopolsce (1832–1875). (Die Genießer des Aufführens, das heißt polnische Liebhabertheater in Großpolen [Wielkopolska], 1832–1875). Poznań u. Kalisz, Inst. Pedagog.-Artystyczny Uniw. A. Mickiewicza, 2000, 232 p.

6726. POIGER (Uta G.). Jazz, rock and rebels. Cold War politics and American culture in a divided Germany. Berkeley a. Los Angeles, University of California Press, 2000, XIII-333 p. (Studies on the history of society and culture, 35).

6727. POWELL (John S.). Music and theatre in France 1600–1680. Oxford, Clarendon Press, 2000, 598 p.

6728. QVIST (Per Olof). Guide to the cinema of Sweden and Finland. London, Westport, 2000, VI-308 p.

6729. Richard Wagner im Dritten Reich. Ein Schloß Elmau-Symposion. Hrsg. v. Jörn RÜSEN und Saul FRIEDLÄNDER. München, C.H. Beck, 2000, 372 p.

6730. RICKITT (Richard). Special effects: the history and technology. London, Virgin Books, 2000, 318 p. (ill.).

6731. RIMONDI (Francesca). Alle origini del noir: cronaca e letteratura popolare del cinema francese degli anni '10. Roma, Associazione italiana per le ricerche di storia del cinema, 2000, 269 p.

6732. RINGDAL (Nils Johan). Nationaltheatrets historie 1899–1999. (History of the Norwegian National Theatre 1899–1999). Oslo, Gyldendal, 2000, 678 p. (ill.).

6733. Romantische Drama (Das). Hrsg. v. Uwe JAPP, Stefan SCHERER und Claudia STOCKINGER. Tübingen, Max Niemeyer Verlag, 2000, IX-287 p. (Untersuchungen zur deutschen Literaturgeschichte, 103).

6734. SCOTT (Virginia). Molière: a theatrical life. Cambridge, Cambridge U. P., 2000, 323 p.

6735. SILBER (Karl Bernhard). Die dramatischen Werke Sigmund von Birkens (1626–1681). Tübingen, Narr, 2000, 501 p. (Mannheimer Beiträge zur Sprach- und Literaturwissenschaft, 44).

6736. SLOTTEN (Hugh R.). Radio and television regulation: broadcast technology in the United States, 1920–1960. Baltimore, Johns Hopkins U. P., 2000, XV-308 p.

6737. SMALLMAN (Basil). Schütz. Oxford, Oxford U. P., 2000, XVII-218 p.

6738. STOCKINGER (Claudia). Das dramatische Werk Friedrich de la Motte Fouqués. Tübingen, Max Niemeyer Verlag, 2000, VI-372 p. (Studien zur deutschen Literatur, 158).

6739. TAWA (Nicholas E.). High-minded and lowdown: music in the lives of Americans 1800–1861. Boston, Northeastern U. P., 2000, XIII-350 p.

6740. Teatri barocchi. Tragedie, commedie pastorali nella drammaturgia europea fra '500 e '600. A cura di Silvia CARANDINI. Roma, Bulzoni, 2000, 602 p.

6741. TEDESCO (Antonio). Underground e trasgressione. Il cinema dell'altra America in due generazioni. Roma, Castelvecchi, 2000, 187 p.

6742. THORAVAL (Yves). Le cinéma du Moyen-Orient: Iran, Egypte, Turquie (1896–2000). Paris, Seguier, 2000, 321 p.

6743. TRIVERO (Paola). Tragiche donne. Tipologie femminili nel teatro italiano del Settecento. Alessandria, Edizioni dell'Orso, 2000, 156 p.

6744. TROTT (David). Théâtre du XVIIIe siècle. Jeux, écritures, regards. Essai sur les spectacles en France de 1700 à 1789. Montpellier, Editions Espaces 34, 2000, 304 p. (ill.).

6745. TUCKER (Sherrie). Swing shift: "All-girl" bands of the 1940s. Durham, Duke U. P., 2000, IX-413 p.

6746. USAI CHERCHI (Paolo). Silent cinema: an introduction. Berkeley a. Los Angeles, University of California Press, 2000, 229 p. (ill.).

6747. VAN DER HOEYEN (Roland). Le Théâtre de la Monnaie au XIXe siècle. Contraintes d'exploitation d'un théâtre lyrique (1830–1914). Bruxelles, Université libre de Bruxelles, 2000, 412 p.

6748. VEIDLINGER (Jeffrey). The Moscow state Yiddish theater: Jewish culture on the Soviet stage. Bloomington, Indiana U. P., 2000, IX-356 p. (Jewish literature and culture; Indiana-Michigan series in Russian and East European studies).

6749. VEYRAT-MASSON (Isabelle). Quand la télévision explore le temps. L'histoire au petit écran. Paris, Fayard, 2000, 567 p.

6750. Visible nations: Latin American cinema and video. Ed. by Chon A. Noriega. Minneapolis, University of Minnesota Press, 2000, XXV-305 p.

6751. VUILLERMOZ (Marc). Le système des objets dans le théâtre français des années 1625–1650: Corneille, Mairet, Rotrou, Scudéry. Genève, Droz, 2000, 333 p. (Travaux du Grand Siècle, 17).

6752. WEIMANN (Robert). Author's pen and actor's voice: playing and writing in Shakespeare theatre. Cambridge, Cambridge U. P., 2000, XIV-298 p. (Cambridge studies in Renaissance literature and culture, 39).

6753. WIENS (Birgit). Grammatik der Schauspielkunst. Die Inszenierung der Geschlechter in Goethes klassischem Theater. Tübingen, Max Niemeyer Verlag, 2000, XII-256 p. (Theatron, 31).

6754. Word and image in Japanese cinema. Ed. by Dennis WASHBURNT and Carole CAVANAUGH. Cambridge, Cambridge U. P., 2000, 384 p.

Cf. n°s 5430, 6070, 8419

N

MODERN ECONOMIC AND SOCIAL HISTORY

§ 1. General. 6755-6819. – § 2. Political economy. 6820-6841.– § 3. Industry, mining and transportation. 6842-6913. – § 4. Trade. 6914-6956. – § 5. Agriculture and agricultural problems. 6957-6989. – § 6. Money and finance. 6990-7040. – § 7. Demography and urban history. 7041-7111. – § 8. Social history. 7112-7298. – § 9. Working-class movement and socialism. 7299-7366.

§ 1. General.

* 6755. DENG (Kent G.). A critical survey of recent research in Chinese economic history. *Economic history review*, 2000, 53, 1, p. 1-28.

* 6756. ROSS (Duncan M.). British business history: a review of the periodical literature for 1998. *Business history*, 2000, 42, 2, p. 1-16.

** 6757. BOYNS (R. E.), BOYNS (T.), EDWARDS (J. R.). Historical accounting records: a guide for archivists and researchers. London, Society of Archivists, 2000, X-109 p.

** 6758. Prodovol'stvennaja bezopasnost' Urala v XX veke, 1900–1984 gg.: Dokumenty i materialy. (Food independence of Urals in the 20th century, 1900–1984: Documents and materials). Vol. 1. 1900–1928 gg. Vol 2. 1929–1984 gg. Ekaterinburg, Akademkniga, 2000, 552, 455 p.

** 6759. Quellen zur deutschen Wirtschafts- und Sozialgeschichte in der Zeit des Nationalsozialismus. Teilband 1. 1933–1939. Teilband 2. Die Kriegswirtschaft. Hrsg. v. Walter STEITZ. Darmstadt, Wissenschaftliche Buchgesellschaft, 2000, 2 vol., XIV-294 p., XV-425 p. (Ausgewählte Quellen zur Geschichte der Neuzeit. Freiherr vom Stein-Gedächtnisausgabe, 39).

6760. AHRENS (Ralf). Gegenseitige Wirtschaftshilfe? Die DDR im RGW – Strukturen und handelspolitische Strategien 1963–1976. Köln, Weimar u. Wien, Böhlau, 2000, 386 p. (Schriften des Hannah-Arendt-Instituts für Totalitarismusforschung, 15).

6761. Asia Pacific dynamism, 1550–2000. Ed. by A. J. H. LATHAM and Heita KAWAKATSU. London a. New York, Routledge, 2000, XIV-281 p.

6762. BERNANKE (Ben S.). Essays on the Great depression. Princeton, Princeton U. P., 2000, VII-310 p.

6763. BULUT (Mehmet). Ottoman-Dutch economic relations in the early modern period 1571–1699 = Ottomaans Nederlandse economische betrekkingen in de vroeg-moderne periode 1571–1699. Utrecht, M. Bulut, 2000, XIII-223 p.

6764. Business institutions and behaviour in Australia. Ed. by David T. MERRETT. *Business history*, 2000, 42, 3, p. 1-152. [Cf. n^{os} <choice> 6816, 7340.]

6765. CHURCH (Roy). Ossified or dynamic? Structure, markets and the competitive process in the British business system of the nineteenth century. *Business history*, 2000, 42, 1, p. 1-20.

6766. CIRIACONO (Salvatore). La Rivoluzione industriale. Dalla protoindustrializzazione alla produzione possibile. Milano, Bruno Mondadori, 2000, 214 p.

6767. DE WIT (O.), VAN DEN ENDE (J.). The emergence of a new regime: business management and office mechanisation in the Dutch financial sector in the 1920s. *Business history*, 2000, 42, 2, p. 87-118.

6768. DRABBLE (John H.). An economic history of Malaysia, c.1800–1990. London, Macmillan Press a. New York, St. Martin's Press, 2000, XXIII-320 p.

6769. Early modern Atlantic economy (The). Ed. by John J. MAC CUSKER and Kenneth MORGAN. Cambridge, Cambridge U. P., 2000, 369 p.

6770. Economic change and the national question in twentieth-century Europe. Ed. by Alice TEICHOVA, Herbert MATHIS and Jaroslav PÁTEK. Cambridge, Cambridge U. P., 2000, XV-433 p.

6771. Economic history (An) of twentieth-century Latin America. Vol. 1. The export age: the Latin American economies in the nineteenth and early twentieth centuries. Vol. 2. Latin America in the 1930s: the role of the periphery in world crisis. Vol. 3. Industrialization and the state in Latin America: the postwar years. Ed.

by Enrique CÁRDENAS, José Antonio OCAMPO and Rosemary THORP. New York, Palgrave, 2000, 3 vol., 329 p., 297 p., 345 p.

6772. Economic transition in historical perspective: lessons from the history of economics. Ed. by Charles M. A. CLARK and Janina ROSICKA. Burlington, Ashgate, 2000, XII-165 p.

6773. ENGERMAN (David C.). Modernization from the other shore: American observers and the costs of Soviet economic development. *American historical review*, 2000, 105, 2, p. 383-416.

6774. Equità e sviluppo: considerazioni di lungo periodo. (Seminario, Roma, 14 gennaio 2000). *Rivista di storia economica*, 2000, 16, 2, p. 189-266. [Cf. n^{os} <scelta> 6795, 6815.]

6775. FEDERICO (Giovanni), O'ROURKE (Kevin). A social accounting matrix for Italy, 1911. *Rivista di storia economica*, 2000, 16, 1, p. 3-36.

6776. First World War (The) and the international economy. Ed. by Chris WRIGLEY. Cheltenham, Edward Elgar, 2000, X-221 p.

6777. FORSBERG (Aaron). America and the Japanese miracle: the Cold War context of Japan's postwar economic revival, 1950–1960. Chapel Hill a. London, University of North Carolina Press, 2000, XVIII-332 p. (Luther Hartwell Hodges series on business, society, and the state).

6778. GENÇ (Mehmet). Osmanlı İmparatorluğunda devlet ve ekonomi. (State and economy in the Ottoman Empire). İstanbul, Ötüken yayınları, 2000, 368 p.

6779. GERŠLOVÁ (Jana). Die wirtschaftliche Vergangenheit der böhmischen Länder (1870–1914). Industrie, Handel und Banken. *Vierteljahrschrift für Sozial- und Wirtschaftsgeschichte*, 2000, 87, p. 308-321.

6780. GORNIG (Martin). Gesamtwirtschaftliche Leitsektoren und regionaler Strukturwandel. Eine theoretische und empirische Analyse der sektoralen und regionalen Wirtschaftsentwicklung in Deutschland 1895–1987. Berlin, Duncker & Humblot, 2000, 291 p. (Schriften zur Wirtschafts- und Sozialgeschichte, 59).

6781. GROSSMAN (P. Z.). The dynamics of Hungarian hyperinflation 1945–1946: a new perspective. *Journal of European economic history*, 2000, 2-3, p. 405-430.

6782. HACKENBERG (Gerd R.). Wirtschaftlicher Wiederaufbau in Sachsen 1945–1949/50. Köln, Weimar u. Wien, Böhlau, 2000, 416 p. (Geschichte und Politik in Sachsen, 11).

6783. HARLEY (C. Knick), CRAFTS (N. F. R.). Simulating the two views of the British industrial revolution. *Journal of economic history*, 2000, 60, 3, p. 819-841.

6784. History (A) of Japanese trade and industry policy. Ed. by Mikio SUMIYA. Oxford, Oxford U. P., 2000, VIII-662 p.

6785. HODNE (Fritz), GRYTTEN (Ola Honningdal). Norsk økonomi i det nittende århundre. (Norwegian economy in the 19th century). Bergen, Fagbokforlaget, 2000, 318 p. (ill.).

6786. HONNINGDAL GRYTTEN (Ola). Differences in the standard of living in inter-war Norway: a comparative analysis. *Scandinavian economic history review*, 2000, 48, 3, p. 22-41.

6787. Istorija predprinimate'stva v Rossii. (A history of enterprise in Russia). Vol. 1. Ot srednevekov'ja do serediny XIX v. (From the Middle Ages to the middle of the 19th century). Ed. Anna V. SEMENOVA. Vol. 2. Vtoraja polovina XIX–nachalo XX veka. (The 2nd half of the 19th and the early 20th century). Ed. Ju. A. PETROV. Moskva, ROSSPEN, 2000, 2 vol., 480 p., 575 p. (ill.). [Eng. summary]

6788. KEISTER (Lisa). Wealth in America. Cambridge, Cambridge U. P., 2000, X-307 p.

6789. KOMLOS (J.). The Industrial Revolution as the escape from the Malthusian trap. *Journal of European economic history*, 2000, 2-3, p. 307-334.

6790. LACHMANN (Richard). Capitalists in spite of themselves: elite conflict and economic transitions in early modern Europe. Oxford, Oxford U. P., 2000, 314 p.

6791. LLOYD-JONES (R.), LEWIS (M. J.). The long wave and turning points in British industrial capitalism: a new-Schumpeterian approach. *Journal of European economic history*, 2000, 2-3, p. 359-404.

6792. LONDÁK (Miroslav). K niektorým problémom ekonomického vývoja na Slovensku na prelome 50. a 60. rokov. (Zu einigen Problemen der ökonomischen Entwicklung in der Slowakei an der Wende der 50er und 60er Jahre). *Historický časopis*, 2000, 48, 1, p. 99-112. Deutsche Zfassung].

6793. LUCCHINI (Cristina). Industrialismo y nacionalidad en Argentina y el Brasil (1890–1950). Buenos Aires, Ediciones del Signo, 2000, 153 p. (Colección MERCOSUR, 2).

6794. MAGNUSSON (Lars). An economic history of Sweden. London a. New York, Routledge, 2000, XVII-305 p.

6795. MALANIMA (Paolo). Crescita e ineguaglianza nell'Europa preindustriale. *In*: Equità e sviluppo: considerazioni di lungo periodo [Cf. n° 6774], p. 189-212.

6796. Mediterranean response (The) to globalization before 1950. Ed. by Şevket PAMUK and Jeffrey WILLIAMSON. London a. New York, Routledge, 2000, XVII-430 p.

6797. MIDDLETON (Roger). The British economy since 1945. Basingstoke, Macmillan Press, 2000, XIX-198 p.

6798. NEAL (Larry). A shocking view of economic history. *Journal of economic history*, 2000, 60, 2, p. 317-334.

6799. NICHOLAS (Tom). Wealth making in the nineteenth and early twentieth century: the Rubinstein hypothesys revisited. *Business history*, 2000, 42, 2, p. 155-168. [Cf. n° 6806.]

6800. NORTH (Michael). Kommunikation, Handel, Geld und Banken in der frühen Neuzeit. München, Oldenbourg, 2000, X-134 p. (Enzyklopädie deutscher Geschichte, 59).

6801. Occupation (L'), l'Etat français et les entreprises. Ed. par Olivier DARD, Jean-Claude DAUMAS et François MARCOT. Paris, ADHE, 2000, 487 p.

6802. PÉAN (Leslie Jean-Robert). Economie politique de la corruption: de Saint-Domingue à Haïti, 1791–1870. Port-au-Prince, Editions Mémoire, 2000, 523 p.

6803. PRECHEL (Harland). Big business and the state: historical transitions and corporate transformation, 1880s–1990s. Albany, State University of New York Press, 2000, XVI-317 p.

6804. Rise (The) and fall of state-owned enterprise in the Western World. Ed. by Pier Angelo TONINELLI. Cambridge, Cambridge U. P., 2000, IX-370 p.

6805. Rolle (Die) der Wirtschaftsbeziehungen zu Deutschland in der Entwicklung Bulgariens: von den 1930er Jahren bis zur Gegenwart. Hrsg. v. Tsvetana TODOROVA. Sofia, GAL-IKO, 2000, 251 p. (Collegium Germania, 3).

6806. RUBINSTEIN (W. D.). Wealth making in the late nineteenth and early twentieth centuries: a response. *Business history*, 2000, 42, 2, p. 141-154. [Cf. n° 6799.]

6807. SCHLARP (Karl-Heinz). Zwischen Konfrontation und Kooperation: die Anfangsjahre der deutsch-sowjetischen Wirtschaftsbeziehungen in der Ära Adenauer. Hamburg, Lit, 2000, 405 p. (Osteuropa: Geschichte, Wirtschaft, Politik, 28).

6808. SCHÖN (Lennart). En modern svensk ekonomisk historia. Tillväxt och omvandling under två sekel. (Une histoire économique moderne de la Suède. Croissance et mutation aux 19ᵉ et 20ᵉ siècles). Stockholm, SNS förlag, 2000, 560 p.

6809. SÖDERBERG (Johan). Controversial consumption in Sweden, 1914–1945. *Scandinavian economic history review*, 2000, 48, 3, p. 5-21.

6810. STANCIU (Ion Gh.). Afaceri noi în lumea veche: relațiile economice ale Statelor Unite ale Americii cu țările din centrul și sud-estul Europei pâna în 1939. (Nuovi affari nel vecchio mondo. le relazioni economiche degli Stati Uniti d'America con i paesi del centro e del sud-est europeo fino al 1939). București, Silex, 2000, 204 p.

6811. TIRATSOO (Nick). The United States technical assistance programme in Japan, 1955–1962. *Business history*, 2000, 42, 4, p. 117-136 p.

6812. TORTELLA (Gabriel). The development of modern Spain. An economic history of the nineteenth and twentieth centuries. Cambridge, Harvard U. P., 2000, XVI-528 p.

6813. TRIVELLATO (F.). Fondamenta dei vetrai. Lavoro, tecnologia e mercato a Venezia tra Sei e Settecento. Roma, Donzelli, 2000, VIII-343 p.

6814. Unternehmenskommunikation im 19. und 20. Jahrhundert. Neue Wege der Unternehmensgeschichte. Hrsg. v. Clemens WISCHERMANN, Peter BORSCHEID und Karl-Peter ELLERBROCK. Dortmund, Gesellschaft für Westfälische Wirtschaftsgeschichte e. V., 2000, 263 p. (Untersuchungen zur Wirtschafts-, Sozial- und Technikgeschichte, 19).

6815. VECCHI (Giovanni). Quali lezioni dall'analisi della distribuzione della spesa familiare in Italia, 1881–1961? *In*: Equità e sviluppo: considerazioni di lungo periodo [Cf. n° 6774], p.247-260.

6816. VILLE (Simon), MERRETT (David T.). The development of large scale enterprise in Australia, 1910–1964. *In*: Business institutions and behaviour in Australia [Cf. n° 6764], p. 13-46.

6817. WINTLE (Michael). An economic and social history of the Netherlands, 1800–1920: demographic, economic and social transformation. Cambridge, Cambridge U. P., 2000, XV-399 p.

6818. WRIGHT (Christopher). From shop floor to boardroom: the historical evolution of Australian management consulting, 1940s to 1980s. *Business history*, 2000, 42, 1, p. 85-106.

6819. YAMAMOTO (Tadashi). Jentoruman de arukoto: Sono henyô to Igirisu kindai. (To be a gentleman: Its change and modern Britain). Tokyo, Tosui Shobo, 2000, 256p.

§ 2. Political economy.

6820. ACCATI (Luisa). Volontà e autorità. Francisco Suarez e la naturale privazione dell'etica. *In*: Etiche economiche [Cf. n° 6826], p. 623-654.

6821. Adam Smith across nations: translations and receptions of The Wealth of Nations. Ed. by Chengchung LAI. Oxford, Oxford U. P., 2000, XXXIII-403 p.

6822. BECKMANN (Ulf). Von Löwe bis Leontief: Pioniere der Konjunkturforschung am Kieler Institut fur Weltwirtschaft. Marburg, Metropolis, 2000, 544 p. (Beiträge zur Geschichte der deutschsprachigen Ökonomie, 15).

6823. BENSEL (Richard Franklin). The political economy of American industrialization, 1877–1900. Cambridge, Cambridge U. P., 2000, XXIII-549 p. (ill., maps).

6824. CREMASCHI (Sergio). Legge di natura e scienza economica. *In*: Etiche economiche [Cf. n° 6826], p. 697-730.

6825. Development (The) of economics in western Europe since 1945. Ed. by A. W. Bob COATS. London a. New York, Routledge, 2000, XIV-262 p.

6826. Etiche economiche. A cura di Maria Luisa PESANTE. *Quaderni storici*, 2000, 35, 104, p. 573-766. [Cf. n^os <scelta> 3889, 6820, 6824, 6833, 6838.]

6827. FINKELSTEIN (Andrea). Harmony and the balance: an intellectual history of seventeenth-century English economic thought. Ann Arbor, University of Michigan Press, 2000, VIII-381 p.

6828. "From our Italian correspondent": Luigi Einaudi's articles in The Economist, 1908–1946. Ed. by Roberto MARCHIONATTI. Firenze, Olschki, 2000, 2 vol., LXVIII-834 p. (Studi / Fondazione Luigi Einaudi, 36).

6829. HERSTAD (John). I helstatens grep. Kornmonopolet 1735–1788. (The monopoly of grain 1735–1788). Oslo, Tano Aschehoug, 2000, 440 p. (ill.). (Skriftserie / Riksarkivaren, 8).

6830. Historicity (The) of economics. Continuities and discontinuities of historical thought in 19th and 20th century economics. Ed. by H. H. NAU and B. SCHEFOLD. Berlin a. Heidelberg, Springer Verlag, 2000, VIII-245 p.

6831. IRIGOIN (María Alejandra). Finance, politics and economics in Buenos Aires, 1820s–1860s: the political economy of currency stabilisation. London, [s. n.], 2000, 333 p. [Thesis (Ph. D.)London, LSE, 2000]

6832. LACINA (Vlastislav). Zlatá léta československého hospodářství 1918–1929. (The golden years of Czechoslovak economy, 1918–1929). Praha, Historický ústav AV ČR, 2000, 254 p. (photogr.). (Práce HÚ AV ČR. Řada A Monographia, 17).

6833. MACCABELLI (Terenzio), MORATO (Erica). Il «bisognevole» e il «superfluo»: occupazioni e distribuzione della ricchezza in Giammaria Ortes. *In*: Etiche economiche [Cf. n° 6826], p. 731-766.

6834. MORNATI (Fiorenzo). Pareto observateur du libéralisme économique suisse et vaudois fin de siècle. D'une vision mythique à une ré-vision critique. *Schweizerische Zeitschrift für Geschichte*, 2000, 50, p. 403-420.

6835. OMICCIOLI (Massimo). Einaudi e l'inflazione del 1946–1947: un riesame. *Rivista di storia economica*, 2000, 16, 1, p. 37-64.

6836. Pensamiento y politica económica en la Edad Moderna. Ed. por L. A. RIBOT GARCIA y L. DE ROSA. Madrid, Instituto Universitario de Historia Simancas y Actas Editorial, 2000, 253 p.

6837. PERELMAN (Michael). The invention of capitalism: classical political economy and the secret history of primitive accumulation. Durham, Duke U. P., 2000, 405 p.

6838. PESANTE (Maria Luisa). Il commercio nella repubblica. *In*: Etiche economiche [Cf. n° 6826], p. 655-696.

6839. TADMAN (Michael). The demographic Cost of Sugar: debates on slave societies and natural increase in the Americas. *American historical review*, 2000, 105, 5, p. 1534-1575.

6840. TRIBE (Keith). "The Price is Right": Léon Walras and economic justice. *Schweizerische Zeitschrift für Geschichte*, 2000, 50, p. 388-402.

6841. WHATMORE (Richard). 'L'amitié de grands Etats est leur plus sûr appui'. The small state dilemma in Genevan political economy, 1762–1798. *Schweizerische Zeitschrift für Geschichte*, 2000, 50, p. 353-371.

§ 3. Industry, mining and transportation.

6842. AMDAM (Rolv Petter). Norwegian Industrial Committee in New York 1943–1945. *Historisk tidsskrift*, 2000, 79, 2, p. 3-21.

6843. ANDERSEN (Dan H.), VOTH (Hans-Joachim). The grapes of war: neutrality and Mediterranean shipping under the Danish flag, 1747–1807. *Scandinavian economic history review*, 2000, 48, 1, p. 5-27.

6844. ANDERSSON-SKOG (Lena). National patterns in the regulation of railways and telephony in the Nordic Countries to 1950. *Scandinavian economic history review*, 2000, 48, 2, p. 30-46.

6845. ANTILA (Kimmo), NENONEN (Marko). Relief works, the World Bank and private business: building highways in Finland. *Scandinavian economic history review*, 2000, 48, 2, p. 47-71.

6846. BAMBERG (James). British petroleum and global oil, 1950–1957: the challenge of nationalism. Cambridge, Cambridge U. P., 2000, XXVIII-637 p.

6847. BANKEN (Ralf). Die Industrialisierung der Saarregion 1815–1914. Band 1. Die Frühindustrialisierung 1815–1850. Stuttgart, Steiner, 2000, 481 p. (Regionale Industrialisierung, 1).

6848. BERGERON (Louis), MAIULLARI-PONTOIS (Maria Teresa). Le patrimoine industriel des Etats-Unis. Paris, Hoëbeke, 2000, 288 p.

6849. BLACK (Brian). Petrolia: the landscape of America's first oil boom. Baltimore, Johns Hopkins U. P., 2000, XII-235 p. (Creating the North American landscape).

6850. BOYCE (Robert W.). Imperial dreams and national realities: Britain, Canada and the struggle for a Pacific telegraph cable, 1879–1902. *English historical review*, 2000, 115, 460, p. 39-70.

6851. CAN (Bilmez Bülent). Demiryolundan petrole Chester projesi (1908–1923). (From railroad to oil: the Chester project) Ed. by Osman KÖKER. İstanbul, Türkiye Ekonomik ve Toplumsal Tarih Vakfı, 2000, 409 p.

6852. CHANUT (Jean-Marie), HEFFER (Jean), MAIRESSE (Jacques), POSTEL-VINAY (Gilles). L'industrie française au milieu du XIXe siècle. Les enquêtes de la Statistique générale de la France. Paris, Ed. de l'EHESS, 2000, 215 p.

6853. CHASTAGNARET (Gérard). L'Espagne, puissance minière dans l'Europe du XIXe siècle. Madrid, Casa de Velázquez, 2000, XV-1170 p.

6854. CIRONISOVÁ (Eva). Český plzeňský pivovar Světovar 1910–1933. (Die Pilsner tschechische Bierbrauerei Světovar, 1910–1933). *Západočeský historický sborník*, 2000, 6, p. 233-263.

6855. Créateurs et créations d'entreprises de la révolution industrielle à nos jours. Ed. par Jacques MARSEILLE. Paris, ADHE, 2000, 751 p.

6856. Culture and technology in modern Japan. Ed. by Ian INKSTER and Fumihiko SATOFUKA. London a. New York, I. B. Tauris, 2000, VII-169 p.

6857. DALSGAARD LARSEN (Keld). Dansk papirindustri: mennesker, teknologi og produktion 1829–1999. (L'industre papetière danoise: hommes, technologie, production). Silkeborg, Silkeborg museum, 2000, 272 p. (ill.).

6858. D'ANGIO (Agnès). Schneider et Cie et la naissance de l'ingénierie. Des pratiques internes à l'aventure internationale 1836–1949. Préf. de Henri LACHMANN. Paris, CNRS Editions, 2000, 320 p.

6859. Dansk søfartshistorie. Band 6. Damp og Diesel. 1920–1960 (Histoire de la marine danoise. Tome 6. Vapeur et Diesel. 1920–1960). Ed par Frank A. RASMUSSEN, Bent VEDSTED RØNNE et Hans Chr. JOHANSEN. København, Gyldendal, 2000, 251 p. (diagr., ill.).

6860. DAVIES (R. E. G.). The birth of commercial aviation in the United States. *In*: National paths to the sky [Cf. n° 6888], p. 993-1008.

6861. DEUSTUA (José R.). The bewitchment of silver: the social economy of mining in nineteenth-century Peru. Athens, Ohio University Center for International Studies, 2000, XVI-290 p. (Monographs in international studies. Latin American series, 31).

6862. DIENEL (Hans-Liudiger), SCHIEFELBUSCH (Martin). German commercial air transport until 1945. *In*: National paths to the sky [Cf. n° 6888], p. 945-968.

6863. DIERIKX (Marc). Routes versus revenue: pioneering commercial aviation in Holland, 1919–1940. *In*: National paths to the sky [Cf. n° 6888], p. 889-912.

6864. ELTIS (David), ENGERMAN (Stanley L.). The importance of slavery and the slave trade to industrializing Britain. *Journal of economic history*, 2000, 60, 1, p. 123-144.

6865. FENOALTEA (Stefano). The growth of Italy's wool industry, 1861–1913: a statistical reconstruction. *Rivista di storia economica*, 2000, 16, 2, p. 119-146.

6866. GOHM (Lilian). Technologietransfer deutscher Unternehmen in die USA 1870–1939. St. Katharinen, [s. n.], 2000, 480 p. (Studien zur Wirtschafts- und Sozialgeschichte, 20).

6867. HAYCRAFT (William R.). Yellow steel: the story of the earthmoving equipment industry. Urbana a. Chicago, University of Illinois Press, 2000, XVI-465 p.

6868. HAYES (Peter). Industry and ideology: IG Farben in the Nazi era. New York, Cambridge U. P., 2000, [s. p.].

6869. HIGGINS (David), TOMS (Steven). Public subsidy and private investment: the Lancashire cotton textile industry, c.1950–c.1965. *Business history*, 2000, 42, 1, p. 59-84.

6870. HONEYMAN (Katrina). Women, gender and industrialisation in England, 1700–1870. New York, St. Martin's Press, 2000, VIII-204 p.

6871. Industria y época moderna. Ed. por L. A. RIBOT GARCIA y L. DE ROSA. Madrid, Actas Editorial y Napoli, Istituto italiano per gli Studi Filosofici, 2000, 402 p.

6872. Industrial Revolution (The) in comparative perspective. Ed. by Christine RIDER and Micheal THOMPSON. Malabar, Krieger, 2000, X-268 p.

6873. KENNEDY (Sean). The Croix de Feu, the Parti Social Français, and the politics of aviation, 1931–1939. *French historical studies*, 2000, 2, p. 373-399.

6874. KIESEWETTER (Hubert). Region und Industrie in Europa 1815–1995. Stuttgart, Steiner, 2000, 224 p. (Grundzüge der modernen Wirtschaftsgeschichte, 2).

6875. Kommunen und Unternehmen im 20. Jahrhundert. Wechselwirkungen zwischen öffentlicher und privater Wirtschaft. Hrsg. v. Matthias FRESE und Burkhard ZEPPENFELD. Essen, Klartext, 2000, 264 p. (Bochumer Schriften zur Unternehmens- und Industriegeschichte, 7).

6876. KOMULAINEN (Arvo). Suojattu merenkulku elinehtomme. Miinoitukset ja miinanraivauksen vaikutukset meriliikenteeseen vuosina 1939–1950. (Protection of merchant fleet in the Baltic sea, 1939–1950 by laying mine and mine clearance. Covering the history of Nobel mine laying before the Crimean war to the mine clearance in the Baltic Sea, 1950). Helsinki, Suomi merellä –säätiö, 2000, 355 p.

6877. KRANTZ (Olle). The transport and communications sector in economic development: views from the historical national accounts. *Scandinavian economic history review*, 2000, 48, 2, p. 5-29.

6878. LAMOREAUX (Naomi R.), SOKOLOFF (Kenneth L.). The geography of invention in the American glass industry, 1870–1925. *Journal of economic history*, 2000, 60, 3, p. 700-755.

6879. LYTH (Peter). The empire's airways: British civil aviation from 1919. *In*: National paths to the sky [Cf. n° 6888], p. 865-888.

6880. MÄKINEN (Tuomo), RITARANTA (Eino). Finnish civil aircraft 1926–1999. Helsinki, Koala-Kustannus, 2000, 170 p.

6881. MARKOVIC (Sacha). Le rôle de l'Etat dans la naissance de l'aviation commerciale française (1918–

1933). *In*: National paths to the sky [Cf. n° 6888], p. 969-992.

6882. MATSUSHITA (T.). Railway policy during the second Okuma Cabinet. *Journal of Japanese history*, 2000, 460, p. 1-28.

6883. MIERZEJEWSKI (Alfred C.). The most valuable asset of the Reich. A history of the German national railway. Vol. 2. 1933–1945. Chapel Hill a. London, University of North Carolina Press, 2000, XXI-248 p.

6884. MITCHELL (Allan). The great train race: railways and the Franco-German rivalry, 1815–1914. New York, Berghahn Books, 2000, XV-328 p.

6885. MOLÀ (L.). The silk industry of Renaissance Venice. Baltimore a. London, Johns Hopkins U. P., 2000, XIX-457 p.

6886. MORE (Charles). Understanding the industrial revolution. London a. New York, Routledge, 2000, IX-188 p.

6887. MÜLLER (Uwe). Infrastrukturpolitik in der Industrialisierung. Der Chauseebau in der preußischen Provinz Sachsen und dem Herzogtum Braunschweig vom Ende des 18. Jahrhunderts bis in die siebziger Jahre des 19. Jahrhunderts. Berlin, Duncker & Humblot, 2000, 585 p. (Schriften zur Sozial- und Wirtschaftsgeschichte, 57).

6888. National paths to the sky. The origins of commercial air transport in Western Europe and the United States (1919–1939). Intr. by Guy VANTHEMSCHE. *Revue belge de philologie et d'histoire*, 2000, 78, 3-4, p. 853-1022. [Cf. n[os] <choice> 6860, 6862, 6863, 6879, 6881, 6909.]

6889. NORDBY (Trond). I politikkens sentrum. Variasjoner i Stortingets makt 1814 til 2000. (Variations of power in the Norwegian Parliament from 1814 to 2000). Oslo, Universitetsforlaget, 2000, 346 p. (ill.).

6890. OLIEN (Roger M.), DAVIDS OLIEN (Diana). Oil and ideology: the cultural creation of the American petroleum industry. Chapell Hill, University of North Carolina Press, 2000, XVIII-305 p. (Luther Hartwell Hodges series on business, society, and the state).

6891. ÖZYÜKSEL (Murat). Hicaz demiryolu. (Hedjaz railroad) Ed. by Ali BERKTAY. İstanbul, Türkiye Ekonomik ve Toplumsal Tarih Vakfı, 2000, 312 p.

6892. PAQUIER (S.). Le mythe de l'industrie électrique nazionale helvétique (1875–années vingt). *Relations internationales*, 2000, 101, p. 5-20.

6893. POLLARD (Sidney). Essays on the Industrial Revolution in Britain. Aldershot, Ashgate, 2000, 320 p. (Variorum collected studies series).

6894. Quel modèle productif? Trajectoires et modèles industriels des constructeurs automobiles mondiaux. Dir. par Michel FREYSSENET, Andrew MAIR, Koïchi SHIMIZU et Giuseppe VOLPATO. Paris, La Découverte, 2000, 525 p.

6895. ROEMER (Klaus). Geschichte der Papiermühlen in Westpreußen und Danzig, nebst einem Anhang für den Netzedistrikt. Münster, Copernicus-Verlag, 2000, 397 p. (Quellen und Darstellungen zur Geschichte Westpreußens, 30).

6896. RÖNDINGS (Uwe). Globarisierung und europäische Integration. Der Strukturwandel des Energiesektors und die Politik der Montaunion, 1952–1962. Baden-Baden, Nomos, 2000, 472 p.

6897. ROSE (Mary B.). Firms, networks and business values: the British and American cotton industries since 1750. Cambridge, Cambridge U. P., 2000, XII-352 p. (Cambridge studies in modern economic history, 8).

6898. SCHMITZ (Christopher J.). The world copper industry: geology, mining techniques and corporate growth, 1870–1939. *Journal of European economic history*, 2000, 1, p. 77-108.

6899. SPEHR (Michael). Maschinensturm. Protest und Widerstand gegen technische Neuerungen am Anfang der Industrialisierung. Münster, Westfälisches Dampfboot, 2000, 225 p. (Theorie und Geschichte der bürgerlichen Gesellschaft, 18).

6900. STANCIU (Laura). Free-standing companies in the oil sector in Romania and Poland before 1948: typologies and competencies. *Business history*, 2000, 42, 4, p. 27-66.

6901. STIER (Bernhard). Die neue Elektrizitätsgeschichte zwischen kulturhistorischer Erweiterung und kommunikationspolitischer Instrumentalisierung. Anmerkungen zum Forschungsstand am Ende des "langen 20. Jahrhunderts der Elektrizität". *Vierteljahrschrift für Sozial- und Wirtschaftsgeschichte*, 2000, 87, p. 477-487.

6902. STOKES (Raymond G.). Constructing socialism: technology and change in East Germany 1945–1990. Baltimore, Johns Hopkins U. P., 2000, XI-260 p. (Johns Hopkins studies in the history of technology).

6903. Svenska skärgårdsflottan (Den). (La Flotte suédoise de l'archipel aux 18e et 19e siècles). Ed. par Hans NORMAN. Lund, Historiska media, 2000, 414 p. (ill., planches).

6904. TEICH (Mikulás). Bier, Wissenschaft und Wirtschaft in Deutschland, 1800–1914. Ein Beitrag zur deutschen Industrialisierungsgeschichte. Köln, Weimar u. Wien, Böhlau, 2000, 355 p.

6905. THOMSEN (Steen). The survival of foreign subsidiaries and domestically-owned companies in Denmark, 1895–1995. *Scandinavian economic history review*, 2000, 48, 2, p. 72-87.

6906. THORBURN (Thomas). Economics of transport: the Swedish case 1780–1980. Södertalje, Almqvist and Wicksell International, 2000, 589 p.

6907. UNGER (Stefan). Eisen und Stahl für den Sozialismus. Modernisierungs- und Innovationsstrategien der Schwarzmetallurgie in der DDR von 1949 bis 1971.

Berlin, Duncker & Humblot, 2000, 431 p. (Schriften zur Wirtschafts- und Sozialgeschichte, 61).

6908. VALENTA (Vladimír). Organizační struktura uranového průmyslu v Příbrami. (The organizational structure of the Uranium industry in Příbram). *Podbrdsko*, 2000, 7, p. 162-225.

6909. VANTHEMSCHE (Guy). The birth of commercial air transport in Belgium, 1919–1923. *In*: National paths to the sky [Cf. n° 6888], p. 913-944.

6910. Vedvarende energi i Danmark: en krønike om 25 opvækstår 1975–2000. (Les énergies renouvelables au Danemark: chronique de 25 années de croissance, 1975–2000). Ed. par Ejvin BEUSE. Århus, OVE, 2000, 367 p. (ill.).

6911. VIGO (G.). Nel cuore della crisi. Politica economica e metamorfosi industriale nella Lombardia del Seicento. Pavia, Università di Pavia, 2000, 171 p.

6912. WALSH (Margaret). Making connections: the long-distance bus industry in the USA. Aldershot a. Burlington, Ashgate Publishing Limited, 2000, XVII-245 p.

6913. WHITE (Nicola). Reconstructing Italian fashion: America and the development of the Italian fashion industry. New York, Berg, 2000, XVII-181 p. (Dress, body, culture).

Cf. n°s 893, 5083, 7180

§ 4. Trade.

6914. Advertising and the European city: historical perspectives. Ed. by Clemens WISCHERMANN and Elliott SHORE. Aldershot, Ashgate, 2000, 225 p. (Historical urban studies).

6915. BARTON (Jonathan R.). Struggling against decline: British business in Chile, 1919–1933. *Journal of Latin American studies*, 2000, 32, 1, p. 235-264.

6916. BEATTY (Edward). The impact of foreign trade on the Mexican economy: terms of trade and the rise industry, 1880–1923. *Journal of Latin American studies*, 2000, 32, 2, p. 399-434.

6917. Business and society. Entrepreneurs, politics and networks in a historical perspective. Proceedings of the third European business history association conference "Business and society", September 24–26, 1999, Rotterdam. Ed. by Anne-Marie KUIJLAARS, Kim PRUDON and Joop VISSER. Rotterdam, Centre of Business History, 2000, 548 p.

6918. CHURCH (Roy). Advertising consumer goods in nineteenth-century Britain. *Economic history review*, 2000, 53, 4, p. 621-645.

6919. Consumption studies and the history of the Ottoman empire, 1550–1922: an introduction. Ed. by Donald QUARTER. Albany, State University of New York Press, 2000, VII-358 p.

6920. COX (Howard). The global cigarette: origins and evolution of British American tobacco 1880–1945. Oxford, Oxford U. P., 2000, XXII-401 p.

6921. CRESPO SOLANA (Ana). El comercio marítimo entre Amsterdam y Cádiz (1713–1778). Madrid, Banco de España, 2000, 164 p. (Estudios de historia económica, 40).

6922. EWALD (Janet J.). Crossers of the sea: slaves, freedman, and other migrants in northwestern Indian Ocean, c. 1750–1914. *American historical review*, 2000, 105, 1, p. 69-91.

6923. FOREMAN-PECK (James), PEPELASIS MINOGLOU (Ioanna). Entrepreneurship and convergence. Greek businessmen in the later nineteenth century. *Rivista di storia economica*, 2000, 16, 3, p. 279-304.

6924. GAZIŃSKI (Jarosław). Handel morski Szczecina w latach 1720–1805. (Der Seehandel von Szczecin [Stettin] in den Jahren 1720–1805). Szczecin, Wydaw. Nauk. Uniw. Szczecińskiego, 2000, 380 p. (Rozprawy i Sprawozdania, 32).

6925. GÓRCZYŃSKA-PRZYBYŁOWICZ (Bożena). Polityka handlowa Trzeciej Rzeszy wobec państw Europy Południowo-Wschodniej, Środkowej i Wschodniej w latach 1933–1939. (Die Handelspolitik des Dritten Reiches gegen die Länder des Süd-Ost-, Mittel-, und Ost-Europas in den Jahren 1933–1939). Poznań, Inst. Historii Uniw. A. Mickiewicza, 2000, 311 p. (Publikacje Inst. Historii Uniw. A. Mickiewicza, 30).

6926. HAGMARK (Hanna), BARNARD (Michaela). Åland shipping and Hull's Baltic timber trade. *Sjöhist. Årsskrift för Åland*, 1999-2000, 12, p. 92-109. (ill., tables).

6927. Handel (Der) im Kurfürstentum / Königreich Hannover (1780–1850). Hrsg. v. Karl Heinrich KAUFHOLD und Markus A. DENZEL. Stuttgart, Steiner, 2000, [s. p.].

6928. HARRIS (Ron). Industrializing English law: entrepreneurship and business organization, 1720–1844. Cambridge, Cambridge U. P., 2000, XVI-331 p. (Political economy of institutions and decisions).

6929. HIGMAN (B. W.). The sugar revolution. *Economic history review*, 2000, 53, 2, p. 213-236.

6930. IVANOV (M.). Foreign trade in the economic policy of the popular bloc in Bulgaria (1931–1934). *Bulgarian historical review*, 2000, 28, 3-4, p. 148-179.

6931. JONES (Geoffrey). Merchants to multinationals: British trading companies in the nineteenth and twentieth centuries. Oxford, Oxford U. P., 2000, X-404 p.

6932. KAGOYA (Naoto). Ajia kokusai tsûshô chitsujo to kindai Nihon. (International commercial order in Asia and modern Japan). Nagoya, The University of Nagoya Press, 2000, 517 p.

6933. KHAN (B. Zorina). Commerce and cooperation: litigation and settlement of civil disputes on the

Australian frontier, 1860–1900. *Journal of economic history*, 2000, 60, 4, p. 1088-1119.

6934. KLINE (Ronald R.). Consumers in the country: technology and social change in rural America. Baltimore, Johns Hopkins U. P., 2000, XII-372 p. (Revisiting rural America).

6935. KÖHLER (Meike). Die Narvafahrt. Mittel- und westeuropäischer Rußlandhandel 1558–1581. Hamburg, Kovač, 2000, 173 p. (Hamburger Beiträge zur Geschichte des östlichen Europa, 6).

6936. KÖNIG (Wolfgang). Geschichte der Konsumgesellschaft. Stuttgart, Steiner, 2000, 509 p. (Vierteljahrschrift für Sozial- und Wirtschaftsgeschichte, 154).

6937. KÜMIN (Beat), RADEFF (Anne). Markt-Wirtschaft. Handelsinfrastruktur und Gastgewerbe im alten Bern. *Schweizerische Zeitschrift für Geschichte*, 2000, 50, p. 1-19.

6938. LAMBERTY (Christiane). Reklame in Deutschland 1890–1914. Wahrnehmung, Professionalisierung und Kritik der Wirtschaftswerbung. Berlin, Duncker & Humblot, 2000, 535 p. (Beiträge zur Verhaltensforschung, 38).

6939. LAMPEN (Angelika). Stadt und Fisch: Konsum, Produktion und Handel im Hanseraum der Frühzeit. *Vierteljahrschrift für Sozial- und Wirtschaftsgeschichte*, 2000, 87, p. 281-307.

6940. LEPORE (Amedeo). Mercato e impresa in Europa. L'azienda Gonzáles de la Sierra nel commercio gaditano tra XVIII e XIX secolo. Bari, Cacucci Editore, 2000, 429 p.

6941. LINEBAUGH (Peter), REDIKER (Marcus). The many headed hydra: sailors, slaves, commoners, and the hidden history of the revolutionary Atlantic. Boston, Beacon, 2000, 433 p.

6942. MARKOVITS (Claude). The global world of Indian merchants, 1750–1947: traders of Sind from Bukhara to Panama. Cambridge, Cambridge U. P., 2000, XV-327 p. (Cambridge studies in Indian history and society, 6).

6943. MOTONO (Eiichi). Conflict and cooperation in Sino-British business, 1860–1911: the impact of pro-British commercial network in Shanghai. Basingstoke, Macmillan, 2000, XIII-229 p. (St. Antony's series).

6944. PFISTER (Ulrich). Vom Kiepenkerl zu Karstadt. Einzelhandel und Warenkultur im 19. und frühen 20. Jahrhundert. *Vierteljahrschrift für Sozial- und Wirtschaftsgeschichte*, 2000, 87, p. 38-66.

6945. PIERENKEMPER (Toni). Unternehmensgeschichte. Eine Einführung in ihre Methoden und Ergebnisse. Stuttgart, Steiner, 2000, 328 p. (Grundzüge der modernen Wirtschaftsgeschichte, 1).

6946. Region and strategy in Britain and Japan: business in Lancashire and Kansai, 1890–1990. Ed. by Douglas A. FARNIE [et al.]. New York, Routledge, 2000, XVIII-322 p. (Routledge International Studies in Business History, 7).

6947. RENDERS FOLMER (Corrie). Van Prooijen: van Goederenkandel naar Slavenkandel. De Middelburgse Commercie compagnie 1720–1755. Middelbourg, [s. n.], 2000, 222 p. (Werken uitgegeven door het Koninklijk Zeeuwsch Genootschap der Wetenschappen, 10).

6948. SEYF (Ahmad). Foreign firms and local merchants in nineteenth-century Iran. *Middle eastern studies*, 2000, 36, 4, p. 137-155.

6949. ŠTEFÁNIKOVÁ (Antónia). O niektorých aspektoch zahraničného obchodu Slovenskej republiky v období rokov 1939–1945. (Einige Aspekte des auswärtigen Handels der slowakischen Republik in den Jahren 1939–1945). *Historický časopis*, 2000, 48, 3, p. 462-477. [Deutsche Zfassung].

6950. STEIN (Stanley J.), STEIN (Barbara H.). Silver, trade, and war: Spain and America in the making of Early Modern Europe. Baltimore, Johns Hopkins U. P., 2000, IX-351 p.

6951. STYLES (John). Product innovation in early modern London. *Past and present*, 2000, 168, p. 124-169.

6952. THOMPSON (Victoria E.). The virtuous marketplace: women and men, money and politics in Paris, 1830–1870. Baltimore, Johns Hopkins U. P., 2000, VIII-229 p.

6953. TOPOLSKI (Jerzy). Przełom gospodarczy w Polsce XVI wieku i jego następstwa. (Der wirtschaftliche Umbruch in Polen des XVI. Jahrhunderts und dessen Auswirkungen). Poznań, Wydaw. Poznańskie, 2000, 172 p.

6954. WEINER (Richard). Battle for survival: Porfirian views of the international marketplace. *Journal of Latin American studies*, 2000, 32, 3, p. 645-670.

6955. WHITE (Nicholas J.). The business and the politics of decolonization: the British experience in the twentieth century. *Economic history review*, 2000, 53, 3, p. 544-564.

6956. WILLETT (Julie A.). Permanent waves: the making of the American beauty shop. New York, New York U. P., 2000, XII-249 p.

Cf. n[os] *6860, 6862, 6863, 6879, 6881, 6888, 6909, 7120*

§ 5. Agriculture and agricultural problems.

** 6957. BORISOVA (L). [et al.]. Sovetskaja derevnja glazami VChK – OGPU – NKVD, 1918–1939: dokumenty i materialy. (Les campagnes soviétiques vues par la TCHEKA – O.G.P.U. – N.K.V.D., 1918–1939: documents and matériaux). Ed. A. BERELOVICH, V.P. DANILOV. RAN, In-t rossijskoj istorii; Maison des sciences de l'homme, etc. Vol. 2. 1923–1929. Moskva, ROSSPEN, 2000, 1167 p. (tables; bibl. incl.; ind. p. 1109-1137). [Cf. Bibl. 98, n° 6822.]

5. AGRICULTURE AND AGRICULTURAL PROBLEMS

** 6958. Tragedija sovetskoj derevni: Kollektivizatsija i raskulachivanie: Dokumenty i materialy, 1929–1939. (The tragedy of Soviet village: the collectivization and the dispossession of Kulaks: Documents and Materials, 1929–1939). Vol. 2. November 1929–December 1930. Ed. V. P. DANILOV, R. Th. MANNING, L. VIOLA, N. IVNITSKIJ. Moskva, ROSSPEN, 2000, 927 p.

6959. Agrarian history (The) of England and Wales. Vol. VII. 1850–1914. Ed. by E. J. T. COLLINS. Cambridge, Cambridge U. P., 2000, XL-2277 p.

6960. Agriculture and politics in England, 1815–1939. Ed. by J. R. WORDIE. Basingstoke, Macmillan, 2000, VII-260 p.

6961. ALDENHOFF-HÜBINGER (Rita). "Les nations anciennes, écrasées ...". Agrarprotektionismus in Deutschland und Frankreich, 1880–1914. *Geschichte und Gesellschaft*, 2000, 26, 3, p. 439-470.

6962. BÉAUR (Gérard). Histoire agraire de la France au XVIIIe siècle. Paris, Sedes, 2000, 320 p. (Regards sur l'histoire-histoire moderne).

6963. BJØRN (Claus), FONNEBESCH-WULFF (Benedicte). Mark og mennesker: studier i Danmarks historie 1500–1800 - tilegnet Karl-Erik Frandsen. (Terre et hommes: études d'histoire danoise offertes à K.-E. Frandsen). Ebeltoft, Skippershoved, 2000, 309 p. (ill.).

6964. BORGES (Emília Salvado). Homens, fazenda e poder no alentejo de setecentos: o caso de Cuba. Lisboa, Colibri, 2000, 513 p. (ill., maps). (Colibri história, 28).

6965. DUMONT (Dora M.). 'Strange and exorbitant demands': rural labour in nineteenth-century Bologna. *European history quarterly*, 2000, 30, 4, p. 467-492.

6966. GADD (Carl Johan). Det svenska jordbrukets historia. Band 3. (Histoire de l'agriculture suédoise. Tome 3). Den agrara revolutionen 1700–1870. (La révolution agricole en Suède de 1700 à 1870). Stockholm, Natur och Kultur & LTs förlag, 2000, 415 p.

6967. GARAVAGLIA (Juan Carlos). Les hommes de la pampa. Une histoire agraire de la campagne de Buenos Aires (1700–1830). Paris, Editions de l'EHESS et Editions de la MSH, 2000, 465 p.

6968. HEYDE (Jürgen). Bauer, Gutshof und Königsmacht. Die estnischen Bauern in Livland unter polnischer und schwedischer Herrschaft 1561–1650. Köln, Weimar u. Wien, Böhlau, 2000, 377 p. (Quellen und Studien zur baltischen Geschichte, 16).

6969. HOPCROFT (R. L.), EMIGH (R. J.). Divergent paths of agrarian change. Eastern England and Tuscany compared. *Journal of European economic history*, 2000, 1, p. 9-52.

6970. KULIKOFF (Allan). From British peasants to colonial American farmers. Chapel Hill, University of North Carolina Press, 2000, XIII-484 p.

6971. Landesgeschichte und historische Demographie. Hrsg. v. Michael MATHEUS und Walter G. RÖDEL. Stuttgart, Steiner, 2000, VII-194 p. (Geschichtliche Landeskunde, 50).

6972. Lange Abschied (Der) vom Agrarland. Agrarpolitik, Landwirtschaft und ländliche Gesellschaft zwischen Weimar und Bonn. Hrsg. v. Daniela MÜNKEL. Göttingen, Wallstein, 2000, 316 p. (Veröffentlichungen des Arbeitskreises Geschichte des Landes Niedersachsen, 16).

6973. LAUCK (Jon). American agriculture and the problem of monopoly: the political economy of grain belt farming, 1953–1980. Lincoln, University of Nebraska Press, 2000, XIV-254 p.

6974. LI (Lillian). Integration and disintegration in north China's grain markets, 1738–1911. *Journal of economic history*, 2000, 60, 3, p. 665-699.

6975. MARTIN (John). The development of modern agriculture: British farming since 1931. Foreword by Tom BLUNDELL. New York, St. Martin's Press, 2000, XVII-236 p.

6976. MARTINI (Manuela). Fedeli alla terra: scelte economiche e attività pubbliche di una famiglia nobile Bolognese nell'Ottocento. Bologna, Mulino, 2000, 434 p. (Collana di storia dell'economia e del credito, 8).

6977. NARO (Nancy Priscilla). A slave's place, a master's world: fashioning dependency in rural Brazil. New York, Continuum, 2000, XI-212 p. (Black Atlantic).

6978. PAZZAGLI (Carlo). La mezzadria senese in età moderna. *Archivio storico italiano*, 2000, 158, 586, p. 751-786.

6979. Politická a stavovská zemědělská hnutí ve 20. století. Sborník příspěvků. (Politische und Standeslandwirtschaftliche Bewegungen im 20ten Jahrhundert). Ed. Blanka RAŠTICOVÁ. Uherské Hradiště, Slovácké muzeum, 2000, 319 p. (Studie Slováckého muzea, 5).

6980. SCHÖNE (Jens). Landwirtschaftliches Genossenschaftswesen und Agrarpolitik in der SBZ/DDR 1945–1950/51. Stuttgart, [s. n.], 2000, 97 p.

6981. SCOTT JENKINS (Virginia). Bananas: an American history. Washington, Smithsonian Institution, 2000, XIII-210 p.

6982. Small town and rural economic development: a case studies approach. Ed. by Peter V. SCHAEFFER and Scott LOVERIDGE. Westport, Praeger Publishers, 2000, XVI-293 p.

6983. SPUFFORD (Margaret). Figures in the landscape: rural society in England, 1500–1700. Aldershot, Ashgate, 2000, X-364 p.

6984. TURNER (Michael). Corporate strategy or individual priority? Land management, income and tenure on Oxbridge agricultural land in the mid-nineteenth century. *Business history*, 2000, 42, 4, p. 1-26.

6985. VAN CRUYNINGEN (P. J.). Bekous maar buigzaam. Boeren in West-Zeeuws-Vlaanderen 1650–1850. Wageningen, Afdeling Agrarische Geschiedenis Landboux Universiteit, 2000, 486 p. (A. A. G., 40).

6986. WALKER (Melissa). All we knew was to farm: rural women in the upcountry South, 1919–1941. Baltimore, Johns Hopkins U. P., 2000, XVII-341 p. (Revisited Rural America).

6987. WHITED (Tamara L.). Forests and peasant politics in modern France. New Haven a. London, Yale U. P., 2000, XII-274 p. (Yale agrarian studies series).

6988. WHITTLE (Jane). The development of agrarian capitalism: land and labour in Norfolk, 1440–1580. Oxford, Oxford U. P., 2000, XII-361 p.

6989. Zemědělské družstevnictví. Kolektivizace zemědělství – vznik JZD 1952. [Agricultural cooperatives. The collectivization of agriculture and the establishment of the Standard Farming Co-operative (JZD), 1952]. Ed. Jana PŠENIČKOVÁ. Praha, Státní ústřední archiv, 2000, 322 p. (Edice dokumentů z fondů Státního ústředního archivu v Praze).

Cf. n°s 5616, 7180, 7249

§ 6. Money and finance.

6990. BALZANI (Roberto). Il forziere della città. La Cassa dei risparmi e la società forlivese dalle origini al secondo dopoguerra. Bologna, Il Mulino, 2000, 300 p.

6991. Banche e reti di banche nell'Italia postunitaria. A cura di Giuseppe CONTI e Salvatore LA FRANCESCA. Bologna, Il Mulino, 2000, 2 vol., 872 p. (Collana di storia dell'economia e del credito, 9).

6992. Banques et entreprises industrielles en Europe de l'Ouest, XIXe–XXe siècles: aspects nationaux et régionaux. Ed. par Philippe MARGUERAT, Laurent TISSOT et Yves FROIDEVAUX. Genève, Droz, 2000, 270 p.

6993. BODENHORN (Howard). A history of banking in antebellum America. Financial markets and economic development in an nation-building. Cambridge, Cambridge U. P., 2000, 260 p.

6994. BONELLI (Franco), CERRITO (Elio). L'emergere di una funzione pubblica di controllo monetario. La Banca d'Italia dal 1894 al 1913 (II). *Rivista di storia economica*, 2000, 16, 1, p. 65-110.

6995. BRETON (Stéphane). Le monde de la dette. *In*: La monnaie des sciences sociales. *Annales*, 2000, 55, 6, p. 1361-1366.

6996. CARDARELLI (S.), MARTANO (R.). I nazisti e l'oro della Banca d'Italia. Sottrazione e recupero, 1943–1958. Roma e Bari, Laterza, 2000, 186 p.

6997. CESARANO (Filippo). Gli accordi di Bretton Woods. La costruzione di un ordine monetario internazionale. Presentazione di Marcello DE CECCO. Roma e Bari, Laterza, 2000, 166 p.

6998. Contributi-Ricerche per la storia della Banca d'Italia. Vol. 7. Stabilità e sviluppo negli anni Cinquanta. 1. L'Italia nel contesto internazionale. A cura di Franco COTULA. Roma e Bari, Laterza, 2000, 518 p.

6999. COWEN (David J.). The first bank of the United States and the securities market crash of 1792. *Journal of economic history*, 2000, 60, 4, p. 1041-1060. – IDEM. The origins and economic impact of the first bank of the United States, 1791–1797. New York, Garland Publishing, 2000, XXIX-323 p.

7000. Currency crises. Ed. by Paul KRUGMAN. Chicago, University of Chicago Press, 2000, 356 p.

7001. DE SANTIAGO FERNANDEZ (J.). Política monetaria en Castilla durante el siglo XVII. Valladolid, Junta de Castilla y León, 2000, 298 p.

7002. Dějin (Z) českého bankovnictví v 19. a 20. století. (On the history of the Czech banking system in the 19th and 20th centuries). Ed. Praha, Karolinum, 2000, 209 p. (Acta Universitatis Carolinae. Philosophica et Historica. Studia historica, 47).

7003. DESZERI (K.). Introducting currency convertibility in Western Europe in the 1950s. *Journal of European economic history*, 2000, 1, p. 131-172.

7004. Development (The) of London as a financial centre. Ed. by R. C. MICHIE. London, I. B. Tauris, 4 vol., XXXVII-298 p., X-356 p., X-262 p., XIII-433 p.

7005. EINAUDI (Luca). From the franc to the 'Europe': the attempted transformation of the Latin Monetary Union into a European Monetary Union, 1865–1873. *Economic history review*, 2000, 53, 2, p. 284-298.

7006. ELDEM (Edhem). Osmanlı Bankası Tarihi. (The history of the Otoman Bank). İstanbul, Tarih Vakfı, 2000, 587 p.

7007. ERMER (Matthias). Von der Reichsmark zur Deutschen Mark der Deutschen Notenbank: zum Binnenwährungsumtausch in der Sowjetischen Besatzungszone Deutschlands (Juni/Juli 1948). Stuttgart, In Kommission bei Steiner, 2000, 222 p. (ill.). (Beiträge zur Wirtschafts- und Sozialgeschichte, 91).

7008. FERGUSON (Niall), GRANVILLE (Brigitte). "Weimar on the Volga": causes and consequences of inflation in 1990s Russia compared with 1920s Germany. *Journal of economic history*, 2000, 60, 4, p. 1061-1087.

7009. GREGORY (Paul R.), TIKHONOV (Aleksei). Central planning and unintended consequences: creating the Soviet financial system, 1930–1939. *Journal of economic history*, 2000, 60, 4, p. 1017-1040.

7010. GRENIER (Jean-Yves). Penser la monnaie autrement (note critique). *In*: La monnaie des sciences sociales. *Annales*, 2000, 55, 6, p. 1335-1342.

7011. HALLON (Ľudovít). Expanzia slovenského finančného kapitálu v účastninných podnikoch v rokoch 1939–1944. (Der Expansion des slowakischen Finanzkapitals in den Aktiengesellschaften in den Jahren 1939–1944). *Historický časopis*, 2000, 48, 1, p. 75-98. [Deutsche Zfassung].

7012. High politics and low finance. Symposium. *Journal of economic history*, 2000, 60, 2, p. 414-496. [Contents: NORTH (Douglass C.), WEINGAST (Barry R.). Introduction: institutional analysis and economic history (p. 414-417). – WELLS (John), WILLS (Douglas). Revolution, restoration, and debt repudiation: the Jacobite threat to England's institutions and economic growth (p. 418-441). – SUSSMAN (Nathan), YAFEH (Yishay). Institutions, reforms, and country risk: lessons from Japanese government debt in the Meiji era (p. 442-467). – FREY (Bruno S.), KUCHER (Marcel). History as reflected in capital markets: the case of World War II (p. 468-496].

7013. HOFFMAN (Philip T.), POSTEL-VINAY (Gilles), ROSENTHAL (Jean-Laurent). Priceless markets: the political economy of credit in Paris, 1660–1870. Chicago, University of Chicago Press, 2000, X-350 p.

7014. İhsaiyat-ı maliye: maliye istatistikleri 1885–1909. (Financial statistics 1885–1909). Ankara, Maliye Bakanlığı Araştırma, Planlama ve Koordinasyon Kurulu Başkanlığı, 2000, 450 p.

7015. JOHNMAN (Lewis), MURPHY (Hugh). Scotland, 'A Dead Loss'. The industrial and commercial finance corporation and the Scottish banks, 1945–1965. *Scottish economic and social history*, 2000, 20, 2, p.153-175.

7016. KASERER (Christoph). Der Fall der Herstatt-Bank 25 Jahre danach. Überlegungen zur Rationalität regulierungspolitischer Reaktionen unter besonderer Berücksichtigung der Einlagensicherung. *Vierteljahrschrift für Sozial- und Wirtschaftsgeschichte*, 2000, 87, p. 166-193.

7017. KAUFHOLD (Karl Heinrich). Hauptstadtfinanzierung – ein (weithin) neues Forschungsgebiet. *Vierteljahrschrift für Sozial- und Wirtschaftsgeschichte*, 2000, 87, p. 468-476.

7018. KWASS (Michael). Privilege and the politics of taxation in eighteenth-century France: liberté, égalité, fiscalité. Cambridge a. New York, Cambridge U. P., 2000, XVII-353 p. (ill.).

7019. LAZAREVIČ (Žarko), PRINČIČ (Jože). Zgodovina slovenskega bančništva. (The history of Slovene banking). Ljubljana, ZBS – Združenje bank Slovenije, 2000, 438 p.

7020. LORDON (Frédéric). La légitimité au regard du fait monétaire. *In*: La monnaie des sciences sociales. *Annales*, 2000, 55, 6, p. 1343-1360.

7021. Mercado (El) del tabaco en España durante el siglo XVIII: fiscalidad y consumo. Ed. por Santiago DE LUXÁN MELÉNDEZ, Sergio SOLBES FERRI y Juan José LAFORET. Las Palmas, Fundación Altadis, Universidad de Las Palmas de Gran Canaria y Real Sociedad Económica de Amigos del País de Gran Canaria, 2000, VI-311 p.

7022. NOTERMANS (Ton). Money, markets, and the state: social democratic economic policies since 1918. Cambridge, Cambridge U. P., 2000, XIX-302 p.

7023. Osmanlı bütçeleri 1909–1918. (Ottoman budgets 1909–1918). Ankara, Maliye Bakanlığı Araştırma, Planlama ve Koordinasyon Kurulu Başkanlığı, 2000, 601 p.

7024. Osmanlı Maliyesi Hakkında İngiliz Raporları (1861–1892). (British reports on Ottoman finance 1861–1892). Ed. by Nezih VARCAN. Ankara, Maliye Bakanlığı Araştırma, Planlama ve Koordinasyon Kurulu Başkanlığı, 2000, 185 p.

7025. PAMUK (Şevket). A monetary history of the Ottoman empire. Cambridge, Cambridge U. P., 2000, XXVI-276 p.

7026. POITRAS (Geoffrey). The early history of financial economics, 1478–1776. Cheltenham and Northampton, 2000, X-522 p.

7027. POTTER (Mark). Good offices: intermediation by corporate bodies in early modern French public finance. *Journal of economic history*, 2000, 60, 3, p. 599-626.

7028. PRACHOWNY (Martin F. J.). The Kennedy-Johnson tax cut: a revisionist history. Cheltenham a. Nothampton, Edward Elgar, 2000, IX-227 p.

7029. REDISH (Angela). Bimetallism: an economic and historical analysis. Cambridge, Cambridge U. P., 2000, XII-276 p.

7030. REITMAYER (Morten). Bankiers im Kaiserreich: Sozialprofil und Habitus der Deutschen Hochfinanz. Göttingen, Vandenhoeck und Ruprecht, 2000, 428 p. (Kritische Studien zur Geschichtswissenschaft, 136).

7031. ROSELLI (A.). Il governatore Vincenzo Azzolini, 1931–1944. Roma e Bari, Laterza, 2000, XV-380 p.

7032. SCATAMACCHIA (Rosanna). Istituzioni finanziarie e prospettive di ricerca: la Banca d'Italia e le donne. *Quaderni storici*, 2000, 35, 103, p. 163-200.

7033. SEGRETO (Luciano). Finanza, industria e relazioni internazionali nella Ricostruzione. Il prestito dell'Eximbank all'Italia. *Passato e presente*, 2000, 51, p. 67-93.

7034. Staatsfinanzen – Staatsverschuldung – Staatsbankrotte in der europäischen Staaten- und Rechtsgeschichte. Hrsg. v. Gerhard LINGELBACH. Köln, Weimar u. Wien, Böhlau, 2000, X-386 p.

7035. Storia della Cassa Depositi e Prestiti. A cura di Marcello DE CECCO e Gianni TONIOLO. Presentazione di Maria Teresa SALVEMINI. Roma e Bari, Laterza, 2000, 452 p.

7036. TRINER (Gail D.). Banking and economic development: Brazil, 1189–1930. New York, Palgrave, 2000, XV-333 p.

7037. WEICHEL (Thomas). Gontard & MetallBank. Die Banken der Frankfurter Familien Gontard und Merton. Stuttgart, Thorbecke, 2000, 269 p.

7038. WILSON (Ted). Battles for the standard: bimetallism and the spread of the gold standard in the nineteenth century. Aldershot, Ashgate, 2000, XI-200 p.

7039. WORMELL (Jeremy). The management of the national debt of the United Kingdom, 1900–1932. London a. New York, Routledge, 2000, XXX-805 p.

7040. YETKİN (Sabri). İzmir Esnaf ve Ahali Bankası'ndan Egebank'a (1928–2000). (From İzmir craftsmen and community bank to Egebank). İstanbul, Egebank, 2000, 256 p.

§ 7. Demography and urban history.

7041. ALVAREZ MORA (Alfonso). Centro e periferia nella formazione della città moderna: Roma e Madrid a confronto. *Storia urbana*, 2000, 24, 92, p. 143-168.

7042. BACKOUCHE (Isabelle). La trace du fleuve. La Seine et Paris (1750–1850). Paris, Ed. de l'EHESS, 2000, 430 p.

7043. BAXANDALL (Rosalyn), EWEN (Elizabeth). Picture windows: how the suburbs happened. New York, Basic Books, 2000, XXII-298 p.

7044. Biographieforschung und Stadtgeschichte. Lemgo in der Spätphase der Hexenverfolgung. Hrsg. v. Gisela WILBERTZ und Jürgen SCHEFFLER. Bielefeld, Verlag für Regionalgeschichte, 2000, 486 p. (Studien zur Regionalgeschichte, 13. Beiträge zur Geschichte der Stadt Lemgo, 5).

7045. BOLOVAN (Ioan). Transilvania între Revoluția de la 1848 și Unirea din 1918. Contribuții demografice. Cluj-Napoca, CST, 2000, XXXI-267 p.

7046. BORSAY (Peter). The image of Georgian Bath, 1700–2000: towns, heritage, and history. New York, Oxford U. P., 2000, XI-434 p.

7047. BOURDIEU (Jérôme), POSTEL-VINAY (Gilles), ROSENTAL (Paul-André), SUWA-EISENMANN (Akiko). Migrations et transmissions inter-générationnelles dans la France du XIXe et du début du XXe siècle. *In*: Migrations, réseaux, patrimoine: renouveler les perspectives. *Annales*, 2000, 55, 4, p. 749-790.

7048. BURG (Peter). Saarbrücken 1789–1860. Von der Residenzstadt zum Industrienzentrum. Blieskastel, Gollenstein, 2000, 512 p. (Saarland-Bibliothek, 14).

7049. Cambridge urban history (The) of Britain. Vol. 2. 1540–1840. Ed. by Peter CLARK. Vol. 3. 1840–1950. Ed. by Martin DAUNTON. Cambridge, Cambridge U. P., 2000, 2 vol., 906 p., XXVI-944 p.

7050. COHN (Raymond L.). Nativism and the end of the mass migration of the 1840s and 1850s. *Journal of economic history*, 2000, 60, 2, p. 361-383.

7051. COLE (Joshua). The power of large numbers: population, politics, and gender in nineteenth-century France. Ithaca, Cornell U. P., 2000, XI-252 p.

7052. DARQUES (Régis). Salonique au XXe siècle. De la cité ottomane à la métropole grecque. Paris, CNRS Editions, 2000, 390 p.

7053. DAVIS (Mike). Magical urbanism: latinos reinvent the U. S. city. London a. New York, Verso, 2000, XVIII-172 p.

7054. DE GRAZIA (Victoria). Die Radikalisierung der Bevölkerungspolitik im faschistischen Italien: Mussolinis "Rassenstaat". *Geschichte und Gesellschaft*, 2000, 26, 2, p. 219-254.

7055. DEL RIO BARREDO (María José). Madrid, Urbs Regia: la capital ceremonial de la Monarquia Catolica. Prol. de Peter BURKE. Madrid, Marcial Pons, 2000, VIII-258 p.

7056. DOYLE (Shane). Population decline and delayed recovery in Bunyoro, 1860–1960. *Journal of African history*, 2000, 41, 3, p. 429-458.

7057. FOLIN (Marco). Sui criteri di classificazione degli insediamenti urbani nell'Italia centro-settentrionale, secoli XIV–XVIII. *Storia urbana*, 2000, 24, 92, p. 5-24.

7058. FRIEDRICHS (Christopher R.). Urban politics in early modern Europe. London a. New York, Routledge, 2000, XIV-87 p.

7059. FURUSE (N.), KAMIKAWA (M.), OHTOMO (K.), SUMITOMO (A.). Theme: for the creation of discussion on the metropolis. *Journal of Japanese history*, 2000, 458, p. 4-22.

7060. GALLMAN (J. Matthew). Receiving Erin's children: Philadelphia, Liverpool, and the Irish famine migration, 1845–1855. Chapell Hill, University of North Carolina Press, 2000, XII-306 p.

7061. GARRETSON (Peter P.). A history of Addis Abäba from its foundation in 1886 to 1910. Wiesbaden, Harrassowitz, 2000, XXI-226 p. (Äthiopistische Forschungen, 49).

7062. GEHRMANN (Rolf). Bevölkerungsgeschichte Norddeutschlands zwischen Aufklärung und Vormärz. Berlin, Berlin Verlag A. Spitz, 2000, 499 p. (Schriftenreihe des Forschungsinstituts für die Geschichte Preußens, 1).

7063. HOPPE (Bert). Auf den Trümmern von Königsberg. Kaliningrad 1946–1970. München, Oldenbourg, 2000, 166 p. (Schriftenreihe der Vierteljahrshefte für Zeitgeschichte, 80).

7064. İPEK (Nedim). Mübadele ve Samsun. (Population exchange [between Greece and Turkey after the Lausanne Treaty] and Samsun) Ankara, Türk Tarih Kurumu, 2000, 194 p.

7065. JACOBSON (Charles David). Ties that bind: economic and political dilemmas of urban utility net-

7. DEMOGRAPHY AND URBAN HISTORY

works, 1800–1990. Pittsburgh, University of Pittsburgh Press, 2000, XI-282 p.

7066. JOHN (Michael). Bevölkerung in der Stadt: "Einheimische" und "Fremde" in Linz (19. und 20. Jahrhundert). Linz, Archiv der Stadt Linz, 2000 568 p. (ill.). (Linzer Forschungen, 7).

7067. Kassel im 18. Jahrhundert. Residenz und Stadt. Hrsg. v. Heide WUNDER, Christina VANJA und Karl-Hermann WEGNER. Kassel, Euregio, 2000, 384 p. (Kasseler Semesterbücher, Studia Casselana, 10).

7068. KLEINFELD (Martin). Die wirtschaftliche Entwicklung der Stadt Lauenburg/Elbe vom 18. bis zum 20. Jahrhundert. Hamburg, Kovač, 2000, 498 p. (Studien zur Geschichtsforschung der Neuzeit, 17).

7069. KAOCH (Lene). Tvangssterilisation i Danmark 1929–1967. (Les stérilisations forcées au Danemark de 1929 à 1967). København, Gyldendal, 2000, 414 p.

7070. KOZINA (Valerija V.). Naselenie Tsentral'nogo Kazakhstana. (The population of Central Kazakhstan). Vol. 1. Konets XIX–30-e gody XX v. (From the late 19th century to the 1930s.). In-t istorii i etnologii im. Ch. Ch. Valikhanova. Almaaty, Orkeniet, 2000, 143 p.

7071. Krakau, Prag und Wien: Funktionen von Metropolen im frühmodernen Staat. Hrsg. v. Marina DMITRIEVA und Karen LAMBRECHT. Stuttgart, Steiner, 2000, 432 p. (Forschungen zur Geschichte und Kultur des östlichen Mitteleuropa, 10).

7072. LINDBERG (Kirsten). Sirenernes stad. By- og bygningshistorie for 1728. Bd 5, 6 (Copenhague, la ville des sirènes. Histoire de la ville et de son urbanisme avant 1728. Tomes 5, 6). Ebeltoft, Skipperhoved, 590 p., 385 p. (ill.). [English summary].

7073. LITTLE (J. I.). From the Isle of Arran to inverness township: a case study of Highland emigration and North American settlement, 1829–1934. *Scottish economic and social history*, 2000, 20, 1, p. 3-30.

7074. LOVELL (William George), LUTZ (Christopher H.). Demografía e imperio: guía para la historia de la población de la América Central española, 1500–1821. [S. l.], Universidad de San Carlos de Guatemala, Editorial Universitaria, 2000, XVII-258 p. (Colección Monografías).

7075. MAINARDI (Roberto). Lo spazio europeo in un contesto mondiale: appunti e definizioni. *Storia urbana*, 2000, 24, 93, p. 115-132.

7076. Malthus, medicine and morality: Malthusianism after 1798. Ed. by Brian DOLAN. Amsterdam a. Atlanta, Rodopi, 2000, 232 p.

7077. Material London, ca. 1600. Ed. by Lena C. ORLIN. Phiadelphia, University of Pennsylvania Press, 2000, X-393 p.

7078. MATTL (Siegfried). Wien im 20. Jahrhundert. Wien, Pichler, 2000, 192 p. (ill.). (Geschichte Wiens, 6).

7079. MELOSI (Martin C.). The sanitary city: urban infrastructure in America from colonial times to the present. Baltimore, Johns Hopkins U. P., 2000, XII-578 p. (Creating the North American landscape).

7080. Metropolie Europy Środkowo-Wschodniej w XV i XVI wieku. (Die Metropolen des Mittel-Ost-Europas im XV. und XVI. Jahrhundert). Hrsg. v. Leszek BELZYT und Jan PIROŻYŃSKI. Kraków, PAU, 2000, 141 p. (Prace Komisji Środkowoeuropejskiej, 8).

7081. MISKELL (Louise), KENEFICK (William). 'A Flourishing Seaport': Dundee harbour and the making of the industrial town, c.1815–1850. *Scottish economic and social history*, 2000, 20, 2, p. 176-198.

7082. NAQUIN (Susan). Peking: temples and city life, 1400–1900. Berkeley a. Los Angeles, University of California Press, 2000, XXXIV-816 p.

7083. Naselenie Rossii v XX veke: Istoricheskie ocherki. (The population of Russia in the 20th century: historical essays). Vol. 1. 1900–1939. Ed. Ju. A. POLJAKOV; V. B. ZHIROMSKAJA. RAN, Nauchnyj sovet po istoricheskoj demografii i istoricheskoj geografii; In-t rossijskoj istorii. Moskva, ROSSPEN, 2000, 462 p.

7084. NEAD (Lynda). Victorian Babylon: people, street and images in Nineteenth century London. New Haven a. London, Yale U. P., 2000, 251 p. (ill.)

7085. Nouvelles approches de la documentation notarile et histoire urbaine. Le cas italien (XVIIe–XIXe siècle). Table ronde organisée par l'Ecole française de Rome, l'Università degli studi di Milano et le Centre Roland Mousnier. *Mélanges de l'Ecole française de Rome. Italie et Mediterranée*, 2000, 112, 1, 504 p.

7086. OUTTES (Joel). Disciplining society through the city? The birth of urbanismo (city planning) in Brazil (1916–1941). Oxford, [s. n.], 2000, XXXVII-517 p. [Thesis (D. Phil.). University of Oxford. Faculty of Anthropology and Geography].

7087. PÉROTIN-DUMON (Anne). La ville aux îles. La ville dans l'île. Basse-Terre et Pointe-à-Pitre, Guadeloupe, 1650–1820. Paris, Karthala, 2000, 990 p.

7088. Population history (A) of North America. Ed. by Michael R. HAINES and Richard H. STECKEL. Cambridge, Cambridge U. P., 2000, 736 p.

7089. PROKOP (Radim). K populačním přeměnám v české části Těšínska v meziválečném období a po roce 1945. (On population transformations in the Czech part of Těšín in the interwar period and after the year 1945). *Slezský sborník*, 2000, 98, 1/2, p. 46-62.

7090. Regimi dei suoli in transizione sulle coste del Pacifico: Hong Kong, Shanghai, Hanoi, Manila. *Storia urbana*, 2000, 24, 91, p. 5-180.

7091. REIS FILHO (Nestor Goulart), PICCOLOTTO SIQUEIRA BUENO (Beatriz), BRUNA (Paulo Julio Valentino). Imagens de vilas e ciudades do Brasil colonial. Sao Paulo, Editora da Universidade de Sao Paulo, 2000, 411 p. (ill.).

7092. Remaking the Chinese city: modernity and national identity, 1900–1950. Ed. by Joseph W. ESHERICK. Honolulu, University of Hawai'i Press, 2000, X-278 p.

7093. SÁNCHEZ-ALONSO (Blanca). European emigration in the late nineteenth century: the paradoxical case of Spain. *Economic history review*, 2000, 53, 2, p. 309-330.

7094. SCOTT (Peter). The state, internal migration, and the growth of new industrial communities in interwar Britain. *English historical review*, 2000, 115, 461, p. 329-353.

7095. STAPLETON (Kristin). Civilizing Chengdu: Chinese urban reform, 1895–1937. Cambridge, Harvard U. P., 2000, XII-341 p. (Harvard east monographs, 186).

7096. STEINWASCHER (Gerd). Osnabrück und der Westfälische Frieden. Die Geschichte der Verhandlungsstadt 1641–1650. Osnabrück, Selbstverlag des Vereins, 2000, X-416 p. (Osnabrücker Geschichtsquellen und Forschungen, 42).

7097. Storia di Torino. 6. La città nel Risorgimento (1798–1864). A cura di Umberto LEVRA. Torino, Einaudi, 2000, CLX-896 p.

7098. SWIERENGA (Robert P.). Faith and family: Dutch immigration and settlement in the United States, 1820–1920. New York, Holmes and Meyer, 2000, XX-362 p. (Ellis Island series).

7099. Tenements (From) to the Taylor homes: in search of an urban housing policy in twentieth-century America. Ed. by John F. BAUMAN, Roger BILES and Kristin SZYLVIAN. University Park, Pennsylvania State U. P., 2000, XIII-288 p.

7100. TIPPACH (Thomas). Koblenz als preussische Garnison- und Festungsstadt. Wirtschaft, Infrastruktur und Städtebau. Köln, Weimar u. Wien, Böhlau, 2000, VI-336 p. (Städteforschung, Rh. A: Darstellungen, 53).

7101. Urban fortunes: property and inheritance in the town, 1700–1900. Ed. by Jon STOBART and Alastair OWENS. Aldershot, Ashgate, 2000, XIV-241 p.

7102. VAN DER WOUDE (Ad), VAN POPPEL (Frans). De rol van de migratie in het demografisch systeem van de Republiek. *In*: VAN DER WOUDE (A.). Leven met geschiedenis. Theorie, praktijk en toepassing van historische kennis. Amsterdam, Uitgeverij Balans, 2000, p. 225-246; p. 454-455.

7103. Venezuela: città e territorio in una prospettiva storica. *Storia urbana*, 2000, 24, 90, p. 5-98.

7104. WALSH (Brendan M.). Urbanization and the regional distribution of population in post-famine Ireland. *Journal of European economic history*, 2000, 1, p. 109-130.

7105. WEIL (François). Histoire de New York. Paris, Fayard, 2000, 377 p.

7106. WERMEIL (Sara E.). The fireproof building: technology and public safety in the nineteenth-century American city. Baltimore, Johns Hopkins U. P., 2000, VIII-301 p. (Studies in Industry and Society).

7107. WHITING YOUNG (Biloine), FOWLER (Melvin L.). Cahokia: the great native American metropolis. Urbana a. Chicago, University of Illinois Press, 2000, XI-366 p.

7108. WOODS (Robert). The demography of Victorian England and Wales. Cambridge, Cambridge U. P., 2000, XXV-447 p. (Cambridge studies in population, economy and society in past time, 35).

7109. WRIGHT (R.). Historical underdosing: pop demography and the crisis in Canadian history *Canadian historical review*, 2000, 81, 4, p. 646-667.

7110. XU (Yinong). The Chinese city in space and time: the development of urban form in Suzhou. Honolulu, Universiti of Hawai'i Press, 2000, X-361 p.

7111. YEATES (Pádraig). Lockout: Dublin 1913. New York, Palgrave, 2000, XXX-670 p.

Cf. n°s 787, 5737, 7242

§ 8. Social history.

* 7112. Studentenproteste (Die) der 60er Jahre. Archivführer, Chronik, Bibliographie. Hrsg. v. Thomas P. BECKER und Ute NEUMANN. Köln, Böhlau, 2000, 381 p.

* 7113. TOPIK (Steven C.). Coffee anyone? Recent research on Latin American coffee societies. *Hispanic American historical review*, 2000, 80, 2, p. 225-266.

7114. ÅBERG (Alf). Kvinnorna i Nya Sverige. (Les femmes dans la colonie suédoise de la Nouvelle Suède). Stockholm, Natur och Kultur, 179 p. (ill.).

7115. Adel und Bürgertum in Deutschland. Entwicklungslinien und Wendepunkte im 19. Jahrhundert. Hrsg. v. Heinz REIF. Berlin, Akademie, 2000, 355 p. (Elitenwandel in der Moderne, 1).

7116. ALLYN (David). Make love not war: the sexual revolution; an unfettered history. New York, Little, Brown, 2000, XI-381 p.

7117. Alter (Das) im Spiel der Generationen. Historische und Sozialwissenschaftliche Beiträge. Hrsg. v. Josef EHMER und Peter GUTSCHNER. Köln, Weimar u. Wien, Böhlau, 2000, 437 p.

7118. Amministrazione, formazione e professione: gli ingegneri in Italia tra Sette e Ottocento. A cura di L. BLANCO. Bologna, Il Mulino, 2000, 540 p.

7119. ANDERSON (Bonnie S.). Joyous greetings: the first international women's movement, 1830–1860. Oxford, Oxford U. P., 2000, XII-288 p.

7120. ANDERSON (Claire). Convicts in the Indian Ocean: transportation from South Asia to Mauritius, 1815–1853. New York, St. Martin's Press, 2000, XII-192 p.

7121. ARCHER (John E.). Social unrest and popular protest in England, 1780–1840. Cambridge, Cambridge U. P., 2000, VI-110 p.

7122. ASSAYAG (Jackie). En quête de classe moyenne en Inde. Grandeur, recomposition, forfaiture. *In*: Catégories sociales de l'espace politique. *Annales*, 2000, 55, 6, p. 1229-1254.

7123. BARBAGLI (Marzio). Sotto lo stesso tetto: mutamenti della famiglia in Italia dal XV. al XX. secolo. Bologna, Il Mulino, 2000, 533 p. (Biblioteca).

7124. BARBAZZA (Marie-Catherine). La société paysanne en Nouvelle-Castille: famille, mariage et transmission des biens à Pozuelo de Aravaca (1580–1640). Madrid, Casa de Velázquez, 2000, XVII-361 p. (Bibliothèque de la Casa de Velázquez, 15).

7125. BAY (Mia). The white image in the black mind: African-American ideas about white people, 1830–1925. Oxford, Oxford U. P., 2000, VIII-288 p.

7126. BENNINGHAUS (Christina). Mothers' toil and daughters' leisure: working-class girls and time in 1920s Germany. *History workshop*, 2000, 50, p. 45-72.

7127. BEREBITSKY (Julie). Like our very own: adoption and the changing culture of motherhood, 1851–1950. Lawrence, University Press of Kansas, 2000, VIII-248 p.

7128. BERRY (Chad). Southern migrants, northern exiles. Urbana a. Chicago, University of Illinois Press, 2000, XIII-236 p.

7129. BERSELLI (Edmondo). Canzoni. Storie dell'Italia leggera. Bologna, Il Mulino, 2000, 186 p.

7130. BERTOLOTTO (Sabrina), CEVASCO (Roberta). Fonti osservazionali e fonti testuali. Le «consegne dei boschi» e il sistema dell'«alnocoltura» nell'Appennino ligure orientale (1822). *In*: Pratiche del territorio [Cf. n° 7483], p. 87-108.

7131. Beyond kinship. Social and material reproduction in house societies. Ed. By Rosemary A. JOYCE and Susan D. GILLESPIE. Forew. by Clark E. CUNNINGHAM. Philadelphia, University of Pennsylvania Press, 2000, 269 p.

7132. BLAMONT (Jacques). Le lion et le moucheron. Histoire des marranes de Toulouse. Paris, O. Jacob, 2000, 464 p.

7133. BLANC-CHALÉARD (Marie-Claude). Les Italiens dans l'Est parisien: une histoire d'intégration (1880–1960). Roma, Ecole française de Rome, 2000, 803 p.

7134. BLICKLE (Peter). Kommunalismus. Skizzen einer gesellschaftlichen Organisationsform. Band 1. Oberdeutschland. Band 2. Europa. München, Oldenbourg, 2000, 2 vol., IX-196 p., IX-422 p.

7135. BØGE PEDERSEN (Merete). Den reglementerede prostitution i København fra 1874 til 1906: en undersøgelse af prostitutionsmiljøet og de prostitueredes livsvilkår. (La prostitution réglementaire à Copenhague entre 1874 et 1906: étude du milieu et des conditions de vie des prostituées). København, Museum Tusculanum, 2000, 191 p. (ill.).

7136. BOULTON (Jeremy). Food prices and the standard of living in London in the 'century of revolution', 1580–1700. *Economic history review*, 2000, 53, 3, p. 455-492.

7137. BOYER (Christopher R.). The threads of class at La Virgen: misrepresentation and identity at a Mexican textile mill, 1918–1935. *American historical review*, 2000, 105, 5, p. 1576-1598.

7138. BREWER (Priscilla J.). From fireplace to cookstove: technology and the domestic ideal in America. Syracuse, Syracuse U. P., 2000, XIX-338 p.

7139. BUDDE (Gunilla-Friederike). Der Körper der "sozialistischen Frauenpersönlichkeit". Weiblichkeits-Vorstellungen in der SBZ und frühen DDR. *Geschichte und Gesellschaft*, 2000, 26, 4, p. 602-628.

7140. Bulgarskoto obshtestvo XV–XVIII vek. (La société bulgare XVe–XVIIIe siècles). Sustavitel Mariia RADEVA. Sofiia, UI "Sv. Kl. Okhridski", 2000, 203 p.

7141. BUREŠOVÁ (Jana). Zemská organizace Pokrokových žen moravských jako představitelka a pokračovatelka liberalizačních a demokratizačních snah žen na Moravě za první ČSR. Pohled do dvacátých let 20. století. (The Land Organisation of Progressive Moravian Women as a representative and successor of liberal and democratic efforts of women in Moravia in the first Czechoslovak Republic. The view in the 1920's). *Acta Universitatis Palackianae Olomucensis, Fac. Philosophica. Historica 29 Sborník prací historických*, 2000, 17, p. 143-156.

7142. Bürgerliche Familien. Lebenswege im 19. und 20. Jahrhundert. Hrsg. v. Hannes STEKL. Köln, Weimar u. Wien, Böhlau, 2000, 298 p. (Bürgertum in der Habsburgermonarchie, 8).

7143. BUSH (M. L.). Servitude in modern times. Oxford, Blackwell, 2000, XII-292 p.

7144. Celem nauki jest człowiek ... Studia z historii społecznej ofiarowane Helenie Madurowicz-Urbańskiej. (Wissenschaft menschenorientiert ... Helena Madurowicz-Urbańska gewidmete Studien zur Gesellschaftsgeschichte). Hrsg. v. Piotr FRANASZEK. Kraków, Wydaw. Uniw. Jagiellońskiego, 2000, 403 p.

7145. CERNOVODEANU (Paul). The structure of the Wallachian nobility in the fifteenth-seventeenth centuries: Byzantine and Western Influences. *In*: Nouvelles Études d'Histoire. X. Publiées à l'occasion du XIXe Congrès International des Sciences Historiques. Oslo, 2000. București, Editura Academiei Române, 2000, p. 103-113.

7146. CHOJNACKI (Stanley). Women and men in Renaissance Venice. Twelve essays on Patrician society. Baltimore a. London. Johns Hopkins U. P., 2000, X-370 p.

7147. CONZE (Eckart). Von deutschen Adel. Die Grafen von Bernstorff im zwanzigsten Jahrhundert. Stuttgart u. München, Deutsche Verlags-Anstalt, 2000, 560 p.

7148. COOPER (Frederick), HOLT (Thomas C.), SCOTT (Rebecca J.). Beyond slavery. Explorations of race, labor, and citizenship in postemancipation societies. Chapel Hill a. London, University of North Carolina Press, 2000, XII-198 p.

7149. COTT (Nancy F.). Public vows: a history of marriage and the nation. Cambridge, Harvard U. P., 2000, V-297 p.

7150. CRANE (Diana). Fashion and its social agendas: class, gender, and identity in clothing. Chicago, University of Chicago Press, 2000, X-294 p.

7151. CRAWFORD (Patricia). Women's dreams in early Modern England. *History workshop*, 2000, 49, p. 129-141.

7152. DALGÅRD (Sune). Poul Laxmands sag: dyk i dansk historie omkring år 1500. (Le procès contre Poul Laxmand: plongée dans l'histoire du Danemark autour de 1500). København, Reitzel, 2000, 233 p. (Det Kongelige Danske Videnskabernes Selskab 79).

7153. DE ROSA (L.). The Balkan minorities (Slavs and Albanians) in South Italy. *Journal of European economic history*, 2000, 2-3, p. 249-270.

7154. DELATTRE (Simone). Les douzes heures noires. La nuit à Paris au XIXe siècle. Paris, Albin Michel, 2000, 679 p. (L'évolution de l'humanité).

7155. DRĂGAN (Ioan). Nobilimea românească din Transilvania între anii 1440–1514, Bucureşti, Editura Enciclopedică, 2000, 466 p.

7156. DUTTON (Paul V.). An overlooked source of social reform: family policy in French agriculture, 1936–1945. *Journal of modern history*, 2000, 72, 2, p. 375-412.

7157. EISENBERG (Christiane). Rival interpretations of football hooliganism: figurational sociology, social history and anthropology. *In*: Representations of emotional excess. Ed. by Jürgen SCHLAEGER. Tübingen, [s. n.], 2000, p. 297-306. (Yearbook of Research in English and American Literature, 16). – EADEM. Von England in die Welt: Entstehung und Verbreitung des modernen Fußballs. *In*: Über Fußball. Ein Lesebuch zur wichtigsten Nebensache der Welt. Hrsg. v. Wolfgang SCHLICHT und Werner LANG. Schorndorf, Hofmann Verlag, 2000, p. 59-84.

7158. EKSTRÖM (Anders). Dödens exempel. (L'Exemple de la mort. Regards sur le suicide au 19e siècle). Stockholm, Atlantis, 2000, 325 p.

7159. ERNST (Christoph). Den Wald entwickeln. Ein Politik- und Konfliktfeld in Hunsrück und Eifel im 18. Jahrhundert. München, Oldenbourg, 2000, VII-408 p. (Ancien Régime, Aufklärung und Revolution, 32).

7160. FARRY (Michael). The aftermath of revolution: Sligo 1921–1923. Dublin, University College Dublin Press, 2000, XV-270 p.

7161. FAUVE-CHAMOUX (Antoinette). Veuvages et remariage en France pre-industrielle. *In*: Femmes (Les) dans la société européenne [Cf. n° 7163], p. 281-302.

7162. FELDSTEIN (Ruth). Motherhood in black and white: race and sex in American liberalism 1930–1965. Ithaca, Cornell U. P., 2000, 241 p.

7163. Femmes (Les) dans la société européenne = Die Frauen in der europäischen Gesellschaft. 8e Congrès des Historiennes suisses = 8. Schweizerische Historikerinnentagung. Hrsg. v. Anne-Lise HEAD-KÖNIG und Liliane MOTTU-WEBER. Genève, Société d'histoire et archéologie de Genève, 2000, VIII-336 p. (ill.). [Cf. n° <sélection> 7161.]

7164. Femmes (Les) et la guerre. Intr. de Chantal ANTIER et Marianne WALLE. *Guerres mondiales et conflits contemporaine*, 2000, 50, 198, p. 7-150. [Cf. n^{os} <sélection> 4424, 7174, 7246, 7291.]

7165. FERNÁNDEZ PÉREZ (P.). Tolerance and endogamy: entrepreneurial strategies in XVIIIth century Spain. *Journal of European economic history*, 2000, 2-3, p. 271-294.

7166. FILIPPONE (Mario), GUASCONI (Giovanni B.), PUCCI (Silvio). Una signoria nella Toscana moderna. Il vescovado di Murlo (Siena) nelle carte del secolo XVIII. Siena, Università degli Studi di Siena, Dipartimento di Scienze Storiche, Giuridiche, Politiche e Sociali, 2000, XXXII-409 p.

7167. FLÜGEL (Axel). Bürgerliche Rittergüter: Sozialer Wandel und politische Reform in Kursachsen, 1680–1844. Göttingen, Vandenhoeck & Ruprecht, 2000, 304 p. (Bürgertum: Beiträge zur Gesellschaftsgeschichte, 16).

7168. FÖRHAMMAR (Staffan). Med känsla eller förnuft: svensk debatt om filantropi 1870–1914. (Bons sentiments ou bon sens: le débat sur la philanthropie en Suède entre 1870 et 1914). Stockholm, Almqvist & Wiksell International, 2000, 259 p. (Stockholm studies in History. 59).

7169. Frauen und dem Weg zur Elite. Hrsg. v. Günther SCHULZ. München, Oldenbourg, 2000, 220 p. (Deutsche Führungsschichten in der Neuzeit, 23. Büdinger Forschungen zur Sozialgeschichte, 1998).

7170. FRENCH (H. R.). Social status, localism and the 'Middle sort of people' in England 1620–1750. *Past and present*, 2000, 166, p. 66-99.

7171. FRIEDEBURG (Robert von). Welche Wegscheide in die Neuzeit? Widerstandsrecht, "Gemeiner Mann" und konfessioneller Landespatriotismus zwischen "Münster" und "Magdeburg". *Historische Zeitschrift*, 2000, 270, 3, p. 561-616.

7172. Fritz K. Ein deutsches Leben in zwanzigsten Jahrhundert. Hrsg. v. Hartmut BERGHOFF und Cornelia

RAUH-KÜHNE. Stuttgart u. München, Deutsche Verlags-Anstalt, 2000, 447 p.

7173. GALLANT (Thomas W.). Honor, masculinity, and ritual knife fighting in nineteenth-century Greece. *American historical review*, 2000, 105, 2, p. 359-382.

7174. GAUJAC (Paul). Des corps féminins aux AFAT. Afrique du Nord, 1943–1944. *In*: Femmes (Les) et la guerre [Cf. n° 7164], p. 109-122.

7175. GAZZINI (Marina). Patriziati urbani e spazi confraternali in età rinascimentale: l'esempio di Milano. *Archivio storico italiano*, 2000, 158, 585, p. 491-514.

7176. Gender and the southern body politic. Ed. by Nancy BERCAW. Jackson, University of Mississippi Press, 2000, XIX-259 p.

7177. GERSON (Stéphane). Town, nation, or humanity? Festive delineations of place and past in northern France, ca. 1825–1865. *Journal of modern history*, 2000, 72, 3, p. 628-682.

7178. GIESBERG (Judith Ann). Civil war sisterhood: the U. S. sanitary commission and women's politics in transition. Boston, Northeastern U. P., 2000, XIV-239 p.

7179. GLASSMAN (Jonathan). Sorting out the tribes: the creation of racial identities in Colonial Zanzibar's newspaper wars. *Journal of African history*, 2000, 41, 3, p. 395-428.

7180. GOLIKOVA (Svetlana V.), MINENKO (Nina A.), POBEREZHNIKOV (Igor' V.). Gornozavodskie tsentry i agrarnaja sreda v Rossii: vzaimodejstvie i protivorechija, XVIII–pervaja polovina XIX v. (Mining centers and their agrarian surroundings in Russia: interaction and contradictions, the 18th and the 1st half of the 19th century). RAN, Ural'skoe otd., In-t istorii i arkheologii. Moskva, Nauka, 2000, 261 p.

7181. GOODHEW (David). Working-class respectability: the example of the Western areas of Johannesburg, 1930–1955. *In*: Bad behaviour in urban South Africa [Cf. n° 5125], p. 241-266.

7182. GORSUCH (Anne E.). Youth in Revolutionary Russia: enthusiasts, bohemians, delinquents. Bloomington, Indiana U. P., 2000, X-274 p. (Indiana-Michigan series in Russian and East European studies).

7183. Greffe humaine (La). (In)certitudes éthiques: du don de soi à la tolérance de l'autre. Dir. par Robert CARVAIS et Marilyne SASPORTES. Paris, Presses Universitaires de France, 2000, 1000 p. (Science, histoire et société).

7184. Großbürger und Unternehmer. Die deutsche Wirtschaftselite im 20. Jahrhundert. Hrsg. v. Dieter ZIEGLER. Göttingen, Vandenhoeck & Ruprecht, 2000, 282 p. (Bürgertum. Beiträge zur europäischen Gesellschaftsgeschichte, 17).

7185. GRYTTEN (Ola Honningdal). Differences in the standards of living in inter-war Norway. A comparative analysis. *Scandinavian Economic History Review*, 2000, 48, 3, 22-41.

7186. GUILLAUME (Pierre). Les Hospices de Bordeaux au XIXe siècle (1796–1855). Bordeaux, Les Etudes hospitalières Editions, 2000, 277 p.

7187. HAAVE (Per). Sterilisering av tatere 1934–1977. En historisk undersøkelse av lov og Praksis. (The sterilization of gipsies during 1934–1977. An historical investigation of law and practice). Oslo, Norges forskningsråd, Området for kultur og samfunn, 2000, 427 p.

7188. HABERMAS (Rebekka). Frauen und Männer des Bürgertums. Eine Familiengeschichte (1750–1850). Göttingen, Vandenhoeck & Ruprecht, 2000, VIII-456 p. (Bürgertum, 14).

7189. HARDACH-PINKE (Irene). Bleichsucht und Blütenträume. Junge Mädchen 1750–1850. Frankfurt am Main u. New York, Campus, 2000, 238 p. (Geschichte und Geschlechter).

7190. HEDENBORG (Susanna). The world is full of sorrow: infant mortality in Stockholm, 1754–1850. *Scandinavian economic history review*, 2000, 48, 1, p. 64-80.

7191. Hidden histories of gender and the state in Latin America. Ed. by Elizabeth DORE and Maxine MOLYNEAUX. Durham, Duke U. P., 2000, XIII-381 p.

7192. HILTON (Matthew). Smoking in British popular culture 1800–2000: perfect pleasures. Manchester, Manchester U. P., 2000, XII-284 p. (Studies in popular culture).

7193. HINDLE (Steve). The state and society in early modern England, c. 1550–1640. London, Macmillan Press a. New York, St. Martin's Press, 2000, XI-338 p. (Early modern history: society and culture).

7194. HOERDER (Dirk). Metropolitan migration in the past: labour markets, commerce, and cultural interaction in Europe, 1600–1914. *Journal of international migration and integration*, 2000, 1, p. 39-58.

7195. HOFFMANN (Stefan-Ludwig). Die Politik der Geselligkeit. Freimaurerlogen in der deutschen Bürgergesellschaft 1840–1918. Göttingen, Vandenhoeck & Ruprecht, 2000, 425 p. (Kritische Studien zur Geschichtswissenschaft, 141).

7196. HOLMAN (Andrew C.). A sense of their duty: middle-class formation in Victorian Ontario towns. Montreal, McGill-Queen's U. P., 2000, XI-243 p.

7197. HOMMEN (Tanja). Körperdefinition und Körpererfahrung. "Notzucht" und "unzüchtige Handlungen an Kindern" im Kaiserreich. *Geschichte und Gesellschaft*, 2000, 26, 4, p. 577-601.

7198. HOPKINS (Eric). Industrialisation and society: a social history, 1830–1951. London a. New York, Routledge, 2000, X-308 p.

7199. Household strategies for survival 1600–1000: fission, faction and cooperation. Ed. by Laurence FON-

TAINE and Jürgen SCHLUMBOHM. Cambridge, Cambridge U. P., 2000, 196 p. (International review of social history, 8).

7200. HUNEFELDT (Christine). Liberalism in the bedroom: quarreling spouses in nineteenth-century Lima. University Park, Pennsylvania State U. P., 2000, XIX-388 p.

7201. HUSBAND (William B.). "Godless communists": atheism and society in Soviet Russia 1917–1932. DeKalb, Northern Illinois U. P., 2000, XVIII-241 p.

7202. ISENBERG (Andrew C.). The destruction of the bison: an environmental history, 1750–1920. Cambridge, Cambridge U. P., 2000, XII-206 p. (Studies in environment and history).

7203. JÄCKEL (Hartmut). Menschen in Berlin. Das letzte Telefonbuch der alten Reichshaupstadt 1941. Stuttgart u. München, DVA, 2000, 400 p.

7204. JAHAN (Sébastien). Profession, parenté, identité sociale. Les notaires de Poitiers aux temps modernes (1515–1815). Toulouse, Presses universitaires du Mirail, 2000, 384 p.

7205. JUPP (Peter), MAGENNIS (Eoin). Crowds in Ireland, c. 1720–1920. New York, St. Martin's Press, 2000, XII-277 p.

7206. KAHK (Juhan). Bauer und Baron im Baltikum. Versuch einer historisch-phänomenologischen Studie zum Thema "Gutsherrschaft in den Ostseeprovinzen". Tallin, von Wistinghausen, 2000, 199 p.

7207. KASCHKE (Lars). Nichts als "Bettelgelder"? Wert und Wertschätzung der Alters und Invalidenrenten im Kaiserreich. *Historische Zeitschrift*, 2000, 270, 2, p. 345-388.

7208. KASHANI-SABET (Firoozeh). Hallmarks of humanism: hygiene and love of homeland in Qajar Iran. *American historical review*, 2000, 105, 4, p. 1171-1203.

7209. KIRKLEY (Evelyn A.). Rational mothers and infidel gentlemen: gender and American atheism, 1865–1915. Syracuse, Syracuse U. P., 2000, XVIII-198 p. (Women and Gender in North American Religions).

7210. Kleinstadtbürgertum in der Habsburgermonarchie, 1861–1914. Hrsg. v. Peter URBANITSCH und Hannes STEKL. Köln, Weimar u. Wien, Böhlau, 2000, 516 p. (Bürgertum in der habsburgermonarchie, 9).

7211. KLING (Gudrun). Frauen im öffentlichen Dienst des Großherzogtums Baden. Von den Anfängen bis zum Ersten Weltkrieg. Stuttgart, Kohlhammer, 2000, XLI-250 p. (Veröffentlichungen der Kommission für geschichtliche Landeskunde Baden-Württemberg, Rh. B: Forschungen, 142).

7212. KOIVUNIEMI (Jussi). Tehtaan pillin tahdissa. Nokian tehdasyhdyskunnan sosiaalinen järjestys 1870–1939. (Social order in the factory community of Nokia, 1870–1939). Helsinki, SKS, 2000, 263 p. (ill., maps). (Bibiliotheca historica, 64).

7213. KÖNIG (Wolfgang). Bahnen und Berge. Verkehrstechnik, Tourismus und Naturschutz in den Schweizer Alpen 1870–1939. Frankfurt am Main, Campus, 2000, 242 p. (Deutsches Museum. Beiträge zur Historischen Verkehrsforschung, 2).

7214. KOPEČEK (Lubomír). Konfliktní linie slovenské společnosti v první polovině 20. století. (Cleavages of Slovak society in the first half of the 20th century). *Politologická revue*, 2000, 6, 1, p. 41-59.

7215. KRAUSMAN BEN-AMOS (Ilana). Gifts and favors: informal support in early modern England. *Journal of modern history*, 2000, 72, 2, p. 295-338.

7216. KUHN (Bärbel). Familienstand: ledig. Ehelose Frauen und Männer im Bürgertum (1850–1914). Köln, Weimar u. Wien, Böhlau, 2000, X-488 p. (L'homme, Schriften, 5).

7217. KURTOĞLU (Ayşenur). [et al.]. Osmanlı'dan Cumhuriyet'e kadının tarihi dönüşümü. (Transformation of women from the Ottomans to the republic). Ed. by Yıldız RAMAZANOĞLU. İstanbul, Pınar yayınları, 2000, 236 p.

7218. KÜTÜKOĞLU (Mübahat S.). XV ve XVI. asırlarda İzmir kazasının sosyal ve iktisâdî yapısı. (Social and economic structure of the İzmir District in the XVth and XVIth centuries). Ed. by Mustafa ÖZTURANLİ. İzmir, İzmir Büyükşehir Belediyesi, 2000, 326 p.

7219. LA MARCA (Nicola). La nobiltà romana e i suoi strumenti di perpetuazione del potere. Roma, Bulzoni, 2000, 3 vol., IX-1202 p.

7220. LAHIRI (Shompa). Indians in Britain: Anglo-Indian encounters, race and identity 1880–1930. Portland, Frank Cass, 2000, XVIII-249 p. (Colonial legacy in Britain, 1).

7221. LAKE (Marilyn). The ambiguities for feminists of national belonging: race and gender in the imagined Australian community. *In*: Gendered nations/nationalisms in the long 19th century Europe and beyond [Cf. n° 4064], p. 159-76.

7222. LASLETT (John H. M.). Colliers across the sea: a comparative study of class formation in Scotland and American Midwest, 1830–1924. Urbana a. Chicago, University of Illinois Press, 2000, XIV-314 p. (Working class in American history).

7223. LATHAM (Angela J.). Posing a threat: flappers, chorus girls, and other brazen performers of the American 1920s. Middletown, Wesleyan U. P. a. Hanover, University Press of New England, 2000, XI-203 p.

7224. LENEMAN (Leah). 'A natural foundation in equity': marriage and divorce in eighteenth and nineteenth-century Scotland. *Scottish economic and social history*, 2000, 20, 2, p. 199-215.

7225. LITWICKI (Ellen M.). America's public holidays 1865–1920. Washington, D.C., Smithsonian Institution, 2000, IX-293 p.

7226. Livre (Le) des vies coupables. Autobiographies de criminels (1896–1909). Ed. par Philippe ARTIÈRES. Paris, A. Michel, 2000, 427 p.

7227. LOEHLIN (Jennifer A.). From rugs to riches: housework, consumption and modernity in modern Germany. New York, Berg, 2000, IX-250 p.

7228. LUCASSEN (Leo). Sekse en nationaliteit als ordenend principe. De uitsluiting van vrouwen en vreemdelingen op de Nederlandse arbeidsmarkt (1904–1997). In: De grote lijn in de sociale geschiedenis van Nederland in de 20e eeuw. Ed. C. VAN EIJL, L. Heerma VAN VOSS en P. DE ROOY. Amsterdam, Spinhuis/Stichting Beheer IISG, 2000, p. 87-107. – IDEM. Zigeuner im frühneuzeitlichen Deutschland: neue Forschungsergebnisse, Probleme, und Vorschläge. In: Policey und frühneuzeitliche Gesellschaft [Cf. n° 7481], p. 235-262.

7229. LÜDTKE (Alf). People working: everyday life and German fascism. History workshop, 2000, 50, p. 74-92.

7230. MAC CROSSEN (Alexis). Holy day, holiday: the American Sunday. Ithaca, Cornell U. P., 2000, XIII-209 p.

7231. MAC ENANEY (Laura). Civil defense begins at home: militarization meets everyday life in the fifties. Princeton, Princeton U. P., 2000, X-213 p. (Politics and society in twentieth-century America).

7232. MAIFREDA (Germano). Gli ebrei e l'economia milanese. L'Ottocento. Milano, Franco Angeli, 2000, 336 p.

7233. MALCOMSON (Scott L.). One drop of blood: the American misadventure of race. New York, Farrar, Straus and Giroux, 2000, VIII-584 p.

7234. MARLING (Karal Ann). Merry Christmas! Celebrating America's greatest holiday. Cambridge, Harvard U. P., 2000, XIII-442 p.

7235. MÉTAYER (Christine). Au tombeau des secrets. Les écrivains publics du Paris populaire. Cimetière des Saints-Innocents, XVIIe–XVIIIe siècle. Paris, Albin Michel, 2000, 480 p. (Bibliothèque histoire).

7236. Mutations d'identités en Méditerranée. Moyen Age et époque contemporaine. Dir. par Henri BRESC et Christian VEAUVY. Avec la participation d'Eliane DUPUY. Saint-Denis, Editions Bouchène, 2000, 302 p.

7237. NASSIET (Michel). Parenté, noblesse et Etats dynastiques, XVe–XVIe siècles. Paris, Ed. de l'EHESS, 2000, 376 p. (Recherches d'histoire et de sciences sociales, 90).

7238. NELSON LIMERICK (Patricia). Something in the soil: legacies and reckonings in the New West. New York, W. W. Norton, 2000, 384 p.

7239. Neue Geschichte der Sexualität. Beispiele aus Ostasien und Zentraleuropa 1700–2000. Hrsg. v. Franz X. EDER und Sabine FRÜHSTÜCK. Wien, Turia-Kant, 2000, 301 p. (Querschnitte, 3).

7240. New biography (The): performing femininity in nineteenth-century France. Ed. by Jo BURR MARGADANT. Berkeley a. Los Angeles, University of California Press, 2000, X-298 p. (Studies on the history of society and culture, 38).

7241. NØKLEBY (Berit), HJELTNES (Guri). Barn under krigen. (Children during the Second World War). Oslo, Aschehoug, 2000, 272 p. (ill.).

7242. Normy i tsennosti povsednevnoj zhizni: stanovlenie sotsialisticheskogo obraza zhizni v Rossii, 1920–1930-e gg. (Everyday norms and values: formation of the Socialist way of living in Russia, 1920s and 1930s: [Articles]). Akademija nauk Finljandii, In-t Finljandii v Sankt-Peterburge. Sankt-Peterburg, Zhurnal 'Neva', 2000, 478 p.

7243. OGILVIE (Sheilagh), EDWARDS (Jeremy). Women and the "Second Serfdom": evidence from early modern Bohemia. Journal of economic history, 2000, 60, 4, p. 961-994.

7244. ORTAYLI (İlber). Osmanlı toplumunda aile. (Family in the Ottoman society). İstanbul, Pan yayınları, 2000, 184 p.

7245. Partnerka, matka, opiekunka. Status kobiety w dziejach nowożytnych od XVI do XX wieku. (Partnerin, Mutter, Betreuerin. Die Lage der Frau in der Neuzeit vom XVI. bis zum. XX. Jahrhundert. Hrsg. v. Krzysztof JAKUBIK. Bydgoszcz, Wydaw. Uczelniane WSP, 585 p.

7246. PASTEUR (Paul). Violences et viols des vainqueurs: les femmes à Vienne et en Basse-Autriche, avril–août 1945. In: Femmes (Les) et la guerre [Cf. n° 7164], p. 123-136.

7247. PICKERING (Paul A.), TYRRELL (Alex). The people's bread: a history of the Anti-Corn Law League. New York, Continuum, 2000, X-304 p.

7248. PLECK (Elizabeth H.). Celebrating the family: ethnicity, consumer culture, and family rituals. Cambridge, Harvard U. P., 2000, IX-328 p.

7249. PORSHNEVA (Ol'ga S.). Mentalitet i sotsial'noe povedenie rabochikh, krest'jan i soldat Rossii v period Pervoj mirovoj vojny. (Mentality and social behaviour of Russian workers, peasants and soldiers during the First World War). Ekaterinburg, Izd-vo Ural'skogo otdelenija RAN, 2000, 415 p. (ill.; bibl. p. 379-386). – EADEM. Sotsial'noe povedenie rossijskogo krest'janstva v gody Pervoj mirovoj vojny, 1914–fevral' 1917 g. (Social behaviour of Russian peasantry in the years of the First World War, 1914–February 1917). In: Sotsial'naja istorija: Ezhegodnik [Cf. n° 7277], p. 57-82. [Eng. summary]

7250. PRASHAD (Vijay). Untouchable freedom: a social history of a Dalit community. Oxford, Oxford U. P., 2000, XX-176 p.

7251. PRESTA (Ana María). Encomienda, familia y negocios en Charcas colonial. Los encomenderos de La Plata, 1550–1600. Lima, Institutos de Estudios Peruanos, Banco Central de Reserva del Perú, 2000, 308 p.

7252. RANGASWAMY (Padma). Namasté America: Indian immigrants in an American metropolis. University Park, Pennsylvania State U. P., 2000, XVIII-386 p.

7253. RANSEL (David L.). Village mothers: three generations of change in Russia and Tataria. Bloomington, Indiana U. P., 2000, VIII-314 p. (Indiana-Michigan series in Russian and East European studies).

7254. RAPPAPORT (Erika Diane). Shopping for pleasure: women in the making of London's West End. Princeton, Princeton U. P., 2000, XIII-323 p.

7255. Reconstructing criminality in Latin America. Ed. by Carlos A. AGUIRRE and Robert BUFFINGTON. Wilmington, Scholarly Resources, 2000, XIX-254 p. (Jaguar books on Latin America, 19).

7256. REDDY (William M.). Sentimentalism and its erasure: the role of emotions in the era of the French Revolution. *Journal of modern history*, 2000, 72, 1, p. 109-152.

7257. Regole (Le) dei mestieri e delle professioni, secoli XV–XIX. A cura di M. MERIGGI e A. PASTORE. Milano, Franco Angeli, 2000, 378 p.

7258. RHEINHEIMER (Martin). Arme, Bettler, und Vaganten: Überleben in der Not 1450–1850. Frankfurt, Fischer Taschenbuch, 2000, 252 p. (Europäische Geschichte).

7259. RHEUBOTTOM (David). Age, marriage, and politics in fifteenth-century Ragusa. Oxford a. New York, Oxford U. P., 2000, X-220 p. (Oxford studies in social and cultural anthropology).

7260. RODRÍGUEZ SÁENZ (Eugenia). Hijas, novias y esposas: familia, matrimonio y violencia doméstica en el Valle Central de Costa Rica (1750–1850). Heredia, Editorial Universidad Nacional, 2000, 178 p.

7261. ROYNETTE (Odile). «Bons pour le service». L'expérience de la caserne en France à la fin du XIXe siècle. Paris, Belin, 2000, 458 p.

7262. SAFLEY (Th. Max). Bankruptcy: family and finance in early modern Augsburg. *Journal of European economic history*, 2000, 1, p. 53-76.

7263. SAHLINS (Peter). La nationalité avant la lettre. Les pratiques de naturalisation en France sous l'Ancien Régime. *In*: Bannir au Moyen Age, naturaliser à l'époque moderne. *Annales*, 2000, 55, 5, p. 1081-1108.

7264. SALAZAR VERGARA (Gabriel). Labradores, peones y proletarios: formación y crisis de la sociedad popular chilena del siglo XIX. Santiago, LOM, 2000, 334 p. (ill.). (Historia).

7265. SAPPINEN (Eero). Arkielämän murros 1960-ja 1970-luvuilla. Tutkimus suomalaisen työväestön elämäntavoista ja niiden paikallisista raumalaisista piirteistä (Change in everyday life in the 1960s and 1970s. A study of workers' ways of life in Finland and local Rauma traits). Helsinki, SMY,02, in-4, 509 p. (Kansat. ark., 46).

7266. SCHÄFER (Axel R.). American progressives and German social reform, 1875–1920: social ethics, moral control, and the regulatory state in a transatlantic context. Stuttgart, Franz Steiner Verlag, 2000, 252 p. (USA-Studien, 12).

7267. SCHNEIDER (Arnd). Futures lost. Nostalgia and identity among Italian immigrants in Argentina. Bern, P. Lang, 2000, 343 p.

7268. SCHULZ (Günther). Die deutschen Angestellten seit dem 19. Jahrhundert. München, Oldenbourg, 2000, 152 p. (Enzyklopädie deutscher Geschichte, 54).

7269. SHEPARD (Alexandra). Manhood, credit and patriarchy in early modern England c. 1580–1640. *Past and present*, 2000, 167, p. 75-106.

7270. SHOEMAKER (Robert B.). The decline of public insult in London 1660–1800. *Past and present*, 2000, 169, p. 97-131.

7271. SHOVLIN (John). Toward a reinterpretation of revolutionary antinobilism: the political economy of honor in the Old Regime. *Journal of modern history*, 2000, 72, 1, p. 35-66.

7272. SMITH (Jay M.). Social categories, the language of patriotism and the origins of the French Revolution: the debate over noblesse commerçante. *Journal of modern history*, 2000, 72, 2, p. 339-374.

7273. SMITH (Timothy B.). The plight of able-bodied poor and the unemployed in urban France, 1880–1914. *European history quarterly*, 2000, 30, 2, p. 147-184.

7274. SOÉNIUS (Ulrich S.). Wirtschaftsbürgertum im 19. und 20. Jahrhundert. Die Familie Scheidt in Kettwig 1848–1925. Köln, Selbstverlag Stiftung Rheinisch-Westfälisches Wirtschaftsarchiv zu Köln, 2000, 799 p. (Schriften zur rheinisch-westfälischen Wirtschaftsgeschichte, 40).

7275. SØLAND (Birgitte). Becoming modern: young women and the reconstruction of womanhood in the 1920s. Princeton, Princeton U. P., 2000, X-249 p.

7276. SOMMER (Elisabeth W.). Serving two masters: Moravian Brethren in Germany and North Carolina, 1727–1801. Lexington, University Press of Kentucky, 2000, XVII-234 p.

7277. Sotsial'naja istorija: Ezhegodnik, 2000. (Social history: annual, 2000). Ed. Kirill M. ANDERSON. RAN, In-t vseobshchej istorii. Moskva, ROSSPEN, 2000, 351 p. [Eng. summaries and table of contents]. [Cf. n^{os} <choice> 1159, 5616, 7249.]

7278. STENSETH (Bodil). Eilert Sundt og det Norge han fant. (Social conditions in Norway during the 19th century). Oslo, Gyldendal, 2000, 335 p. (ill.).

7279. STRUPP (Christoph). Erbe und Auftrag. Bürgerliche Revolutionserinnerung im Kaiserreich. *Historische Zeitschrift*, 2000, 270, 2, p. 309-344.

7280. TAÏEB (Jacques). Sociétés juives du Maghreb moderne (1500–1900). Un monde en mouvement. Paris, Editions Maisonneuve et Larose, 2000, 223 p.

7281. THER (Philipp). Die einheimische Bevölkerung des Oppelner Schlesiens nach dem Zweiten Weltkrieg: die Entstehung einer deutschen Minderheit. *Geschichte und Gesellschaft*, 2000, 26, 3, p. 407-438.

7282. TOSH (John). Maschilità e genere nell'Inghilterra vittoriana. *Quaderni storici*, 2000, 35, 104, p. 803-822.

7283. TRAMA (Luciana). Un'Opera Pia nell'Italia unita. Il 'Suor Orsola Benincasa' dall'Unità alla nascita del Magistero. Introduzione di Giuseppe GALASSO, intervento di Piero CRAVERI. Napoli, Editoriale scientifica, 2000, 225 p.

7284. TYGIEL (Jules). Past time: Baseball as history. New York, Oxford U. P., 2000, XIII-258 p.

7285. VALE (Lawrence J.). From the puritans to the projects: public housing and public neighbors. Cambridge, Harvard U. P., 2000, XIII-460 p.

7286. VAN POPPEL (F.). Children in one-parent families: survival as an indicator of the role of the parents. *Journal of family history*, 2000, 25, 3, p. 269-290.

7287. VAN SOLINGE (Hanna), WALHOUT (Evelien), VAN POPPEL (Frans). Determinants of institutionalization of orphans in a 19th century Dutch town. *Continuity and change*, 2000, 15, 1, p. 139-166.

7288. VANGELISTA (Chiara). Strategie di frontiera in un territorio tribale: il Brasile occidentale tra Otto e Novecento. *In*: Pratiche del territorio [Cf. n° 7483], p. 109-138.

7289. Ville promise (La). Mobilité et accueil à Paris (fin XVIIe début XIXe siècle). Sous la dir. de Daniel ROCHE. Paris, Fayard, 2000, 440 p.

7290. WALASZEK (Adam). Polskie imigrantki w miastach USA. (Polish immigrant women in the United States of America). *Przegląd Polonijny*, 2000, 2, p. 25-41. – IDEM. Polscy imigranci w USA i ich ojczyzna (1914–1923). (Polish immigrants in the U.S.A. and their homeland). *In*: Idea niepodległości i suwerenności narodowej w działalności Polaków w kraju i na obczyźnie (1918–1998). (Idea of independence and sovereignty in the activity of Poles abroad 1918–1998). Ed. by Marek SZCZERBIŃSKI. Gorzów Wielkopolski, Magnum s.c., 2000, p. 137-153.

7291. WALLE (Marianne). Vichy ou la féminité imposée. *In*: Femmes (Les) et la guerre [Cf. n° 7164], p. 99-108.

7292. WEIDNER (Marcus). Landadel in Münster 1600–1760. Stadtverfassung, Standesbehauptung und Fürstenhof. Münster, Aschendorff, 2000, 2 vol., XIII-1268 p. (Quellen und Forschungen zur Geschichte der Stadt Münster, 18/1, 2).

7293. WELLS (Robert W.). Facing the «King of Terrors». Death and society in an American community, 1750–1990. Cambridge, Cambridge U. P., 2000, 301 p.

7294. WELSKOPP (Thomas). Sattelzeitgenosse. Freiherr Karl vom Stein zwischen Bergbauverwaltung und gesellschaftlicher Reform in Preußen. *Historische Zeitschrift*, 2000, 271, 2, p. 347-372.

7295. WHYTE (Ian D.). Migration and society in Britain, 1550–1830. Basingstoke, Macmillan, 2000, X-198 p.

7296. Women in Renaissance and Early Modern Europe. Ed. by Christine MEEK. Portland, Four Courts Press, 2000, 230 p.

7297. ZÁRATE TOSCANO (Verónica). Los nobles ante la muerte en México. Actitudes, ceremonias y memoria (1750–1850). Mexico, El Colegio de México/Instituto Mora, 2000, 298 p.

7298. ZUBKOVA (Elena Ju.). Poslevoennoe sovetskoe obshchestvo: politika i povsednevnost', 1945–1953. (Soviet society after the war: politics and everyday life, 1945–1953). RAN, In-t rossijskoj istorii. Moskva, ROSSPEN, 2000, 230 p. [Cf. Bibl. 93, n° 6740.]

Cf. n°s 114-133, 893, 895, 1159, 4050, 4586, 5044, 5045, 5061, 5072, 5082, 5758, 6919, 6936, 7021, 7342

§ 9. Working-class movement and socialism.

* 7299. MEYER FILARDO (Peter). Labor history bibliography, 1999. *Labor history*, 2000, 41, 4, p. 465-500.

* 7300. SZMRECSÁNYI (Tamás). Novas contribuçes à historiografia do trabalho no Estado de São Paulo. *In*: Presente e futuro das relaçes de trabalho: estudios em homenagem a Roberto Araujo de Oliveira Santos. Ed. Georgenor DE SOUSA FRANCO FILHO. São Paulo, Editora LTR, 2000, p. 37-65.

** 7301. Quellensammlung zur Geschichte der deutschen Sozialpolitik 1867 bis 1914. Abt. 1. Von der Reichsgründungszeit bis zur kaiserlichen Sozialbotschaft (1867–1881). Band 7. Halbbde. 2. Armengesetzgebung und Freizügigkeit, bearb. v. Christoph SACHßE, Floria TENNSTEDT und Elmar ROEDER. Darmstadt, Wissenschaftliche Buchgesellschaft, 2000, XLVIII-986 p.

** 7302. Rabochee dvizhenie v Rossii, 1895–fevral' 1917 g.: Khronika. (Working-class movement in Russia, 1895–February 1917: a chronicle). Ed. Jurij I. KIR'JANOV. Gos. arkhivnaja sluzhba RF; RAN, In-t rossijskoj istorii. Vol. 7. 1901 god. (Year 1901). Ed. Irina M. PUSHKAREVA. Sankt-Peterburg, BLITs, 2000, 607 p. [Eng. summary]. [Cf. Bibl. 99, n° 7037.]

7303. Ausschuß für die Reform der Sozialversicherung/für Sozialversicherung (1934 bis 1944). Versorgungswerk und Gesundheitswerk des Deutschen Volkes (1940–1942). Hrsg. und mit einer Einleitung versehen v. Werner SCHUBERT. Frankfurt am Main, Lang, 2000, XXXIV-619 p. (Akademie für Deutsches Recht 1933–1945. Protokolle der Ausschüsse, 10).

7304. BEECHER (Jonathan). Victor Considerant and the rise and fall of French romantic socialism. Berkeley and Los Angeles, University of California Press, 2000, XVI-584 p.

7305. BEITO (David T.). From mutual aid to the welfare state: fraternal societies and social services, 1890–1967. Chapel Hill, University of North Carolina Press, 2000, XIV-320 p.

7306. BIX (Amy Sue). Inventing ourselves out of jobs? America's debate over technological unemployment 1929–1981. Baltimore, Johns Hopkins U. P., with the assistance of the Hagley Museum and Library, 2000, X-376 p. (Studies in Industry and Society).

7307. BLASZAK (Barbara J.). The matriarchs of England's co-operative movement: a study in gender politics and female leadership, 1883–1921. Westport a. London, Greenwood Press, 2000, X-209 p.

7308. BORTZ (Jeffrey). The revolution, the labour regime and conditions of works in the cotton textile industry in Mexico, 1910–1927. *Journal of Latin American studies*, 2000, 32, 3, p. 671-704.

7309. BROMM (Boris Franz Leo). Die Entstehungsgeschichte des Berufs des Handelsvertreters. Unter besonderer Berücksichtigung der Sozialgesetzgebung in den Jahren 1871–1933. Frankfurt am Main, Berlin, Bern, Bruxelles, New York u. Wien, Lang, 2000, 152 p.

7310. BROŽEK (Marek). Činnost odborových organizací a závodního výboru v Měšťanském pivovaru v Plzni 1918–1938. Příspěvek k sociálním dějinám 20. století. (Die Tätigkeit der Gewerkschaftsorganisationen und des Gewerkschaftsausschusses bei der Bürgerlichen Bierbrauerei in Pilsen in den Jahren von 1918 bis 1938). *Studie k sociálním dějinám*, 2000, 5, p. 231-272. – IDEM. Dělníci Měšťanského pivovaru v Plzni v letech 1918–1938. Příspěvek k sociálním dějinám Plzně. (Die Arbeiter des Bürgerlichen Brauhauses in Plzeň/Pilsen in den Jahren 1918-1938: Ein Beitrag zur Sozialgeschichte Pilsens). *Západočeský historický sborník*, 2000, 6, p. 265-298.

7311. Bund. 100 lat historii 1897–1997. ("Bund" [Allgemeiner Jüdischer Arbeiterbund]. 100 Jahre Geschichte 1897–1997). Hrsg. v. Feliks TYCH und Jürgen HENSEL. Warszawa, Volumen, 2000, 352 p.

7312. CADÉ (Michel). L'écran bleu. La représentation des ouvriers dans le cinéma français. Perpignan, Presses universitaires de Perpignan, 2000, 272 p. (Etudes).

7313. CHASE (Malcolm). Early trade unionism: fraternity, skill and the politics of labour. Aldershot, Ashgate, 2000, VI-294 p.

7314. CLEVERLEY (John). In the lap of tigers: the Communist Labor University in Jangxi province. Lanham, Rowman & Littlefield, 2000, XIV-249 p.

7315. CUNNINGHAM (Hugh). The decline of child labour: labour markets and family economies in Europe and North America since 1830. *Economic history review*, 2000, 53, 3, p. 409-428.

7316. DE GIORGI (Alvaro), DOMINZAÍN (Susana). Respuestas sindicales en Chile y Uruguay bajo las dictaduras y en los inicios de la democratización. Directora del proyecto Lucía SALA DE TOURÓN. Montevideo, Centro de Estudios Interdisciplinarios Latinoamericanos, Facultad de Humanidades y Ciencias de la Educación, Universidad de la República, 2000, 171 p.

7317. DERICKSON (Alan). "Take health from the list of luxuries": labor and the right to health care, 1915–1949. *Labor history*, 2000, 41, 2, p. 171-188.

7318. DESLIPPE (Dennis A.). "Rights, not roses": unions and the rise of working-class feminism, 1945–1980. Urbana a. Chicago, University of Illinois Press, 2000, X-259 p. (Working class in American history).

7319. Dock workers: international explorations in comparative labour history, 1790–1970. Ed. by Sam DAVIES, Colin J. DAVIS, David DE VRIES, Lex HEERMA VAN VOSS, Lidewij HESSELINK and Klaus WEINHAUER. Aldershot, Ashgate, 2000, 2 vol., XIV-863 p.

7320. DOLLINGER (Sol), JOHNSON DOLLINGER (Genora). Not automatic: women and the Left in the forging of the Auto Worker's Union. New York, Monthly Review Press, 2000, XV-214 p.

7321. DUBOFSKY (Melvyn). Hard work: the making of labor history. Urbana a. Chicago, University of Illinois Press, 2000, X-249 p.

7322. EGGERT (Gerald G.). Making iron on the bad eagle: Ronald Curtin's ironworks and workers' community. University Park, Pennsylvania State U. P., 2000, XVI-189 p.

7323. FARNSWORTH-ALVEAR (Ann). Dulcinea in the factory: myths, morals, men, and women in Colombia's industrial experiment, 1905–1960. Durham, Duke U. P., 2000, XVI-303 p. (Comparative and international working-class history).

7324. GRYPA (Dietmar). Die katholische Arbeiterbewegung in Bayern nach dem Zweiten Weltkrieg (1945–1963). Paderborn, München u. Wien, Schöningh, 2000, 594 p. (Veröffentlichungen der Kommission für Zeitgeschichte, Rh. B: Forschungen, 91).

7325. HÁJEK (Miloš), MEJDROVÁ (Hana). Vznik Třetí internacionály. (The creation of the Third Internationale). Praha, Karolinum, 2000, 491 p.

7326. HOLTWICK (Bernd). Der zerstrittene Berufsstand. Handwerker und ihre Organisationen in Ostwestfalen-Lippe 1929–1953. Paderborn, München, Wien u. Zürich, Schöningh, 2000, IX-463 p. (Forschungen zur Regionalgeschichte, 36).

7327. HOWLETT (Peter). Evidence of the existence of an internal labour market in the Great Eastern Railway Company, 1875–1905. *Business history*, 2000, 42, 1, p. 21-40.

7328. HUTTULA (Tapio). Nauloilla laadittu laki. Työväentalojen sulkemiset 1929–1932. (A law nailed down with force. The shutting down of Labour Halls in 1929–1932). (Helsinki, SKS, 2000, 443 p. (ill., maps). (Bibliotheca historica, 55).

7329. IRONS (Janet). Testing the New Deal: the general textile strike of 1934 in the American South. Champaign, University of Illinois Press, 2000, IX-262 p. (The Working Class in American history).

7330. JAFFE (James A.). Striking a bargain: work and industrial relations in England 1815–1865. Manchester, Manchester U. P., 2000, IX-273 p.

7331. JONES (William P.). Black workers and CIO's turn toward racial liberalism: operation dixie and the North Carolina Lumber industry, 1946–1953. *Labor history*, 2000, 41, 3, p. 279-306.

7332. KAUFMAN (Bruce E.). The case for the company union. *Labor history*, 2000, 41, 3, p. 321-350.

7333. KING (Steven). Poverty and welfare in England, 1700–1850. Manchester, Manchester U. P., 2000, X-294 p.

7334. KOCKA (Jürgen). Arbeit früher, heute, morgen: Zur Neuartigkeit der Gegenwart. *In*: Geschichte und Zukunft der Arbeit [Cf. n° 1144], p. 476-492.

7335. KOSAK (Hadassa). Cultures of opposition: Jewish immigrant workers, New York City, 1881–1905. Albany, State University of New York Press, 2000, X-220 p. (SUNY Series in American Labor History).

7336. Labouring the Canadian millennium: writings on work and workers, history and historiography. Ed. by Bryan D. PALMER. St. John's, Canadian Committee on Labour History, 2000, 486 p.

7337. LAUGHLIN (Kathleen A.). Women's work and public policy: a history of the women's bureau, U.S. Departmen of Labor 1945–1970. Boston, Northeastern U. P., 2000, X-172 p.

7338. LIM (Jie-Hyun). From labour emancipation to labour mobilisation: socialism as a way of anti-western modernisation in underdeveloped countries. *In*: Die Arbeiterbewegung – Ein gescheitertes Projekt der Moderne? 35 Linzer Konferenz der ITH. Leipzig, Akademische Verlagsanstalt, 2000, [s. p.].

7339. MAC RAILD (Donald M.), MARTIN (David E.). Labour in British society, 1830–1914. Basingstoke, Macmillan, 2000, XI-214 p.

7340. MERRETT (David T.), SELTZER (Andrew). Work in the financial services industry and worker monitoring: a study of the Union Bank of Australia in the 1920s. *In*: Business institutions and behaviour in Australia [Cf. n° 6764], p. 133-152.

7341. METZGAR (Jack). Striking steel: solidarity remembered. Philadelphia, Temple U. P., 2000, VIII-264 p. (Critical Perspectives on the Past).

7342. MITCHISON (Rosalind). The old poor law in Scotland: the experience of poverty, 1574–1845. Edinburgh, Edinburgh U. P., 2000, 246 p.

7343. MULCAHY (Richard P.). A social contract for the coal fields: the rise and fall of the united mine workers of America welfare and retirement fund. Knoxville, University of Tennessee Press, 2000, XIII-274 p.

7344. MURPHY (Michelle). Toxicity in the details: the history of the women's office worker movement and occupational health in the late-capitalist office. *Labor history*, 2000, 41, 2, p. 189-214.

7345. Nonunions employee representation: history, contemporary practice and policy. Ed. by Bruce E. KAUFMAN and Gottlieb TARAS DAPHNE. New York a. London, Sharpe, 2000, 576 p.

7346. OBADELE-STARKS (Ernest). Black unionism in the industrial south. College Station, Texas A&M U. P., 2000, XXII-183 p.

7347. PARODI (Maurice), LANGEVIN (Philippe), OPPENHEIM (Jean-Pierre), RICHEZ-BATTESTI (Nadine). La question sociale en France depuis 1945. Paris, Armand Colin, 2000, 233 p. (ill.). (Collection U).

7348. PECK (Gunther). Reinventing free labor: padrones and immigrant workers in the North American West, 1880–1930. Cambridge, Cambridge U. P., 2000, 308 p.

7349. PEDERSEN (Vernon L.). George Mink, the marine workers industrial union, and the Comintern in America. *Labor history*, 2000, 41, 3, p. 307-320.

7350. PERRY (Matt). Bread and work: the experience of unemployment 1918–1939. London, Pluto Press, 2000, XV-244 p.

7351. PHILLIPS (Laura L.). Bolsheviks and the bottle: drink and worker culture in St. Petersburg, 1900–1929. DeKalb, Northern Illinois U. P., 2000, VIII-212 p.

7352. PIERCE (Michael). The populist president of the American Federation on Labor: the career of John Mac Bride, 1880–1895. *Labor history*, 2000, 41, 1, p. 5-24.

7353. PIETTE (Valérie). Domestiques et servantes. Des vies sous condition. Essai sur le travail domestique en Belgique au 19e siècle. Bruxelles, Académie royale de Belgique, 2000, 521 p.

7354. PILBEAM (Pamela M.). French socialists before Marx: workers, women and the social question in France. Montreal a. London, McGill-Queen's U. P., 2000, X-259 p. (ill.).

7355. Politics (The) of retribution in Europe: World War II and its aftermath. Ed. by István DEÁK, Jan T. GROSS and Tony JUDT. Princeton, Princeton U. P., 2000, XII-337 p.

7356. PRYWES (Ruth W.). The United States labor force: a descriptive analysis. Westport, Quorum Books, XIV-393 p.

7357. ROBERTSON (David Brian). Capital, labor, and state: the battle for American labor markets from the Civil War to the New Deal. Lanham, Rowman & Littlefield, 2000, XXII-297 p.

7358. ROCKEL (Stephen J.). 'A nation of porters': the Nyamwezi and the labour market in nineteenth-century Tanzania. *Journal of African history*, 2000, 41, 2, p. 173-196.

7359. ROTH (Karl Heinz). Facetten des Terrors. Der Geheimdienst der "Deutschen Arbeitsfront" und die Zerstörung der Arbeiterbewegung 1933 bis 1938. Bremen, Temmen, 2000, 280 p.

7360. SILVERMAN (Victor). Imagining internationalism in American and British labor, 1939–1949. Urbana a. Chicago, University of Illinois Press, 2000, XIV-298 p. (The working class in American history).

7361. STORRS LANDON (R. Y.). Civilizing capitalism: the National Consumers' League and the politics of labor standards in the New Deal era. Chapel Hill, University of North Carolina Press, 2000, [s. p.]. (Gender & American culture).

7362. THATCHER (Ian D.). Leon Trotsky and World War One: August 1914–February 1917. New York, St. Martin's, 2000, VII-262 p.

7363. URILOV (Il'ja Kh.). Isotrija rossijskoj sotsial-demokratii (men'shevizma). (A history of Russian Social Democracy [Menshevism]). Part 1. Istochnikovedenie (Study of sources). Moskva, Raritet, 2000, 286 p.

7364. VOTH (Hans Joachim). Time and work in England, 1750–1830. Oxford, Oxford U. P., 2000, 304 p.

7365. WARREN (Christian). Brush with death: a social history of lead poisoning. Baltimore, Johns Hopkins U. P., 2000, XIV-362 p.

7366. ZWEIG (Michael). The working class majority: America's best kept secret. Ithaca, Cornell U. P., 2000, VIII-198 p.

Cf. n°s 4763, 5124, 5136, 7249

O

MODERN LEGAL AND CONSTITUTIONAL HISTORY

§ 1. General. 7367-7395. – § 2. History of constitutional law. 7396-7437. – § 3. Public law and institutions. 7438-7494. – § 4. Civil and penal law. 7495-7565. – § 5. International law. 7566-7575.

§ 1. General.

* 7367. Catalogo del fondo Filippo Vassalli. A cura di Sandro BULGARELLI e Alessandra CASAMASSIMA. Prefazione di Aldo MASULLO, introduzione di Gian Battista FERRI e Ugo PETRONIO. Roma, Senato della Repubblica, 2000, XLIII-291 p.

* 7368. DUYNSTEE (M.), FEENSTRA (R.), WAELKENS (L.). Repertorium bibliographicum institutorum et sodalitatum iuris historiae, iussu societatis c. n. Association internationale d'histoire du droit et des institutions, Editio tertia. Groeninghe, Kortrijk, 2000, XXXVIII-482 p.

** 7369. SAVIGNY (Friedrich Karl von). Politik und neuere Legislationen. Materialen zum "Geist der Gesetzgebung". Aus den Nachlassmaterialen hrsg. v. Hidetake AKAMATSU und Joachim RÜCKERT. Frankfurt am Main, Klostermann, 2000, LXIX-314 p. (Savignyana 5 = Ius Commune Sonderheft, 135).

7370. BARGAGLI (Roberta). Bartolomeo Sozzini giurista e politico (1436–1506). Milano, Giuffrè, 2000, XVI-255 p.

7371. BECCHI (Paolo). Giuristi e principi. Genova, Compagnia dei Librai, 2000, 166 p.

7372. CALDWELL (Peter C.), SCHEUERMAN (William E.). From liberal democracy to fascism: legal and political thought in the Weimar Republic. Boston, Humanities Press, 2000, VII-165 p. (Studies in Central European histories).

7373. Cesare Beccaria. La pratica dei Lumi. A cura di Vincenzo FERRONE e Gianni FRANCIONI. Firenze, Olschki, 2000, 182 p.

7374. CLÉMENT (Jean-Paul). Aux sources du libéralisme français: Boissy d'Anglas, Daunou, Lanjuinais. Paris, L.G.D.J., 2000, 310 p. (Bibliothèque constitutionnelle et de science politique, 95).

7375. FINLAY (J.). Men of law in pre-Reformation Scotland. East Lothian, Tuckwell Press, 2000, X-253 p. (Scottish historical review monograph, 9).

7376. GROSSO (G.). Scritti storico-giuridici. Tomo 1. Storia – Diritto – Società. Torino, Giappichelli, 2000, LII-1000 p.

7377. GRUNERT (Frank). Normbegründung und politische Legitimität: zur Rechts- und Staatsphilosophie der deutschen Frühaufklärung. Tübingen, Niemeyer, 2000, VII-310 p. (Frühe Neuzeit, 57).

7378. HAALAND (Torstein). Byråkrati, politikk og situasjonisme. Norsk statshistorie 1814–1900. (Bureaucracy, politics and situationalism. Norwegian state history 1814–1900). *Historisk tidsskrift*, 2000, 79, 1, p. 53-95.

7379. HOCHSTRASSER (Tim J.). Natural law theories in the early Enlightenment. Cambridge, Cambridge U. P., 2000, XIII-246 p. (Ideas in context, 58).

7380. Human rights and revolutions. Ed. by Jeffrey N. WASSERSTROM, Lynn HUNT and Marilyn B. YOUNG. Lanham, Rowman and Littlefield, 2000, XII-253 p.

7381. İNALCIK (Halil). Osmanlı'da devlet, hukuk, adâlet. (State, law, justice in the Ottoman State). İstanbul, Eren yayınları, 2000, 208 p.

7382. KINKLEY (Jeffrey C.). Chinese justice, the fiction: law and literature in modern China. Stanford, Stanford U. P., 2000, XI-497 p.

7383. Kodifikationsgedanke (Der) und das Modell des Bürgerlichen Gesetzbuches (BGB). Hrsg. v. Okko BEHRENDS und Wolfgang SELLERT. Göttingen, Vandenhoeck & Ruprecht, 2000, 229 p. (Abhandlungen der Akademie der Wissenschaften in Göttingen, Phil.-Hist. Klasse, Dritte Folge, 236).

7384. Latinskaja Amerika i Kariby: Politicheskie instituty i protsessy. (Latin America and the Caribbean: political institutions and processes: [essays on formation of political systems of the countries]). RAN, In-t

Latinskoj Ameriki. Moskva, Nauka, 2000, 448 p. (tables). [Enlish and Spanish summaries and tables of content].

7385. LEMMINGS (David). Professors of the law: barristers and English legal culture in the eighteenth century. Oxford, Oxford U. P., XIV-399 p.

7386. Lexicon Juridicum Romano-Teutonicum. Hrsg. v. Samuel OBERLÄNDER. 4 aufl. 1753. Hrsg. v. Rainer POLLEY. Köln, Weimar u. Wien, Böhlau, 2000, XXXII-X-748 p.

7387. MOSCATI (Laura). Italienische Reise. Savigny e la scienza giuridica della restaurazione. Roma, Viella, 2000, 200 p. («Ius nostrum». Studi e testi pubblicati dall'Istituto di Storia del diritto italiano dell'Università di Roma «La Sapienza», 26).

7388. Rechtskultur, Rechtswissenschaft, Rechtsberufe im 19. Jahrhundert. Professionalisierung und Verrechtilichung in Deutschland und Italien. Hrsg. v. Christof DIPPER. Berlin, Duncker & Humblot, 2000, 167 p. (Schriften zur europäischen Rechts- und Verfassungsgeschichte, 35).

7389. ROSENBERG (Mathias). Friedrich Karl von Savigny (1779–1861) im Urteil seiner Zeit. Frankfurt am Main, Lang, 2000, XII-187 p. (Rechtshistorische Reihe, 215).

7390. SCHMIDT-RECLA (Adrian). Theorien zur Schuldfähigkeit. Psychowissenschaftliche Konzepte zur Beurteilung strafrechtlicher Verantwortlichkeit im 19. und 20. Jahrhundert. Eine Anleitung zur juristischen Verwertbarkeit. Leipzig, Leipziger Universitätsverlag, 2000, 334 p. (Leipziger Juristische Studien, Strafrechtliche Abteilung, 4).

7391. SCHRAMM (Nils-Eberhard). Die Vereinigung demokratischer Juristen (1949–1999). Frankfurt am Main, Lang, 2000, 405 p. (Rechtshistorische Reihe, 222).

7392. SVAREZ (Carl Gottlieb). Gesammelte Schriften. Band 4. Die Kronprinzenvorlesungen. Teil 1. Staatsrecht. Teil 2. Das positive preußische Recht. Stuttgart, Frommann-Holzboog, 2000, XC-948 p.

7393. VANO (Cristina). 'Il nostro autentico Gaio', strategie della Scuola Storica alle origini della romanistica moderna. Napoli, Editoriale scientifica, 2000, XIX-389 p. (Pubblicazioni del Dipartimento di diritto romano e storia della scienza romanistica dell'Università degli studi di Napoli 'Federico II', 16).

7394. ZARKA (Yves Charles). L'autre voie de la subjectivité: six études sur le sujet et le droit naturel au XVIIe siècle. Paris, Beauchesne, 2000, VIII-132 p. (Grenier à sel).

7395. Zwische Romanistik und Germanistik. Carl Georg von Waechter (1797–1880). Hrsg. v. Bernd-Rüdiger KERN. Berlin, Duncker & Humblot, 2000, 172 p. (Schriften zur Rechtsgeschichte, 81).

Cf. n°s 1085-1116, 3486, 3523, 5904, 6010

§ 2. History of constitutional law.

** 7396. End (The) of empire: dependencies since 1948: select documents on the constitutional history of the British Empire and Commonwealth. Part 1. The West Indies, British Honduras, Hong Kong, Fiji, Cyprus, Gibraltar, and the Falklands. Ed. by Frederick MADDEN. Westport a. London, Greenwood Press, 2000, XXXVI-555 p. (Documents in imperial history, 8).

7397. ALJOVÍN DE LOSADA (Cristóbal). Caudillos y Constituciones: Perú: 1821–1845. Lima, Pontificia Universidad Católica del Perú, Instituto Riva-Agüero, Fondo de Cultura Economica, 2000, 354 p.

7398. ALVAREZ (Silvia T.), [et al.]. Los territorios nacionales australes en la historia constitucional argentina: el caso de Tierra del Fuego. Coordinador Alberto David LEIVA. Bahía Blanca, Publicaciones de la Cátedra de Historia Constitucional Universidad Nacional del Sur, 2000, 143 p

7399. BIGAUT (Christian). Le réformisme constitutionnel en France (1789–2000). Préface de Didier MAUS. Paris, Documentation francaise, 2000, 256 p. (Les études de la Documentation française).

7400. BJÖRNER (Ulf). Die Verfassungsgerichtsbarkeit im Norddeutschen Bund und Deutschen Reich (1867–1918). Eine rechtshistorische Untersuchung über Gerichtsbarkeit im Spannungsfeld von Politik und Recht innerhalb der von Bismarck geschaffenen deutschen Bundesstaaten. Frankfurt am Main, Lang, 2000, XIX-183 p. (Rechtshistorische Reihe, 214).

7401. BODINEAU (Pierre), VERPEAUX (Michel). Histoire constitutionnelle de la France. Paris, Presses universitaires de France, 2000, 127 p. (Que sais-je? 3547).

7402. Challenge (The) to Westminster: sovereignty, devolution and independence. Ed. by H. T. DICKINSON and Michael LYNCH. Linton, Tuckwell Press, 2000, IX-174 p.

7403. CLAVERO (Bartolomé). Ama Llunku, Abya Yala: Constituyencia indígena y código ladino por América. Madrid, Centro de Estudios Políticos y Constitucionales, 2000, 481 p.

7404. Constitution (La) du 4 novembre 1848: l'ambition d'une république démocratique. Dijon, Publications de l'université de Bourgogne, Editions universitaires de Dijon, 2000, 463 p.

7405. Costituzione politica della monarchia spagnola. A cura e con introduzione di Andrea ROMANO. Nota bibliografica di Bartolomé CLAVERO. Messina, Rubettino, 2000, [s. p.]. (Materiali per una storia delle istituzioni giuridiche e politiche medievali, moderne e contemporanee).

7406. DE GUILLENCHMIDT (Michel). Histoire constitutionnelle de la France depuis 1789. Paris, Economica, 2000, 302 p.

7407. DURANTI (Francesco). Le origini della giustizia costituzionale in Francia e la nascita del Conseil Constitutionnel. *Politica del diritto*, 2000, 1, p. 167-198.

7408. FENSKE (Hans). Der moderne Verfassungsstaat. Eine vergleichende Geschichte von der Entstehung bis zum 20. Jahrhundert. Paderborn, Schöningh, 2000, XIII-577 p.

7409. GERHARDT (Michael J.). The federal appointment process: a constitutional and historical analysis. Durham, Duke U. P., 2000, IX-400 p. (Constitutional conflicts).

7410. GHISALBERTI (Carlo). Venezia e il costituzionalismo del '48–'49. *Clio*, 2000, 1, p. 21-34.

7411. GRONSKÝ (Ján). Dokumenty k ústavnímu vývoji Československa. Díl 3, 1968–1989. (Documents on Czechoslovak constitutional development. Tomo 3, 1968–1989). Praha, Karolinum, 2000, 308 p.

7412. Grundlovens Danmark: samfund og kultur før og efter 1849. (Le Danmark et sa Constitution: société et culture avant et après 1849). Ed. par Søren BITSCH CHRISTENSEN. København, National Museet: Århus, Historisk Institut, Aarhus universitet, 2000, 87 p. (ill.).

7413. HUTCHINS (Francis G.). Tribes and the American Constitution. Brookline, Amarta, 2000, 273 p.

7414. KLOEVEKORN (Andreas). Die irische Verfassung von 1937. Entstehung und Rezeption. Stuttgart, Steiner, 2000, 199 p. (Historische Mitteilungen, 39).

7415. KOTULLA (Michael). Die Entstehung der Kriegsverfassung des Deutschen Bundes vor dem Hintergrund verfassungsrechtlicher und verfassungspolitischer Kontroversen in der Bundesversammlung 1816–1823. *Zeitschrift der Savigny-Stiftung für Rechtsgeschichte. Germanistische Abteilung*, 2000, 117, p. 122-237.

7416. LINCK (Stephan). Der Ordnung verpflichtet: Deutsche Polizei 1933–1949. Der Fall Flensburg. Paderborn, München, Wien u. Zürich, Schöningh, 2000, 368 p.

7417. MAC GRATH (Charles Ivar). The making of the eighteenth-century Irish constitution: government, parliament and the revenue, 1692–1714. Dublin, Four Courts Press, 2000, 314 p. (ill.).

7418. MARKO (Joseph), [et al.]. Revolution und Recht: Systemtransformation und Verfassungsentwicklung in der Tschechischen und Slowakischen Republik. Frankfurt am Main u. New York, P. Lang, 2000, 502 p.

7419. Modelos constitucionales en la historia comparada. Ed. por Joaquín VARELA SUANZES. Oviedo, Junta General del Principado de Asturias, Oviedo, 2000, 696 p. (Fundamentos. Cuadernos monográficos de teoría del estado, derecho público e historia constitucional, 2/2000).

7420. MÜLLER (Oda). Die Verfassungsbeschwerde nach der bayerischen Verfassung von 1818 (1818–1848). Ein Beitrag zur Geschichte der Verfassungs- und Verwaltungsgerichtsbarkeit in Deutschland. Frankfurt am Main, Lang, 2000, 221 p. (Europäische Hochschulschriften 2, 2845).

7421. Nascita (La) delle costituzioni europee del secondo dopoguerra: Torino, 25–26 ottobre 1996. Padova, Cedam, 2000, 283 p.

7422. NOE (Manfred). Berufsständische Elemente in den deutschen Staatsverfassungen des 19. und 20. Jahrhunderts. Frankfurt am Main, Neue Wissenschaft, 2000, 223 p.

7423. PELLERITI (Enza). 1812–1848. La Sicilia fra due costituzioni: con un'appendice di testi. Milano, Giuffre, 2000, CXXIV-229 p. (Studi storico giuridici / Università degli studi di Messina, Facoltà di scienze politiche, 7).

7424. PURCELL (Edward A., Jr.). Brandeis and the progressive constitution: Erie, the judicial power, and the politics of the Federal Courts in twentieth-century America. New Haven a. London, Yale U. P., 2000, X-417 p.

7425. Razvoj slovenskega parlamentarizma: kolokvij ob 10. obletnici parlamentarizma v Sloveniji: zbornik referatov, koreferatov in razprav. (The development of Slovene parliamentarianism: collection of papers on the 10th anniversary of parlamentarism in Slovenia). Zbrala in uredila Tatjana KRAŠOVEC. Ljubljana, Državni zbor Republike Slovenije, 2000, 215 p. (Zbirka Poslanske pisarne / Državni zbor Republike Slovenije).

7426. (Re)parlamentarizacija v Sloveniji: kronologija dogodkov: (ob 10. obletnici prvih demokratičnih volitev v Sloveniji). (Parliamentarianism in Slovenia: chronology of events. On the occasion of the 1st anniversary of the first democratic elections in Slovenia). Zbrali in uredili Tatjana KRAŠOVEC [et al.]. Ljubljana, Državni zbor Republike Slovenije, 2000, 59 p. (Zbirka Poslanske pisarne / Državni zbor Republike Slovenije).

7427. Reforming the constitution: debates in twentieth-century Britain. Ed. by Peter CATTERALL, Wolfram KAISER and Ulrike WALTON-JORDAN. London, Frank Cass, 2000, XIII-304 p. (Cass series, British politics and society).

7428. RUIZ IBÁÑEZ (José Javier). Théories et pratiques de la souveraineté dans la Monarchie hispanique: un conflit de jurisdictions à Cambrai. *In*: Conflits politiques d'Ancien Régime (France et monarchie hispanique). *Annales*, 2000, 55, 3, p. 623-644.

7429. SCATURRO (Frank J.). The Supreme Court's retreat from Reconstruction: a distortion of constitutional jurisprudence. Westport a. London, Greenwood Press, 2000, XII-305 p (Contributions in legal studies, 91).

7430. SHIBUTANI (Akira). Kinsei Doitsu teikoku kokusei-shi kenkyû: Tôzoku-sei gikai to teikoku kuraisu. (Constitutional history of early modern imperial Germany: Imperial Diet and provinces). Kyoto, Minerva Shobo, 2000, 220 p.

7431. STIPLOVŠEK (Miroslav). Slovenski parlamentarizem: 1927–1929: avtonomistična prizadevanja skupščin ljubljanske in mariborske oblasti za ekonomskosocialni in prosvetno-kulturni razvoj Slovenije ter za udejanjenje parlamentarizma. (Slovene parliamentarianism 1927–1929: the struggle of the Ljubljana and Maribor district assemblies for economic, social, educational and cultural development of Slovenia and the Realization of parliamentaranism). Ljubljana, Znanstveni inštitut Filozofske fakultete, 2000, 497 p. (Razprave Filozofske fakultete).

7432. Unintended consequences of constitutional amendment. Ed. by David E. KYVIG. Athens, University of Georgia Press, 2000, 260 p.

7433. Verfassungsgeschichte und Staatsrechtslehre. Griechisch-deutsche Wechselwirkungen. Hrsg. v. Georg KASSIMATIS und Michael STOLLEIS. Frankfurt am Main, Klostermann, 2000, VIII-290 p. (Ius Commune Sonderheft, 140).

7434. VOLPE (Giuseppe). Il costituzionalismo del Novecento. Roma e Bari, Laterza, 2000, 327 p.

7435. WAGNER (Norbert Berthold). Der Deutsche Kaiser und König von Preußen. Zugleich ein Beitrag zur Regentschaft und zur Regierungsstellvertretung. *Zeitschrift der Savigny-Stiftung für Rechtsgeschichte. Germanistische Abteilung*, 2000, 117, p. 450-495.

7436. WHITE (G. Edward). The Constitution and the New Deal. Cambridge a. London, Harvard U. P., 2000, X-385 p.

§ 2. Addenda 1999.

7437. GANCI (Massimo). Costituzionalisti e costituzioni in Sicilia e a Napoli dal 1812 al 1848. (Numero monografico. Atti del Convegno di studi 150° anniversario della Rivoluzione del 1848 in Sicilia). *Archivio storico siciliano*, 99, p. 7-179.

Cf. n[os] 1024-1084, 4181, 5304

§ 3. Public law and institutions.

* 7438. Bibliografia di storia delle istituzioni contemporanee. A cura di Carla ABBAMONDI e Laura LANZA. *Le carte e la storia*, 2000, 6, 1, p. 77-90. – Bibliografia di storia delle istituzioni contemporanee. Monografie italiane – 1999. A cura di Carla ABBAMONDI e Laura LANZA. *Le carte e la storia*, 2000, 6, 2, p. 74-84.

** 7439. Inventar der Akten des Reichskammergerichts 1495–1806. Frankfurter Bestand. Bearb. v. Inge KALTWASSER. Frankfurt am Main, Kramer, 2000, 1280 p. (Veröffentlichungen der Frankfurter Historischen Kommission, 21).

7440. 150 Jahre Erfurter Unionsparlament (1850–2000). Hrsg. v. Thüringer Landtag Erfurt, red. v. Harald MITTELSDORF. Weimar, Wartburg Verlag, 2000, 194 p. (Schriften zur Geschichte des Parlamentarismus in Thüringen, 15).

7441. ABABAY (Feridun). Çıldır'ın Yönetsel Örgüt Süreci: Kuzeydoğu Anadolunun Tarihi Coğrafyası İle Osmanlı Taşra Örgütü. (The evolution of administrative organisation in Çıldır: historical geography of Northwestern Anatolia and Ottoman provincial administration). Ankara, 2000, 513 p.

7442. AIMO (Piero). La giustizia nell'amministrazione dall'Ottocento a oggi. Roma e Bari, Laterza, 2000, 147 p. (Istituzioni dell'Europa contemporanea. Libri del tempo Laterza, 306).

7443. AMSON (Daniel). La responsabilité politique et pénale des ministres de 1789 à 1958. *Pouvoirs*, 2000, 1, p. 31-60.

7444. ANGELOZZI (Giancarlo), CASANOVA (Cesarina). Diventare cittadini. La cittadinanza ex privilegio a Bologna (secoli XVI–XVIII). Bologna, Comune di Bologna, 2000, 535 p. (Biblioteca dell'Archiginnasio, 3, 1).

7445. AQUARONE (Alberto). Le riforme fasciste e gli apparati centrali dello Stato. *Nuova storia contemporanea*, 2000, 6, p. 17-24.

7446. BAHL (Peter). Der Hof des Großen Kurfürsten. Studien zur höheren Amtsträgerschaft Brandenburg-Preußen. Köln, Weimar u. Wien, Böhlau, 2000, VIII-777 p. (Veröffentlichungen aus den Archiven preußischer Kulturbesitz beiheft, 8).

7447. Biographisches Handbuch der Reichsrätekongresse 1918/1919. Bearb. v. Sabine ROß. Düsseldorf, Droste, 2000, 295 p. (Handbücher zur Geschichte des Parlamentarismus und der politischen Parteien, 11).

7448. Burocrazia a scuola. Per una storia della formazione del personale pubblico nell'Otto-Novecento. A cura di Guido MELIS e Angelo VARNI. Torino, Rosenberg & Sellier, 2000, 286 p.

7449. CLARK (Linda L.). The rise of professional women in France: gender and public administration since 1830. New York, Cambridge U. P., 2000, XIV-324 p.

7450. CLINQUART (Jean). L'administration des douanes en France de 1914 à 1940. Paris, Comité pour l'histoire économique et financière de la France, 2000, 482 p.

7451. COLLIN (Peter). "Wächter der Gesetze" oder "Organ der Staatsregierung"? Konzipierung, Einrichtung und Anleitung der Staatsanwaltschaft durch das preußische Justizministerium von den Anfängen bis 1860. Frankfurt am Main, Klostermann, 2000, XX-452 p. (Rechtsprechung, 16).

7452. CORCUFF (Philippe), SANIER (Max). Politique publique et action stratégique en contexte de décentralisation. Aperçus d'un processus décisionnel «après la bataille». *In*: Comment se prend une décision publique aujourd'hui? *Annales*, 2000, 55, 4, p. 845-870.

7453. Dansk forvaltningshistorie: stat, forvaltning og samfund. Bd. 3. Stat, forvaltning og samfund efter 1950. (Histoire de l'administration danoise: Etat, administration et société. Tome 3. Etat, administration et société après 1950). Ed. par Peter BOGASON. København, Jurist- og Økonomforbundets forlag, 2000, 258 p. (ill.).

7454. Dinámicas de Antiguo Régimen y orden constitucional. Representación, justicia, y administración en Iberoamérica. Siglos XVIII–XIX. Coord. por Marco BELLINGERI. Torino, Otto Editore, 2000, 509 p. (Nova americana).

7455. DOYLE (William). La vénalité. Paris, Presses Universitaires de France, 2000, 127 p. (Que sais-je? 3557).

7456. Erfurter Union (Die) und das Erfurter Unionsparlament 1850. Hrsg. v. Gunther MAI. Köln, Weimar u. Wien, Böhlau, 2000, 470 p.

7457. Europapolitik und Bundesstaatsprinzip: die Europafähigkeit Deutschlands und seiner Länder im Vergleich mit anderen Föderalstaaten. Hrsg. v. Rudolf HRBEK. Baden-Baden, Nomos, 2000, 203 p. (ill.). (Schriftenreihe des Europäischen Zentrums für Föderalismus-Forschung, 17).

7458. FAHRMEIER (Andreas). Paßwesen und Staatsbildung im Deutschland des 19. Jahrhunderts. *Historische Zeitschrift*, 2000, 271, 1, p. 57-92. – IDEM. Citizens and aliens: foreigners and the law in Britain and the German states, 1789–1870. New York, Berghahn Books, 2000, XIII-258 p. (Monographs in German History, 5).

7459. FEOLA (Raffaele). Istituzioni e cultura giuridica. Percorsi. Napoli, Edizioni Scientifiche Italiane, 2000, XVII-628 p.

7460. FIMPEL (Martin). Reichsjustiz und Territorialstaat. Württemberg als Kommissar von Kaiser und Reich im Schwäbischen Kreis (1648–1806). Tübingen, Bibliotheca academica, 2000, 347 p. (Frühneuzeit-Forschungen, 6).

7461. GIANA (Luca). Pratica delle istituzioni: procedure e ambiti giurisdizionali a Spihno nella prima metà del XVII secolo. *In*: Pratiche del territorio [Cf. n° 7483], p. 11-48.

7462. GIANNI (Andrea). Stato e chiesa in Croazia: un caso di laicità dello Stato alla prova della storia. Padova, Cedam, 2000, IX-180 p.

7463. HENNESSY (Peter). The Prime Minister: the office and its holders since 1945. London, Allen Lane, 2000, 685 p.

7464. History of suffrage, 1760–1867. Ed. by Sarah RICHARDSON. London, Pickering & chatto, 2000, 6 vol., XL-351 p., 320 p., 348 p., 396 p., 341 p., 346 p.

7465. HOUTE (Arnaud-Dominique). Gendarmes et gendarmerie dans le département du Nord (1814–1852). Paris, Phénix Editions, 2000, 244 p.

7466. Istituzione parlamentare (L') nel XIX secolo. Una prospettiva comparata = Die parlamentarische Institution im 19. Jahrhundert. Eine Perspektive im Vergleich. A cura di Anna Gianna MANCA e Wilhelm BRAUNEDER. Bologna, Il Mulino e Berlin, Duncker & Humblot, 2000, 449 p. (Annali dell'Istituto storico italogermanico in Trento. Contributi, 10).

7467. KALIFA (Dominique). Naissance de la police privée. Détectives et agences de recherches en France 1832–1942. Paris, Plon, 2000, 334 p. (Civilisations et mentalités).

7468. KITSON (Simon). Les policiers Marseillais et le front populaire (1936–1938). *XX siècle*, 2000, 65, p. 47-57.

7469. KOHL (Gerald). Die Anfänge der modernen Gerichtsorganisation in Niederösterreich. Verlauf und Bedeutung der Organisierungsarbeiten 1849–1854. Sankt Pölten, Niederösterreichisches Institut für Landeskunde, 2000, 339 p.

7470. LANDWEHR (Achim). Policey im Alltag. Die Implementation frühneuzeitlicher Policeyordnungen in Leonberg. Frankfurt am Main, Klostermann, 2000, X-430 p.

7471. Legitimidad, representación y alternancia en España y América Latina: las reformas electorales, 1880–1930. Coor. de Carlos MALAMUD. México, Fondo de Cultura Económica, El Colegio de México, 2000, 318 p (Serie Ensayos. Fideicomiso Historia de las Américas).

7472. LONDREGAN (John). Legislative institutions and ideology in Chile. Cambridge, Cambridge U. P., 2000, 260 p. (ill.). (Political economy of institutions and decisions).

7473. M. d. B. Volksvertretung im Wiederaufbau 1946–1961. Bundestagskandidaten und Mitglieder der westzonalen Vorparlamente. Eine biographische Dokumentation. Hrsg. v. Martin SCHUMACHER. Düsseldorf, Droste, 2000, 574 p.

7474. MAY (Johanna). Vom obrigkeitlichen Stadtregiment zur bürgerlichen Kommunalpolitik. Entwicklungslinien der hannoverschen Stadtpolitik von 1699 bis 1824. Hannover, Hahnsche Buchhandlung, 2000, 541 p. (Veröffentlichungen der Historischen Kommission für Niedersachsen und Bremen, 198).

7475. MELIS (Guido). Uomini e scrivanie. Personaggi e luoghi della pubblica amministrazione. Roma, Editori Riuniti, 2000, 239 p.

7476. MIELONEN (Eevamaria). From cadres to civil servants. Transition in China's public administration. *In*: China @com. Näkökulmia 2000-luvun Itä-Aasiaan. Toim. Eevamaria MIELONEN. Turku, Turun yliopisto, 2000, p. 152-182.

7477. MOUYSSET (Sylvie). Le pouvoir dans la bonne ville. Les consuls de Rodez sous l'Ancien Régime. Toulouse, Société des lettres, sciences et arts de l'Aveyron et CNRS-Université de Toulouse-Le Mirail, 2000, 644 p.

7478. Obligations of citizenship and demands of faith: religious accommodation in pluralist democracies. edited by Nancy L. ROSENBLUM. Princeton, Princeton U. P., 2000, VI-438 p.

7479. OLECHOWSKI-HRDLICKA (Karin). Die gemeinsamen Angelegenheiten der Österreichisch-Ungarischen Monarchie. Vorgeschichte – Ausgleich 1867 – Staatsrechtliche Kontroversen. Frankfurt am Main, Lang, 2000, 552 p. (Rechtshistorische Reihe, 232).

7480. ORTAYLI (İlber). Tanzimat devrinde Osmanlı mahallî idareleri (1840–1880). (Ottoman provincial government in the Tanzimat period). Ankara, Türk Tarih Kurumu, 2000, 265 p.

7481. Policey und frühneuzeitliche Gesellschaft. Hrsg. v. Karl HÄRTER. Frankfurt am Main, Klostermann, 2000, XIII-625 p. (Ius Commune, Sonderhefte: Studien zur Europäischen Rechtsgeschichte, 129). [Cf. n° <Auswahl> 7228.]

7482. POSADA-CARBÓ (Eduardo). Electoral juggling: a comparative history of the corruption of suffrage in Latin America, 1830–1930. *Journal of Latin American studies*, 2000, 32, 3, p. 611-644.

7483. Pratiche del territorio. A cura di Angelo TORRE. *Quaderni storici*, 2000, 35, 103, p. 3-138. [Cf. n^{os} <scelta> 7130, 7288, 7461, 7542.]

7484. Praxis (Die) des Bundesgerichtshofes im deutschen Rechtsleben. Festgabe aus der Wissenschaft zum 50-jährigen Bestehen des Bundesgerichtshofes. Band 1. Bürgerliches Recht. Hrsg. v. Claus Wilhelm CANARIS und Andreas HELDRICH. Band 2. Handels- und Wirtschaftsrecht Europäisches und Internationales Recht. Hrsg. v. Andreas HELDRICH und Klaus J. HOPT. Band 3. Zivilprozeßrecht, Insolvenzenrecht, Öffentliches Recht. Hrsg. v. Karsten SCHMIDT. Band 4. Strafrecht, Strafprozeßrecht. Hrsg. v. Claus ROXIN und Gunter WIDMAIER. München, Beck, 2000, 4 vol., 950 p., 950 p., 950 p., 950 p.

7485. PRÖVE (Ralf). Stadtgemeindlicher Republikanismus und die "Macht des Volkes". Civile Ordungsformationen und kommunale Leitbilder politischer Partizipation in den deutschen Staaten vom Ende des 18. bis zur Mitte des 19. Jahrhunderts. Göttingen, Vandenhoeck & Ruprecht, 2000, 580 p. (Veröffentlichungen des Max-Planck-Instituts für Geschichte, 159).

7486. RÖSSLER (Ruth-Kristin). Justizpolitik in der SBZ/DDR 1945–1956. Frankfurt am Main, Klostermann, 2000, IX-315 p. (Ius Commune, Sonderhefte: Studien zur Europäischen Rechtsgeschichte, 136).

7487. ROUBAN (Luc). Les préfets sous la Cinquième République 1958–1997. *La revue administrative*, 2000, 317, p. 522-534.

7488. RUSCO (Elmer R.). A fateful time: the background and legislative history of the Indian Reorganization Act. Reno, University of Nevada Press, 2000, XV-363 p. (Wilbur S. Shepperson series in history and humanities).

7489. STENVALL (Jari). Käskyläisestä toimitsijaksi. Valtion keskushallinnon virkamiehistön pätevyyden arvostusten kehitys suuriruhtinaskunnan ajan alusta 2000-luvulle. (From subordinate to initiator. Development of the esteem of central administration officials' competence from the early days Finnish Grand Duchy to the 21st century). Tampere, 2000, 290 p. (Acta Univ. Tamperensis, 759).

7490. TORPEY (John). The invention of the passport: surveillance, citizenship and the state. New York, Cambridge U. P., 2000, XI-211 p. (Cambridge Studies in Law and History).

7491. TUCKER (P.). The early history of the Court of Chancery: a comparative study. *English historical review*, 2000, 115, 463, p. 791-811.

7492. VILLARD (Pierre). Histoire des institutions publiques de la France (de 1789 à nos jours). Paris, Dalloz, 2000, 244 p.

7493. YARBROUGH (Tinsley E.). The Rehnquist Court and the Constitution. Oxford, Oxford U. P., 2000, XII-306 p.

§ 3. Addenda 1998.

7494. SODDU (Francesco). The Italian Senate in the era of Giolitti and the House of Lords: some comparative insights. *Parliamentary, estates of representation*, 98, p. 103-133.

Cf. n°s 1024-1084, 4154, 4492, 4586, 7384

§ 4. Civil and penal law.

** 7495. Codex Fridericianus Marchicus. T. 1. Project des Codicis Fridericiani Marchici, 1748. T. 2. Plan du Roi, 1748. Anhang zum Codice Fridericiano, 1761. Milano, Giuffrè, 2000, 2 vol., XXXIV-576 p., XXXVIII-764 p. (Testi e documenti per la storia del processo, 3).

** 7496. Dokumentation des NS-Strafrechts. Hrsg. v. Heribert OSTENDORF. Baden-Baden, Nomos, 2000, 363 p.

** 7497. GOTTLOB MORGENBESSER (Ernst). Beiträge zum republikanischen Gesetzbuche enthalten in Anmerkungen zum allgemeinen Landrechte und zur allgemeinen Gerichtsordnung für die preußischen Staaten. Königsberg 1798. Hrsg. v. Wolfgang SCHILD. Freiburg, Haufe, 2000, 344 p.

** 7498. Libros registros – cedularios del Tucuman y Paraguay (1573–1716). Catalogo. Buenos Aires, Instituto de Investigaciones de historia del derecho, 2000, 414 p.

7499. ÅGREN (Maria). Contracts for the old or gifts for the young? On the use of wills in early modern Sweden. *Scandinavian journal of history*, 2000, 25, 3, p. 197-218.

7500. ALPA (Guido). La cultura delle regole. Storia del diritto civile italiano. Roma e Bari, Laterza, 2000, XV-486 p.

7501. ALVAZZI DEL FRATE (Paolo). L'interpretazione autentica nel XVIII secolo. Divieto di interpretatio e «riferimento al legislatore» nell'illuminismo giuridico. Torino, Giappichelli, 2000, VI-181 p.

7502. BARBACETTO (Stefano). "Tanto del ricco quanto del povero". Proprietà collettive ed usi civici in Carnia tra Antico regime ed età contemporanea. Pasian di Prato, Coordinamento Circoli Culturali della Carnia, 2000, 405 p.

7503. Beni culturali a Napoli nell'Ottocento. Atti del Convegno di studi, Napoli 5–6 novembre 1997. Roma, Ministero per i Beni e le Attività Culturali, 2000, 287 p.

7504. BERKOWITZ (Alan J.). Patterns of disengagement: the practice and portrayal of reclusion in early medieval China. Stanford, Stanford U. P., 2000, XII-296 p.

7505. Bürgerliche Gesetzbuch und seine Richter. Zur Reaktion der Rechtsprechung auf die Kodifikation des deutschen Privatrechts (1896–1914). Hrsg. v. Ulrich FALK und Heinz MOHNHAUPT. Frankfurt am Main, Klostermann, 2000, XV-676 p. (Rechtsprechung, Materialen und Studien, Veröffentlichungen des Max-Planck-Instituts für Europäische Rechtsgeschichte, 14).

7506. CASALNUOVO (Mario). L'avvocato penale nel Novecento. Soveria Mannelli, Rubbettino, 2000, 202 p.

7507. CAVANNA (Adriano), VANZELLI (Gianfrancesco). Il primo progetto di codice penale per la Lombardia napoleonica (1801–1802). Padova, CEDAM, 2000, 342 p. (Casi, fonti e studi per il diritto penale, raccolti da S. Vinciguerra, Serie III, Gli Studi, Storia del diritto penale, 15).

7508. Civil rights since 1787: a reader on the black struggle. Ed. by Jonathan BIRNBAUM and Clarence TAYLOR. New York, New York U. P., 2000, XXI-935 p.

7509. Code civil français (Le). Evolution de textes depuis 1804. Sous la dir. de Ph. BIHR. Paris, Dalloz, 2000, XVII-653 p.

7510. COGROSSI (Cornelia). La criminalistica italiana del XVIII secolo sulla «certezza morale» antesignana del libero convincimento del giudice: note. *Rivista di storia del diritto italiano*, 2000, 73, p. 121-236.

7511. COLAO (Floriana). Progetti di codificazione civile nella Toscana della restaurazione. Bologna, Monduzzi, 2000, 204 p.

7512. DA PASSANO (Mario). Emendare o intimidire? La codificazione del diritto penale in Francia e in Italia durante la Rivoluzione e l'Impero. Torino, Giappichelli, 2000, 377 p.

7513. DEL BAGNO (Ileana). Il collegio napoletano dei dottori. Privilegi, decreti, decisioni. Napoli, Jovene, 2000, 310 p.

7514. DEZZA (Ettore). Lezioni di storia della codificazione civile: il Code Civil (1804) e l'Allgemeines Burgerliches Gesetzbuch (ABGB, 1812). Torino, Giappichelli, 2000, VIII-172 p.

7515. Eigentum im internationalen Vergleich 18.–20. Jahrhundert. Hrsg. v. Hannes SIEGRIST und David SUGARMAN. Göttingen, Vandenhoeck & Ruprecht, 2000, 294 p. (Kritische Studien zur Geschichtswissenschaft, 130).

7516. EWING (K. D.), GEARTY (C. A.). The struggle for civil liberties: political freedom and the rule of law in Britain, 1914–1945. Oxford, Oxford U. P., 2000, XVII-451 p.

7517. FALK (Ulrich). Zur Geschichte der Strafverteidigung. Aktuelle Beobachtungen und rechtshistorische Grundlagen. *Zeitschrift der Savigny-Stiftung für Rechtsgeschichte. Germanistische Abteilung*, 2000, 117, p. 395-449.

7518. FUCHS (Rachel). Seduction, paternity, and the law in Fin de Siècle France. *Journal of modern history*, 2000, 72, 4, p. 944-989.

7519. GALOUNOV (T.). La loi pénale concernant les responsables de la catastrophe nationale en Bulgarie de l'an 1919. *Bulgarian historical review*, 2000, 28, 3-4, p. 139-147.

7520. GASKILL (Malcolm). Crime and mentalities in early modern England. Cambridge, Cambridge U. P., 2000, 377 p.

7521. GIGLIO (Francesco). Condictio proprietaria und europäisches Bereicherungsrecht. Eine Untersuchung auf rechtshistorischer und rechtsvergleichender Basis mit besonderer Berücksichtigung des deutschen und italienischen Rechts. Berlin, Duncker & Humblot, 2000, 352 p. (Untersuchungen zum Europäischen Privatrecht, 2).

7522. GÓMEZ GONZÁLEZ (Inés). La justicia en almoneda. La venta de oficios en la Chancillería de Granada (1505–1834). Prol. de Antonio DOMÍNGUEZ ORTIZ. Granada, Editorial Comares, 2000, 258 p.

7523. HEIN (Oliver). Von Rohen zum Hohen. Öffentliches Strafrecht im Spiegel der Strafrechtsgeschichtsschreibung des 19. Jahrhunderts. Köln, Weimar u. Wien, Böhlau, 2000, XIV-418 p. (Konflikt, Verbrechen und Sanktion in der Gesellschaft Alteuropas, Symposien und Synthesen, 3).

7524. HELGI (Gunnlaugsson), GALLIHER (John F.). Wayward Icelanders: punishment, boundary maintenance, and the creation of crime. Madison, University of Wisconsin Press, 2000, XIII-170 p.

7525. HOFER (Sibylle). Drittschutz und Zeitgeist. Ein Beitrag zur privatrechtlichen Zeitgeschichte. *Zeitschrift der Savigny-Stiftung für Rechtsgeschichte. Germanistische Abteilung*, 2000, 117, p. 377-394.

7526. HOHOFF (Ute). An den Grenzen des Rechtsbeugungstatbestandes. Eine Studie zu den Strafverfahren gegen DDR-Juristen. Berlin, Berlin Verlag, 2000, XVII-237 p. (BJU, Strafrecht, 9).

7527. Ideologie e patrimonio storico-culturale nell'età rivoluzionaria e napoleonica: a proposito del trattato di Tolentino. Atti del convegno, Tolentino, 18–21 settembre 1997. Roma, Ministero per i Beni e le Attività Culturali, 2000, 654 p.

7528. KEBBEDIES (Frank). Außer Kontrolle. Jugendkriminalpolitik in der NS-Zeit und der frühen Nachkriegszeit. Essen, Klartext, 2000, 307 p. (Düsseldorfer Schriften zur Neueren Landesgeschichte und zur Geschichte Nordrhein-Westfalens, 54).

7529. KING (Peter). Crime, justice and discretion in England, 1740–1820. Oxford, Oxford U. P., 2000, XIII-383 p.

7530. KROGH (Tyge). Oplysningstiden og det magiske: Henrettelser og korporlige straffe i 1700-tallets første halvdel. (The Enlightenment and the magical: executions and corporal punishments in Denmark in the first half of the eighteenth-century). København, Samleren, 2000, 608 p.

7531. LABARDI (Andrea). La Facoltà giuridica senese e la Restaurazione, con il testo delle Istituzioni civili di Pietro Capei. Milano, Giuffrè, 2000, 284 p. (Quaderni di "Studi Senesi", 93).

7532. MAAS (Martin Jürgen). Die Geschichte des Eigentumsvorbehalts, insbesondere im 18. und 19. Jahrhundert. Frankfurt am Main, Lang, 2000, 356 p. (Rechtshistorische Reihe, 224).

7533. MARTSCHUKAT (Jürgen). Inszeniertes Töten. Eine Geschichte der Todesstrafe vom 17. bis zum 19. Jahrhundert. Köln, Weimar u. Wien, Böhlau, 2000, VIII-365 p.

7534. MELIK (Jelka). V imenu njegovega veličanstva kralja!: kazensko sodstvo v jugoslovanski Sloveniji v letih 1930–1941. (In the name of his majesty the king: Slovene penal judiciary system in Yugoslav Slovenia 1930–1941). Ljubljana, Arhiv Republike Slovenije, 2000, 163 p.

7535. MIERSCH (Matthias). Der sogenannte référé législatif. Eine Untersuchung zum Verhältnis Gesetzgeber, Gesetz und Richteramt seit dem 18. Jahrhundert. Baden-Baden, Nomos, 2000, 323 p. (Fundamenta juridica, 36).

7536. NESCHWARA (Christian). Österreichs Notariatsrecht in Mittel- und Osteuropa. Zur Geltung und Ausstrahlung des österreichischen Notariats. Wien, Manz, 2000, XVI-102 p.

7537. NOBIS (Frank). Die Strafprozessgesetzgebung der späten Weimarer Republik (1930–1932), insbesondere die Notverordnung vom 14. Juni 1932. Baden-Baden, Nomos, 2000, XV-190 p. (Juristische Zeitgeschichte 3, Beiträge zur modernen deutschen Strafgesetzgebung, 6).

7538. NOVARESE (Daniela). Costituzione e codificazione nella Sicilia dell'Ottocento. Il «Progetto di Codice penale" del 1813. Milano, Giuffrè, 2000, 346 p. (Università degli Studi di Messina, Facoltà di Scienze Politiche, Serie Studi storico giuridici, 10). – EADEM. Istituzioni e processo di codificazione nel regno delle Due Sicilie. Le "Leggi penali" del 1819. Milano, Giuffrè, 2000, 303 p. (Università degli Studi di Messina, Facoltà di Scienze Politiche, Serie Studi storico giuridici, 9).

7539. OBERWITTLER (Dietrich). Von der Strafe zur Erziehung? Jugendkriminalpolitik in England und Deutschland (1850–1920). Frankfurt am Main u. New York, 2000, 382 p. (Campus Forschung, 799).

7540. Officium advocati. Hrsg. v. Laurent MAYALI, Antonio PADOA SCHIOPPA und Dieter SIMON. Frankfurt am Main, Klostermann, 2000, X-419 p. (Rechtsprechung, Materialen und Studien, 15).

7541. PACE (Giacomo). Il discernimento dei fanciulli. Ricerche sulla imputabilità dei minori nella cultura giuridica moderna. Torino, Giappichelli, 2000, 237 p.

7542. PALMERO (Beatrice). Regole e registrazione del possesso in età moderna. Modalità di costruzione del territorio in alta Val Tanaro. In: Pratiche del territorio [Cf. n° 7483], p. 49-86.

7543. PRAVILOVA (E. A.). Zakonnost' i prava lichnosti: Administrativnaja justitsija v Rossii, vtoraja polovina XIX v.–oktjabr' 1917 g. (Administrative justice in Russia, from the 2nd half of the 19th century to October 1917). Sankt-Peterburg, SZAGS a. Obrazovanie-Kul'tura, 2000, 287 p.

7544. Pravo, zgodovina, arhivi. 1, Prispevki za zgodovino pravosodja. (Law, history, archives: 1. Papers on the history of jurisdiction). Glavni urednik Jože ŽONTAR. Ljubljana, Arhiv Republike Slovenije, 2000, 299 p.

7545. RASCHKA (Johannes). Justizpolitik im SED-Staat. Anpassung und Wandel des Strafrechts während der Amtszeit Honeckers. Köln, Weimar u. Wien, Böhlau, 2000, 375 p. (Schriften des Hannah-Arendt-Instituts für Totalitarismusforschung, 13).

7546. RÓDENAS VILAR (Rafael). El hombre de la capa verde. Historia de un error judicial en la España de Felipe II. Alicante, Ed. Univers. de Alicante, 2000, 330 p.

7547. RÜFNER (Th.). Vertretbare Sachen? Die Geschichte des res, quae pondere numero mensura constant. Berlin, Duncker & Humblot, 2000, 172 p. (Schriften zur europäischen Rechts- und Verfassungsgeschichte, 31).

7548. SANDMANN (Hendrik). Die Entwicklung vom Begriff und Inhalt des Wirtschaftsrechts durch die Rechtswissenschaft in der Weimarer Republik. Frankfurt am Main, Lang, 2000, 215 p. (Rechtshistorische Reihe, 230).

7549. SCHERMAIER (M. J.). Die Bestimmung des wesentlichen Irrtums von den Glossatoren bis zum

BGB. Köln, Weimar u. Wien, Böhlau, 2000, 789 p. (Forschungen zur Neueren Privatrechtsgeschichte, 29).

7550. SCHÖNIGER-HEKELE (Bernhard). Die österreichische Zivilprozessreform 1895. Wirkung im Inland bis zum Ausbruch des Ersten Weltkrieges 1914. Ausstrahlung ins Ausland. Frankfurt am Main, Lang, 2000, 204 p. (Rechts- und sozialwissenschaftliche Reihe, 26).

7551. SCIUMÈ (Alberto). Organizzare l'economia. Le Camere di commercio nell'Italia contemporanea fra diritto commerciale e diritto amministrativo. Brescia, Promodis Italia, 2000, 99 p. (Università degli Studi di Brescia. Quaderni brevi del Dipartimento di Scienze giuridiche).

7552. SOMMA (Alessandro). Autonomia privata e struttura del consenso contrattuale. Aspetti storico-comparativi di una vicenda concettuale. Milano, Giuffrè, 2000, 466 p. (Problemi di diritto comparato, 4).

7553. Strafjustiz und DDR-Unrecht. Dokumentation. Band 1. Wahlfälschung. Hrsg. v. Klaus MARXEN und Gerhard WERLE unter Mitarbeit von Jan MÜLLER und Petra SCHÄFTER. Berlin, De Gruyter, 2000, XLVIII-528 p.

7554. STÜBINGER (Stephan). Schuld, Strafrecht und Geschichte. Die Entstehung der Schuldzurechnung in der deutschen Strafrechtshistorie. Köln, Weimar u. Wien, Böhlau, 2000, 460 p. (Konflikt, Verbrechen und Sanktion in der Gesellschaft Alteuropas, Symposien und Synthesen, 4).

7555. SUFAJ (Fehmi). Historia e burgjeve të Shqipërisë: gjatë shek. XX. (La storia delle prigioni in Albania nel XX secolo). Tiranë, Albin, 2000, 182 p. (ill).

7556. SYDOW (Gernot). Die Verwaltungsgerichtsbarkeit des ausgehenden 19. Jahrhunderts. Eine Quellenstudie zu Baden, Württemberg und Bayern mit einem Anhang archivalischer und parlamentarischer Quellen. Heidelberg, C. F. Müller, 2000, XIX-273 p. (Freiburger Rechts- und Staatswissenschaftlichen Abhandlungen, 66).

7557. TAVILLA (Carmelo Elio). Riforme e giustizia nel Settecento estense. Il Supremo Consiglio di Giustizia (1761–1796). Milano, Giuffrè, 2000, 563 p.

7558. TESSITORE (Giovanni). Fascismo e pena di morte. Consenso e informazione. Milano, F. Angeli, 2000, 598 p. (Collana di diritto e società, 24).

7559. THIEMRODT (Ivo). Strafjustiz und DDR-Spionage. Zur Strafverfolgung ehemaliger DDR-Bürger wegen Spionage gegen die Bundesrepublik. Berlin, Berlin Verlag, 2000, XVIII-376 p. (BJU, Strafrecht, 7).

7560. TITA (Massimo). Sentenze senza motivi. Documenti sull'opposizione delle magistrature napoletane ai dispacci del 1774. napoli, Jovene, 2000, XX-226 p. (Storia e diritto. Collana di Studi e Testi, Studi, 45).

7561. TRETVIK (Aud Mikkelsen). Tretter, ting og tillitsmenn. En undersøkelse av konflikthåndtering i det norske bygdesamfunnet på 1700-tallet. (The handling of conflicts in rural society during the 18th century). Trondheim, Historisk institutt, Norges teknisk-naturvitenskapelige universitet, 2000, 456 p. (ill.). (Skriftserie fra Historisk institutt, 32).

7562. TROITSKIJ (Nikolaj A.). Advokatura v Rossii i politicheskie protsessy 1866–1904 gg. (The bar in Russia and the political trials of 1866–1904). Tula, Avtograf, 2000, 454 p. (ill.).

7563. WICKE (H.). Respondeat Superior: Haftung fü Verrichtungsgehilfen im römischen, römisch-holländischen, englischen und südafrikanischen Recht. Berlin, Duncker & Humblot, 2000, 483 p. (Schriften zur Europäischen Rechts- und Verfassungsgeschichte, 32).

7564. YLIKANGAS (Heikki). Aikansa rikos. Historiallisen kehityksen valaisijana. (Crime of its time. Social impacts on the understanding crime and criminals in Finland during the 17th and 19th centuries). Helsinki, WS, 2000, 365 p. (ill.).

7565. YRTTIAHO (Eero). Tuomarimentaliteetti ja kansan oikeustaju. Oikeudenkäytön representaatiot suomalaisessa kaunokirjallisuudessa ja mediassa käytävässä oikeudenmukaisuuskeskustelussa. (The mentality of the judge and the sense of right and wrong among people. The representations of processing in court concerning the discussion of justice in literature and media). Rovaniemi, [s. n], 2000, 385 p. (Acta Universitatis Lapponiensis, 34).

Cf. n°s 182, 5038, 5071, 6957

§ 5. International law.

7566. ALLSTADT (Markus). Die offene Völkerrechtslage im deutsch-tschechischen Verhältnis und ihre Bedeutung für die Osterweiterung der Europäischen Union: zugleich Überlegungen zu den Auswirkungen zwingenden Völkerrechts in der Vetriebenenfrage. Frankfurt, Lang, 2000, 208 p. (Schriften zum Staats- und Völkerrecht, 86).

7567. BARKAN (Alazar). Guilt of nations: restitution and negotiating historical injustices. New York, W. W. Norton, 2000, XLI-414 p.

7568. BELL (Christine). Peace agreements and human rights. Oxford a. New York, Oxford U. P., 2000, X-409 p.

7569. BOROUMAND (Ladan). Emigration and the Rights of Man: French revolutionary legislators equivocate. *Journal of modern history*, 2000, 72, 1, p. 67-108.

7570. GREWE (Wilhelm G.). The epochs of international law. Berlin a. New York, de Gruyter, 2000, XXII-780 p.

7571. HIGURASHI (Yoshinobu). Gassyûkoku to tainichi senpan saiban no syûketsu. (The United States of America and the conclusion of the trials of Japanese war criminals in 1946–1948). *Shigaku Zasshi*, 2000, 109, 11, p. 1-34. [English summary].

7572. KÉVONIAN (D.). Représentation, enjeux politiques et codification juridique: les réfugiés des années vingt. *Relations internationales*, 2000, 101, p. 21-39.

7573. MÜLLER (Sibylle). Gibt es Menschenrechte bei Samuel Pufendorf? Frankfurt am Main, Lang, 2000, XLII-166 p. (Rechtshistorische Reihe, 231).

7574. SIMMONS (Beth A.). International law and state behavior: commitment and compliance in international monetary affairs. *American political science review*, 2000, 4, p. 819-836.

7575. WALTON-JORDAN (Ulrike). Die britische Gerichtsbarkeit in Nordwestdeutschland 1945–1949. *Zeitschrift der Savigny-Stiftung für Rechtsgeschichte. Germanistische Abteilung*, 2000, 117, p. 362-376.

P

HISTORY OF INTERNATIONAL RELATIONS

§ 1. General. 7576-7641. – § 2. History of colonization and decolonization (a. General; b. Asia; c. Africa; d. America; e. Oceania). 7642-7758. – § 3. From 1500 to 1789 (a. General; b. 1500–1648; c. 1648–1789). 7759-7851. – § 4. From 1789 to 1815. 7852-7871. – § 5. From 1815 to 1910. 7872-7952. – § 6. From 1910 to 1935. The First World War. 7953-8054. – § 7. From 1935 to 1945. The Second World War (a. General; b. Diplomacy. Economy; c. Military operations; d. Resistance). 8055-8235. – § 8. From 1945. 8236-8473.

§ 1. General.

** 7576. Iran in the Persian Gulf 1843–1966. Research editors A. L. P. BURDETT and A. SEAY. Slough, Archive Editions, 2000, 6 vol., [s. p.]. (Arabian geopolitics).

7577. Alien justice: wartime internment in Australia and North America. Ed. by Kay SAUNDERS and Roger DANIELS. St. Lucia, University of Queensland Press, 2000, XIX-323 p.

7578. Anglo-American attitudes: from revolution to partnership. Ed. by Fred M. LEVENTHAL and Roland QUINAULT. Aldershot, Ashgate, 2000, XI-313 p. (ill.).

7579. Anglo-French relations in the twentieth century. Ed. by Glyn STONE and Alan SHARP. New York, Routledge, 2000, XI-335 p.

7580. Anglo-German dialogue (An): the Munich lectures on the history of international relations. Ed. by Adolf M. BIRKE, Magnus BRECHTKEN and Alaric SEARLE. München, Saur, 2000, 286 p.

7581. Argentina – Estados Unidos: acuerdos bilaterales: 1853–2000. Buenos Aires, Consejo Argentino para las Relaciones Internacionales, Centro de Estudios de Politica Exterior, 2000, XIII-895 p.

7582. Balkans (Les) dans les relations internationales. Part 1, 2. Hommage à Raymond Poidevin par P GUILLEN; introduction de P. DU BOIS. *Relations internationales*, 2000, 103, p. 269-379; 104, p. 397-523. [Cf. n^{os} <sélection> 7618, 7621, 7886, 7905, 7959, 7968, 7998, 8018, 8042, 8276, 8315, 8340, 8388, 8446.]

7583. BAUMFALK (Gerhard). Tatsachen zur Kriegsschuldfrage: Diplomatie, Politik, Hintergrund, 1871–1939. Tubingen, Grabert, 2000, 750 p. (ill.). (Veröffentlichungen des Institutes für Deutsche Nachkriegsgeschichte, 26).

7584. Between empire and nation: Australia's external relations from Federation to the Second World War. Ed. by Carl BRIDGE and Bernard ATTARD. Melbourne, Australian Scholary Publishing, 2000, XI-258 p.

7585. Biographisches Handbuch des deutschen Auswärtigen Dienstes: 1871–1945. Band 1. A–F. Bearb. v. Johannes HÜRTER [et al.]. Paderborn, Schoningh, 2000, XLVIII-633 p.

7586. BLITZ (Amy). The contested state: American foreign policy and regime change in the Philippines. Lanham, Rowman & Littlefield Publishers, 2000, [s. p.].

7587. BOCCIA ROMANACH (Alfredo). Paraguay y Brasil: crónica de sus conflictos. Asunción, Editorial El Lector, 2000, 243 p. (Gran enciclopedia de la cultura paraguaya).

7588. Bridging the divide: 400 years the Netherlands-Japan. Ed. by Leonard BLUSSÉ, Willem REMMELINK and Ivo SMITS. Leiden, Hotei, 2000, 288 p. (ill.).

7589. CARTER (John J.). Covert operations as a tool of presidential foreign policy in American history from 1800 to 1920: foreign policy in the shadows. Lewiston, Edwin Mellen Press, 2000, V-223 p. (Studies in American history, 26).

7590. Caucasia between the Ottoman Empire and Iran, 1555–1914. Hrsg. v. Raoul MOTIKA und Michael URSINUS. Wiesbaden, Reichert, 2000, 222 p. (ill., maps). (Kaukasienstudien, 2 = Caucasian studies, 2).

7591. CERVO (Amado Luiz), CALVET DE MAGALHAES (José). Depois das caravelas: as relaçoes entre Portugal e Brasil, 1808–2000. Organizaçao e apresentaçao de Dário Moreira de Castro Alves. Brasília, Editora Universidade de Brasília, 2000, 397 p.

7592. CHKNAVEROVA (A. A.). K istorii russko-iaponskikh otnoschenii. (History of Russian-japanese relations). Moskva, Kompaniia Sputnik, 2000, 183 p.

7593. CLOGG (Richard). Anglo-Greek attitudes: studies in history. Basingstoke, Macmillan a. Oxford, St. Antony's College, 2000, VIII-207 p. (St. Antony's series).

7594. Conflits et opérations en Asie orientale, 1930–1950. Intr. de J.-C. ALLAIN. *Guerres mondiales et conflits contemporaine*, 2000, 50, 199, p. 3-80. [Cf. n^{os} <sélection> 8156, 8166, 8170, 8197, 8208.]

7595. Deutsche-russische Kriege und der Export der Revolution. Hrsg. v. Nikolaus LOBKOWICZ. Köln, Weimar u. Wien, Böhlau, 2000, 330 p. (Forum für osteuropäische Ideen- und Zeitgeschichte, 2000, 1).

7596. Deutschland und Frankreich: vom Konflikt zur Aussöhnung; die Gestaltung der westeuropäischen Sicherheit, 1914–1963. Hrsg. V. Stephen A. SCHUKER; unter Mitarbeit von Elisabeth MÜLLER-LUCKNER. München, Oldenbourg, 2000, XIX-280 p. (Schriften des Historischen Kollegs. Kolloquien, 46).

7597. Deutschland, Frankreich, Russland: Begegnungen und Konfrontationen = La France et l'Allemagne face à la Russie. Im Auftrag des Deutsch-Französischen Historikerkomitees herausgegeben von Ilja MIECK und Pierre GUILLEN. München, Oldenbourg, 2000, 383 p. (ill.). [Cf. n° <Auswahl> 7986.]

7598. DOS SANTOS ALVEZ (Jorge Manuel). Portugal e a China: Conferencias no III curso livre de historia das relaççoes entre Portugal e a China (seculos XVI–XIX). Lisboa, Fundacao Oriente, 2000, 167 p.

7599. EMMES (Manfred). Die Aussenpolitiken der USA, Japans und Deutschlands im wechselseitigen Einfluss von der Mitte des 19. bis Ende des 20. Jahrhunderts. Münster, Lit, 2000, 256 p. (Studien zur Politikwissenschaft. Abteilung B, 91).

7600. FORMIGONI (Guido). Storia della politica internazionale nell'età contemporanea. Bologna, Il Mulino, 2000, 554 p.

7601. GALLICCHIO (Marc S.). The African American encounter with Japan and China: Black internationalism in Asia, 1895–1945. Chapel Hill a. London, University of North Carolina Press, 2000, X-262 p. (ill., ports).

7602. GEBHART (Jan). Proměny válečné propagandy ve světových konfliktech 20. století. (Changes in wartime propaganda in the world conflicts of the 20th century). *Historie a vojenství*, 2000, 49, 1, p. 3-13.

7603. GILDERHUS (Mark Theodore). The second century: U.S.-Latin American relations since 1889. Wilmington, Scholarly Resources, 2000, XVI-282 p.

7604. GILQUIN (Michel). D'Antioche au Hatay: l'histoire oubliée du Sandjak d'Alexandrette: nationalisme turc contre nationalisme arabe, la France, arbitre? Paris, L'Harmattan, 2000, 220 p. (Comprendre le Moyen-Orient).

7605. Grenzen und Grenzräume in der deutschen und polnischen Geschichte: Scheidelinie oder Begegnungsraum? Hrsg. v. Georg STÖBER und Robert MAIER. Hannover, Hahn, 2000, 326 p. (ill., maps). (Studien zur internationalen Schulbuchforschung, 104).

7606. Guerre (La) de course en Méditerranée (1515–1830). Actes du Colloque de Bonifacio, 1999 présentés et réunis par Michel VERGÉ-FRANCESCHI et Antoine-Marie GRAZIANI. Paris, Presses de l'Univ. Paris-Sorbonne et Ajaccio, Ed. Alin Piazzola, 2000, 277 p. [Cf. n^{os} <sélection> 7785, 7795.]

7607. HALE (William Mathew). Turkish foreign policy, 1774–2000. London, Frank Cass, 2000, IX-375 p.

7608. HIRATA (Masahiro). Igirisu teikoku to sekai sisutemu. (The British Empire and the world system). Tokyo, Koyo Shobo, 2000, 275 p.

7609. History (The) of Anglo-Japanese relations. Vol. 1. The political-diplomatic dimension, 1600–1930. Ed. by Ian NISH and Yoichi KIBATA. Assisted by Tadashi KURAMATSU. Foreword by Chihiro HOSOYA and Ian NISH. New York, St. Martin's Press, 2000, XIII-282 p. (The history of Anglo-Japanese relations, 1600–2000).

7610. HOLDEN (Robert H.), ZOLOV (Eric). Latin America and the United States: a documentary history. New York a. Oxford, Oxford U. P., 2000, 363 p.

7611. Hongrie (La) dans les conflits du XXe siècle. *Guerres mondiales et conflits contemporaine*, 2000, 50, 200, p. 3-138. [Cf. n^{os} <sélection> 4738, 4739, 8049, 8094, 8282, 8377.]

7612. Internationale Geschichte. Themen – Ergebnisse – Aussichten. Hrsg. v. Wilfried LOTH und Jürgen OSTERHAMMEL. München, Oldenbourg, 2000, XIV-415 p. (Studien zur Internationalen Geschichte, 10).

7613. Irish foreign policy, 1919–1966: from independence to internationalism. Ed. by Michael KENNEDY and Joseph Morrison SKELLY. Dublin, Four Courts Press, 2000, 350 p.

7614. KALLEY (Jacqueline Audrey). South Africa's treaties in theory and practice, 1806–1998. Lanham, Scarecrow Press, 2000, [s. p.].

7615. KASTNER (Quido). Některé historické souvislosti česko-německých vztahů na lokální úrovni. Zpráva z výzkumné sondy o názorech starousedlíků na česko-německé vztahy. (Some historical contexts of Czech-German relations on the local level). *In*: Vytváření přeshraničního společenství na česko-německé hranici. Ed. František ZICH. Ústí nad Labem, 2000, p. 154-225.

7616. KHAN (Zubeda K.). International politics and India's foreign policy under Nehru. Lucknow, New Royal Book Co., 2000, 122 p.

7617. LAFONT (Jean-Marie). Indika: essays in Indo-French relations, 1630–1976. New Delhi, Manohar, 2000, 492 p.

7618. LIEBICH (A.). Les conflits dans les Balkans: deux sources, aucune solution. *In*: Balkans (Les) dans les relations internationales, Part 2 [Cf. n° 7582], p. 507-518.

7619. LIEVEN (Dominic). Empire: the Russian empire and its rivals. London, John Murray, 2000, XXIII-486 p. (ill., maps).

7620. MĄCZAK (Antoni). Nieformalne systemy władzy w Polsce i Europie XVI–XVIII w. (Die unformellen Machtsysteme in Polen und in Europa XVI.–XVIII. Jhdt). Warszawa, Semper, 2000, 355 p.

7621. MICHEL (B.). Les enjeux des Grandes Puissances dans les Balkans, depuis le Congrès de Berlin jusqu'à nos jours. *In*: Balkans (Les) dans les relations internationales, Part 1 [Cf. n° 7582], p. 279-288.

7622. MOLINARI (Maurizio). L'interesse nazionale: dieci storie dell'Italia nel mondo. Roma, Editori Laterza, 2000, XI-128 p. (Saggi tascabili Laterza, 243).

7623. MONNET (Sylvie). La politique extérieure de la France depuis 1870. Paris, A. Colin, 2000, 224 p.

7624. Nations et Saint-Siège au XXe siècle. Ed. par Hélène CARRÈRE D'ENCAUSSE et Philippe LAVILLAIN. Paris, Fayard, 2000, 448 p.

7625. Ours (L') et le coq: trois siècles de relations franco-russes: essais en l'honneur de Michel Cadot. Textes réunis par Francine-Dominique LIECHTENHAN. Paris, Presses de la Sorbonne Nouvelle, 2000, 286 p.

7626. PAZ (Abel). La cuestión de Marruecos y la República española. Madrid, Fundacion de Estudios Libertarios Anselmo Lorenzo, 2000, 238 p. (ill.).

7627. PFEIFFER (Rolf). Neuseeland und die Wahrnehmung seiner aussenpolitischen Interessen 1856–1972: das Phänomen der Unabhängigkeit in der Anhängigkeit. Frankfurt am Main u. Oxford, P. Lang, 2000, 2 vol., [s. p.].

7628. Polen und Österreich im 18. Jahrhundert. Hrsg. v. Walter LEITSCH und Stanisław TRAWKOWSKI unter Mitwirkung v. Wojciech KRIEGSEISEN. Warszawa, Semper, 2000, 152 p. (Poln. Akad. der Wissenschaften. Tadeusz-Manteuffel-Inst. für Geschichte).

7629. RHODES (Carolyn). Pivotal decisions: selected cases in twentieth century international politics. Fort Worth a. London, Harcourt College Publishers, 2000, XII-179 p.

7630. RICCARDS (Michael P.). The presidency and the middle kingdom: China, the United States, and executive leadership. Lanham, Lexington Books, 2000, [s. p.].

7631. Rossiia i Britaniia vypusk 2: chteniia pamiati N. A. Erofeeva = Russia and Britain volume 2: readings in memory of N. A. Erofeev. Otv. red. V. G. TRUKHANOVSKII. Moskva, Rossiiskaia Akademiia Nauk. Institut vseobshchei istorii, 2000, 186 p.

7632. SEELY (Robert). Russo-Chechen political relations, 1800–2000: a deadly embrace. London, Frank Cass, 2000, XI-333 p. (ill.). (Cass series on Soviet military experience, 6).

7633. SHISHOV (A. V.). Rossiia i Iaponiia: istoriia voennykh konfliktov. Moskva, Veche, 2000, 572 p. (Voennye tainy XX veka).

7634. SMITH (Tony). Foreign attachments: the power of ethnic groups in the making of American foreign policy. Cambridge a. London, Harvard U. P., 2000, X-190 p.

7635. SUGARMAN (Sidney). The unrelenting conflict: Britain, Balfour and betrayal. Lewes, Book Guild, 2000, 282 p, (ill.).

7636. Tolérance et solidarités dans les pays pyrénéens. Actes du Colloque de Foix 18–20 septembre 1998, publiés par Claudine PAILHÈS avec la collab. de Philippe DE ROBERT. Foix, Archives départementales de l'Ariège, 2000, 565 p. [Cf. n° <sélection> 7812.]

7637. TOWLE (Philip). Democracy and peace making: negotiations and debates, 1815–1973. London a. New York, Routledge, 2000, VI-212 p.

7638. TUMINEZ (Astrid S.). Russian nationalism since 1856: ideology and the making of foreign policy. Lanham a. Oxford, Rowman & Littlefield, 2000, X-339 p.

7639. VÁZQUEZ CARRIZOSA (Alfredo). Relatos de historia diplomática de Colombia, siglo XX. T. 1. Literatura y política. T. 2. Estados Unidos, San Andrés y Providencia y América Latina. T. 3. La paz internacional, relaciones con la Santa Sede. T. 4. La exportación de café, los intercambios internacionales. Santafé de Bogotá, Centro Editorial Javeriano, 2000, 4 vol., 1040 p.

7640. Vestiges of war: the Philippine-American War and the aftermath of an imperial dream, 1899–1999. Ed. by Angel VELASCO SHAW and Luis H. FRANCIA. New York, New York U. P., 2000, [s. p.].

7641. VINOGRADOV (Vladilen N.). Rossija i Balkany: Ot Ekateriny Velikoj do Pervoj mirovoj vojny. (Russia and the Balkans: from Catherine the Great till World War I). Lewinston, Queenston a. Lampeter, Edwin Mellen Press, 2000, 392 p. (Rossijskie issledovanija po mirovoj istorii i kul'ture, 4). [Eng. summary and table of contents].

Cf. n[os] *931, 964, 1932, 4035-4121, 8295*

§ 2. History of colonization and decolonization.

a. General

** 7642. SCHMIDT (Nelly). Abolitionnistes de l'esclavage et réformateurs des colonies 1820–1851. Analyse et documents. Paris, Karthala, 2000, 1196 p.

7643. ALEXANDRE (Valentim). Velho Brasil, novas Áfricas: Portugal e o Império (1808–1975). Lisboa,

Edicoes Afrontamento, 2000, 244 p. (Colecçao Biblioteca das ciências do homem. História, 19).

7644. BRANDAO FERREIRA (Joao José). A evoluçao do conceito estratégico ultramarino português: da conquista de Ceuta à conferência de Berlim. S.Pedro do Estoril, Atena, 2000, 436 p.

7645. DARWIN (John). Diplomacy and decolonization. *In*: International diplomacy and colonial retreat [Cf. n° 7650], p. 5-24.

7646. DUFOUR (Jean-Louis). Les décolonisation de 1945 à nos jours. Paris, A. Colin, 2000, 95 p.

7647. ELTIS (David), [et al.]. The Trans-Atlantic slave trade: a database on CD-ROM. Cambridge, Cambridge U. P., 2000, CD-ROM.

7648. ETEMAD (Bouda). La possession du monde: poids et mesures de la colonisation (XVIIIe–XXe siècles). Paris et Bruxelles, Ed. Complexe, 2000, 351 p. (Questions à l'histoire).

7649. GROSSE (Pascal). Kolonialismus, Eugenik und bürgerliche Gesellschaft in Deutschland 1850–1918. Frankfurt a. New York, Campus Verlag, 2000, 266 p. (Campus Forschung, 815).

7650. International diplomacy and colonial retreat. Ed. by Kent FEDOROWICH and Martin THOMAS. *Journal of imperial and Commonwealth history*, 2000, 28, 3, p. 1-252. [Cf. n^{os} <choice> 7645, 7665, 7673, 7678, 7679, 7693, 7705, 7721, 7725, 8347, 8353, 8359.]

7651. KIELSTRA (Paul Michael). The politics of slave trade suppression in Britain and France, 1814–1848: diplomacy, morality and economics. Basingstoke, Macmillan, 2000, XIV-388 p.

7652. KIRK-GREENE (Anthony). Britain's imperial administrators, 1858–1866. London, Macmillan, 2000, XVI-347 p.

7653. KUMAR (Dharma). Colonialism, property and the state. Delhi a. Oxford, Oxford U. P., 2000, 393 p.

7654. LENMAN (Bruce P.). Britain's colonial wars c. 1550–1688. Harlow, Longman, 2000, 304 p. (ill.). (Modern wars in perspective). – IDEM. Britain's colonial wars, 1688–1783. Harlow, Longman, 2000, 304 p. (ill.). (Modern wars in perspective).

7655. LEWIS (J. E.). 'Tropical east ends' and the Second World War: some contradictions in colonial office welfare initiatives. *Journal of imperial and Commonwealth history*, 2000, 28, 2, p. 42-66.

7656. *Vacat*.

7657. ORIZIO (Riccardo). Lost white tribes: the end of privilege and the last colonials in Sri Lanka, Jamaica, Brazil, Haiti, Namibia and Guadeloupe. London, Secker & Warburg, 2000, XII-270 p. (ill.).

7658. THOMPSON (Andrew S.). Imperial Britain: the empire in British politics, c. 1880–1932. Harlow, Longman, 2000, XVII-219 p.

7659. THORN (Gary). End of empires: European decolonisation, 1919–1980. London, Hodder & Stoughton Educational, 2000, VIII-132 p. (ill., maps). (Access to history).

7660. TORRÓ (Josep). Jérusalem ou Valence: la première colonie d'Occident. *In*: Contre les stéréotypes. Etudes sur la colonisation et l'esclavage. *Annales*, 2000, 55, 5, p. 983-1008.

Cf. n°s 6955, 8086

b. Asia

7661. AMES (Glenn J.). Renascent empire? The House of Braganza and the quest for stability in Portuguese Monsoon Asia, c. 1640–1683. Amsterdam, Amsterdam U. P., 2000, 262 p.

7662. Armée française (L') dans la guerre d'Indochine (1946–1954): adaptation ou inadaptation? Sous la direction de Maurice VAÏSSE. Bruxelles, Complexe, 2000, 363 p. (Interventions).

7663. BLYTH (Robert J.). Britain versus India in the Persian Gulf: the struggle for political control, c. 1928–1948. *Journal of imperial and Commonwealth history*, 2000, 28, 1, p. 90-111.

7664. BODIN (Michel). Les africains dans la guerre d'Indochine, 1947–1954. Paris, l'Harmattan, 2000, 240 p.

7665. FEDOROWICH (Kent). Decolonization deferred? The re-establishment of colonial rule in Hong Kong, 1942–1945. *In*: International diplomacy and colonial retreat [Cf. n° 7650], p. 25-50.

7666. GHOSH (Parimal). Brave men of the hills: resistance and rebellion in Burma, 1825–1932. Honolulu, Universiti of Hawai'i Press, 2000, 197 p.

7667. IMADA (Shusaku). Pakusu-Buritanika to shokuminchi Indo: Igirisu-Indo keizai-shi no "sôkan haaku". (Pax Britannica and colonial India: the mutuality of economic history of Britain and India). Kyoto, Kyoto University Academic Press, 2000, 458p.

7668. KNIGHT (G. R.). Narratives of colonialism: sugar, Java, and the Dutch. Huntington, Nova Science Publishers, 2000, XX-191 p. (Horizons in post-colonial studies).

7669. MAHAJAN (Sucheta). Independence and partition: the erosion of colonial power in India. New Delhi a. London, Sage Publications, 2000, 425 p. (Sage series in modern Indian history, 1).

7670. MANN (Michael). Bengalen im Umbruch. Die Herausbildung des britischen Kolonialstaates 1754–1793. Stuttgart, Steiner, 2000, 469 p. (Beiträge zur Kolonial- und Überseegeschichte, 78).

7671. MARTIN (Virginia). Law and custom in the steppe: the Kazakhs of the Middle Horde and Russian colonialism in the nineteenth century. Richmond, Curzon, 2000, 264 p. (ill., map).

7672. MATHUR (Laxman Prasad). Protest movements of Bhils under the British "Raj". Jaipur, Publication Scheme, 2000, 166 p.

7673. MOORE (Bob). Decolonization by default: Suriname and the Dutch retreat from empire. *In*: International diplomacy and colonial retreat [Cf. n° 7650], p. 228-251.

7674. PERNAU-REIFELD (Margrit). The passing of patrimonialism: politics and political culture in Hyderabad, 1911–1948. New Delhi, Manohar, 2000, X-395 p.

7675. SEN (Satadru). Disciplining punishment: colonialism and convict society in the Andaman Islands. Oxford, Oxford U. P., 2000, VI-283 p.

7676. SHAMIR (Ronen). The colonies of law: colonialism, zionism and law in early mandate Palestine. Cambridge, Cambridge U. P., 2000, XIII-216 p. (Cambridge studies in law and society).

7677. SHEPHERD (Naomi). Ploughing sand: British rule in Palestine 1917–1948. New Brunswick, N.J., Rutgers U. P., 2000, X-290 p.

7678. SUBRITZKY (John). Britain, Konfrontasi, and the end of empire in Southeast Asia, 1961–1965. *In*: International diplomacy and colonial retreat [Cf. n° 7650], p. 209-227.

7679. THOMAS (Martin). Divisive decolonization: the Anglo-French withdrawal from Syria and Lebanon, 1944–1946. *In*: International diplomacy and colonial retreat [Cf. n° 7650], p. 71-93.

7680. THOMPSON (Elizabeth). Colonial citizens: republican rights, paternal privilege, and gender in French Syria and Lebanon. New York a. Chichester, Columbia U. P., 2000, XVII-402 p. (ill., map). (History and society of the modern Middle East series).

7681. VAN DEN DOEL (Wim). Afscheid van Indië: de val van het Nederlands imperium in Azië. Amsterdam, Prometheus, 2000, 418 p.

7682. VAN VEEN (Ernst). Decay or defeat? An inquiry into the Portuguese decline in Asia 1580–1645. Leiden, Research School of Asian, African, and Amerindian Studies, Leiden University, 2000, IV-306 p. (CNWS publications, 96. Studies in overseas history, 1).

7683. VAN VIÊT (Dang). De la RC4 à la N4, la campagne des frontières. [S. l.], Lectoure Le Capucin, 2000, 140 p.

Cf. n°s 8474-8785

c. Africa

** 7684. ALLÈS (Jean-François). Commandos de chasse Gendarmerie. Algérie, 1959–1962, récits et témoignages. [S. l.], Atlante Rditions, 2000, 174 p.

** 7685. FAIVRE (Maurice). Les archives inédites de la politique algérienne: 1958–1962. Paris, L'Harmattan, 2000, 431 p. (ill.). (Collection Histoire et perspectives méditerranéennes).

** 7686. GATTARI (Nicola). La strada per Addis Abeba: lettere di un camionista dall'Impero (1936–1941). Compilazione e introduzione a cura di Sergio LUZZATTO. Torino, Paravia scriptorium, 2000, 194 p. (ill., ports.).

** 7687. KLEINKNECHT (Charles). Administrateur civil au Sahara: une vie au service de l'Algérie et des Territoires du Sud, 1942–1962. Paris, l'Harmattan, 2000, 350 p.

** 7688. WILLIAMSON (Thora). Gold Coast diaries: chronicles of political officers in West Africa, 1900–1919. Ed. by Anthony KIRK-GREENE. London, Radcliffe, 2000, XVI-419 p. (ill., maps, ports.).

7689. AFONSO (Aniceto), DE MATOS GOMES (Carlos). Guerra colonial. Lisboa, Editorial Notícias, 2000, 635 p. (ill., some col., maps).

7690. ALTHABE (Gérard). Anthropologie politique d'une décolonisation. Paris, L'Harmattan, 2000, 329 p.

7691. ANDERSON (David M.). Master and servant in colonial Kenya, 1895–1939. *Journal of African history*, 2000, 41, 3, p. 459-486.

7692. BOTTE (Roger). L'esclavage africain après l'abolition de 1848. Servitude et droit du sol. *In*: Contre les stéréotypes. Etudes sur la colonisation et l'esclavage. *Annales*, 2000, 55, 5, p. 1009-1038.

7693. BUTLER (L. J.). Britain, the United States, and the demise of Central Africa Federation, 1959–1963. *In*: International diplomacy and colonial retreat [Cf. n° 7650], p. 131-151.

7694. CARRASCO GARCÍA (Antonio), DE MESA GUTIÉRREZ (José Luis). Las tropas de África en las campañas de Marruecos. Madrid, Almena Ediciones, 2000, 96 p. (Especial Serga, 1).

7695. Colonialismo (Un), due sponde del Mediterraneo: atti del seminario di studi italo-iberici (Siena-Pistoia, 13–14 gennaio 2000). A cura di Nicola LABANCA e Pierluigi VENUTA. Pistoia, CRT, 2000, 169 p. (Collana studi e ricerche. Istituto storico provinciale della Resistenza di Pistoia, 4).

7696. DABBAB (Mohamed). Commisaires et commissaires généraux de gouvernment: cinquante ans de mainmise française progressive sur les jurisdictions indigènes et l'administration de la justice tunisienne dans la Régence (1881–fin de la 1ère moitié du XXe siècle) *Revue d'histoire maghrébine*, 2000, 27, 99-100, p. 255-290.

7697. EL MACHAT (Samya). Le Groupement de Tunisie du Rassemblement du Peuple français ou la tentation de l'immobilisme (1950–1951). *Revue d'histoire maghrébine*, 2000, 27, 99-100, p. 327-342.

7698. ELKINS (Caroline). The struggle for Mau Mau rehabilitation in late colonial Kenya. *International journal of African historical studies*, 2000, 33, 1, p. 25-58.

7699. ELTIS (David). The rise of African slavery in the Americas. Cambridge, Cambridge U. P., 2000, XVII-353 p.

7700. Frühe Kolonialgeschichte Namibias, 1880–1930. Hrsg. v. Wilhelm J. G. MÖHLIG. Köln, R. Köppe, 2000, 207 p. (ill.). (History, cultural traditions, and innovations in Southern Africa, 9).

7701. GERSHOVICH (Moshe). French military rule in Morocco: colonialism and its consequences. London, Frank Cass, 2000, XVII-238 p. (Cass series: history and society in the Islamic world).

7702. Guerre (La) d'Algérie au miroir des décolonisations françaises: en l'honneur de Charles-Robert Ageron: actes du colloque international, Paris, Sorbonne (23, 24, 25 novembre 2000). Paris, Société française d'histoire d'outre-mer et Luisant, Distique, 2000, 683 p

7703. HUBER (Hansjörg Michael). Koloniale Selbstverwaltung in Deutsch-Südwestafrika: Entstehung, Kodifizierung und Umsetzung. Frankfurt am Main u. New York, P. Lang, 2000, 325 p. (Rechtshistorische Reihe, 213).

7704. Imperialism, decolonization, and Africa: studies presented to John Hargreaves. Ed. by Roy BRIDGES. London, Macmillan Press, 2000, XI-213 p. (Cambridge Commonwealth series).

7705. KELLY (Saul). Britain, the United States, and the end of the Italian empire in Africa, 1940–1952. In: International diplomacy and colonial retreat [Cf. n° 7650], p. 51-70. – IDEM. Cold War in the desert: Britain, the United States and the Italian colonies, 1945–1952. Basingstoke, Macmillan, 2000, 207 p. (Cold War history series).

7706. Lesson learned? Development experiences in the late colonial period. *Journal of African history*, 2000, 41, 1, p. 29-130. [Contents: VAN BEUSEKOM (Monica M.), HODGSON (Dorothy L.). Lesson learned? Development experiences in the late colonial period (p. 29-34). – FAIRHEAD (James), LEACH (Melissa). Desiccation and domination: science and struggles over environment and development in colonial Guinea (p. 35-54). – HODGSON (Dorothy L.). Taking stock: state control, ethnic identity and pastoralist development in Tanganika, 1948–1958 (p. 55-78). – VAN BEUSEKOM (Monica M.). Disjunctures in theory and practice: making sense of change in agricultural development at the Office Du Niger, 1920–1960 (p. 79-100). – WORBY (Eric). 'Discipline without oppression': sequence, timing and marginality in souther Rhodesia's post-war development regime (p. 101-126). – BERRY (Sara). Afterword (p. 127-130)].

7707. LOVEJOY (Paul). Transformations in slavery: a history of slavery in Africa. Cambridge, Cambridge U. P., 2000, XXII-367 p. (African studies series, 36).

7708. MONNERET (Jean). La phase finale de la guerre d'Algérie. Paris, L'Harmattan, 2000, 399 p. (Collection Histoire et perspectives méditerranéennes).

7709. MOREIRA (Adriano), [et al.]. Estudos sobre as campanhas de África (1961–1974). Lisboa, Ediçoes Atena, Instituto de Altos Estudos Militares, 2000, 356 p. (ill., maps).

7710. MOUILLEAU (Elisabeth). Fonctionnaires de la République et artisans de l'empire: le cas des contrôleurs civils en Tunisie, (1881–1956). Préface de Annie REY-GOLDZEIGUER. Paris, L'Harmattan, 2000, 432 p. (Collection Histoire et perspectives méditerranéennes).

7711. NAYLOR (Phillip Chiviges). France and Algeria: a history of decolonization and transformation. Gainesville, University Press of Florida, 2000, XVIII-457 p.

7712. NUHN (Walter). Feind überall. Der große Nama-Aufstand (Hottentottenaufstand) 1904–1908 in Deutsch-Südwestafrika (Namibia). Der erste Partisanenkrieg in der Geschichte der deutschen Armee. Bonn, Bernard & Graefe, 2000, 344 p.

7713. NYAMBARA (Pius S.). Colonial policy and peasant cotton agriculture in southern Rhodesia, 1904–1953. *International journal of African historical studies*, 2000, 33, 1, p. 81-112.

7714. OSAGIE (Iyunolu Folayan). The Amistad revolt: memory, slavery, and the politics of identity in the United States and Sierra Leone. Athens a. London, University of Georgia Press, 2000, 180 p, (ill.).

7715. PATANÈ (Pietro). Eroi dimenticati: la contrastata sistemazione dei caduti della campagna 1935–1936 in Africa orientale: la normalizzaione dei rapporti con l'Etiopia: Somalia, Eritrea, Etiopia. Napoli, Istituto grafico editoriale italiano, 2000, 136 p. (ill.). (Itinera).

7716. RATELBAND (Klaas). Nederlanders in West-Afrika 1600–1650: Angola, Kongo en São Tomé. Ed. by René BAESJOU. Zutphen, Walburg Pers, 2000, 319 p.

7717. RATHBONE (Richard). The transfer of power and colonial civil servants in Ghana. *Journal of imperial and Commonwealth history*, 2000, 28, 2, p. 67-84.

7718. Route (La) des esclaves, système servile et traite dans l'Est malgache: actes du colloque international de Toamasina, 20–22 septembre 1999. Textes réunis et présentés par Rakoto IGNACE; préface Mangalaza EUGÈNE. Paris, L'Harmattan, 2000, 348 p. (ill.).

7719. SCHMUHL (Hans-Walter). Deutsche Kolonialherrschaft und Ethnogenese in Ruanda, 1897–1916. [Diskussionsforum]. *Geschichte und Gesellschaft*, 2000, 26, 2, p. 307-334.

7720. SCHNEIDER (Gabriele). Mussolini in Afrika: die faschistische Rassenpolitik in den italienischen Kolonien; 1936–1941. Köln, SH-Verlag, 2000, 315 p. (Italien in der Moderne, 8).

7721. SMITH (Simon C.). Revolution and reaction: South Arabia in the aftermath of the Yemeni revolution. In: International diplomacy and colonial retreat [Cf. n° 7650], p. 193-208.

7722. South African war reappraised (The). Ed. by Donald LOWRY. Manchester, Manchester U. P., 2000, XII-258 p.

7723. STAPLETON (Timothy J.). Faku: rulership and colonialism in the Mpondo Kingdom (c. 1760–1867). Waterloo, Wilfrid Laurier U. P., 2000, XVIII-198 p.

7724. STOCKWELL (Sarah). The business of decolonization: British business strategies in the Gold Coast. Oxford, Clarendon Press a. New York, Oxford U. P., 2000, IX-265 p. (ill., map). (Oxford historical monographs).

7725. STONE (Glyn). Britain and the Portuguese Africa, 1961–1965. *In*: International diplomacy and colonial retreat [Cf. n° 7650], p. 169-192.

7726. THOMAS (Martin). The French North African crisis: colonial breakdown and Anglo-French relations, 1945–1962. Basingstoke, Macmillan a. New York, St. Martin's Press, 2000, XV-287 p. (Studies in military and strategic history).

7727. TSCHAPEK (Rolf Peter). Bausteine eines zukünftigen deutschen Mittelafrika: deutscher Imperialismus und die portugiesischen Kolonien; deutsches Interesse an den südafrikanischen Kolonien Portugals vom ausgehenden 19. Jahrhundert bis zum ersten Weltkrieg. Stuttgart, F. Steiner, 2000, 475 p. (Beiträge zur Kolonial- und Überseegeschichte, 77).

Cf. n°s 4932, 5143, 6864, 7984, 8786-8809

d. America

** 7728. Colonial lives: Documents on Latin American history, 1550–1850. Ed. by Richard BOYER and Geoffrey SPURLING. New York a. Oxford, Oxford U. P., 2000, XVIII-352 p. (ill., maps).

** 7729. COPELAND (David A.). Debating the issues in colonial newspapers: primary documents on events of the period. Westport a. London, Greenwood Press, 2000, XVII-397 p. (ill.).

7730. BALBOA NAVARRO (Imilcy). Los brazos necesarios: inmigración, colonización y trabajo libre en Cuba, 1878–1898. Valencia, Centro Francisco Tomas y Valiente UNED Alzira-Valencia, Fundacion Instituto de Historia Social, 2000, 267 p. (Biblioteca Historia social, 6).

7731. BASKES (Jeremy). Indians, merchants, and markets: a reinterpretation of the Repartimiento and Spanish-Indian economic relations in colonial Oaxaca, 1750–1821. Stanford, Stanford U. P., 2000, 305 p.

7732. BRADLEY (Peter T.), CAHILL (David). Habsburg Peru: images, imagination and memory. Liverpool, Liverpool U. P., 2000, XII-167 p. (Liverpool Latin American studies, 2).

7733. CASANOVAS (Joan). O pan, o plomo! Los trabajadores urbanos y el colonialismo español en Cuba, 1850–1898. Prólogo de Nicolás SÁNCHEZ-ALBORNOZ. Madrid, Siglo XXI, 2000, XIX-326 p. (Historia).

7734. DÍAZ (María Elena). The Virgin, the king, and the royal slaves of El Cobre: negotiating freedom in colonial Cuba, 1670–1780. Stanford, Stanford U. P., 2000, XVIII-440 p. (ill., maps). (Cultural sitings).

7735. DOMINGUES (Ângela). Quando os índios eram vassalos: colonizaçao e relaçoes de poder no norte do Brasil na segunda metade do século XVIII. Lisboa, Comissao Nacional Comemoracoes dos Descobrimentos Portugueses, 2000, 388 p. (Colecçao Outras margens).

7736. DOS SANTOS (Ricardo Evaristo). Carlos V, Portugal y Brasil. Madrid, Brand Editorial, 2000, 306 p. (ill.). (Colección Auryn ensayo).

7737. Estudios coloniales. Encuentro de Historia Colonial (1998: Universidad Nacional Andrés Bello). Ed. por Julio RETAMAL AVILA. Providencia, RIL editores y Santiago de Chile, Universidad Andres Bello, 2000, [s. p.]. (Serie monografías).

7738. FABEL (Robin F. A.). Colonial challenges: Britons, native Americans, and Caribs, 1759–1775. Gainesville, University Press of Florida, 2000, X-282 p.

7739. FISHER (John R.). The royalist regime in the viceroyalty of Peru, 1820–1824. *Journal of Latin American studies*, 2000, 32, 1, p. 55-84.

7740. GONÇALVES DE SOUZA (Miguel Augusto). O descobrimento e a colonizacao portuguesa no Brasil. Belo Horizonte, Editora Itatiaia, 2000, 944 p. (Coleçao Reconquista do Brasil, 220).

7741. GRØNGAARD JESPERSEN (Torben). Dannebrog på den amerikanske prærie: et dansk koloniprojekt i 1870'erne – landkøb, bygrundlæggelse og integration. (Le Drapeau danois flotte sur la prairie américaine: un projet de colonie danoise dans les années 1870 - acquisitions foncières, établissement et intégration). Odense, Odense Bys Museer, 2000, 278 p. (ill.).

7742. HAMNETT (Brian R.). La política contrarrevolucionaria del virrey Abascal: Perú, 1806–1816. Lima, IEP, 2000, 16 p. (Documento de trabajo, 112. Serie historia, 18).

7743. Insurrection (L') des esclaves de Saint-Domingue (22–23 août 1791): actes de la table ronde internationale de Port-au-Prince (8 au 10 décembre 1997). Sous la direction de Laënnec HURBON. Paris, Karthala, 2000, 271 p.

7744. MONTIEL ARGÜELLO (Alejandro). Nicaragua colonial. Managua, Banco Central de Nicaragua, 2000, 417 p.

7745. ORDHAL KUPPERMAN (Karen). Indians and English: facing off in early America. Ithaca, Cornell U. P., 2000, XI-297 p.

7746. O'SHAUGHNESSY (Andrew Jackson). An empire divided: the American revolution and the British

Caribbean. Philadelphia, University of Pennsylvania Press, 2000, XVI-357 p. (Early American studies).

7747. PETROV (Aleksandr Ju.). Obrazovanie Rossijsko-Amerikanskoj kompanii. (The formation of the Russian-American company, [1781–1799]). Ed. N. N. BOLKHOVITINOV. RAN, In-t vseobshchej istorii. Moskva, Nauka, 2000, 153 p. [Eng. summary]. [Cf. Bibl. 99, n° 7374.]

7748. RUSSELL-WOOD (A. J. R.). 'Acts of grace': Portuguese monarchs and thei subjects of African descent in eighteenth-century Brazil. *Journal of Latin American studies*, 2000, 32, 2, p. 307-332.

7749. SCHOFIELD SAEGER (James). The Chaco mission frontier: the Guaycuruan experience. Tucson, University of Arizona Press, 2000, XVIII-266 p.

7750. SCHRÖTER (Bernd). Volksbewegungen in den lateinamerikanischen Unabhängigkeitsrevolutionen von 1810–1826: Mexiko, Paraguay und Uruguay im Vergleich. Leipzig, Leipziger Universitätsverlag, 2000, 251 p.

7751. SCOTT (Rebecca J.). Slave emancipation in Cuba: the transition to free labor, 1860–1899. Pittsburgh, University of Pittsburgh Press, 2000, XVI-327 p.

7752. SHANNON (Timothy J.). Indians and colonists at the crossroads of empire: the Albany Congress of 1754. Ithaca a. London, Cornell U. P., 2000, XV-268 p.

7753. VIDAL (Laurent). La présence française dans le Brésil colonial au XVIe siècle. *Cahiers des Amériques latines*, 2000, 2, p. 17-38.

7754. Wars (The) of independence in Spanish America. Ed. by Christon I. ARCHER. Wilmington, Scholarly Resources, 2000, XIII-325 p.

d. America. Addenda 1999

7755. SANTOS PÉREZ (José Manuel). Élites, poder local y régimen colonial: el cabildo y los regidores de Santiago de Guatemala, 1700–1787. Cádiz, Servicio de Publicaciones de la Universidad de Cádiz y Vermont, Plumsock Mesoamerican Studies y Miami, Centro de Investigaciones Regionales de Mesoamérica, 99, XXII-416 p.

Cf. n°s 1469, 4234, 7816, 7827, 8810-8820

e. Oceania

7756. CLARK (Jeffery J.). Steel to stone: a chronicle of colonialism in the Southern Highlands of Papua New Guinea. Ed. by Chris BALLARD and Michael NIHILL. Oxford, Oxford U. P., 2000, XXX-187 p. (ill.). (Oxford studies in social and cultural anthropology).

7757. HASSAM (Andrew). Through Australian eyes: colonial perceptions of imperial Britain. Brighton a. Portland, Sussex Academic Press, 2000, IX-220 p. (ill.).

7758. BARKO (Ivan). The Cobar incident, its antecedents and sequels. The warfare between Earl Beauchamp, Governor of New South Wales, and Georges Biard d'Aunet, Consul General de France a 1899–1900. *Journal of the Royal Australian Historical Society*, 2000, 86, p. 134-158.

Cf. n°s 8821-8826

§ 3. From 1500 to 1789

a. General

** 7759. Asnad-i ravabit-i dawlat-i Safavi ba hukumat'ha-yi Italiya. (Documents on relations between Safavid government and Italian governments). Bih kushish-i Muhammad Hasan KAVUSI `IRAQI. Tihran, Vizarat-i Umur-i Kharijah, Markaz-i Chap va Intisharat, 2000, 402 p. (ill., facsims). (Asnad, 69).

** 7760. Letters from the Pyrenees: Don Luis Méndez de Haro's correspondence to Philip IV of Spain, July to November 1659. Ed. by Lynn WILLIAMS. Exeter, Exeter U. P., 2000, XXVII-147 p. (Exeter Hispanic texts, 57).

7761. ANDRETTA (Stefano). La repubblica inquieta: Venezia nel Seicento tra Italia ed Europa. Roma, Carocci editore, 2000, 260 p. (Ricerche / Università degli studi Roma Tre, Dipartimento di studi storici geografici antropologici, 1).

7762. BÉRENGER (J.). Ferdinand III et la France de Mazarin. *In*: Europe (L') des Traités de Westphalie [Cf. n° 7828], p. 163-180.

7763. BJÖRKMAN (S.). Diplomates suédois au XVIIe siècle. *In*: Europe (L') des Traités de Westphalie [Cf. n° 7828], p. 503-516.

7764. DEAKIN (Quentin). Expansion, war and rebellion: Europe, 1598–1661. Cambridge, Cambridge U. P., 2000, IV-178 p. (Cambridge perspectives in history).

7765. France (La) et la Pologne. Histoire, mythes, représentations. Textes réunis et présentés par François LAVOCAT. Actes du colloque des 16–18 septembre 1998 à l'Université Lumière-Lyon II. Lyon, Presses Universitaires de Lyon, 2000, 385 p.

7766. HAAN (Bertrand). La médiation pontificale entre la France et la Savoie de la paix de Vervins [Aisne] à la paix de Lyon (1598–1601). *Cahiers René de Lucinge*, 2000, 4, 34, p. 5-20.

7767. HUGON (Alain). Des Habsbourg aux Bourbons: le combat espagnol pour la conservation de l'hégémonie européenne milieu XVIe–fin XVIIe siècle. *Bulletin de la Société d'histoire moderne et contemporaine*, 2000, 3-4, p. 34-55.

7768. IANOVSKII (O. A.), BOBYSHEV (V. I.). Diplomatiia i diplomaty Rossii: ot Ivana III do Petra Velikogo. Minsk, BGU, 2000, 91 p. (ill.).

7769. KOLODZIEJCZYK (Dariusz). Ottoman-Polish diplomatic relations (15th–18th century): an annotated edition of `ahdnames and other documents. Leiden, Brill, 2000, XVIII-721 p. (Ottoman Empire and its heritage, 18).

7770. KOTOSHIKHIN (Grigorii) [et al.]. Moskoviia i Evropa. Sostavlenie A. LIBERMAN, S. SHOKAREV. Moskva, Fond Sergeia Dubova, 2000, 618 p. (Istoriia Rossii i Doma Romanovykh v memuarakh sovremennikov XVII-XX).

7771. MURDOCH (Steve). Britain, Denmark-Norway and the House of Stuart, 1603–1660: a diplomatic and military analysis. East Linton, Tuckwell Press, 2000, XII-318 p. (ill.).

7772. PÉREZ-BUSTAMANTE (Rogelio). El gobierno del imperio español: los Austrias (1517–1700). Madrid, Comunidad de Madrid, Consejeria de Educacion, 2000, 586 p.

7773. Politics and diplomacy in early modern Italy: the structure of diplomatic practice, 1450–1800. Ed. by Daniela FRIGO. Cambridge, Cambridge U. P., 2000, V-262 p. (Cambridge studies in Italian history and culture).

7774. Russkaia i ukrainskaia diplomatiia v Evrazii: 50-e gody XVII veka. (Russian and Ukraine diplomaci in Eurasia). Otv. redaktor B. N. FLORIA. Moskva, In-t slavianovedeniia RAN, 2000, 222 p.

7775. SCOTT (Jonathan). England's troubles: seventeenth-century English political instability in European context. Cambridge, Cambridge U. P., 2000, XII-546 p.

7776. VALLADARES (Rafael). Portugal y la monarquía hispánica, 1580–1668. Madrid, Arco/Libros, 2000, 63 p. (Cuadernos de historia, 74).

b. 1500–1648

* 7777. NEUHAUS (Helmut). Westfälischer Frieden und Dreißigjähriger Krieg. *Archiv für Kulturgeschichte*, 2000, 82, p. 455-476.

** 7778. Due ambasciatori veneziani nella Spagna di fine Cinquecento: i diari dei viaggi di Antonio Tiepolo (1571–1572) e Francesco Vendramin (1592–1593). A cura di Luigi MONGA. Torino, Centro interuniversitario di ricerche sul viaggio in Italia, 2000, 295 p. (Dimensioni del viaggio = Dimensions du voyage, 11).

** 7779. WIDMARCKTER (Obristen Caspar von). Söldnerleben am Vorabend des Dreissigjährigen Krieges. Lebenslauf und Kriegstagebuch 1617 des hessischen Obristen Caspar von Widmarckter. Hrsg. u. bearb. v. Holger Th. GRÄF. Marburg an der Lahn, Trautvetter & Fischer, 2000, 168 p. (ill.). (Beiträge zur hessischen Geschichte, 16).

7780. ALLEN (Paul C.). Philip III and the Pax Hispanica, 1598–1621: the failure of grand strategy. New Haven a. London, Yale U. P., 2000, XVI-335 p.

7781. ASCH (Ronald G.). 'Wo der soldat hinkömbt, da ist alles sein': military violence and atrocities in the thirty years war re-examined. *German history*, 2000, 18, 3, p. 291-309.

7782. AUBIN (Jean). Le latin et l'astrolabe. Recherches sur le Portugal de la Renaissance, son expansion en Asie et les relations internationales. Vol. 2. Paris, Fondation Calouste Gulbekian, 2000, 590 p.

7783. BARBICHE (B.), DE DAINVILLE-BARBICHE (S.). La diplomatie pontificale de la paix de Vervins aux traitée de Westphalie (1598–1648), permanences et ruptures. *In*: Europe (L') des Traités de Westphalie [Cf. n° 7828], p. 555-566.

7784. BARZEGAR (Karim Najafi). Mughal-Iranian relations: during sixteenth century. Delhi, Indian Bibliographies Bureau, 2000, XVIII-288 p.

7785. BÉRENGER (Jean). La politique française en Méditerranée au XVIe siècle et l'alliance ottomane. *In*: Guerre (La) de course en Méditerranée [Cf. n° 7606], p. 9-26.

7786. BORROMEO (Agostino). Clément VIII, la diplomatie pontificale et la paix de Vervins. *In*: Traité (Le) de Vervins [Cf. n° 7819], p. 323-344.

7787. CABEZA (Antonio). La diplomatie espagnole en Italie devant la paix de Vervins. *In*: Traité (Le) de Vervins [Cf. n° 7819], p. 283-296.

7788. CĂZAN (Ileana). Habsburgii şi otomanii la linia Dunării. Tratate şi negocieri de pace. 1526–1576. Bucureşti, Editura Oscar Print, 2000, 297 p.

7789. CRISTIAN (Luca). Participarea Transilvaniei la Războiul de treizeci de ani în timpul domniei lui Gabriel Bethlen (1613–1629). *Istros*, 2000, 10, p. 425-434.

7790. DE LUCINGE (René). Les occurences de la Paix de Lyon (1601). Tome 2. Des œuvres. Texte établi et annoté par Alain DUFOUR. Genève, Droz, 2000, 117 p.

7791. DORAN (Susan). Elizabeth I and foreign policy, 1558–1603. London, Routledge, 2000, XX-75 p. (ill., maps). (Lancaster pamphlets).

7792. DRECOLL (Volker Henning). Der Passauer Vertrag (1552). Einleitung und Edition. Berlin u. New York, de Gruyter, 2000, XII-382 p.

7793. EXTERNBRINK (Sven). "Le cœur du monde" et la "liberté de l'Italie". Aspects de la politique italienne de Richelieu, 1624–1642. *Revue d'histoire diplomatique*, 2000, 114, 3, p. 181-208.

7794. GARCÍA HERNÁN (Enrique). Irlanda y el rey prudente. Madrid, Ediciones del Laberinto, 2000, 286 p. (ill.). (Colección Hermes, 2). – IDEM. La acción diplomática de Francisco de Borja al servicio del Pontificado, 1571–1572. Valencia, generalitat Valenciana, 2000, 562 p.

7795. GRAZIANI (Antoine-Marie). Les ouvrages de défense en Corse contre les Turcs (1530–1650). *In*: Guerre (La) de course en Méditerranée [Cf. n° 7606], p. 73-158.

7796. HOUSSIAU (Jean). Les ambassadeurs des Pays-Bas à Vervins: prémices d'une diplomatie «belge»? *In*: Traité (Le) de Vervins [Cf. n° 7819], p. 267-281.

7797. ISRAEL (Jonathan Irvine). El Brasil y la política holandesa en el Nuevo Mundo, 1618-1648. *In*: Acuarela de Brasil, 500 anos después [Cf. n° 4225], [s. p.].

7798. KĄKOLEWSKI (Igor). Nadużycia władzy i korupcja w Prusach Książęcych w połowie XVI wieku. Narodziny państwa wczesnonowożytnego. (Machtmissbrauch und Bestechlichkeit in dem Herzogtum Preußen um die Mitte des XVI. Jahrhunderts. Das Zur-Welt-Kommen des frühneuzeitlichen Staates). Warszawa, Trio, 2000, 243 p.

7799. KOLLER (A.). Le rôle du Saint-Siège au début de la guerre de Trente Ans, les objectifs de la politique allemande de Grégoire XV (1621–1623). *In*: Europe (L') des Traités de Westphalie [Cf. n° 7828], p. 123-134.

7800. Krieg (Der) vor den Toren: Hamburg im Dreissigjährigen Krieg, 1618–1648. Hrsg. v. Martin KNAUER und Sven TODE; unter Mitarbeit von Niels WIECKER. Hamburg, Verlag Verein für Hamburgische Geschichte, 2000, 504 p. (Beiträge zur Geschichte Hamburgs, 60).

7801. LABOURDETTE (Jean-François). L'importance du traité de Vervins dans l'histoire de l'Europe. *In*: Traité (Le) de Vervins [Cf. n° 7819], p. 15-26.

7802. LOUREIRO (Rui Manuel). Fidalgos, missionários e mandarins: Portugal e China no século XVI. Lisboa, Fundaçao Oriente, 2000, 736 p. (Orientalia, 1).

7803. MAJOROS (Ferenc). Karl V. Habsburg als Weltmacht. Graz, Wien u. Köln, Styria, 2000, 255 p.

7804. MALETTKE (Klaus). Le concept de sécurité collective de Richelieu et les traités de paix de Westphalie. *In*: Europe (L') des Traités de Westphalie [Cf. n° 7828], p. 55-66. – IDEM. Le traité de Vervins et ses conséquences pour l'empereur et pour l'empire. *In*: Traité (Le) de Vervins [Cf. n° 7819], p. 493-512.

7805. MAROŃ (Jerzy). Militarne aspekty wojny trzydziestoletniej na Śląsku. (Der dreißigjährige Krieg in Schlesien in militärischer Hinsicht). Wrocław, Wydaw. Uniw. Wrocławskiego, 2000, 277 p. (Acta Universitatis Wratislaviensis, 2201).

7806. MAROSZEK (Józef). Pogranicze Litwy i Korony w planach króla Zygmunta Augusta. Z historii dziejów realizacji myśli monarszej między Niemnem a Narwią. (Grenzgebiete von Litauen und Krone in Königs Sigismund August [Zygmunt August] Vorsätzen. Zur Geschichte der Umsetzung des Gedankens des Monarchen zwischen Flüssen Memel und Narew). Białystok, Wydaw. Uniw. W Białymstoku, 2000, 646 p.

7807. MARQUES DE OLIVEIRA (Julieta Teixeira). Veneza e Portugal no século XVI: subsídios para a sua história. Lisboa, Comissao Nacional para as Comemoraçoes dos Descobrimentos Portugueses, Imprensa Nacional-Casa da Moeda, 2000, 382 p. (Mare liberum).

7808. MITCHELL (Colin Paul). Sir Thomas Roe and the Mughal Empire. Karachi, Area Study Centre for Europe, 2000, XXI-258 p.

7809. OCHOA BRUN (M. A.). La diplomatie espagnole dans la première moitié du XVIIe siècle. *In*: Europe (L') des Traités de Westphalie [Cf. n° 7828], p. 537-554.

7810. OSBORNE (Toby). Abbot Scaglia, the Duke of Buckingham and Anglo-Savoyard relations during the 1620s. *European history quarterly*, 2000, 30, 1, p. 5-32.

7811. PILLORGET (R.). Alliance, déceptions, ruptures: les relations franco-bavaroises au cours de la guerre de Trente ans (1618–1648). *In*: Europe (L') des Traités de Westphalie [Cf. n° 7828], p. 517-526.

7812. POUJADE (Patrice). Conflictualité, solidarités et relations frontalières dans les Pyrénées (v. 1550–v. 1650). *In*: Tolérance et solidarités dans les pays pyrénéens [Cf. n° 7636], p. 431-447.

7813. POUSSOU (Jean-Pierre). La politique extérieure d'Elisabeth Ière et la paix de Vervins. *In*: Traité (Le) de Vervins [Cf. n° 7819], p. 247-263.

7814. SERWAŃSKI (M.). La neutralité de la Pologne pendant la guerre de Trente ans. *In*: Europe (L') des Traités de Westphalie [Cf. n° 7828], p. 135-146.

7815. SILKE (John J.). Kinsale: the Spanish intervention in Ireland at the end of the Elizabethan wars. Dublin, Four Courts, 2000, 208 p.

7816. STELLA (Roseli Santaella). Brasil durante el gobierno español, 1580–1640. Madrid, Fundacion Historica Tavera, 2000, XXXIII-298 p. (ill.).

7817. Sterbzeiten: der Dreissigjährige Krieg im Herzogtum Westfalen: eine Dokumentation. Hrsg. v. Horst CONRAD und Gunnar TESKE. Münster, Landschaftsverband Westfalen-Lippe, Westfälisches Archivamt, 2000, 568 p. (ill., maps). (Westfälische Quellen und Archivpublikationen, 23).

7818. TALLON (A.). Les puissances catholiques face à la tolérance religieuse en France au XVIe siècle. *In*: Europe (L') des Traités de Westphalie [Cf. n° 7828], p. 21-30.

7819. Traité (Le) de Vervins [Aisne]. Actes du Colloque de Vervins, 1er–3 mai 1998, réunis par Jean-François LABOURDETTE, Jean-Pierre POUSSOU et Marie-Catherine VIGNAL. Paris, Presses de l'Univ. de Paris-Sorbonne, 2000, 572 p. [Cf. n^{os} <sélection> 7786, 7787, 7796, 7801, 7804, 7813, 7821.].

7820. TYGIELSKI (Wojciech). Dyplomacja, informacja, propaganda: podróz Luigiego Bevilacqua, posla toskanskiego na dwory europejskie w 1609 r. Warszawa, Wydawn. Naukowe Semper, 2000, 108 p.

7821. VÁZQUEZ DE PRADA (Valentin). Philippe II et la France: de Cateau-Cambrésis à Vervins. Quelques réflexions, quelques préscisions. *In*: Traité (Le) de Vervins [Cf. n° 7819], p. 135-157.

Cf. n°s 7834, 7840

c. *1648–1789*

** 7822. CIOBANU (Veniamin). Informaţii documentare privind politica orientală a Angliei (1786–1791). *Anuarul Institutului de Istorie «A. D. Xenopol» Iaşi*, 2000, 37, p. 223-247.

** 7823. DOBSON (David). American data from the records of the High Court of the Admiralty of Scotland, 1675–1800. Baltimore, Clearfield, 2000, 129 p. (David Dobson publications).

** 7824. Russko-mongol'skie otnosheniia: 1685–1691. Sbornik dokumentov. (Russian-Mongolian relations: 1685–1691). Sostavitel' G. I. SLESARCHUK; otvetstvennyi redaktor N. F. DEMIDOVA. Moskva, Vostochnaia lit-ra RAN, 2000, 487 p. (Materialy po istorii russko-mongol'skikh otnoshenii).

7825. ANDERSON (Fred). Crucible of war: the Seven Years' War and the fate of empire in British North America 1754–1766. New York, Alfred A. Knopf, 2000, XXV-862 p.

7826. CHAUSSINAND-NOGARET (Guy). Le Cardinal Dubois, 1656–1723, ou, une certaine idée de l'Europe. Paris, Perrin, 2000, 257 p.

7827. CONWAY (Stephen). The British Isles and the War of American Independence. Oxford, Oxford U. P., VII-407 p. (ill., map).

7828. Europe (L') des Traités de Westphalie. Esprit de la diplomatie et diplomatie de l'esprit. Sous la dir. de L. BÉLY, avec le concours d'I. RICHEFORT. Paris, Presses Universitaires de France, 2000, 612 p. [Cf. n^{os} <sélection> 7762, 7763, 7783, 7799, 7804, 7809, 7811, 7814, 7818, 7830, 7832, 7834, 7840, 7841.]

7829. FOUCHÉ (Nicole). Benjamin Franklin et Thomas Jefferson: aux sources de l'amitié franco-américaine 1776–1808. Préface par Claude FOHLEN. Paris, M. Houdiard, 2000, 102 p. (Biographies américaines).

7830. GIRY-DELOISON (C.). Westphalie 1648: l'Angleterre en marge de l'Europe. *In*: Europe (L') des Traités de Westphalie [Cf. n° 7828], p. 401-410.

7831. HERRERO SÁNCHEZ (Manuel). El acercamiento hispano-neerlandés (1648–1678). Madrid, Consejo Superior de Investigaciones Científicas, 2000, 427 p. (Biblioteca de historia, 39).

7832. HILDESHEIMER (F.). Guerre et paix selon Richelieu. *In*: Europe (L') des Traités de Westphalie [Cf. n° 7828], p. 31-54.

7833. KRASNOV (Nikolaj A.). SShA i Frantsija: Diplomaticheskie otnoshenija, 1775–1801 gg. (The USA and France: diplomatic relations, 1775–1801). RAN, In-t vseobshchej istorii, Moskva, PAIMS, 2000, 423 p.

7834. LABOURDETTE (J.-F.). La diplomatie française à Liège (1646–1650): l'action du président de Lumbres. *In*: Europe (L') des Traités de Westphalie [Cf. n° 7828], p. 567-578.

7835. LAMBOTTE (Miguel). La bataille de Rocourt (1746). Liège, Centre d'éditions, 2000, 80 p.

7836. MAQUART (Marie-Françoise). L'Espagne de Charles II et la France, 1665–1700. Toulouse, Presses universitaires du Mirail, 2000, 185 p. (Amphi 7. Histoire).

7837. MATVEEV (V.). Summit diplomacy of the seventeenth century: William III and Peter I in Utrecht and London, 1697–1698. *Diplomacy and statecraft*, 2000, 11, 3, p. 29-48.

7838. MATVEYEV (Vladimir). The Karlowitz congress and the debut of Russia's multilateral diplomacy (1698–1699). Leicester, University of Leicester, Centre for the Study of Diplomacy, 2000, 15 p. (Diplomatic Studies Programme discussion papers, 61).

7839. PETSCHEL (Dorit). Sächsische Aussenpolitik unter Friedrich August I.: zwischen Rétablissement, Rheinbund und Restauration. Köln, Bohlau, 2000, IX-355 p. (Dresdner historische Studien, 4).

7840. PONCET (O.). Les contradictions d'une diplomatie, le Saint-Siège face aux demandes indultaires des souverains catholiques (Espagne, France, Portugal), de 1640 à 1668. *In*: Europe (L') des Traités de Westphalie [Cf. n° 7828], p. 253-266.

7841. POUMARÈDE (G.). La question d'orient au temps de Westphalie. *In*: Europe (L') des Traités de Westphalie [Cf. n° 7828], p. 363-390.

7842. PRITCHARD (Earl H.). The crucial years of early Anglo-Chinese relations, 1750–1800. London, Routledge, 2000, 442 p. (ill., map). (Britain and the China trade, 1635–1842, 6).

7843. ROWLANDS (Guy). Louis XIV, Vittorio Amadeo II and French military failure in Italy, 1689–1696. *English historical review*, 2000, 115, 462, p. 534-569.

7844. SELLIN (Volker). Le Palatinat après les Traités de Westphalie. *Revue d'histoire diplomatique*, 2000, 114, 3, p. 229-250.

7845. SÉRÉ (Daniel). Les difficultés d'exécution d'un traité: le cas du Traité des Pyrénées. *Revue d'histoire diplomatique*, 2000, 114, 3, p. 209-228.

7846. STORRS (Christopher). War, diplomacy and the rise of Savoy, 1690–1720. Cambridge, Cambridge U. P., 2000, XIV-345 p. (Cambridge studies in Italian history and culture).

7847. Strukturwandel Internationaler Beziehungen. Zum Verhältnis von Staat und internationalem System seit dem Westfälischen Frieden. Hrsg. v. Jens SIEGELBERG und Klaus SCHLICHTE. Wiesbaden, Westdeutscher Verlag, 2000, 437 p.

7848. Vek Ekateriny II: Dela balkanskie. (The age of Catherine II: the affairs of the Balkans: [Articles]).

Ed. Vladilen N. VINOGRADOV. RAN, In-t slavjanovedenija. Moskva, Nauka, 2000, 295 p. (ill.).

7849. WILLS (Rebecca). The Jacobites and Russia, 1715–1750. East Linton, Tuckwell Press, 2002, 253 p.

7850. WINDLER (Christian). Tributes and presents in Franco-Tunisian diplomacy. *Journal of early modern history*, 2000, 4, 2, p. 168-199.

7851. WŁODARSKI (Józef). Polityka pruska elektora brandenburskiego Fryderyka Wilhelma I w latach 1640–1660. Studium z dziejów dyplomacji. (Die preußische Politik des Kurfürsten von Brandenburg Friedrich Wilhelm I. in den Jahren 1640–1660. Studie zur Geschichte der Staatskunst). Gdańsk, Wydaw. Uniw. Gdańskiego, 2000, 233 p.

Cf. n°s 4670, 5043, 5862, 7859, 7860, 7861, 7641

§ 4. From 1789 to 1815.

7852. ABALIKHIN (Boris S.). 1812 god: Aktual'nye problemy istorii. (The year 1812: some questions of [military] history). Intr. by Vladlen G. SIROTKIN. Elista, Dzhangar, 2000, 171 p.

7853. ASSERETO (G.). La seconda Repubblica ligure: 1800–1805: dal 18 brumaio genovese all'annessione alla Francia. Milano, Selene, 2000, 202 p. (Strumenti).

7854. AUSTIN (Paul Britten). 1812: Napoléon's invasion of Russia. London, Greenhill Books, 2000, 1120 p.

7855. BURROWS (Simon). French exile journalism and European politics, 1792–1814. Suffolk a. Rochester, Boydell Press, 2000, XVI-272 p. (The Royal Historical Society studies in history. New series).

7856. COSSERON (Serge). Les grandes campagnes de Napoléon: des campagnes d'Italie à la campagne de France. Paris, Auzou, 2000, 128 p. (ill.).

7857. DODOLEV (Mikhail A.). Venskij kongress v istoriografii XIX-XX vekov. (The Congress in Vienna in the nineteenth- and the twentieth-century historiography). RAN, In-t vseobshchej istorii. Moskva, [s. n.], 2000, 253 p.

7858. EMSLEY (Clive). Britain and the French Revolution. Harlow, Longman, 2000, 134 p. (ill.). (Seminar studies in history).

7859. GVOSDEV (Nikolas K.). Imperial policies and perspectives towards Georgia, 1760–1819. Basingstoke, Macmillan a. Oxford, St. Antony's College, 2000, XXI-197 p. (ill., maps). (St. Antony's series).

7860. HEPPNER (Harald). Austria și Principatele Dunarene (1774–1812): o contribuție la politica Sud-Est Europeana a Habsburgilor. Cluj, Presa Universitara Clujeana, 2000, 297 p.

7861. HOCHEDLINGER (Michael). Krise und Wiederherstellung: österreichische Grossmachtpolitik zwischen Türkenkrieg und "Zweiter Diplomatischer Revolution" 1787–1791. Berlin, Duncker und Humblot, 2000, 520 p. (Historische Forschungen, 65).

7862. JOFFRIN (Laurent). Les batailles de Napoléon. Paris, Ed. du Seuil, 2000, 238 p. (ill.).

7863. JOOR (Johan). De adelaar en het lam: onrust, opruiing en onwilligheid in Nederland ten tijde van het Koninkrijk Holland en de inlijving bij het Franse keizerrijk (1806–1813). Amsterdam, Bataafsche Leeuw, 2000, 864 p. (ill.).

7864. *Vacat*.

7865. KUCHARSKI (Maciej). Dzialalnosc dyplomacji polskiej w Berlinie w latach 1788–1792. Katowice, Wydawn. Uniwersytetu Slaskiego, 2000, 147 p. (Prace naukowe Uniwersytetu Slaskiego w Katowicach, 1881).

7866. LAMAR (Glenn J.). Jérôme Bonaparte: the war years, 1800–1815. Westport, Greenwood, 2000, XII-136 p. (Contributions in military studies, 189).

7867. RILEY (J. P.). Napoleon and the world war of 1813: lessons in coalition warfighting. London, Frank Cass, 2000, XIV-480 p. (ill., maps).

7868. SIROTKIN (V. G.). Napoleon i Rossii. Moskva, Olma-Press, 2000, 379 p. (ill.). (Serii "Istoricheskoe dos'e").

7869. TERENAS (Gabriela Gândara). O Portugal da Guerra Peninsular: a visao dos militares británicos (1808–1812). Lisboa, Ediçoes Colibri, 2000, 196 p. (ill., map). (Colecçao Estudos / Faculdade de Ciências Sociais e Humanas da Universidade Nova de Lisboa, 7).

7870. TISCHLER (Ulrike). Die Habsburgische Politik gegenüber den Serbien und Montenegrinern, 1791–1822: Förderung oder Vereinnahmung? München, R. Oldenbourg, 2000, 405 p. (Südosteuropäische Arbeiten, 108).

7871. VOVELLE (Michel). Les républiques-soeurs sous le regard de la Grande Nation (1795–1803): de l'Italie aux portes de l'Empire ottoman, l'impact du modèle républicain français. Paris, L'Harmattan, 2000, 350 p.

Cf. n°s 4670, 7641, 7773, 7823, 7829, 7833, 7842, 7848, 7897, 7941

§ 5. From 1815 to 1910.

** 7872. Documenti diplomatici italiani (I). Serie 2. 1870–1896. 27. 1 aprile 1895–9 marzo 1896. Serie 3. 1896–1907. 7. 1 luglio 1902–2 novembre 1903. Roma, Istituto poligrafico e Zecca dello Stato, Libreria dello Stato, 2000, 2 vol., LXXIII-738 p., LXIX-661 p.

** 7873. Documenti diplomatici italiani sull'Armenia. Serie II. 1891–1916. Commissione per la pubblicazione dei documenti italiani sull'Armenia. Vol. 3. 1 settembre–31 dicembre 1895. A cura di Lorenzo MECHI. Firenze, Commissione per la pubblicazione dei documenti italiani sull'Armenia, 2000, XI-517 p.

** 7874. ELIAV (Mordechai). Österreich und das Heilige Land: ausgewählte Konsulatsdokumente aus Jerusalem 1849-1917. Unter Mitarbeit von Barbara HAI-

DER. Wien, Verlag der Österreichischen Akademie der Wissenschaften, 2000, 617 p. (ill.). (Fontes rerum Austriacarum. Zweite Abteilung, Diplomataria et acta, 91).

** 7875. Letters from the 49th parallel, 1857–1873: selected correspondence of Joseph Harris and Samuel Anderson. Ed. with an introduction by C. Ian JACKSON and with a technical appendix by Louis M. SEBERT. Toronto, Champlain Society, 2000, CVIII-477-XX p. (Publications of the Champlain Society, 63).

** 7876. POPOV (Radoslav). Rusiia protiv Stambolov, ili, Stambolov protiv Rusiia? Sofiia, A.I. "Prof. Marin Drinov", 2000, 424 p.

7877. ADIYEKE (Ayşe Nükhet). Osmanlı İmparatorluğu ve Girit bunalımı (1896–1908). (The Ottoman Empire and the Crete crisis). Ankara, Türk Tarih Kurumu, 2000, 367 p.

7878. AGØY (Nils Ivar). "It will serve to increase our union difficulties". Norway, Sweden and the Hague peace conference of 1899. *Historisk tidsskrift*, 2000, 79, 1, p. 181-208.

7879. AGSTNER (Rudolf). Österreich im Kaukasus 1849–1918 = Austria in the Caucasus 1849–1918. Wien, Diplomatische Akademie, 2000, 104 p. (Occasional papers / Diplomatische Akademie Wien, 3/1999).

7880. ARBOIT (Gérald). Aux sources de la politique arabe de la France: le Second Empire au Machrek. Préface de Jacques FRÉMEAUX. Paris, L'Harmattan, 2000, 336 p. (Collection "Comprendre le Moyen-Orient").

7881. ARCINIEGAS DUARTE (Orlando). Los conflictos de intereses en la negociaciones para el reconocimiento de Venezuela por España, 1834–1845. Valencia, Consejo de Desarrollo Científico y Humanístico de la Universidad de Carabobo, 2000, 332 p.

7882. ARNOULET (François). Le faux problème de l'espionite: la mission Seignette (1874). *Revue d'histoire maghrébine*, 2000, 27, 97-98, p. 31-37.

7883. AUSLIN (Michael Robert). Negotiating with imperialism: Japan and the unequal treaty regime, 1858–1872. Ann Arbor, UMI, 2000, XI-307 p. (ill.).

7884. BENSIMON (Fabrice). Les Britanniques face à la Révolution française de 1848. Paris, L'Harmattan, 2000, 451 p. (Aire anglophone).

7885. BLACKETT (R. J. M.). Divided hearts: Britain and the American Civil War. Baton Rouge, Louisiana State U. P., XIII-273 p.

7886. BLED (J.-P.). L'Autriche-Hongrie et les Balkans du Congres de Berlin à la Première Guerre mondiale (1878–1914). *In*: Balkans (Les) dans les relations internationales, Part 1 [Cf. n° 7582], p. 289-295.

7887. BLYTH (Robert J.). Redrawing the boundary between India and Britain: the succession crisis at Zanzibar, 1870–1873. *International history review*, 2000, 22, 4, p. 785-805.

7888. British envoys to Germany, 1816–1866. Vol. 1. 1816–1829. Ed. by Sabine FREITAG and Peter WENDE. Cambridge, Cambridge University Press for the Royal Historical Society in association with the German Historical Institute London, 2000, XXI-592 p (Camden fifth series, 15).

7889. BUZPINAR (Ş. Tufan). The repercussion of the British occupation of Egypt on Syria, 1882–1883. *Middle eastern studies*, 2000, 36, 1, p. 82-91.

7890. CELOZZI BALDELLI (Pia G.). L'Italia e la crisi balcanica: 1876–1879. Galatina, Congedo, 2000, 157 p. (Dipartimento di studi storici dal Medioevo all'età contemporanea / Università di Lecce, 46).

7891. COELHO (Maria Teresa Pinto). A Agulha de Cleópatra: Jaime Batalha Reis e as relações diplomáticas e culturais Luso-Britânicas. Lisboa, Cosmos, 2000, 349 p. (Cosmos Literatura, 51).

7892. CONTE-HELM (Marie), OHTA (Akiko), WESTE (John L.). The Iwakura Mission to Britain. Durham, Department of East Asian Studies, University of Durham, 2000, 46 p. (Durham East Asian papers, 13).

7893. CRAPOL (Edward P.). James G. Blaine: architect of empire. Wilmington, Scholarly Resources, 2000, XX-157 p. (Biographies in American foreign policy, 4).

7894. DASQUE (I.). A la recherche de Monsier de Norpois: prosopographie des ambassadeurs et des ministres plénipotentiaires sous la Troisème République (1871–1914). *Revue d'histoire diplomatique*, 2000, 114, 4, p. 261-288.

7895. Deutsches Ottocento: die deutsche Wahrnehmung Italiens im Risorgimento. Hrsg. v. Arnold ESCH und Jens PETERSEN. Tübingen, M. Niemeyer, 2000, IX-356 p. (ill., maps). (Bibliothek des Deutschen Historischen Instituts in Rom, 94).

7896. DUVAL HERNÁNDEZ (Dolores). Luis de la Rosa y el paso interoceánico en Tehuantepec, 1849–1852. San Juan, Instituto Mora, 2000, 66 p. (Perfiles. Historia diplomática).

7897. EARLE (Rebecca A.). Spain and the independence of Colombia, 1810–1825. Exeter, University of Exeter Press, 2000, IX-254 p.

7898. España y la república Romana de 1849. Ed. por Manuel ESPADAS BURGOS. Roma, Consejo Superior de Investigaciones Científicas, Escuela Española de Historia y Arqueología en Roma, 2000, 143 p. (Serie historica, 1).

7899. FOSTER (Anne L.). Prohibition as superiority: policing opium in South-East Asia, 1898–1925. *International history review*, 2000, 22, 2, p. 253-273.

7900. GARCÍA MARTÍNEZ (José Ramón). Méndez Núñez (1824–1869) y la campana del Pacífico (1862–1869). Vol. 1. Estudio histórico. Vol. 2. Colección documental. Santiago de Compostela, Xunta de Galicia, 2000, 2 vol., [s. p.] (ill.).

7901. GEORGE (Marcus Sebastian). The Cyprus Convention policy and its impact upon the integrity of the Ottoman Empire. London, Centre of Near and Middle Eastern Studies, University of London, 2000, 50 p. (Occasional paper, 18).

7902. German empire (The) and Britain's Pacific dominions, 1871–1919: essays on the role of Australia and New Zealand in world politics in the age of imperialism. Ed. by John A. MOSES and Christopher PUGSLEY. Claremont, Regina Books, 2000, XVIII-576 p.

7903. GONZÁLEZ BARRIOS (René), ESPLUGAS VALDÉS (Héctor). El ejército español en Cuba, 1868–1878. La Habana, Ediciones Verde Olivo, 2000, 255 p. (ill.).

7904. GOODLAD (Graham D.). British foreign and imperial policy, 1865–1919. London, Routledge, 2000, 118 p. (Questions and analysis in history).

7905. GRANGE (D.-J.). L'Italie et les Balkans du Congrès de Berlin à 1915. In: Balkans (Les) dans les relations internationales, Part 1 [Cf. n° 7582], p. 297-305.

7906. GUTIÉRREZ (Harim B.). Una alianza fallida: México y Nicaragua contra Estados Unidos, 1909–1910. San Juan Mixcoac, Instituto Mora, 2000, 86 p. (Perfiles. Historia diplomática).

7907. HERREN (Madeleine). Hintertüren zur Macht: Internationalismus und modernisierungsorientierte Aussenpolitik in Belgien, der Schweiz und den USA 1865–1914. München, Oldenbourg, 2000, VI-551 p. (ill.). (Studien zur internationalen Geschichte, 9).

7908. HUGHES (Michael). Diplomacy before the Russian Revolution: Britain, Russia and the old diplomacy, 1894–1917. Basingstoke, Macmillan, 2000, XII-222 p. (Studies in diplomacy).

7909. IGNAT'JEV (Anatolij V.). Vneshnjaja politika Rossii 1907–1914 gg.: Tendentsii, ljudi, sobytija. (The foreign policy of Russia in 1907–1914: tendencies, people, events). RAN, In-t vseobshchej istorii. Moskva, Nauka, 2000, 233 p.

7910. Inszenierungen des Nationalstaats: politische Feiern in Italien und Deutschland seit 1860/1871. Hrsg. v. Sabine BEHRENBECK und Alexander NÜTZENADEL. Köln, SH-Verlag, 2000, 247 p. (ill.). (Kölner Beiträge zur Nationsforschung, 7).

7911. Internal rivalries and foreign threats, 1869–1879. Ed by Sven RUBENSON; co-editors, Amsalu AKLILU, Merid WOLDE AREGAY and Samuel RUBENSON. Addis Ababa, Addis Ababa U. P., 2000, XXVII-349 p. (Acta aethiopica, 3).

7912. ITO (Yukio). Rikken kokka to Nichiro sensô: Gaisei to naisei 1898–1905. (The constitutional state and the Russo-Japanese War: Foreign and home affairs 1898–1905). Tokyo, Bokutakusha, 2000, 425 p.

7913. JACOBSON (Matthew Frye). Barbarian virtues: the United States encounters foreign peoples at home and abroad, 1876–1917. New York, Hill and Wang, 2000, XII-324 p.

7914. KANIA (Jan). Pius IX a wojna partyzancka Polaków z Rosją. Powstanie styczniowe (1863–1864). (Pius IX. und Partisanenkampf der Polen gegen Russland. Januaraufstand, 1863–1864). Kraków, Spes, 2000, 311 p. (Studia z Dziejów Stosunków Polsko-Watykańskich).

7915. KATO (Masafumi). Hara Takashi to Mantetsu: Tôsei kakuchô to mammô seisaku no yûgô. (Hara Takashi and the Manchurian Railroad Company: The extention of party and the Manchuria-Mongolia policy). In: Kindai Nihon to Mantetsu. Ed. by Hideo KOBAYASHI. Tokyo, Yoshikawa Kobunkan, 2000, p. 31-63.

7916. KOMISSAROV (Boris N.), BOZHKOVA (Svetlana G.). Pervyj russkij poslannik v Brazilii F. F. Borel'. (The first Russian Ambassador in Brazilia: F. F. Borel). Sankt-Peterburg, Izd-vo Sank-Peterburgskogo un-ta, 2000, 279 p. [Portuguese Summary]

7917. KUTOLOWSKI (John F.). The West and Poland: essays on governmental and public responses to the Polish national movement, 1861–1864. Boulder, East European Monographs, 2000, XVII-338 p. (East European monographs, 548).

7918. LE MOAL (Frédéric). Diplomates et diplomatie en France entre 1900 et 1914. Revue d'histoire diplomatique, 2000, 114, 4, p. 289-330.

7919. MAC ALLISTER LINN (Brian). The Philippine War, 1899–1902. Lawrence, University Press of Kansas, 2000, XIV-427 p, (ill.). (Modern war studies).

7920. MAC BETH (Brian Stuart). Gunboats, corruption, and claims: foreign intervention in Venezuela, 1899–1908. Westport a. London, Greenwood Press, 2000, XII-307 p. (Contributions in Latin American studies, 20).

7921. MAC MILLAN (John). From 'separate democratic peace' to 'liberal pacificism': Anglo-Boer relations, 1880–1902. Keele, School of Politics, International Relations and the Environment, Keele University, 2000, 39 p. (Keele research paper, 25. Keele research papers, 25).

7922. MAISCH (C. J.). The Falkland/Malvinas Islands clash of 1831–1832: U.S. and British diplomacy in the South Atlantic. Diplomatic history, 2000, 24, 2, p. 185-209.

7923. Makedonija vo bilateralnite i multilateralni dogovori na Balkanskite drzavi 1861–1913: dokumenti. (Macedonia in the bilateral and multilateral agreements of the Balkan states 1861–1913). Redakcija i komentar Mihajlo MINOSKI. Skopje, Arhiv na Makedonija, Filozofski fakultet, Matica makedonska, 2000, 456 p.

7924. MELLANO (Maria Franca). I principi Maria Clotilde e Amedeo di Savoia e il Vaticano (1870–1890): attraverso la corrispondenza diplomatica della Santa Sede ed altri documenti. Torino, Centro studi piemontesi, 2000, 213 p.

7925. Memoirs of Giambattista Scala: consul of his Italian majesty in Lagos in Guinea (1862). Ed. by Robert SMITH. Oxford, Oxford U. P., 2000, XXVIII-155 p.

7926. MITCHELL (Leslie). Britain's reaction to the revolutions. In: Revolutions (The) in Europe, 1848–1849: from reform to reaction. Oxford, Oxford U. P., 2000, p. 83-98.

7927. MONDOLFI GUDAT (Edgardo). El águila y el león: el presidente Benjamin Harrison y la mediación de los Estados Unidos en la controversia de límites entre Venezuela y Gran Bretana. Presentación de Simón Alberto CONSALVI. Caracas, Academia Nacional de la Historia, 2000, 336 p. (Biblioteca de la Academia Nacional de la Historia. Estudios, monografías y ensayos, 180).

7928. MÜHLHALN (Klaus). Herrschaft und Widerstand in der «Musterkolonie» Kiatschou. Interaktionen zwischen China und Deutschland, 1897–1914. München, Oldenbourg, 2000, 474 p. (Studien zur Internationalen Geschichte, 8).

7929. MURRAY (Scott W.). Liberal diplomacy and German unification: the early career of Robert Morier. Westport a. London, Praeger, 2000, XXII-277 p.

7930. OTTE (T. G.). A question of leadership: lord Salisbury, the Unionist cabinet and foreign policy making, 1895–1900. *Contemporary British history*, 2000, 14, 4, p. 1-27.

7931. PAVEL (Teodor). Între Berlin şi Sankt-Petersburg. Vol. I. Românii în relaţiile germano-ruse din secolul al XIX-lea. Cluj-Napoca, Editura Presa Universitară Clujeană, 2000, 398 p.

7932. PAWLIK (Georg). Tegetthoff und das Seegefecht vor Helgoland: 9. Mai 1864. Wien, Verlag Österreich, 2000, 176 p. (ill.).

7933. PELLISTRANDI (Benoît). Les relations internationales de 1815 à 1870. Paris, A: Colin, 2000, 192 p.

7934. PSILOS (Christopher). The young Turk revolution and the Macedonian question 1908–1912, Leeds, [s. n.], 2000, XIII-253 p. [Thesis (Ph. D.), University of Leeds, Department of Russian and Slavonic Studies].

7935. PULIDO LLANO (Gabriela). Desde Cuba: escenas de la diplomacia porfirista, 1887–1901. San Juan Mixcoac, Instituto Mora, 2000, 80 p. (Perfiles. Historia diplomática).

7936. ROBERTS (Timothy M.), HOWE (Daniel W.). The United States and the revolutions of 1848. In: Revolutions (The) in Europe, 1848–1849: from reform to reaction. Oxford, Oxford U. P., 2000, p. 157-179.

7937. Russen und Rußland aus deutscher Sicht. 4. 19./20. Jahrhundert: von der Bismarckzeit bis zum Ersten Weltkrieg. Hrsg. v. Mechthild KELLER. München, Fink, 2000, 1160 p. (West-östliche Spiegelungen / A, 4).

7938. ŞAŞMAZ (Musa). British policy and the application of reforms for the Armenians in Eastern Anatolia, 1877–1897. Ankara, Turkish Historical Society Printing House, 2000, XXIII-306 p. (Publications of Turkish Historical Society. Serial. VII, 170. Türk Tarih Kurumu yayinlarindan. VII. dizi, 170).

7939. SCHÖLLGEN (Gregor). Imperialismus und Gleichgewicht: Deutschland, England und die orientalische Frage 1871–1914. München, Oldenbourg, 2000, XIV-501 p.

7940. SERRA (Enrico). Ammiragliato britannico e politica estera alla fine del XIX secolo. Roma, Rivista marittima, 2000, 87 p. (ill.). (Supplemento alla Rivista marittima, 2000, 7).

7941. SHAUMIAN (Tatiana). Tibet: the great game and Tsarist Russia. New Delhi a. Oxford, Oxford U. P., 2000, XII-223 p.

7942. STEVENSON (David). Armaments and the coming of the War: Europe, 1904–1914. Oxford, Oxford U. P., 2000, 470 p.

7943. TERRAZAS Y BASANTE (María Marcela). Inversiones, especulación y diplomacia: las relaciones entre México y los Estados Unidos durante la dictadura santannista. México, Universidad Nacional Autónoma de México, 2000, 292 p. (Serie de historia moderna y contemporánea, 35).

7944. TOUSSAINT RIBOT (Mónica). La política exterior de Estados Unidos hacia Guatemala, 1881–1885. México, Instituto Mora, 2000, 202 p.

7945. UÇAROL (Rifat). 1878 Cyprus dispute and Ottoman-English agreement: handover of the island to the British. Lefkosa, Rustem, 2000, III-175 p. (ill.).

7946. UNNO (Fukuju). Kankoku-heigôshi no kenkyû. (A study on the history of the annexation of Korea). Tokyo, Iwanami Shoten, 2000, 442 p.

7947. VALE (Brian). A war between Englishmen: Brazil against Argentina on the River Plate, 1825–1830. London, I.B. Tauris, 2000, XI-275 p.

7948. VÁZQUEZ (Josefina Zoraida), DEL REFUGIO GONZÁLEZ (María). Tratados de México, soberanía y territorio 1821–1910. México, Secretaría de Relaciones Exteriores, 2000, 291 p.

7949. VILLASENOR BELLO (José Miguel). Labor informativa de la legacion mexicana en Washington, 1822–1844. San Juan Mixcoac, Instituto Mora, 2000, 92 p. (Perfiles. Historia diplomática).

7950. VUCHKOV (Aleksandur). Srubsko-bulgarska voina 1885. (Serbo-Bulgarian War, 1885). Sofiia, Anzhela, 2000, 78 p. (Voinite za obedinenie na Bulgariia).

7951. ZACHS (Fruma). 'Novice' or 'Heaven-born' diplomat? Lord Dufferin and his plan for a 'Province of Syria': Beirut, 1860–1861. *Middle eastern studies*, 2000, 36, 3, p. 160-176.

7952. ZBIERSKI (Andrzej), CIEŚLAK (Marek), TRAWICKI (Lech). Udział Polaków w wojnie rosyjsko-japońskiej na morzu w latach 1904-1905. (Die Anteilnahme der Polen am russisch-japanischen Krieg in den Jahren 1904–1905). Gdańsk, CMM, 2000, 328 p. (Materiały Centralnego Muzeum Morskiego, 1).

Cf. n°s 7641, 7857, 7859, 7870, 7971, 7975, 8000, 8022, 8027, 8035

§ 6. From 1910 to 1935.
The First World War.

** 7953. Documents on Irish foreign policy. Vol. 2. 1923–1926. Ed. by Ronan FANNING, Michael KENNEDY, Dermot KEOGH and Eunan O'HALPIN. Dublin, Royal Irish Academy, 2000, XLV-596 p.

** 7954. Inside Stalin's Russia: the diaries of Reader Bullard, 1930–1934. Ed. by Julian BULLARD and Margaret BULLARD. Charlbury, Day Books, 2000, 310 p.

** 7955. Sandino y los U.S. Marines: reportes de los agregados militares y comandantes marines en acción. Compiladores, R. R. ISAGUIRRE y Adrían MARTÍNEZ RODRÍGUEZ. Tegucigalpa, Omni Editores, 2000, 490 p. (ill.). (Informantes de la historia).

** 7956. Sovetsko-rumynskie otnoshenija: Dokumenty i materialy. (Soviet-Roumanian Relations: documents and materials). Vol. 1. 1917–1934. Vol. 2. 1935–1941. Ed. A. A. AVDEEV, M. R. UNGURJANU. Ministerstvo inostrannykh del Rossijskoj Federatsii; Ministerstvo inostrannykh del Rumynii. Moskva, Mezhdunarodnye otnoshenija, 2000, 455, 543 p. (ill., portr.).

** 7957. Sultanate of Oman (The), 1918–1939. Ed. by Raghid EL-SOLH. External affairs. Reading, Ithaca, 2000, XXII-219 p.

7958. AEBISCHER (Tullio). La Commissione tecnica italo-vaticana ed i confini del territorio vaticano 1929–1933. Studi romani, 2000, 1/2, p. 104-117.

7959. ALLAIN (J.-C.). La France et les Balkans pendant l'entre-deux guerres (1920–1938). In: Balkans (Les) dans les relations internationales, Part 1 [Cf. n° 7582], p. 351-359.

7960. ALVAREZ (José E.). The betrothed of death: the Spanish Foreign Legion during the Rif Rebellion, 1920–1927. Westport a. London, Greenwood Press, 2000, 282 p. (ill., maps). (Contributions in comparative colonial studies, 40).

7961. ANDERSON (Mark Cronlund). Pancho Villa's revolution by headlines. Norman, University of Oklahoma Press, 2000, X-301 p.

7962. ANGELOW (Jürgen). Kalkül und Prestige: der Zweibund am Vorabend des Ersten Weltkrieges. Köln, Böhlau, 2000, VIII-530 p.

7963. ARNOLD (Georg). Gustav Stresemann und die Problematik der deutschen Ostgrenzen. Frankfurt am Main u. Oxford, Peter Lang, 2000, 148 p. (ill.). (Europäische Hochschulschriften. Reihe 3, Geschichte und ihre Hilfswissenschaften = Publications universitaires européennes. Série III, Histoire, sciences auxiliaires de l'histoire = European university studies, Series III, History and allied sciences, 870).

7964. ASHTON (Nigel J.), HELLEMA (Duco). Hanging the Kaiser: Anglo-Dutch relations and the fate of Wilhelm II, 1918–1920. Diplomacy and statecraft, 2000, 11, 2, p. 53-77.

7965. AUDOIN-ROUZEAU (Stéphane), BECKER (Annette). 14–18, retrouver la guerre. Paris, Gallimard, 2000, 272 p. (Bibliothèque des histoires).

7966. AUDOIN-ROUZEAU (Stéphane). Corps perdus, corps retrouvés. Trois exemples de deuils de guerre. In: Corps (Le) dans la Première Guerre Mondiale [Cf. n° 7981], p. 47-72.

7967. BAKHTURINA (Aleksandra Ju.). Politika Rossijskoj Imperii v Vostochnoj Galitsii v gody Pervoj mirovoj vojny. (The policy of the Russian Empire in Eastern Galicia during the First World War). Intr. by Valentin V. SHELOKHAEV. Assotsiatsija issledovatelej rossijskogo obshchestva XX v. Moskva, AIRO-XX, 2000, 250 p. (ill., maps). (Pervaja monografija).

7968. BARYÉTY (J.). La France et la naissance du Royaume des Serbs, Croates et Slovènes, 1914–1919. In: Balkans (Les) dans les relations internationales, Part 1 [Cf. n° 7582], p. 307-327.

7969. BAUER (Wolfgang). Tsingtau 1914 bis 1931: Japanische Herrschaft, wirtschaftliche Entwicklung und die Rückkehr der deutschen Kaufleute. München, Indicium, 2000, 247 p. (ill., map).

7970. BECKER (Annette). Guerre totale et troubles mentaux. In: Corps (Le) dans la Première Guerre Mondiale [Cf. n° 7981], p. 135-152.

7971. BENGTSSON (Rikard). Trust, threat, and stable peace: Swedish great power perceptions 1905–1939. Lund, Lund University, 2000, 214 p. – IDEM. Uncertainty, risk, and trust: the Swedish-German general staff negotiations of 1910. Diplomacy and statecraft, 2000, 11, 2, p. 37-52.

7972. BERDAH (Jean-François). La démocratie assassinée: la République espagnole et les grandes puissances 1931–1939. Paris, Berg International, 2000, 451 p. (Ecritures de l'histoire).

7973. BOBROFF (Ronald). Sergei D. Sazonov and the future of Poland, 1910–1916. International history review, 2000, 22, 2, p. 505-528.

7974. BODEN (Ragna). Die Weimarer Nationalversammlung und die deutsche Aussenpolitik: Waffenstillstand, Friedensverhandlungen und internationale Beziehungen in den Debatten von Februar bis August 1919. Frankfurt am Main u. Oxford, P. Lang, 2000, X-191 p. (Europäische Hochschulschriften. Reihe III, Geschichte und ihre Hilfswissenschaften, 884 = Publications universitaires européennes. Série III, Histoire, sciences auxiliaires de l'histoire, 884 = European university studies. Series III, History and allied studies, 884).

7975. BROOKS (Barbara J.). Japan's imperial diplomacy: consuls, treaty ports, and war in China 1895–1938. Honolulu, Universiti of Hawai'i Press, 2000, XI-296 p. (Studies on East Asan Institute, Columbia University).

7976. BURGARD (Oliver). Das gemeinsame Europa-von der politischen Utopie zum aussenpolitischen Programm: Meinungsaustausch und Zusammenarbeit proeuropäischer Verbände in Deutschland und Frankreich, 1924–1933. Frankfurt am Main, Verlag Neue Wissenschaft, 2000, 292 p.

7977. CACCAMO (Francesco). L'Italia e la nuova Europa: il confronto sull'Europa orientale alla conferenza di pace di Parigi, 1919–1920. Prefazione di Francesco PERFETTI. Milano, Luni, 2000, 346 p. (Biblioteca di storia contemporanea, 16).

7978. CAROLI (Giuliano). Rapporti militari fra Italia e Romania dal 1918 al 1945. Le carte dell'Ufficio Storico. Roma, Stato Maggiore dell'Esercito, Ufficio Storico, 2000, 420 p.

7979. CLOAREC (Vincent). Raymond Poincaré et la diplomatie française en Méditerranée orientale à la veille de la Premiére Guerre mondiale: une préfiguration de la politique mandataire? *Revue d'histoire diplomatique*, 2000, 114, 1, p. 25-52.

7980. COLLOTTI (Enzo). Fascismo e politica di potenza: politica estera 1922–1939. con la collaborazione di Nicola LABANCA e Teodoro SALA. Firenze, La Nuova Italia, 2000, VII-494 p. (Biblioteca di storia, 80).

7981. Corps (Le) dans la Première Guerre Mondiale. Près. de Stéphane AUDOIN-ROUZEAU et Annette BECKER. *Annales*, 2000, 55, 1, 43-152. [Cf. n^{os} <sélection> 7966, 7970, 8003, 8036.]

7982. CORRIGAN (Gordon). Sepoys in the trenches: the Indian corps on the Western Front 1914–1915. Staplehurst, Spellmout, 2000, XIV-274 p.

7983. DEJMEK (Jindřich). Masaryk, Beneš a zahraniční politika Československa v první polovině dvacátých let (1919–1924). (The foreign policy of the first President of the CR T. G. Masaryk and the Foreign Minister E. Beneš in the 1920s). *In*: Společnost v přerodu. Češi ve 20. století. Praha, 2000, p. 268–287. – IDEM. Politické vztahy Československa a Velké Británie v době první labouristické vlády (leden–listopad 1924). (Political relations between Czechoslovakia and Great Britain during the period first government of Labour Party, January–November 1924). *In*: Evropa mezi Německem a Ruskem. Ed. Miroslav ŠESTÁK. Praha, 2000, p. 321-337. – IDEM. Velká Británie a Československo v době jednání o Západní garanční pakt (leden–prosinec 1925). (Gross-Britannien und die Tschechoslowakei während der Verhandlungen über den Westlichen Garantie-Pakt, Januar–Dezember 1925). *Český časopis historický*, 2000, 98, 4, p. 775-806.

7984. DENÉCHÈRE (Yves). La campagne française de «pacification» dans le Sud marocain: la question de la coopération militaire espagnole (1931–1934). *Guerres mondiales et conflits contemporaine*, 2000, 50, 199, p. 93-110.

7985. DOERRIES (Reinhard R.). Prelude to the Easter Rising: Sir Roger Casement in imperial Germany. London, F. Cass, 2000, XIV-233 p. (ill., ports). (Cass series – studies in intelligence).

7986. DULLIN (Sabine). Le rôle de l'Allemagne dans le rapprochement franco-soviétique 1932–1935. *In*: Deutschland, Frankreich, Russland: Begegnungen und Konfrontationen = La France et l'Allemagne face à la Russie [Cf. n° 7597], p. 245-262.

7987. FARAH (Caesar E.). The politics of interventionism in Ottoman Lebanon, 1830–1861. London, I. B. Tauris, 2000, XXV-816 p.

7988. FARLEY (Brigit). Ethnic conflict and European affairs revisited: the Serb-Croat quarrel and French diplomacy 1929–1935. Seattle, Henry M. Jackson School of International Studies, University of Washington, 2000, 57 p. (Donald W. Treadgold papers in Russian, East European and Central Asian Studies, 25).

7989. FERRO (Marc). Octobre 1017: révolution dans les relations internationales. Les trois leviers de la politique étrangère soviètique. *Politique étrangère*, 2000, 65, 3-4, p. 855-861.

7990. FRANZ (Corinna). Fernand de Brinon nd die deutsch-französischen Beziehungen 1918–1945. Bonn, Bouvier, 2000, XII-413 p. (Pariser Historischer Studien, 54).

7991. FRENCH (David). Raising Churchill's army: the British Army and the war against Germany 1919–1945. Oxford, Oxford U. P., 2000, XII-319 p.

7992. GABELLINI (Andrea). L'Italia e l'assetto della Palestina: 1916–1924. Firenze, SeSaMO, 2000, IX-221 p.

7993. GIL-HAR (Yitzhak). Boundaries delimitation: Palestine and Trans-Jordan. *Middle eastern studies*, 2000, 36, 1, p. 68-81.

7994. GIORDANO (Giancarlo). Storia diplomatica del patto a quattro. Milano, F. Angeli, 2000, 208 p.

7995. GOEMANS (H. E.). War and punishment: the causes of war termination and the First World War. Princcton, Princeton U. P., 2000, 355 p.

7996. GOL (Ayla). The place of foreign policy in the transition to modernity: Turkish policy towards the south Caucasus, 1918–1921. London, [s. n.], 2000, 298 p. [Thesis (Ph. D.). – LSE, 2000].

7997. GREENHALGH (Elizabeth). Technology development on coalition: the case of the First World War tank. *International history review*, 2000, 22, 4, p. 806-836.

7998. GRUMEL-JACQUIGNON (F.). La France et la Yougoslavie de 1920 à 1940. *In*: Balkans (Les) dans les relations internationales, Part 1 [Cf. n° 7582], p. 361-373.

7999. HALL (Richard C.). The Balkan Wars, 1912–1913: prelude to the First World War. London, Routledge, 2000, 176 p. (Warfare and history).

8000. HEWITSON (Mark). Germany and France before the First World War: a reassessment of Wilhelmine foreign policy. *English historical review*, 2000, 115, 462, p. 570-606.

8001. HEYDE (Philipp). Frankreich und das Ende der Reparationen. Das Scheitern der französischen Stabilisierungskonzepte in der Weltwirtschaftskrise 1930–1932. *Vierteljahrshefte für Zeitgeschichte*, 2000, 48, 1, p. 37-74.

8002. HORNE (John), KRAMER (Alan). German atrocities, 1914. A history of denial. New Haven a. London, Yale U. P., 2000, 608 p.

8003. HORNE (John). Corps, lieux et nation: la France et l'invasion de 1914. *In*: Corps (Le) dans la Première Guerre Mondiale [Cf. n° 7981], p. 73-110.

8004. HOUSDEN (Martyn). Ewald Ammende and the organization of national minorities in inter-war Europe. *German history*, 2000, 18, 4, p. 439-460.

8005. HUGHES (Michael). The virtues of specialization: British and American diplomatic reporting on Russia, 1921–1939. *Diplomacy and statecraft*, 2000, 11, 2, p. 79-104.

8006. HURST (James W.). The Villista prisoners of 1916–1917. Las Cruces, Yucca Tree Press, 2000, X-112 p.

8007. HUTCHISON (John F.). Disaster as the international order: eartqakes, humanitarians, and the Ciraolo project. *International history review*, 2000, 22, 1, p. 1-36.

8008. ISNENGHI (Mario), ROCHAT (Giorgio). La Grande Guerra 1914–1918. Firenze, La Nuova Italia, 2000, 562 p.

8009. JOHNSON (G.). Curzon, Lloyd George and the control of British foreign policy, 1919–1922: a reassessment. *Diplomacy and statecraft*, 2000, 11, 3, p. 49-71.

8010. KASSIM (Mahmoud). Die diplomatischen Beziehungen Deutschlands zu Ägypten, 1919–1936, Hamburg, LIT, 2000, 393 p. (ill.). (Studien zur Zeitgeschichte des Nahen Ostens und Nordafrikas, 6).

8011. KEN (O. N.), RUPASOV (A. I.). Politbjuro TsK VKP(b) i otnoshenija SSSR s zapadnymi sosednimi gosudarstvami (konets 1920–1930-e gody): Problemy, dokumenty, opyt kommentarija. (The Politburo of the Central Committee of the VKP(b) and the Relations of the USSR with its western neighbouring countries, the Late 1920s and the 1930s: problems, documents and an attempt at commentary). Part 1. December 1928–June 1934. Sankt-Peterburg, Evropejskij dom, 2000, 703 p. (ill., portr., maps).

8012. KOVÁČ (Dušan). Štefánikovo talianske memorandum z apríla 1916. (Štefánik's Italian Memorandum on April 1916). *Historický časopis*, 2000, 48, 3, p. 517-533.

8013. KÜPPERS (Heinrich). Zwischen londoner Ultimatum und Rapallo. Joseph Wirth und die deutsche Außenpolitik 1921/1922. *Historische Mitteilungen*, 2000, 13, p. 150-175.

8014. KURAEV (Oleksyj). Der Verband "Freie Ukraine" im Kontext der deutschen Ukraine-Politik des Ersten Weltkriegs. München, Osteuropa-Institut München, 2000, 47 p. (Mitteilungen / Osteuropa-Institut München, 35).

8015. LATOUR (Francis). La Saint-Siège et les Etats-Unis à la recherche de la paix pendant la Première Guerre mondiale: la collaboration impossible. *Guerres mondiales et conflits contemporaine*, 2000, 50, 199, p. 81-92.

8016. LIULEVICIUS (Vejas G.). War land on the Eastern Front: culture, national identity and German occupation in World War I. Cambridge a. New York, Cambridge U. P., 2000, VIII-309 p. (Studies in the social and cultural history of modern warfare, 9).

8017. MESKANK (Timo). Kultur besteht, Reich vergeht: Tschechen und Sorben (Wenden) 1914–1945. Berlin, Mensch & Buch, 2000, 264 p.

8018. MILZA (Pierre). L'Italie fasciste et les Balkans (1922–début des années 30). *In*: Balkans (Les) dans les relations internationales, Part 2 [Cf. n° 7582], p. 397-411.

8019. MINNITI (Fortunato). Fino alla guerra: strategie e conflitto nella politica di potenza di Mussolini: 1923–1940. Napoli, Edizioni scientifiche italiane, 2000, 238 p. (Quaderni di Clio, 2). – IDEM. Il Piave. Bologna, Il Mulino, 2000, 148 p.

8020. MLADENOVIC (Bozica). Grad u austrougarskoj okupacionoj zoni u Srbiji od 1916. do 1918. godine. Beograd, Cigoja stampa, 2000, 256 p.

8021. NOVAK TALAVERA (Fabián). Las conversaciones entre Perú y Chile para la ejecución del tratado de 1929. San Miguel, Pontificia Universidad Católica del Perú, Instituto de Estudios Internacionales, Fondo Editorial, 2000, 223 p.

8022. OZGHANBAI (Omirzaq). Gosudarstvennaia Duma Rossii i Kazakhstan: 1905–1917. Almaty, Izd-vo "Arys", 2000, 280 p.

8023. PAGE (Melvin E.). The Chiwaya war: Malawians and the First World War. Boulder, Westview, 2000, XVI-276 p. (History and Warfare).

8024. PASZTOR (Maria H.). Un lobby en voie d'affaiblissement. L'activité du groupe franco-polonaise à l'Assemblée nationale (1921–1936). *Revue d'histoire diplomatique*, 2000, 114, 1, p. 53-68.

8025. PORTZ (Thomas). Grossindustrie, Kriegszielbewegung und OHL, Siegfrieden und Kanzlersturz: Carl Duisberg und die deutsche Aussenpolitik im Ersten Weltkrieg. Lauf a.d. Pegnitz, Europaforum-Verlag,

2000, 503 p. (Subsidia academia. Reihe A, Neuere und neueste Geschichte, 1).

8026. PROKŠ (Petr). Česká politika a válečné hospodářství 1914–1918. (Czech policy and the war economy in 1914–18). *Slovanský přehled*, 2000, 86, 3, p. 309–332.

8027. Recueil des traités, conventions, protocoles, arrangements et déclarations signés entre l'Empire ottoman et les puissances étrangers, 1903–1922. Vol. 1. 1903–1916. Vol. 2. 1917–1922. Recuillis et publiés par Sinan KUNERALP. İstanbul, Editions Isis, 2000, 2 vol., 647 p. (Studies in Ottoman diplomatic history, 10).

8028. RIBEIRO DE MENESES (Filipe). 'All of us are looking forward to leaving': the censored correspondence of the Portuguese Expeditionary Corps in France, 1917–1918. *European history quarterly*, 2000, 30, 3, p. 333-356.

8029. ROSSI (Ernesto). Il manganello e l'aspersorio. [La collusione fra il Vaticano e il regime fascista nel ventennio]. Milano, Kaos, 2000, 370 p. (Libertaria).

8030. ROSSINI (Daniela). Il mito americano nell'Italia della grande guerra. Roma, Laterza, 2000, X-284 p. (ill.). (Quadrante Laterza, 106).

8031. RUPASOV (A. I.), CHISTIKOV (Aleksandr Nikolaevich). Sovetsko-finliandskaia granitsa, 1918–1938 gg. Sankt-Peterburg, Evropeiskii dom, 2000, 163 p.

8032. ŠEDIVÝ (Ivan). Rakousko-uherská válečná propaganda 1914–1918. (Austro-Hungarian wartime propaganda 1914–18). *Historie a vojenství*, 2000, 49, 1, p. 38-55.

8033. ŠILPOCH (Karel). Etiopie a Československo mezi válkami 1918–1938. (Ethiopia and Czechoslovakia between the two world wars, 1918–38). *Moderní dějiny*, 2000, 8, p. 53-82.

8034. SLÁDEK (Zdeněk). Malá dohoda 1919–1938. Její hospodářské, politické a vojenské komponenty. (Die Kleine Entente 1919–1938. Ihre ekonomische, politische und militärische Komponenten). Praha, Karolinum, 2000, 297 p.

8035. SMITH (Jeremy). Britain and Ireland: from home rule to independence. Harlow, Longman, 2000, X-163 p. (Seminar studies in history).

8036. SMITH (Leonard V.). Le corps et la survie d'une identité dans les écrits de guerre français. *In*: Corps (Le) dans la Première Guerre Mondiale [Cf. n° 7981], p. 111-134.

8037. STAPLES (Anne) [et al.]. Diplomacia y revolución: homenaje a Berta Ulloa. Pedegral de Santa Teresa, Colegio de México, Centro de Estudios Históricos, 2000, 311 p. (ill.).

8038. SUCHOPLES (Jaroslaw). Finland and the United States, 1917–1919. Early years of mutual relations. Helsinki, SKS, 2000, 221 s. (Bibliotheca historica, 56).

8039. SUNDBÄCK (Esa). Finland in the British Baltic policy. British political and economic interests regarding Finland in the aftermath of the First World War, 1918–1925. Helsinki, Finnish Academy of Science and Letters, 2000, 392 p. (A. Acad. Sci. Fennicae, Hum., 315).

8040. SZCZEPAŃSKI (Janusz). Społeczeństwo Polski w walce z najazdem bolszewickim 1920 roku. (Polnische Gesellschaft im Kampf gegen bolschewistischen Überfall 1920). Warszawa u. Pułtusk, Oficyna Wydawnicza Tow. Opieki nad Zabytkami, 2000, 551 p.

8041. TANAKA (Ryûichi). Mansyûkoku chigaihouken teppai to Mantetsu. (The abolition of extraterritoriality in Manchukuo and the Manchurian Railroad Company). *In*: Kindai Nihon to Mantetsu. Ed. Hideo KOBAYASHI. Tokyo, Yoshikawa Kobunkan, 2000, p. 187-218.

8042. THOBIE (J.). La politique balkanique de la Turquie kémaliste. *In*: Balkans (Les) dans les relations internationales, Part 1 [Cf. n° 7582], p. 329-349.

8043. Traités (les) de paix de Versailles, Saint-Germain-en-Laye, Neuilly, Trianon, Sèvres, 1919–1920. Actes du colloque international de Saint-Germain-en-Laye. *Bulletin des Amis du Vieux Saint-Germain*, 2000 (numéro spécial), 230 p.

8044. Trianon: a magyar béküldöttség tevékenysége 1920-ban: válogatás 'A magyar béketárgyalások: jelentés a magyar béküldöttség muködéserol Neuilly-sur-Seine-ben' I-II. kötetébol (térképmelléklet: III/B. kötet) Budapest, 1920–1921. [The Hungarian peace negotiations: an account of the work of the Hungarian peace delegation at Neuilly s/S, from January to March, 1920]. A bevezeto tanulmányt írta és az iratokat válogatta POMOGÁTS Béla; a forrásokat szerkesztette és sajtó alá rendezte ÁDÁM Magda és CHOLNOKY Gyozo. Budapest, Lucidus Kiadó, 2000, 543 p. (Kisebbségkutatás Könyvek).

8045. URSS (L') et l'Europe dans les années 20: actes du colloque organisé à Moscou les 2 et 3 octobre 1997 par l'Institut d'histoire universelle de l'Académie des sciences de Russie; sous la direction de Mikhail NARINSKI, Elisabeth DU RÉAU, Georges-Henri SOUTOU et Alexandre TCHOUBARIAN. Paris, Presses de l'Université de Paris-Sorbonne, 2000, 184 p. (Mondes contemporains). [Cf. n° <sélection> 8067.]

8046. VARGAS GARCIA (Eugênio). O Brasil e a Liga das Naçoes (1919–1926): vencer ou nao perder. Porto Alegre, Editora da Universidade, Universidade Federal do Rio Grande do Sul, 2000, 167 p. (Relaçoes internacionais e integraçao).

8047. VELTER (Tiberiu). Relații româno-britanice între anii 1914 1924. Cluj-Napoca, Presa Universitară Clujeană, 2000, XIV-242 p.

8048. VESELÝ (Zdeněk). Československá zahraniční politika 1914–1945. Dokumenty. (Czechoslovak foreign policy, 1914–1945. Documents). Praha, Vysoká škola ekonomická, 2000, 319 p. (mp.).

8049. VRAIN (Cécile). La politique de la France en Hongrie entre 1921 et 1931. *In*: Hongrie (La) dans les conflits du XXe siècle [Cf. n° 7611], p. 53-66.

8050. WELCH (David). Germany, propaganda, and total war, 1914–1918: the sins of omission. New Brunswick, Rutgers U. P., 2000, IX-335 p.

8051. WYSZCZELSKI (Lech). Bitwa na przedpolach Warszawy. (Die Schlacht am Rande Warschaus). Warszawa, Bellona, 2000, 549 p. (O Wolność i Niepodleglość).

8052. ZABLOUDILOVÁ (Jitka). Příspěvek k tematice propagandy v čs. vojsku v Rusku v letech 1914–1920. (On the theme of propaganda among the Czech forces in Russia 1914–1920). *Historie a vojenství*, 2000, 49, 1, p. 56-66.

8053. ZIEGER (Robert H.). America's great war: World War I and the American experience. Oxford, Rowman & Littlefield, 2000, XII-275 p.

8054. Zwischen Tradition und Revolution. Determinanten und Strukturen sowjetischer Außenpolitik 1917–1941. Hrsg. v. Ludmila THOMAS und Viktor KNOLL. Stuttgart, Steiner, 2000, 441 p. (Quellen und Studien zur Geschichte des östlichen Europa, 59).

Cf. n°s 4738, 4763, 5050, 5261, 6776, 7249, 7874, 7879, 7886, 7894, 7899, 7902, 7904, 7905, 7907, 7908, 7909, 7913, 7918, 7923, 7928, 7934, 7937, 7939, 7941, 7942, 8055, 8061, 8067, 8077, 8078, 8079, 8083, 8084, 8098, 8102, 8103, 8122, 8126, 8133, 8144, 8145, 8153, 8169, 8176

§ 7. From 1935 to 1945.
The Second World War

a. General

** 8055. Times (The) and appeasement: the journals of A. L. Kennedy, 1932–1939. Ed. by Gordon MARTEL. Cambridge, Cambridge U. P. for the Royal Historical Society, 2000, XVII-312 p. (Camden fifth series, 16).

8056. ALBORD (Maurice). L'armée française et les États du Levant, 1936–1946. Préface de Pierre MESSMER. Paris, CNRS Editions, 2000, 336 p. (ill.). (Moyen-Orient).

8057. ALDRICH (Richard J.). Intelligence and war against Japan: Britain, America and the politics of secret service. Cambridge, Cambridge U. P., 2000, XXIV-500 p.

8058. BATTEL (Franco). "Wo es hell ist, dort ist die Schweiz": Flüchtlinge und Fluchthilfen an der Schaffhauser Grenze zur Zeit des Nationalsozialismus. Zürich, Chronos, 2000, 375 p, (ill.). (Schaffhauser Beiträge zur Geschichte, 77).

8059. BILLINGER (Robert D.). Hitler's soldiers in the sunshine state. German POWs in Florida. Gainesville, University Press of Florida, 2000, XIX-262 p.

8060. BORÁK (Mečislav). Druhá fáze delimitace hranic mezi Československem a Polskem na Těšínsku v listopadu 1938. Výběr dokumentů. (Die zweite Etappe der Delimitation zwischen der Tschechoslowakei und Polen im November 1938: Auszug aus den Dokumenten). *Časopis Slezského zemského muzea. Ser. B Vědy historické*, 2000, 49, 1, p. 51-94.

8061. BOWEN (Wayne H.). Spaniards and Nazi Germany: collaboration in the new order. Columbia a. London, University of Missouri Press, 2000, XII-250 p.

8062. BYSTRICKÝ (Valerián). Zahraničnopolitické koncepcie politických strán na Slovensku koncom tridsiatych rokov 20. storočia. (Die aussenpolitische Konzeption der politischen Parteien in der Slowakei Ende der 30er Jahre). *Historický časopis*, 2000, 48, 2, p. 257-281. [Deutsche Zfassung].

8063. Camps (Les) d'internement français: 1939–1942: témoignages d'un dessinateur autrichien [Bil Spira]. Ed. par Claude WINKLER-BESSONE et Jean-Marie WINKLER. Mont-Saint-Aignan, Publ. de l'Université de Rouen, 2000, 142 p.

8064. CEPREGANOV (Todor). Sudirot na britanskite i germanskite interesi na balkanot vo tekot na vtorata svetska vojna. (Clash of the British and German interests in the Balkans in the Second World War). Skopje, Institut za nacionalna istorija, 2000, 266 p.

8065. CHARLES (D. M.). Informing FDR: FBI political surveillance and the isolationist-interventionist foreign policy debate, 1939–1945. *Diplomatic history*, 2000, 24, 2, p. 211-232.

8066. COGAN (Frances B.). Captured: the Japanese internment of American civilians in the Philippines, 1941–1945. Athens, University of Georgia Press, 2000, XI-357 p.

8067. DULLIN (Sabine). Litvinov, les diplomates soviétiques et l'Europe au seuil des années 1930. *In:* URSS (L') et l'Europe dans les années 20 [Cf. n° 8045], p. 151-166.

8068. Frankreich und Deutschland im Krieg (November 1942–Herbst 1944), Okkupation, Kollaboration, Résistance: Akten des deutsch-französischen Kolloquiums La France et l'Allemagne en guerre (novembre 1942–automne 1944), occupation, collaboration, résistance. Von Stefan MARTENS und Maurice VAÏSSE. Deutsches Historisches Institut (Paris, France); Centre d'études d'histoire de la défense (France); Institut für Zeitgeschichte (München, Deutschland); Institut d'histoire du temps présent (France). Bonn: Bouvier, 2000, XVIII-944 p. (Pariser historische Studien, 55).

8069. FRIJTAG (Geraldien von), KÜNZEL (Drabbe). Rechtpolitik im Reichkommissariat. Zum Einsatz deutscher Strafrichter in den Niederlanden und in Norwegen 1940–1944. *Vierteljahrshefte für Zeitgeschichte*, 2000, 48, 3, p. 461-490.

8070. GELVIN (James L.). Zionism and the representation of 'Jewish Palestine' at the New York world's fair, 1939–1940. *International history review*, 2000, 22, 1, p. 1-36.

8071. GROSS (Jan Tomasz). Sąsiedzi. Historia zaglady żydowskiego miasteczka. (Nachbarn. Die Geschichte der Vernichtung einer jüdischen Kleinstadt). Sejny, Pogranicze, 2000, 17 p. (Biblioteka Krasnogrudy).

8072. HAYNES (Rebecca). Romanian policy towards Germany, 1936–1940. Basingstoke, Macmillan in association with School of Slavonic and East European Studies, University College London, 2000, VIII-205 p. (Macmillan Studies in Russia and East Europe).

8073. HEALY (Michael S.). Colour, climate, and combat: the Caribbean regiment in the Second World War. *International history review*, 2000, 22, 1, p. 65-85.

8074. ICHINOSE (Toshiya). Gunji engo to Jûgo hôkô kai. (Military support and the 'society of service' in home front in World War II). *Nihon Rekishi*, 2000, no. 627, p.71-87.

8075. Ireland and the Second World War: politics, society and remembrance. Ed. by Brian GIRVIN and Geoffrey ROBERTS. Dublin, Four Courts Press, 2000, 186 p.

8076. JACKSON (Peter). Stratégie et idéologie: le haut commandement français et la guerre civile espagnole. *Guerres mondiales et conflits contemporaine*, 2000, 50, 199, p. 111-134.

8077. KALLIS (Aristotle A.). Fascist ideology: territory and expansionism in Italy and Germany, 1922–1945. London a. New York, Routledge, 2000, IX-286 p.

8078. KHLEVNIUK (O. V.). The reasons for the 'Great Terror': the foreign-political aspect. *In*: Russia in the age of wars, 1914–1945 [Cf. n° 5077], p. 159-169.

8079. KNOX (MacGregor). Common destiny: dictatorship, foreign policy, and war in Fascist Italy and Nazi Germany. Cambridge, Cambridge U. P., 2000, XIV-262 p. – IDEM. Hitler's Italian allies. Royal armed forces, fascist regime, and the war of 1940–1943. Cambridge, New York a. Melbourne, Cambridge U. P., 2000, XIV-207 p.

8080. KOTOWSKI (Albert S.). Hitlers Bewegung im Urteil der polnischen Nationaldemokratie. Wiesbaden, Harrassowitz, 2000, VIII-298 p. (Studien der Forschungsstelle Ostmitteleuropa an der Universität Dortmund, 28).

8081. Latvija otraja pasaules kara: starptautiskas konferences materiali 1999. gada 14.–15. junijs, Riga = Latvia in World War II: materials of an international conference, 14–15 June 1999, Riga. Riga: Latvijas vestures instituta apgads, 2000, 389 p. (Latvijas Vesturnieku komisijas raksti, 1).

8082. LAWSON (Fred. H.). Westphalian sovereignity and the emergence of the Arab states system: the case of Syria. *International history review*, 2000, 22, 3, p. 529-556.

8083. LUCONI (Stefano). La "diplomazia parellela": il regime fascista e la mobilitazione politica degli italo-americani. Milano, F. Angeli, 2000, 157 p. (Temi di storia, 14). – IDEM. Le fascisme et la campagne des Italo-Américains contres les modifications de la loi sur la neutralité aux Etats-Unis en 1939. *Guerres mondiales et conflits contemporaine*, 2000, 50, 199, p. 135-146.

8084. MARTÍNEZ SÁNCHEZ (Antonio), MARTINEZ GÓMEZ (Rosario). Los franceses y la IIa República espanola: estudios de prensa 1931–1936. Granada, Grupo Editorial Universitario, 2000, 149 p.

8085. MAY (Ernest R.). Strange victory: Hitler's conquest of France. New York, Farrar, Straus & Giroux, 2000, 594 p.

8086. METZGER (C.). L'Empire colonial français dans la stratégie du Troisième Reich (1936–1945). *Relations internationales*, 2000, 101, p. 41-55.

8087. MICCOLI (Giovanni). I dilemmi e i silenzi di Pio XII (Vaticano, Seconda guerra mondiale e Shoah). Milano, Rizzoli, 2000, 570 p.

8088. Misión de Luis I. Rodríguez en Francia: la protección de los refugiados españoles, julio a diciembre de 1940. Prólogo de Rafael SEGOVIA y Fernando SERRANO. México, El Colegio de México, Secretaría de Relaciones Exteriores, Consejo Nacional de Ciencia y Tecnología, 2000, XVI-604 p.

8089. Nachalo Vtoroj mirovoj vojny (1939 god): Sovremennoe sostojanie problemy: Kruglyj stol. (The beginning of the Second World War, 1939: contemporary state of the question: a 'Round Table'). Ed. S.Z. Sluch. *Slavjanovedenie*, Moskva, 2000, [36], 6, p. 58-82.

8090. NICOLL (Peter H.). Englands Krieg gegen Deutschland: Ursachen, Methoden und Folgen des Zweiten Weltkriegs. Tübingen, Grabert, 2000, 572 p. (Veröffentlichungen des Instituts für Deutsche Nachkriegsgeschichte, 2).

8091. OVERMANS (Rüdiger). Deutsche militärische Verluste im Zweiten Weltkrieg. München, Oldenbourg, 2000, XIV-367 p. (Beiträge zur Militärgeschichte, 46).

8092. PONS (Silvio). In the aftermath of the age of wars: the impact of World War II on Soviet security policy. *In*: Russia in the age of wars, 1914–1945 [Cf. n° 5077], p. 277-307.

8093. PRAZMOWSKA (Anita). Eastern Europe and the origins of the Second World War. London, Macmillan, 2000, X-278 p. (Making of the 20th century).

8094. PRITZ (Pál). La crise de guerre internationale et la Hongrie (1938–1941). *In*: Hongrie (La) dans les conflits du XXe siècle [Cf. n° 7611], p. 67-82.

8095. RALPH (Barry). They passed this way: the United States of America, the States of Australia and World War II. Sydney, Kangaroo Press, 2000, 310 p. (ill).

8096. RAY (Roland). Annäherung an Frankreich im Dienste Hitlers? Otto Abetz und die deutsche Frankreichpolitik, 1930–1942. München, R. Oldenburg, 2000, 419 p. (Studien zur Zeitgeschichte, 59).

8097. Réfugiés civils (Les) et la frontière genevoise durant la Deuxième Guerre mondiale: fichiers et archives. Ed. par Pierre FLÜCKIGER et Gérard BAGNOUD, sous la direction de Catherine SANTSCHI, avec la collaboration de Joëlle DROUX, Ruth FIVAZ-SILBERMANN et Roger ROSSET, avant-propos de Robert CRAMER, préface de Jean-Claude FAVEZ. Genève, Archives d'État de Genève, 2000, 176 p., (ill.).

8098. ROBERTS (Geoffrey Charles). The Fascist war threat and Soviet politics in the 1930s. *In*: Russia in the age of wars, 1914–1945 [Cf. n° 5077], p. 147-158.

8099. ROMSICS (Ignác). La politique des grandes puissances et la Hongrie pendant et après la Deuxième Guerre mondiale. 83-98.

8100. RYCHLAK (Ronald J). Hitler, the war, and the pope. Huntington, Our Sunday Visitor, 2000, XIV-470 p.

8101. SEMIRJAGA (Mikhail I.). Kollaboratsionism: priroda, tipologija i projavlenija v gody Vtoroj mirovoj vojny. (Collaborationism: its nature, typology and manifestation during the Second World War). Moskva, ROSSPEN, 2000, 863 p. (ill.; portr.; maps; schemes; tables; bibl. incl.).

8102. SIMONI (Marcella). At the roots of division: a new perspective on Arabs and Jews, 1930–1939. *Middle eastern studies*, 2000, 36, 3, p. 52-92.

8103. SIROIS (Herbert). Zwischen Illusion und Krieg: Deutschland und die USA, 1933–1941. Paderborn, Schöningh, 2000, 317 p. (Sammlung Schöningh zur Geschichte und Gegenwart).

8104. Slovensko a druhá svetová vojna. Zborník príspevkov z medzinárodnej vedeckej konferencie v Bratislave 29.–31. mája 2000 organizovanej Vojenským historickým ústavom a Historickým ústavom Slovenskej akadémie vied. (Slovakia and the Second World War. Contributions Miscellany from the International Scientific Conference in Bratislava, 29.–31. May 2000, organized by the Institute of Military History and the Historical Institute of SAS). Ed. by František CSÉFALVAY and Miloslav PÚČIK. Bratislava, Vojenský historický ústav Ministerstva obrany, 2000, 427 p.

8105. SLUTSCH (Sergej). 17. September 1939: der Eintritt der Sowjetunion in den Zweiten Weltkrieg. Eine historische und völkerrechtliche Bewertung. *Vierteljahrshefte für Zeitgeschichte*, 2000, 48, 2, p. 219-254.

8106. STROBL (Gerwin). The Germanic isle: Nazi perceptions of Britain. Cambridge a. New York, Cambridge U. P., 2000, X-274 p. (ill.).

8107. TESAŘ (Jan). Mnichovský komplex. Jeho příčiny a důsledky. (The Munich complex. Its causes and consequences). Praha, Prostor, 2000, 255 p. (Obzor, 28).

8108. THOM (Françoise). Les Mémoires de Chepilov et les relations internationales. *Revue d'histoire diplomatique*, 2000, 114, 4, p. 331-350.

8109. TIERNEY (Dominic). The belated partisan: Franklin D. Roosevelt and the Spanish Civil War, 1936–1939. Oxford, [s. n.], 2000, 128 p. [Thesis (M. Phil.) – University of Oxford, 2000].

8110. TOWNSEND (Kenneth William). World War II and the American Indian. Albuquerque, University of New Mexico Press, 2000, X-272 p.

8111. VAÏSSE (Maurice), MARTENS (Stefan). Frankreich und Deutschland im Krieg (November 1942–Herbst 1944), Okkupation, Kollaboration, Résistance: Akten des deutsch-französischen Kolloquiums La France et l'Allemagne en Guerre (novembre 1942–automne 1944), Occupation, Collaboration, Résistance. Bonn, Bouvier, 2000, XVIII-944 p. (ill.). (Pariser historische Studien, 55).

8112. VALLAUD (Pierre). L'exode, mai–juin 1940. Paris, Perrin, 2000, 112 p. (ill.).

8113. VENNER (Dominique). Histoire de la collaboration: suivi des dictionnaires, des acteurs, partis et journaux. Paris, Pygmalion-Gérard Waelet, 2000, 766 p.

8114. VIHAVAINEN (Timo). Stalin i finny. Sankt-Peterburg, Zhurnal Neva, 2000, 285 p.

8115. WEGNER (Bernd). Hitler, der Zweite Weltkrieg und die Choreographie des Untergangs. [Diskussionsforum]. *Geschichte und Gesellschaft*, 2000, 26, 3, p. 493-518.

8116. WILKINSON (Sarah). Perceptions of public opinion: British foreign policy decisions about Nazi Germany, 1933–1938. Oxford, [s. n.], 2000, IX-II-350 p. [Thesis (D. Phil.) – University of Oxford, 2000].

8117. WINGEATE PIKE (David). Spaniards in the Holocaust: Mauthausen, the horror on the Danube. London a. New York, Routledge, 2000, 445 p. [Cañada Clanch Studies on Contemporary Spain].

8118. WOLTON (Suke). Lord Hailey, the Colonial Office and the politics of race and empire in the Second World War: the loss of white prestige. Basingstoke, Macmillan Press, 2000, XII-221 p.

8119. YAMAMOTO (Masahiro). Nanking: anatomy of an atrocity. Westport, Praeger, 2000, XV-352 p.

8120. ŽÁČEK (Rudolf). Těšínsko v československo-polských vztazích v letech 1939–1945. (Gebiet Teschin in den tschechoslowakisch-polnischen Beziehungen in Jahren 1939–1945). Český Těšín, Muzeum Těšínska, 2000, 120 p. (Studie o Těšínsku, 16).

8121. ZIMA (Veniamin F.). Mentalitet narodov Rossii v vojne 1941–1945 godov. (The war of 1941–1945 and mentalities of Russia's peoples). RAN, In-t rossijskoj istorii. Moskva, IRI RAN, 2000, 279 p.

8122. ZÖLLNER (Hans-Bernd). Der Feind meines Feindes ist mein Freund: Subhas Chandra Bose und das zeitgenössische Deutschland unter dem Nationalsozialismus, 1933–1943. Munster, Lit, 2000, III-57 p. (Geschichte, 25).

Cf. n^{os} *4577, 4692, 5050, 7959, 7971, 7975, 7978, 7998, 8004, 8017, 8019, 8024, 8029, 8031, 8033, 8042, 8048, 8220, 8242, 8316, 8333, 8344, 8357, 8375, 8384, 8455*

b. *Diplomacy. Economy*

* 8123. WELLS (Anne Sharp). The Anglo-American special relationship during the Second World War: a selective guide to materials in the British Library. London, Eccles Centre for American Studies, British Library, 2000, 44 p.

** 8124. BERRY (Burton Yost). Romanian diaries 1944–1947. Ed. by Cornelia BODEA. Iaşi a. Oxford, Center for Romanian Studies, 2000, 715 p.

** 8125. Dokumenty vneshnej politiki. 22 ijunja 1941–1 janvarja 1942. (Documents of the Foreign Policy [of the USSR], 22 June 1941–1 January 1942). Vol. 24. Ed. E. P. GUSAROV. Ministerstvo inostrannykh del RF. Moskva, Mezhdunarodnye otnoshenija, 2000, 630 p. (tabl.). [Cf. Bibl. 98, n° 7888.]

** 8126. Indo-Russian relations, 1917–1947: select documents from the archives of the Russian Federation. Part 2. 1929–1947. Edited and compiled by Purabi ROY, Sobhanlal DATTA GUPTA and Hari VASUDEVAN, Calcutta, Asiatic Society, 2000, [s. p.].

** 8127. LUNGU (Corneliu Mihail), NEGREANU (Ioana Alexandra). România în jocul Marilor Puteri: 1939–1940: documente, 1938–1941. Cu o introducere de Dinu C. GIURESCU. Bucureşti, Curtea Veche, 2000, 556 p.

** 8128. Nuovo ordine Europeo (Nel): documenti sulla Republica di Salò sotto il Terzo Reich. Con una guida delle fonti tedesche presso l'Istituto veneto per la storia della Resistenza e dell'età contemporanea. A cura di Monica FIORAVANZO. Padova, Cleup-Istituto veneto per la storia della Resistenza e dell'età contemporanea, 2000, 196 p.

** 8129. Quellen zur deutschen Aussenpolitik 1933–1939. Hrsg. v. Friedrich KIESSLING; mit einem Vorwort von Gregor SCHÖLLGEN. Darmstadt, Wissenschaftliche Buchgesellschaft, 2000, XXXI-318 p. (Ausgewählte Quellen zur deutschen Geschichte der Neuzeit, 34).

** 8130. Russko-kitajskie otnoshenija v XX veke. (Russian-Chinese relations in the 20th century). Ed. S. L. TIKHVINSKIJ. Compiled by A. M. LEDOVSKIJ, R. A. MIROVITSKAJA. V. S. MJASNIKOV. Vol. 4. Sovetsko-kitajskie otnoshenija (Soviet-Chinese relations), 1937–1945. Book 1. 1937–1944. Book 2. 1945. Moskva, Pamjatniki istoricheskoj mysli, 2000, 2 vol., 870 p., 704 p.

** 8131. SCURTU (Ioan). România şi Marile Puteri: documente. Vol. 2. 1933–1940. Bucureşti, Editura Fundaţiei România de Mâine, 2000, [s. p.].

** 8132. Transilvanskij vopros. Vengersko-Rumynskij territorial'nyj spor i SSSR, 1940–1946 gg.: Dokumenty rossijskikh arkhivov. (The question of Transilvania: the USSR and the territorial conflict between Hungary and Roumania, 1940–1946: documents from Russian archives). Ed. Tofik M. ISLAMOV. RAN, In-t slavjanovedenija; Feder. arkhivnaja sluzhba Rossii. Moskva, ROSSPEN, 2000, 456 p.

8133. ALVAREZ (David J.). Secret messages: codebreaking and American diplomacy, 1930–1945. Lawrence, University Press of Kansas, 2000, XI-292 p. (ill.). (Modern war studies).

8134. BARNETT (Enid). The war budget of September 1939: Keynes comes to Canada. Kingston, Harbinger House, 2000, 64 p.

8135. BOHN (Robert). Reichkommissariat Norwegen. «Nationalsozialistische Neuordnung» und Kriegswirtschaft. München, Oldenbourg, 2000, XII-508 p. (Beiträge zur Militärgeschichte, 54).

8136. BOYER DE SAINTE-SUZANNE (Raymond). Une politique étrangère: Le Quai d'Orsay et Saint-John Perse à l'épreuve d'un regard: journal novembre 1938–juin 1940. Présenté et annoté par Henriette LEVILLAIN et Philippe LEVILLAIN. Paris, V. Hamy, 2000, 349 p.

8137. BRODY (J. Kenneth). The avoidable war. Vol. 2. Pierre Laval and the politics of reality, 1935–1936. New Brunswick a. London, Transaction Publishers, X-368 p.

8138. BYRNES (Mark). Unfinished business: the United States and Franco's Spain, 1944–1947. *Diplomacy and statecraft*, 2000, 11, 1, p. 128-160.

8139. ČAPLOVIČ (Miloslav). Tri dokumenty k slovensko-poľským vzťahom z jari 1938. (Three documents on Slovak-Polish relations in Spring of 1938). *Historický časopis*, 2000, 48, 2, p. 340-348.

8140. CAPUTI (Robert J.). Neville Chamberlain and appeasement. Selinsgrove a. London, Susquehanna U. P. a. Associated University Presses, 2000, 271 p.

8141. Československá zahraniční politika v roce 1938. Sv. 1, 1. leden–30. červen 1938. (Czechoslovak foreign policy in the year 1938. Tomo 1). Ed. Jindřich DEJMEK, collab. Jan NĚMEČEK, Helena NOVÁČKOVÁ and Ivan ŠŤOVÍČEK. Praha, Ústav mezinárodních vztahů, 2000, 591 p. (Dokumenty československé zahraniční politiky).

8142. CHADANI (S.). Political process of the withdrawal from the League of Nations. *Journal of Japanese history*, 2000, 457, p. 1-27.

8143. CHIPER (Ioan). România şi Germania nazistă: relaţiile româno-germane între comandamente politice şi interese economice: ianuarie 1933–martie 1938. Bucureşti, Editura Elion, 2000, 279 p.

8144. CORSI (Francisco Luiz). Estado novo: política externa e projeto nacional. Sao Paulo, Editora UNESP, FAPESP, 2000, 304 p. (Coleçao Prismas).

8145. CRAFT (S. G.). Saving the League: V. K. Wellington Koo, the League of Nations and Sino-Japanese conflict, 1931–1939. *Diplomacy and statecraft*, 2000, 11, 3, p. 91-112.

8146. DE RISIO (Carlo). La clessidra di Mussolini: i nove mesi di non belligeranza. 1 Settembre 1939–10 giugno 1940. Roma, Settimo sigillo, 2000, 286 p. (ill.). (Saggi, 53).

8147. DEMIDOV (Sergej V.). Anglo-frantsuzskie otnoshenija nakanune Vtoroj mirovoj vojny, 1936–1939 gg. (The Anglo-French relations on the eve of the Second World War, 1936–1939). Rjazanskij gos. ped. un-t im. S.A. Esenina. Rjazan', [s. n.], 2000, 180 p.

8148. Deutsch-polnische Beziehungen: 1939–1945–1949; eine Einführung. Hrsg. v. Włodzimierz BORODZIEJ. Osnabrück, Fibre, 2000, 348 p. (Einzelveröffentlichungen des Deutschen Historischen Instituts Warschau, 5).

8149. DOENECKE (Justus D.). Storm on the horizon: the challenge to American intervention, 1939–1941. Lanham a. Oxford, Rowman & Littlefield Publishers, 2000, XIX-551 p.

8150. DRYBURGH (Marjorie). North China and Japanese expansion, 1933–1937: regional power and the national interest. Richmond, Curzon Press, 2000, VIII-249 p. (Curzon studies in East Asia).

8151. EMANUILOV (Emanuil Georgiev). Bulgariia v politikata na velikite sili 1939-1947. (La Bulgaria nella politica delle grandi potenze). V. Turnovo, "PAN-VT", 2000, 397 p.

8152. FOLLY (Martin H.). Churchill, Whitehall and the Soviet Union, 1940–1945. New York, St. Martin's Press, 2000, XI-237 p. (Cold War history series).

8153. FORBES (Neil). Doing business with the Nazis: Britain's economic and financial relations with Germany, 1931–1939. With a foreword by Richard OVERY. London, Frank Cass, 2000, XVIII-250 p. (ill.).

8154. GOGOL (Bogusław). "Czerwony Sztandar". Rzecz o sowietyzacji ziem Małopolski Wschodniej, wrzesień 1939–czerwiec 1941. ("Rote Fahne". Ein Wort über die Sowjetisierung der Gebiete des Ost-Kleinpolens [Małopolska Wschodnia]: September 1939–Juni 1941). Gdańsk, Wydaw. Uniw. Gdańskiego, 2000, 365 p.

8155. HRYCIUK (Grzegorz). Polacy we Lwowie 1939–1944. Życie codzienne. (Polen in Lwow 1939–1944. Alltagsleben). Warszawa, Książka i Wiedza, 2000, 430 p.

8156. HSIN CHIH (Chen). La réponse chinoise à l'occupation française des îles Spratly en 1933. In: Conflits et opérations en Asie orientale, 1930–1950 [Cf. n° 7594], p. 5-24.

8157. JACKSON (Peter). France and the Nazi menace: intelligence and policy making, 1933–1939. Oxford, Oxford U. P., 2000, XII-446 p.

8158. JAYNE (Catherine E.). Oil, war, and Anglo-American relations: American and British reactions to Mexico's expropriation of foreign oil properties, 1937–1941. Westport, Greenwood Press, 2000, XII-210 p. (ill.). (Contributions in Latin American studies, 19).

8159. JINDRA (Zdeněk). Dva nové zbrojní závody Kruppova koncernu ve Slezsku za 2. světové války. (Zwei neue Waffenfabriken des Krupp-Konzernes in Schlesien während des 2. Weltkrieges). In: Evropa mezi Německem a Ruskem. Ed. Miroslav ŠESTÁK. Praha, 2000, p. 421-440.

8160. JONJIĆ (Tomislav). Hrvatska vanjska politika: 1939.–1942. (Croatian foreign policy, 1939–1942). Zagreb, Libar, 2000, 942 p.

8161. KOCHANKOV (Nikolai). Bulgariia i Nezavisimata Hurvatska durzhava, 1941–1944: politicheski i diplomaticheski otnosheniia. Sofiia, Kheron pres, 2000, 204 p.

8162. KREIS (Georg). La Suisse pendant la deuxième guerre mondiale: ses réponses aus défis de l'époque. Zürich, Pro Helvetia, 2000, 150 p. (Information).

8163. KUBŮ (Eduard). Československo-francouzská spojenecká smlouva a Německo. 1-2. (The Czechoslovak-French alliance and Germany. 1-2). Historie a vojenství, 2000, 49, 2-3, p. 266-293; p. 555-580.

8164. LEFEBVRE D'OVIDIO (Francesco). La questione etiopica nei negoziati italo-franco-britannici del 1935. Roma, Pixel Press, 2000, 109 p. (Streghe).

8165. LEITZ (Christian). Nazi Germany and neutral Europe during the second world war: sympathy for the devil? Manchester, Manchester U. P., 2000, VII-213 p.

8166. LESOURD (Emmanuel). La bataille de la production des porte-avions avant et pendant la Seconde Guerre mondiale. In: Conflits et opérations en Asie orientale, 1930–1950 [Cf. n° 7594], p. 41-48.

8167. LOEWENSTEIN (Andrew B.). The veiled protectorate of Kowait': liberalized imperialism and British efforts to influence Kuwaiti domestic policy during the reign of Sheikh Ahmad al-Jaber, 1938–1950. Middle eastern studies, 2000, 36, 2, p. 103-123.

8168. LOUÇA (António). Hitler e Salazar: comércio em tempos de guerra, 1940–1944. Lisboa, Terramar, 2000, 261 p. (ill.). (Arquivos do século XX, 4).

8169. MALLETT (Robert). Fascist foreign policy and official italian views of Anthony Eden in 1930s. Historical journal, 2000, 43, 1, p. 157-189.

8170. MERCIER-BERNADETTE (Fabienne). Le rôle du Japon dans le conflit franco-thaïlandais (juin 1940–mai 1941). In: Conflits et opérations en Asie orientale, 1930–1950 [Cf. n° 7594], p. 25-40.

8171. MÖCKLI (Daniel). Neutralität, Solidarität, Sonderfall: die Konzeptionierung der schweizerischen Aussenpolitik der Nachkriegszeit, 1943–1947. Zürich, ETH, 2000, 336 p. (Zürcher Beiträge zur Sicherheitspolitik und Konfliktforschung, 55).

8172. NĚMEČEK (Jan). K slovensko-jugoslávským vztahům 1939–1941. (The Slovak-Yugoslav relations, 1939–1941). In: Evropa mezi Německem a Ruskem. Ed. Miroslav ŠESTÁK. Praha, 2000, p. 385-398.

8173. NEVILLE (Peter). Appeasing Hitler: the diplomacy of Sir Nevile Henderson, 1937–1939. Basingstoke, Macmillan, 2000, XV-237 p. (Studies in diplomacy).

8174. ODDATI (Nicola). Dalla guerra alla pace: Italia ed alleati, 1943–1946. Salerno, Edizioni del Paguro, 2000, 118 p. (Gli uomini e il tempo, 3).

8175. PARKER (R. A. C.). Churchill and appeasement. London, Macmillan, 2000, XI-290 p.

8176. PETROV (Liudmil). Bulgariia i Turtsiia: 1931–1941. Sofiia, Ivrai, 2000, 278 p. (ill.). (Bulgarska istoricheska biblioteka, 2).

8177. PETRUF (Pavol). Taliansko-etiópska vojna v rokoch 1935–1936. Príčiny, priebeh, dôsledky. (Italian-Ethiopian War in the Years 1935–1936. Reasons, course, consequences). Bratislava, Univerzita Komenského, 2000,182 p. – IDEM. Vichystické Francúzsko a diplomatické uznanie Slovenskej republiky. (The Vichy France and the diplomatic recognition of the Slovak Republic). *Historický časopis*, 2000, 48, 1, p. 131-152.

8178. RINZEMA (Win). Java, het laatste front: de sociale gevolgen van de Japanse bezetting op Centraal-Java voor de Indonesiërs en de Europeanen. Zutphen, Walburg, 2000, 202 p. (ill., maps).

8179. RODRÍGUEZ (Luis I.). Misión de Luis I. Rodríguez en Francia: la protección de los refugiados españoles, julio a diciembre de 1940. Prólogo de Rafael SEGOVIA y Fernando SERRANO. México, El Colegio de México, Secretaría de Relaciones Exteriores, Consejo Nacional de Ciencia y Tecnología, 2000, XVI-604 p.

8180. ROTH (Andreas). Mr Bewley in Berlin: aspects of the career of an Irish diplomat, 1933–1939. Dublin, Four Courts Press, 2000, 119 p.

8181. ŠAMBERGER (Zdeněk). Mnichov 1938 v řeči archivních dokumentů. (Munich 1938 in the archive documents). Praha, Státní ústřední archiv, 2000, 166 p.

8182. SILVA SEITENFUS (Ricardo Antônio). A entrada do Brasil na Segunda Guerra Mundial. Porto Alegre, EDIPUCRS, 2000, 378 p. (Coleçao História, 33).

8183. SIPOLS (Vilnis Ja.). Velikaja pobeda i diplomatija, 1941–1945. (The Great Victory and diplomacy, 1941–1945). Moskva, Novina, 2000, 382 p.

8184. STINNETT (Robert B.). Day of deceit: the truth about FDR [Franklin Delano Roosevelt] and Pearl Harbor. London, Constable, 2000, XIV-386 p. (ill., maps).

8185. STOLER (Mark A.). Allies and adversaries: the Joint Chiefs of Staff, the Grand Alliance, and U.S. strategy in World War II. Chapel Hill a. London, University of North Carolina Press, 2000, XXII-380 p.

8186. Switzerland and the Second World War. Ed. by Georg KREIS. London, F. Cass, 2000, XVII-378 p.

8187. TELO (António José). A neutralidade portuguesa e o ouro Nazi. Lisboa, Quetzal, 2000, 384 p. (ill.).

8188. TOURETTE (Florence). Yalta et ses mythes. *Revue d'histoire diplomatique*, 2000, 114, 2, p. 101-114.

8189. Victory in Europe 1945: from World War to Cold War. Ed. by Arnold A. OFFNER and Theodore A. WILSON. Lawrence, University Press of Kansas, 2000, X-308 p.

Cf. n^{os} *4412, 4692, 7956, 7972, 7980, 7990, 8005, 8010, 8011, 8034, 8054, 8255, 8262, 8265, 8295, 8301, 8314, 8364, 8392, 8408, 8423, 8434, 8448*

c. Military operations

** 8190. FORMENTO (Ettore). Kai Bandera: Etiopia 1936–1941, una banda irregolare. Introduzione di Angelo DEL BOCA. Milano, Mursia, 2000, XII-290 p. (ill., maps). (Testimonianze fra cronaca e storia. Seconda guerra mondiale).

** 8191. Fra neutralitet til besaettelse: Viceadmiral H. Rechnitzers dagbog om fladens virke 1939–1940. Udgivet og kommenteret af Hans Christian BJERG. Kobenhaven, Selskabet for Udgivelse af Kilder til Dansk Historie, 2000, 376 p.

8192. Belorussia 1944: the Soviet General Staff study. Translated and edited by David M. GLANTZ and Harold S. ORENSTEIN. London a. Portland, Frank Cass, 2000, XIV-337 p. (Cass series on the Soviet study of war, 12).

8193. CHAIX (Bruno). En mai 1940, fallait-il entrer en Belgique? Décisions stratégiques et plans opérationnels de la campagne de France. Paris, Economica, 2000, XII-349 p. (Campagnes & stratégies. Les grande batailles, 35).

8194. DUNN (Walter Scott). Soviet blitzkrieg: the battle for White Russia, 1944. Boulder a. London, Lynne Rienner, 2000, XII-249 p. (ill., maps).

8195. DUTU (Alesandru). Între Wehrmacht şi armata roşie: relaţii de comandament româno-germane şi româno-sovietice, 1941–1945. Bucureşti, Editura Enciclopedică, 2000, 356 p. (ill.). (Colecţia "Biblioteca de istorie contemporană a României").

8196. FOLLY (Martin). The Red Air Force in Italy, 1944: a case study in the formation of British policy to the Soviet Union in World War II. *Diplomacy and statecraft*, 2000, 11, 2, p. 105-136.

8197. GARÇON (François). La guerre du Pacifique: le minage des voies maritimes japonaise (octobre 1942–août 1945). *In*: Conflits et opérations en Asie orientale, 1930–1950 [Cf. n° 7594], p. 49-56.

8198. HARTMANN (Christian), ZARUSKY (Jürgen). Stalins "Fackelmänner-Befehl" vom November 1941. Ein verfälschtes Dokument. *Vierteljahrshefte für Zeitgeschichte*, 2000, 48, 4, p. 667-674.

8199. HILLS (Alice). Britain and the occupation of Austria, 1943–1945. Basingstoke, Macmillan a. London, King's College, 2000, XI-222 p. (Studies in military and strategic history).

8200. HÜRTER (Johannes). "Es herrschen Sitten und Gebräuche, genauso wie im 30-jährigen Krieg". Das erste Jahr des deutsch-sowjetischen Krieges in Dokumenten des Generals Gotthard Heinrici. *Vierteljahrshefte für Zeitgeschichte*, 2000, 48, 2, p. 329-403.

8201. JOHNSTON (Mark). Fighting the enemy: Australian soldiers and their adversaries in World War II. Cambridge, Cambridge U. P., 2000, XI-206 p.

8202. MASSON (Philippe). La marine française et la guerre, 1939–1945. Paris, Tallandier, 2000, 547 p. (Approches).

8203. MURRAY (Williamson), MILLETT (Allan R.). A war to be won: fighting the Second World War. Cambridge, Harvard U. P., 2000, XIV-656 p.

8204. PALONIEMI (Jarmo). Yhdysvaltojen ja Ison-Britannian sodanjohdon yhteistyö Normandian maihinnousun valmistelussa 1941–1944. (From Sledgehammer to Overlord: the co-operation between the American and British war leaders during the preparations for the Normandy invasion, 1941–1944) Oulu, Oulun yliopisto, 2000, 224 [6] p. (Acta Universitatis Ouluensis. Series B. Humaniora, 38).

8205. Präventivkrieg? Der deutsche Angriff auf die Sowjetunion. Hrsg. v. Bianka PIETROW-ENNKER. Frankfurt am Main, Fischer-Taschenbuch Verlag, 2000, 218 p. (Geschichte: die Zeit des Nationalsozialismus).

8206. ROGERS (Anthony). Battle over Malta: aircraft losses and crash sites, 1940–1942. Stroud, Sutton, 2000, 224 p. (ill., map).

8207. ŠTEFANSKÝ (Václav). Slovenskí vojaci v Taliansku 1943–1945. (Slovak Soldiers in Italy 1943–1945). Bratislava, Ministerstvo obrany Slovenskej republiky vo Vydavateľskej a informačnej agentúre, 2000, 114 p.

8208. SURLEAU (Jean-Claude). La task force de porte-avions, bras armé de l'Amérique pendant la guerre du Pacifique. *In*: Conflits et opérations en Asie orientale, 1930–1950 [Cf. n° 7594], p. 57-66.

8209. WOODMAN (Richard). Malta convoys, 1940–1943. London, John Murray, 2000, XX-532 p.

Cf. n°s 7984, 7991

d. Resistance

** 8210. PANAGIOTAKES (Georgios I.). Dokoumenta apo te mache kai ten antistase tes Kretes, 1941–1945 = Documents from the battle of and the resistance of Crete = Dokumente zur Schlacht und zum Widerstand auf Kreta. Herakleio, [s. n.], 2000, 430 p. (ill.).

8211. Atlante storico della Resistenza italiana. A cura di Luca BALDISSARA. Milano, Istituto nazionale per la storia del Movimento di Liberazione in Italia e Bruno Mondadori, 2000, 159 p.

8212. BÉDARIDA (François). La résistance de Jean Moulin: l'Etat et l'éthique. *In*: Jean Moulin face à l'histoire [Cf. n° 8226], p. 9-13.

8213. BELOT (Robert). Jean Moulin et Henri Frenay: les enjeux d'un affrontement. *In*: Jean Moulin face à l'histoire [Cf. n° 8226], p. 163-183.

8214. BOURSIER (Jean-Yves). Chroniques du maquis (1943–1944): FTP du camp Jean Pierson et d'ailleurs. Paris, l'Harmatan, 2000, 350 p.

8215. BURRIN (Philippe). L'unité de la Résistance dans une perspective internationale. *In*: Jean Moulin face à l'histoire [Cf. n° 8226], p. 213-220.

8216. COLLIN (Claude). Carmagnolc et Liberté. Les étrangers dans la Résistance en Rhône-Alpes. Grenoble, PUG, 2000, [s. p.]. (Résistances).

8217. CORDIER (Daniel). Le délégué général. *In*: Jean Moulin face à l'histoire [Cf. n° 8226], p. 126-138. – IDEM. Les rencontres décisives de Londres. *In*: Jean Moulin face à l'histoire [Cf. n° 8226], p. 111-125.

8218. CRÉMIEUX-BRILHAC (Jean-Louis). De Gaulle et la mort de Jean Moulin. *In*: Jean Moulin face à l'histoire [Cf. n° 8226], p. 195-208.

8219. ČRNUGELJ (Franc). Na zahodnih mejah – 1944. Briško-beneški odred. Kazali Knj. 3. (On western borders – 1944 detachment from Brda and Venetian Slavia). Ljubljana, Društvo piscev zgodovine NOB Slovenije, 2000, [s. p.]. (Zbirka Partizanski knjižni klub, 6, 31).

8220. DAVIES (Peter Jonathan). France and the Second World War: occupation collaboration and resistance. London, Routledge, 2000, 145 p. (Introductions to history).

8221. Dizionario della Resistenza. Vol. 1. Storia e geografia della Liberazione. A cura di Enzo COLLOTTI, Renato SANDRI e Frediano SESSI. Torino, Einaudi, 2000, 617 p.

8222. DUBICKI (Tadeusz). Bazy wojskowej łączności zagranicznej ZWZ-AK w latach 1939–1945. Studia i materiały. (Stützpunkte des ausländischen militärischen Fernmeldedienstes des Verbandes des Bewaffneten Kampfes der Landesarmee [ZWZ-AK] in den Jahren 1939–1945. Studien und Materialien). Częstochowa, Wydaw. WSP, 2000, 269 p.

8223. EMKJÆR (Stefan). Stikkerdrab: modstandsbevægelsernes likvidering af danskere under besæ ttelsen. Oslo, Ascheoug, 2000, 223 p.

8224. FLEUTOT (François-Marin). Des royalistes dans la Résistance. Paris, Flammarion, 2000, 514 p.

8225. GUÉRIN (Alain). Chronique de la Résistance. Paris, Omnibus, 2000, 1806 p.

8226. Jean Moulin face à l'histoire. Actes du Colloque organisé les 10 et 11 juin 1999, à Paris. Sous la dir. de Jean-Pierre AZÉMA. Paris, Flammarion, 2000, 418 p. [Cf. n^{os} <sélection> 8212, 8213, 8215, 8217, 8218.]

8227. JESPERSEN (Knud J. V.). Med hjælp fra England: Special Operations Executive og den danske modstandskamp 1940–1945. Bd. 2. Den væbnede kamp 1943–1945. (Avec l'aide de l'Angleterre: le SOE et la résistance danoise, 1940–1945. Tome 2. La Lutte armée, 1943–1945). Odense, Odense Universitets forlag, 2000, 493 p. (ill.). (Odense University Studies in History and Social Sciences. 228).

8228. KOMOROWSKI (Krzysztof). Polityka i walka. Konspiracja zbrojna ruchu narodowego 1939–1945. (Politik und Kampf. Bewaffnete Geheimorganisationen der Nationalbewegung 1939–1945). Warszawa, Rytm, 2000, 614 p.

8229. KUKLÍK (Jan). Kontrapropaganda v ilegálním tisku v letech 1939–1941. (Counter-propaganda in the illegal press, 1939–1941). *Historie a vojenství*, 2000, 49, 1, p. 100-113.

8230. LEVINE (Ellen). Darkness over Denmark: the Danish resistance and the rescue of the Jews. New York, Holiday House, 2000, X-164 p.

8231. MARIJAN (Davor). Koliko je u Drugom svjetskom ratu bilo partizanskih divizija iz Hrvatske? (How many Partisan Divisions were there from Croatia in the WWII?). *Časopis za suvremenu povijest*, 2000, 32, 3, p. 517-525.

8232. MONTAGNON (Pierre). Les maquis de la Libération, 1942–1944. Paris, Pygmalion, 2000, 411 p. (ill., maps).

8233. Resistance in Western Europe. Ed. by Bob MOORE. Oxford, Berg, 2000, X, 256 p. (ill.).

8234. STENTON (Michael). Radio London and resistance in occupied Europe: British political warfare, 1939–1943. Oxford a. New York, Oxford U. P., 2000, XVI-423 p.

8235. TAYLOR (Lynne). Between Resistance and collaboration: popular protest in Northern France, 1940–1945. New York, St. Martin's Press, 2000, V-195 p.

Cf. n°s 4364, 4537, 4825, 8068, 8111, 8265

§ 8. From 1945.

* 8236. GIENOW-HECHT (J. C. E.). Shame on U.S.? Academics, cultural transfer, and the Cold War: a critical review. *Diplomatic history*, 2000, 24, 3, p. 465-494.

* 8237. Italia (L') e il Consiglio d'Europa. Bibliografia 1949–1999. A cura di Gisella BOCHICCIO, Rosanna DE LONGIS, Fabrizio DOLCI e Patrizia RUSCIANI. Saggio introduttivo di Andrea CHITI BATELLI. Roma, Carocci, 2000, 125 p.

* 8238. Reviewing the Cold War. Approaches, interpretations, theory. Ed. by Odd Arne WESTAD. London a. Portland, Cass, 2000, 382 p.

* 8239. TINGVOLD PETERSEN (Tore). Crossing the Rubicon? Britain's withdrawal from the Middle East, 1964–1968: a bibliographical review. *International history review*, 2000, 22, 2, p. 318-340.

* 8240. WESTAD (O. A.). The new international history of the Cold War: three (possible) paradigms. *Diplomatic history*, 2000, 24, 4, p. 551-565.

** 8241. Akten zur auswärtigen Politik der Bundesrepublik Deutschland. Herausgegeben im Auftrag des Auswärtigen Amts vom Institut für Zeitgeschichte; Hauptherausgeber Hans-Peter SCHWARZ; Mitherausgeber Helga HAFTENDORN [et al.]. 1952. 1. Januar bis 31. Dezember 1952. Wissenschaftlicher Leiter Rainer A. BLASIUS; Bearbeiter Martin KOOPMANN und Joachim WINTZER. 1969, 1. 1. Januar bis 30. Juni 1969. 1969, 2. 1. Juli bis 31. Dezember 1969. Wissenschaftlicher Leiter Rainer A. BLASIUS; Bearbeiter Franz EIBL und Hubert ZIMMERMANN. München, Oldenbourg, 2000, 3 vol., LVIII-842 p., LXXXIII-750 p., 850 p.

** 8242. American-British-Canadian intelligence relations, 1939–2000. Ed. by David STAFFORD and Rhodri JEFFREYS-JONES. London, Frank Cass, 2000, 270 p. (Studies in intelligence).

** 8243. Australia and the Indonesian incorporation of Portuguese Timor, 1974–1976. Editor, Wendy WAY, assistant editors, Damien BROWNE and Vivianne JOHNSON. Carlton, Melbourne U. P., 2000, XXIX-885 p. (Documents on Australian foreign policy).

** 8244. Beziehungen (Die) der Bundesrepublik Deutschland zum Heiligen Stuhl 1949–1966: aus den Vatikanakten des Auswärtigen Amts: eine Dokumentation. Hrsg. v. Michael F. FELDKAMP. Köln, Böhlau, 2000, 541 p. (Bonner Beiträge zur Kirchengeschichte, 21).

** 8245. BINNS (Jack R.). The United States in Honduras, 1980–1981: an ambassador's memoir. Jefferson a. London, McFarland & Co, 2000, X-397 p. (ill.).

** 8246. Britain and Hungary in the post-war years, 1945–1951: a parallel history in narrative and documents. Vol. 2. The documents. Compiled by Eva HARASZTI-TAYLOR. Nottingham, Astra, 2000, IX-344 p.

** 8247. British documents on foreign affairs: reports and papers from the Foreign Office confidential print. General editors Paul PRESTON and Michael PARTRIDGE. Part 4. From 1946 through 1950. Series B. Near and Middle East. Vol. 3. Afghanistan, Persia and Turkey: January 1947–December 1947. Vol. 4. Eastern affairs: January 1947–December 1947. Ed. by Malcolm YAPP. Bethesda, University Publications of America, 2000, 2 vol., XX-532 p., XX-486 p.

** 8248. British documents on foreign affairs: reports and papers from the Foreign Office confidential print. General editors Paul PRESTON and Michael PARTRIDGE. Part 4. From 1946 through 1950. Series D. Latin America. Vol. 1. Latin America: January 1946–June 1946. Vol. 2. Latin America: July 1946–December 1946. Ed. by James DUNKERLEY. Bethesda, University Publications of America, 2000, 2 vol., XIX-568 p., XIX-577 p.

** 8249. British documents on foreign affairs: reports and papers from the Foreign Office confidential print. General editors Paul PRESTON and Michael PARTRIDGE. Part 4. From 1946 through 1950. Series E. Asia. Vol. 1. Far Eastern affairs: January 1946–June 1946. Vol. 2. Far Eastern affairs: July 1946–December 1946. Ed. by Anthony BEST. Bethesda, University Publications of America, 2000, 2 vol., XIX-578 p., XIX-450 p.

** 8250. British documents on foreign affairs: reports and papers from the Foreign Office confidential print. General editors Paul PRESTON and Michael PARTRIDGE. Part 4. From 1946 through 1950. Series F. Europe. Vol. 1. Central Europe: January 1946–June 1946. Vol. 2. Central Europe: July 1946–December 1946. Vol. 3. Western Europe: January 1946–June 1946. Vol. 4. Western Europe: October 1946–December 1946 and Scandinavia 1946. Vol. 5. South-Eastern Europe: January 1946–June 1946. Vol. 6. South-Eastern Europe: July 1946–December 1946. Ed. by Denis SMYTH. Bethesda, University Publications of America, 2000, 6 vol., XXV-502 p., XXV, 481 p., XXV-469 p., XXV-313 p., XXV-536 p., XXV-385 p.

** 8251. British documents on foreign affairs: reports and papers from the Foreign Office confidential print. General editors Paul PRESTON and Michael PARTRIDGE. Part 4. From 1946 through 1950. Series G. Africa. Vol. 1. Africa: January 1946–June 1946. Vol. 2. Africa: July 1946–December 1946. Ed. by Peter WOODWARD. Bethesda, University Publications of America, 2000, 2 vol., XX-428 p., XX-362 p.

** 8252. British documents on foreign affairs: reports and papers from the Foreign Office confidential print. General editors Paul PRESTON and Michael PARTRIDGE. Part 4. From 1946 through 1950. Series M. International organizations, Commonwealth affairs and general. Vol. 1. Commonwealth affairs, 1945–1946 and General, January 1946–March 1946. Vol. 2. General: April 1946–December 1946. Ed. by M. L. DOCKRILL. Bethesda, University Publications of America, 2000, 2 vol., XVIII-414 p., XVIII-514 p.

** 8253. British documents on the end of Empire. Series A. Vol. 4. The Conservative government and the end of Empire 1957–1964. General ed. S. R. ASHTON; eds. Ronald HYAM and Wm. Roger LOUIS. Pt. 1. High policy, political and constitutional change. Pt. 2. Economics, international relations, and the Commonwealth. London, Stationery Office, 2000, 2 vol., CIX-825 p., XXXVII-811 p.

** 8254. Documenti diplomatici italiani (I). Serie 10. 1943–1948. 7. 15 dicembre 1947–7 maggio 1948. Roma, Istituto poligrafico e Zecca dello Stato, Libreria dello Stato, 2000, LV-891 p.

** 8255. Documents diplomatiques français. 23. 1963. Tome 1. 1er janvier–30 juin. Ministère des affaires étrangères. Commission de publication des documents diplomatiques. Paris, Imprimerie nationale, 2000, XLIII-728 p.

** 8256. Documents on Israeli-Soviet relations, 1941–1953. Israel Ministry of Foreign Affairs [et al.]. Part 1. 1941–May 1949. Part 2. May 1949–1953. London, Frank Cass, 2 vol., LIX-467 p., LIX-467 p. (The Cummings Center series).

** 8257. FERAOUN (Mouloud). Journal, 1955–1962: reflections on the French-Algerian War. Edited and with an introduction by James D. LE SUEUR. Lincoln a. London, University of Nebraska Press, 2000, LI-340 p.

** 8258. Foreign relations of the United States. 1964–1968. General editor David S. PATTERSON. 16. Cyprus; Greece; Turkey. Editor James E. MILLER. 18. Arab-Israeli dispute, 1964–1967. Editor Harriet Dashiell SCHWAR. 21. Near east region, Arabian peninsula. Editor Nina DAVIS HOWLAND. 25. South Asia. Editors Gabrielle S. MALLON and Louis J. SMITH. 27. Mainland Southeast Asia; Regional affairs. Editor Edward C. KEEFER. 29. 1. Korea. Editor Karen L. GATZ. Washington, United States Government Printing Office, 2000, 6 vol., XXXVII-796 p., XXXI-853 p., XXXI-919 p., XXXVIII-1106 p., XLI-946 p., XXXI-829 p.

** 8259. Front lines (On the) of the Cold War: documents on the intelligence war in Berlin, 1946 to 1961. Ed. by Donald P. STEURY. Washington, CIA History Staff, Center for the Study of Intelligence, 2000, XVI-575 p. (ill.). (CIA Cold War records).

** 8260. KAUL (Triloki Nath). A diplomat's diary, 1947–1999: China, India, and USA, the tantalising triangle. Delhi, Macmillan India, 2000, XIV-273 p. (ill.).

** 8261. Kennedys (The) and Cuba: the declassified documentary history. Edited with commentary by Mark J. WHITE. Chicago, Ivan R. Dee, XVIII-356p.

** 8262. KYNIN (G. P.), LAUFER (J. P.). SSSR i germanskij vopros, 1941–1949: Dokumenty iz arkhiva vneshnej politiki Rossijskoj Federatsii. (The USSR and the question of Germany, 1941–1949: documents from the Foreign Policy Archive of the Russian Federaion). Vol. 2. 9 May 1945–3 October 1946. Ministerstvo inostrannykh del Rossijskoj Federatsii. Moskva, Mezhdunarodnye otnoshenija, 2000, 879 p.

** 8263. PELIKÁN (Jan). Jugoslávie a okupace Československa v srpnu 1968. Pět dokumentů k srpnovým událostem. (Yugoslavia and the Soviet Occupation of Czechoslovakia. Five documents on the events of August 1968). *Soudobé dějiny. Suppl. Archiv soudobých dějin*, 2000, 7, 4, p. 594-618.

** 8264. REIMAN (Michal), LUŇÁK (Petr). Studená válka 1954–1964. Sovětské dokumenty v českých archivech. (Cold War 1954–1964. Soviet documents in the Czech archives). Brno, Doplněk, 2000, 439 p.

** 8265. Sovetsko-izrail'skie otnoshenija: Sb. dokumentov. (The relations between the USSR and Israel: documents). Ed. B. L. KOLOKOLOV, E. BENTSUR. Federal'naja arkhivnaja sluzhba Rossii, etc. Vol. 1. 1941–1953. Book 1. 1941–May 1949. Book 2. May 1949–1953. Moskva, Mezhdunarodnye otnoshenija, 2000, 2 vol., 554 p., 560 p.

** 8266. Vneshnjaja politika Rossii: Sb. dokumentov. (The foreign policy of Russia: documents). 1993, part 1. January–May. 1993, part 1. June–December. 1994, part 1. January–May. 1993, part 1. June–December. 1995. Ministerstvo inostrannykh del Rossijskoj Federatsii. Moskva, Mezhdunarodnye otnoshenija, 2000, 5 vol., 536 p., 536 p., 408 p., 544 p., 504 p.

** 8267. WINKLER (Allan M.). The Cold War: a history in documents. New York, Oxford U. P., 2000, 159 p. (Pages from history).

** 8268. YOHANNES (Petros). The Ethiopia-Eritrea conflict: selected documents of Ethiopian foreign policy. London, Centre of Ethiopian Studies, XIII-153 p. (Ethiopian studies: international relation and Ethiopian foreign policy series).

8269. AIJAZUDDIN (F. S.). From a head, through a head, to a head: the secret channel between the US and China through Pakistan. Oxford, Oxford U. P., 2000, XXIV-163 p.

8270. ALLEY (Roderic Martin). The domestic politics of international relations: cases from Australia, New Zealand and Oceania. Aldershot, Ashgate, 2000, XII-275 p.

8271. ALTER (Peter). The German question and the structure of Europe: a history. London: Arnold, 2000, 2000, X-173 p. (ill., maps).

8272. ALZUGARAY TRETO (Carlos). Crónica de un fracaso imperial: la administración de Eisenhower y el derrocamiento de la dictadura de Batista. La Habana, Editorial de Ciencias Sociales, 2000, 205 p.

8273. AMIN (Shahid M.). Pakistan's foreign policy: a reappraisal. Oxford, Oxford U. P., 2000, 327 p.

8274. ANDERSON (Sheldon R). A Cold War in the Soviet Bloc: Polish-East German relations: 1945–1962. Boulder, Westview Press, 2000, [s. p.].

8275. ANTUNES (José Freire). Portugal na guerra do petróleo: os Açores e as vitórias de Israel, 1973. Carnaxide, Edeline, 2000, 270 p. (ill., facsims., ports). (Obras de José Freire Antunes, 1).

8276. ARCIDIACONO (Bruno). Les Balkans et les origines de la guerre froide: grandes puissances et «facteur local». In: Balkans (Les) dans les relations internationales, Part 2 [Cf. n° 7582], p. 413-432.

8277. ARUNOVA (Marianna Rubenovna). Afganskaia politika SShA v 1945–1999 gg.: (kratkii ocherk). Moskva, Institut izucheniia Izrailia i Blizhnego Vostoka, Institut vostokovedeniia RAN, 2000, 127 p.

8278. AYDÍN (Mustafa). Determinants of Turkish foreign policy: changing patterns and conjunctures during the Cold War. Middle eastern studies, 2000, 36, 1, p. 103-139.

8279. BANGE (Oliver). The EEC crisis of 1963: Kennedy, Macmillan, De Gaulle, and Adenauer in conflict. Foreword by Peter CATTERALL. New York, St. Martin's, 2000, XV-291 p. (Contemporary History in Context).

8280. BATTAGLIA (Rosario). Gaetano Martino e la politica estera italiana, 1954–1964. Messina, Confcommercio, EDAS, 2000, 315 p.

8281. BEJARANO (Jesús Antonio). El proceso de paz en Colombia y la política exterior de los Estados Unidos. Washington, Latin American Program, Woodrow Wilson International Center for Scholars, 2000, 94 p. (Documento de trabajo, 247).

8282. BÉKÈS (Csaba). Retour en Europe. In: Hongrie (La) dans les conflits du XXe siècle [Cf. n° 7611], p. 115-138.

8283. Belgique (La) et l'Afrique centrale: de 1960 à nos jours. Sous la direction de Olivier Lanotte, Claude Roosens et Caty Clément. Bruxelles, GRIP et Complexe, 2000, 380 p. (Les livres du GRIP, 243-245).

8284. BERENDT (Grzegorz). Żydzi na gdańskim rozdrożu (1945–1950) [Juden am Scheideweg. Geschehnisse in Gdańsk [Danzig], 1945–1950). Gdańsk, Uniw. Gdański. Zakł. Historii Kultury i Myśli Polit., 2000, 211 p.

8285. BERNKOPF TUCKER (Nancy). China confidential: American diplomats and Sino-American relations since 1945. New York, Columbia U. P., 2000, XXV-569 p. (ill., maps).

8286. BÍLEK (Jiří). Vojenské aspekty čs.-polského sporu o Těšínsko v roce 1945. Od ozbrojených záměslů k politickému řešení. (Military aspects of the Czechoslovak-Polish dispute over the Těšín region in 1945). Historie a vojenství, 2000, 49, 2, p. 235-265.

8287. BLACKWELL (S.). Britain, the United States and the Syrian crisis, 1957. Diplomacy and statecraft, 2000, 11, 3, p. 139-158.

8288. BONORA-WAISMAN (Camille). France and the Algerian conflict: issues in democracy and political stability, 1988–1995. Aldershot a. Burlington, Ashgate, 2000, XII-233 p. (Leeds studies in democratization).

8289. BONWETSCH (Bernd), FILITOV (Alexei). Chruschtschow und der Mauerbau. Die Gipfelkonferenz der Warschauer-Pakt-Staaten vom 3.–5. August 1961. Vierteljahrshefte für Zeitgeschichte, 2000, 48, 1, p. 155-198.

8290. BORCHARD (Michael). Die deutschen Kriegsgefangenen in der Sowjetunion: zur politischen Bedeutung der Kriegsgefangenenfrage 1949–1955. Düsseldorf, Droste, 2000, 349 p. (Forschungen und Quellen zur Zeitgeschichte, 35).

8291. BORODZIEJ (Wlodzimierz), KOCHANOWSKI (Jerzy), SCHÄFER (Bernd). Grenzen der Freundschaft: zur Kooperation der Sicherheitsorgane der DDR und der Volksrepublik Polen zwischen 1956 und 1989. Dresden, Hannah-Arendt-Institut für Totalitarismusforschung e. V. an der Technischen Universität Dresden, 2000, 90 p. (Berichte und Studien, 30).

8292. BORSTELMANN (T.). 'Hedging our bets and buying time': John Kennedy and racial revolution in the American South and Southern Africa. *Diplomatic history*, 2000, 24, 3, p. 435-463.

8293. BOTH (Norbert). From indifference to entrapment: the Netherlands and the Yugoslav crisis, 1990–1995. Amsterdam, Amsterdam U. P., 2000, 267 p.

8294. BRENCHLEY (Thomas Frank). Britain and the 1967 Arab-Israeli War. Oxford, [s. n.], 2000, VII-409 p. [Thesis (D. Phil.) – University of Oxford, 2000].

8295. CARLTON (David). Churchill and the Soviet Union. New York, Manchester U. P., 2000, 234 p.

8296. CARY (Noel D.). Reassessing Germany's Ostpolitik. Part 1. From détente to refreeze. Part 2. From refreeze to reunification. *Central European history*, 2000, 33, 2, p. 235-262; 2000, 33, 3, p. 369-390.

8297. CHAITANI (Youssef). Dissension among allies: Ernest Bevin's Palestine policy between Whitehall and the White House, 1945–1947. London, Saqi, 2000, 156 p.

8298. CHI-KWAN (Mark). A reward for good behaviour in the Cold War: bargaining over the defence of Hong Kong, 1949–1957. *International history review*, 2000, 22, 4, p. 837-861.

8299. Chinese perspectives on Sino-American relations, 1950-2000. Ed. by Elizabeth VAN WIE DAVIS. Lewiston a. Lampeter, E. Mellen Press, 2000, IX-259 p. (Chinese studies, 12).

8300. CHRISTISON (Kathleen). Perceptions of Palestine: their influence on U.S. Middle East policy. Berkeley a. London, University of California Press, 2000, IX-390 p.

8301. CHUNG (Henry). Korea and the United States through war and peace: 1943–1960. Seoul, Yonsei U. P., 2000, XXXIV-538 p. (Institute for Modern Korean Studies historical materials series, 4).

8302. Cold War (The) – reassessments. Ed. by Arthur L. ROSENBAUM and Chae-Jin LEE. Claremont, Keck Center for International and Strategic Studies, 2000, 214 p. (Monograph series, 11).

8303. Cold War respite: the Geneva Summit of 1955. Ed. by Günther BISCHOF and Saki DOCKRILL. Baton Rouge, Louisian State U. P., 2000, XII-319 p.

8304. Conflit (Du) d'Indochine aux conflits indochinois. Sous la dir. de Pierre BROCHEUX. Bruxelles, Ed. Complexe, 2000, 179 p.

8305. CORALLUZZO (Valter). La politica estera dell'Italia repubblicana: 1946–1992: modello di analisi e studio di casi. Milano, F. Angeli, 2000, 398 p. (Collana Gioele Solari / Dipartimento di studi politici dell'Università di Torino, 34).

8306. CORDOVEZ (Diego). Nuestra propuesta inconclusiva: Ecuador-Perú del inmovilismo al acuerdo de Brasilia. Quito, Centro Andino de Estudios Internacionales, Universidad Andina Simon Bolivar, 2000, 271 p. (Serie Estudios internacionales, 1).

8307. CYGAŃSKI (Mirosław). Niemieckie mniejszości narodowe w państwach Europy Środkowo-Wschodniej w latach 1945–1995. (Deutsche nationale Minderheiten in den Ländern des Mittel-Ost-Europas in den Jahren 1945–1995). Opole, Wydaw. Inst. Śląskiego, 2000, 136 p. (Stowarzyszenie Inst. Śląskiego).

8308. DAMODARAN (A. K.). Beyond autonomy, India's foreign policy. Mumbai, Somaiya Publications, 2000, XV-246 p.

8309. Denmark's policy towards Europe after 1945: history, theory and options. Ed. by Hans BRANNER and Morten KELSTRUP. Odense, Odense U. P., 2000, 441 p.

8310. DEROCHE (Andrew). Black, white and chrome: the United States and Zimbabwe, 1953 to 1998. Trenton, Africa World, 2000, 300 p.

8311. Deutsch-französischen Beziehungen (Die): Chronologie und Dokumente 1948–1999 = Les relations franco-allemandes: chronologie et documents 1948–1999. Bonn, Europa Union Verlag, 2000, 199 p. (ill.).

8312. DOCKRILL (Saki). Britain's power and influence: dealing with three roles and the Wilson government's defence debate at Chequers in November 1964. *Diplomacy and statecraft*, 2000, 11, 1, p. 210-239. – IDEM. Dealing with soviet power and influence: Eisenhower's management of U.S. national security. *Diplomatic history*, 2000, 24, 2, p. 345-352.

8313. DOLHAR (Erik). Prelomnost Osimskih sporazumov. (The Osimo agreement turning point). Trieste, Krožek za družbena vprašanja Virgil Šček, 2000, 310 p. (Krožek za družbena vprašanja Virgil Šček, 33).

8314. DRUKS (Herbert). The uncertain friendship: the U.S. and Israel from Roosevelt to Kennedy. Westport a. London, Greenwood Press, 2000, XII-243 p. (Contributions to the study of world history, 80).

8315. DU BOIS (P.). L'Union européenne et le naufrage de la Yougoslavie (1991–1995). *In*: Balkans (Les) dans les relations internationales, Part 2 [Cf. n° 7582], p. 469-485.

8316. DUNTHORN (David J.). Britain and the Spanish anti-Franco opposition, 1940–1950. Houndmills a. New York, Palgrave, 2000, IX-236 p. – IDEM. The Prieto-Gil-Robles meeting of October 1947: Britain and the failure of the Spanish anti-Franco coalition, 1945–1950. *European history quarterly*, 2000, 30, 1, p. 49-78.

8317. DURAND (Pierre-Michel). Le pacte de Bagdad, pacte oublié de la Guerre Froide. *Revue d'histoire diplomatique*, 2000, 114, 2, p. 115-150.

8318. ELLISON (James). Threatening Europe: Britain and the Creation of the European Community, 1955–1958. New York, St. Martin's, 2000, XIII-310 p. (Contemporary History in Context).

8319. Empire and revolution: the United States and the Third World since 1945. Ed. by Peter L. HAHN and Mary Ann HEISS. Columbus, Ohio State U. P., 2000, VII-295 p.

8320. ENGEL (Ulf). Die Afrikapolitik der Bundesrepublik Deutschland 1949–1999: Rollen und Identität. Munster, Lit, 2000, 334 p.

8321. EUBANK (Keith). The missile crisis in Cuba. Malabar, Krieger Pub, 2000, VIII-235 p

8322. EVERTS (Steven). Adaptation in foreign policy: French and British reactions to German unification. Oxford, [s. n.], 2000, 280 p. [Thesis (D. Phil.) – University of Oxford, 2000].

8323. FANZUN (Jon A.), LEHMANN (Patrick). Die Schweiz und die Welt: Aussen- und sichereitspolitische Beiträge der Schweiz zur Frieden, Sicherheit und Stabilität, 1945–2000. Hrsg. v. Kurt SPILLMANN und Andreas WENGER. Zürich, Forschungsstelle für Sicherheitspolitik und Konfliktanalyse, ETH-Zentrum, cop. 2000, 362 p. (Zürcher Beiträge zur Sicherheitspolitik und Konfliktforschung, 57).

8324. FISCH (Bernhard). Stalin und die Oder-Neisse-Grenze: ein europäisches Problem. Berlin, Gesellschaftswissenschaftliches Forum, 2000, 64 p. (Hefte zur DDR-Geschichte, 64).

8325. FITZGERALD (Frances). Way out there in the blue: Reagan, Star Wars, and the end of the Cold War. New York a. London, Simon & Schuster, 2000, 592 p.

8326. FRASER (Cary). Crossing the color line in Little Rock: the Eisenhower administration and the dilemma of race for U.S. foreign policy. *Diplomatic history*, 2000, 24, 2, p. 233-264. – IDEM. The 'New Frontier' of Empire in the Caribbean: the transfer of power in British Guiana, 1961–1964. *International history review*, 2000, 22, 3, p. 583-610.

8327. FREEDMAN (Lawrence). Kennedy's wars: Berlin, Cuba, Laos, and Vietnam. New York, Oxford U. P., 2000, [s. p.]

8328. FREUDING (Christian). Deutschland in der Weltpolitik: die Bundesrepublik Deutschland als nichtständiges Mitglied im Sicherheitsrat der Vereinten Nationen in den Jahren 1977/78, 1987/88 und 1995/6. Baden-Baden, Nomos, 2000, 543 p. (Nomos Universitätsschriften, Politik, 113).

8329. GARDET (Claudie). Les relations de la République populaire de Chine et de la République démocratique allemande (1949–1989). Préface de Marie-Claire BERGÈRE. Bern et Oxford, Peter Lang, 2000, XVII-711 p. (Schweizer asiatische Studien. Monografien. Etudes asiatiques suisses. Monographies, 36).

8330. GAWRYCH (George Walter). The albatross of decisive victory: war and policy between Egypt and Israel in the 1967 and 1973 Arab-Israeli wars. Westport a. London, Greenwood Press, 2000, XV-281 p. (Contributions in military studies, 188).

8331. GINAT (Rami). The Soviet Union and the Syrian Ba'th regime: from hesitation to Rapprochement. *Middle eastern studies*, 2000, 36, 2, p. 150-171.

8332. GIOVAGNOLI (Agostino). L'Italia nel nuovo ordine mondiale: politica ed economia dal 1945 al 1947. Milano, Vita e pensiero, 2000, VIII-215 p. (Cultura e storia, 17).

8333. GIŻEJEWSKA (Małgorzata). Kołyma 1944–1956 we wspomnieniach polskich więźniów. (Kolyma 1944–1956 in Erinnerungen der polnischen Gefangenen). Warszawa, ISP PAN, 2000, 257 p. (Biblioteka Ziem Wschodnich).

8334. Globaler Wandel und schweizerische Aussenpolitik: Informationsbeschaffung und Entscheidungsfindung der Schweizerischen Bundesverwaltung. Hrsg. v. Thomas BERNAUER und Dieter RULOFF. Chur, Rüegger, 2000, 368 p. (ill.). (NFPNR, 42).

8335. GOLAN (Shimon). Gevul ham, milhamah karah: hitgabshut mediniyut ha-bitahon shel Yisra'el, 1949–1953. [Hot border, cold war]. Tel Aviv, Tsahal, Hotsa'at "Ma`arakhot", Misrad ha-bitahon, 2000, 570 p.

8336. Grandes potencias, el 9 de Abril y la violencia. Ed. por Gonzalo SÁNCHEZ G. Santafé de Bogotá, Planeta Columbiana Editorial S. A., 2000, 362 p. (ill.). (Colección La Línea del horizonte).

8337. *Vacat.*

8338. GUDERZO (Massimiliano). Interesse nazionale e responsabilità globale: gli Stati Uniti, l'Alleanza atlantica e l'integrazione europea negli anni di Johnson, 1963–1969. Firenze, Aida, 2000, XXI-611 p. (Il maestrale, 6).

8339. HALL (Michael R.). Sugar and power in the Dominican Republic: Eisenhower, Kennedy, and the Trujillos. Westport a. London, Greenwood Press, 2000, XII-163 p. (Contributions in Latin American studies, 13).

8340. HAMMER (M.). La Chine et les Balkans, 1960–1978. *In*: Balkans (Les) dans les relations internationales, Part 2 [Cf. n° 7582], p. 455-467.

8341. HEADLEY (James Henry). The Russian Federation and the conflicts in former Yugoslavia, 1992–1995. London, [s. n.], 2000, 334 p. [Thesis: (Ph D) – University of London 2000].

8342. HENKE (Holger). Between self-determination and dependency: Jamaica's foreign relations, 1972–1989. Kingston, University of the West Indies Press, 2000, XV-221 p.

8343. HEPPERLE (Sabine). Die SPD und Israel: von der grossen Koalition 1966 bis zur Wende 1982. Frankfurt am Main, Lang, 2000, 499 p. (Europäische Hochschulschriften: Reihe 3, Geschichte und ihre Hilfswissenschaften, 861).

8344. HILGER (Andreas). Deutsche Kriegsgefangene in der Sowjetunion, 1941–1956: Kriegsgefangenpolitik, Lageralltag und Erinnerung. Essen, Klartext Verlag, 2000, 486 p. (Schriften der Bibliothek für Zeitgeschichte, 11).

8345. HIPPEL (Karin, von). Democracy by force: U.S. military intervention in the post-Cold War world. Cambridge, Cambridge U. P., 2000, XII-224 p. (ill., maps). (LSE monographs in international studies).

8346. HÖNIG (Patrick). Der Kaschmirkonflikt und das Recht der Völker auf Selbstbestimmung. Berlin, Duncker & Humblot, 2000, 405 p. (Schriften zum Völkerrecht, 138).

8347. HUGHES (Philip). Division and discord: British policy, Indochina, and the origins of the Vietnam war, 1954–1956. In: International diplomacy and colonial retreat [Cf. n° 7650], p. 94-112.

8348. ILSAAS PHARO (Per Fredrik). A procondition for peace: transparency and Test-Ban negotiations, 1958–1963. International history review, 2000, 22, 3, p. 557-582.

8349. ISSA-SALWE (Abdisalam M.). Cold War fallout: boundary politics and conflict in the Horn of Africa. London, HAAN, 2000, XIV-175 p.

8350. JACKSON (I.). 'The limits of international leadership': the Eisenhower administration, East-West trade and the Cold War, 1953–1954. Diplomacy and statecraft, 2000, 11, 3, p. 113-138.

8351. JACKSON (Robert). United States Air Force in Britain: its aircraft, bases and strategy since 1948. Shrewsbury, Airlife, 2000, 144 p. (ill.).

8352. JAIN (Satya Bhusan). India's foreign policy and non-alignment. New Delhi, Anamika Publishers, 2000, VIII-349 p.

8353. JAMES (Alan). Britain, the Cold War and the Congo crisis, 1960–1963. In: International diplomacy and colonial retreat [Cf. n° 7650], p. 152-168.

8354. Japan and Russia: the tortuous path to normalization, 1949–1999. Ed. by Gilbert ROZMAN. Basingstoke, Macmillan, 2000, 389 p.

8355. JAUVERT (Vincent). L'Amérique contre De Gaulle: histoire secrète (1961–1969). Paris, Ed. du Seuil, 2000, 284 p. (Histoire immédiate).

8356. JEFFERY (Judith S.). Ambiguous commitments and uncertain policies: the Truman Doctrine in Greece, 1947–1952. Lanham, Lexington Books, 2000, XII-342 p.

8357. JENSEN (Bent). Soviet occupation of a new type: the long liberation of the Danish island of Bornholm 1944–1946. Scandinavian journal of history, 2000, 25, 3, p. 219-238.

8358. JESPERSEN (T. C.). The bitter end and the lost chance in Vietnam: Congress, the Ford administration, and the battle over Vietnam, 1975–1976. Diplomatic history, 2000, 24, 2, p. 265-293.

8359. JOHNSON (Edward). Britain and the Cyprus problem at the United Nations, 1954–1958. In: International diplomacy and colonial retreat [Cf. n° 7650], p. 113-130. – IDEM. Keeping Cyprus off the Agenda: British and American Relations at the United Nations, 1954–1958. Diplomacy and statecraft, 2000, 11, 3, p. 227-255.

8360. JOO (Seung-Ho). Gorbachev's foreign policy toward the Korean peninsula, 1985–1991: power and reform. Lewiston, Edwin Mellen Press, 2000, XVI-248 p. (Studies in Russian history, 5).

8361. KALAWOUN (Nasser M.). The struggle for Lebanon: a modern history of Lebanese-Egyptian relations. London, I. B. Tauris, 2000, XVI-224 p. (Library of international relations, 14).

8362. KAPLAN (Karel). Jednání československé vlády s Vatikánem v letech 1963–1967. (The negotiation of the Czechoslovak government with Vatican in the year 1963–1967). Historický časopis, 2000, 48, 2, p. 283-316.

8363. KAUFMAN (Victor S.). 'Chirep': the Anglo-American dispute over Chinese representation in the United Nations, 1950–1971. English historical review, 2000, 115, 461, p. 354-377.

8364. KENDALL MOORE (J.). Between expediency and principle: U.S. repatriation policy toward Russian nationals, 1944–1949. Diplomatic history, 2000, 24, 3, p. 381-404.

8365. KNIGHT (Robert). Ethnicity and identity in the Cold War: the Carinthian border dispute, 1945–1949. International history review, 2000, 22, 2, p. 274-303.

8366. KOOPMANN (Martin). Das schwierige Bündnis: die deutsch-französischen Beziehungen und die Aussenpolitik der Bundesrepublik Deutschland 1958–1965. Baden-Baden, Nomos, 2000, 330 p.

8367. KOSSDORFF (Felix). Die Republik Irland: ein europäischer Kleinstaat und seine aussenpolitischen Strategien als Mitglied der EU. Wien, WUV, 2000, 291 p. (ill.). (Dissertationen der Universität Wien, 68).

8368. KÖSTER (Klaus). Bundesrepublik Deutschland und Vereinte Nationen 1949 bis 1963. Frankfurt am Main u. Oxford, P. Lang, 2000, 272 p. (Europäische Hochschulschriften. Reihe III, Geschichte und ihre Hilfswissenschaften, 879 = Publications universitaires européennes. Série III, Histoire, sciences auxiliaires de l'histoire, 879 = European university studies. Series III, History and allied studies, 879).

8369. KROUCK (Bernard). La France face à la crise du Tibet. Revue d'histoire diplomatique, 2000, 114, 2, p. 151-176.

8370. KÜSTERS (Hanns Jürgen). Der Integrationsfriede: Viermächte-Verhandlungen über die Friedensregelung mit Deutschland 1945–1990. München, Oldenbourg, 2000, 1026 p. (Dokumente zur Deutschlandpolitik. Studien, 9).

8371. LANGGUTH (A. J.). Our Vietnam: the war, 1954-1975. New York a. London, Simon & Schuster, 2000, 766 p. (ill., maps, ports).

8372. LANUS (Juan Archibaldo). De Chapultepec al Beagle: politica exterior argentina: 1945-1980. Buenos Aires, Emece, 2000, 571 p.

8373. LARONCE (Cécile). Nkrumah, le panafricanisme et les États-Unis. Paris, Éditions Karthala, 2000, 325 p. (Hommes et sociétés. Insatisfaits).

8374. LATHAM (Michael E.). Modernization as ideology: American social science and "nation building" in the Kennedy era. Chapel Hill a. London, University of North Carolina Press, 2000, XII-288 p. (New Cold War history).

8375. LAYTON (Azza Salama). International politics and civil rights policies in the United States, 1941-1960. Cambridge a. New York, Cambridge U. P., 2000, XI-217 p.

8376. LEE (David Tawei). The making of the Taiwan Relations Act: twenty years in retrospect. Oxford, Oxford U. P., 2000, 217 p. (Studies on contemporary Taiwan).

8377. LITVÁN (György). 1956: crise de Hongrie, crise de Suez. *In*: Hongrie (La) dans les conflits du XXe siècle [Cf. n° 7611], p. 99-114.

8378. LUU (Van Loi). Fifty years of Vietnamese diplomacy, 1945-1995. Vol. 1. 1945-1975. Hanoi, Thê Gioi Publishers, 2000, [s. p.].

8379. MAC EACHIN (Douglas J.). US intelligence and the Polish crisis: 1980-1981. Washington, Center for the Study of Intelligence, 2000, XII-211 p.

8380. MAC ELRATH (Karen). Unsafe haven: the United States, the IRA, and political prisoners. London a. Sterling, Pluto Press, 2000, XII-164 p.

8381. MAC EVOY-LEVY (Siobhán). American exceptionalism and US foreign policy: public diplomacy at the end of the Cold War. New York, Palgrave, 2000, IX-265 p.

8382. MADHOK (Balraj). Kashmir, Kargil and Indo-Pak relations. New Delhi, D.A.V., 2000, 175 p.

8383. MARGOLIS (Eric S.). War at the top of the world: the struggle for Afghanistan, Kashmir, and Tibet. New York, Routledge, 2000, XII-250 p.

8384. MASSOT I MUNTANER (Josep). De la guerra i de l'exili: Mallorca, Montserrat, França, Mèxico, 1936-1975. Barcelona, Publicacions de l'Abadia de Montserrat, 2000, 331 p. (Biblioteca Serra d'or, 247).

8385. MAYERS (D.). JFK's ambassadors and the Cold War. *Diplomacy and statecraft*, 2000, 11, 3, p. 183-211.

8386. MAZZA (Jacqueline). Don't disturb the neighbors: United States and democracy in Mexico, 1980-1995. London a. New York, Routledge, 2000, [s. p.].

8387. MEIER (Edward Franklin). Libya's foreign policy in Africa: from destabilization to diplomacy. 2000, Oxford, [s. n], 98 p. [Thesis (M. Phil.) – University of Oxford, 2000].

8388. MELANDRI (P.). Les Etats-Unis et les premières guerres de l'ex-Yougoslavie (1991-1995). *In*: Balkans (Les) dans les relations internationales, Part 2 [Cf. n° 7582], p. 487-506.

8389. MENON (Anand). France, NATO and the limits of independence, 1981-1997: the politics of ambivalence. Basingstoke, Macmillan Press a. New York, St. Martin's Press, 2000, XII-258 p.

8390. METHOL FERRÉ (Alberto). Perón y la alianza argentino-brasilena con textos complementarios. Córdoba, Ediciones del Corredor Austral, 2000, 89 p. (Colección Eje de la integración).

8391. MILLER (Rory). Divided against Zion: anti-Zionist opposition in Britain to a Jewish state in Palestine, 1945-1948. London, Frank Cass, 2000, XII-275 p. (Cass series – Israeli history, politics, and society, 11).

8392. MINION (Mark). The Fabian Society and Europe during the 1940s: the search for a 'Socialist foreign policy'. *European history quarterly*, 2000, 30, 2, p. 237-270.

8393. MITROVICH (Gregory). Undermining the Kremlin: America's strategy to subvert the Soviet Bloc, 1947-1956. Ithaca a. London, Cornell U. P., 2000, X-235 p. (Cornell studies in security affairs).

8394. MUTH (Ingrid). Die DDR-Aussenpolitik 1949-1972: Inhalte, Strukturen, Mechanismen. Berlin, Ch. Links, 2000, 318 p. (Forschungen zur DDR-Gesellschaft).

8395. Nationale Außen- und Bündnispolitik der NATO-Mitgliedstaaten. Im Autrag des MGFA hrsg. v. Norbert WIGGERHAUS und Winfried HEINEMANN. München, Oldenbourg, 2000, XVIII-352 p. (Entstehung und Probleme des Atlantischen Bündnisses bis 1956, 2).

8396. NEGASH (Tekeste), TRONVOLL (Kjetil). Brothers at war: making sense of the Eritrean-Ethiopian war. Oxford, James Currey, 2000, XI-179 p. (Eastern African studies).

8397. NEUSS (Beate). Geburtshelfer Europas? Die Rolle der Vereinigten Staaten im europäischen Integrationsprozeß 1945-1958. Baden-Baden, Nomos-Verl.-Ges., 2000, 388 p.

8398. Neutralen (Die) und die europäische Integration, 1945-1995 = The neutrals and the European integration, 1945-1995. Hrsg. v. Michael GEHLER und Rolf STEININGER. Wien, Böhlau, 2000, 800 p. (Historische Forschungen, Veröffentlichungen / Institut für Zeitgeschichte der Universität Innsbruck, Arbeitskreis Europäische Integration, 3).

8399. NEWMAN (Kathleen Paula). Britain and the Soviet Union: the search for an interim agreement on

West Berlin, November 1958–May 1960. London, [s. n.], 2000, 334 p. [Thesis (Ph. D.) – London, LSE, 2000]

8400. Norway and national liberation in Southern Africa. Ed. by Tore Linne ERIKSEN. Uppsala, Nordiska Afrikainstitutet, 2000, 416 p.

8401. OSGOOD (K. A.). Form before substance: Eisenhower's commitment to psychological warfare and negotiations with the enemy. *Diplomatic history*, 2000, 24, 3, p. 405-433.

8402. Österreich im frühen Kalten Krieg 1945–1958: Spione, Partisanen, Kriegspläne. Hrsg. v. Erwin A. SCHMIDL. Wien, Böhlau, 2000, 175 p.

8403. O'SULLIVAN (Donal). Das amerikanischen Venona-Projekt. Die Enttarnung der sowjetischen Auslandsspionage in den vierziger Jahren. *Vierteljahrshefte für Zeitgeschichte*, 2000, 48, 4, p. 603-630.

8404. OTTOSSON (Sten). Den (o)moraliska neutraliteten. Tre politikers och tre tidningars moraliska värdering av svensk utrikespolitik 1945–1953. (La neutralité (im)morale. L'évaluation morale de la politique étrangère suédoise de 1945 à 1953 par trois personnalités politiques et par trois journaux). Stockholm, Santérus förlag, 2000, 277 p. (Sverige under det kalla kriget. 7).

8405. PAGEDAS (Constantine A.). Anglo-American strategic relations and the French problem, 1960–1963: a troubled partnership. London a. Portland, Frank Cass, 2000, 308 p.

8406. PAPE (Matthias). Die Deutschlandinitiative des österreichischen Bundeskanzler Julius Raab im Frühjahr 1958. *Vierteljahrshefte für Zeitgeschichte*, 2000, 48, 2, p. 281-318. – IDEM. Ungleiche Brüder: Österreich und Deutschland 1945-1965. Köln, Böhlau, 2000, 713 p.

8407. PARSONS (Michael). The Falklands War. Stroud, Sutton, 2000, X-118 p. (Sutton pocket histories).

8408. PAUL (Septimus H.). Nuclear rivals: Anglo-American atomic relations, 1941–1952. Columbus, Ohio State U. P., 2000, IX-266 p.

8409. PELIKÁN (Jan). Jugoslávie a státy sovětské zájmové sféry ve druhé polovině roku 1955. (Yugoslavia and the states of the Soviet sphere of influence in the second half of 1955). *Slovanské historické studie*, 2000, 26, p. 285-326.

8410. PERNES (Jiří). Československý rok 1956. K dějinám destalinizace v Československu. (Czechoslovak in 1956. Concerning the history of de-Stalinization in Czechoslovakia). *Soudobé dějiny*, 2000, 7, 4, p. 594-618.

8411. PESTKOWSKA (Maria). Za kulisami rządu polskiego na emigracji. (Hinter den Kulissen der polnischen Exil-Regierung). Warszawa, Rytm, 2000 329 p.

8412. PETERSEN (Tore T.). The Middle East between the Great Powers: Anglo-American conflict and cooperation, 1952–1957. New York, St. Martin's, 2000, XIII-170 p.

8413. PINTEV (S.). Nachaloto na konflikta Tito-Stalin i otnoshenieto mu vurkhu amerikanskite politicheski planove kum Bulgariia (1948–1949 g.). (Le début du conflit Tito-Stalin et son attitude sur les plans politiques américains envers la Bulgarie, 1948–1949). Mezhdunarodni otnosheniia, 2000, 6, p. 71-86.

8414. PLOETZ (Michael). Wie die Sowjetunion den Kalten Krieg verlor: von der Nachrüstung zum Mauerfall. Berlin, Propylaen, 2000, 456 p.

8415. POIRIER (Bernard W). Witness to the end: Cold War revelations, 1959–1969. Lanham, University Press of America, 2000, [s. p.].

8416. POTEAT (Eugene). The use and abuse of intelligence: an intelligence provider's perspective. *Diplomacy and statecraft*, 2000, 11, 2, p. 1-16.

8417. POWASKI (Ronald E.). Return to Armageddon: the United States and the nuclear arms race, 1981–1999. New York, Oxford U. P., 2000, XI-294 p.

8418. PSOMIADES (Harry J.). The Eastern question: the last phase: a study in Greek-Turkish diplomacy. With an introduction by Van COUFOUDAKIS. New York, Pella Pub. Co, 2000, XI-139 p. (Modern Greek research series, 9).

8419. PUDDINGTON (Arch). Broadcasting freedom: the Cold War triumph of Radio Free Europe and Radio Liberty. Lexington, University Press of Kentucky, 2000, XIX-382 p. (ill.).

8420. PUTENSEN (Dörte). Im Konfliktfeld zwischen Ost und West: Finnland, der Kalte Krieg und die deutsche Frage (1947–1973). Berlin, Berlin-Verlag A. Spitz, 2000, 457 p. (ill.). (Schriftenreihe der Deutsch-Finnischen Gesellschaft e. V., 3).

8421. QUINTERO TORRES (José Gilberto). Venezuela-U.S.A.: estrategia y seguridad en lo regional y en lo bilateral, 1952–1958. Caracas, Fondo Editorial Nacional, 2000, 239 p.

8422. RANGEL (Domingo Alberto). El Vietnam colombiano. Caracas, Universidad Central de Venezuela, Ediciones de la Biblioteca-EBUC, 2000, 152 p. (Colección Temas, 87).

8423. Relaciones exteriores (Las) de la Argentina subordinada, 1943–1989. T. 11. Las relaciones económicas externas, 1943–1989. T. 12. La diplomacia de Malvinas, 1945–1989. T. 13. Las relaciones políticas, 1943–1966. T. 14. Las relaciones políticas, 1966–1989. Con la colaboración específica de Ana MARGHERITIS [et al.]; obra dirigida por Andrés CISNEROS y Carlos ESCUDÉ. [S. l.], Centro de Estudios de Política Exterior, Consejo Argentino para las Relaciones Internacionales y Buenos Aires, Nuevohacer, Grupo Editor Latinoamericano, 1999–2000, 4 vol., [s. p.]. (ill., maps). (Historia general de las relaciones exteriores de la República Argentina, 3. Colección Estudios internacionales).

8424. RETEGAN (Mihai). In the shadow of the Prague spring: Romanian foreign policy and the crisis in Czechoslovakia, 1968. Iaşi, Romania a. Oxford, Center for Romanian Studies, 2000, 248 p.

8425. Revisiting the Yom Kippur war. Ed. by P. R. KUMARASWAMY. Portland, Frank Cass, 2000, 249 p. (Israeli history, politics and society).

8426. RIZAS (Soteres Ch.). Henose dichotomese anexartesia: hoi Henomenes Politeies kai he Vretania sten anazetese lyses gia to Kypriako, 1963–1967. (Unione, divisione, indipendenza: le Nazioni Unite e la Gran Bretagna nella richiesta di risoluzioni per Cipro). Athena, Vivliorama, 2000, 247 p.

8427. ROMEO (Giuseppe). La politica estera italiana nell'era Andreotti (1972–1992). Soveria Mannelli, Rubbettino, 2000, 186 p. (Saggi, 69).

8428. Roots (The) of the Vietnam War. Edited with introductions by Walter L. HIXSON. New York, Garland, 2000, XVI-330 p. (The United States and the Vietnam War: significant scholarly articles, 1).

8429. ROTTER (A. J.). Christians, Muslims, and Hindus: religion and U.S.-South Asian relations, 1947–1954. *Diplomatic history*, 2000, 24, 4, p. 593-613.

8430. ROTTER (Andrew Jon). Comrades at odds: the United States and India, 1947–1964. Ithaca a. London, Cornell U. P., 2000, XXIX-337 p. (ill.).

8431. RUANE (Kevin). The rise and fall of the European defense community: Anglo-American relations and the crisis of European defense, 1950–1955. New York, Palgrave, 2000, IX-252 p. (Cold War history series).

8432. RYDEL (Jan). "Polska okupacja" w północnowschodnich Niemczech 1945–1948. Nieznany rozdział stosunków polsko-niemieckich. ("Die polnische Besatzung" in Nord-Ost-Deutschland 1945–1948. Ein unbekanntes Kapitel der polnisch-deutschen Verhältnisse). Kraków, Fundacja CDCN u. Księgarnia Akademicka, 2000, 351 p. (Biblioteka Centrum Dokumentacji Czynu Niepodległościowego, 6).

8433. SAHA (Rekha). India-Bangladesh relations. Calcutta, Minerva Associates, 2000, 241 p.

8434. SARANTAKES (Nicholas Evan). Keystone: the American occupation of Okinawa and U.S.-Japanese relations. College Station, Texas A&M University, 2000, XXIII-264 p. (ill.). (Foreign relations and the presidency, 6).

8435. Satelliten nach Stalins Tod. der "Neue Kurs"; 17. Juni 1953 in der DDR; ungarische Revolution 1956. Hrsg. v. András B. HEGEDÜS. Berlin, Akad.-Verl., 2000, 316 p. (Studien der Forschungsverbundes SED-Staat an der Freien Universität Berlin).

8436. SCHAAD (Martin P. C.). Bullying Bonn: Anglo-German diplomacy on European integration, 1955–1961. Basingstoke, MacMillan, 2000, VIII-243 p. (St. Antony's/Macmillan series).

8437. SCHOFIELD (Victoria). Kashmir in conflict: India, Pakistan and the unfinished war. London, I. B. Tauris, 2000, XVIII-286 p.

8438. SELIKTAR (Ofira). Failing the crystal ball test: the Carter administration and the fundamentalist revolution in Iran. Westport, Praeger, 2000, XXI-245 p.

8439. SHIRAEV (Eric), ZUBOK (V. M.). Anti-Americanism in Russia: from Stalin to Putin. New York a. Basingstoke, Palgrave, 2000, 182 p. – IIDEM. Anti-Americanism in Russia after Perestroika. Basingstoke, Macmillan, 2000, 320 p.

8440. SHLAIM (Avi). The iron wall: Israel and the Arab world. London, Allen Lane, 2000, XXV-670 p. (ill.; maps).

8441. SHORE (Peter). Separate ways: the heart of Europe. London, Duckworth, 2000, XI-244 p.

8442. SMITH (T.). New bottles for new wine: a pericentric framework for the study of the Cold War. *Diplomatic history*, 2000, 24, 4, p. 567-591.

8443. SONYEL (Salâhi Ramadan). Why did the intercommunal talks on Cyprus (1968–1971) fail? Revelations in the light of British official documents released recently. Lefkosa, CYREP, 2000, 41 p.

8444. SORBY (Karol). Suezská kríza 1955–1956. (The Suez crisis 1955–1956). *Historický časopis*, 2000, 48, 4, p. 650-672.

8445. SOULET (Jean-François). L'empire Stalinien. L'U.R.S.S. et les pays de l'Est depuis 1945. Paris, Rèférence, 2000, 253 p. (Le livre de poche).

8446. SOUTOU (George-Henri). La politique française envers la Yougoslavie, 1945-1956. *In*: Balkans (Les) dans les relations internationales, Part 2 [Cf. n° 7582], p. 433-454.

8447. SPÄTI (Christoph). Die Schweiz und die Tschechoslowakei, 1945–1953: wirtschaftliche, politische und kulturelle Beziehungen im Polarisationsfeld des Ost-West-Konflikts. Zürich, Chronos, 2000, 762 p. (Schweiz und der Osten Europas, 6).

8448. SREBRAKOWSKI (Aleksander). Polacy w Litewskiej SRR 1944–1989. Torun, Wyd. A. Marszalek, 2000, 392 p.

8449. STAMENOVA (M.). Makedonskijat vupros v bulgaro-jugoslavskite otnoshenija (1953–1963 g.). (Le problème macédonien dans les relations bulgaro-yougoslaves, 1953–1963). Izvestija na durzhavnite arkhivi, 2000, 74, p. 50-68.

8450. STREETER (Stephen M.). Managing the counterrevolution: the United States and Guatemala, 1954–1961. Athens, Ohio University Center for International Studies, 2000, XV-384 p. (Research in international studies. Latin America series, 34).

8451. STYAN (David A.). Franco-Iraqi relations and Fifth Republic foreign policy 1958-1990. London, [s. n.], 2000, 264 p. [Thesis (Ph. D.) – London, LSE, 2000].

8452. SUSSER (Asher). Jordan: case study of a pivotal state. Washington, Washington Institute for Near East Policy, 2000, XIV-134 p. (Policy papers, 53).

8453. TAKEYH (Ray). The origins of the Eisenhower doctrine: the US, Britain and Nasser's Egypt, 1953–1957. Basingstoke, Macmillan, 2000, XIX-216 p. (St Antony's series).

8454. TAL (David). Symbol not substance? Israel's campaign to acquire Hawk missiles, 1960–1962. *International history review*, 2000, 22, 2, p. 304-317.

8455. TAN (Patricia S. M.). Idea factories: American policies for German higher education and reorientation, 1944–1949. Oxford, [s. n.], 2000, XV-324 p. [Thesis (D. Phil.) – Oxford, Faculty of modern history].

8456. TAREKE (Gebru). The Ethiopia-Somalia war of 1977 revisited. *International journal of African historical studies*, 2000, 33, 3, p. 635-667.

8457. TOMARU (Junko). The postwar rapprochement of Malaya and Japan, 1945–1961: the roles of Britain and Japan in South-East Asia. New York, St. Martin's, in association with St. Antony's College, Oxford, 2000, XIV-317 p. (St. Antony's Series).

8458. TORKUNOV (Anatolij V.). Zagadochnaja vojna: Korejskij konflikt 1950–1953 godov. (A mysterious war: the Korean conflict of 1950–1953). Moskva, ROSSPEN, 2000, 310 p. (ill.).

8459. TORREIRA CRESPO (Ramón), BUAJASÁN MARRAWI (José). Operación Peter Pan: un caso de guerra psicológica contra Cuba. La Habana, Editorial Politica, 2000, 443 p.

8460. TRIMBUR (Dominique). De la Shoah à la réconciliation? La question des relations RFA-Israël, 1949–1956. Paris, CNRS, 2000, 447 p. (Cahiers du CRFJ. Hommes et sociétés, 7).

8461. Truman (Von) bis Harmel: die Bundesrepublik Deutschland im Spannungsfeld von NATO und europäischer Integration. Hrsg. v. Hans-Joachim HARDER. München, Oldenbourg, 2000, XIV-238 p. (Militärgeschichte seit 1945, 11).

8462. Uneasy allies: British-German relations and European integration since 1945. Ed. by Klaus LARRES with Elizabeth MEEHAN. Oxford a. New York, Oxford U. P., 2000, XVI-344 p. (ill.).

8463. USLU (Nasuh). Türk-Amerikan ilişkilerinde Kıbrıs. (Cyprus in Turkish-American relations). Ed. by Mustafa EVERDI. Ankara, 21. Yüzyıl yayınları, 2000, 405 p.

8464. VAN DEN DOEL (Wim). Afschied van Indië: die geschiedenis van Nederlands-Indië, 1945–1949. Amsterdam, Prometheus, 2000, 418 p.

8465. VANDER LIPPE (John M.). The forgotten brigade of the forgotten war: Turkey's participation in the Korean war. *Middle eastern studies*, 2000, 36, 1, p. 92-102.

8466. VOGT (Timothy R.). Denazification in Soviet-occupied Germany: Brandenburg, 1945–1948. Cambridge a. London, Harvard U. P., 2000, XII-314 p. (ill.). (Harvard historical studies, 137).

8467. WAGENSOHN (Tanja). Von Gorbatschow zu Jelzin: Moskaus Deutschlandpolitik (1985–1995) im Wandel. Baden-Baden, Nomos, 2000, 366 p.

8468. WEITZ (Yechiam). Ben-Gurions Weg zum "Andere Deutschland" 1952–1963. *Vierteljahrshefte für Zeitgeschichte*, 2000, 48, 2, p. 255-280.

8469. Whitehall and the Suez crisis. Ed. by Saul KELLY and Anthony GORST. London,: Frank Cass, 2000, 250 p. (British foreign and colonial policy, 1).

8470. YOFRE (Juan Bautista). Misión argentina en Chile (1970–1973). Santiago de Chile, Editorial Sudamericana, 2000, 467 p. (Colección Crónicas y testimonios).

8471. ZELIKOW (P.). American policy and Cuba, 1961–1963. *Diplomatic history*, 2000, 24, 2, p. 317-334.

8472. ZHAI (Qiang). China and the Vietnam Wars 1950–1975. Chapel Hill: University of North Carolina Press, 2000, XII-304 p. (The new Cold War history).

§ 8. Addenda 1999.

8473. MILZA (Pierre). L'année 1947 dans les combats culturels de la Guerre froide. *In*: Année (L') 1947 [Cf. n° Bibl. 99, n° 4298.], p. 411-436.

Cf. n°ˢ 5082, 5297, 5390, 8056, 8082, 8099, 8108, 8124, 8126, 8132, 8138, 8148, 8151, 8167, 8171, 8174, 8189

R

ASIA

§ 1. General. 8474-8479. – § 2. Western and central Asia. 8480-8495. – § 3. South Asia and Southeast Asia. 8496-8503. – § 4. China. 8504-8738. – § 5. Japan (before 1868). 8739-8781. – § 6. Korea. 8782-8785.

§ 1. General.

8474. Asian nationalism. Ed. by Michael LEIFER. London a. New York, Routledge, 2000, VI-203 p.

8475. Columbia chronologies of Asian history and culture. Ed. by John S. BOWMAN. New York, Columbia U. P., 2000, XVI-751 p. (map).

8476. MASON (Colin). A short history of Asia: Stone Age to 2000 AD. Basingstoke, Macmillan a. New York, St. Martin's Press, 2000, VIII-316 p.

8477. MURPHEY (Rhoads). A history of Asia. New York a. Harlow, Longman, 2000, XXVII-468 p. (ill., maps, ports.).

8478. Nation work: Asian elites and national identities. Ed. by Timothy BROOK and Andre SCHMID. Ann Arbor, University of Michigan Press, 2000, 270 p.

8479. ROGOZIŃSKI (Jan). Honor among thieves: Captan Kidd, Henry Every, and the pirate democracy in the Indian ocean. Mechanicsburg, Stackpole Books, 2000, XXII-298 p.

Cf. n°s 1025, 1126, 7661-7683

§ 2. Western and central Asia.

8480. AKASAKA (Tsuneaki). 14 seiki chuyô–16 seiki hajime ni okeru "Uzubeku": Isuramuka go no Juchi-Urusu no sôsho. ("Özbeg": from the middle of fourteenth century to the beginning of the sixteenth century: as a general term for Ulūs-i Jūjī). *Shigaku Zasshi*, 2000, 109, 3, p.1-39. [Eng. Summary].

8481. HANEDA (Tadashi). Mittsu no "isurâmu-kokka". (Three "Islamic States": the Ottoman, Safavids, and Mogul dynasties). *In*: Iwanami kôza sekai rekishi 14: isurâmu·kan-indoyou sekai. Tokyo, Iwanami Shoten, 2000, p. 3-90. (Iwanami World History, 14. The Islamic and Indian Ocean world).

8482. IBRAGIMOV (B. I.). Srednevekovyj gorod Kiran. (Kiran: a medieval city [Azerbaidjan]). Baku a. Moskva, [s. n.], 2000, 176 p. (ill.). [Eng. summary]

8483. ISHIMARU (Yumi). Kotoba to identity: Şemseddin Sâmi no identity wo motomete. (Terminology and national identity in the writings of Şemseddin Sâmi). *Nihon Ch$\bar{u}$tou Gakkai Nempou*, 2000, 15, p. 205-223. [Eng. Summary].

8484. IWATAKE (Akio). Mongoru no Isuramuka no shosô. (Some notes on the Islamization of the Mongols in the Il-khanids of the thirteenth century). *Historical Studies of Kansei Gakuin*, 2000, 27, p. 71-100.

8485. KAWAMOTO (Masatomo). Chûo-ajia no "Tuman" naru chiiki-kubun ni tsuite. ("Tuman" as regional district in Central Asia). *Seinan-ajia kenkyû*, 2000, 53, p. 24-60.

8486. KONDO (Nobuaki). Iran·turan·hindo: Perushiago bunka-ken no hatten to henyou. (Iran, Turan, and Hind: the development and change of the world of Persian language). *In*: Iwanami kôza sekai rekishi 14: isurâmu·kan-indoyou sekai. Tokyo, Iwanami Shoten, 2000, p. 93-114.

8487. NAKAMACHI (Nobutaka). Iru-han koku kara mamurûku-cho ni ryunyu shita boumei gunji shudan: Torai no haikei to toraigo no keireki wo chushin ni. (The military refugees from the Ilkhanid to the Mamluk sultanate: The backgroud of their flight and later careers). *Shigaku Zasshi*, 2000, 109, 4, p. 1-42. [Eng. Summary].

8488. OTSUKA (Kazuo). Kindai·isurâmu no jinruigaku. (An anthropological approach to the modern and Islam). Tokyo, University of Tokyo Press, 2000, 287 p.

8489. PIGULEVSKAJA (Nina V.). Sirijskaja srednevekovaja istoriografija: Issledovanija i perevody. (Historical writing in Medieval Syria: studies and translations). Intr. by Elena N. MESHCHERSKAJA. RAN, In-t vostokovedenija. Sankt-Peterburg, Dmitrij Bulanin, 2000, 760 p.

8490. SAKAI (Hiroki). Tyuruku eiyû-joji-shi no chiiki-teki tokuchô: "Chora Batir" no bunrui wo motoni. (The regional peculiarities of Turkic heroic epics: with special reference to the classification of the "Chora Batir"). *Chiiki Kenkyu Ronshu*, 2000, 3, 2, p. 95-122. [Eng. Summary].

8491. SATO (Kentaro). Slave elites and the Saqāliba in al-Andalus in the Umayyad period. *In*: Slave elites in the Middle East and Africa. Ed. by Toru MIURA and John Edward PHILIPS. London a. New York, Kegan Paul International, 2000, p. 25-40.

8492. SHIMIZU (Kazuhiro). Ghulâm no shosô: Abbasu-chou ni okeru ie to gunji-ryoku. (Ghulâms in households and battlefields during the Abbâsid period). *Seinan-ajia kenkyû*, 2000, 52, p. 38-58.

8493. Shinpan Sekai kakkoku-shi 4: Chûo-yûrashia shi. (A history of Central Eurasia. History of Countries, 4). Ed. by Hisao KOMATSU. Tokyo, Yamakawa Shuppansha, 2000, 551 p.

8494. UYAMA (Tomohiko). Chûo-ajia no rekishi to genzai. (Past and present of Central Asia). Tokyo, Toyo Shoten, 2000, 64 p.

8495. YAMAGUCHI (Akihiko). Osuman Kenchi-chou ni miru 18-seiki shotou iran no chiiki-shakai (1): iran naibu arudarān chihou no nouson to yûbokumin-shakai. (Early 18th century Iranian local societies as seen from Ottoman Tahrir Registers: peasants and nomads in Ardalan province). *Touyou-bunka kinkyû-jyo kiyou*, 2000, 140, p. 208-264.

Cf. n°s 1208, 4763

§ 3. South Asia and Southeast Asia.

8496. Imin kara shimin he: sekai no indo-kei komyunithi. (From migrant to citizen: studies on South Asian communities overseas). Ed. by Masanori KOGA, Masao NAITO and Norio HAMAGUCHI. Tokyo, University of Tokyo, 2000, 328 p.

8497. ITO (Toshikatsu). Shoyû no teitô-baikyaku: Koubaun-chô zenki Biruma ni okeru chû-shô-dajî no jirei. (Mortgaging and selling Myo or Ywa by local Thugyi in Myanmar during the early Koubaung period). *Toyo Gakuho*, 2000, 82, 2, p. 97-129.

8498. KITAGAWA (Takako). Suiô no keifu: Surei-Sânto ôken-si. (History of the Water Kings in Srei Santhor). *Southeast Asian Studies*, 2000, 38, 1, p. 50-73. [Eng. Summary].

8499. MIKI (Sayako). 18 seiki matsu–19 seiki zenhan ni okeru Bengaru no kokumotsu-ryûtsû shisutemu: kokumotsu-kôeki wo meguru indo syônin to Igirisu Higashi Indo Gaisha. (The English East India Company and indigenous trading system: a case study of the grain trade in late eighteenth and early nineteenth century Bengal). *Socio-Economic History*, 2000, 66, 1, p. 67-84.

8500. Nærbilder av India. Samfunn, politikk og utvikling. (Close-ups of India. Society, politics and development). Ed. by Kathinka FRØYSTAD, Eldrid MAGELI and Arild ENGELSEN RUUD. Oslo, Cappelen akademisk forlag, 2000, 353 p. (ill.).

8501. NAMBA (Chizuru). Vishî-ki furansu no teikoku-teki ketsugô-seisaku to Indoshina no "fukken". (The unionist policy of the French colonial empire and "the rehabilitation" of Indo-China in the Vichy period). *Mita Gakkai Zasshi*, 2000, 93, 2, p. 127-148.

8502. TAN (Tai Yong), KUDAISYA (Gyanesh). The aftermath of partition in South Asia. London a. New York, Routledge, 2000, XIV-322 p. (Routledge series in the modern history of Asia, 3).

§ 3. Addenda 1997.

8503. KIKUCHI (Yoko). Takêku no tatakai: Râo-Isara undô. (The battle of Thakhek in 1946: a phase of the Lao-Issara movement). *Historica*, 97, 50, p. 31-48.

§ 4. China.

8504. A (Feng). Ming Qing shiqi Huizhou funü zai tudi mimai zhongde quanli yu diwei. (Rights and status of women in land transactions in Huizhou during the Ming Qing dynasties). *Lishi yanjiu*, 2000, 1, p.73-85.

8505. AO (Wenwei). 1927–1937 nian Zhongguo baoxianye fazhan jiannan zhi yuangyin. (The difficult development of Chinese insurance, 1927–1937). *Minguo dang'an*, 2000, 2, p. 71-76.

8506. BAI (Chun). Taiwan guangfu hou de minzhong xintai yu er erba shijian. (Taiwanese popular feelings after the recovery of the Taiwan and the February 28th incident). *Minguo dang'an*, 2000, 3, p. 102-108.

8507. BIAN (Li). Mingdai Huizhou de minsi jiufen yu minsi susong. (Civil disputes and law suits in Huizhou during the Ming dynasty). *Lishi yanjiu*, 2000, 1, p. 94-105.

8508. BUOYE (Thomas M.). Manslaughter, markets, and moral economy: violence disputes over property rights in eighteenth-century China. New York, Cambridge U. P., 2000, XVI-283 p. (Cambridge studies in Chinese history, literature, and institutions).

8509. CAI (Shuangquan), YANG (Xiulin). Wang Jingwei panguo toudi xinli tanjiu. (Wang Jingwei's psichology and his defection to the enemy). *Minguo dang'an*, 2000, 4, p. 97-101.

8510. CAO (Shuji), LIU (Rentuan). Qingdai qianqi ding de shizhi. (On the nature of ding in early Qing). *Zhongguo shi yanjiu*, 2000, 4, p. 137-147.

8511. Changjiang wenming shi (A cultural history of the Yangze river). Ed. by YE Shuzong. Shanghai, Jiaoyu chubanshe, 2000, 496 p.

8512. CHAO (Xiaohong). Qingdai qian zhongqi Shenxi renkou shuzi pingxi. (An evaluation of population figures in Shenxi during the early and mid Qing). *Qingshi yanjiu*, 2000, 2, p. 55-63.

8513. CHEN (Feng). Qingdai qianqi zouxiao zhidu yu zhengce yanbian. (The system for the submission of expense account during the first half of the Qing dynasty and its evolution). *Lishi yanjiu*, 2000, 2, p. 63-74.

8514. CHEN (Hongmin), LUO (Junqing). Buomin zhengfu yi er liangjie li fayuan zucheng fenxi. (An analysis of the formation of the first and second legislative yuan under the Nationalist government). *Minguo dang'an*, 2000, 2, p. 65-70.

8515. CHEN (Jinjin). Dongbei jun yu zhongyuan dazhan. (The Manchurian army and the China's civil war in 1930). *Jindaishi yanjiu*, 2000, 5, p. 1-34.

8516. CHEN (Mingguang), MAO (Lei). Tang Song yilai de yaren yu tianzhai diandang maimai (Real estates: the business and brokers since the Tang and Song dynasties). *Zhongguo shi yanjiu*, 2000, 4, p. 63-72.

8517. CHEN (Weiming). Ming Qing shiqi Lingnan shaoshu minzu de hunsu wenhua. (Wedding institutions among the ethnic minorities in Lingnan during the Ming and the Qing). *Zhongguo shi yanjiu*, 2000, 4, p. 148-159.

8518. CHENG (Hua). Guanyu Zhong Ri jindai mian fangzhi pin maoyi de kaocha. (An investigation on modern cotton textile trade between China and Japan). *Qingshi yanjiu*, 2000, 2, p. 64-72.

8519. CHENG (Weiming). Ming Qing Yue Min haishang de goucheng yu tedian. (The organization and characteristics of businessmen in foreign trade in Guangdong and Fujian during the Ming and Qing dynasties). *Lishi dang'an*, 2000, 2, p. 80-88.

8520. CHENG (Zhaoyun). Kangzhan chuqi de nanmin qianyi. (Refugees' migration to the interior in early days of the resistance war). *Kang Ri zhanzheng yanjiu*, 2000, 2, p. 79-97.

8521. CHI (Yunfei). Qingji zhuzhang lixian de guanyuan dui xianzheng de tiren. (The concept of constitutional government of officials supporting the constitution at the end of the Qing). *Qingshi yanjiu*, 2000, 1, p. 14-22.

8522. China: empire and civilization. Ed. by Edward L. SHAUGHNESSY. Oxford a. New York, Oxford U. P., 2000, 256 p. (ill., maps).

8523. CHIU-DUKE (Joséphine). To rebuild the empire: Lu Chih's Confucian pragmatism approach to the Mid-Tang predicament. Albany, State University of New York Press, 2000, XIII-311 p.

8524. CHUNG (Dooeum). Élitist fascism: Chiang Kaishek's Blueshirts in 1930s China. Aldershot, Ashgate, 2000, XVI-366 p.

8525. COCHRAN (Sheran). Encountering Chinese networks: Western, Japanese, and Chinese corporations in China, 1880–1937. Berkeley a. Los Angeles, University of California Press, 2000, XII-257 p.

8526. Cong Yapian zhanzheng dao Baguo lianjun 1840–1900. (From the Opium war to the Eight powers allied army 1840–1900). Edited by ZHE Fu. Tianjin, Renmin chubanshe, 2000, 288 p.

8527. DAI (Jianguo). Tang "Kaiyuan ershiwu nian ling – tian ling" yanjiu. (A study onf the laws of the 25th year of the Kaiyuan reign period: Tang dynasty laws on the land). *Lishi yanjiu*, 2000, 2, p. 36-50.

8528. DENG (Ye). Guomindang liujie er zhongquanhui yanjiu. (A study on the second conference of the 6th central committee of the Nationalist Party). *Lishi yanjiu*, 2000, 1, p. 3-20.

8529. DENG (Yibing). Qingdai qianqi shanpin liutong de yundao. (The circulation of commodities in early Qing). *Lishi dang'an*, 2000, 1, p. 99-106.

8530. Dentô chûgoku no chiikizô. (Images of regions in traditional China). Ed. by Eishi YAMAMOTO. Tokyo, University of Keio Press, 2000, 386 p.

8531. DIAMANT (Neil J.). Revolutionizing the family: politics, love, and divorce in urban and rural China, 1949–1968. Berkeley a. Los Angeles, University of California Press, 2000, XVIII-440 p.

8532. DING (Zeqin). Lun Baituan dazhan hou Riben dui Huabei de zhengce. (On the Japanese policy in Northern China after the One-hundred regiment campaigns). *Kang Ri zhanzheng yanjiu*, 2000, 2, p. 1-28.

8533. DU (Xuncheng). Zhongguo jindai liangzhong jinrong zhidu de bijiao. (A comparison of two kinds of financial systems in modern China). *Zhongguo shehui kexue*, 2000, 2, p. 178-190.

8534. DUAN (Tanli). Tangdai funü diwei yanjiu. (Women's position in Tang period). Beijing, Renmin chubanshe, 2000, 316 p.

8535. ELMAN (Benjamin A.). A cultural history of civil examination in late imperial China. Berkeley a. Los Angeles, University of California Press, 2000, XLII-847 p.

8536. EMURA (Haruki). Shunjû sengoku shinkan jidai syutsudo moji shiryô no kenkyû. (Studies on inscribed artifacts unearthed from Spring and Autumn, warring states, Qin and Han period archaeological sites). Tokyo, Kyuko Shoin, 2000, 788 p.

8537. FANG (Biao). Beijing shidafu. (Beijing literati). Beijing, Jinghua chubanshe, 2000, 508 p.

8538. FANG (He). Jiang Jieshi he tade gaoji muliao. (Chiang Kai-shek and his high bureaucrats). Zhengzhou, Henan renmin chubanshe, 2000, 543 p.

8539. FENG (Wei). Riben zai Taiwan tuixing zhimin tongzhi de benzhi tezheng. (The essential features of Japanese colonial rule in Taiwan). *Kang Ri zhanzheng yanjiu*, 2000, 3, p. 1-33.

8540. FOGEL (Joshua A.). The Nanjing massacre in history and historiography. Berkeley a. Los Angeles, University of California Press, 2000, XVI-248 p. (Asia: local studies/global themes, 2).

8541. FORÊT (Philippe). Mapping Chengde: the Qing landscape enterprise. Honolulu, University of Hawai'i Press, 2000, XVIII-209 p.

8542. FOX (Josephine). Common sense in Shanghai: the General Chamber of Commerce and political legitimacy in Republican China. *History workshop*, 2000, 50, p. 22-44.

8543. FUKAZAWA (Hideo). Bojutsu henpô undôshi no kenkyû. (Studies on the history of the reform movement of 1898). Tokyo, Kokusyo Kankôkai, 2000, 710 p.

8544. FUNG (Edmund S. K.). In search of Chinese democracy: civil opposition in Nationalist China, 1929–1949. New York, Cambridge U. P., 2000, XVIII-407 p. (Cambridge Modern China Series).

8545. FURUMATSU (Takashi). Gendai katô enchi shimbyôhi kenkyû josetsu. (Introduction to He-Dong Salt-Lake-God Temple's inscriptions of the Yuan Dynasty). *Toho Gakuho*, 2000, 72, p. 347-380.

8546. FURUTA (Kazuko). Shanhai nettowâku to kindai higashi ajia. (Shanghai network: the economic order in late nineteenth-century East Asia). Tokyo, Tokyo U. P., 2000, 237 p.

8547. Gansu geming wenhua shiliao cuibian. (Selected documents in revolutionary culture in Gansu). Ed. by QIAO Nan. Lanzhou, Gansu wenhua chubanshe, 2000, 510 p.

8548. GAO (Wangling). Qingdai youguan nongmin kangzu de falü he zhengfu zhengming. (Laws and regulations pertaining peasant rent resistance during the Qing period). *Qingshi yanjiu*, 2000, 3, p. 41-49.

8549. GAO (Xiang). Lun Qing qianqi Zhongguo shehui de jindaihua qushi. (Modernizing trends in Chinese society in early Qing). *Zhongguo shehui kexue*, 2000, 4, p. 178-189.

8550. GE (Fuping). Lun Yihetuan yundong shiqi de Faguo dui Hua waijiao. (French diplomatic policy towards China during the Boxer movement). *Jindaishi yanjiu*, 2000, 2, pp.136-149.

8551. GE (Zhaoguang). Luoyang yu Bianliang: wenhua zhongxin yu zhengzhe zhongxin de fenli. (Luoyang and Bianyang: the separation of cultural center from political center). *Lishi yanjiu*, 2000, 5, p. 24-37.

8552. Gender and Manhood in Chinese history. [AHR Forum]. *American historical review*, 2000, 105, 5, p. 1599-1666. [Contents: MANN (Susan). The male bond in Chinese history and culture. – KUTCHER (Norman). The fifth relationship: dangerous friendship in the Confucian context. – DAVIS (Adrian). Fraternity and fratricide in late imperial China. – MAC ISAAC (Lee). "Righteous fraternities" and honorable men: sworn brotherhoods in wartime Chongqing. – NYE (Ronert A.). Kinship, male bonds, and masculinity in comparative perspective].

8553. GOTTSCHANG (Thomas R.), LARY (Diana). Swallows and settlers: the great migration from North China to Manchuria. Ann Arbor, Center for Chinese Studies, University of Michigan, 2000, XVII-231 p.

8554. GREGOR (A. James). A place in the sun: marxism and fascism in China's long revolution, Boulder, Westview, 2000, XV-231 p.

8555. GUAN (Xiaohong). Qingmo zhongyang jiaoyu hui shulun. (A review of late Qing central conference on education). *Jindaishi yanjiu*, 2000, 4, p. 116-140. – IDEM. Zhang Zhidong yu wan Qing xuebu. (Zhang Zhidong and the late Qing board of education). *Lishi yanjiu*, 2000, 3, p. 80-91.

8556. GUO (Chengkang). Ye tan Manzu hanhua. (On Manchu sinification). *Qingshi yanjiu*, 2000, 2, p. 24-35.

8557. GUO (Hongmao). Dongbei lunxian shiqi de Mantie tielu Zhongguo gongren zhuangkuang. (Chinese laborers in South Manchuria Railway Company in occupied Manchuria). *Kang Ri zhanzheng yanjiu*, 2000, 1, p. 75-104.

8558. GUO (Songyi). Qingdai 403 zongmin xing anlie zhongde sitong xingwei kaocha. (An examination of extramarital affairs on the basis of 403 civil and criminal cases in the Qing period). *Lishi yanjiu*, 2000, 3, p. 51-67.

8559. GUO (Weidong). Zhaohui yu Zhongguo waijiao wenshu jindai fanshi de chugou. (The diplomatic note the the initial format of modern diplomatic documents in China). *Lishi yanjiu*, 2000, 3, p. 92-102.

8560. HAN (Dongping). The unknown cultural revolution: educational reforms and their impact on China's rural development. New York a. London, Garland Pub, 2000, XII-195 p. (East Asia).

8561. HAN (Shufeng). Hedong Liu shi zai Nanchao de dute fazhan licheng. (The unique development of Liu clan in Hedong during the Southern Dynasties). *Zhongguo shi yanjiu*, 2000, 1, p. 45-58.

8562. HARRISON (Henrietta). The making of the Republican citizen: political ceremonies and symbols in China, 1911–1929. Oxford, Oxford U. P., 2000, VIII-270 p. (Studies on contemporary China).

8563. HE (Baogang), GUO (Yingjie). Nationalism, national identity and democratization in China. Aldershot a. Brookfield, Ashgate, 2000, XVIII-240 p.

8564. HOU (Jie). Wan Qing shehui wenhua yu minjian shiguan. (Late Qing social culture and historical conceptions). *Qingshi yanjiu*, 2000, 3, p. 63-70.

8565. HOU (Yangfang). Minguo shiqi quanguo renkou tongji shuzi de laiyuan. (Sources on the national population statistics in the republican period). *Lishi yanjiu*, 2000, 4, p. 3-16.

8566. HU (Cheng). Jindai Jiangnan nongcun de gongjia ji qi yingxiang – jianlun xiaomin yu jingyingshi nongchang shuaibai de guanxi. (Wages of rural workers in modern Jiangnan and their impact – on the relationship between small peasants and the decline of agricultural enterprises). *Lishi yanjiu*, 2000, 6, p. 56-71.

8567. HUANG (Juyan). Kangzhan shiqi Guangdong sheng yinggongye de sunshi yu chongjian. (Loss and reconstruction of business and industry in Guangdong province during the resistance war). *Minguo dang'an*, 2000, 2, p. 95-100.

8568. HUANG (Lingjun). 30–40 niandai Zhongguo sixiangjie de "jihua jingji" sichao. (The ideological trend towards a planned economy in Chinese intellectual circles in 1930s and 1940s). *Jindaishi yanjiu*, 2000, 2, pp.150-176.

8569. HUTCHINGS (Graham). Modern China: a companion to a rising power. London, Allen Lane, 2000, XXIII-530 p.

8570. IIJIMA (Wataru). Pesuto to kinadai Chûgoku: Eisei no seidoka to shakai henyô. (The plague and modern China: hygienic institutionalization and social change). Tokyo, Kembun Shuppan, 2000, 380p.

8571. INOUE (Toru). Chûgoku no Sôzoku to kokka no reisei. (Kinship groups and state rituals in China). Tokyo, Kembun Shuppan, 2000, 522p.

8572. ISAEVA (Marina V.). Predstavlenija o mire i gosudarstve v Kitae v III–VI vekakh n.e.: po dannym "normativnykh istoriopisanij". (The concepts of the social world and of the state in China betweeb the 3rd and the 6th centruries A.D.: based on the Zheng sih standard histories). RAN, In-t vostokovedenija. Moskva, [s. n.], 2000, 263 p. [Eng. summary]

8573. IWAI (Shigeki). Shindai no hanto junsôhô to sono shûhen. (The Pantu-Shunzhuang register system and tax collection in the Qing Period). *Toho Gakuho*, 2000, 72, p. 381-450.

8574. IWAMA (Kazuhiro). Chûgoku kyûsai fujukai no katsudô to ronri: Minkokuki shanhai ni okeru minkan jitsugyôka no shakai rinri. (Activity and philosophy of the anti-kidnapping society: a case study of the social ethics of merchants in Republican Shanghai). *Shigaku Zasshi*, 2000, 109, 10, p. 65-90.

8575. JI (Peng). 1927–1935 nian Guomin zhengfu jin yan shuping. (An account of Nationalist government's opium suppression activities, 1927–1935). *Minguo dang'an*, 2000, 1, p. 77-82.

8576. JIA (Erqiang). Lun Tangdai de Huashan xinying. (Worship at Mt. Hua during the Tang). *Zhongguo shi yanjiu*, 2000, 2, p. 90-99.

8577. JIANG (Pei). Huabei kang Ri genjudi de shehui bianqian pingxi. (Society change in Northern China base areas). *Kang Ri zhanzheng yanjiu*, 2000, 2, p. 98-118.

8578. JIANG (Dachun). Sun Zhongshan minsheng shiguan xilun. (On historical perspectives in Sun Zhongshan's people's welfare). *Zhongguo shehui kexue*, 2000, 1, p. 191-204.

8579. JIN (Guangyao). Guomindang zai Meiguo de youshuo huodong - yi Gu Weijun wei zhongxin de taolun. (Nationalist Party lobbying in America – A case-study of Wellington Koo). *Lishi yanjiu*, 2000, 4, p. 61-72.

8580. KE (Dawei), LIU (Zhiwei). Zongzu yu difang shehui de guojia rentong – Ming Qing Huanan diqu zongzu fazhan de yishi xingtai jichu. (Clans and State representation in local society – The ideological foundation of the development of clans in Southern China during the Ming and Qing periods). *Lishi yanjiu*, 2000, 3, p. 3-14.

8581. KIM (Yung Sik). The natural philosophy of Chu Hsi (1130–1200). Philadelphia, American Philosophical Society, 2000, XII-380 p. (Memoirs of the American Philosophical Society Held at Philadelphia for Promoting Useful Knowledge, 235).

8582. KIN (Seikei). Sôdai no seihoku mondai to iminzoku seisaku. (The northwest problems and policies toward barbarians in the Song period). Tokyo, Kyuko Shoin, 2000, 366p.

8583. KONG (Xiangji). Jiawu zhanzheng zhong Beiyang shuishi shangceng renwu de xintai – yingwuchu zongban Luo Fenglu jiashu jiedu. (The mood of the leaders of the Northern Navy during the 1894–1895 war: interpreting the family correspondance of general Luo Fenglu). *Jindaishi yanjiu*, 2000, 6, p. 140-160.

8584. LI (Bozhong). "Xuanjing", "jicui" yu "Songdai Jiangnan nongye geming" – dui chuantong jingji shi yanjiu fangfa de jiantao. (On "picking up the best one" and "the agricultural revolution in Jiangnan during the Song" – a discussion on research methodology about the history of traditional economy). *Zhongguo shehui kexue*, 2000, 1, p. 177-192.

8585. LI (Jinzheng). Lun 1938–1949 nian Huabei kangri genjudi, jiefangqu de nongdai. (Agricultural loans in anti-Japanese bases and liberated areas in Northern China, 1938–1949). *Jindaishi yanjiu*, 2000, 4, p. 178-212.

8586. LI (Yinzi). 19shiji houbanqi Lu xi minjian zongjiao jieshe yu quanhui de dongxiang. (The trends of religious sects and boxers in western Shandong during the second half of the nineteenth century). *Jindaishi yanjiu*, 2000, 6, p. 161-183.

8587. LIANG (Mancang). Han Tang jian zhengzhi yu wenhua tansuo. (Researches on politics and culture from the Han to the Tang). Guiyang, Guizhou, 2000, 418 p.

8588. LIAO (Chi-yang). Nagasaki kashô to higashi ajia kôekimô no keisei. (Chinese business in Nagasaki and the formation of East Asian business networks). Tokyo, Kyuko Shoin, 2000, 428 p.

8589. LIN (Wenhai), ZHAO (Xiaohua). Zhongguo zheyang zoujin ershi shiji – Yihetuan yundong houde

Zhongguo zhengju. (China's entry in the twentieth century Chinese politics after the Boxer movement). *Qingshi yanjiu*, 2000, 4, p. 1-8.

8590. LINDERT (Peter H.). Shifting ground: the changing agricultural soils of China and Indonesia. Cambridge, MIT Press, 2000, 351 p.

8591. LIU (Fusheng). Ru Shu Liao ren de minsu tezheng yu yuyan yicun. (Folklore and linguistics relics of Liao immigrants in Sichuan: a reconsideration). *Zhongguo shi yanjiu*, 2000, 2, p. 49-62.

8592. LIU (Haifeng). Keju zhi de qiyuan yu jinshi ke de qishi. (The origins of the imperial examination systems and the beginning of the examination for selected talents). *Lishi yanjiu*, 2000, 6, p. 3-16. – IDEM. Tangdai *junshi* ke bianxi. (An analysis of the junshi degree examination during the Tang). *Zhongguo shi yanjiu*, 2000, 2, p. 157-167.

8593. LIU (Hong). Xinjiapo Zhonghua zongshanghui yu Yazhou Huashang wangluo de zhiduhua. (The Singapore Chinese chamber of commerci and industry and the institutionalization of Chinese business network in Asia). *Lishi yanjiu*, 2000, 1, p. 106-118.

8594. LIU (Jiafeng). Jidujiao yu jindai nongye keji chuanbo: yi Jinling daxue nonglinke wei zhongxin yanjiu. (Christianity and the spread of modern agronomy: the case of Nanjing university college of agriculture and forestry). *Jindaishi yanjiu*, 2000, 2, p. 177-204.

8595. LIU (Jinzhong), WANG (Shucai). Shilun Feng Yuxiang ji guominjun zai 1925–1927 nian de zhengzhe taidu. (The political attitude of Feng Yuxiang and the National Army). *Lishi yanjiu*, 2000, 5, p. 100-110.

8596. LIU (Pujiang). Liaochao de touxia zhidu yu touxia junzhou. (The touxia system and the touxia junzhou in the Liao dynasty). *Zhongguo shi yanjiu*, 2000, 3, p. 86-101.

8597. LIU (Qiugen). Ming Qing gaoli dai ziben. (High interest loan capitals in Ming and Qing period). Beijing, Kexue wenxian, 2000, 300 p.

8598. LIU (Xiao). Yuandai shouyang zhidu yanjiu. (Adoption during the Yuan dynasty). *Zhongguo shi yanjiu*, 2000, 3, p. 113-124.

8599. LÜ (Xiaobo). Cadres and corruption: the organizational involution of the Chinese Communist Party. Stanford, Stanford U. P., 2000, XVIII-368 p. (Studies of the East Asian Institute, Columbia University).

8600. LU (Yu). Zhongguo renkou tongshi. (A general history of Chinese population). Jinan, Shandong renmin chubanshe, 2000, 1244 p.

8601. LUO (Min). Kangzhan shiqi de Zhongguo Guomindang yu Yuenan duli yundong. (The Nationalist party and the independence movement in Vietnam during the resistance war against Japan). *Kang Ri zhanzheng yanjiu*, 2000, 4, p. 27-53.

8602. LUO (Zhitian). Gujin yu zhongwau de shikong hudong: xin wanhua yundong shiqi guanyu zhengli guogu de sixiang lunzheng. (Space-time interactions between the ancient and modern, Chinese and foreign: intellectual controversies about reorganizing the national heritage). *Jindaishi yanjiu*, 2000, 6, p. 56-106. – IDEM. Zouxiang guoxue yu shixue de "sai xiansheng" – wusi qianhou Zhongguoren xinmu zhongde "kexue" yili. ("Mr. Science" 's turn towards national studies and history: an example of science as seen by Chinese during the May fourth period). *Jindaishi yanjiu*, 2000, 3, p.59-94.

8603. Luoyang wenwu yu kaogu: Han Wei Luoyang gucheng yanjiu. (Relics and archeology in Luoyang: researches on the old city from Han to Wei). Edited by the Committee for Luoyang relics and archeology. Beijing, Kexue chubanshe, 2000, 1016 p.

8604. MA (Chunli). Zhongguo wan Qing wenxue geming shi. (A history of the revolution of literature in late Qing China). Dalian, Liaoning daxue, 2000, 273 p.

8605. MA (Junjie). Lun "Beiyang haijun zhangcheng". (The regulation of the Beiyang Navy). *Lishi dang'an*, 2000, 4, p. 102-110.

8606. MA (Lie). Cong "Jiang Jieshi nianpu chugao" kan Deng Yanda yu Jiang Jieshi de guanxi. (The relationship between Deng Yanda e Jiang Jieshi as seen from "Draft for a Jiang Jieshi biography"). *Minguo dang'an*, 2000, 4, p. 65-72.

8607. MATSUMOTO (Toshiro). Manshûkoku kara shinchûgoku e. (From Manchukuo to Communist China: The industrial reconstruction of northeast China observed through the iron and steel industry in Anshan, 1940–1954). Nagoya, University of Nagoya Press, 2000, 374 p.

8608. MIN (Jie). Lun Qing mo caipiao. (Lotteries in the late Qing). *Jindaishi yanjiu*, 2000, 4, pp.1-52.

8609. Ming Qing liangchao Shenzhen dang'an wenxian yanyi. (Ming and Qing archival documents on Shenzhen). Edited by the Archives of Shenzhen. Shenzhen, Huacheng chubanshe, 2000, 2134 p.

8610. MITTER (Rana). The Manchurian myth: nationalism, resistance, and collaboration in modern China. Berkeley a. Los Angeles, University of California Press, 2000, XI-295 p.

8611. NAKAJIMA (Gakusho). Mindai no soshô seido to rôjinsei: Esso mondai to chôbatsuken wo megutte. (The Community Elder System and administration of justice in the early Ming Dynasty: Yue-su problem and the right of Sanction). *Chugoku*, 2000, vol.15, p. 135-59.

8612. NIIMURA (Yoko). Ahen bôeki ronsô: Igirisu to chûgoku. (Controversy over the opium trade: Britain and China). Tokyo, Kyuko Shoin, 2000, 421 p.

8613. PANTSOV (Alexander). The Bolsheviks and the Chinese Revolution 1919–1927. Honolulu, University of Hawai'i Press, 2000, XII-324 p. (Chinese Worlds).

8614. PENG (Houwen). Shanghai zaoqi de waishang zhengquan shichang. (On foreign businessmen securities market in early years in Shanghai). *Lishi dang'an*, 2000, 3, p. 97-103.

8615. QIAN (Jing). Zhang Xueliang yu Dongbei yizhi xinshi. (New explications about Zhang Xueliang and Manchuria's adherence to Natioanlist government). *Minguo dang'an*, 2000, 4, p. 79-84.

8616. Qingchu wushi Dalai Lama dang'an shiliao xuanbian. (A selection of Qing archival documents on the first five Dalai Lama). Ed. by LI Pengnian. Beijing, Zhongguo Zangxue, 2000, 208 p.

8617. QU (Wenjun). Yuandai de baiguan jiyi. (The practice of officers' meetings and discussions during the Yuan). *Zhongguo shi yanjiu*, 2000, 2, p. 127-134.

8618. QUAN (Renrong). Cong Qimen xian "Xie shi fenzheng" kan Ming mo Huizhou de ditu zhangliang yu lijia zhi. (Land measuring in Huizhou toward the end of the Ming period and the lijia system as seen from the internal disputes of the Xie clan in Qimen county). *Lishi yanjiu*, 2000, 1, pp.86-93.

8619. Reappraising Republican China. Ed. by Frederic E. WAKEMAN and Richard Louis EDMONDS. Oxford, Oxford U. P., 2000, VIII-209 p. (Studies on contemporary China).

8620. REED (Bradly Ward). Talons and teeth: county clerks and runners in the Qing Dynasty. Stanford, Stanford U. P. a. Cambridge, Cambridge U. P., 2000, XXIII-318 p. (Law, society, and culture in China).

8621. REN (Donglai). 1934–1936 nianjian Zhong Mei guanxi zhongde baiyin waijiao. (Success from disaster: the Sino-US silver deal 1934–1936). *Lishi yanjiu*, 2000, 3, p. 103-115.

8622. RHOADS (Edward J. M.). Manchus and Han: ethnic relations and political power in late Qing and early republican China. Seattle, University of Washington Press, 2000, X-394 p. (Studies on ethnic groups in China).

8623. RI (Kaigen). Kan teikoku no seiritsu to ryûhô shûdan. (The establishment of the Hang Empire and the group of Liu Bang: studies on the military meritocracy). Tokyo, Kembun Shuppan, 2000, 220 p.

8624. SAKAKURA (Atsuhide). Min ôchô chûô tôchi kikô no kenkyû. (Studies on the organization of central government of the Ming Dynasty). Tokyo, Kyuko Shoin, 2000, 347p.

8625. SANG (Bing). Xiamen daxue guoxueyuan fengpo. (The break-up of the Institute of Chinese Studies of Xiamen University in 1927). *Jindaishi yanjiu*, 2000, 5, p. 71-93.

8626. SASAKI (Yo). Shimmatsu chûgoku ni okeru nihonkan to seiyôkan. (Chinese views of Japan and the West in the late Qing Period). Tokyo, Tokyo U. P., 2000, 314 p.

8627. SHAO (Dongfang). "Jinben zhisu jinian" Zhou Wu wang, Cheng wang jipu pailie wenti zai fenxi. (Re-examination of the chronicle of King Wu and King Cheng of Zhou in the "Modern text of Bamboo Annals"). *Zhongguo shi yanjiu*, 2000, 1, p. 19-28.

8628. SHAO (Yong). Du Yuesheng yu Shanghai kang Ri qiuwang yundong. (Du Yuesheng and the patriotic movement in Shanghai). *Kang Ri zhanzheng yanjiu*, 2000, 2, p. 118-159.

8629. SHEN (Xiaoyun). Si yier qianhou de Jiang Jieshi yu lieqiang. (On the relations between Chiang Kai-shek and imperialist powers around the time of the 12 April massacre). *Lishi yanjiu*, 2000, 6, p. 96-106.

8630. SHENG (Yichang). Taiping Tianguo sishi nian. (Forty years of the Heavenly Kingdom). Beijing, Xuelin, 2000, 323 p.

8631. Shiji huimou – ershi shiji de Zhongguo. (Looking back at century – China in the twentieth century). Ed. by WANG Hongjiang. Tianjin, Renmin chubanshe, 2000, 370 p.

8632. SMITH (S. A.). A road is made: Communism in Shanghai 1920–1927. Honolulu, University of Hawai'i Press, 2000, X-315 p. (Chinese Worlds).

8633. SOMMER (Matthew H.). Sex, law, and society in late imperial China. Stanford, Stanford U. P., 2000, XVI-413 p. (Law, society, and culture in China).

8634. SONG (Jiong). Liang Song juyang zhidu de fazhan – Songdai guanban cishan shiye chutan. (The juyang system: a tentative discussion on the government-run philantrophy in the Song). *Zhongguo shi yanjiu*, 2000, 4, p. 73-82.

8635. Songdai lishi wenhua yanjiu. (Research on Song history and culture). Ed. by ZHANG Qifan. Beijing, Renmin chubanshe, 2000, 440 p.

8636. SU (Lizhi). Yihetuan yanjiu yibai nian. (One hundred years of research on the Boxers). Jinan, Jilu shushe, 2000, 834 p.

8637. SUN (Longji). Qingjie minzuzhuyi yu Huangdi chongbai de faming. (Late Qing nationalism and the invention of the cult of the Yellow Emperor). *Lishi yanjiu*, 2000, 3, p. 68-79.

8638. Tai Gang Zhonggong dangshi. Zhongguo xiandaishi yanjiu pingxi. (History of the Chinese Communist Party in Taiwan and Hong Kong). Ed. by GUO Tieqiang. Guangzhou, Guangdong renmin chubanshe, 2000, 490 p.

8639. TAKASHIMA (Ko). Shindai no fueki zensho. (The complete books of taxes and labor services in the Qing period). *Toho Gakuho*, 2000, vol.72, p. 451-502.

8640. TANG (Baolin). Chong ping Gongchang Guoji zhidao Zhongguo da geming de luxian. (A re-evaluation of the basic line of the Communist international in directing the revolution in China). *Lishi yanjiu*, 2000, 2, p. 75-96.

8641. TANG (Lixing). Cong beike kan Ming Qing yilai Suzhou shehui de bianqian. (Social change in Suzhou since the Ming Qing dynasties as seen from the table inscriptions). *Lishi yanjiu*, 2000, 1, p. 61-73.

8642. TANG (Xuefeng). Zhongguo kongjun kangzhan shi. (History of Chinese military aircraft during the resistance war). Chengdu, Sichuan daxue chubanshe, 320 p.

8643. TANG (Zhijun). Qiu Shuyuan yu Kang Youwei. (Qiu Shuyuan and Kang Youwei). *Jindaishi yanjiu*, 2000, 3, p. 218-228.

8644. TAYLOR (Jay). The Generalissimo's son: Chiang Ching-Kuo and the revolutions in China and Taiwan. Cambridge, Harvard U. P., 2000, XIV-520 p.

8645. TIAN (Baoguo). 30 niandai de Zhong Su wenhua jiaoliu. (Sino-Soviet cultural exchange in the thirties). *Minguo dang'an*, 2000, 2, p. 83-88.

8646. TIAN (Tao). 19 shiji xiabann qi Zhongguo zhishijie de guojifa guannian. (Chinese intellectuals' perception of international law in the second half of the nineteenth century). *indaishi yanjiu*, 2000, 2, p. 102-135.

8647. TIAN (Xuan). Zhanhou zhonggong "heping, minzhu, tuanjie" zong fangzhen de queding ji qi zhuanbian. (The establishment of the Chinese Communist Party policy of "peace, democracy and unity" and its changes). *Jindaishi yanjiu*, 2000, 4, p. 141-177.

8648. TIAN (Yuqing). Daibei diqu Tuoba yu Wuhuan de gongsheng guanxi – Wei Shu Xu Ji youguan shiliao jiexi. (The coesistence of the Tuoba and the Wuhuan in the Daibei area – an analytical study of the related historical records in Xu Ji of Wei Shu). *Zhongguo shi yanjiu*, 2000, 3, p. 58-85; 4, p. 17-45.

8649. WANG (Chaoguang). 1946 nian zaochun Zhongguo minzhuhua jincheng de dunzuo – yi Zhengxie huiyi ji Guo Gong guanzi wei zhongxin de yanjiu. (The setback of democratizaion in China in early 1946 – the Political Consultive Conference and the relationship between the Nationalist and the Communist parties). *Lishi yanjiu*, 2000, 6, p. 107-119.

8650. WANG (Di). Cong xiangsi diwei bianhua kan Songdai xiangcun guanli tizhi de zhuanbian. (The transformation of the countryside administration system as seen from the changing status of village officers during the Song). *Zhongguo shi yanjiu*, 2000, 1, p. 82-93.

8651. WANG (Fangwen). Jindai Xiamen shehui lüeying. (A sketch of modern Xiamen society). Xiamen, Xiamen Daxue, 2000, 200 p.

8652. WANG (Fan-sen). Fu Ssu-sien: a life in Chinese history and politics. Cambridge, Cambridge U. P., 2000, XVI-261 p. (Cambridge studies in Chinese history, literature and institutions).

8653. WANG (Hui). Xi Zhou chunqiu Wu du qiantu kao. (The movement of the Wu Capital during the Western Zhou and the Spring and Autumn period). *Lishi yanjiu*, 2000, 5, p. 63-78.

8654. WANG (Hui). Zhou chu gaizhi kao. (The institutional reform of early Zhou). *Zhongguo shi yanjiu*, 2000, 2, p. 19-34.

8655. WANG (Lingling). Song zhengfu de kuangchan shoumai cuoshi ji qi xiaoguo. (The purchase of minerals by Song goverments and its consequences). *Zhongguo shi yanjiu*, 2000, 2, p. 100-111.

8656. WANG (Lixi). Zhongguo jindai minzuzhuyi de xingqi yu dizhi Mei huo huodong. (The rise of modern Chinese nationalism and the 1905 boycott against American goods). *Lishi yanjiu*, 2000, 1, pp. 21-33.

8657. WANG (Qifa). Xunzi yu Ru Mo Dao Fa ming zhujia. (Xunzi and other intellectual schools on his time). *Zhongguo shi yanjiu*, 2000, 3, p. 40-57.

8658. WANG (Qingcheng). Taiping Tianguo Youtian wang, Gan wang deng weikan gongci zhong de xin shiliao ji bianzheng. (Unpublished new historical materials in the confessions of the Young Heavenly King, King Gan and others of the Taiping Heavenly Kingdom). *Lishi yanjiu*, 2000, 5, p. 79-99.

8659. WANG (Qisheng). Lun Guomindang gaizu hou de shehui goucheng yu jiceng zuzhi. (The social composition and grassroot organization of the Nationalist party after the reorganization). *Jindaishi yanjiu*, 2000, 2, p. 40-80.

8660. WANG (Yinhuan). Minguo shiqi de renlichefu fenxi. (An analysis of rickshaw men in republican period). *Jindaishi yanjiu*, 2000, 3, pp.193-217.

8661. WANG (Yuesheng). Qingdai zhongqi hunyin xingwei fenxi – lizu yu 1781–1791 nian de kaocha. (Marriage during the mid-Qing dynasty – based on the analysis of 1781–1791 decade). *Lishi yanjiu*, 2000, 6, p. 44-55. – IDEM. Shiba shiji zhonghou de Zhongguo jiating jiegou. (Family structure in mid and late eighteenth century China). *Zhongguo shehui kexue*, 2000, 2, p. 167-177.

8662. WANG (Yunlai). Shilun Jiangsu gaodeng jiaoyu zhutu de jindaihua. (The modernization of higher education in Jiangsu). *Minguo dang'an*, 2000, 1, p. 96-102.

8663. WANG (Yuxin). Yinren baoyu, yong yu ji duiyu wenhua yanjiu de jidian qishi. (Yin people's honoring and using jade, and some ideas on the jade culture). *Zhongguo shi yanjiu*, 2000, 1, p. 3-18.

8664. WANG (Zhen). Lun Riben qin Hua qijian xueruo Zhongguo guoli de jingji zhanlüe. (Japanese economic strategy for weaking Chinese national force during the occupation). *Minguo dang'an*, 2000, 3, p. 73-83.

8665. WU (Jingping). Jiangsu jian Shanghai caizheng weiyuanhui shulun. (The Jiangsu and Shanghai financial committee in 1927). *Jindaishi yanjiu*, 2000, 1, p. 20-49.

8666. WU (Jingping). Yingguo yu Zhongguo de fabi pingzhun qijin. (Great Britain and the stabilization fund of the Chinese national currency). *Lishi yanjiu*, 2000, 1, pp.34-50.

8667. WU (Limei). Shixi Liu Yan licai de gongting beijing – jianlun Tang houqi caizheng shizhi yu chenguang guanxi. (An analysis of the court politics influencing Liu Yan's financial refork – the relationship between financial commissioners and eunuchs in the late Tang). *Zhongguo shi yanjiu*, 2000, 1, p. 68-81.

8668. WU (Yanhong). Mingdai liuxing kao. (The system of banishment in the Ming dynasty). *Lishi yanjiu*, 2000, 6, p. 33-43.

8669. WU (Yixiong). Yi ming zhi zheng yu zaoqi de "Shengjing" zhong yi. (The terminology dispute and the early translations of the Bible into Chinese). *Jindaishi yanjiu*, 2000, 2, pp.205-232.

8670. XIA (Mingfang). Kangzhan shiqi Zhongguo de niehuang yu renkou qianyi. (Famine and migrations in China during the resistance war). *Kang Ri zhanzheng yanjiu*, 2000, 2, p. 59-78.

8671. XIAO (Jianxin). Songchao shenji jiegou de yanbian. (Evolution of the auditing system in the Song dynasty). *Zhongguo shi yanjiu*, 2000, 2, p. 112-126.

8672. XIAO-PLANES (Xiaohong). La construction du politique dans la Chine du début du XXe siècle. L'action des élites locales du Jiangsu. *In*: Catégories sociales de l'espace politique. *Annales*, 2000, 55, 6, p. 1201-1228.

8673. XIE (Xiaopeng). Beiyang zhengfu fei yue waijiao shulun. (On Beiyang goverment's diplomacy on the abrogation of unequal treaties). *Minguo dang' an*, 2000, 2, p. 59-64.

8674. XING (Long). Jindai Huabei nongcun renkou xiaochang ji qi liudong. (Population change and mobility in modern rural Northern China). *Lishi yanjiu*, 2000, 4, p. 17-25.

8675. XIONG (Victor Cunrui). Sui-Tang Chang'an: a study in the urban history of Medieval China. Ann Arbor, Center for Chinese Studies, University of Michigan Press, 2000, XIV-370 p. (Michigan Monographs in Chinese Studies, 85).

8676. XIONG (Yuanbao). Shindai minkoku jiki ni okeru pekin no mizu baibaigyô to suidôro. (The water supply business and 'waterways' in Beijing, 1644–1949). *Shakai Keizai Shigaku*, 2000, 66, 2, p. 47-67.

8677. XU (Hao). Lun Qingdai Huabei nongye de zufang jingying. (The management of extensive agriculture in Northern China during the Qing). *Qingshi yanjiu*, 2000, 1, p. 50-60.

8678. XU (Min). Shilun Qingdai qianqi pu shang huji wenti – jianlun Qingdai shangji (The census register of merchants in the early Qing and the shang ji). *Zhongguo shi yanjiu*, 2000, 3, p. 140-154.

8679. XU (Shiduan). Shizi Yuandai funü zai falü zhong de diwei. (Women's legal status in the Yuan dynasty). *Zhongguo shi yanjiu*, 2000, 4, p. 103-115.

8680. XU (Tan). Ming Qing shiqi chengxiang shichang wangluo tixi de xingcheng ji yiyi. (The formation and significance of urban rural market network in Ming and Qing period). *Zhongguo shehui kexue*, 2000, 3, p. 191-202.

8681. YAMAMOTO (Susumu). Shindai kôki kônan ni okeru zatsuen to zendô. (Miscellaneous contribution and benevolent halls in Jiangnan during the late Qing period). *Shirin*, 2000, vol.83, no.6, p. 126-47.

8682. YAN (Buke). Wei Jin de chaoban, guanpin he weijie. (The official grades and sub-grades in the Wei and Jin dynasties). *Zhongguo shi yanjiu*, 2000, 4, p. 46-62.

8683. YAN (Guangwen). Zhang Juzheng zhi Yue fanglüe pingxi. (A reappraisal of Zhang Juzheng's plan for governing Guangdong). *Lishi dang'an*, 2000, 4, p. 82-87.

8684. YAN (Jun). Hu Shi Qingdai sixiang shi yanjiu qianyi. (An analysis of Hu Shi's study of the intellectual history of the Qing dynasty). *Jindaishi yanjiu*, 2000, 1, pp. 154-174.

8685. YAN (Xiaojun). Liang Han "gushi" lunkao. (The use of "precedents" in the Han dynasty). *Zhongguo shi yanjiu*, 2000, 1, p. 29-36.

8686. YANG (Guisong). Jiang Jieshi kang Ri taidu zhi yanjiu. (Chiang Kai-shek's attitude toward the war of resistance against Japan). *Kang Ri zhanzheng yanjiu*, 2000, 4, p. 54-95.

8687. YANG (Tianhong). Guomindang yu shanhou huiyi guanxi kaoxi. (A study on the relations between the Nationalist party and the Reconstruction conference in 1925). *Jindaishi yanjiu*, 2000, 3, p. 95-116.

8688. YANG (Tianshi). "Yuefa" zhi zheng yu Jiang Jieshi ruanjin Hu Hanmin shijian. (The fight about Constitution and Chiang Kai-shek soft arrest of Hu Hanmin). *Zhongguo shehui kexue*, 2000, 1, p. 193-203.

8689. YANG (Tianshi). Guanyu Sun Zhongshan "san da zhengce" gainian de xingcheng yu tichu. (The formation and presentation of Sun Yat-sen's "Three great policies" concept). *Jindaishi yanjiu*, 2000, 1, p. 1-19.

8690. YANG (Tieshi). Jiang Jieshi yu Hanguo duli yundong. (Chiang Kai-shek and the independence movement in Korea). *Kang Ri zhanzheng yanjiu*, 2000, 4, p. 1-26.

8691. YANG (Xingmei). Cong quandao dao jinfa: Qing jie Sichuan fan zhouzu nuli shulüe. (From persuasion to punishment: an account of the anti footbinding movement in Sichuan during the Qing). *Lishi yanjiu*, 2000, 6, p. 80-95. – IDEM. Guannian yu shehui: nuzi xiaojiao de meichou yu jindai Zhongguo de liangge shijie. (Ideas and society: the aesthetics of bound feet and the two worlds of modern China). *Jindaishi yanjiu*, 2000, 4, p. 53-86.

8692. YANG (Yuqing). Guominjun yu Egong (bu) zhongyang zhengzhiju Zhongguo weiyuanhui. (The Nationalist army and the China committee of the Central Politburo of the Russian Communist Party). *Jindaishi yanjiu*, 2000, 3, p. 117-133.

8693. YE (Hanming). Mingdai zhonghouqi Lingnan de difang shehui yu jiazu wenhua. (Local society and clan culture in Lingnan during the mid and late Ming). *Lishi yanjiu*, 2000, 3, pp.15-30.

8694. YEH (Wen-hsin). Becoming Chinese: passages to modernity and beyond. Berkeley a. Los Angeles, University of California Press, 2000, X-435 p. (Studies on China, 23).

8695. YIN (Weiguang). Mingdai Zangzu shi yanjiu. (History of the Tibetans in the Ming period). Beijing, Minzu chubanshe, 2000, 250 p.

8696. YU (Dahua). Dong Zhi dufu yu Yihetuan yundong de xingqi. (The governors of Zhili and Shandong and the rise of the Boxer movement). *Qingshi yanjiu*, 2000, 4, p. 18-24.

8697. YU (Dahua). Wan Qing wenhua baoshou sichao yanjiu. (Cultural conservative thinking in late Qing). Beijing, Renmin chubanshe, 2000, 295 p.

8698. YU (Guilin). Songdai maimai funü xianxiang chutan. (A preliminary enquiry on the commercial transactions about women in the Song dynasty). *Zhongguo shi yanjiu*, 2000, 3, p. 102-112.

8699. YU (Heping). Qusi yundong yu shangren waijiao. (Merchant diplomacy and the May fourth movement). *Jindaishi yanjiu*, 2000, 2, pp. 81-102.

8700. YU (Kunqi). Yinwan Han mu jiandu yu Xi Han guanzhi tanxi. (Han bamboo scriptures at Yinwan and their implication of the official institution of the Western Han). *Zhongguo shi yanjiu*, 2000, 2, pp. 35-48.

8701. YU (Li), TIAN (Ziyu). Chen Cheng yu Hubei kangzhan. (Chen Cheng and the resistance war in Hubei). *Kang Ri zhanzheng yanjiu*, 2000, 3, p. 129-168.

8702. YUAN (Chengyi). Zhongguo jindai dui Ri zhanzheng peikuan shulun. (On Chinese modern war indemnities to Japan). *Lishi dang'an*, 2000, 1, p. 106-113.

8703. ZANG (Rong). Sui Tang Wudai shilun. (Essays on the Sui, Tang and the Five dynasties). Shijiazhuang, Hebei jiaoyu chubanshe, 2000, 405 p.

8704. ZENG (Yeying). Jiang Jieshi 1929 nian tao Gui zhanzheng zhong de junshi moulüe. (Chiang Kaishek's military strategy in the war against the Gui clique in 1929). *Jindaishi yanjiu*, 2000, 2, p. 1-39.

8705. ZHANG (Yongjiang). Lun Qingdai Xizang xingzheng tizhi de yanbian ji qi tedian. (The development and characteristics of Qing administration of Tibet). *Qingshi yanjiu*, 2000, 3, p. 31-43.

8706. ZHANG (Baoming). Chen Duxiu de zuihou shinian. (The last ten years of Chen Duxiu). Zhengzhou, Henan renmin chubanshe, 2000, 375 p.

8707. ZHANG (Fuyun). 1927–1937 nian Nanjing qianzhuangye de xing shuai. (The rise and fall of private banks in Nanjing 1927–1937). *Minguo dang'an*, 2000, 1, p. 82-88.

8708. ZHANG (Gefu). Kangzhan shiqi Zhejiang sheng de renkou qianyi yu diyu fenbu. (Population migration and distribution in Zhejiang during the Sino-Japanese War). *Lishi yanjiu*, 2000, 4, p. 26-38.

8709. ZHANG (Hao). 1932–1933 nian Luo Wengan chuzhang waijiao yu Guomin zhengfu dui Ri zhengce. (Luo Wengan and the Nationalist government's diplomatic policy to Japan in 1932–1933). *Kang Ri zhanzheng yanjiu*, 2000, 3, p. 34-57.

8710. ZHANG (Kaiyuan). Zhongguo jindaishi shangde guan shen shangxue. (The commercial expertise of the officials and the gentry in modern China). Wuhan, Hubei renmin chubanshe, 2000, 852 p.

8711. ZHANG (Lei). Sun Zhongshan zhuan. (A biography of Sun Yat-sen). Guangzhou, Guangzhou chubanshe, 2000, 500 p.

8712. ZHANG (Lianhong). Nanjing guomin zhengfu fabi zhengce de shishi yu ge sheng difang zhengfu de fanying. (The implementation of national currency policy under the Nanjing Nationalist government and its repercussions for provincial governments). *Minguo dang'an*, 2000, 2, p. 76-82.

8713. ZHANG (Shengzhen). 1931–1937 nianjian Mantie tujian gongcheng zhong shiying Huabei minfu de jiben zhuangkuang. (The state of Northern China laborers used by South Manchuria Railway Company in civil engineering works 1931–1937). *Kang Ri zhanzheng yanjiu*, 2000, 1, p. 52-74.

8714. ZHANG (Xianwen). Minguo lishi dang'an de yanjiu yu yongli. (Research and use of historical archives of the Republican period). *Minguo dang'an*, 2000, 1, p. 63-70.

8715. ZHANG (Xiaoye). Qingdai yanzheng zhongde jisi wenti. (The problem of anti-smuggling in Qing salt admininstration). *Qingshi yanjiu*, 2000, 1, p. 32-41.

8716. ZHANG (Xin). Social transformation in modern China: the state and local elites in Henan, 1900–1937. Cambridge, Cambridge U. P., 2000, XVI-320 p. (ill., maps). (Cambridge modern China series).

8717. ZHANG (Yan). Bao Shichen yu jindai qianye de "haiyun nancao" gaige. (Bao Shichen and the reform of the transportation of tribute in pre-modern times). *Jindaishi yanjiu*, 2000, 1, pp.154-174.

8718. ZHANG (Yan). Qingdai shehui de man bianliang – cong Qingdai jiceng shehui zuzhi kan Zhong guo fengjian shehui jiegou yu jingji jiegou de yanbian qishi. (Slow variables in Qing society – Changing trends in social and economic structures of feudal China as seen from the basic social organization in Qing period). Taiyuan, Shanxi remin chubanshe, 2000, 393 p.

8719. ZHANG (Yan). Qingdai zhonghouqi Zhongguo jiceng shehui zuzhi de congheng yilai yu xianghu lianxi. (Mutual communication and interdependency among basic social organization in China during the early and mid Qing). *Qingshi yanjiu*, 2000, 2, p. 79-91.

8720. ZHANG (Zhuangqiang). Guangxi jindai yuan Yue kang Fa zhanzheng. (Guangxi support of Vietnam in the war of resistance against France). Xiamen, Xiamen daxue chubanshe, 289 p.

8721. ZHENG (Genan). Mao Zedong zuji ji jiashi kaoshu. (An investigation on Mao Zedong's ancestral home and family). Beijing, Zhongyang wenxian, 2000, 210 p.

8722. ZHENG (Qin). Kangxi xianxing zelie kao. (An analysis of regulations during the reign of Kangxi). *Lishi dang'an*, 2000, 3, p. 87-92.

8723. Zhongguo fantan shi. (A history of fight against corruption in China). Ed. by WANG Chunyu. Chengdu, Sichuan daxue, 2000, 1254 p.

8724. Zhongguo gaige tongshi. (A history of reforms in China). Ed. by QI Xia. Shijiazhuang, Hebei jiaoyu chubanshe, 2000, 7160 p.

8725. Zhongguo Gongchandang lishi ziliao congshu: Zhonggong Zhongyang beifangju. Beifang quwei shiqi juan. (Historical documents of the Chinese Communist Party: the Northern department of Central Committee. The period of the Northern area commission). Edited by the Committee for the historical documents on the Chinese Communist Party). Beijing, Zhonggong dangshi chubanshe, 2000, 675 p.

8726. Zhongguo Gongchandang Ningxia shi. (History of the Chinese Communist Party in Ningxia province). Edited by the Committee for the history of the Chinese Communist Party in Ningxia province. Yinchuan, Ningxia renmin chubanshe, 2000, 364 p.

8727. Zhongguo jin xiandai jishu shi. (The history of technology in modern and contemporary China). Ed. by WU Xijin. Beijing, Kexue chubanshe, 2000, 1512 p.

8728. Zhongguo kexue jishu shi. Jixie juan. (History of science and technology in China. Mechanics). Ed. by LU Jinyan. Beijing, Kexue chubanshe, 2000, 436 p.

8729. ZHOU (Qiuguang). Wan Qing shiqi de Zhongguo hongshizi hui shulun. (The Chinese Red Cross in late Qing). *Jindaishi yanjiu*, 2000, 3, p. 134-192.

8730. ZHOU (Qiuwang). Minguo Beijing zhengfu shiqi Zhongguo hongshizi hui de cishan jiuhu yu zhenji huodong. (The philantropic and relief activities of the Chinese Red Cross during the Republican Beijing government). *Jindaishi yanjiu*, 2000, 6, p. 107-139.

8731. ZHOU (Shaoquan). Huizhou wenshu yu Hui xue. (The Huizhou documents and the studies of Huizhou). *Lishi yanjiu*, 2000, 1, p. 51-60.

8732. ZHOU (Yumin). Jihai jianchu yu Yihetuan yundong. (The attempt to establish a crown prince in 1899 and the Boxer movement). *Qingshi yanjiu*, 2000, 4, p. 8-17.

8733. ZHU (Qingbao), CAO (Dachen). Riben dui Hainandao diaocha zhi pingshu. (Japan's survey of Hainandao). *Minguo dang'an*, 2000, 3, p. 65-72.

8734. ZHU (Ying). Jindai Zhongguo guanggao de chansheng fazhan ji qi yingxiang. (The emergence, development and influence of advertising in modern China). *Jindaishi yanjiu*, 2000, 4, p. 87-115.

8735. ZHU (Zongchen). Jiangsu dudu Cheng Dequan anfu huidang zhengce de shibai. (The failure of Jiangsu provincial military governor Cheng Dequan's policy of pacifying secret societies). *Minguo dang'an*, 2000, 1, p. 71-76.

8736. ZHUANG (Weimin). Jindai Shandong xingzhan ziben de fazhan yu qi yingxiang. (The development and influence of brokers' storehouses in modern Shandong). *Jindaishi yanjiu*, 2000, 5, p. 35-70.

8737. ZOU (Xiaozhan). Zhang Shizhao "Jiayin" shiqi ziyouzhuyi zhengzhi sixiang pingxi. (Zhang Shizhao's political liberalism during the period of the "Jiayin" monthly). *Jindaishi yanjiu*, 2000, 1, pp. 68-126.

8738. ZOU (Zhenhuan). Wan Qing xifang dilixue zai Zhongguo – yi 1815 dao 1911 nian xifang dilixue yizhu de chuanbo yu yingxiang wei zhongxin. (Western geography in late Qing China – the diffusion and impact of western geography translations from 1815 to 1911). Shanghai, Guji chubanshe, 2000, 445 p.

Cf. n°s 1243, 7092, 8130

§ 5. Japan (before 1868).

8739. 17 Seiki no Nihon to Higashi-Ajia. (Japan and East Asia in the seventeenth century). Ed. by Satoru FUJITA. Tokyo, Yamakawa Shuppansha, 2000, 228 p.

8740. ADOLPHSON (Mikael S.). The gates of power: monks, courtiers, and warriors in premodern Japan. Honolulu, University of Hawai'i Press, 2000, XVII-456 p.

8741. AOYAMA (Tadamasa). Meiji-Ishin to kokkakeisei. (The Meiji Restoration and Japanese state formation). Tokyo, Yoshikawa Kobunkan, 2000, 311 p.

8742. EBISAWA (Tadashi). Shôen-kôryô-sei to chûsei-sonraku. (Proprietary land institutions and the village in medieval Japan). Tokyo, Azekura Shobo, 2000, 544 p.

8743. Geinô no chûsei. (Performing arts in medieval Japan). Ed. by Fumihiko GOMI. Tokyo, Yoshikawa Kobunkan, 2000, 316 p.

8744. GOMI (Fumihiko). Meigetsuki no shiryô-gaku. (Historical analysis of the Meigetsuki Diary). Tokyo, Seishi shuppan, 2000, 348 p.

8745. HAYAKAWA (Shohachi). Nihon kodai no zaisei seido. (The financial system in ancient Japan). Tokyo, Meicho kanko kai, 2000, 335 p. – IDEM. Tennou to kodai kokka. (Emperors and state in ancient Japan). Tokyo, Kodansha, 2000, 301 p.

8746. HAYASHI (Reiko). Kinsei no shijô-kôzô to ryûtsû (Market-structure and distribution in early modern Japan). Tokyo, Yoshikawa Kobunkan, 2000, 328 p.

8747. HIRAKAWA (Minami). Bokusho doki no kenkyû. (A study on the earthenware with notes). Tokyo, Yoshikawa Kobunkan, 2000, 552 p.

8748. HOJO (Hideki). Nihon kodai kokka no chiiki shihai. (The regional governance by the ancient Japanese state). Tokyo, Yoshikawa Kobunkan, 2000, 307p.

8749. HOSOKAWA (Shigeo). Kamakura-seiken Tokusô-sensei-ron. (The Kamakura-Bakufu and Tokusô dictatorship). Tokyo, Yoshikawa Kobunkan, 2000, 557 p.

8750. IIDA (Mizuho). IIDA Mizuho chosaku shu. 1. Shôtoku-Taishi no kenkyu. 2-4. Kodai shiseki no kenkyu. (The collected works of Iida Mizuho. Vol. 1. A study on Shôtoku-Taishi. Vol. 2-4. A study on ancient history books). Tokyo, Yoshikawa Kobunkan, 2000, 373 p.

8751. IMATANI (Akira). Muromachi-jidai seiji-shi-ron. (Political history of the muromachi period in Japan). Tokyo, Hanawa Shobo, 2000, 378 p.

8752. ITO (Kiyoo). Chûsei-nihon no kokka to jiin. (The state and temples in medieval Japan). Tokyo, Koshi Shoin, 2000, 400 p.

8753. KATSUURA (Reiko). Nihon kodai no souni to shakai. (The position of monks and nuns in the ancient Japanese society). Tokyo, Yoshikawa Kobunkan, 2000, 440 p.

8754. KEIICHI (Eguchi). 'Concentric Circle', 'Ring' and 'Radiation' structures in the Japanese colonial empire. *Journal of Japanese history*, 2000, 453, p. 85.

8755. KITO (Kiyoaki). Kodai mokkan to tojô no kenkyû. (A study on wooden tablets and capitals in ancient Japan). Tokyo, Haniwa shobo, 2000, 410 p.

8756. KOBAYASHI (Shoji). Nihon kodai no sonraku to noumin shihai. (Villages and the governance over peasantry in ancient Japan). Tokyo, Haniwa shobo, 2000, 468 p.

8757. KURODA (Hideo). Chûsei-shôen-ezu no kaishaku-gaku. (Iconology of estate maps in medieval Japan). Tokyo, University of Tokyo Press, 2000, 423 p.

8758. Kyoto Reizencho monjo bekkan: Kaidai, shiryô-mokuroku, sankô-shiryô (Documents of Kyoto Reizencho: Bibliographical introduction, catalogue and reference materials). Ed. Kyoto Reizencho Monjo Kenkyukai. Kyoto, Shibunkaku Shuppan, 2000, 466 p.

8759. MIZUNO (Shoji). Nihon-chûsei no sonraku to shôen-sei. (The medieval village and the manor system in Japan). Tokyo, Azekura Shobo, 2000, 510 p.

8760. MOMOSE (Kesao). Kôan-shosatsu-rei no kenkyû. (Court-prescribed protocols for letters in medieval Japan). Tokyo, University of Tokyo Press, 2000, 338 p.

8761. MORI (Kimiyuki). Kodai Gunji seido no kenkyû. (A study on the system of Country Local Officers in ancient Japan). Tokyo, Yoshikawa Kobunkan, 2000, 379 p. – IDEM. Nagaya ôke mokkan no kisoteki kenkyû. (The fundamental study on wooden tablets from Prince Nagaya's palace). Tokyo, Yoshikawa Kobunkan, 2000, 388 p.

8762. NAGAMURA (Makoto). Chûsei-jiin shiryô-ron. (Historical sources of medieval temples and monasteries). Tokyo, Yoshikawa Kobunkan, 2000, 414 p.

8763. NAKAMURA (Tadashi). Kinsei taigai-kôshô-shi ron. (History of diplomatic negotiations in early modern Japan). Tokyo, Yoshikawa Kobunkan, 2000, 255 p.

8764. Nihon kinsei shiryô-gaku kenkyû. (Studies of historical documents in early modern Japan). Ed. by Shunsuke TAKAGI and Kouichi WATANABE. Hokkaido, Hokkaido U. P., 2000, 546 p.

8765. Oke to taru: Wakiyaku no nihonshi. (Japanese history of goods: The case of tubs and barrels). Ed. by Kazuko KOIZUMI. Tokyo, Hosei U. P., 2000, 459 p.

8766. SAKAEHARA (Towao). Nara jidai no shakyô to dairi. (Transcriptions of sutras and the Imperial Court in Nara period Japan). Tokyo, Haniwa shobo, 2000, 428 p.

8767. SATO (Kenji). Chûsei kenmon no seiritsu to kasei. (The formation of the medieval aristocracy and its household). Tokyo, Yoshikawa Kobunkan, 2000, 318 p.

8768. SHIRANE (Yasuhiro). Chûsei no ôchoô-shakai to insei. (Medieval court society and government by the retired emperor). Tokyo, Yoshikawa Kobunkan, 2000, 294 p.

8769. Shirîzu kinsei no mibun-teki shûen. 1. Minkan ni ikiru shûkyôsha. (Marginal social status in early modern Japan. Vol. 1. Man of religion in society). Ed. by Toshihiko TAKANO. 2. Geinô, bunka no sekai. (Vol. 2. Entertainment and culture). Ed. by Fuyuhiko YOKOTA. 3. Shokunin, oyakata, nakama. (Vol. 3. Craftsman, master and fellowship). Ed. by Takashi TSUKADA. 4. Akinai no ba to shakai. (Vol. 4. Market and society). Ed. by Nobuyuki YOSHIDA. 5. Shihai wo sasaeru hitobito. (Vol. 5. People who sustained the regime). Ed. by Hiroshi KURUSHIMA. 6. Mibun wo toinaosu. (Vol. 6. Reconsidering the social status). Ed. by Hiroshi KURUSHIMA, Toshihiko TAKANO, Takashi TSUKADA, Fuyuhiko YOKOTA and Nobuyuki YOSHIDA. Tokyo, Yoshikawa Kobunkan, 2000, 6 vol., 272 p., 312 p., 268 p., 277 p., 258 p., 216 p.

8770. SONE (Masato). Kodai bukkyô kai to ôchô shakai. (The Buddhistic circle and society under the Imperial dynasty in ancient Japan). Tokyo, Yoshikawa Kobunkan, 2000, 304 p.

8771. TAKAHASHI (H.). Secret imperial ordinance of overthrowing the Shogunate and the imperial message of putting it off. *Journal of Japanese history*, 2000, 457, p. 28-42.

8772. TAKAHASHI (Hidenao). Tôbaku no micchoku to miawase satasyo, (Secret Imperial Ordinance to overthrow the Shogunate and the Imperial Message of putting it off). *Nihonshi Kenkyu*, 2000, 457, p. 28-42.

8773. TAKAHASHI (Osamu). Chûsei-bushidan to chi-iki-shakai. (Japanese medieval warrior bands and regional society). Osaka, Seibundo Shuppan, 2000, 329 p.

8774. TAKAMAKI (Minoru). Kinsei no toshi to sairei (The city and festival in early modern Japan). Tokyo, Yoshikawa Kobunkan, 2000, 404 p.

8775. TAMAI (Chikara). Heian jidai no kizoku to tennou. (The aristocracy and Emperors in Heian Japan). Tokyo, Iwanami shoten, 2000, 444 p.

8776. TORAO (Toshiya). Engi-shiki, jô: Yakuchû nihon shiryou. (Engi-shiki, the First volume: translation and annotation for Japanese historical materials). Tokyo, Shueisha, 2000, 1139 p.

8777. TSUKADA (Takashi). Mibun-ron kara rekishi-gaku wo kangaeru (A historical study viewed from social status). Tokyo, Azekura Shobo, 2000, 324 p.

8778. YAMADA (Yasuhiro). Sengoku-ki Muromachi-bakufu to shôgun. (Muromachi-bakufu and shôgun in the Japanese civil war). Tokyo, Yoshikawa Kobunkan, 2000, 271 p.

8779. YOSHIDA (Nobuyuki). Kyodai jôkamachi Edo no Bunsetsu-kôzô (Edo as the great castle town: an analysis). Tokyo, Yamakawa Shuppansha, 2000, 390 p.

8780. YOSHIE (Akiko). Nihon kodai keihu youshiki ron. (A study on the styles of genealogies). Tokyo, Yoshikawa Kobunkan, 2000, 288 p.

§ 5. Addenda 1999.

8781. IKEGAMI (Hiroko). Sengoku-jidai shakai-kôzô no kenkyû. (Studies on the social structure in the Japanese civil war). Tokyo, Azekura Shobo, 99, 626 p.

§ 6. Korea.

8782. CHO (Wan-je). A cultural history of modern Korea: a history of Korean civilization. Ed. with an intr. by Hongkyu A. CHOE. Elizabeth, Hollym, 2000, XXVIII-876 p. (ill., maps, ports.).

8783. LIM (Jie-Hyun). Domination and hegemony in the nationalist discourse of the Korean peninsular. *Contemporary Criticism*, 2000, 10, [s. p.]. – IDEM. Junjigoojuk Gundaesunggwa Minjokjooui. (The global modernity and nationalism). *Critical Studies on Modern Korean History*, 2000, 4, [s. p.]. – IDEM. Minjokdamronui Spektrum. (Spectrums of nationalist discourse). *In/Outside: English Studies in Korea*, 2000, 8, [s. p.].

8784. SIN (Yong-ha). Modern Korean history and nationalism. Seoul, Jimoondang, 2000, VIII-293 p. (Korean studies series, 16).

8785. Sources of Korean tradition. Vol. 2: from the sixteenth to the twentieth centuries. Ed. by Yong-ho CH'OE, Peter H. LEE and Wm. Theodore DE BARY. New York a. Chichester, Columbia U. P., 2000, XX-487 p. (Introduction to Asian civilizations).

Cf. n° 8458

S

AFRICA
(esp. to its colonization)

---

* 8786. SPEAR (T.). Swahili history and society to 1900: A classified bibliography. *History in Africa*, 2000, 27, p. 339-373.

8787. African (The) origin of civilisation and the destiny of Africa: proceedings of a conference held in Windhoek, 24–25 May 1999. Ed. by N. ANGULA and B. F. BANKIE. Windhoek, Gamsberg Macmillan, 2000, VII-208 p. (ill.).

8788. ALEXANDER (Jocelyn), MAC GREGOR (JoAnn), RANGER (Terence). Violence and memory: one hundred years in the "Dark Forests" of Metabeleland. Portsmouth, Heinemann, 2000, XIV-291 p. (Social history of Africa).

8789. ALLMAN (Jean), TASHJIAN (Victoria). "I will not eat stone": a women's history of colonial Asante. Portsmouth, Heinemann, 2000, XLVI-255 p. (Social history of Africa).

8790. BLYDEN (Nemata Amelia). West Indians in West Africa, 1808–1880: the African diaspora in reverse. Rochester, University of Rochester Press, 2000, XI-258 p. (Rochester studies in African history and the diaspra, 8).

8791. CHRÉTIEN (Jean-Pierre). L'Afrique de Grands Lacs. Deux mille ans d'histoire. Paris, Aubier, 2000, 411 p. (Collection historique).

8792. Coptica Hermitagiana: Sbornik materialov k 100-letiju koptskoj kollektsii Ermitazha. (Articles and materials to the 100th anniversary of the Coptic Collection of the Hermitage, [Saint-Petersburg]). Ed. Aleksandr Ja. KAKOVKIN. Gosudarstvennyj Ermitazh. Sankt-Peterburg, Izd-vo Gos. Ermitazha, 2000, 183 p. (ill.). [Eng. summary]

8793. CRUMMEY (Donald). Land and society in the Christian kingdom of Ethiopia: from the thirteenth to the twentieth century. Urbana a. Chicago, University of Illinois Press, 2000, XV-373 p.

8794. DE HESSE (Paul). Nouvelle Jérusalem de Lalibela – Part 1. Une Maison de Dieu. Recherche sur l'axiologie des sanctuaires médiévaux monolithes de Roha, Éthiopie. [S. l.], Paul de Hesse, 2000, 269 p. (ill.).

8795. FABIAN (Johannes). Out of our minds: reason and madness in the exploration of Central Africa. Berkeley a. Los Angeles, University of California Press, 2000, XV-230 p. (Ad. E. Jensen lectures at the Frobenius-Institut University of Frankfurt).

8796. GLASER (Clive). Bo-Tsotsi: the youth gangs of Soweto, 1935–1976. Portsmouth, Heinemann, 2000, XVI-214 p. (Social history of Africa).

8797. History (The) of Islam in Africa. Ed. by Nehemia LEVTZION and Randall L. POUWELS. Athens, Ohio U. P., 2000, X-591 p.

8798. LENTZ (C.). Of hunters, goats and earth-shrines: settlement histories and the politics of oral tradition in northern Ghana. *History in Africa*, 2000, 27, p. 193-214.

8799. MAC CASKIE (T. C.). Asante identities: history and modernity in an African village 1850–1950. Edinburgh, Edinburgh U. P., 2000, IX-277 p. (International African library, 25).

8800. MASONEN (Pekka). The negroland revisited: discovery and invention of the Sudanese Middle Ages. Helsinki, Finnish Academy of Science and Letters, 2000, 599 p.

8801. ODUWOBI (T.). Oral historical traditions and political integration in Ijebu. *History in Africa*, 2000, 27, p. 249-259.

8802. PARKER (John). Making the town: Gã state and society in early colonial Accra. Portsmouth, Heinemann a. Oxford, James Currey a. Cape Town, David Philip, 2000, XXXIII-264 p. (Social history of Africa series).

8803. PEEL (J. D. Y.). Religious encounter and the making of the Yoruba. Bloomington, Indiana U. P., 2000, XI-420 p. (African systems of thought).

8804. ROBERTSHAW (P.). Sibling rivalry? The intersection of archeology and history. *History in Africa*, 2000, 27, p. 261-286.

8805. ROBERTSON (J. H.), BRADLEY (R.). A new paradigm: the African early Iron Age without Bantu migrations. *History in Africa*, 2000, 27, p. 287-323.

8806. SPAULDING (J.). The chronology of Sudanese Arabic genealogical tradition. *History in Africa*, 2000, 27, p. 325-337.

8807. SPEAR (Thomas). Early Swahili history reconsidered. *International journal of African historical studies*, 2000, 33, 2, p. 257-290.

8808. TAKAKI (Keiko). Kita-afurika no isurâmu seija-shinkou: Chunijia·Sedada-mura no rekishi-minzokushi. (The faith in the Islamic Saints of North-Africa: the historical ethnography at Sedada in Tunisia). Tokyo, Tosui Shobou, 2000, 414 p.

8809. Traditsionnye kul'tury afrikanskikh narodov: Proshloe i nastojashchee. (Traditional cultures of African peoples: past and present). RAN, In-t Afriki. Ed. Roza N. ISMAGILOVA. Moskva, Vostochnaja literatura, 2000, 271 p.

Cf. n°s 1025, 1524, 7684-7727

T

AMERICA
(esp. to its colonization)

8810. BOONE (Elizabeth Hill). Stories in red and black: pictorial histories of the Aztecs and Mixtecs. Austin, University of Texas Press, 2000, XIV-296 p. (ill.).

8811. GARFIELD (Seth). Where the earth touches the sky: the Xavante Indians' struggle for land in Brazil, 1951–1979. *Hispanic American historical review*, 2000, 80, 3, p. 537-564.

8812. Indian (The) in Latin American history: resistance, resilience, and acculturation. Ed. by John E. KICZA. Wilmington, SR Books, 2000, XXVIII-296 p. (ill.).

8813. Indians and the greater Southeast: historical archaeology and ethnohistory. Ed. by Bonnie G. MAC EWAN. Gainesville, University Press of Florida, 2000, XVI-336 p.

8814. JULIEN (Catherine J.). Reading Inca history. Iowa City, University of Iowa Press, 2000, XI-338 p.

8815. MACERA DALL'ORSO (Pablo), FORNS (Santiago). Nueva crónica del Perú, siglo XX. Lima, Fondo Editorial del Congreso del Perú, 2000, 591 p.

8816. PIEPER (Renate). Die Vermittlung einer neuen Welt. Amerika im Nachrichtennetz des Habsburgischen Imperiums 1493–1598. Mainz, von Zabern, 2000, XI-354 p. (Veröffentlichungen des Instituts für europäische Geschichte Mainz, Abt. für Universalgeschichte, 163).

8817. Possible pasts: becoming colonial in early America. Ed. by Robert BLAIR ST. GEORGE. Ithaca, Cornell U. P. a. Philadelphia, McNeil Center for early American studies a. Williamsburg, Omohundro Institute of early American history and culture, 2000, XII-417 p.

8818. Razvitie tsivilizatsii v Novom Svete: Sb. st. po materialam [Pervykh] Knorozovskikh chtenij (20–21 oktjabrja 1999 g.). (The development of civilizations in the New World: Proceedings of the [First] Conference dedicated to the memory of Jurij V. Knorozov). Ed. Aleksandr P. LOGUNOV, Dmitrij D. BELJAEV [et al.]. Ros. gos. gumanit. un-t. Moskva, [s. n.], 2000, 190 p. (ill.). [Eng. summaries]

8819. STAVIG (Ward). Ambiguous visions: nature, law, and culture in indigenous-Spanish land relations in colonial Peru. *Hispanic American historical review*, 2000, 80, 1, p. 77-112.

8820. VALLE (Perla). Ordeñanza del señor Cuauhtémoc. Paleografía y traducción del náhuati, Rafael TENA. Ciudad de México, MAG Ediciones y Gobierno del Distrito Federal, 2000, 168 p. (ill.).

Cf. n°s 424, 969, 1025, 7728-7755

U

OCEANIA
(esp. to its colonization)

* 8821 ANGLEVIEL (F.). Bibliographie. Question identitaire et identités océaniennes. *Nouvelle Revue du Pacifique*, 2000, 1, 1, p. 210-224. – IDEM. Bibliographie des principaux ouvrages d'histoire religieuse consacrés récemment à l'Océanie, 1975-1980. *In*: Religion et sacré en Océanie [Cf. n° 8826], p. 11-24; p. 121-146; p. 293-300.

** 8822. ZARAGOZA (Justo). Historia del descubrimiento de las regiones austriales: hecho por el general Pedro Fernández de Quirós: el Pacífico hispano y la búsqueda de la "Terra Australis". Madrid, Dove, 2000, 1126 p. (ill., maps). (Colección Mundus novus, 7).

8823. ANGLEVIEL (F.). Les écrits historiques calédoniens au coeur du débat identitaire. Approche multiculturelle ou discours contradictoires? *In*: L'extraordinaire et le quotidien. Variations anthropologiques. Hommages au professeur Pierre Vérin. Paris, Karthala, 2000, p. 203-216.

8824. ARKLEY (Lindsey). The hated protector: the story of Charles Wightman Sievwright, protector of Aborigines, 1839–1842. Mentone, Orbit Press, 2000, X-509 p. (ill.).

8825. DENOON (Donald), MEIN-SMITH (Philippa), WYNDHAM (Marivic). A history of Australia, New Zealand, and the Pacific. Oxford a. Malden, Blackwell Pub., 2000, XIV-523 p. (ill., maps).

8826. Religion et sacré en Océanie. Sous la dir. de F. ANGLEVIEL. Paris, L'Harmattan-UNC, XIIe colloque Corail, Paris, 2000, 306 p. [Cf. n° <sélection> 8821.]

Cf. n°s 7756-7758

INDEX OF NAMES

A

A (Feng), 8504.
Aachen (Hans von), 6591.
AARTS (Kees), 4936.
AARUP-KRISTENSEN (Jørn), 957.
ABABAY (Feridun), 7441.
ABADIE-REYNAL (Catherine), 1746.
ABALIKHIN (Boris S.), 7852.
Abascal y Sousa (José Fernando, marqués de la Concordia), 7742.
ABASHEV (Vladimir V.), 5737.
ABBAMONDI (Carla), 7438.
ABBEY (Ruth), 6117.
ABBRI (Ferdinando), 6210.
ABDEL-MALEK (Kamal), 1473.
Abderhalden (Emil), 6251.
ABDO (Geneive), 4393.
Abdülhamid II, sultan of the Turks, 5248.
Abelardus (Petrus), 3869.
ABENDSTERN (Michele), 5944.
ÅBERG (Alf), 7114.
ABERNETHY (David R.), 4037.
Abetz (Otto Friedrich), 8096.
ABKAI-KHAVARI (Manijeh), 1887.
ABOUBAKER ALWAN (Daoud), 902.
Abraham Zakut, 3867.
ABRAMENKO (A.), 2046.
ABRAMOV (Andrei), 2355.
ABRAMZON (Mikaïl), 2355.
ABŪ MA'ŠAR, 3846.
ABULAFIA (David), 3530.
Accame (Silvio), 642.
ACCARDO (S.), 2871.
ACCATI (Luisa), 6820.
Acestes, king of Sicily, 2721.
ACEVES PASTRANA (Patricia Elena), 6216.

ACHAM (Karl), 5795.
ACHARD (G.), 2720.
Achebe (Chinua), 6477.
Acton (John Emerich Edward), 643.
Adalbertus, Sanctus, 3966.
ÁDÁM (Magda), 8044.
ADAMCZEWSKI (Marek), 134.
ADAMS (David K.), 609.
Adams (John), 5308, 5367.
ADAMS (Steve), 6211.
ADAMS (Walter Randolph), 1329.
ADAMS (Willi Paul), 5278.
Adenauer (Konrad), 4511, 4523, 6807, 8279.
ADERT (Laurent), 6356.
ADHIKARI (Mohamed), 5134.
ADIBEKOV (Grant M.), 5040.
ADIEGO (Ignasi-Xavier), 1888.
ADIYEKE (Ayşe Nükhet), 7877.
ADKIN (Neil), 2721.
ADKINS (Gregory Matthew), 6118.
ADLER (Alfred), 833.
ADLER (Glenn), 5128, 5136.
ADOLPHSON (Mikael S.), 8740.
Adorno (Theodor Wiesengrund), 6135.
ADOVASIO (J. M.), 1536.
Adrianus VI, Papa, 5804.
AEBISCHER (Tullio), 4802, 7958.
AERS (David), 3687, 3727.
Aeschines, 1971.
Aeschylus, 1895, 2195, 2223, 2298.
AFFRIFAH (Kofi), 4640.
AFONSO (Aniceto), 7689.
AFYONCU (Erhan), 5251.
AGAPITOS (P. A.), 3236.
Agatocles, 2055.
AGBESE (Dan), 4942.
Ageron (Charles-Robert), 7702.

AGIER (Michel), 834.
AGNOLETTI (M.), 1141.
AGOSTI (Aldo), 4055, 5046.
AGOSTINIANI (L.), 2408.
AGØY (Nils Ivar), 7878.
AGRELL (Wilhelm), 5207.
ÅGREN (Maria), 7499.
AGRIKOLIANSKY (Eric), 4420.
Agrippa (Marcus Vipsanius), 1854.
Agrippina (Vipsania), maior, 2623.
AGSTNER (Rudolf), 7879.
AGUADÉ NIETO (Santiago), 3596.
AGUD (Ana), 1889.
AGUILAR-RIVERA (José Antonio), 4917.
AGUIRRE (Carlos A.), 7255.
AGUIRRE ROJAS (Carlos Antonio), 410.
AGULHON (Maurice), 4421.
Ahmad al-Jaber, Sheikh of Kuwait, 8167.
AHMIDA (Ali Abdullatif), 427.
AHRENS (Ralf), 6760.
AHSMANN (Margreet), 5904.
AHTOKARI (Reijo), 4405.
AIGLE (Denise), 3972.
AIGRAIN (René), 3955.
AIJAZUDDIN (F. S.), 8269.
AILES (Marianne), 411.
AIMO (Piero), 7442.
AINSWORTH (Peter), 270, 3531.
AIRES (M.), *IX*.
AIRÒ (Anna), 3532.
AIZPURU (Mikel), 5138.
AIZPURUA (Paul), 3020.
AKAMATSU (Hidetake), 7369.
AKASAKA (Tsuneaki), 8480.
AKBAR KHAN (H.), 2460.
AKINBODE (Ralımon O.), *XII*.
AKIYAMA (T.), 4866.
AKLILU (Amsalu), 7911.

---

1. The Slavonic and in particular the Russian names are given in their national form transliterated following the usual methods and are classified accordingly. Characters with diacritics, for instance ć, ś, č, š are considered as if ordinary c, s. the German modified vowels ä, ö, ø, ü are considered as if a, o, u. The names of Classical authors, Saints and Popes are indexed in their Latin form. Authors' names are given in capital letters.

AKSENOVA (Elena P.), 412.
AKŞİN (Sina), 5246.
AKTAŞ (Necati), 3256, 3257, 5241.
AKURGAL (Ekrem), 1943.
AKYÜREK ŞAHIN (N. E.), 1639.
ALABE (Françoise), 2354.
ALABRUDZIŃSKA (Elżbieta), 5666.
ALAKOM (Rohat), 903.
ALAMO MARTELL (María Dolores), 5139.
ALARCÓN COSTTA (César), 904.
AL-BAKHIT (M. A.), 1210.
ALBANESE (Gabriella), 669.
ALBERS-SCHÖNBERG (Heinz), 5224.
ALBERT (Pierre), 6049.
ALBERTI (Giovanni), 660.
Alberti (L. B.), 2167.
ALBERTI (M. Emanuela), 1944.
ALBERTINI (Luigi), 4794.
ALBERTUS (Bohemus), 3259.
Albinana Sanz (José María), 5168.
ALBINI (Pierluigi), 2390.
ALBINUS (L.), 2294.
ALBORD (Maurice), 8056.
ALBRECHT (Christian), 5622.
ALBRECHT (Dieter), 4524.
ALBRECHT (Willy), 4522.
ALBRECTSEN (Esbern), 937.
ALDA MEJÍAS (Sonia), 4721.
ALDENHOFF-HÜBINGER (Rita), 6961.
ALDERINK (Larry J.), 1330.
ALDGATE (Anthony), 5894.
ALDRICH (Richard J.), 8057.
ALDRIGHI (Clara), 5371.
ALEMÁN (Inmaculada), 1483.
ALEMDAR (S.), 2314.
Aleramo (Sibilla), 6452.
ALESSANDRONE PERONA (Ersilia), 385.
ALESSE (F.), 2234.
ALEXANDER (Jocelyn), 8788.
ALEXANDER (Martin S.), 4457.
ALEXANDER (Matthias), 4525.
ALEXANDER (Peter), 5124.
ALEXANDRE (Valentim), 7643.
Alexandros III, ho Megas, re di Macedonia, 965, 1620, 2046, 2047, 2086, 2120, 2267, 2451.
ALEXANDROVSKY (A. L.), 1559.
Alexandru cel Bun, 1146.
ALEXEIEVA (Ekaterina), 2355.
ALFIERI (Luigi), 4829.
ALFIERI (N.), 285.
ALFÖLDY (Géza), 2414, 2419, 2535, 2674, 2695.
Alfonso el Magnànimo, rey de Aragona, 3433.
Alfonso X el Sabio, rey de Castilla y de Léon, 414, 3488, 3549, 3833.

Alfred the Great, king of England, 3750.
al-Gazali (Abu Hamid Muhammad ibn Muhammad), 5700.
ALGAZI (Gadi), 3453.
AL-HINAI (Abdulmalik Abdullah), 4954.
ALI (Azra Asghar), 5679.
ALJOVÍN DE LOSADA (Cristóbal), 7397.
ALLAIN (Jean-Claude), 4465, 7594, 7959.
ALLAN (W.), 2159.
Allart (Hortense), 5888.
ALLEN (Calvin H.), 4955.
ALLEN (Danielle S.), 2092.
ALLEN (David), 336.
ALLEN (Paul C.), 7780.
ALLEN (Roger M. A.), 1435.
ALLÈS (Elisabeth), 5680.
ALLÈS (Jean-François), 7684.
ALLEY (Roderic Martin), 8270.
ALLGOR (Catherine), 5279.
ALLIBERT (Claude), 850.
ALLIEVI (Stefano), 5398.
ALLINSON (Mark), 4526.
ALLISON (Robert J.), 5283.
ALLMAN (Eileen), 6635.
ALLMAN (Jean), 8789.
ALLMAND (Christopher), 3371.
ALLSTADT (Markus), 7566.
ALLYN (David), 7116.
Almosnino (Moisés), 1871.
ALON (Yoav Zvi), 4885.
ALONGE (Roberto), 1291.
ALONSO (Natàlia), 1541.
ALONSO (Paula), 4143.
ALONSO-NUNEZ (José Manuel), 728.
ALPA (Guido), 7500.
AL-RAWI (Farouk), 1734.
AL-RAWWAS (Issam), 4953.
Alsted (Johann Heinrich), 5644.
ALT (Peter André), 6431.
ALTER (Peter), 8271.
ALTERMATT (Urs), 5466.
ALTHABE (Gérard), 7690.
ALTHANN (Robert), 1322, 2968.
ALTHOFF (Gerd), 3377.
ALTMAN (Ida), 5140.
ALTRIPP (Michael), 3172.
ALTSCHULER (Glenn C.), 5280.
ALTURO PERUCHO (Jesus), 43.
ÁLVAREZ (David J.), 8133.
ÁLVAREZ (José E.), 7960.
ÁLVAREZ (Silvia T.), 7398.
ÁLVAREZ LLANOS (Jaime), 4287.
ÁLVAREZ MARQUEZ (Maria del Carmen), 44.
ÁLVAREZ MORA (Alfonso), 7041.

ÁLVAREZ PINEDO (Francisco Javier), 320.
Alvaro (Corrado), 6497.
ALVAZZI DEL FRATE (Paolo), 7501.
ALZATI (Cesare), 5416.
ALZUGARAY TRETO (Carlos), 8272.
AMADASI GUZZO (Maria Giulia), 1806.
AMALADASS (Anand), 5578.
AMANN (Petra), 2391.
Amaral (Azevedo), 4240.
AMATA (Biagio), 2972.
AMAYA BANEGAS (Jorge Alberto), 5681.
AMBAGLIO (D.), 1985.
Ambrosius, episcopus Mediolanensis, Sanctus, 2970, 2983, Sanctus, 3111.
AMBRUSTER (Barbara), 1500.
AMDAM (Rolv Petter), 6842.
Amedeo I, rey de España, 7924.
AMERINGER (Charles D.), 4310.
AMERUZES DE TREBISONDA (Jeorge), 3981.
AMES (Glenn J.), 7661.
AMES-LEWIS (Francis), 6501.
AMIC (Sylvain), 291.
AMIN (Shahid M.), 8273.
AMIRKHANOV (H. A.), 1536.
AMIRKHANOV (Khizri A.), 1501.
AMIR-MOEZZI (Mohammad Ali), 1383.
Amiroutzes (Georgios), 3128.
AMIT (D.), 1831.
Ammende (Ewald), 8004.
Ammianus Marcellinus, 464, 2437, 2438, 2748.
AMORES (Juan Bosco), 4311.
AMORY (Hugh), 87.
AMOURETTI (M. C.), 2048.
Ampère (André Marie), 6236.
AMSON (Daniel), 7443.
AMTOWER (Laurel), 45.
ANADOL (Ayşen), 4735.
ANAGNOSTOU-CANAS (Barbara), 1679.
ANASTASIADIS (Vasilis I.), 2415.
Anastasius I, Byzantine emperor, 3154.
Anaximenes, 2178.
ANCEAU (Eric), 4422.
ANDALORO (Maria), 292.
ANDENMATTEN (Bernard), 3407.
ANDERMANN (Kurt), 5738.
ANDERSEN (Dan H.), 6843.
ANDERSEN (Knud-Erik), 3476.
ANDERSON (Bonnie S.), 7119.
ANDERSON (Carolyn Bernadette), 415.

INDEX OF NAMES

ANDERSON (Claire), 7120.
ANDERSON (D. G.), 1502.
ANDERSON (David M.), 7691.
ANDERSON (Fred), 7825.
ANDERSON (G.), 1914.
ANDERSON (Greg), 2049.
ANDERSON (Kirill M.), 5040, 7277.
ANDERSON (Margaret Lavinia), 4527.
ANDERSON (Mark Cronlund), 7961.
ANDERSON (Robin), 5505.
ANDERSON (S.), 1141.
Anderson (Samuel), 7875.
ANDERSON (Sheldon R), 8274.
ANDERSON (Stanford), 6541.
ANDERSON (Trevor), 920.
ANDERSSON (Lars M.), 5682.
ANDERSSON (Theodore M.), 3298, 3475.
ANDERSSON-SKOG (Lena), 6844.
ANDO (C.), 2536.
ANDÒ (V.), 2006.
ANDOR (Eszter), 5473.
ANDRADE-EEKHOFF (Katharine), 4398.
ANDREA (Alfred J.), 3299.
ANDRÉANI (Roland), 6016.
ANDREEV (Andrej Ju.), 5961.
ANDRÉN (Åke), 5673.
Andreotti (Giulio), 8427.
ANDRÉS SANTOS (F. J.), 3173.
ANDRÉ-SALVINI (Beatrice), 1807.
ANDRESS (David), 4423.
ANDRETTA (Stefano), 7761.
ANDREWS (Dee E.), 5623.
ANDREWS (Gill), 1477.
ANDREWS (Malcolm), 6502.
ANDREWS (Robert), 3849.
ANDREWS (Stuart), 6050.
ANDRIĆ (Stanko), 3895, 3956.
ANGELERI (Sandra), 4288.
ANGELETTI (Luciana Rita), 5905.
ANGELOW (Jürgen), 657, 7962.
ANGELOZZI (Giancarlo), 7444.
ANGLEVIEL (F.), 8821, 8823, 8826.
ANGUERA (Pere), 5141.
ANGULA (N.), 8787.
Anjou (Robert d'), 3436.
ANKUDINOV (I. Ju.), 4024.
Anna Komnena, 3174.
ANNAS (J.), 2235.
Anne of Cleves, 4708.
Anonimo di Gall, 673.
Anonimo romano, 644.
Anonymus Notary, 3297.
ANSANI (Michele), 34.
ANSELMI (Gian Mario), 3631.
ANSHELM (Jonas), 5208.
ANTES (Peter), 1332.

ANTHONY (Daid W.), 1565.
ANTIER (Chantal), 4424, 7164.
Antigonus, re di Macedonia, 1095.
ANTILA (Kimmo), 6845.
ANTIN (Kirsti), IV.
Antiochos I, re di Commagene, 1660.
ANTIPOV (Il'ja V.), 3773.
ANTOHI (Sorin), 4044, 5962, 6003.
Antoine (André), 6648.
ANTOINE (François), 4432.
ANTOMARCHI (Véronique), 4425.
ANTÓN (Susan C.), 1537.
ANTONACCI (Nicola), 4803.
Antonescu (Ion), 5023, 5032.
Antoninus Pius (T. Aurelius Fulvus Boionius Arrius), 2591.
ANTONOPOULOU (Theodora), 3115.
ANTONOV (S.), 135.
ANTONSEN (Jack), 154.
ANTOSHCHENKO (A. V.), 1109.
ANTTONEN (Veikk), 1333.
ANTUNES (José Freire), 8275.
AO (Wenwei), 8505.
AOYAMA (Tadamasa), 8741.
APOLINÁRIO (António), 1448.
Apollonius Pergaeus, 2191.
APPEL (W.), 2838.
APPELQVIST (Örjan), 5209.
Appianus, 645, 2439, 2530.
APPLEBAUM (Wilbur), 1200.
APPLEBY (Joyce), 5285.
APPLEBY (R. Scott), 1335.
APPUHN (Karl), 4804.
APPUHN-RADTKE (Sibylle), 6573.
APPY (Christian G.), 5297.
APRILE (Sylvie), 4426.
Apuleius (Lucius), 2440-2444, 2722, 2743, 2825, 2851, 2856.
AQUARONE (Alberto), 7445.
ARA (Angelo), 729.
ARACIL (Rafael), 5181.
Aratus, 1973, 1974, 2822.
Araujo de Oliveira (Roberto), 7300.
ARBAIZA VILALLONGA (Mercedes), 5155.
ARBOIT (Gérald), 7880.
ARCANGELI (Alessandro), 6636.
ARCHER (Caroline), 6051.
ARCHER (Christon I.), 7754.
ARCHER (John E.), 7121.
ARCHER-STRAW (Petrina), 5740.
Archestratus, 1975.
ARCHETTI GIAMPAOLINI (Elisabetta), 3896.
ARCHIBALD (Z. H.), 2131.
Archilocus, 2186, 2222.
ARCIDIACONO (Bruno), 8276.
ARCINIEGAS DUARTE (Orlando), 7881.

ARDOUIN (Claude Daniel), 386.
ARECCO (Sergio), 6637.
ARENA (A.), 1680.
ARENA (Gabriella), 4021.
ARENDS OLSEN (L.), 2673.
Arendt (Hannah), 6123.
ARES QUEIJA (Berta), 4085.
ARGUINDEGUY (Diego L.), 4149.
ARIAS CABAL (Pablo), 1566.
ARINZE (Emmanuel), 386.
Ariosto (Ludovico), 6391.
Aristophanes, 1940, 1976-1979, 1984, 2212, 2226.
Aristoteles, 1980-1984, 2240, 2250, 2251, 2253, 2257, 2259, 2267, 2273, 2282, 2286, 2287, 2292, 3847, 3850.
ARKLEY (Lindsey), 8824.
ARLIGHAUS (Franz-Josef), 3533.
ARMANDO (David), 4805.
ARMANI (Giuseppe), 5494.
ARMITAGE (D.), 1027.
ARMITAGE (David), 5741.
ARMSTRONG (Adrian), 6374.
ARN (Mary-Jo), 3697.
ARNALL I JUAN (M. Josepa), 3283.
ARNAU DE VILANOVA, 3848.
ARNAUD (Daniel), 1832.
ARNAUD (Jean-Luc), 3460.
ARNAUD (Vicente Guillermo), 222.
Arnobius, 2972.
ARNOLD (Dana), 6542.
ARNOLD (David), 6212.
ARNOLD (Edward J.), 4445.
ARNOLD (Georg), 7963.
ARNOLD (John), 748.
ARNOTT (W. G.), 2020.
ARNOULD (D.), 2114.
ARNOULET (François), 7882.
ARNOUX (Mathieu), 3931.
ARON (Raymond), 831, 1028.
ARONSON (Arnold), 6638.
ARREGUI ZAMORANO (Pilar), 5142.
Arrianus, 1985, 1986.
ARSLAN (Murat), 1641.
ARTEMOV (Viktor A.), 4556, 5047.
ARTIÈRES (Philippe), 7226.
ARTIFONI (Enrico), 3631.
ARUNOVA (Marianna Rubenovna), 8277.
ARVANITAKES (Demetres), 4/96.
ARVIDSSON (Stefan), 418.
ARWAS (Victor), 6615.
ARZAKANJAN (Marina Ts.), 4493.
ARZE CUADROS (Eduardo), 4213.
ASBRIDGE (Thomas S.), 3378.
ASCH (Ronald G.), 7781.
ASCHE (Matthias), 5906.
ASCHERI (Mario), 3490.
ASCHTON (Gail), 3957.

Ashby (Thomas), 646.
ASHDOWN (Paddy), 4646.
ASHMANN (Margreet J. A. M.), 5907.
ASHTON (Nigel J.), 7964.
ASHTON (S. R.), 8253.
ASIAMAH (Alfred Effah Aduenum), 4641.
ASİLTÜRK (Bâki), 223.
ASIMOV (M. S.), 1208, 1210.
Asinius Pollio (Gaius), 2780.
ASKÉNAZI-SEKNADJE (Enrique), 6639.
ASMUTH (Christoph), 6193.
ASOV (Aleksandr Igorevich), 4.
ASSAF (Antoine-Joseph), 4895.
ASSAYAG (Jackie), 316, 7122.
ASSERETO (G.), 7853.
Aššurbanipal, king of Assyria, 1769.
Aššurnasirpal II, king of Assyria, 1748.
ASTON (Elaine), 1265.
ASTON (Nigel), 5399, 5400.
ASTVALDSSON (Astvaldur), 4214.
Atatürk v. Kemal (Mustafa 'Atatürk).
Atilius Regulus (Marcus), 2512.
ATKINS (E. M.), 2973.
ATSMA (Hartmut), 3288.
ATTANASIO (Agostino), 323.
ATTARD (Bernard), 7584.
Attlee (Clement Richard), 4651.
ATTRIDGE (Derek), 6432.
AUBERT (Roger), 1350.
AUBIN (Jean), 7782.
AUBREY (Elizabeth), 3828.
AUDOIN-ROUZEAU (Stéphane), 7965, 7966, 7981.
AUDRING (Gert), 695.
AUERBERGER (Janick), 2160.
AUGÉ (Jean Louis), 6503.
AUGENTI (Andrea), 3820.
Augustinus Aurelius, Sanctus, 1928, 2973-2980, 2990, 3009, 3066, 3072, 3865, 6421.
Augustinus Cantuariensis, 3058.
AUGUSTINUS DE FERRARIA, 3849.
Augustus (Gaius Iulius Caesar Octavianus), Roman emperor, 2530, 2556, 2577, 2589, 2597, 2604, 2608, 2618, 2626, 2665, 2711, 2764, 2822, 2883, 2905, 2926, 2930.
AULICH (James), 6574.
AUNESLUOMA (Juhana), 6213.
AUPERT (Pierre), 2316.
AURELL (Martin), 3383.
AURELL I CARDONA (J.), 3620.
AURIEMMA (Rita), 2872.
Auriol (Peter), 3886.

AUSLIN (Michael Robert), 7883.
Ausonius (Decimus Magnus), 2981, 2982.
AUSTIN (Paul Britten), 7854.
AUYERO (Javier), 4144.
AVDEEV (A. A.), 7956.
Avdeeva (Klaudia D.), 769.
AVELINE (J.), 2518.
AVENARIUS (A.), 3175.
Averroes, 3850, 3866.
Avicenna, 3877.
Avigad (Nahman), 1851.
AVILOVA (L. I.), 1484.
AX (W.), 2161.
AXTMANN (Roland), 1222.
AYALA MARTÍNEZ (Carlos de), 3265.
AYDİN (Mustafa), 8278.
AYERS (Edward L.), 5287.
AYKUT (Şevki Nezihi), 155.
AYMARD (Maurice), 804.
AYNSLEY (Jeremy), 6616.
Azariah (V. S.), 5590.
AZÉMA (Jean-Pierre), 4427, 8226.
AZEVEDO SANTOS (Maria José), 6.
AZIMI (Vida), 4428.
Azzolini (Vincenzo), 7031.
Azzolino (Decio), 4856.

B

BABELON (Jean-Pierre), 4477.
BABEROWSKI (Jörg), 4038.
BABINI (Valeria), 5963.
BABIUCH (Jolanta), 5558.
BÄBLER (B.), 1992.
Bacchelli (Riccardo), 6486.
Bacchus, Sanctus, 3247.
BACCI (Michele), 293.
BACHLEITNER (Norbert), 6434.
BACHORSKI (Hans-Jürgen), 3690.
BÄCK (A.), 2814.
BACKES (Uwe), 4529.
BACKHOUSE (Janet), 3535.
BACKOUCHE (Isabelle), 7042.
BADAL'JANETS (Jurij S.), 2317.
BADE (Klaus J.), 4039.
BADER (N. O.), 1532.
BADIA (Gilbert), 4530.
BADRAWI (Malak), 4394.
BAESJOU (René), 7716.
Baffier (Jean), 6596.
BAGG (Ariel M.), 1735.
BAGH (Peter von), 6641.
Baglivi (Giorgio), 5905.
BAGNOUD (Gérard), 8097.
BAGOLAN (M.), 2367.
BAHARAL (D.), 2675.
BAHL (Peter), 7446.
BAHL (Vinay), 492.
BAHN (Paul G.), 1503.
BAI (Chun), 8506.

BAILLIE (James), 1412.
Bainville (Jacques), 647.
BAIONI (Massimo), 395.
Bairoch (Paul), 648.
BAJOHR (Frank), 4531.
BAJONI (M. G.), 2722.
BAKER (Denise N.), 3639.
BAKER (Fiona), 6617.
BAKER (John H.), 1085.
BAKER (Kevin), 5307.
BAKER (Malcolm), 6575.
BAKER (Peter S.), 3300.
BAKHOUCHE (B.), 2839.
BAKHTADZE (R. A.), 792.
BAKHTURINA (Aleksandra Ju.), 7967.
BAKKEN (D.), 1527.
BAKKER (Paul J. J. M.), 3882.
BALAKIN (Vasilij D.), 3379.
BALAKRISHNAN (Gopal), 6119.
BALARD (Michel), 397.
BALASHOV (E. M.), 5072.
BALBOA NAVARRO (Imilcy), 7730.
BALCEDO (Antonio), 4145.
BAŁCZEWSKI (Marian), 5964.
BALDAN (Sergio), 5506.
BALDASSARRI (Stefano U.), 3667.
BALDINI (Antonio), 2050.
BALDINI (Massimo), 1186.
BALDISSARA (Luca), 8211.
BALDWIN (John W.), 3691.
Balfour (Arthur James, Earl of), 7635.
BALFOUR (Sebastiano), 5143.
BALL (Christine), 1542.
BALL (Stuart), 4651.
BALL (W.), 2537.
BALLARD (Chris), 7756.
BALLERIO (R.), 2030.
BALLETTO (Laura), 3280.
BALLINA (Augustín), 3784.
BALLINI (Pier Luigi), 4100, 4864.
BALMUTH (Daniel), 6052.
BALTUSSEN (H.), 2236.
BALTZER (Rebecca A.), 3903.
BALUKIEWICZ (Małgorzata), 5965.
BALZANI (Roberto), 6990.
BAMBERG (James), 6846.
BAMBERGER-STEMMANN (Sabine), 4040.
Bamford (Samuel), 4649.
BANDELLI (G.), 2388.
BANDHAUER-SCHÖFFMANN (Irene), 1123.
BANDINI (M.), 2999.
BANERJEE (Mukulika), 4957.
BĂNESCU (N.), 3164.
BANGE (Oliver), 8279.
BANKEN (Ralf), 6847.
BANKIE (B. F.), 8787.
BANNET TAVOR (Eve), 6400.

BANNISTER (Mark), 4429.
BANNO (Junji), 4867.
BANNON (C. J.), 2723.
BANTI (Alberto M.), 420, 4806.
Bao (Shichen), 8717.
BARADAT (Leon P.), 1029.
BARAKO (T. J.), 2116.
BARANOVA (Jurate), 1421.
BARANSKI (Zygmunt G.), 3692.
BARBACETTO (Stefano), 7502.
BARBAGALLO (Francesco), 732.
BARBAGLI (Marzio), 7123.
Barbarigo (Gregorio), Cardinale, 5550.
BARBAZZA (Marie-Catherine), 7124.
BARBER (John), 5083.
BARBERI (Alessandro), 750.
BARBERI (Claudio), 28.
BARBERIS (Peter), 905.
BARBERO (Alessandro), 3325, 3372, 3631.
BARBICHE (B.), 7783.
Barbusse (Henri), 5770.
BARCELÓ CRESPÍ (Maria), 3621.
Barcelos (João Batista), 5594.
BARCLAY (GORDON), 1558.
BARCZEWSKI (Stephanie L.), 421.
BARDAVÍO (Joaquín), 906.
BARDEN DOWLING (Melissa), 2538.
BARDET (Jean-Pierre), 4453.
BARDON (Jonathan), 907.
Bareš (Gustav), 5826.
BARGAGLI (Roberta), 7370.
BARJAU I RICO (M. E.), 1833.
BARKAI (R.), 1873.
BARKAN (Alazar), 7567.
BARKER (Graeme), 2392.
BARKER (Hannah), 6053.
BARKER (Peter), 4555.
BARKO (Ivan), 7758.
Barlaam, 1878.
BARLOW (Frank), 3380.
BARMON (Pascale), 3319.
BARNARD (Alan), 836.
BARNARD (Michaela), 6926.
BARNETT (Betty), 6472.
BARNETT (Enid), 8134.
BARNWELL (P. S.), 3491.
BAROFFIO (Giacomo), 3829, 3838, 3842.
BARON (Beth), 4071.
BARONI (Maria Franca), 35.
BARONOV (David), 4226.
BARRAGÁN MORIANA (Antonio), 5144.
BARRAQUÉ (J.-P.), 3536.
BARRATT (G.), 1890.
BARRÉ (Jean-Luc), 4415.
BARREAU (Jean-Michel), 5966.

BARRELL (A. D. M.), 3326.
BARRELL (John), 6401.
Barrès (Maurice), 4504.
BARRET (Sébastien), 3288.
BARRETT (Anthony A.), 2539, 2724.
BARRETT (John C.), 1477.
BARRETT-GAINES (K.), 5264.
BARROT (Olivier), 6642.
BARRY (Catherine), 1834.
BARTELS (Klaus), 2416.
BARTHELEMY (Tiphaine), 180.
BARTKY (Ian R.), 97.
BARTLETT (Robert), 3381.
BARTLETT (W. B.), 3176.
BARTLEY (Paula), 6214.
BARTNIK (Paweł), 4093.
BARTOLI LANGELI (Attilio), 5, 181.
BARTOLINI (Stefano), 4041.
BARTOLONI (Piero), 2368.
BARTON (Jonathan R.), 6915.
BARTON (Monika), 2989.
BARTON (Simon), 3318.
BARTOŠ (Josef), 422.
BARTOSZEWICZ (Agnieszka), 3323.
BARTOV (Omer), 478, 4042.
BARTSCH (Karl), 3675.
BARUCH (Olivier), 4492.
BARVÍKOVÁ (Hana), 4353.
Barwis (Richard), 4697.
BARYÉTY (J.), 7968.
BARZANÒ (A.), 1988.
BARZEGAR (Karim Najafi), 7784.
BARZMAN (Karen), 5908.
BAS MARTÍN (Nicolás), 321.
BAŞARAN (Sait), 1642.
BASCHET (Jerôme), 3622.
BAŞGELEN (Nezih), 1641, 2901.
BASHA (Ali Musa), 3454.
BASHFORD (Christina), 6715.
Basilius Magnus, episcopus Caesariensis, Sanctus, 2983, 3070.
Basilius Minimus, episcopus Caesareae Cappadociae, 3116.
BASILOTTA (Nicoletta), 5576.
BASKES (Jeremy), 7731.
BASS (Harold Franklin), 908.
BAST (Rainer A.), 6120.
BASTEA (Eleni), 6543.
BATALHA (Claudio H. M.), 4227.
Batalha Reis (Jaime), 7891.
Bateman (William), bishop of Norwich, 3292.
BATES (D.), 6215.
BATHMANN (Torsten), 618.
Batista y Zaldívar (Fulgencio), 4314, 8272.
BATLLÓ ORTIZ (Josep), 6340.
BATTAGLIA (Rosario), 8280.

BATTEL (Franco), 8058.
BATTINI (Michele), 694.
BATTISTINI (Andrea), 5742.
BATTY (Roger), 2492.
BAUBÉROT (Jean), 5402.
BAUDINO (Isabelle), 6054.
Baudoin IV, King of Jerusalem, 3388.
BAUDRY (G. H.), 1343.
BAUDRY (Marie-Pierre), 3755.
BAUER (Dieter R.), 3962.
BAUER (Erika), 3717.
BAUER (Franz J.), 5967.
BAUER (Hermann), 6576.
Bauer (Wilhelm), 649.
BAUER (Wolfgang), 7969.
BAUMAN (John F.), 7099.
BAUMAN (R. A.), 2638.
BAUMBACH (M.), 1915, 2836.
BAUMFALK (Gerhard), 7583.
BAUMGARTEN (Albert I.), 1334.
Baumgarten (Alexander Gottlieb), 6207.
BAXANDALL (Rosalyn), 7043.
BAY (Mia), 7125.
BAYERLEIN (Bernhard H.), 5041.
BAYKARA (Tuncer), 3537.
Bayle (Pierre), 5626.
BAYLISS (Richard), 3177.
BAYNHAM (E. J.), 2047.
BAYROU (François), 5624.
BAZÁN DÍAZ (Iñaki), 1187.
BAZIN (L.), 1210.
BAZZANA (André), 3364, 3550.
BEAMISH (Matt), 1479.
BEAN (Simon C.), 156.
BEARMAN (P. J.), 1351.
BEATTIE (Kirk J.), 4395.
BEATTY (Edward), 6916.
BEAUCAMP (J.), 3178.
Beauchamp (Earl), 7758.
BEAUJARD (Brigitte), 3104, 3894.
Beaumarchais (Pierre Augustine Caron de), 6402.
BÉAUR (Gérard), 6962.
BÉCARES (V.), 1926.
Beccaria (Cesare), 7373.
BECCHI (Paolo), 7371.
BECHER (Ursula A. J.), 813.
BECHT (Michael), 5403.
BECK (M.), 2493.
BECK (Roger), 2840.
BECKER (Annette), 7965, 7970, 7981.
BECKER (Cornelia), 1614.
BECKER (Thomas P.), 7112.
BECKER-CANTARINO (Barbara), 6433.
Beckett (Samuel), 6657.
BECKMANN (Ulf), 6822.
BÉDARIDA (François), 8212.

Bédier (Joseph), 739.
BEDNÁŘ (Miloslav), 601.
BEDOS-REZAK (Brigitte Miriam), 423.
BEDOUELLE (Guy), 3039.
BEDUHN (Jason David), 1336.
BEECHER (Jonathan), 7304.
BEETZ (Manfred), 6155.
Behaim (Albert), 3259.
BEHLMER (George K.), 595.
BEHMER (Markus), 6062.
BEHRENBECK (Sabine), 7910.
BEHRENDS (Okko), 7383.
Behrens (Peter), 6541.
BEHROOZ (Maziar), 4760.
BEIHAMMER (Alexander Daniel), 3117, 3179.
BEITO (David T.), 7305.
BEJARANO (Jesús Antonio), 8281.
BEJARANO ESCANILLA (Ingrid), 1835.
BÉKÈS (Csaba), 8282.
BELAVIN (Andrej M.), 3538.
BELAYCHE (N.), 2841.
Belcher (Edith), 4316.
Belcher (Taylor G.), 4316.
BELCHER (William), 1514.
BELFIORE (E. S.), 2162.
BELHOCINE (Mabrouk), 4133.
BELINKI (Karmela), 5683.
BELINOV (M. K.), 654.
BELINSKIJ (Andrej B.), 1480.
BELJAEV (Dmitrij D.), 8818.
BELJAEV (Leonid A.), 3776.
BELKE (K.), 3167.
BELL (Bill), 6088.
BELL (Christine), 7568.
BELLAMY (Richard), 658.
Bellarmino (Roberto), 5553.
BELLASSAI (Sandro), 4807.
BELL-FIALKOFF (Andrew), 880.
BELLIN (Joshua David), 6350.
BELLINGERI (Marco), 7454.
BELLINI (Piero), 5404.
BELLINTANI (P.), 2369.
BELLITTO (Christopher M.), 5441.
BELMONTE MARÍN (Juan A.), 1643.
BELOKONSKI (Petko), 4249.
BELOT (Robert), 8213.
BELOUSOV (Lev S.), 4808.
BELOVA (Ol'ga V.), 3623.
BÉLY (L.), 7828.
BELZYT (Leszek), 7080.
Ben Gurion (David), 4788, 8468.
BEN-AMOS (Avner), 4430.
BENARDETE (S.), 2163.
BENAVIDES (Gustavo), 1337.
BENDA (Patrik), 4327.
BENDALL (A. Sarah), 5909.
BENDL (Regine), 1123.

BEN-DROR (Graciela), 5684.
BENEDICT (Philip), 5665.
BENEDIKTSON (D. Th.), 2318.
Beneš (Edvard), 4332, 7983.
BENESCELLI (Alberto), 6644.
BEN-GHIAT (Ruth), 5743.
BENGTSSON (Rikard), 7971.
BENICHOU (Lucien D.), 4745.
BENIGNI (Mario), 5507.
BENINI CLEMENTI (Enrica), 5625.
BENITEZ (José Antonio), 6055.
BENITO MARTIN (Félix), 4005.
BENITO RUANO (Eloy), 3327.
BENJAMIN (Thomas), 424.
Benjamin (Walter), 808.
BENNASSAR (Bartolomé), 4228.
BENNETT (D.), 3180.
BENNETT (G. H.), 5288.
BENNETT (Lerone), 5289.
BENNETT (Martyn), 909.
BENNETT (Philip E.), 3548.
BENNINGHAUS (Christina), 7126.
BENOIT-DUSAUSOY (Annick), 1436, 6358.
BENRATH (Gustav Adolf), 5564.
BENROSE (Stephen), 3693.
BENSA (Alban), 6544.
BENSCH (Stephen Paul), 3382.
BENSEL (Richard Franklin), 6823.
BENSER (Günter), 4532.
BENSIMON (Fabrice), 7884.
BENSON (H. H.), 2237.
BENTHIEN (Claudia), 1136.
Ben-Tor (Amnon), 1849, 1879.
BENTSUR (E.), 8265.
BENUCCI (Elisabetta), 6435.
BENZ (Wolfgang), 4533.
BENZO D'ALESSANDRIA, 3851.
Beolco (Angelo, detto Ruzante), 6679.
BERCAW (Nancy), 7176.
BERCÉ (Yves-Marie), 5145.
BERDAH (Jean-François), 7972.
Berdjaev (Nikolaj Aleksandrovič), 6147.
BEREBITSKY (Julie), 7127.
BERELOVICH (A.), 6957.
BERENDT (Grzegorz), 8284.
BÉRENGER (Jean), 5405, 7762, 7785.
BERENGUER DE PUIGPARDINES, 3301.
BERESFORD (Andrew M.), 3735.
BERG (NICOLAS), 425.
BERGDOLL (Barry), 6545.
BERGER (Albrecht), 3181.
BERGER (Elisabeth), 182.
BERGER (Harry), 6375.
BERGER WALDENEGG (Georg Christoph), 6121.
BERGÈRE (Marie-Claire), 8329.

BERGERON (Louis), 948, 6546, 6848.
BERGES (Dietrich), 1644.
BERGFELD (Christoph), 1087.
BERGGREN (J. L.), 2035.
BERGGREN (Lars), 910.
BERGHAHN (Klaus L.), 5463.
BERGHOFF (Hartmut), 7172.
Bergman (Ingmar), 6637.
BERGMANN (B.), 2115.
BERGSTEIN (Mary), 6577.
BERINDEI (Dan), 5025.
BERKOWITZ (Alan J.), 7504.
BERKOWITZ (Michael), 5685.
BERKTAY (Ali), 6891.
BERLIOZ (Jacques), 3993.
BERMAN (Constance Hoffmann), 3928.
BERNANKE (Ben S.), 6762.
Bernard (Francis), 5347.
BERNARD (Jacques-Emmanuel), 689.
Bernard de Montfaucon, 740.
BERNARDINI (Aldo), 6662.
BERNARDINI (Paola A.), 2079.
Bernardinus Senensis, Sanctus, 3443, 4851.
Bernardus, Abbas Claraevallensis, Sanctus, 3873, 3875.
BERNAUER (Thomas), 8334.
BERNAYS (L.), 2164.
BERNDT (Rainer), 5855.
BERNER (Ulrich), 1338.
BERNET (J.), 4446.
Bernhard (Thomas), 6464.
Bernier (François), 6336.
BERNIER (Olivier), 4043.
BERNKOPF TUCKER (Nancy), 8285.
BERNSTEIN (Deborah), 4963.
BERRA (Claudia), 6391.
BERRY (Burton Yost), 8124.
BERRY (Chad), 7128.
BERRY (J.), 2553.
BERRY (Sara), 7706.
BERSCHIN (Walter), 3939.
BERSELLI (Edmondo), 7129.
BERSTEIN (Serge), 4480.
BERTAUD (Jean Paul), 6056.
BERTÉ (M.), 2459.
BERTHIAU (Denis), 1088.
BERTOLOTTO (Sabrina), 7130.
BERTRAND (Giles), 221.
BERTRAND (Michael T.), 6645.
BERZINA (S. Ja.), 1655.
BESANÇON (Alain), 5406.
BESIER (Gerhard), 5464.
BEßLICH (Barbara), 5744.
BESSMERTNYJ (Jurij L.), 787, 1126.
BESSONE (L.), 2540.
BEST (Anthony), 8249.

BESTE (H.-J.), 2873.
BETH (Williamson), 3774.
Bethlen (Gabriel), 7789.
Bethmann Hollweg (Theobald von), 4622.
BETRI (Maria Luisa), 5773.
BETT (R.), 2051.
BETTINI (Maurizio), 838.
BETTONI (M.), *IX*.
BETTS (Paul), 426.
BETZ (Hans Dieter), 1319.
BEULLENS (Pieter), 3847.
BEUSE (Ejvin), 6910.
BEUTIN (Wolfgang), 5424.
Bevilacqua (Luigi), 7820.
Bevin (Ernest), 8297.
BEWELL (Alan), 6217.
Bewley (Charles), 8180.
BEYER (Andreas), 6504.
BEYER (Renate), 5407.
BEYLIE (Claude), 6694.
BEYOĞLU (Ağacan), 839.
BEYRAU (Dietrich), 4045.
BEYRU (Rauf), 840.
Beza (Theodore), 5654.
BEZRUCHENKO (Igor), 2355.
BHATTACHARYA (Gouriswar), 294.
BIAGETTI (Stefania), 5626.
BIAGIONI (Mario), 5523.
BIALAS (Wolfgang), 5815.
BIAN (Li), 8507.
BIANCHI (Serge), 183.
Bianchi Bandinelli (Ranuccio), 650.
BIANQUIS (T.), 1351.
BICHLER (Reinhold), 682, 2165.
BICHO (Nino), 1538.
BICKERS (Robert), 4086.
BIDEAULT (Marise), 1255.
BIDOUZE (Frédéric), 4431.
BIDUSSA (David), 4046, 5398.
BIELENBERG (Andy), 4773.
BIEŃKOWSKI (Piotr), 934, 1615.
BIEŃKOWSKI (Wiesław), *XIV*.
BIERBRODT (Johannes), 6218.
BIERNACKI (Richard), 761.
BIFFI (Nicola), 1986.
Biget (Jean-Louis), 3915.
BIGLARI (Fereydoun), 1504.
BIGLIONE DI VIARIGI (Luigi Amedeo), 4845.
BIGSBY (Christopher), 1266.
BIHL (Wolfdieter), *I*.
BIHR (Ph.), 7509.
BIKAI (P. M.), 1752.
BÍLEK (Jiří), 8286.
BILES (Roger), 7099.
BILIKTÜ (Betül), 223.
BILLAULT (A.), 2024, 2166.
BILLÉ (Louis-Marie), 3058.
BILLER (Peter), 3859.

BILLERBECK (M.), 2167.
BILLINGER (Robert D.), 8059.
BILLINGTON HARPER (Susan), 5590.
BILLOTTE (Denis), 3860.
BINCHI (Carmela), 319.
BINDER (Gerhard), 2725.
BINGEN (R. James), 4907.
BINGGELI (A.), 3121.
BINGHAM (D. J.), 3008.
BINNS (Jack R.), 8245.
BIOCCA (Dario), 4809.
BIRASCHI (Anna Maria), 278.
BIRD (Margaret), 4272.
BIRD (Stephen), 428.
BIRD (T. A.), 2726.
BIRKE (Adolf M.), 7580.
BIRKELUND (Peter), 4364.
Birken (Sigmund von), 6735.
BIRLEY (Anthony R.), 723, 2490, 2541.
BIRNBAUM (Jonathan), 7508.
BISCHOF (Günter), 4185, 8303.
BISCONTI (Fabrizio), 315.
Bismarck (Otto von), 4524, 4626, 7937.
BISNAUTH (D. A.), 4724.
BISPHAM (E. H.), 2370.
BISPHAM (E.), 2864.
BITSCH CHRISTENSEN (Søren), 7412.
BIX (Amy Sue), 7306.
BIZEUL (Yves), 1068.
BIZZARRI (Hugo O.), 911.
BIZZARRO (F. C.), 2168.
BIZZOCCHI (Roberto), 429.
BJERG (Hans Christian), 8191.
BJÖRK (Ragnar), 4947.
BJÖRKMAN (S.), 7763.
BJØRN (Claus), 4365, 6963.
BJÖRN (Ismo), 1124.
BJÖRNER (Ulf), 7400.
BLAAS (P. B. M.), 430.
BLACK (Alistair), 348.
BLACK (Brian), 6849.
BLACK (Jeremy), 1734.
BLACK (Robert), 3861.
BLACKETT (R. J. M.), 7885.
BLACKMAN (D. J.), 1481.
BLACKMUN VISONÀ (Monica), 12/4.
BLACKWELL (S.), 8287.
BLAIKIE (A.), 1117.
Blaine (James G.), 7893.
BLAIR ST. GEORGE (Robert), 8817.
BLAMONT (Jacques), 7132.
BLANC-CHALÉARD (Marie-Claude), 7133.
BLANCO (L.), 7118.
BLAND (Kalman P.), 1262.
BLÄNSDORF (Jürgen), 737.

BLASCHKE (Olaf), 5408, 5432, 5466.
BLASCO (Mònica), 1505.
BLASIUS (Dirk), 6646.
BLASIUS (Rainer A.), 8241.
BLASZAK (Barbara J.), 7307.
BLASZCZYK LEE (Regina), 6618.
BLAUERT (Andreas), 1151, 3492.
BLÁZQUEZ (José María), 1482.
BLÁZQUEZ MARTÍNEZ (José M.), 1567.
BLEANEY (C. H.), 1324.
BLECKMANN (B.), 2417.
BLED (J.-P.), 7886.
BLEIBER (Helmut), 4545, 4546.
BLEKER (Johanna), 6219.
BLICKLE (Peter), 7134.
BLITZ (Amy), 7586.
BLITZ (Hans M.), 5745.
BLJUM (Arlen V.), 6057.
BLOCH (Herbert), 2433.
BLOCH (MARC), 651.
BLOCH (Olivier), 6403.
BLOCKLEY (S. P. E.), 1506.
BLODIG (Vojtěch), 4328.
BLOK (Frans F.), 5746.
BLOM (Ida), 4064.
BLOMKVIST (Vermund), 3025.
BLONDÉ (Francine), 2319, 2336.
BLUCHE (Frédéric), 4485.
BLÜCKERT (Kjell), 5409.
BLÜDNIKOW (Bent), 6647.
BLUM (Alain), 6220.
Blum (Léon), 4474.
BLUM (Paul Richard), 751.
BLUME (Dieter), 3624.
Blume (H.-D.), 2194.
Blumenberg (Hans), 6123.
BLUMIN (Stuart M.), 5280.
BLUNDELL (Tom), 6975.
BLUSSÉ (Leonard), 7588.
BLYDEN (Nemata Amelia), 8790.
BLYTH (Robert J.), 7663, 7887.
Boardman (John), 2363.
BOATWRIGHT (Mary T.), 2542.
Bobbio (Norberto), 1043, 6116.
BOBRÍK (Miroslav), 5092.
BOBROFF (Ronald), 7973.
BOBYSHEV (V. I.), 7768.
Boccaccio (Giovanni), 3749.
BOCCIA ROMANACH (Alfredo), 7587.
BOCHICCIO (Gisella), 8237.
BOCHNAK (Władysław), 5527.
BÖCK (Barbara), 1736, 1737.
BÖCKER (Manfred), 5146.
Bocola (Sandro), 6505.
BOCQUET-APPEL (Jean-Pierre), 1507.
BODEA (Cornelia), 8124.
BODEMANN (Ulrike), 20.

BODEN (Ragna), 7974.
BODENHORN (Howard), 6993.
BODIN (Michel), 7664.
BODIN (Per Arne), 1188.
BODINEAU (Pierre), 7401.
BODINIER (Bernard), 4432.
BOECKL (Christine M.), 295.
BOEHM (Laetitia), 5747.
BOENIG (Robert), 3726, 3911.
BOESPFLUG (François), 296.
Boethius (Anicius Manlius Torquatus Severinus), 2775, 3118, 3860.
BOGAERT (R.), 1681.
BOGASON (Peter), 7453.
BOGATYREV (Sergei), 3418.
BØGE PEDERSEN (Merete), 7135.
Bognetti (Gian Piero), 652.
BOHN (Robert), 8135.
BOHRN (Patricia), 6648.
BOIA (Lucian), 1030.
BOÏADJIEV (D.), 2543.
BOIS (Guy), 3539.
BOISITS (Barbara), 6649.
BOISSON (Didier), 5627.
Boissy d'Anglas (François-Antoine, comte de), 7374.
BOITEAUX (M.), 2676.
BOJADŽIEVA (E.), 5910.
BOJE (Lars Mortensen), 3862.
BOJE MORTENSEN (Lars), 431.
BOJTSOV (Mikhail A.), 787, 1126, 3302.
BOLBECHER (Siglinde), 1437.
BOLENS (Guillemette), 1189.
Bolívar (Simón), 4219.
BOLKHOVITINOV (N. N.), 7747.
Bollata (Issa J.), 1473.
BOLÒS (Jordi), 4031.
BOLÒS I MASCLANS (Jordi), 912.
BOLOVAN (Ioan), 7045.
BÖLSKER-SCHLICHT (Franz), 5981.
BOLT (Marvin), 6331.
Bolton (Whitney F.), 3726.
BÖMER (F.), 2727.
BOMMELAER (Jean-François), 2320.
BOMPAIRE (Marc), 157, 158.
BONA CASTELLOTTI (Marco), 6629.
BONACASA (Nicola), 2321.
BONACINI (Pierpaolo), 3540.
BONANI (G.), 1495.
Bonaparte (Jérôme), 7866.
BONARIA URBAN (Maria), 4006.
BONATZ (Dominik), 1616, 1617, 1738.
BONDARENKO (D. M.), 1025.
BONDÉELLE-SOUCHIER (Anne), 49.
BONECHI (Simone, 5528.
BONELLI (Franco), 6994.

BONGARD-LEVIN (Grigorij M.), 1633.
BONGHI JOVINO (Maria), 2393.
BONGIOVANNI (Bruno), 6122.
Bonham Carter (Violet), 4648.
BONIAS (Zissis), 1958.
Bonifatius VIII, Papa, 3918.
BONNASSIÉ (Pierre), 3541.
BONNAUD (Robert), 752.
BONNET (Corinne), 1836.
BONNEY (Richard), 1161.
BONORA-WAISMAN (Camille), 8288.
BONVICINI (M.), 2728.
BONWETSCH (Bernd), 8289.
BOOBBYER (Philip), 5048.
BOOCKMANN (Hartmut), 3328.
BOONE (Elizabeth Hill), 8810.
BOONE (R.), 432.
BOOTH (Alan Rundlett), 913.
BOOTH (Robert R.), 4162.
BORÁK (Mečislav), 4329, 8060.
BORBEIN (Adolf), 2904.
BORCA (Federico), 184, 2544.
BORCHARD (Michael), 8290.
BORCHARDT (Knut), 736.
Borchardt (Ludwig), 653.
BORCHARDT (Peter), 373.
BORDES-BENAYOUN (Chantal), 5707.
BORDREUIL (Pierre), 1739.
Borel (Franz), 7916.
BORGEAUD (Philippe), 722.
BORGERSRUD (Lars), 914, 4946.
BORGES (Emília Salvado), 6964.
BORGETTO (Michel), 4433.
BORGOLTE (Michael), 1172.
BORISOVA (L), 6957.
BORKOPP (B.), 3169.
BORNSCHLEGEL (Franz-Albrecht), 3255.
BORODZIEJ (Włodzimierz), 8148, 8291.
BOROUMAND (Ladan), 7569.
BOROWIEC (Andrew), 4317.
Borozdin (Ilya Nikolaevich), 654.
BOROZDINA (Polina A.), 654.
BORRA (Antonello), 3694.
BORRELL (Mònica), 1505.
BORRELLI (Antonio), 6222.
BORRIS (Kenneth), 6376.
BORROMEO (Agostino), 7786.
Borromeo (Carlo), cardinale, 5524.
BORRUSO (Paolo), 5410.
BORSAY (Peter), 7046.
BORSCHEID (Peter), 6814.
BORSOOK (Eve), 3812.
BORSTELMANN (T.), 8292.
BORTZ (Jeffrey), 7308.
BOSBACH (Franz), 1026, 6023.
BOSCH (Joseph), 1505.

BOSCH (Lynette M. F.), 3983.
BOSCHIAN (G.), 1508.
BOSCO (Anna), 4047.
Bose (Subhas Chandra), 8122.
BOSSHARD-NEPUSTIL (Erich), 1837.
BOSSIER (Fernand), 3847.
BOSSINA (L.), 3136.
BOST POUDERON (Cécile), 2169.
BOSWORTH (A. B.), 2047, 2170.
BOSWORTH (C. E.), 1208, 1351.
BOSWORTH (R. J. B.), 4810.
BOTELLA (Miguel C.), 1483.
BOTH (Norbert), 8293.
BOTHWELL (James), 3600.
BOTTE (Roger), 7692.
BOTTI (Alfonso), 5560.
BOTTICINI (Maristella), 3542.
BOUCHER (David), 1032.
BOUCHERON (Patrick), 3754, 3915.
BOUDON (Jacques-Olivier), 4434.
BOUDON (Véronique), 2001.
BOUDOU (Bénédicte), 5748.
BOUGUESSA (Kamel), 4134.
BOULARD (Claire), 6058.
BOULHOL (P.), 3105.
BOULTON (Jeremy), 7136.
BOUNEGRU (O.), 2587.
BOUNEGRU (Octavian), 1645.
BOUQUET BOYER (Marie-Therese), 3827.
BOURDIEU (Jérôme), 7047.
BOURDIN (Philippe), 5529.
BOUREAU (Alain), 3166, 3493, 3761.
BOURGAIN (Pascale), 3734.
BOURGOING (Jacqueline de), 98.
BOURIN (Monique), 1165, 3543.
BOURSIER (Jean-Yves), 8214.
BOUSMANNE (Bernard), 361.
BOUTET (Dominique), 3354.
BOUTRY (Philippe), 667, 5439, 5491.
BOUWSMA (William J.), 5749.
BOUYSSY (Maïté), 667.
BOVE (Roger), 4132.
BOVESSE (Jean), II.
BOVEY (M.), 2519.
BOWEN (Merle), 4929.
BOWEN (Wayne H.), 8061.
BOWERSOCK (Glen W.), 656, 2944.
BOWIE (Fiona), 1339.
BOWLING (Kenneth R.), 5346.
BOWMAN (Alan K.), 2548.
BOWMAN (John S.), 8475.
BOYCE (Robert W.), 6850.
BOYD (Brian), 1838.
BOYDE (Patrick), 3695.
BOYER (Christopher R.), 7137.
BOYER (Richard), 7728.
BOYER DE SAINTE-SUZANNE (Raymond), 8136.

BOYLE (David), 967.
BOYLE (Frank), 6404.
BOYLE (Leonard E., O.P.), 3863.
Boyle (Robert), 6304.
BOYNS (R. E.), 6757.
BOYNS (T.), 6757.
BOZHKOVA (Svetlana G.), 7916.
BRACCESI (L.), 2371.
BRACCINI (Massimo), 4795.
BRACHET (J.-P.), 2729.
BRACKE (W.), 2449.
BRADDICK (Michael J.), 4653.
BRADFORD (Helen), 802.
BRADLEY (G. J.), 2370.
BRADLEY (G.), 2372.
BRADLEY (Keith), 2677.
BRADLEY (Peter T.), 7732.
BRADLEY (R.), 8805.
BRADLEY (Richard), 1542, 2874, 5290.
BRADSHAW (Ann), 6223.
BRADY (David W.), 5298.
BRAHAM (Randolph), 5503.
Brahe (Tycho), 6230.
BRAIDA (Ludovica), 6059.
BRAKE (Laurel), 6088.
BRAKENSIEK (Stefan), 569.
BRAKMANN (H.), 3047.
BRAMBILLA (Elena), 5530.
BRAMON (Dolors), 3455.
BRANCACCI (A.), 2815.
Brancacci (Aldo), 2820.
Brancati (Francesco), 237.
BRAND (Charles M.), 3182.
BRAND (Helmut), 2117.
BRAND (Paul), 3439.
BRANDAO FERREIRA (Joao José), 7644.
BRANDES (W.), 3183.
BRANDHERM (Dirk), 1568.
BRANDON-JONES (Douglas), 1690.
BRANDSCH (Juliane), 4515.
BRANDT (Willy), 4513.
BRANIGAN (K.), 2587.
BRANNER (Hans), 8309.
Brasillach (Robert), 5821.
BRATOŽ (Rajko), 888.
Braudel (Fernand), 470.
BRAUN (Bernd), 4512.
BRAUN (Emily), 6506.
BRAUN (Maximilian), 2781.
BRAUN (Willi), 1303.
BRAUNEDER (Wilhelm), 4534, 7466.
BRAUNSTEIN (Philippe), 780.
BRAVO (Gian Mario), 6097.
BRAVO VALDIVIESO (Germán), 4275.
BRAYER (John), 1525.
BRAZAITIS (Tom), 5296.
BRAZZINI (Gianfranco), 6192.

BREARTON (Fran), 6436.
Brecht (Bertolt), 6652.
BRECHTKEN (Magnus), 7580.
BRÉCOULAKI (Hariclia), 2322.
BREDOHL (Thomas M.), 4535.
BREGEL (Yuri), 241.
BREMER (Thomas), 5401.
BRENCHLEY (Thomas Frank), 8294.
BRENDECKE (Arndt), 99.
BRENDON (Piers), 4048.
BRENK (Lan), 968.
BRENNAN (James P.), 4159.
BRENNAN (T. C.), 2639.
BRENNECKE (H. Ch.), 3048.
BRENNER (Louis), 1340.
BRENOVÁ (Vera), *III*.
BRENTJES (Burchard), 4124.
BRENTJES (Helga), 4124.
BRESC (Henri), 1142, 7236.
BRESSON (A.), 2118.
Bresson (Robert), 6643.
BRETON (Stéphane), 6995.
BRETONE (Mario), 753.
BREVAGLIERI (Sabina), 1263.
BREWER (Priscilla J.), 7138.
BREYER (Francis A. K.), 1682.
BREZIANU (Andrei), 915.
Briard d'Aunet (Georges), 7758.
BRIDGE (Carl), 7584.
BRIDGES (Roy), 7704.
Bridges Adams (Mary), 6009.
BRIENT (Elizabeth), 6123.
BRIESEMEISTER (Dietrich), 5933.
BRILLI (Attilio), 6437.
BRILLIANT (R.), 2545.
BRILLIANT (Richard), 2323.
BRINCKEN (A.-D. Von den), 100.
BRINK (Claudia), 6578.
Brinner (William M.), 5706.
BRIOIS (François), 2339.
BRIQUEL (D.), 2394, 2469.
BRISSON (Luc), 2171, 2233, 2238.
BRISTON (Adrian), 5968.
BRITNELL (Jennifer), 6394.
BRITNELL (Richard), 6394.
BRITO (E.), 1297.
BRIVATI (Brian), 4675.
BRIZZI (Gian Paolo), 5953.
Broch (Hermann), 6201.
BROCHEUX (Pierre), 8304.
BROCK (E. A. G.), 3040.
BROCK (Michael G.), 5929.
BROCKE (Michael), 5699.
BRODERSON (Kai), 697.
BRODY (J. Kenneth), 8137.
BRODY (Judit), 5969.
BROGIOLO (Gian Pietro), 2964, 3267.
BROISE (Henri), 3754.
BROKLOVÁ (Eva), 4330.

BROMM (Boris Franz Leo), 7309.
BRON (François), 1839.
BRONZINI (Giovanni Battista), 841.
Bronzino (Agnolo), 6602.
BROOK (Timothy), 4088, 8478.
BROOKE (Christopher Nugent Lawrence), 5909.
BROOKS (Barbara J.), 7975.
BROOKS (Douglas A.), 6377.
BROOKS (Jeanice), 6651.
BROOKS (Jeffrey), 5750.
BROOKS (Nicholas), 3363.
Broszat (Martin), 655.
BROTHERS (A. J.), 2524.
BROTTON (Jerry), 6625.
BROUCEK (Peter), 434.
BROWN (A. G.), 2875.
BROWN (Andrew), 3544.
BROWN (Dorcas R.), 1565.
BROWN (J. P.), 1840.
BROWN (Katharine Reynolds), 3775.
BROWN (Keith M.), 5628.
BROWN (Michael), 1229.
BROWN (Peter), 1440.
BROWN (V. Max), 1694.
BROWN (William O.), 5291.
BROWNE (Damien), 8243.
BROWNING (Christopher R.), 4536.
BROWNING (Don S.), 1319.
BROŽEK (Marek), 7310.
Brügel (Pieter), 6587.
BRUHNS (Hinnerk), 736.
BRUMETT (Palmira), 6060.
BRUN (J.-P.), 2119.
BRUNA (Paulo Julio Valentino), 7091.
Brunelleschi (Filippo), 6557.
BRUNET (François), 6619.
BRUNETTA (G. Piero), 1290.
BRUNGARDT (Maurice P.), 959.
BRUNI (Domenico Maria), 6061.
BRUNI (Leonardo), 3667, 5812.
BRUNNTHALER (Adolf), 4171.
Bruno (Giordano), 5567, 6191, 6203.
BRUSH (Stephen G.), 687.
BRUUN (Christer), 2546, 2612.
BRUZEK (J.), 1604.
BRYANT-BERTAIL (Sarah), 6652.
BRYSAC (Shareen Blair), 4537.
BUAJASÁN MARRAWI (José), 8459.
Buarque de Holanda (Sérgio), 4240.
BUCCI (Onorato), 1089.
BUCHER (G. S.), 645.
Bücher (Karl), 531.
BUCHI (Ezio), 2385.
BÜCHSENSCHÜTZ (Ulrich), 4250.
BUCKLAND (Gail), 5307.
BUCK-MORSS (Susan), 5751.

BUDDE (Gunilla-Friederike), 7139.
BUDICK (Sanford), 754.
BUELL (Denise Kimber), 1341.
BUENO MADURGA (Jesús Ignacio), 5147.
BUFFET (Marc), 6674.
BUFFINGTON (Robert), 7255.
BÜHL (G.), 3184.
BÜHNEMANN (Gudrun), 297.
BUIKE (Bruno), 4168.
BUJOSA (Francesc), 1224.
BUKEY (Evan Burr), 4172.
BULATOVA (Angara G), 842.
BULEI (Ion), 5026.
Bulgakov (Vladimir), 5618.
BULGARELLI (Sandro), 7367.
BULKIN (Val. A.), 3773.
BULLARD (Alice), 1191.
BULLARD (Julian), 7954.
BULLARD (Margaret), 7954.
Bullard (Reader W.), 7954.
BULLÓN-FERNÁNDEZ (María), 3545.
Bullough (Donald A.), 2947.
BULUT (Mehmet), 6763.
BUNTING (M. J.), 1553.
Buñuel (Luis), 6678.
BÜNZ (E.), 720.
Buonaiuti (Ernesto), 5521.
Buonarroti (Michelangelo), 6600, 6607, 6612.
BUOYE (Thomas M.), 8508.
BURCHARD (Christoph), 3028.
Burchard von Worms, Bishof, 3897.
Burckhardt (Jacob), 656, 6159.
BURDEKIN (Richard C. K.), 5291.
BURDETT (A. L. P.), 7576.
BURDIEL (Isabel), 5175.
BUREŠOVÁ (Jana), 7141.
BURG (Peter), 7048.
BURGARD (Oliver), 7976.
BURGAT (François), 917.
BURGESS (Miranda J.), 6405.
BURGHARTZ (Susanna), 719.
BURGIO (Alberto), 6124.
BURGUIÈRE (André), 6224.
BURGUIÈRE (P.), 2042.
BURIAN (Peter), 2172.
BURK (Kathleen), 726.
BURKARD (Dominik), 5411.
Burke (Edmund), 657, 1027.
BURKE (J.), 3185.
BURKE (Peter), 5757, 7055.
BURKERT (Martin), 436.
Burkert (Walter), 1330.
BURKHARD (Bud), 6125.
BURKHARDT (Johannes), 1143.
BURKLE-YOUNG (Francis), 5508.
BURLEIGH (Michael), 4538.
BURLEY (Walter), 3852.

BURLING (William J.), 6653.
BURLINGAME (Michael), 5272, 5277, 5337.
BURNETT (Charles), 3846.
BURNETT (D. Graham), 225.
BURNEY (Jan), 6225.
BURNS (Robert M.), 807.
BURR MARGADANT (Jo), 7240.
BURRIN (Philippe), 8215.
BURROUGHS (Catherine), 6654.
BURROW (J. A.), 3645.
BURROW (John Wyon), 6126.
BURROWS (Simon), 7855.
BURSCHEL (Peter), 876.
BURTON (G. P.), 2547.
BURTON (Nick), 1590.
BURTON (P.), 3023.
BURTON (Paul J.), 2470.
BUSCHFORT (Wolfgang), 4539.
BUSH (M. L.), 7143.
BUSINO (Giovanni), 676, 729.
BUSQUETA (Joan J.), 4031.
BUSSFELD (Christina), 4654.
BUSUIOC VON HASSELBACH (Dan Nicolae), 3929.
BUSYGINA (Irina M.), 4540.
BUTLER (Jon), 5292.
BUTLER (L. J.), 7693.
BUTLER (Michael), 5227.
BUTLER (Virginia), 1569.
BUVIT (Ian), 1573.
BUZAGLO (E.), 1689.
BUZPINAR (Ş. Tufan), 7889.
Buzzi (Paolo), 6497.
BYNUM (Caroline Walker), 3984.
BYRNES (Mark S.), 5293, 8138.
Byron (George Gordon), 5982.
BYRSKOG (Samuel), 3049.
BYSTRICKÝ (Valerián), 8062.
Byulek (Muhammad), 40.

C

CABADA (Ladislav), 4331, 5753.
CABALLERO (Manuel), 5377.
CABANA I VANCELLS (Francesc), 5196.
CABANEL (Patrick), 5629, 5707.
CABAÑERO SUBIZA (Bernabé), 3777.
CABANES (Pierre), 2324.
CABEZA (Antonio), 7787.
CABEZA SÁNCHEZ-ALBORNOZ (María Cruz), 349.
CABOURET (B.), 2842.
CABRITA (Joao M.), 4930.
CABY (Cécile), 3930.
CACCAMO (Francesco), 7977.
CACCIATORE (Giuseppe), 683, 6155.
CADÉ (Michel), 7312.
CADOGAN (Jean K.), 3778.

Cadot (Michel), 7625.
Caesar (Gaius Iulius), 2446, 2610, 2620, 2895.
CAESAR (Michael), 6468.
CAETANO (Gerardo), 5376.
CAFFERA (Hugo R.), 5970.
CAFFIERO (Marina), 5412.
CAFIERO (Antonio Francisco), 4141.
ÇAGLAR (Gazi), 5247.
CAGLIOTI (Francesco), 6579.
CAHILL (David), 7732.
CAHN (H. A.), 2120.
CAI (Shuangquan), 8509.
CAIE (Graham D.), 3680.
CAILLAVET (Chantal), 843.
CALABRIA (P.), 2679.
CALAFATE (Pedro), 6157.
CALAFETEANU (Ion), 5027.
CALAME (Claude), 2173, 2295.
CALDELLI (Maria Letizia), 2419, 2434.
CALDER (W. M.), 698.
Calderón de la Barca (Pedro), 6719.
CALDERS I ARTÍS (T.), 1833.
CALDWELL (Peter C.), 7372.
CALI (Maria), 6580.
CALÌ (Vincenzo), 5816.
Caligula (Gaius Iulius Caesar Germanicus), Roman emperor, 2539, 2573, 2636, 2876.
ÇALIK (Ramazan), 5248.
CALLAHAN (William J.), 5467.
CALLEBAT (Louis), 3888.
CALLEGARIN (Laurent), 159.
CALLEN (Anthea), 6581.
CALLENDER (Dexter E. Jr.), 1841.
Callicles, 2244.
Callimachus, 1987, 2450.
CALLOW (John), 4655.
CALONACI (Stefano), 4811.
CALORE (Antonello), 2640.
CALVESI (Maurizio), 6547.
CALVET DE MAGALHAES (José), 7591.
Calvia Crispinilla, 2594.
Calvin (Jean), 5403, 5621, 5635, 5657.
ÇAM (Nusret), 1264.
CAMARERO (Hernán), 4160.
CAMASSA (Giorgio), 2052.
CAMBIANO (Giuseppe), 1033.
CAMBOU (Pierre), 6406.
Camerarius (Joachim), 5650.
CAMERON (A.), 2945.
CAMERON (Ewan), 3898.
CAMERON (Keith), 5397.
CAMILLE (Michael), 3626.
CAMODECA (G.), 2418.
CAMPANILE (M. D.), 2549.

CAMPBELL (Brian), 2436.
CAMPBELL (Bruce M. S.), 3546.
Campbell (James), 3346.
CAMPBELL (Lyle), 185.
CAMPBELL (Michelle), 1542.
CÂMPEANU (Remus), 5028.
Campion (Edmund), 5830.
CAMPIONE (Ada), 3050.
CAMPOREALE (Giovannangelo), 2395.
CAMPOS Y FERNÁNDEZ DE SEVILLA (Francisco J.), 3935.
CAMPS (Guiu), 1740.
CAMUNAS MADERA (Ricardo R.), 5019.
CAMURRI (Renato), 4814.
Camus (Albert), 6201.
CAN (Bilmez Bülent), 6851.
CANAL (Josep), 4027.
CANALI (Mauro), 4809.
CANARIS (Claus Wilhem), 7484.
CANART (Paul), 56.
CANCIK (H.), 1934.
CANCIK-LINDEMAIER (H.), 2843.
CANDAU MORÓN (José María), 2174.
CANELLIS (A.), 3002.
CANESSA (ANDREW), 5532.
CANEVA (Kenneth L.), 687.
CANFORA (Luciano), 732, 3134.
CANGUILHELM (Georges), 830.
Canguilhem (Georges), 6270.
Canidius Crassus (Publius), 2567.
CANIM (Rıdvan), 1439.
CANNADINE (David), 755.
CANNON (Joanna), 3774.
Canova (Antonio), 6572.
CANTARELLA (Eva), 2094.
CANTERA MONTENEGRO (Santiago), 3899.
CANTRELLE (Sylvier), 963.
CANZIANI (Guido), 6184.
CAO (Dachen), 8733.
CAO (Shuji), 8510.
CAPANNELLI (Emilio), 331.
CAPASSO (Riccardo), 101.
CAPATA (Alessandro), 6378.
CAPATTI (Alberto), 1192.
Capci (Pietro), 7531.
Capella (Martianus), 2786.
CAPITANI (Ovidio), 672.
CAPLAN (Jane), 478, 1253.
CAPLAN (Jay), 5754.
ČAPLOVIČ (Dušan), 5093.
ČAPLOVIČ (Miloslav), 8139.
CAPO (Lidia), 5951.
CAPOGROSSI BOLOGNESI (Luigi), 735.
CAPONE (Alfredo), 701.
CAPONETTO (Salvatore), 5630.
CAPRETTINI (Gian Paolo), 845.

CAPUA (Michelangelo), 6655.
Capuana (Luigi), 6478.
CAPUTI (Robert J.), 8140.
CAPUTO (Riccardo), 2325.
CARABINE (Deidre), 3864.
CARANDINI (Silvia), 6740.
CARASSI (Marco), 318.
CARATOZZOLO (Vittorio), 3732.
Caravia (Alessandro), 5625.
CARBASSE (Jean-Marie), 1090.
Cardano (Girolamo), 6257.
CARDARELLI (S.), 6996.
CARDENAL TELLERÍA (Marco A.), 4939.
CÁRDENAS (Enrique), 6771.
CARDOSO RUÍZ (Patricio), 4276.
Carducci (Giosuè), 6478.
CARDY (Michael), 6640.
CAREY (C.), 1971.
Carías Andino (Tiburcio), 4732.
CÂRJA (Ion), 5029.
CARL (Horst), 3419.
CARLEBACH (E.), 5687.
CARLIER (Omar), 1193.
Carlos II, rey de España, 7836.
Carlos IV, rey de España, 5198.
CARLSSON (Bo Göran), 1194.
CARLTON (David), 8295.
CARMICHAEL (Cathie), 5117.
CARNEMOLLA (Piero Antonio), 5533.
Carnesecchi (Pietro), 5462.
CARNEY (Elizabeth Donnelly), 2095.
CARNEY (Ray), 6656.
Caro (Heinrich), 6307.
CAROCCI (Sandro), 3919.
CAROL (Harrison), 3865.
CAROLI (Giuliano), 7978.
CARPENTER (D. A.), 3547.
CARRARA (Paolo), 2992.
CARRASCO (Juan), 3293.
CARRASCO GARCÍA (Antonio), 7694.
CARRERAS MONFORT (C.), 2680.
CARRÈRE (Isabelle), 2339.
CARRÈRE D'ENCAUSSE (Hélène), 7624.
CARRILERO MARTÍNEZ (Ramón), 322.
CARRILLO CÁZARES (Alberto), 4908.
CARRINO (Annastella), 4812.
CARRITHERS (David W.), 1061.
CARROBLES SANTOS (Jesus), 1543.
CARROLL (Brett E.), 1342.
CARROLL (M.), 2877.
CARROUÉ (François), 1741.
CARRUBA (O.), 1646.
CARSANA (C.), 2550.
CARTER (J. A.), 1518.

CARTER (J. C.), 2121.
Carter (James Earl), 8438.
CARTER (John J.), 7589.
CARTER (William), 6438.
CARTIER (Stephan), 437.
CARTLEDGE (P.), 2053.
CARTOCCI (Roberto), 4865.
CARTON (Benedict), 5126.
CARVAIS (Robert), 7183.
CARY (Noel D.), 8296.
CASADEI (Alberto), 6439.
CASALNUOVO (Mario), 7506.
CASAMASSIMA (Alessandra), 7367.
CASANOVA (Cesarina), 7444.
CASANOVAS (Joan), 7733.
CASANOVAS MIRÓ (Jorge), 1842.
CASASSAS (Jordi), 5196.
CASCETTA (Annamaria), 6657.
CASCIONE (C.), 2641.
CASELLA (Francesco), 5971.
Casement (Roger, sir), 7985.
CASERTANO (G.), 2239.
CASHDOLLAR (Charles D.), 5631.
CASINI (Simona), 699.
CASSAGNE (Jean-Marie), 186.
CASSAGNES-BROUQUET (Sophie), 3329.
CASSANDRO (Cristiana), 69.
CASSAR (Carmel), 918.
CASSELMAN (Bill), 187.
CASSESE (Michele), 5468.
CASSIO (Albio Cesare), 1959, 2122.
Cassiodorus, 2990.
Cassirer (Ernst), 6120.
Cassius Dio, 698, 1988, 2447.
CASSUTO (Ph.), 1872.
CASTÁN LANASPA (Guillermo), 3549.
Castaña (Juan Bautista), cardinale, 5524.
CASTELLANO (Juan Luis), 5188.
CASTELLI (C.), 2175.
CASTELLI (Rosario), 6467.
CASTELLS (Irene), 5148.
Castiglione (Baldassarre), 6385.
CASTILLEJO GORRÁIZ (Miguel), 3866.
CASTILLO (Santiago), 5149.
CASTONGUAY (René), 4264.
CASTOR (Helen), 3420.
CASTORIO (J.-N.), 2878.
CASTRO LÓPEZ (Marcelo), 1517.
Catharina, Sancta, 3967.
Catilina (Lucius Sergius), 2540.
Cato (Marcus Porcius), Censor, 2448.
Cattaneo (Carlo), 6478.
CATTANEO (Massimo), 4805.
CATTARUZZA (Alejandro), 438.
CATTERALL (Peter), 6090, 7427, 8279.

CATTINI (Marco), 662.
Catullus (Gaius Valerius), 1987, 2449, 2450, 2723, 2728, 2739, 2755, 2779, 2794.
CAUDHARI (Amita), 4746.
CAVALIERE (Paola), 1843.
CAVALIERI (M.), 3051.
CAVALIERI MANASSE (G.), 2642.
CAVALLO (Guglielmo), 7.
CAVANAUGH (Carole), 6754.
CAVANAUGH (Jan), 6507.
CAVANNA (Adriano), 7507.
CAVARZERE (Alberto), 2730.
Cavour (Camillo Benso, conte di), 4823.
CAYLA-VARDHAN (Fabienne), 439.
CAYRAC-BLANCHARD (Françoise), 4756.
CĂZAN (Ileana), 7788.
CAZES (G.), 286.
CAZORLA SÁNCHEZ (Antonio), 5150.
Ceauşescu (Nicolae), 5031, 5033, 5962.
CEAUŞU (Mihai-Ştefan), 5030.
CÉBEILLAC-GERVASONI (Mireille), 2561.
CECCARELLI (Maria Grazia), 369.
CECCOPIERI (Isabella), 144.
ČECHURA (Jaroslav), 4332.
ČECHUROVÁ (Jana), 4332, 4333.
Celdon, Sanctus, 3960.
Céline (Louis-Ferdinand), 6201, 6354.
ČELOVSKÝ (Bořivoj), 4334.
CELOZZI BALDELLI (Pia G.), 7890.
CELS SAINT-HILAIRE (J.), 2643.
Celsus (Aulus Cornelius), 2821.
CENA (Juan Carlos), 4146.
CENTRONE (B.), 2816.
CEPEDA FANDIÑO (Antonio), 345.
CEPREGANOV (Todor), 8064.
CERASI (Laura), 5755.
CERETTI (Marinella), 733.
CERISOLA (Pier Luigi), 6440.
CERNOVODEANU (Paul), 7145.
CERNUSCHI (Alain), 6658.
CERRATO (Rocco), 5560.
CERRITO (Elio), 6994.
Cervantes (Miguel de), 6379.
CERVELLÓ SÁNCHEZ (Josep), 5006.
CERVO (Amado Luiz), 7591.
CESARANO (Filippo), 6997.
CEVASCO (Roberta), 7130.
ÇEVİKEL (Nuri), 4318.
CHABÁS (José), 3867.
CHABAUD (G.), 286.
CHABOT (Hugues), 6226.
CHADANI (S.), 8142.
CHAITANI (Youssef), 8297.

CHAIX (Bruno), 8193.
CHAKRABARTY (Dipesh), 756.
CHALINE (J. P.), 3350.
CHAMBERLAIN (Muriel E.), 919.
Chamberlain (Neville), 8140.
CHAMBOREDON (Jean-Claude), 831.
CHAMPEAU (Sthéphanie), 6441.
CHAN (Sewell), 5294.
CHANCE (Jane), 3696.
CHANDE (M. B.), 6127.
CHANDEZON (Christophe), 2296.
CHANDLER (Alfred D. Jr.), 5846.
CHANG (Claudia), 1603.
CHANIOTIS (A.), 1970.
CHANKOWSKI (Véronique), 2326.
CHANSON (H.), 2879.
CHANUT (Jean-Marie), 6852.
CHAO (Xiaohong), 8512.
CHAPA BRUNET (Teresa), 1602.
CHAPLIN (J. D.), 689.
CHAPMAN (Henry P.), 1509.
CHAPMAN (James), 5894.
CHAPUIS (Oscar), 5389.
CHARIER (Alain), 5378.
CHARIM (Isolde), 4188.
CHARLES (D.), 2240, 8065.
Charles d'Orléans, 3697, 3700.
Charles IV, re di Boemia, 3821.
Charles of Flanders, 3515.
Charles V, roi de France, 322, 3286.
CHARLES-EDWARDS (T. M.), 3900.
CHARLET (J. L.), 2455.
CHARNLEY (Joy), 5227.
CHARON (Annie), 96.
Charvaz (Andrea), 6027.
CHASE (Malcolm), 7313.
CHASIN (José), 4230.
CHASKALSON (Matthew), 5135.
CHASTAGNARET (Gérard), 6853.
Chateaubriand (François René de), 6720.
CHATELAIN (Jean-Marc), 71.
CHATELAIN (Violaine), 6659.
CHÂTELET-LANGE (Liliane), 6548.
CHÂTELLIER (Louis), 5490.
CHATTOPADHYAYA (D.P.), 1211.
CHAUBET (F.), 5911.
Chaucer (Geoffrey), 1440, 3664, 3698, 3771.
CHAUNU (Pierre), 5452.
CHAUSSINAND-NOGARET (Guy), 948, 7826.
CHAUSSON (François), 2551.
CHAUVAUD (Frédéric), 6227.
CHAUVIN (Charles), 1344.
Chávez Frías (Hugo), 5377, 5380.
CHAZAN (Michael), 1510.
CHAZAN (Mireille), 441.
CHAZAN (Robert), 3440.

CHEGAI (Andrea), 6660.
CHELKOWSKI (Peter J.), 4761.
CHEMELLO (Adriana), 6407.
Chen (Cheng), 8701.
Chen (Duxiu), 8706.
CHEN (Feng), 8513.
CHEN (Hongmin), 8514.
CHEN (Jinjin), 8515.
CHEN (Mingguang), 8516.
CHEN (Wei-chun), 1556.
CHEN (Weiming), 8517.
CHÈNE (Olivier), 2054.
CHENEY (C. R.), 953.
Cheng (Dequan), 8735.
CHENG (Hua), 8518.
CHENG (Weiming), 8519.
CHENG (Zhaoyun), 8520.
CHENTSOV (Viktor Vasil'evich), 5269.
Chepilov (Dimitri), 8108.
CHERIX (Pierre), 1844.
CHERKASOV (Petr P.), 4102.
CHERNYJ (E. N.), 1484.
CHERTINA (Zoja S.), 5295.
CHEVALLIER (Raymond), 228.
CHI (Yunfei), 8521.
Chiang (Ching-Kuo), 8644.
Chiang (Kai-shek) v. Jiang (Jieshi).
Chiari (Pietro), 6418.
Chiaromonte (Nicola), 5849.
Chiaromonti (Barnaba) v. Pius VII, Papa.
CHIBNALL (Marjorie), 3330.
CHICHAGOVA (O. A.), 1559.
CHIDESTER (David), 1345.
CHIELLINO (Carmine), 1451.
CHIESA (Paolo), 705.
CHIFFOLEAU (Jacques), 3915.
CHIGLINTSEV (Evgenij A.), 1916.
CHIH (Rachida), 5688.
CHI-KWAN (Mark), 8298.
CHILDRESS (Diana), 3698.
CHILDS (Nick), 4435.
CHILTON (Bruce), 5704.
CHIOFFI (Laura), 2419.
CHIPER (Ioan), 8143.
CHIPPINDALE (Christopher), 1591, 1917.
CHIREAU (Yvonne), 5686.
CHIRIBIRI (Alessandro), 4335.
CHISTIKOV (Aleksandr Nikolaevich), 5945, 8031.
CHISTJAKOV (A. N.), 5072.
CHITI BATELLI (Andrea), 8237.
CHITTY (Dennis), 6229.
CHIU-DUKE (Joséphine), 8523.
CHKNAVEROVA (A. A.), 7592.
CHLEVNJUK (Oleg), 5051.
CHO (Wan-je), 8782.
CHOCANO MENA (Magdalena), 5756.

INDEX OF NAMES

CHOE (Hongkyu A.), 8782.
CH'OE (Yong-ho), 8785.
Choibalsan (Khorlogiin), 5080.
CHOJNACKI (Stanley), 7146.
CHOLNOKY (Gyozo), 8044.
CHORHERR (Thomas), 4173.
CHRÉTIEN (Jean-Pierre), 8791.
CHRIST (Karl), 656.
CHRISTE (Yves), 3779.
CHRISTENSON (David M.), 2489.
Christian VII, konge av Danmark, 4369.
CHRISTIANSON (John Robert), 6230.
CHRISTIE (Kenneth), 5127.
CHRISTIE (N.), 2964.
CHRISTIE (Nancy), 4265.
CHRISTIE (Yves), 306.
CHRISTIEN (J.), 2048.
CHRISTINGER (Raymond), 844.
CHRISTISON (Kathleen), 8300.
Christodoros, 3119.
Chruščëv (Nikita Sergeevič), 5083, 8289.
Chu Hsi, 8581.
CHUBAR'JAN (Aleksandr O.), 4493, 5075.
CHUNG (Dooeum), 8524.
CHUNG (Henry), 8301.
CHURCH (Roy), 6765, 6918.
CHURCHILL (J. Bradford), 2448, 2499.
Churchill (Winston Leonard Spencer sir), 4651, 7991, 8152, 8175, 8295.
CHWALBA (Andrzej), 4984.
CIAMPANI (Andrea), 5536.
CIAPPELLI (Giovanni), 417.
CIAPPI (M.), 2844.
CIARDI (Roberto Paolo), 6537.
CIBULKA (Pavel), 4326.
ČIČAJ (Viliam), 5093.
CICCARESE (Maria Pia), 3045.
CICCOLELLA (F.), 3120.
Cicero (Marcus Tullius), 656, 2452-2454, 2638, 2771, 2798, 2818, 5812.
CIEŚLAK (Marek), 7952.
CIFARELLI (Francesco), 2845.
CIFUENTES COMAMALA (Lluís), 3868.
CIGLENEČKI (Slavko), 4010.
CINGARI (Salvatore), 658.
CINGOLANI (Stefano Maria), 3699.
Cinna (Helvius), 2739.
CIOBANU (Veniamin), 7822.
CIPAIANU (George), 580.
CIPRIANI (Fernando), 6442.
CIRANNA (Simonetta), 3780.
Ciraolo (Giovanni), 8007.
CIRAVOĞLU (Öner), 6078.
CIRIACONO (Salvatore), 6766.

CIRONISOVÁ (Eva), 6854.
CISNEROS (Andrés), 8423.
CISSOKO (S. M.), 1210.
CITRON (Marcia J.), 6663.
CITRONI MARCHETTI (S.), 2731.
CITTI (F.), 2461.
CIURTIN (Eugen), 1346.
CIVIL (Miguel), 1743.
ÇİZAKÇA (Murat), 1195.
CLAASSENS (Geert H. M.), 3709, 3721.
CLACKSON (J.), 1683.
CLAEYS (V.), 286.
CLAMAN (Henry N.), 3441.
CLANCHY (Michael), 3869.
CLARAMUNT (Salvador), 3762.
CLARK (Alan), 4647.
CLARK (Charles E.), 5972.
CLARK (Charles M. A.), 6772.
CLARK (Christopher), 4541.
CLARK (Dymphna), 1035.
CLARK (Gillian), 1919.
CLARK (Gregory T.), 3781.
CLARK (James G.), 5632.
CLARK (Jeffery J.), 7756.
CLARK (Linda L.), 7449.
CLARK (Manning), 1035.
CLARK (Peter), 5758, 7049.
CLARKE (M.), 2176.
CLARYSSE (W.), 1684.
CLASSEN (Carl Joachim), 3024.
CLAUDE (Sandrine), 3782.
Claudianus (Claudius), 2455, 2456.
Claudius Galenus, 2001, 2002, 2003, 2280.
Claudius Nero Germanicus (Tiberius), Roman emperor, 2501.
Claudius Ptolemaeus, 2035, 2036.
CLAVERO (Bartolomé), 7403, 7405.
Clemens I, Papa, Sanctus, 3102.
Clemens VIII, Papa, 7786.
CLÉMENT (Caty), 8283.
CLÉMENT (Jean-Paul), 7374.
CLÉMENT (S.), 2732.
CLEMENTE (Josep Carles), 5151.
Cleopatra VII, queen of Egypt, 1730.
CLERICI (Luca), 6042.
CLEVERLEY (John), 7314.
CLIFF (Andrew D.), 6231.
Clifford (Martin), 6200.
CLIFT (Eleanor), 5296.
Clift (Montgomery), 6655.
CLINGAN (C. Edmund), 4542.
CLINQUART (Jean), 7450.
Clisthenes, 2052, 2097.
CLOAREC (Vincent), 7979.
Clodia, 2572.
Clodius Macer, 2594.

Clodius Pulcher (Publius), 2572.
CLOGG (Richard), 7593.
CLOONEY (Francis X.), 5578.
CLOSE (Anthony), 6379.
CLOSSON (Marianne), 6380.
CLOUZOT (M.), 3627.
CLUNIES ROSS (Margaret), 3731.
CLUSE (Christoph), 3442.
COARELLI (Filippo), 2373, 2552, 2880.
COATES (Simon), 2946.
COATS (A. W. Bob), 6825.
COBO ROMERO (Francisco), 5203.
COCHET (A.), 2681.
COCHRAN (Sheran), 8525.
COCKSHAW (Pierre), 3552.
COCO (Antonio), 5912.
COELHO (Maria Teresa Pinto), 7891.
COENEN (Marc), 1685.
CŒURÉ (Sophie), 741.
COFFEY (John), 5633.
COGAN (Frances B.), 8066.
COGAN (John F.), 5298.
COGROSSI (Cornelia), 7510.
COHEN (Adam S.), 3870.
COHEN (B.), 2358.
COHEN (E. E.), 2096.
COHEN (Eran), 1744.
COHEN (Esther), 3871.
COHEN (Jeffrey Jerome), 3357.
COHEN (R.), 1678.
COHEN (Robin), 4822.
COHEN (Sarah), 6509.
COHEN (Susan Sarah), 896.
COHEN (William B.), 443.
COHEN-SOLAL (Annie), 6582.
COHN (Norman), 3901.
COHN (Raymond L.), 7050.
COHN (Samuel K.), 3421.
COINTET (Jean-Paul), 935.
COINTET (Michèle), 935.
Cola di Rienzo, 292.
COLAO (Floriana), 7511.
COLARIZI (Simona), 4813.
COLAS (Gérard), 8.
COLBOW (G.), 1745.
COLDIRON (A. E. B.), 3700.
COLE (Ann), 188.
COLE (Joshua), 7051.
COLE (Laurence), 4174.
COLE (Stephanie), 5370.
COLEMAN (Janet), 1036.
COLEMAN (Patrick), 6188, 6369.
Coleridge (Samuel Taylor), 805.
COLIN (Andrew), 5634.
COLINET (A.), 1972.
COLLARD (Patrick), 5759.
COLLATZ (Christian-Friedrich), 2219.
COLLE (Enrico), 6628.

COLLEDGE (Sue), 1531.
COLLEY (Linda), 444.
COLLIN (Claude), 8216.
CÖLLIN (Jan), 3688.
COLLIN (Peter), 7451.
Collingwood (Robin George), 805.
COLLINI (Stefan), 5810.
COLLINS (E. J. T.), 6959.
COLLINS (Hugh E.L.), 3422.
COLLINS (J. J.), 1845.
COLLINS (Minta), 3783.
COLLINSON (Patrick), 5909.
COLLON (Dominique), 1794.
COLLOTTI (Enzo), 460, 4793, 7980, 8221.
COLONOMOS (Ariel), 5414.
COLPAS GUTIÉRREZ (Jaime), 4287.
COLPE (C.), 3047.
COLTON (Robert E.), 2733.
COMBA (Rinaldo), 3534.
COMBEAU (Yvan), 4436.
Comestor (Peter), 3742.
COMFORT (Anthony), 1746.
Commodus (Marcus Aurelius), imperatore romano, 608.
COMOTH (Katharina), 2241.
COMOTTI (G.), 2030.
COMPAGNON (Daniel), 5395.
CONCONI (Bruna), 6381.
CONDE (Rafael), 3762.
CONDETTE (Jean-François), 5761, 5913.
CONGAR (Yves), 5470.
CONGOURDEAU (M.-H.), 2990.
CONLON (Pierre M.), 5730.
CONNELL (William J.), 3423, 3426.
CONNELLY (John), 5914.
CONNELLY (Matthew), 446.
CONNOLLY (Sean J.), 4781.
CONNON (Derek), 6640.
CONNOR (Aileen), 1592.
CONNOR (Ian), 4543.
CONRAD (Christoph), 757.
CONRAD (Horst), 7817.
Conrad (Joseph), 6447.
CONRAD (Sebastian), 757.
CONSALVI (Simón Alberto), 7927.
Considerant (Victor), 7304.
CONSOLINO (F. E.), 2764.
CONSOLO LANGHER (Sebastiana Nerina), 2055.
CONSTABLE (Giles), 3932.
Constans (Flavius Iulius), Roman emperor, 2468.
Constantinus (Lucius Flavius Valerius), Roman emperor, 292, 1933, 2590, 2876, 3055, 3097.
Constantinus V Copronymus, Byzantine emperor, 3243.
Constantinus VII Porphyrogenetus, Byzantine emperor, 3242.

Constantius (Flavius Iulius), Roman emperor, 2468.
CONTAMINE (Philippe), 1020, 3367.
CONTE (Domenico), 758.
CONTE-HELM (Marie), 7892.
CONTI (Fulvio), 4815.
CONTI (Giuseppe), 6991.
CONTRERAS (Carlos A.), 4732.
CONWAY (Stephen), 7827.
CONYBEARE (C.), 3014.
CONZE (Eckart), 7147.
COOK (Brad L.), 2177.
COOK (Chris), 924.
COOK (Michael), 5690.
COOLEY (A. E.), 2424.
COOMANS (Thomas), 3785.
COOPER (Darius), 6664.
COOPER (Frederick), 7148.
COOPER (Tristan), 4956.
COOTE (Lesley A.), 3629.
COPELAND (David A.), 7729.
Copernicus (Nicolaus), 1200.
COPLEY (A. R. H.), 5691.
COPPENS (E. C.), 1106.
COPPER (John Franklin), 925.
COPSEY (Nigel), 4656.
CORALLUZZO (Valter), 8305.
CORBELLARI (Alain), 739.
CORBET (Patrick), 5583.
CORBIER (M.), 2413.
CORBRIDGE (Stuart), 1038, 4747.
CORCUFF (Philippe), 7452.
CORDIER (Daniel), 8217.
CORDOVEZ (Diego), 8306.
CORIAT (P.), 2644.
Corinna, 2496.
CORLEY (Brigitte), 3786.
CORMACK (Robin), 3160.
CORN (Wanda), 6510.
Corneille (Pierre), 6751.
CORNELIẞEN (Christoph), 775.
Cornelius Nepos, 2807.
CORNELL (T. J.), 2645.
CORNETTE (Joël), 4438, 4476.
CORNI (Gustavo), 5816.
CORNISH (Alison), 3701.
CORNWALL (Mark), 4734.
CORONEO (Roberto), 3787.
CORRADINI (Richard), 447.
CORREIA DE OLIVEIRA ANDRADE (Manuel), 4229.
CORRIGAN (Gordon), 7982.
CORS I MEYA (Jordi), 1846.
CORSI (Francisco Luiz), 8144.
CORTADA (James W.), 5846.
CORTEN (André), 4725.
CORTÉS (Valeria), 768.
CORTÉS TOVAR (R.), 1926.
CORTESE (Delia), 57.
CORTESI (Mariarosa), 3748.

CORTIJO OCANA (Antonio), 759.
CORTINI (Maria Antonietta), 6382.
CORVALÁN MARQUÉZ (Luis), 4277.
COSANDEY (Fanny), 5762.
COSENTINO (S.), 3188.
Cosimo de' Medici, 5822.
COSMAN (Madeleine Pelner), 3553.
COSMANN (Peggy), 760.
COSS (Peter), 1107.
COSSERON (Serge), 7856.
COSTA (Barbara), 5734.
COSTA (Maria), 3268.
COSTA (Pietro), 1037.
Costa (Stefano), 3509.
COSTA I BOU (Joan), 5415.
COSTA PINTO (Antonio), 5005, 5017.
COSTANTINI (Massimo), 1011.
COTLER (Julio), 4973.
COTT (Nancy F.), 7149.
COTTICA (Daniela), 1647.
COTTON (Hannah M.), 698.
Cotugno (Domenico), 6222.
COTULA (Franco), 6998.
COUÉGNAS (Daniel), 6371.
COUFOUDAKIS (Van), 8418.
COULET (Noël), 3593.
COULIÉ (Anne), 2328.
COULIE (Bernard), 2995, 3037.
COULSON (Frank T.), 3702.
Cournot (Augustin), 470.
COURTENAY (William J.), 3770.
COURTNEY (Cecil P.), 699.
COURVILLE (Serge), 229.
COUSSY (Denise), 6443.
COVA (Anne), 5537.
COWDREY (H.E.J.), 3920.
COWEN (David J.), 6999.
COX (Howard), 6920.
COX (Michael), 4770.
COX (Philip), 1442.
COZZOLI (Umberto), 2056.
CRABB (Ann), 115.
CRACA (C.), 2734.
CRAEYBECKX (Jan), 4210.
CRAFT (S. G.), 8145.
CRAFTS (N. F. R.), 6783.
CRAGOE (Matthew), 5399.
CRAIG (Douglas B.), 6665.
CRAINZ (Guido), 4816.
CRAMER (Robert), 8097.
CRAMPTON (R. J.), 926.
CRANE (Diana), 7150.
CRANE (Eva), 1485.
CRANE (Susan A.), 387.
CRANSTON (Jody), 6583.
CRAPOL (Edward P.), 7893.
Cratinus, 2190.
CRAVERI (Piero), 7283.

CRAWFORD (M.), 2121.
CRAWFORD (Patricia), 7151.
Creagh (Richard), 5482.
CREMASCHI (Sergio), 6824.
CRÉMIEUX-BRILHAC (Jean-Louis), 4415, 8218.
CREMONINI (Giorgio), 6666.
CRÉPIN (Thierry), 1275.
CRÈPON (Marc), 656.
CRESPO SOLANA (Ana), 6921.
CRESSIER (Patrice), 3474.
CREUZBERGER (Stefan), 627.
CRIMI (Carmelo), 2998.
CRISCUOLO (Eduardo Luis), 898.
CRISCUOLO (Ugo), 477, 3189.
CRISP (Roger), 1980.
CRISTIAN (Luca), 3554, 7789.
CRISTIANI (Emilio), 718.
CRITTENDEN (Camille), 6667.
ČRNUGELJ (Franc), 8219.
Croce (Benedetto), 658, 6204.
CROCKER (Richard L.), 3832.
CROCKETT (Clayton), 1347.
Croesus, re di Lidia, 1653.
CROFT (Paul), 1531.
CROFT (Sharon), 1542.
CROISILLE (Christian), 6430.
CROIZY-NAQUET (Catherine), 448, 639.
Croker (John), 690.
CROMEY (R. D.), 2097.
Cromwell (Oliver), 4680.
CRONE (Patricia), 3456.
CRONIN (Mike), 4772.
CRONIN (Stephanie), 4759.
CROOK (John), 3958.
CROPP (M. J.), 1993.
CROPPER (Elizabeth), 6584.
CROSBY (Everett U.), 3362.
CROSSLAND (Zoë), 1838.
CROUCH (David J.F.), 3384, 3555.
CROUZET (Denis), 5225, 5635.
CROWLEY (David), 6634.
CRUBÉZY (E.), 1604.
CRUMMEY (Donald), 8793.
CRUNDEN (Robert M.), 5763.
CRUSELLS (Magi), 6668.
CRUZ (Jesus), 5154.
CRUZ ARTACHO (Salvador), 5203.
CRUZ HERNANDEZ (Miguel), 3872.
CRUZ INFANTE (José Abigail), 4385.
CSÉFALVAY (František), 8104.
Ctesias, 2022.
CUBELLI (V.), 2123.
CUBITT (Geoffrey), 961.
CUENCA TORIBIO (D. José Manuel), 708, 5189.
CULLA I CLARÀ (Joan B.), 5158.
CUMBER (Janey), 188.
CUMINETTI (Benvenuto), 6661.

CUMONT (Franz), 659, 2846.
CUNCHILLOS (Jesús-Luis), 1747.
CUNHA (Norberto Ferreira da), 5915.
CÜNNEN (Janina), 3704.
CUNNINGHAM (Andrew), 5639.
CUNNINGHAM (Clark E.), 7131.
CUNNINGHAM (Hugh), 7315.
CUNNINGHAM (Martin G.), 3833.
CUNZ (Reiner), 160.
CURCIO (Giovanna), 6568.
CURRAN (J.), 3054.
CURRAN (Stuart), 6472.
CURRIE (Elizabeth J.), 1593.
CURRY (Anne), 3417, 3424, 3628.
CURTHOYS (Mark C.), 5929.
CURTI (Danilo), 3838.
CURTIN (Philip D.), 4052.
Curtin (Ronald), 7322.
CURTIS (Bruce), 6233.
CURTIS (J.), 1781.
CURTIS (Penelope), 6585.
CURTIS (Sarah A.), 5973.
CURVERS (Hans H.), 1668, 1795.
Curzon of Kedleston (George Nathaniel, sir), 8009.
Cusin (Fabio), 660.
Cuthbert, Sanctus, 3969.
CUTINO (M.), 3009.
CUTLER (William W.), 5974.
Cvetic (Matthew), 5335.
CYGAŃSKI (Mirosław), 8307.
Cyprianus, episcopus Carthaginiensis, Sanctus, 3036, 3068.
Cyrillus Alexandrinus, Sanctus, 2984-2986.
CZERNETZKY (Günter), 5054.
CZERNIAKOWSKA (Małgorzata), 5916.

D

D'AGOSTINO (Salvo), 6234.
D'Agoult (Marie), 5888.
D'AIUTO (Francesco), 3999.
D'ALFONSO (Lorenzo), 1808.
D'AMBROSIO (Gaetano), 85.
D'ANGIO (Agnès), 6858.
D'ANNA (G.), 3003.
D'ARIENZO (Luisa), 1302.
D'ARMS (J. H.), 2420, 2882.
D'AVRAY (Anthony), 4402.
D'INTINO (Franco), 6468.
D'IPPOLITO (Federico M.), 2647.
D'MONTE (Rebecca), 5780.
D'ORAZI FLAVONI (Francesco), 4748.
D'ORSI (Angelo), 5774, 6116.
D'URSEL (Caroline), 3485.
DA COSTA (Cláudio), 6669.
DA PASSANO (Mario), 1179, 7512.

DA PAZ (Alfredo), 6511.
DA SILVA ROQUETTE LOPREATO (Christina), 4232.
DABASHI (Hamid), 4761.
DABBAB (Mohamed), 7696.
DAGER ALVA (Joseph), 709.
DAGRADA (Elena), 6670.
DAGRON (G.), 3121.
Dagron (Gilbert), 3166, 3170.
DAHL (Nils Alstrup), 3025.
DAHL (Per), 280.
DAHLBERG (Hans), 5210.
DAHLHEIM (Werner), 2554.
DAI (Jianguo), 8527.
DAILEADER (Philip), 3332.
DAIN (Phyllis), 350.
DAINTREE (D.), 2736.
DAL BONI (Diego), 4971.
DALBY (A.), 2682.
DALENA (Pietro), 3556.
DALFEN (J.), 2817.
DALGÅRD (Sune), 7152.
DALL'AGLIO (Pier Luigi), 285.
DALLA CASA (Brunella), 4817.
Dalle Mole (Riccardo), 4814.
DALLIN (Alexander), 5042.
DALONGEVILLE (Rémi), 1522.
DALSGAARD LARSEN (Keld), 6857.
DAMAMME (Dominique), 651.
Damasceno (Giovanni), 3053.
DAMINA-GRINT (Peter), 449.
DAMODARAN (A. K.), 8308.
Damon of Chaeronea, 2069.
DAMSHOLT (Tine), 4366.
DANDRAU (Alain), 1945.
DANEK (G.), 3142.
DANIEL (Elton L.), 4762.
DANIEL (Pete), 5299.
DANIEL (Thomas), 6235.
Daniele, profeta, 306.
DANIELS (Roger), 288, 7577.
DANILEVSKIJ (Igor' N.), 3333.
DANILOV (V. P.), 6957, 6958.
DANNA (Bianca), 6444.
DANOWSKA (Bogumiła), 4985.
Dantal (Charles), 5890.
Dante Alighieri, 1443, 3692, 3693, 3695, 3701, 3737, 3740.
DARBO-PESCHANSKI (Catherine), 2242.
DARBYSHIRE (Gareth), 1648.
DARD (Olivier), 6801.
DARDAINE (S.), 2413.
Darius I, king of Persia, 1892, 1908.
DARK (Petra), 1511.
Darlan (François), 4441.
DARLAS (Andréas), 2329.
Darmesteter (James), 1357.
DARNTON (Robert), 5765.
DARQUES (Régis), 7052.

DARRIGOL (Olivier), 6236.
DARROW (Margaret M.), 5766.
Darwin (Charles), 6252, 6334.
DARWIN (John), 7645.
DASQUE (I.), 7894.
DASSMANN (Ernst), 1388, 3047.
DAUBEN (Joseph W.), 1184.
DAUMAS (Jean-Claude), 6801.
Daumier (Honoré), 6595.
Daunou (Pierre Claude François), 7374.
DAUNTON (Martin), 7049.
DAUPHIN (Cécile), 5767.
DAUVOIS (Nathalie), 6383.
DAVAZ (Kemal Özcan), 5249.
DAVICO BONINO (Guido), 1291.
DAVID (Bruno), 1525.
David (Jacques-Louis), 6605.
DAVID (Jean-Michel), 2555.
David (Louis), 6606.
DAVID (Robert G.), 230.
DAVID-FOX (Michael), 5903.
DAVIDOFF (Leonore), 471.
DAVIDS OLIEN (Diana), 6890.
DAVIDSON (Clifford), 3831.
DAVIDSON (Pamela), 1468.
DAVIE (Grace), 5417.
DAVIES (Adrian), 5538.
DAVIES (G.), 1649.
DAVIES (Horton), 5692.
DAVIES (J.), 2131.
DAVIES (Joan), 4657.
DAVIES (Marie-Hélène), 5692.
DAVIES (P. J. E.), 2883.
DAVIES (Peter Jonathan), 8220.
DAVIES (Peter), 5768.
DAVIES (R. E. G.), 6860.
DAVIES (Sam), 7319.
DAVIS (Adrian), 8552.
DAVIS (Belinda J.), 4544.
DAVIS (Colin J.), 7319.
DAVIS (David Brion), 1129.
DAVIS (Derek H.), 5471.
DAVIS (Lisa Fagin), 3834.
DAVIS (Michael T.), 4694.
DAVIS (Mike), 7053.
DAVIS (Natalie Z.), 895.
DAVIS (Simon), 1007.
DAVIS (Tracy C.), 6671.
DAVIS HOWLAND (Nina), 8258.
DAVISON (Graeme), 450.
DAVRIL (A., O.S.B.), 3854.
DAVRIL (Anselme), 3933.
DAWSON (S.), 678.
DE ALENCASTRO (Luiz Felipe), 4233.
De Amicis (Edmondo), 6444.
DE BARY (Wm. Theodore), 8785.
DE BEAUREPAIRE (Fr.), 3350.
De Béthune (Evrard), 3876.
DE BLOIS (L.), 2683.

DE BOER (Esther), 3106.
De Borja (Francisco), 7794.
De Brinon (Fernand), 7990.
DE CARLOS MORALES (Carlos Javier), 5153.
DE CAROLIS (E.), 2847.
De Carturia (Angelo), 3155, 3270.
DE CASTELNAU-L'ESTOILE (Charlotte), 5593.
DE CAZANOVE (O.), 2124, 2421.
DE CECCO (Marcello), 6997, 7035.
De Certeau (Michel), 798.
DE CESARE (Giovanni Battista), 3705.
DE CLERCQ (Jan), 5988.
DE DAINVILLE-BARBICHE (S.), 7783.
DE DREUILLE (Christophe), 3058.
DE DURAND (G. M.), 3004.
De Felice (Franco), 4843.
De Felice (Renzo), 661.
DE FILIPPIS CAPPAI (Ch.), 3011.
DE FLEURQUIN (L.), 1297.
DE GAULLE (Charles), 4415, 4421, 4461, 4478, 4493, 8218, 8279, 8355.
DE GAULLE (Philippe), 4416.
DE GIORGI (Alvaro), 7316.
De Girardin (Delphine Gay), 5888.
DE GIUSTI (Luciano), 6643.
De Goncourt (Edmund et Jules), 6441.
DE GRAND (Alexander), 4818.
DE GRAZIA (Victoria), 7054.
DE GROSSI MAZZORIN (Jacopo), 2684.
DE GUILLENCHMIDT (Michel), 7406.
DE GUIO (A.), 2374.
DE HEMPTINNE (Thérèse), 39.
DE HESSE (Paul), 8794.
De Joanes (Joan), 6586.
DE JOINVILLE (Jean), 3953.
DE JONG (Albert), 4905.
DE JONGH (Joané), 1591.
DE KEUNING (M.), XI.
De L'Orme (Philibert), 6562.
DE LA BARRERA (José Luis), 2884.
DE LA CERDA (María Soledad), 4278.
DE LA CRUZ PALMA (Óscar), 3128, 3981.
DE LA FUENTE (Ariel), 451, 4147.
DE LA FUENTE COBOS (Concepción), 3282.
DE LA FUENTE COLLELL (Pere), 6340.
DE LA FUENTE MONGE (Gregorio), 5156.
DE LA GRANJA SAINZ (José Luis), 5157.

De la Houssaye (Amelot), 600.
De la Rosa (Luis), 7896.
DE LA TORRE (Carlos), 4389.
DE LA TORRE GÓMEZ (Hipólito), 5006.
DE LAMARTINE (Alphonse), 6430, 6430.
De Lapouge (Valcher), 6264.
De Léry (Jean), 889, 6381.
DE LIGT (L.), 2685.
DE LONGIS (Rosanna), 8237.
DE LUCA (Vittorio), 5509.
DE LUCINGE (René), 7790.
De Lumbres (Antoine), 7834.
DE LUMLEY (Henry), 2329.
DE LUXÁN MELÉNDEZ (Santiago), 7021.
De Maddalena (Aldo), 662.
De Maistre (Joseph), 6180.
De Maistre (Xavier), 6435.
DE MARTINO (Stefano), 1783, 1809.
DE MATOS GOMES (Carlos), 7689.
DE MESA GUTIÉRREZ (José Luis), 7694.
De Miranda (Francisco), 146.
DE MIRANDA (Girolamo), 6384.
De Morkeba (Guillelmi), 3847.
DE NADAÏ (J. C.), 2737.
DE NOBILI (Roberto), 5578.
De Novais (Paulo Dias), 4139.
DE OYUELA (Leticia), 4731.
DE PLANHOL (Xavier), 927.
DE POLIGNAC (François), 2885.
DE POMMEROL (Marie-Henriette Jullien), 366.
De Porcioles (Josep Maria), 5180.
DE RIDDER (P.), 324.
DE RIQUER (Borja), 5158.
DE RISIO (Carlo), 8146.
De Rivo (François), 663.
DE RIZ (Liliana), 4148.
DE ROBERT (Philippe), 7636.
DE ROJAS (Fernando), 6408.
DE ROOY (P.), 7228.
DE ROSA (Gabriele), 5535.
DE ROSA (L.), 6836, 6871, 7153.
DE SALAS (Jaime), 5933.
De Sanctis (Francesco), 656, 6478.
DE SANTIAGO FERNANDEZ (J.), 7001.
De Scudéry (Madeleine), 6410.
De Seyssel (Claude), 432.
DE SOUSA (Marcelo Rebelo), 5007.
DE SOUSA FRANCO FILHO (Georgenor), 7300.
DE SPIRITO (Angelomichele), 5539.
DE TITTO (Ricardo), 4149.
DE VINCENTI (Giorgio), 6672.

DE VOGÜÉ (Adalbert), 2994.
DE VRIES (David), 7319.
DE VRIES ZANUCCOLI (Luciana), 5540.
DE WAELE (Michel), 4439.
DE WAILLY (Henri), 4440.
De Wette (W.L.M.), 6159.
DE WIT (O.), 6767.
DE WITTE (Ludo), 4298.
DE ZAVALA MORENCOS (Ignacio), 1589.
De' Medici (Cosimo), 5822.
DEÁK (István), 7355.
DEAKIN (Quentin), 7764.
DEAN (James M.), 3739.
DEAN (Martin), 4202.
DEAN (Trevor), 4032.
DEAN ANDERSON (R.), 2178.
DEARNE (M. J.), 2587.
DEBBY (N. B-A.), 3443.
DEBORD (André), 3364.
DECAUX (Alain), 4441.
DECHERF (Dominique), 647.
DECLERCQ (Georges), 102.
DÉCULTOT (Elisabeth), 738.
DEDIEU (Jean Pierre), 5188.
DEE (J. H.), 2179.
DEFORGE (Bernard), 2180.
DEFRANCE (Corinne), 5917.
Degas (Edgar), 6611.
DEGIEL (Rafał), 5612.
DEGLER-SPENGLER (Brigitte), 5460.
DEGN (Ole), 4363.
DEHAUDT (C.), 4442.
DEHEE (Yannick), 6673.
DEHN-NIELSEN (Henning), 4369.
DEISSLER (J.), 2098.
DEJMEK (Jindřich), 7983, 8141.
DEKKER (Ton), 881.
DEL BAGNO (Ileana), 7513.
DEL BOCA (Angelo), 8190.
DEL BONO (Gianna), 351.
DEL LUCCHESE (A.), 1512.
DEL LUNGO (Stefano), 3457.
DEL MASTRO (G.), 2422.
DEL MONACO (Lavinio), 1960.
Del Pulgar (Fernando), 493.
DEL REFUGIO GONZÁLEZ (María), 7948.
Del Rio (Martin), 5519.
DEL RÍO ALDAZ (Ramón), 5159.
DEL RIO BARREDO (María José), 7055.
DEL TACCA (MARIO), 5918.
Del Treppo (Mario), 3347.
DELACAMPAGNE (Christian), 6128.
DELANCEY (Mark Wakeman), 928.
DELARUE (F.), 2738.
DELATTRE (Simone), 7154.
DELCOURT (Thierry), 3707.

DELEIS (Mónica), 4149.
DELFOSSE (Heinrich P.), 6114.
DELIBES DE CASTRO (Germán), 1513.
Délicieux (Brother Bernard), 3986.
DELLA VALLE (Antimo), 5613.
DELON (Michel), 5769, 5887.
DELPAR (Helen), 959.
DELSALLE (Paul), 325, 4056.
DELUMEAU (Jean), 1206, 1304, 5396.
DELUMEAU (Jean-Pierre), 3341.
DELUSSU (Fabrizio), 2886.
DELUZ (Christian), 3310.
DEMAND (N.), 2587.
DEMANTOWSKY (Marko), 399.
DEMARIA (Enrico), 3827.
DEMARS (Pierre Yves), 1507.
DEMEL (Walter), 4053.
Demetrius Poliorcetes, 2058.
DEMICHELI (A. M.), 3190.
DEMIDOV (Sergej V.), 8147.
DEMIDOVA (N. F.), 7824.
DEMIR (Remzi), 1197.
DEMİRKAN (Tarık), 4735.
DEMM (Eberhard), 5770.
Democritus, 2265.
DEMONT (Paul), 2298.
Demosthenes, 1989, 1990, 2110, 2177, 2224.
DEMPSEY (Charles), 6584.
DEMPSEY (Corinne), 1348.
DENÉCHÈRE (Yves), 7984.
Deneckes (Ludwig), 66.
DENEMARKOVÁ (Radka), 6500.
DENG (Kent G.), 6755.
Deng (Yanda), 8606.
DENG (Ye), 8528.
DENG (Yibing), 8529.
Denina (Carlo), 664.
DENIS (Benoît), 6352.
DENİZ (Bekir), 1130.
DENKER (Alfred), 6193.
DENLEY (Peter), 5930.
DENNIS (Mike), 4548.
DENNISON (Lynda), 3805.
DENOIX (Sylvie), 3460.
Denon (Dominique-Vivant), 394, 6535.
DENOON (Donald), 8825.
DENTAN (Paul-Emile), 5636.
DENTI (Mario), 2330.
DENZEL (Markus A.), 6927.
DEOL (Harnik), 5418.
DEPAEPE (Marc), 5975.
DEPAULE (Jean-Charles), 262.
DEPEW (M.), 2181.
DEPTUŁA (Czesław), 673.
DER MÜLBE (Wolf-Christian von), 6576.
DEREGNAUCOURT (Gilles), 5448.

DEREMETZ (A.), 2848.
DERICKSON (Alan), 7317.
DEROCHE (Andrew), 8310.
DÉROCHE (François), 58.
DERORI (Ze'ev), 4787.
DEROUEN (Karl R., Jr), 964.
DEROUX (Carl), 2739, 2806.
Derrida (Jacques), 6174.
DERVILLE (Alain), 3557.
DESA WIGGINS (Peter), 6385.
DESBORDES (Olivier), 3888.
Descartes (René), 1413, 6129, 6130, 6197.
DESCLOS (Marie-Laurence), 879.
DESFORGES (Michel), 4444.
DESLIPPE (Dennis A.), 7318.
DESSE (Jean), 1650, 2887.
DESSE-BERSET (Nathalie), 1650, 2887.
DESZERI (K.), 7003.
DETIENNE (Marcel), 762.
DETTENHOFER (Maria H.), 2556.
DETTI (Tommaso), 4054.
DEUSTUA (José R.), 6861.
DEUTSCH (Nathaniel), 5686.
DEUTSCH (Robert), 137.
DEUVE (J.), 4893.
DEVILLERS (O.), 2511.
DEVINE (T. M.), 5637.
DEVROEY (Jean-Pierre), 3558.
DEWEY (Scott Hamilton), 1132.
DEYERMOND (Alan), 3735, 3736.
DEZEM (Rogério), 4224.
DEZZA (Ettore), 7514.
Di Banco (Nanni), 6577.
DI BENEDETTO (Arnaldo), 6445.
DI BERARDINO (Angelo), 3053.
DI BIAGIO (Anna), 5049.
DI COSTANZO (Giuseppe), 683.
DI DONATO (Riccardo), 697, 732.
DI FAZIO (M.), 2396.
DI GENNARO (F.), 2375, 2918.
DI GIOVINE (C.), 2476.
DI GIUSEPPE (H.), 2918.
DI GIUSEPPE (Riccardo), 2500.
DI MARCO (Massimo), 2183.
DI SALVO (Lucia), 2982.
DI SIMONE (Maria Rosa), 5951.
DI SIMPLICIO (Oscar), 6238.
DI ZIO (Tiziana), 319.
DIAFANI (Laura), 6446.
Diakonoff[Igor (M.), 1633.
DIAMANT (Neil J.), 8531.
DIANTEILL (Erwan), 5693.
DÍAZ (María Elena), 7734.
Díaz (Porfirio), 4915.
DÍAZ DE CERIO DÍEZ (M.), 2027.
DÍAZ NIEVA (José), 4279.
DIAZ-MAS (Paloma), 6408.
DICHTBURN (David), 3365.
DICKINSON (Gary), 3902.

DICKINSON (H. T.), 7402.
Diderot (Denis), 6674.
DIDIER (Hugues), 5609.
DIDI-HUBERMAN (Georges), 1268.
DIEBOLD (William J.), 3788.
DIEHL (Gerhard), 452.
DIENEL (Hans-Liudiger), 6862.
DIERICHS (A.), 2888.
DIERIKX (Marc), 6863.
DIERKENS (Alain), 233.
DIETRICH (Manfried), 1651.
DIETZ (G.), 2184.
DIETZ (K.), 2557.
DÍEZ MERINO (Luis), 1847.
DIGGINS (John P.), 5300.
DIHLE (A.), 3047.
DIJKSTRA (Meindert), 1652.
DILG (Peter), 6239.
DILLON (Janette), 6675.
DILLON (S.), 2889.
DILTHEY (Wilhelm), 656, 789, 6131.
DIMITROV (Georgi), 5041, 5042.
DIMITROV (S.), 400.
DIMOU (Élefthéria), 1946.
DIMUNDO (R.), 2740.
DINER (Hasia R.), 5694.
DING (Zeqin), 8532.
Dingane, second Zulu king, 802.
DINGERS (Martin), 1177.
DINGWALL (Helen M.), 6240.
Dio Chrysostomus, 1991, 1992, 2169.
Diocletianus (Gaius Aurelius Valerius), Roman emperor, 2693, 2702.
Diodorus Siculus, 635.
Diogenes Oenoandensis, 2279, 2288.
DION (Emmanuel), 948.
DION (P.-E.), 1748.
Dionysius Halicarnassensis, 665, 2469, 2641.
Dionysius I, 2063.
DIOUF (Eu.), 2648.
DIPPELREITER (Michael), 4175.
DIPPER (Christof), 763, 1138, 7388.
DIRLIK (Arif), 492, 764.
DITHMAR (Christiane), 5695.
DIU (Isabelle), 71.
DİVİTÇİOĞLU (Sencer), 3373.
DIXIT (Jyotindra Nath), 4122.
DIXON (Chris), 4726.
DIXON (Karen R.), 2962.
DIXSAUT (M.), 2243.
DLUBEK (Rolf), 4545.
DMITRIEVA (Marina), 7071.
DO AMARAL (Ilídio), 4139.
DOBIAS-LALOU (Catherine), 1961.
DOBOSZ (Józef), 3349, 5577.

DOBSON (B.), 2695.
DOBSON (Barrie), 944.
DOBSON (David), 7823.
DOCKRAY-MILLER (Mary), 3559.
DOCKRILL (M. L.), 8252.
DOCKRILL (Saki), 8303, 8312.
DOCTER (Roald F.), 2943.
DODARO (R. J.), 2973.
DODD (Diane), 454.
DODDE (Nan L.), 1213.
DODOLEV (Mikhail A.), 7857.
DODWELL (Charles Reginald), 3708.
DOENECKE (Justus D.), 8149.
DOERRIES (Reinhard R.), 7985.
DOGAN (Ismail), 9.
DOĞER (Lale), 2331.
DOGLIO (Maria Luisa), 5772.
DOGNINI (C.), 2501.
DOĞRU (Halime), 1133.
DOĞRUEL (Fatma), 1134.
DOLAN (Brian), 234, 7076.
DOLBEAU (François), 207.
DOLCI (Fabrizio), 8237.
DOLCINI (Carlo), 1066.
DOLGOVA (S. R.), 5043.
DOLHAR (Erik), 8313.
DOLL (Peter M.), 4658.
DOLLINGER (Marc), 5301.
DOLLINGER (Sol), 7320.
DOLUKHANOV (Pavel M.), 1486.
DOMANSKI (Marian), 1487.
DOMBRADI (Eva), 1749.
DOMÉNECH (F. Benito), 6586.
DOMÈNECH FAUS (Elisa María), 1544.
DOMINGUES (Ângela), 7735.
DOMÍNGUEZ ORTIZ (Antonio), 7522.
DOMÍNGUEZ SÁNCHEZ (Santiago), 3266.
DOMINICZAK (Henryk), 4986.
DOMINIK (William J.), 2741.
DOMINZAÍN (Susana), 7316.
DOMPNIER (Bernard), 92.
DONAHUE (R. E.), 1506.
Donatello (Donato di Betto Bardi), 6579.
DONATO (Maria Monica), 3815.
DONATO (Maria Pia), 4805, 5919.
DONCHEV (Doncho), 116.
DONGA-SYLVESTER (Eva), 5054.
DONIGER (W.), 2185.
Donne (John), 6385.
DOOLEY (Brendan), 455.
DÖPP (Siegmar), 1444.
DORAN (Susan), 7791.
DORANDI (Tiziano), 10.
DORATI (M.), 2004.
DORE (Elizabeth), 7191.
DORÉ (Joseph), 5574.

DÖRING (Detlef), 5924.
DORIVAL (G.), 2850, 2863.
DORMEYER (Detlev), 5433.
DORN (Matthias), 6241.
DOROSZEWSKI (Jerzy), 4987.
DORSCH (Klaus-Dieter), 2890.
DOS SANTOS (Ricardo Evaristo), 7736.
DOS SANTOS ALVEZ (Jorge Manuel), 7598.
DOSENRODE (Søren), 4367.
DOSSE (François), 765.
DOSSI (Eugenia), 940.
DOTOLI (Giovanni), 6386.
DOUGLASS (John Aubrey), 5976.
DOUWES (Dick), 5232.
DOVE (Alfred), 711.
DOVERT (Stéphane), 4756.
DOWDEN (Ken), 1305.
DOYLE (Shane), 7056.
DOYLE (William), 5541, 7455.
DRABBLE (John H.), 6768.
Dracontius (Blossius Aemilius), 2988.
DRAELANTS (Isabelle), 3656.
DRĂGAN (Ioan), 7155.
DRÄGER (Michael), 2558.
DRAKE (H. A.), 3055.
DRÁPALA (Milan), 6063.
DRAPER (J. A.), 2987.
DRAYTON (Richard), 6242.
DRECOLL (Carsten), 235.
DRECOLL (Volker Henning), 7792.
DREW (Allison), 5129.
DREWS (Robert), 1750.
DREYER (Boris), 2058.
DREYFUS (Michel), 4106.
DREYFUS-ARMAND (Geneviève), 5739.
DRGRELL (Ole Peter), 5639.
DRIESSEN (J. M.), 1576.
DRINKWATER (J. F.), 2559.
DRINOT DE ECHAVE (Paulo), 4974.
DRIVER (Toby), 1594.
DRNOVŠEK (Darinka), 5114.
DROBNER (Hubertus R.), 2974, 2975, 3000.
DRORY (Rina), 3453.
DROUX (Joëlle), 8097.
DRUKS (Herbert), 8314.
DRUMMOND (Andrew), 2560.
DRUMMOND (Diane K.), 5934.
DRWESKI (Bruno), 4900.
DRYBURGH (Marjorie), 8150.
DRYDEN (Linda), 6447.
DRYSDALE (John Gordon Stewart), 5122.
DU (Xuncheng), 8533.
Du (Yuesheng), 8628.
DU BOIS (P.), 7582, 8315.
DU RÉAU (Elisabeth), 8045.

DUAN (Tanli), 8534.
DUARTE (Angel), 5160.
DUARTE (Carlos F.), 1269.
Duarte (Nestor), 4240.
DUARTE CASANUEVA (Felipe R.), 4150.
DUBICKI (Tadeusz), 8222.
DUBOFSKY (Melvyn), 7321.
Dubois (Guillaume), 7826.
DUBOW (Saul), 6321.
DUBROVIN (Gennadij E.), 3560.
DUBROVSKIJ (I. V.), 3335.
DUBUISSON (Michel), 1199, 2743.
DUBY (Georges), 224, 666, 3272, 3789.
DUCATÉ (S.), 2891.
DUCCINI (Hélène), 4447.
DUCHESNE (Sylvie), 2339.
DUCHET (Claude), 899.
DUCHHARDT (Heinz), 975, 5455.
DUCK (Ian), 6243.
DUCLERT (Baruch), 4492.
DUCLERT (Vincent), 326.
DUCREUX (Marie Elisabeth), 781.
DUCZMAL (Małgorzata), 4736.
DUDZIAK (Mary L.), 5302.
DUECK (D.), 2043.
DUE-NIELSEN (Carsten), 937.
Dufferin and Ava (Frederick Temple Blackwood, Marquis of), 7951.
DUFFIN (Martin J.), 3841.
DUFFY (Michael), 4659.
DUFFY (Seán), 938, 4018.
DUFOUR (Alain), 7790.
DUFOUR (Jean-Louis), 7646.
DUGGAN (Anne J.), 3592.
DUHAMEL (Éric), 4448, 4449.
DUHOUX (Y.), 1947.
Duichev (Ivan), 21.
DUIJZINGS (Ger), 5419.
Duisberg (Carl), 8025.
DULFER (Kurt), 12.
DULLIN (Sabine), 5050, 7986, 8067.
DULOVIČ (Erik), 5094.
DUMAN (Hasan), 6041.
DUMAS (Françoise), 157.
DUMÉZIL (Georges), 1352, 3478.
DUMINUCO (Vincent J.), 5580.
DUMOLYN (J.), 3496.
DUMONT (Dora M.), 6965.
DUMOULIN (Olivier), 640, 651.
DUNKERLEY (James), 8248.
DUNN (Charles W.), 5303.
DUNN (Marylin), 3934.
DUNN (Walter Scott), 8194.
DUNTHORN (David J.), 8316.
DÜNZL (Franz), 3056.
DUPÂQUIER (Jacques), 205.
DUPLUOY (A.), 1653.

DUPRÉ (X.), 2927.
DUPRÉ RAVENTÓS (X.), 2714.
Dupront (Alphonse), 667.
DUPUY (Eliane), 7236.
DURAN (M.), 2186.
DURAND (Bernard), 3498.
DURAND (Frédéric), 4756.
DURAND (Jean-Marie), 1637, 1751.
DURAND (Jorge), 298.
DURAND (Pierre-Michel), 8317.
DURAND (Robert), 3458, 3543.
DURAND (Yves), 4453, 5145.
DURANTI (Francesco), 7407.
ĎURČANSKÝ (Marek), 4320.
Dürer (Albrecht), 6551.
Durfey (Thomas), 6706.
DURRILL (Wayne K.), 5920.
DÜRRING (Norbert), 653.
DUŠANIĆ (S.), 2686.
DUSO (Giuseppe), 6172.
DUSSART (Françoise), 847.
DUSTDAR (Farah), 6132.
DUTTON (Paul V.), 7156.
DUTU (Alesandru), 8195.
DUVAL (Frédéric), 3319.
DUVAL (Sophie), 6353.
DUVAL (Y.), 3057.
Duval (Yvette), 3091.
DUVAL HERNÁNDEZ (Dolores), 7896.
DUYNSTEE (M.), 7368.
DVOŘÁKOVÁ (Daniela), 5095.
DYBDAHL (Audun), 1496.
DYBIEC (Julian), 5921.
DYER (Christopher), 1135, 3561.
DYSON (M.), 2187.
DZHUROVA (Aksiniia), 21, 1217.

E

EAGLES (R. D. E.), 5775.
EARLE (Rebecca A.), 7897.
EASTWOOD (David), 727.
EATON (Richard Maxwell), 5696.
EBENBAUER (Alfred), 3671.
EBENSTEIN (Alan), 1040.
EBENSTEIN (William), 1040.
EBISAWA (Tadashi), 8742.
Eccelus Lucanus, 1415.
ECHAZÚ (Carlos), 4215.
ECK (W.), 2423, 2695.
ECKART (Otto), 3034.
ECKERT (Hans-Wilhelm), 4450.
EDELHEIT (Hershel), 939.
EDELSTEIN (Bruce L.), 1270.
Eden (Anthony), 8169.
EDENS (C.), 1636, 1890.
EDER (Franz X.), 7239.
EDGERTON (Robert Breckenridge), 4361.
EDLING (Max M.), 5304.

EDMONDS (Richard Louis), 8619.
EDMONDSON (J.), 2105.
EDOUARD (Sylvène), 5192.
EDROIU (Nicolae), 117.
EDSFORTH (Ronald), 5305.
Edward I, king of England, 3408.
Edward II, king of England, 3408.
Edward III, king of England, 3409.
Edward VI, king of England, 5653.
EDWARDS (A. S. G.), 60.
EDWARDS (C.), 2516.
EDWARDS (Clive), 6620.
EDWARDS (Elizabeth), 5979.
EDWARDS (J. R.), 6757.
EDWARDS (Jeremy), 7243.
EDWARDS (M. J.), 2188.
EFFE (Bernd), 2823.
EFFENDI (M. Y.), 4959.
EFIMOV (Alexei), 2355.
EGAN (V.), 1752.
EGANA BARAONA (María Loreto), 5980.
EGELHAAF-GAISER (Ulrike), 2851.
EGGER (Christoph), 13.
EGGERT (Gerald G.), 7322.
EGGERT (Katherine), 6387.
Egk (Werner), 6698.
EHLERS (Joachim), 3385.
EHMER (Josef), 7117.
EHRENFREUND (Jacques), 4550.
EIBL (Franz), 8241.
EICH (Armin), 2744.
Eichhorn (Wolfgang), 774.
EICHLER (Joachim), 4512.
EIDE (Oyvind), 5697.
EIDE (T.), 1687.
EIGENMANN (Eric), 6356.
EIGLER (U.), 2818.
EINAUDI (Luca), 7005.
Einaudi (Luigi), 6828, 6835.
Einstein (Albert), 6236, 6244.
EISENBERG (Christiane), 7157.
Eisenhower (Dwight David), 8272, 8312, 8326, 8339, 8350, 8401, 8453.
EISENLOHR (Erika), 14.
EISNER (Marc Allen), 5306.
Ekaterina II, empress of Russia, 4120, 7848.
EKLOF (Ben), 5065.
EKONOMOU (A. J.), 3191.
EKREM (Inger), 431.
ËKSHTUT (Semen A.), 456.
EKSTRÖM (Anders), 7158.
EL GAMMAL (Jean), 4475.
EL MACHAT (Samya), 7697.
El'tsin (Boris Nikolaevich) 8467.
ELDEM (Edhem), 7006.
ELDUROV (Svetlozar), 4248.
Eleonora d'Aragona, 1270.
Eleonora di Toledo, 1270.

ELEUTERI (Paolo), 3143.
ELEY (Geoff), 478.
Eliade (Mircea), 1346, 1352, 1390, 1397.
ELIÆSON (Sven), 4947.
Elias Synkellos, 3120.
ELIASSON (Per), 6245.
ELIAV (Mordechai), 7874.
Eliot (George), 523.
Elissa, queen of Carthago, 1852.
Elizabeth I, Queen of England and Ireland, 7791, 7813.
EL-KHAZEN (Farid), 4896.
ELKINS (Caroline), 7698.
ELLEDGE (Paul), 5982.
ELLERBROCK (Karl-Peter), 6814.
ELLI (Maria), 662.
Elliott (John), 668.
ELLIOTT (David W. P.), 5390.
ELLISON (James), 8318.
ELM (Kaspar), 3927.
ELM (S.), 3082.
ELMAN (Benjamin A.), 8535.
Elmham (Thomas), 240.
ELORZA (Antonio), 5172.
ELORZA MAIZTEGUI (Javier), 327.
EL-SABBAN (Sherif), 103.
ELSAESSER (Thomas), 6676.
ELSAGHE (Yahya), 6448.
ELSNER (Jaś), 2562.
EL-SOLH (Raghid), 4953, 7957.
ELTIS (David), 6864, 7647, 7699.
ELZE (Reinhard), 710.
EMANUILOV (Emanuil Georgiev), 8151.
EMCHENKO (Elena B.), 3486.
EMERSON (Everett), 6449.
EMERY (Anthony), 3790.
EMERY (Elizabeth), 457.
EMIGH (R. J.), 6969.
EMKJÆR (Stefan), 8223.
EMMANUEL (Steven), 1413.
EMMES (Manfred), 7599.
Empedocles, 2853.
EMPEREUR (Jean-Yves), 2332.
EMPTOZ (Gérard), 6216.
Ems (Rudolf von), 3742.
EMSLEY (Clive), 7858.
EMURA (Haruki), 8536.
ENCUENTRA ORTEGA (A.), 3016.
Endelechius, 2989.
ENDERE (María Luz), 1488.
ENDERLE-BURCEL (Gertrude), 4170.
ENDERS (Ulrich), 4517.
ENGEL (David), 4551.
ENGEL (Ulf), 8320.
ENGEL (Ute), 6550.
Engelbert von Admont, 3892.
ENGELS (Jens Ivo), 4452.
ENGELSEN RUUD (Arild), 8500.

ENGELSKIRCHEN (Lutz), 6344.
ENGELSTEIN (Laura), 1246.
ENGEMANN (J.), 3047.
ENGERMAN (David C.), 6773.
ENGERMAN (Stanley L.), 1125, 1129, 6864.
ENGLUND (Peter), 5211.
ENGMAN (Max), 5212.
Engracia, Sancta, 3078.
ENGSTROM (Eric J.), 766.
ENNABLI (Liliane), 3791.
Ennius (Quintus), 2457, 2458.
Ennodius (Magnus Felix), 2991, 3069.
Enrique II, rey de Castilla, 493.
Enrique III, rey de Castilla, 493.
ENSALACO (Mark), 4280.
ENSENYAT I PUJOL (Gabriel), 3621.
ENSOLI (Serena), 2321, 4004.
Epicurus, 2265, 2280, 2834.
Epifanio di Salamina, 175.
EPSTEIN (S. R.), 1137.
Erasistratus, 2280.
Erasmus Roterodamus (Desiderius), 5403, 5651, 5675, 5804, 6168.
ERBE (Michael), 4176.
ERCOLANI (A.), 2189.
ERDEM (Ayşe), 5984.
ERDEMGIL (Selahattin), 1645.
ERDENEBAATAR (D.), 1604.
ERGEÇ (Rifat), 1746.
ERICKSON (Peter), 6512.
Erik VII. Pomeranski, 3430.
ERIKSEN (Tore Linne), 8400.
ERIKSSON SKOOG (Gun), 5234.
ERKANAL-ÖKTÜ (Armağan), 1753.
ERLER (M.), 2819, 2832.
ERMANN (Joachim), 2649.
Ermenter, Sanctus, 3960.
ERMER (Matthias), 7007.
Erofeev (Nikolai Aleksandrovich), 7631.
Erpingham (Thomas, sir), 3417.
ERRERA (Andrea), 5542.
ERTUĞ (Füsun), 1654.
ERVINE (Roberta R.), 175.
ESCH (Arnold), 686, 7895.
ESCUDÉ (Carlos), 8423.
ESCUDERO (José Antonio), 5139.
ESCUDERO ANDÚJAR (Fuensanta), 5162.
ESDERS (Michael), 6135.
ESDERS (Stefan), 3905.
ESHERICK (Joseph W.), 7092.
ESLER (Philip F.), 3042.
ESMONDE CLEARY (A. S.), 2563.
ESPADAS BURGOS (Manuel), 5161, 7898.

ESPAGNE (Michel), 5776.
ESPARZA VALDIVIA (Ricardo Cuauhtémoc), 4910.
ESPEJEL LÓPEZ (Laura), 4911.
ESPLUGAS VALDÉS (Héctor), 7903.
Estienne (Henri), 5748.
ETABA OTOA (Didier), 4263.
ETEMAD (Bouda), 7648.
ETHELBERG (Per), 1489.
ÉTIENNE (Roland), 2333, 2687.
ÈTINGOF (Ol'ga E.), 299.
EUBANK (Keith), 8321.
EUGÈNE (Mangalaza), 7718.
EUJANIAN (Alejandro), 438.
Eunomius of Cyzicus, 3101.
Euripides, 1993-1999, 2159, 2180, 2187, 2206, 2220.
Eusebia Augusta, 2438.
Eusebius Caesareensis, 2468, 2992, 3003.
Eustathius Thessalonicensis, 3126, 3127.
Evagrius Scholasticus, 2993.
EVANS (Eric J.), 967, 4661.
EVANS (G. R.), 3873.
EVANS (Gwenogvryn), 3676.
EVANS (Harold), 5307.
EVANS (J. A. S.), 3192.
EVENDEN (Doreen), 6246.
EVERDI (Mustafa), 8463.
EVERETT-HEATH (John), 189.
EVERSON (Paul), 300.
EVERTS (Steven), 8322.
Every (Henry), 8479.
ÉVIEUX (Pierre), 3007.
EWALD (Janet J.), 6922.
EWANS (Martin), 4125.
EWEN (Elizabeth), 7043.
EWERT (Ulf Christian), 3562.
EWIGLEBEN (Cornelia), 2678.
EWING (K. D.), 7516.
EXTERNBRINK (Sven), 7793.
EZE (E. C.), 6136.
Ezechiele, profeta, 306, 1868.
Ezpeleta (José de), 4311.

F

FABBRINI (Fabrizio), 642.
FABEL (Robin F. A.), 7738.
FABI (Lucio), 384.
FABIAN (Ann), 459.
FABIAN (Claudia), 209.
FABIAN (Johannes), 8795.
FABRE (Daniel), 846.
FABRE (Giorgio), 2425.
FABRE (Ginette), 4941.
FABRE (Pierre-Antoine), 5491.
FABRE (Thierry), 271.
FÁBREGAS VALCARCE (Ramón), 1554.
FABRE-SERRIS (Jacqueline), 2868.

FABRIZIO (Daniela), 4964.
FACCHETTI (Giulio M.), 2397.
FACCHINI (Fiorenzo), 1377.
Facio (Bartolomeo), 669.
FAHNENSCHMIDT (Willi), 4552.
FAHRMEIER (Andreas), 7458.
FAIRCLOUGH (H. R.), 2533.
FAIRHEAD (James), 7706.
FAIVRE (Maurice), 7685.
FAIX (Gerhard), 4180.
Faku, king of Mpondo, 7723.
FALCETTO (Bruno), 6042.
FALCO (Raphael), 6677.
FALCONE (Luigi), 3193.
FALDBORG (Henne), 4370.
FALES (F. M.), 1783.
FALK (Harry), 1619.
FALK (Ulrich), 7505, 7517.
FALLER (S.), 1620.
FAŁOWSKI (Janusz), 4988.
FAŁOWSKI (Wojciech), 3323.
FALTIN (Thomas), 6247.
FALZONE (Maria Teresa), 5544.
FAMERÉE (J.), 1297.
FANG (Biao), 8537.
FANG (He), 8538.
FANNING (Ronan), 7953.
FANTALKIN (A.), 1595, 1848.
FANTI (Mario), 190.
FANTINI TERZI (L.), 1759.
FANZUN (Jon A.), 8323.
Faqir of Ipi, 4754.
FARA (Giovanni Maria), 6551.
FARAH (Caesar E.), 7987.
FARASSINO (Alberto), 6678.
FAREWELL (A), 4770.
FARIA (Cristina Isabel), 5008.
FARIA (Telmo), 5009.
FARÍAS (Víctor), 4273, 4281.
FARIOLI (Marcella), 2190.
FARLEY (Brigit), 7988.
FARMER (Sharon), 3940.
FARNIE (Douglas A.), 6946.
FARNSWORTH-ALVEAR (Ann), 7323.
FARR (James R.), 1139.
FARRELL (Sean), 4771.
FARRUGIA (Edward G.), 936.
FARRY (Michael), 7160.
FASANO GUARINI (Elena), 694.
FASSIN (Didier), 6248.
FASSLER (Margot E.), 3903.
FATOUROS (G.), 3194.
FATTORI (Marta), 6137.
Faucon (Jean-Claude), 3716.
FAUGHT (M. K.), 1502.
FAULKNER (N.), 2565.
Faulkner (William), 6471.
FAUPEL-DREVS (Kirstin), 3874.
Faustina (Annia Galeria), minor, Roman empress, 2675.

FAUVE-CHAMOUX (Antoinette), 7161.
FAVEZ (Jean-Claude), 8097.
FAVRHOLDT (David), 1008.
FAWCUS (Peter), 4222.
FAYANT (M. C.), 3138.
FAYE (Cheikh Faty), 5089.
FAYET-SCRIBE (Sylvie), 328.
FAZLHASHEMI (Mohammad), 5779.
FEAR (A. T.), 2688.
FEATHERSTONE (Jeffrey Michael), 3158.
FEATHERSTONE (M.), 3121.
FEBRER Y ROMAGUERA (Manuel Vicente), 3564.
Febvre (Lucien), 651.
FEDERICO (Giovanni), 6775.
FEDERMAYER (Frederik), 143.
FEDERSPIEL (Michel), 2191.
FEDOROWICH (Kent), 7650, 7665.
FEDOSOV (N. F.), 654.
FEENSTRA (R.), 7368.
FEHRENBACH (Heide), 5884.
FEICHTLBAUER (Hubert), 4177.
FEIN (Susanna), 33.
FEINGOLD (Mordechai), 5930.
FEINMAN (Gary M.), 1596.
FEISSEL (D.), 2426.
FELCMAN (Ondřej), 4336.
FELDBÆK (Ole), 937.
FELDKAMP (Michael F.), 5511, 8244.
FELDMANN (R.), 2598.
FELDNER (Josef), 65.
FELDSTEIN (Ruth), 7162.
FELICE (Costantino), 1011.
FELICE (Domenico), 6138.
FELICI (Lucia), 5525.
Felipe II, rey de España, 5163, 5182, 7546, 7794, 7821.
Felipe III, rey de España, 5164, 7780.
Felipe IV, rey de España, 7760.
Felipe V, rey de España, 5179.
FELIU (Lluís), 1754.
FELL (Ulrike), 6249.
FELMY (Karl Christian), 5614.
FELTEN (F.), 2587.
FENG (Wei), 8539.
Feng (Yuxiang), 8595.
FENGER (Ole), 3499.
FENIELLO (A.), 3565.
FENOALTEA (Stefano), 6865.
FENSKE (Hans), 7408.
FENTON (Paul B.), 1353.
FENTRESS (James), 4819.
FEOLA (Raffaele), 7459.
FERAOUN (Mouloud), 8257.
Ferdinand III, röm.-deutscher Kaiser, 7762.

Ferdinando de' Medici, 4811.
FERENC (Tone), 5113, 5115.
FERGUSON (Niall), 7008.
FERGUSON (Ronnie), 6679.
FERLING (John), 5308.
FERNANDES (Bernardo Mançano), 4235.
FERNÁNDEZ ALVAREZ (Manuel Angel), 1429.
FERNÁNDEZ CONTI (Santiago), 5153.
FERNÁNDEZ CORTE (J. C.), 1926.
Fernandez de Quirós (Pedro), 8822.
FERNÁNDEZ MANZANO (Julio), 1610.
FERNÁNDEZ PÉREZ (P.), 7165.
FERNÁNDEZ POSSE (M. Dolores), 1610.
FERNÁNDEZ TERRICABRAS (Ignasi), 5163.
FERNÁNDEZ-ORDÓÑEZ (Inés), 414.
FERNGREN (Gary B.), 1207.
FERNIE (Eric), 3793.
FEROS (Antonio), 5164.
FERRANDI (Giuseppe), 5816.
FERRANTE (D.), 2017.
FERRARI (Ada), 6713.
FERRARI (G. R. F.), 2025.
FERRARI TONIOLO (Chiara), 2984.
FERRARY (Jean-Louis), 1962, 2566.
FERRAZZI (Marialuisa), 6680.
FERREIRA (Fernando), 4152.
FERREIRA BICHO (Nino), 1514.
FERRER (André), 4056.
FERRER I MALLOL (Maria Teresa), 3497.
FERRETTI (Paolo), 2650.
FERREYROLLES (Gérard), 1457.
FERRI (Gian Battista), 7367.
FERRI (Mauro), 674.
FERRIÈS (M.-C.), 2567.
FERRIS (David S.), 6450.
FERRO (Marc), 7989.
FERRONE (Vincenzo), 7373.
FERTIG (Georg), 4059.
FEUCHTWANGER-SARIG (Naomi), 6065.
FEYEL (Christophe), 1963.
FEYEL (Gilles), 6066.
Fichte (Johann Gottlieb), 6172, 6187, 6193, 6198.
FIEDROWICZ (Michael), 3060.
Fielding (Henry), 6416.
FIERS (Stefaan), 4205.
FIGGE (Valerie), 3794.
FIGUEIRA (Daurius), 5238.
FILIP (Václav V.), 138.
FILIPOW (Krzysztof), 173.
FILIPPONE (Mario), 7166.

FILIPPOV (I. S.), 3567.
FILITOV (Alexei), 8289.
FILMER-SANKEY (William), 6023.
FILONI (Marco), 6139.
FIMPEL (Martin), 7460.
FINALDI (Gabriele), 301.
FINCH (Alison), 6451.
FINE (John V. A.), 3195.
FINI (Carla), 2991.
FINK (Jørgen), 4371.
FINKELSTEIN (Andrea), 6827.
FINKELSTEIN (David), 6088.
FINKELSTEIN (I.), 1849.
FINLAY (J.), 7375.
FINLAYSON (Bill), 1528.
Finley (M. I.), 670.
FINO (Antonio), 4820.
FIORANI (Luigi), 5518.
FIORAVANTI (Gigliola), 5900.
FIORAVANZO (Monica), 8128.
FIORE (Fabio), 5421.
FIORE-MAROCHETTI (Elisa), 1686.
FIORINA (Morris P.), 5298.
FIREY (A.), 3500.
FIRPO (Massimo), 5462, 5545, 5620.
FIRSOV (Fridrich I.), 5042.
FISCH (Bernhard), 8324.
FISCHBACH (Michael R.), 4886.
FISCHER (Claudia), 1621.
FISCHER (Didier), 5983.
FISCHER (Joachim), 5784.
FISCHER (Klaus-Dietrich), 1183.
FISCHER (Thomas E.), 461.
FISCHER (Wolfram), 5943.
FISHER (John R.), 7739.
FISHER (Kate), 6250.
FISHWICK (D.), 2892.
FISSORE (Gian Giacomo), 7.
FITSCHEN (Klaus), 3019.
FITZGERALD (Frances), 8325.
FITZGERALD (Timothy), 1308.
FITZGERALD (William), 2745.
FITZMYER (J. A.), 1850.
FITZPATRICK (Sheila), 5084.
FIVAZ-SILBERMANN (Ruth), 8097.
FLACHOWSKY (Sören), 353.
FLAMARION CARDOSO (Ciro), 809.
FLANAGAN (Clare), 5736.
Flandin (Pierre-Etienne), 4474.
FLASCH (Kurt), 5785.
FLASHAR (H.), 2040.
Flaubert (Gustave), 817, 6489.
Flavius Josephus, 671.
FLEET (Kate), 3425.
FLEMING (K. E.), 462.
FLEMMING (R.), 2821.
FLENSTED JENSEN (P.), 2078.
FLETCHER (Alan), 6681.
FLETCHER (Richard), 3318.
FLETCHER (Stella), 945.

FLEUTOT (François-Marin), 8224.
FLIGGE (Jörg), 373.
FLIKSCHUH (Katrin), 6142.
FLINT (Kate), 6513.
FLOOD (Christopher), 5764.
FLOOD (Josephine), 1591.
FLORACK (Arnd), 5828.
FLORES (E.), 2457.
FLORES ARROYUELO (Francisco José), 851.
FLORES RUBIO (J. A.), 3147.
FLORIA (B. N.), 7774.
Florus (Lucius Annaeus), 2459, 3018.
FLORY (David A.), 3985.
FLOWER (Harriet I.), 2427.
FLÜCKIGER (Pierre), 8097.
FLÜGEL (Axel), 569, 1140, 7167.
FLÜGEL (Christof), 2614.
FLUSIN (B.), 3121.
FOEHR-JANSSENS (Yasmina), 3710.
FOFI (Goffredo), 5849.
FOGEL (Joshua A.), 543, 8540.
FÖGEN (Th.), 2746.
FOHLEN (Claude), 7829.
FOL (Aleksandur), 916.
Folengo (Teofilo), 6378.
FOLIN (Marco), 7057.
FOLLAIN (Antoine), 1122.
FOLLET (Simone), 2099.
FOLLI (Anna), 6452.
FOLLIET (Georges), 2977.
FOLLINI (Marco), 4821.
FOLLY (Martin H.), 8152, 8196.
Fomenko (Anatoly T.), 503.
FONNEBESCH-WULFF (Benedicte), 6963.
Fonseca Amador (Carlos), 4940.
FONTAINE (Guy), 6358.
FONTAINE (Jacques), 3633.
FONTAINE (Laurence), 7199.
FONTANA (Benedetto), 6143.
FONTANA (F.), 463.
FONTANA (S.), 2918.
Fontanella (Donato), 3155, 3270.
Fontanelle (Bernard de), 6118.
FOOT (Sarah), 3906.
FORABOSCHI (Daniele), 2123.
FORBES (Neil), 8153.
FORBES MUNRO (J.), 5954.
FORCADE (Olivier), 4448.
Ford (Gerald R.), 8358.
FORD (Patrick K.), 3676.
FOREMAN-PECK (James), 6923.
FORÊT (Philippe), 8541.
FÖRHAMMAR (Staffan), 7168.
FORMENTO (Ettore), 8190.
FORMIGONI (Guido), 7600.
FORMOZOV (Alexandr A.), 6453.
FORNACA (Remo), 5546.

FORNBERG (Tord), 3025.
FORNS (Santiago), 8815.
FORSBERG (Aaron), 6777.
FÖRSTER (Hans), 3061.
FORTE (Isabel), 6067.
FORTE (Ricardo), 4909.
FORTESCUE (William), 4454.
Fortunatus (Venantius Honorius Clementianus), 2946.
FOSCHIA (L.), 3062.
FÖßEL (Amalie), 3366.
FOSSI (Gloria), 1261.
FOSSIER (Robert), 3529, 3569.
FOSTER (Anne L.), 7899.
FOTIOU (A.), 3129.
FOUACHE (Éric), 2326, 2356.
Foucault (Michel), 750, 793, 6174.
FOUCHÉ (Nicole), 7829.
FOUCHER (A.), 2747.
FOUCHER (Antoine), 464.
FOUNTOULAKIS (A.), 2299.
Fouqué (Friedrich de la Motte), 6738.
FOURNET (Jean-Luc), 2127.
FOURRIER (Sabine), 2316.
FOUSEK (John), 5786.
FOWKES (Ben), 4060.
FOWLER (Melvin L.), 7107.
FOWLER (Robert Louis), 465.
FOWLER (Will), 4912.
FOX (Adam), 5787.
FOX (Jo), 6682.
FOX (Josephine), 8542.
FOX (Mary-Jane), 5123.
Fox Morcillo (Sebastián), 759.
FRAGNITO (Gigliola), 5547.
Fragonard (Jean-Honoré), 6606.
FRAHM (Eckart), 1755.
FRAISSE (Philippe), 2354.
FRAKES (R. M.), 2748.
FRANASZEK (Piotr), 7144.
FRANCESCHI (Franco), 1261.
FRANCESIO (M.), 2192.
FRANCH BENAVENT (Ricardo), 5165.
FRANCIA (Luis H.), 7640.
FRANCIONI (Gianni), 7373.
FRANCISCO (Ronald A.), 1042.
Franciscus Assisiensis, Sanctus, 5, 3973.
FRANCK (Georgia), 3063.
FRANCO (Artur Henrique), 5594.
FRANCO MASIDE (Susana), 1554.
Franco y Bahamonde (Francisco), 906, 5014, 5150, 5183, 8138.
François (Autrand), 3354.
François (Etienne), 641.
FRANCOVICH (Riccardo), 4008.
Frandsen (Karl-Erik), 6963.
FRANDSEN (P. J.), 1710.
FRÄNGSMYR (Tore), 5788.

Frank (Hermann Karl), 4323.
FRANKEL (David), 1570.
FRANKEL (J.), 6068.
FRANKEL (Jonathan), 5705.
Franklin (Benjamin), 5273, 7829.
FRANTZ (John B.), 4105.
FRANZ (Corinna), 7990.
FRANZ GÄRTNER (Kurt), 80.
FRASCA-SPADA (Marina), 1190.
FRASCHETTI (Augusto), 292, 2568.
FRASCINA (Francis), 6514.
FRASER (Cary), 8326.
FRASER (P. M.), 199, 1921.
FRATEANTONIO (Christa), 2601.
Frauendorfer (Max), 4615.
FRAZER (William O.), 3610.
FRAZIK (Wojciech), XIV.
FRÉDÉRIC II DE HOHENSTAUFEN, 3634.
FREEDMAN (Lawrence), 8327.
FREEDMAN (Paul), 1165, 3984.
FREEDMAN (Richard), 6683.
FREEDMAN (Robert Owen), 4790.
FREELY (John), 5984.
FREEMAN (Jo), 5309.
FREGUGLIA (Paolo), 3769.
FREI (Norbert), 4528, 4627.
FREIBERG (M.), 104.
FREIBERGER (Oliver), 5698.
FREIFELD (Alice), 4737.
FREITAG (K.), 2334.
FREITAG (Sabine), 7888.
FREITÄGER (Andreas), 5789.
FRÉMEAUX (Jacques), 7880.
Frenay (Henri), 8213.
FRENCH (David), 7991.
FRENCH (H. R.), 7170.
FRENCH (Roger Kenneth), 1201.
FRENDO (D.), 3129.
FRENZ (Thomas), 3259.
FRÈRE (Marie-Soleil), 4211.
FRESE (Matthias), 6875.
FRESNEDA PADILLA (Eduardo), 1583.
Freud (Sigmund), 808.
FREUDING (Christian), 8328.
FREUND (S.), 3036.
FREVERT (Ute), 545, 5790.
FREWER (Andreas), 6251.
FREY (Bruno S.), 7012.
FREY (Jörg), 3027.
FREYBURGER (G.), 2293.
FREYDANK (Helmut), 1622.
FREYSSENET (Michel), 6894.
FRIAS GARCIA (Maria del Carmen), 5548.
FRICKE (Karl Wilhelm), 4554.
FRIEDBERG (Aaron L.), 5310.
FRIEDE (Susanne), 3688.
FRIEDEBURG (Robert von), 7171.
FRIEDEL (A.), 354.

FRIEDER (Ludwig), 5422.
FRIEDLANDER (Alan), 3986.
FRIEDLÄNDER (Saul), 6729.
FRIEDMAN (Anna Felicity), 6331.
FRIEDMAN (Isaiah), 4965.
FRIEDMAN (John Block), 287.
FRIEDRICH (Hans-Edwin), 6454.
FRIEDRICH (Karin), 4061, 5782.
Friedrich I. von Sachsen-Gotha und Altenburg, 4515.
Friedrich II der Großen, König von Preußen, 4120, 5890.
Friedrich II von Hohenstaufen, röm.-deutscher Kaiser, 3397.
Friedrich III, röm.-deutscher Kaiser, 3291, 3302.
Friedrich Wilhelm I, Kurfürsten von Brandenburg, 7851.
FRIEDRICHS (Christopher R.), 7058.
FRIEDRICHS (J.), 2045.
FRIEDRICHS (Jörg), 728.
FRIGO (Daniela), 7773.
FRIISBERG (Claus), 4372.
FRIJTAG (Geraldien von), 8069.
FRITZ (Jean-Marie), 3836.
FROESCHLÉ-CHOPARD (Marie-Hélène), 92.
FROGLEY (M. R.), 1948.
FRÖHLICH (Uwe), 2512.
FROIDEVAUX (Yves), 6992.
Froissart (Jean), 3531.
FROST (Dan R.), 5922.
FROT (Y.), 2971.
FRØYSTAD (Kathinka), 8500.
Frugoni (Arsenio), 672.
FRÜHMORGEN-VOSS (Hella), 20.
FRÜHSTÜCK (Sabine), 7239.
FRUYT (Michèle), 2735.
FRYDE (E.), 3196.
Fu (Ssu-sien), 8652.
FUBINI (Riccardo), 656.
FUCHS (A.), 2193.
FUCHS (Markus E.), 2601.
FUCHS (Rachel), 7518.
FUDGE (Erica), 6221.
FUGAZZOLA DELPINO (M. A.), 1545.
Fujimori (Alberto), 4973, 4975.
FUJITA (Satoru), 8739.
FUKAZAWA (Hideo), 8543.
FULCHER (Jane F.), 6684
FULLER (Robert C.), 5474.
FULLER (Steve), 687.
Fumagalli (Vito), 3599.
FUMAROLI (Marc), 1450.
FUNARO (Liana Elda), 329.
FUNDÁRKOVÁ (Anna), 5096.
FUNG (Edmund S. K.), 8544.
FUNK (Wolf-Peter), 1834.
FUNKE (S.), 2059.

FUREIX (Emmanuel), 466.
Furius Camillus (Marcus), 2546.
FURUMATSU (Takashi), 8545.
FURUSE (N.), 7059.
FURUTA (Kazuko), 8546.
FUSI (A.), 2477.
FUSI AIZPURÚA (Juan Pablo), 5166.
FUTRELL (Alison), 2690.
FYFE (Gordon), 1272.

G

GABACCIA (Donna R.), 4822.
GABANYI (Anneli Ute), 5031.
GABBA (Emilio), 467, 1273, 2569.
GABDRAKHMANOV (Pavel Sh.), 1126.
GABELENKO (Oleg L.), 1932.
GABELLINI (Andrea), 7992.
GABERSCEK (Carlo), 6685.
GÄBLER (Ulrich), 5564.
GABRIEL (Pere), 5160.
GABRIELLI (Patrizia), 5791.
GABRIELSEN (V.), 2131.
Gadamer (Hans Georg), 808.
GADD (Carl Johan), 6966.
GADE (Kari Ellen), 3298, 3475.
GAFFNEY (V.), 2918.
GAIGER (Jason), 6144.
GAILLARD (Jean Michel), 5985.
GAL (Stéphane), 5225.
GALÁN (José M.), 1688.
GALANIDOU (N.), 1948.
GALASSO (Giuseppe), 770, 7283.
GALCHEV (Ilia), 4255.
GALDERISI (Claudio), 3711.
GALE (M. R.), 2749.
Galenus (Claudius), 2821.
GALERA (Andrés), 6252.
GALERA PEDROSA (Andreu), 3960.
GALIBIN (V. A.), 792.
GALIL (Gershon), 279.
Galilei (Galileo), 6241.
GALIMBERTI (A.), 1988.
GALIMBERTI (Umberto), 5423.
GALL (Lothar), 4521.
Gall Anonim, 673.
GALLABARES (Giorgos), 61.
GALLAGHER (Gary W.), 541, 5355.
GALLANT (Thomas W.), 7173.
GALLAZZI (C.), 1684.
GALLICCHIO (Marc S.), 7601.
Gallienus (Publius Licinius Egnatius), Roman emperor, 2691.
GALLIHER (John F.), 7524.
GALLMAN (J. Matthew), 7060.
GALLMAN (Robert E.), 1125.
GALLO (Luigi), 6552.
GALLOWAY (Jess), 2411.
GALMARINI (Hugo R.), 4151.

GALMÉS DE FUENTES (Álvaro), 191.
GALOUNOV (T.), 7519.
GALPERIN (Charles), 6253.
GAMBERALE (Leopoldo), 692.
GAMBLE (Andrew), 4706.
Gamble (C. S.), 1506.
GAMESON (Richard), 62.
GANCHOU (Th.), 3197.
GANCI (Massimo), 7437.
Gandev (Khristo), 4254.
GANGL (Manfred), 5815.
GANTET (C.), 3635.
GANZERT (Joachim), 2893.
GAO (Wangling), 8548.
GAO (Xiang), 8549.
GAPOSCHKIN (M. Cecilia), 302.
GARAVAGLIA (Juan Carlos), 6967.
GARBARINO (Osvaldo), 3339.
GARCIA (José Manuel), 232.
GARCÍA (Juan Manuel), 4386.
GARCIA (Patrick), 468.
GARCÍA BAZÁN (Francisco), 1356.
GARCÍA COLÓN (Pablo), 5020.
GARCÍA HERNÁN (Enrique), 7794.
GARCÍA MARSILLA (Juan Vicente), 3795.
GARCÍA MARTÍNEZ (José Ramón), 7900.
GARCÍA RECIO (Jesús), 1756.
GARCÍA ROMERO (F. A.), 3146.
GARCÍA SCHMIDT (Armando), 5167.
GARCIA-ARENAL (Mercedes), 3474.
GARCÍA-BELLIDO (Ma. Paz), 159.
GARCIN (Jean-Claude), 3459, 3460.
GARÇON (François), 8197.
GARDAZ (Michel), 1357.
GARDET (Claudie), 8329.
GARDIES (André), 6686.
GARDINER (Juliet), 996.
GARFF (Joakim), 6145.
GARFIELD (Seth), 8811.
GARGARELLA (Roberto), 5311.
GARNIER (Claudia), 3386.
GARNSEY (Peter), 2548.
Garosci (Aldo), 674.
GARRARD (John G.), 4058.
GARRETSON (Peter P.), 7061.
GARRISON (D. H.), 2128.
Gärtner (Hans Armin), 2756.
GÄRTNER (T.), 2750.
GARZYA (Antonio), 3022, 3198.
GASCOU (J.), 2428, 2651.
GĄSIOROWSKI (Antoni), 3765.
GĄSIOROWSKI (Stefan), XIV.
GASKILL (Malcolm), 7520.
GASPARRI (Stefano), 652.
GASTALDELLI (Ferruccio), 3712.

GASTALDI (Silvia), 2244.
GATHERCOLE (Patricia May), 3796.
GATTARI (Nicola), 7686.
GATTO (Romano), 5923.
GATTO TROCCHI (Cecilia), 1358.
GATTONI (Maurizio), 5512.
GATZ (Erwin), 5476.
GATZ (Karen L.), 8258.
GAUER (W.), 2300, 2398.
GAUER (Werner), 1359.
GAUGER (J.-D.), 1891.
GAUJAC (Paul), 7174.
GAUKROGER (Stephen), 6130.
GAULTHIER (N.), 2964.
GAUNT (Peter), 4660.
GAUTHIER (N.), 3099.
GAUTHIER (Nancy), 3894.
GAUTHIER (Philippe), 1964.
GAUVARD (Claude), 3520.
GAUVIN (Gilles), 5022.
GAVILANES (José Luis), 1448.
GAVRISHINA (Oksana V.), 469.
GAWRECKI (Dan), 4063.
GAWRYCH (George Walter), 8330.
GAZI (Effi), 6254.
GAZIAUX (É.), 1297.
GAZIŃSKI (Jarosław), 6924.
GAZZINI (Marina), 7175.
GE (Fuping), 8550.
GE (Zhaoguang), 8551.
GEAREY (Benjamin R.), 1509.
GEARTY (C. A.), 7516.
GEARY (Laurence M.), 4782.
GEBHARDT (Gabriele), 6131.
GEBHART (Jan), 7602.
GEDE (F. J.), 1524.
GEE (Austin), *VII*.
GEE (Emma), 2822.
GEERLINGS (Wilhelm), 1444.
GEERTZ (Armin W.), 1309, 1317, 1360.
GEHLER (Michael), 8398.
GEHRING (Roger W.), 3064.
GEHRKE (H.-J.), 2100.
GEHRMANN (Rolf), 7062.
GEISSLER (Gert), 5986.
GEJ (Aleksandr N.), 1571.
GELICHI (Sauro), 3758.
GELLING (Margaret), 188.
Gellius (Aulus), 2445.
GELVIN (James L.), 8070.
GEMEINHARDT (Peter), 5828.
GEMELLI (Giuliana), 470.
GEMENNE (Louis), 3683.
GENÇ (Mehmet), 6778.
GENEVIÈVE (Vincent), 161.
GENEVRAY (Françoise), 6455.
GENGENBACH (Heidi), 473.
GENICOT (Luc-Francis), 3485.
GENIS (Vladimir L.), 4763.

GENOVESE (Michael A.), 5334.
GENTA (Enrico), 4823.
GENTILE (Emilio), 4824.
Gentile (Giovanni), 734, 5937.
GENTILE MESSINA (R.), 3130.
GENTILI (Bruno), 2208.
Genucius (Gaius), 2670.
GEORG (Odile), 458.
GEORGE (Marcus Sebastian), 7901.
GEORGELIN (Christine), 3319.
GEORGES (Martin), 489.
GEORGES (Pericles B.), 1892.
GEORGIEV (Ivo), 4250.
GEORGIEV (Velichko), 5792.
Georgius Grammaticus, 3120.
GERA (Deborah Levine), 2000.
GÉRARD (Jean-Louis), 946.
GERARDO MAGNO, 3855.
GERASIMOVA (K.), 5045.
GERBER (Christoph), 1671.
Gerbert de Montreuil, 3691.
GERBOD (Paul), 5793.
Gerdes (Daniel), 5626.
GERDING (Henrik), 2852.
GERETTO (Paola), 6330.
GERGELY (András), 966.
GERHARDT (Michael J.), 7409.
GERLACH (Irene), 1104.
GERMAIN (Lucienne), 4663.
GERMANN (Martin), 6069.
GERMIVNIK (Franciscus, c. m.), 3501.
GERRING (Britta), 139.
GERRITSEN (F. A.), 1795.
GERSHOVICH (Moshe), 7701.
GERŠLOVÁ (Jana), 6779.
GERSON (Stéphane), 7177.
GERVAIS (Raymond R.), 4262.
GERVASONI (Marco), 5794.
GERVERS (Michael), 38.
GEUENICH (Dieter), 3939.
GEVA (Hillel), 1851.
GEWALD (Jan-Bart), 4931.
GEYER-KORDESCH (Johanna), 6255.
GHAI (Raj Krishan), 122.
GHELARDI (Maurizio), 656.
GHIATI (Cl.), *V*.
Ghirlandaio (Domenico), 3778, 6592.
GHISALBERTI (Carlo), 734, 7410.
GHISLOTTI (Stefano), 6661.
GHOSH (Parimal), 7666.
GIACOPINI (Vittorio), 5849.
GIALLOMBARDO (Laura), 144.
GIAMBONINI (Francesco), 6397.
GIANA (Luca), 7461.
GIANNI (Alessandra), 3977.
GIANNI (Andrea), 7462.
GIANNINI (Pietro), 3045.
Giannone (Pietro), 6190.

GIANOTTI (G. F.), 2440.
GIARDINA (Andrea), 476, 2611.
GIBBINS (David), 1757.
GIBERTI (Gian Matteo), 5549.
GIBSON (Jeremy Sumner Wycherley), 118.
GIBSON (John C. L.), 1810.
GIBSON (Marion), 5420.
GIBSON (McGuire), 1758.
GIBSON (Walter S.), 6587.
GIEBEL (Marion), 2031.
GIENOW-HECHT (J. C. E.), 8236.
GIESBERG (Judith Ann), 7178.
GIESE (Wolfgang), 3502.
GIESECKE (Annette Lucia), 2824.
GIESEKE (Jens), 4558.
GIESS BEVILACQUA (Valérie), 2335.
Gieysztor (Aleksander), 675.
GIGANTE LANZARA (Valeria), 2016.
GIGLIO (Francesco), 7521.
GIGLIONI (Guido), 6256.
GIL PECHARROMÁN (Julio), 5168.
GILBERT (Jane), 3743.
GILBERT (Ruth), 6221.
GILDENHARD (I.), 2478.
GILDENHARD (Ingo), 2751, 2752.
GILDERHUS (Mark Theodore), 7603.
Giles (Mary E.), 3911.
GIL-HAR (Yitzhak), 7993.
GILJE (Nils), 1416.
GILL (David W. J.), 693, 1656, 1917.
GILLESPIE (Susan D.), 7131.
GILLESPIE (Vincent), 60.
GILLINGHAM (John), 3387.
GILMAN (Sander), 5796.
GILQUIN (Michel), 7604.
Gil-Robles y Quiñones (José María), 8316.
GILROY (Amanda), 272.
GIMÉNEZ RODRÍGUEZ (Alejandro), 5372.
GINAT (Rami), 8331.
GINATEMPO (Maria), 4008.
GINGELL (John), 1060.
GINIO (Ruth), 5987.
GINZBURG (Carlo), 676, 750, 776.
Gioacchino da Fiore, 3875, 3884.
GIOFFREDI SUPERBI (Fiorella), 3812.
Giolitti (Giovanni), 4818, 5563, 7494.
GIOLLÁIN (Diarmuid Ó), 852.
GIORDANO (Giancarlo), 7994.
GIORGIERI (Mauro), 1811.
GIOSTRA (Caterina), 3756.
GIOVAGNOLI (Agostino), 5526, 8332.

Giovanni Dominici, 3443.
GIOVANNINI (A.), 2652.
GIOVANNINI (Adalberto), 722.
GIOVANNUCCI (Pierluigi), 5550.
GIOVIALE (Fernando), 6456.
GIPPIUS (Aleksej A.), 3305.
GIRARD (Pascale), 5596.
Girard (René), 1400.
GIRBAL (Christian), 1812.
GIROU SWIDERSKI (Marie-Laure), 5781.
GIRVIN (Brian), 8075.
GIRY-DELOISON (C.), 7830.
GISCARD (P. H.), 1604.
GIUA (M. A.), 2520.
GIULIANO (Antonio), 2894.
GIUNTA (R.), 1893.
GIUNTELLA (Maria Cristina), 5551.
GIURESCU (Dinu C.), 8127.
GIUSTINO VITOLO (Angela), 6147.
GIUSTO (Gaia), 6713.
GIUSTOZZI (Antonio), 4126.
GIUZELEV (Vasil), 3278.
GIŻEJEWSKA (Małgorzata), 8333.
Glabrio (Manius Acilius), 2448.
GLANTZ (David M.), 8192.
GLASER (Clive), 8796.
GLASER (Maria), 3255.
GLASS (Joseph B), 4966.
GLASSMAN (Jonathan), 7179.
GLASSNER (Christine), 19.
GLAUDE (Eddie S.), 5313.
GLEADLE (Kathryn), 4712.
GLEASON (Elisabeth), 5552.
GLEI (Reinhold F.), 2823.
GLENN (Susan A.), 6687.
GLIOZZI (Giuliano), 853.
GLORIA (Glenda M.), 4981.
GLUSHKOV (Igor' G.), 1572.
GNECCO (Cristóbal), 533.
GNILKA (Christian), 3017.
GNOLI (Tommaso), 2653.
GØBEL (Erik), 4373.
Gobetti (Piero), 5794.
GÖBL (Robert), 2691.
GOCKEL (Michael), 3269.
GOCKING (Roger), 4642.
GODA (Norman J.), 4559.
GÖDDE (S.), 2194, 2195.
GODDEN (D.), 2877.
GODDING (Rober), 3955.
GODMAN (Peter), 5553.
GOEBEL (Ted), 1573.
GOEMANS (H. E.), 7995.
Goethe (Johann Wolfgang), 1434, 5424, 6413, 6419, 6420, 6627, 6753.
GOETSCHEL (Roland), 1353.
GOETTE (Hans Rupprecht), 2129.
GOETZ (Hans-Werner), 192, 413.
GOETZ (Helmut), 5925.

GOEZ (Werner), 3921.
GOGOL (Bogusław), 8154.
GOHM (Lilian), 6866.
GÖKBEL (Ahmet), 5250.
GOL (Ayla), 7996.
GOLAN (Shimon), 8335.
GOLDBERG (P. J. P.), 3600.
GOLDBERG (Sylvie Anne), 105.
GOLDEN (Peter B.), 201.
Goldhagen (Daniel Jonah), 478.
GOLDHILL (Simon), 2060, 2196.
GOLDING (John), 6588.
GOLDNER (Loren), 5010.
Goldoni (Carlo), 6418.
GOLDSTEIN (Bernhard R.), 3867.
GOLDSTEIN (Claudine), 4458.
Goldstein (Eugen), 6273.
GOLDSTEIN (Robert Justin), 6111.
Goldziher (Ignaz), 5714.
GOLDZINK (Jean), 6688.
GOLIKOVA (Svetlana V.), 7180.
GOLINELLI (Paolo), 3976.
GOLOUBEVA (Maria), 4178.
GOLSAN (Richard Joseph), 479.
GOLVERS (N.), 237.
GOMES (Nilma Lino), 835.
GÓMEZ (Carlos Alarico), 5379.
GÓMEZ DE VALENZUELA (Manuel), 3275.
GÓMEZ GONZÁLEZ (Inés), 7522.
GÓMEZ SIERRA (Esther), 3669.
GÓMEZ VILLEGAS (N.), 2997.
GOMI (Fumihiko), 8743, 8744.
GONÇALVES DE SOUZA (Miguel Augusto), 7740.
Goncourt (Edmond et Jules de), 6441.
GONEN (Jay Y.), 4560.
Gonzaga (Federico), 6613.
GONZÁLEZ BARRIOS (René), 7903.
GONZÁLEZ BLANCO (Antonino), 3199.
GONZÁLEZ CHAMORRO (Ever), 4287.
GONZÁLEZ CUEVAS (Pedro Carlos), 947.
GONZÁLEZ MÉNDEZ (Matilde), 1490.
GONZÁLEZ MORALES (Manuel R.), 1546.
GONZÁLEZ SAINZ (César), 1566.
GONZÁLEZ TORRE (Yolotl), 1362.
Gonzalo de Berceo, 3961.
GONZALO SÁNCHEZ-MOLERO (J. L.), 355.
GOOCH (Todd A.), 6148.
GOODHEW (David), 7181.
GOODLAD (Graham D.), 7904.
GOODMAN (Joyce), 6040.
GOODY (Jack), 119.
GOONERATNE (John), 5204.

GOPAL (S.), 4744.
GOPHER (A.), 1873.
GOPHNA (R.), 1689.
Gorbachev (Mikhail Sergeevič), 8360, 8467.
GÓRCZYŃSKA-PRZYBYŁOWICZ (Bożena), 6925.
GORDIENKO (Elisa A.), 3814.
GORDILLO (José M.), 4216.
GORDON (Bruce), 1240.
GÖRGEMANNS (H.), 2836.
GORINOV (M. M.), 5044.
GORJAEVA (Tat'jana M.), 6070.
GÖRKE (Susanne), 1760.
GORLOV (Youri), 2355, 2356.
GORNENSKY (Ioann), 854.
GORNER (Paul), 6149.
GORNIG (Martin), 6780.
GORST (Anthony), 8469.
GORSUCH (Anne E.), 7182.
GORYS (Andrea), 2920.
GOSLAN (Richard J.), 534.
GOSSEL-RAECK (Berthild), 2920.
GOSSMAN (Lionel), 656.
GOSWAMY (B. N.), 1277.
GOTSHALK (R.), 2012.
GOTT (Richard), 5380.
GOTTERI (Nicole), 330.
GOTTLIEB (Anthony), 1417.
GOTTLIEB (Julie V.), 4664.
GOTTLOB MORGENBESSER (Ernst), 7497.
GOTTSCHANG (Thomas R.), 8553.
GOTTSCHILD DIXON (Brenda), 6689.
GOTZMANN (Andreas), 4575.
GOUDINEAU (CH.), 2895.
GOUDRIAN (K.), 3277.
GOUDSMIT (Jaap), 1690.
GOULD (Eliga H.), 4665.
GOULD (John), 682.
GOULD (Warwick), 3884.
GOULET (R.), 1415.
GOULET-CAZÉ (Marie-Odile), 1196.
GOUMA-PETERSON (T.), 3174.
GOUNAROPOULOU (L.), 1966.
GOUREVITCH (D.), 2042.
GOURINAT (J. B.), 2246.
GOURON (André), 3498, 3503.
GOUTTEBROZE (Jean-Guy), 3714.
GOW (Ian), 4868.
GOW (James), 5117.
GOW (Robert James), 5634.
GOWER (John), 3545, 3670.
GOY (Corinne), 963.
Goya (Francisco José), 6505.
GOYARD-FABRE (Simone), 1041.
Gozzi (Carlo), 6418.
GOZZI (Marco), 3838.
GOZZINI (Giovanni), 4054.

GOZZOLI (Roberto B.), 1691.
GRABARCZYK (Tadeusz), 3428.
GRABER (Klaus), 5829.
GRABSKI (Andrzej Feliks), 480.
GRACIA CÁRCAMO (Juan), 5169.
GRADMANN (Christoph), 5926.
GRAESER (A.), 2832.
GRAF (Friedrich Wilhelm), 767.
GRÄF (Holger Th.), 7779.
GRAFENSTEIN GAREIS (Johanna von), 4387.
GRAFTON (Anthony), 6257.
GRAHAM (Jenny), 4666.
GRAHAM (Lisa Jane), 4459.
GRAINGER (J. D.), 2061.
Gramsci (Antonio), 1059, 1073, 1075, 6143, 6170.
GRAN (Peter), 492.
GRANA (Daniela), 42.
GRANADA (Miguel A.), 6184.
GRANDE QUEJIGO (Francisco Javier), 3961.
GRANDIN (Greg), 4722.
GRANDJEAN (Yves), 2336.
GRANGE (D.-J.), 7905.
GRANT (Alfred), 6071.
GRANT (M.), 2003, 2502.
GRANT (Susan-Mary), 5314.
Grant (Ulysses S.), 5275.
GRANTLEY (Darryl), 3625, 5989.
GRANVILLE (Brigitte), 7008.
GRAS (Michel), 777.
GRASMAN (Edward), 481.
GRATWICK (A. S.), 2515.
GRAY (Floyd), 6388.
GRAY (Jeremy), 6258.
GRAY (Ronald), 193.
GRAZIANI (Antoine-Marie), 7606, 7795.
GRAZIOSI (Andrea), 5051.
GRAZZINI (Giovanni), 5899.
GRAZZINI (S.), 2451.
GRBIC (Dusica), 63.
GREATEX (G.), 2948.
GREBE (S.), 2753.
GRECI (Roberto), 3324.
GRECO (Emanuele), 2337.
GRECO (Tommaso), 1043.
GREEN (Anthony), 1623.
GREEN (Christopher), 6515.
GREEN (Ian M.), 5640.
GREEN (Monika H.), 3636.
GREENBAUM (Fred), 5315.
GREENE (Jack P.), 922, 5316.
GREENE (John C.), 6690.
GREENE (Kevin), 670.
GREENE (Molly), 4065.
GREENEWALT (C. H. Jr.), 1657.
GREENGRASS (Mark), 5397.
GREENHALGH (Elizabeth), 7997.
GREENSTEIN (Fred I.), 5317.

GREENWOOD (Keith M.), 5990.
GREGOIRE (Reginald), 3959.
GREGOR (A. James), 8554.
Gregorius I Magnus, Papa, Sanctus, 2994, 3058, 3072.
Gregorius Nazianzenus, Sanctus, 2995-2997, 3037, 3070.
Gregorius Nyssenus, Sanctus, 2990, 2999-3001, 3038, 3070.
Gregorius VII, Papa, Sanctus, 3921.
Gregorius XIII, Papa, 4811.
Gregorius XV, Papa, 7799.
GREGORY (Brad), 5598.
GREGORY (Jeremy), 950, 5641.
GREGORY (Paul R.), 7009.
GREINER (Bernhard), 6691.
GREINER (Frank), 6409.
GREISCH (Jean), 1310.
GRELL (Chantal), 5797.
GRELL (Ole Peter), 5453, 5883.
Grendi (Edoardo), 677.
GRENIER (Jean-Yves), 7010.
GRESCHAT (Martin), 4562, 5431.
GRÉSILLON (Almuth), 86.
GREVELHÖRSTER (Ludger), 4563.
GREWE (Wilhelm G.), 7570.
GREYERZ (Kaspar), 5425.
GRIESE (Sabine), 6064.
GRIESSMER (Axel), 4564.
GRIFFEL (Frank), 5700.
GRIFFIN (J.), 2008.
GRIFFITH (Th.), 2025.
Griffiths (Jeremy), 60.
GRIFONI CREMONESI (R.), 1515.
GRIGOR'EV (A. P.), 40.
GRIGSBY (John L.), 3715.
GRILLI (A.), 1976.
GRILLON (Louis), 3276.
GRILLONE (A.), 2988.
Grimm (Jacob), 5733.
Grimm (Wilhelm), 5733.
Grimm Wild (Dorothea), 5733.
GRIMOULT (Cédric), 6259.
GRIMSTAD (Kaaren), 3484.
GRINDER-HANSEN (Keld), 3571.
GRINER (Massimiliano), 4825.
GRISSOM (C. A.), 1547.
GRIX (Jonathan), 4565.
GRMEK (Mirko Dražen), 1202, 6269.
GRODDEK (Detlev), 1813.
GROEN (Gerrit), 4887.
GROMADA (Barbara), 5991.
GROMPONE (Romeo), 4973.
GRONDEUX (Anne), 3876.
GRONEMAN (Carol), 6260.
GRØNGAARD JESPERSEN (Torben), 7741.
GRONSKÝ (Ján), 7411.
GROPPO (Bruno), 4106.

INDEX OF NAMES

GROSS (Jan Tomasz), 7355, 8071.
GROSSE (Pascal), 7649.
GROSSENBACHER (Maya), 306.
Grosseteste (Robert), 3879.
GROSSI (Paolo), 1093.
GRÖSSING (Helmuth), 5798.
GROSSMAN (P. Z.), 6781.
GROSSMANN (Atina), 478.
GROSSO (G.), 7376.
Grote (George), 678.
GROTHE (Ewald), 5733.
Grotius (Hugo), 1084.
GROTUM (Thomas), 4627.
GROVER (Verinder), 4960, 5237.
GROßE (Jürgen), 656.
GRUCA (Anna), *XIV*.
GRUMEL-JACQUIGNON (F.), 7998.
GRÜNBAUM (Robert), 4566.
GRUNCHAROV (Mikhail), 4252.
GRÜNER (Pia), 1008.
GRUNER (Wolf), 4179, 4567.
GRUNERT (Frank), 1044, 7377.
GRÜNEWALD (Dietrich), 6621.
GRUNTOVÁ (Jitka), 4337.
GRYPA (Dietmar), 7324.
GRYSON (Roger), 3032.
GRYTTEN (Ola Honningdal), 6785, 7185.
GSELL (Danièle), 339.
Gu (Weijun), 8579.
GUAJARDO (Guillermo), 4909.
GUALDO ROSA (Lucia), 73.
GUAN (Xiaohong), 8555.
GUARDASOLE (A.), 2247.
GUASCONI (Giovanni B.), 7166.
GUDERZO (Massimiliano), 8338.
GUEDEA (Virginia), 4914.
GUELKE (Adrian), 4770.
GUENIFFEY (Patrice), 4460.
GUÉRIN (Alain), 8225.
GUERRA (Alessandro), 5579.
GUERRINI (Roberto), 3798.
GUEST (Harriet), 5992.
Guevara (Ernesto 'Che'), 4221.
GUEX (Sophie), 2497.
GUGGISBERG (Daniel), 5799.
Gui (Bernard), 679.
GUICHARD (Jean Pierre), 4461.
GUICHARD (V.), 2895.
GUIDETTI (Massimo), 3340.
GUIDI (A.), 2375.
GUIDI BRUSCOLI (Francesco), 5513.
GUIDO (Francesco), 162.
Guido d'Arezzo, 3842.
GUIDOT (Bernard), 3685.
GUIDOT (Raymond), 6622.
GUILAINE (Jean), 2339.
GUILD (Rollins), 3894.
GUILLAUME (Pierre), 7186.
Guillaume d'Orange, 3683.

GUILLELMUS DURANTIUS, 3854.
GUILLEN (Pierre), 7582, 7597.
GUILLERÉ (Christian), 3284.
GUILLET (François), 238.
Guittone d'Arezzo, 3694.
GÜLSOY (Ufuk), 5251.
GÜLTEPE (Necati), 3256, 3257.
GUNDERT (B.), 2248.
GUNDLE (Stephen), 5800.
GUNN (Simon), 5801.
GUNNAR (Karlsson), 4742.
GÜNTERT (Georges), 3732.
GÜNTHER (R.), 2572.
GUO (Chengkang), 8556.
GUO (Hongmao), 8557.
GUO (Songyi), 8558.
GUO (Tieqiang), 8638.
GUO (Weidong), 8559.
GUO (Yingjie), 8563.
GUREVIC (Aron Jakovlevic), 3335, 3637.
GURY (F.), 2573.
GUSAROV (E. P.), 8125.
GUSTAFSSON (Harald), 4068.
GUSTAFSSON (Karl-Erik), 6072.
GUSY (Christoph), 4547.
GUT (Waltraud), 140.
GUTAS (Dimitri), 3461.
GUTHMÜLLER (Bodo), 5777.
GUTIÉRREZ (Harim B.), 7906.
GUTIÉRREZ MASSON (L.), 2654.
GUTIÉRREZ SÁNCHEZ (Tomás), 4975.
GUTOT-BACHY (Isabelle), 3306.
GUTSCHNER (Peter), 7117.
GUY (Josephine M.), 6457.
GUYARD (Marius-François), 4415.
GUYER (Paul), 6150.
GUYON (Jean), 3065.
GUYOT-BACHY (Isabelle), 685.
GUYOTJEANNIN (Olivier), 3262.
GUZMAN (Gregory G.), 287.
GUZZETTI (Luca), 1204.
GUZZO (P. G.), 2376.
GVOSDEV (Nikolas K.), 7859.
GYSELEN (R.), 1894.

H

HAALAND (Torstein), 7378.
HAAN (Bertrand), 7766.
HAAR (Ingo), 482.
HAARDT DE LA BAUME (Caroline), 6589.
HAAS (Stefan), 1220.
HAAS (Volkert), 1814.
HAAVE (Per), 7187.
Habermas (Jürgen), 797, 1067, 6174.
HABERMAS (Rebekka), 7188.
HACHEM (Lamys), 1516.
HACHMANN (E.), 2503.

HACKENBERG (Gerd R.), 6782.
HACKETT (Helen), 6389.
HACOHEN (Malachi Haim), 6151.
HADJISAVVAS (Sophoklis), 2340.
HADJÚ (Tibor), 4738.
HADLEY (Dawm M.), 3477, 3479.
Hadrianus (Publius Aelius), Roman emperor, 2541, 2542, 2558, 2607, 2827.
HAECKEL (Ilse), 356.
HAEGEMANS (Karen), 1852.
HAEHLING (Raban von), 3089.
HAENTJENS (A. M. E.), 2130.
HAERS (J.), 1297.
HAFEZ-ERGAUT (Agnès), 6354.
HAFTENDORN (Helga), 8241.
HAGEDORN (Dieter), 3033.
HAGEDORN (Ursula), 3033.
HAGELWEIDE (Gert), 6043.
HAGEMANN (Karen), 4064.
HÄGG (T.), 1687, 2949.
HAGGETT (Peter), 6231.
HAGMARK (Hanna), 6926.
HAHN (Peter L.), 8319.
HAHN (Thomas), 1242.
HÄHNER-ROMBACH (Sylvelyn), 1205.
HAIDER (Barbara), 7874.
Hailey (William Malcolm Hailey, Baron), 8118.
HAINES (Michael R.), 7088.
HAINSWORTH (Paul), 4755.
HAJDU (P.), 2456.
HÁJEK (Miloš), 7325.
HALASZ (Katalin), 3666.
Halbwachs (Maurice), 680.
HALDON (J. F.), 3122.
HALDON (J.), 3168.
HALE (William Mathew), 7607.
HALÉVI (Elie), 741.
HALÉVI (Florence), 741.
HÁLFDANARSON (Guðmundur), 4743.
HALL (Allan), 1598.
HALL (Catherine), 4064, 4667.
HALL (David), 87, 1491.
HALL (E.), 1994.
HALL (John R.), 761.
HALL (Michael R.), 8339.
HALL (Nina), 6295.
HALL (Patrik), 5213.
HALL (Richard C.), 7999.
HALL STERNBERG (Rachel), 2197.
HALLAGER (B. P.), 2338.
HALLAGER (E.), 2338.
HALLAQ (Wael), 1473.
HALLENSLEBEN (Markus), 6458.
Haller (Heinrich), 3717.
HALLER (Rudolf), 6196.
HALLINGBERG (Gunnar), 6692.
HALLOF (K.), 2574.

HALLON (Ľudovít), 7011.
HALPIN (Andrew), 3289.
HALTENHOFF (Andreas), 2756, 2781.
HALTERN (Utz), 1045.
HAMAGUCHI (Norio), 8496.
HAMANN (Matthias), 3799.
HAMBURGER (Jeffrey F.), 3800.
HAMER (Richard), 3978.
HAMERNIK (Gottfried), 1692.
HAMES (Harvey J.), 3907.
Hamilton (Alexander), 5353.
HAMILTON (Bernard), 3388, 3987.
HAMILTON (Marybeth), 6693.
HAMILTON (Mike), 1594.
HAMILTON (R.), 2101.
HAMLIN (William M.), 6152.
HAMMAN (A.-G.), 2971.
HAMMANN (Konrad), 5642.
HAMMEL-KIESOW (Rolf), 3572.
HAMMER (M.), 8340.
HAMMER (Olav), 837.
HAMMERSTEIN (Notker), 5879, 5927.
HAMMOND (M.), 2008.
HAMMOND (Nicholas G. L.), 2062.
Hammurabi, king of Babylonia, 1792.
HAMNETT (Brian R.), 7742.
HAMZA (Mahmoud), 1693.
HAN (Dongping), 8560.
HAN (Shufeng), 8561.
HANAWALT (Barbara A.), 4019.
HANČIČ (Damjan), 952.
HANCOCK (Ian), 4163.
Handke (Peter), 6464.
HANEDA (Tadashi), 8481.
HANIOGLU (M. Sükrü), 5252.
HANISCH (Ludmilla), 5714.
HANKINS (James), 5858.
HANKS (David A.), 6623.
HANNA (Ralph), 60.
HANNAH (Matthew G.), 5318.
Hannibal, 2549.
HÄNNINEN (Marja-Leena), 2692.
HANSCHMIDT (Alwin), 5981.
HANSEN (Inge Lyse), 2953.
HANSEN (Lars Ivar), 3563.
HANSEN (M. H.), 1128.
HANSEN (Peer Erik), 4374.
HANSEN (William), 2301.
HANUŠ (Jiří), 5450.
HANUSCHEK (Svan), 6372.
HANZAL (Josef), 4325.
HAOUR (Anne), 1492.
HAQUIN (A.), 1297.
Hara (Takashi), 7915.
HARADA (Keiichi), 4869.
HARAN (Alexandre Yali), 4462.
HARARI (Maurizio), 2377, 2381, 2399.

HARASKO (Alois), 5091.
HARASZTI-TAYLOR (Éva), 8246.
HARBISON (Robert), 6516.
HARBUĽOVÁ (Ľubica), 5097.
HARDACH-PINKE (Irene), 7189.
Hardenburg (Karl August von), 347.
HARDER (Hans-Joachim), 8461.
HARDIN (R. E.), 2198, 3200.
HARDT (Michael), 1046.
HARDWICK (L.), 1922.
HARDY (Lyda Mary), 956.
Hare (Francis), 4683.
HAREN (Michael), 3573.
HARGREAVES (John), 5170, 7704.
HARITONOW (Alexandr), 5994.
HARITONOW (Berit), 5994.
HÄRKE (Heinrich), 1480.
HARLEY (C. Knick), 6783.
HARLINE (Craig), 5477.
HÄRMÄNMAA (Marja), 5802.
HARMSEN (Theodorus Hendrikus Bernardus Maria), 681.
HARNA (Josef), 4338.
Harnack (Mildred), 4537.
HAROOTUNIAN (Harry), 483.
HARRELL (James H.), 1694.
HARRIGAN (Jane), 4906.
HARRIS (Jay M.), 3452.
Harris (Joseph S.), 7875.
HARRIS (Michael), 362.
HARRIS (Ron), 6928.
HARRIS (Roy), 15.
HARRIS (Tim), 4690.
HARRISON (A.), 2918.
Harrison (Benjamin), 7927.
HARRISON (Brian), 5757.
HARRISON (C.), 3066.
HARRISON (Dick), 778, 855.
HARRISON (George W. M.), 2803.
HARRISON (Henrietta), 6073, 8562.
HARRISON (Mark), 5083.
HARRISON (S. G.), 2825.
HARRISON (Simon), 1034.
HARRISON (Stephen H.), 1229.
HARRISON (T. P.), 1636.
HARRISON (T.), 1895.
HARRISON (T.), 2302.
HARRISS (John), 4747.
HARROP (Sylvia), 6040.
HART (A. R.), 3504.
HART (Mitchell B.), 6261.
HART (Peter), 4668.
HARTENSTEIN (Judith), 3067.
HARTER (Gabriele), 2897.
HÄRTER (Karl), 7481.
HARTMANN (Christian), 8196.
Hartmann (Karl Amadeus), 6698.
Hartmann (Martin), 5714.
HARTMANN (Wilfried), 3897.
HARTOG (François), 484, 779.

HARTSOCK (John C.), 6074.
HARTWIG (Angela), 5945.
Harvey (Barbara), 1201.
HARVEY (P. B., jr.), 2898.
Harvey (William), 6215.
HARY (Benjamin H.), 5706.
HARZER (Friedmann), 6459.
HASAN (Burhanuddin), 4958.
HASAN (Mushirul), 4749.
HASE (F.-W. von), 2400.
HASIL (Jan), 5827.
HASKELL (Francis), 3175.
HASSAM (Andrew), 7757.
HASSE (Dag Nikolaus), 3877.
HASSELINK (A. W.), 3277.
Hassenpflug (Amalie), 5733.
Hassenpflug (Ludwig), 5733.
Hassenpflug Grimm (Charlotte), 5733.
HASSLER (Uta), 508.
HATCHER (Richard W. III), 5350.
HATHAWAY (Heather), 6460.
HATINA (Meir), 4396.
HATSCHER (Christoph R.), 2575.
HATZOPOULOS (M.), 1966.
HAUDE (Sigrun), 5643.
HAUE (Henry), 957.
HAUG (Eldbjørg), 3429.
Haultin (Pierre), 6108.
HAUNER (Milan), 4320.
HAUROLD (Johannes), 2199.
HAUSBERGER (Bernd), 5599.
HAUSER (Claude), 4419.
HAUSMANN (Frank-Rutger), 485.
HÄUSSLING (Roger), 6153.
HAUTMANN (Hans), 4169.
HAVEL (Ivan M.), 921.
Havel (Vaclav), 5885.
HAVELANGE (I.), V.
HAVERALS (M.), 1301.
HAVERKAMP (A.), 3988.
HAVERLING (Gerd), 2754.
HAVLÍK (Bohumil), 4339.
HAWKINS (Thomas), 6262.
HAWS (Jonathan), 1514.
HAWTHORNE (J. W. J.), 2370.
Hay (John), 5337.
HAYAKAWA (Shohachi), 8745.
HAYASHI (Reiko), 8746.
HAYCOCK (D.), 6314.
HAYCRAFT (William R.), 6867.
HAYE (L.), 1604.
HAYES (Kevin J.), 6461.
HAYES (Peter), 6868.
HAYNES (Rebecca), 8072.
HAYNES (S.), 2401.
HAYWARD (Rhodri), 5803.
HAZELGROVE (Jenny), 5426.
HAZEN (Crai James), 6263.
HAZRA (Kanai Lal), 5236.
HAZZARD (R. A.), 1695.

HE (Baogang), 8563.
HEAD (Thomas), 3971.
HEAD-KÖNIG (Anne-Lise), 7163.
Headlam (Cuthbert Morley) 4651.
HEADLEY (James Henry), 8341.
HEADON (David), 1035.
HEALY (John F.), 2826.
HEALY (Michael S.), 8073.
HEAP (Ruby), 486.
Hearne (Thomas), 681.
HEATHORN (Stephen J.), 5993.
HEBER-SUFFRIN (François), 3943.
HECHT (Jennifer Michael), 6264.
HECK (V. Kilian), 472.
HECKER (Karl), 1761.
HECTOR (Michel), 4727.
HEDELER (Wladislaw), 5041.
HEDENBORG (Susanna), 7190.
HEDRICK (Charles W.), 487.
HEDWIG (Andreas), 4518.
HEERICH (Thomas), 1047, 6154.
HEERMA VAN VOSS (Lex), 7319.
HEFFER (Jean), 6852.
HEFFERNAN (Richard), 4675.
HEGEDÜS (András B.), 8435.
Hegel (Georg Wilhelm Friedrich), 796, 1422, 6164, 6172, 6187, 6193.
HEIBERG (Steffen), 958.
HEIDECKER (Karl), 36.
Heidegger (Martin), 6179, 6201, 6205.
HEIL (M.), 2574.
HEIMPEL (W.), 1762.
HEIN (Laura), 440.
HEIN (Oliver), 7523.
HEIN (Rudolf Branko), 5804.
HEINEMANN (Manfred), 5994.
HEINEMANN (Winfried), 8395.
HEINIG (Paul-Joachim), 3359.
HEINRICH VON DEM TÜRLIN, 3671.
HEINRICHS (W. P.), 1351.
HEINRICHS (Wolfgang E.), 5702.
Heinrici (Gotthard), 8200.
HEINTZE (Beatrix), 871.
Heinz (Friedrich Wilhelm), 4593.
HEINZ (Hürten), 1048.
HEINZE (T.), 2194.
HEINZELMANN (Michael), 2899.
HEIRBAUT (D.), 1094.
HEIRBAUT (Dirk), 3505.
HEISE (Joachim S.), 6075.
HEISER (Richard R.), 3411.
HEISS (Mary Ann), 8319.
HEISSIG (Walther), 856.
HEITZ (Carol), 3943.
HELDRICH (Andreas), 7484.
HELFRICH (Cornelia), 6355.
HELFT-MALZ (Véronique), 4463.
HELGERSON (Richard), 6517.

HELGI (Gunnlaugsson), 7524.
Heliodorus Emesenus, 3142.
HELLEMA (Duco), 7964.
HELLER (Geneviève), 6278.
HELLER (Joseph), 4788.
HELLHOLM (David), 3025.
HELLMANN (M.), 955.
HELLMANN (Martin), 17.
HELLMANN (O.), 1949.
HELLMUTH (Thomas), 1221.
HELLY (Bruno), 2325.
HELMS (Hadwig), 2219.
HELSLOOT (John), 881.
HELSTRUP (Søren), 4375.
HELTZER (Michael), 1896.
HELVE (Helena), 1363.
HEMPHILL (P.), 2402.
HEN (Yitzhek), 611.
HENDERSON (J.), 1977.
HENDERSON (James D.), 959.
Henderson (Nevile, Sir), 8173.
HENDERSON (Peter V. N.), 4915.
HENIG (Robin Marantz), 6265.
HENKE (Holger), 8342.
HENKE (Josef), 4517.
HENKE (Rainer), 2982.
HENNESSY (Peter), 7463.
HENNING (Eckart), 141.
HENNING (Joseph M.), 5805.
HENRAD (Nadine), 3683.
Henri de Flandre et de Hainaut, empereur latin de Constantinople, 3252.
Henri I, Duc de Montmorency, 4657.
Henri IV, roi de France, 4439.
HENRICKSON (Robert C.), 1612.
HENRIET (Patrick), 3963.
HENRIOT (Christian), 4086.
Henrique O Navegador, infante de Portugal, 3438.
Henry IV, king of England, 4707.
Henry V, king of England, 3417.
HENSEL (Jürgen), 7311.
HENSELLECK (Werner), 2979.
HENTSCHEL (Frank), 3837.
HENZE (Paul B.), 960.
HEPLER (Alison L.), 6266.
HEPPERLE (Sabine), 8343.
HEPPNER (Harald), 7860.
HERBER (Mark D), 120.
HERBERS (Klaus), 3962.
HERBERT (Robert L.), 6614.
HERCIGONJA (Eduard), 1009.
HERDE (Peter), 3259.
Herder (Johann Gottfried von), 6429.
HÉRICHÉ (Sandrine), 3686.
HERKLOTZ (Ingo), 3922.
HERMAN (Arthur), 5319.
HERMANN (Christian), 5011.

HERMANN DE FRANCESCHI (Sylvio), 5514.
HERMANSON (Lars), 3389.
HERMARY (A.), 2297.
HERMARY (Antoine), 2316.
Herodes I, king of Iudaea, 1854, 2902.
Herodotus, 484, 682, 2004, 2005, 2091, 2165, 2213, 2302, 5748.
Herophilus, 2280.
HERREN (Madeleine), 7907.
HERRERA HERNÁNDEZ (María Teresa), 343.
HERRERO SÁNCHEZ (Manuel), 7831.
HERRIN (Judith), 3201.
HERRING (E.), 1913.
HERRING (Susan C.), 217.
HERŠAK (Emil), 857.
HERSTAD (John), 6829.
HERZBERG (Guntolf), 6156.
HERZIG (Arno), 5478.
Hesiodus, 2012, 2207.
HESK (J.), 2102.
HESS (Andreas), 1049.
HESSELINK (Lidewij), 7319.
HESSENBRUCH (Arne), 1241.
Hetepheres I, queen of Egypt, 1711.
HETTLING (Manfred), 806, 5752.
HEULLANT-DONAT (Isabelle), 3341, 3631.
HEUSS (Anja), 6518.
HEWITSON (Mark), 5806, 8000.
HEWITT (Martin), 4649.
HEWITT (Steve), 5928.
HEWLETT (Nick), 5764.
HEWSEN (Robert H.), 239.
HEXELSCHNEIDER (Erhard), 5807.
HEY (David), 121.
HEYDARI (Saman), 1504.
HEYDE (Jürgen), 6968.
HEYDE (Philipp), 8001.
HEYDEMANN (Steven), 4116.
HEYWOOD (Linda Marinda), 4140.
HEYWOOD (T.), 2479.
HIATT (A.), 240.
HIBNER KOBLITZ (Ann), 6267.
HIEBL (Ewald), 1221.
Hieronymus (Sophronius Eusebius), Sanctus, 2721, 3002, 3003.
HIERY (Hermann), 6023.
HIGGINS (David), 6869.
HIGGIT (John), 3638.
HIGHAM (Charles), 1493.
HIGMAN (B. W.), 6929.
HIGUCHI (Hidemi), 4870.
HIGURASHI (Yoshinobu), 7571.
HILAIRE-PÉREZ (Liliane), 6268.
Hilarius Pictaviensis, 3004, 3104.
Hilbert (David), 6258.

HILD (F.), 3167.
Hildegard von Bingen, 3988.
HILDERMEIER (Manfred), 5052.
HILDESHEIMER (Françoise), 4464, 7832.
HILGER (Andreas), 8344.
HILKEN (Charles, FSC), 3964.
HILL (Clive), 5813.
HILL (D. E.), 2482.
HILL (John M.), 3718.
HILL (Roland), 643.
HILLEBRECHT (Frauke), 5808.
HILLERBRAND (Hans Joachim), 962.
HILLGARTH (Jocelyn N.), 5171.
HILLS (Alice), 8199.
HILLYAR (Anna), 5053.
HILMI BAŞ (Mustafa), 1230.
HILTON (Mary), 6021.
HILTON (Matthew), 7192.
HINARD (François), 2576.
HINDE (John R.), 656.
Hindemith (Paul), 6698.
HINDLE (Steve), 7193.
HINE (H. M.), 2504.
HINGLEY (R.), 2900.
HINOJOSA MONTALVO (José), 3574.
HINRICHS (Ernst), 4069.
HINSKE (Norbert), 6114.
HINTERBERGER (M.), 3202.
HINTERSTOISSER (Hermann), 4182.
HINTIKKA (Kaisamari), 5615.
Hintze (Otto), 683.
HINZ (Hans-Martin), 3336.
HIPPEL (Karin, von), 8345.
HIPPEL (Wolfgang von), 163.
Hippocrates, 1201, 2006, 2007.
HIRAKAWA (Minami), 8747.
HIRATA (Helena), 1039.
HIRATA (Masahiro), 7608.
HIRSCH (Pam), 6021.
HIRSCHBIEGEL (Jan), 3562, 3570.
HIRSCHFELD (N.), 1956.
HIRSCHFELD (Yizhar), 1831, 1853.
HIRST (John), 4164.
HIRVONEN (Ari), 1095.
HISCOCK (Nigel), 3801.
HITCHCOCK (L. A.), 1950.
Hitler (Adolf), 496, 521, 4045, 4172, 4559, 4560, 4577, 4593, 4594, 4597, 4617, 5014, 8085, 8100, 8115, 8168.
HIXSON (Walter L.), 8428.
HJELDE (Sigurd), 1376.
HJELTNES (Guri), 7241.
HO (Cynthia), 3703.
HOAERAU -DODINAU (Jacqueline), 3519.
Hobbes (Thomas), 782, 1047, 1077, 1082, 1084, 6154.

HOBBS (A.), 2249.
HOCHADEL (Oliver), 1100.
HOCHEDLINGER (Michael), 7861.
HOCHLEITNER (Janusz), 5427.
HOCHSTRASSER (Tim J.), 7379.
HOCKETT (Bryan), 1514.
HODGE (Jonathan), 6270.
HODGES (Richard), 646, 4013.
HODGKISS (Andrew), 6271.
HODGSON (Dorothy L.), 7706.
HODKINSON (S.), 2132.
HÖDL (Günther), I.
HODNE (Fritz), 6785.
HOEGEN (Saskia von), 493.
HOEGES (Dirk), 6158.
HOENEN (Maarten J. F. M.), 3882.
HOENSCH (Jörg Konrad), 5099.
HOERDER (Dirk), 7194.
HOFER (Marc), 2063.
HOFER (Sibylle), 7525.
HOFFMAN (Philip T.), 7013.
HOFFMANN (Andreas), 3068.
HOFFMANN (David L.), 5078.
HOFFMANN (Dierk), 4569, 4584.
HOFFMANN (F.), 1696.
HOFFMANN (Hartmut), 3313.
HOFFMANN (Roland J.), 5091.
HOFFMANN (Stefan-Ludwig), 4570, 5752, 7195.
HOFFMANNOVÁ (Jaroslava), 334.
HOFMAN (Rijcklof), 3855.
HOFMANN (Andreas R.), 4989.
HOFMANN (Catherine), 242.
HOFMANN (Daniel), 4571.
HOFMANN (Hasso), 1096.
HOFSTETTER (Michael J.), 5932.
HOFTIJZER (Jacob), 1763.
HOGAN (Michael J.), 551.
HOGAN (Patrick), 6359.
HÖGEMANN (Peter), 1658.
HOHEISEL (K.), 3047.
HOHLFELD (Rainer), 5943.
HOHLFELDER (Robert L.), 1854.
HOHLS (Rüdiger), 618.
HOHOFF (Ute), 7526.
HOJO (Hideki), 8748.
HOLDEN (Robert H.), 7610.
HOLDEN (T. G.), 1553.
Hölderlin (Friedrich), 6178, 6205.
HÖLKESKAMP (K. J.), 2577, 2655.
HOLLEDGE (Julie), 6695.
HOLLMANN (Michael), 4516.
HOLLOWAY (R. R.), 2341.
HOLLSTEIN (W.), 2704.
HOLMAN (Andrew C.), 7196.
HOLMAN (Valerie), 4456.
HOLMES (Frederic L.), 6275.
HOLMES (Heather), 5978.
HOLO (Joshua), 3203.
HÖLSCHER (Tonio), 2570, 2904.
HOLT (Thomas C.), 7148.

HOLTON PIERCE (R.), 1687.
HOLTWICK (Bernd), 7326.
HOLTZ (Bärbel), 4520.
HOLTZMANN (Bernard), 2342.
HOLZBERG (N.), 2755.
HOLZEM (Andreas), 5479.
HOLZHAUER (Heinz), 1097.
HOLZHAUSEN (J.), 1984, 2827.
Homerus, 484, 1928, 2008, 2009, 2010, 2012, 2160, 2176, 2179, 2184, 2199, 2204, 2228, 2775.
HOMMEN (Tanja), 7197.
HOMZA (Lu Ann), 5428.
Honecker (Erich), 7545.
HONEMANN (Volker), 6064.
HONEYMAN (Katrina), 6870.
HONG (Yong-pyo), 4891.
HÖNIG (Patrick), 8346.
HONNINGDAL GRYTTEN (Ola), 6786.
HONORÉ (Jean-Paul), 206.
Honorius Augustidunensis, 3891.
HOOPER (R. W.), 965.
HOOREBEECK (Céline), 361.
HOPCROFT (R. L.), 6969.
HOPE (V. M.), 2125.
HOPKINS (Andrew), 6554.
HOPKINS (David), 6519.
HOPKINS (Eric), 7198.
HOPKINSON (N.), 2481.
HOPPE (Bert), 7063.
HOPPIT (Julian), 4669.
HOPT (Klaus J.), 7484.
HORÁK (Antonín), 6696.
Horatius Flaccus (Quintus), 2460-2464, 2736, 2834.
HORBAS (Claudia), 6624.
HORCAJO (Arturo), 858.
HORCAJO (Carlos), 858.
HORCÁKOVÁ (Václava), III.
HORDEN (P.), 1923.
Hormizd, sacerdote, 1902.
HORN (Angelica), 6129.
HORNBLOWER (S.), 1924, 1925.
HORNE (John), 8002, 8003.
HÖRNIGK (Therese), 6372.
HORNOS MATA (Francisca), 1517.
HORNUNG (Klaus), 1051.
HOROWITZ (Wayne), 1764.
HORROCKS (M.), 1518.
HORSFALL (N.), 2532.
HORSKY (Vilen), 5270.
HORSMAN (Reginald), 5320.
HORSMANN (Gerhard), 2064.
HOSE (Mertin), 2775.
HOSOKAWA (Shigeo), 8749.
HOSOYA (Chihiro), 7609.
HOTSON (Howard), 5644.
HOU (Jie), 8564.
HOU (Yangfang), 8565.
HOURIHANE (Colum), 3802.

HOURMANT (François), 5811.
HOURS (Bernard), 5555.
HOUSDEN (Martyn), 8004.
Housley (R. A.), 1506.
HOUSSIAU (Jean), 7796.
HOUTE (Arnaud-Dominique), 7465.
Hovius (Mathias), 5477.
HOWARD (Angela M.), 954.
HOWARD (Deborah), 3803.
HOWARD (Don), 6244.
HOWARD (Thomas A.), 6159.
HOWE (Daniel W.), 7936.
HOWE (Stephen), 494.
HOWKINS (Richard A.), 1118.
HOWLETT (David), 3757.
HOWLETT (Peter), 7327.
HOY (Anne), 6623.
HRBEK (Rufolf), 7457.
HRYCIUK (Grzegorz), 8155.
HSIN CHIH (Chen), 8156.
HU (Cheng), 8566.
Hu (Hanmin), 8688.
Hu (Shi), 8684.
HUANG (H. T.), 6272.
HUANG (Juyan), 8567.
HUANG (Lingjun), 8568.
HUBER (Christoph), 3713.
HUBER (Hansjörg Michael), 7703.
HUBER (Martin), 859.
HUBER (Peter J.), 106.
HUBER (Sandrine), 2362.
HUBERT (Etienne), 3550, 3603.
HUBERT (Marie-Clotilde), 31, 107.
HUBERT (Olivier), 5600.
HUBNER (Gert), 3719.
HÜBNER (Klaus), 6273.
HUDEMANN-SIMON (Calixte), 6274.
HUDSON (D.), 5645.
HUET (V.), 2881.
Hufeland (Christoph Wilhelm), 6302.
HUFFMANN (Joseph P.), 3390.
Hugh of Wells, bishop of Lincoln, 3258.
HUGHES (Michael), 7908, 8005.
HUGHES (Philip), 8347.
HUGON (Alain), 7767.
HUHTAMIES (Mikko), 5214.
Huizinga (Johan), 684.
HÜLDEN (O.), 1659.
HULSE (Clark), 6512.
Hume (David), 826, 1412, 1418, 6136, 6165, 6176.
HUMM (M.), 2578.
HUMMEL (Pascale), 495.
HUMPHREY (E. M.), 1855.
HUNEEUS (Carlos), 4282.
HUNEFELDT (Christine), 7200.
HUNT (Lynn), 7380.

HUNTER (F. Robert), 4397.
HUNTER (I.), 6161.
HUNTER (V.), 2105.
HURBON (Laënnec), 7743.
HURD (Madeleine), 4070.
HUREL (Daniel-Odon), 740.
HURLET (Frédéric), 2579.
HURST (James W.), 8006.
HÜRTER (Johannes), 7585, 8200.
Husák (Gustáv), 4322.
HUSBAND (William B.), 7201.
HUSKINSON (J.), 2564.
HUSSON (Edouard), 496.
HUTCHINGS (Graham), 8569.
HUTCHINS (Francis G.), 7413.
HUTCHISON (John F.), 8007.
HUTNER (Gordon), 309.
HUTTNER (Ulrich), 2580.
HUTTULA (Tapio), 7328.
HUVELIN (H.), 2693.
HUWS (Daniel), 18.
Huygen (Christiaan), 6348.
HUYGENS (R. B. C.), 3684.
Huysmans (Joris Karl), 6354.
HVASS (Steen), 1015.
HYAM (Ronald), 8253.
HYAMS (P.R.), 3506.
HYMPÁNOVÁ (Ingrid), 6323.
Hypereides, 2013.

I

Iacovleff (Alexandre), 6589.
IADEVAIA (F.), 3113.
Iamblichus, 3142.
IANCU (Carol), 5023.
IANOVSKII (O. A.), 7768.
IANZITI (Gary), 5812.
IAPPELLI (F.), 5941.
IBAÑEZ ROJO (Enrique), 4217.
IBARRA GUITART (Jorge Renato), 4312.
IBELINGS (B. J.), 3277.
IBLER (Mladen), 3430.
İBN KEMAL (Kemalpaşazâde), 5241.
IBORRA (Joan), 3301.
IBRAGIMOV (B. I.), 8482.
IBRAHIME (Mahmoud), 4297.
ICHINOSE (Toshiya), 8074.
IDDENG (Jon W.), 747.
IDOWU (Stephen Babatunde), 4932.
Iesus Christus, 301, 305, 3061, 3064, 3075, 3077, 3084, 3087, 3090, 3103.
IGARASHI (Yoshiyuki), 498.
IGGERS (Georg C.), 784, 827.
IGLESIA FERREIRÓS (Aquilino), 1092.
IGLESIAS (Francisco), 499.
IGLESIAS ZOIDO (J. C.), 2200.

IGNACE (Rakoto), 7718.
Ignacio de Loyola, Sanctus, 5585.
IGNAT'JEV (Anatolij V.), 7909.
Ignatius Diaconus, 3120.
IGOUNET (Valérie), 500.
İHSANOĞLU (Ekmeleddin), 3471.
IIDA (Mizuho), 8750.
IIJIMA (Wataru), 8570.
IKEGAMI (Hiroko), 8781.
IKEGAMI (Shunnichi), 3804.
IKEMA (Makoto), 5931.
IL'IN (Pavel V.), 5039.
IL'INA (Irina N.), 5055.
ILBERT (Robert), 262, 271.
İLDEM (Arzu Etensel), 243.
Ilg (Alfred), 4401.
ILHAN (M. Mehdi), 5243.
ILIESCU (Octavian), 1146.
ILIFFE (Rob), 6162.
ILKOSZ (Jerzy), 6553.
ILLARIONOVA (L. I.), 64.
ILLINGWORTH (J. S.), 1536.
ILSAAS PHARO (Per Fredrik), 8348.
IMADA (Shusaku), 7667.
IMAŃSKA (Iwona), 88.
IMATANI (Akira), 8751.
IMAZAWA (Koji), 5242.
Imbriani (Vittorio), 6478.
IMÍZCOZ (J. M. F.), 2018.
IMORDE (J.), 1276.
IMPEY (Oliver), 389.
IMSEN (Steinar), 3391.
İNALCIK (Halil), 7381.
İNAN (Jale), 2901.
INCH (Marcela), 357.
INEDA (Masahiro), 4871.
INGELMAN-SUNDBERG (Catharina), 3480.
INGEMARK (Dominic), 2852.
INGERFLOM (Claudio), 4106.
INGLEBY (Jonathan C.), 5601.
Ingres (Jean Auguste Dominique), 6606.
INIGO FERNÁNDEZ (Luis), 5173.
INKSTER (Ian), 6856.
INNES (Matthew), 3374.
INNES (Matthew), 611.
Innocentius III, Papa, 3875.
INOUE (Toru), 8571.
INSABATO (Elisabetta), 331.
IOANID (Radu), 5032.
Iohannes Chrysostomus, Sanctus, 3005, 3079.
Iohannes Gazaeus, 3120.
Iohannes IV Comnenus, 3197.
Iohannes Kameniates, 3129.
Iohannes Kapistrans, Sanctus, 3956.
Iohannes Kinnamos, 3130.
Iohannes Lydus, 2796.
Iohannes Malalas, 3131.

Iohannes Oxeites, 3249.
Iohannes, Apostolus, Evangelista, Sanctus, 3032, 3139.
IOLY ZORATTINI (Pier Cesare), 5703.
Iosephus Genesios, 3208.
Iosephus, Aseneth, 1855.
Iovianus (Flavius Claudius), Roman emperor, 2586.
İPEK (Nedim), 7064.
IRACE (Erminia), 502.
IRACHETA CENECORTA (Alfonso X.), 4913.
IRIGOIN (María Alejandra), 6831.
Irnerio, 3514.
IRONS (Janet), 7329.
IRUROZQUI VICTORIANO (Marta), 4218, 4220.
Irving (David), 407.
IRWIN (Arthur), 1519.
IRWIN G. (Domingo), 5381.
ISAAC (Peter Charles Gordon), 6046.
ISAEVA (Marina V.), 8572.
ISAGER (J.), 2849.
ISAGUIRRE (R. R.), 7955.
Isaiah, 3008.
ISENBERG (Andrew C.), 7202.
ISHIMARU (Yumi), 8483.
Isidorus Pelusiota, Sanctus, 3007.
Isidorus, 3006, 3633, 3673.
ISLAMOV (Tofik M.), 8132.
ISMAGILOVA (Roza N.), 8809.
ISNENGHI (Mario), 8008.
Isocrates, 2014, 2197, 2224.
ISRAEL (Fred L.), 5276.
ISRAEL (Jonathan Irvine), 4225, 7797.
ISSA-SALWE (Abdisalam M.), 8349.
ISSERMAN (Maurice), 5321.
ISTANBULI (Yasin), 3462.
Ister, 2015.
ISUSOV (Mito), 4256.
ITO (Kiyoo), 8752.
ITO (Toshikatsu), 8497.
ITO (YUKIO), 7912.
ITON (Richard), 5322.
Iulianus (Flavius Claudius), Roman emperor, 2426, 2465, 2557, 2606, 2842.
Iulianus Severus (Marcus Didius), 2551.
Iustinianus I (Flavius), Byzantine emperor, 965, 3173, 3190, 3192, 3222, 3229.
Iustinus Martyr, 3008.
Iuvenalis (Decimus Iunius), 1415, 2466, 2721, 2741.
Ivan VI Frankapan, 3430.
IVANIČ (Martin), 968.

IVANOV (M.), 6930.
IVES (Eric William), 5934.
IVETIC (E.), 4826.
IVISON (Eric), 3204.
IVNITSKIJ (N.), 6958.
IWAI (Jun), 4645.
IWAI (Shigeki), 8573.
Iwakura (Tomomi), 7892.
IWAMA (Kazuhiro), 8574.
IWATAKE (Akio), 8484.
IYANGA PENDI (Augusto), 1214.
Izabela Jagiellonka, Königin Ungarns, 4736.
IZBICKI (M.), 5441.
IZENBERG (Gerald N.), 6522.
IZZET (Vedia), 2403.

J

JACHEE (Nancy), 6523.
JÄCKEL (Hartmut), 7203.
JACKSON (Adam), 1531.
JACKSON (C. Ian), 7875.
JACKSON (I.), 8350.
JACKSON (Peter), 8076, 8157.
JACKSON (Richard A.), 3369.
JACKSON (Robert H.), 5603.
JACKSON (Robert), 8351.
JACKSON (S.), 2015.
JACKSON (Tat'jana N.), 3307, 3392.
JACKSON (W. H.), 3689.
JACOB (Robert), 3520, 3575.
JACOBI (R. M.), 1520.
JACOBITTI (Edmund E.), 445.
JACOBS (Bruno), 1660.
JACOBS (Emil), 698.
JACOBSEN (Roswitha), 4515.
JACOBSEN BUCKLEY (Jorunn), 1765.
JACOBSON (Charles David), 7065.
JACOBSON (Matthew Frye), 7913.
Jacobus, Sanctus, 3970, 3975.
JACOBY (D.), 3205.
JACOBY (Joachim), 6591.
JACQUEMIN (Anne), 2202.
JAEGER (Achim), 3444.
JAFARI (Reza), 4764.
JAFFA (Harry V.), 5323.
JAFFE (James A.), 7330.
JAFFRELOT (Christophe), 994.
JAFFRO (Laurent), 6195.
JAGGI (Om Prakash), 6276.
JAHAN (Sébastien), 948, 7204.
JAHN (Bernhard), 472.
JAHNKE (Carsten), 3576.
JAHNS (Sigrid), 3359.
JAIN (Satya Bhusan), 8352.
Jakeš (Miloš), 4322.
JAKOVLEV (Aleksandr N.), 5038.
JAKOVLEV (Nikolaj N.), 4670.
JAKUBIK (Krzysztof), 7245.

JALALZAI (Musa Khan), 4127.
JAMES (A.), 2037.
JAMES (Alan), 8353.
JAMES (Daniel), 4153.
JAMES (Harold), 4572.
James (Henry), 6471, 6483.
JAMES (Scott Curtis), 5324.
James II, king of Great Britain, 4655.
JAMES-RAOUL (Danièle), 3588.
JAMZADEH (P.), 1897.
JAN (Libor), 4338, 5413.
JANÁK (Dušan), 4342.
JANAN (M.), 2494.
JANCHUK (I. I.), 969.
JANIN (Enrico), 164.
JANIN (Valentin L.), 991, 3279, 3814.
JANKOVIC (V.), 6277.
JANKOWSKI (Paul), 4466.
JANKULAK (Karen), 3965.
JANOS (Andrew C.), 4073.
JANOWSKI (Berndt), 1319.
JANSEN (Christian), 4573.
JANSEN (Katherine Ludwig), 3989.
JANSEN (Trine S.), 4376.
JANSEN-WINKELN (K.), 1697.
JANSSEN (Jac. J.), 1698.
JANSSENS (Emiel), 728.
JANSSON (Maija), 4652.
JANVIER (Y.), 2581.
JANZ (Oliver), 4121.
JAPP (Sarah), 2902.
JAPP (Uwe), 6733.
JARAUSCH (Konrad H.), 618.
JARDINE (Lisa), 6625.
JARDINE (Nick), 1190.
JARMOLJUK (S. F.), 6048.
JARRICK (Arne), 786, 1147, 1238.
JASKOT (Paul B.), 6555.
Jaspers (Karl), 6201.
JAUVERT (Vincent), 8355.
JAWOR (Grzegorz), 4014.
JAYNE (Catherine E.), 8158.
Jean de Meun, 3860.
Jean de Saint-Victor, 441, 685, 3306.
Jean II le Bon, roi de France, 3286.
JEAN-BAPTISTE (Chantal), 1298.
JEAN-MARIE (Laurence), 4015.
JEANMONOD (Gilles), 6278.
JEANNENEY (Jean-Noël), 505.
JEAN-PIERRE (Jean Reynold), 4728.
JEDAN (Christoph), 2250.
JEDIN (Hubert), 3043.
JĘDRYCHOWSKA (Barbara), 5056.
Jefferson (Thomas), 5260, 5308, 5348, 5353, 5367, 7829.
JEFFERY (Judith S.), 8356.
JEFFREY (Elizabeth), 3160.
JEFFREY (Robin), 6076.

JEFFREYS (D.), 1733.
JEFFREYS-JONES (Rhodri), 8242.
JEISMANN (Michael), 5735.
Jelinek (Elfriede), 6464.
JENDORFF (Alexander), 5480.
JENKINS (Alice), 5859.
JENNINGS (Francis), 5325.
JENNINGS (Lawrence C.), 5818.
JENSEN (Bent), 8357.
JENSSEN (Dag), 716.
JESPERSEN (Knud J.V.), 8227.
JESPERSEN (Leon), 4098, 4368.
JESPERSEN (T. C.), 8358.
JESSEE (W. Scott), 3393.
JESSENNE (Jean-Pierre), 4446.
Jesus Christus v. Iesus Christus.
JEWERS (Caroline A.), 1453.
JHALDIYAL (RICHA), 1580.
JI (Peng), 8575.
JIA (Erqiang), 8576.
JIANG (Dachun), 8578.
Jiang (Jieshi), 8524, 8538, 8606, 8629, 8686, 8688, 8690, 8704.
JIANG (Pei), 8577.
JIMÉNEZ (Sylvia A.), 1483.
JIMENO MARTÍNEZ (Alfredo), 1597.
JIN (Guangyao), 8579.
JINDRA (Zdeněk), 8159.
JIRÁSEK (Miroslav), 4321.
JOACHIMIDES (Alexis), 376.
JOANNÈS (F.), 1920.
JOANNÈS (Francis), 1638.
JOBST (Kerstin S.), 715.
JODOCK (Darrell), 5469.
JOFFRIN (Laurent), 7862.
JOHANEK (Peter), 603, 3345.
Johannes Cincinnius von Lippstadt, 5789.
Johannes Philoponos, 3152.
Johannes XXIII, Papa, 5507, 5509.
JOHANSEN (Hans Chr.), 6859.
JOHANSEN (Øystein Kock), 1574.
JOHN (Juliet), 5859.
JOHN (Michael), 7066.
JOHNMAN (Lewis), 7015.
JOHNSON (Charlotte Nicola), 5373.
JOHNSON (David F.), 3709, 3721.
JOHNSON (Edward), 8359.
JOHNSON (G.), 8009.
Johnson (Lyndon Baines), 5290, 7028, 8338.
JOHNSON (Odai), 4671.
JOHNSON (Paul), 1560.
JOHNSON (Vivianne), 8243.
JOHNSON (W. R.), 2758.
JOHNSON (William A.), 2133.
JOHNSON DOLLINGER (Genora), 7320.
JOHNSTON (M.), 3059.
JOHNSTON (Mark), 8201.

JOLIVET (J.-Ch.), 2853.
JONAS (Raymond), 5819.
JONES (A.), 2035.
JONES (Ann Rosalind), 1215.
JONES (B. W.), 2517.
JONES (Colin), 6279.
JONES (Eric Lionel), 1148.
JONES (Geoffrey), 6931.
JONES (H.), 2656.
JONES (Larry Eugene), 4574.
JONES (M. D.), 1518.
JONES (Michael), 953, 3434, 3577.
JONES (Steven E.), 6462.
JONES (Vivien), 6427.
JONES (Whitney R. D.), 4672.
JONES (William P.), 7331.
JONJIĆ (Tomislav), 8160.
JOO (Seung-Ho), 8360.
JOOR (Johan), 7863.
JORDAN (B.), 2134.
JORDAN (Stefan), 829.
JORDÁN FLÓREZ (Fernando), 4285.
JØRGENSEN (John Gunnar), 3674.
JORIO (Mario), 970.
JÖRN (Nils), 3507.
Josaphat, 1878.
JOSÉ DE SIGÜENZA, 3935.
JOSEPHSON (Paul R.), 5057.
JOUANNA (Jacques), 2007.
JOUHAUD (Christian), 6360.
JOURDAN (Annie), 4467.
JOY (Morny), 1311.
JOYAL (M.), 2026.
Joyce (James), 6432.
JOYCE (Rosemary A.), 7131.
Juan I, rey de Castilla, 493.
JUDD COLLINS (Cristle), 6697.
JUDICE LISBOA (Lourdes), 5609.
JUDSON (Pieter), 478.
JUDT (Matthias), 5994.
JUDT (Tony), 7355.
JULIA (Dominique), 667, 5491.
JULIEN (Catherine J.), 8814.
JULLIEN (François), 1216.
JUNCO (José Alvarez), 5198.
JUNG (Peter), 4182.
JUNGEL (Eberhard), 1319.
Jünger (Ernst), 5892.
JUNILA (Marianne), 4406.
JUNKELMANN (M.), 2582, 2694.
JUNYENT (Emili), 1541.
JUPP (Peter), 7205.
JURIST (Elliot L.), 6164.
JURSA (Michael), 1856.
JUSSEN (Bernhard), 3578, 3595.
JUZARTE (Teotônio José), 244.

K

KAASCH (Joachim), 5935.
KAASCH (Michael), 5935.
KABACALI (Alpay), 6078.

KACIN-WOHNIZ (Milica), 5118.
KACUNOV (V.), 4257.
KADIKOV (Boris Kh.), 1548.
KADMON (Naftali), 194.
KAEGBEIN (Paul), 897.
KAENEL (Gilbert), 2410.
KAEUPER (Richard W.), 3619.
Kafka (Franz), 6492.
KAGAN (Richard), 6556.
KAGOYA (Naoto), 6932.
KAHK (Juhan), 7206.
KÂHYA (Esin), 6280.
KAISER (Thomas), 5820.
KAISER (Wolfram), 7427.
KĄKOLEWSKI (Igor), 7798.
KAKOVKIN (Aleksandr Ja.), 8792.
KAKRIDIS (F. I.), 1927.
KALAWOUN (Nasser M.), 8361.
KALBERG (Stephen), 1365.
KALIFA (Dominique), 7467.
KALIMTZIS (K.), 2251.
KALININA (Tat'jana M.), 3334.
Kallai (Zecharia), 279.
KALLEY (Jacqueline Audrey), 358, 7614.
KALLIS (Aristotle A.), 1053, 8077.
Kallistos I, patriarch of Constantinople, 3209.
KALMYKOV (Alexej A.), 1480.
KALOUDIS (George Stergiou), 4717.
KALPAXIS (Thanassis), 2337.
KALTWASSER (Inge), 7439.
KAMEN (Henry Arthur Francis), 971.
KAMENEC (Ivan), 5098, 5100.
KAMIKAWA (M.), 7059.
KAMIŃSKI (Łukasz), 4990.
KAMMEN (Carol), 972.
Kamp (Norbert), 686.
KAMPITS (Peter), 1419.
Kampitsis (I. S.), 1927.
KANARFOGEL (E.), 3447.
Kandaules Myrsilos, 1888.
Kandinsky (Vasily), 6522, 6588.
KANE (Anne), 761.
KANE (S.), 2370.
KANELLOS (Nicolás), 6044.
KANG (Yoonhee), 5058.
Kang (Youwei), 8643.
KANGASPURO (Markku), 5059.
Kangxi, emperor of China, 8722.
KANIA (Jan), 7913.
KANIOR (Marian), 5581.
Kanji (Kato), 4868.
KANSU (Aykut), 5253.
Kant (Immanuel), 808, 6114, 6132, 6142, 6150, 6172, 6178, 6195, 6209.
KAOCH (Lene), 7069.
KAPITALNIAK (Tomasz), 4673.

KAPLAN (Alice), 5821.
KAPLAN (Jeffrey), 5465.
KAPLAN (Karel), 4343, 8362.
KAPLAN (Mustafa), 3256, 3257.
KAPLAN (Yosef), 5708.
KAPLONY-HECKEL (Ursula), 1699.
KAPSTEIN (Matthew), 5709.
KARAAGAC (John), 5326.
KARAGEORGHIS (V.), 1956, 2343.
KARAMAN (Igor), 4303.
KARAMBOULA (D.), 3206.
KARAVANIC (Ivor), 1521.
KARIPSIADES (Giorgos), 4718.
Karl I der Große, röm-deutscher Kaiser, König der Franken, 3372.
Karl V, röm.-deutscher Kaiser, 355, 4194, 5153, 5677, 6567, 7736, 7803.
Karl X, king of Sweden, 5211.
Karl XII, king of Sweden, 5217.
KÁRNÍK (Zdeněk), 4344.
KARP (S. Ja.), 5862.
KARPAT (Kemal H.), 548.
KARPIŃSKI (Andrzej), 4991.
KARSH (Efraim), 506, 4789.
KARTOUS (Peter), 142.
KASCHKE (Lars), 7207.
KASEKAMP (Andres), 4400.
KASERER (Christoph), 7016.
KASFIR LETTELFIELD (Sidney), 6524.
KASHANI-SABET (Firoozeh), 7208.
KASHTANOV (Sergej M.), 32.
KASPER (Walter), 1370.
KASSIM (Mahmoud), 8010.
KASSIMATIS (Georg), 7433.
KÄSTNER (Ingrid), 6292.
KASTNER (Quido), 7615.
KATAINEN (Elina), 4407.
KATER (Michael H.), 6698.
KATO (Masafumi), 7915.
KATOUZIAN (Homa), 4765.
KATSIAOUNES (Rolandos), 4319.
KATSUURA (Reiko), 8753.
KATZ (Joshua T.), 2203.
KAUDERS (Anthony), 4576.
KAUFFMAN (Kyle D.), 1121.
KAUFHOLD (Karl Heinrich), 6927, 7017.
KAUFHOLD (Martin), 3394, 3579.
KAUFMAN (Bruce E.), 7332, 7345.
KAUFMAN (Victor S.), 8363.
KAUL (Triloki Nath), 8260.
KAVENIK (Frances M.), 954.
KAVUSI `IRAQI (Muhammad Hasan), 7759.
KAWAKATSU (Heita), 6761.
KAWAMOTO (Masatomo), 8485.
KAY (Lily E.), 6281.
KAYA (Mehmet Ali), 1661, 5936.

KAYANAKIS (Nicolas), 4136.
KAYE (Joel), 3580.
KAZANSKI (M.), 3187.
KAZIN (Michael), 5321.
KE (Dawei), 8580.
KEARNS (G.), 246.
KEATS-ROHAN (K. S. B.), 208.
KEAY (S. J.), 2918.
KEBBEDIES (Frank), 7528.
KEBEDE (John Admassu), 5235.
KECKS (Ronald G.), 6592.
Keddie (Nikki R.), 4071.
KEDOURIE (Sylvia), 5259.
KEE (K. H.), 2187.
KEE (Robert), 4774.
KEEFER (Edward C.), 8258.
KEENAN (Jeremy), 1898.
KEIICHI (Eguchi), 8754.
KEISTER (Lisa), 6788.
KEITA (Maghan), 507.
KEITH (A. M.), 2759.
KEITH DIX (T.), 2760.
KEITSCH (Christine), 6079.
Kekaumenos, 3132.
KELLER (Hildegard Elisabeth), 3640.
KELLER (Mechthild), 7937.
KELLERHALS-MAEDER (Andreas), 337.
KELLERMANN (Karina), 3720.
KELLEY (Donald R.), 788.
KELLY (Catriona), 6372.
KELLY (Debra), 4456.
KELLY (James), 5554.
KELLY (S. E.), 3263.
KELLY (Saul), 7705, 8469.
KELM-HANSEN (Christian), 4377.
KELSTRUP (Morten), 8309.
Kemal (Mustafa 'Atatürk'), 5240, 5249, 5257, 5260.
KEMP (Barry), 1693, 1733, 1766.
KEMP (C.), 6165.
KEMP (Martin), 1278.
KEMPENEERS (Paul), 195.
KEN (O. N.), 8011.
KENDALL (Harry H.), 5080.
KENDALL MOORE (J.), 8364.
KENDEROVA (Stoyanka), 359.
KENEFICK (William), 7081.
Kennedy (Aubrey Leo), 8055.
Kennedy (John Fitzgerald), 5290, 7028, 8261, 8279, 8292, 8327, 8339, 8374, 8385.
KENNEDY (Michael J.), 4468, 4775.
KENNEDY (Michael), 7613, 7953.
KENNEDY (Nick), 25.
Kennedy (Robert Francis), 8261.
KENNEDY (Sean), 6873.
KENNELL (S. A. H.), 3069.
KENNON (Donald R.), 5346.

KENNY (A.), 2252.
KENNY (Kevin), 973.
KENT (Dale), 5822.
KENWARD (Harry), 1598.
KEOGH (Daire), 5554.
KEOGH (Dermot), 7953.
KEOHANE (Dan), 4674.
KÉRIVEN (B.), V.
KERN (Bernd-Rüdiger), 7395.
KERSH (Rogan), 5327.
KERSHAW (Ian), 4577.
KESKİN (Mustafa), 5244.
KEB (Bettina), 6525.
KESSEL (Jürgen), 5068.
KESSENER (P.), 2903.
KESSLER (Herbert L.), 3581, 3806.
KESTNER (Joseph A.), 6463.
KETOLA (Mikko), 5647.
KETTLER (David), 4490.
KETTUNEN-HUJANEN (Eija), 4266.
KEUM (Jang-tae), 5710.
KEUNECKE (H.-O.), 360.
KEUTH (Herbert), 6166.
KÉVONIAN (D.), 7572.
KEY (Newton E.), 3308.
Keynes (John Maynard), 8134.
KEYSSAR (Alexander), 5328.
KHADIAGALA (L.), 5264.
KHALIL (Jehanzeb), 4959.
KHAN (B. Zorina), 6933.
KHAN (Iftikhar Ali), 122.
KHAN (Shaharyar M.), 4750.
KHAN (Zubeda K.), 7616.
KHARITONOV (Khristo), 165.
KHLEVNIUK (O. V.), 8078.
KHODZHASH (S. I.), 1655.
Khomeini (Ruhollah), 4767.
KHRISTOVA (Boriana), 21, 1217.
KHROMOV (Oleg R.), 6045.
KHVOSTOVA (Ksenija V.), 819.
KIBATA (Yoichi), 7609.
KICZA (John), 8812.
KIDD (Dafydd), 3775.
Kidd (William), 8479.
Kiehn (Fritz), 7172.
KIELSTRA (Paul Michael), 7651.
KIELT COSTELLO (Sarah), 1767.
KIENZLE (Beverly Mayne), 3295.
KIERDORF (Alexander), 508.
Kierkegaard (Søren), 6145, 6201.
KIESER (Hans-Lukas), 5254.
KIESEWETTER (Hubert), 6874.
KIESSLING (Friedrich), 8129.
KIEVAL (Hillel J.), 5711.
KIEVEN (Elisabeth), 6568.
KIKUCHI (Yoko), 8503.
KIM (Hyonim), 3582.
KIM (In-Kyuoung), 899.
KIM (J.), 2204.
KIM (Keechang), 3508.
KIM (Young Tae), 4578.

KIM (Yung Sik), 8581.
KIMMERLING (Baruch), 4967.
KIMURA (Shozaburo), 3764.
KIN (Seikei), 8582.
KINANE (Vincent), 352.
KINDT (Tom), 789.
KING (Charles), 4924.
KING (Desmond), 5329.
KING (Edmund), 3395.
KING (Margaret L.), 1218.
KING (Pamela M.), 3831.
KING (Peter), 7529.
KING (Ross), 6557.
KING (Steven), 7333.
KINGSTON (Beverley), 4165.
KINKLEY (Jeffrey C.), 7382.
KINNA (Ruth), 6526.
KINTZINGER (Martin), 3368.
KIOURTZIAN (G.), 3133.
KIPKE (Rüdiger), 5101.
KIPPENBERG (Hans G.), 1366.
KIR'JAK (Dikova, Margarita A.), 1494.
KIR'JANOV (Jurij I.), 7302.
KIRCH (Patrick V.), 1537.
KIRCHBERGER (Ulrike), 6282.
KIRIŞCI (Kemal), 5255.
KIRJUSHIN (Jurij F.), 1548.
KIRK-GREENE (Anthony), 7652, 7688.
KIRKHAM (Victoria), 1279.
KIRKLEY (Evelyn A.), 7209.
KIRSCH (Fritz Peter), 6361.
KIRSTEIN (Robert), 698.
KISELEV (A. S.), 5044.
KISSI (Edward), 4403.
KISTEREV (S. N.), 3312.
KITAGAWA (Takako), 8498.
KITO (Kiyoaki), 8755.
KITSON (Simon), 7468.
KITTELSON (James), 5481.
KLAPISCH-ZUBER (Christiane), 123.
KLAPP-LEHRMANN (Astrid), 1431.
KLARÉN (Peter F.), 4976.
KLAUCK (H. J.), 1992.
KLAUß (Jochen), 6627.
KLEIN (Friz), 401.
KLEIN (Michael), 4568.
KLEIN (Natalie), 5823.
KLEIN (Richard), 3070.
KLEINFELD (Martin), 7068.
KLFINKNECHT (Charles), 7687.
KLEINMAN (Mark L.), 5330.
KLEINSCHMIDT (Harald), 3641.
Kleist (Heinrich von), 6691.
KLEMPERER (Victor), 196.
KLENER (Pavel), 7002.
KLENGEL (Evelyn), 1624.
KLENGEL (Horst), 1624.
Klenze (Leo von), 6559.

KLEPSCH (Michael), 5824.
KLESOW (Rainer M.), 749.
KLEßMANN (Eckart), 4469.
KLIER (Dzh. D.), 5712.
KLIMEK (Antonín), 4345.
KLINE (Ronald R.), 6934.
KLING (Gudrun), 7211.
KLINKHAMMER (Lutz), 1138, 4793.
KLIPPEL (Diethelm), 1098.
KLJASHTONYJ (S. G.), 974.
KLOCZOWSKI (Jerzy), 1312, 4062.
KLOEVEKORN (Andreas), 7414.
KLONDER (Andrzej), 860.
KLOSS (B. M.), 3312.
KLOTZ (Heinrich), 1280.
KLÖTZER (Wolfgang), 5825.
KLUG (W.), 2828.
KLUGE (Bernd), 179.
KLYNNE (Allan), 2905.
KNAPÍK (Jiří), 5826.
KNAPP (Fritz Peter), 3671.
KNAPPETT (Carl), 1951.
KNATZ (Christian), 4579.
KNAUER (Martin), 7800.
KNAUF (E. A.), 1575.
KNELL (H.), 2344.
KNIGHT (Franklin W.), 4729.
KNIGHT (G. R.), 7668.
KNIGHT (Robert), 8365.
KNIGHTLEY (Phillip), 4166.
KNIRK (James E.), 1460.
KNITTLER (H.), 1149.
KNOLL (Viktor), 8054.
KNOPF (Thomas), 2906.
Knorozov (Jurij V.), 8818.
KNOX (MacGregor), 8079.
KNUDSEN (Tim), 4368, 4381.
KOBAYASHI (Hideo), 7915, 8041.
KOBAYASHI (Shoji), 8756.
KOBER (Michael), 2530.
KOBI (Silvia), 5226.
KOBIALKA (Michal), 4019.
KOBUSCH (Th.), 1426.
KOCATÜRK (Kenan), 6078.
KOCH (Christian), 4892.
KOCH (G.), 3186.
KOCH (Johannes), 1768.
Koch (Pietro), 4825.
KOCH (Robert), 5926.
KOCH (Ursula E.), 6049, 6062.
KOCH (Walter), 3255.
KOCH-WESTENHOLZ (Ulla), 1769.
KOCKA (Jürgen), 509, 790, 1144, 4520, 7334.
KOCZERSKA (Maria), 675.
KODER (J.), 3167.
KOEPPEL (Philippe), 5516.

KOFF (Leonard Michael), 3706.
KOGA (Masanori), 8496.
KOHEN (Pirhiyah), 4791.
KOHL (Gerald), 7469.
KÖHLER (H.), 2829, 2836.
KÖHLER (Meike), 6935.
KOHLMEYER (Kay), 1770.
KÖHNE (Eckart), 2678.
KOHUT (Andrew), 5331.
KOIVUNIEMI (Jussi), 7212.
KOIZUMI (Kazuko), 8765.
KOJÈVE (Alexandre), 1055.
KÖKER (Osman), 6851.
KOKURIN (Aleksandr I.), 5038.
KOLA (Paulin), 4130.
KOLB (Anne), 2696.
KOLBABA (Tia M.), 3207.
KOLCHIN (Peter), 1129.
KOLENEKO (Vadim A.), 4267.
KOLEV (V.), 4258.
KOLL (Beatrix), 65.
KOLLER (A.), 7799.
KOLMER (Lothar), 1228.
KOLODZIEJCZYK (Dariusz), 7769.
KOLOKOLOV (B. L.), 8265.
KOLSRUP (Inge-Lise), 3936.
KOLTA (K. S.), 1700.
KÖLZER (Theo), 41.
KOMATSU (Hisao), 8493.
KOMISSAROV (Boris Nikolaevich), 7916.
KOMLOS (J.), 6789.
Komnena (Anna), 3174.
KOMOLOVA (Nelli P.), 5817.
KOMOROWSKI (Krzysztof), 8228.
KOMOROWSKI (Manfred), 5829.
KOMULAINEN (Arvo), 6876.
KONDO (Kazuhiko), 1023.
KONDO (Nobuaki), 8486.
KONDOLEON (C.), 2115.
KONG (Xiangji), 8583.
KONIAS (Andrzej), 247.
KÖNIG (I.), 2583.
KÖNIG (Wolfgang), 6936, 7213.
KÖNNEKER (Carsten), 5828.
KONONENKO (E. I.), 1655.
KONOVALOVA (Irina G.), 3342.
Konrad II, röm.-deutscher Kaiser, 3415.
KONSTAM (Angus), 248.
KONSTAN (David), 2253.
KONSTANTINOV (Aleksander V.), 1573.
KONSTANTINOV (Mikhail V.), 1573.
KONUK (Neval), 1130.
KONZETT (Matthias), 1447, 6464.
Koo (Wellington), 8579.
KOOI (Christine), 5648.
KOOPMANN (Martin), 8241, 8366.
KOPEČEK (Lubomír), 7214.

KORANJAK (M.), 2135.
KORBEN (A.), 1420.
KORENEVSKIJ (Sergej N.), 1480.
KORN (Hans-Enno), 12.
KORNBLUH (Mark Lawrence), 5332.
KÖRNER (Axel), 4036.
KÖRNER (Martin), 4011.
KOROTAEV (A. V.), 1025.
KORSAK (Mariola), 186.
KORTE (Barbara), 249.
KORUNIĆ (Petar), 4074.
KORZILIUS (Jean Loup), 6593.
KORZUN (Valentina P.), 510.
KOSAK (Hadassa), 7335.
KOSCH (Wilhelm), 1433.
KÖSE (Ensar), 901.
KOSELLECK (Reinhart), 791.
KOSHAR (Rudy), 250, 511.
KOSHELENKO (Gennadij A.), 1282.
KOSHELEV (Aleksej D.), 503.
KOSHELEV (Vjacheslav A.), 6465.
KOSHMAN (Lidija V.), 5851.
KOSIŃSKI (Krzysztof), 5995.
KOSLOFSKY (Craig), 5649.
KOSMALA (Beate), 5003.
KOSONEN (Katariina), 251.
KOSSDORFF (Felix), 8367.
KOSSMANN (Ernst Heinrich), 4933.
KÖSTER (Klaus), 8368.
KOSTINER (Joseph), 4081.
Köstlin (Konrad), 894.
KOSTYRJA (G. V.), 1549.
Kotelawala (John Lionel), 5205.
KOTOSHIKHIN (Grigorii), 7770.
KOTOWSKI (Albert S.), 8080.
KOTSONIS (Yanni), 5078.
KOTTSIEPER (I.), 1771.
KOTULLA (Michael), 7415.
Kotzebue (August von), 5893.
KOUAMÉ (Nathalie), 22.
KOUDELKA (František), 4322.
KOUKOLI-CHRYSSANTHAKI (Haïdo), 1522.
KOUNTOURA (E.), 3208.
KOURINOU (E.), 2345.
KOUVIBIDILA (Gaston-Jonas), 4299.
KOUWENBERG (N. J. C.), 1772.
KOVÁČ (Dušan), 596, 4075, 4357, 5093, 8012.
KOVALEVSKAJA (Vera B.), 792.
KOWALCZUK (Ilko-Sascha), 4554.
KOWALIK (Jill), 6188, 6369.
KOWALSKÁ (Eva), 5098.
KOZINA (Valerija V.), 7070.
KOZŁOWSKI (Kazimierz), 4093.
KRADIN (N. N.), 1025.
Krafft-Ebing (Richard), 6299.
KRAG (Claus), 3396.
KRAGELUND (P.), 2521.

KRAH (Adelheid), 3375.
KRAHMALKOV (Charles R.), 1857.
KRAJNC (Viktor), 5120.
KRAMER (Alan), 8002.
KRAMER (Claudia), 2419.
KRANTZ (Olle), 6877.
KRAPAUSKAS (Virgil), 512.
KRASNOV (Nikolaj A.), 7833.
KRAŠOVEC (Tatjana), 7425, 7426.
KRATOCHVIL (Jan), 4328.
KRAUSE (Jens-Uwe), 2419.
KRAUSE (M.), 1625.
KRAUSMAN BEN-AMOS (Ilana), 7215.
KRAUTSCHICK (Stefan), 2065.
KRAYE (Jill), 6160.
KRECH (Volkhard), 1367.
KREILE (Michael), 4827.
KREIS (Georg), 8162, 8186.
KREJČOVÁ (Helena), 4346.
KRENKEL (Werner A.), 2527, 2528.
KŘESŤAN (Jiří), 338.
KRESTEN (Otto), 3123, 3124, 3210.
KREUTZER (Heike), 4580.
KRIEGER (Wolfgang), 4523.
KRIEGSEISEN (Wojciech), 4092, 7628.
Kristensen (W. Brede), 1376.
Kristina, queen of Sweden, 4856, 5216, 5746.
KRIVUSHIN (I. V.), 769.
KROBB (Florian), 6466.
KROEN (Sheryl), 4470.
KROGH (Tyge), 7530.
KROLL (Stefan), 4082.
KROSTENKO (Brian A.), 2452.
KROUCK (Bernard), 8369.
KRUCHTEN (Jean-Marie), 1701.
KRUEGER (Roberta L.), 1438.
KRÜGER (Kersten), 4082.
KRÜGER (Klaus), 6521.
KRUGLER (David F.), 6699.
KRUGMAN (Paul), 7000.
KRUIT (N.), 1684.
KRUMMACHER (Hans-Henril), 80.
KRUMREICH (Gerd), 4561.
KRUSCHWITZ (Peter), 2419.
KRYDER (Daniel), 5333.
Krzyżaniakowa (Jadwiga), 3349.
KUBŮ (Eduard), 8163.
KUCHARSKI (Maciej), 7865.
KUCHER (Marcel), 7012.
KUDAISYA (Gyanesh), 8502.
KUDELIĆ (Zlatko), 5520.
KUHLEMANN (Frank M.), 806.
KÜHLMANN (Wilhelm), 5777.
KUHN (Bärbel), 7216.
KUHN (Rudolf), 6594.
Kuhn (Thomas), 687.
KÜHN (W.), 2254.

KÜHNE (Thomas), 620.
KUIJLAARS (Anne-Marie), 6917.
KUJALA (Antti), 4408.
KUKKONEN (T.), 3463.
KUKLÍK (Jan), 5827, 8229.
KULEMANN-OSSEN (Sabina), 1773.
KULIKOFF (Allan), 6970.
KULIKOWSKI (Michael), 2467.
KÜLZER (A.), 513.
Kumaichirô (Okuno), 4869.
KUMAR (Dharma), 7653.
KUMAR (Nita), 5996.
KUMAR (Ravinder), 4744.
KUMARASWAMY (P. R.), 8425.
KÜMIN (Beat), 6937.
Kundera (Milan), 6201.
KUNÉ (C.), 305.
KUNERALP (Sinan), 8027.
KUNGUROVA (Natal'ja Ju.), 1548.
KUNKLER (Stephan), 5650.
KUNST (Ch.), 2571.
KUNZ (Georg), 514.
KUNZE (Johannes), 5651.
KUNZE (Rolf-Ulrich), 4934.
KUNZE (Thomas), 5033.
KÜNZEL (Drabbe), 8069.
KUPIAINEN (Jari), 861.
KUPPER (Jean-Louis), 233.
KÜPPERS (Heinrich), 8013.
KURAEV (Oleksyj), 8014.
KURAMATSU (Tadashi), 7609.
KURKE (L.), 2205.
KURKINEN (Marjaana), 6700.
KURNAZ (Cemâl), 1454.
KURODA (Hideo), 8757.
KURT (Yılmaz), 5245.
KURTÉN-LINDBERG (Birgitta), 3343.
KURTOĞLU (Ayşenur), 7217.
KURUNMÄKI (Jussi), 5215.
KURUSHIMA (Hiroshi), 8769.
KURZ (Gerhard), 532.
KURZE (Wilhelm), 3397.
KÜSTERS (Hanns Jürgen), 8370.
KUTCHER (Norman), 8552.
KUTOLOWSKI (John F.), 7917.
KÜTÜKOĞLU (Mübahat S.), 5997, 7218.
KUTZ (Martin), 5778.
KUZMICS (Helmut), 1222.
KUZMIN (Yaroslav V.), 1550.
KUZNETSOV (Igor' Nikolaevich), 4203.
KUZNETSOV (Vladimir D.), 2346, 2355.
KUZNIAR (Alice), 6701.
KWASS (Michael), 7018.
KYNIN (G. P.), 8262.
KYNOCH (Gary), 5130.
Kynoch (George), 6051.
KYVIG (David E.), 7432.

L

L'HOMME (M.-L.), 2103.
L'VOVA (Z. A.), 792.
La Bruyère (Jean de), 6415.
La Fayette (Marie-Madeleine Pioche de la Vergne, Madame de), 6421.
LA FRANCESCA (Salvatore), 6991.
LA MANNA (Federica), 5760.
LA MARCA (Nicola), 7219.
LA PENNA (A.), 2697.
La Pira (Giorgio), 5533.
LA PORTE (Pablo), 5143.
LA ROCCA (Eugenio), 4004.
La Rochefoucauld (François de), 6421.
LAAKKONEN (Simo), 1152.
LAAMANEN (Hilkka), 5998.
LABAHN (Michael), 3030.
LABANCA (Nicola), 7695, 7980.
LABARDI (Andrea), 7531.
LABBÉ (A.), 3716.
LABIANO ILUNDAIN (Juan Miguel), 1978.
LABOURDETTE (Jean-François), 7801, 7819, 7834.
LABSVIRS (Janis), 4894.
LABUDA (Gerard), 3966.
LACAM (Jean-Patrice), 4471.
LACAPRA (Dominick), 793.
LACHAUD (Frédérique), 488.
LACHMANN (Henri), 6858.
LACHMANN (Richard), 6790.
LACINA (Vlastislav), 6832.
LACROIX (D. W.), 3716.
LADERO QUESADA (Miguel Ángel), 3431.
LADEWIG PETERSEN (E.), 4368.
LAFFI (U.), 2569.
LAFONT (Jean-Marie), 7617.
LAFONT (S.), 1774.
LAFORE (Robert), 4433.
LAFORET (Juan José), 7021.
LAFOZ RABAZA (Herminio), 3264.
LAFUENTE (Ángel), 1541.
LAGERLUNF (Henrik), 3878.
Laget (Mireille), 6016.
LAGROU (Pieter), 515.
LAHIRI (Shompa), 7220.
LAI (Cheng-chung), 6821.
LAIGNEAU (S.), 2761.
LAIOU (A.), 3210.
LAKE (Marilyn), 7221.
LAKE (Peter), 5830.
LALKOÙ (Ihar), 4900.
LALLEMAND (Marie-Gabrielle), 6410.
LALU (Premesh), 802.
LAMAR (Glenn J.), 7866.
LAMARRIGUE (Anne-Marie), 679.
Lamb (Winifred), 1656.

LAMBDIN (Laura Cooner), 1446.
LAMBDIN (Robert Thomas), 1446.
LAMBERIGTS (M.), 1297.
LAMBERT (Thomas A.), 5621.
LAMBERTINI (Roberto), 3937.
LAMBERTY (Christiane), 6938.
LAMBERT-ZAZULAK (Patricia), 1702.
LAMBOLEY (Jean-Luc), 2324.
LAMBOTTE (Miguel), 7835.
LAMBRECHT (Karen), 7071.
LAMBROPOULOU (A.), 3211.
LAMMERS (William W.), 5334.
LAMOREAUX (Naomi R.), 6878.
LAMPEN (Angelika), 6939.
LANA (Italo), 5952.
LANATA (Giuliana), 2963.
LANCASTER (Joanna), 335.
LANCASTER (L.), 2907.
Lancisi (Giovanni Maria), 5905.
LANDOLFI (L.), 2762.
LANDOLFI (M.), 2378.
LANDOLFI (Maurizio), 2366.
LANDUCCI (Sergio), 6167.
LANDWEHR (Achim), 1091, 7470.
LANDWEHR (Christa), 2908.
LANDY (Marcia), 6702.
LANE (Melissa), 1034.
LANE-POOLE (Stanley), 124.
LANFRANCHI (G.), 1783.
LANG (Werner), 7157.
LANGE (Hans-Jürgen), 4625.
LANGE (Tadeusz Wojciech), 4992.
LANGENSIEPEN (Bernd), 5256.
LANGER (Stefanie), 1099.
LANGEVIN (Philippe), 7347.
LANGEWIESCHE (Dieter), 4553.
LANGFORD (Paul), 1223.
LANGGUTH (A. J.), 8371.
LANGLEY (Leanne), 6715.
LANGLOIS (Charles-Victor), 688.
LANGMUIR (Erika), 1281.
LANGSLOW (D. R.), 2830.
LANGUE (Frédérique), 5382.
Lanjuinais (Jean-Denis, comte de), 7374.
LANOTTE (Olivier), 8283.
LANSING (Richard), 1443.
LANUS (Juan Archibaldo), 8372.
LANZA (Laura), 7438.
LANZI (S.), 1928.
LANZILLOTTA (Eugenio), 2104.
LAPIDGEÒ (Michael), 944.
LAPTEVA (T. A.), 5043.
LARA (Lúcio), 4138.
LARCHER (P.), 1872.
LARDIN (Philippe), 4034.
LARGUIER (Gilbert), 4437.
LARIZZA (Vincenzo), 4828.
LAROCHE-TRAUNECKER (Françoise), 1703.

LARONCE (Cécile), 8373.
LARONDE (André), 917.
LAROSA (Michael), 4289.
Larrea Sagarminaga (María Angeles), 5169.
LARRES (Klaus), 8462.
LARSON (Pier M.), 516.
LARSSON (Lars), 1551.
LARY (Diana), 8553.
LASCARIDES (Celia), 5999.
LASERNA (Mario), 794.
Lasker Schüler (Else), 6458.
Laski (Jan), 5429.
Lasky (Melvin J.), 5872.
LASLETT (John H. M.), 7222.
LASSANDRO (Domenico), 2950.
LÄSSIG (Simone), 197.
LASSNER (Jacob), 517.
LASZKIEWICZ (Hubert Mikolaj), 4062.
LASZLO (Pierre), 6283.
LATHAM (A. J. H.), 6761.
LATHAM (Angela J.), 7223.
LATHAM (Michael E.), 8374.
LATOUR (Francis), 8015.
LATTEUR (Nicolas), 4206.
LAUBE (Adolf), 5638.
LAUCK (Jon), 6973.
LAUER (Gerhard), 859.
LAUER (Josh), 1245.
LAUFER (J. P.), 8262.
LAUFER (Ulrike), 6000.
LAUGHLIN (Kathleen A.), 7337.
LAURA (Ernesto G.), 6703.
Laura Battiferra degli Ammannati, 1279.
LAURENCE (R.), 2553.
LAURENT (Natacha), 6704.
LAUTERBACH (Ansgar), 4581.
LAUX (Claire), 5604.
LAVAGNE (Henri), 659, 2909.
Laval (Pierre), 4441, 8137.
LAVEN (David), 4084.
LAVENCY (M.), 2763.
LAVILLAIN (Philippe), 7624.
LAVOCAT (François), 7765.
LAWES (Kim), 4677.
LAWSON (E. Thomas), 1368.
LAWSON (Fred. H.), 8082.
LAWSON (I. T.), 1948.
Laxmand (Paul), 7152.
LAYBOURN (Keith), 4678.
LAYTON (Azza Salama), 8375.
LAYTON (Bentley), 1704.
LAZAREV (Viktor Nikitic), 3807.
LAZAREVIČ (Žarko), 7019.
LÁZARO (José), 1224.
LAZZARINI (Isabella), 3260.
LAZZARINI (M. Letizia), 2106.
LE BÉGUEC (Gilles), 4451, 4475.
LE BIÉVE (Daniel), 3583.

LE BOHEC (Yann), 2585, 2698.
LE BOT (Yvon), 5378.
LE CORNEC (Jacques), 4362.
LE DIVIDICH (Aude), 71.
LE FLEM (Jean-Paul), 5145.
LE GALL (J.-M.), 6168.
LE GOFF (Jacques), 3661.
LE GUILLOU (Louis), 696.
LE HOUÉROU (Fabienne), 4404.
LE MOAL (Frédéric), 7918.
LE ROUX (P.), 2413, 2699.
LE ROY LADURIE (Emmanuel), 4509.
LE SUEUR (James D.), 8257.
LEAB (Daniel J.), 5335.
LEACH (Melissa), 7706.
LEACH (Michael), 4167.
LEAMAN (Oliver), 1775.
LEASE (Gary), 1313.
LEBOUTTE (René), 1155.
LEBRAVE (Jean-Louis), 86.
LECIEJEWICZ (Lech), 2910.
LECLANT (J.), 1705.
LECLERC (Gérard), 1225.
LEDDEROSE (Lothar), 6626.
LEDENTU (M.), 2720.
LEDGER (Sally), 5783.
LEDIN (Per), 6080.
Ledoux (Claude-Nicolas), 6564.
LEDOVSKIJ (A. M.), 8130.
LEE (A. D.), 2951.
LEE (Chae-Jin), 8302.
LEE (David Tawei), 8376.
LEE (K.), 2037.
LEE (Nicholas), 5534.
LEE (Peter H.), 8785.
LEE DOWNS (Laura), 6001.
LEE RUBIN (Patricia), 417.
LEE VAN SCOTT (Donna), 4290.
LEFEBURE (Leo D.), 1369.
Lefebvre (Henri), 6125.
LEFEBVRE D'OVIDIO (Francesco), 8164.
LEFÈVRE (Marianne), 4472.
LEGASSICK (Martin), 802.
Leguay (Jean-Pierre), 4034.
LEHMAN (P.), 2121.
LEHMANN (Axel), 4582.
LEHMANN (Hartmut), 491, 4103, 4105, 4561.
Lehmann (Orla), 4372.
LEHMANN (Patrick), 8323.
LEHMKUHL (Ursula), 5881.
LEHTONEN (Tuomas M.S.), 490.
LEHUU (Isabelle), 6081.
Leibniz (Gottfried Wilhelm), 6192.
LEIFER (Michael), 8474.
LEIGH (Matthew), 2473.
Leigh (Mike), 6656.
LEIKERT (Jozef), 4347.
LEINKAUF (Thomas), 6284.

LEINO-KAUKIAINEN (Pirkko), 1153.
LEINWEBER (Luise), 6558.
LEITSCH (Walter), 4092, 7628.
LEITZ (Christian), 8165.
LEIVA (Alberto David), 7398.
LEIVERS (Matt), 1594.
LEKEBY (Kjell), 5216.
LEL'CHUK (Vitalij S.), 5082.
LELLOUCH (B.), 3250.
LEMA PUEYO (José Angel), 3607.
LEMAIRE (André), 137, 1858.
LEMARIE (Gérard-Georges), 6527.
LEMBERG (Hans), 4066, 5099.
Lemieux (Rodolphe), 4264.
LEMMINGS (David), 7385.
LEŃ (Kazimierz), 6002.
LENEMAN (Leah), 7224.
LENFANT (Dominique), 2022.
LENGEMANN (Jochen), 4583.
LENGWILER (Martin), 6285.
Lenin (Vladimir Ilič Uljanov), 5074, 5081, 5087.
LENMAN (Bruce P.), 920, 7654.
LENNER (Andrew), 1056.
LENNON (Colm), 5482.
LENOIR (Frédéric), 1307.
LENSKY (N.), 2586.
Lentulus Niger, 2867.
LENTZ (C.), 8798.
LENZI (Mauro), 3584.
LENZIN (Rene), 4643.
Leo Magistros, 3120.
Leo VI, Byzantine emperor, 3115.
Leo X, Papa, 5512, 5947.
León de la Barra (Francisco), 4915.
LEONARDI (Claudio), 3321, 3748.
LEONARDI (G.), 2367.
LEONARDI (Lino), 3668.
Leone (frate), 5.
LEONOVA (Lira S.), 5831.
LEONT'EVA (T. G.), 5616.
Leopardi (Giacomo), 6437, 6446, 6468.
Leopardi (Paolina), 6435.
Leopold I, röm.-deutscher Kaiser, 4178.
LEPORE (Amedeo), 6940.
LEPORE (Giuseppe), 3808.
LEPORE (P.), 2437.
LEPPIN (Hartmut), 3071.
LEPSCHY (Giulio), 200.
LERA (Pétrika), 2347.
LERI (Jean-Marc), 378.
LERNOULD (Alain), 2304.
LEROY (Béatrice), 3432, 3585.
LESCENT-GILLES (Isabelle), 488.
LESLIE (Julia), 5602.
LESOURD (Emmanuel), 8166.
LESPEZ (Laurent), 1522.
LESSING (Hans-Ulrich), 6131.

LESTRIGANT (Frank), 6390.
LESZCZYŃSKI (Adam), 6082.
LETTIERI (Alberto Rodolfo), 4154.
LETZ (Róbert), 5102.
LEUKER (Maria-Theresia), 3345.
LEUWERS (Hervé), 4446.
LEV (Daniel S.), 4757.
LEVEAU (Philippe), 795.
LEVENTHAL (Fred M.), 595, 7578.
LÉVÊQUE (Guillaume), 948.
LEVERE (Trevor H.), 6275.
LEVILLAIN (Henriette), 8136.
LEVILLAIN (Philippe), 8136.
LEVIN (Carole), 3338.
LEVIN (Leonid I.), 5060.
LEVINE (Ellen), 8230.
LEVINGER (Matthew), 4585.
LEVINSON (Kirill A.), 4586.
LEVITT (Jan), 4650.
LEVRA (Umberto), 7097.
Levski (Vassil), 4252.
LEVTZION (Nehemia), 8797.
LÉVY (Bernard Henri), 6169.
LEVY (Carl), 6170.
LÉVY (Edmond), 1618, 1952, 2305.
LEVY (Lia), 6171.
LÉVY (Paule H.), 4463.
LEVYKIN (Konstantin G.), 379.
LEW (Rolando), 4106.
LEWIS (Albert C.), 1184.
LEWIS (J. E.), 7655.
LEWIS (Jayne), 6188, 6369.
LEWIS (Jill), 4183.
LEWIS (Joanna), 4889.
LEWIS (John S. C.), 1477.
LEWIS (Katherin J.), 3967.
LEWIS (M. J.), 6791.
LEWIS (Michael), 4872.
LEWY (Günter), 4587.
LEYS (Ruth), 6286.
LEYSER (Conrad), 3072.
LI (Bozhong), 8584.
LI (Jinzheng), 8585.
LI (Lillian), 6974.
LI (Pengnian), 8616.
LI (Yinzi), 8586.
LIANG (Mancang), 8587.
LIANG (Qichao), 1057.
LIAO (Chi-yang), 8588.
Libanius, 2142, 2192, 2468.
LIBERMAN (A.), 7770.
Licentius, 3009.
LICHNER (Vlastimil Patrick), 5103.
LICHTENHAM (Francine-Dominique), 518.
LIDOV (A. M.), 304.
LIDOV (Aleksej M.), 377.
Lie (Sophus), 6335.
LIEBERMAN (Robbie), 5336.
Lieberwirth (Rolf), 1111.

LIEBICH (A.), 7618.
LIEBIG (Michael), 1776.
LIECHTENHAN (Francine-Dominique), 7625.
LIEDTKE (Rainer), 4575.
LIETZMANN (Hans J.), 1054.
LIEVEN (Dominic), 7619.
LIGHT (Timothy), 1371.
LIICEANU (Gabriel), 6003.
LILJEGREN (Bengt), 5217.
LILJENSTOLPE (Peter), 2905.
LILLIOS (Karina T.), 1552.
LIM (Jie-Hyun), 519, 4076, 4993, 8783.
LIN (Wenhai), 8589.
LINCK (Stephan), 7416.
Lincoln (Abraham), 589, 5272, 5277, 5300, 5323, 5337, 5344.
LINCOLN (B.), 2306.
LINCOLN (Bruce), 1372.
LINDBERG (Kirsten), 7072.
LINDEMANN (Andreas), 3029.
LINDEMANN (Thomas), 4077.
LINDEMANN (Uwe), 6362.
LINDER (Jan), 5218.
LINDERT (Peter H.), 8590.
LINDH (Björn), 5206.
LINDHAL (Carl), 867.
LINDOW (John), 867.
LINEBAUGH (Peter), 6941.
LINEHAN (Thomas), 4679.
LINGELBACH (Gerhard), 7034.
LINGUITI (A.), 2028.
LINI (Gabriella), 306.
LINIGER-GOUMAZ (Max), 978.
LINK (Eugene Perry), 1455.
LINK (Stefan), 2066.
LINKE (B.), 2596.
LINNÉR (S.), 3149, 3150.
LINTHOE NÆSHAGEN (Ferdinand), 3990, 4948.
LIONNET (Jean-François), 4941.
LIOU (Bernard), 2429.
LIPHSCHITZ (N.), 1495.
LIPIŃSKI (Edward), 1859.
LIPPI (E.), 51.
LIPPIELLO (Armando), 3953.
LIPSET (Seymour Martin), 5338.
Lipstadt (Deborah), 407.
LIPTÁK (Ľubomír), 5093, 5098, 5104.
LISBERG JENSEN (Ebba), 6245.
LISITSYN (Nikolaj F.), 1523.
LITAVRIN (Gennadij G.), 3134, 3161.
LITERA (Bohuslav), 4115.
LITRI (Marina), 5046.
LITTLE (Adrian), 1060.
LITTLE (Charles T.), 3775.
LITTLE (J. I.), 7073.
Little (Lester K.), 3940.

LITTLEWOOD (A. R.), 3212.
LITVÁN (György), 8377.
Litvinov (Maksim Maksimovič), 5050, 8067.
LITWICKI (Ellen M.), 7225.
LIU (Fusheng), 8591.
LIU (Haifeng), 8592.
LIU (Hong), 8593.
LIU (Jiafeng), 8594.
LIU (Jinzhong), 8595.
LIU (Pujiang), 8596.
LIU (Qiugen), 8597.
LIU (Rentuan), 8510.
LIU (Xiao), 8598.
Liu (Yan), 8667.
LIU (Zhiwei), 8580.
LIULEVICIUS (Vejas G.), 8016.
Liutprando, vescovo di Cremona, 3183.
LIUZZA (R.M.), 3723.
LIVERMORE (Harold V.), 979.
LIVI (Angelo), 4829.
Livius (Titus), 689, 2469-2472, 2795.
LIVREA (E.), 3139.
LIZZI TESTA (Rita), 3073.
LJUBENOVA (L.), 435.
LJUBIN (V. P.), 1524.
LLANQUE (Marcus), 5832.
Lloyd George (David), 8009.
LLOYD-JONES (R.), 6791.
LO CASCIO (Elio), 2111, 2700, 2703.
LOBATO (Mirta Zaida), 254.
LOBE (M.), 2765.
LOBERA (Francisco J.), 6408.
LOBKOWICZ (Nikolaus), 7595.
Locke (John), 6134, 6195.
LODI (Enzo), 3968.
LOEHLIN (Jennifer A.), 7227.
LOESDAU (Alfred), 474.
LOEWENSTEIN (Andrew B.), 8167.
LOGGI (Saturnino), 55.
LOGUNOV (Aleksandr P.), 8818.
LOHRMANN (Dietrich), 3759.
LOHRMANN (Klaus), 4184.
LOMBARDI (Giuseppe), 67.
LOMBARDO (S.), 2009.
Lomellini (Napoleone), 3274.
LONDÁK (Miroslav), 6792.
LONDON (Louise), 5713.
LONDREGAN (John), 7472.
LONE (Stewart), 4873.
LONG (D. J.), 1553.
LONG (Eugene Thomas), 6173.
LONGHITANO (Adolfo), 5912.
LONGLEY (David), 980.
Longley (Michael), 6436.
LONGO (Gisella), 5937.
LONGONI (Ana), 6528.
LÖNNQVIST (K.), 2911.

LÖNNQVIST (Minna Angelina), 863.
LONZA (Nella), 4309.
LOOSER (Devoney), 520.
LOPES MATOS (Paulo), 5609.
LÓPEZ (Jesús), 1706.
LÓPEZ (Joan B.), 1541.
LÓPEZ CRESPÍ (Miquel), 5833.
López de Ayala (Pero), 493.
LÓPEZ FÉREZ (J. A.), 2182, 2255.
LÓPEZ GARCÍA (Bernabé), 4926.
LÓPEZ GARCÍA (Pilar), 1589.
LÓPEZ PÉREZ (Miguel), 6287.
LÓPEZ RAMOS (Sergio), 849.
LÓPEZ SÁEZ (José Antonio), 1589.
LÓPEZ-ALVES (Fernando), 5375.
LÓPEZ-CORDÓN (María Victoria), 5176, 5188.
LÓPEZ-DAVALILLO (Larrea, Julio), 255.
LORA CAM (Jorge), 4977.
LORBERBAUM (Menachem), 1052.
LORBERBAUM (Yair), 1052.
Lorca (F. García), 2223.
LORDON (Frédéric), 7020.
LORENZ (Chris), 761.
LORENZ (Einhart), 4513.
LORENZ (Maren), 1226.
LORENZ (Richard), 1017.
LORENZ (Stefan), 521.
LORENZANI (M.), 3083.
LORETO (Luigi), 2657.
LORETZ (Oswald), 1651.
ŁOŚ (Andrzej), 2701.
LÖSCHE (Peter), 4588.
LOST (Christine), 6004.
LOTH (Wilfried), 7612.
LOTTIN (Alain), 5448, 5483.
LOTTO (Lorenzo), 5552.
LOUÇA (António), 8168.
LOUD (G. A.), 3938.
LOUD (Graham), 3398.
LOUIS (Cameron), 3844.
LOUIS (Wm. Roger), 8253.
Louis II de Bourbon, Prince de Condé, 4429.
Louis XIII, roy de France, 5582.
Louis XIV, roi de France, 313, 6563, 7843.
Louis XV, roi de France, 4459.
Louis XVIII, roi de France, 4500.
LOUREIRO (Rui Manuel), 7802.
LOURENÇO (Frederico), 2206.
LOVATT (Marie), 3273.
LOVEJOY (Paul), 7707.
LOVELL (Stephen), 89, 6372.
LOVELL (William George), 7074.
LOVERIDGE (Scott), 6982.
LOVISI (Claire), 2658.
LÖWE (Heinz-Dietrich), 4083.
LOWE (K. J. P.), 5834.

LOWE (Roy), 1209.
LOWRY (Donald), 7722.
LOYER (E.), 5911.
LOYN (H. R.), 3908.
LOYRETTE (Henri), 6595.
LÖYTÖNEN (Markku), 256.
Lozek (Gerhard), 474.
LOZOVSKY (Natalia), 257.
LU (Jinyan), 8728.
LÜ (Xiaobo), 8599.
LU (Yu), 8600.
LUALDI (Jackson), 5662.
LUCA (Santo), 56.
Lucanus (Marcus Annaeus), 2473-2475, 2737, 6425.
LUCAS LAWSON (Ann), 6469.
LUCASSEN (Leo), 7228.
LUCCHESI (E.), 3151.
LUCCHESI (Marzia), 3509.
LUCCHINI (Cristina), 6793.
LUCHTERHANDT (Martin), 4589.
LUCIANI (S.), 2831.
Lucianus, 2167, 2262.
Lucilius (Gaius), 2741.
Lucilius Gamala (Publius), 2882.
LUCK (Georg), 1626.
LÜCK (Heiner), 1111.
LUCKEN (Christopher), 501.
LUCKHURST (Roger), 5783.
LUCONI (Stefano), 8083.
Lucretius Carus (Titus), 2265, 2734, 2749, 2758, 2799, 2829, 2831, 2848.
Lucullus, 2760.
LUDINGTON (Charles C.), 522.
LUDMERER (Kenneth M.), 6288.
LÜDTKE (Alf), 7229.
LUDWIG (Frieder), 5606.
Ludwig II, König von Bayern, 4524.
LÜHE (Marion), 4830.
LUKAČKA (Ján), 5093.
LUKIN (Pavel V.), 5061.
Lukonin (V. G.), 1781.
LULL (Vicente), 1599.
LULLUS (Raimundus), 3856.
LUMINI (Antonella), 5461.
Lumumba (Patrice), 4298, 4301.
LUNA (Félix), 4149, 4155.
LUNA (María), 4909.
LUŇÁK (Petr), 8264.
LUND (Hakon), 1227.
LUND (Niels), 3481.
LUNDGREN-NIELSEN (Flemming), 1008.
LUNDIN (Ingvar), 5219.
LUNDSTRÖM (Steven), 1777.
LUNGER KNOPPERS (Laura), 4680.
LUNGU (Corneliu Mihail), 8127.
Luo (Fenglu), 8583.
Luo (Junqing), 8514.

LUO (Min), 8601.
Luo (Wengan), 8709.
LUO (Zhitian), 8602.
LUONGO (Gennaro), 5543.
LUPI (M.), 2107.
LUPO (Salvatore), 4831.
LUPOI (Maurizio), 1101.
LÜSEBRINK (Hans-Jürgen), 6095.
LUSSE (Jackie), 5583.
LUST (J.), 1297.
LUSTOSA (Isabel), 6083.
Luther (Martin), 584, 5403, 5651, 5671, 5675.
Luthuli (Albert John), 802.
LUTZ (Christopher H.), 7074.
LUTZKA (Carolina), 2969.
LUU (Van Loi), 8378.
LUXMOORE (Jonathan), 5558.
LUZZATTO (Sergio), 4832, 7686.
Lycophron, 2016.
LYKOVA (L. A), 5040.
LYNCH (Michael), 7402.
LYNDE-RECCHIA (Molly), 3724.
LYNSHA (V. A.), 1025.
Lysander, 2174.
Lysias, 2017-2019, 2138, 2229.
Lysicrates, 2314.
LYTH (Peter), 6879.

M

MA (Chunli), 8604.
MA (John), 1662.
MA (Junjie), 8605.
MA (Lie), 8606.
MAAS (M.), 2954.
MAAS (Martin Jürgen), 7532.
MAC ALLISTER (William B.), 6289.
MAC ALLISTER LINN (Brian), 7919.
MAC BETH (Brian Stuart), 7920.
Mac Bride (John), 7352.
MAC CANN (Sean), 6470.
MAC CARGO (Duncan), 4874.
MAC CARNEY (Joe), 796.
MAC CARNEY (Joseph), 1422.
MAC CARTHY (Conor), 5835.
Mac Carthy (Joseph Raymond), 5319.
MAC CARTHY (K.), 2767.
MAC CARTNEY (Carol), 1531.
MAC CASKIE (T. C.), 8799.
MAC CAULEY (Robert N.), 1373.
MAC CAW (Neil), 523.
MAC CLANCY (Jeremy), 5177.
MAC CLELLAND (Keith), 471, 4667.
MAC CLOSKEY (Stephen), 4755.
MAC CONNELL (Curt), 258.
MAC CONNELL (Louise), 1458.
MAC CORMACK (J. A.), 1795.
MAC CROSSEN (Alexis), 7230.

MAC CULLOCH (Diarmand), 5653.
MAC CUMBER (John), 6174.
MAC CUSKER (John J.), 6769.
MAC CUTCHEON (Russell T.), 1303, 1317, 1360, 1374.
MAC DERMID (Jane), 5053.
MAC DONALD (Lee Martin), 3074.
MAC DONALD (Robert A.), 3488.
MAC DONOUGH (Richard), 335.
MAC DOWELL (D. M.), 1989.
MAC EACHIN (Douglas J.), 8379.
MAC ELRATH (Karen), 8380.
MAC ENANEY (Laura), 7231.
MAC EVOY (James), 3879.
MAC EVOY-LEVY (Siobhán), 8381.
MAC EWAN (Bonnie G.), 8813.
MAC FARLANE (John), 2257.
MAC GEE (Mary), 5602.
MAC GIBBON (Ian), 992.
MAC GOWAN (Margaret M.), 524.
MAC GRATH (Charles Ivar), 7417.
MAC GRATH (Michael), 6005.
MAC GREGOR (Arthur), 389, 1600.
MAC GREGOR (JoAnn), 6290, 8788.
MAC HUGH (John), 905.
MAC INERNEY (J.), 2067.
MAC ISAAC (Lee), 8552.
MAC KAY (Barry), 6046.
Mac Kay (Claude), 6460.
MAC KAY (I.), 4268.
MAC KEE (Patricia), 6471.
MAC KEE (Sally), 3344.
MAC KENZIE (Steven L.), 574.
Mac Kie (Euan), 1558.
MAC KILLOP (James), 864.
MAC KINNON (James), 3991.
MAC KITTERICK (Rosamond), 525, 3309.
MAC LAUGHLIN (Jim), 4776.
MAC LEAN (Roderick R.), 4621.
MAC LEOD (Roy), 363.
MAC MILLAN (John), 7921.
MAC MILLAN (Norman), 6024.
MAC MULLEN (Haynes), 364.
MAC NAIRN (Jeffrey), 4269.
MAC NAMARA (John), 867.
MAC NEILL (J. R.), 1154.
MAC NELLY (Theodore), 4875.
MAC NIVEN (Ian J.), 1525.
MAC PHERRAN (M. L.), 2258.
MAC QUEEN (E. I.), 2005.
MAC RAILD (Donald M.), 7339.
MAC VAUGH (Michael), 3848.
MAC VEAGH (John), 6706.
MAC WILLIAM (Neil), 6596.
MAC WILLIAMS (Wilson C.), 5339.
Macarius Aegyptius, Sanctus, 3109.
MACARRO VERA (José Manuel), 5178.

Macaulay (Thomas), 690.
MACCABELLI (Terenzio), 6833.
MACDONALD (Sharon), 585.
MACÉ (Laurent), 3399.
MACERA DALL'ORSO (Pablo), 8815.
MACGILLIVRAY (J. A.), 1576.
MACGREGOR MORRIS (Ian), 2068.
Machiavelli (Niccolò), 1081, 6121, 6158.
MACHIN (Ian), 4681.
MACIOTTI (Maria Immacolata), 5434.
MACK (Phyllis), 5435.
MACKAY (Christopher S.), 2069, 2588.
Mackintosh (Charles Rennie), 6615.
MACKLEY (Alan), 6570.
MACLEAN (Fitzroy), 981.
MACLURE (Stuart), 6006.
Macmillan (Harold), 8279.
MACMULLEN (R.), 2589.
Macnamara (E.), 1913.
MACONIE (Stuart), 865.
MACRAE (Clare), 1375.
Macrianus, Roman emperor, 2691.
MĄCZAK (Antoni), 7620.
MADDEN (Frederick), 7396.
MADDICOTT (J. R.), 3346.
MADDOX (Donald), 3725.
MADER (G.), 2768, 2976.
MADER (Gottfried), 671.
MADHOK (Balraj), 8382.
Madison (James), 5274, 5353.
MADRAZO (Santos), 5179.
MADRIGAL BELINCHÓN (Antonio), 1602.
MADRIGNANI (Carlo Alberto), 6411.
Madurowicz-Urbańska (Helena), 7144.
MAEIR (A. M.), 1899.
MAEKAWA (Kumiko), 3809.
MAFFI (Davide), 668.
MAGDALENA NOM DE DÉU (J. Ramón), 1860.
MAGDALINO (Paul), 3213.
MAGELI (Eldrid), 8500.
MAGENNIS (Eoin), 4777, 7205.
MAGNALDI (G.), 2440, 2769.
MAGNALDI (J.), 2002.
MAGNANI (Paolo), 1377.
MAGNELLO (Eileen), 166.
Magnentius (Flavius), 2559.
MAGNESS (J.), 1831.
MAGNO (P.), 2462.
MAGNUSSON (Lars), 6794.
MAH (Harold), 797.
MAHAJAN (Sucheta), 7669.
MAHAL (Bhupinder Singh), 866.

MAHÉ (J. P.), 3214.
MÄHRLE (Wolfgang), 5938.
MAI (Gunther), 7456.
MAIER (Bernhard), 982.
MAIER (Charles S.), 526.
MAIER (Christoph T.), 3630.
MAIER (Robert), 7605.
MAIER (Wilhelm), 3797.
MAIFREDA (Germano), 7232.
MAIGNAN (Jean-Claude), 4941.
MAIGRET (Eric), 798.
MAIN (Steven J.), 5062.
MAINARDI (Roberto), 7075.
MAIORCA (Bruno), 6007.
Maiorianus (Iulius Valerius), Roman emperor, 2510.
MAIORINI (Maria Grazia), 4799.
MAIR (Andrew), 6894.
MAIRE VIGUEUR (Jean-Claude), 3518.
MAIRESSE (Jacques), 6852.
Mairet (Jean), 6751.
MAISANO (Riccardo), 477.
MAISCH (C. J.), 7922.
MAISONNEUVE (Sophie), 6707.
MAÎTRE (Jacques), 1314.
MAIULLARI (Franco), 2136.
MAIULLARI PONTOIS (Maria Teresa), 6546, 6848.
MAJD (Mohammad Gholi), 4766.
MAJOROS (Ferenc), 7803.
MAKDISI (Ussama), 5715.
MÄKINEN (Anne), 6560.
MÄKINEN (Tuomo), 6880.
MAKRIS (G.), 3171.
MAKSAKOVA (Lidija V.), 504.
MAKSIMOVIĆ (L.), 3215.
MAKTASH (Muhammad), 1758.
MAKUMBE (John), 5395.
MALAMUD (Carlos), 7471.
MALAMUT (E.), 3216.
MALANDRINO (Corrado), 6097.
MALANIMA (Paolo), 6795.
MALATO (Enrico), 1432.
MALCOMSON (Scott L.), 7233.
MALDINI CHIARITO (Daniela), 5773.
MALEK (Roman), 5607.
MALENDE (Christine), 6373.
MALERBA (Jurandir), 809, 4236.
MALETTKE (Klaus), 7804.
MALEUVRE (J.-Y.), 2770.
Malevich (Kasimir), 6588.
MALHERBE (A. J.), 3031.
MALHOTRA (Bimal), 4751.
Malik al-Afdal al-`Abbas ibn `Ali, 201.
MALINAS (Y.), 2042.
MALINOVSKIJ (I. V.), 469.
MALITZ (Jürgen), 1912.
MALIYAMKONO (T. L.), 5394.

MALLETT (Robert), 8169.
MALLMANN (Klaus-Michael), 4557.
MALLON (Gabrielle S.), 8258.
MALLORY (J. P.), 2213.
MALLY (Lynn), 6708.
MALOSSE (P. L.), 2468.
MALSBERGER (John W.), 5340.
Malthus (Thomas Robert), 7076.
MALVERN (Sue), 380.
MALYCHA (Andrea), 4590.
MALYSHEV (Alexei), 2355.
MALYSHEVA (Svetlana Ju.), 5063.
MAMINA (Ion), 5034.
MANCA (Anna Gianna), 7466.
MANCA (Joseph), 6597.
MANČEV (Krastjo), 4078.
MANDÉ (Issiaka), 4262.
Mandela (Winnie), 5132.
MANDELBROTE (Giles), 362.
MANDEVILLE (Jean de), 3310.
MANE (Perrine), 3658.
MANETSCH (Scott M.), 5654.
MANGAN (James Anthony), 6008.
MANGO (Cyril), 3217.
MANGOLD (M.), 2348.
MANIACI (Marilena), 47.
MANIATIS (George C.), 3218.
MANIKOWSKA (Halina), 3323.
Manilius (Marcus), 2799.
MANISCALCHI (Monia), 3815.
MANN (Alastair J.), 6084.
MANN (Michael), 7670.
MANN (Susan), 8552.
Mann (Thomas), 6448, 6493, 6494, 6522.
MANN (Vivian B.), 3446.
MANN (W. R.), 2259.
MANNELLI (Goggioli, Marina), 365.
MANNELLI (Maria Francesca), 3810.
Mannerheim (Gustaf), 4413.
MANNING (Brian), 528.
MANNING (R. Th.), 6958.
MANNOVÁ (Elena), 923, 5098.
MANNTEUFEL (Ingo), 627.
MANOLOVA (Mariia G.), 4259.
MANRIQUE (María Elena), 6598.
MANSON (Enrique), 4156.
MANTOVANI (Dario), 2659.
Manuel I Comnenus, Byzantine emperor, 3194, 3212.
MANUWALD (G.), 2757.
MANZI (F.), 3075.
MAO (Lei), 8516.
Mao (Zedong), 5836, 6030, 8721.
MAQUART (Marie-Françoise), 7836.
MARANGONI (C.), 2442.
MARANGOU (Antigone), 2316.

MARC (Jean-Yves), 2349.
MARCADÉ (Jacques), 5011.
MARCELLESI (Marie-Christine), 1627.
Marcellina, Sancta, 3111.
Marcellus (Marcus Claudius), 2795.
MARCHESI (Henri), 2912.
MARCHETTI (Patrick), 2350, 2359.
MARCHETTI (S. C.), 2771.
MARCHETTI-LAKAKI (Maria), 2359.
MARCHEVA (Iliiana), 4260.
MARCHIONATTI (Roberto), 6828.
MARCO SIMÓN (Francisco), 2855.
MARCONE (A.), 2913.
MARCONE (Arnaldo), 529, 664, 2590, 2955.
MARCONI (Silvio), 3464.
MARCOS (Sylvia), 1361.
MARCOT (François), 6801.
MARCOTTE (Didier), 236, 2033.
MARCUS (Kenneth H.), 4591.
Marcus Aurelius Antoninus, Roman emperor, 2593, 2817, 2881, 2883.
MARÉCHAUX (Pierre), 3005.
MAREČKOVÁ (Sylvie), 5105.
MAREK (Doris), 1430.
MAREK (Pavel), 4348.
MARENBON (John), 3880.
MARGALIT (Baruch), 1778.
MARGETIĆ (Lujo), 3909.
MARGHERITIS (Ana), 8423.
MĂRGINEANU-CÂRSTOIU (Monica), 2351.
MARGOLIAN (Howard), 4270.
MARGOLIS (Eric S.), 8383.
Margrete (1353–1412), queen of Denmark, Norway and Sweden, 3429.
Margrethe II, queen of Denmark, 1008, 1015.
MARGRY (P. J.), 5436.
MARGUERAT (Philippe), 6992.
MARGUERON (J.), 1779.
MARGUERON (Jean-Claude), 1628.
Maria Clotilde di Savoia, 7924.
Maria Magdalena, 3040, 3106.
Maria Teresa, imperatrice d'Austria, 4120.
Mariette (Pierre-Jean), 561.
MARIJAN (Davor), 8231.
MARÍN (Manuela), 3465.
MARIN (Richard), 4228.
MARÍN I CORBERA (Martí), 5180.
MARINATOS (N.), 2307.
Marinetti (Filippo Tommaso), 5802, 6497.
MARINHO (José da Silva), 5012.
Marino (Giambattista), 6397.

MARINO (Giuseppe Carlo), 4079.
MARINOVICH (Ljudmila P.), 562, 1282, 1953.
MARIOTTE (Jean-Yves), 339.
Mariotti (Scevola), 692.
Marius Gratidianus (Marcus), 2855.
MARKKANEN (Tapio), 259.
MARKKOLA (Pirjo), 5475.
MARKO (Joseph), 7418.
MARKOE (G. E.), 1861.
MARKOPOULOS (Athanasios), 3114.
MARKOVIC (Sacha), 6881.
MARKOVIĆ (Zvezdan), 5120.
MARKOVITS (Claude), 6942.
MARKS (Gary), 5338.
MARKS (Richard), 3331.
MARKUS (Tomislav), 4304.
MARKWICK (Roger D.), 530.
MARLING (Karal Ann), 7234.
MARNEF (Guido), 5665.
MARNER (Dominic), 3969.
MARNOT (Bruno), 4473.
MAROŃ (Jerzy), 7805.
MAROSZEK (Józef), 7806.
MAROTTA (Valerio), 2660.
MARQUES DE OLIVEIRA (Julieta Teixeira), 7807.
MÁRQUEZ MORA (Belén), 1606.
MÁRQUEZ ROWE (Ignacio), 1736.
MARQUILHAS (Rita), 5939.
MARQUIS (J. P.), 6291.
MARRAMAO (Giacomo), 1058.
Marretti (Lelio), 4833.
MARRO (Catherine), 1640.
MARROCU (Luciano), 4055.
MARSCH (Robert M.), 1102.
MARSDEN (William E.), 799.
MARSEILLE (Jacques), 6855.
MARSHALL (E.), 2125.
Marshall (John Hubert), 693.
MARSHALL (P.-K.), 2772.
Marshall (Paule), 6460.
MARSHALL (Peter), 1240.
MARSILIO (Maria S.), 2207.
Marsilio da Padova, 1076.
MARSILIUS VON INGHEN, 3857, 3882.
MARSTON SPEIGHT (R.), 1862.
MARTANO (R.), 6996.
MARTEL (Gordon), 8055.
MARTELLONE (Anna Maria), 4833.
MARTENS (Stefan), 8068, 8111.
Martialis (Marcus Valerius), 2476, 2477.
MARTIN (A.), 1708.
MARTIN (Catherine), 5582.
MARTIN (David E.), 7339.
MARTIN (David W.), 4316.
MARTIN (Dieter), 6412.

MARTIN (H.), 1604.
MARTIN (Henri-Jean), 71.
MARTIN (Jane), 6009.
MARTIN (Jean-Marie), 3303.
MARTIN (John), 1016, 6975.
MARTÍN (Jorgelina), 4282.
MARTIN (Paul M.), 2446.
MARTIN (Raymond), 6175.
MARTIN (Terry), 5051.
MARTIN (Vanessa), 4767.
MARTIN (Virginia), 7671.
MARTIN (Wallraff), 8071.
Martin du Gard (Roger), 368.
MARTÍN MONTERO (Marcelino), 1583.
MARTÍN VISO (Iñaki), 4017.
MARTINCIC (Lorena), 3811.
MARTINELLI (Vittorio), 6709.
MARTINENGO (Marirí), 3642.
Martínez (Jusepe), 6598.
MARTINEZ (Marc), 6353.
MARTÍNEZ CORTIZAS (Antonio), 1554.
MARTÍNEZ DE SAS (María Teresa), 5176.
MARTINEZ GÓMEZ (Rosario), 8084.
MARTÍNEZ LACY (Ricardo), 1663.
MARTÍNEZ MARTÍNEZ (María), 3586.
MARTÍNEZ MATA (Emilio), 1429.
MARTÍNEZ MILLÁN (José), 5153.
MARTÍNEZ NAVARRETE (M. Isabel), 1589.
MARTÍNEZ RODRÍGEZ (Adrián), 7955.
MARTÍNEZ SÁNCHEZ (Antonio), 8084.
MARTÍNEZ VALLE (Adolfo), 4916.
MARTÍNEZ VALLE (Rafael), 1601.
MARTINI (Manuela), 6976.
Martini (Martino), 5607.
MARTINI (S.), 1512.
Martino (Gaetano), 8280.
MARTINS (Manuel Gonçalves), 5013.
MARTON (E.), 2598.
MARTSCHUKAT (Jürgen), 7533.
MARWICK (Arthur), 4682, 5894.
Marx (Karl), 1423, 6122, 6124.
MARX (Roland), 4663.
MARXEN (Klaus), 7553.
Masaryk (Tomáš Garrigue), 4330, 4357, 7983.
MASCANZONI (Leardo), 3970.
MASIÁ (Andrés), 2458.
MASIA (K.), 1900.
MASIER (A.), 2591.
MASLENNIKOV (Alexandre), 2355.
MASON (Colin), 8476.
MASONEN (Pekka), 8800.

MASROORI (C.), 6176.
MASS (Jeffrey P.), 1103.
MASSABÒ RICCI (Isabella), 318.
MASSEAU (Didier), 6177.
MASSER (Achim), 3925.
MASSEY (Irving), 5716.
MASSON (O.), 1930.
MASSON (Philippe), 8202.
MASSOT I MUNTANER (Josep), 8384.
MASTELLONE (Salvo), 4834.
MASTROGREGORI (Massimo), *VIII*, 398.
MASULLO (Aldo), 7367.
MATCHABELI (K.), 307.
MATEER (David), 6351.
MÁTHÉ (Gábor), 966.
MATHEUS (Michael), 6971.
MATHIEU-CASTELLANI (Gisèle), 6363.
MATHIS (Herbert), 6770.
MATHON (G.), 1343.
MATHUR (Laxman Prasad), 7672.
MATHY (Jean-Philippe), 5837.
MATIJEVIĆ (Zlatko), 4305.
Matilde di Canossa, 3921.
MATRAVERS (Derek), 1081.
MATSCHKE (K.-P.), 3219.
MATSUMOTO (Toshiro), 8607.
MATSUSHITA (T.), 6882.
MATSUZAWA (Yûsaku), 4876.
MATTEINI (Maurizio), 147.
MATTERN (David B.), 5274.
Matthaeus, Evangelista, Sanctus, 3090.
MATTHÄUS (H.), 2379.
MATTHÄUS VON KRAKAU, 3489.
MATTHEE (Rudolph P.), 4071.
Matthei (Caspar), 5520.
MATTHEW (Colin), 4687.
MATTHEW (Elizabeth), 3628.
MATTHEWS (E.), 199, 1921, 1924.
MATTHEWS-GRECO (Sara F.), 1267.
MATTHIESEN (Helge), 4592.
MATTILA (Raija), 1780.
MATTINGLY (D. J.), 2126.
MATTINGLY (Harold B.), 2070.
MATTL (Siegfried), 7078.
MATTONE (Antonello), 1179.
MATUSCHEK (Maria), 5041.
MATVEEV (V.), 7837.
MATVEYEV (Vladimir), 7838.
MATZ (D.), 1931.
MATZ (Jean-Michel), 3593.
MATZ (Siegel), 136.
MAU (Rudolf), 5655.
MAUELSHAGEN (Franz), 5838.
MAUER (Benedikt), 5838.
MAUERSBERGER (Arno), 2219.
MAUGER-PLICHON (B.), 2523.

MAURACH (G.), 2506.
MAURER (Michael), 983.
MAUS (Didier), 7399.
Mauss (Marcel), 694.
MAWBY (Arthur Andrew), 5131.
MAWDSLEY (Evan), 5064.
MAXFIELD (St.), 3220.
MAXIA (C.), 2773.
Maximus Chrysopolitanus (Confessor), 2990.
Maximus Tauriniensis, Sanctus, 3011.
MAXON (Robert M.), 984.
MAXWELL (Kathleen), 3221.
MAXWELL-STUART (P.G.), 5519.
MAY (Ernest R.), 8085.
MAY (Gerhard), 5455.
MAY (Johanna), 7474.
MAY (Lary), 6710.
MAYAFFRE (Damon), 4474.
MAYALI (Laurent), 3498, 7540.
MAYER (Alicia), 5595.
MAYER (Arno), 4080.
MAYER (Thomas F.), 5559.
MAYERS (D.), 8385.
MAYET (F.), 2687.
MAYEUR (Jean-Marie), 1364.
MAYHEW (Robert J.), 260.
MAYOR (Adrienne), 2137.
MAYORAL HERRERA (Victorino), 1602.
MAZDON (Lucy), 6711.
Maždrakov (Panajot D.), 133.
MAZETS (Valentin Genrikhovich), 4203.
MAZOWER (Mark), 4716.
MAZUY (R.), 286.
MAZZA (Jacqueline), 8386.
MAZZA (Mario), 531.
MAZZARELLA (Eugenio), 800.
Mazzarino (Giulio Raimondo), 7762.
Mazzini (Giuseppe), 4832, 4834.
MAZZIO (Carla), 5809.
MAZZOCCHINI (P.), 2774.
MAZZONI (Vieri), 3400.
MEADOWS (I.), 2875.
MECHI (Lorenzo), 7873.
MEDICI (Rita), 1059.
MEDINA (Joao), 5014.
MEDWID (L. M.), 2352.
MEDYNTSEVA (Al'bina A.), 3285.
MEEHAN (Elizabeth M.), 8462.
MEEK (Christine), 7296.
MEEKS (Nigel D.), 1794.
MEGAW (Vincent), 1577.
MÉGIER (Elisabeth), 704, 742.
MEHLMAN (Jeffrey), 5839.
MEHR KIAN (J.), 1901.
MEHTONEN (Paivi), 490.
MEI (Mauro), 55.

MEIER (Edward Franklin), 8387.
MEIER (Helmut), 474.
MEIER (J. P.), 3077.
MEIER (Thomas Dominik), 381.
MEIGHOO (Kirk Peter), 5239.
MEIKLE (Scott), 1423.
MEILMAN (Patricia), 6599.
MEIN (Georg), 6178.
MEINECK (P.), 1979, 2041.
MEINEL (Cristoph), 1212.
MEINL (Susanne), 4593.
MEIN-SMITH (Philippa), 8825.
MEISSNER (Burkhard), 2071, 2702.
MEIßNER (Jochen), 4921.
MEJDROVÁ (Hana), 7325.
MEJER (Jørgen), 1424.
Mejía (Hipólito), 4385.
MEL'NIKOVA (Elena A.), 706.
MELAMED (Diego), 5717.
MELANDER (Torben), 2327.
MELANDRI (P.), 8388.
Melantone (Filippo), 5630.
MELE (Franca), 1179.
MELE (G.), 3083.
MELEGODA (Nayani Samarasinghe), 5205.
Meletius Antiochensis, episcopus, Sanctus, 3056.
MELIK (Jelka), 7534.
MELIS (François), 6085.
MELIS (Guido), 7448, 7475.
MELLANO (Maria Franca), 7924.
MELLONI (Alberto), 5574.
MELLOR (Anne K.), 6473.
MELLUS (M.), 3222.
MELO CAVALCANTI (Sandra), 4229.
MELOSI (Martin C.), 7079.
MELTZER (David J.), 1526.
Menander, 2020, 2021.
MENASSE (Robert), 4186.
Mende (Durandus von), 3874.
Mendel (Gregor), 6252, 6265, 6300.
Mendeleyev (Dmitry Ivanovich), 6333.
Mendes da Costa (Emanuel), 6314.
Méndez de Haro y Guzmán (Luis), 7760.
Méndez-Núnez (Casto), 7900.
Menelik II, Negus of Ethiopia, 4401.
MENESTRINA (Giovanni), 3144.
MENGOZZI (Marino), 5517.
MENIGHETTI (Romolo), 4835.
MENK (Gerhard), 5652.
MENOCAL (María Rosa), 3722.
MENON (Anand), 8389.
MENOZZI (Daniele), 5510, 5515.
Menshikov (Alexander D.), 5043.
MENSING (Hans Peter), 4511.

MENTZER (Alf), 1470.
MENU (M.), 2138.
MENUGE (Noël), 3510.
MERCATTO (Dario), 5462.
MERCIER-BERNADETTE (Fabienne), 8170.
MERÇİL (Erdoğan), 4020.
MERGNAC (Marie-Odile), 205.
MERİÇ (Nevin), 1230.
MERIGGI (M.), 7257.
MERIGGI (Marco), 4836.
MERKEL (Gerhard), 5902.
MERKEL (Kerstin), 5771.
MERKER (G. S.), 2353.
MERLI (E.), 2776.
MERLO (Grado Giovanni), 35.
MERLOTTI (Andrea), 4837.
MERRETT (David T.), 6764, 6816, 7340.
MERSIOWSKY (Mark), 3345.
MERVAUD (Christiane), 5887.
MESA SANZ (J. F.), 3078.
MESHCHERSKAJA (Elena N.), 8489.
MESKANK (Timo), 8017.
MESSINEO (Gaetano), 2914.
MESTMAN (Mariano), 6528.
Metastasio (Pietro), 6644, 6660.
MÉTAYER (Christine), 7235.
Metge (Bernat), 3699.
METHOL FERRÉ (Alberto), 8390.
MÉTHY (N.), 2491, 2856.
MÉTHY (Nicole), 2777.
METREVELI (Helene), 2996.
METS (David R.), 4049.
METZ (Bernhard), 339.
METZDORF (Jens), 4683.
METZGAR (Jack), 7341.
METZGER (C.), 8086.
METZLER (Gabriele), 6293.
MEULDER (M.), 2857.
MEUNIER (F.), 3223.
MEURANT (Alain), 2858.
MEUTEN (Ludger), 3511.
MEVES (Uwe), 66.
MEYER (Andreas), 3512.
MEYER (C.), 3224.
MEYER (Christoph H. F.), 3513.
Meyer (Eduard), 531, 695.
MEYER (Frank), 547.
MEYER (Horst), 84.
MEYER (Hugo), 2915.
MEYER (Jan-Waalke), 1629.
MEYER FILARDO (Peter), 7299.
MEYERHOFER (Ursula), 5228.
MEYER-STOLL (Cornelia), 736.
MEYNEN (Alain), 4210.
MEZ'ER (Avgusta V.), 6045.
MEZA DORTA (Norelky), 5383.
MIARI (Monica), 2404.
MIBRATHU (Yohanis), 902.

MICCOLI (Giovanni), 8087.
MICELI (Sergio), 6712.
MICH (Włodzimierz), 4994.
Michael Psellus, 3135.
MICHÁLEK (Slavomír), 5103.
MICHAŁOWSKA (Anna), 5718.
MICHALSKY (Tanja), 308.
MICHEL (Andreas), 6179.
MICHEL (B.), 7621.
MICHEL (Franck), 263, 4758.
MICHEL (Henri), 6016.
Michelet (Jules), 696.
MICHELINI (Ann N.), 2260.
MICHELS (T.), 6086.
MICHIE (R. C.), 7004.
MIDDLETON (Catherine M.), 5341.
MIDDLETON (Roger), 6797.
MIDGLEY (David), 6474.
MIECK (Ilja), 7597.
MIELONEN (Eevamaria), 5836, 7476.
MIERAU (Heike Johanna), 3905.
MIERSCH (Matthias), 7535.
MIERZEJEWSKI (Alfred C.), 6883.
MIETHKE (Jürgen), 3770.
MIEZA Y MIEG (Rafael), 5169.
MIGLUS (Peter A.), 1782.
MIGNOLO (Walter D.), 1231.
Mihnea (Radu), 3554.
MIJATOVIĆ (Anđelko), 985.
Mikhail, Metropolitan of Russia, 40.
MIKHAL'CHENKO (Sergej I.), 999, 6010.
MIKHEEVA (Galina V.), 5037.
MIKI (Sayako), 8499.
MIKRUT (Jan), 4189.
MIL'SKAJA (L. T.), 562.
MILANO (Ernesto), 3746.
MILANO (L.), 1783.
MILBURN (P.), 1553.
MILEVSKI (I.), 1863.
Mill (John Stuart), 1081.
MILLAR (Eileen Anne), 4841.
MILLARD (Alan), 934, 1615.
MILLER (Dean D. A.), 2209.
MILLER (Elaine R.), 3449.
MILLER (James E.), 8258.
MILLER (John F.), 2592.
MILLER (John), 4684.
MILLER (Maureen Catherine), 3647, 3813.
MILLER (N. F.), 1795.
MILLER (Peter N.), 707.
MILLER (Rory), 8391.
MILLER (William Ian), 1232.
MILLER ROSEN (Arlene), 1603.
MILLER-ANTONIO (S.), 1527.
MILLET ALBÀ (Adelina), 1784.
MILLETT (Allan R.), 8203.
MILLETT (M.), 2918.

MILLIS (Ludo), 3311.
MILLMAN (Brock), 4685.
MILLS (Sophie), 2210.
MILNER (John), 6529.
MILNER (N. P.), 1664.
Miltiades, 2066.
Milton (John), 5440, 6376, 6387.
Milutin, Serb King, 3216.
MILZA (Pierre), 8018, 8473.
MIN (Jie), 8608.
MINAHAN (James), 987.
MINAS (Martina), 1709.
MINAULT-GOUT (A.), 1705.
MINENKO (Nina A.), 7180.
MINEO (B.), 2471.
MINEVRINI (Laura), 3304.
MINION (Mark), 8392.
Mink (George), 7349.
MINKLEY (Gary), 802.
MINKOVA (Milena), 202.
MINNIS (A. J.), 3331.
MINNITI (Fortunato), 8019.
MINOIS (Georges), 1233.
MINOSKI (Mihajlo), 7923.
Minucius Felix (Marcus), 3036.
MIQUEL (Bastien), 6180.
MIRHADY (D. C.), 2014.
MIRON (Dolores), 2108.
MIRONOV (Boris Nikolaevich), 5065.
MIROVITSKAJA (R. A.), 8130.
MISELLI (Walter), 167.
MISHRA (Shreegovind), 4752.
MISKELL (Louise), 7081.
MÍŠKOVÁ (Alena), 4323.
MISSAGGIA (Maria Giovanna), 4838.
MITCHELL (Allan), 6884.
MITCHELL (Colin Paul), 7808.
MITCHELL (Gordon R.), 5342.
MITCHELL (L. G.), 2072.
MITCHELL (Leslie), 7926.
MITCHELL (Margaret M.), 3079.
MITCHELL (Rosemary), 535.
MITCHELL (S.), 2948.
MITCHELL (Stephen), 1648.
MITCHISON (Rosalind), 7342.
MITFORD (Tim), 2211.
MITHEN (Steven), 1528, 1529.
Mithridates IV Eupator, king of Pontus, 2566.
MITROVICH (Gregory), 8393.
MITTAG (P. F.), 1665.
MITTELSDORF (Harald), 7440.
MITTER (Rana), 8610.
MITTERAUER (Michael), 868.
MITTHOF (Fritz), 2419.
MITTLMEIER (Christine), 6087.
MIURA (Toru), 8491.
MIX (York-Gothart), 90, 6475.
MIYAKOSHI (Eiichi), 6011.

MIZUNO (Shoji), 8759.
MJAGKOV (German P.), 536.
MJASNIKOV (V. S.), 8130.
MKHIZE (Sibongiseni), 802.
MLADENOVIC (Bozica), 8020.
MOADDEL (Mansoor), 5689.
MOATTI (Claudia), 1920, 2661, 2705.
MÖCKLI (Daniel), 8171.
MÓCSY (A.), 2598.
MÖDE (Erwin), 1378.
MOENNIGHOFF (Burkhard), 6413.
MOES (M.), 2261.
MOGROBEJO (Endika), 125.
Mohammadan, 124.
MOHNHAUPT (Heinz), 1105, 7505.
MÖHRING (Hannes), 3648.
MOLÀ (L.), 6885.
MOLA (Luca), 1166.
MOLAS RIBALTA (Pere), *XVII.*
MOLCHANOV (Arkadij A.), 1953.
MOLENDIJK (Arie L.), 1379.
Molière (Jean -Baptiste Poquelin, dit), 6403, 6724, 6734.
MOLINA (Iván), 6012.
MOLINA (Manuel), 1785.
MOLINA GONZÁLEZ (Fernando), 1583.
MOLINARI (Maurizio), 7622.
MOLINER (Antonio), 5148.
Molkenbuhr (Hermann), 4512.
MÖLLE (Herbert), 5433.
MÖLLENDORFF (P. von), 2262.
MÖLLER (A.), 2139.
MÖLLER (Renate), 6624.
MØLLER JENSEN (Elisabeth), 1459.
MOLLIER (Jean-Yves), 6095.
MOLLO (S.), 2706.
MOLYNEAUX (Maxine), 7191.
MOLYVIATI (O.), 2778.
MOMIGLIANO (Arnaldo), 697.
MOMMSEN (Hans), 4057, 4594.
Mommsen (Theodor), 698.
MOMMSEN (Wolfgang J.), 736, 5840.
MOMOSE (Kesao), 8760.
MONACCHIA (Paola), 5572.
MONACI CASTAGNO (Adele), 3012.
MONDOLFI GUDAT (Edgardo), 7927.
Mondrian (Piet), 6588.
MONGA (Luigi), 7778.
MONNERET (Jean), 7708.
MONNET (Sylvie), 7623.
MONOSON (S. Sara), 2263.
MONRAD PETERSEN (Andreas), 4379.
MONSON (Jamie), 537.
MONTAGNON (Pierre), 8232.
Montaigne (Michel Eyquem de), 6152, 6415.

MONTANARI (Massimo), 1179, 1192, 3599.
MONTBARBUT (Johnny), 203.
MONTECCHI (Giorgio), 77.
MONTERO FENOLLÓS (Jean Luis), 1786.
MONTESQUIEU (Charles Louis de Secondat, baron de la Brède et de), 699, 1024, 1061, 6138.
Montessori (Maria), 5963.
MONTI (Enrico), 2593.
MONTIEL ARGÜELLO (Alejandro), 7744.
MONTIGLIO (Silvia), 2264.
MONZALI (Luciano), 4794.
MOOGK (Peter N.), 5841.
MOORE (Barrington), 1234.
MOORE (Bob), 7673, 8233.
MOORE (Nina M.), 5343.
MOORE (R. I.), 3401.
MOORE (Rosemary), 5656.
MOORMAN VAN KAPPEN (O.), 1106.
MOORMANN (Eric M.), 2943.
MORADIELLOS (Enrique), 801, 5183.
MORAES (Antonio Carlos Robert), 264.
MORALES (H.), 2201.
MORAN (Bruce T.), 6294.
MORAN (Jeffrey P.), 6013.
Morandi (Carlo), 700.
MORATO (Erica), 6833.
Moraw (Peter), 3359.
MORAWETZ (T.), 2140.
MORDECHAI (Eliav), 6014.
MORE (Charles), 6886.
More (Thomas), 5804.
MOREAU (A.), 2854.
MOREAU (Philippe), 2662.
MOREIRA (Adriano), 7709.
MOREIRA (Isabel), 3650.
MOREIRA DE CASTRO ALVES (Dário), 7591.
MOREL (Ch.), 3004.
MOREL (Lucas E.), 5344.
MOREL (P. M.), 1981.
MOREL (Pierre-Marie), 2265.
MORELLI (A. M.), 2707, 2779.
MORELLI (Serena), 402.
MORELLO (Giovanni), 3999.
MORENO FRAGINALS (Manuel), 4313.
MORENO MENGIBAR (Andrés), 1187.
MORENZ (Ludwig D.), 1837.
MORESCHINI (Claudio), 2998, 3188.
MORETTI (Jean-Charles), 2308, 2354.
MORETTI (Mauro), 651, 5900.
MORETTI (P. F.), 2970.
MOREY (James H.), 3910.

MORGAN (David), 1380.
MORGAN (Graham), 1577.
MORGAN (Gwyn), 2594.
MORGAN (K. A.), 2266.
MORGAN (Kenneth), 6769.
MORGAN (Llewelyn), 2780.
MORI (Jennifer), 4686.
MORI (Kimiyuki), 8761.
Mori (Orinori), 4883.
Morier (Robert), 7929.
Moritz (Karl Phillip), 6178.
Moritz (Landgraf von Hesse-Kassel), 5652.
MORLEY (Neville), 2595.
MORLICCHIO (Elda), 198.
MORNATI (Fiorenzo), 6834.
MORONI (Andrea), 340.
MOROVICS (Miroslav Tibor), 6323.
MOROZOVA (Ljudmila E.), 5066.
MORPURGO (Piero), 3651.
Morris (William), 6526.
MORRISON (Jennifer Klein), 4652.
Morrison (Toni), 6471.
MORSE (David), 5842.
MORSEL (Joseph), 3589.
MORTIER (Roland), 5843.
MORTON (Patricia A.), 382.
MORWOOD (J.), 1994.
MOSCATI (Italo), 6714.
MOSCATI (Laura), 3514, 7387.
Moscati (Ruggero), 701.
MOSCOVICH (M. J.), 2447.
MOSES (John Anthony), 7902.
MOSHER (J. C.), 4237.
MOSHER (Michael A.), 1061.
MOSIG-WALBURG (K.), 1902.
MOSS (Michael), 5954.
MOSSLER FIGG (Kristen), 287.
MOST (Glenn W.), 697.
MOTA (Carlos), 6408.
MOTIKA (Raoul), 7590.
MOTONO (Eiichi), 6943.
MOTTU-WEBER (Liliane), 7163.
MOUBAYED (Sami M.), 5233.
MOUILLEAU (Elisabeth), 7710.
Moulin (Jean), 8212, 8213, 8218, 8226.
MOULINET (Daniel), 1299.
MOULINIER (J.-C.), 3107.
MOURE ROMANILLO (Alfonso), 1566.
MOURY FERNANDES (Eliane), 4229.
MOUYSSET (Sylvie), 7477.
MOXÓ Y ORTIZ DE VILLAJOS (Salvador de), 3590.
MOYANO LAISSUÉ (Miguel Angel), 4142.
MREVLJE (Božidar), 5119.
MROZEK (S.), 2430, 2708.
MROZOWICZ (Wojciech), 24.

MUCHEMBLED (Robert), 1235, 5473.
MÚCSKA (Vincent), 3297.
MUELLER (Joan), 3973.
MUELLER (Reinhold C.), 1166.
MUELLER-GOLDINGEN (Chr.), 2309.
MÜHLHALN (Klaus), 7928.
MUIR (Bernard J.), 25.
MULAS (Luisa), 6382.
MULCAHY (Richard P.), 7343.
MULHOLLAND (Marc), 4778.
MULLEN (Anne), 4839.
MULLER (Arthur), 2336.
MÜLLER (C. W.), 1995.
MÜLLER (Christel), 2355, 2356.
MÜLLER (Frank-Bernhard), 1256.
MÜLLER (G.), 1401.
MÜLLER (H.-P.), 1815.
MÜLLER (Hans H.), 789.
MÜLLER (Jan), 7553.
MÜLLER (Jan-Werner), 5844.
Müller (Klaus E.), 883.
MÜLLER (Michael), 6015.
MÜLLER (Oda), 7420.
MULLER (Richard Alfred), 5657.
MÜLLER (Sibylle), 7573.
MÜLLER (Siegfried), 391.
MÜLLER (Uwe), 6887.
MÜLLER (Wolfgang P.), 1108.
MÜLLER-CELKA (Sylvie), 2357.
MÜLLER-LUCKNER (Elisabeth), 1086, 7596.
MULLIEZ (Dominique), 2336.
MUMFORD (Eric), 6561.
MÜNCH (Hans-Hubertus), 1711.
MUNCK (Thomas), 5845.
MUNDELL MANGO (Marlia), 3225.
MUNDUBELTZ (G.), 2597.
MUNIER (Claudine), 963.
MUNIER (Gerald), 538.
MÜNKEL (Daniela), 4513, 6972.
MÜNKLER (Herfried), 977.
MÜNKLER (Marina), 977.
MUNN (M.), 2073.
MUNNICHS (Geert), 4935.
Muñoz (Juan Bautista), 321.
MUÑOZ IBÁÑEZ (Francisco Javier), 1578.
MUNSLOW (Alun), 539.
MUNTING (Roger), 6232.
MUNZI (Luigi), 73.
MURAIL (P.), 1604.
MURASE (Shinichi), 4877.
MURDOCH (Steve), 7771.
MURDOCK (Graeme), 5658.
MUREDDU (P.), 2212.
MUREȘAN (Camil), 540.
MURGATROYD (P.), 2485.
MURNAGHAN (S.), 2009.
MURPHEY (Rhoads), 8477.
MURPHY (Caroline P.), 1284.

MURPHY (E. M.), 2213.
MURPHY (Hugh), 7015.
MURPHY (Michelle), 7344.
MURPHY (Robert), 6650.
MURRAY (Alan V.), 3402, 3515.
MURRAY (Alan), 3320.
MURRAY (Alexander Callander), 2956.
MURRAY (Alexander), 3652.
MURRAY (Jane), 1530.
MURRAY (Mary Anne), 1531.
MURRAY (Scott W.), 7929.
MURRAY (Williamson), 8203.
Muršili II, king of Hittites, 1818.
MUSAEV (V. I.), 5072.
MUSERE (Jonathan), 204.
MUSGROVE (Margaret Worsham), 2480.
MUSI (A.), 4840.
Musil (Robert), 6201.
Mussolini (Benito), 4808, 4817, 7054, 7720, 8019, 8146.
MUSTÉ (Marcello), 703.
MUSTI (Domenico), 692, 1933, 2122, 2141, 2214.
MUTGÉ VIVES (Josefina), 3433, 3497.
MUTH (Ingrid), 8394.
MUTH (S.), 2859.
MUTSCHLER (Fritz-Heiner), 2756, 2781.
MÜTTER (Bernd), 773.
MUZERELLE (Denis), 70.
MYERS (Robin), 362.
MYHRE (Jan E.), 547.
Myrdal (Gunnar), 5209.
MYSZOR (Wincenty), 3080.

N

NA'AMAN (N.), 1787.
NADEAU (Y.), 2782.
NAGAMURA (Makoto), 8762.
NAGEL (Alexander), 6600.
NAGEL (Anne Christine), 5901.
NAGLE (B.), 2267.
NAGY (Piroska), 3653.
NAITO (Masao), 8496.
NAKAJIMA (Gakusho), 8611.
NAKAMACHI (Nobutaka), 8487.
NAKAMURA (Tadashi), 8763.
NAKKEN (Alfhild), 4949.
NAMAZOVA (Alla S.), 4072.
NAMBA (Chizuru), 8501.
NAMER (Gérard), 680.
Namier (Lewis), 702.
NANNI (Stefania), 5561.
NANZKA (Martin), 4595.
Napoléon Ier, empereur de France, 590, 4467, 4469, 4500, 5505, 6535, 7854, 7856, 7862, 7867, 7868.

NAQUIN (Susan), 7082.
NARDI (E.), 3226.
NARINSKII (Mikhail Matveevich), 8045.
Narmer, 1732.
NARO (Nancy Priscilla), 6977.
NASKA (Kaliopi), 4131.
NASO (Paolo), 5398.
NASS (I. A.), 4899.
Nasser (Gamal Abd el-), 8453.
NASSIET (Michel), 7237.
NATALI (Antonio), 6537.
NATALINI (Terzo), 341.
NATOLI (Claudio), 4842.
NAU (H. H.), 6830.
NAVRÁTIL (Jaromír), 4359.
NAYAK (Anand), 1389.
NAYLOR (Phillip Chiviges), 7711.
NAZARENKO (Aleksandr V.), 3403.
NDLOVU (Sifiso Mxolisi), 802.
NEAD (Lynda), 7084.
NEAL (Larry), 6798.
NEBBIAI DALLA GUARDA (Donatella), 67.
NEBELSICK (Louis Daniel), 1605.
Necho, king of Egypt, 1691.
NECHUTOVA (Jana), 3728.
NECIPOĞLU (N.), 3227.
NECK (Rudolf), 4170.
NEDASHKOVSKIJ (Leonard F.), 4022.
NEDELCHEV (Mikhail), 4261.
NEDERMAN (Carl J.), 4002.
NEDKVITNE (Arnved), 869, 4023.
NEEDHAM (Joseph), 1236.
NEEMANN (Andreas), 4596.
Neera (Anna Radius Zuccari), 6452.
Neferefre, king of Egypt, 1720.
NEGASH (Tekeste), 8396.
NEGREANU (Ioana Alexandra), 8127.
NEGRETTO (Gabriel L.), 4917.
Negri (Ada), 6452.
NEGRI (Antonio), 1046.
NEGRI (M.), 1973, 1974.
NEGRI (Mario), 3.
NEGRINO (F.), 1512.
Nehru (Jawaharlal), 4744, 7616.
NEIBERG (Michael S.), 5345.
NEJKOVA (A.), 342.
NEKIPELOV (Aleksandr D.), 4112.
NELIS-CLÉMENT (J.), 2709.
NELLES (Paul), 5847.
NELSON (Janet L.), 544, 2959, 3729.
NELSON LIMERICK (Patricia), 7238.
NĚMEČEK (Jan), 8141, 8172.
NENONEN (Marko), 6845.
NERALIĆ (Jadranka), 900.

NERCESSIAN (A.), 3187.
NERDINGER (Winfried), 6559.
Neriglissar, king of Babylonia, 1804.
Nero (Claudius Caesar), Roman emperor, 2491, 2501, 2521.
NESCHWARA (Christian), 7536.
Nestor the Chronicler, 3317.
NETHERCOTT (Frances), 5848.
NETTIS (A. V.), 2663.
NEUGEBAUER (Wolfgang), 4520.
NEUHAUS (Helmut), 7777.
NEUMANN (Gerhard), 862.
NEUMANN (Iver B.), 4945.
NEUMANN (Klaus), 546.
NEUMANN (Ute), 7112.
NEUMAYR (Anton), 6364.
NEUMEISTER (Christoff), 2791.
NEUSNER (Jacob), 5704.
NEUSS (Beate), 8397.
NEVEROV (O. Ja.), 1655.
NEVILLE (Ann), 2629.
NEVILLE (Leonora), 3228.
NEVILLE (Peter), 8173.
NEWELL (W. R.), 2268.
NEWHAUSER (Richard), 3654.
NEWLYN (Lucy), 6476.
Newman (Barnett), 6588.
NEWMAN (Kathleen Paula), 8399.
Newton (Isaac), 1200.
NEWTON (William Ritchey), 4477.
Ney (Frederick James), 4272.
NEZHAD (Saeid Edalat), 1381.
NÍ AOLÁIN (Fionnuala), 4779.
NICASTRO (Franco), 4835.
Nicetas Choniata, 3136.
NICHOLAS (Linda M.), 1596.
NICHOLAS (Tom), 6799.
NICHOLLS (Christine Stephanie), 5940.
NICHOLLS (David), 870, 4597.
NICHOLS (Ann Eljenholm), 3805.
NICHOLSON (Paul T.), 1733.
NICKLAS (Thomas), 4478.
Nicola della Cicogna, Sanctus, 3964.
Nicolai (Friedrich), 93.
NICOLAJ (Giovanna), 7.
NICOLAS (Christian), 2735.
NICOLÁS CHECA (Elena), 1606.
Nicolaus Damascenus, 2022, 2530.
Nicolay (John G.), 5272.
NICOLET (Claude), 262, 2710, 4479.
NICOLL (Peter H.), 8090.
NICOLOTTI (A.), 3137.
NICOLSON (Colin), 988, 5347.
NICOSIA (F.), 2408.
NIEBEL (Wilhelm F.), 6129.
Niebuhr (Reinhold), 5330.
NIEDDU (G. F.), 2212.

NIEDERMAN (Carry J.), 3655.
NIELSEN (B. E.), 1712.
NIELSEN (T. H.), 2057, 2078.
NIEMANN (H. M.), 1864.
NIEMANN (Mario), 4598.
NIEMI (Marjaana), 6296.
NIESNER (Manuela), 3671.
NIETHAMMER (Ortrun), 6414.
NIETO HERNÁNDEZ (Pura), 2215.
NIETO SORIA (José Manuel), 3516.
Nietzsche (Friedrich Wilhelm), 800, 1413, 1425, 6153, 6164, 6182.
NIEZEN (Ronald), 5719.
NIGHTINGALE (Pamela), 3591.
NIHILL (Michael), 7756.
NIIMURA (Yoko), 8612.
NIIMURA (Yuichiro), 1935.
NIKOLAENKO (G.), 2121.
NIKOLIC (Kosta), 5392.
NILSSON (Ingemar), 1237, 6297.
NINCI (M.), 2029.
NIPPEL (Wilfried), 736, 1018.
NIPPERT (Klaus), 4599.
NIQUET (Heike), 2419, 2431.
NISH (Ian), 7609.
NISHIKAWA (Makoto), 4878.
NITSCHKE (Bernadetta), 4995.
NITSCHKE (Peter), 1054, 1062, 1104.
NIVAT (Georges), 5270.
NIVET (Philippe), 4436.
Nixon (Richard Milhous), 5290.
NIŽŇANSKÝ (Eduard), 5112.
Nkosi (Johannes), 802.
Nkrumah (Kwame), 8373.
NOACK (Christian), 2023, 5720.
NOAKES (J.), 4520.
NOBIS (Frank), 7537.
NOE (Manfred), 7422.
NOËL (Bernard), 989.
NOERGAARD (Asbjoern Sonne), 4380.
NOESKE (Hans-Christoph.), 170.
NOETHLICHS (K. L.), 3229.
NOGUÈRES (Louis), 4417.
Noica (Constantin), 5962.
NOIREL-SCHUTZ (Claudine), 1522.
NOKANDEH (Gabriel), 1504.
NØKLEBY (Berit), 7241.
NOLAN (Alan T.), 541.
NOLTE (Paul), 806, 3594.
NOLZEN (Armin), 655.
NONNU (Monica), 5849.
Nonnus Panopolitanus, 3138, 3139.
NONOSE (Koji), 5659.
NONOT, (Jean-Jacques), 3677.
NOODT (Birgit), 3992.
NORDBLOM (Pia), 6089.
NORDBY (Trond), 6889.

NORDHOF (Anton Wilhelm), 5068.
NORDIN (Jonas), 5220.
NORIEGA (Chon A.), 6750.
NORMAN (A. F.), 2142.
NORMAN (Hans), 6903.
NORMAND (Lawrence), 5457.
NORRIS (Jim), 5608.
NORSENG (Per), 4023.
NORTH (Douglass C.), 7012.
NORTH (Michael), 1131, 3507, 6800.
NORTH (Robert), 1327.
Northcliffe (Alfred Charles William Harmsworth, viscount of), 6090, 6104.
NORTIER (Michel), 3286.
NOSKOVÁ (Helena), 4349, 6017.
NOTERMANS (Ton), 7022.
NÖTZOLDT (Peter), 5943.
NOUWEN (R.), 2587.
NOVA (Alessandro), 6521.
NOVÁČKOVÁ (Helena), 8141.
NOVÁK (Mirko), 1616, 1631, 1760, 1773, 1788.
NOVAK TALAVERA (Fabián), 8021.
Novalis (Friedrich von Hardenberg), 808.
NOVARA (A.), 2860.
NOVARA (Paola), 3758, 3996.
NOVARESE (Daniela), 7538.
NOVARO-LEFÈVRE (Daniela), 2310.
Novatianus, antipapa, Arnobius, 3036.
NOVELLI (Claudio), 4844.
NOVELLI (Edoardo), 310.
NOVELO (Victoria), 849.
Novoseltsev (Anatoly P.), 3334.
NOWAK (Kurt), 5924.
NOWAKOWSKI (Jacek), 1579.
NOY (David), 2916.
NOYA (Manuel Santos), 3857.
NUDING (Matthias), 3489.
NUHN (Walter), 7712.
Numa Pompilius, king of Rome, 2783.
NÚÑEZ FLORENCIO (Rafael), 5184.
NUNN (Astrid), 1865.
NÜTZENADEL (Alexander), 1138, 7910.
NUVOLONE (Flavio G.), 3568.
NYAMBARA (Pius S.), 7713.
NYBERG (Tore), 3941.
NYE (Malory), 1382.
NYE (Ronert A.), 8552.

O

Ó CIARDHA (Éamonn), 4780.
O'BALLANCE (Edgar), 4300.
O'CONNELL (Michael), 5437.
O'CONNOR (Erin), 6298.

O'GORMAN (Ellen), 723.
O'HALPIN (Eunan), 7953.
O'HANLON (Michael), 872.
O'MALLEY (John W.), 5484.
O'MALLEY (Tom), 6091.
O'NEILL (Kerill), 2495.
O'Neill (Terence), 4778.
O'ROURKE (Kevin), 6775.
O'ROURKE (Shane), 5069.
O'SHAUGHNESSY (Andrew Jackson), 7746.
O'SULLIVAN (Donal), 8403.
OBADELE-STARKS (Ernest), 7346.
OBBINK (D.), 2181.
OBERG (Barbara B.), 5273.
OBERG (Eberhard), 2487.
OBERLÄNDER (Samuel), 7386.
Oberman (Heiko A.), 5634.
OBERWITTLER (Dietrich), 7539.
OBINATA (Sumio), 4879.
OBMANN (Jürgen), 2917.
OBOLENSKAJA (Svetlana V.), 3335, 5850.
OBREGÓN (Clotilde María), 4302.
ÖÇAL (Şamil), 6183.
OCAMPO (José Antonio), 6771.
OCCHIPINTI (Elisa), 3351.
OCEGUERA RAMOS (Rafael), 4918.
OCHIENG' (William Robert), 4888.
OCHOA (Enrique C.), 4919.
OCHOA BRUN (M. A.), 7809.
OCITTI (Jim), 5265.
Octavius (Gaius), 2934.
ODDATI (Nicola), 8174.
ODED (Arye), 4890.
ODONGO (Onyango), 5266.
ODUWOBI (T.), 8801.
OELDEMANN (Johannes), 5617.
OETTEL (Andreas), 1616, 1630.
OEXLE (Otto G.), 1126.
OFCANSKY (Thomas P.), 984.
OFFE (Claus), 1144.
OFFEN (Karen M.), 4087.
OFFLER (H.S.), 3912.
OFFNER (Arnold A.), 8189.
OGAWA (Hiroshi), 3730.
OGEDE (Ode), 6477.
OGILVIE (Sheilagh), 7243.
OGOT (Bethwell A.), 4888.
OHLMEYER (Jane), 5856.
OHTA (Akiko), 7892.
OHTOMO (K.), 7059.
OKA (M.), 2216.
OKEY (Robin), 4187.
ÖKSE (A. Tuba), 1816.
Okuma (Shigenobu), 6882.
ÓLAFUR HALLDÓRSSON, 3679.
OLCOTT (Jocelyn Harrison), 4920.
OLECHOWSKI-HRDLICKA (Karin), 7479.
Olga, Sancta, 3254.

OLIEN (Roger M.), 6890.
OLIVA (A.), 1985.
OLIVA (Juan C.), 1789.
OLIVEIRA (Jorge), 1478.
OLIVEIRA E COSTA (Joao Paulo), 231.
OLIVER (G. J.), 2131.
Olivieri (Benvenuto), 5513.
OLIVOVÁ (Věra), 4350.
Ollenhauer (Erich), 4522.
OLSEN (Jørgen), 957.
OLSEN (Rikke Agnete), 990.
OLSHAUSEN (E.), 2587.
Olson (Carl), 1390.
OLSON (Roberta), 6601.
OLSON (S. D.), 1975.
OLUMOROTI (Oluranti), *XII*.
Olympiodorus Thebanus, 2050.
OMICCIOLI (Massimo), 6835.
Omodeo (Adolfo), 703.
ONSLOW (Barbara), 6092.
ONTAÑON PEREDO (Roberto), 1566.
ONUF (Peter S.), 5348.
OOSTERHUIS (Harry), 6299.
OPHER-COHN (Liliane), 585.
OPINEL (Annick), 311.
OPITZ (Peter J.), 1063.
Opočenský (Jan), 4320.
OPPEN VON (Karoline), 6093.
OPPENHEIM (Jean-Pierre), 7347.
OPPERMANN (Irene), 2453.
Ordericus Vitalis, 704, 742.
ORDHAL KUPPERMAN (Karen), 7745.
OREJAS (Almudena), 1610.
ORENSTEIN (Harold S.), 8192.
OREŠKOVIĆ (Luc), 4306.
Orff (Carl), 6698.
ORIGAS (Jean-Jacques), 1461.
Origenes, 3012, 3013.
ORIGONE (S.), 3230.
ORIZIO (Riccardo), 7657.
ORLANDI (Silvia), 2419.
Orlando di Lasso, 6683.
ORLIK (Igor' I.), 4112.
ORLIN (Lena C.), 7077.
ORLOVA (Lyubov A.), 1550.
ORLOVSKAIA (L. B.), 1484.
ORLOW (Dietrich), 4089.
ORME (Nicholas), 3974.
ORMOS (Mária), 4739.
ORMROD (W. M.), 3600.
ORNATO (Ezio), 72.
OROFINO (Giulia), 47, 3772.
Orosius (Paulus), 743.
ORPHANIDES (Andreas G.), 4316.
ORR (Michael T.), 3805.
ORTAYLI (İlber), 7244, 7480.
ORTEGA (Pascual), 3466.
ORTEGA CANADELL (Rosa), *XVII*.
Ortes (Giammaria), 6833.

ORTI GOST (Pere), 3597.
ORTIZ (A. M.), 2032.
ORTIZ (David), 5185.
ORTIZ MESA (Luis Javier), 542.
OSAGIE (Iyunolu Folayan), 7714.
OSBORN (Emily Lynn), 4723.
OSBORNE (Ken), 803.
OSBORNE (R.), 2074.
OSBORNE (Toby), 7810.
OSEI-TUTU (John Kwadwo), 4644.
OSGOOD (K. A.), 8401.
OSGOOD (Robert L.), 6018.
OSHCHEPKOVA (M. M.), 3322.
OSLER (Margaret), 6310.
OSTENDORF (Heribert), 7496.
OSTERHAMMEL (Jürgen), 1156, 7612.
OSTI GUERRAZZI (Amedeo), 4846.
OSTRANDER (Rick), 5660.
OSTROWSKI (Jerzy), 245.
OSTWALD (Martin), 2109.
OTSUKA (Kazuo), 8488.
OTT (Norbert H.), 20, 312.
OTTE (T. G.), 7930.
OTTEN-FROUX (Catherine), 3274.
OTTMANN (Henning), 1425.
OTTO (Adelheid), 1866.
OTTO (Cl.), 2861.
OTTO (E.), 1867.
OTTO (Ingeborg), 1119.
Otto (Rudolf), 6148.
OTTOMANO (C.), 1512.
OTTOSSON (Sten), 8404.
OUDIJK (Michel R.), 549.
OUÉDRAOGO (Arouna P.), 1239.
OURY (G.-M.), 145.
OUTTES (Joel), 7086.
OUZOULIAS (P.), 2716.
Overberg (Bernard), 5981.
ØVERLAND (Orm), 5349.
OVERMANS (Rüdiger), 8091.
OVERY (Richard), 8153.
Ovidius Naso (Publius), 463, 2477-2484, 2727, 2728, 2731, 2740, 2751, 2752, 2761, 2762, 2771, 2776, 2782, 2788, 2810, 2812, 2822, 2844, 2866, 3702.
OWEN (David W. D.), 1418.
OWEN (Sara), 2075.
OWENS (Alastair), 7101.
OWENS (William M.), 2488.
OYEWOLE (A.), 993.
ØYSTEIN (Rian), 4950.
ÖZBURUN (Serkan), 1230.
ÖZCAN (Abdülkadir), 416.
ÖZGAN (Ramazan), 1790.
OZGHANBAI (Omirzaq), 8022.
ÖZKİLİNÇ (Ahmet), 3256, 3257.
OŻÓG (Krzysztof), 3356.
ÖZTURANLI (Mustafa), 1943, 7218.
ÖZYÜKSEL (Murat), 6891.

P

PACE (Enzo), 1384.
PACE (Giacomo), 7541.
PACE (Valentino), 3816.
PACIOCCO (Roberto), 3769.
PADDAYYA (K.), 1580.
PADEN (William E.), 1316, 1385.
PADILLA (Mark), 2217.
PADOA SCHIOPPA (Antonio), 7540.
PADRÓ (Josep), 1713.
PADUANO (Guido), 1996.
PAECH (Anne), 6717.
PAGÁN (Victoria E.), 2514.
PAGANO (Emanuele), 4847.
PAGE (Mark), 3406.
PAGE (Melvin E.), 8023.
PAGEDAS (Constantine A.), 8405.
PAGLIARULO (Giovanni), 3812.
PAI (Hyung Il), 550.
PAIANO (Maria), 5562.
PAIGE (John Rhodes), 6019.
PAILHÈS (Claudine), 7636.
PALACIOS MARTÍN (Bonifacio), 3265.
PALAZZO (Éric), 3913, 3933.
PALEČEK (Pavel), 4324.
PALERMO (Antonio), 6478.
PALERNE (J. C.), 2270.
PALESE (Salvatore), 332.
PALKA (Joel W.), 995.
PALKIJ (Henryk), 4996.
PALLAVICINO (Eleonora), 3281.
PALLISER (David M.), 3346, 4007.
PALMER (Bryan D.), 7336.
PALMER (Peter), 5393.
PALMER (Rog), 1491, 1592.
PALMER (Steven), 6012.
PALMERINI (P.), 1936.
PALMERO (Beatrice), 7542.
PALMUCCI (A.), 2405.
PALONIEMI (Jarmo), 8204.
Pammenes, 2099.
PAMUK (Şevket), 6796, 7025.
PANAGIOTAKES (Georgios I.), 8210.
PANAYI (Panikos), 4601.
PANCIERA (Silvio), 2419, 2432.
PANEJAKH (Viktor M.), 713.
PÁNEK (Jaroslav), *III*.
PANERO (Francesco), 651, 3534.
PANGRAZIO (Miguel Ángel), 4972.
PANI ERMINI (Letizia), 3052, 4009.
PANIZZA (Giorgio), 5734.
PANIZZA (Letizia), 5895.
PANKHURST (Richard), 4402.
PANTAZZI (Michael), 6595.
PANTSOV (Alexander), 8613.
PANVINI ROSATI (Francesco), 168, 2380, 2711.
PAOLINO (Laura), 3681.

Paolo Diacono, 705.
PAPAGNO (Giuseppe), 804.
PAPAIOANNOU (E.), 3135.
PAPANEK (Hanna), 4490.
PAPASOTERIOU (Ch.), 3231.
PAPE (Matthias), 8406.
PAPENFUSS (Dietrich), 4549.
PAPI (Emanuele), 2406.
Papon (Maurice), 534.
PAQUIER (S.), 6892.
PARAMONOVA (M. Ju.), 3335.
PARANAGUÁ (Paulo Antonio), 6718.
PARATORE (Emanuele), 5951.
PARAVICINI (Werner), 3570.
PARAVICINI BAGLIANI (Agostino), 3407, 3551.
PARCER (Jan), 4627.
PARCERO OUBIÑA (César), 1607.
PARDEE (Denis), 1739.
PARDI (Renzo), 3817.
PARENTE (U.), 5941.
Pareto (Vilfredo), 6834.
PARINET (Elisabeth), 96.
Parini (Giuseppe), 6399.
PARIS (Michael), 5852.
PARISE (Nicola Franco), 2143.
PARISELLA (Antonio), 4848.
PARISH (Helen L.), 5661.
PARISI (Giovanni), 6300.
PARISINOU (Eva), 2144.
PARK (Mungo), 266.
PARKER (Alison M.), 5370.
PARKER (Christopher), 805.
PARKER (Deborah), 6602.
PARKER (Holt N.), 2463.
PARKER (John), 8802.
PARKER (R. A. C.), 8175.
PARKINSON (Stephen), 3830.
PARLATO (Giuseppe), 4849.
PARMEGIANI (Neda), 1817.
PARMENTIER (Bérengère), 6415.
PARODI (Maurice), 7347.
PARRA (Iván Darío), 146.
PARRA DÁVILA (Alvaro), 4219.
PARRONI (P.), 2802.
PARSONS (David N.), 220.
PARSONS (Laila Helen), 4792.
PARSONS (Michael), 8407.
PÄRSSINEN (Leena), 283.
PARTOENS (G.), 3018.
PARTRIDGE (Michael), 8247, 8248, 8249, 8250, 8251, 8252.
PARUSHEV (Parashkev), 5257.
PAS (N.), 4481.
Pascal (Blaise), 6352.
PASCHIDIS (P.), 1966.
Pascoe (Louis), 5441.
Pascoli (Giovanni), 6440.
Pashuto (Vladimir T.), 706.
PASLER (Ralf G.), 66.

PASPALAS (S. A.), 1903.
PASQUALI (Roberto), 5549.
PASQUINI (Elisabetta), 3840.
PAŠTEKA (Július), 5557.
PASTENA (Carlo), 27.
PASTEUR (Paul), 7246.
PASTOR (María Alba), 5595.
PASTORE (A.), 7257.
PASZKIEWICZ (Borys), 172.
PASZTOR (Edith), 3914.
PASZTOR (Maria H.), 8024.
PASZYN (Danuta), 4090.
PATANÈ (Pietro), 7715.
PÁTEK (Jaroslav), 6770.
PATIN (V.), 286.
PATLAGEAN (Evelyne), 3170, 3232.
Patočka (Jan), 4341, 5885.
PATRIARCA (Fátima), 5015.
Patrizi (Francesco), 751.
PATRIZI (Giorgio), 6365.
PATROUCH (Joseph F.), 5485.
PATRY (Sylvie), 291.
PATTERSON (David S.), 8258.
PATTERSON (H.), 2918.
PATTERSON (John R.), 2600.
PATTISON (David G.), 3747.
PATZOLD (Steffen), 3942.
PAUK (Marcin, Rafał), 3598.
PAUL (Gerhard), 4557.
PAUL (Septimus H.), 8408.
Paulinus Nolanus, Sanctus, 3014, 3015.
PAULMANN (Johannes), 4091.
PAULO (Heloisa), 4238.
PAULSON (Ronald), 6416.
PAULUS (Anne), 3634.
Paulus III, Papa, 5513.
Paulus V, Papa, 4852.
Paulus, Apostolus, Sanctus, 3020, 3064, 3079, 3098.
Pausanias, 2202.
PAUVERT (Jean-Jacques), 6366.
PAVANELLO (Giuseppe), 6572.
PAVEAU (Marie-Anne), 206.
PAVEL (Teodor), 7931.
PAVESIO (Monica), 6719.
PAVIOT (Jacques), 3367.
PAVLIČ (Slavica), 6020.
PAVLOVA (Tat'jana A.), 1315.
PAVLOVICH (P.), 1065.
PAVLOVICH GRIGORIEV (Fedor), 1603.
PAWLIK (Georg), 7932.
Payne (Robert O.), 3718.
PAYNE (Stanley G.), 5186.
PAZ (Abel), 7626.
PAZZAGLI (Carlo), 6978.
PAZZAGLIA (Luciano), 5563.
PÉAN (Leslie Jean-Robert), 6802.
PEARCE (M.), 1913.

PEARCE (Mark), 2381.
PEARD (Julyan G.), 6301.
PEARSALL (Derek), 26, 3727.
PEBALL (Kurt), 434.
PÉBARTHE (Ch.), 2145.
PECCHIOLI DADDI (Franca), 1818.
PECH (Thierry), 6417.
PECK (Gunther), 7348.
PECORELLA (Paolo Emilio), 1819.
PEDEN (G. C.), 4688.
PEDERSEN (Andreas Monrad), 4382.
PEDERSEN (Vernon L.), 7349.
PEDERSON (William D.), 5312.
PEDIO (Alessia), 4850.
PEDRAZA JIMÉNEZ (Felipe B.), 1462.
PEDREGAL (Amparo), 3108.
PEDREIRA (J. M.), 4239.
Pedro I, rey de Castilla, 493.
PEEL (J. D. Y.), 8803.
PEGLAU (M.), 2783.
PEIKOLA (Matti), 3994.
Peiresc (Nicolas-Claude Fabri de), 707.
PEKAŘ (Josef), 4325.
PELAEZ ALVAREZ (Victor Manuel), 6337.
PELÁEZ Manuel (J.), 3526.
PÉLAQUIER (Elie), 6016.
PELAZ LÓPEZ (José-Vidal), 5187.
PELIKÁN (Jan), 8263, 8409.
PELINKA (Anton), 4185, 4190.
PELIPAS' (Mikhail Ja.), 5284.
PELIZZARI (Maria Rosaria), 5853.
PELLAND (G.), 3004.
PELLEGRINI (Alfredo), 323.
PELLEGRINI (E.), 1533.
PELLEGRINI (Luigi), 3769, 3995.
PELLEGRINO (C.), 2522.
PELLEGRINO SOARES (Gabriela), 4978.
PELLERITI (Enza), 7423.
PELLETIER (Marcel R.), 4271.
PELLETIER (Réjean), 4271.
PELLING (C. B. R.), 553.
PELLING (Christopher), 2076.
PELLISTRANDI (Benoît), 7933.
Pellizzi (Camillo), 5937.
PELON (Olivier), 1946.
PELTENBURG (Edgar), 1531.
PELTONEN (Matti), 4409.
PELTOVUORI (Risto), 4410.
PELÚAS (Daniel), 5374.
PENDER (E. E.), 2271.
PENDER (Malcolm), 5227.
PENDERGAST (Tom), 5854.
PENELLA (R. J.), 2044.
PENG (Houwen), 8614.
PENNELL (C. R.), 4927.
PENNER (T.), 2077.

PENNETIER (Claude), 4106.
PENTER (Tanja), 5070.
PENZEN STADLER (Franz), 6480.
PEPELASIS MINOGLOU (Ioanna), 6923.
PEPPIS (Paul), 6481.
PERÄLÄ (Anna), 91.
PERALTA RUIZ (Víctor), 4220, 4979.
PEREA (Alicia), 1500.
PEREGUDOVA (Zinaida I.), 5071.
PEREIRA SIESO (Juan), 602, 1543.
PERELMAN (Michael), 6837.
PÉREZ (Joseph), 3596.
PÉREZ (Luis Carlos), 1610.
PÉREZ (Sebastián Celestino), 1581.
Pérez de Guzman (Fernán), 493, 3432.
PÉREZ FERREIRO (Elvira), 5584.
PÉREZ LECUNA (Roberto), 5384.
PÉREZ LEDESMA (Manuel), 5175.
PÉREZ SAMPER (María Angeles), 5176.
PÉREZ VEGA (Ana), 2483.
PÉREZ VILATELA (Luciano), 873.
PÉREZ-BUSTAMANTE (Rogelio), 7772.
Pérez-Embid (Florentino), 708.
PERFETTI (Francesco), 7977.
PERGAMI (Federico), 2664.
PÉRICARD-MÉA (Denise), 3975.
Pericles, 728.
PÉRIÈS (Gabriel), 206.
PERKHAVKO (Valerij B.), 3342.
PERKOWSKA (Urszula), 5942.
PERNAU-REIFELD (Margrit), 7674.
PERNES (Jiří), 4351, 8410.
PERNOT (L.), 2218, 2293.
PERNOUD (Régine), 4895.
Perón (Evita), 4144.
Perón (Juan Domingo), 8390.
PERONNET (M.), 554.
PÉROT (Nicolas), 6720.
PÉROTIN-DUMON (Anne), 7087.
PEROTTI (P. A.), 2784.
PÉROUAS (Louis), 5486.
PÉROUSE DE MONTCLOS (Jean-Marie), 6562.
Perrault (Claude), 6563.
PERRIN (Dominique), 4968.
PERROT (Charles), 3084.
PERROT (Jean-Pierre), 3677.
PERRY (J. P.), 5438.
PERRY (Matt), 7350.
Persius (Flaccus Aulus), 17.
PERTICI (Roberto), 555, 703.
PERUSINO (Franca), 2208.
PERUTELLI (A.), 2785.
PESANTE (Maria Luisa), 6826, 6838.
PESARINO (ASTRID), 3085.

PESCHANSKI (Denis), 4451.
PESCHOT (Bernard), 4482.
PESCOSOLIDO (Guido), 714.
PEŠEK (Jan), 5106, 5107.
PEŠEK (Jiří), 408.
PESTKOWSKA (Maria), 8411.
Pétain (Henri-Philippe), 4441, 5987.
PÉTERI (György), 5903.
PETERS (Christine), 3657.
PETERS (Frederic H.), 1386.
PETERS (Shawn Francis), 5487.
PETERS STONE (Julie), 6721.
PETERSEN (Jens), 7895.
PETERSEN (Klaus), 4383.
PETERSEN (Nikolaï), 937.
PETERSEN (Tore T.), 8412.
PETERSON (Rick), 1594.
PETERSSON (Hans-Inge), 6297.
PETIT (Th.), 1954.
PETIT-AUPERT (Catherine), 2316.
Petitt (P.), 1506.
PETOLETTI (Marco), 3851.
Pëtr I Velikij [le Grand], empereur de Russie, 7837.
PETRAGLIA (M.D.), 1580.
PETRARCA (Francesco), 2459, 3681, 3732.
PETRIE (Duncan), 6722.
Petroc, Sanctus, 3965.
PETRONE (Karen), 5073.
PETRONIO (Ugo), 7367.
Petronius Arbiter, 2485, 2486, 2787.
Petronius, Sanctus, 3968.
PETROPOULOS (Jonathan), 6530.
PETROV (Aleksandr Ju.), 7747.
PETROV (E. V.), 556.
PETROV (Ju. A.), 6787.
PETROV (Liudmil), 8176.
PETROV (Nikolaj V.), 5038.
PETROVA (L. I.), 4024.
PETROVA (Marina S.), 2786.
PETRUCCIANI (Stefano), 1067.
PETRUF (Pavol), 8177.
PETRUKHIN (Vladimir Ja.), 3355.
Petrus Canisius SJ, 5855.
Petrus Damianus, Sanctus, 3896.
Petrus, Apostolus, Sanctus, 3040.
PETRŮV (Helena), 4352.
PETSAS (P. H. M.), 1966.
PETSCHEL (Dorit), 7839.
PETTI BALBI (Giovanna), 126.
PETTIFER (James), 4719, 4902.
PETTITT (P. B.), 1520, 1532.
PETY (Dominique), 899.
PETZET (Michael), 6563.
PETZINA (Dietmar), 1157.
PETZOLD (Joachim), 557.
PEVNY (Olenka Z.), 3140.
PEYLET (Gérard), 6479.

PEYRACHE- LEBORGNE (Dominique), 6371.
PFATSCHBACHER (Klaus), 6482.
PFEIFER (Klaus), 6302.
PFEIFFER (Rolf), 7627.
PFISTER (Ulrich), 6944.
Pfitzner (Hans), 6698.
Pfitzner (Josef), 4323.
PFREPPER (Regine), 6292.
Phaedrus, 2487.
Phalaris, 2063.
PHATOUROU (A. A.), 4718.
PHAYER (Michael), 5488.
PHELPS (M. K.), 2587.
Pherekydes Atheniensis, 2082.
PHILETICUS (Martinus), 3682.
Philipp den Guten, Herzog von Burgund, 3781.
PHILIPPA-TOUCHAIS (Anna), 2359, 2362.
Philippus (Marcus Julius), imperatore romano, 608.
PHILIPS (John Edward), 1168, 8491.
PHILLIPS (David), 5977.
PHILLIPS (Laura L.), 7351.
PHILLIPS (Mark), 558.
PHILLIPS (Quitman E.), 6603.
PHILLIPS (Steve), 5074.
PHILLIPS (Tim), 1542, 1608.
Philo Alexandrinus, 2023, 2833.
Philo of Biblo, 1846.
Philostorgius, 3097.
Philostratus, 2024.
Philotheos Kokkinos, patriarch of Constantinople, 3209.
Photius, patriarch of Constantinople, 3141-3144, 3152.
PIANETTO (Ofelia), 4159.
PIAZZONI (Ambrogio M.), 3999.
PIBIRI (Eva), 3407.
PICCARI (P.), 101.
PICCININI (Chiara), 3818.
Piccolomini (Enea Silvio) v. Pius II, Papa.
PICCOLOTTO SIQUEIRA BUENO (Beatriz), 7091.
PICK (Hella), 4191.
PICKERING (Paul A.), 7247.
PICKSTONE (John), 6303.
PICON (Maurice), 2319.
PIEKARSKI (Stanisław), 6723.
PIEPER (Renate), 4921, 8816.
PIER (Jean-Paul), 6237.
PIERALLINI (Sibilla), 1820.
PIÉRART (Marcel), 2110, 2359.
PIERCE (Michael), 7352.
PIERENKEMPER (Toni), 6945.
PIERONI (Geraldo), 5489.
Pierre II de Savoie, 3407.

PIERRI (Bruno), 4689.
Pierson (Jean), 8214.
PIETRI (Charles), 1364.
PIETRI (Luce), 1364, 3044.
PIETROW-ENNKER (Bianka), 8205.
PIETSCHMANN (Horst), 4921.
PIETTE (Valérie), 7353.
PIGEAUD (A. J.), 2837.
PIGULEVSKAJA (Nina V.), 8489.
PIKE (Jon), 1081.
PILBEAM (Pamela M.), 4483, 7354.
PILE (John F.), 6630.
PİLEHVARİAN (Nuran Kara), 1285.
PILGER (John), 4755.
PILLORGET (R.), 7811.
PIMENTEL (Irene Flunser), 5016.
Pimentel (famiglia), 128.
PINA POLO (Francisco), 2855.
PINCELLI (Maria Agata), 3682.
Pindarus, 3127.
PINDEMONTE (Ippolito), 6398.
PINEAU (Joseph), 6724.
PINELLI (Antonio), 6531.
PINELLI (Lucia), 3321.
PINET (Patrice), 6304.
PINO ITURRIETA (Elías), 5385.
Pinochet Ugarte (Augusto), 4280, 4282.
PIÑOL ALABART (Daniel), 3517.
PINON (Laurent), 71.
PINTEV (S.), 435, 8413.
PINTO (Giuliano), 718.
PINXTEN (Rik), 1387.
PIOTROWSKA (Magdalena), 6725.
PIPERNO (M.), 1533.
PIPONNIER (Françoise), 3658.
PIPPIN (Robert B.), 6483.
PIRART (Éric), 1904.
Pirenne (Henri), 651.
PIRENNE DELFORGE (V.), 2303.
PIRILLO (Nestore), 710.
PIRJEVEC (Jože), 5118.
PIROŻYŃSKI (Jan), 7080.
Pisistratus, 2083.
PISTON (William Gerret), 5350.
PITCHER (Edward W. R.), 267.
PITSAKIS (C. G.), 3233.
Pitt (William), the Younger, 4659.
PITTAU (Massimo), 2407.
Pius II, Papa, 6396.
Pius IX, Papa, 7914.
Pius V, Papa, 4811, 5462.
Pius VI, Papa, 5517.
Pius VII, Papa, 5505, 5506, 5517.
Pius XII, Papa, 5511, 8087, 8100.
PIUZ (Anne M.), 648.
PIVA (Luiz Guilherme), 4240.
PIZZAMAGLIO (Gilberto), 6398.
PIZZORUSSO (G.), *IX*.
Planck (Max Karl Ernst Ludwig), 6243.

PLANERT (Ute), 6305.
PLANT (Alisa), 4652.
Plato, 1040, 1715, 2025, 2026, 2027, 2171, 2212, 2236-2238, 2241, 2243, 2244, 2249, 2254, 2258, 2261, 2263, 2266, 2270, 2271, 2278, 2281, 2285, 2289, 5848.
PLATON (Alexandru-Florin), 874.
PLATON (Gheorghe), 559.
Plautus (Titus Maccius), 463, 2488, 2489, 2731, 2767.
PLAX (Julie Anne), 6604.
PLAZA (Elena), 5386.
PLAZA (M.), 2787.
PLECK (Elizabeth H.), 7248.
Pleistarchos, Athenian politician, 1659.
PLEKET (H. W.), 1970.
Plessner (Helmuth), 884.
PLETERSKI (Janko), 5121.
PLETNEVA (Svetlana A.), 3287.
Plinius Caecilius Secundus (Gaius), minor, 2490.
Plinius Secundus (Gaius), senior, 2491, 2648, 2809, 2826.
PLISIECKI (Piotr), 4062.
PLOETZ (Michael), 8414.
PLOKHII (Serhii), 560.
PLONGERON (Bernard), 5401.
Plotinus, 2028, 2029, 2274.
PLOTZ (John), 1463.
PLOUX (F.), 4484.
Plutarchus, 2030-2032, 2174, 2177, 2221, 2277.
POBEREZHNIKOV (Igor' V.), 7180.
POBST (Phyllis E.), 3292.
PODANÝ (Václav), 4320, 4353.
PODOSINOV (Aleksandr V.), 268, 3086.
PODSKALSKY (Gerhard), 3883.
Poe (Edgar Allan), 6461.
POE (Elizabeth W.), 3733.
POE (Marshall), 875.
Poelzig (Hans), 6553.
POGGI (Vincenzo), 5565.
POGREBOVA (M. N.), 1655.
POHJONEN (Juha), 4411.
POHL (Dieter), 4602.
POHL (Nicole), 5780.
POHL (Walter), 2602, 2965.
POHLANDT-MAC CORMICK (Helena), 802, 5132.
Poidevin (Raymond), 7582.
POIGER (Uta G.), 5884, 6726.
Poincaré (Raymond), 4489, 7979.
POIRIER (Bernard W), 8415.
POIRIER (Paul-Hubert), 1834.
POISSON (Jean-Michel), 3364.
POITRAS (Geoffrey), 7026.
POLANSKÝ (Lubos), *III*.

Pole (J. R.), 5286.
POLE (Jack R.), 922.
Pole (Reginald), 5559.
POLECRITTI (Cynthia L.), 4851.
POLENBERG (Richard), 5351.
POLGÁR (Lászlo), 5576.
POLICHETTI (Antonio), 743.
POLJAKOV (Ju. A.), 7083.
POLLACK (Detlef), 4603.
POLLARD (A. J.), 3435.
POLLARD (A. M.), 1506.
POLLARD (Sidney), 6893.
POLLEY (Martin), 997.
POLLEY (Rainer), 7386.
Pollock (Jackson), 6588.
Polo (José Toribio), 709.
POLONI-SIMARD (Jacques), 4390.
Polybius, 2219, 2869.
POMBENI (Paolo), 710.
POMERANZ (Kenneth), 1158.
POMEYROLS (Catherine), 4419.
POMIAN (Krzysztof), 561, 712.
POMOGÁTS (Béla), 8044.
Pompeius Magnus (Cneus), 2451.
Pompeius Trogus, 635.
Pompidou (Georges), 6659.
Pomponius Mela, 2492.
PONCE MURIEL (Alvaro), 4291.
PONCELET (Christian), 4428.
PONCET (O.), 7840.
PONESSA-SALATHÉ (J.), 2862.
PONOMAREV (A. N.), 5044.
PONS (Anaclet), 676.
PONS (Silvio), 4843, 5077, 8092.
Pontius Pilatus, 2911.
Poole (Edward), 1727.
POOLE (Kristen), 5440.
POOLE (Robert), 4649.
POOLE (Steve), 4691.
POPE (Daniel), 5282.
POPE (Gregory A.), 1497.
POPESCU (J.), 998.
POPOV (Radoslav), 7876.
POPOV (V. A.), 4024.
POPOVA (O. S.), 3234.
POPOVITCH (Miroslav), 5270.
Popper (Karl), 6151, 6166.
PORCIANI (Ilaria), 5900.
PORCIANI (Leone), 665.
PORDOMINGO (F.), 1926.
POROTOV (Alexei), 2356.
PORRET (Michel), 1024.
Porsenna, 2396.
PORSHNEVA (Ol'ga S.), 7249.
PORTE (D.), 2529.
PORTER (Brian A.), 4997.
PORTER (Pamela), 3370.
PORTER (Roy), 5453, 5857, 5883.
PORTER (Stanley E.), 3026, 3074, 3087.
PORTZ (Thomas), 8025.

PORZECANSKI (Teresa), 5371.
POSADA-CARBÓ (Eduardo), 7482.
Posidonius, 635.
POSTEL-VINAY (Gilles), 6852, 7013, 7047.
POSTOLEC (Genevieve), 454.
POTEAT (Eugene), 8416.
POTTER (Mark), 7027.
POTTER (P.), 2272.
POTTER (Simon James), 6094.
POTTER (T. W.), 2919.
POTTLE (Mark), 4648.
PÖTZSCH (Hansjörg), 4604.
POUCET (J.), 2603.
POUJADE (Patrice), 7812.
POULLE (Emmanuel), 31.
POULOT (Dominique), 390.
POUMARÈDE (G.), 7841.
POUNDS (N. J. G.), 1318.
Poussin (Nicolas), 6584, 6606.
POUSSOU (Jean-Pierre), 7813, 7819.
POUWELS (Randall L.), 8797.
POWASKI (Ronald E.), 8417.
POWE (Lucas A., jr.), 5352.
POWELL (John S.), 6727.
POWER (Helen), 6296.
POWER (James E.), 3494.
POZDEEVA (L. V.), 4692.
POZHARSKAJA (Svetlana P.), 4072.
POZZI (Pablo A.), 4158, 4160.
PRACHE (Anne), 3819.
PRACHOWNY (Martin F. J.), 7028.
PRADA (M. Esmeralda), 4292.
PRAG (A. J. N. W.), 2363.
PRALON (D.), 2863.
PRANDI (Anna), 77.
PRASAD (H. Y. Sharada), 4744.
PRASHAD (Vijay), 7250.
PRASLOV (N. D.), 1536.
PRATO (Giancarlo), 23.
PRAVILOVA (E. A.), 7543.
PRAWITZ (Dag), 786.
Praxagoras, 2468.
PRAYON (Friedhelm), 2389.
PRAŽÁKOVÁ (Jana), 334.
PRAZMOWSKA (Anita), 8093.
PREBISH (Charles), 1296.
PRECHEL (Harland), 6803.
PREISENDÖRFER (Bruno), 4605.
PREISER (C.), 1997.
PRENDI (Frano), 2347.
Presley (Elvis), 6645.
PRESS (G. A.), 2289.
PRESTA (Ana María), 7251.
PRESTON (Paul), 8247, 8248, 8249, 8250, 8251, 8252.
PRESTON (Peter Wallace), 4880.
PRÊTRE (Clarisse), 1967.
PREVENIER (Walter), 39.
PRÉVOT (F.), 3091.
Prezzolini (Giuseppe), 5521.

PRICOCO (Salvatore), 3010.
PRIDDAT (Birger P.), 1143.
PRIDHAM (G.), 4519.
PRIETO (Víctor Manuel), 6022.
Prieto y Tuero (Indalecio), 8316.
Prieur (Jean-Louis), 6605.
PRINČIČ (Jože), 7019.
PRINGLE (Denys), 4025.
PRITCHARD (Earl H.), 7842.
PRITCHARD (Gareth), 4606.
PRITZ (Pál), 8094.
PRJAKHIN (A.D.), 3337.
PROCACCI (Giuliano), 4094.
PROCHASKA (Frank), 4693.
PROCHASSON (Christophe), 563.
Procopius Caesariensis, 3146, 3147, 3148, 3149, 3150.
Prodi (Paolo), 710, 1110.
PRÖGER (Susanne), 1430.
PROKOP (Radim), 7089.
PROKŠ (Petr), 8026.
PRONIN (Aleksandr A.), 564.
Propertius (Aurelius Sextus), 2493, 2494, 2495, 2789, 2860.
Prosdocimi (Luigi), 5416.
PROSPERI (Adriano), 5518, 5545, 5566.
PROSS (Harry), 6096.
Proust (Marcel), 6438.
PRÖVE (Ralf), 7485.
Prudentius Clemens (Aurelius), 3016, 3017, 3018.
PRUDON (Kim), 6917.
PRUNETI (Luigi), 4829.
PRUSS (Alexander), 1631, 1632.
PRYDS (Darleen N.), 3436.
PRYWES (Ruth W.), 7356.
PŠENIČKOVÁ (Jana), 6989.
Pseudo-Augustinus, 2990.
Pseudo-Claudianus, 2497.
Pseudo-Macarius, 3019.
Pseudo-Quintilianus, 2498.
Pseudo-Scymnus, 2033.
Pseudo-Zeno, 2034.
PSILOS (Christopher), 7934.
PSOMIADES (Harry J.), 8418.
Ptahhotep, vizir egiziano, 1682.
PUCCI (Francesco), 5523.
PUCCI (Silvio), 7166.
PUCCINI DELBEY (G.), 2788.
Pucheu (Pierre Firmin), 4441.
PÚČIK (Miloslav), 8104.
PUDAL (Bernard), 4106.
PUDDINGTON (Arch), 8419.
Pufendorf (Samuel), 7573.
PUGACHEVA (M. G.), 6048.
PUGH (Martin), 4095.
PUGSLEY (Christopher), 7902.
PUIG AGUILAR (Roser), 6340.
PUIG I TARRECH (Roser), 3261.
PULIDO LLANO (Gabriela), 7935.

INDEX OF NAMES

PULLAN (Brian), 5944.
PULLEYN (S.), 2010.
PUNTONI (Gabriella), 3810.
PURCELL (Edward A., Jr.), 7424.
PURCELL (N.), 1923.
PUŞCAŞ (Vasile), 4096.
PUSCH (C. M.), 1535.
PUSHKAREVA (Irina M.), 7302.
PUSHKAREVA (Natal'ja L.), 1159.
Pushkin (Aleksandr Sergeevich), 6453, 6465.
PUT (Eddy), 5477.
PUTENSEN (Dörte), 8420.
PUTTER (Ad), 3743.
PUTTFARKEN (Thomas), 1286.
PÜTZ (P. Horst), 3671.
PUYMÈGE (G.), 286.
PYLE (Andrew), 1414.
Pyrrho, 2051.

Q

Qabus Bin Said, Sultan of Oman, 4955.
QI (Xia), 8724.
QIAN (Jing), 8615.
QIAO (Nan), 8547.
Qiu (Shuyuan), 8643.
QU (Wenjun), 8617.
QUACK (Joachim Friedrich), 1714.
QUAGLIONI (Diego), 710, 3495.
QUAN (Renrong), 8618.
QUANTIN (Patrick), 4211.
QUARETTI (Jérôme), 4437.
QUART (Leonard), 6656.
QUARTER (Donald), 6919.
QUARTHAL (Franz), 4180.
Quecke SJ (P. Hans), 1625.
QUÉREL (D.), 3716.
QUERZOLI (Serena), 2665.
QUESTIER (Michael), 5830.
Quiccheberg (Samuel), 1260.
Quietus, Roman emperor, 2691.
QUIJADA (Mónica), 877.
QUINAULT (Roland), 7578.
QUINN (Seán E.), 210.
QUINN (Sholeh Alysia), 565.
QUINN (Stephanie), 2811.
QUINTAVALLE (Arturo Carlo), 3617.
QUINTERO TORRES (José Gilberto), 8421.
Quintilianus (Marcus Fabius), 2178, 2499, 2753, 2801.
Quintus Smyrnaeus, 2037.
QVARSELL (Roger), 527.
QVIST (Per Olof), 6728.

R

Raab (Julis), 8406.
RAAFLAUB (K. A.), 2080.
RABAN (Sandra), 3408.

RABASA (José), 566.
RABAU (Sophie), 6367.
RABEHL (Bernd), 4608.
RABEMANANJARA (Raymond William), 4903.
RABINOVICI (Doron), 4188.
RABINOWITZ (J.), 2789.
RABOTTI (Giuseppe), 7.
RABREAU (Daniel), 6564.
RACCANELLI (R.), 2790.
Racine (Jean), 6421.
RACZKOWSKI (Wlodzimierz), 1579.
RADEFF (Anne), 6937.
Radek (Karl), 5047.
RADELIĆ (Zdenko), 4307.
RADER (Olaf B.), 878.
RADEV (Ivan), 4246.
RADEVA (Mariia), 4251, 7140.
RADICI COLACE (P.), 2256.
RADKAU (Joachim), 1160.
RADMILLI (A. M.), 1515.
RADNOTI ALFÖLDI (M.), 2146, 2957.
RADVANOVSKÝ (Zdeněk), 4340.
RADY (Martyn), 3601.
RADZIK (Ryszard), 4204.
RAEBURN (David), 2220.
RAECK (Wulf), 2791, 2920.
RAFFA (Guy P.), 3737.
RAFFAELE (Silvana), 5912.
RAGGI (Barbara), 5726.
RAGGIO (Osvaldo), 677.
RAHE (Paul A.), 1061.
RAHMAN (A.), 1211.
RAINER (J. Michael), 2646.
RAINERI (Osvaldo), 54.
RAINVILLE (Lynn), 1582.
RAITH (Dirk), 4490.
RAITHEL (Thomas), 4609.
RAITT (Suzanne), 6484.
RAJAONARIMANANA (Narivelo), 850.
RAJSFUS (Maurice), 4418.
RAJŠP (Vincenc), 276.
RALPH (Barry), 8095.
RAMAGE (Edwin S.), 2604.
RAMAZANOĞLU (Yıldız), 7217.
RAMAZANOV (Sergej P.), 567.
RAMBALDI (Enrico I.), 6185.
RAMETTA (Gaetano), 6172.
RAMONDETTI (Paola), 721.
RAMOS ROVI (María José), 5189.
RAMOS-HORTA (José), 4755.
RANAWAKE (S.A.), 3689.
RANCATI (Fiorano), 6705.
RANDLES (W. G. L.), 269.
RANGASWAMY (Padma), 7252.
RANGEL (Domingo Alberto), 8422.
RANGER (Terence), 6290, 8788.
Ranke (Leopold von), 711, 820.

RANSEL (David L.), 7253.
RANZATO SANTIN (Federica), 6368.
RAPHAEL (Lutz), 568.
RAPONE (Leonardo), 4055.
RAPPAPORT (Erika Diane), 7254.
RAPPORT (Michael), 4486.
RASCHKA (Johannes), 7545.
RASHID (Ahmed), 4128.
RASMUSSEN (Frank A.), 6859.
RASMUSSEN (Tom), 2392.
RAŠTICOVÁ (Blanka), 6979.
RASTODER (Serbo), 4925.
RATELBAND (Klaas), 7716.
RATH (Wolfgang), 1464.
RATHBONE (D. W.), 2111.
RATHBONE (Dominic), 2548.
RATHBONE (Richard), 7717.
RATHMELL (Andrew), 5021.
RATTÉ (C.), 1669.
RAUBER (André), 5229.
RAUCH (Angelika), 286, 808.
RAUDVERE (Catharina), 837.
RAUH-KÜHNE (Cornelia), 7172.
RAUKAR (Tomislav), 3602.
RAUTMAN (M. L.), 1657.
RAUX (Paul), 1534.
RAVEN (J. E.), 2147.
RAWLINSON (Mark), 6485.
RAWLS (John), 1427.
RAY (Roland), 8096.
Ray (Satyajit), 6664.
RAYCHAUDHURY (Ladli Mohon), 4753.
RAYMENT-PICKARD (Hugh), 807.
RAZAFINDRANALY (Jacques), 4904.
REA (R.), 2921.
READ (James H.), 5353.
Reagan (Ronald Wilson), 5326, 8325.
REBENICH (Stefan), 698.
REBÉRIOUX (Madeleine), 744, 4207.
REBILLARD (É.), 3082.
REBOIRAS (Fernando Domínguez), 3856.
Rechnitzer (Hjalmar), 8191.
REDDÉ (M.), 2895.
REDDY (William M.), 7256.
REDFIELD (Peter), 6306.
REDFORD (Scott), 3467.
REDHEAD (David), 6631.
REDIKER (Marcus), 6941.
REDISH (Angela), 7029.
REDSLOB (Beate), 1144.
RĘDZIŃSKI (Kazimierz), 6025.
REED (Bradly Ward), 8620.
REES JONES (Sarah), 3331.
REEVES (Marjorie), 3884.
Regalianus (Publius Gaius), 2691.
REGAN (John M.), 4772.

REGEN (F.), 2443.
Rehnquist (William), 7493.
REICHL (Karl), 3839.
REID (Susan E.), 6634.
Reid (Thomas), 6208.
REIF (Heinz), 7115.
REIMAN (Michal), 8264.
REIMITZ (Helmut), 2965.
Reinach (Salomon), 659.
REINALTER (Helmut), 1071.
REINELT (Janette), 1265.
REINHARD SEELIGER (Hans), 2890.
REINHARDT (Carsten), 6307.
REINHARDT (Nicole), 4852.
REINLE (Christine), 3604.
REINSCH (D. R.), 3236.
REIS FILHO (Daniel Aarao), 4241.
REIS FILHO (Nestor Goulart), 7091.
REISER (M.), 3088.
Reitemeiers (Johann Friedrich), 2098.
REITER (Reimond), 571.
REITMAYER (Morten), 7030.
REJ (Krzysztof Jan), 5666.
REMBOLD (Elfie), 572.
REMMELINK (Willem), 7588.
REMOLÀ (J.-A.), 2714.
RÉMY (Bernard), 2605.
REMY (Johannes), 6026.
REN (Donglai), 8621.
Renan (Ernest), 1344.
Renart (Jean), 3691.
RENDALL (Jane), 4667.
RENDELI (M.), 2918.
RENDERS FOLMER (Corrie), 6947.
RENEVEY (Denis), 4003.
RENNEVILLE (Marc), 6308.
RENNIE (Bryan S.), 1390.
Renoir (Pierre Auguste), 6614.
RENTON (Dave), 4695.
RENUCCI (P.), 2606.
RENZI (Chiara), 6632.
RENZI (Giovanni), 6632.
REPP (Kevin), 6309.
RESINA (Joan Ramon), 453.
RETAMAL AVILA (Julio), 7737.
RETEGAN (Mihai), 8424.
REUST (Hans Rudolf), 381.
REUTHER (Thomas), 5860.
REVIAKINA (L.), 435.
REVIRIEGO (Bernard), 3276.
REXOVÁ (Kristina), *III*.
REY (Roselyne), 6189.
REY-GOLDZEIGUER (Annie), 7710.
REYNARD (Jean), 1715.
REYNOLDS (J. M.), 2607.
REYNOLDS (Larry J.), 309.
Rhee (Syngman), 4891.
RHEINHEIMER (Martin), 7258.
RHEUBOTTOM (David), 3605, 7259.

RHOADS (Edward J. M.), 8622.
RHODES (Carolyn), 7629.
RHOR VIO (Francesca), 2608.
RI (Kaigen), 8623.
RIALL (Lucy), 4084.
RIAÑO RUFILANCHAS (D.), 1968.
RIBARD (Dinah), 575.
RIBEIRO DE MENESES (Filipe), 8028.
RIBERA FLORIT (Josep), 1868.
RIBICHINI (Sergio), 1869.
RIBOT GARCÍA (Luis A.), 5182, 6836, 6871.
RICALDONE (Luisa), 6407.
RICCARDI (Andrea), 5492.
RICCARDS (Michael P.), 7630.
RICCI (C.), 2609.
RICCI (Cecilia), 2419.
RICCI (Saverio), 5567.
RICCÒ (Laura), 6418.
RICHARD (Thibault), 4487.
Richard I Lionheart, king of England, 3411.
Richard, Earl of Cornwall, 3406.
RICHARDS (Edward Graham), 108.
RICHARDS (Leonard L.), 5354.
RICHARDSON (J. S.), 2439.
RICHARDSON (L.), 2922.
RICHARDSON (Roger Charles), 404.
Richardson (Samuel), 6420.
RICHARDSON (Sarah), 4712, 7464.
RICHE (Denyse), 3944.
RICHE (Denyse), 663.
RICHEFORT (I.), 7828.
Richelieu (Armand-Jean du Plessis, cardinal de), 346, 1080, 4464, 7793, 7804, 7832.
RICHER (N.), 2081.
RICHER VON SAINT-REMI, 3313.
RICHEZ-BATTESTI (Nadine), 7347.
RICHTER (Reinhard), 5568.
RICHTER (Steffi), 109.
RICKITT (Richard), 6730.
RICO (Francisco), 6408.
RICO MORENO (Javier), 576.
RICŒUR (Paul), 765, 810, 6115.
RICOTTILLI (L.), 2792.
RICUPERATI (Giuseppe), 570, 6190.
RIDDY (Felicity), 3358.
RIDER (Christine), 6872.
RIDGWAY (B. S.), 2361.
RIDGWAY (D.), 1913, 2382.
RIDLEY (Ronald T.), 1791, 2610.
RIEDE (Judith), 1113.
RIEDEL (Volker), 1465.
RIEDENAUER (Markus), 2273.
RIEGER (Dietmar), 367.
RIEKE-MÜLLER (Annelore), 391.
RIEMENSCHNEIDER (Rainer), 813.

RIEMER (Ellen), 2923.
RIEMER (U.), 2571.
RIES (Julien), 1352.
RIESEBRODT (martin), 1391.
RIEU (Josiane), 6390.
RIFFLET (Jacques), 1392.
RIGGIO (Andrea), 4021.
RIGGS (Christina), 1716.
RIGHETTI TOSTI-CROCE (Marina), 3918.
RIGSBEE (W. Lynn), 4955.
RILEY (J. P.), 7867.
RIMONDI (Francesca), 6731.
RINEHART (Michael), 1255.
RINGDAL (Nils Johan), 6732.
RINGSHAUSEN (Gerhard), 1064.
RINZEMA (Win), 8178.
RIO (Joseph), 577.
Ripka (Hubert), 4324.
RIPPER (Susan), 1479.
RITARANTA (Eino), 6880.
RITNER (Robert K.), 1717.
RITTER (Werner H.), 1393.
RIVA (Silvia), 6370.
RIVOAL (Isabelle), 5721.
RIX (Helmut), 2383.
RIZAKIS (Yvonne), 2359.
RIZAS (Soteres Ch.), 8426.
RIZO-PATRÓN BOYLAN (Paul), 4980.
RO'I (Yaacov), 5722.
Robbe-Grillet (Alain), 6491.
ROBBINS (Ruth), 1474.
ROBERGE (Michel), 1870.
Robert the Burgundian, 3393.
Roberto d'Altavilla 'Il Guiscardo', duca di Puglia, di Calabria e di Sicilia, 3398.
ROBERTS (David D.), 578.
ROBERTS (Gareth), 5457.
ROBERTS (Geoffrey Charles), 8098.
ROBERTS (Geoffrey), 8075.
ROBERTS (Harry E.), 6311.
ROBERTS (Jon H.), 5946.
ROBERTS (Julia), 1594.
ROBERTS (Lois L.), 4391.
ROBERTS (P.), 2918.
ROBERTS (Penny), 5397.
ROBERTS (Richard), 579.
ROBERTS (Timothy M.), 7936.
ROBERTS (Warren), 6605.
ROBERTSHAW (P.), 8804.
ROBERTSHAW (Peter), 5267.
ROBERTSON (Anne Strachan), 174.
ROBERTSON (D.), 1466.
ROBERTSON (David Brian), 7357.
ROBERTSON (J. H.), 8805.
ROBEY (David), 3740.
ROBIC (Marie-Claire), 281.
ROBINE (Nicole), 6098.

ROBINSON (B.), 2919.
ROBINSON (Chase F.), 3468.
ROBINSON (David), 4907, 5723.
ROBINSON (John Martin), 152.
ROBINSON (Philip), 6402.
ROBINSON (Richard D.), 5960.
ROCCHI (Jean), 6191.
ROCH (Jean-Louis), 4034.
ROCHAT (Giorgio), 4855, 8008.
ROCHE (Daniel), 573, 1206, 7289.
ROCK (David), 5375, 5443.
ROCKEL (Stephen J.), 7358.
RODAS CHAVES (Germán), 4392.
RÖDEL (Walter G.), 6971.
RODÉN (Marie-Louise), 4856.
RÓDENAS VILAR (Rafael), 7546.
RÖDER (Katrin), 211.
RODRÍGUEZ (Luis I.), 8088, 8179.
RODRÍGUEZ ALCALDE (Ángel L.), 1589.
RODRÍGUEZ ARIZA (M. Oliva), 1555, 1583.
RODRIGUEZ BESNE (Jose Ramon), 5569.
RODRÍGUEZ CÁCERES (Milagros), 1462.
RODRÍGUEZ DE DIEGO (José Luis), 320.
RODRÍGUEZ JIMÉNEZ (José Luis), 5190.
RODRÍGUEZ MÉNDEZ (Jesús), 1606.
RODRÍGUEZ NOZAL (Raúl), 6312.
RODRÍGUEZ O. (Jaime E.), 4101.
RODRÍGUEZ SÁENZ (Eugenia), 7260.
RODRÍGUEZ-ALMEIDA (Emilio), 2666, 2924.
Roe (Thomas), 7808.
ROEDER (Elmar), 7301.
ROEMER (Klaus), 6895.
ROEMER (N.), 110.
ROEST (Bert), 3766.
ROFFE (David), 3294.
ROFFIN (Françoise), 212.
ROFFIN (Raymond), 212.
ROGERS (Anthony), 8206.
ROGERS (Clifford J.), 3409.
ROGERS (G. A. J.), 782.
ROGERS (Joel), 5364.
ROGERS (Rebecca), 458.
ROGOZIŃSKI (Jan), 8479.
ROHBECK (Johannes), 811.
ROHLÍKOVÁ (Slavena), *III*.
ROHR (Christian), 1228.
ROHWEDER (Ch.), 2793.
ROIG (Alanyà), 4026.
ROKÉAH (Zefira Entin), 3448.
ROLANDO (Daniele), 5521.
Roldán (María), 4153.
ROLETT (Barry V.), 1556.
RÖLL (Wolfgang), 4610.

Rolland (Romain), 5824.
ROLLASON (David), 3296.
ROLLEY (Claude), 2222.
RÖLLIG (Wolfgang), 2389.
ROLLINGER (Robert), 682.
ROLLO (David), 3659.
ROMANI (Marzio A.), 662.
ROMANIELLO (Giuseppe), 3090.
ROMANO (A.), 3082, 5953, 7405.
ROMANO (Andrea), 5077.
ROMANO (Dennis), 1016.
Romano (Ruggiero), 712.
ROMANO (Serena), 292.
Romanov (Boris Aleksandrovich), 713.
ROMEK (Zbigniew), 4983.
ROMEO (Giuseppe), 8427.
Romeo (Rosario), 701, 714.
RÖMER (M.), 1718.
RÖMER (Thomas), 574.
ROMERO (Federico), 497.
ROMERO (Saul Jerónimo), 768.
ROMERO MAURA (Joaquín), 5191.
ROMEU FERRÉ (Pilar), 1871.
ROMSICS (Ignác), 8099.
Romulus, re di Roma, 965, 2577.
RONCHEY (Silvia), 3092, 3237.
RONCHI DE MICHELIS (Laura), 5667.
RONCROFFI (Stefania), 3826.
RÖNDINGS (Uwe), 6896.
RONZÓN LEÓN (José A.), 768.
ROOSENS (Claude), 8283.
Roosevelt (Franklin Delano), 5351, 8109, 8184.
ROPA (Rossella), 4798.
ROPER (G. J.), 1324.
ROPER (Lyndal), 5493.
ROPER (Michael), 581.
ROQUES (Denis), 3022.
Rorty (Richard), 6174.
ROSANVALLON (Pierre), 4488.
ROSARIO (Iva), 3821.
ROSATI (Massimo), 5861.
ROSE (Mary B.), 6897.
ROSE (Pamela), 1733.
ROSE (Peter), 4784.
ROSE (R. S.), 4242.
ROSELLI (A.), 7031.
ROSEMBLATT (Karin Alejandra), 4283.
ROSEN (Jean), 1275.
ROSENBAND (Leonard N.), 6313.
ROSENBAUM (Arthur L.), 8302.
ROSENBERG (Mathias), 7389.
ROSENBERG (Pierre), 6606.
ROSENBERG (Rafael), 6607.
ROSENBERGER (Sieglinde), 4190.
ROSENBLUM (Nancy L.), 7478.
RÖSENER (Werner), 409, 1010, 1219.

ROSENFELD (Gavriel D.), 582.
ROSENSTONE (Robert A.), 812.
ROSENTAL (Paul-André), 7047.
ROSENTHAL (Jean-Laurent), 7013.
ROSENWEIN (Barbara H.), 3940.
ROSICKA (Janina), 6772.
ROSIVACH (Vincent J.), 1969, 2148.
ROSKAM (G.), 2794.
ROSLAVTSEVA (Lidija I.), 882.
ROSS (Corey), 4611.
ROSS (Duncan M.), 6756.
ROSS (Friso), 1091.
ROß (Sabine), 7447.
ROSSATANGA-RIGNAULT (Guy), 4510.
RÖSSEL (Jörg), 4612.
Rossellini (Roberto), 6639.
ROSSET (Roger), 8097.
ROSSETTI (Gabriella), 3347.
ROSSETTO (Nicola), 6027.
ROSSI (Andrea), 4857.
ROSSI (Andreola), 2474, 2795.
ROSSI (E.), 2507.
ROSSI (Ernesto), 5494, 8029.
ROSSI (Franco), 3824.
ROSSI (L.), 2450.
ROSSI (Laura), 1987.
ROSSI (Luigi Enrico), 2122.
ROSSI-DORIA (Anna), 4858.
ROSSINI (Daniela), 8030.
ROSSINI (Giulio), 6705.
RÖSSLER (Ruth-Kristin), 7486.
ROSSLYN (F.), 2223.
ROSSO (Paolo), 3521.
ROSTGAARD (Marianne), 4378.
Rostovcev (Michajl Ivanovic), 531.
ROTH (Andreas), 1097, 8180.
ROTH (François), 4475, 4489.
ROTH (Harriet), 1260.
ROTH (Karl Heinz), 7359.
ROTH (M. T.), 1792.
ROTHSCHILD (Jean-Pierre), 3319.
Rotrou (Jean), 6751.
ROTTER (Andrew Jon), 583, 8429, 8430.
ROTTLÄNDER (Rolf C.A.), 171.
ROTTOLI (M.), 1557.
ROUBAN (Luc), 7487.
ROUBAUD-BÉNICHOU (Sylvia), 3741.
ROUCHE (Michel), 3644.
ROUDOMETOF (Victor), 4901.
ROUSE (Mary A.), 74.
ROUSE (Richard H.), 74.
ROUSSEAU (George Sebastian), 6314.
Rousseau (Jean-Jacques), 6199, 6364, 6420, 6426.
ROUSSEAU (Ph.), 2510, 2949.
Rousseau (Théodore), 6610.

ROWE (Christopher), 1034.
ROWE (Galen O.), 2224.
ROWE (Greg), 722.
ROWELL (S. C.), 3437.
ROWLANDS (Guy), 7843.
ROXIN (Claus), 7484.
ROY (Bruno), 3702.
ROY (J.), 2057.
ROY (Martin), 584.
ROYLE (Edward), 4696.
ROYNETTE (Odile), 7261.
ROZELL (Mark J.), 5312.
ROZENBLIT (Marsha L.), 4192.
ROZMAN (Gilbert), 8354.
ROZZO (Ugo), 5570, 5663.
RUANE (Kevin), 8431.
RUBENSON (Samuel), 7911.
RUBENSON (Sven), 7911.
RUBIÉS (Joan-Pau), 273.
RUBIN (Rachel), 1467.
RUBIN LEE (Patricia), 6532.
RUBINSTEIN (L.), 2078.
RUBINSTEIN (Nina), 4490.
RUBINSTEIN (W. D.), 6799, 6806.
RUBIO Y LLUCH (Antoni), 3271.
RÜBSAMEN (Dieter), 3291.
Ruck (Peter), 14.
RÜCKERT (Joachim), 7369.
RUDAVSKY (T.M.), 3451.
RUDEN (S.), 2486.
RUDERMAN (David B.), 5724.
RUDGERS (David F.), 5356.
RUDNEVA (Svetlana E.), 5076.
RÜFNER (Th.), 7547.
RUFOLO (Scott), 1591.
Rufus de Shotep, 3151.
RUGGIU (François-Joseph), 488.
RUGGLES (Clive), 1558.
RUGGLES (D. Fairchild), 3469.
RUHE (Ernstpeter), 3858.
RUIZ (Néstor), 4152.
RUIZ ARZALLUZ (Inigo), 6408.
RUIZ IBÁÑEZ (José Javier), 7428.
RUIZ TABOADA (Arturo), 1543.
Rullus (Publius Servilius), Roman tribune of the people, 2560.
RULOFF (Dieter), 8334.
RUMJANTSEV (S. Ju.), 5865.
RUMMEL (Erika), 5668.
RUMPLER (Helmut), 4181.
RUMSCHEID (J.), 2615.
RUNASOV (A. M.), 5072.
RUNBLOM (Harald), 5004.
RUNIA (David T.), 2833.
RÜNITZ (Lone), 4384.
RUPASOV (A. I.), 8011, 8031.
RUPKE (Nicolaas A.), 261.
RUPPRECHT (Hermann), 2513.
RÜRUP (Reinhard), 4607.
RUSAN (Romolus), 5024.
RUSBROCHIUS (Ioannes), 3855.

RUSCHENBUSCH (E.), 2082.
RUSCIANI (Patrizia), 8237.
RUSCO (Elmer R.), 7488.
RUSCONI (Roberto), 3875.
RÜSEN (Jörn), 585, 813, 883, 6729.
RUSSELL (Colin A.), 6228.
RUSSELL (F. S.), 2112.
RUSSELL (Norman), 2985.
RUSSELL (Peter), 3438.
RUSSELL (Vida), 3978.
RUSSELL-WOOD (A. J. R.), 7748.
Russolo (Luigi), 6497.
RUSTER (Thomas), 5433.
RUTGERS (L. V.), 2925.
RUTHERFORD (Ian), 1719.
RUTLAND (Robert Allen), 442.
RUTTER (N. K.), 2149.
RUZÉ (F.), 2048.
RYAN (Mary P.), 5863.
RYAN (Ray), 5897.
RYAN (Thomas F.), 3885.
RYCHLAK (Ronald J), 8100.
RYCHLÍK (Jan), 4354.
RYDEL (Jan), 8432.
RYDÉN (Per), 6072.
RYHOLT (K.), 1710.
RZEPKA (Slawomir), 1720.

S

SAADALLAH (Aboul-Kassem), 4137.
SAALER (Sven), 4882.
SAAR (Stefan Chr.), 1097.
SAARD (Riho), 5669.
SABATÉ CURULL (Flocel), 3614.
SABATIER (Gérard), 313, 5192.
SABBAN (Françoise), 1243.
SABETTI (Filippo), 4859.
SABIN (Philip), 2616.
SABROW (Martin), 475.
SABW KANYANG (Jean-Anatole), 3093.
SACCENTI (Mario), 6486.
Sacco (Catone), 3521.
SACHßE (Christoph), 7301.
SACK (Friz), 1177.
SACKETT (L. H.), 1576.
Sacy (Isaac Le Maistre de), 6421.
Sadat (Anwar), 4395.
SADEH (M.), 1609.
SADHU (Shyam Lal), 213.
SADOUN (Marc), 4443.
Sadruddin Sadar-i-Jahan, Sheikh, 122.
SÁEZ (Carlos), 37.
SÁEZ (Emilio), 37.
SAFDAR (Mahmood), 4961.
SAFFREY (H. D.), 2273.
SAFLEY (Th. Max), 7262.
SAFRAN (Janina M.), 3470.
SAGÁS (Ernesto), 4388.

SAGATELJAN (Garij Sh.), 5082.
SAGEL-GRANDE (Irene), 5907.
SAHA (Rekha), 8433.
ŞAHIN (S.), 1639.
SAHLINS (Peter), 7263.
SAHLSTRÖM (André), 4412.
Said (E.W.), 715.
SAINE (Harri), 6028.
Sainte-Beuve (Charles Augustin), 5847.
Saint-John Perse (Alexis Saint-Léger Léger, dit), 8136.
SAKAEHARA (Towao), 8766.
SAKAI (Hiroki), 8490.
SAKAKURA (Atsuhide), 8624.
SAKHAROV (Andrej N.), 3523.
SALA (Teodoro), 7980.
SALA DE TOURÓN (Lucía), 7316.
SALADIN (Jean Christophe), 5947.
Saladin the Ayyubid, 3471.
SALAH MASOUD (Masoud), 1694.
SALAMONE (Frank), 1329.
Salazar (António de Oliveira), 5009, 5014, 5015, 6067, 8168.
SALAZAR (C. F.), 2275.
SALAZAR VERGARA (Gabriel), 7264.
SALEMME (G.), 2475.
SALER (Benson), 1320.
SALETTI (C.), 2926.
SALEWSKI (Michael), 1000.
SALGADO (Carlos), 4292.
Salgari (Emilio), 6469.
SALIMOVA (Kadriya), 1213.
Salisbury (Robert Arthur Talbot), 7930.
SALITOT (Michelle), 274.
SALLABERGER (Walther), 1793.
Sallustius Crispus (Gaius), 464, 2500.
SALMANN (Martin), 5564.
SALMERÓN GIMÉNEZ (Francisco Javier), 5193.
SALMIERI (Giovanni), 278.
SALMON (J.), 2126.
SALMON (John Hearsey McMillan), 5864.
SALOMON (Christine), 6315.
SALOMON (Frank), 6565.
Saloninus (Publius Licinius Cornelius), Roman emperor, 2691.
SALVEMINI (Maria Teresa), 7035.
SALVIAT (François), 2336.
SALVINI (Mirjo), 1807, 1821.
ŠAMBERGER (Zdeněk), 8181.
SAMONS IL (L. J.), 2150.
SAMSONOWICZ (Henryk), 3323, 3416.
SAMUELSON (Lennart), 5079.
San Millán de la Cogolla, Sanctus, 3961.

Sánchez Albornoz (Claudio), 3596.
SÁNCHEZ G. (Gonzalo), 8336.
SÁNCHEZ GONZÁLEZ DE HERRERO (María Nieves), 343.
SÁNCHEZ LLAMA (Inigo), 6099.
SÁNCHEZ MARTÍN (José María), 3006, 3673.
SÁNCHEZ PALENCIA (F.-Javier), 1610.
SÁNCHEZ-ALBORNOZ (Nicolás), 7733.
SÁNCHEZ-ALONSO (Blanca), 7093.
SANCISI-WEERDENBURG (H.), 2083.
Sand (George), 5888, 6455.
SANDAG (Shagdariin), 5080.
SANDBERG (K.), 2667.
SANDELOWSKI (Margarete), 6316.
SANDER (Rudolf), 4355.
SANDIFORD (Keith A.), 1469.
SANDIN (Bengt), 527.
Sandino (Augusto César), 7955.
SANDL (Marcus), 6317.
SANDLER (Stephanie), 1246.
SANDMANN (Hendrik), 7548.
SANDMO (Erling), 1162.
SANDRI (Renato), 8221.
SANER (Turgut), 1666.
SANG (Bing), 8625.
SANI (Chiara), 2998.
SANIER (Max), 7452.
SANMARTÍN (Joaquín), 1822.
SANSTERRE (Jean-Marie), 233.
Santa Anna (Antonio López de), 4912, 7943.
Santarosa (Teodoro di), 4823.
SANTELIA (Stefania), 2613.
SANTOMASSIMO (Gianpasquale), 4860.
SANTONI (Francesca), 7.
SANTORELLI (Paola), 3035.
SANTOS MOLANO (Enrique), 4286.
SANTOS PÉREZ (José Manuel), 4225, 7755.
SANTSCHI (Catherine), 8097.
SANTUCCI (Antonio), 6141.
Sappho, 2755.
SAPPINEN (Eero), 7265.
SARANTAKES (Nicholas Evan), 8434.
SARASOHN (Lisa T.), 1072.
SARDIELLO (R.), 2465.
SARFATTI (Michele), 4861.
SARIANIDI (V.), 1905.
SARINAY (Yusuf), 3256, 3257, 5241.
SARKOWICZ (Hans), 1470.
SARONIO (P.), 1584.
Sars (Ernst), 716.
Sartre (Jean Paul), 6169, 6201, 6352, 6354.

SASAKI (Yo), 8626.
SASHI (Akihiro), 4645.
ŞAŞMAZ (Musa), 5258, 7938.
SASPORTES (Marilyne), 7183.
SASSO (Gennaro), 734.
SASSOON (Anne Showstack), 1073.
SASTRE (Inés), 1610.
SATO (Kenji), 8767.
SATO (Kentaro), 8491.
SATOFUKA (Fumihiko), 6856.
SAUER (Thomas), 5431.
SAUERMANN (Eberhard), 5866.
SAUGE (André), 1937.
SAUL (Nigel), 1001, 3427.
SAUNDERS (Christopher), 1002.
SAUNDERS (Kay), 7577.
SAUNDERS (T. J.), 2233.
SAUTON (Gilles), 2617.
SAUVÉ (Rachel), 6487.
SAVIGNY (Friedrich Karl von), 7369, 7387, 7389.
SAVOSTINA (Elena), 2355.
SAVVIDES (A. G. C.), 3238.
SAWYER (Birgit), 3482.
SAX (Margaret), 1794.
SAYER (Karen), 1163.
Sazonov (Sergei D.), 7973.
SBARDELLA (Livio), 2039.
Scaglia (Allessandro Cesare), 7810.
Scala (Giambattista), 7925.
SCALABRIN (C.), 1409.
SCARAVELLI (Luigi), 6192.
SCARRE (Chris), 1534.
SCATAMACCHIA (Rosanna), 7032.
SCATURRO (Frank J.), 7429.
SCHAAB (Meinrad), 4568.
SCHAAD (Martin P. C.), 8436.
SCHABEL (Chris), 3886.
SCHACHT (Richard), 6182.
SCHAEFFER (Peter V.), 6982.
SCHÄFER (Annette), 4613.
SCHÄFER (Axel R.), 7266.
SCHÄFER (Bernd), 8291.
SCHÄFER (Melsene), 2219.
SCHÄFER (N.), 2618.
SCHÄFKE (Werner), 6520.
SCHÄFTER (Petra), 7553.
SCHALHORN (Andreas), 6608.
SCHAMP (J.), 2276, 2796, 3145, 3152.
SCHARBAUM (Heike), 724.
SCHARF (Claus), 5068.
SCHATZ (J.), 1074.
SCHAUB (Jean-Frédéric), 4491.
SCHEER (Tanja S.), 2311.
SCHEFFLER (Jürgen), 7044.
SCHEFOLD (B.), 6830.
SCHEID (John), 814, 1383, 2881.
SCHEINDLIN (Raymond P.), 3722.
SCHEITHAUER (A.), 2797.

SCHEITHAUER (Andrea), 2419.
Schelling (Friedrich Wilhelm Joseph), 6172, 6187, 6193.
SCHEMBRA (R.), 3153.
SCHEMMANN (Ulrike), 3767.
SCHEPARTZ (L. A.), 1527.
SCHERER (Stefan), 6733.
Scherer (Wilhelm), 789.
SCHERMAIER (M. J.), 7549.
SCHEUCH (Manfred), 275.
SCHEUERMAN (William E.), 7372.
SCHEUERMANN (Martin), 4614.
SCHIAMONE (Pietro), 5585.
SCHIAVONE (A.), 2619.
SCHIEDER (Wolfgang), 1138, 4549.
SCHIEFELBUSCH (Martin), 6862.
SCHIERA (Pierangelo), 4121.
SCHIEVENIN (R.), 2798.
SCHIKEDANZ (Hans-Joachim), 5867.
SCHILBRACK (Kevin), 1394.
SCHILD (Wolfgang), 7497.
SCHILDGEN (Brenda Deen), 3706.
SCHILDT (Bernd), 1111.
Schiller (Johann Christoph Friedrich von), 820, 6144, 6178, 6364, 6431.
SCHILLING (Peter), 2979.
SCHIMDT (Jürgen Michael), 5495.
SCHIMMEL (Annemarie), 6533.
SCHINDLER (C.), 2799.
SCHINGO (G.), 2928.
SCHIPKE (Renate), 29, 76.
SCHIRMACHER (Wolfang), 6146.
SCHIRRMACHER (Freimut), 884.
SCHLAEGER (Jürgen), 7157.
SCHLAGER (Neil), 1245.
SCHLARP (Karl-Heinz), 6807.
SCHLEGEL (C.), 2464.
Schlegel (Friedrich), 6135.
SCHLEGEL (Oliver), 2929.
SCHLEIER (Hans), 815.
SCHLEIFER (Ronald), 5868.
SCHLEMMER (Thomas), 4615.
SCHLICH (Jutta), 5814.
SCHLICHT (Wolfgang), 7157.
SCHLICHTE (Klaus), 7847.
SCHLICK (Werner), 6419.
SCHLIEWENZ (Birgit), 5041.
SCHLOBACH (Jochen), 368.
SCHLOSSER (Marianne), 3887.
SCHLUMBOHM (Jürgen), 7199.
SCHMALE (Wolfgang), 1003.
SCHMID (Andre), 8478.
SCHMID (Hans Heinrich), 574.
SCHMID (Martin), 2316.
SCHMID (Wolfgang), 3797.
SCHMIDL (Erwin A.), 8402.
SCHMIDT (Christoph), 5670.
SCHMIDT (E. A.), 2800.
SCHMIDT (Georg), 4553.

SCHMIDT (Hans-Joachim), 3359.
SCHMIDT (Hartwig), 586.
SCHMIDT (Jens), 5869.
SCHMIDT (Karsten), 7484.
SCHMIDT (Klaus), 1667.
SCHMIDT (Leigh Eric), 6318.
SCHMIDT (Manfred G.), 2419.
SCHMIDT (Nelly), 7642.
SCHMIDT (Peer), 4921.
SCHMIDT (Peter Lebrecht), 1471.
SCHMIDT (Regin), 5357.
SCHMIDT (Th. S.), 3116.
SCHMIDT (Tilmann), 5945.
SCHMIDT (Walter), 4545.
SCHMIDTCHEN (Dietrich), 1395.
SCHMIDT-RECLA (Adrian), 7390.
SCHMIECHEN-ACKERMANN (Detlef), 4616.
SCHMIFT (Walter), 4546.
SCHMITT (Anne), 1946.
Schmitt (Carl), 6119.
SCHMITZ (Christine), 2466.
SCHMITZ (Christopher J.), 6898.
SCHMITZ (Thomas A.), 2225.
SCHMITZ-BERNING (Cornelia), 1004.
SCHMITZ-BURGARD (Sylvia), 6420.
SCHMITZER (Ulrich), 731.
SCHMOECKEL (Mathias), 1114.
SCHMÖLDERS (Claudia), 4617, 5796.
SCHMUHL (Hans-Walter), 7719.
SCHMUTZ (Jürg), 3768.
SCHNÄDELBACH (Herbert), 6129.
SCHNAPP (Alain), 2337.
SCHNAPP (Jeffrey T.), 4797.
SCHNEEBERGER (Paul), 587.
SCHNEIDER (Alejandro), 4158, 4160.
SCHNEIDER (Arnd), 7267.
SCHNEIDER (Barbara), 6029.
SCHNEIDER (C.), 2498.
SCHNEIDER (Gabriele), 7720.
SCHNEIDER (H.), 1934.
SCHNEIDER (Ivo), 6319.
SCHNEIDER (Wolfgang Christian), 75.
SCHNEIDMÜLLER (Bernd), 3410.
SCHÖBER (Peter), 1164.
SCHOBER (Richard), 4193.
SCHOELLER (Wilfried F.), 790.
Schoenberg (Arnold), 6698.
SCHOEP (Ilse), 1951.
SCHOFIELD (M.), 2084.
SCHOFIELD (Malcolm), 1034.
SCHOFIELD (Victoria), 8437.
SCHOFIELD SAEGER (James), 7749.
SCHÖLLGEN (Gregor), 7939, 8129.
SCHOLL-SEIBERT (Jutta), 4518.
SCHOLTEN (C.), 3094.

SCHOLTEN (J. B.), 2085.
SCHOLTZ (Gunter), 785.
SCHOLZ (Birgit), 588.
SCHOLZ (C.), 3171.
SCHOLZ (M.), 1535.
SCHOMBURG-SCHERIFF (Sylvia M.), 871.
SCHÖN (Lennart), 6808.
SCHÖNBERGER (Otto), 2981.
SCHÖNE (Jens), 6980.
SCHÖNEMANN (Bernd), 773.
SCHÖNHAGEN (Anne Renate), 5221.
SCHÖNIGER-HEKELE (Bernhard), 7550.
SCHÖNLE (Andreas), 1472.
SCHOONOVER (Thomas D.), 5870.
Schopenauer (Arthur), 6206.
SCHOPPA (R. Keith), 1005.
SCHORN-SCHÜTTE (Luise), 4194.
SCHØSLER (Lene), 217.
SCHÖTTLER (Peter), 641.
SCHRADER (Wolfgang H.), 6198.
SCHRAMM (G.), 955.
SCHRAMM (Jan Melissa), 6488.
SCHRAMM (Nils-Eberhard), 7391.
SCHREINER (Peter), 3162.
SCHRENCK-NOTZIG (Caspar von), 1078.
SCHRÖDER (St.), 2508.
SCHROLL (Heike), 344.
SCHRÖTER (Bernd), 7750.
SCHRÖTER (Harm G.), 3606.
SCHUBERT (Ernst), 5948.
SCHUBERT (Werner), 7303.
SCHUBRING (Gert), 1244.
SCHUCHARD (Christiane), 3923.
SCHUH (Anna Marie), 5358.
SCHUKER (Stephen A.), 7596.
SCHÜLER (Barbara), 4618.
SCHÜLLER (Karin), 1006.
SCHULTE-KLÖCKER (Ursula), 2978.
SCHULZ (Andreas), 6100.
SCHULZ (Günther), 7169, 7268.
SCHULZ (Jindřich), 422.
SCHULZ (Monika), 885.
SCHULZ (R.), 2712.
SCHULZ (Raimund), 2620.
SCHULZE (Kirsten), 5021.
Schumacher (Kurt), 4522.
SCHUMACHER (Martin), 7473.
Schuschnigg (Kurt), 4170.
SCHUSTER (John), 6130.
SCHUTKOWSKI (Holger), 886.
SCHÜTT (Hans W.), 6320.
SCHÜTZ (Bernhard), 6566.
Schütz (Heinrich), 6737.
SCHWAR (Harriet Dashiell), 8258.
SCHWARCZ (Lilia K. Moritz), 835.
SCHWARTZ (Barry), 589.

SCHWARTZ (Glenn M.), 1668, 1795.
SCHWARTZ (Richard Alan), 5871.
SCHWARZ (Hans-Peter), 4104, 5917, 8241.
SCHWARZ (Kathryn), 6392.
SCHWARZ (Leonard D.), 5934.
SCHWARZ (Michael Viktor), 3797.
SCHWARZMAIER (Hansmartin), 4568.
SCHWARZMANN-SCHAFHAUSER (D.), 1700.
Schwedt (Herman H.), 5531.
SCHWEIGARD (Jörg), 5949.
SCHWERHOFF (Gerd), 1151.
SCHWICKER (François), 339.
SCHWINDT (Jürgen Paul), 2801.
SCHWINGES (Reiner Christoph), 3359.
Sciascia (Leonardo), 4839, 6467.
SCIOLLA (Gianni Carlo), 1288.
SCIUMÈ (Alberto), 7551.
Sclopis (Federico), 3514.
SCOLZ (Michael F.), 4619.
SCOTT (David), 4697.
SCOTT (Jonathan), 7775.
SCOTT (Kathleen L.), 3805.
SCOTT (Peter), 7094.
SCOTT (R. T.), 2930.
SCOTT (R.), 3185.
SCOTT (Rebecca J.), 1129, 7148, 7751.
SCOTT (S.), 2960.
SCOTT (Tom), 270.
SCOTT (Virginia), 6734.
SCOTT JENKINS (Virginia), 6981.
SCOTT-SMITH (Giles), 5872.
Scotus Eriugena (Johannes), 3864.
SCRAGG (Donald), 1456.
SCREECH (Timon), 1289.
SCREEN (John Ernest Oliver), 4413.
SCROCCARO (Mauro), 4862.
Scudéry (Georges de), 6751.
SCURTU (Ioan), 8131.
SEARLE (Alaric), 7580.
SEARS (Olivia E.), 4797.
SEAY (A.), 7576.
SEBASTIAN (Anton), 1198.
SEBE (Andrei), 2351.
SEBERT (L. M.), 7875.
SECCI (E.), 2865.
SECO SERRANO (Carlos), 5194.
SECORD (James A.), 6322.
SĘCZYS (Elżbieta), 131.
SEDDA (Lidia), 1075.
SEDGWICK (Sally), 6187.
ŠEDIVÝ (Ivan), 8032.
SEDLEY (D.), 2269.
SEDLIAKOVÁ (Alžbeta), XVI.
SEE (Klaus von), 5950.

SEEBERG (Vilma), 6030.
SEEKINGS (Jeremy), 5133.
SEELY (Robert), 7632.
SEGAL (Ch.), 2866.
SEGAL (Daniel A.), 816.
SEGAL (Robert A.), 1321.
SÉGINGER (Gisèle), 817, 6489.
SEGOVIA (Rafael), 8088, 8179.
SEGRE RUTZ (Vera), 3853.
SEGRETO (Luciano), 7033.
SÉGUENNY (André), 5444.
SEGURA (Antoni), 5181.
SÉGUY (Mireille), 501.
SEIBT (Ferdinand), 4620.
SEIBT (Gustav), 644.
SEIDEL MENCHI (Silvana), 3495.
SEIDLER (Eduard), 6324.
SEIDMAN (Michael), 5195.
SEIFERT (Katharina), 5445.
SEIFERT (Siegfried), 1434.
Seignette (Napoléon), 7882.
Selassie (Hailé), 5410.
SELDEN (Mark), 440.
SELF (Robert), 4698.
SELIGMANN (Matthew S.), 4621.
SELIKTAR (Ofira), 8438.
SELLERT (Wolfgang), 7383.
SELLÉS I QUINTANA (Magda), 5196.
SELLIER (Philippe), 6421.
SELLIN (Volker), 7844.
SELLS (Michael), 3722.
SELTZER (Andrew), 7340.
SELWYN (Pamela E.), 93.
SEMENOVA (Anna V.), 6787.
SEMIRJAGA (Mikhail I.), 8101.
SEMKOV (Milen), 4254.
SEMMEL (Stuart), 590.
SEMOTANOVÁ (Eva), 253.
SEN (Satadru), 7675.
Senanayake (Don Stephen), 5205.
Senanayake (Dudley), 5205.
SENDEROVICH (S. Ja.), 3314.
Seneca (Lucius Annaeus), 737, 2501-2509, 2721, 2768, 2773, 2800, 2802, 2803, 2804, 2865, 3020.
SENS (A.), 1975.
Septimius Severus (Lucius), Roman emperor, 2679.
SERBIN (Kenneth P.), 4243.
SÉRÉ (Daniel), 7845.
SERENI (Emilio), 717.
SERÉS (Guillermo), 6408.
SERGHIDOU (S.), 1938.
Sergius, Sanctus, 3247.
SERIO (A.), 2834.
SERNA ALONSO (Justo), 676.
SEROV (Vadim V.), 3154.
SERPICO (MARGARET), 1721.
SERRA (Enrico), 7940.

SERRA RIDGWAY (F.), 1913.
SERRAI (Alfredo), 369.
SERRANO (Fernando), 8088, 8179.
SERRANO CANTARÍN (R.), 2027.
SERRANO RAMOS (Encarnacion), 2931.
SERRAO (Joel), 930.
Sertorius (Quintus), 2583.
SERVICE (Robert), 5081.
Servius Honoratus (Maurus), 2772.
SERWAŃSKI (M.), 7814.
ŞEŞEN (Ramazan), 3471.
SESMERO CUTANDA (Enriqueta), 5197.
SESSI (Frediano), 8221.
ŠESTÁK (Miroslav), 4063, 4330, 4356, 7983, 8159, 8172.
Sestan (Ernesto), 718.
SETAIOLI (Aldo), 2804.
SETH (Catriona), 5887.
SETTIA (Aldo), 3608.
SETTIPANI (Christian), 127, 208.
SEVERIANO TEIXEIRA (Nuno), 5017.
Severus (Alexander), Roman emperor, 2693.
SEVOST'JANOV (G. N.), 562.
Sextus Empiricus, 2038, 2284.
SEY (Yıldız), 1248.
SEYF (Ahmad), 6948.
SEYMOUR-URE (Colin), 6090.
SGARD (Jean), 6422.
SHACKLETON-BAILEY (D. R.), 2526.
Shakespeare (William), 1458, 1466, 5440, 6387, 6752.
Shakhmatov (Aleksey), 3314.
SHALIMOV (Oleg A.), 2622.
SHAMIR (Ronen), 7676.
SHANNON (Timothy J.), 7752.
SHAO (Dongfang), 8627.
SHAO (Yong), 8628.
SHAPIRO (Barbara J.), 592.
SHAPIRO (Paul A.), 5032.
SHARKANSKY (Ira), 5446.
SHARMA (Arvind), 5447.
SHARP (Alan), 7579.
SHARPE (Kevin), 94, 4699.
SHARROCK (A.), 2201.
SHATTUCK (Gardiner H.), 5496.
SHATZMILLER (Joseph), 3609.
SHAUGHNESSY (Edward L.), 8522.
SHAUMIAN (Tatiana), 7941.
SHAVIT (A.), 1585.
SHAVIT (Jacob), 593.
SHAW (Christine), 4863.
SHAW (I.), 1722.
SHAW (Prue), 200.
SHEARD (Sally), 6296.
SHEEHAN (James J.), 392.
SHELDON (Garrett Ward), 5260.

SHELL (COLIN A.), 1586.
SHELLER (Mimi), 4730.
Shelley (Mary), 6472.
SHELOKHAEV (Valentin V.), 7967.
SHEN (Xiaoyun), 8629.
SHENG (Yichang), 8630.
SHEPARD (Alexandra), 7269.
SHEPHERD (Naomi), 7677.
SHER (Richard B.), 6325.
SHERWOOD-SMITH (Maria C.), 3742.
SHIBUTANI (Akira), 7430.
SHIELDS (Sarah), 4768.
SHIGAKI (Yoshio), 4494.
SHIMELMITZ (R.), 1873.
SHIMIZU (Kazuhiro), 8492.
SHIMIZU (Koïchi), 6894.
SHIPLEY (G.), 2086.
SHIRAEV (Eric), 8439.
SHIRANE (Yasuhiro), 8768.
SHIRINIAN (M. E.), 2034.
SHIROKOVA (Nadezhda S.), 2835.
SHISHKIN (V. A.), 5072.
SHISHLINA (N, I.), 1559.
SHISHOV (A. V.), 7633.
SHLAIM (Avi), 8440.
SHLYKOV (Vitalii), 5079.
SHOEMAKER (Robert B.), 7270.
SHOKAREV (S.), 7770.
SHORE (Elliott), 6914.
SHORE (Peter), 8441.
SHORTLAND (Andrew), 1723.
SHOSTAKOVSKIJ (Vjacheslav N.), 5038.
SHOTTER (David C. A.), 2623.
SHOVLIN (John), 7271.
SHUBERT (Adrian), 5199.
SHUL'GINA (Emilija V.), 30.
SHULMAN (Robert), 6490.
SIBERRY (Elizabeth), 594.
SICARD (D.), 2990.
SICARD (Patrice), 3319.
SICKER-AKMAN (Martina), 1823.
Siculo (Giorgio), 5545, 5566.
SIDLAUSKAS (Susan), 6609.
Sidney (Philip), 6376.
Sidonius Apollinaris, 2510, 2733.
SIEG (Uwe), 4622.
SIEGEL (Jonah), 6534.
SIEGELBERG (Jens), 7847.
SIEGRIST (Hannes), 4121, 7515.
SIEVERS (S.), 2895.
SIEVERT (James), 1167.
Sievwright (Charles Wightman), 8824.
SIFAKIS (G. M.), 1927.
Sigebertus Gemblacensis, 441.
Sigismund II Augustus, king of Poland, 7806.
Sigismund, König von Luxemburg, 5095, 5109.

Sigismund, röm.-deutscher Kaiser, 3368.
SIGMIREAN (Cornel), 5873.
SIGNES CODOÑER (J.), 3132, 3148, 3239.
SIHLER (Andrew L.), 214.
SILAEVA (T. V.), 4024.
SILBER (Karl Bernhard), 6735.
Silius Italicus (Tiberius Catius Asconius), 2511, 2512.
SILK (M. S.), 2226.
SILKE (John J.), 7815.
SILLIÈRES (P.), 2425.
Silone (Ignazio), 4809, 5849.
ŠILPOCH (Karel), 8033.
SILVA SEITENFUS (Ricardo Antônio), 8182.
Silvanus, 2417.
SILVER (Marie-France), 5781.
SILVERMAN (Victor), 7360.
SIMA (Alexander), 1906.
SIMBULA (Pinuccia F.), 1179.
ŠIMEKOVÁ (Slávka), 5108.
SIMI BONINI (Eleonora), 3825.
SIMMONS (Beth A.), 7574.
SIMMONS (Clare A.), 5874.
SIMON (Anne), 5732.
SIMON (Dieter), 749, 7540.
SIMON (Jacques), 4135.
SIMON (John Y.), 5275.
SIMONCINI (Giorgio), 6549.
SIMONETTA (Stefano), 1076.
SIMONETTI (Manlio), 3010.
SIMONI (Marcella), 8102.
Simonides, 2039.
SIMONIN (Anne), 6491.
SIMONINI (Gian Luca), 5875.
SIMONTON (Deborah), 6031.
SIMPSON (C. J.), 2713.
ŠIMŮNEK (Robert), 253.
SIN (Yong-ha), 8784.
Sinclair (Mary), 6484.
SINEUX (P.), 2048.
SINGH (Kavita), 1277.
SINISCALCO (Paolo), 3052.
SINNREICH-LEVI (Deborah M.), 3718.
SINOVA (Justino), 906.
SINTON (John M.), 1556.
SION-JENKINS (Karin), 2624.
SIPAHI (Tunç), 1824.
SIPOLS (Vilnis Ja.), 8183.
Siret (L.), 1568.
SIRGES (Thomas), 5701.
SIRINELLI (Jean), 2277.
SIRINIAN (Anna), 3037.
SIROIS (Herbert), 8103.
Sironi (Mario), 6506.
SIRONNEAU (Jean-Pierre), 1396.
SIROT (Stéphane), 4495.
SIROTKIN (Vladlen G.), 7852, 7868.

SITHOLE (Jabulani), 802.
SJÖBLOM (Tom), 887.
SKEATES (Robin), 393, 1498.
SKEB (Matthias), 3015.
SKEGRO (Ante), 2961.
SKELLY (Joseph Morrison), 7613.
SKINNER (Alexander), 3240.
SKIRBEKK (Gunnar), 1416.
SKOCZYLAS POWNALL (F.), 2227.
SKOGOREV (Aleksandr P.), 3021.
SKOVGAARD-PETERSEN (Karen), 431.
SKRŽINSKAJA (E. Ch.), 3241.
SLAATTÉ (H. A.), 2278.
SLACK (Paul), 5757.
SLÁDEK (Zdeněk), 8034.
SLATER (Terry R.), 1175.
SLESARCHUK (G. I.), 7824.
ŚLESZYŃSKI (Wojciech), 4998.
SLOMP (Gabriella), 1077.
SLOTTEN (Hugh R.), 6736.
SLUTSCH (Sergej), 8105.
SMAJE (Chris), 1169.
SMALL (Jan), 6457.
SMALLMAN (Basil), 6737.
SMALLMAN-RAYNOR (Matthew R.), 6231.
SMELSER (Ronald), 4624.
SMERDOU ALTOLAGUIRRE (Luis), 5198.
SMITH (Adrian), 6090, 6821.
SMITH (Angela K.), 5876.
SMITH (Anthony D.), 597.
SMITH (C.), 2864.
SMITH (David M.), 3258.
SMITH (Erin A.), 6633.
SMITH (Graeme), 25.
SMITH (Jay M.), 7272.
SMITH (Jeremy), 4700, 8035.
Smith (John), 719.
SMITH (Jonathan Z.), 1397.
SMITH (Joseph), 1007.
SMITH (Julia M. H.), 2947.
SMITH (Leonard V.), 8036.
SMITH (Louis J.), 8258.
SMITH (Marc H.), 31.
SMITH (Martin Ferguson), 2279.
SMITH (Paul Julian), 5877.
SMITH (R. R. R.), 1669.
SMITH (Robert), 7925.
SMITH (S. A.), 8632.
SMITH (Shawn), 4652.
SMITH (Simon C.), 7721.
SMITH (T.), 8442.
SMITH (Timothy B.), 7273.
SMITH (Tony), 7634.
SMITS (Ivo), 7588.
SMOLANDER (Jyrki), 4414.
SMOUT (T. C.), 1170.
SMULDERS (P.), 3004.
SMYSER (W. R.), 4623.

SMYTH (Denis), 8250.
SMYTHE (D. C.), 3246.
SNODGRASS (A. M.), 2363.
SNORRI STURLUSON, 3315.
SNYDER (H. G.), 1874.
SOCAS GAVILÁN (Francisco), 2483.
Socrates, 2073, 2077, 2260.
SODDU (Francesco), 7494.
SÖDERBERG (Johan), 5222, 6809.
SOÉNIUS (Ulrich S.), 7274.
SOFFER (O.), 1536.
SOHN (Andreas), 1145.
SOKOLOFF (Kenneth L.), 6878.
SOKOLOVSKII (Vladimir), 111.
SOLAK (Zbigniew), XIV.
SOLANA PUJALTE (J.), 2805.
SØLAND (Birgitte), 7275.
Solari (Gioele), 6116.
SOLARO (Giuseppe), 1428.
SOLBERG (Bergljot), 1611.
SOLBES FERRI (Sergio), 7021.
SOLDANI (Simonetta), 718.
SOLDI RONDININI (Gigliola), 598, 3348.
SOLER (Leticia), 599.
SOLEY (Clive), 6091.
SOLL (Jacob), 600.
SOLNTSEVA (L. P.), 5865.
SOLOMON (J.), 2036.
Solomon, 1864.
Solon, 2064, 2103.
SOLOV'EV (S. L.), 2315.
SOLOV'EVA (Tat'jana S.), 4701.
Soloviev (Alexandr), 5618.
SOMERVILLE (Christopher), 6326.
SOMMA (Alessandro), 7552.
SOMMER (Elisabeth W.), 7276.
SOMMER (Matthew H.), 8633.
SOMMER (Michael), 1670.
SOMMERS (Mary Catherine), 3852.
SONE (Masato), 8770.
SONG (Jiong), 8634.
SONNENBERG-STERN (Karina), 5725.
SONNET (M.), V.
Sonnino (Sidney), 4864.
SONYEL (Salâhi Ramadan), 5261, 8443.
Sophocles, 2040, 2041, 2155.
SORABJI (Richard), 3095.
Soranus Ephesinus, 2042.
SORBY (Karol), 8444.
SORDI (Marta), 2087.
SORELL (Tom), 782, 6197.
SØRENSEN (Jakob), 4374.
SORLIN (I.), 3242.
SOTNIKOVA (Svetlana I.), 818.
SOULET (Jean-François), 8445.
SOULHOL (H.), 2895.
SOURIS (George A.), 2415.

SOUSA (Bernardo Vasconcelos), 128.
SOUSTAL (P.), 3167.
SOUTHERN (Eileen), 314.
SOUTHERN (Pat), 2962.
SOUTHWICK (Michael), 129.
SOUTOU (George-Henri), 8045, 8446.
SOVERINI (P.), 2509.
SOWERWINE (Charles), 4496.
SOWIŃSKI (Paweł), 4999.
SOYSAL (Oğuz), 1825.
SOZINA (S. A.), 969.
Sozzini (Bartolomeo), 7370.
SPACCAPELO (N.), 3083.
SPADONI (Maria Carla), 2384.
SPAETH (Donald A.), 5497.
SPANG (Rebecca L.), 5878.
SPARY (E. C.), 6327.
SPÄTH (Th.), 2689.
SPÄTI (Christoph), 8447.
SPAULDING (J.), 8806.
SPAWFORTH (T.), 1925.
SPEAKE (G.), 1939.
SPEAR (T.), 8786.
SPEAR (Thomas), 8807.
SPECK (P.), 3243.
SPECTOR (Scott), 6492.
SPÈDE (Raphaël), 3485.
SPEHNJAK (Katarina), 4308.
SPEHR (Michael), 6899.
SPEIDEL (M. A.), 2715.
Spenser (Edmund), 6376, 6387.
SPERA (Lucinda), 6423.
SPETH (Rudolf), 4107.
SPEYER (W.), 3047.
SPICKERMANN (Wolfgang), 3096.
SPIEGEL (Régis), 394, 6535.
SPIESER (Cathie), 1724.
SPILLANE (Joseph F.), 6328.
SPILLMANN (Kurt), 8323.
SPINA (L.), 2151.
SPINELLI (E.), 2038.
SPINELLI (Maria Estela), 4161.
SPINETO (Natale), 1352.
SPINI (Giorgio), 5449.
Spinoza (Baruch), 1047, 6154, 6171.
Spira (Bil), 8063.
SPISAREVSKA (Ioanna D.), 4253.
SPITZ (Jean-Fabien), 4497.
SPITZ (Richard), 5135.
SPREE (Ulrike), 6101.
SPUFFORD (Margaret), 6983.
SPURLING (Geoffrey), 7728.
SPURR (John), 4702.
SQUATRITI (Paolo), 3760.
SREBRAKOWSKI (Aleksander), 8448.
SRIVASTAVA (Kamal Shankar), 602.

STAATZ (John M.), 4907.
STACHEL (John), 6244.
STACK (Annetta), 6032.
STACKMANN (Karl), 3675.
STADEN (H. von), 2280.
STADLER (Friedrich), 6133.
STAHL (Alan M.), 3155, 3270, 3611.
STAHNCKE (Holmer), 4881.
STAIKOS (K.), 370.
Stalin (Iosif Visarionovič Džugašvili), 4045, 5042, 5050, 5073, 5080, 5083, 5084, 5750, 6704, 8114, 8198, 8324, 8413.
STALLYBRASS (Peter), 1215.
STALMANN (Volker), 4626.
Stambolov (Stefan), 7876.
STAMENOVA (M.), 8449.
STÄMPFLI (Regula), 5230.
STAMPINO (Maria G.), 4797.
STAMSØ (May Britt), 226.
STANCIU (Ion Gh.), 6810.
STANCIU (Laura), 6900.
STANCO (E.), 2932.
Stanislaw August, king of Poland, 5916.
STANLEY (Brian), 5592.
STANZIANI (Alessandro), 6329.
STAPLES (Anne), 8037.
STAPLETON (Kristin), 7095.
STAPLETON (Timothy J.), 7723.
STÄRK (Ekkehard), 2742.
STARR (Rebecca), 5286.
STARY (Giovanni), 130.
STATELOVA (Elena), 4247.
Statilius Ammianus, 1708.
Statius (Publius Papinius), 2513, 2514, 2738.
STAUFFER (D.), 2281.
STAVIG (Ward), 8819.
Stavisky (Alexandre), 4466.
STAYER (James M.), 5671.
STEADMAN (David W.), 1537.
STEARNS (Peter N.), 942.
STECKEL (Richard H.), 7088.
STEED (Christopher), 5451.
STEEN (Inken), 6493.
ŞTEFĂNESCU (Ştefan), 604.
STEFANI (Piero), 1384.
ŠTEFÁNIK (Martin), 5109.
Štefánik (Milan Ratislav), 8012.
ŠTEFÁNIKOVÁ (Antónia), 6949.
ŠTEFANSKÝ (Václav), 8207.
STEIN (Barbara H.), 6950.
Stein (Karl, Freiherr vom), 7294.
STEIN (S. A.), 6102.
STEIN (Stanley J.), 6950.
STEINBACHER (Sybille), 4528, 4627, 4628.
STEINBERG (Jonny), 5128.
STEINBERGER (Peter J.), 1070.

STEINBY (Eve Margarete), 252, 2433.
STEINER (Claudia), 4293.
STEINER (Richard C.), 1875.
STEINER (Uwe), 6424.
STEINHAUS (Hubert), 5981.
STEIN-HÖLKESKAMP (E.), 2577.
STEINHÜBEL (Ján), 5110.
STEININGER (Alexander), 627.
STEININGER (Rolf), 8398.
STEINKOHL (Franz), 889.
STEINSDORFER (Helmut), 4629.
STEINSLAND (Gro), 3916.
STEINWASCHER (Gerd), 7096.
STEITZ (Walter), 6759.
STEKL (Hannes), 7142, 7210.
STELLA (Alessandro), 1171, 4085.
STELLA (Roseli Santaella), 7816.
STEMMLER (Michael), 2596, 2625.
STENGERS (Jean), 4208.
STENSETH (Bodil), 7278.
STENTON (Michael), 8234.
STENVALL (Jari), 7489.
STEPHAN (Alexander), 5359.
STEPHEN (Fiona), 4770.
Stephen of Blois, Count of Mortain and Boulogne, 3395.
Stephen, king of England, 3384.
STEPHENS (Sonya), 6357.
STEPHENSON (Bruce), 6331.
STEPHENSON (P.), 3244.
STEPPAN (Th.), 3169.
STERN (Yaakov), 2668.
STERNBERG (Myriam), 1725.
STERNHELL (Zeev), 4498.
Stesichorus, 2309.
STEUER (Heiko), 3939.
STEURY (Donald P.), 8259.
STEVENS (Annick), 2282.
STEVENSON (Barbara), 3703.
STEVENSON (David), 7942.
STEVENSON (John), 950.
STEWART (Alexander), 6134.
STEWART (H. M.), 1726.
STEWART (Mary Lynn), 6033.
STEWART STOKES (Hamish I.), 277.
STEYN (Frances Caroline), 3843.
STICHEL (Rudolf H. W.), 3245, 3247.
STIEBEROVÁ (Mária), 5090.
STIEFEL (Dieter), 4185.
STIELDORF (A.), 148.
STIER (Bernhard), 6901.
STIGLMAYR (Cristina María), 6567.
Still (Clyfford), 6588.
STINNETT (Robert B.), 8184.
STIPLOVŠEK (Miroslav), 7431.
STIVERS (Camilla), 5360.
STOBART (Jon), 7101.

STÖBER (Georg), 7605.
STÖBER (Gunda), 4630.
STÖBER (Rudolf), 6103.
STOCKER (David), 300.
STOCKERT (Harald), 4631.
STOCKING (Rachel L.), 3376.
STOCKINGER (Claudia), 6733, 6738.
STOCKWELL (Sarah), 7724.
STOCLET (Alain J.), 3522.
STODDARD (William Osborn), 5277.
STODDART (S.), 2918.
STOIANOV (Valeri), 215.
STOICHITA (Victor), 6536.
STOJANOV (L.), 435.
STOK (Fabio), 73.
STOKES (Geoff), 4167.
STOKES (Raymond G.), 6902.
STOKLUND (Bjarne), 1173.
STOL (M.), 1876.
STOLBERG (Michael), 6332.
STOLER (Mark A.), 8185.
STOLJAROVA (Lubov' V.), 32.
STOLLBERG-RILINGER (Barbara), 4108.
Stolle (Conrad), 720.
STOLLEIS (Michael), 7433.
STÖLLNER (Thomas), 1577.
STONE (David R.), 5085.
STONE (Glyn), 7579, 7725.
STONE (M. E.), 2034.
STONE (M. W. F.), 6160.
STONE (Michael Edward), 175.
STOPANI (Renato), 3997.
STOPAR (Ivan), 4028.
STORA (Benjamin), 605.
Storer (Johann Christoph), 6573.
STOREY (I. C.), 1979.
STORRS (Christopher), 7846.
STORRS LANDON (R. Y.), 7361.
STÖRTKUHL (Beate), 6553.
STOTZ (Peter), 216.
STOUFF (Louis), 4029.
ŠŤOVÍČEK (Ivan), 8141.
STOY (Manfred), 649.
Strabo, 278, 2043.
STRAFFORD (David), 8242.
STRAKA (Tomás), 5387.
STRATHERN (Paul), 6333.
STRAUS (Lawrence G.), 1546.
STRAUS (Lawrence Guy), 1538.
Strauss (Johann), 6667.
Strauss (Richard), 6698.
STRČIĆ (Petar), 3430.
STRECK (Bernhard), 848.
STRECK (Michael), 1796.
STREET (M.), 1536.
STREETER (Stephen M.), 8450.
STRENGE (Irene), 4609.
Stresemann (Gustav), 7963.

STŘÍBRNÝ (Jan), 5450.
STRICK (James Edgar), 6334.
Strindberg (August), 6364.
STROBEL (Jochen), 6494.
STROBEL (Karl), 1671.
STROBL (Gerwin), 8106.
STROPPA (A.), 1988.
STROTHMANN (M.), 2626.
STROUD (R. S.), 1970.
STROUTHOUS (Andrew), 5361.
Strozzi (Filippo), 3579.
STRUBBE (J. H. M.), 1970.
STRUBBINGS (Derek), 193.
STRUPP (Christoph), 684, 7279.
STRZELCZYK (Jerzy), 3349.
STUART (Barbara), 1668.
STUBHAUG (Arild), 6335.
STÜBINGER (Stephan), 7554.
STUCHTEY (Benedikt), 433.
STUDSTILL (Randall), 1398.
STUNT (Timothy C. F.), 5672.
STURGIS (James), 4272.
ŠTURM (Lovro), 5116.
STÜRMER (Michael), 4632.
STUURMAN (Siep), 6336.
STYAN (David A.), 8451.
STYLES (John), 6951.
STYLES (Tania), 220.
SU (Lizhi), 8636.
Suarez (Francisco), 6820.
SUÁREZ CORTINA (Manuel), 4051, 5200.
SUÁREZ DE LA TORRE (E.), 2303.
SUÁREZ PIÑEIRO (A. M.), 2669.
Subleyras (Pierre), 6608.
SUBRITZKY (John), 7678.
SUC (Jean-Pierre), 1522.
SUCHMIEL (Jadwiga), 6034.
SUCHOPLES (Jaroslaw), 8038.
SUCIU (Dumitru), 5035.
SUDA (Max Josef), 2977.
Suetonius Tranquillus (Gaius), 2515-2517.
SUFAJ (Fehmi), 7555.
SUGARMAN (David), 7515.
SUGARMAN (Sidney), 7635.
ŠULC (Michal), 4320.
ŠULC (Zdislav), 5880.
Šulgi, king of Ur, 1741.
Sulla (Lucius Cornelius), 2174, 2538, 2560, 2588, 2610.
SULLIVAN (Lawrence E.), 1399.
SULLIVAN (S. D.), 1998.
Sulpicia, 2793.
SULTANOV (T. I.), 974.
SUMITOMO (A.), 7059.
SUMIYA (Mikio), 6784.
SUMMERS (Geoffrey D.), 1672.
SUMMERS (Mark Wahlgren), 5362.
SUMMIT (Jennifer), 3744.
SUN (Longji), 8637.

Sun (Yat-sen) v. Sun (Zhongshan).
Sun (Zhongshan), 8578, 8689, 8711.
SUNDBÄCK (Esa), 8039.
SUNDKLER BENGT (Gustaf Malcolm), 5451.
SUNSERI (Thaddeus), 606.
SURCHAT (P. L.), XVIII.
SURIANO (Juan), 254.
SURIKOV (Igor' E.), 2088.
SURLEAU (Jean-Claude), 8208.
SÜSLÜ (Azmi), 5240.
SUSSER (Asher), 8452.
SUSSMAN (Nathan), 7012.
SÜßMANN (Johannes), 820.
ŠUSTEK (Vojtech), 4323.
SUTER (Claudia), 1826.
SUTTON (D. G.), 1518.
SUTTON (John), 6130.
SUTTON (Marilyn), 3664.
SUWA-EISENMANN (Akiko), 7047.
SUZIKI (Seiichi), 4030.
SVANBERG (Fredrik), 3483.
SVANIDZE (Ada A.), 4012, 4050.
SVAREZ (Carl Gottlieb), 7392.
SVENSHON (Helge), 3247.
SVENTSITSKAJA (N. S.), 3021.
Svetonio, 721.
SWAHN (Jan-Öjvind), 1247.
SWAIN (S.), 1991.
SWALE (Alistair), 4883.
SWAN (Mary), 3738.
Swedenborg (Emanuel), 6190.
SWENSON (James), 6199.
SWIERENGA (Robert P.), 7098.
Swift (Jonathan), 6404.
SWIGGERS (Pierre), 5988.
SWITZER (Les), 5134.
SYDOW (Gernot), 7556.
SYDRAC LE PHILOSOPHE, 3858.
SYKES (Patricia Lee), 5363.
SYLVESTROVA (Marta), 6574.
Syme (Ronald), 722.
SYMEON OF DURHAM, 3296.
Symmachus (Quintus Aurelius), 2981, 3018.
Symmachus, Papa, Sanctus, 3083.
Synesius Cyrenaicus, episcospus, 3022.
SYRING (Enrico), 4624.
SZABÓ (A.), 2283.
SZABÓ (Anikó), 6035.
SZABÓ (Lajos), 2627.
SZANIAWSKA (Lucyna), 245.
SZCZEPAŃSKI (Janusz), 8040.
SZCZERBIŃSKI (Marek), 7290.
SZCZUR (Stanisław), 3356.
Szeberényi (Ludwig Sigismund), 4740.
SZILÁGYI (M.), 2598.
SZLACHTA (Bogdan), 5000.

SZMRECSÁNYI (Tamás), 6232, 6337, 6338, 7300.
SZOCKI (Józef), 371.
SZOMOLÁNYI (Soňa), 5106.
SZONYI (György E.), 303.
SZPILMAN (Andrzej), 4982.
SZPILMAN (Władysław), 4982.
SZWAGRZYK (Krzysztof), 5001.
SZYLVIAN (Kristin), 7099.

T

TABACCO (Giovanni), 3360.
TABERNER (Stuart), 5736.
TABOR (Richard), 1560.
TACCOLINI (Mario), 5498.
Tacitus (Publius Cornelius), 600, 723, 2518, 2519, 2521, 2522, 2828.
TACKETT (Timothy), 4499.
Taddeo di Bartolo, 3798.
TADDEY (Gerhard), 4568.
TADMAN (Michael), 6839.
TAFFIN (Dominique), 383.
TAFLA (Bairu), 4401.
TAGLIAFERRI (Teodoro), 702, 725.
TAHTAH (M.), 4928.
TAÏEB (Jacques), 7280.
Taine (Hippolyte), 656.
TAJTÁK (Ladislav), 5111.
TAKAGI (Shunsuke), 8764.
TAKAHASHI (H.), 8771.
TAKAHASHI (Hidenao), 8772.
TAKAHASHI (Osamu), 8773.
TAKAKI (Keiko), 8808.
TAKAMAKI (Minoru), 8774.
TAKANO (Toshihiko), 8769.
TAKAOĞLU (Turan), 1587.
TAKASHIMA (Ko), 8639.
TAKEUCHI (Keiichi), 282.
TAKEYH (Ray), 8453.
TAL (Abraham), 1877.
TAL (David), 8454.
TALALAY (L. E.), 2312.
TALATTOF (Kamran), 5689.
TALENTI (Simona), 6569.
TALHA (Naureen), 4962.
TALLON (Alain, 5499, 7818.
Talmon (Jacob), 1051.
TAMAI (Chikara), 8775.
TAMBURRI (Pascual), 3293.
TAMELANDER (Michael), 4951.
TAMMEN (Björn R.), 3845.
TAN (Patricia S. M.), 8455.
TAN (Tai Yong), 8502.
Tanagli (Caterina), 3579.
TANAKA (R.), 2284.
TANAKA (Ryûichi), 8041.
TANASESCU (Florian), 5036.
TANDETER (Enrique), 4109, 4157.
TANG (Baolin), 8640.
TANG (Lixing), 8641.

TANG (Xuefeng), 8642.
TANG (Zhijun), 8643.
TANKUTER (Korkut), 3373.
TANNER (Duncan), 4676.
TANNER (Jeremy), 2628.
TANSEY (Patrick), 2867.
TANTILLO (I.), 3097.
TANUCCI (Bernardo), 4799.
TANZARELLA (Sergio), 1400.
TANZINI (Lorenzo), 3316.
TAPLIN (Oliver), 1929, 2228.
TARACHA (Piotr), 1827.
TARADEL (Ruggero), 5726.
TARAGNA (A. M.), 3156.
TARANOV (E. V.), 5044.
TARANTINO (G.), 6200.
TARAS DAPHNE (Gottlieb), 7345.
TÂRAU (Virgiliu), 580.
TARDAN-MASQUELIER (Ysé), 1307.
Tardieu (André), 4474.
TAREKE (Gebru), 8456.
TARIC-ZUMSTEG (Fabienne), 5500.
Tarō (Katsura), 4873.
TARRANT (H.), 2285.
TARRÊTE (Alexandre), 6390.
TARTAGLIA (A.), 3157.
TARTERA (Enric), 1541.
TASHJIAN (Victoria), 8789.
TASSINARI (G.), 149.
TATON (René), 1249.
TAUBER (Christine), 656.
Täubler (Eugen), 724.
TAUNTON (Nina), 3625.
TAUSCH (Harald), 6538.
TAVERES DA SILVA (C.), 2687.
TAVILLA (Carmelo Elio), 7557.
TAWA (Nicholas E.), 6739.
Tawney (R. H.), 725.
Taylor (A. J. P.), 726.
TAYLOR (Bruce), 5586.
Taylor (Charles), 6117.
TAYLOR (Clarence), 7508.
TAYLOR (David), 5267.
TAYLOR (Jay), 8644.
TAYLOR (John H.), 1727.
TAYLOR (Lynne), 8235.
TAYLOR (Miles), 4110.
TAYLOR (Stephen), 5438.
TCHOUBARIAN (Alexandre), 8045.
TEDESCHI (John A.), 5731.
TEDESCHI (John), 5620.
TEDESCO (Antonio), 6741.
Tegetthoff (Wilhelm), 7932.
TEICH (Mikuláš), 6904.
TEICHERT (Carsten), 4633.
TEICHNER (Felix), 2629.
TEICHOVA (Alice), 6770.
TEIXEIRA (Ruy A.), 5364.
TEJCHMAN (Miroslav), 4115.
TEJCHMANOVÁ (Ladislava), 4356.

TELLECHEA IDIGORAS (J. I.), 5524.
TELO (António José), 8187.
TEMPERÁN VILLAVERDE (Elisardo), 345.
TENA (Rafael), 8820.
Tenca (Carlo), 6478.
TENENTI (Alberto), 1079.
TENNSTEDT (Floria), 7301.
Teotochi Albrizzi (Isabella), 6398.
TERÁN (Oscar), 5882.
TERENAS (Gabriela Gândara), 7869.
Terentius Afer (Publius), 2515, 2523, 2524.
Teresa d'Avila, 6355.
TERHOEVEN (Petra), 4549.
TERMES (Josep), 5201.
TERNAUX (Jean-Claude), 6425.
TERRAZAS Y BASANTE (María Marcela), 7943.
TERRIEN (Marie-Pierre), 3894.
TERRY (Arthur), 3745.
Tertullianus (Quintus Septimius Florens), 2990, 3036.
TESAŘ (Jan), 8107.
TESKE (Gunnar), 7817.
TESSITORE (Fulvio), 822.
TESSITORE (Giovanni), 7558.
TESTEN (David), 1797.
TESTI (Arnaldo), 5365.
TEXIER (Pascal), 3519.
TEYSSIER (Arnaud), 4500.
TEYSSIER (Eric), 4432.
THACKERAY (H.St. J.), 671.
THANE (Pat), 1012, 1174, 4676.
THAPAR (Romila), 823.
THATCHER (Ian D.), 7362.
THAYER (Katharine), 5662.
THÉBERT (Yvon), 3754.
THEMELES (P. G.), 3235.
Themistius, 2044.
Theoderic, Ostrogothic king, 3048.
THEODORESCU (Răzvan), 2933.
Theodoricus Magnus, Ostrogothic king v. Theoderic.
Theodorus Monachus, 3862.
Theodorus II Ducas Lascaris, Byzantine emperor, 3157.
Theodorus Metochites, 3158
Theognis, 2731.
Theophanes Confessor, 3159.
Theophrastus, 2177, 2236.
Theopompus, 2177.
THER (Philipp), 7281.
THÉRIAULT (Michel), 3501.
THEUWS (Frans), 2959.
THIBODEAU (T. M.), 3854.
THIEMRODT (Ivo), 7559.
THOBIE (J.), 8042.
THODY (Philip Malcolm Waller), 4111.

THOM (Françoise), 8108.
THOMAS (Greg), 6610.
THOMAS (Helen), 6495.
Thomas (Keith), 5757.
THOMAS (Ludmila), 8054.
THOMAS (Martin), 7650, 7679, 7726.
THOMAS (Michael L.), 2411.
THOMAS (Ronald R.), 6496.
THOMAS (Rosalind), 682.
THOMAS (T. K.), 1728.
THOMAS (Terence), 1402.
THOMAS (William), 690.
Thomas Aquinas, Sanctus, 3863, 3885, 3889, 3893.
Thomasius (Christian), 6161.
THOMASSEN (Jacques), 4936.
THOMASSET (Claude), 3588.
THOME (Gabriele), 2630.
Thomis (Malcolm Ian), 4694.
THOMMEN (Lukas), 2089.
THOMPSON (Alastair P.), 4634.
THOMPSON (Andrew S.), 7658.
Thompson (E. P.), 727.
THOMPSON (Elizabeth), 7680.
THOMPSON (J. Lee), 6104.
THOMPSON (Michéal), 6872.
THOMPSON (Victoria E.), 6952.
THOMPSON (Willie), 824.
THOMSEN (Steen), 6905.
THOMSON (Robert W.), 3487.
Thonet (Michael), 6632.
THORAU (Peter), 1907.
THORAVAL (Yves), 6742.
THORBURN (Thomas), 6906.
Thorez (Maurice), 4474, 4495.
THORN (Gary), 7659.
THORP (Rosemary), 6771.
THORPE (Andrew), 4703.
THOSARAT (Rachanie), 1493.
THRAEDE (K.), 3047.
THUAU (Étienne), 1080.
Thucydides, 728, 2045, 2054, 2091, 2170, 2200.
THURÉN (Lauri), 3098.
THURLOW (Richard C.), 4704.
THURN (I.), 3131.
THÝFNER (Margit), 6105.
TIAN (Baoguo), 8645.
TIAN (Tao), 8646.
TIAN (Xuan), 8647.
TIAN (Yuqing), 8648.
TIAN (Ziyu), 8701.
Tiberius (Julius Alexander), governor of Judaea and Egypt, 721.
Tiberius Claudius Nero, Roman emperor, 731.
TIBET (Aksel), 1640.
Tibullus (Albius), 2777.
TICHY (Christiane), 5674.
TICKNER (Lisa), 6539.

Tiepolo (Antonio), 7778.
TIERNEY (Dominic), 8109.
TIHON (Anne), 3656.
TIKHONOV (Aleksei), 7009.
TIKHVINSKIJ (S. L.), 8130.
TILBURY (Alan), 4222.
TILKOVSZKY (Loránt), 4740, 4741.
TILLION (Germaine), 890.
Timaeus, 1852.
TIMELLI (Maria Colombo), 3672.
TIMMER (Karsten), 4635.
TIMONEN (Asko), 608.
TIMOSHINA (L. A.), 3312.
TINGVOLD PETERSEN (Tore), 8239.
TINNEFELD (F.), 3248.
TINNEY (Steve), 1798.
Tintoretto (Jacopo Robusti), 1263.
TIPPACH (Thomas), 7100.
TIPPETT-SPIRTOU (Sandy), 5501.
TIPPING (R.), 1553.
TIRATSOO (Nick), 4676, 6811.
TISCHLER (Ulrike), 7870.
TISMANEANU (Vladimir), 4044.
TISSONI (Francesco), 3119.
TISSOT (Laurent), 284, 286, 6992.
TITA (Massimo), 7560.
Tito (Josip Broz), 8413.
Tiziano Vecellio, 6613.
TJULENEV (V. M.), 769.
TKHAMOKOVA (Irina Kh.), 891.
TOCANCIPÁ (Jairo), 4294.
TOCK (Benoît-Michel), 3311.
Tocqueville (Charles Alexis Henri Morice Clérel de), 793, 1035.
TODA (Satoshi), 3109.
TODD (Malcolm), 2631.
TODD (S. C.), 2019.
TODE (Sven), 7800.
TODESCHINI (Giacomo), 3889.
TODOROVA (Tsvetana), 6805.
Togliatti (Palmiro), 5046.
TOGNETTI (Sergio), 3613.
TOGNOTTI (Eugenia), 6339.
TOKO (H.), 3249.
TOLLET (Daniel), 5452.
TOLSTIKOV (Vladimir), 2355.
TOLSTOJ (N. I.), 3317.
TOLZ (Vera), 4058.
TOMA (Hildegard), 5054.
TOMARU (Junko), 8457.
TOMASELLO (Dario), 6497.
TOMBER (Roberta), 1729.
TOMEI (Maria Antonietta), 2934.
TOMPKINS (Joanne), 6695.
TOMS (Steven), 6869.
TOMUSK (Voldemar), 6036.
TONINELLI (Pier Angelo), 6804.
TONIOLO (Gianni), 7035.
TONTSCH (Günther H.), 4083.
TOO (Y. L.), 2014, 2152.
TOPIK (Steven C.), 4922, 7113.

TOPOLSKI (Jerzy), 6953.
TORA (Miura), 1168.
TORAL-NIEHOFF (Isabel). 1878.
TORAO (Toshiya), 8776.
TORELLI (Mario), 2409, 2670.
TORI (Giorgio), 4800.
TORKE (Hans-Joachim), 5067.
TORKUNOV (Anatolij V.), 8458.
Tornel y Mendívil (José María), 4912.
TÖRÖK (L.), 1687.
TORPEY (John), 7490.
TORRE (Angelo), 7483.
TORREIRA CRESPO (Ramón), 8459.
TORRES DEL RÍO (César), 4295.
TORRES SÁNCHEZ (Concha), 5587.
TORRÓ (Josep), 7660.
TORSTENDAHL (Rolf), 419, 746.
TORT (O.), 4501.
TORTELLA (Gabriel), 6812.
TORZINI (Roberto), 5675.
TOSCANO (Tobia R.), 6393.
TOSCHI (Alberto), 2441.
TOSH (John), 7282.
TOSHKOVA (Vitka), 435.
TOTH (Istvan György), 95, 5473.
TOUATI (François-Olivier), 1019.
TOUATI (Houari), 79, 3472.
TOUCHAIS (Gilles), 2347, 2359, 2362.
TOUGHER (Shaun), 2438.
TOULOUMAKOS (I.), 1927.
TOURETTE (Florence), 8188.
Toussaint Louverture (Pierre), 4728.
TOUSSAINT RIBOT (Mónica), 7944.
TOWLE (Philip), 7637.
TOWNSEND (Kenneth William), 8110.
TOWNSON (Nigel), 5202.
TRACY (James D.), 1127.
Traianus (Marcus Ulpius), Roman emperor, 2595, 2777, 2880, 2907.
TRAILL (David A.), 1673.
TRAINA (A), 2496.
TRAINOR (Richard H.), 5954.
Trakl (Georg), 6364.
TRAMA (Luciana), 7283.
TRAMONTANA (Salvatore), 3361.
TRAMPUS (Antonio), 5588.
TRANFAGLIA (Nicola), 5952, 6106.
TRANVOUEZ (Yvon), 4502.
TRAPL (Miloš), 422, 5454.
Trasibulus, 2087.
TRASK (Robert Lawrence), 218.
TRAVERSO (M.), 2671.
TRAWICKI (Lech), 7952.
TRAWKOWSKI (Stanisław), 4092, 7628.

TRAWNY (Peter), 6205.
TREADGOLD (Warren T.), 1013.
TRÉDÉ (M.), 1999.
TREECE (Dave), 4244.
TREGIDGA (Garry), 4705.
TREHARNE (Elaine M.), 3738.
TREIBER (Gerhard), 6201.
TREICHEL (Eckhardt), 4521.
TRELLISÓ CARREÑO (L.), 1535.
TRELOGAN (J.), 2121.
TREMBLAY (Manon), 4271.
TRÉMOUILLE (Marie-Claude), 1828.
TRENCHER (Susan R.), 892.
TŘEŠTÍK (Dušan), 921.
TRETVIK (Aud Mikkelsen), 7561.
TREUIL (René), 1522.
TREVARTHEN (David), 1542.
TREVOR (Douglas), 5809.
TREWIN (Ion), 4647.
TRÉZINY (H.), 2297.
TRIBE (Keith), 6840.
TRIESSNIG (Simon), 4195.
TRIFOGLI (Cecilia), 3890.
TRIMBLE (J.), 2935.
TRIMBUR (Dominique), 8460.
TRINER (Gail D.), 7036.
TRIPP (Aili Mari), 5268.
TRIPP (Charles), 4769.
TRIVELLATO (F.), 6813.
TRIVERO (Paola), 6743.
TROEBST (Stefan), 4083.
Troeltschs (Ernst), 767.
TROISI (F. F.), 2153.
TROITSKIJ (Nikolaj A.), 7562.
TROMPF (Garry W.), 610.
TRONVOLL (Kjetil), 8396.
TROTT (David), 6744.
TROTTER (David), 6498.
Trotzkij (Lev Davidovič pseud. Di Lejba Bronštejn), 7362.
TROUVÉ (Susan), 6037.
TRUBAROV (T. M.), 654.
TRUHART (Peter), 1014.
Trujillo y Molina (Rafael Leónidas), 8339.
TRUKHANOVSKII (V. G.), 7631.
Truman (Harry Shippe), 5293, 8356.
TSAGARAKIS (O.), 1927, 2011.
TSAN (Tsong-Sheng), 1829.
TSCHAPEK (Rolf Peter), 7727.
TSETSKHLADZE (G. R.), 2363.
TSIRTSONI (Zoï), 1561.
TSUCHIYA (Hiroshi), 1403.
TSUKADA (Takashi), 8769, 8777.
TSUZUKI (Chushichi), 4884.
TUBBS (J.W.), 3524.
TUCCI CARNEIRO (Maria Luiza), 4224.
TUCK (S.), 2936.

TUCKER (Avezier), 5885.
TUCKER (P.), 7491.
TUCKER (Richard P.), 1176.
TUCKER (Sherrie), 6745.
TUCKER (Spencer C.), 943.
Tukhachevsky (Mikhail Nikolaevich), 5079.
TUKSAR (Stanislav), 1292.
TULLIO-ALTAN (Carlo), 4865.
TULUM (Mertol), 821.
TUMBLETY (Joan), 4503.
TUMINEZ (Astrid S.), 7638.
TUNA (Numan), 1644.
TUPLIN (C.), 2807.
Tura (Cosmè), 6597.
TURBANTI (Giovanni), 5571.
TURBAT (T.), 1604.
TURCUŞ (Şerban), 3945, 4033.
TURCUŞ (Veronica), 3822, 3946.
TUREK-KWIATKOWSKA (Lucyna), 5002.
TURFAN (Mehmed Naim), 5262.
TURI (Gabriele), 5959.
TÜRK (Monika), 3891.
TURLEY (D.), 2154.
TURNER (James), 5946.
TURNER (Jane), 1287.
TURNER (John D.), 1834.
TURNER (Michael), 6984.
TURNER (Ralph V.), 3411.
TURNER (Thomas), 4301.
TURNEY (Caroline), 1203.
TURPIN (J.-C.), 2854.
Tuthalija, king of Hittites, 1827.
Twain (Mark), 6449.
TWERSKY (Isadore), 3452.
TYCH (Feliks), 7311.
TYERMAN (Christopher), 6038.
TYGIEL (Jules), 7284.
TYGIELSKI (Wojciech), 7820.
TYLDESLEY (Mike), 905.
TYLER Elizabeth (M.), 3615.
TYRREL (Andrew), 3610.
TYRRELL (Alex), 7247.
TYSSENS (Madeleine), 3683.
Tzachas, Turkish emir of Smyrna, 3238.
TZEDAKIS (P. C.), 1948.

U

UBIÑA (José Fernández), 3100.
UBL (Karl), 3892.
UÇANKUŞ (Hasan Tahsin), 1674.
UÇAROL (Rifat), 7945.
UEBERSCHÄR (Gerd. R.), 4600.
UFFELMANN (Uwe), 773.
UGOLINI (Gherardo), 2155.
UHDE (Karsten), 12.
UHLÍŘ (Jan B.), 4358.
UKOLOVA (Viktorija I.), 6396.
ULATE SEGURA (Bodil), 5206.

ULLMANN (Ernst), 1256.
Ulloa (Berta), 8037.
Ulpianus (Domitius), 2660.
Ulrich (Anton), 5060.
ULUÇAM (Abdüsselam), 3823.
UNGER (Stefan), 6907.
UNGERN-STERNBERG (J. von), 2632.
UNGURJANU (M. R.), 7956.
UNNO (Fukuju), 7946.
UNSER (Jutta), 627.
UNTERMANN (J.), 2386.
UNZUETA (Patxo), 5172.
URAZMANOVA (Raufa K.), 893.
URBAIN (Jean-Didier), 263.
URBAN (Hugh B.), 1404.
URBAŃCZYK (Przemysław), 3404.
URBANITSCH (Peter), 4181, 7210.
URÍA (Jorge), 5174.
URIBE URÁN (Víctor Manuel), 542, 4296.
URILOV (Il'ja Kh.), 7363.
UROFSKY (Melvin I.), 5281.
URQUIJO GARCÍA (José Ignacio), 5388.
URQUIZU SARASUA (Patricio), 1449.
URSINUS (Michael), 7590.
URSO (Carmela), 3616.
USAI CHERCHI (Paolo), 6746.
USLU (Nasuh), 8463.
USSISHKIN (D.), 1879.
UTTERSTRÖM (Gudrun), 219.
UTZ TREMP (Kathrin), 3998.
UVAROV (Pavel Ju.), 3660, 4012.
UYAMA (Tomohiko), 8494.
UZEL (İlter), 1675.

V

VAAHTERA (J. E.), 2869.
VÁCHOVÁ (Jana), 4349.
VAGENHEIM (Ginette), 612.
VAGGIONE (Richard Paul), 3101.
VAGI (David L.), 176.
VAINFAS (Ronaldo), 4234.
VAINIO (R.), 2808.
VAÏSSE (Maurice), 931, 7662, 8068, 8111.
VAJDA (Sarah), 4504.
VAJNBERG (B. I.), 1655.
VAKALOUDI (Anastasia), 3251.
VALADE (Jean-Michel), 4505.
VALAKH (Yehudah), 4791.
VALANTASIS (R.), 2958.
VALDERAS (José), 730.
VALDEZ DEL ALAMO (Elizabeth), 3646.
VALE (Brian), 7947.
VALE (Lawrence J.), 7285.
VALENTA (Vladimír), 6908.
VALENTI (Filippo), 42.
VALENTIN (Hellwig), 4196.

VALENTIN (Joachim), 6163.
VALENTINI (Filippo), 5525.
Valerianus (Publius Licinius), Roman emperor, 2691.
Valerius Flaccus Setinus Balbus (Gaius), 2525.
Valerius Maximus, 53, 2526.
VALERO GARCÍA (Pilar), 5955.
Valiani (Leo), 729.
VALKEAPÄÄ (Leena), 613.
VALLACQUA GUARIENTO (Maria Luisa), 81.
VALLADARES (Rafael), 7776.
VALLAUD (Pierre), 8112.
VALLE (Perla), 8820.
VALLIERE (Paul), 5618.
VAN BELLE (G.), 1297.
VAN BEUSEKOM (Monica M.), 7706.
VAN BRUAENE (Anne L.), 6107.
VAN CAENEGEM (R. C.), 3525.
VAN CRUYNINGEN (P. J.), 6985.
VAN DEN ABEELE (Baudoin), 3551, 3634, 3656.
VAN DEN DOEL (Wim), 7681, 8464.
VAN DEN ENDE (J.), 6767.
VAN DER HOEYEN (Roland), 6747.
VAN DER KLIS (Jolande), 1271.
VAN DER KOLK (Henk), 4936.
VAN DER PLAAT (G. N.), XI.
VAN DER PLICHT (J.), 1559.
VAN DER STOCKT (L.), 2221.
VAN DER STRAETEN (Joseph), 3110, 3112.
VAN DER WOUDE (Ad), 7102.
VAN DONZEL (E.), 1351.
VAN DÜLMEN (Richard), 614.
VAN EENOO (Romain), II.
VAN EFFENTERRE (H.), 1955.
VAN EFFENTERRE (M.), 1955.
VAN EIJL (C.), 7228.
VAN GELDER (K.), 2364.
VAN HARTESVELDT (Fred R.), 405.
Van Helmont (Jan Baptiste), 6256.
VAN HERWIJNEN (G.), XI.
VAN HOURS (Elisabeth), 3405.
VAN KESSEL (Ineke), 5137.
VAN LIEFFERINGE (C.), 2156.
VAN LOKEREN (Sven), 1588.
Van Maerlant (Jacob), 3742.
VAN MINNEN (P.), 1730.
VAN NIEROP (Henk), 5665.
VAN ORDEN (Kate), 6716.
VAN OSSEL (P.), 2716.
VAN POPPEL (Frans), 7102, 7286, 7287.
VAN PRAET (J.), 286.
VAN RAHDEN (Till), 4575, 5502.
VAN RENEEN (Pieter), 217.
VAN RIEL (Arthur), 4937.

Van Ruisdael (Jacob), 6587.
VAN RULER (Han), 6202.
VAN SANT (John E.), 288.
Van Seters (John), 574.
VAN SOLINGE (Hanna), 7287.
VAN STEEN (G. A. H.), 1940.
VAN TRICHT (F.), 3252.
VAN VEEN (Ernst), 7682.
VAN VELTEN (A. A.), 1115.
VAN VIÊT (Dang), 7683.
VAN VLIET (C.), XI.
VAN VOSS (L. Heerma), 7228.
VAN WIE DAVIS (Elizabeth), 8299.
VAN ZANDEN (J. L.), 4937.
VANDENHOUTE (Thierry), 4209.
VANDENSTEENDAM (Gh.), 2157.
VANDER LIPPE (John M.), 8465.
VANDERPUTTEN (Steven), 615.
VÁNDOR (Jaime), 1880.
VANGELISTA (Chiara), 7288.
VANJA (Christina), 7067.
VANO (Cristina), 7393.
VANSINA (F. D.), 6115.
VANSINA (Jan), 616.
VANTHEMSCHE (Guy), 6888, 6909.
VANZELLI (Gianfrancesco), 7507.
VAPORIS (Nomikos Michael), 5619.
VAR'JASH (Ol'ga I.), 4012.
VARCAN (Nezih), 7024.
VARDAR (Levent), 1648.
VARDI (Amiel D.), 2445.
VARELA SUANZES (Joaquín), 7419.
Vargas (Getúlio), 4242.
VARGAS GARCIA (Eugênio), 8046.
VARGYAS (P.), 1908.
VARIKAS (Eleni), 471.
VARNER (E. R.), 2876.
VARNI (Angelo), 7448.
VARONE (Antonio), 2937.
Varro (Marcus Terentius), 2527, 2528, 2753, 2783.
Varro Atacinus (Publius Terentius), 2529.
Vasari (Giorgio), 481.
VAŠEK (František), 4337.
VASILIKOU (Dora), 150.
VASINA (Augusto), 3599.
VATER (Michael), 6193.
VATIN (Nicolas), 3948.
VAUCHEZ (André), 476, 617, 944, 1364.
VAUDAGNA (Maurizio), 609.
VAUGHAN (Olufemi), 4943.
VÁZQUEZ (Josefina Zoraida), 7948.
VÁZQUEZ CARRIZOSA (Alfredo), 7639.
VÁZQUEZ DE PRADA (Valentin), 7821.
VÁZQUEZ GARCÍA (Francisco), 1187.

VÁZQUEZ-VIANA (Humberto), 4221.
VEAUVY (Christian), 7236.
VECCHI (Giovanni), 6815.
VECCHIOTTI (Icilio), 6203.
Vecellio (Tiziano), 6599.
VEDJUSHKIN (Vladimir A.), 4050.
VEDSTED RØNNE (Bent), 6859.
VEGETTI (Mario), 2286.
Vegio (Maffeo), 2805.
VEGLIA (Marco), 3749.
VEIDLINGER (Jeffrey), 6748.
VEIT (Patrice), 975.
VEKOV (M.), 177.
VELASCO SHAW (Angel), 7640.
VELCIC-CANIVEZ (M.), 666.
Vélez de Arciniega (Francisco), 730.
VELICHKOVA (G.), 406.
VELINOVA (Vasia), 21, 1217.
Velleius Paterculus, 731, 2530.
VELTER (Tiberiu), 8047.
Venantius Fortunatus, 2981, 3035.
VENARD (Marc), 1364, 5573, 5665.
Vendramin (Francesco), 7778.
VENNER (Dominique), 8113.
VENTURA (António), 5018.
VENUTA (Pierluigi), 7695.
Vercingetorix, chief of the Arverni, 2895.
VERDICCHIO (Massimo), 6204.
VERDUGO (Patricia), 4274.
VERGÉ-FRANCESCHI (Michel), 7606.
VERGER (Jacques), 3354.
VERGER (Stéphane), 1941.
Vergerio (Pier Paolo), 5663.
VERGIER (Jacques), 3367.
Vergilius Maro (Publius), 463, 2532, 2533, 2721, 2749, 2765, 2770, 2774, 2782, 2799, 2811, 2812, 2813, 2853, 3036.
VERHEY (Jeffrey), 4636.
VERHEYDEN (J.), 1297.
Vérin (Pierre), 850, 8823.
Vernadsky (Vladimir I.), 5831.
VERNANT (Jean-Pierre), 732.
Verne (Jules), 6482.
VERNET (Joan), 1881.
VERNOIT (Stephen), 374.
VERPEAUX (Michel), 7401.
Verri (Alessandro), 733.
VERRIÈRE (Jacques), 4506.
VERSTRAETE (J.), 1636.
VERTECCHI (Giulia), 1178.
VERVENIOTI (Tasoula), 4720.
VERVLIET (Hendrik D. L.), 6108.
VERZÁR-BASS (M.), 2388.
VESELÝ (Zdeněk), 8048.

INDEX OF NAMES

Vespasianus (Titus Flavius),
  Roman emperor, 2517.
VÉSTEINSSON (Orri), 4000.
Vetranius, 2559.
VEYNE (Paul), 2313, 2717.
VEYRAT-MASSON (Isabelle), 6749.
VEZIN (Jean), 3288.
Viana (Oliveira), 4240.
VIARD (Georges), 5583.
VIARRE (S.), 2484.
VICARIO (Federico), 3982.
VICENT GARCÍA (Juan M.), 1589.
VICIANO (Albert), 3000.
VICKERS (Michael), 4944.
VICKERS (Rhiannon), 4706.
Vico (Giovanni Battista), 788.
Victor Massiliensis, Sanctus, 3107.
VIDAL (Laurent), 7753.
Vidal de La Blache (Paul), 281.
VIDAL NAQUET (Pierre), 744,
  2090.
VIDIC (Marko), 968.
VIELBERG (Meinolf), 3102.
VIELLIARD (Jeanne), 366.
VIGANÒ (Lorenzo), 1799.
VIGARELLO (Georges), 1250.
VIGGIANI (Maria Carmen), 3111.
VIGNAL (Marie-Catherine), 346,
  7819.
VIGNE (Jean-Denis), 2339.
VIGO (G.), 6911.
VIGORELLI (Amedeo), 6140.
VIHAVAINEN (Timo), 8114.
VILADRICH (Mercè), 1882.
VILAR (Pierre), 5158.
VILARDO (M.), 2021.
Villa (Pancho), 7961.
Villani (Filippo), 3316.
VILLAR (Francisco), 1731.
VILLARD (Pierre), 7492.
VILLARI (Rosario), *IX*, 986.
VILLASENOR BELLO (José Miguel),
  7949.
VILLE (Simon), 6816.
VILLERS (Caroline), 3792.
Villiers (George), Duke of
  Buckingham, 7810.
VILLSTRAND (Nils Erik), 5223.
VIN (Jurij Ja.), 3618.
VINCENT (Andrew), 1032.
VINCENT (C.), 4001.
VINICELLI (Maria), 2998.
Vinković (Benedikt), 5520.
Vinnikov (A. V.), 3337.
VINOGRADOV (Pavel Gavrilovich),
  1109, 3526.
VINOGRADOV (Vladilen N.), 7641,
  7848.
VINZENT (Markus), 3046.
VIOLA (L.), 6958.
VIOLA (Paolo), 4113.

VIOLLET (Pierre-Louis), 1634.
VIOTTI DA COSTA (Emilia), 4245.
Virtanen (Olga), 4407.
Visconti (Ottone), arcivescovo di
  Milano, 35.
VISMARA (C.), 2434.
VISOCCHI (Paola), 4021.
VISSER (J. C.), 3277.
VISSER (Joop), 6917.
VISSIERE (Isabelle), 5605.
VISSIERE (Jean-Louis), 5605.
VITA (Juan-Pablo), 1800.
VITALI (Daniele), 2410.
VITOLO. PISA (Giovanni), 3347.
VITTORIA (Albertina), 6106.
Vittorio Amedeo II di Savoia,
  7843.
Vittorio Emanuele II, 6027.
VITUCCI (G.), *IX*.
VITUG (Marites Danguilan), 4981.
VITUKHNOVSKAJA (M.), 5045.
VIVARELLI (Roberto), 4801.
VIVIERS (Didier), 2337.
VOCELKA (Karl), 4197.
VODICKA (Karel), 5101.
VOGEL (Lothar), 3979.
VOGLER (Werner), 3943.
VOGT (Lino), 3412.
VOGT (Marion), 6611.
VOGT (Timothy R.), 8466.
VOGTHERR (Thomas), 3949.
VOGT-SPIRA (Gregor), 2742.
VOIGT (Mary M.), 1612.
VOIGT (Rainer), 1883.
VOISENAT (Claudie), 846.
VOISENET (Jacques), 3661.
VOITHOFER (Richard), 4198.
VÖLCKER (Lars), 5886.
Volckertsz Coornhert (Dirk), 5676.
VOLDMAN (Danièle), 619.
VOLK (Katharina), 2203.
VOLK (Konrad), 1801.
VOLKOV (A. I.), 6048.
VOLKOV (Vladimir K.), 4114.
VOLMER (Annette), 6109.
VOLPATO (Giuseppe), 6894.
Volpe (Antonio), 6632.
VOLPE (Giuseppe), 714, 734, 7434.
VOLPE (Rita), 2938.
Voltaire (François-Marie Arouet
  de, dit), 428, 5887, 5896, 6406.
VONDROVÁ (Jitka), 4359.
VONKOVÁ (Erika), 4341.
VONS (J.), 2809.
VOOGT (Gerrit), 5676.
VORDERSTEMANN (Jürgen), 3678.
VOSS (Rüdiger von), 1064.
Vossius (Isaac), 5746.
VOTH (Hans-Joachim), 6843,
  7364.
VOVELLE (Michel), 7871.

VOX (Onofrio), 2229.
VRAIN (Cécile), 8049.
VREKAJ (Bashkim), 2324.
VRTEĽ (Ladislav), 142.
VUCHKOV (Aleksandur), 7950.
VUGRINEC (Jože), 5664.
VUILLERMOZ (Marc), 6751.
VUILLEUMIER LAURENS
  (Florence), 6395.
VYKOUKAL (Jiří), 4115.

W

WACHINGER (Burghart), 3713.
WADDEL (Chrysogonus, OCSO),
  3924.
WADE (Rex A.), 5086.
WADHAM (John), 4779.
Waechter (Carl Georg), 7395.
WAELKENS (L.), 7368.
WAGENSOHN (Tanja), 8467.
WAGNER (Anthony Richard), 151.
WAGNER (Anthony, Sir), 151.
WAGNER (Bernd C.), 4528, 4627,
  4637.
WAGNER (Jacques), 5430, 6077.
WAGNER (Michael F.), 4378.
WAGNER (Norbert Berthold), 7435.
Wagner (Richard), 6729.
WAGNER-HASEL (B.), 2689.
WAHL (Johannes), 132.
WAINWRIGHT (Geoffrey), 1499,
  1562.
WAITE (Gary K.), 5677.
WAITE (Greg), 3750.
WAKABAYASHI (Bob Tadashi),
  4088.
WAKEMAN (Frederic E.), 8619.
WALA (Michael), 5881.
WALASZEK (Adam), 5004, 7290.
WALDHERR (G. H.), 2633.
WALDHERR (G.), 2587.
WALHOUT (Evelien), 7287.
WALIA (Shelley), 825.
WALKER (Brian Mercer), 4785.
WALKER (J.), 1942.
WALKER (Melissa), 6986.
WALKER (Penelope), 1485.
WALKER (Simon), 4707.
Wallace (Henry Agard), 5330.
WALLACE (Paul W.), 4316.
WALLE (Marianne), 7164, 7291.
Walras (Léon), 6840.
WALSER (Gerold), 1909.
WALSH (Anne), 352.
WALSH (Brendan M.), 7104.
WALSH (Margaret), 6912.
WALSH (P. G.), 2454.
WALSHAM (Alexandra), 6110.
WALTER (Axel), 5829.
WALTER (Franz), 4588.
WALTER (Peter), 5531.

WALTERSKIRCHEN (Gudula), 4199.
WALTERSKIRCHEN (Helene), 1180.
WALTHER (Gerrit), 5879.
WALTON (Whitney), 5888.
WALTON-JORDAN (Ulrike), 7427, 7575.
WALZ (Robin), 5889.
WALZER (Michael), 1052.
WAMSER (Ludwig), 2614.
WANG (Chaoguang), 8649.
WANG (Chunyu), 8723.
WANG (Di), 8650.
WANG (Fangwen), 8651.
WANG (Fan-sen), 8652.
WANG (Hongjiang), 8631.
WANG (Hui), 8653.
WANG (Hui), 8654.
Wang (Jingwei), 8509.
WANG (Lingling), 8655.
WANG (Lixi), 8656.
WANG (Qifa), 8657.
WANG (Qingcheng), 8658.
WANG (Qisheng), 8659.
WANG (Shucai), 8595.
WANG (Victor), 178.
WANG (Yinhuan), 8660.
WANG (Yuesheng), 8661.
WANG (Yunlai), 8662.
WANG (Yuxin), 8663.
WANG (Zhen), 8664.
WAQUET (Françoise), 5890.
WAQUET (Jean-Claude), 458, 5891.
Warburg (Aby), 735.
WARBURTON (Nigel), 1081.
WARD (Aengus), 607.
WARD (Ian), 4167.
WARD (Joseph P.), 3308.
WARD (Keith), 5456.
WARD (Kevin), 5592.
WARDEN (P. Gregory), 2411.
WARD-PERKINS (Bryan), 1251, 2945.
WARDY (Robert), 2287.
WARNER (Carolyn M.), 5575.
WARNICKE (Retha M.), 4708.
WARREN (Alan), 4754.
WARREN (Allen), 961.
WARREN (Christian), 7365.
Warren (Earl), 5352.
WARREN (James), 2288.
WARREN (Michelle R.), 3413.
WARRENDER (Howard), 1082.
WARRING (Anette), 1150.
WARTELLE (André), 2230.
WASHBURNT (Dennis), 6754.
Washington (George), 5308, 5312.
WASSERMAN (Mark), 4923.
WASSERMANN (Heinz P.), 4200.
WASSERSTROM (Jeffrey N.), 7380.
WASSERSTROM (Steven M.), 1405.

WASSON (Ellis), 4709.
WATANABE (Kouichi), 8764.
WATANABE-O'KELLY (Helen), 5732.
WATERS (Michael R.), 1573.
Watson (Aaron), 1608.
WATSON (Andrew G.), 59.
WATSON (Nigel), 5956.
WATSON (Wilfred G. E.), 1802.
WATSON (William), 1293.
WATSON-TREUMANN (Brigitte), 1884.
WATT (Isabella M.), 5621.
Watteau (Jean-Antoine), 6604, 6606.
WATTS (J. F.), 5276.
WAUQUELIN (Jehan), 3686.
WAWN (Andrew), 621.
WAY (Wendy), 8243.
WAYLEN (Georgina), 4284.
WEAR (Andrew), 6342.
WEATHERALL (Mark), 6343.
WEBB (Jennifer M.), 1570.
WEBB (John A.), 1487.
WEBBY (Elizabeth), 1475.
WEBER (G.), 2966.
WEBER (Hermann), 1145.
WEBER (J. A.), 1795.
Weber (Max), 736, 1102, 2575.
WEBER (Philippe), 3485.
WEBER (Wolfhard), 6344.
WEBER-FAS (Rudolf), 1083.
WEBSTER (Eddie), 5136.
WEBSTER (Jill Rosemary), 3950.
WECHSLER (Diana Beatriz), 6590.
Wedekind (Frank), 6522.
WEDIN (M. V.), 1982.
WEFERS (Sabine), 3359.
WEGEHAUPT (Heinz), 6047.
WEGMANN (Nikolaus), 372.
WEGNER (Bernd), 976, 8115.
WEGNER (Karl-Hermann), 7067.
WEHDE (Susanne), 6112.
WEHRLE (William T.), 2741.
WEI (Ian P.), 3645.
WEICHEL (Thomas), 7037.
WEIDLING (Tor), 4952.
WEIDNER (Marcus), 7292.
WEIGEL (Sigrid), 862.
WEIGL (Herwig), 13.
WEIGLEY (Russell Frank), 5366.
WEIL (Brigitte), 339.
Weil (Eric), 6139.
WEIL (Françoise), 699, 7105.
Weill (Kurt), 6698.
WEIMANN (Robert), 6752.
WEINBERG (Carole), 1456.
WEINBERG (J.), 113.
WEINBERGER (Leon J.), 3445.
WEINCKE (M. H.), 1613.
WEINDLING (Paul Julian), 4117.

WEINER (Richard), 6954.
WEINER (Thomas), 6206.
WEINFELD (Moshe), 279.
WEINGAST (Barry R.), 7012.
WEINHAUER (Klaus), 7319.
WEISBERGER (Bernard A.), 5367.
WEISBROD (Bernd), 5892.
WEISER (Johanna), 347.
WEIß (Bardo), 3917.
WEIß (Dieter J.), 5472.
WEISS (P.), 2718.
WEIß (Ulman), 5638.
WEISSER (Bernhard), 179.
WEITZ (Yechiam), 8468.
Weizsäcker (Richard von), 737.
WELCH (David), 8050.
WELCH (Robert), 1441.
WELDON (Richard N.), 959.
WELLENREUTHER (Hermann), 4105.
Wellington Koo (V. K.), 8145.
WELLS (Anne Sharp), 8123.
WELLS (John), 7012.
WELLS (Robert W.), 7293.
WELS (Volkhard), 6039.
WELSCH (Robert L.), 872.
WELSKOPP (Thomas), 4638, 7294.
WELWEI (K.-W.), 2672.
WENBAN-SMITH (Francis F.), 1539.
WEN-CHAO (Li), 5611.
WENDE (Peter), 433, 951, 4067, 7888.
WENDEL (Saskia), 6163.
WENGER (Andreas), 8323.
WENTA (Jaroslaw), 3951.
Wenz (Eugene), 6247.
WENZERUL (Rosemary), 114.
WERLE (Gerhard), 7553.
WERMEIL (Sara E.), 7106.
WERNER (Wilfried), 3926.
WERTZ (Spencer K.), 826.
WESCH-KLEIN (Gabriele), 2419.
WESSEL (Carola), 4105.
WESSEL (Susan), 2986.
WEST (Nancy Martha), 622.
WESTAD (Odd Arne), 8238, 8240.
WESTBROOK (R.), 1678.
WESTE (John L.), 7892.
WESTGATE (R.), 2939.
WESTREM (Scott D.), 287.
WESZELI (Michaela), 1856.
WETTENGL (Kurt), 771.
WETTMANN (Andrea), 5957.
WETZELL (Richard F.), 6345.
WEYLAND (Petra), 5778.
WHALEN (Brett E.), 3299.
WHALEY (Leigh Ann), 4507.
WHATMORE (Richard), 5810, 6841.
WHEATLEY (Edward), 3771.

WHEELER (Bonnie), 3643.
WHEELER (S. M.), 2810.
WHELAN (Bernadette), 4786.
WHITBY (Kenny J.), 5368.
WHITBY (M.), 2945, 2993.
WHITE (G. Edward), 7436.
WHITE (George W.), 4118.
WHITE (Graeme J.), 3414.
WHITE (Hayden V.), 750, 784, 827.
WHITE (Hugh), 3751.
WHITE (James D.), 5087.
WHITE (Luise), 802, 1476.
WHITE (Mark J.), 8261.
WHITE (Nicholas J.), 6955.
WHITE (Nicola), 6913.
WHITE (Ralph), 4058.
WHITE (Raymond), 1721.
WHITE (Stephen), 5064.
WHITED (Tamara L.), 6987.
WHITEHEAD (Christiania), 4003.
WHITEHEAD (D.), 2013.
WHITEHOUSE (R. D.), 1913.
WHITING (Richard), 4710.
WHITING YOUNG (Biloine), 7107.
WHITMAN (Jon), 1452.
WHITNEY (Robert), 4314.
WHITTLE (Alasdair), 1563, 1564.
WHITTLE (Jane), 6988.
WHORTON (James C.), 6346.
WHYTE (Ian D.), 7295.
WICHKAM (Chris), 2953.
WICKE (H.), 7563.
WICKHAM (Chris), 3527.
WIDDER (Ellen), 3345.
WIDENHEIM (Cecilia), 6540.
WIDMAIER (Gunter), 7484.
WIDMARCKTER (Obristen Caspar von), 7779.
WIEBE (Don), 1407.
WIECKER (Niels), 7800.
WIECZOREK (Alfried), 3336.
WIEGELS (R.), 2435.
WIELL (Stine), 623.
WIENS (Birgit), 6753.
WIESEL (Elie), 5032.
WIESER (Marie Therese), 2980.
WIEVORKA (Olivier), 4427.
WIGAL (Donald), 289.
WIGGERHAUS (Norbert), 8395.
WIJERS (Carla), 881.
WIJSMAN (II. J. W.), 2525.
WILBERTZ (Gisela), 7044.
Wilde (Oscar), 6457.
WILDFAG (R. L.), 2849.
WILHELM (Susanne), 1635.
Wilhelm II, Dt. Kaiser u. König von Preussen, 4541, 7964.
WILKE (Jürgen), 1181.
WILKES (J. J.), 2719.
WILKINS (J. B.), 1913.
WILKINS (J.), 2231.

WILKINSON (Endymion Porter), 1021.
WILKINSON (Sarah), 8116.
WILKINSON (T. J.), 1636, 1890.
WILKINSON (Toby A. H.), 1677, 1732.
WILLCOX (George), 2339.
WILLETT (Julie A.), 6956.
William II, King of England, 3380.
William III, king of England, 7837.
William of Jumièges, 3862.
WILLIAMS (Frank J.), 5312.
WILLIAMS (John), 1035.
WILLIAMS (Lynn), 7760.
WILLIAMSON (George S.), 5893.
WILLIAMSON (Jeffrey), 6796.
WILLIAMSON (Philip), 4711.
WILLIAMSON (Thora), 7688.
WILLING (Matthias), 624.
WILLOWEIT (Dietmar), 1086.
WILLS (Douglas), 7012.
WILLS (Rebecca), 7849.
WILMETH (Don B.), 1266.
WILSON (Andrew), 5271.
Wilson (Harold), 8312.
Wilson (James), 5353.
WILSON (John F.), 5946.
WILSON (L. M.), 1252.
WILSON (M.), 1983.
WILSON (Penelope), 1733.
WILSON (Peter H.), 4119.
WILSON (Peter), 2113.
WILSON (R. J. A.), 2940.
WILSON (Renate), 4105, 6347.
WILSON (Richard), 6570.
WILSON (Sephen), 3662.
WILSON (Ted), 7038.
WILSON (Theodore A.), 8189.
WILSON JONES (M.), 2365, 2941.
WINCH (Christopher), 1060.
Winckelmann (Johann Joachim), 561, 738.
WIND (Edgar), 6612.
WINDLER (Christian), 7850.
WINEGARDNER (Ann C.), 1538.
WINGEATE PIKE (David), 8117.
WINGROVE (Elizabeth Rose), 6426.
WINKLER (Allan M.), 8267.
WINKLER (Eduard), 4201.
WINKLER (Heinrich August), 1022.
WINKLER (Jean-Marie), 8063.
WINKLER-BESSONE (Claude), 8063.
WINROTH (Anders), 3528.
WINSTEAD (Karen A.), 3952.
WINTLE (Michael), 6817.
WINTON (Harold R.), 4049.
WINTON (R.), 2091.
WINTZER (Joachim), 8241.
WIPPEL (John F.), 3893.
WIRSCHING (Andreas), 4639.

WIRTH (John), 1182.
Wirth (Joseph), 8013.
WIRTH (Peter), 3126.
WIRTY (Emeline), 396.
WISCART (Jean-Marie), 948.
WISCHERMANN (Clemens), 1220, 6814, 6914.
WISEMAN (Susan), 6221.
WISEMAN (T. P.), 2870.
WISSELGREN (Per), 625.
Wissowa (Georg), 695.
WITCHER (R.), 2918.
WITNEY (Kenneth), 3612.
WITSCHEL (Christian), 2419, 2634.
WITT (Ronald G.), 626.
WITTE (Egbert), 6207.
WITTE (Els), 4210.
WITTENBURG (Andreas), 697.
WITZEL (Carsten), 1803.
WLODARCZYK (Marta Anna), 2290.
WŁODARSKI (Józef), 7851.
WODRUFF (P.), 2041.
WÖHRLE (G.), 2245.
WOJCIECHOWSKIEGO (P.), 2838.
WOLDE AREGAY (Merid), 7911.
WOLF (Kirsten), 3954.
WOLFENSBERGER (Donald R.), 5369.
WOLFRAM (A. X.), 2232.
WOLFRAM (Herwig), 3415.
WOLFREYS (Julian), 1474.
WOLFRUM (Edgar), 628.
WOLFTHAL (Diane), 3353.
WOLIKOW (Serge), 4106.
WOLLOCH (Nathaniel), 6348.
WOLTERS (R.), 2635.
WOLTERS (Wolfgang), 6571.
WOLTERSTORFF (Nicholas), 6208.
WOLTON (Suke), 8118.
WOOD (Elisabeth Jean), 4399.
WOOD (Ian), 2965.
WOOD (Paul), 6194.
WOODCOCK (Thomas), 152.
WOODFIELD (Richard), 735.
WOODING (Lucy E. C.), 5458.
WOODMAN (Richard), 8209.
WOODS (D.), 2636.
WOODS (Robert), 7108.
WOODWARD (Peter), 8251.
WOOLF (D. R.), 629.
WOOTTON (David), 5896.
WORBOYS (Michael), 6349.
WORBY (Eric), 7706.
WORDIE (J. R.), 6960.
WORLEY (Matthhew), 4713.
WORM (Peter), 14.
WORMELL (Jeremy), 7039.
WORP (K. A.), 2587.
WORTHINGTON (I.), 1990.
WORTMAN (Richard S.), 5088.

WOYTEK (E.), 2812.
WREDE (Martin), 630.
WRIGHT (Alison)., 6532.
WRIGHT (Beth S.), 631.
WRIGHT (Christopher), 6818.
WRIGHT (Donald), 632.
WRIGHT (G. R. H.), 3253.
WRIGHT (J. P.), 2272.
WRIGHT (Josephine), 314.
WRIGHT (R.), 7109.
WRIGLEY (Chris), 6776.
WU (Jingping), 8665.
WU (Jingping), 8666.
WU (Limei), 8667.
WU (Xijin), 8727.
WU (Yanhong), 8668.
WU (Yixiong), 8669.
WULFRAM (Hartmut), 3688.
WUNDER (Heide), 5771, 7067.
WUNDERLICH (Dieter), 4120.
WUNN (Ina), 1408.
WUNSCH (Cornelia), 1804.
WÜNSCH (Thomas), 3980.
WUTHENOW (Ralph-Rainer), 6428.
WYATT (N.), 1805.
WYLIN (K.), 2412.
WYMER (Rowland), 303.
WYNDHAM (Marivic), 8825.
WYRWA (Andrzej Marek), 5577.
WYRWA (Ulrich), 5727.
WYSOCKI (Michael), 1564.
WYSZCZELSKI (Lech), 8051.

X

XELLA (Paolo), 1885.
Xenophon, 2211, 2214, 2227.
XIA (Mingfang), 8670.
XIAO (Jianxin), 8671.
XIAO-PLANES (Xiaohong), 8672.
XIE (Xiaopeng), 8673.
XING (Long), 8674.
XIONG (Victor Cunrui), 8675.
XIONG (Yuanbao), 8676.
XU (Hao), 8677.
XU (Min), 8678.
XU (Shiduan), 8679.
XU (Tan), 8680.
XU (Yinong), 7110.

Y

YAFEH (Yishay), 7012.
YAGIL (Limoré), 4508.
YALÇIKLI (Derya), 1676.
YAMADA (Yasuhiro), 8778.
YAMAGUCHI (Akihiko), 8495.
YAMAMOTO (Dorothy), 3752.
YAMAMOTO (Eishi), 8530.
YAMAMOTO (Keiji), 3846.
YAMAMOTO (Masahiro), 633, 8119.
YAMAMOTO (Susumu), 8681.
YAMAMOTO (Tadashi), 6819.

YAN (Buke), 8682.
YAN (Guangwen), 8683.
YAN (Jun), 8684.
YAN (Xiaojun), 8685.
YANG (Guisong), 8686.
YANG (Tianhong), 8687.
YANG (Tianshi), 8688, 8689.
YANG (Tieshi), 8690.
YANG (Xingmei), 8691.
YANG (Xiulin), 8509.
YANG (Yuqing), 8692.
YANNOPOULOS (P.), 3159.
YAPP (Malcolm), 8247.
YARBROUGH (Tinsley E.), 7493.
YARDENI (Myriam), 634.
YARDLEY (J. C.), 2472.
Yarim-Lim, king of Syria, 1789.
YARROW (Liv Mariah), 635.
YAVUZ (Celalettin), 5263.
YAZBAK (Mahmoud), 4969.
YE (Hanming), 8693.
YE (Shuzong), 8511.
YEATES (Pádraig), 7111.
Yeats (William Butler), 6436.
YEE (Jennifer), 6499.
YEH (Wen-hsin), 8694.
Yeltsin (Boris Nikolaevich) v.
   El'tsin (Boris Nikolaevich).
YENER (K. A.), 1636.
YERASIMOS (S.), 3250.
YETKİN (Sabri), 7040.
YILDIRIM (Hacı Osman), 5241.
YILDIRIM (Tayfun), 1830.
YILDİZ (Gültekin), 5251.
YILDIZ (Nuray), 82.
YIN (Weiguang), 8695.
YLIKANGAS (Heikki), 7564.
YOFRE (Juan Bautista), 8470.
YOHANNES (Petros), 8268.
YOKOTA (Fuyuhiko), 8769.
YON (M.), 1956.
YOSHIDA (Nobuyuki), 8769, 8779.
YOSHIE (Akiko), 8780.
YOUNG (Brian), 5810.
YOUNG (Crawford), 4301.
YOUNG (Marilyn B.), 7380.
YOUNIS (Mona), 4970.
YRTTIAHO (Eero), 7565.
YU (Dahua), 8696, 8697.
YU (Guilin), 8698.
YU (Heping), 8699.
YU (Kunqi), 8700.
YU (Li), 8701.
YUAN (Chengyi), 8702.
YU-JOSE (Lydia N), 5898.
YUZO (Mizoguchi), 828.

Z

ZABLOUDILOVÁ (Jitka), 8052.
ŽÁČEK (Pavel), 4360.
ŽÁČEK (Rudolf), 8120.

ZACHARIAS (Johanna), 3581.
Zacharias, 1847.
ZACHS (Fruma), 7951.
ZADOK (R.), 1787.
ZAFRA DE LA TORRE (Narciso), 1517.
ZAGDOUN (M. A.), 2291.
ZAGORIN (Perez), 1084.
ZAHARIADE (M.), 2587.
Zahn-Harnack (Agnes von), 698.
ZAKHOZHAJA (Tat'jana M.), 1572.
ZALIZNJAK (Anatolij A.), 3279.
Zamboni (Anteo), 4817.
ZAMBRANO (Marta), 533.
ZAMIR (Meir), 4897.
ZAMORA (K.), 1752.
ZANATTONI (Marzio), 6140.
ZANCARINI (Jean-Claude), 1116.
ZANCHI (Goffredo), 5507.
ZANG (Rong), 8703.
ZANGEMEISTER (Karl), 698.
ZANINI (A.), 2387.
ZANKER (Graham), 2292.
ZANKER (Paul), 2904.
Zanni Rosiello (Isabella), 319.
ZAPRJANOVA (A.), 133.
ZARAGOZA (Justo), 8822.
ZÁRATE TOSCANO (Verónica), 7297.
ZARET (David), 4714.
ZARETSKIJ (Jurij P.), 6396.
ZARKA (Yves Charles), 6184, 7394.
ZARRI (Gabriella), 1267, 5459.
ZARUSKY (Jürgen), 8198.
ZATLOUKAL (Klaus), 3671.
ZAVOIKINE (Alexei), 2355.
ZAVOS (John), 5728.
ZBIERSKI (Andrzej), 7952.
ZEDINGER (Renate), 4938.
ZEHNDER (Frank. G.), 6520.
ZEIT (Lisa), 6613.
ZELIKOW (P.), 8471.
ZELKA (Luan), 6113.
ZELLER (Joachim), 636.
ZELNICK-ABRAMOVITZ (R.), 2158.
ZENG (Yeying), 8704.
Zeno, 1684.
ZEPPENFELD (Burkhard), 6875.
Zerbold von Zutphen (Gerald), 3717.
ZERNACK (K.), 955.
ZETTERLING (Niklas), 4951.
ZEUCH (Ulrike), 6429.
ZEUSKE (Michael), 4315.
ZEVOLINO (G.), 2942.
ZHAI (Qiang), 8472.
ZHANG (Baoming), 8706.
ZHANG (Fuyun), 8707.
ZHANG (Gefu), 8708.
ZHANG (Hao), 8709.

Zhang (Juzheng), 8683.
ZHANG (Kaiyuan), 8710.
ZHANG (Lei), 8711.
ZHANG (Lianhong), 8712.
ZHANG (Qifan), 8635.
ZHANG (Shengzhen), 8713.
Zhang (Shizhao), 8737.
ZHANG (Xianwen), 8714.
ZHANG (Xiaoye), 8715.
ZHANG (Xin), 8716.
Zhang (Xueliang), 8615.
ZHANG (Yan), 8717, 8718, 8719.
ZHANG (Yongjiang), 8705.
Zhang (Zhidong), 8555.
ZHANG (Zhuangqiang), 8720.
ZHAO (Xiaohua), 8589.
ZHE (Fu), 8526.
ZHELIAZKOVA (Antonina), 4129.
ZHENG (Genan), 8721.
ZHENG (Qin), 8722.
ZHIROMSKAJA (V. B.), 7083.
ZHIVOV (Vladimir M.), 3663.
ZHMODIKOV (A.), 2637.
ZHOU (Qiuguang), 8729, 8730.
ZHOU (Shaoquan), 8731.
ZHOU (Yumin), 8732.
Zhou Cheng (King Cheng of Zhou), 8627.
Zhou Wu (King Wu of Zhou), 8627.
ZHU (Qingbao), 8733.
ZHU (Ying), 8734.

ZHU (Zongchen), 8735.
ZHUANG (Weimin), 8736.
ZICH (František), 7615.
ZICHE (Paul), 6181.
ZIEGAUS (Bernward), 2614.
ZIEGER (Robert H.), 8053.
ZIEGLER (Charlotte), 68.
ZIEGLER (Dieter), 7184.
ZIEGLER (Hans-Joachim), 3713.
ZIEMANN (Benjamin), 620.
ZIFF (Larzer), 290.
ZILHÃO (Joâo), 1540.
ZILSEL (Edgar), 1254.
ZIMA (Veniamin F.), 8121.
ZIMMER (Oliver), 637.
ZIMMERMAN (M.), 2444.
ZIMMERMANN (Bernhard), 1294.
ZIMMERMANN (Harald), 3290.
ZIMMERMANN (Hubert), 8241.
ZIMMERMANN (Matilde), 4940.
ZIMMERMANN (Susanne), 5958.
ZINGERLE (Arnold), 5607.
ZINK (Anne), 4509.
Zinzendorf (Nikolaus Ludwig Graf), 5695.
ZISSER (Eyal), 4898.
ZISSOS (Andrew), 2478, 2751, 2752.
Živkov (Todor Christov), 4260.
Zix (Benjamin), 6535.
ZLINSZKY (J.), 3103.

ZNAMIEROWSKI (Alfred), 153.
ZOHAR (Noam), 1052.
Zola (Emile), 5761.
ZÖLLNER (Hans-Bernd), 8122.
ZOLOV (Eric), 7610.
ZOMEÑO (Amalia), 3473.
ZONTA (M.), 1886.
ŽONTAR (Jože), 7544.
ZORZI (Andrea), 3426.
Zosimus (Q. Cornelius), 2428.
ZOTTA (Franco), 6209.
ZOU (Xiaozhan), 8737.
ZOU (Zhenhuan), 8738.
ZOURNATZI (A.), 1910.
ZUB (Alexandru), 638.
ZUBER (Valentine), 5402, 5678.
ZUBKOVA (Elena Ju.), 7298.
ZUBLER (C.), 2167.
ZUBOK (V. M.), 8439.
ZUCCOTTI (Susan), 5504.
ZUCKERMAN (C.), 3187, 3254.
ZUCKERMANN (Moshe), 772.
ZULIC (Arif), 5391.
ZUMBO (A.), 2256.
Zweibrücken (Johann Casimir von), 6548.
ZWEIG (Michael), 7366.
ZWEINIGER-BARGIELOWSKA (Ina), 4715.
ZWIERLEIN (O.), 2812.
ZYSBERG (André), 4448.

GEOGRAPHICAL INDEX

A

Aberdeenshire, 1542.
Abruzzo, 1011.
Accra, 4644, 8802.
Acquitaine, 158.
Actium, 2555.
Acy-Romance, 1941.
Adamello, 384.
Addis Ababa, 7061.
Adriaticum (Mare), 2366, 2377, 2378.
Aegaeum (Mare), 2394.
Aegyptus, 1677-1733, 1759, 3205.
Afghanistan, 4122-4128, 8247, 8277, 8383.
Africa proconsularis, 2579.
Afrique, 204, 266, 833, 834, 1168, 1274, 1340, 1476, 2604, 2634, 2698, 3057, 5422, 5451, 5606, 6359, 6443, 6524, 6686, 7684-7727, 8251, 8320, 8786-8809. – A. centrale, 8283, 8795. – A. méridionale, 6321, 8292. – A. occidentale, 386, 5987, 8790. – A. orientale, 7715. – A. septentrionale, 3473, 7174, 7726, 8808. – A. sud-occidentale, 7703.
Agia Aikaterini Square Kastelli (Khania), 2338.
Agincourt, 3417, 3424.
ᶜAin Ghazal, 1547.
Ainos, Enez, 1642.
Aix, 5582.
Alalah, 1789, 1800.
Albania v. Shqiperi.
Alcántara, 3265.
Aleppo, 1770.
Alexandrette, 7604.
Alexandria, 363, 1690, 2332.
Algérie, 443, 446, 605, 946, 4132-4137, 7684, 7685, 7687, 7702, 7708, 7711, 8257, 8288.
Alicante, 3574.
Almería, 2429.
Alpes Cottiae, 2605.
Alpes Graiae, 2605.
Alpes Maritimae, 2434, 2605.
Alpes Poeninae, 2605.
Alpes, 2614.
Alpes-de-Haute-Provence, 3782.
Altai, 1548.
Altinum, 2388.
Amarna (el-Hagg Qandil), 1693.
Amarna, 1678.
Amathus, 2316.
Amérique, 185, 290, 314, 364, 593, 870, 892, 954, 973, 1049, 1056, 1342, 4099, 4101, 4105, 5414, 5595, 5686, 6839, 6950, 7403, 7699, 7728-7755, 7823, 8810-8820. – A. centrale, 5870, 7074. – A. latine, 410, 542, 959, 4109, 5597, 6718, 6750, 6771, 7191, 7255, 7384, 7454, 7471, 7482, 7603, 7610, 7639, 7728, 8248, 8812. – A. méridionale, 6055, 8292. – A. septentrionale, 1182, 4658, 5719, 7079, 7088, 7315, 7825.
Amsterdam, 5725, 6921.
Amuq Valley, 1636.
Ana Manuku, 1537.
Anatolie, 155, 1587, 1632, 1640, 1641, 1661, 1674, 1675, 1676, 3467, 5261, 7441, 7938.
Ancon, 2371.
Ancona, 3808.
Andalucía, 3431, 3465, 3470, 3473, 3474, 3722, 3872, 5178, 5189, 8491.
Andaman Islands, 7675.
Andes, 4214.
Angers, 3593.
Angkor, 1493.
Angola, 4138-4140, 7716.
Anshan, 8607.
Antiochia, 2142, 3209, 3378.
Antium, 2885.
Aphrodisias, 1669, 2607.
Apollonia (Illyria, Shqiperi), 2324.
Appennino, 7130.
Appenninus (Mons), 2369, 2381.
Aqht, 1778.
Aquileia, 2926.
Aquitania, 2425.
Arab countries, 8258, 8294, 8425, 8440.
Arab Gulf, 4956.
Arabie, 1839, 1890, 1906, 8258.
Aragona, 3264, 3275, 3282, 3433, 3795, 3868.
Arbeca, 1541.
Arcadia, 2057.
Arctic, 230.
Ardalan (province), 8495.
Argentina, 254, 438, 451, 877, 1050, 4141-4161, 4284, 4917, 5375, 5684, 5717, 5970, 6528, 6590, 6793, 7267, 7398, 7581, 7947, 8372, 8390, 8407, 8423, 8470.
Argos, 2157, 2350, 2359.
Arles, 4029.
Armenija, 175, 239, 3214, 7873.
Asante, 8789, 8799.
Asie, 1126, 1821, 1895, 2718, 6761, 7661-7683, 7782, 8249, 8474-8785. – A. centrale, 241, 1208, 8480-8495. – A. du sud-est, 4086, 7678, 7899, 8258, 8457, 8496-8503. – A. méridionale, 7120, 8258, 8429, 8496-8503. – A. minor, 278, 1648, 1662, 1666, 2558, 2566. – A. occidentale, 8480-8495. – A. orientale, 7239, 7594, 8739.
Aspendus, 2903.
Assisi, 147.
Asti, 7.
Atapuerca (Burgos), 1606.
Athenae, 1845, 2040, 2058, 2064, 2072, 2073, 2088, 2092, 2099, 2102, 2105, 2106, 2110, 2140, 2150, 2155, 2158, 2308, 2344, 2348, 6543.
Atlantique (océan), 4233.
Atria, 2366.
Atthica, 2049.
Augsburg, 7262.
Augusta Emerita, 2884.
Auschwitz v. Oświęcim.

Australia, 450, 1475, 1591, 4162-4167, 6764, 6816, 6933, 7340, 7584, 7757, 7902, 8095, 8201, 8243, 8270, 8822, 8825.
Austria v. Österreich.
Avebury, 1590.
Avranches, 3888.
Ayalon, 1585.
Azerbaidjan, 8482.
Azores, 8275.

B

Babylonia, 106, 1670, 1768, 1769, 1771, 1776, 1876, 1920.
Bactria, 1905.
Baden, 7211, 7556.
Baden-Württemberg, 4568.
Bahia, 834.
Bălgarija, Bulgaria, 21, 135, 165, 202, 342, 400, 406, 435, 916, 926, 998, 3134, 3278, 3883, 4246-4261, 5792, 5910, 6805, 6930, 7519, 7876, 7950, 8151, 8161, 8176, 8413, 8449.
Balkans, 462, 3195, 3244, 4078, 7582, 7618, 7621, 7641, 7848, 7886, 7890, 7905, 7959, 7999, 8018, 8042, 8064, 8276, 8340.
Baltique (mer, pays), 6876, 7206.
Banat, 5873.
Bangladesh, 5249, 8433.
Barcelona, 3382, 3597, 3620, 3762.
Bari, 4803.
Barranquilla (Colombia), 4287.
Basel, 656, 6069.
Basilica di San Prospero (Reggio Emilia), 3826.
Basilicata, 3050.
Basse-Terre (Guadeloupe), 7087.
Bath, 7046.
Bayern, 6000, 6566, 7324, 7556, 7811.
Beersheba, 1848.
Beijing, 7082, 8537, 8676, 8730.
Beirut, 7951.
Balarus v. Belorussija.
Belfast, 6690.
Belgique, 361, 2980, 4205-4210, 4938, 5975, 6909, 7353, 7907, 8283.
Belorussija, Belarus, 173, 4062, 4202-4204, 8192, 8194.
Benares/Varanasi, 5996.
Benevento, 3938.
Bengal, 7670, 8499.
Benin, 4211.
Benizaa, 549.
Bergè, 1958.
Berlin, 344, 353, 4544, 4550, 5926, 6273, 7203, 8259, 8327, 8399.

Bern, 6069.
Bettola (Piacenza), 1584.
Beycesultan (Asia Minor), 3253.
Bharatpur (Princely State), 4746.
Białystok, 4998.
Bianyang, 8551.
Bir Umm Fawakhir (Aegyptus), 3224.
Birkenau, 4589.
Birmingham, 5934, 6296.
Bisignano (Cosenza), 3193.
Bjørvika, 4023.
Black Sea, 268, 2315.
Bobbio, 3568.
Bogotá, 6022.
Bolivia, 4212-4221, 4290, 5532.
Bologna, 190, 1284, 3503, 3768, 3968, 4852, 6558, 6965, 6976, 7444.
Bomarzo, 6547.
Bordeaux, 7186.
Bornholm, 8357.
Bosna-Hercegovina, 5391.
Bosphorus, 2355.
Boston, 6018.
Botswana, 4222.
Bourgogne, 361, 3368, 3548, 3577, 3627, 3799.
Brabant, 3785.
Bracciano, 1545.
Brandenburg, 7446, 7851.
Brasil, 231, 232, 264, 499, 889, 4223-4245, 5489, 5593, 5594, 6083, 6301, 6669, 6793, 6977, 7036, 7091, 7288, 7587, 7591, 7657, 7735, 7736, 7740, 7748, 7753, 7797, 7816, 7916, 7947, 8046, 8144, 8182, 8390, 8811.
Bratsberg County, 4950.
Braunschweig, 4604, 6887.
Brda, 8219.
Brescia, 4845.
Breslau, 5502, 6553.
Bretagne, 577.
Brihuega, 5140.
Britannia, 2434, 2563, 2565, 2680, 2960.
British Honduras, 7396.
British Isles, 4682.
Brixia, 2706.
Bruges, 3544.
Bruttium, 2871.
Bruxelles, 324, 6105.
Bryn yr Hen Bobl (Llanedwen, Anglesey), 1594.
Buchenwald, 4610.
Bucovina, 5030.
Buenos Aires, 898, 5882, 6831, 6967.
Bulgaria v. Bălgarija.
Bunyoro, 7056.

Burkina Faso, 4262.
Burma, 7666.
Bützow, 5906.
Büyüknefes (Provinz Yozgat, Anatolie), 1671.
Byzantion, Empire byzantin, 233, 299, 513, 1013, 3113-3254, 3663.

C

Cadiz, 6921.
Caen, 4015.
Caere, 2400, 2403, 2670.
Caesarea (Mauretania), 2908.
Cagliari, 162, 4006.
Cahokia, 7107.
Calabria, 2871.
Calcedonia, 3053.
California, 5976.
Cambodia, 5390.
Cambrai, 70.
Cambridge, 193, 5909, 6343.
Cameroun, 928, 4263.
Campanaio (Sicilia), 2940.
Campania, 2382, 2701.
Campos, 5142.
Canada, 187, 486, 632, 803, 4264-4272, 5841, 5928, 6850, 7109, 7336, 7577, 8134, 8242.
Canarias, 5139.
Cancho Roano (Zalamea de la Serena, Badajoz), 1567, 1581.
Caporetto, 384.
Caracas, 5387.
Caria, 1646, 1669.
Caribbean area, 542, 1469, 6359, 6460, 7738, 7746, 8326.
Carinthia, 8365.
Carnia, 7502.
Carpates (Bassin de), 1605.
Carthago, 1852, 1885, 2055, 2441, 2633, 2652, 3791.
Cassino, 3772.
Castelfranco Veneto, 3824.
Castellón Alto (Galera, Granada), 1583.
Castilla, 3432, 3488, 3494, 3516, 3549, 3590, 4005, 5142, 7001.
Cataluña Vella, 1882.
Cataluña, 43, 912, 1092, 3261, 3271, 3466, 3614, 3950, 5141, 5170, 5201.
Catania, 5912.
Catherinenburg (Drachenbronn-Birlenbach, France), 6548.
Čechy, 334, 408, 4352, 4354, 5711, 5753, 5827, 6500, 6508, 6779, 7002, 7243.
Československo, Czechoslovak Republik, 333, 4320-4360, 5091, 5880, 5885, 5914, 6063,

6341, 6832, 7411, 7418, 7615, 7983, 8017, 8026, 8033, 8048, 8060, 8120, 8141, 8163, 8229, 8263, 8286, 8362, 8410, 8424.
Chaco, 7749.
Charcas, 7251.
Chechnia, 7632.
Chengde, 8541.
Chengdu, 7095.
Chersonesus, 2121.
Chiapas, 424.
Chile, 4273-4284, 5980, 6915, 7264, 7316, 7472, 7900, 8021, 8470.
China, 237, 925, 1005, 1021, 1057, 1063, 1102, 1158, 1236, 1293, 1455, 2287, 4088, 5596, 5607, 5611, 5680, 5811, 5836, 6030, 6073, 6272, 6626, 6755, 6943, 6974, 7095, 7382, 7476, 7504, 7598, 7601, 7630, 7802, 7842, 7928, 7975, 8130, 8145, 8150, 8156, 8260, 8269, 8285, 8299, 8329, 8340, 8363, 8472, 8501, 8504-8738.
Chios, 1965.
Chongqing, 8552.
Chulim-Yenisey (region), 1523.
Çıldır, 7441.
Cilicia, 2566.
Città del Vaticano, 7624, 7924, 7958, 8015, 8029, 8087, 8244, 8362.
Civitella Cesi, 2402.
Claros, 1962.
Cluny, 3288, 3577.
Clusium, 2396.
Cochabamba, 4216.
Colombia, 4285-4296, 7323, 7639, 7897, 8281, 8336, 8422.
Commagene, 1660.
Comores, 4297.
Compostela, 3975.
Congo, 4298-4301, 8353.
Congo-Kinshasa, 6370.
Constantinopolis, 2997, 3124, 3176, 3177, 3181, 3209, 3213, 3217, 3218, 3219, 3225, 3250, 3254.
Córdoba, 5144.
Corduba, 2892.
Corinthus, 2334, 2353.
Cornwall, 3974.
Corrèze, 4505.
Corse, 4472, 7795.
Cortona, 2407, 2408.
Çorum, 1830.
Costa Rica, 4302, 6012.
Côte d'Ivoire, 1524.
Crawford, 1600.
Cremna, 1649.

Crna Gora, Montenegro, 4925, 7870.
Croatia v. Hrvatska.
Cuba, 4310-4315, 5811, 6964, 7730, 7733, 7734, 7751, 7903, 7935, 8261, 8321, 8327, 8459, 8471.
Cuenca, 4390.
Cumae, 2376.
Cyclades (insulae), 3133.
Cyprus v. Kypros.
Cyrene, 1961, 2321.
Czechoslovak Republik v. Československo.

D

Dacia, 2627.
Dadong Cave, (China), 1527.
Dahomey, 4361, 4362.
Daibei, 8648.
Dakar, 5089.
Dalmacija, Dalmatia, 1521, 2719.
Dalmatia v. Dalmacija.
Danmark, 937, 957, 958, 990, 1227, 3389, 3571, 3936, 4363-4384, 6647, 6843, 6857, 6859, 6905, 7069, 7412, 7453, 7741, 7771, 8191, 8223, 8227, 8230, 8309.
Danube (region), 3342.
Dead sea, 1850.
Deir el-Bahri, 1716.
Deitschland, 425.
Delos, 1779, 1963, 1965, 1967, 2119, 2202, 2349, 2354.
Delphi, 2156, 2300, 2320.
Deutschland, 93, 148, 160, 178, 376, 392, 397, 399, 426, 436, 437, 440, 475, 478, 482, 485, 496, 511, 546, 557, 582, 584, 586, 618, 624, 628, 636, 766, 815, 832, 1022, 1064, 1068, 1131, 1451, 1470, 3368, 3374, 3390, 3394, 3492, 3526, 3594, 3635, 3640, 3689, 3894, 3905, 4089, 4121, 4511-4639, 4741, 5047, 5060, 5091, 5221, 5411, 5431, 5443, 5445, 5503, 5511, 5562, 5636, 5648, 5649, 5668, 5687, 5699, 5702, 5716, 5744, 5760, 5768, 5771, 5782, 5784, 5785, 5796, 5806, 5813, 5814, 5840, 5844, 5850, 5862, 5886, 5914, 5917, 5924, 5932, 5933, 5967, 5977, 5986, 5994, 6004, 6014, 6023, 6029, 6043, 6047, 6049, 6062, 6093, 6096, 6101, 6103, 6146, 6149, 6156, 6172, 6247, 6251, 6271, 6282, 6285, 6292, 6293, 6309, 6319, 6344, 6345, 6412, 6434, 6466, 6474,

6494, 6530, 6555, 6593, 6616, 6676, 6701, 6726, 6759, 6760, 6780, 6805, 6807, 6862, 6868, 6883, 6884, 6902, 6904, 6907, 6925, 6938, 6961, 6980, 7007, 7008, 7030, 7062, 7115, 7126, 7134, 7139, 7195, 7227, 7229, 7266, 7276, 7301, 7359, 7372, 7388, 7400, 7415, 7416, 7422, 7430, 7433, 7435, 7458, 7485, 7486, 7526, 7528, 7537, 7539, 7559, 7566, 7575, 7580, 7583, 7585, 7596, 7597, 7599, 7605, 7615, 7649, 7727, 7799, 7865, 7888, 7895, 7902, 7910, 7928, 7929, 7931, 7939, 7963, 7971, 7974, 7976, 7985, 7986, 7990, 7991, 8000, 8002, 8010, 8013, 8014, 8016, 8017, 8025, 8050, 8061, 8064, 8068, 8072, 8077, 8079, 8080, 8086, 8090, 8091, 8096, 8103, 8106, 8111, 8116, 8122, 8129, 8135, 8143, 8148, 8153, 8157, 8163, 8165, 8168, 8173, 8175, 8180, 8187, 8195, 8200, 8205, 8241, 8244, 8262, 8271, 8274, 8289, 8290, 8291, 8296, 8311, 8320, 8322, 8328, 8329, 8343, 8344, 8366, 8368, 8370, 8394, 8395, 8406, 8420, 8432, 8435, 8436, 8455, 8460, 8461, 8462, 8466, 8467, 8468. – v. Germaniae.
Dikili Tash (Macedonia, Graecia), 1522.
Dipoldsau, 4171.
Djibouti, 902.
Dominican Republic v. República Dominicana.
Dongbei, Manchuria, 7915, 8041, 8515, 8553, 8557, 8615.
Dordogne, 3276.
Dublin, 352, 3289, 4018, 7111.
Dubrovnik, 4253, 4309, 7259.
Dundee, 7081.

E

East Central Europe, 5903.
East Midlands, 1479.
Ebla, 1799, 1809.
Ecbatana, 1904.
Ecuador, 843, 904, 4389-4392, 8306.
Edessa, 2653.
Edinburgh, 6240.
Egyin Gol valley (Mongolia), 1604.
Egypt, 103, 170, 3445, 4393-4397, 5688, 6742, 7889, 8010, 8330, 8361, 8453, 8469. – v. Aegyptus.

Eibar, 327.
Eichstätt, 354.
El Cobre, 7734.
El Salvador, 4090, 4398-4399.
El-Amarna, 1800.
Elbląg, 860.
Eleutherna, 3235.
Elmina, 4642.
Elymaide, 1901.
Emar, 1808.
Emecik, 1644.
Emilia Romagna, 3324.
ᶜEn Besor, 1689.
ᶜEn Gedi, 1831, 1853.
England, 60, 62, 94, 188, 220, 246, 249, 303, 528, 535, 592, 629, 983, 1001, 1012, 1076, 1170, 1174, 1272, 1318, 1414, 3273, 3292, 3294, 3329, 3330, 3368, 3381, 3387, 3390, 3406, 3408, 3409, 3414, 3417, 3420, 3422, 3424, 3427, 3435, 3439, 3448, 3477, 3493, 3526, 3546, 3547, 3559, 3561, 3573, 3591, 3600, 3629, 3638, 3639, 3657, 3680, 3687, 3697, 3698, 3730, 3738, 3743, 3744, 3750, 3752, 3790, 3793, 3805, 3906, 3908, 3910, 3952, 3967, 3978, 3994, 4003, 4669, 4691, 4699, 4702, 4708, 4714, 5420, 5435, 5437, 5538, 5633, 5787, 5801, 5830, 5864, 5932, 5989, 6011, 6023, 6031, 6040, 6053, 6058, 6110, 6134, 6214, 6225, 6342, 6353, 6376, 6377, 6389, 6392, 6394, 6473, 6512, 6565, 6570, 6677, 6827, 6870, 6928, 6959, 6960, 6969, 6983, 7012, 7108, 7121, 7151, 7157, 7170, 7193, 7215, 7269, 7282, 7307, 7330, 7333, 7364, 7385, 7520, 7529, 7539, 7609, 7771, 7810, 7822, 7830.
Equatorial Guinea, 978.
Erfurt, 7440, 7456.
Eritrea, 439, 4402, 4404, 7715, 8268, 8396.
Erlangen, 356.
España, Spain, 255, 320, 355, 414, 493, 668, 906, 947, 1224, 1429, 1462, 1482, 1713, 3376, 3449, 3450, 3469, 3585, 3705, 3735, 3736, 3741, 3784, 3899, 3970, 3985, 4047, 5010, 5011, 5138-5203, 5428, 5467, 5524, 5548, 5586, 5703, 5759, 5877, 5933, 6287, 6312, 6355, 6503, 6668, 6812, 6853, 6950, 7021, 7093, 7165, 7405, 7428, 7471, 7546, 7626, 7760, 7767, 7772, 7776, 7778, 7780, 7794, 7809, 7815, 7816, 7821, 7831, 7836, 7840, 7881, 7897, 7898, 7900, 7903, 7960, 7972, 7984, 8061, 8076, 8084, 8088, 8109, 8138, 8179, 8316, 8384. – v. Hispania.
Esshu-tefnut, 1692.
Estland v. Ëstonija.
Estonie v. Ëstonija.
Ëstonija, Estland, Estonie, 897, 4400, 4412, 5075, 5465, 5647, 6968.
Ethiopia, 960, 4401-4404, 5410, 7686, 7705, 7715, 7911, 8033, 8164, 8177, 8190, 8268, 8396, 8456, 8793.
Etruria, 2389, 2398, 2402, 2404, 2405, 2406.
Euphrate, 1742.
Euphrates (flumen), 1746.
Eurasie, 792, 974, 3337, 7774, 8493.
Europe, 119, 189, 223, 248, 373, 389, 545, 569, 585, 609, 668, 756, 874, 919, 921, 942, 945, 971, 986, 987, 988, 997, 1000, 1024, 1026, 1037, 1066, 1101, 1112, 1126, 1137, 1138, 1139, 1149, 1158, 1161, 1203, 1204, 1214, 1240, 1294, 1305, 1383, 1436, 1450, 2554, 3080, 3232, 3331, 3359, 3368, 3401, 3405, 3481, 3526, 3553, 3587, 3592, 3642, 3655, 3660, 3662, 3703, 3760, 3764, 3792, 3819, 3898, 3901, 3912, 3928, 3941, 4002, 4012, 4036, 4039, 4040, 4053, 4057, 4058, 4064, 4072, 4084, 4091, 4105, 4108, 4111, 4254, 4670, 5119, 5405, 5414, 5416, 5417, 5421, 5425, 5439, 5449, 5452, 5453, 5491, 5558, 5559, 5598, 5703, 5732, 5777, 5778, 5782, 5793, 5804, 5867, 5881, 5883, 5887, 6059, 6111, 6126, 6137, 6274, 6358, 6511, 6531, 6545, 6574, 6670, 6721, 6740, 6770, 6790, 6795, 6874, 6896, 6940, 6950, 7058, 7134, 7296, 7315, 7355, 7457, 7620, 7761, 7764, 7767, 7770, 7801, 7826, 7942, 7976, 7977, 8004, 8045, 8165, 8237, 8250, 8271, 8309, 8315, 8318, 8338, 8392, 8397, 8398, 8431, 8441, 8461, 8462. – E. centrale, 95, 781, 4096, 4097, 4112, 4114, 4119, 5454, 5577, 6810, 6925, 7080, 7239, 8250. – E. centrale-orientale, 4066, 4073, 4083, 8307. – E. du sud-est, 4118, 6810, 6925, 8250. – E. méridionale, 4051. –
E. occidentale, 515, 6825, 6888, 6992, 7003, 8233, 8250, 5884. – E. orientale, 268, 706, 781, 3334, 4045, 4060, 4112, 4114, 4117, 6036, 6634, 6925, 7080, 8093, 8445.

F

Falkland Islands, 7922, 7936, 8407.
Ferrara, 6597.
Fiji, 7396.
Filipinas, Philippine, 4981, 7586, 7640, 7919, 8066.
Finland v. Suomi.
Firenze, 365, 417, 1261, 3316, 3400, 3412, 3421, 3579, 3613, 3667, 3804, 3818, 3840, 5755, 5822, 5908, 5959, 6532, 6592, 6601.
Flanders, 3496, 3515, 3781, 5477, 5483, 6985.
Florida, 566, 8059.
Folsom, 1526.
Forlì, 6990.
France, 49, 67, 70, 71, 157, 180, 183, 205, 242, 281, 302, 311, 316, 326, 328, 390, 397, 403, 410, 411, 457, 468, 479, 500, 524, 573, 575, 617, 619, 631, 640, 657, 895, 1003, 1068, 1275, 1354, 1445, 3311, 3332, 3364, 3368, 3369, 3371, 3385, 3503, 3526, 3557, 3583, 3616, 3639, 3650, 3691, 3711, 3715, 3725, 3796, 3862, 3904, 3985, 3986, 4080, 4098, 4102, 4135, 4415-4509, 5145, 5192, 5396, 5400, 5430, 5448, 5501, 5516, 5529, 5537, 5541, 5555, 5573, 5575, 5587, 5624, 5654, 5665, 5674, 5723, 5754, 5762, 5764, 5766, 5770, 5797, 5806, 5813, 5818, 5819, 5820, 5864, 5874, 5886, 5888, 5891, 5966, 5973, 5983, 6015, 6016, 6033, 6050, 6098, 6125, 6189, 6224, 6227, 6249, 6259, 6271, 6313, 6327, 6353, 6355, 6357, 6361, 6371, 6374, 6380, 6386, 6388, 6390, 6394, 6409, 6422, 6425, 6442, 6451, 6499, 6509, 6515, 6518, 6529, 6552, 6569, 6596, 6604, 6610, 6640, 6642, 6659, 6673, 6683, 6684, 6694, 6709, 6711, 6719, 6727, 6731, 6744, 6751, 6801, 6852, 6873, 6881, 6884, 6961, 6962, 6987, 7018, 7027, 7047, 7051, 7161, 7177, 7240, 7256, 7261, 7263, 7273, 7291, 7304, 7347, 7354, 7374, 7401,

7404, 7406, 7407, 7449, 7450, 7465, 7467, 7487, 7492, 7509, 7512, 7518, 7569, 7579, 7596, 7597, 7604, 7617, 7623, 7625, 7651, 7662, 7679, 7690, 7701, 7702, 7708, 7710, 7711, 7726, 7760, 7762, 7765, 7766, 7785, 7811, 7818, 7821, 7829, 7833, 7834, 7836, 7840, 7850, 7852, 7853, 7858, 7863, 7868, 7871, 7880, 7884, 7894, 7918, 7959, 7968, 7976, 7979, 7984, 7986, 7988, 7990, 7998, 8001, 8003, 8028, 8049, 8056, 8063, 8068, 8076, 8084, 8085, 8088, 8096, 8111, 8113, 8136, 8147, 8156, 8157, 8163, 8164, 8170, 8177, 8179, 8193, 8202, 8220, 8224, 8226, 8232, 8235, 8255, 8257, 8288, 8311, 8322, 8355, 8366, 8369, 8384, 8389, 8405, 8446, 8451, 8720.
Franconie, 3589.
Frankfurt am Main, 5825, 6146, 7439.
Freiburg (Schweiz), 3998.
Fujian, 8519.

G

Gabon, 4510.
Galizien, 6025, 7967.
Gallia Cisalpina, 2926.
Gallia Narbonensis, 2434, 2909.
Gallia, 2434, 2529, 2681, 2716, 2946, 3058, 3099, 3104.
Gansu, 8547.
Garigliano (fiume), 2898.
Gatemala, 4090.
Gavá (Barcelona), 1505.
Gdańsk, 860, 4985, 6895, 8284.
Gebel Manzal el-Seyl, 1694.
Genève, 6841.
Genova, 3280, 3281, 7853.
Gent, 6107.
Georgia, 307, 7859.
Germaniae, 2434, 2635.
Ghana, 4640-4644, 7717, 8373, 8798.
Gibraltar, 7396.
Gilan, 4763.
Girnavaz-Höyük, 1753.
Girona, 3283, 3284, 4027, 5149.
Giza, 1711.
Glasgow, 5954.
Göbekli Tepe, 1667.
Gold Coast, 7688, 7724.
Gollion, 5500.
Gordion (Phrygia), 1612.
Gortyna, 1952, 1955.
Göteborg, 5808, 6296.
Gradus, 2872.

Graecia, 168, 199, 879, 1965, 2080, 2100, 2109, 2112, 2122, 2128, 2136, 2147, 2264, 2318, 2330, 2362, 2859, 3062.
Granada, 6567, 7522.
Grandford (Cambridgeshire), 2919.
Great Britain, 114, 120, 152, 234, 260, 348, 421, 488, 558, 572, 595, 905, 909, 924, 950, 967, 996, 1222, 1251, 1463, 3413, 3610, 4007, 4088, 4110, 4665-4715, 5205, 5399, 5426, 5438, 5440, 5458, 5563, 5631, 5656, 5661, 5672, 5713, 5724, 5741, 5758, 5775, 5810, 5813, 5842, 5852, 5857, 5886, 6050, 6071, 6090, 6092, 6094, 6101, 6141, 6228, 6242, 6250, 6271, 6277, 6282, 6349, 6405, 6427, 6434, 6481, 6485, 6539, 6650, 6654, 6671, 6715, 6765, 6783, 6797, 6819, 6846, 6850, 6864, 6879, 6893, 6897, 6915, 6918, 6931, 6943, 6946, 6975, 7039, 7049, 7094, 7192, 7220, 7295, 7339, 7396, 7402, 7427, 7458, 7463, 7494, 7578, 7579, 7580, 7593, 7608, 7609, 7631, 7635, 7651, 7652, 7654, 7658, 7663, 7667, 7677, 7678, 7679, 7693, 7705, 7724, 7725, 7726, 7775, 7791, 7808, 7827, 7842, 7858, 7869, 7884, 7885, 7887, 7888, 7889, 7891, 7892, 7904, 7908, 7921, 7922, 7926, 7927, 7930, 7938, 7939, 7940, 7945, 7964, 7983, 7991, 8005, 8009, 8035, 8039, 8047, 8057, 8064, 8077, 8090, 8106, 8116, 8118, 8123, 8140, 8147, 8152, 8153, 8158, 8164, 8167, 8173, 8175, 8196, 8199, 8204, 8227, 8234, 8239, 8242, 8246, 8247, 8248, 8249, 8250, 8251, 8252, 8253, 8287, 8294, 8297, 8312, 8316, 8318, 8322, 8347, 8351, 8353, 8359, 8363, 8380, 8391, 8399, 8405, 8407, 8408, 8426, 8431, 8436, 8453, 8457, 8462, 8469, 8612, 8666.
Great Hale, 300.
Greece v. Hellas.
Grenoble, 5225, 5582.
Grossraming, 4171.
Grotta del Pino (Sassano, Salerno), 1533.
Grotta Patrizi (Sasso di Furbara, Cerveteri, Roma), 1508, 1515.
Gruzija, 6705.
Guadelupe, 7657.
Guangdong, 8519, 8567, 8683.

Guangxi, 8720.
Guatemala, 4721, 4722, 7944, 8450.
Guimaraes, 5012.
Guinea, 4723, 7706.
Guipúzcoa, 3607, 5138.
Gürcütepe, 1667.
Guyana, 4724, 6306, 8326.

H

Hacienda Zuleta (Ecuador), 1593.
Hainandao, 8733.
Haiti, 4725-4730, 6802, 7657.
Hamburg, 4070, 7800.
Hammat al-Qa, 1890.
Hannover, 4599, 6075, 6927, 7474.
Hanoi, 7090.
Har Haruvim (Israel), 1873.
Harrow on the Hill, 6038.
Ha-Shophtim Street (Qiryat 'Ata), 1595.
Hattuša, 1820.
Hawaii, 288.
Hawara, 1686.
Hazor, 1849.
Hedjaz, 6891.
Hedong, 8561.
Heidelberg, 2929, 5949.
Helgoland, 7932.
Hellas, Greece, 243, 656, 949, 1264, 1282, 1840, 1911-2365, 4716-4720, 6254, 6923, 7064, 7173, 7433, 7593, 8258, 8356, 8418. – v. Graecia.
Helsinki, 1152.
Henan, 8716.
Heraclea apud Latmum, 1659.
Herculaneum, 2422, 2889, 2922, 2935.
Hermione, 2070.
Hessen, 4604.
Hessen-Nassau, 4604.
Hesses, 4518.
Hiberia, 1538.
Hierapolis (Türkiye), 1647.
Hierusalem, 3089.
Hispania Superior, 2535.
Hispania, 2699.
Holon (Israel), 1510.
Honduras, 4731-4733, 5681, 8245.
Hong Kong v. Xianggang.
Horn of Africa, 8349.
Horvat Qidmit, 1837.
Hrvatska, Croatia, 900, 985, 1292, 3430, 3909, 4303-4309, 7462, 7988, 8160, 8161, 8231.
Huashan, 8576.
Hubei, 8701.
Huizhou, 8504, 8507, 8618, 8731.
Hungary v. Magyarország.
Hüseyindede Tepesi, 1824.

Hyderabad (Princely State), 4745, 7674.

I

Ibérique (peninsula), 979, 1171, 3867, 4017.
Iceland, 3298, 3307, 3475, 3731, 4000, 4742, 4743, 7524.
Igel (Germania), 2913.
India, 8, 273, 602, 1211, 1277, 1619, 1889, 1985, 2501, 4744-4754, 5418, 5447, 5590, 5601, 5602, 5605, 5645, 5679, 5691, 5696, 5728, 6076, 6127, 6212, 6276, 6359, 7616, 7617, 7663, 7667, 7669, 7672, 7784, 7887, 7982, 8122, 8126, 8260, 8308, 8352, 8382, 8430, 8433, 8437, 8500, 8501.
Indian (océan), 6922, 8479.
Indochine, 7662, 7664, 7683, 8304, 8347.
Indonesia, 4755-4758, 8178, 8464, 8590.
Ionia, 1892.
Ionian islands, 4718.
Ipsos, 2058.
Iran, 1781, 1887-1910, 4071, 4759-4767, 5779, 6742, 6948, 7208, 7576, 7590, 7759, 8247, 8438.
Iraq, 4768, 4769, 8451.
Ireland, 210, 494, 522, 905, 907, 909, 938, 1441, 3504, 3900, 4700, 4770-4786, 5784, 5835, 5897, 6005, 6024, 6032, 6434, 6436, 6681, 7104, 7205, 7417, 7613, 7794, 7815, 7953, 7985, 8035, 8075, 8180, 8367, 8380.
Irtysh (region), 1572.
Isampur (Karnataka, India), 1580.
Israel, 494, 506, 1840, 4787-4792, 5721, 8256, 8258, 8265, 8275, 8294, 8314, 8330, 8335, 8343, 8425, 8440, 8454, 8460, 8468.
İstanbul, 1285.
Istria, 4826.
Italia, 7, 67, 181, 221, 293, 308, 332, 351, 393, 397, 497, 651, 664, 686, 845, 1011, 1079, 1093, 1166, 1167, 1192, 1913, 2372, 2375, 2380, 2417, 2424, 2430, 2864, 2923, 2935, 3051, 3188, 3198, 3241, 3303, 3339, 3341, 3351, 3379, 3397, 3398, 3490, 3495, 3512, 3518, 3534, 3542, 3556, 3576, 3608, 3631, 3647, 3774, 3930, 3970, 3995, 3997, 4032, 4047, 4121, 4121, 4793-4865, 5118, 5412, 5468, 5492, 5505, 5506, 5509, 5510, 5512, 5516, 5518, 5521, 5524, 5528, 5530, 5536, 5539, 5540, 5546, 5551, 5552, 5560, 5563, 5572, 5575, 5620, 5625, 5626, 5630, 5731, 5760, 5800, 5817, 5834, 5853, 5861, 5875, 5895, 5953, 5971, 6042, 6065, 6106, 6140, 6330, 6339, 6396, 6411, 6423, 6452, 6478, 6506, 6551, 6568, 6578, 6583, 6590, 6594, 6628, 6662, 6702, 6743, 6775, 6815, 6865, 6913, 6991, 6994, 6996, 6998, 7032, 7033, 7054, 7057, 7085, 7118, 7123, 7129, 7153, 7283, 7388, 7494, 7512, 7538, 7551, 7622, 7705, 7715, 7720, 7759, 7761, 7773, 7787, 7793, 7871, 7872, 7873, 7890, 7895, 7905, 7910, 7958, 7977, 7978, 7980, 7992, 8018, 8019, 8029, 8030, 8079, 8083, 8128, 8164, 8169, 8174, 8177, 8196, 8207, 8211, 8237, 8254, 8280, 8305, 8427.
Itanos (Krētē), 2337.
Ivry-sur-Seine, 6001.
İzmir, 840, 7040.

J

Jabbul Plain (Syria), 1795.
Jamaica, 4730, 7657, 8342.
Jangsu, 8672.
Jangxi, 7314.
Japan, 22, 282, 440, 483, 498, 1103, 1289, 1354, 1403, 1461, 3703, 4088, 4866-4884, 5805, 5884, 5931, 6603, 6700, 6754, 6777, 6784, 6811, 6856, 6882, 6946, 7012, 7571, 7588, 7592, 7599, 7601, 7609, 7633, 7883, 7892, 7912, 7952, 7969, 7975, 8057, 8145, 8150, 8170, 8178, 8197, 8354, 8434, 8457, 8518, 8532, 8539, 8626, 8664, 8702, 8709, 8733, 8739-8781.
Java, 7668, 8178, 8178.
Jedwabne, 8071.
Jena, 5958.
Jerusalem, 1575, 1845, 1851, 3388, 3402, 7660. – v. Hierusalem.
Jiangnan, 8566, 8584.
Jiangsu, 8662, 8665.
Johannesburg, 7181.
Jordan, 1528, 1752, 4885, 4886, 7993, 8452.
Jugoslavija, Yugoslavia, 5391-5393, 7534, 7968, 7988, 7998, 8172, 8263, 8293, 8315, 8341, 8388, 8409, 8413, 8446, 8449.
Jutland, 1489.

K

Kabardino-Balkaria, 891.
Kainuu, 4266.
Kalamakia (Aréopolis, Peloponnesus), 2329.
Kaliningrad, 7063.
Kalmykia, 1559.
Kano, 1492.
Kansai, 6946.
Karatepe, 1806.
Karatepe-Aslantaş, 1823.
Karelia, 4266, 5059.
Kargaly (Orenburg, Rossija), 1589.
Kargil, 8382.
Kärnten, 4195, 4196.
Kashmir, 213, 8346, 8382, 8383, 8437.
Kassel, 7067.
Kastritsa (Graecia), 1948.
Kavkaz, 1480, 1549, 1742, 7590, 7879, 7996.
Kayseri, 5244.
Kazakhstan, 7070, 7671, 8022.
Kenya, 984, 4887-4890, 4905, 7691, 7698.
Kerkenes Dağ, 1672.
Kiatschou, 7928.
Kiran, 8482.
Kirghizstan, 832.
København, 4949, 7072, 7135.
Koblenz, 7100.
Köln, 3786.
Kolyma, 8333.
Kongo, 7716.
Konigsberg, 66.
Korea, 550, 943, 4891, 5710, 7946, 8258, 8301, 8360, 8458, 8465, 8690, 8782-8785.
Kosovo, 4130, 5419.
Kozluca (district), 1133.
Kraków, 7071.
Krētē, 693, 3155, 3203, 3270, 3344, 7877, 8210.
Krym, 1454, 2121.
Kufan Kanawa, 1492.
Kursachsen, 6109.
Kuşaklı, 1816.
Kuwait, 4892, 8167.
Kypros, Cyprus, 1531, 1570, 1954, 2340, 2343, 3274, 4310-4319, 7396, 7901, 7945, 8258, 8359, 8426, 8443, 8463.

L

La Ferté-sur-Grosne (abbaye), 3272.
La Garma (Cantabria), 1566.
La Marmotta, 1557.
La Plata, 7251.
La Rioja, 4147.
La Virgen, 7137.

Ladenburg am Neckar, 2435.
Ladenburg, 2929.
Lagos, 7925.
Lancashire, 6869, 6946.
Laos, 4893, 8327.
L'Aquila, 5941.
Las Herrerías (Cuevas de Almanzora, Almería), 1568.
Las Médulas (León), 1610.
Lascaux, 1519.
Latium, 2418.
Latvia, Lettland, Lettonie, 173, 897, 4894, 5075, 8081.
Lauenburg, 7068.
Lebanon, 1884, 4895-4898, 5715, 7679, 7680, 7987, 8361.
Leiden, 5904.
Lemgo, 7044.
Leningrad, 5045, 5058.
León, 3266, 3549, 4005, 5142.
Leonberg, 7470.
Lerna (Argos), 1613.
Lettland v. Latvia.
Lettonie v. Latvia.
Leukopetra, 1966.
Liberia, 4899.
Libya, 917, 7705, 8387.
Lietuva, Lithuania, Lituanen, 173, 512, 897, 1421, 3437, 4062, 4900, 5075, 7806, 8448.
Liguria, 717, 3230.
Lille, 5483, 5913.
Lima, 4980, 7200.
Limes Farm (Landbeach, Cambridgeshire), 1592.
Limousin, 5486.
Lincolnshire, 300.
Lingnan, 8517, 8693.
Linz, 7066.
Lippe, 7326.
Lithuania v. Lietuva.
Lituanen v. Lietuva.
Liverpool, 7060.
Livland, 5670, 6968.
Ljubljana, 5115, 7431.
Lombardia, 78, 7507.
London, 5956, 6054, 6246, 6542, 6653, 6675, 6951, 7004, 7077, 7084, 7136, 7254, 7270.
Lopodunum, 2435.
Los Castellones de Céal (Hinojares, Jaén), 1602.
Löwenstein-Wertheim-Freudenberg, 4631.
Lübeck, 3992.
Lubelszczyzna, 4987.
Ludanice, 5095.
Lugdunum, 2550.
Luoyang, 8551, 8603.
Lusitania, 873, 2629, 2687.
Luxembourg, 2980.

Lwow, 5965, 8155.
Lyon, 5582.

M

Macedonia v. Makedonija.
Madagascar, 516, 4903, 4904, 7690, 7718.
Madrid, 5154, 7041, 7055.
Magdeburg, 3511, 7171.
Maghreb, 427, 3474, 7280.
Magna Graecia, 2149.
Magyarország, Hungary, 966, 1563, 3601, 3666, 4182, 4734-4741, 5096, 5108, 5658, 6781, 7479, 7611, 7886, 8044, 8049, 8094, 8099, 8132, 8246, 8282, 8377.
Mainz, 2897, 5480, 5949.
Majorca, 8384.
Makedonija, Macedonia, 1561, 1903, 1966, 2062, 2095, 2108, 2322, 3185, 4255, 4901, 4902, 7923, 7934, 8064, 8449.
Malawi, 4905, 4906, 8023.
Malaya, 8457.
Malaysia, 6768.
Mali, 4907.
Malia, 1944, 1946, 2357.
Mallorca, 3621.
Małopolska Wschodnia, 169, 8154.
Malta, 918, 8206, 8209.
Malvinas, 222.
Manahat, 1863.
Manchester, 1702, 5944.
Manchuria v. Dongbei.
Mangaia (Cook Islands), 1537.
Manhattan, 5839.
Manila, 7090.
Mannheim, 2929.
Mantova, 6613.
Marburg, 5957.
Marča (diocese), 5520.
Marengo, 3608.
Mari, 1762, 1779, 1784.
Maribor, 7431.
Marki Alonia, 1570.
Maroc, Morocco, 605, 4926-4928, 5143, 7626, 7694, 7701, 7984.
Marroquíes Bajos (Jaén), 1517.
Marseille, 7468.
Maşat Höyük, 1816.
Matabeleland (Zimbabwe), 6290.
Mauritania, 5723.
Mauritius, 7120.
Mauthausen, 8117.
Mecklenburg, 4598.
Medinet Habu, 1750.
Méditerranée (mer), 159, 271, 2146, 3340, 3458, 3460, 3530, 3959, 4065, 7236, 7606, 7785, 7979. – v. Mediterraneum (mare).

Méditerranée (pays), 1165.
Mediterraneum (mare), 1482, 1663, 1821, 1876, 1913, 1923.
Megara, 1921.
Megarid, 199.
Megiddo, 1826, 1864.
Mekong Delta, 5390.
Melanesia, 872.
Memphis, 1733.
Merina, 516.
Mesoamerica, 995.
Mesopotamia, 1582, 1638, 1734-1805, 3468.
Messene, 1964.
Metabeleland, 8788.
México, 298, 424, 576, 1362, 4908-4923, 5599, 5756, 6916, 6954, 7297, 7308, 7691, 7750, 7906, 7935, 7943, 7948, 7949, 8037, 8158, 8179, 8384, 8386.
Michaelbeuern, 65.
Middelbourg, 6947.
Milano, 35, 3111, 3260, 6713, 7175, 7232.
Mindanao, 4981.
Minturnae, 2898.
Mirón (Ramales de la Victoria, Cantabria), 1546.
Misenum, 2420.
Mitanni, 1809, 1821.
Moesia superior, 2686.
Moesia, 2543.
Mogul Empire, 7784, 7808.
Moldavia v. Moldova.
Moldova, Moldavia, 915, 1146, 4924.
Monferrato, 3348.
Mongolia, 856, 5080, 7824, 7915.
Monopoli, 4812.
Monte Penide Redondela (Pontevedra), 1554.
Monte Sirai, 2368.
Montecassino, 3938.
Montenegro v. Crna Gora.
Monteprandone, 55.
Montpellier, 5582.
Mont-Saint-Michel (baie du), 274.
Montserrat, 8384.
Móra d'Ebre (Tarragona), 1842.
Moravia, 4338, 4339, 4352, 4354, 7141.
Morella, 4026.
Morocco v. Maroc.
Mortola Superiore (Ventimiglia, Imperia), 1512.
Moskva, 377, 379, 3418, 3486, 5044, 5068, 5961, 6748.
Mosul, 4768.
Moulins, 6015.
Mozambique, 473, 4929, 4930.
Mpondo, 7723.

München, 4576, 5910.
Münster, 5479, 5643, 5981, 7171, 7292.
Murcia, 3586, 5162, 5193.
Murlo (Siena), 7166.
Musoleo (Corsica), 2912.
Myanmar, 8497.

N

N'Gaous, 1869.
Nagasaki, 8588.
Namibia, 4931-4932, 7657, 7700, 7712.
Nanjing, 543, 633, 8119, 8540, 8594, 8707, 8712.
Napoli, 402, 2418, 3613, 5923, 6384, 6393, 6504, 7437, 7503, 7513, 7560.
Naukratis, 2139.
Navarra, 4431, 5159.
Neapolis (Nabeul, Tunisie), 1725.
Nederland, Netherlands, Pays Bas, 1115, 1354, 1379, 2980, 3271, 3442, 4089, 4933-4938, 5436, 5448, 5540, 5665, 5677, 6763, 6767, 6817, 6863, 7287, 7588, 7673, 7681, 7716, 7796, 7797, 7831, 7863, 7964, 8069, 8293, 8464.
Nemrud Daği, 1660.
Nereditsa, 3814.
Netherlands v. Nederland.
New Mexico, 566, 5608.
New York, 350, 5911, 6582, 7105, 7335.
New Zealand, 933, 992, 7627, 7902, 8270, 8825.
Newclare, 5130.
Nicaragua, 4090, 4939, 4940, 7744, 7906, 7955.
Niger, 1492, 4211, 4941, 7706.
Nigeria, 993, 4942-4944.
Nilus (flumen), 1687, 1731.
Ningxia, 8726.
Ninive, 1813.
Nocera Superiore, 2942.
Noedic countries, 6844.
Nokia, 7212.
Nordic Countries, 1162, 5475.
Norfolk, 6988.
Norge, Noeway, 431, 914, 1574, 1611, 3315, 3391, 3392, 3396, 3475, 3576, 3862, 3990, 4373, 4945-4952, 6732, 6786, 6889, 7185, 7278, 7378, 7771, 8069, 8135, 8400.
Normandie, 238, 3286, 3330, 3350.
North Carolina, 7276, 7331.
Norway v. Norge.
Novgorod, 3279, 3560, 3814, 4024.
Novi Sad, 63.
Nubia, 1687.
Numancia, 1597.
Nürnberg, 391.
Nusaybin/Mardin, 1753.
Nya Sverige, 7114.

O

Oaxaca (Mexico), 1596, 7731.
Oberschwaben, 6566.
Océanie, 7756-7758, 8270, 8821-8826.
Oder-Neisse (line), 8324.
Odessa, 5070.
Oenoanda, 1664.
Oeta (Mons), 2507.
Okinawa, 8434.
Olbia (Sardinia), 1843.
Olympia, 1960, 2398.
Oman, 4749, 4953-4962, 7957.
Ontario, 7196.
Oppius (Mons), 2938.
Orkney, 1558.
Oronsay (Hebrides insulae), 1529.
Osnabrück, 7096.
Ost Preußen, 5829.
Österreich, Austria, 275, 434, 478, 1222, 1419, 1437, 1577, 4092, 4168-4201, 4534, 4734, 4938, 5112, 5121, 5485, 5588, 5795, 5866, 7210, 7246, 7469, 7479, 7536, 7550, 7628, 7772, 7860, 7861, 7870, 7874, 7879, 7886, 7967, 8199, 8402, 8406.
Ostia, 2899.
Oświęcim, Auschwitz, 4337, 4627, 4628, 4637.
Oxbridge, 6984.
Oxford, 59, 1685, 1692, 5929, 5940.

P

Pacifique (océan), 277, 1191, 8825.
Padova, 69.
Paestum, 2119.
País Vasco, 1187, 1449, 5155, 5157.
Pakistan, 994, 8269, 8273, 8382, 8437.
Palaestina, 1839.
Palaikastro, 1576.
Palatinat, 7844.
Palatinus (Mons), 2885, 2934, 3191.
Palencia, 5187.
Palestine, 494, 4025, 4792, 4963-4970, 7635, 7676, 7677, 7874, 7992, 7993, 8070, 8297, 8300, 8391. – v. Palaestina.
Palmyra, 2653.
Palomar de Pintado (Villafranca de los Caballeros, Toledo), 1543.
Panama, 4971.
Papua New Guinea, 7756.
Paraguay, 4972, 7498, 7587, 7750.
Paris, 74, 302, 378, 466, 1191, 3638, 3886, 4435, 4436, 5582, 5740, 5765, 5878, 6015, 6279, 6572, 6582, 6952, 7013, 7042, 7133, 7154, 7235, 7289.
Parnassus (mons), 2067.
Pau de Santa Maria, 1833.
Pays Bas v. Nederland.
Pearl Harbor, 8184.
Peloponnesus, 2145, 3211.
Perachora, 2296, 2310.
Pergamum, 1645, 2141, 2920.
Perm, 5737.
Pernabuco, 4237.
Persian Gulf, 7576, 7663.
Peru, 969, 4973-4980, 6861, 7397, 7732, 7739, 7742, 8021, 8306, 8815, 8819.
Pesaro, 3808.
Pescia, 4795.
Petanjci, 5664.
Petrograd, 5072.
Phaistos, 1947.
Philadelphia, 7060.
Philippine v. Filipinas.
Phoenicia, 1896.
Phokis, 2067.
Piave, 8019.
Picareiro Cave (Portugal), 1514.
Picenum, 2366.
Piemonte, 7, 4837.
Pisa, 331, 5918.
Pisaurum, 2373.
Pisidia, 1649.
Pistiros (Bălgarija), 2326.
Pistoia, 3423.
Plataea, 2039.
Plzeň, 7310.
Poggio Colla, 2411.
Pointe-à-Pitre (Guadeloupe), 7087.
Poitiers, 7204.
Poland v. Polska.
Polesine, 2369.
Polinesia, 5604.
Polska, Poland, 131, 173, 480, 1068, 1312, 2952, 3356, 3404, 3416, 3428, 4061, 4062, 4092, 4982-5004, 5429, 5670, 5914, 6010, 6507, 6900, 6953, 6968, 7290, 7605, 7620, 7628, 7765, 7769, 7814, 7820, 7865, 7914, 7917, 7952, 7967, 7973, 8040, 8060, 8080, 8120, 8139, 8148, 8274, 8286, 8291, 8379, 8432, 8447, 8448.
Pommern, 4592.

Pomorze Zachodnie, 4093.
Pompeii, 2474, 2847, 2877, 2922, 2937, 2939.
Port-Royale, 6421.
Portugal, 128, 255, 930, 1448, 1503, 1552, 3438, 4047, 4238, 4239, 5005-5018, 5609, 5834, 5915, 5939, 6157, 6503, 7591, 7598, 7643, 7644, 7661, 7682, 7689, 7725, 7727, 7736, 7740, 7776, 7782, 7802, 7807, 7840, 7869, 7891, 8028, 8168, 8187, 8243, 8275.
Potosí, 357.
Pouerua (Northland, New Zealand), 1518.
Poznań, 5718.
Pozuelo de Aravaca, 7124.
Praha, 4323, 6492, 7071.
pre-Urals (region), 3538.
Preußen, 347, 3951, 4061, 4520, 4585, 4881, 5957, 6255, 6895, 7294, 7392, 7446, 7798, 7851.
Příbram, 6908.
Priene, 1968.
Puebla, 5140.
Puerto Rico, 5019, 5020.
Punjab, 866.
Pyrénées, 3536, 7636, 7812.
Pyrgi, 2407.

Q

Qara Qūzāq, 1835.
Qasr Ibrim, 1733.
Qatar, 5021.
Qimen, 8618.
Qiryat ʿAta, 1609.
Québec, 229, 5600.

R

Ragusa, 3605.
Ramat Hanadiv, 1853.
Rauma, 7265.
Ravenna, 7, 149, 3996.
Red Barns, 1539.
República Dominicana, Dominican Republic, 4385-4388.
Reunion, 5022.
Rhein, 6520.
Rhodesia, 7706, 7713.
Rhodus, 2317, 3203, 3948.
Rhône-Alpes, 8216.
Richmond, 5355.
Rocourt, 7835.
Rodez, 7477.
Roha, 8794.
Roma, 168, 176, 252, 292, 467, 476, 524, 646, 2043, 2094, 2318, 2366-2967, 3052, 3054, 3089, 3191, 3581, 3584, 3780, 3816, 3820, 4004, 4009, 4805, 4852, 4856, 5951, 6565, 6572, 7041, 7219, 7898, 7956, 8132.
România, 559, 580, 1346, 4033, 4924, 5023-5036, 5962, 6900, 7931, 7956, 7978, 8047, 8072, 8124, 8127, 8131, 8132, 8143, 8195, 8424.
Rossija, Russia, 4, 30, 32, 64, 89, 377, 379, 412, 435, 456, 469, 504, 510, 530, 536, 556, 562, 564, 567, 654, 713, 832, 842, 875, 882, 891, 893, 955, 980, 991, 1159, 1188, 1246, 1468, 1480, 1494, 3134, 3241, 3285, 3305, 3307, 3312, 3314, 3317, 3333, 3334, 3342, 3355, 3392, 3403, 3486, 3523, 3663, 3773, 3776, 3807, 4080, 4102, 4353, 4692, 4763, 4924, 5037-5088, 5212, 5465, 5616, 5618, 5667, 5712, 5720, 5722, 5737, 5807, 5817, 5831, 5848, 5850, 5851, 5862, 5865, 5903, 5961, 5972, 6010, 6026, 6045, 6048, 6052, 6057, 6070, 6109, 6267, 6292, 6329, 6372, 6453, 6455, 6465, 6680, 6758, 6787, 6935, 6957, 6958, 7008, 7070, 7083, 7180, 7182, 7201, 7242, 7249, 7253, 7277, 7298, 7302, 7363, 7543, 7562, 7592, 7597, 7619, 7625, 7631, 7632, 7633, 7638, 7641, 7671, 7747, 7768, 7770, 7774, 7824, 7838, 7848, 7849, 7852, 7854, 7868, 7876, 7908, 7909, 7912, 7914, 7916, 7931, 7941, 7952, 7956, 7967, 8005, 8011, 8022, 8052, 8067, 8078, 8092, 8098, 8121, 8125, 8130, 8132, 8183, 8262, 8265, 8266, 8341, 8354, 8360, 8364, 8439, 8692. – v. SSSR.
Rostock, 5906, 5945.
Ruanda, 7719.
Ruś Czerwona, 4014.
Russia v. Rossija.
Rwanda, 616.

S

Saar, 6847.
Saarbrücken, 7048.
Sachsen, 88, 1256, 3511, 4596, 4604, 5776, 5807, 6782, 6887, 7167, 7839.
Saguntum, 2414.
Sahara, 7687.
Saint Petersburg, 1764.
Saint-Gall (abbaye), 3943.
Saint-Jean-de-Garguier, 2428.
Saint-Maur, 145.
Saint-Vanne, 145.
Saint-Yved de Braine (abbaye), 3262.
Sais, 1733.
Saka-Wusun (Kazakhstan), 1603.
Salamanca, 5955.
Salamis, 2365.
Salerno (provincia), 85.
Salomó ha-Leví, 1833.
Salonique v. Thessaloniki.
Salto (valle del), 3603.
Salzburg, 4198.
Samerina, 1787.
Samokov, 359.
Samsun, 7064.
Samus, 1965.
San Andrés y Providencia, 7639.
San Benedetto Polirone, 3540.
San Nicolo in Carpi (Convento), 77.
San Salvador, 4398.
Santa Maria di Tricesimo, 3982.
Santa Maria in Trastevere (Roma), 3825.
Santa Severa (Roma), 2891.
Santiago de Guatemala, 7755.
Santo Domingo, 6802, 7743.
São Tomé, 7716.
Saqqara, 1690.
Sardegna, 162, 3787.
Sardis, 1657.
Savo, 4266.
Savoie, 7766.
Savoy, 7810.
Scandinavie, 1120, 3187, 3430, 3478, 3481, 3482, 4098, 4619, 8250.
Schaffhauser, 8058.
Schlesien, 172, 247, 4063, 4628, 4989, 7281, 7805, 8159.
Schleswig-Holstein, 76, 4543.
Schweiz, Suisse, Svizzera, Switzerland, 284, 637, 844, 970, 5224-5231, 5460, 5659, 5672, 6285, 6834, 6892, 7907, 8058, 8097, 8162, 8171, 8186, 8323, 8334, 8447.
Schweizer Alpen, 7213.
Scotland, 981, 1170, 1530, 1608, 3326, 3365, 3757, 4650, 4662, 4673, 5457, 5637, 5978, 6084, 6325, 6394, 6722, 7015, 7073, 7222, 7224, 7342, 7375, 7823.
Sedada, 8808.
Sedano (Burgos), 1513.
Seine, 7042.
Seine-et-Oise, 4487.
Senegal, 5089, 5723.
Serbia v. Srbija.
Seriphos, 2114.
Sevastopol, 560.
Sevilla, 44.

Shandong, 8586, 8696, 8736.
Shanghai, 237, 6943, 7090, 8542, 8546, 8574, 8614, 8628, 8632, 8665.
Shenxi, 8512.
Shenzhen, 8609.
Shillourokambos (Parekklisha, Kypros), 2339.
Shqiperi, Albania, 3454, 4129-4131, 4248, 7555.
Shuqba Cave, 1838.
Sibir', 1550, 5056, 5284.
Sichuan, 8591, 8691.
Sicilia, 235, 2063, 2149, 2341, 2597, 3464, 4819, 5544, 6467, 6497, 7423, 7437, 7538.
Siena, 3798, 6238, 6978, 7531.
Sierra Leone, 7714.
Signia, 2845.
Sigwells (Somerset, England), 1560.
Sind, 6942.
Singapore, 8593.
Sirya, 3471.
Skåne, 3483.
Slavonija, 3895.
Sligo (county), 7160.
Slovakia v. Slovensko.
Slovenija, 276, 888, 952, 968, 4028, 5113-5121, 6020, 7019, 7425, 7426, 7431, 7534.
Slovensko, Slovakia, 142, 143, 596, 923, 4354, 5090-5112, 6792, 6949, 7011, 7418, 8062, 8104, 8139, 8172, 8177.
Słuck, 5612.
Solomon Islands, 861.
Somalia, 5122, 5123, 7715, 8456.
Soraluze/Placentia, 327.
Soudan, 1705.
South Africa, 358, 812, 901, 1002, 4399, 4970, 5124-5137, 5920, 7614, 7722, 7921, 8400.
Sovjan (Shqiperi), 2347.
Soweto, 802, 5132, 8796.
Spain v. España.
Sparta, 1935, 1959, 2081, 2089, 2107, 2132, 2345.
Spigno, 7461.
Spina, 2366.
Split, 3602.
Spratly (îles), 8156.
Srbija, Serbia, 3215, 3883, 7870, 7950, 7988, 8020.
Sri Lanka, 1620, 5204, 5205, 7657.
SSSR, 286, 741, 4090, 4692, 4703, 5050, 5082, 5083, 5750, 5811, 5865, 6057, 6070, 6518, 6704, 6708, 6773, 6807, 6957, 6958, 7007, 7009, 7298, 7954, 7986, 7989, 8011, 8031, 8045, 8054, 8078, 8092, 8098, 8105, 8114, 8125, 8126, 8130, 8152, 8194, 8195, 8196, 8200, 8205, 8256, 8262, 8264, 8265, 8290, 8295, 8331, 8344, 8360, 8393, 8399, 8403, 8413, 8414, 8445, 8467, 8645.
St. Petersburg, 7351.
Stabiae, 2922.
Star Carr (North Yorkshire), 1511.
Stockholm, 4070, 7190.
Stonehenge, 1562.
Strasbourg, 339, 5481.
Sudan, 8800, 8806.
Sudeten, 5091, 6576, 7566.
Sudtirolo, 4862.
Suez, 8377, 8444, 8469.
Suisse v. Schweiz.
Sungchiang, 237.
Sungir, 1532.
Suomi, Finland, 91, 251, 613, 1153, 4405-4414, 5212, 5465, 5683, 6028, 6213, 6560, 6641, 6728, 6845, 7265, 7564, 8031, 8038, 8039, 8114, 8420.
Suriname, 7673.
Suzhou, 7110, 8641.
Sverige, 219, 280, 625, 910, 1147, 1247, 1551, 4947, 5206-5223, 5682, 5788, 5969, 6072, 6540, 6692, 6728, 6794, 6808, 6809, 6903, 6906, 6966, 6968, 7168, 7499, 7763, 7971, 8404.
Svizzera v. Schweiz.
Swarzędz, 5718.
Swaziland, 913.
Switzerland v. Schweiz.
Syme, 2054.
Syria, 1839, 3251, 5232, 5233, 7679, 7680, 7889, 7951, 8082, 8287, 8331, 8489.
Szczecin, 5002, 6924.

T

T. Istaba (Israel), 1899.
Taiwan, 925, 1556, 8376, 8506, 8539, 8638.
Tall Šēh Hamad, 1738, 1760.
Tamam (Rossija), 2356.
Tanganika, 7706.
Tanzania, 537, 4905, 5234, 5235, 7358.
Taprobane, 1620.
Taranto, 3532.
Tarquinia, 2393.
Tarragona, 3517.
Tarsus, 2169.
Tebtunis, 1875.
Tehuantepec, 7896.
Tekel, 1134.
Tektas Burnu (Türkiye), 1757.
Tel Hadid, 1787.
Tel Jezreel, 1879.
Tell el-Amarna, 1733, 1766.
Tell Hamoukar (Syria), 1758.
Tell Umm el-Marra (Syria), 1668.
Ternberg, 4171.
Těšín, 7089, 8286.
Thailand, 5236, 5237, 8170.
Thakhek, 8503.
Thasos, 2075, 2336.
Thebae, 2079, 2298.
Thermopylae, 2068.
Thessalia, 199, 1921, 2325.
Thessaloniki, 7052.
Thracia, 1910, 1969, 2075, 2543.
Thüringen, 6109.
Tiberis (flumen), 2918.
Tibet, 5609, 5709, 7941, 8369, 8383, 8705.
Tibur, 2889.
Tierra del Fuego, 7398.
Timor, 4755, 8243.
Tinje, 4010.
Tiranë, 6113.
Tiriolo, 2421.
Tirol, 4174, 4193.
Tirreno, mare, 3457.
Tjaiharpata, 1692.
Tlos, 1639.
Toledo, 3983.
Torcello, 2910, 3234.
Torino, 3827, 5774, 5952, 7097.
Toscana, 331, 2387, 3426, 3527, 3977, 4008, 6061, 6537, 6969, 7511, 7820.
Toul (Meurthe-et-Moselle), 2878.
Toulouse, 161, 3399, 7132.
Tours, 17.
Toyama, 4872.
Transbaikal (Sibir'), 1573.
Transcaucasia, 1632.
Transilvania, 5035, 3945, 5028, 5029, 5873, 7045, 7155, 7789, 8132.
Transvaal, 358.
Trenčín, 143.
Trentino, 2385, 4862.
Treviso, 51.
Trieste, 4201.
Trinidad and Tobago, 5238, 5239.
Troezen, 2070.
Troia, 2348.
Tucuman, 7498.
Tunisie, 7696, 7697, 7710, 7850.
Turano (valle del), 3603.
Turkey v. Türkiye.
Türkiye, Turkey, 9, 243, 548, 1282, 1654, 3256, 3257, 3373, 3425, 3537, 3823, 4020, 5240-

GEOGRAPHICAL INDEX

5263, 5619, 5820, 5964, 6060, 6078, 6280, 6742, 6763, 6778, 6919, 7006, 7025, 7064, 7217, 7244, 7381, 7480, 7590, 7604, 7607, 7769, 7871, 7877, 7901, 7934, 7945, 7987, 7996, 8027, 8042, 8176, 8247, 8258, 8278, 8418, 8463, 8465.
Tusculum, 2927.
Tyrus, 1832.

U

U.S.A., 87, 97, 203, 288, 298, 440, 442, 459, 478, 551, 556, 609, 908, 922, 956, 964, 1125, 1132, 1176, 1290, 1354, 4403, 4910, 5103, 5260, 5272-5370, 5603, 5623, 5631, 5660, 5786, 5805, 5846, 5854, 5860, 5881, 5922, 5960, 5974, 5976, 6044, 6071, 6074, 6081, 6263, 6266, 6288, 6318, 6328, 6347, 6350, 6470, 6471, 6490, 6510, 6514, 6546, 6582, 6638, 6665, 6689, 6710, 6736, 6741, 6777, 6788, 6810, 6811, 6823, 6848, 6849, 6860, 6866, 6878, 6888, 6890, 6897, 6912, 6934, 6970, 6973, 6981, 6993, 6999, 7053, 7098, 7099, 7106, 7138, 7162, 7178, 7209, 7222, 7225, 7230, 7231, 7234, 7252, 7266, 7290, 7306, 7337, 7343, 7346, 7348, 7356, 7357, 7361, 7366, 7413, 7424, 7429, 7571, 7577, 7578, 7581, 7586, 7589, 7599, 7601, 7603, 7610, 7630, 7634, 7639, 7640, 7693, 7705, 7714, 7747, 7827, 7829, 7833, 7885, 7893, 7906, 7907, 7913, 7922, 7927, 7936, 7943, 7944, 7949, 7955, 8005, 8015, 8030, 8038, 8053, 8057, 8065, 8083, 8095, 8103, 8109, 8123, 8133, 8138, 8149, 8158, 8184, 8185, 8204, 8208, 8236, 8242, 8245, 8258, 8260, 8261, 8269, 8277, 8281, 8285, 8287, 8292, 8297, 8299, 8300, 8301, 8310, 8312, 8314, 8319, 8325, 8326, 8327, 8336, 8338, 8339, 8345, 8350, 8351, 8355, 8356, 8358, 8359, 8363, 8364, 8371, 8373, 8374, 8375, 8379, 8380, 8381, 8386, 8388, 8393, 8397, 8401, 8403, 8405, 8408, 8413, 8417, 8421, 8428, 8429, 8430, 8431, 8434 8438, 8439, 8450, 8453, 8455, 8463, 8471, 8579, 8621, 8656, 8813.
Uganda, 5264-5268, 6019.
Ugarit, 1805, 1807, 1822.
Ukek, 4022.
Ukraine, 173, 882, 891, 2121, 4062, 4202, 5269-5271, 7774, 7967, 8014.
Ulster, 4770, 4771, 4778, 4779.
Umbria, 2372, 2932, 3817.
Umm el-Marra, 1795.
Uphill Quarry (Somerset), 1520.
Upton Lovell, 1586.
Urals, 6758.
Uruguay, 599, 5371-5376, 7316, 7750.
Uruk, 1791.

V

Val Canale, 4862.
Val Tanaro, 7542.
Valachia, 7145.
Valencia, 349, 3081, 3301, 3564, 3745, 5165, 7660.
Valle Central de Costa Rica, 7260.
Valle d'Aosta, 81, 3268, 3325.
Valle del Côa (Portugal), 1540.
Valltorta-Gasulla (Castellón), 1601.
Van (lake), 3823.
Vandée, 186.
Vaticano, 341.
Vaud (canton de), 6278.
Veliko Turnovo, 4246.
Venetia, 2367, 2374, 2388.
Veneto, 4814.
Venezia, 1016, 1263, 3344, 3554, 3611, 3803, 4804, 4830, 5109, 6065, 6554, 6571, 6572, 6599, 6813, 6885, 7146, 7410, 7761, 7807, 8219.
Venezuela, 146, 929, 1269, 5377-5388, 7103, 7881, 7920, 7927, 8421, 8422.
Verona, 2642.
Versilia, 3810.
Vetulonia, 2395.
Vicenza, 69.

Vietnam, 4481, 5389, 5390, 6291, 8327, 8371, 8378, 8428, 8472, 8601, 8720.
Virginia, 719.
Vizcaya, 5197.
Volterra, 2399.

W

Wadi en-Natuf (Judaea), 1838.
Wales, 3258, 3790, 6959, 7108.
Wallonie, 3485.
Warmia, 5427.
Warszawa, 8051.
Washington, 5279.
Waziristan, 4754.
West Indies, 7396.
Westphalia, 7326, 7817.
Wielkopolska, 134, 1579, 6725.
Wien, 1178, 6151, 6632, 6667, 7071, 7078, 7246.
Wilson's Creek, 5350.
Worcester, 6550.
Wrocław, 5527.
Württemberg, 163, 1205, 4613, 7460, 7556.
Würzburg, 5949, 6525.

X

Xiamen, 8625, 8651.
Xi'an (Shaanxi Sheng), 8675.
Xianggang, Hong Kong, 7090, 7396, 7665, 8298, 8638.

Y

Yemen, 7721.
Yinwan, 8700.
York, 3273.
Yorkshire, 3555.
Yörüklü/Hüseyindede, 1830.
Yugoslavia v. Jugoslavija.

Z

Zagros Mountains, 1504.
Zakynthos, 4796.
Zanzibar, 5394, 7179, 7887.
Zaragoza, 3078, 3777, 5147, 6598.
Zaraisk, 1501.
Zeugma, 1746.
Zhejiang, 8708.
Zhili, 8696.
Zimbabwe, 5395, 8310.
Zürich, 6069.